THE 80x86 IBM PC AND COMPATIBLE COMPUTERS

VOLUMES I & II

Assembly Language, Design, and Interfacing

Fourth Edition

Muhammad Ali Mazidi
Janice Gillispie Mazidi

Upper Saddle River, New Jersey
Columbus, Ohio

Editor in Chief: Stephen Helba
Assistant Vice President and Publisher: Charles E. Stewart, Jr.
Production Editor: Alexandrina Benedicto Wolf
Design Coordinator: Diane Ernsberger
Cover Designer: Jeff Vanik
Cover image: Digital Images
Production Manager: Matthew Ottenweller
Marketing Manager: Ben Leonard

This book was set in Times Roman by Janice Mazidi. It was printed and bound by Courier/Kendallville. The cover was printed by Phoenix Color Corp.

Pearson Education Ltd.
Pearson Education Australia Pty. Limited
Pearson Education Singapore Pte. Ltd.
Pearson Education North Asia Ltd.
Pearson Education Canada, Ltd.
Pearson Educación de Mexico, S.A. de C.V.
Pearson Education—Japan
Pearson Education Malaysia Pte. Ltd.
Pearson Education, *Upper Saddle River, New Jersey*

10 9 8 7 6 5 4 3 2
ISBN 0-13-061775-X

THE 80x86 IBM PC
AND COMPATIBLE COMPUTERS

VOLUMES I & II

Assembly Language, Design, and Interfacing

Fourth Edition

*Regard man as a mine
rich in gems of inestimable value.
Education can, alone,
cause it to reveal its treasures,
and enable mankind
to benefit therefrom.*

Baha'u'llah

DEDICATIONS

This book is dedicated to the memory of Muhammad Ali's parents, who raised 10 children and persevered through more than 50 years of hardship together with dignity and faith.

We feel especially blessed to have the support, love, and encouragement of Janice's parents whose kindness, wisdom, and sense of humor have been the bond that has welded us into a family.

In addition, we must also mention our two most important collaborations: our sons Robert Nabil and Michael Jamal who have taught us the meaning of love and patience.

We would also like to honor the memory of a dear friend, Kamran Lotfi.

CONTENTS AT A GLANCE

Assembly Language Programming on the IBM PC, PS, and Compatibles

Design and Interfacing of the IBM PC, PS, and Compatibles

APPENDICES

CONTENTS

PREFACE TO VOLUMES I AND II

Purpose

This combined volume is intended for use in college-level courses in which both Assembly language programming and 80x86 PC interfacing are discussed. It not only builds the foundation of Assembly language programming, but also provides a comprehensive treatment of 80x86 PC design and interfacing for students in engineering and computer science disciplines. This volume is intended for those who wish to gain an in-depth understanding of the internal working of the IBM PC, PS, and 80x86 compatible computers. It builds a foundation for the design and interfacing of microprocessor-based systems using the real-world example of the 80x86 IBM PC. In addition, it can also be used by practicing technicians, hardware engineers, computer scientists, and hobbyists who want to do PC interfacing and data acquisition.

Prerequisites

Readers should have a minimal familiarity with the IBM PC and the DOS operating system in addition to having had an introductory digital course. Knowledge of other programming languages would be helpful, but is not necessary.

Although a vast majority of current PCs use 386, 486, or Pentium microprocessors, their design is based on the IBM PC/AT, an 80286 microprocessor system introduced in 1984. A good portion of PC/AT features, hence its limitations, are based on the original IBM PC, an 8088 microprocessor system, introduced in 1981. In other words, one cannot expect to understand fully the architectural philosophy of the 80x86 PC and its expansion slot signals unless the 80286 PC/AT and its subset, the IBM PC/XT, are first understood. For this reason, we describe the 8088 and 80286 microprocessors in Chapter 9.

Contents of Volume I

A systematic, step-by-step approach has been used in covering various aspects of Assembly language programming. Many examples and sample programs are given to clarify concepts and provide students an opportunity to learn by doing. Review questions are provided at the end of each section to reinforce the main points of the section. We feel that one of the functions of a textbook is to familiarize the student with terminology used in technical literature and in industry, so we have followed that guideline in this text.

Chapter 0 covers concepts in number systems (binary, decimal, and hex) and computer architecture. Most students will have learned these concepts in previous courses, but Chapter 0 provides a quick overview for those students who have not learned these concepts, or who may need to refresh their memory.

Chapter 1 provides a brief history of the evolution of x86 microprocessors and an overview of the internal workings of the 8086 as a basis of all x86 processors. Chapter 1 should be used in conjunction with Appendix A (a tutorial introduction to DEBUG) so that the student can experiment with concepts being learned on the PC. The order of topics in Appendix A has been designed to correspond to the order of topics presented in Chapter 1. Thus, the student can begin programming with DEBUG without having to learn how to use an assembler.

Chapter 2 explains the use of assemblers to create programs. Although the programs in the book were developed and tested with Microsoft's MASM assembler, any Intel-compatible assembler such as Borland's TASM may be used.

Chapter 3 introduces the bulk of the logic and arithmetic instructions for unsigned numbers, plus bitwise operations in C.

Chapter 4 introduces DOS and BIOS interrupts. Programs in Assembly and C allow the student to get input from the keyboard and send output to the monitor. In addition, interrupt programming in C is described, as well as how to put Assembly language code in C programs.

Chapter 5 describes how to use macros to develop Assembly language programs in a more time-efficient and structured manner. We also cover INT 33H mouse function calls and mouse programming.

Chapter 6 covers arithmetic and logic instructions for signed numbers as well as string processing instructions.

Chapter 7 discusses modular programming and how to develop larger Assembly language programs by breaking them into smaller modules to be coded and tested separately. In addition, linking Assembly language modules with C programs is thoroughly explained.

Chapter 8 introduces some 32-bit concepts of 80386 and 80486 programming. Although this book emphasizes 16-bit programming, the 386/486 is introduced to help the student appreciate the power of 32-bit CPUs. Several programs are run across the 80x86 family to show the dramatic improvement in clock cycles with the newer CPUs.

Contents of Volume II

Chapter 9 describes the 8088 and 286 microprocessors and supporting chips in detail and shows how they are used in the original IBM PC/XT/AT. In addition, the origin and function of the address, data, and control signals of the ISA expansion slot are described.

Chapter 10 provides an introduction to various types of RAM and ROM memories, their interfacing to the microprocessor, the memory map of the 80x86 PC, the timing issue in interfacing memory to the ISA bus, and the checksum byte and parity bit techniques of ensuring data integrity in RAM and ROM.

Chapter 11 is dedicated to the interfacing of I/O ports, the use of IN and OUT instructions in the 80x86, and interfacing and programming of the 8255 programmable peripheral chip. We describe I/O programming in several languages, as well.

Chapter 12 covers the PC Interface Trainer and Bus Extender, which are used to interface PCs to devices for data acquisition such as LCDs, stepper motors, ADC, DAC, and sensors.

Chapter 13 discusses the use of the 8253/54 timer chip in the 80x86 PC, as well as how to generate music and time delays.

Chapter 14 is dedicated to the explanation of hardware and software interrupts, the use of the 8259 interrupt controller, the origin and assignment of IRQ signals on the expansion slots of the ISA bus, and exception interrupts in 80x86 microprocessors.

Chapter 15 is dedicated to direct memory access (DMA) concepts, the use of the 8237 DMA chip in the 80x86 PC, and DMA channels and associated signals on the ISA bus.

Chapter 16 covers the basics of video monitors and various video modes and adapters of the PC, in addition to the memory requirements of various video boards in graphics mode.

Chapter 17 discusses serial communication principles, the interfacing and programming of National Semiconductor's 8250/16450/16550 UART chip, Intel's 8251 USART chip, and verifying data integrity using the CRC method.

Chapter 18 covers the interfacing and programming of the keyboard in the 80x86 PC, in addition to printer port interfacing and programming. In addition, a discussion of various types of parallel ports such as EPP and ECP is included.

Chapter 19 discusses both floppy and hard disk storage organization and terminology. We also show how to write Assembly language programs to access files using INT 21H DOS function calls.

Chapter 20 examines the 80x87 math coprocessor, its programming and interfacing, and IEEE single and double precision floating point data types.

Chapter 21 explores the programming and hardware of the 386 microprocessor, contrasts and explains real and protected modes, and discusses the implementation of virtual memory.

Chapter 22 is dedicated to the interfacing of high-speed memories and describes various types of DRAM, including EDO and SDRAM, and examines cache memory and various cache organizations and terminology in detail.

In Chapter 23 we describe the main features of the 486, Pentium and Pentium Pro and compare these microprocessors with the RISC processors. Chapter 23 also provides a discussion of MMX technology and how to write programs to detect which CPU a PC has.

Chapter 24 describes the MS DOS structure and the role of CONFIG.SYS and batch files in the 80x86 PC, the writing of TSR (terminate and stay resident) programs and device drivers.

Chapter 25 explains 80x86 PC memory terminology, such as conventional memory, expanded memory, upper memory block, high memory area, as well as MS DOS memory management.

Chapter 26 provides an overview of the IC technology including the recent advances in the IC fabrication, describes IC interfacing and system design issues, and covers error detection and correction.

Chapter 27 is dedicated to the discussion of the various types of PC buses, such as ISA, EISA, USB, their performance comparisons, the local bus and features of the PCI local bus.

In Chapter 28 we show how to use C language to access DOS function calls, BIOS interrupts, memory, input/output ports, and CMOS RAM of the 80x86.

Appendices

The appendices have been designed to provide all reference material required for the topics covered in this combined volume so that no additional references should be necessary.

Appendix A provides a tutorial introduction to DEBUG. Appendix B provides a listing of Intel's 8086 instruction set along with clock cycles for 80x86 microprocessors. Appendix C describes assembler directives with examples of their use. Appendix D lists some commonly used DOS 21H function calls and INT 33H mouse functions. Appendix E lists the function calls for various BIOS interrupts. Appendix F provides a table of ASCII codes. Appendix G lists the I/O map of 80x86-based ISA computers. Appendix H provides a description of the BIOS data area. Appendix I contains data sheets for various IC chips.

Lab Manual

The lab manual for this series is available on the following web site:

www.microdigitaled.com

Acknowledgments

This book is the result of the dedication, work and love of many individuals. Our sincere and heartfelt appreciation goes out to all of them. First, we must thank the original reviewers who provided valuable suggestions and encouragement: Mr. William H. Shannon of the University of Maryland, Mr. Howard W. Atwell of Fullerton College, Mr. David G. Delker of Kansas State University, Mr. Michael Chen of Duchess Community College, Mr. Yusuf Motiwala of Prairie View A&M University, and Mr. Donald T. Coston of ITT Technical Institute. We were truly amazed by the depth and breadth of their knowledge of microprocessor-based system design in general and 80x86 PC architecture in particular. We sincerely appreciate their comments and suggestions.

Thanks also must go to the many students whose comments have helped shape this book, especially Daniel Woods, Sam Oparah, Herbert Sendeki, Greg Boyle, Philip Fitzer, Adnan Hindi, Kent Keeter, Mark Ford, Shannon Looper, Mitch Johnson, Carol Killelea, Michael Madden, Douglas McAlister, David Simmons, Dwight Brown, Clifton Snyder, Phillip Boatright, Wilfrid Lowe, Robert Schabel, John Berry, Clyde Knight, Robert Jones (all of DeVry Institute of Technolgy), Lynnette Garetz (Heald College), Peter Woof (Southern Sydney Institute, Lidcombe College of Tafe), M. Soleimanzadeh, Mark Lessley, Snehal Amin, Travis Erck, Gary Hudson, Nathan Noel, Dan Bent, and Frank Fortman.

A word must also be said of our colleagues, especially the late Mr. Allan Escher, whose encouragement set the making of this series into motion. For the last 25 years, his dedication and love of microprocessor education were a source of inspiration to many. A special thanks goes to Mr. James Vignali for his enthusiasm in discussing the internal intricacies of the 80x86 PC and his readiness to keep current with the ever-changing world of the PC.

In addition, we offer our appreciation for the dedicated professionals at Prentice Hall. Many thanks to Charles Stewart for his continued support and guidance of this series.

Finally, we would like to sincerely thank the following professors from some outstanding engineering schools whose enthusism for the book, suggestions, and kind words have been encouraging to us and made us think we are on the right track: Dr. Michael Chwialkowski (Electrical Engineering Dept., University of Texas at Arlington), Dr. Roger S. Walker (Computer Science Engineering Dept., University of Texas at Arlington), Dr. Behbood Zoghi (Electronics Engineering Technology, Texas A&M University).

ABOUT THE AUTHORS

Muhammad Ali Mazidi holds Master's degrees from both Southern Methodist University and the University of Texas at Dallas, and currently is a.b.d. on his Ph.D. in the Electrical Engineering Department of Southern Methodist University. He is a co-founder and chief researcher of Microprocessor Education Group, a company dedicated to bringing knowledge of microprocessors to the widest possible audience. He also teaches microprocessor-based system design at DeVry Institute of Technology in Dallas, Texas.

Janice Gillispie Mazidi has a Master of Science degree in Computer Science from the University of North Texas. After several years experience as a software engineer in Dallas, she co-founded Microprocessor Education Group, where she is the chief technical writer, production manager, and is responsible for software development and testing.

The Mazidis have been married since 1985 and have two sons, Robert Nabil and Michael Jamal.

The authors can be contacted at the following address if you have any comments, suggestions, or if you find any errors.

Microprocessor Education Group
P.O. Box 381970
Duncanville, TX 75138

email: mazidi@mail.dal.devry.edu
   or: profmazidi@yahoo.com

The web site www.microdigitaled.com provides much support for this book.

CHAPTER 0

INTRODUCTION TO COMPUTING

OBJECTIVES

Upon completion of this chapter, you will be able to:

» **Convert any number from base 2, base 10, or base 16 to any of the other two bases**
» **Count in binary and hex**
» **Add and subtract hex numbers**
» **Add binary numbers**
» **Represent any binary number in 2's complement**
» **Represent an alphanumeric string in ASCII code**
» **Explain the difference between a bit, a nibble, a byte, and a word**
» **Give precise mathematical definitions of the terms *kilobyte*, *megabyte*, *terabyte*, and *gigabyte***
» **Explain the difference between RAM and ROM and describe their use**
» **List the major components of a computer system and describe their functions**
» **List the three types of buses found in computers and describe the purpose of each type of bus**
» **Describe the role of the CPU in computer systems**
» **List the major components of the CPU and describe the purpose of each**
» **Trace the evolution of computers from vacuum tubes to transistors to IC chips**
» **State the differences between the RISC and CISC design philosophies**

To understand the software and hardware of the computer, one must first master some very basic concepts underlying computer design. In this chapter (which in the tradition of digital computers can be called Chapter 0), the fundamentals of numbering and coding systems are presented. Then an introduction to the workings of the inside of the computer is given. Finally, in the last section we give a brief history of CPU architecture. Although some readers may have an adequate background in many of the topics of this chapter, it is recommended that the material be scanned, however briefly.

SECTION 0.1: NUMBERING AND CODING SYSTEMS

Whereas human beings use base 10 (*decimal*) arithmetic, computers use the base 2 (*binary*) system. In this section we explain how to convert from the decimal system to the binary system, and vice versa. The convenient representation of binary numbers called *hexadecimal* also is covered. Finally, the binary format of the alphanumeric code, called *ASCII*, is explored.

Decimal and binary number systems

Although there has been speculation that the origin of the base 10 system is the fact that human beings have 10 fingers, there is absolutely no speculation about the reason behind the use of the binary system in computers. The binary system is used in computers because 1 and 0 represent the two voltage levels of on and off. Whereas in base 10 there are 10 distinct symbols, 0, 1, 2, ..., 9, in base 2 there are only two, 0 and 1, with which to generate numbers. Base 10 contains digits 0 through 9; binary contains digits 0 and 1 only. These two binary digits, 0 and 1, are commonly referred to as *bits*.

Converting from decimal to binary

One method of converting from decimal to binary is to divide the decimal number by 2 repeatedly, keeping track of the remainders. This process continues until the quotient becomes zero. The remainders are then written in reverse order to obtain the binary number. This is demonstrated in Example 0-1.

Example 0-1

Convert 25_{10} to binary.

Solution:

		Quotient	Remainder	
25/2	=	12	1	LSB (least significant bit)
12/2	=	6	0	
6/2	=	3	0	
3/2	=	1	1	
1/2	=	0	1	MSB (most significant bit)

Therefore, $25_{10} = 11001_2$.

Converting from binary to decimal

To convert from binary to decimal, it is important to understand the concept of weight associated with each digit position. First, as an analogy, recall the weight of numbers in the base 10 system:

$$740683_{10} =$$

$$
\begin{array}{rcr}
3\times10^{0} & = & 3 \\
8\times10^{1} & = & 80 \\
6\times10^{2} & = & 600 \\
0\times10^{3} & = & 0000 \\
4\times10^{4} & = & 40000 \\
7\times10^{5} & = & \underline{700000} \\
 & & 740683
\end{array}
$$

By the same token, each digit position in a number in base 2 has a weight associated with it:

$$110101_2 =$$

					Decimal	Binary
1×2^0	=	1×1	=		1	1
0×2^1	=	0×2	=		0	00
1×2^2	=	1×4	=		4	100
0×2^3	=	0×8	=		0	0000
1×2^4	=	1×16	=		16	10000
1×2^5	=	1×32	=		$\underline{32}$	$\underline{100000}$
					53	110101

Knowing the weight of each bit in a binary number makes it simple to add them together to get its decimal equivalent, as shown in Example 0-2.

Example 0-2

Convert 11001_2 to decimal.

Solution:

Weight:	16	8	4	2	1
Digits:	1	1	0	0	1
Sum:	16 +	8 +	0 +	0 +	$1 = 25_{10}$

Knowing the weight associated with each binary bit position allows one to convert a decimal number to binary directly instead of going through the process of repeated division. This is shown in Example 0-3.

Example 0-3

Use the concept of weight to convert 39_{10} to binary.

Solution:

Weight:	32	16	8	4	2	1
	1	0	0	1	1	1
	32 +	0 +	0 +	4 +	2 +	$1 = 39$

Therefore, $39_{10} = 100111_2$.

Hexadecimal system

Base 16, the *hexadecimal* system as it is called in computer literature, is used as a convenient representation of binary numbers. For example, it is much easier for a human being to represent a string of 0s and 1s such as 100010010110 as its hexadecimal equivalent of 896H. The binary system has 2 digits, 0 and 1. The base 10 system has 10 digits, 0 through 9. The hexadecimal (base 16) system must have 16 digits. In base 16, the first 10 digits, 0 to 9, are the same as in decimal, and for the remaining six digits, the letters A, B, C, D, E, and F are used. Table 0-1 shows the equivalent binary, decimal, and hexadecimal representations for 0 to 15.

Converting between binary and hex

To represent a binary number as its equivalent hexadecimal number, start from the right and group 4 bits at a time, replacing each 4-bit binary number with its hex equivalent shown in Table 0-1. To convert from hex to binary, each hex digit is replaced with its 4-bit binary equivalent. Converting between binary and hex is shown in Examples 0-4 and 0-5.

Converting from decimal to hex

Converting from decimal to hex could be approached in two ways:
1. Convert to binary first and then convert to hex. Experimenting with this method is left to the reader.
2. Convert directly from decimal to hex by the method of repeated division, keeping track of the remainders. Example 0-6 demonstrates this method of converting decimal to hex.

Converting from hex to decimal

Conversion from hex to decimal can also be approached in two ways:
1. Convert from hex to binary and then to decimal.
2. Convert directly from hex to decimal by summing the weight of all digits. Example 0-7 demonstrates the second method of converting from hex to decimal.

Table 0-1: Decimal, Binary, and Hex

Decimal	Binary	Hexadecimal
0	0000	0
1	0001	1
2	0010	2
3	0011	3
4	0100	4
5	0101	5
6	0110	6
7	0111	7
8	1000	8
9	1001	9
10	1010	A
11	1011	B
12	1100	C
13	1101	D
14	1110	E
15	1111	F

Example 0-4

Represent binary 100111110101 in hex.

Solution:

First the number is grouped into sets of 4 bits: 1001 1111 0101
Then each group of 4 bits is replaced with its hex equivalent:

$$\begin{array}{ccc} 1001 & 1111 & 0101 \\ 9 & F & 5 \end{array}$$

Therefore, 100111110101_2 = 9F5 hexadecimal.

Example 0-5

Convert hex 29B to binary.

Solution:

$$\begin{array}{cccc} & 2 & 9 & B \\ = & 0010 & 1001 & 1011 \end{array}$$

Dropping the leading zeros gives 1010011011.

Example 0-6

(a) Convert 45_{10} to hex.

Solution:

		Quotient	Remainder	
45/16	=	2	13 (hex D)	(least significant digit)
2/16	=	0	2	(most significant digit)

Therefore, $45_{10} = 2D_{16}$.

(b) Convert decimal 629 to hexadecimal.

Solution:

		Quotient	Remainder	
629/16	=	39	5	(least significant digit)
39/16	=	2	7	
2/16	=	0	2	(most significant digit)

Therefore, $629_{10} = 275_{16}$.

(c) Convert 1714 base 10 to hex.

Solution:

		Quotient	Remainder	
1714/16	=	107	2	(least significant digit)
107/16	=	6	11	(hex B)
6/16	=	0	6	(most significant digit)

Therefore, $1714_{10} = 6B2_{16}$.

Example 0-7

Convert the following hexadecimal numbers to decimal.

(a) $6B2_{16}$

Solution:

6B2 hexadecimal =

$$
\begin{aligned}
2 \times 16^0 &= 2 \times 1 &&= &2 \\
11 \times 16^1 &= 11 \times 16 &&= &176 \\
6 \times 16^2 &= 6 \times 256 &&= &\underline{1536} \\
& && &1714
\end{aligned}
$$

Therefore, $6B2_{16} = 1714_{10}$.

(b) $9F2D_{16}$

Solution:

9F2D hexadecimal =

$$
\begin{aligned}
13 \times 16^0 &= 13 \times 1 &&= &13 \\
2 \times 16^1 &= 2 \times 16 &&= &32 \\
15 \times 16^2 &= 15 \times 256 &&= &3840 \\
9 \times 16^3 &= 9 \times 4096 &&= &\underline{36864} \\
& && &40749
\end{aligned}
$$

Therefore, $9F2D_{16} = 40749_{10}$.

SECTION 0.1: NUMBERING AND CODING SYSTEMS

Counting in bases 10, 2, and 16

To show the relationship between all three bases, in Figure 0-1 we show the sequence of numbers from 0 to 31 in decimal, along with the equivalent binary and hex numbers. Notice in each base that when one more is added to the highest digit, that digit becomes zero and a 1 is carried to the next-highest digit position. For example, in decimal, $9 + 1 = 0$ with a carry to the next-highest position. In binary, $1 + 1 = 0$ with a carry; similarly, in hex, $F + 1 = 0$ with a carry.

Table 0-2: Binary Addition

A + B	Carry	Sum
0 + 0	0	0
0 + 1	0	1
1 + 0	0	1
1 + 1	1	0

Addition of binary and hex numbers

The addition of binary numbers is a very straightforward process. Table 0-2 shows the addition of two bits. The discussion of subtraction of binary numbers is bypassed since all computers use the addition process to implement subtraction. Although computers have adder circuitry, there is no separate circuitry for subtractors. Instead, adders are used in conjunction with *2's complement* circuitry to perform subtraction. In other words, to implement "$x - y$", the computer takes the 2's complement of y and adds it to x. The concept of 2's complement is reviewed next, but the process of subtraction of two binary numbers using 2's complement is shown in detail in Chapter 3. Example 0-8 shows the addition of binary numbers.

Decimal	Binary	Hex
0	00000	0
1	00001	1
2	00010	2
3	00011	3
4	00100	4
5	00101	5
6	00110	6
7	00111	7
8	01000	8
9	01001	9
10	01010	A
11	01011	B
12	01100	C
13	01101	D
14	01110	E
15	01111	F
16	10000	10
17	10001	11
18	10010	12
19	10011	13
20	10100	14
21	10101	15
22	10110	16
23	10111	17
24	11000	18
25	11001	19
26	11010	1A
27	11011	1B
28	11100	1C
29	11101	1D
30	11110	1E
31	11111	1F

Figure 0-1. Counting in 3 Bases

Example 0-8

Add the following binary numbers. Check against their decimal equivalents.

Solution:

	Binary	Decimal
	1101	13
	1001	9
+	10110	22
	101100	44

2's complement

To get the 2's complement of a binary number, invert all the bits and then add 1 to the result. Inverting the bits is simply a matter of changing all 0s to 1s and 1s to 0s. This is called the *1's complement*. See Example 0-9.

CHAPTER 0: INTRODUCTION TO COMPUTING

Example 0-9

Take the 2's complement of 10011101.

Solution:

10011101	binary number
01100010	1's complement
+ 1	
01100011	2's complement

Addition and subtraction of hex numbers

In studying issues related to software and hardware of computers, it is often necessary to add or subtract hex numbers. Mastery of these techniques is essential. Hex addition and subtraction are discussed separately below.

Addition of hex numbers

This section describes the process of adding hex numbers. Starting with the least significant digits, the digits are added together. If the result is less than 16, write that digit as the sum for that position. If it is greater than 16, subtract 16 from it to get the digit and carry 1 to the next digit. The best way to explain this is by example, as shown in Example 0-10.

Example 0-10

Perform hex addition: 23D9 + 94BE.

Solution:

```
  23D9
+ 94BE
  B897
```

LSD: 9 + 14	=	23	23 − 16 = 7 with a carry to next digit
1 + 13 + 11	=	25	25 − 16 = 9 with a carry to next digit
1 + 3 + 4	=	8	
MSD: 2 + 9 = B			

Subtraction of hex numbers

In subtracting two hex numbers, if the second digit is greater than the first, borrow 16 from the preceding digit. See Example 0-11.

Example 0-11

Perform hex subtraction: 59F − 2B8.

Solution:

```
  59F
− 2B8
  2E7
```

LSD:	8 from 15 = 7
	11 from 25 (9 + 16) = 14, which is E
MSD:	2 from 4 (5 − 1) = 2

ASCII code

The discussion so far has revolved around the representation of number systems. Since all information in the computer must be represented by 0s and 1s, binary patterns must be assigned to letters and other characters. In the 1960s a standard representation called *ASCII* (American Standard Code for Information Interchange) was established. The ASCII (pronounced "ask-E") code assigns binary patterns for numbers 0 to 9, all the letters of the English alphabet, both uppercase (capital) and lowercase, and many control codes and punctuation marks. The great advantage of this system is that it is used by most computers, so that information can be shared among computers. The ASCII system uses a total of 7 bits to represent each code. For example, 100 0001 is assigned to the uppercase letter "A" and 110 0001 is for the lowercase "a". Often, a zero is placed in the most significant bit position to make it an 8-bit code. Figure 0-2 shows selected ASCII codes. A complete list of ASCII codes is given in Appendix F. The use of ASCII is not only standard for keyboards used in the United States and many other countries but also provides a standard for printing and displaying characters by output devices such as printers and monitors.

The pattern of ASCII codes was designed to allow for easy manipulation of ASCII data. For example, digits 0 through 9 are represented by ASCII codes 30 through 39. This enables a program to easily convert ASCII to decimal by masking off the "3" in the upper nibble. As another example, notice in the codes listed below that there is a relationship between the uppercase and lowercase letters. Namely, uppercase letters are represented by ASCII codes 41 through 5A while lowercase letters are represented by ASCII codes 61 through 7A. Looking at the binary code, the only bit that is different between uppercase "A" and lowercase "a" is bit 5. Therefore conversion between uppercase and lowercase is as simple as changing bit 5 of the ASCII code.

Hex	Symbol	Hex	Symbol
41	A	61	a
42	B	62	b
43	C	63	c
44	D	64	d
45	E	65	e
46	F	66	f
47	G	67	g
48	H	68	h
49	I	69	i
4A	J	6A	j
4B	K	6B	k
4C	L	6C	l
4D	M	6D	m
4E	N	6E	n
4F	O	6F	o
50	P	70	p
51	Q	71	q
52	R	72	r
53	S	73	s
54	T	74	t
55	U	75	u
56	V	76	v
57	W	77	w
58	X	78	x
59	Y	79	y
5A	Z	7A	z

Figure 0-2. Alphanumeric ASCII Codes

Review Questions

1. Why do computers use the binary number system instead of the decimal system?
2. Convert 34_{10} to binary and hex.
3. Convert 1101011_2 to hex and decimal.
4. Perform binary addition: $101100 + 101$.
5. Convert 101100_2 to its 2's complement representation.
6. Add 36BH + F6H.
7. Subtract 36BH − F6H.
8. Write "80x86 CPUs" in its ASCII code (in hex form).

SECTION 0.2: INSIDE THE COMPUTER

In this section we provide an introduction to the organization and internal working of computers. The model used is generic, but the concepts discussed are applicable to all computers, including the IBM PC, PS/2, and compatibles. Before embarking on this subject, it will be helpful to review definitions of some of the most widely used terminology in computer literature, such as *K, mega, giga, byte, ROM, RAM,* and so on.

Some important terminology

One of the most important features of a computer is how much memory it has. Next we review terms used to describe amounts of memory in IBM PCs and compatibles. Recall from the discussion above that a *bit* is a binary digit that can have the value 0 or 1. A *byte* is defined as 8 bits. A *nibble* is half a byte, or 4 bits. A *word* is two bytes, or 16 bits. The following display is intended to show the relative size of these units. Of course, they could all be composed of any combination of zeros and ones.

```
Bit                                        0
Nibble                                  0000
Byte                          0000 0000
Word           0000 0000 0000 0000
```

A *kilobyte* is 2^{10} bytes, which is 1024 bytes. The abbreviation K is often used. For example, some floppy disks hold 356K bytes of data. A *megabyte*, or meg as some call it, is 2^{20} bytes. That is a little over 1 million bytes; it is exactly 1,048,576. Moving rapidly up the scale in size, a *gigabyte* is 2^{30} bytes (over 1 billion), and a *terabyte* is 2^{40} bytes (over 1 trillion). As an example of how some of these terms are used, suppose that a given computer has 16 megabytes of memory. That would be 16×2^{20}, or $2^4 \times 2^{20}$, which is 2^{24}. Therefore 16 megabytes is 2^{24} bytes.

Two types of memory commonly used in microcomputers are *RAM,* which stands for random access memory (sometimes called *read/write memory*), and *ROM,* which stands for read-only memory. RAM is used by the computer for temporary storage of programs that it is running. That data is lost when the computer is turned off. For this reason, RAM is sometimes called *volatile memory*. ROM contains programs and information essential to operation of the computer. The information in ROM is permanent, cannot be changed by the user, and is not lost when the power is turned off. Therefore, it is called *nonvolatile memory*.

Internal organization of computers

The internal working of every computer can be broken down into three parts: *CPU* (central processing unit), *memory* , and *I/O* (input/output) devices (see Figure 0-3). The function of the CPU is to execute (process) information stored in memory. The function of I/O devices such as the keyboard and video monitor is to provide a means of communicating with the CPU. The CPU is connected to memory

---

and I/O through strips of wire called a *bus*. The bus inside a computer carries information from place to place just as a street bus carries people from place to place. In every computer there are three types of buses: *address bus*, *data bus*, and *control bus*.

For a device (memory or I/O) to be recognized by the CPU, it must be assigned an *address*. The address assigned to a given device must be unique; no two devices are allowed to have the same address. The CPU puts the address (of course, in binary) on the address bus, and the decoding circuitry finds the device. Then the CPU uses the data bus either to get data from that device or to send data to it. The control buses are used to provide read or write signals to the device to indicate if the CPU is asking for information or sending it information. Of the three buses, the address bus and data bus determine the capability of a given CPU.

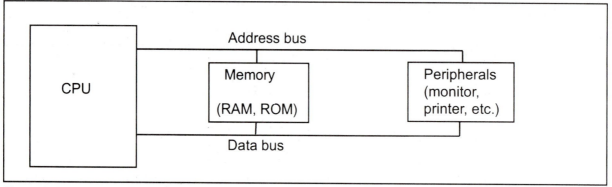

Figure 0-3. Inside the Computer

More about the data bus

Since data buses are used to carry information in and out of a CPU, the more data buses available, the better the CPU. If one thinks of data buses as highway lanes, it is clear that more lanes provide a better pathway between the CPU and its external devices (such as printers, RAM, ROM, etc.; see Figure 0-4). By the same token, that increase in the number of lanes increases the cost of construction. More data buses mean a more expensive CPU and computer. The average size of data buses in CPUs varies between 8 and 64. Early computers such as Apple 2 used an 8-bit data bus, while supercomputers such as Cray use a 64-bit data bus. Data buses are bidirectional, since the CPU must use them either to receive or to send data. The processing power of a computer is related to the size of its buses, since an 8-bit bus can send out 1 byte a time, but a 16-bit bus can send out 2 bytes at a time, which is twice as fast.

More about the address bus

Since the address bus is used to identify the devices and memory connected to the CPU, the more address buses available, the larger the number of devices that can be addressed. In other words, the number of address buses for a CPU determines the number of locations with which it can communicate. The number of locations is always equal to 2^x, where x is the number of address lines, regardless of the size of the data bus. For example, a CPU with 16 address lines can provide a total of 65,536 (2^{16}) or 64K bytes of addressable memory. Each location can have a maximum of 1 byte of data. This is due to the fact that all general-purpose microprocessor CPUs are what is called *byte addressable*. As another example, the IBM PC AT uses a CPU with 24 address lines and 16 data lines. In this case the total accessible memory is 16 megabytes (2^{24} = 16 megabytes). In this example there would be 2^{24} locations, and since each location is one byte, there would be 16 megabytes of memory. The address bus is a *unidirectional* bus, which means that the CPU uses the address bus only to send out addresses. To summarize: The total number of memory locations addressable by a given CPU is always equal to 2^x where x is the number of address bits, regardless of the size of the data bus.

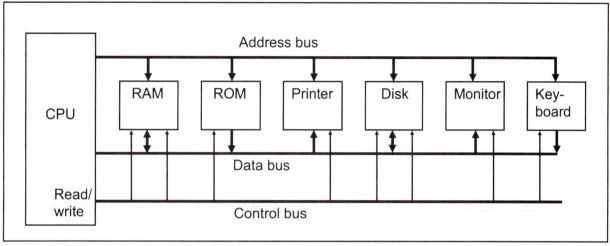

Figure 0-4. Internal Organization of Computers

CPU and its relation to RAM and ROM

For the CPU to process information, the data must be stored in RAM or ROM. The function of ROM in computers is to provide information that is fixed and permanent. This is information such as tables for character patterns to be displayed on the video monitor, or programs that are essential to the working of the computer, such as programs for testing and finding the total amount of RAM installed on the system, or programs to display information on the video monitor. In contrast, RAM is used to store information that is not permanent and can change with time, such as various versions of the operating system and application packages such as word processing or tax calculation packages. These programs are loaded into RAM to be processed by the CPU. The CPU cannot get the information from the disk directly since the disk is too slow. In other words, the CPU gets the information to be processed, first from RAM (or ROM). Only if it is not there does the CPU seek it from a mass storage device such as a disk, and then it transfers the information to RAM. For this reason, RAM and ROM are sometimes referred to as *primary memory* and disks are called *secondary memory*. Figure 0-4 shows a block diagram of the internal organization of the PC.

Inside CPUs

A program stored in memory provides instructions to the CPU to perform an action. The action can simply be adding data such as payroll data or controlling a machine such as a robot. It is the function of the CPU to fetch these instructions from memory and execute them. To perform the actions of fetch and execute, all CPUs are equipped with resources such as the following:

1. Foremost among the resources at the disposal of the CPU are a number of *registers*. The CPU uses registers to store information temporarily. The information could be two values to be processed, or the address of the value needed to be fetched from memory. Registers inside the CPU can be 8-bit, 16-bit, 32-bit, or even 64-bit registers, depending on the CPU. In general, the more and bigger the registers, the better the CPU. The disadvantage of more and bigger registers is the increased cost of such a CPU.

2. The CPU also has what is called the *ALU* (arithmetic/logic unit). The ALU section of the CPU is responsible for performing arithmetic functions such as add, subtract, multiply, and divide, and logic functions such as AND, OR, and NOT.

3. Every CPU has what is called a *program counter*. The function of the program counter is to point to the address of the next instruction to be executed. As each instruction is executed, the program counter is incremented to point to the address of the next instruction to be executed. It is the contents of the program counter that are placed on the address bus to find and fetch the desired instruction. In the IBM PC, the program counter is a register called IP, or the instruction pointer.

4. The function of the *instruction decoder* is to interpret the instruction fetched into the CPU. One can think of the instruction decoder as a kind of dictionary, storing the meaning of each instruction and what steps the CPU should take upon receiving a given instruction. Just as a dictionary requires more pages the more words it defines, a CPU capable of understanding more instructions requires more transistors to design.

Internal working of computers

To demonstrate some of the concepts discussed above, a step-by-step analysis of the process a CPU would go through to add three numbers is given next. Assume that an imaginary CPU has registers called A, B, C, and D. It has an 8-bit data bus and a 16-bit address bus. Therefore, the CPU can access memory from addresses 0000 to FFFFH (for a total of 10000H locations). The action to be performed by the CPU is to put hexadecimal value 21 into register A, and then add to register A values 42H and 12H. Assume that the code for the CPU to move a value to register A is 1011 0000 (B0H) and the code for adding a value to register A is 0000 0100 (04H). The necessary steps and code to perform them are as follows.

Action	Code	Data
Move value 21H into register A	B0H	21H
Add value 42H to register A	04H	42H
Add value 12H to register A	04H	12H

If the program to perform the actions listed above is stored in memory locations starting at 1400H, the following would represent the contents for each memory address location:

Memory address	Contents of memory address	
1400	(B0)	the code for moving a value to register A
1401	(21)	the value to be moved
1402	(04)	the code for adding a value to register A
1403	(42)	the value to be added
1404	(04)	the code for adding a value to register A
1405	(12)	the value to be added
1406	(F4)	the code for halt

The actions performed by the CPU to run the program above would be as follows:

1. The CPU's program counter can have a value between 0000 and FFFFH. The program counter must be set to the value 1400H, indicating the address of the first instruction code to be executed. After the program counter has been loaded with the address of the first instruction, the CPU is ready to execute.

2. The CPU puts 1400H on the address bus and sends it out. The memory circuitry finds the location while the CPU activates the READ signal, indicating to memory that it wants the byte at location 1400H. This causes the contents of memory location 1400H, which is B0, to be put on the data bus and brought into the CPU.

3. The CPU decodes the instruction B0 with the help of its instruction decoder dictionary. When it finds the definition for that instruction it knows it must bring into register A of the CPU the byte in the next memory location. Therefore, it commands its controller circuitry to do exactly that. When it brings in value 21H from memory location 1401, it makes sure that the doors of all registers are closed except register A. Therefore, when value 21H comes into the CPU it will go directly into register A. After completing one instruction, the program counter points to the address of the next instruction to be executed, which in this case is 1402H. Address 1402 is sent out on the address bus to fetch the next instruction.

4. From memory location 1402H it fetches code 04H. After decoding, the CPU knows that it must add to the contents of register A the byte sitting at the next address (1403). After it brings the value (in this case 42H) into the CPU, it provides the contents of

register A along with this value to the ALU to perform the addition. It then takes the result of the addition from the ALU's output and puts it in register A. Meanwhile the program counter becomes 1404, the address of the next instruction.

5. Address 1404H is put on the address bus and the code is fetched into the CPU, decoded, and executed. This code is again adding a value to register A. The program counter is updated to 1406H.

6. Finally, the contents of address 1406 are fetched in and executed. This HALT instruction tells the CPU to stop incrementing the program counter and asking for the next instruction. In the absence of the HALT, the CPU would continue updating the program counter and fetching instructions.

Now suppose that address 1403H contained value 04 instead of 42H. How would the CPU distinguish between data 04 to be added and code 04? Remember that code 04 for this CPU means move the next value into register A. Therefore, the CPU will not try to decode the next value. It simply moves the contents of the following memory location into register A, regardless of its value.

Review Questions

1. How many bytes is 24 kilobytes?
2. What does "RAM" stand for? How is it used in computer systems?
3. What does "ROM" stand for? How is it used in computer systems?
4. Why is RAM called volatile memory?
5. List the three major components of a computer system.
6. What does "CPU" stand for? Explain its function in a computer.
7. List the three types of buses found in computer systems and state briefly the purpose of each type of bus.
8. State which of the following is unidirectional and which is bidirectional.
   (a) data bus   (b) address bus
9. If an address bus for a given computer has 16 lines, then what is the maximum amount of memory it can access?
10. What does "ALU" stand for? What is its purpose?
11. How are registers used in computer systems?
12. What is the purpose of the program counter?
13. What is the purpose of the instruction decoder?

SECTION 0.3: BRIEF HISTORY OF THE CPU

In the 1940s, CPUs were designed using vacuum tubes. The vacuum tube was bulky and consumed a lot of electricity. For example, the first large-scale digital computer, ENIAC, consumed 130,000 watts of power and occupied 1500 square feet. The invention of transistors changed all of that. In the 1950s, transistors replaced vacuum tubes in the design of computers. Then in 1959, the first IC (integrated circuit) was invented. This set into motion what many people believe is the second industrial revolution. In the 1960s the use of IC chips in the design of CPU boards became common. It was not until the 1970s that the entire CPU was put on a single IC chip. The first working CPU on a chip was invented by Intel in 1971. This CPU was called a *microprocessor*. The first microprocessor, the 4004, had a 4-bit data bus and was made of 2300 transistors. It was designed primarily for the hand-held calculator but soon came to be used in applications such as traffic-light controllers. The advances in IC fabrication made during the 1970s made it possible to design microprocessors with an 8-bit data bus and a 16-bit address bus. By the late 1970s, the Intel 8080/85 was one of the most widely used microprocessors, appearing in everything from microwave ovens to homemade computers. Meanwhile, many other companies joined in the race for faster and better microprocessors. Notable among them was Motorola with its 6800 and 68000 microprocessors. Apple's Macintosh computers use the 68000 series microprocessors. Figure 0-5 shows a block diagram of the internal structure of a CPU.

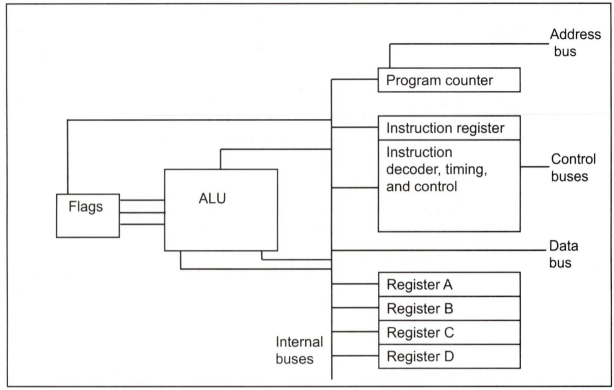

Figure 0-5. Internal Block Diagram of a CPU

CISC vs. RISC

Until the early 1980s, all CPUs, whether single-chip or whole-board, followed the *CISC* (complex instruction set computer) design philosophy. CISC refers to CPUs with hundreds of instructions designed for every possible situation. To design CPUs with so many instructions consumed not only hundreds of thousands of transistors, but also made the design very complicated, time-consuming, and expensive. In the early 1980s, a new CPU design philosophy called *RISC* (reduced instruction set computer) was developed. The proponents of RISC argued that no one was using all the instructions etched into the brain of CISC-type CPUs. Why not streamline the instructions by simplifying and reducing them from hundreds to around 40 or so and use all the transistors that are saved to enhance the power of the CPU? Although the RISC concept had been explored by computer scientists at IBM as early as the 1970s, the first working single-chip RISC microprocessor was implemented by a group of researchers at the University of California at Berkeley in 1980. Today the RISC design philosophy is no longer an experiment limited to research laboratories. Since the late 1980s, many companies designing new CPUs (either single-chip or whole-board) have used the RISC philosophy. It appears that eventually the only CISC microprocessors remaining in use will be members of the 80x86 family (8086, 8088, 80286, 80386, 80486, 80586, etc.) and the 680x0 family (68000, 68010, 68020, 68030, 68040, 68050, etc.). The 80x86 will be kept alive by the huge base of IBM PC, PS, and compatible computers, and the Apple Macintosh is prolonging the life of 680x0 microprocessors.

Review Questions

1. What is a microprocessor?
2. Describe briefly how advances in technology have affected the size, cost, and availability of computer systems.
3. Explain the major difference between CISC and RISC computers.

SUMMARY

The binary number system represents all numbers with a combination of the two binary digits, 0 and 1. The use of binary systems is necessary in digital computers because only two states can be represented: on or off. Any binary number can be coded directly into its hexadecimal equivalent for the convenience of humans. Converting from binary/hex to decimal, and vice versa, is a straightforward process that becomes easy with practice. The ASCII code is a binary code used to represent alphanumeric data internally in the computer. It is frequently used in peripheral devices for input and/or output.

The major components of any computer system are the CPU, memory, and I/O devices. "Memory" refers to temporary or permanent storage of data. In most systems, memory can be accessed as bytes or words. The terms *kilobyte, megabyte, gigabyte,* and *terabyte* are used to refer to large numbers of bytes. There are two main types of memory in computer systems: RAM and ROM. RAM (random access memory) is used for temporary storage of programs and data. ROM (read-only memory) is used for permanent storage of programs and data that the computer system must have in order to function. All components of the computer system are under the control of the CPU. Peripheral devices such as I/O (input/output) devices allow the CPU to communicate with humans or other computer systems. There are three types of buses in computers: address, control, and data. Control buses are used by the CPU to direct other devices. The address bus is used by the CPU to locate a device or a memory location. Data buses are used to send information back and forth between the CPU and other devices.

As changes in technology were incorporated into the design of computers, their cost and size were reduced dramatically. The earliest computers were as large as an average home and were available only to a select group of scientists. The invention of transistors and subsequent advances in their design have made the computer commonly available. As the limits of hardware innovation have been approached, computer designers are looking at new design techniques, such as RISC architecture, to enhance computer performance.

PROBLEMS

1. Convert the following decimal numbers to binary.
    (a) 12   (b) 123   (c) 63   (d) 128   (e) 1000
2. Convert the following binary numbers to decimal.
    (a) 100100   (b) 1000001   (c) 11101   (d) 1010   (e) 00100010
3. Convert the values in Problem 2 to hexadecimal.
4. Convert the following hex numbers to binary and decimal.
    (a) 2B9H   (b) F44H   (c) 912H   (d) 2BH   (e) FFFFH
5. Convert the values in Problem 1 to hex.
6. Find the 2's complement of the following binary numbers.
    (a) 1001010   (b) 111001   (c) 10000010   (d) 111110001
7. Add the following hex values.
    (a) 2CH+3FH   (b) F34H+5D6H   (c) 20000H+12FFH   (d) FFFFH+2222H
8. Perform hex subtraction for the following.
    (a) 24FH–129H   (b) FE9H–5CCH   (c) 2FFFFH–FFFFFH   (d) 9FF25H–4DD99H
9. Show the ASCII codes for numbers 0, 1, 2, 3, ..., 9 in both hex and binary.
10. Show the ASCII code (in hex) for the following string:
    "U.S.A. is a country" CR,LF
    "in North America" CR,LF
    CR is carriage return
    LF is line feed

---

11. Answer the following:
    (a) How many nibbles are 16 bits?
    (b) How many bytes are 32 bits?
    (c) If a word is defined as 16 bits, how many words is a 64-bit data item?
    (d) What is the exact value (in decimal) of 1 meg?
    (e) How many K is 1 meg?
    (f) What is the exact value (in decimal) of giga?
    (g) How many K is 1 giga?
    (h) How many meg is 1 giga?
    (i) If a given computer has a total of 8 megabytes of memory, how many bytes (in decimal) is this? How many kilobytes is this?
12. A given mass storage device such as a hard disk can store 2 gigabytes of information. Assuming that each page of text has 25 rows and each row has 80 columns of ASCII characters (each character = 1 byte), approximately how many pages of information can this disk store?
13. In a given byte-addressable computer, memory locations 10000H to 9FFFFH are available for user programs. The first location is 10000H and the last location is 9FFFFH. Calculate the following:
    (a) The total number of bytes available (in decimal)
    (b) The total number of kilobytes (in decimal)
14. A given computer has a 32-bit data bus. What is the largest number that can be carried into the CPU at a time?
15. Below are listed several computers with their data bus widths. For each computer, list the maximum value that can be brought into the CPU at a time (in both hex and decimal).
    (a) Apple 2 with an 8-bit data bus
    (b) IBM PS/2 with a 16-bit data bus
    (c) IBM PS/2 model 80 with a 32-bit data bus
    (d) CRAY supercomputer with a 64-bit data bus
16. Find the total amount of memory, in the units requested, for each of the following CPUs, given the size of the address buses.
    (a) 16-bit address bus (in K)
    (b) 24-bit address bus (in meg)
    (c) 32-bit address bus (in megabytes and gigabytes)
    (d) 48-bit address bus (in megabytes, gigabytes and terabytes)
17. Regarding the data bus and address bus, which is unidirectional and which is bi-directional?
18. Which register of the CPU holds the address of the instruction to be fetched?
19. Which section of the CPU is responsible for performing addition?
20. Which type of CPU (CISC or RISC) has the greater variety of instructions?

ANSWERS TO REVIEW QUESTIONS

SECTION 0.1: NUMBERING AND CODING SYSTEMS

1. Computers use the binary system because each bit can have one of two voltage levels: on and off.
2. $34_{10} = 100010_2 = 22_{16}$
3. $110101_2 = 35_{16} = 53_{10}$
4. 1110001
5. 010100
6. 461
7. 275
8. 38 30 78 38 36 20 43 50 55 73

SECTION 0.2: INSIDE THE COMPUTER

1. 24,576
2. random access memory; it is used for temporary storage of programs that the CPU is running, such as the operating system, word processing programs, etc.
3. read-only memory; it is used for permanent programs such as those that control the keyboard, etc.
4. the contents of RAM are lost when the computer is powered off
5. the CPU, memory, and I/O devices
6. central processing unit; it can be considered the "brain" of the computer, it executes the programs and controls all other devices in the computer
7. the address bus carries the location (address) needed by the CPU; the data bus carries information in and out of the CPU; the control bus is used by the CPU to send signals controlling I/O devices
8. (a) bidirectional   (b) unidirectional
9. 64K, or 65,536 bytes
10. arithmetic/logic unit; it performs all arithmetic and logic operations
11. for temporary storage of information
12. it holds the address of the next instruction to be executed
13. it tells the CPU what steps to perform for each instruction

SECTION 0.3: BRIEF HISTORY OF THE CPU

1. a CPU on a single chip
2. The transition from vacuum tubes to transistors to ICs reduced the size and cost of computers and therefore made them more widely available.
3. CISC computers use many instructions whereas RISC computers use a small set of instructions.

CHAPTER 1

THE 80x86 MICROPROCESSOR

OBJECTIVES

Upon completion of this chapter, you will be able to:

» Describe the Intel family of microprocessors from the 8085 to the 80486 in terms of bus size, physical memory, and special features

» Explain the function of the EU (execution unit) and BIU (bus interface unit)

» Describe pipelining and how it enables the CPU to work faster

» List the registers of the 8086

» Code simple MOV and ADD instructions and describe the effect of these instructions on their operands

» State the purpose of the code segment, data segment, stack segment, and extra segment

» Explain the difference between a logical address and a physical address

» Describe the "little endian" storage convention of 80x86 microprocessors

» State the purpose of the stack

» Explain the function of PUSH and POP instructions

» List the bits of the flag register and briefly state the purpose of each bit

» Demonstrate the effect of ADD instructions on the flag register

» List the addressing modes of the 8086 and recognize examples of each mode

This chapter begins with a history of the evolution of Intel's family of microprocessors. The second section is an overview of the internal workings of 80x86 microprocessors. An introduction to 80x86 Assembly language programming is given in the third section. The fourth and fifth sections cover segments of Assembly language programs and how physical addresses are generated. Finally, the last section describes in detail the addressing modes of the 80x86.

SECTION 1.1: BRIEF HISTORY OF THE 80x86 FAMILY

In this section we trace the evolution of Intel's family of microprocessors from the late 1970s, when the personal computer had not yet found widespread acceptance, to the powerful microcomputers widely in use today.

Evolution from 8080/8085 to 8086

In 1978, Intel Corporation introduced a 16-bit microprocessor called the 8086. This processor was a major improvement over the previous generation 8080/8085 series Intel microprocessors in several ways. First, the 8086's capacity of 1 megabyte of memory exceeded the 8080/8085's capability of handling a maximum of 64K bytes of memory. Second, the 8080/8085 was an 8-bit system, meaning that the microprocessor could work on only 8 bits of data at a time. Data larger than 8 bits had to be broken into 8-bit pieces to be processed by the CPU. In contrast, the 8086 is a 16-bit microprocessor. Third, the 8086 was a pipelined processor, as opposed to the nonpipelined 8080/8085. In a system with pipelining, the data and address buses are busy transferring data while the CPU is processing information, thereby increasing the effective processing power of the microprocessor. Although pipelining was a common feature of mini- and mainframe computers, Intel was a pioneer in putting pipelining on a single-chip microprocessor. Pipelining is discussed further in Section 1.2.

Evolution from 8086 to 8088

The 8086 is a microprocessor with a 16-bit data bus internally and externally, meaning that all registers are 16 bits wide and there is a 16-bit data bus to transfer data in and out of the CPU. Although the introduction of the 8086 marked a great advancement over the previous generation of microprocessors, there was still some resistance in using the 16-bit external data bus since at that time all peripherals were designed around an 8-bit microprocessor. In addition, a printed circuit board with a 16-bit data bus was much more expensive. Therefore, Intel came out with the 8088 version. It is identical to the 8086 as far as programming is concerned, but externally it has an 8-bit data bus instead of a 16-bit bus. It has the same memory capacity, 1 megabyte.

Success of the 8088

In 1981, Intel's fortunes changed forever when IBM picked up the 8088 as their microprocessor of choice in designing the IBM PC. The 8088-based IBM PC was an enormous success, largely because IBM and Microsoft (the developer of the MS-DOS operating system) made it an *open system*, meaning that all documentation and specifications of the hardware and software of the PC were made public. This made it possible for many other vendors to clone the hardware successfully and thus spawned a major growth in both hardware and software designs based on the IBM PC. This is in contrast with the Apple computer, which was a closed system, blocking any attempt at cloning by other manufacturers, both domestically and overseas.

Other microprocessors: the 80286, 80386, and 80486

With a major victory behind Intel and a need from PC users for a more powerful microprocessor, Intel introduced the 80286 in 1982. Its features included 16-bit internal and external data buses; 24 address lines, which give 16 megabytes of memory (2^{24} = 16 megabytes); and most significantly, *virtual memory*. The

80286 can operate in one of two modes: real mode or protected mode. *Real mode* is simply a faster 8088/8086 with the same maximum of 1 megabyte of memory. *Protected mode* allows for 16M of memory but is also capable of protecting the operating system and programs from accidental or deliberate destruction by a user, a feature that is absent in the single-user 8088/8086. Virtual memory is a way of fooling the microprocessor into thinking that it has access to an almost unlimited amount of memory by swapping data between disk storage and RAM. IBM picked up the 80286 for the design of the IBM PC AT, and the clone makers followed IBM's lead.

With users demanding even more powerful systems, in 1985 Intel introduced the 80386 (sometimes called 80386DX), internally and externally a 32-bit microprocessor with a 32-bit address bus. It is capable of handling physical memory of up to 4 gigabytes (2^{32}). Virtual memory was increased to 64 terabytes (2^{46}). All microprocessors discussed so far were general-purpose microprocessors and could not handle mathematical calculations rapidly. For this reason, Intel introduced numeric data processing chips, called math coprocessors, such as the 8087, 80287, and 80387. Later Intel introduced the 386SX, which is internally identical to the 80386 but has a 16-bit external data bus and a 24-bit address bus which gives a capacity of 16 megabytes (2^{24}) of memory. This makes the 386SX system much cheaper. With the introduction of the 80486 in 1989, Intel put a greatly enhanced version of the 80386 and the math coprocessor on a single chip plus additional features such as *cache memory*. Cache memory is static RAM with a very fast access time. Table 1-1 summarizes the evolution of Intel's microprocessors. It must be noted that all programs written for the 8086/88 will run on 286, 386, and 486 computers. The advances made in the Pentium and Pentium Pro are summarized in Chapter 9.

Table 1-1: Evolution of Intel's Microprocessors

Product	8080	8085	8086	8088	80286	80386	80486
Year introduced	1974	1976	1978	1979	1982	1985	1989
Clock rate (MHz)	2 - 3	3 - 8	5 - 10	5 - 8	6 - 16	16 - 33	25 - 50
No. transistors	4500	6500	29,000	29,000	130,000	275,000	1.2 million
Physical memory	64K	64K	1M	1M	16M	4G	4G
Internal data bus	8	8	16	16	16	32	32
External data bus	8	8	16	8	16	32	32
Address bus	16	16	20	20	24	32	32
Data type (bits)	8	8	8, 16	8, 16	8, 16	8, 16, 32	8, 16, 32

Notes:
1. The 80386SX architecture is the same as the 80386 except that the external data bus is 16 bits in the SX as opposed to 32 bits, and the address bus is 24 bits instead of 32; therefore, physical memory is 16MB.
2. Clock rates range from the rates when the product was introduced to current rates; some rates have risen during this time.

Review Questions

1. Name three features of the 8086 that were improvements over the 8080/8085.
2. What is the major difference between 8088 and 8086 microprocessors?
3. Give the size of the address bus and physical memory capacity of the following:
   (a) 8086   (b) 80286   (c) 80386
4. The 80286 is a _____-bit microprocessor, whereas the 80386 is a _____-bit microprocessor.
5. State the major difference between the 80386 and the 80386SX.
6. List additional features introduced with the 80286 that were not present in the 8086.
7. List additional features of the 80486 that were not present in the 80386.

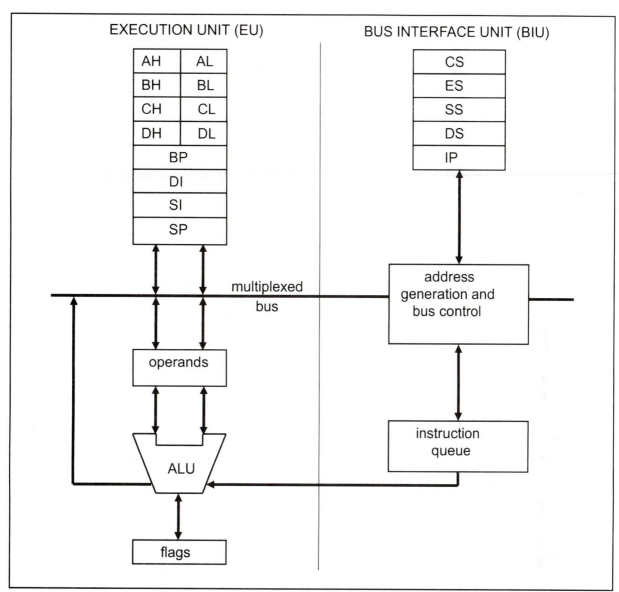

EXECUTION UNIT (EU)

AH	AL
BH	BL
CH	CL
DH	DL
BP	
DI	
SI	
SP	

multiplexed bus

operands

ALU

flags

BUS INTERFACE UNIT (BIU)

CS
ES
SS
DS
IP

address generation and bus control

instruction queue

Figure 1-1. Internal Block Diagram of the 8088/86 CPU
(Reprinted by permission of Intel Corporation, Copyright Intel Corp. 1989)

SECTION 1.2: INSIDE THE 8088/8086

In this section we explore concepts important to the internal operation of the 8088/86, such as pipelining and registers. See the block diagram in Figure 1-1.

Pipelining

There are two ways to make the CPU process information faster: increase the working frequency or change the internal architecture of the CPU. The first option is technology dependent, meaning that the designer must use whatever technology is available at the time, with consideration for cost. The technology and materials used in making ICs (*integrated circuits*) determine the working frequency, power consumption, and the number of transistors packed into a single-chip microprocessor. A detailed discussion of IC technology is beyond the scope of this book. It is sufficient for the purpose at hand to say that designers can make the CPU work faster by increasing the frequency under which it runs if technology and cost allow. The second option for improving the processing power of the CPU has to do with the internal working of the CPU. In the 8085 microprocessor, the CPU could either fetch or execute at a given time. In other words, the CPU had to fetch an instruction from memory, then execute it and then fetch again, execute it, and so on. The idea of *pipelining* in its simplest form is to allow the CPU to fetch and execute at the same time as shown in Figure 1-2. It is important to point out that Figure 1-2 is not meant to imply that the amount of time for fetch and execute are equal.

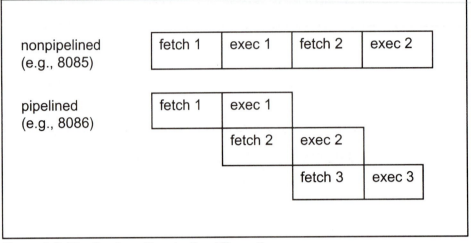

Figure 1-2. Pipelined vs. Nonpipelined Execution

Intel implemented the concept of pipelining in the 8088/86 by splitting the internal structure of the microprocessor into two sections: the *execution unit* (EU) and the *bus interface unit* (BIU). These two sections work simultaneously. The BIU accesses memory and peripherals while the EU executes instructions previously fetched. This works only if the BIU keeps ahead of the EU; thus the BIU of the 8088/86 has a buffer, or queue (see Figure 1-1). The buffer is 4 bytes long in the 8088 and 6 bytes in the 8086. If any instruction takes too long to execute, the queue is filled to its maximum capacity and the buses will sit idle. The BIU fetches a new instruction whenever the queue has room for 2 bytes in the 6-byte 8086 queue, and for 1 byte in the 4-byte 8088 queue. In some circumstances, the microprocessor must flush out the queue. For example, when a jump instruction is executed, the BIU starts to fetch information from the new location in memory and information in the queue that was fetched previously is discarded. In this situation the EU must wait until the BIU fetches the new instruction. This is referred to in computer science terminology as a *branch penalty*. In a pipelined CPU, this means that too much jumping around reduces the efficiency of a program. Pipelining in the 8088/86 has two stages: fetch and execute, but in more powerful computers pipelining can have many stages. The concept of pipelining combined with an increased number of data bus pins has, in recent years, led to the design of very powerful microprocessors.

Registers

In the CPU, registers are used to store information temporarily. That information could be one or two bytes of data to be processed or the address of data. The registers of the 8088/86 fall into the six categories outlined in Table 1-2. The general-purpose registers in 8088/86 microprocessors can be accessed as either

AX 16-bit register	
AH 8-bit reg.	AL 8-bit reg.

16-bit or 8-bit registers. All other registers can be accessed only as the full 16 bits. In the 8088/86, data types are either 8 or 16 bits. To access 12-bit data, for example, a 16-bit register must be used with the highest 4 bits set to 0. The bits of a register are numbered in descending order, as shown below.

8-bit register:

| D7 | D6 | D5 | D4 | D3 | D2 | D1 | D0 |

16-bit register:

| D15 | D14 | D13 | D12 | D11 | D10 | D9 | D8 | D7 | D6 | D5 | D4 | D3 | D2 | D1 | D0 |

Different registers in the 8088/86 are used for different functions, and since some instructions use only specific registers to perform their tasks, the use of registers will be described in the context of instructions and their application in a given program. The first letter of each general register indicates its use. AX is used for the accumulator, BX as a base addressing register, CX is used as a counter in loop operations, and DX is used to point to data in I/O operations.

Table 1-2: Registers of the 8086/286 by Category

Category	Bits	Register Names
General	16	AX, BX, CX, DX
	8	AH, AL, BH, BL, CH, CL, DH, DL
Pointer	16	SP (stack pointer), BP (base pointer)
Index	16	SI (source index), DI (destination index)
Segment	16	CS (code segment), DS (data segment), SS (stack segment), ES (extra segment)
Instruction	16	IP (instruction pointer)
Flag	16	FR (flag register)

Note:
The general registers can be accessed as the full 16 bits (such as AX), or as the high byte only (AH) or low byte only (AL).

Review Questions

1. Explain the functions of the EU and the BIU.
2. What is pipelining, and how does it make the CPU execute faster?
3. Registers of the 8086 are either _____ bits or _____ bits in length.
4. List the 16-bit registers of the 8086.

SECTION 1.3: INTRODUCTION TO ASSEMBLY PROGRAMMING

While the CPU can work only in binary, it can do so at very high speeds. However, it is quite tedious and slow for humans to deal with 0s and 1s in order to program the computer. A program that consists of 0s and 1s is called *machine language,* and in the early days of the computer, programmers actually coded programs in machine language. Although the hexadecimal system was used as a more efficient way to represent binary numbers, the process of working in machine code was still cumbersome for humans. Eventually, *Assembly languages* were developed, which provided mnemonics for the machine code instructions, plus other features that made programming faster and less prone to error. The term *mnemonic* is frequently used in computer science and engineering literature to refer to codes and abbreviations that are relatively easy to remember. Assembly language programs must be translated into machine code by a program called an *assembler.* Assembly language is referred to as a *low-level language* because it deals directly with the internal structure of the CPU. To program in Assembly language, the programmer must know the number of registers and their size, as well as other details of the CPU.

Today, one can use many different programming languages, such as Pascal, BASIC, C, and numerous others. These languages are called *high-level languages* because the programmer does not have to be concerned with the internal details of the CPU. Whereas an assembler is used to translate an Assembly language program into machine code (sometimes called *object code*), high-level languages are translated into machine code by a program called a *compiler*. For instance, to write a program in C, one must use a C compiler to translate the program into machine language.

There are numerous assemblers available for translating 80x86 Assembly language programs into machine code. One of the most commonly used assemblers, MASM by Microsoft, is introduced in Chapter 2. The present chapter is designed to correspond to Appendix A: DEBUG programming. The program in this chapter can be entered and run with the use of the DEBUG program. If you are not familiar with DEBUG, refer to Appendix A for a tutorial introduction. The DEBUG utility is provided with the DOS operating system and therefore is widely accessible.

Assembly language programming

An Assembly language program consists of, among other things, a series of lines of Assembly language instructions. An Assembly language instruction consists of a mnemonic, optionally followed by one or two operands. The operands are the data items being manipulated, and the mnemonics are the commands to the CPU, telling it what to do with those items. We introduce Assembly language programming with two widely used instructions: the move and add instructions.

MOV instruction

Simply stated, the MOV instruction copies data from one location to another. It has the following format:

```
MOV    destination,source      ;copy source operand to destination
```

This instruction tells the CPU to move (in reality, copy) the source operand to the destination operand. For example, the instruction "MOV DX,CX" copies the contents of register CX to register DX. After this instruction is executed, register DX will have the same value as register CX. The MOV instruction does not affect the source operand. The following program first loads CL with value 55H, then moves this value around to various registers inside the CPU.

```
MOV    CL,55H      ;move 55H into register CL
MOV    DL,CL       ;copy the contents of CL into DL (now DL=CL=55H)
MOV    AH,DL       ;copy the contents of DL into AH (now AH=DL=55H)
MOV    AL,AH       ;copy the contents of AH into AL (now AL=AH=55H)
MOV    BH,CL       ;copy the contents of CL into BH (now BH=CL=55H)
MOV    CH,BH       ;copy the contents of BH into CH (now CH=BH=55H)
```

The use of 16-bit registers is demonstrated below.

```
MOV    CX,468FH    ;move 468FH into CX (now CH=46,CL=8F)
MOV    AX,CX       ;copy contents of CX to AX (now AX=CX=468FH)
MOV    DX,AX       ;copy contents of AX to DX (now DX=AX=468FH)
MOV    BX,DX       ;copy contents of DX to BX (now BX=DX=468FH)
MOV    DI,BX       ;now DI=BX=468FH
MOV    SI,DI       ;now SI=DI=468FH
MOV    DS,SI       ;now DS=SI=468FH
MOV    BP,DI       ;now BP=DI=468FH
```

In the 8086 CPU, data can be moved among all the registers shown in Table 1-2 (except the flag register) as long as the source and destination registers match in size. Code such as "MOV AL,DX" will cause an error, since one cannot move the contents of a 16-bit register into an 8-bit register. The exception of the flag register means that there is no such instruction as "MOV FR,AX". Loading the flag register is done through other means, discussed in later chapters.

If data can be moved among all registers including the segment registers, can data be moved directly into all registers? The answer is no. Data can be moved directly into nonsegment registers only, using the MOV instruction.   For example, look at the following instructions to see which are legal and which are illegal.

```
MOV    AX,58FCH      ;move 58FCH into AX    (LEGAL)
MOV    DX,6678H      ;move 6678H into DX    (LEGAL)
MOV    SI,924BH      ;move 924B  into SI    (LEGAL)
MOV    BP,2459H      ;move 2459H into BP    (LEGAL)
MOV    DS,2341H      ;move 2341H into DS    (ILLEGAL)
MOV    CX,8876H      ;move 8876H into CX    (LEGAL)
MOV    CS,3F47H      ;move 3F47H into CS    (ILLEGAL)
MOV    BH,99H        ;move 99H into BH      (LEGAL)
```

From the discussion above, note the following three points:

1. Values cannot be loaded directly into any segment register (CS, DS, ES, or SS). To load a value into a segment register, first load it to a nonsegment register and then move it to the segment register, as shown next.

```
MOV    AX,2345H      ;load 2345H into AX
MOV    DS,AX         ;then load the value of AX into DS

MOV    DI,1400H      ;load 1400H into DI
MOV    ES,DI         ;then move it into ES, now ES=DI=1400
```

2. If a value less than FFH is moved into a 16-bit register, the rest of the bits are assumed to be all zeros. For example, in "MOV BX,5" the result will be BX = 0005; that is, BH = 00 and BL = 05.

3. Moving a value that is too large into a register will cause an error.

```
MOV    BL,7F2H       ;ILLEGAL: 7F2H is larger than 8 bits
MOV    AX,2FE456H    ;ILLEGAL: the value is larger than AX
```

ADD instruction

The ADD instruction has the following format:

```
ADD    destination,source      ;ADD the source operand to the destination
```

The ADD instruction tells the CPU to add the source and the destination operands and put the result in the destination. To add two numbers such as 25H and 34H, each can be moved to a register and then added together:

```
MOV    AL,25H        ;move 25 into AL
MOV    BL,34H        ;move 34 into BL
ADD    AL,BL         ;AL = AL + BL
```

Executing the program above results in AL = 59H (25H + 34H = 59H) and BL = 34H. Notice that the contents of BL do not change. The program above can be written in many ways, depending on the registers used. Another way might be:

```
MOV     DH,25H          ;move 25 into DH
MOV     CL,34H          ;move 34 into CL
ADD     DH,CL           ;add CL to DH: DH = DH + CL
```

The program above results in DH = 59H and CL = 34H. There are always many ways to write the same program. One question that might come to mind after looking at the program above is whether it is necessary to move both data items into registers before adding them together. The answer is no, it is not necessary. Look at the following variation of the same program:

```
MOV     DH,25H          ;load one operand into DH
ADD     DH,34H          ;add the second operand to DH
```

In the case above, while one register contained one value, the second value followed the instruction as an operand. This is called an *immediate operand*. The examples shown so far for the ADD and MOV instructions show that the source operand can be either a register or immediate data. In the examples above, the destination operand has always been a register. The format for Assembly language instructions, descriptions of their use, and a listing of legal operand types are provided in Appendix B.

The largest number that an 8-bit register can hold is FFH. To use numbers larger than FFH (255 decimal), 16-bit registers such as AX, BX, CX, or DX must be used. For example, to add two numbers such as 34EH and 6A5H, the following program can be used:

```
MOV     AX,34EH         ;move 34EH into AX
MOV     DX,6A5H         ;move 6A5H into DX
ADD     DX,AX           ;add AX to DX: DX = DX + AX
```

Running the program above gives DX = 9F3H (34E + 6A5 = 9F3) and AX = 34E. Again, any 16-bit nonsegment registers could have been used to perform the action above:

```
MOV     CX,34EH         ;load 34EH into CX
ADD     CX,6A5H         ;add 6A5H to CX (now CX=9F3H)
```

The general-purpose registers are typically used in arithmetic operations. Register AX is sometimes referred to as the accumulator.

Review Questions

1. Write the Assembly language instruction to move value 1234H into register BX.
2. Write the Assembly language instructions to add the values 16H and ABH. Place the result in register AX.
3. No value can be moved directly into which registers?
4. What is the largest hex value that can be moved into a 16-bit register? Into an 8-bit register? What are the decimal equivalents of these hex values?

SECTION 1.4: INTRODUCTION TO PROGRAM SEGMENTS

A typical Assembly language program consists of at least three segments: a code segment, a data segment, and a stack segment. The *code segment* contains the Assembly language instructions that perform the tasks that the program was designed to accomplish. The *data segment* is used to store information (data) that needs to to be processed by the instructions in the code segment. The *stack* is used to store information temporarily. In this section we describe the code and data segments of a program in the context of some examples and discuss the way data is stored in memory. The stack segment is covered in Section 1.5.

Origin and definition of the segment

A segment is an area of memory that includes up to 64K bytes and begins on an address evenly divisible by 16 (such an address ends in 0H). The segment size of 64K bytes came about because the 8085 microprocessor could address a maximum of 64K bytes of physical memory since it had only 16 pins for the address lines ($2^{16} = 64K$). This limitation was carried into the design of the 8088/86 to ensure compatibility. Whereas in the 8085 there was only 64K bytes of memory for all code, data, and stack information, in the 8088/86 there can be up to 64K bytes of memory assigned to each category. Within an Assembly language program, these categories are called the code segment, data segment, and stack segment. For this reason, the 8088/86 can only handle a maximum of 64K bytes of code and 64K bytes of data and 64K bytes of stack at any given time, although it has a range of 1 megabyte of memory because of its 20 address pins ($2^{20} = 1$ megabyte). How to move this window of 64K bytes to cover all 1 megabyte of memory is discussed below, after we discuss logical address and physical address.

Logical address and physical address

In Intel literature concerning the 8086, there are three types of addresses mentioned frequently: the physical address, the offset address, and the logical address. The *physical address* is the 20-bit address that is actually put on the address pins of the 8086 microprocessor and decoded by the memory interfacing circuitry. This address can have a range of 00000H to FFFFFH for the 8086 and real-mode 286, 386, and 486 CPUs. This is an actual physical location in RAM or ROM within the 1 megabyte memory range. The *offset address* is a location within a 64K-byte segment range. Therefore, an offset address can range from 0000H to FFFFH. The *logical address* consists of a segment value and an offset address. The differences among these addresses and the process of converting from one to another is best understood in the context of some examples, as shown next.

Code segment

To execute a program, the 8086 fetches the instructions (opcodes and operands) from the code segment. The logical address of an instruction always

consists of a CS (code segment) and an IP (instruction pointer), shown in CS:IP format. The physical address for the location of the instruction is generated by shifting the CS left one hex digit and then adding it to the IP. IP contains the offset address. The resulting 20-bit address is called the physical address since it is put on the external physical address bus pins to be decoded by the memory decoding circuitry. To clarify this important concept, assume values in CS and IP as shown in the diagram. The offset address is contained in IP; in this case it is 95F3H. The logical address is CS:IP, or 2500:95F3H. The physical address will be 25000 + 95F3 = 2E5F3H. The physical address of an instruction can be calculated as follows:

1. Start with CS.

2. Shift left CS.

3. Add IP.

The microprocessor will retrieve the instruction from memory locations starting at 2E5F3. Since IP can have a minimum value of 0000H and a maximum of FFFFH, the logical address range in this example is 2500:0000 to 2500:FFFF. This means that the lowest memory location of the code segment above will be 25000H (25000 + 0000) and the highest memory location will be 34FFFH (25000 + FFFF). What happens if the desired instructions are located beyond these two limits? The answer is that the value of CS must be changed to access those instructions. See Example 1-1.

Example 1-1

If CS = 24F6H and IP = 634AH, show:
(a) The logical address
(b) The offset address
 and calculate:
(c) The physical address
(d) The lower range
(e) The upper range of the code segment

Solution:

(a) 24F6:634A   (b) 634A
(c) 2B2AA (24F60 + 634A)   (d) 24F60 (24F60 + 0000)
(e) 34F5F (24F60 + FFFF)

Logical address vs. physical address in the code segment

In the code segment, CS and IP hold the logical address of the instructions to be executed. The following Assembly language instructions have been assembled (translated into machine code) and stored in memory. The three columns show the logical address of CS:IP, the machine code stored at that address and the corresponding Assembly language code. This information can easily be generated by the DEBUG program using the Unassemble command.

Logical address CS:IP	Machine language opcode and operand	Assembly language mnemonics and operand
1132:0100	B057	MOV AL,57
1132:0102	B686	MOV DH,86
1132:0104	B272	MOV DL,72
1132:0106	89D1	MOV CX,DX
1132:0108	88C7	MOV BH,AL
1132:010A	B39F	MOV BL,9F
1132:010C	B420	MOV AH,20
1132:010E	01D0	ADD AX,DX
1132:0110	01D9	ADD CX,BX
1132:0112	05351F	ADD AX,1F35

The program above shows that the byte at address 1132:0100 contains B0, which is the opcode for moving a value into register AL, and address 1132:0101 contains the operand (in this case 57) to be moved to AL. Therefore, the instruction "MOV AL,57" has a machine code of B057, where B0 is the opcode and 57 is the operand. Similarly, the machine code B686 is located in memory locations 1132:0102 and 1132:0103 and represents the opcode and the operand for the instruction "MOV DH,86". The physical address is an actual location within RAM (or even ROM). The following are the physical addresses and the contents of each location for the program above. Remember that it is the physical address that is put on the address bus by the 8086 CPU to be decoded by the memory circuitry:

Logical address	Physical address	Machine code contents
1132:0100	11420	B0
1132:0101	11421	57
1132:0102	11422	B6
1132:0103	11423	86
1132:0104	11424	B2
1132:0105	11425	72
1132:0106	11426	89
1132:0107	11427	D1
1132:0108	11428	88
1132:0109	11429	C7
1132:010A	1142A	B3
1132:010B	1142B	9F
1132:010C	1142C	B4
1132:010D	1142D	20
1132:010E	1142E	01
1132:010F	1142F	D0
1132:0110	11430	01
1132:0111	11431	D9
1132:0112	11432	05
1132:0113	11433	35
1132:0114	11434	1F

Data segment

Assume that a program is being written to add 5 bytes of data, such as 25H, 12H, 15H, 1FH, and 2BH, where each byte represents a person's daily overtime pay. One way to add them is as follows:

```
MOV   AL,00H      ;initialize AL
ADD   AL,25H      ;add 25H to AL
ADD   AL,12H      ;add 12H to AL
ADD   AL,15H      ;add 15H to AL
ADD   AL,1FH      ;add 1FH to AL
ADD   AL,2BH      ;add 2BH to AL
```

In the program above, the data and code are mixed together in the instructions. The problem with writing the program this way is that if the data changes, the code must be searched for every place the data is included, and the data retyped. For this reason, the idea arose to set aside an area of memory strictly for data. In 80x86 microprocessors, the area of memory set aside for data is called the data segment. Just as the code segment is associated with CS and IP as its segment register and offset, the data segment uses register DS and an offset value.

The following demonstrates how data can be stored in the data segment and the program rewritten so that it can be used for any set of data. Assume that the offset for the data segment begins at 200H. The data is placed in memory locations:

```
DS:0200 = 25
DS:0201 = 12
DS:0202 = 15
DS:0203 = 1F
DS:0204 = 2B
```

and the program can be rewritten as follows:

```
MOV   AL,0          ;clear AL
ADD   AL,[0200]     ;add the contents of DS:200 to AL
ADD   AL,[0201]     ;add the contents of DS:201 to AL
ADD   AL,[0202]     ;add the contents of DS:202 to AL
ADD   AL,[0203]     ;add the contents of DS:203 to AL
ADD   AL,[0204]     ;add the contents of DS:204 to AL
```

SECTION 1.4: INTRODUCTION TO PROGRAM SEGMENTS

Notice that the offset address is enclosed in brackets. The brackets indicate that the operand represents the address of the data and not the data itself. If the brackets were not included, as in "MOV AL,0200", the CPU would attempt to move 200 into AL instead of the contents of offset address 200. Keep in mind that there is one important difference in the format of code for MASM and DEBUG in that DEBUG assumes that all numbers are in hex (no "H" suffix is required), whereas MASM assumes that they are in decimal and the "H" must be included for hex data.

This program will run with any set of data. Changing the data has no effect on the code. Although this program is an improvement over the preceding one, it can be improved even further. If the data had to be stored at a different offset address, say 450H, the program would have to be rewritten. One way to solve this problem would be to use a register to hold the offset address, and before each ADD, to increment the register to access the next byte. Next a decision must be made as to which register to use. The 8086/88 allows only the use of registers BX, SI, and DI as offset registers for the data segment. In other words, while CS uses only the IP register as an offset, DS uses only BX, DI, and SI to hold the offset address of the data. The term *pointer* is often used for a register holding an offset address. In the following example, BX is used as a pointer:

```
MOV    AL,0          ;initialize AL
MOV    BX,0200H      ;BX points to the offset addr of first byte
ADD    AL,[BX]       ;add the first byte to AL
INC    BX            ;increment BX to point to the next byte
ADD    AL,[BX]       ;add the next byte to AL
INC    BX            ;increment the pointer
ADD    AL,[BX]       ;add the next byte to AL
INC    BX            ;increment the pointer
ADD    AL,[BX]       ;add the last byte to AL
```

The "INC" instruction adds 1 to (increments) its operand. "INC BX" achieves the same result as "ADD BX,1". For the program above, if the offset address where data is located is changed, only one instruction will need to be modified and the rest of the program will be unaffected. Examining the program above shows that there is a pattern of two instructions being repeated. This leads to the idea of using a loop to repeat certain instructions. Implementing a loop requires familiarity with the flag register, discussed later in this chapter.

Logical address and physical address in the data segment

The physical address for data is calculated using the same rules as for the code segment. That is, the physical address of data is calculated by shifting DS left one hex digit and adding the offset value, as shown in Examples 1-2, 1-3, and 1-4.

Example 1-2

Assume that DS is 5000 and the offset is 1950. Calculate the physical address of the byte.

Solution:

| DS | : | offset |
| 5 | 0 | 0 | 0 | : | 1 | 9 | 5 | 0 |

The physical address will be 50000 + 1950 = 51950.

1. Start with DS.

| 5 | 0 | 0 | 0 |

2. Shift DS left.

| 5 | 0 | 0 | 0 | 0 |

3. Add the offset.

| 5 | 1 | 9 | 5 | 0 |

Example 1-3

If DS = 7FA2H and the offset is 438EH,

(a) Calculate the physical address.   (b) Calculate the lower range.

(c) Calculate the upper range of the data segment.   (d) Show the logical address.

Solution:

(a) 83DAE (7FA20 + 438E)   (b) 7FA20 (7FA20 + 0000)

(c) 8FA1F (7FA20 + FFFF)   (d) 7FA2:438E

Example 1-4

Assume that the DS register is 578C. To access a given byte of data at physical memory location 67F66, does the data segment cover the range where the data is located? If not, what changes need to be made?

Solution:

No, since the range is 578C0 to 678BF, location 67F66 is not included in this range. To access that byte, DS must be changed so that its range will include that byte.

Little endian convention

Previous examples used 8-bit or 1-byte data. In this case the bytes are stored one after another in memory. What happens when 16-bit data is used? For example:

```
MOV    AX,35F3H      ;load 35F3H into AX
MOV    [1500],AX     ;copy the contents of AX to offset 1500H
```

In cases like this, the low byte goes to the low memory location and the high byte goes to the high memory address. In the example above, memory location DS:1500 contains F3H and memory location DS:1501 contains 35H.

DS:1500 = F3   DS:1501 = 35

This convention is called little endian versus big endian. The origin of the terms *big endian* and *little endian* is from a *Gulliver's Travels* story about how an egg should be opened: from the little end or the big end. In the big endian method, the high byte goes to the low address, whereas in the little endian method, the high byte goes to the high address and the low byte to the low address. See Example 1-5. All Intel microprocessors and many minicomputers, notably the Digital VAX, use the little endian convention. Motorola microprocessors (used in the Macintosh),

Example 1-5

Assume memory locations with the following contents: DS:6826 = 48 and DS:6827 = 22.
Show the contents of register BX in the instruction "MOV BX,[6826]".

Solution:

According to the little endian convention used in all 80x86 microprocessors, register BL should contain the value from the low offset address 6826 and register BH the value from offset address 6827, giving BL = 48H and BH = 22H.

DS:6826 = 48

DS:6827 = 22

BH	BL
22	48

---

along with some mainframes, use big endian. This difference might seem as trivial as whether to break an egg from the big end or little end, but it is a nuisance in converting software from one camp to be run on a computer of the other camp.

Extra segment (ES)

ES is a segment register used as an extra data segment. Although in many normal programs this segment is not used, its use is absolutely essential for string operations and is discussed in detail in Chapter 6.

Memory map of the IBM PC

For a program to be executed on the PC, DOS must first load it into RAM. Where in RAM will it be loaded? To answer that question, we must first explain some very important concepts concerning memory in the PC. The 20-bit address of the 8088/86 allows a total of 1 megabyte (1024K bytes) of memory space with the address range 00000 - FFFFF. During the design phase of the first IBM PC, engineers had to decide on the allocation of the 1-megabyte memory space to various sections of the PC. This memory allocation is called a *memory map*. The memory map of the IBM PC is shown in Figure 1-3. Of this 1 megabyte, 640K bytes from addresses 00000 - 9FFFFH were set aside for RAM. The 128K bytes from A0000H to BFFFFH were allocated for video memory. The remaining 256K bytes from C0000H to FFFFFH were set aside for ROM.

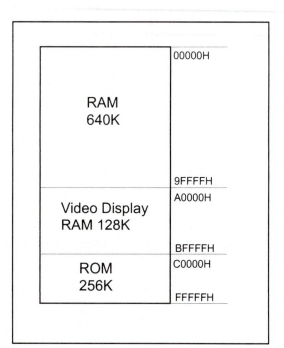

Figure 1-3. Memory Allocation in the PC

More about RAM

In the early 1980s, most PCs came with only 64K to 256K bytes of RAM memory, which was considered more than adequate at the time. Users had to buy memory expansion boards to expand memory up to 640K if they needed additional memory. The need for expansion depends on the DOS version being used and the memory needs of the application software being run. The DOS operating system first allocates the available RAM on the PC for its own use and then lets the rest be used for applications such as word processors. The complicated task of managing RAM memory is left to DOS since the amount of memory used by DOS varies among its various versions and since different computers have different amounts of RAM, plus the fact that the memory needs of application packages vary. For this reason we do not assign any values for the CS, DS, and SS registers since such an assignment means specifying an exact physical address in the range 00000 - 9FFFFH, and this is beyond the knowledge of the user. Another reason is that assigning a physical address might work on a given PC but it might not work on a PC with a different DOS version and RAM size. In other words, the program would not be portable to another PC. Therefore, memory management is one of the most important functions of the DOS operating system and should be left to DOS. This is very important to remember because in many examples in this book we have values for the segment registers CS, DS, and SS that will be different from the values that readers will get on their PCs. Therefore, do not try to assign the value to the segment registers to comply with the values in this book.

CHAPTER 1: THE 80x86 MICROPROCESSOR

Video RAM

From A0000H to BFFFFH is set aside for video. The amount used and the location varies depending on the video board installed on the PC. Table E-2 of Appendix E lists the starting addresses for video boards.

More about ROM

From C0000H to FFFFFH is set aside for ROM. Not all the memory space in this range is used by the PC's ROM. Of this 256K bytes, only the 64K bytes from location F0000H - FFFFFH are used by BIOS (basic input/output system) ROM. Some of the remaining space is used by various adapter cards (such as cards for hard disks), and the rest is free. In recent years, newer versions of DOS have gained some very powerful memory management capabilities and can put to good use all the unused memory space beyond 640. The 640K-byte memory space from 00000 to 9FFFFH is referred to as *conventional memory*, while the 384K bytes from A0000H to FFFFFH are called the UMB (*upper memory block)* in DOS 5 literature. A complete discussion of the various memory terminology and configurations such as expanded and extended memory appears in Chapter 25.

Function of BIOS ROM

Since the CPU can only execute programs that are stored in memory, there must be some permanent (nonvolatile) memory to hold the programs telling the CPU what to do when the power is turned on. This collection of programs held by ROM is referred to as BIOS in the PC literature. BIOS, which stands for *basic input-output system*, contains programs to test RAM and other components connected to the CPU. It also contains programs that allow DOS to communicate with peripheral devices such as the keyboard, video, printer, and disk. It is the function of BIOS to test all the devices connected to the PC when the computer is turned on and to report any errors. For example, if the keyboard is disconnected from the PC before the computer is turned on, BIOS will report an error on the screen, indicating that condition. It is only after testing and setting up the peripherals that BIOS will load DOS from disk into RAM and hand over control of the PC to DOS. Although there are occasions when either DOS or applications programs need to use programs in BIOS ROM (as will be seen in Chapter 4), DOS always controls the PC once it is loaded.

Review Questions

1. A segment is an area of memory that includes up to ____ bytes.
2. How large is a segment in the 8086? Can the physical address 346E0 be the starting address for a segment? Why or why not?
3. State the difference between the physical and logical addresses.
4. A physical address is a ____-bit address; an offset address is a ____-bit address.
5. Which register is used as the offset register with segment register CS?
6. If BX = 1234H and the instruction "MOV [2400],BX" were executed, what would be the contents of memory locations at offsets 2400 and 2401?

SECTION 1.5: MORE ABOUT SEGMENTS IN THE 80x86

In this section we examine the concept of the stack, its use in 80x86 microprocessors, and its implementation in the stack segment. Then more advanced concepts relating to segments are discussed, such as overlapping segments.

What is a stack, and why is it needed?

The *stack* is a section of read/write memory (RAM) used by the CPU to store information temporarily. The CPU needs this storage area since there are only

---

a limited number of registers. There must be some place for the CPU to store information safely and temporarily. Now one might ask why not design a CPU with more registers? The reason is that in the design of the CPU, every transistor is precious and not enough of them are available to build hundreds of registers. In addition, how many registers should a CPU have to satisfy every possible program and application? All applications and programming techniques are not the same. In a similar manner, it would be too costly in terms of real estate and construction costs to build a 50-room house to hold everything one might possibly buy throughout his or her lifetime. Instead, one builds or rents a shed for storage.

Having looked at the advantages of having a stack, what are the disadvantages? The main disadvantage of the stack is its access time. Since the stack is in RAM, it takes much longer to access compared to the access time of registers. After all, the registers are inside the CPU and RAM is outside. This is the reason that some very powerful (and consequently, expensive) computers do not have a stack; the CPU has a large number of registers to work with.

How stacks are accessed

If the stack is a section of RAM, there must be registers inside the CPU to point to it. The two main registers used to access the stack are the SS (stack segment) register and the SP (stack pointer) register. These registers must be loaded before any instructions accessing the stack are used. Every register inside the 80x86 (except segment registers and SP) can be stored in the stack and brought back into the CPU from the stack memory. The storing of a CPU register in the stack is called a *push,* and loading the contents of the stack into the CPU register is called a *pop.* In other words, a register is pushed onto the stack to store it and popped off the stack to retrieve it. The job of the SP is very critical when push and pop are performed. In the 80x86, the stack pointer register (SP) points at the current memory location used for the top of the stack and as data is pushed onto the stack it is decremented. It is incremented as data is popped off the stack into the CPU. When an instruction pushes or pops a general-purpose register, it must be the entire 16-bit register. In other words, one must code "PUSH AX"; there are no instructions such as "PUSH AL" or "PUSH AH". The reason that the SP is decremented after the push is to make sure that the stack is growing downward from upper addresses to lower addresses. This is the opposite of the IP (instruction pointer). As was seen in the preceding section, the IP points to the next instruction to be executed and is incremented as each instruction is executed. To ensure that the code section and stack section of the program never write over each other, they are located at opposite ends of the RAM memory set aside for the program and they grow toward each other but must not meet. If they meet, the program will crash. To see how the stack grows, look at the following examples.

Pushing onto the stack

Notice in Example 1-6 that as each PUSH is executed, the contents of the register are saved on the stack and SP is decremented by 2. For every byte of data saved on the stack, SP is decremented once, and since push is saving the contents of a 16-bit register, it is decremented twice. Notice also how the data is stored on the stack. In the 80x86, the lower byte is always stored in the memory location with the lower address. That is the reason that 24H, the contents of AH, is saved in memory location with address 1235 and AL in location 1234.

Popping the stack

Popping the contents of the stack back into the 80x86 CPU is the opposite process of pushing. With every pop, the top 2 bytes of the stack are copied to the register specified by the instruction and the stack pointer is incremented twice. Although the data actually remains in memory, it is not accessible since the stack pointer is beyond that point. Example 1-7 demonstrates the POP instruction.

Example 1-6

Assuming that SP = 1236, AX = 24B6, DI = 85C2, and DX = 5F93, show the contents of the stack as each of the following instructions is executed:

```
PUSH  AX
PUSH  DI
PUSH  DX
```

Solution:

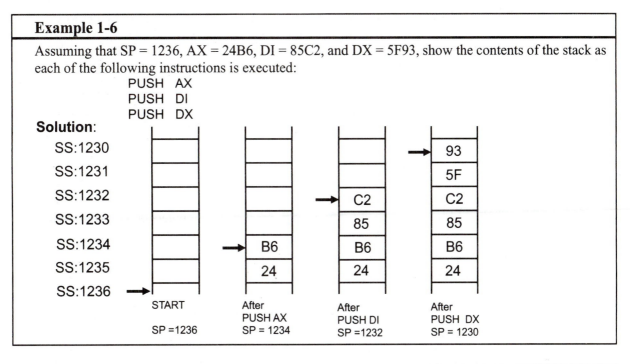

Example 1-7

Assuming that the stack is as shown below, and SP = 18FA, show the contents of the stack and registers as each of the following instructions is executed:

```
POP   CX
POP   DX
POP   BX
```

Solution:

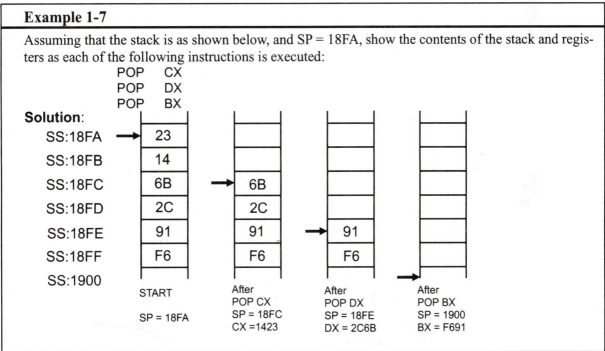

Logical address vs. physical address for the stack

Now one might ask, what is the exact physical location of the stack? That depends on the value of the stack segment (SS) register and SP, the stack pointer. To compute physical addresses for the stack, the same principle is applied as was used for the code and data segments. The method is to shift left SS and then add offset SP, the stack pointer register. This is demonstrated in Example 1-8.

What values are assigned to the SP and SS, and who assigns them? It is the job of the DOS operating system to assign the values for the SP and SS since memory management is the responsibility of the operating system. Before leaving the discussion of the stack, two points must be made. First, in the 80x86 literature, the top of the stack is the last stack location occupied. This is different from other CPUs.

Second, BP is another register that can be used as an offset into the stack, but it has very special applications and is widely used to access parameters passed between Assembly language programs and high-level language programs such as C. This is discussed in Chapter 7.

Example 1-8

If SS = 3500H and the SP is FFFEH,
(a) Calculate the physical address of the stack.   (b) Calculate the lower range.
(c) Calculate the upper range of the stack segment.   (d) Show the logical address of the stack.

Solution:

(a) 44FFE (35000 + FFFE)   (b) 35000 (35000 + 0000)
(c) 44FFF (35000 + FFFF)   (d) 3500:FFFE

A few more words about segments in the 80x86

Can a single physical address belong to many different logical addresses? Yes, look at the case of a physical address value of 15020H. There are many possible logical addresses that represent this single physical address:

Logical address (hex)	Physical address (hex)
1000:5020	15020
1500:0020	15020
1502:0000	15020
1400:1020	15020
1302:2000	15020

This shows the dynamic behavior of the segment and offset concept in the 8086 CPU. One last point that must be clarified is the case when adding the offset to the shifted segment register results in an address beyond the maximum allowed range of FFFFFH. In that situation, wrap-around will occur. This is shown in Example 1-9.

Example 1-9

What is the range of physical addresses if CS = FF59?

Solution:

The low range is FF590 (FF590 + 0000). The range goes to FFFFF and wraps around, from 00000 to 0F58F (FF590 + FFFF = 0F58F), which is illustrated below.

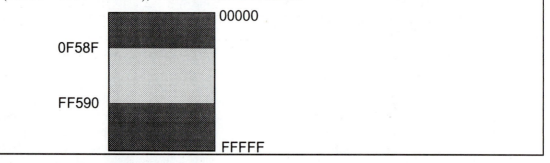

Overlapping

In calculating the physical address, it is possible that two segments can overlap, which is desirable in some circumstances. For example, overlapping is used in COM files, as will be seen in Chapter 2. Figure 1-4 illustrates overlapping and nonoverlapping segments.

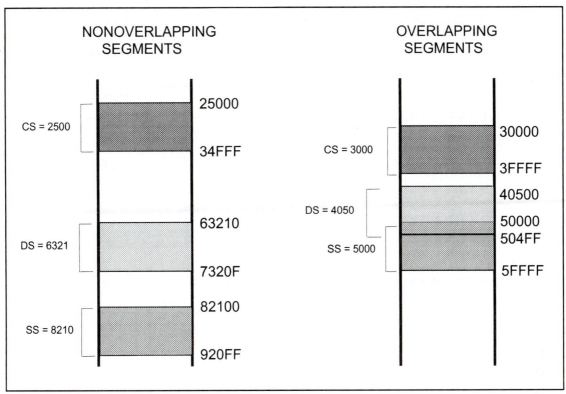

Figure 1-4. Nonoverlapping vs. Overlapping Segments

Flag register

The flag register is a 16-bit register sometimes referred to as the *status register.* Although the register is 16 bits wide, only some of the bits are used. The rest are either undefined or reserved by Intel. Six of the flags are called conditional flags, meaning that they indicate some condition that resulted after an instruction was executed. These six are CF, PF, AF, ZF, SF, and OF. The three remaining flags are sometimes called control flags since they are used to control the operation of instructions before they are executed. A diagram of the flag register is shown in Figure 1-5.

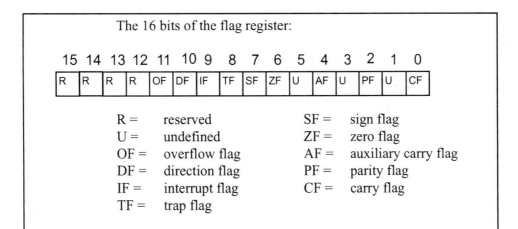

Figure 1-5. Flag Register

Bits of the flag register

Below are listed the bits of the flag register that are used in 80x86 Assembly language programming. A brief explanation of each bit is given. How these flag bits are used will be seen in programming examples throughout the textbook.

CF, the Carry Flag . This flag is set whenever there is a carry out, either from d7 after an 8-bit operation, or from d15 after a 16-bit data operation.

PF, the Parity Flag. After certain operations, the parity of the result's low-order byte is checked. If the byte has an even number of 1s, the parity flag is set to 1; otherwise, it is cleared.

AF, Auxiliary Carry Flag. If there is a carry from d3 to d4 of an operation, this bit is set; otherwise, it is cleared (set equal to zero). This flag is used by the instructions that perform BCD (binary coded decimal) arithmetic.

ZF, the Zero Flag. The zero flag is set to 1 if the result of an arithmetic or logical operation is zero; otherwise, it is cleared.

SF, the Sign Flag. Binary representation of signed numbers uses the most significant bit as the sign bit. After arithmetic or logic operations, the status of this sign bit is copied into the SF, thereby indicating the sign of the result.

TF, the Trap Flag. When this flag is set it allows the program to single-step, meaning to execute one instruction at a time. Single-stepping is used for debugging purposes.

IF, Interrupt Enable Flag. This bit is set or cleared to enable or disable only the external maskable interrupt requests.

DF, the Direction Flag. This bit is used to control the direction of string operations, which are described in Chapter 6.

OF, the Overflow Flag. This flag is set whenever the result of a signed number operation is too large, causing the high-order bit to overflow into the sign bit. In general, the carry flag is used to detect errors in unsigned arithmetic operations. The overflow flag is only used to detect errors in signed arithmetic operations.

Flag register and ADD instruction

In this section we examine the impact of the ADD instruction on the flag register as an example of the use of the flag bits. The flag bits affected by the ADD instruction are CF (carry flag), PF (parity flag), AF (auxiliary carry flag), ZF (zero flag), SF (sign flag), and OF (overflow flag). The overflow flag will be covered in Chapter 6, since it relates only to signed number arithmetic. To understand how each of these flag bits is affected, look at Examples 1-10 and 1-11.

Example 1-10

Show how the flag register is affected by the addition of 38H and 2FH.

Solution:

```
MOV    BH,38H       ;BH= 38H
ADD    BH,2FH       ;add 2F to BH, now BH=67H

       38           0011    1000
+      2F           0010    1111
       67           0110    0111
```

CF = 0 since there is no carry beyond d7
PF = 0 since there is an odd number of 1s in the result
AF = 1 since there is a carry from d3 to d4
ZF = 0 since the result is not zero
SF = 0 since d7 of the result is zero

Example 1-11

Show how the flag register is affected by

```
MOV    AL,9CH      ;AL=9CH
MOV    DH,64H      ;DH=64H
ADD    AL,DH       ;now AL=0
```

Solution:

```
        9C          1001    1100
  +     64          0110    0100
        00          0000    0000
```

CF=1 since there is a carry beyond d7
PF=1 since there is an even number of 1s in the result
AF=1 since there is a carry from d3 to d4
ZF=1 since the result is zero
SF=0 since d7 of the result is zero

The same concepts apply for 16-bit addition, as shown in Examples 1-12 and 1-13. It is important to notice the differences between 8-bit and 16-bit operations in terms of their impact on the flag bits. The parity bit only counts the lower 8-bits of the result and is set accordingly. Also notice the CF bit. The carry flag is set if there is a carry beyond bit d15 instead of bit d7.

Example 1-12

Show how the flag register is affected by

```
MOV    AX,34F5H    ;AX= 34F5H
ADD    AX,95EBH    ;now  AX= CAE0H
```

Solution:

```
        34F5        0011    0100    1111    0101
  +     95EB        1001    0101    1110    1011
        CAE0        1100    1010    1110    0000
```

CF = 0 since there is no carry beyond d15
PF = 0 since there is an odd number of 1s in the lower byte
AF = 1 since there is a carry from d3 to d4
ZF = 0 since the result is not zero
SF = 1 since d15 of the result is one

Example 1-13

Show how the flag register is affected by

```
MOV    BX,AAAAH    ;BX= AAAAH
ADD    BX,5556H    ;now  BX= 0000H
```

Solution:

```
        AAAA        1010    1010    1010    1010
  +     5556        0101    0101    0101    0110
        0000        0000    0000    0000    0000
```

CF = 1 since there is a carry beyond d15
PF = 1 since there is an even number of 1s in the lower byte
AF = 1 since there is a carry from d3 to d4
ZF = 1 since the result is zero
SF = 0 since d15 of the result is zero

SECTION 1.5: MORE ABOUT SEGMENTS IN THE 80x86

Notice the zero flag (ZF) status after the execution of the ADD instruction. Since the result of the entire 16-bit operation is zero (meaning the contents of BX), ZF is set to high. Do all instructions affect the flag bits? The answer is no; some instructions such as data transfers (MOV) affect no flags. As an exercise, run these examples on DEBUG to see the effect of various instructions on the flag register.

Example 1-14

Show how the flag register is affected by

```
        MOV    AX,94C2H      ;AX=94C2H
        MOV    BX,323EH      ;BX=323EH
        ADD    AX,BX         ;now AX=C700H
        MOV    DX,AX         ;now DX=C700H
      . MOV    CX,DX         ;now CX=C700H
```

Solution:

```
      94C2        1001  0100  1100  0010
   +  323E        0011  0010  0011  1110
      C700        1100  0111  0000  0000
```

After the ADD operation, the following are the flag bits:
CF = 0 since there is no carry beyond d15
PF = 1 since there is an even number of 1s in the lower byte
AF = 1 since there is a carry from d3 to d4
ZF = 0 since the result is not zero
SF = 1 since d15 of the result is 1

Running the instructions in Example 1-14 in DEBUG will verify that MOV instructions have no effect on the flag. How these flag bits are used in programming is discussed in future chapters in the context of many applications. In Appendix B we give additional information about the effect of various instructions on the flags.

Use of the zero flag for looping

One of the most widely used applications of the flag register is the use of the zero flag to implement program loops. The term *loop* refers to a set of instructions that is repeated a number of times. For example, to add 5 bytes of data, a counter can be used to keep track of how many times the loop needs to be repeated. Each time the addition is performed the counter is decremented and the zero flag is checked. When the counter becomes zero, the zero flag is set (ZF = 1) and the loop is stopped. The following shows the implementation of the looping concept in the program, which adds 5 bytes of data. Register CX is used to hold the counter and BX is the offset pointer (SI or DI could have been used instead). AL is initialized before the start of the loop. In each iteration, ZF is checked by the JNZ instruction. JNZ stands for "Jump Not Zero" meaning that if ZF = 0, jump to a new address. If ZF = 1, the jump is not performed and the instruction below the jump will be executed. Notice that the JNZ instruction must come immediately after the instruction that decrements CX since JNZ needs to check the affect of "DEC CX" on the zero flag. If any instruction were placed between them, that instruction might affect the zero flag.

```
              MOV    CX,05       ;CX holds the loop count
              MOV    BX,0200H    ;BX holds the offset data address
              MOV    AL,00       ;initialize AL
   ADD_LP:    ADD    AL,[BX]     ;add the next byte to AL
              INC    BX          ;increment the data pointer
              DEC    CX          ;decrement the loop counter
              JNZ    ADD_LP      ;jump to next iteration if counter not zero
```

Review Questions

1. Which registers are used to access the stack?
2. With each PUSH instruction, the stack pointer register SP is (circle one) incremented/decremented by 2.
3. With each POP instruction, SP is (circle one) incremented/decremented by 2.
4. List three possible logical addresses corresponding to physical address 143F0.
5. The ADD instruction can affect which bits of the flag register?
6. The carry flag will be set to 1 in an 8-bit ADD if there is a carry out from bit ___.
7. CF will be set to 1 in a 16-bit ADD if there is a carry out from bit ____.

SECTION 1.6: 80x86 ADDRESSING MODES

The CPU can access operands (data) in various ways, called addressing modes. The number of addressing modes is determined when the microprocessor is designed and cannot be changed. The 80x86 provides a total of seven distinct addressing modes:

1. register
2. immediate
3. direct
4. register indirect
5. based relative
6. indexed relative
7. based indexed relative

Each addressing mode is explained below, and application examples are given in later chapters as the reader understands Assembly language programming in greater detail. Since the reader is now familiar with ADD and MOV instructions, these are used below to explain addressing modes.

Register addressing mode

The register addressing mode involves the use of registers to hold the data to be manipulated. Memory is not accessed when this addressing mode is executed; therefore, it is relatively fast. Examples of register addressing mode follow:

```
MOV    BX,DX          ;copy the contents of DX into BX
MOV    ES,AX          ;copy the contents of AX into ES
ADD    AL,BH          ;add the contents of BH to contents of AL
```

It should be noted that the source and destination registers must match in size. In other words coding "MOV CL,AX" will give an error, since the source is a 16-bit register and the destination is an 8-bit register.

Immediate addressing mode

In the immediate addressing mode, the source operand is a constant. In immediate addressing mode, as the name implies, when the instruction is assembled, the operand comes immediately after the opcode. For this reason, this addressing mode executes quickly. However, in programming it has limited use. Immediate addressing mode can be used to load information into any of the registers except the segment registers and flag registers. Examples:

```
MOV    AX,2550H       ;move 2550H into AX
MOV    CX,625         ;load the decimal value 625 into CX
MOV    BL,40H         ;load 40H into BL
```

To move information to the segment registers, the data must first be moved to a general-purpose register and then to the segment register. Example:

```
MOV    AX,2550H
MOV    DS,AX
```

In other words, the following would produce an error:

```
MOV DS,0123H          ;illegal!!
```

In the first two addressing modes, the operands are either inside the microprocessor or tagged along with the instruction. In most programs, the data to be processed is often in some memory location outside the CPU. There are many ways of accessing the data in the data segment. The following describes those different methods.

Direct addressing mode

In the direct addressing mode the data is in some memory location(s) and the address of the data in memory comes immediately after the instruction. Note that in immediate addressing, the operand itself is provided with the instruction, whereas in direct addressing mode, the address of the operand is provided with the instruction. This address is the offset address and one can calculate the physical address by shifting left the DS register and adding it to the offset as follows:

```
MOV    DL,[2400]       ;move contents of DS:2400H into DL
```

In this case the physical address is calculated by combining the contents of offset location 2400 with DS, the data segment register. Notice the bracket around the address. In the absence of this bracket it will give an error since it is interpreted to move the value 2400 (16-bit data) into register DL, an 8-bit register. Example 1-15 gives another example of direct addressing.

Example 1-15

Find the physical address of the memory location and its contents after the execution of the following, assuming that DS = 1512H.
```
MOV    AL,99H
MOV    [3518],AL
```

Solution:
First AL is initialized to 99H, then in line two, the contents of AL are moved to logical address DS:3518 which is 1512:3518. Shifting DS left and adding it to the offset gives the physical address of 18638H (15120H + 3518H = 18638H). That means after the execution of the second instruction, the memory location with address 18638H will contain the value 99H.

Register indirect addressing mode

In the register indirect addressing mode, the address of the memory location where the operand resides is held by a register. The registers used for this purpose are SI, DI, and BX. If these three registers are used as pointers, that is, if they hold the offset of the memory location, they must be combined with DS in order to generate the 20-bit physical address. For example:

```
MOV    AL,[BX]         ;moves into AL the contents of the memory location
                       ;pointed to by DS:BX.
```

Notice that BX is in brackets. In the absence of brackets, it is interpreted as an instruction moving the contents of register BX to AL (which gives an error because source and destination do not match) instead of the contents of the memory location whose offset address is in BX. The physical address is calculated by shifting DS left one hex position and adding BX to it. The same rules apply when using register SI or DI.

```
MOV    CL,[SI]        ;move contents of DS:SI into CL
MOV    [DI],AH        ;move contents of AH into DS:DI
```

In the examples above, the data moved is byte sized. Example 1-16 shows 16-bit operands.

Example 1-16

Assume that DS = 1120, SI = 2498, and AX = 17FE. Show the contents of memory locations after the execution of

```
MOV [SI],AX
```

Solution:

The contents of AX are moved into memory locations with logical address DS:SI and DS:SI + 1; therefore, the physical address starts at DS (shifted left) + SI = 13698. According to the little endian convention, low address 13698H contains FE, the low byte, and high address 13699H will contain 17, the high byte.

Based relative addressing mode

In the based relative addressing mode, base registers BX and BP, as well as a displacement value, are used to calculate what is called the effective address. The default segments used for the calculation of the physical address (PA) are DS for BX and SS for BP. For example:

```
MOV    CX,[BX]+10     ;move DS:BX+10 and DS:BX+10+1 into CX
                      ;PA = DS (shifted left) + BX + 10
```

Alternative codings are "MOV CX,[BX+10]" or "MOV CX,10[BX]". Again the low address contents will go into CL and the high address contents into CH. In the case of the BP register,

```
MOV    AL,[BP]+5      ;PA = SS (shifted left) + BP + 5
```

Again, alternative codings are "MOV AL,[BP+5]" or "MOV AL,5[BP]". A brief mention should be made of the terminology *effective address* used in Intel literature. In "MOV AL,[BP]+5", BP+5 is called the effective address since the fifth byte from the beginning of the offset BP is moved to register AL. Similarly in "MOV CX,[BX]+10", BX+10 is called the effective address.

Indexed relative addressing mode

The indexed relative addressing mode works the same as the based relative addressing mode, except that registers DI and SI hold the offset address. Examples:

```
MOV    DX,[SI]+5      ;PA = DS (shifted left) + SI + 5
MOV    CL,[DI]+20     ;PA = DS (shifted left) + DI + 20
```

Example 1-17 gives further examples of indexed relative addressing mode.

Example 1-17

Assume that DS = 4500, SS = 2000, BX = 2100, SI = 1486, DI = 8500, BP = 7814, and AX = 2512. Show the exact physical memory location where AX is stored in each of the following. All values are in hex.

(a) MOV [BX]+20,AX   (b) MOV [SI]+10,AX

(c) MOV [DI]+4,AX   (d) MOV [BP]+12,AX

Solution:

In each case PA = segment register (shifted left) + offset register + displacement.

(a) DS:BX+20   location 47120 = (12) and 47121 = (25)

(b) DS:SI+10   location 46496 = (12) and 46497 = (25)

(c) DS:DI+4   location 4D504 = (12) and 4D505 = (25)

(d) SS:BP+12   location 27826 = (12) and 27827 = (25)

Based indexed addressing mode

By combining based and indexed addressing modes, a new addressing mode is derived called the based indexed addressing mode. In this mode, one base register and one index register are used. Examples:

```
MOV    CL,[BX][DI]+8        ;PA = DS (shifted left) + BX + DI + 8
MOV    CH,[BX][SI]+20       ;PA = DS (shifted left) + BX + SI + 20
MOV    AH,[BP][DI]+12       ;PA = SS (shifted left) + BP + DI + 12
MOV    AH,[BP][SI]+29       ;PA = SS (shifted left) + BP + SI + 29
```

The coding of the instructions above can vary; for example, the last example could have been written

```
MOV    AH,[BP+SI+29]
or
MOV    AH,[SI+BP+29]  ;the register order does not matter.
```

Note that "MOV AX,[SI][DI]+displacement" is illegal.

In many of the examples above, the MOV instruction was used for the sake of clarity, even though one can use any instruction as long as that instruction supports the addressing mode. For example, the instruction "ADD DL,[BX]" would add the contents of the memory location pointed at by DS:BX to the contents of register DL.

Table 1-3: Offset Registers for Various Segments

Segment register:	CS	DS	ES	SS
Offset register(s):	IP	SI, DI, BX	SI, DI, BX	SP, BP

Segment overrides

Table 1-3 provides a summary of the offset registers that can be used with the four segment registers of the 80x86. The 80x86 CPU allows the program to override the default segment and use any segment register. To do that, specify the segment in the code. For example, in "MOV AL,[BX]", the physical address of the operand to be moved into AL is DS:BX, as was shown earlier since DS is the default segment for pointer BX. To override that default, specify the desired segment in the instruction as "MOV AL,ES:[BX]". Now the address of the operand being moved to AL is ES:BX instead of DS:BX. Extensive use of all these addressing modes is shown in future chapters in the context of program examples. Table 1-4 shows more examples of segment overrides shown next to the default address in the absence of the override. Table 1-5 summarizes addressing modes of the 8086/88.

Table 1-4: Sample Segment Overrides

Instruction	Segment Used	Default Segment
MOV AX,CS:[BP]	CS:BP	SS:BP
MOV DX,SS:[SI]	SS:SI	DS:SI
MOV AX,DS:[BP]	DS:BP	SS:BP
MOV CX,ES:[BX]+12	ES:BX+12	DS:BX+12
MOV SS:[BX][DI]+32,AX	SS:BX+DI+32	DS:BX+DI+32

Table 1-5: Summary of 80x86 Addressing Modes

Addressing Mode	Operand	Default Segment
Register	reg	none
Immediate	data	none
Direct	[offset]	DS
Register indirect	[BX]	DS
	[SI]	DS
	[DI]	DS
Based relative	[BX]+disp	DS
	[BP]+disp	SS
Indexed relative	[DI]+disp	DS
	[SI]+disp	DS
Based indexed relative	[BX][SI]+disp	DS
	[BX][DI]+disp	DS
	[BP][SI]+ disp	SS
	[BP][DI]+ disp	SS

SUMMARY

Intel's 80x86 family of microprocessors are used in all IBM PC, PS, and compatible computers. The 8088 was the microprocessor used by IBM in the first PCs, which revolutionized the computing industry in the early 1980s. Each generation of Intel microprocessors brought improvements in speed and processing power.

A typical Assembly language program consists of at least three segments. The code segment contains the Assembly language instructions to be executed. The data segment is used to store data needed by the program. The stack segment is used for temporary storage of data. Memory within each segment is accessed by combining a segment register and an offset register. The flag register is used to indicate certain conditions after the execution of an instruction such as carry, overflow, or zero result.

Assembly language instructions can use one of seven addressing modes. An addressing mode is simply a method by which the programmer tells the CPU where to find the operand for that instruction.

PROBLEMS

1. Which microprocessor, the 8088 or the 8086, was released first?
2. If the 80286 and 80386SX both have 16-bit external data buses, what is the difference between them?
3. What does "16-bit" or "32-bit" microprocessor mean? Does it refer to the internal or external data path?
4. Do programs written for the 88/86 run on 80286-, 80386-, and 80486-based CPUs?
5. What does the term *upward compatibility* mean?
6. Name a major difference between the 8088 and 8086.
7. Which has the larger queue, the 8088 or 8086?
8. State another way to increase the processing power of the CPU other than increasing the frequency.
9. What do "BIU" and "EU" stand for, and what are their functions?
10. Name the general-purpose registers of the 8088/86.
    (a) 8-bit   (b) 16-bit
11. Which of the following registers cannot be split into high and low bytes?
    (a) CS   (b) AX   (c) DS
    (d) SS   (e) BX   (f) DX
    (g) CX   (h) SI   (i) DI
12. Which of the following instructions cannot be coded in 8088/86 Assembly language? Give the reason why not, if any. To verify your answer, code each in DEBUG. Assume that all numbers are in hex.
    (a) MOV AX,27   (b) MOV AL,97F   (c) MOV DS,9BF2
    (d) MOV CX,397   (e) MOV SI,9516   (f) MOV CS,3490
    (g) MOV DS,BX   (h) MOV BX,CS   (i) MOV CH,AX
    (j) MOV AX,23FB9   (k) MOV CS,BH   (l) MOV AX,DL
13. Name the segment registers and their functions in the 8088/86.
14. If CS = 3499H and IP = 2500H, find:
    (a) The logical address
    (b) The physical address
    (c) The lower and upper ranges of the code segment
15. Repeat Problem 14 with CS = 1296H and IP = 100H.
16. If DS = 3499H and the offset = 3FB9H, find:
    (a) The physical address
    (b) The logical address of the data being fetched
    (c) The lower and upper range addresses of the data segment
17. Repeat Problem 16 using DS = 1298H and the offset = 7CC8H.
18. Assume that the physical address for a location is 0046CH. Suggest a possible logical address.
19. If an instruction that needs to be fetched is in physical memory location 389F2 and CS = 2700, does the code segment range include it or not? If not, what value should be assigned to CS if the IP must be = 1282?
20. Using DEBUG, assemble and unassemble the following program and provide the logical address, physical address, and the content of each address location. The CS value is decided by DOS, but use IP = 170H.
    MOV AL,76H
    MOV BH,8FH
    ADD BH,AL
    ADD BH,7BH
    MOV BL,BH
    ADD BL,AL

21. Repeat Problem 20 for the following program from page 29.

```
MOV AL,0              ;clear AL
ADD AL,[0200]         ;add the contents of DS:200 to AL
ADD AL,[0201]         ;add the contents of DS:201 to AL
ADD AL,[0202]         ;add the contents of DS:202 to AL
ADD AL,[0203]         ;add the contents of DS:203 to AL
ADD AL,[0204]         ;add the contents of DS:204 to AL
```

22. The stack is:
    (a) A section of ROM
    (b) A section of RAM used for temporary storage
    (c) A 16-bit register inside the CPU
    (d) Some memory inside the CPU
23. In problem 22, choose the correct answer for the stack pointer.
24. When data is pushed onto the stack, the stack pointer is _____, but when data is popped off the stack, the stack pointer is _____.
25. Choose the correct answer:
    (a) The stack segment and code segment start at the same point of read/write memory and grow upward.
    (b) The stack segment and code segment start at opposite points of read/write memory and grow toward each other.
    (c) There will be no problem if the stack and code segments meet each other.
26. What is the main disadvantage of the stack as temporary storage compared to having a large number of registers inside the CPU?
27. If SS = 2000 and SP = 4578, find:
    (a) The physical address
    (b) The logical address
    (c) The lower range of the stack segment
    (d) The upper range of the stack segment
28. If SP = 24FC, what is the offset address of the first location of the stack that is available to push data into?
29. Assume that SP = FF2EH, AX = 3291H, BX = F43CH, and CX = 09. Find the content of the stack and stack pointer after the execution of each of the following instructions.

```
PUSH AX
PUSH BX
PUSH CX
```

30. In order for each register to get back their original values in Problem 29, show the sequence of instructions that needs to be executed. Show the content of the SP at each point.
31. The following registers are used as offsets. Assuming that the default segment is used to get the logical address, give the segment register associated with each offset.
    (a) BP   (b) DI   (c) IP
    (d) SI   (e) SP   (f) BX
32. Show the override segment register and the default segment register used (if there were no override) in each of the following cases.
    (a) MOV SS:[BX],AX   (b) MOV SS:[DI],BX
    (c) MOV DX,DS:[BP+6]
33. Find the status of the CF, PF, AF, ZF, and SF for the following operations.
    (a)MOV BL,9FH   (b) MOV AL,23H   (c) MOV DX,10FFH
       ADD BL,61H   ADD AL,97H   ADD DX,1
34. Assume that the registers have the following values (all in hex) and that CS = 1000, DS = 2000, SS = 3000, SI = 4000, DI = 5000, BX = 6080, BP = 7000, AX = 25FF, CX = 8791, and DX = 1299. Calculate the physical address of the memory where the operand is stored and the contents of the memory locations in each of the following addressing examples.
    (a) MOV [SI],AL   (b) MOV [SI+BX+8],AH
    (c) MOV [BX],AX   (d) MOV [DI+6],BX
    (e) MOV [DI][BX]+28,CX   (f) MOV [BP][SI]+10,DX

PROBLEMS

        (g) MOV [3600],AX   (h) MOV [BX]+30,DX
        (i) MOV [BP]+200,AX   (j) MOV [BP+SI+100],BX
        (k) MOV [SI]+50,AH   (l) MOV [DI+BP+100],AX.
    35. Give the addressing mode for each of the following:
        (a) MOV AX,DS   (b) MOV BX,5678H
        (c) MOV CX,[3000]   (d) MOV AL,CH
        (e) MOV [DI],BX   (f) MOV AL,[BX]
        (g) MOV DX,[BP+DI+4]   (h) MOV CX,DS
        (i) MOV [BP+6],AL   (j) MOV AH,[BX+SI+50]
        (k) MOV BL,[SI]+10   (l) MOV [BP][SI]+12,AX
    36. Show the contents of the memory locations after the execution of each instruction.
        (a) MOV BX,129FH   (b) MOV DX,8C63H
            MOV [1450],BX   MOV [2348],DX
            DS:1450 ....   DS:2348 ....
            DS:1451 ....   DS:2349 ....

ANSWERS TO REVIEW QUESTIONS

SECTION 1.1: BRIEF HISTORY OF THE 80X86 FAMILY
1. (1) increased memory capacity from 64K to 1 megabyte; (2) the 8086 is a 16-bit microprocessor instead of an 8-bit microprocessor; (3) the 8086 was a pipelined processor
2. the 8088 has an 8-bit external data bus whereas the 8086 has a 16-bit data bus
3. (a) 20-bit, 1 megabyte   (b) 24-bit, 16 megabytes   (c) 32-bit, 4 gigabytes
4. 16, 32
5. the 80386 has 32-bit address and data buses, whereas the 80386SX has a 24-bit address bus and a 16-bit external data bus
6. virtual memory, protected mode
7. math coprocessor on the CPU chip, cache memory and controller

SECTION 1.2: INSIDE THE 8088/8086
1. the execution unit executes instructions; the bus interface unit fetches instructions
2. pipelining divides the microprocessor into two sections: the execution unit and the bus interface unit; this allows the CPU to perform these two functions simultaneously; that is, the BIU can fetch instructions while the EU executes the instructions previously fetched
3. 8, 16
4. AX, BX, CX, DX, SP, BP, SI, DI, CS, DS, SS, ES, IP, FR

SECTION 1.3: INTRODUCTION TO ASSEMBLY PROGRAMMING
1. MOV BX,1234H
2. MOV AX,16H
   ADD AX,ABH
3. the segment registers CS, DS, ES, and SS
4. FFFFH = 65535_{10}, FFH = 255_{10}

SECTION 1.4: INTRODUCTION TO PROGRAM SEGMENTS
1. 64K
2. a segment contains 64K bytes; yes because 346E0H is evenly divisible by 16
3. the physical address is the 20-bit address that is put on the address bus to locate a byte; the logical address is the address in the form xxxx:yyyy, where xxxx is the segment address and yyyy is the offset into the segment
4. 20, 16
5. IP
6. 2400 would contain 34 and 2401 would contain 12

SECTION 1.5: MORE ABOUT SEGMENTS IN THE 80X86
1. SS is the segment register; SP and BP are used as pointers into the stack
2. decremented
3. incremented
4. 143F:0000, 1000:43F0, 1410:02F0
5. CF, PF, AF, ZF, SF, and OF
6. 7
7. 15

CHAPTER 2

ASSEMBLY LANGUAGE PROGRAMMING

OBJECTIVES

Upon completion of this chapter, you will be able to:

» Explain the difference between Assembly language instructions and pseudo-instructions

» Identify the segments of an Assembly language program

» Code simple Assembly language instructions

» Assemble, link, and run a simple Assembly language program

» Code control transfer instructions such as conditional and unconditional jumps and call instructions

» Code Assembly language data directives for binary, hex, decimal, or ASCII data

» Write an Assembly language program using either the full segment definition or the simplified segment definition

» Explain the difference between COM and EXE files and list the advantages of each

This chapter is an introduction to Assembly language programming with the 80x86. First the basic form of a program is explained, followed by the steps required to edit, assemble, link, and run a program. Next, control transfer instructions such as jump and call are discussed and data types and data directives in 80x86 Assembly language are explained. Then the full segment definition is discussed. Finally, the differences between ".exe" and ".com" files are explained. The programs in this chapter and following ones can be assembled and run on any IBM PC, PS and compatible computer with an 8088/86 or higher microprocessor.

SECTION 2.1: DIRECTIVES AND A SAMPLE PROGRAM

In this section we explain the components of a simple Assembly language program to be assembled by the assembler. A given Assembly language program (see Figure 2-1) is a series of statements, or lines, which are either Assembly language instructions such as ADD and MOV, or statements called directives. *Directives* (also called *pseudo-instructions*) give directions to the assembler about how it should translate the Assembly language instructions into machine code. An Assembly language instruction consists of four fields:

[label:]   mnemonic   [operands]   [;comment]

Brackets indicate that the field is optional. Do not type in the brackets.

1. The label field allows the program to refer to a line of code by name. The label field cannot exceed 31 characters. Labels for directives do not need to end with a colon. A label must end with a colon when it refers to an opcode generating instruction; the colon indicates to the assembler that this refers to code within this code segment. Appendix C, Section 2 gives more information about labels.

2,3. The Assembly language mnemonic (instruction) and operand(s) fields together perform the real work of the program and accomplish the tasks for which the program was written. In Assembly language statements such as

```
ADD    AL,BL
MOV    AX,6764
```

ADD and MOV are the mnemonic opcodes and "AL,BL" and "AX,6764" are the operands. Instead of a mnemonic and operand, these two fields could contain assembler pseudo-instructions, or directives. They are used by the assembler to organize the program as well as other output files. Directives do not generate any machine code and are used only by the assembler as opposed to instructions, which are translated into machine code for the CPU to execute. In Figure 2-1 the commands DB, END, and ENDP are examples of directives.

4. The comment field begins with a ";". Comments may be at the end of a line or on a line by themselves. The assembler ignores comments, but they are indispensable to programmers. Comments are optional, but are highly recommended to make it easier for someone to read and understand the program.

Model definition

The first statement in Figure 2-1 after the comments is the MODEL directive. This directive selects the size of the memory model. Among the options for the memory model are SMALL, MEDIUM, COMPACT, and LARGE.

```
.MODEL SMALL           ;this directive defines the model as small
```

SMALL is one of the most widely used memory models for Assembly language programs and is sufficient for the programs in this book. The small model uses a maximum of 64K bytes of memory for code and another 64K bytes for data. The other models are defined as follows:

```
.MODEL MEDIUM      ;the data must fit into 64K bytes
                   ;but the code can exceed 64K bytes of memory
.MODEL COMPACT     ;the data can exceed 64K bytes
                   ;but the code cannot exceed 64K bytes
.MODEL LARGE       ;both data and code can exceed 64K
                   ;but no single set of data should exceed 64K
.MODEL HUGE        ;both code and data can exceed 64K
                   ;data items (such as arrays) can exceed 64K
.MODEL TINY        ;used with COM files in which data and code
                   ;must fit into 64K bytes
```

Notice in the above list that MEDIUM and COMPACT are opposites. Also note that the TINY model cannot be used with the simplified segment definition described in this section.

Segment definition

As mentioned in Chapter 1, the 80x86 CPU has four segment registers: CS (code segment), DS (data segment), SS (stack segment), and ES (extra segment). Every line of an Assembly language program must correspond to one of these segments. The simplified segment definition format uses three simple directives: ".CODE", ".DATA", and ".STACK", which correspond to the CS, DS, and SS registers, respectively. There is another segment definition style called the *full segment definition*, which is described in Section 2.6.

Segments of a program

Although one can write an Assembly language program that uses only one segment, normally a program consists of at least three segments: the stack segment, the data segment, and the code segment.

```
.STACK      ;marks the beginning of the stack segment
.DATA       ;marks the beginning of the data segment
.CODE       ;marks the beginning of the code segment
```

Assembly language statements are grouped into segments in order to be recognized by the assembler and consequently by the CPU. The stack segment defines storage for the stack, the data segment defines the data that the program will use, and the code segment contains the Assembly language instructions. In Chapter 1 we gave an overview of how these segments were stored in memory. In the following pages we describe the stack, data, and code segments as they are defined in Assembly language programming.

Stack segment

The following directive reserves 64 bytes of memory for the stack:

```
.STACK  64
```

Data segment

The data segment in the program of Figure 2-1 defines three data items: DATA1, DATA2, and SUM. Each is defined as DB (define byte). The DB directive is used by the assembler to allocate memory in byte-sized chunks. Memory can be allocated in different sizes, such as 2 bytes, which has the directive DW (define word). More of these pseudo-instructions are discussed in detail in Section 2.5. The data items defined in the data segment will be accessed in the code segment by their labels. DATA1 and DATA2 are given initial values in the data section. SUM is not given an initial value, but storage is set aside for it.

```
;THE FORM OF AN ASSEMBLY LANGUAGE PROGRAM
;NOTE: USING SIMPLIFIED SEGMENT DEFINITION
            .MODEL SMALL
            .STACK 64
            .DATA
DATA1       DB      52H
DATA2       DB      29H
SUM         DB      ?
            .CODE
MAIN        PROC FAR            ;this is the program entry point
            MOV     AX,@DATA    ;load the data segment address
            MOV     DS,AX       ;assign value to DS
            MOV     AL,DATA1    ;get the first operand
            MOV     BL,DATA2    ;get the second operand
            ADD     AL,BL       ;add the operands
            MOV     SUM,AL      ;store the result in location SUM
            MOV     AH,4CH      ;set up to return to DOS
            INT     21H         ;
MAIN        ENDP
            END     MAIN        ;this is the program exit point
```

Figure 2-1. Simple Assembly Language Program

Code segment definition

The last segment of the program in Figure 2-1 is the code segment. The first line of the segment after the .CODE directive is the PROC directive. A *procedure* is a group of instructions designed to accomplish a specific function. A code segment may consist of only one procedure, but usually is organized into several small procedures in order to make the program more structured. Every procedure must have a name defined by the PROC directive, followed by the assembly language instructions and closed by the ENDP directive. The PROC and ENDP statements must have the same label. The PROC directive may have the option FAR or NEAR. The operating system that controls the computer must be directed to the beginning of the program in order to execute it. DOS requires that the entry point to the user program be a FAR procedure. From then on, either FAR or NEAR can be used. The differences between a FAR and a NEAR procedure, as well as where and why each is used, are explained later in this chapter. For now, just remember that in order to run a program, FAR must be used at the program entry point.

A good question to ask at this point is: What value is actually assigned to the CS, DS, and SS registers for execution of the program? The DOS operating system must pass control to the program so that it may execute, but before it does that it assigns values for the segment registers. The operating system must do this because it knows how much memory is installed in the computer, how much of it is used by the system, and how much is available. In the IBM PC, the operating system first finds out how many kilobytes of RAM memory are installed, allocates some for its own use, and then allows the user program to use the portions that it needs. Various DOS versions require different amounts of memory, and since the user program must be able to run across different versions, one cannot tell DOS to give the program a specific area of memory, say from 25FFF to 289E2. Therefore, it is the job of DOS to assign exact values for the segment registers. When the program begins executing, of the three segment registers, only CS and SS have the proper values. The DS value (and ES, if used) must be initialized by the program. This is done as follows:

```
MOV AX,@DATA          ;DATA refers to the start of the data segment
MOV DS,AX
```

Remember from Chapter 1 that no segment register can be loaded directly. That is the reason the two lines of code above are needed. You cannot code "MOV DS,@DATA".

After these housekeeping chores, the Assembly language program instructions can be written to perform the desired tasks. In Figure 2-1, the program loads AL and BL with DATA1 and DATA2, respectively, ADDs them together, and stores the result in SUM.

```
MOV  AL,DATA1
MOV  BL,DATA2
ADD  AL,BL
MOV  SUM,AL
```

The two last instructions in the shell are:

```
MOV   AH,4CH
INT   21H
```

Their purpose is to return control to the operating system. The last two lines end the procedure and the program, respectively. Note that the label for ENDP (MAIN) matches the label for PROC. The END pseudo-instruction ends the entire program by indicating to DOS that the entry point MAIN has ended. For this reason the labels for the entry point and END must match.

Figure 2-2 shows a sample shell of an Assembly language program. When writing your first few programs, it is handy to keep a copy of this shell on your disk and simply fill it in with the instructions and data for your program.

```
;THE FORM OF AN ASSEMBLY LANGUAGE PROGRAM
; USING SIMPLIFIED SEGMENT DEFINITION
          .MODEL SMALL
          .STACK 64
          .DATA
          ;
          ;place data definitions here
          ;
          .CODE
MAIN      PROC  FAR            ;this is the program entry point
          MOV   AX,@DATA       ;load the data segment address
          MOV   DS,AX          ;assign value to DS
          ;
          ;place code here
          ;
          MOV   AH,4CH         ;set up to
          INT   21H            ;return to DOS
MAIN      ENDP
          END   MAIN           ;this is the program exit point
```

Figure 2-2. Shell of an Assembly Language Program

SECTION 2.1: DIRECTIVES AND A SAMPLE PROGRAM

Review Questions

1. What is the purpose of pseudo-instructions?
2. _____ are translated by the assembler into machine code, whereas _____ are not.
3. Write an Assembly language program with the following characteristics:
   (a) A data item named HIGH_DAT, which contains 95
   (b) Instructions that move HIGH_DAT to registers AH, BH, and DL
   (c) A program entry point named START
4. Find the errors in the following:

```
        .MODEL ENORMOUS
        .STACK
        .CODE
        .DATA
MAIN    PROC  FAR
        MOV   AX,DATA
        MOV   DS,@DATA
        MOV   AL,34H
        ADD   AL,4FH
        MOV   DATA1,AL
START ENDP
        END
```

SECTION 2.2: ASSEMBLE, LINK, AND RUN A PROGRAM

Now that the basic form of an Assembly language program has been given, the next question is: How is it created and assembled? The three steps to create an executable Assembly language program are outlined as follows:

Step	Input	Program	Output
1. Edit the program	keyboard	editor	myfile.asm
2. Assemble the program	myfile.asm	MASM or TASM	myfile.obj
3. Link the program	myfile.obj	LINK or TLINK	myfile.exe

The MASM and LINK programs are the assembler and linker programs for Microsoft's MASM assembler. If you are using another assembler, such as Borland's TASM, consult the manual for the procedure to assemble and link a program. Many excellent editors or word processors are available that can be used to create and/or edit the program. The editor must be able to produce an ASCII file. Although filenames follow the usual DOS conventions, the source file must end in ".asm" for the assembler used in this book. This ".asm" source file is assembled by an assembler, such as Microsoft's MASM, or Borland's TASM. The assembler will produce an object file and a list file, along with other

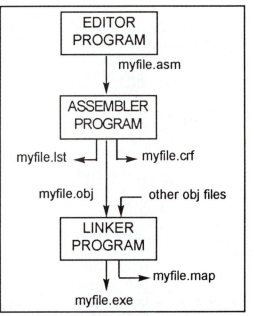

Figure 2-3. Steps to Create a Program

files that may be useful to the programmer. The extension for the object file must be ".obj". This object file is input to the LINK program, which produces the executable program that ends in ".exe". The ".exe" file can be executed by the microprocessor. Before feeding the ".obj" file into LINK, all syntax errors produced by the assembler must be corrected. Of course, fixing these errors will not guarantee that the program will work as intended since the program may contain conceptual errors. Figure 2-3 shows the steps in producing an executable file.

Figure 2-4 shows how an executable program is created by following the steps outlined above, and then run under DEBUG. The portions in bold indicate what the user would type in to perform these steps. Figure 2-4 assumes that the MASM, LINK, and DEBUG programs are on drive C and the Assembly language program is on drive A. The drives used will vary depending on how the system is set up.

```
C>MASM A:MYFILE.ASM <enter>

Microsoft (R) Macro Assembler  Version  5.10
Copyright (C) Microsoft Corp 1981, 1988.  All rights reserved.

Object filename [A:MYFILE.OBJ]: A: <enter>
Source listing  [NUL.LST]:A:MYFILE.LST  <enter>
Cross-reference [NUL.CRF]: <enter>

    47962 + 413345 Bytes symbol space free

        0 Warning Errors
        0 Severe  Errors


C>LINK A:MYFILE.OBJ <enter>

Microsoft (R) Overlay Linker  Version 3.64
Copyright (C) Microsoft Corp 1983-1988.  All rights reserved.

Run File [A:MYFILE.EXE]:A:<enter>
List File [NUL.MAP]: <enter>
Libraries [.LIB]:<enter>
LINK : warning L4021: no stack segment

 C>DEBUG A:MYFILE.EXE <enter>
 -U CS:0 1 <enter>
 1064:0000 B86610        MOV     AX,1066
 -D 1066:0 F <enter>
 1066:0000 52 29 00 00 00 00 00 00-00 00 00 00 00 00 00 00  R)..............
-G <enter>
Program terminated normally
-D 1066:0 F <enter>
1066:0000 52 29 7B 00 00 00 00 00-00 00 00 00 00 00 00 00  R){.............
-Q <enter>
C>
```

Figure 2-4. Creating and Running the .exe File
Note: The parts you type in are printed in **bold**.

.asm and .obj files

The ".asm" file (the source file) is the file created with a word processor or line editor. The MASM (or other) assembler converts the .asm file's Assembly language instructions into machine language (the ".obj" object file). In addition to creating the object program, MASM also creates the ".lst" list file.

.lst file

The ".lst" file, which is optional, is very useful to the programmer because it lists all the opcodes and offset addresses as well as errors that MASM detected. MASM assumes that the list file is not wanted (NUL.LST indicates no list). To get a list file, type in a filename after the prompt. This file can be displayed on the monitor or sent to the printer. The programmer uses it to help debug the program. It is only after fixing all the errors indicated in the ".lst" file that the ".obj" file can be input to the LINK program to create the executable program.

One way to look at the list file is to use the following command at the DOS level. This command will print myfile.lst to the monitor, one sceen at a time.

C>type myfile.lst | more

Another way to look at the list file is to bring it into a word processor. Then you can read it or print it. There are two assembler directives that can be used to make the ".lst" file more readable: PAGE and TITLE.

PAGE and TITLE directives

The format of the PAGE directive is

PAGE [lines],[columns]

and its function is to tell the printer how the list should be printed. In the default mode, meaning that the PAGE directive is coded with no numbers coming after it, the output will have 66 lines per page with a maximum of 80 characters per line. In this book, programs will change the default settings to 60 and 132 as follows:

PAGE 60,132

The range for number of lines is 10 to 255 and for columns is 60 to 132. When the list is printed and it is more than one page, the assembler can be instructed to print the title of the program on top of each page. What comes after the TITLE pseudo-instruction is up to the programmer, but it is common practice to put the name of the program as stored on the disk immediately after the TITLE pseudo-instruction and then a brief description of the function of the program. The text after the TITLE pseudo-instruction cannot be more than 60 ASCII characters.

.crf file

MASM produces another optional file, the cross-reference, which has the extension ".crf". It provides an alphabetical list of all symbols and labels used in the program as well as the program line numbers in which they are referenced. This can be a great help in large programs with many data segments and code segments.

CHAPTER 2: ASSEMBLY LANGUAGE PROGRAMMING

LINKing the program

The assembler (MASM) creates the opcodes, operands, and offset addresses under the ".obj" file. It is the LINK program that produces the ready-to-run version of a program that has the ".exe" (EXEcutable) extension. The LINK program sets up the file so that it can be loaded by DOS and executed.

In Figure 2-4 we used DEBUG to execute the program in Figure 2-1 and analyze the result. In the program in Figure 2-1, three data items are defined in the data segment. Before running the program, one could look at the data in the data segment by dumping the contents of DS:offset as shown in Figure 2-4. Now what is the value for the DS register? This can vary from PC to PC and from DOS to DOS. For this reason it is important to look at the value in "MOV AX,xxxx" as was shown and use that number. The result of the program can be verified after it is run as shown in Figure 2-4. When the program is working successfully, it can be run at the DOS level. To execute myfile.exe, simply type in

C>A:myfile

However, since this program produces no output, there would be no way to verify the results. When the program name is typed in at the DOS level, as shown above, DOS loads the program in memory. This is sometimes referred to as *mapping*, which means that the program is mapped into the physical memory of the PC.

.map file

When there are many segments for code or data, there is a need to see where each is located and how many bytes are used by each. This is provided by the map file. This file, which is optional, gives the name of each segment, where it starts, where it stops, and its size in bytes. In Chapter 7 the importance of the map will be seen when many separate subroutines (modules) are assembled separately and then linked together.

Review Questions

1. (a) The input file to the MASM assembler program has the extension _____.
   (b) The input file to the LINK program has the extension _____.
2. Select all the file types from the second column that are the output of the program in the first column.

   _____ Editor   (a) .obj (b) .asm

   _____ Assembler   (c) .exe (d) .lst

SECTION 2.3: MORE SAMPLE PROGRAMS

      _____ Linker   (e) .crf (f) .map

Now that some familiarity with Assembly language programming in the IBM PC has been achieved, in this section we look at more example programs in order to allow the reader to master the basic features of Assembly programming. The following pages show Program 2-1 and the list file generated when the program was assembled. After the program was assembled and linked, DEBUG was used to dump the code segment to see what value is assigned to the DS register. Precisely where DOS loads a program into RAM depends on many factors, including the amount of RAM on the system and the version of DOS used. Therefore, remember that the value you get could be different for "MOV AX,xxxx" as well as for CS in the program examples. Do not attempt to modify the segment register contents to conform to those in the examples, or your system may crash!

Write, run, and analyze a program that adds 5 bytes of data and saves the result. The data should be the following hex numbers: 25, 12, 15, 1F, and 2B.

```
PAGE         60,132
TITLE        PROG2-1  (EXE)   PURPOSE: ADDS 5 BYTES OF DATA
             .MODEL SMALL
             .STACK 64
;————————————————————————
             .DATA
DATA_IN      DB     25H,12H,15H,1FH,2BH
SUM          DB              ?
;————————————————————————
             .CODE
MAIN         PROC FAR
             MOV    AX,@DATA
             MOV    DS,AX
             MOV    CX,05                 ;set up loop counter CX=5
             MOV    BX,OFFSET DATA_IN     ;set up data pointer BX
             MOV    AL,0                  ;initialize AL
AGAIN:       ADD    AL,[BX]               ;add next data item to AL
             INC    BX                    ;make BX point to next data item
             DEC    CX                    ;decrement loop counter
             JNZ    AGAIN                 ;jump if loop counter not zero
             MOV    SUM,AL                ;load result into sum
             MOV    AH,4CH                ;set up return
             INT    21H                   ;return to DOS
MAIN         ENDP
             END    MAIN
```

After the program was assembled and linked, it was run using DEBUG:

```
C>debug prog2-1.exe
-u cs:0 19
1067:0000 B86610      MOV    AX,1066
1067:0003 8ED8        MOV    DS,AX
1067:0005 B90500      MOV    CX,0005
1067:0008 BB0000      MOV    BX,0000
1067:000D 0207        ADD    AL,[BX]
1067:000F 43          INC    BX
1067:0010 49          DEC    CX
1067:0013 A20500      MOV    [0005],AL
1067:0016 B44C        MOV    AH,4C
1067:0018 CD21        INT    21
-d 1066:0 f
1066:0000  25 12 15 1F 2B 00 00 00-00 00 00 00 00 00 00 00 %...+.........
-g

Program terminated normally
-d 1066:0 f
1066:0000  25 12 15 1F 2B 96 00 00-00 00 00 00 00 00 00 00 %...+.........
-q
C>
```

Program 2-1

Analysis of Program 2-1

The DEBUG program is explained thoroughly in Appendix A. The commands used in running Program 2-1 were (1) u, to unassemble the code from cs:0 for 19 bytes; (2) d, to dump the contents of memory from 1066:0 for the next F bytes; and (3) g, to go, that is, run the program.

Notice in Program 2-1 that when the program was run in DEBUG, the contents of the data segment memory were dumped before and after execution of the program to verify that the program worked as planned. Normally, it is not necessary to unassemble this much code, but it was done here because in later sec-

```
 1   0000                              .MODEL SMALL
 2   0000                              .STACK 64
 3                              ;_____
 4   0000                              .DATA
 5   0000  25 12 15 1F 2B   DATA_IN   DB      25H,12H,15H,1FH,2BH
 6   0005  ??              SUM       DB      ?
 7                              ;_____
 8   0006                              .CODE
 9   0000              MAIN      PROC    FAR
10   0000  B8 0000s              MOV     AX,@DATA
11   0003  8E D8                 MOV     DS,AX
12   0005  B9 0005               MOV     CX,05          ;set up loop counter CX=5
13   0008  BB 0000r              MOV     BX,OFFSET DATA_IN   ;set up data
14   000B  B0 00                 MOV     AL,0            ;initialize AL
15   000D  02 07    AGAIN:       ADD     AL,[BX]         ;add next data item  to AL
16   000F  43                    INC     BX              ;make BX point to next
17   0010  49                    DEC     CX              ;decrement loop counter
18   0011  75 FA                 JNZ     AGAIN           ;jump if counter not zero
19   0013  A2 0005r              MOV     SUM,AL          ;load result into sum
20   0016  B4 4C                 MOV     AH,4CH          ;set up return
21   0018  CD 21                 INT     21H             ;return to DOS
22   001A          MAIN      ENDP
23                              END     MAIN
```

Symbol Name	Type	Value
??DATE	Text	"06/25/99"
??FILENAME	Text	"test21 "
??TIME	Text	"12:05:32"
??VERSION	Number	0300
@32BIT	Text	0
@CODE	Text	_TEXT
@CODESIZE	Text	0
@CPU	Text	0101H
@CURSEG	Text	_TEXT
@DATA	Text	DGROUP
@DATASIZE	Text	0
@FILENAME	Text	TEST21
@INTERFACE	Text	00H
@MODEL	Text	2
@STACK	Text	DGROUP
@WORDSIZE	Text	2
AGAIN	Near	_TEXT:000D
DATA_IN	Byte	DGROUP:0000
MAIN	Far	_TEXT:0000
SUM	Byte	DGROUP:0005

Groups & Segments	Bit Size Align Combine Class
DGROUP	Group
STACK	16 0040 Para Stack STACK
_DATA	16 0006 Word Public DATA
_TEXT	16 001A Word Public CODE

List File for Program 2-1

tions of the chapter we examine the jump instruction in this program. Also notice that the first 5 bytes dumped above are the data items defined in the data segment of the program and the sixth item is the sum of those five items, so it appears that the program worked correctly (25H + 12H + 15H + 1FH + 2BH = 96H). Program 2-1 is explained below, instruction by instruction.

"MOV CX,05" will load the value 05 into the CX register. This register is used by the program as a counter for iteration (looping).

"MOV BX,OFFSET DATA_IN" will load into BX the offset address assigned to DATA. The assembler starts at offset 0000 and uses memory for the data and then assigns the next available offset memory for SUM (in this case, 0005).

"ADD AL,[BX]" adds the contents of the memory location pointed at by the register BX to AL. Note that [BX] is a pointer to a memory location.

"INC BX" simply increments the pointer by adding 1 to register BX. This will cause BX to point to the next data item, that is, the next byte.

"DEC CX" will decrement (subtract 1 from) the CX counter and will set the zero flag high if CX becomes zero.

"JNZ AGAIN" will jump back to the label AGAIN as long as the zero flag is indicating that CX is not zero. "JNZ AGAIN" will not jump (that is, execution will resume with the next instruction after the JNZ instruction) only after the zero flag has been set high by the "DEC CX" instruction (that is, CX becomes zero). When CX becomes zero, this means that the loop is completed and all five numbers have been added to AL.

Various approaches to Program 2-1

There are many ways in which any program may be written. The method shown for Program 2-1 defined one field of data and used pointer [BX] to access data elements. In the method used below, a name is assigned to each data item that will be accessed in the program. Variations of Program 2-1 are shown below to clarify the use of addressing modes in the context of a real program and also to show that the 80x86 can use any general-purpose register to do arithmetic and logic operations. In earlier-generation CPUs, the accumulator had to be the destination of all arithmetic and logic operations, but in the 80x86 this is not the case. Since the purpose of these examples is to show different ways of accessing operands, it is left to the reader to run and analyze the programs.

```
;from the data segment:
DATA1  DB 25H
DATA2  DB 12H
DATA3  DB 15H
DATA4  DB 1FH
DATA5  DB 2BH
SUM DB  ?
;from the code segment:
MOV  AL,DATA1     ;MOVE DATA1 INTO AL
ADD  AL,DATA2     ;ADD DATA2 TO AL
ADD  AL,DATA3
ADD  AL,DATA4
ADD  AL,DATA5
MOV  SUM,AL       ;SAVE AL IN SUM
```

There is quite a difference between these two methods of writing the same program. While in the first one the register indirect addressing mode was used to access the data, in the second method the direct addressing mode was used.

Write and run a program that adds four words of data and saves the result. The values will be 234DH, 1DE6H, 3BC7H, and 566AH. Use DEBUG to verify the sum is D364.

```
TITLE          PROG2-2  (EXE)  PURPOSE: ADDS 4 WORDS OF DATA
PAGE   60,132
                .MODEL SMALL
                .STACK 64
;_____
                .DATA
DATA_IN         DW     234DH,1DE6H,3BC7H,566AH
                ORG    10H
SUM             DW .   ?
;_____
                .CODE
MAIN            PROC FAR
                MOV    AX,@DATA
                MOV    DS,AX
                MOV    CX,04            ;set up loop counter CX=4
                MOV    DI,OFFSET DATA_IN  ;set up data pointer DI
                MOV    BX,00            ;initialize BX
ADD_LP:         ADD    BX,[DI]          ;add contents pointed at by [DI] to BX
                INC    DI               ;increment DI twice
                INC    DI               ;to point to next word
                DEC    CX               ;decrement loop counter
                JNZ    ADD_LP           ;jump if loop counter not zero
                MOV    SI,OFFSET SUM    ;load pointer for sum
                MOV    [SI],BX          ;store in data segment
                MOV    AH,4CH           ;set up return
                INT    21H              ;return to DOS
MAIN            ENDP
                END    MAIN
```

After the program was assembled and linked, it was run using DEBUG:
```
C>debug a:prog2-2.exe
1068:0000 B86610    MOV    AX,1066
-D 1066:0 1F
1066:0000 4D 23 E6 1D C7 3B 6A 56-00 00 00 00 00 00 00 00 M#f.G;jV........
1066:0010 00 00 00 00 00 00 00 00-00 00 00 00 00 00 00 00 ................
-G

Program terminated normally
-D 1066:0 1F
1066:0000 4D 23 E6 1D C7 3B 6A 56-00 00 00 00 00 00 00 00 M#f.G;jV........
1066:0010 64 D3 00 00 00 00 00 00-00 00 00 00 00 00 00 00 dS.............
-Q
C>
```

Program 2-2

Analysis of Program 2-2

First notice that the 16-bit data (a word) is stored with the low-order byte first. For example, "234D" as defined in the data segment is stored as "4D23", meaning that the lower address, 0000, has the least significant byte, 4D, and the higher address, 0001, has the most significant byte, 23. This is shown in the DEBUG display of the data segment. Similarly, the sum, D364, is stored as 64D3. As discussed in Chapter 1, this method of low byte to low address and high byte to high address operand assignment is referred to in computer literature as "little endian."

Second, note that the address pointer is incremented twice, since the operand being accessed is a word (two bytes). The program could have used "ADD DI,2" instead of using "INC DI" twice. When storing the result of word addition, "MOV SI,OFFSET SUM" was used to load the pointer (in this case 0010, as defined by ORG 0010H) for the memory allocated for the label SUM,

---

SECTION 2.3: MORE SAMPLE PROGRAMS   61

and then "MOV [SI],BX" was used to move the contents of register BX to memory locations with offsets 0010 and 0011. Again, as was done previously, it could have been coded simply as "MOV SUM,BX", using the direct addressing mode.

Program 2-2 uses the ORG directive. In previous programs where ORG was not used, the assembler would start at offset 0000 and use memory for each data item. The ORG directive can be used to set the offset addresses for data items. Although the programmer cannot assign exact physical addresses, one is allowed to assign offset addresses. The ORG directive in Program 2-2 caused SUM to be stored at DS:0010, as can be seen by looking at the DEBUG display of the data segment.

Write and run a program that transfers 6 bytes of data from memory locations with offset of 0010H to memory locations with offset of 0028H.

```
TITLE          PROG2-3 (EXE)  PURPOSE: TRANSFERS 6 BYTES OF DATA
PAGE   60,132
               .MODEL SMALL
               .STACK 64
               .DATA
               ORG    10H
DATA_IN        DB              25H,4FH,85H,1FH,2BH,0C4H
               ORG    28H
COPY           DB              6 DUP(?)
;───────────────────────
               .CODE
MAIN           PROC FAR
               MOV    AX,@DATA
               MOV    DS,AX
               MOV    SI,OFFSET DATA_IN   ;SI points to data to be copied
               MOV    DI,OFFSET COPY      ;DI points to copy of data
               MOV    CX,06H              ;loop counter = 6
MOV_LOOP: MOV  AL,[SI]                    ;move the next byte from DATA area to AL
               MOV    [DI],AL             ;move the next byte to COPY area
               INC    SI                  ;increment DATA pointer
               INC    DI                  ;increment COPY pointer
               DEC    CX                  ;decrement LOOP counter
               JNZ    MOV_LOOP            ;jump if loop counter not zero
               MOV    AH,4CH              ;set up to return
               INT    21H                 ;return to DOS
MAIN           ENDP
               END    MAIN
```

After the program was assembled and linked, it was run using DEBUG:

```
C>debug prog2-3.exe
-u cs:0 1
1069:0000 B86610     MOV    AX,1066
-d 1066:0 2f
1066:0000 00 00 00 00 00 00 00 00-00 00 00 00 00 00 00 00 ................
1066:0010 25 4F 85 1F 2B C4 00 00-00 00 00 00 00 00 00 00 %O..+D..........
1066:0020 00 00 00 00 00 00 00 00-00 00 00 00 00 00 00 00 ................
-g

Program terminated normally
-d 1066:0 2f
1066:0000 00 00 00 00 00 00 00 00-00 00 00 00 00 00 00 00 ................
1066:0010 25 4F 85 1F 2B C4 00 00-00 00 00 00 00 00 00 00 %O..+D..........
1066:0020 00 00 00 00 00 00 00 00-25 4F 85 1F 2B C4 00 00 %O..+D..........
-q
C>
```

Program 2-3

Analysis of Program 2-3

The DEBUG example shows the data segment being dumped before the program was run and after to verify that the data was copied and that the program ran successfully. Notice that C4 was coded in the data segments as 0C4. This is required by the assembler to indicate that C is a hex number and not a letter. This is required if the first digit is a hex digit A through F.

This program uses two registers, SI and DI, as pointers to the data items being manipulated. The first is used as a pointer to the data item to be copied and the second as a pointer to the location the data item is to be copied to. With each iteration of the loop, both data pointers are incremented to point to the next byte.

Stack segment definition revisited

One of the primary functions of the DOS operating system is to determine the total amount of RAM installed on the PC and then manage it properly. DOS uses the portion it needs for the operating system and allocates the rest. Since memory requirements vary for different DOS versions, a program cannot dictate the exact physical memory location for the stack or any segment. Since memory management is the responsibility of DOS, it will map Assembly programs into the memory of the PC with the help of LINK.

Although in the DOS environment a program can have multiple code segments and data segments, it is strongly recommended that it have only one stack segment, to prevent RAM fragmentation by the stack. It is the function of LINK to combine all different code and data segments to create a single executable program with a single stack, which is the stack of the system. Various options for segment definition are discussed in Chapter 7 and many of these concepts are explained there.

Review Questions

1. What is the purpose of the INC instruction?
2. What is the purpose of the DEC instruction?
3. In Program 2-1, why does the label AGAIN have a colon after it, whereas the label MAIN does not?
4. State the difference between the following two instructions:
   ```
   MOV   BX,DATA1
   MOV   BX,OFFSET DATA1
   ```
5. State the difference between the following two instructions:
   ```
   ADD   AX,BX
   ADD   AX,[BX]
   ```

SECTION 2.4: CONTROL TRANSFER INSTRUCTIONS

In the sequence of instructions to be executed, it is often necessary to transfer program control to a different location. There are many instructions in the 80x86 to achieve this. This section covers the control transfer instructions available in the 8086 Assembly language. Before that, however, it is necessary to explain the concept of FAR and NEAR as it applies to jump and call instructions.

FAR and NEAR

If control is transferred to a memory location within the current code segment, it is NEAR. This is sometimes called *intrasegment* (within segment). If control is transferred outside the current code segment, it is a FAR or intersegment (between segments) jump. Since the CS:IP registers always point to the address of the next instruction to be executed, they must be updated when a control transfer instruction is executed. In a NEAR jump, the IP is updated and CS remains the same, since control is still inside the current code segment. In a FAR jump, because control is passing outside the current code segment, both CS and IP have to be updated to the new values. In other words, in any control transfer instruction such as jump or call, the IP must be changed, but only in the FAR case is the CS changed, too.

Conditional jumps

Conditional jumps, summarized in Table 2-1, have mnemonics such as JNZ (jump not zero) and JC (jump if carry). In the conditional jump, control is transferred to a new location if a certain condition is met. The flag register is the one that indicates the current condition. For example, with "JNZ label", the processor looks at the zero flag to see if it is raised. If not, the CPU starts to fetch and execute instructions from the address of the label. If ZF = 1, it will not jump but will execute the next instruction below the JNZ.

Table 2-1: 8086 Conditional Jump Instructions

Mnemonic	Condition Tested	"Jump IF ..."
JA/JNBE	(CF = 0) and (ZF = 0)	above/not below nor zero
JAE/JNB	CF = 0	above or equal/not below
JB/JNAE	CF = 1	below/not above nor equal
JBE/JNA	(CF or ZF) = 1	below or equal/not above
JC	CF = 1	carry
JE/JZ	ZF = 1	equal/zero
JG/JNLE	((SF xor OF) or ZF) = 0	greater/not less nor equal
JGE/JNL	(SF xor OF) = 0	greater or equal/not less
JL/JNGE	(SF xor OR) = 1	less/not greater nor equal
JLE/JNG	((SF xor OF) or ZF) = 1	less or equal/not greater
JNC	CF = 0	not carry
JNE/JNZ	ZF = 0	not equal/not zero
JNO	OF = 0	not overflow
JNP/JPO	PF = 0	not parity/parity odd
JNS	SF = 0	not sign
JO	OF = 1	overflow
JP/JPE	PF = 1	parity/parity equal
JS	SF = 1	sign

Note:
"Above" and "below" refer to the relationship of two unsigned values; "greater" and "less" refer to the relationship of two signed values.
(Reprinted by permission of Intel Corporation, Copyright Intel Corp. 1989)

Short jumps

All conditional jumps are short jumps. In a short jump, the address of the target must be within -128 to +127 bytes of the IP. In other words, the conditional jump is a two-byte instruction: one byte is the opcode of the J condition and the second byte is a value between 00 and FF. An offset range of 00 to FF gives 256 possible addresses; these are split between backward jumps (to -128) and forward jumps (to +127).

In a jump backward, the second byte is the 2's complement of the displacement value. To calculate the target address, the second byte is added to the

IP of the instruction after the jump. To understand this, look at the unassembled code of Program 2-1 for the instruction JNZ AGAIN, repeated below.

```
1067:0000 B86610      MOV   AX,1066
1067:0003 8ED8        MOV   DS,AX
1067:0005 B90500      MOV   CX,0005
1067:0008 BB0000      MOV   BX,0000
1067:000D 0207        ADD   AL,[BX]
1067:000F 43          INC   BX
1067:0010 49          DEC   CX
1067:0011 75FA        JNZ   000D
1067:0013 A20500      MOV   [0005],AL
1067:0016 B44C        MOV   AH,4C
1067:0018 CD21        INT   21
```

The instruction "JNZ AGAIN" was assembled as "JNZ 000D", and 000D is the address of the instruction with the label AGAIN. The instruction "JNZ 000D" has the opcode 75 and the target address FA, which is located at offset addresses 0011 and 0012. This is followed by "MOV SUM,AL", which is located beginning at offset address 0013. The IP value of MOV, 0013, is added to FA to calculate the address of label AGAIN (0013 + FA = 000D) and the carry is dropped. In reality, FA is the 2's complement of -6, meaning that the address of the target is -6 bytes from the IP of the next instruction.

Similarly, the target address for a forward jump is calculated by adding the IP of the following instruction to the operand. In that case the displacement value is positive, as shown next. Below is a portion of a list file showing the opcodes for several conditional jumps.

```
0005   8A 47 02 AGAIN:     MOV   AL,[BX]+2
0008   3C 61               CMP   AL,61H
000A   72 06               JB    NEXT
000C   3C 7A               CMP   AL,7AH
000E   77 02               JA    NEXT
0010   24 DF               AND   AL,0DFH
0012   88 04   NEXT:       MOV   [SI],AL
```

In the program above, "JB NEXT" has the opcode 72 and the target address 06 and is located at IP = 000A and 000B. The jump will be 6 bytes from the next instruction, which is IP = 000C. Adding gives us 000CH + 0006H = 0012H, which is the exact address of the NEXT label. Look also at "JA NEXT", which has 77 and 02 for the opcode and displacement, respectively. The IP of the following instruction, 0010, is added to 02 to get 0012, the address of the target location.

It must be emphasized that regardless of whether the jump is forward or backward, for conditional jumps the address of the target address can never be more than -128 to + 127 bytes away from the IP associated with the instruction following the jump (- for the backward jump and + for the forward jump). If any attempt is made to violate this rule, the assembler will generate a "relative jump out of range" message. These conditional jumps are sometimes referred to as SHORT jumps.

Unconditional jumps

"JMP label" is an unconditional jump in which control is transferred unconditionally to the target location label. The unconditional jump can take the following forms:

---

1. SHORT JUMP, which is specified by the format "JMP SHORT label". This is a jump in which the address of the target location is within −128 to +127 bytes of memory relative to the address of the current IP. In this case, the opcode is EB and the operand is 1 byte in the range 00 to FF. The operand byte is added to the current IP to calculate the target address. If the jump is backward, the operand is in 2's complement. This is exactly like the J condition case. Coding the directive "short" makes the jump more efficient in that it will be assembled into a 2-byte instruction instead of a 3-byte instruction.

2. NEAR JUMP, which is the default, has the format "JMP label". This is a near jump (within the current code segment) and has the opcode E9. The target address can be any of the addressing modes of direct, register, register indirect, or memory indirect:

   (a) Direct JUMP is exactly like the short jump explained earlier, except that the target address can be anywhere in the segment within the range +32767 to −32768 of the current IP.

   (b) Register indirect JUMP; the target address is in a register. For example, in "JMP BX", IP takes the value BX.

   (c) Memory indirect JMP; the target address is the contents of two memory locations pointed at by the register. Example: "JMP [DI]" will replace the IP with the contents of memory locations pointed at by DI and DI+1.

3. FAR JUMP which has the format "JMP FAR PTR label". This is a jump out of the current code segment, meaning that not only the IP but also the CS is replaced with new values.

CALL statements

Another control transfer instruction is the CALL instruction, which is used to call a procedure. CALLs to procedures are used to perform tasks that need to be performed frequently. This makes a program more structured. The target address could be in the current segment, in which case it will be a NEAR call or outside the current CS segment, which is a FAR call. To make sure that after execution of the called subroutine the microprocessor knows where to come back, the microprocessor automatically saves the address of the instruction following the call on the stack. It must be noted that in the NEAR call only the IP is saved on the stack, and in a FAR call both CS and IP are saved. When a subroutine is called, control is transferred to that subroutine and the processor saves the IP (and CS in the case of a FAR call) and begins to fetch instructions from the new location. After finishing execution of the subroutine, for control to be transferred back to the caller, the last instruction in the called subroutine must be RET (return). In the same way that the assembler generates different opcode for FAR and NEAR calls, the opcode for the RET instruction in the case of NEAR and FAR is different, as well. For NEAR calls, the IP is restored; for FAR calls, both CS and IP are restored. This will ensure that control is given back to the caller. As an example, assume that SP = FFFEH and the following code is a portion of the program unassembled in DEBUG:

```
12B0:0200  BB1295  MOV BX,9512
12B0:0203  E8FA00  CALL 0300
12B0:0206  B82F14  MOV AX,142F
```

Since the CALL instruction is a NEAR call, meaning that it is in the same code segment (different IP, same CS), only IP is saved on the stack. In this case, the IP address of the instruction after the call is saved on the stack as shown in Figure 2-5. That IP will be 0206, which belongs to the "MOV AX,142F" instruction.

The last instruction of the called subroutine must be a RET instruction which directs the CPU to POP the top 2 bytes of the stack into the IP and resume executing at offset address 0206. For this reason, the number of PUSH and POP instructions (which alter the SP) must match. In other words, for every PUSH there must be a POP.

```
12B0:0300  53  PUSH BX
12B0:0301  ...  ...... ..
...........  ...  ...... ..
12B0:0309  5B  POP BX
12B0:030A  C3  RET
```

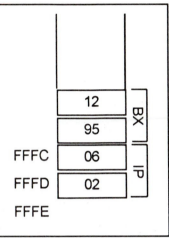

Figure 2-5. IP in the Stack

Assembly language subroutines

In Assembly language programming it is common to have one main program and many subroutines to be called from the main program. This allows you to make each subroutine into a separate module. Each module can be tested separately and then brought together, as will be shown in Chapter 7. The main program is the entry point from DOS and is FAR, as explained earlier, but the subroutines called within the main program can be FAR or NEAR. Remember that NEAR routines are in the same code segment, while FAR routines are outside the current code segment. If there is no specific mention of FAR after the directive PROC, it defaults to NEAR, as shown in Figure 2-6. From now on, all code segments will be written in that format.

Rules for names in Assembly language

By choosing label names that are meaningful, a programmer can make a program much easier to read and maintain. There are several rules that names must follow. First, each label name must be unique. The names used for labels in Assembly language programming consist of alphabetic letters in both upper and lower case, the digits 0 through 9, and the special characters question mark (?), period (.), at (@), underline (_), and dollar sign ($). The first character of the name must be an alphabetic character or special character. It cannot be a digit. The period can only be used as the first character, but this is not recommended since later versions of MASM have several reserved words that begin with a period. Names may be up to 31 characters long. A list of reserved words is given at the end of Appendix C.

Review Questions

1. If control is transferred outside the current code segment, is it NEAR or FAR?
2. If a conditional jump is not taken, what is the next instruction to be executed?
3. In calculating the target address to jump to, a displacement is added to the contents of register _____.
4. What is the advantage in coding the operator "SHORT" in an unconditional jump?
5. A(n) _____ jump is within −128 to +127 bytes of the current IP. A(n) _____ jump is within the current code segment, whereas a(n) _____ jump is outside the current code segment.
6. How does the CPU know where to return to after executing a RET?
7. Describe briefly the function of the RET instruction.
8. State why the following label names are invalid.
   (a) GET.DATA   (b) 1_NUM   (c) TEST-DATA   (d) RET

```
                .CODE
MAIN            PROC  FAR          ;THIS IS THE ENTRY POINT FOR DOS
                MOV   AX,@DATA
                MOV   DS,AX
                CALL  SUBR1
                CALL  SUBR2
                CALL  SUBR3
                MOV   AH,4CH
                INT   21H
MAIN            ENDP
;
SUBR1           PROC
                ...
                ...
                RET
SUBR1           ENDP
;
SUBR2           PROC
                ...
                ...
                RET
SUBR2           ENDP
;
SUBR3           PROC
                ...
                ...
                RET
SUBR3           ENDP
;
                END   MAIN          ;THIS IS THE EXIT POINT
```

Figure 2-6. Shell of Assembly Language Subroutines

SECTION 2.5: DATA TYPES AND DATA DEFINITION

The assembler supports all the various data types of the 80x86 micro-processor by providing data directives that define the data types and set aside memory for them. In this section we study these directives and how they are used to represent different data types of the 80x86. The application of these directives becomes clearer in the context of examples in subsequent chapters.

80x86 data types

The 8088/86 microprocessor supports many data types, but none are longer than 16 bits wide since the size of the registers is 16 bits. It is the job of the programmer to break down data larger than 16 bits (0000 to FFFFH, or 0 to 65535 in decimal) to be processed by the CPU. Many of these programs are shown in Chapter 3. The data types used by the 8088/86 can be 8-bit or 16-bit, positive or negative. If a number is less than 8 bits wide, it still must be coded as an 8-bit register with the higher digits as zero. Similarly, if the number is less than 16 bits wide it must use all 16 bits, with the rest being 0s. For example, the number 5 is only 3 bits wide (101) in binary, but the 8088/86 will accept it as 05 or "0000 0101" in binary. The number 514 is "10 0000 0010" in binary, but the 8088/86 will accept it as "0000 0010 0000 0010" in binary. The discussion of signed numbers is postponed until later chapters since their representation and application are unique.

Assembler data directives

All the assemblers designed for the 80x86 (8088, 8086, 80188, 80186, 80286, 80386, 80386SX, 80486, and Pentium) microprocessors have standard-ized the directives for data representation. The following are some of the data directives used by the 80x86 microprocessor and supported by all software and hardware vendors of IBM PCs and compatibles.

ORG (origin)

ORG is used to indicate the beginning of the offset address. The number that comes after ORG can be either in hex or in decimal. If the number is not followed by H, it is decimal and the assembler will convert it to hex. Although the ORG directive is used extensively in this book in the data segment to separate fields of data to make it more readable for the student, it can also be used for the offset of the code segment (IP).

DB (define byte)

The DB directive is one of the most widely used data directives in the assembler. It allows allocation of memory in byte-sized chunks. This is indeed the smallest allocation unit permitted. DB can be used to define numbers in decimal, binary, hex, and ASCII. For decimal, the D after the decimal number is optional, but using B (binary) and H (hexadecimal) for the others is required. Regardless of which one is used, the assembler will convert them into hex. To indicate ASCII, simply place the string in single quotation marks ('like this'). The assembler will assign the ASCII code for the numbers or characters automatically. DB is the only directive that can be used to define ASCII strings larger than two characters; therefore, it should be used for all ASCII data definitions. Following are some DB examples:

```
DATA1    DB    25              ;DECIMAL
DATA2    DB    10001001B       ;BINARY
DATA3    DB    12H             ;HEX
         ORG   0010H
DATA4    DB    '2591'          ;ASCII NUMBERS
         ORG   0018H
DATA5    DB    ?               ;SET ASIDE A BYTE
         ORG   0020H
DATA6    DB    'My name is Joe' ;ASCII CHARACTERS
```

```
0000  19                    DATA1 DB   25              ;DECIMAL
0001  89                    DATA2 DB   10001001B       ;BINARY
0002  12                    DATA3 DB   12H             ;HEX
0010                              ORG  0010H
0010  32 35 39 31           DATA4 DB   '259'           ;ASCII NUMBERS
0018                              ORG  0018H
0018  00                    DATA5 DB   ?               ;SET ASIDE A BYTE
0020                              ORG  0020H
0020  4D 79 20 6E 61 6D     DATA6 DB   'My name is Joe' ;ASCII CHARACTERS
      65 20 69 73 20 4A
      6F 65
```

List File for DB Examples

Either single or double quotes can be used around ASCII strings. This can be useful for strings which should contain a single quote such as "O'Leary".

DUP (duplicate)

DUP is used to duplicate a given number of characters. This can avoid a lot of typing. For example, contrast the following two methods of filling six memory locations with FFH:

```
                    ORG   0030H
        DATA7  DB     0FFH,0FFH,0FFH,0FFH,0FFH,0FFH  ;FILL 6 BYTES WITH FF
                    ORG   38H
        DATA8  DB     6 DUP(0FFH)    ;FILL 6 BYTES WITH FF
    ; the following reserves 32 bytes of memory with no initial value given
                    ORG   40H
        DATA9  DB     32 DUP (?)     ;SET ASIDE 32 BYTES
    ;DUP can be used inside another DUP
    ; the following fills 10 bytes with 99
        DATA10    DB     5 DUP (2 DUP (99))  ;FILL 10 BYTES WITH 99
```

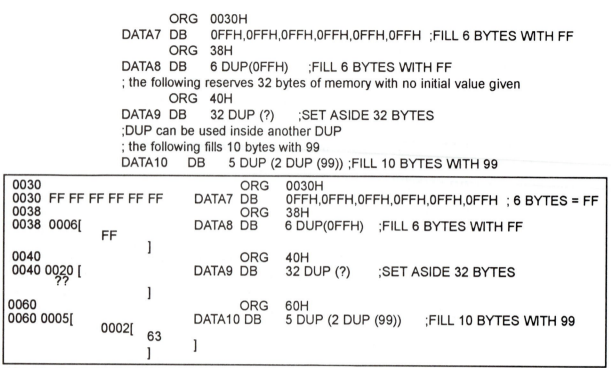

List File for DUP Examples

DW (define word)

DW is used to allocate memory 2 bytes (one word) at a time. DW is used widely in the 8088/8086 and 80286 microprocessors since the registers are 16 bits wide. The following are some examples of DW:

```
              ORG   70H
    DATA11   DW    954                ;DECIMAL
    DATA12   DW    100101010100B      ;BINARY
    DATA13   DW    253FH              ;HEX
              ORG   78H
    DATA14   DW    9,2,7,0CH,00100000B,5,'HI'  ;MISC. DATA
    DATA15   DW    8 DUP (?)              ;SET ASIDE 8 WORDS
```

```
0070                          ORG    70H
0070 03BA                     DATA11   DW    954                ;DECIMAL
0072 0954                     DATA12   DW    100101010100B      ;BINARY
0074 253F                     DATA13   DW    253FH              ;HEX
0078                          ORG    78H
0078 0009 0002 0007 000C      DATA14   DW    9,2,7,0CH,00100000B,5,'HI'  ;MISC. DATA
     0020 0005 4849
0086 0008[                    DATA15   DW    8 DUP (?)              ;SET ASIDE 8 WORDS
        ????                    ]
```

List File for DW Examples

EQU (equate)

This is used to define a constant without occupying a memory location. EQU does not set aside storage for a data item but associates a constant value with a data label so that when the label appears in the program, its constant value will be substituted for the label. EQU can also be used outside the data segment, even in the middle of a code segment. Using EQU for the counter constant in the immediate addressing mode:

COUNT EQU 25

When executing the instructions "MOV CX,COUNT", the register CX will be loaded with the value 25. This is in contrast to using DB:

COUNT DB 25

When executing the same instruction "MOV CX,COUNT" it will be in the direct addressing mode. Now what is the real advantage of EQU? First, note that EQU can also be used in the data segment:

```
COUNT      EQU      25
COUNTER1   DB       COUNT
COUNTER2   DB       COUNT
```

Assume that there is a constant (a fixed value) used in many different places in the data and code segments. By the use of EQU, one can change it once and the assembler will change all of them, rather than making the programmer try to find every location and correct it.

DD (define doubleword)

The DD directive is used to allocate memory locations that are 4 bytes (two words) in size. Again, the data can be in decimal, binary, or hex. In any case the data is converted to hex and placed in memory locations according to the rule of low byte to low address and high byte to high address. DD examples are:

```
           ORG  00A0H
DATA16     DD   1023                       ;DECIMAL
DATA17     DD   10001001011001011100B      ;BINARY
DATA18     DD   5C2A57F2H                   ;HEX
DATA19     DD   23H,34789H,65533
```

```
00A0                                ORG  00A0H
00A0 000003FF       DATA16  DD   1023                       ;DECIMAL
00A4 0008965C       DATA17  DD   10001001011001011100B      ;BINARY
00A8 5C2A57F2       DATA18  DD   5C2A57F2H                   ;HEX
00AC 00000023 00034789   DATA19  DD   23H,34789H,65533
     0000FFFD
```

List File for DD Examples

DQ (define quadword)

DQ is used to allocate memory 8 bytes (four words) in size. This can be used to represent any variable up to 64 bits wide:

```
           ORG  00C0H
DATA20     DQ   4523C2H      ;HEX
DATA21     DQ   'HI'         ;ASCII CHARACTERS
DATA22     DQ   ?            ;NOTHING
```

```
00C0                            ORG  00C0H
00C0 C223450000000000   DATA20  DQ   4523C2H      ;HEX
00C8 4948000000000000   DATA21  DQ   'HI'         ;ASCII CHARACTERS
00D0 0000000000000000   DATA22  DQ   ?            ;NOTHING
```

List File for DQ Examples

SECTION 2.5: DATA TYPES AND DATA DEFINITION 71

DT (define ten bytes)

DT is used for memory allocation of packed BCD numbers. The application of DT will be seen in the multibyte addition of BCD numbers in Chapter 3. For now, observe how they are located in memory. Notice that the "H" after the data is not needed. This allocates 10 bytes, but a maximum of 18 digits can be entered.

```
            ORG  00E0H
DATA23  DT   867943569829    ;BCD
DATA24  DT   ?               ;NOTHING
```

```
00E0                                ORG 00E0H
00E0 299856437986000000  DATA23 DT    867943569829          ;BCD
     00
00EA 000000000000000000  DATA24 DT    ?                     ;NOTHING
     00
```

List File for DT Examples

DT can also be used to allocate 10-byte integers by using the "D" option:

```
DEC  DT  65535d      ;the assembler will convert the decimal
                     ;number to hex and store it
```

Figure 2-7 shows the memory dump of the data section, including all the examples in this section. It is essential to understand the way operands are stored in memory. Looking at the memory dump shows that all of the data directives use the little endian format for storing data, meaning that the least significant byte is located in the memory location of the lower address and the most significant byte resides in the memory location of the higher address. For example, look at the case of "DATA20 DQ 4523C2", residing in memory starting at offset 00C0H. C2, the least significant byte, is in location 00C0, with 23 in 00C1, and 45, the most significant byte, in 00C2. It must also be noted that for ASCII data, only the DB directive can be used to define data of any length, and the use of DD, DQ, or DT directives for ASCII strings of more than 2 bytes gives an assembly error. When DB is used for ASCII numbers, notice how it places them backwards in memory. For example, see "DATA4 DB '2591'" at origin 10H: 32, ASCII for 2, is in memory location 10H; 35, ASCII for 5, is in 11H; and so on.

```
-D 1066:0 100
1066:0000 19 89 12 00 00 00 00 00-00 00 00 00 00 00 00 00  ................
1066:0010 32 35 39 31 00 00 00 00-00 00 00 00 00 00 00 00  2591............
1066:0020 4D 79 20 6E 61 6D 65 20-69 73 20 4A 6F 65 00 00  My name is Joe..
1066:0030 FF FF FF FF FF FF 00 00-FF FF FF FF FF FF 00 00  ................
1066:0040 00 00 00 00 00 00 00 00-00 00 00 00 00 00 00 00  ................
1066:0060 63 63 63 63 63 63 63 63-63 63 00 00 00 00 00 00  cccccccccc......
1066:0070 BA 03 54 09 3F 25 00 00-09 00 02 00 07 00 0C 00  :.T.?%..........
1066:0080 20 00 05 00 4F 48 00 00-00 00 00 00 00 00 00 00  ...OH...........
1066:0090 00 00 00 00 00 00 00 00-00 00 00 00 00 00 00 00  ................
1066:00A0 FF 03 00 00 5C 96 08 00-F2 57 2A 5C 23 00 00 00  ....\...rW*\#...
1066:00B0 89 47 03 00 FD FF 00 00-00 00 00 00 00 00 00 00  B#E......IH.....
1066:00C0 C2 23 45 00 00 00 00 00-49 48 00 00 00 00 00 00  ................
1066:00D0 00 00 00 00 00 00 00 00-00 00 00 00 00 00 00 00  ................
1066:00E0 29 98 56 43 79 86 00 00-00 00 00 00 00 00 00 00  9.VCy6..........
```

Figure 2-7. DEBUG Dump of Data Segment

Review Questions

1. The _____ directive is always used for ASCII strings longer than 2 bytes.
2. How many bytes are defined by the following?

   ```
   DATA_1      DB      6 DUP (4 DUP (0FFH))
   ```
3. Do the following two data segment definitions result in the same storage in bytes at offset 10H and 11H? If not, explain why.

   ```
               ORG  10H          |           ORG  10H
   DATA_1  DB    72              |  DATA_1  DW    7204H
   DATA_2  DB    04H             |
   ```
4. The DD directive is used to allocate memory locations that are ____ bytes in length. The DQ directive is used to allocate memory locations that are _____ bytes in length.
5. State briefly the purpose of the ORG directive.
6. What is the advantage in using the EQU directive to define a constant value?
7. How many bytes are set aside by each of the following directives?

   (a) ASC_DATA DB '1234'   (b) HEX_DATA DW 1234H
8. Does the little endian storage convention apply to the storage of ASCII data?

SECTION 2.6: FULL SEGMENT DEFINITION

The way that segments have been defined in the programs above is a newer definition referred to as *simple segment definition*. It is supported by Microsoft's MASM 5.0 and higher, Borland's TASM version 1 and higher, and many other compatible assemblers. The older, more traditional definition is called the *full segment definition*. Although the simplified segment definition is much easier to understand and use, especially for beginners, it is essential to master full segment definition since many older programs use it.

Segment definition

The "SEGMENT" and "ENDS" directives indicate to the assembler the beginning and ending of a segment and have the following format:

```
label    SEGMENT      [options]
         ;place the statements belonging to this segment here
label    ENDS
```

The label, or name, must follow naming conventions (see the end of Section 2.4) and must be unique. The [options] field gives important information to the assembler for organizing the segment, but is not required. The ENDS label must be the same label as in the SEGMENT directive. In the full segment definition, the ".MODEL" directive is not used. Further, the directives ".STACK", ".DATA", and ".CODE" are replaced by SEGMENT and ENDS directives that surround each segment. Figure 2-8 shows the full segment definition and simplified format, side by side. This is followed by Programs 2-2 and 2-3, rewritten using the full segment definition.

Stack segment definition

The stack segment shown below contains the line: "DB 64 DUP (?)" to reserve 64 bytes of memory for the stack. The following three lines in full segment definition are comparable to ".STACK 64" in simple definition:

```
STSEG       SEGMENT      ;the "SEGMENT" directive begins the segment
            DB 64 DUP (?) ;this segment contains only one line
STSEG       ENDS         ;the "ENDS" segment ends the segment
```

```
;FULL SEGMENT DEFINITION                    ;SIMPLIFIED FORMAT
        ;——— stack segment ———                    .MODEL  SMALL
        name1  SEGMENT                             .STACK   64
                DB      64 DUP (?)        ;
        name1  ENDS                       ;
        ;——— data segment ———            ;————————————————————————
        name2  SEGMENT                               . DATA
        ;data definitions are placed here  ;data definitions are placed here
        name2  ENDS                        ;
        ;——— code segment ———            ;————————————————————————
        name3  SEGMENT                               .CODE
        MAIN   PROC FAR                   MAIN   PROC  FAR
                ASSUME ...                         MOV    AX,@DATA
                MOV   AX,name2                     MOV    DS,AX
                MOV   DS,AX                        ...
                ...                                ...
        MAIN   ENDP                       MAIN   ENDP
        name3  ENDS                                END   MAIN
                END    MAIN
```

Figure 2-8. Full versus Simplified Segment Definition

```
TITLE          PURPOSE: ADDS 4 WORDS OF DATA
PAGE  60,132
STSEG          SEGMENT
               DB     32 DUP (?)
STSEG          ENDS
DTSEG          SEGMENT
DATA_IN        DW     234DH,1DE6H,3BC7H,566AH
               ORG    10H
SUM            DW     ?
DTSEG          ENDS
;————————————————————————
CDSEG          SEGMENT
MAIN           PROC  FAR
               ASSUME CS:CDSEG,DS:DTSEG,SS:STSEG
               MOV   AX,DTSEG
               MOV   DS,AX
               MOV   CX,04              ;set up loop counter CX=4
               MOV   DI,OFFSET DATA_IN  ;set up data pointer DI
               MOV   BX,00              ;initialize BX
ADD_LP:        ADD   BX,[DI]            ;add contents pointed at by [DI] to BX
               INC   DI                 ;increment DI twice
               INC   DI                 ;to point to next word
               DEC   CX                 ;decrement loop counter
               JNZ   ADD_LP             ;jump if loop counter not zero
               MOV   SI,OFFSET SUM      ;load pointer for sum
               MOV   [SI],BX            ;store in data segment
               MOV   AH,4CH             ;set up return
               INT   21H                ;return to DOS
MAIN           ENDP
CDSEG          ENDS
               END    MAIN
```

Program 2-2, rewritten with full segment definition

```
TITLE        PURPOSE: TRANSFERS 6 BYTES OF DATA
PAGE         60,132
STSEG        SEGMENT
             DB              32 DUP (?)
STSEG        ENDS
;─────────────────────
DTSEG        SEGMENT
             ORG    10H
DATA_IN      DB              25H,4FH,85H,1FH,2BH,0C4H
             ORG    28H
COPY         DB              6 DUP(?)
DTSEG        ENDS
;─────────────────────
CDSEG        SEGMENT
MAIN         PROC  FAR
             ASSUME CS:CDSEG,DS:DTSEG,SS:STSEG
             MOV    AX,DTSEG
             MOV    DS,AX
             MOV    SI,OFFSET DATA_IN   ;SI points to data to be copied
             MOV    DI,OFFSET COPY      ;DI points to copy of data
             MOV    CX,06H              ;loop counter = 6
MOV_LOOP:    MOV    AL,[SI]             ;move the next byte from DATA area to AL
             MOV    [DI],AL             ;move the next byte to COPY area
             INC    SI                  ;increment DATA pointer
             INC    DI                  ;increment COPY pointer
             DEC    CX                  ;decrement LOOP counter
             JNZ    MOV_LOOP            ;jump if loop counter not zero
             MOV    AH,4CH              ;set up to return
             INT    21H                 ;return to DOS
MAIN         ENDP
CDSEG        ENDS
             END    MAIN
```

Program 2-3, rewritten with full segment definition

Data segment definition

In full segment definition, the SEGMENT directive names the data segment and must appear before the data. The ENDS segment marks the end of the data segment:

```
DTSEG        SEGMENT      ;the SEGMENT directive begins the segment
             ;define your data here
DTSEG        ENDS         ;the ENDS segment ends the segment
```

Code segment definition

The code segment also begins with a SEGMENT directive and ends with a matching ENDS directive:

```
CDSSEG       SEGMENT      ;the SEGMENT directive begins the segment
             ;your code is here
CDSEG        ENDS         ;the ENDS segment ends the segment
```

In full segment definition, immediately after the PROC directive is the ASSUME directive, which associates segment registers with specific segments by

---

SECTION 2.6: FULL SEGMENT DEFINTION   **75**

assuming that the segment register is equal to the segment labels used in the program. If an extra segment had been used, ES would also be included in the ASSUME statement. The ASSUME statement is needed because a given Assembly language program can have several code segments, one or two or three or more data segments and more than one stack segment, but only one of each can be addressed by the CPU at a given time since there is only one of each of the segment registers available inside the CPU. Therefore, ASSUME tells the assembler which of the segments defined by the SEGMENT directives should be used. It also helps the assembler to calculate the offset addresses from the beginning of that segment. For example, in "MOV AL,[BX]" the BX register is the offset of the data segment.

Upon transfer of control from DOS to the program, of the three segment registers, only CS and SS have the proper values. The DS value (and ES, if used) must be initialized by the program. This is done as follows in full segment definition:

```
MOV AX,DTSEG        ;DTSEG is the label for the data segment
MOV DS,AX
```

SECTION 2.7: EXE VS. COM FILES

All program examples so far were designed to be assembled and linked into EXE files. This section looks at the COM file, which like the EXE file contains the executable machine code and can be run at the DOS level. At the end of this section, the process of conversion from one file to the other is shown.

Why COM files?

There are occasions where, due to a limited amount of memory, one needs to have very compact code. This is the time when the COM file is useful. The fact that the EXE file can be of any size is one of the main reasons that EXE files are used so widely. On the other hand, COM files are used because of their compactness since they cannot be greater than 64K bytes. The reason for the 64K-byte limit is that the COM file must fit into a single segment, and since in the 80x86 the size of a segment is 64K bytes, the COM file cannot be larger than 64K. To limit the size of the file to 64K bytes requires defining the data inside the code segment and also using an area (the end area) of the code segment for the stack. One of the distinguishing features of the COM file program is the fact that in contrast to the EXE file, it has no separate data segment definition. One can summarize the differences between COM and EXE files as shown in Table 2-2.

Table 2-2: EXE vs. COM File Format

EXE File	COM File
unlimited size	maximum size 64K bytes
stack segment is defined	no stack segment definition
data segment is defined	data segment defined in code segment
code, data defined at any offset address	code and data begin at offset 0100H
larger file (takes more memory)	smaller file (takes less memory)

Another reason for the difference in the size of the EXE and COM files is the fact that the COM file does not have a header block. The header block, which occupies 512 bytes of memory, precedes every EXE file and contains information such as size, address location in memory, and stack address of the EXE module.

Program 2-4, written in COM format, adds two words of data and saves the result. This format is very similar to many programs written on the 8080/85 microprocessors, the generation before the 8088/86. This format of first having the code and then the data takes longer to assemble; therefore, it is strongly recommended to put the data first and then the code, but the program must bypass the data area by the use of a JUMP instruction, as shown in Program 2-5.

```
TITLE   PROG2-4 COM PROGRAM TO ADD TWO WORDS
PAGE  60,132
CODSG        SEGMENT
             ORG   100H
             ASSUME  CS:CODSG,DS:CODSG,ES:CODSG
;———THIS IS THE CODE AREA
PROGCODE    PROC  NEAR
             MOV   AX,DATA1      ;move the first word into AX
             MOV   SUM,AX        ;move the sum
             MOV   AH,4CH        ;return to DOS
             INT   21H
PROGCODE    ENDP
;———THIS IS THE DATA AREA
DATA1        DW    2390
DATA2        DW    3456
SUM          DW    ?
;——————————————————
CODSG        ENDS
             END   PROGCODE
```

Program 2-4

```
TITLE        PROG2-5 COM PROGRAM TO ADD TWO WORDS
PAGE         60,132
CODSG        SEGMENT
             ASSUME  CS:CODSG,DS:CODSG,ES:CODSG
             ORG   100H
START:       JMP   PROGCODE     ;go around the data area
;———THIS IS THE DATA AREA
DATA1        DW    2390
DATA2        DW    3456
SUM          DW    ?
;———THIS IS THE CODE AREA
PROGCODE:    MOV   AX,DATA1      ;move the first word into AX
             ADD   AX,DATA1      ;add the second word
             MOV   SUM,AX        ;move the sum
             MOV   AH,4CH
             INT   21H
;------------------------------------
CODSB        ENDS
             END   START
```

Program 2-5

Converting from EXE to COM

For the sake of memory efficiency, it is often desirable to convert an EXE file into a COM file. The source file must be changed to the COM format shown above, then assembled and linked as usual. Then it must be input to a utility program called EXE2BIN that comes with DOS. Its function is to convert the EXE file to a COM file. For example, to convert an EXE file called PROG1.EXE in drive A, assuming that the EXE2BIN utility is in drive C, do the following:

SECTION 2.7: EXE VS. COM FILES

C>EXE2BIN A:PROG1,A:PROG1.COM

Notice that there is no extension of EXE for PROG1 since it is assumed that one is converting an EXE file. Keep in mind that for a program to be converted into a COM file, it must be in the format shown in Programs 2-4 and 2-5.

SUMMARY

An Assembly language program is composed of a series of statements that are either instructions or pseudo-instructions, also called directives. Instructions are translated by the assembler into machine code. Pseudo-instructions are not translated into machine code: They direct the assembler in how to translate the instructions into machine code. The statements of an Assembly language program are grouped into segments. Other pseudo-instructions, often called data directives, are used to define the data in the data segment. Data can be allocated in units ranging in size from byte, word, doubleword, and quadword to 10 bytes at a time. The data can be in binary, hex, decimal, or ASCII.

The flow of a program proceeds sequentially, from instruction to instruction, unless a control transfer instruction is executed. The various types of control transfer instructions in Assembly language include conditional and unconditional jumps, and call instructions.

PROBLEMS

1. Rewrite Program 2-3 to transfer one word at a time instead of one byte.
2. List the steps in getting a ready-to-run program.
3. Which program produces the ".exe" file?
4. Which program produces the ".obj" file?
5. True or false: The ".lst" file is produced by the assembler regardless of whether or not the programmer wants it.
6. The source program file must have the ".asm" extension in some assemblers. such as MASM. Is this true for the assembler you are using?
7. Circle one: The linking process comes (after, before) assembling.
8. In some applications it is common practice to save all registers at the beginning of a subroutine. Assume that SP = 1288H before a subroutine CALL. Show the contents of the stack pointer and the exact memory contents of the stack after PUSHF for the following:

```
1132:0450 CALL PROC1
1132:0453 INC BX

    PROC1    PROC
             PUSH AX
             PUSH BX
             PUSH CX
             PUSH DX
             PUSH SI
             PUSH DI
             PUSHF
             ....
    PROC1    ENDP
```

9. To restore the original information inside the CPU at the end of a CALL to a subroutine, the sequence of POP instructions must follow a certain order. Write the sequence of POP instructions that will restore the information in Problem 8. At each point, show the contents of the SP.

10. When a CALL is executed, how does the CPU know where to return?
11. In a FAR CALL, _____ and _____ are saved on the stack, whereas in a NEAR CALL, _____ is saved on the stack.
12. Compare the number of bytes of stack taken due to NEAR and FAR CALLs.
13. Find the contents of the stack and stack pointer after execution of the CALL instruction shown next.

```
        CS  : IP
        2450:673A CALL SUM
        2450:673D DEC AH
```

SUM is a near procedure. Assume the value SS:1296 right before the execution of CALL.

14. The following is a section of BIOS of the IBM PC which is described in detail in Chapter 3. All the jumps below are short jumps, meaning that the labels are in the range -128 to +127.

```
    IP    Code
    E06C 733F    JNC ERROR1
    ...  ...  ...
    E072 7139    JNO ERROR1
    ...  ...  ...
    E08C 8ED8    C8: MOV DS,AX
    ...  ...  ...
    E0A7 EBE3    JMP C8
    ...  ...  ...
    E0AD F4 ERROR1: HLT
```

Verify the address calculations of:
(a) JNC ERROR1 (b) JNO ERROR1 (c) JMP C8

15. Find the precise offset location of each ASCII character or data in the following:

```
            ORG    20H
    DATA1 DB    '1-800-555-1234'
            ORG    40H
    DATA2 DB    'Name: John Jones'
            ORG    60H
    DATA3 DB    '5956342'
            ORG    70H
    DATA4 DW    2560H,1000000000110B
    DATA5 DW    49
            ORG    80H
    DATA6 DD    25697F6EH
    DATA7 DQ    9E7BA21C99F2H
            ORG    90H
    DATA8 DT    439997924999828
    DATA9 DB    6 DUP (0EEH)
```

16. The following program contains some errors. Fix the errors and make the program run correctly. Verify it through the DEBUG program. This program adds four words and saves the result.

```
    TITLE   PROBLEM (EXE)  PROBLEM 16 PROGRAM
    PAGE    60,132
            .MODEL SMALL
            .STACK 32
    ;---------------------------
            .DATA
    DATA  DW    234DH,DE6H,3BC7H,566AH
            ORG    10H
    SUM   DW    ?
    ;---------------------------
            .CODE
    START: PROC  FAR
            MOV   AX,DATA
            MOV   DS,AX
            MOV   CX,04        ;SET UP LOOP COUNTER CX=4
```

```
         MOV   BX,0              ;INITIALIZE BX TO ZERO
         MOV   DI,OFFSET DATA    ;SET UP DATA POINTER BX
LOOP1:   ADD   BX,[DI] ;ADD CONTENTS POINTED AT BY [DI] TO BX
         INC   DI                ;INCREMENT DI
         JNZ   LOOP1             ;JUMP IF COUNTER NOT ZERO
         MOV   SI,OFFSET RESULT  ;LOAD POINTER FOR  RESULT
         MOV   [SI],BX           ;STORE THE SUM
         MOV   AH,4CH
         INT   21H
START ENDP
         END   STRT
```

ANSWERS TO REVIEW QUESTIONS

SECTION 2.1: DIRECTIVES AND A SAMPLE PROGRAM
1. Pseudo-instructions direct the assembler as to how to assemble the program.
2. Instructions, pseudo-instructions or directives
3.
```
                         .MODEL SMALL
                         .STACK 64
                         .DATA
         HIGH_DAT        DB 95
                         .CODE
                         START  PROC FAR
                         MOV AX,@DATA
                         MOV DS,AX
                         MOV AH,HIGH_DAT
                         MOV BH,AH
                         MOV DL,BH
                         MOV AH,4CH
                         INT 21H
         START           ENDP
                         END START
```
4. (1) there is no ENORMOUS model
   (2) ENDP label does not match label for PROC directive
   (3) .CODE and .DATA directives need to be switched
   (4) "MOV AX,DATA" should be "MOV AX,@DATA"
   (5) "MOV DS,@DATA" should be "MOV DS,AX"
   (6) END must have the entry point label "MAIN"

SECTION 2.2: ASSEMBLE, LINK, AND RUN A PROGRAM
1. (a) MASM must have the ".asm" file as input
   (b) LINK must have the ".obj" file as input
2. Editor outputs : (b) .asm
   Assembler outputs: (a) .obj, (d) .lst, and (e) .crf files
   Linker outputs: (c) .exe and (f) .map files

SECTION 2.3: MORE SAMPLE PROGRAMS
1. increments the operand, that is, it causes 1 to be added to the operand
2. decrements the operand, that is, it causes 1 to be subtracted from the operand
3. a colon is required after labels referring to instructions; colons are not placed after labels for directives
4. the first moves the contents of the word beginning at offset DATA1, the second moves the offset address of DATA1
5. the first adds the contents of BX to AX, the second adds the contents of the memory location at offset BX.

SECTION 2.4: CONTROL TRANSFER INSTRUCTIONS
1. far   2. the instruction right below the jump   3. IP
4. the machine code for the instruction will take up 1 less byte
5. short, near, far
6. the contents of CS and IP were stored on the stack when the call was
    executed
7. it restores the contents of CS:IP and returns control to the instruction
    immediately following the CALL
8. (a) GET.DATA, invalid because "." is only allowed as the first character
    (b) 1_NUM, because the first character cannot be a number
    (c) TEST-DATA, because "-" is not allowed
    (d) RET, is a reserved word

SECTION 2.5: DATA TYPES AND DATA DEFINITION
1. DB   2. 24
3. no because of the little endian storage conventions, which will cause the word
    "7204H" to be stored with the lower byte (04) at offset 10H and the upper byte
    at offset 11H; DB allocates each byte as it is defined
4. 4, 8   5. it is used to assign the offset address
6. if the value is to be changed later, it can be changed in one place instead of
    at every occurrence
7. (a) 4 (b) 2   8. no

CHAPTER 3

ARITHMETIC AND LOGIC INSTRUCTIONS AND PROGRAMS

OBJECTIVES

Upon completion of this chapter, you will be able to:

» Demonstrate how 8-bit and 16-bit unsigned numbers are added in the 80x86
» Convert data to any of the forms: ASCII, packed BCD, or unpacked BCD
» Explain the effect of unsigned arithmetic instructions on the flag register
» Code the following Assembly language unsigned arithmetic instructions:
»   Addition instructions ADD and ADC
»   Subtraction instructions SUB and SBB
»   Multiplication and division instructions MUL and DIV
» Code BCD and ASCII arithmetic instructions:
»   DAA, DAS, AAA, AAS, AAM, and AAD
» Code the following Assembly language logic instructions:
»   AND, OR, and XOR
»   Logical shift instructions SHR and SHL
»   The compare instruction CMP
» Code BCD and ASCII arithmetic instructions
» Code bitwise rotation instructions ROR, ROL, RCR, and RCL
» Demonstrate an ability to use all of the instructions above in Assembly language programs
» Perform bitwise manipulation using the C language

In this chapter, most of the arithmetic and logic instructions are discussed and program examples are given to illustrate the application of these instructions. Unsigned numbers are used in this discussion of arithmetic and logic instructions. Signed numbers are discussed separately in Chapter 6. Unsigned numbers are defined as data in which all the bits are used to represent data and no bits are set aside for the positive or negative sign. This means that the operand can be between 00 and FFH (0 to 255 decimal) for 8-bit data and between 0000 and FFFFH (0 to 65535 decimal) for 16-bit data. The last section of the chapter describes bitwise operations in the C language.

SECTION 3.1: UNSIGNED ADDITION AND SUBTRACTION

Addition of unsigned numbers

The form of the ADD instruction is

ADD   destination,source ;dest. operand = dest. operand + source operand

The instructions ADD and ADC are used to add two operands. The destination operand can be a register or in memory. The source operand can be a register, in memory, or immediate. Remember that memory-to-memory operations are never allowed in 80x86 Assembly language. The instruction could change any of the ZF, SF, AF, CF, or PF bits of the flag register, depending on the operands involved. The effect of the ADD instruction on the overflow flag is discussed in Chapter 6 since it is used in signed number operations. Look at Example 3-1.

Example 3-1

Show how the flag register is affected by

    MOV   AL,0F5H
    ADD   AL,0BH

Solution:

      F5H   1111 0101
    + 0BH   + 0000 1011
    -----   ---------
     100H   0000 0000

After the addition, the AL register (destination) contains 00 and the flags are as follows:
CF = 1 since there is a carry out from D7
SF = 0 the status of D7 of the result
PF = 1 the number of 1s is zero (zero is an even number)
AF = 1 there is a carry from D3 to D4
ZF = 1 the result of the action is zero (for the 8 bits)

In discussing addition, the following two cases will be examined:

1. Addition of individual byte and word data
2. Addition of multibyte data

CASE 1: Addition of individual byte and word data

In Chapter 2 there was a program that added 5 bytes of data. The total sum was purposely kept less than FFH, the maximum value an 8-bit register can hold. To calculate the total sum of any number of operands, the carry flag should be checked after the addition of each operand. Program 3-1a uses AH to accumulate carries as the operands are added to AL.

Write a program to calculate the total sum of 5 bytes of data. Each byte represents the daily wages of a worker. This person does not make more than $255 (FFH) a day. The decimal data is as follows: 125, 235, 197, 91, and 48.

```
TITLE      PROG3-1A (EXE)  ADDING 5 BYTES
PAGE       60,132
.MODEL SMALL
.STACK 64
;────────────────────────────────
           .DATA
COUNT      EQU    05
DATA       DB     125,235,197,91,48
           ORG    0008H
SUM        DW     ?
;────────────────────────────────
           .CODE
MAIN       PROC   FAR
           MOV    AX,@DATA
           MOV    DS,AX
           MOV    CX,COUNT        ;CX is the loop counter
           MOV    SI,OFFSET DATA  ;SI is the data pointer
           MOV    AX,00           ;AX will hold the sum
BACK:      ADD    AL,[SI]         ;add the next byte to AL
           JNC    OVER            ;If no carry, continue
           INC    AH              ;else accumulate carry in AH
OVER:      INC    SI              ;increment data pointer
           DEC    CX              ;decrement loop counter
           JNZ    BACK            ;if not finished, go add next byte
           MOV    SUM,AX          ;store sum
           MOV    AH,4CH
           INT    21H             ;go back to DOS
MAIN       ENDP
           END    MAIN
```

Program 3-1a

Analysis of Program 3-1a

These numbers are converted to hex by the assembler as follows: 125 = 7DH, 235 = 0EBH, 197 = 0C5H, 91 = 5BH, 48 = 30H. Three iterations of the loop are shown below. The tracing of the program is left to the reader as an exercise.

1. In the first iteration of the loop, 7DH is added to AL with CF = 0 and AH = 00. CX = 04 and ZF = 0.
2. In the second iteration of the loop, EBH is added to AL, which results in AL = 68H and CF = 1. Since a carry occurred, AH is incremented. CX = 03 and ZF = 0.
3. In the third iteration, C5H is added to AL, which makes AL = 2DH. Again a carry occurred, so AH is incremented again. CX = 02 and ZF = 0.

This process continues until CX = 00 and the zero flag becomes 1, which will cause JNZ to fall through. Then the result will be saved in the word-sized memory set aside in the data segment. Although this program works correctly, due to pipelining it is strongly recommended that the following lines of the program be replaced:

Replace these lines			With these lines		
BACK:	ADD	AL,[SI]	BACK:	ADD	AL,[SI]
	JNC	OVER		ADC	AH,00 ;add 1 to AH if CF=1
	INC	AH		INC	SI
OVER:	INC	SI			

The "ADC AH,00" instruction in reality means add 00 + AH + CF and place the result in AH. This is much more efficient since the instruction "JNC OVER" has to empty the queue of pipelined instructions and fetch the instructions from the OVER target every time the carry is zero (CF = 0).

The addition of many word operands works the same way. Register AX (or CX or DX or BX) could be used as the accumulator and BX (or any general-purpose 16-bit register) for keeping the carries. Program 3-1b is the same as Program 3-1a, rewritten for word addition.

Write a program to calculate the total sum of five words of data. Each data value represents the yearly wages of a worker. This person does not make more than $65,555 (FFFFH) a year. The decimal data is as follows: 27345, 28521, 29533, 30105, and 32375.

```
TITLE       PROG3-1B  (EXE)   ADDING 5 WORDS
PAGE        60,132
.MODEL SMALL
.STACK 64
;————————————————————
            .DATA
COUNT   EQU     05
DATA    DW      27345,28521,29533,30105,32375
        ORG     0010H
SUM     DW      2 DUP(?)
;————————————————————
            .CODE
MAIN    PROC    FAR
        MOV     AX,@DATA
        MOV     DS,AX
        MOV     CX,COUNT        ;CX is the loop counter
        MOV     SI,OFFSET DATA  ;SI is the data pointer
        MOV     AX,00           ;AX will hold the sum
        MOV     BX,AX           ;BX will hold the carries
BACK:   ADD     AX,[SI]         ;add the next word to AX
        ADC     BX,0            ;add carry to BX
        INC     SI              ;increment data pointer twice
        INC     SI              ;to point to next word
        DEC     CX              ;decrement loop counter
        JNZ     BACK            ;if not finished, continue adding
        MOV     SUM,AX          ;store the sum
        MOV     SUM+2,BX        ;store the carries
        MOV     AH,4CH
        INT     21H             ;go back to DOS
MAIN    ENDP
        END     MAIN
```

Program 3-1b

CASE 2: Addition of multiword numbers

Assume a program is needed that will add the total U. S. budget for the last 100 years or the mass of all the planets in the solar system. In cases like this, the numbers being added could be up to 8 bytes wide or more. Since registers are only 16 bits wide (2 bytes), it is the job of the programmer to write the code to break down these large numbers into smaller chunks to be processed by the CPU. If a 16-bit register is used and the operand is 8 bytes wide, that would take a total of four iterations. However, if an 8-bit register is used, the same operands would require eight iterations. This obviously takes more time for the CPU. This is one reason to have wide registers in the design of the CPU. Large and powerful computers such as the CRAY have registers of 64 bits wide and larger.

Write a program that adds the following two multiword numbers and saves the result:
DATA1 = 548FB9963CE7H and DATA2 = 3FCD4FA23B8DH.

```
TITLE       PROG3-2 (EXE)  MULTIWORD ADDITION
PAGE        60,132
.MODEL SMALL
.STACK 64
;————————————————————————————
            .DATA
DATA1       DQ      548FB9963CE7H
            ORG     0010H
DATA2       DQ      3FCD4FA23B8DH
            ORG     0020H
DATA3       DQ      ?
;————————————————————————————
            .CODE
MAIN        PROC    FAR
            MOV     AX,@DATA
            MOV     DS,AX
            CLC                         ;clear carry before first addition
            MOV     SI,OFFSET DATA1     ;SI is pointer for operand1
            MOV     DI,OFFSET DATA2     ;DI is pointer for operand2
            MOV     BX,OFFSET DATA3     ;BX is pointer for the sum
            MOV     CX,04               ;CX is the loop counter
BACK:       MOV     AX,[SI]             ;move the first operand to AX
            ADC     AX,[DI]             ;add the second operand to AX
            MOV     [BX],AX             ;store the sum
            INC     SI                  ;point to next word of operand1
            INC     SI
            INC     DI                  ;point to next word of operand2
            INC     DI
            INC     BX                  ;point to next word of sum
            INC     BX
            LOOP    BACK                ;if not finished, continue adding
            MOV     AH,4CH
            INT     21H                 ;go back to DOS
MAIN        ENDP
            END     MAIN
```

Program 3-2

Analysis of Program 3-2

In writing this program, the first thing to be decided was the directive used for coding the data in the data segment. DQ was chosen since it can represent data as large as 8 bytes wide. The question is: Which add instruction should be used? In the addition of multibyte (or multiword) numbers, the ADC instruction is always used since the carry must be added to the next-higher byte (or word) in the next iteration. Before executing ADC, the carry flag must be cleared (CF = 0) so that in the first iteration, the carry would not be added. Clearing the carry flag is achieved by the CLC (clear carry) instruction. Three pointers have been used: SI for DATA1, DI for DATA2, and BX for DATA3 where the result is saved. There is a new instruction in that program, "LOOP XXXX", which replaces the often used "DEC CX" and "JNZ XXXX". In other words:

```
LOOP  xxxx   ;is equivalent to the following two instructions

DEC   CX
JNZ   xxxx
```

When the "LOOP xxxx" is executed, CX is decremented automatically, and if CX is not 0, the microprocessor will jump to target address xxxx. If CX is 0, the next instruction (the one below "LOOP xxxx") is executed.

Subtraction of unsigned numbers

SUB dest,source   ;dest = dest − source

In subtraction, the 80x86 microprocessors (indeed, almost all modern CPUs) use the 2's complement method. Although every CPU contains adder circuitry, it would be too cumbersome (and take too many transistors) to design separate subtractor circuitry. For this reason, the 80x86 uses internal adder circuitry to perform the subtraction command. Assuming that the 80x86 is executing simple subtract instructions, one can summarize the steps of the hardware of the CPU in executing the SUB instruction for unsigned numbers, as follows.

1. Take the 2's complement of the subtrahend (source operand).
2. Add it to the minuend (destination operand).
3. Invert the carry.

These three steps are performed for every SUB instruction by the internal hardware of the 80x86 CPU regardless of the source and destination of the operands as long as the addressing mode is supported. It is after these three steps that the result is obtained and the flags are set. Example 3-2 illustrates the three steps.

Example 3-2

Show the steps involved in the following:
```
MOV    AL,3FH         ;load AL=3FH
MOV    BH,23H         ;load BH=23H
SUB    AL,BH          ;subtract BH from AL.  Place result in AL.
```

Solution:

```
  AL    3F   0011 1111           0011 1111
− BH   − 23 − 0010 0011          +1101 1101    (2's complement)
        1C                     1 0001 1100     CF=0 (step 3)
```

The flags would be set as follows: CF = 0, ZF = 0, AF = 0, PF = 0, and SF = 0. The programmer must look at the carry flag (not the sign flag) to determine if the result is positive or negative.

After the execution of SUB, if CF = 0, the result is positive; if CF = 1, the result is negative and the destination has the 2's complement of the result. Normally, the result is left in 2's complement, but the NOT and INC instructions can be used to change it. The NOT instruction performs the 1's complement of the operand; then the operand is incremented to get the 2's complement. See Example 3-3.

Example 3-3

Analyze the following program:
```
;from the data segment:
DATA1   DB  4CH
DATA2   DB  6EH
DATA3   DB  ?
;from the code segment:
        MOV    DH,DATA1       ;load DH with DATA1 value (4CH)
        SUB    DH,DATA2       ;subtract DATA2 (6E) from DH (4CH)
        JNC    NEXT           ;if CF=0 jump to NEXT target
        NOT    DH             ;if CF=1 then take 1's complement
        INC    DH             ;and increment to get 2's complement
NEXT:   MOV    DATA3,DH       ;save DH in DATA3
```

Solution:

Following the three steps for "SUB DH,DATA2":
```
  4C   0100 1100              0100 1100
− 6E   0110 1110   2's comp   +1001 0010
− 22                        0 1101 1110    CF=1 (step 3) the result is negative
```

SECTION 3.1: UNSIGNED ADDITION AND SUBTRACTION   **87**

SBB (subtract with borrow)

This instruction is used for multibyte (multiword) numbers and will take care of the borrow of the lower operand. If the carry flag is 0, SBB works like SUB. If the carry flag is 1, SBB subtracts 1 from the result. Notice the "PTR" operand in Example 3-4. The PTR (pointer) data specifier directive is widely used to specify the size of the operand when it differs from the defined size. In Example 3-4, "WORD PTR" tells the assembler to use a word operand, even though the data is defined as a doubleword.

Example 3-4

Analyze the following program:

```
DATA_A      DD  62562FAH
DATA_B      DD  412963BH
RESULT      DD  ?
...
...   ...
        MOV    AX,WORD PTR DATA_A          ;AX=62FA
        SUB    AX,WORD PTR DATA_B          ;SUB 963B from AX
        MOV    WORD PTR RESULT,AX          ;save the result
        MOV    AX,WORD PTR DATA_A +2       ;AX=0625
        SBB    AX,WORD PTR DATA_B +2       ;SUB 0412 with borrow
        MOV    WORD PTR RESULT+2,AX        ;save the result
```

Solution:

After the SUB, AX = 62FA – 963B = CCBF and the carry flag is set. Since CF = 1, when SBB is executed, AX = 625 – 412 – 1 = 212. Therefore, the value stored in RESULT is 0212CCBF.

Review Questions

1. The ADD instruction that has the syntax "ADD destination, source" replaces the _____ operand with the sum of the two operands.
2. Why is the following ADD instruction illegal?
   ADD DATA_1,DATA_2
3. Rewrite the instruction above in a correct form.
4. The ADC instruction that has the syntax "ADC destination, source" replaces the _____ operand with the sum of _____.
5. The execution of part (a) below results in ZF = 1, whereas the execution of part (b) results in ZF = 0. Explain why.
   (a) MOV BL,04FH   (b) MOV BX,04FH
       ADD BL,0B1H   ADD BX,0B1H
6. The instruction "LOOP ADD_LOOP" is equivalent to what two instructions?
7. Show how the CPU would subtract 05H from 43H.
8. If CF = 1, AL = 95, and BL = 4F prior to the execution of "SBB AL,BL", what will be the contents of AL after the subtraction?

SECTION 3.2: UNSIGNED MULTIPLICATION AND DIVISION

One of the major changes from the 8080/85 microprocessor to the 8086 was inclusion of instructions for multiplication and division. In this section we cover each one with examples. This is multiplication and division of unsigned numbers. Signed numbers are treated in Chapter 6.

In multiplying or dividing two numbers in the 80x86 microprocessor, the use of registers AX, AL, AH, and DX is necessary since these functions assume the use of those registers.

Multiplication of unsigned numbers

In discussing multiplication, the following cases will be examined: (1) byte times byte, (2) word times word, and (3) byte times word.

---

byte × byte: In byte by byte multiplication, one of the operands must be in the AL register and the second operand can be either in a register or in memory as addressed by one of the addressing modes discussed in Chapter 1. After the multiplication, the result is in AX. See the following example:

```
RESULT DW   ?                ;result is defined in the data segment
       ...
       MOV  AL,25H           ;a byte is moved to AL
       MOV  BL,65H           ;immediate data must be in a register
       MUL  BL              ;AL = 25 × 65H
       MOV  RESULT,AX        ;the result is saved
```

In the program above, 25H is multiplied by 65H and the result is saved in word-sized memory named RESULT. In that example, the register addressing mode was used. The next three examples show the register, direct, and register indirect addressing modes.

```
;from the data segment:
DATA1     DB   25H
DATA2     DB   65H
RESULT    DW   ?
;from the code segment:
          MOV  AL,DATA1
          MOV  BL,DATA2
          MUL  BL                     ;register addressing mode
          MOV  RESULT,AX
or
          MOV  AL,DATA1
          MUL  DATA2        ·         ;direct addressing mode
          MOV  RESULT,AX
or
          MOV  AL,DATA1
          MOV  SI,OFFSET DATA2
          MUL  BYTE PTR [SI]          ;register indirect addressing mode
          MOV  RESULT,AX
```

In the register addressing mode example, any 8-bit register could have been used in place of BL. Similarly, in the register indirect example, BX or DI could have been used as pointers. If the register indirect addressing mode is used, the operand size must be specified with the help of the PTR pseudo-instruction. In the absence of the "BYTE PTR" directive in the example above, the assembler could not figure out if it should use a byte or word operand pointed at by SI. This confusion would cause an error.

word × word: In word by word multiplication, one operand must be in AX and the second operand can be in a register or memory. After the multiplication, registers AX and DX will contain the result. Since word × word multiplication can produce a 32-bit result, AX will hold the lower word and DX the higher word. Example:

```
DATA3       DW   2378H
DATA4       DW   2F79H
RESULT1     DW   2 DUP(?)
       .......
       MOV  AX,DATA3         ;load first operand into AX
       MUL  DATA4            ;multiply it by the second operand
       MOV  RESULT1,AX       ;store the lower word result
       MOV  RESULT1+2,DX     ;store the higher word result
```

SECTION 3.2: UNSIGNED MULTIPLICATION AND DIVISION   **89**

word × byte: This is similar to word by word multiplication except that AL contains the byte operand and AH must be set to zero. Example:

```
;from the data segment:
DATA5          DB      6BH
DATA6          DW      12C3H
RESULT3        DW      2 DUP(?)
```

```
;from the code segment:
      ...   ...
      MOV   AL,DATA5            ;AL holds byte operand
      SUB   AH,AH              ;AH must be cleared
      MUL   DATA6              ;byte in AL multiplied by word operand
      MOV   BX,OFFSET RESULT3  ;BX points to storage for product
      MOV   [BX],AX            ;AX holds lower word
      MOV   [BX]+2,DX          ;DX holds higher word
```

Table 3-1 gives a summary of multiplication of unsigned numbers. Using the 80x86 microprocessor to perform multiplication of operands larger than 16-bit size takes some manipulation, although in such cases the 8087 coprocessor is normally used.

Table 3-1: Unsigned Multiplication Summary

Multiplication	Operand 1	Operand 2	Result
byte × byte	AL	register or memory	AX
word × word	AX	register or memory	DX AX
word × byte	AL = byte, AH = 0	register or memory	DX AX

Division of unsigned numbers

In the division of unsigned numbers, the following cases are discussed:

1. Byte over byte
2. Word over word
3. Word over byte
4. Doubleword over word

In divide, there could be cases where the CPU cannot perform the division. In these cases an interrupt is activated. In recent years this is referred to as an *exception*. In what situation can the microprocessor not handle the division and must call an interrupt? They are

1. if the denominator is zero (dividing any number by 00), and
2. if the quotient is too large for the assigned register.

In the IBM PC and compatibles, if either of these cases happens, the PC will display the "divide error" message.

byte/byte: In dividing a byte by a byte, the numerator must be in the AL register and AH must be set to zero. The denominator cannot be immediate but can be in a register or memory as supported by the addressing modes. After the DIV instruction is performed, the quotient is in AL and the remainder is in AH. The following shows the various addressing modes that the denominator can take.

```
DATA7     DB          95
DATA8     DB          10
QOUT1     DB          ?
REMAIN1   DB          ?
```

;using immediate addressing mode will give an error

```
          MOV         AL,DATA7        ;move data into AL
          SUB         AH,AH           ;clear AH
          DIV         10              ;immed. mode not allowed!!
```

;allowable modes include:

;using direct mode

```
          MOV         AL,DATA7        ;AL holds numerator
          SUB         AH,AH           ;AH must be cleared
          DIV         DATA8           ;divide AX by DATA8
          MOV         QOUT1,AL        ;quotient = AL = 09
          MOV         REMAIN1,AH      ;remainder = AH = 05
```

;using register addressing mode

```
          MOV         AL,DATA7        ;AL holds numerator
          SUB         AH,AH           ;AH must be cleared
          MOV         BH,DATA8        ;move denom. to register
          DIV         BH              ;divide AX by BH
          MOV         QOUT1,AL        ;quotient = AL = 09
          MOV         REMAIN1,AH      ;remainder = AH = 05
```

;using register indirect addressing mode

```
          MOV         AL,DATA7            ;AL holds numerator
          SUB         AH,AH               ;AH must be cleared
          MOV         BX,OFFSET DATA8     ;BX holds offset of DATA8
          DIV         BYTE PTR [BX]       ;divide AX by DATA8
          MOV         QOUT2,AX
          MOV         REMAIND2,DX
```

word/word: In this case the numerator is in AX and DX must be cleared. The denominator can be in a register or memory. After the DIV, AX will have the quotient and the remainder will be in DX.

```
          MOV         AX,10050        ;AX holds numerator
          SUB         DX,DX           ;DX must be cleared
          MOV         BX,100          ;BX used for denominator
          DIV         BX
          MOV         QOUT2,AX        ;quotient = AX = 64H = 100
          MOV         REMAIND2,DX     ;remainder = DX = 32H = 50
```

word/byte: Again, the numerator is in AX and the denominator can be in a register or memory. After the DIV instruction, AL will contain the quotient, and AH will contain the remainder. The maximum quotient is FFH. The following program divides AX = 2055 by CL=100. Then AL = 14H (20 decimal) is the quotient and AH = 37H (55 decimal) is the remainder.

```
          MOV         AX,2055         ;AX holds numerator
          MOV         CL,100          ;CL used for denominator
          DIV         CL
          MOV         QUO,AL          ;AL holds quotient
          MOV         REMI,AH         ;AH holds remainder
```

SECTION 3.2: UNSIGNED MULTIPLICATION AND DIVISION   **91**

doubleword/word: The numerator is in AX and DX, with the most significant word in DX and the least significant word in AX. The denominator can be in a register or in memory. After the DIV instruction, the quotient will be in AX, the remainder in DX. The maximum quotient is FFFFH.

```
;from the data segment:
DATA1    DD       105432
DATA2    DW       10000
QUOT     DW       ?
REMAIN   DW       ?
;from the code segment:
         MOV      AX,WORD PTR DATA1        ;AX holds lower word
         MOV      DX,WORD PTR DATA1+2      ;DX higher word of numerator
         DIV      DATA2
         MOV      QUOT,AX                  ;AX holds quotient
         MOV      REMAIN,DX                ;DX holds remainder
```

In the program above, the contents of DX:AX are divided by a word-sized data value, 10000. Now one might ask: How does the CPU know that it must use the doubleword in DX:AX for the numerator? The 8086/88 automatically uses DX:AX as the numerator anytime the denominator is a word in size, as was seen earlier in the case of a word divided by a word. This explains why DX had to be cleared in that case. Notice in the example above that DATA1 is defined as DD but fetched into a word-size register with the help of WORD PTR. In the absence of WORD PTR, the assembler will generate an error. A summary of the results of division of unsigned numbers is given in Table 3-2.

Table 3-2: Unsigned Division Summary

Division	Numerator	Denominator	Quotient	Rem
byte/byte	AL = byte, AH = 0	register or memory	AL [1]	AH
word/word	AX = word, DX = 0	register or memory	AX [2]	DX
word/byte	AX = word	register or memory	AL [1]	AH
doubleword/word	DXAX = doubleword	register or memory	AX [2]	DX

Notes:
1. Divide error interrupt if AL > FFH.
2. Divide error interrupt if AX >FFFFH.

Review Questions

1. In unsigned multiplication of a byte in DATA1 with a byte in AL, the product will be placed in register(s) _____ .
2. In unsigned multiplication of AX with BX, the product is placed in register(s) _____ .
3. In unsigned multiplication of CX with a byte in AL, the product is placed in register(s) _____ .
4. In unsigned division of a byte in AL by a byte in DH, the quotient will be placed in _____ and the remainder in _____ .
5. In unsigned division of a word in AX by a word in DATA1, the quotient will be placed in _____ and the remainder in _____ .
6. In unsigned division of a word in AX by a byte in DATA2, the quotient will be placed in _____ and the remainder in _____ .
7. In unsigned division of a doubleword in DXAX by a word in CX, the quotient will be placed in _____ and the remainder in _____ .

SECTION 3.3: LOGIC INSTRUCTIONS AND SAMPLE PROGRAMS

In this section we discuss the logic instructions AND, OR, XOR, SHIFT, and COMPARE. Instructions are given in the context of examples.

AND

AND destination, source

This instruction will perform a logical AND on the operands and place the result in the destination. The destination operand can be a register or in memory. The source operand can be a register, in memory, or immediate.

X	Y	X AND Y
0	0	0
0	1	0
1	0	0
1	1	1

Example 3-5

Show the results of the following:

```
    MOV     BL,35H
    AND     BL,0FH          ;AND BL with 0FH.  Place the result in BL.
```

Solution:

```
    35H     00110101
    0FH     00001111
    05H     00000101        Flag settings will be:  SF = 0, ZF = 0, PF = 1, CF = OF = 0.
```

AND will automatically change the CF and OF to zero and PF, ZF and SF are set according to the result. The rest of the flags are either undecided or unaffected. As seen in Example 3-5, AND can be used to mask certain bits of the operand. It can also be used to test for a zero operand:

```
    AND     DH,DH
    JZ      XXXX
    ...
XXXX: ...
```

The above will AND DH with itself and set ZF = 1 if the result is zero, making the CPU fetch from the target address XXXX. Otherwise, the instruction below JZ is executed. AND can thus be used to test if a register contains zero.

OR

OR   destination,source

The destination and source operands are ORed and the result is placed in the destination. OR can be used to set certain bits of an operand to 1. The destination operand can be a register or in memory. The source operand can be a register, in memory, or immediate.

X	Y	X OR Y
0	0	0
0	1	1
1	0	1
1	1	1

The flags will be set the same as for the AND instruction. CF and OF will be reset to zero and SF, ZF, and PF will be set according to the result. All other flags are not affected. See Example 3-6.

The OR instruction can also be used to test for a zero operand. For example, "OR BL,0" will OR the register BL with 0 and make ZF = 1 if BL is zero. "OR BL,BL" will achieve the same result.

Example 3-6

Show the results of the following:

```
MOV AX,0504          ;AX = 0504
OR  AX,0DA68H        ;AX = DF6C
```

Solution:

0504H	0000 0101 0000 0100.	
DA68H	1101 1010 0110 1000	Flags will be: SF =1 , ZF = 0, PF = 1, CF = OF = 0.
DF6C	1101 1111 0110 1100	Notice that parity is checked for the lower 8 bits only.

XOR

XOR dest,src

    The XOR instruction will eXclusive-OR the operands and place the result in the destination. XOR sets the result bits to 1 if they are not equal; otherwise, they are reset to 0. The flags are set the same as for the AND instruction. CF = 0 and OF = 0 are set internally and the rest are changed according to the result of the operation. The rules for the operands are the same as in the AND and OR instructions. See Examples 3-7 and 3-8.

X	Y	X XOR Y
0	0	0
0	1	1
1	0	1
1	1	0

Example 3-7

Show the results of the following:

```
MOV    DH,54H
XOR    DH,78H
```

Solution:

54H	0 1 0 1 0 1 0 0	
78H	0 1 1 1 1 0 0 0	
2C	0 0 1 0 1 1 0 0	Flag settings will be: SF = 0, ZF = 0, PF = 0, CF = OF = 0.

Example 3-8

The XOR instruction can be used to clear the contents of a register by XORing it with itself. Show how "XOR AH,AH" clears AH, assuming that AH = 45H.

Solution:

45H	01000101	
45H	01000101	
00	00000000	Flag settings will be: SF = 0, ZF = 1, PF =1 , CF = OF = 0.

    XOR can also be used to see if two registers have the same value. "XOR BX,CX" will make ZF = 1 if both registers have the same value, and if they do, the result (0000) is saved in BX, the destination.

    Another widely used application of XOR is to toggle bits of an operand. For example, to toggle bit 2 of register AL:

```
XOR    AL,04H           ;XOR  AL with 0000 0100
```

    This would cause bit 2 of AL to change to the opposite value; all other bits would remain unchanged.

SHIFT

There are two kinds of shift: logical and arithmetic. The logical shift is for unsigned operands, and the arithmetic shift is for signed operands. Logical shift will be discussed in this section and the discussion of arithmetic shift is postponed to Chapter 6. Using shift instructions shifts the contents of a register or memory location right or left. The number of times (or bits) that the operand is shifted can be specified directly if it is once only, or through the CL register if it is more than once.

SHR: This is the logical shift right. The operand is shifted right bit by bit, and for

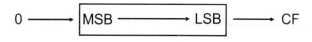

every shift the LSB (least significant bit) will go to the carry flag (CF) and the MSB (most significant bit) is filled with 0. Examples 3-9 and 3-10 should help to clarify SHR.

Example 3-9

Show the result of SHR in the following:

```
MOV     AL,9AH
MOV     CL,3        ;set number of times to shift
SHR     AL,CL
```

Solution:

```
9AH = 10011010
      01001101      CF=0    (shifted once)
      00100110      CF=1    (shifted twice)
      00010011      CF=0    (shifted three times)
```
After three times of shifting right, AL = 13H and CF = 0.

If the operand is to be shifted once only, this is specified in the SHR instruction itself rather than placing 1 in the CL. This saves coding of one instruction:

```
MOV     BX,0FFFFH           ;BX=FFFFH
SHR     BX,1                ;shift right BX once only
```

After the shift above, BX = 7FFFH and CF = 1. Although SHR does affect the OF, SF, PF, and ZF flags, they are not important in this case. The operand to be shifted can be in a register or in memory, but immediate addressing mode is not allowed for shift instructions. For example, "SHR 25,CL" will cause the assembler to give an error.

Example 3-10

Show the results of SHR in the following:
```
;from the data segment:
DATA1   DW 7777H
;from the code segment:
TIMES   EQU     4
MOV     CL,TIMES        ;CL=04
SHR     DATA1,CL        ;shift DATA1 CL times
```

Solution:

After the four shifts, the word at memory location DATA1 will contain 0777. The four LSBs are lost through the carry, one by one, and 0s fill the four MSBs.

SHL: Shift left is also a logical shift. It is the reverse of SHR. After every shift, the LSB is filled with 0 and the MSB goes to CF. All the rules are the same as SHR.

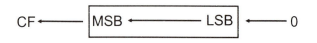

Example 3-11

Show the effects of SHL in the following:

```
MOV     DH,6
MOV     CL,4
SHL     DH,CL
```

Solution:

```
                    00000110
        CF=0        00001100        (shifted left once)
        CF=0        00011000
        CF=0        00110000
        CF=0        01100000        (shifted four times)
```
After the four shifts left, the DH register has 60H and CF = 0.

Example 3-11 could have been coded as

```
MOV     DH,6
SHL     DH,1
SHL     DH,1
SHL     DH,1
SHL     DH,1
```

COMPARE of unsigned numbers

```
CMP     destination,source  ;compare dest and src
```

The CMP instruction compares two operands and changes the flags according to the result of the comparison. The operands themselves remain unchanged. The destination operand can be in a register or in memory and the source operand can be in a register, in memory, or immediate. Although all the CF, AF, SF, PF, ZF, and OF flags reflect the result of the comparison, only the CF and ZF are used, as outlined in Table 3-3.

Table 3-3: Flag Settings for Compare Instruction

Compare operands	CF	ZF
destination > source	0	0
destination = source	0	1
destination < source	1	0

The following demonstrates how the CMP instruction is used:

```
DATA1   DW      235FH
        ....
        MOV     AX,0CCCCH
        CMP     AX,DATA1        ;compare CCCC with 235F
        JNC     OVER            ;jump if CF=0
        SUB     AX,AX
OVER:   INC     DATA1
```

In the program above, AX is greater than the contents of memory location DATA1 (0CCCCH >235FH); therefore, CF = 0 and JNC (jump no carry) will go to target OVER. In contrast, look at the following:

```
        MOV     BX,7888H
        MOV     CX,9FFFH
        CMP     BX,CX           ;compare 7888 with 9FFF
        JNC     NEXT
        ADD     BX,4000H
NEXT:   ADD     CX,250H
```

In the above, BX is smaller than CX (7888H < 9FFFH), which sets CF = 1, making "JNC NEXT" fall through so that "ADD BX,4000H" is executed. In the example above, CX and BX still have their original values (CX = 9FFFH and BX = 7888H) after the execution of "CMP BX,CX". Notice that CF is always checked for cases of greater or smaller than, but for equal, ZF must be used. The next program sample has a variable named TEMP, which is being checked to see if it has reached 99:

```
TEMP        DB ?
            ...
            MOV     AL,TEMP     ;move the TEMP variable into AL
            CMP     AL,99       ;compare AL with 99
            JZ      HOT_HOT     ;if ZF=1 (TEMP = 99) jump to HOT_HOT
            INC     BX          ;otherwise (ZF=0) increment BX
            ...
HOT_HOT:    HLT                 ;halt the system
```

The compare instruction is really a SUBtraction except that the values of the operands do not change. The flags are changed according to the execution of SUB. Although all the flags are affected, the only ones of interest are ZF and CF. It must be emphasized that in CMP instructions, the operands are unaffected regardless of the result of the comparison. Only the flags are affected. This is despite the fact that CMP uses the SUB operation to set or reset the flags. Program 3-3 uses the CMP instruction to search for the highest byte in a series of 5 bytes defined in the data segment. The instruction "CMP AL,[BX]" works as follows, where [BX] is the contents of the memory location pointed at by register BX.

If AL < [BX], then CF = 1 and [BX] becomes the basis of the new comparison.
If AL > [BX], then CF = 0 and AL is the larger of the two values and remains the basis of comparison.

Although JC (jump carry) and JNC (jump no carry) check the carry flag and can be used after a compare instruction, it is recommended that JA (jump above) and JB (jump below) be used for two reasons. One reason is that DEBUG will unassemble JC as JB, and JNC as JA, which may be confusing to beginning programmers. Another reason is that "jump above" and "jump below" are easier to understand than "jump carry" and "jump no carry," since it is more immediately apparent that one number is larger than another, than whether a carry would be generated if the two numbers were subtracted.

Program 3-3 uses the CMP instruction to search through 5 bytes of data to find the highest grade. The program uses register AL to hold the highest grade found so far. AL is given the initial value of 0. A loop is used to compare each of the 5 bytes with the value in AL. If AL contains a higher value, the loop continues to check the next byte. If AL is smaller than the byte being checked, the contents of AL are replaced by that byte and the loop continues.

Assume that there is a class of five people with the following grades: 69, 87, 96, 45, and 75. Find the highest grade.

```
TITLE       PROG3-3 (EXE)  CMP EXAMPLE
PAGE        60,132
.MODEL SMALL
.STACK 64
;————————————————————
            .DATA
GRADES  DB      69,87,96,45,75
        ORG     0008
HIGHEST DB      ?
;————————————————————
            .CODE
MAIN    PROC    FAR
        MOV     AX,@DATA
        MOV     DS,AX
        MOV     CX,5                ;set up loop counter
        MOV     BX,OFFSET GRADES    ;BX points to GRADE data
        SUB     AL,AL               ;AL holds highest grade found so far
AGAIN:  CMP     AL,[BX]             ;compare next grade to highest
        JA      NEXT                ;jump if AL still highest
        MOV     AL,[BX]             ;else AL holds new highest
NEXT:   INC     BX                  ;point to next grade
        LOOP    AGAIN               ;continue search
        MOV     HIGHEST,AL          ;store highest grade
        MOV     AH,4CH
        INT     21H                 ;go back to DOS
MAIN    ENDP
        END     MAIN
```

Program 3-3

Program 3-4 uses the CMP instruction to determine if an ASCII character is uppercase or lowercase. Note that small and capital letters in ASCII have the following values:

Letter	Hex	Binary	Letter	Hex	Binary
A	41	01000001	a	61	01100001
B	42	01000010	b	62	01100010
C	43	01000011	c	63	01100011
...	...	...	...	...	...
Y	59	01011001	y	79	01111001
Z	5A	01011010	z	7A	01111010

As can be seen, there is a relationship between the pattern of lowercase and uppercase letters, as shown below for A and a:

A   0100 0001   41H
a   0110 0001   61H

The only bit that changes is d5. To change from lowercase to uppercase, d5 must be masked. Program 3-4 first detects if the letter is in lowercase, and if it is, it is ANDed with 1101 1111B = DFH. Otherwise, it is simply left alone. To determine if it is a lowercase letter, it is compared with 61H and 7AH to see if it is in the range a to z. Anything above or below this range should be left alone.

```
TITLE        PROG3-4  (EXE)  LOWERCASE TO UPPERCASE CONVERSION
PAGE         60,132
.MODEL SMALL
.STACK 64
;─────────────────────────────────────
             .DATA
DATA1        DB      'mY NAME is jOe'
             ORG     0020H
DATA2        DB      14 DUP(?)
;─────────────────────────────────────
             .CODE
MAIN         PROC    FAR
             MOV     AX,@DATA
             MOV     DS,AX
             MOV     SI,OFFSET DATA1    ;SI points to original data
             MOV     BX,OFFSET DATA2    ;BX points to uppercase data
             MOV     CX,14              ;CX is loop counter
BACK:        MOV     AL,[SI]            ;get next character
             CMP     AL,61H             ;if less than 'a'
             JB      OVER          ;then no need to convert
             CMP     AL,7AH             ;if greater than 'z'
             JA      OVER          ;then no need to convert
             AND     AL,11011111B  ;mask d5 to convert to uppercase
OVER:        MOV     [BX],AL       ;store uppercase character
             INC     SI            ;increment pointer to original
             INC     BX            ;increment pointer to uppercase data
             LOOP    BACK          ;continue looping if CX > 0
             MOV     AH,4CH
             INT     21H                ;go back to DOS
MAIN         ENDP
             END     MAIN
```

Program 3-4

In Program 3-4, 20H could have been subtracted from the lowercase letters instead of ANDing with 1101 1111B. That is what IBM does, as shown next.

IBM BIOS method of converting from lowercase to uppercase

```
              2357          ;—— CONVERT ANY LOWERCASE TO UPPERCASE
              2358
EBFB          2359     K60:                              ;LOWER TO UPPER
EBFB 3C61     2360          CMP   AL,'a'                 ;FIND OUT IF ALPHABETIC
EBFD 7206     2361          JB    K61                    ;NOT_CAPS_STATE
EBFF 3C7A     2362          CMP   AL,'z'
EC01 7702     2363          JA    K61                    ;NOT_CAPS_STATE
EC03 2C20     2364          SUB   AL,'a'-'A'             ;CONVERT TO UPPERCASE
EC05          2365     K61:
EC05 8B1E1C00 2366          MOV   BX,BUFFER_TAIL         ;GET THE END POINTER
                                                         ;TO THE BUFFER
```

(Reprinted by permission from "IBM Technical Reference" c. 1984 by International Business Machines Corporation)

Review Questions

1. Use operands 4FCAH and C237H to perform:
   (a) AND   (b) OR   (c) XOR
2. ANDing a word operand with FFFFH will result in what value for the word operand? To set all bits of an operand to 0, it should be ANDed with _____.
3. To set all bits of an operand to 1, it could be ORed with _____.
4. XORing an operand with itself results in what value for the operand?
5. Show the steps if value A0F2H were shifted left three times. Then show the steps if A0F2H were shifted right three times.
6. The CMP instructions works by performing a(n) _____ operation on the operands and setting the flags accordingly.
7. True or false. The CMP instruction alters the contents of its operands.

BIOS examples of logic instructions

In this section we examine some real-life examples from IBM PC BIOS programs. The purpose is to see the instructions discussed so far in the context of real-life applications.

When the computer is turned on, the CPU starts to execute the programs stored in BIOS in order to set the computer up for DOS. If anything has happened to the BIOS programs, the computer can do nothing. The first subroutine of BIOS is to test the CPU. This involves checking the flag register bit by bit as well as checking all other registers. The BIOS program for testing the flags and registers is given followed by their explanation:

```
              306          ASSUME CS:CODE,DS:NOTHING,ES:NOTHING,SS:NOTHING
E05B          307          ORG    0E05BH
E05B                       ...    ...
E05B          309 START:
E05B FA       310          CLI                         ; DISABLE INTERRUPTS
E05C B4D5     311          MOV    AH,0D5H              ; SET SF, CF, ZF, AND AF FLAGS ON
E05E 9E       312          SAHF
E05F 734C     313          JNC    ERRO1                ; GO TO ERR ROUTINE IF CF NOT SET
E061 754A     314          JNZ    ERRO1                ; GO TO ERR ROUTINE IF ZF NOT SET
E063 7B48     315          JNP    ERRO1                ; GO TO ERR ROUTINE IF PF NOT SET
E065 7946     316          JNS    ERRO1                ; GO TO ERR ROUTINE IF SF NOT SET
E067 9F       317          LAHF                        ; LOAD FLAG IMAGE TO AH
E068 B105     318          MOV    CL,5                 ; LOAD CNT REG WITH SHIFT CNT
E06A D2EC     319          SHR    AH,CL                ; SHIFT AF INTO CARRY BIT POS
E06C 733F     320          JNC    ERRO1                ; GO TO ERR ROUTINE IF AF NOT SET
E06E B040     321          MOV    AL,40H               ; SET THE OF FLAG ON
E070 D0E0     322          SHL    AL,1                 ; SETUP FOR TESTING
E072 7139     323          JNO    ERRO1                ; GO TO ERR ROUTINE IF OF NOT SET
E074 32E4     324          XOR    AH,AH                ; SET AH = 0
E076 9E       325          SAHF                        ; CLEAR SF, CF, ZF, AND PF
E077 7634     326          JBE    ERRO1                ; GO TO ERR ROUTINE IF CF ON
              327                                       ; OR GO TO ERR ROUTINE IF ZF ON
E079 7832     328          JS     ERRO1                ; GO TO ERR ROUTINE IF SF ON
E07B 7A30     329          JP     ERRO1                ; GO TO ERR ROUTINE IF PF ON
E07D 9F       330          LAHF                        ; LOAD FLAG IMAGE TO AH
E07E B105     331          MOV    CL,5                 ; LOAD CNT REG WITH SHIFT CNT
E080 D2EC     332          SHR    AH,CL                ; SHIFT 'AF' INTO CARRY BIT POS
E082 7229     333          JC     ERRO1                ; GO TO ERR ROUTINE IF ON
E084 D0E4     334          SHL    AH,1                 ; CHECK THAT 'OF' IS CLEAR
E086 7025     335          JO     ERRO1                ; GO TO ERR ROUTINE IF ON
              336
              337 ;----- READ/WRITE THE 888 GENERAL AND SEGMENTATION REGISTERS
              338 ;           WITH ALL ONES AND ZEROES.
              339
E088 B8FFFF   340          MOV    AX,0FFFFH            ; SET UP ONE'S PATTERN IN AX
E08B F9       341          STC
E08C          342 C8:
E08C 8ED8     343          MOV    DS,AX                ; WRITE PATTERN TO ALL REGS
E08E 8CDB     344          MOV    BX,DS
E090 8EC3     345          MOV    ES,BX
E092 8CC1     346          MOV    CX,ES
E094 8ED1     347          MOV    SS,CX
E096 8CD2     348          MOV    DX,SS
E098 8BE2     349          MOV    SP,DX
E09A 8BEC     350          MOV    BP,SP
E09C 8BF5     351          MOV    SI,BP
E09E 8BFE     352          MOV    DI,SI
E0A0 7307     353          JNC    C9                   ; TST1A
E0A2 33C7     354          XOR    AX,DI                ; PATTERN MAKE IT THRU ALL REGS
E0A4 7507     355          JNZ    ERRO1                ; NO - GO TO ERR ROUTINE
E0A6 F8       356          CLC
E0A7 EBE3     357          JMP    C8
E0A9          358 C9:
E0A9 0BC7     359          OR     AX,DI                ; ZERO PATTERN MAKE IT THRU?
E0AB 7401     360          JZ     C10                  ; YES - GO TO NEXT TEST
E0AD F4       361 ERRO1: HLT                          ; HALT SYSTEM
```

(Reprinted by permission from "IBM Technical Reference" c. 1984 by International Business Machines Corporation)

Line-by-line explanation:

Line	Explanation

Line   Explanation

310   CLI ensures that no interrupt will occur while the test is being conducted.

311   MOV   AH,0D5H:

```
flag    S Z - AC - P - C
D5H     1 1 0 1  0 1 0 1
```

312   SAHF (store AH into lower byte of the flag register) is one way to move data to flags.
            Another is to use the stack
```
            MOV      AX,00D5H
            PUSH     AX
            POPF
```
            However, there is no RAM available yet to use for the stack because the CPU is tested
            before memory is tested.

313 - 316   Will make the CPU jump to HLT if any flag does not work.

317   LAHF (load AH with the lower byte of flag register) is the opposite of SAHF.

318   Loads CL for five shifts.

319   "SHR AH,CL". By shifting AH five times, AF (auxiliary carry) will be in the CF position.

320   If no AF, there is an error. Lines 317 to 320 are needed because there is no jump condition
            instruction for AF.

321 - 323   Checks the OF flag. This is discussed in Chapter 6 when signed numbers are discussed.

324 - 335   Checks the same flags for zero. Remember that JNZ is the same as JBE.

340   Loads AX with FFFFH.

341   STC (set the carry) makes CF = 1.

343 - 352   Moves the AX value (FFFFH) into every register and ends up with DI = FFFFH if the
            registers are good.

353   Since CF=1 (remember STC) it falls through.

354   Exclusive-ORing AX and DI with both having the same FFFFH value makes AX = 0000
            and ZF = 1 if the registers are good (see lines 343 - 352). If ZF = 0, one of the registers must
            have corrupted the data FFFF, therefore the CPU is bad.

355   If ZF = 0, there is an error.

356   CLC clears the carry flag. This is the opposite of STC.

357   Jumps to C8 and repeats the same process, this time with value 0000. The contents of AX
            are moved around to every register until DI = 0000, and at 353 the JNC C9 will jump since
            CF = 0 by the CLC instruction before it went to the loop.

359   At C9, AX and DI are ORed. If 0000, the contents of AX are copied successfully to all
            registers, DI will be 0000; therefore, ORing will raise the ZF, making ZF = 1.

360   If ZF = 1, the CPU is good and the system can perform the next test. Otherwise, ZF = 0,
            meaning that the CPU is bad and the system should be halted.

SECTION 3.4: BCD AND ASCII OPERANDS AND INSTRUCTIONS

In 80x86 microprocessors, there are many instructions that handle ASCII and BCD numbers. This section covers these instructions with examples.

BCD number system

BCD stands for binary coded decimal. BCD is needed because human beings use the digits 0 to 9 for numbers. Binary representation of 0 to 9 is called BCD (see Figure 3-1). In computer literature one encounters two terms for BCD numbers:

(1) unpacked BCD   (2) packed BCD

Digit	BCD
0	0000
1	0001
2	0010
3	0011
4	0100
5	0101
6	0110
7	0111
8	1000
9	1001

Figure 3-1. BCD Code

Unpacked BCD

In unpacked BCD, the lower 4 bits of the number represent the BCD number and the rest of the bits are 0. Example: "0000 1001" and "0000 0101" are unpacked BCD for 9 and 5, respectively. In the case of unpacked BCD it takes 1 byte of memory location or a register of 8 bits to contain it.

Packed BCD

In the case of packed BCD, a single byte has two BCD numbers in it, one in the lower 4 bits and one in the upper 4 bits. For example, "0101 1001" is packed BCD for 59. It takes only 1 byte of memory to store the packed BCD operands. This is one reason to use packed BCD since it is twice as efficient in storing data.

ASCII numbers

In ASCII keyboards, when key "0" is activated, for example, "011 0000" (30H) is provided to the computer. In the same way, 31H (011 0001) is provided for key "1", and so on, as shown in the following list.

Key	ASCII (hex)	Binary	BCD (unpacked)
0	30	011 0000	0000 0000
1	31	011 0001	0000 0001
2	32	011 0010	0000 0010
3	33	011 0011	0000 0011
4	34	011 0100	0000 0100
5	35	011 0101	0000 0101
6	36	011 0110	0000 0110
7	37	011 0111	0000 0111
8	38	011 1000	0000 1000
9	39	011 1001	0000 1001

It must be noted that although ASCII is standard in the United States (and many other countries), BCD numbers have universal application. Now since the keyboard and printers and monitors are all in ASCII, how does data get converted from ASCII to BCD, and vice versa? These are the subjects covered next.

ASCII to BCD conversion

To process data in BCD, first the ASCII data provided by the keyboard must be converted to BCD. Whether it should be converted to packed or unpacked BCD depends on the instructions to be used. There are instructions that require that data be in unpacked BCD and there are others that must have packed BCD data to work properly. Each is covered separately.

ASCII to unpacked BCD conversion

To convert ASCII data to BCD, the programmer must get rid of the tagged "011" in the higher 4 bits of the ASCII. To do that, each ASCII number is ANDed with "0000 1111" (0FH), as shown in the next example. This example is written in three different ways using different addressing modes. The following three programs show three different methods for converting the 10 ASCII digits to unpacked BCD. All use the same data segment:

```
ASC        DB           '9562481273'
           ORG          0010H
UNPACK     DB           10 DUP(?)
```

In Program 3-5a, notice that although the data was defined as DB, a byte definition directive, it was accessed in word-sized chunks. This is a workable approach; however, using the PTR directive as shown in Program 3-5b makes the code more readable for programmers.

```
          MOV   CX,5
          MOV   BX,OFFSET ASC        ;BX points to ASCII data
          MOV   DI,OFFSET UNPACK     ;DI points to unpacked BCD data
AGAIN:    MOV   AX,[BX]              ;move next 2 ASCII numbers to AX
          AND   AX,0F0FH             ;remove ASCII 3s
          MOV   [DI],AX              ;store unpacked BCD
          ADD   DI,2                 ;point to next unpacked BCD data
          ADD   BX,2                 ;point to next ASCII data
          LOOP  AGAIN
```

Program 3-5a

```
          MOV   CX,5                 ;CX is loop counter
          MOV   BX,OFFSET ASC        ;BX points to ASCII data
          MOV   DI,OFFSET UNPACK     ;DI points to unpacked BCD data
AGAIN:    MOV   AX,WORD PTR [BX]     ;move next 2 ASCII numbers to AX
          AND   AX,0F0FH             ;remove ASCII 3s
          MOV   WORD PTR [DI],AX     ;store unpacked BCD
          ADD   DI,2                 ;point to next unpacked BCD data
          ADD   BX,2                 ;point to next ASCII data
          LOOP  AGAIN
```

Program 3-5b

In both of the solutions so far, registers BX and DI were used as pointers for an array of data. An array is simply a set of data located in consecutive memory locations. Now one might ask: What happens if there are four, five, or six arrays? How can they all be accessed with only three registers as pointers: BX, DI, and SI? Program 3-5c shows how this can be done with a single register used as a pointer to access two arrays. However, to do that, the arrays must be of the same size and defined similarly.

```
          MOV   CX,10               ;load the counter
          SUB   BX,BX               ;clear BX
AGAIN:    MOV   AL,ASC[BX]          ;move to AL content of mem [BX+ASC]
          AND   AL,0FH              ;mask the upper nibble
          MOV   UNPACK[BX],AL       ;move to mem [BX+UNPACK] the AL
          INC   BX                  ;make the pointer to point at next byte
          LOOP  AGAIN               ;loop until it is finished
```

Program 3-5c

Program 3-5c uses the based addressing mode since BX+ASC is used as a pointer. ASC is the displacement added to BX. Either DI or SI could have been used for this purpose. For word-sized operands, "WORD PTR" would be used since the data is defined as DB. This is shown below.

```
          MOV   AX,WORD PTR ASC[BX]
          AND   AX,0F0FH
          MOV   WORD PTR UNPACKED[BX],AX
```

ASCII to packed BCD conversion

To convert ASCII to packed BCD, it is first converted to unpacked BCD (to get rid of the 3) and then combined to make packed BCD. For example, for 9 and 5 the keyboard gives 39 and 35, respectively. The goal is to produce 95H or "1001 0101", which is called packed BCD, as discussed earlier. This process is illustrated in detail below.

Key	ASCII	Unpacked BCD	Packed BCD
4	34	00000100	
7	37	00000111	01000111 or 47H

SECTION 3.4: BCD AND ASCII OPERANDS AND INSTRUCTIONS   **103**

```
            ORG       0010H
VAL_ASC     DB        '47'
VAL_BCD     DB        ?
;reminder: the DB will put 34 in 0010H location and 37 in 0011H.
            MOV       AX,WORD PTR VAL_ASC      ;AH=37,AL=34
            AND       AX,0F0FH                 ;mask 3 to get unpacked BCD
            XCHG      AH,AL                    ;swap AH and AL. :
            MOV       CL,4                     ;CL=04 to shift 4 times
            SHL       AH,CL                    ;shift left AH to get AH=40H
            OR        AL,AH                    ;OR them to get packed BCD
            MOV       VAL_BCD,AL               ;save the result
```

After this conversion, the packed BCD numbers are processed and the result will be in packed BCD format. As will be seen later in this section, there are special instructions, such as DAA and DAS, which require that the data be in packed BCD form and give the result in packed BCD. For the result to be displayed on the monitor or be printed by the printer, it must be in ASCII format. Conversion from packed BCD to ASCII is discussed next.

Packed BCD to ASCII conversion

To convert packed BCD to ASCII, it must first be converted to unpacked and then the unpacked BCD is tagged with 011 0000 (30H). The following shows the process of converting from packed BCD to ASCII.

Packed BCD	Unpacked BCD	ASCII
29H	02H & 09H	32H & 39H
0010 1001	0000 0010 & 0000 1001	011 0010 & 011 1001

```
VAL1_BCD DB 29H
VAL3-ASC DW ?

          ...
          MOV       AL,VAL1_BCD
          MOV       AH,AL          ;copy AL to AH. now AH=29,AL=29H
          AND       AX,0F00FH      ;mask 9 from AH and 2 from AL
          MOV       CL,4           ;CL=04 for shift
          SHR       AH,CL          ;shift right AH to get unpacked BCD
          OR        AX,3030H       ;combine with 30 to get ASCII
          XCHG      AH,AL          ;swap for ASCII storage convention
          MOV       VAL3_ASC,AX    ;store the ASCII
```

BCD addition and subtraction

After learning how to convert ASCII to BCD, the application of BCD numbers is the next step. There are two instructions that deal specifically with BCD numbers: DAA and DAS. Each is discussed separately.

BCD addition and correction

There is a problem with adding BCD numbers, which must be corrected. The problem is that after adding packed BCD numbers, the result is no longer BCD. Look at this example:

```
MOV AL,17H
ADD AL,28H
```

Adding them gives 0011 1111B (3FH), which is not BCD! A BCD number can only have digits from 0000 to 1001 (or 0 to 9). In other words, adding two BCD numbers must give a BCD result. The result above should have been $17 + 28 = 45$ (0100 0101). To correct this problem, the programmer must add 6 (0110) to the low digit:

3F + 06 = 45H. The same problem could have happened in the upper digit (for example, in 52H + 87H = D9H). Again to solve this problem, 6 must be added to the upper digit (D9H +60H =139H), to ensure that the result is BCD (52 + 87 = 139). This problem is so pervasive that all single-chip CISC microprocessors such as the Intel 80x86 and the Motorola 680x0 have an instruction to deal with it. The RISC processors have eliminated this instruction.

DAA

The DAA (decimal adjust for addition) instruction in 80x86 microprocessors is provided exactly for the purpose of correcting the problem associated with BCD addition. DAA will add 6 to the lower nibble or higher nibble if needed; otherwise, it will leave the result alone. The following example will clarify these points:

```
DATA1  DB    47H
DATA2  DB    25H
DATA3  DB    ?

       MOV   AL,DATA1     ;AL holds first BCD operand
       MOV   BL,DATA2     ;BL holds second BCD operand
       ADD   AL,BL        ;BCD addition
       DAA                ;adjust for BCD addition
       MOV   DATA3,AL     ;store result in correct BCD form
```

After the program is executed, the DATA3 field will contain 72H (47 + 25 = 72). Note that DAA works only on AL. In other words, while the source can be an operand of any addressing mode, the destination must be AL in order for DAA to work. It needs to be emphasized that DAA must be used after the addition of BCD operands and that BCD operands can never have any digit greater than 9. In other words, no A - F digit is allowed. It is also important to note that DAA works only after an ADD instruction; it will not work after the INC instruction.

Summary of DAA action

1. If after an ADD or ADC instruction the lower nibble (4 bits) is greater than 9, or if AF = 1, add 0110 to the lower 4 bits.
2. If the upper nibble is greater than 9, or if CF = 1, add 0110 to the upper nibble.

In reality there is no other use for the AF (auxiliary flag) except for BCD addition and correction. For example, adding 29H and 18H will result in 41H, which is incorrect as far as BCD is concerned.

```
   Hex        BCD
   29         0010 1001
 + 18       + 0001 1000
   41         0100 0001  AF = 1
 +  6       +      0110  because AF = 1 DAA will add 6 to the lower nibble
   47         0100 0111  The final result is BCD.
```

Program 3-6 demonstrates the use of DAA after addition of multibyte packed BCD numbers.

BCD subtraction and correction

The problem associated with the addition of packed BCD numbers also shows up in subtraction. Again, there is an instruction (DAS) specifically designed to solve the problem. Therefore, when subtracting packed BCD (single-byte or multibyte) operands, the DAS instruction is put after the SUB or SBB instruction. AL must be used as the destination register to make DAS work.

Two sets of ASCII data have come in from the keyboard. Write and run a program to :
1. Convert from ASCII to packed BCD.
2. Add the multibyte packed BCD and save it.
3. Convert the packed BCD result into ASCII.

```
TITLE      PROG3-6  (EXE)  ASCII TO BCD CONVERSION AND ADDITION
PAGE       60,132
.MODE SMALL
.STACK 64
;————————————————
           .DATA
DATA1_ASC   DB        '0649147816'
            ORG       0010H
DATA2_ASC   DB        '0072687188'
            ORG       0020H
DATA3_BCD   DB        5 DUP (?)
            ORG       0028H
DATA4_BCD   DB        5 DUP (?)
            ORG       0030H
DATA5_ADD   DB        5 DUP (?)
            ORG       0040H
DATA6_ASC   DB        10 DUP (?)
;————————————————
           .CODE
MAIN       PROC    FAR
           MOV     AX,@DATA
           MOV     DS,AX
           MOV     BX,OFFSET DATA1_ASC    ;BX points to first ASCII data
           MOV     DI,OFFSET DATA3_BCD    ;DI points to first BCD data
           MOV     CX,10                  ;CX holds number bytes to convert
           CALL    CONV_BCD               ;convert ASCII to BCD
           MOV     BX,OFFSET DATA2_ASC    ;BX points to second ASCII data
           MOV     DI,OFFSET DATA4_BCD    ;DI points to second BCD data
           MOV     CX,10                  ;CX holds number bytes to convert
           CALL    CONV_BCD               ;convert ASCII to BCD
           CALL    BCD_ADD                ;add the BCD operands
           MOV     SI,OFFSET DATA5_ADD    ;SI points to BCD result
           MOV     DI,OFFSET DATA6_ASC    ;DI points to ASCII result
           MOV     CX,05                  ;CX holds count for convert
           CALL    CONV_ASC               ;convert result to ASCII
           MOV     AH,4CH
           INT     21H                    ;go back to DOS
MAIN       ENDP
;————————————————
;THIS SUBROUTINE CONVERTS ASCII TO PACKED BCD
CONV_BCD  PROC
AGAIN:     MOV     AX,[BX]  ;BX=pointer for ASCII data
           XCHG    AH,AL
           AND     AX,0F0FH        ;mask ASCII 3s
           PUSH    CX              ;save the counter
           MOV     CL,4            ;shift AH left 4 bits
           SHL     AH,CL           ;to get ready for packing
           OR      AL,AH           ;combine to make packed BCD
           MOV     [DI],AL         ;DI=pointer for BCD data
           ADD     BX,2            ;point to next 2 ASCII bytes
           INC     DI              ;point to next BCD data
           POP     CX              ;restore loop counter
           LOOP    AGAIN
           RET
CONV_BCD  ENDP
;————————————————
```

Program 3-6 *(continued on the following page)*

```
;THIS SUBROUTINE ADDS TWO MULTIBYTE PACKED BCD OPERANDS
BCD_ADD  PROC
         MOV    BX,OFFSET DATA3_BCD  ;BX=pointer for operand 1
         MOV    DI,OFFSET DATA4_BCD  ;DI=pointer for operand 2
         MOV    SI,OFFSET DATA5_ADD  ;SI=pointer for sum
         MOV    CX,05
         CLC
BACK:    MOV    AL,[BX]+4            ;get next byte of operand 1
         ADC    AL,[DI]+4            ;add next byte of operand 2
         DAA                         ;correct for BCD addition
         MOV    [SI] +4,AL           ;save sum
         DEC    BX                   ;point to next byte of operand 1
         DEC    DI                   ;point to next byte of operand 2
         DEC    SI                   ;point to next byte of sum
         LOOP   BACK
         RET
BCD_ADD  ENDP
;————————————————————————
;THIS SUBROUTINE CONVERTS FROM PACKED BCD TO ASCII
CONV_ASC  PROC
AGAIN2:  MOV    AL,[SI]              ;SI=pointer for BCD data
         MOV    AH,AL                ;duplicate to unpack
         AND    AX,0F00FH            ;unpack
         PUSH   CX                   ;save counter
         MOV    CL,04                ;shift right 4 bits to unpack
         SHR    AH,CL                ;the upper nibble
         OR     AX,3030H             ;make it ASCII
         XCHG   AH,AL                ;swap for ASCII storage convention
         MOV    [DI],AX              ;store ASCII data
         INC    SI                   ;point to next BCD data
         ADD    DI,2                 ;point to next ASCII data
         POP    CX                   ;restore loop counter
         LOOP   AGAIN2
         RET
CONV_ASC  ENDP
         END    MAIN
```

Program 3-6 *(continued from preceding page)*

Summary of DAS action

1. If after a SUB or SBB instruction the lower nibble is greater than 9, or if AF = 1, subtract 0110 from the lower 4 bits.
2. If the upper nibble is greater than 9, or CF = 1, subtract 0110 from the upper nibble.

Due to the widespread use of BCD numbers, a specific data directive, DT, has been created. DT can be used to represent BCD numbers from 0 to $10^{20}-1$ (that is, twenty 9s). Assume that the following operands represent the budget, the expenses, and the balance, which is the budget minus the expenses.

```
BUDGET     DT    87965141012
EXPENSES   DT    31610640392
BALANCE    DT    ?                          ;balance = budget - expenses

           MOV   CX,10                       ;counter=10
           MOV   BX,00                       ;pointer=0
           CLC                               ;clear carry for the 1st iteration
BACK:      MOV   AL,BYTE PTR BUDGET[BX]      ;get a byte of the BUDGET
           SBB   AL,BYTE PTR EXPENSES[BX]    ;subtract a byte from it
           DAS                               ;correct the result for BCD
           MOV   BYTE PTR BALANCE[BX],AL     ;save it in BALANCE
           INC   BX                          ;increment for the next byte
           LOOP  BACK                        ;continue until CX=0
```

Notice in the code section above that (1) no H (hex) indicator is needed for BCD numbers when using the DT directive, and (2) the use of the based relative addressing mode (BX + displacement) allows access to all three arrays with a single register BX.

In Program 3-7 the DB directive is used to define ASCII values. This makes the LSD (least significant digit) be located at the highest memory location of the array. In VALUE1, 37, the ASCII for 7 is in memory location 0009; therefore, BX must be pointed to that and then decremented. Program 3-7 is repeated, rewritten with the full segment definition.

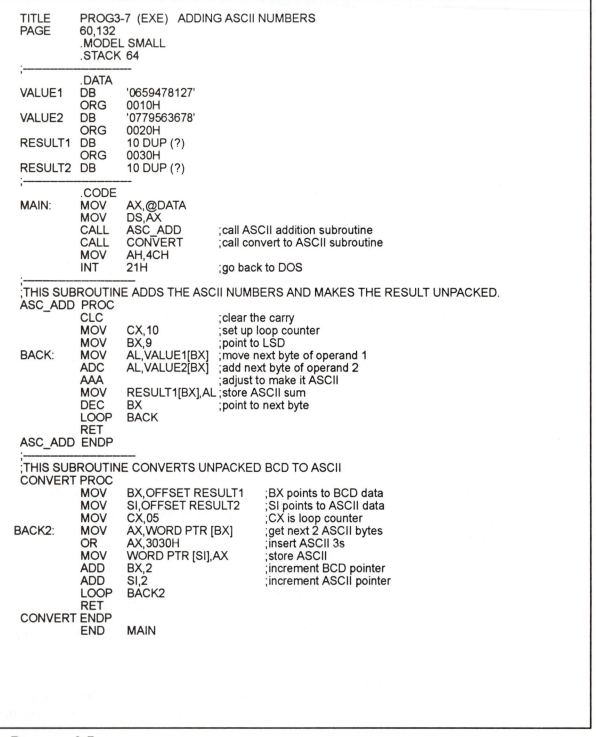

```
TITLE       PROG3-7 (EXE)  ADDING ASCII NUMBERS
PAGE        60,132
            .MODEL SMALL
            .STACK 64
;————————————
            .DATA
VALUE1   DB      '0659478127'
         ORG     0010H
VALUE2   DB      '0779563678'
         ORG     0020H
RESULT1  DB      10 DUP (?)
         ORG     0030H
RESULT2  DB      10 DUP (?)
;————————————
            .CODE
MAIN:    MOV     AX,@DATA
         MOV     DS,AX
         CALL    ASC_ADD         ;call ASCII addition subroutine
         CALL    CONVERT         ;call convert to ASCII subroutine
         MOV     AH,4CH
         INT     21H             ;go back to DOS
;————————————
;THIS SUBROUTINE ADDS THE ASCII NUMBERS AND MAKES THE RESULT UNPACKED.
ASC_ADD PROC
         CLC                     ;clear the carry
         MOV     CX,10           ;set up loop counter
         MOV     BX,9            ;point to LSD
BACK:    MOV     AL,VALUE1[BX]   ;move next byte of operand 1
         ADC     AL,VALUE2[BX]   ;add next byte of operand 2
         AAA                     ;adjust to make it ASCII
         MOV     RESULT1[BX],AL  ;store ASCII sum
         DEC     BX              ;point to next byte
         LOOP    BACK
         RET
ASC_ADD ENDP
;————————————
;THIS SUBROUTINE CONVERTS UNPACKED BCD TO ASCII
CONVERT PROC
         MOV     BX,OFFSET RESULT1   ;BX points to BCD data
         MOV     SI,OFFSET RESULT2   ;SI points to ASCII data
         MOV     CX,05               ;CX is loop counter
BACK2:   MOV     AX,WORD PTR [BX]    ;get next 2 ASCII bytes
         OR      AX,3030H            ;insert ASCII 3s
         MOV     WORD PTR [SI],AX    ;store ASCII
         ADD     BX,2                ;increment BCD pointer
         ADD     SI,2                ;increment ASCII pointer
         LOOP    BACK2
         RET
CONVERT ENDP
         END     MAIN
```

Program 3-7

```
        TITLE     PROG3-7 (EXE) REWRITTEN WITH FULL SEGMENT DEFINITION
        PAGE      60,132
        STSEG     SEGMENT
                  DB      64 DUP (?)
        STSEG     ENDS
;────────────────────
        DTSEG     SEGMENT
        VALUE1    DB      '0659478127'
                  ORG     0010H
        VALUE2    DB      '0779563678'
                  ORG     0020H
        RESULT1   DB      10 DUP (?)
                  ORG     0030H
        RESULT2   DB      10 DUP (?)
        DTSEG     ENDS
;────────────────────
        CDSEG     SEGMENT
        MAIN      PROC    FAR
                  ASSUME CS:CDSEG,DS:DTSEG,SS:STSEG
                  MOV     AX,DTSEG
                  MOV     DS,AX
                  CALL    ASC_ADD            ;call ASCII addition subroutine
                  CALL    CONVERT            ;call convert to ASCII subroutine
                  MOV     AH,4CH
                  INT     21H                ;go back to DOS
        MAIN      ENDP
;────────────────────
        ;THIS SUBROUTINE ADDS THE ASCII NUMBERS AND MAKES THE RESULT UNPACKED.
        ASC_ADD PROC
                  CLC                        ;clear the carry
                  MOV     CX,10              ;set up loop counter
                  MOV     BX,9               ;point to LSD
        BACK:     MOV     AL,VALUE1[BX]      ;move next byte of operand 1
                  ADC     AL,VALUE2[BX]      ;add next byte of operand 2
                  AAA                        ;adjust to make it unpacked BCD
                  MOV     RESULT1[BX],AL     ;store BCD sum
                  DEC     BX                 ;point to next byte
                  LOOP    BACK
                  RET
        ASC_ADD ENDP
;────────────────────
        ;THIS SUBROUTINE CONVERTS UNPACKED BCD TO ASCII
        CONVERT PROC
                  MOV     BX,OFFSET RESULT1  ;BX points to unpacked BCD data
                  MOV     SI,OFFSET RESULT2  ;SI points to ASCII data
                  MOV     CX,05              ;CX is loop counter
        BACK2:    MOV     AX,WORD PTR [BX]   ;get next 2 ASCII bytes
                  OR      AX,3030H           ;insert ASCII 3s
                  MOV     WORD PTR [SI],AX   ;store ASCII
                  ADD     BX,2               ;increment BCD pointer
                  ADD     SI,2               ;increment ASCII pointer
                  LOOP    BACK2
                  RET
        CONVERT ENDP
        CDSEG     ENDS
                  END     MAIN
```

Program 3-7, rewritten with full segment definition

ASCII addition and subtraction

ASCII numbers can be used as operands in add and subtract instructions the way they are, without masking the tagged 011, using instructions AAA and AAS.

```
        MOV     AL,'5'    ;AL=35                          0011  0101
        ADD     AL,'2'    ;add to AL 32 the ASCII for 2   0011  0010
        AAA               ;changes 67H to 07H             0110  0111
        OR      AL,30     ;OR AL with 30H to get ASCII
```

SECTION 3.4: BCD AND ASCII OPERANDS AND INSTRUCTIONS   109

If the addition results in a value of more than 9, AAA will correct it and pass the extra bit to carry and add 1 to AH.

```
SUB      AH,AH        ;AH=00
MOV      AL,'7'       ;AL=37H
MOV      BL,'5'       ;BL=35H
ADD      AL,BL        ;37H+35H=6CH therefore AL=6C.
AAA                   ;changes 6CH to 02 in AL and AH=CF=1
OR       AX,3030H     ;AX=3132 which is the ASCII for 12H.
```

Two facts must be noted. First, AAA and AAS work only on the AL register, and second, the data added can be unpacked BCD rather than ASCII, and AAA and AAS will work fine. The following shows how AAS works on subtraction of unpacked BCD to correct the result into unpacked BCD:

```
MOV      AX,105H      ;AX=0105H unpacked BCD for 15
MOV      CL,06        ;CL=06H
SUB      AL,CL        ;5 - 6 = -1 (FFH)
AAS                   ;FFH in AL is adjusted to 09, and
                      ;AH is decremented, leaving AX = 0009
```

Unpacked BCD multiplication and division

There are two instructions designed specifically for multiplication and division of unpacked BCD operands. They convert the result of the multiplication and division to unpacked BCD.

AAM

The Intel manual says that this mnemonic stands for "ASCII adjust multiplication," but it really is unpacked multiplication correction. If two unpacked BCD numbers are multiplied, the result can be converted back to BCD by AAM.

```
MOV      AL,'7'       ;AL=37H
AND      AL,0FH       ;AL=07 unpacked BCD
MOV      DL,'6'       ;DL=36H
AND      DL,0FH       ;DL=06 unpacked BCD
MUL      DL           ;AX=ALxDL. =07x06=002AH=42
AAM                   ;AX=0402 (7x6=42 unpacked BCD)
OR       AX,3030H     ;AX=3432 result in ASCII
```

The multiplication above is byte by byte and the result is HEX. Using AAM converts it to unpacked BCD to prepare it for tagging with 30H to make it ASCII.

AAD

Again, the Intel manual says that AAD represents "ASCII adjust for division," but that can be misleading since the data must be unpacked BCD for this instruction to work. Before dividing the unpacked BCD by another unpacked BCD, AAD is used to convert it to HEX. By doing that the quotient and remainder are both in unpacked BCD.

```
MOV      AX,3539H     ;AX=3539. ASCII for 59
AND      AX,0F0FH     ;AH=05,AL=09 unpacked BCD data
AAD                   ;AX=003BH hex equivalent of 59
MOV      BH,08H       ;divide by 08
DIV      BH           ;3B / 08 gives AL=07 ,AH=03
OR       AX,3030H     ;AL=37H (quotient) AH=33H (rem)
```

As can be seen in the example above, dividing 59 by 8 gives a quotient of 7 and a remainder of 3. With the help of AAD, the result is converted to unpacked BCD, so it can be tagged with 30H to get the ASCII result. It must be noted that both AAM and AAD work only on AX.

Review Questions

1. For the following decimal numbers, give the packed BCD and unpacked BCD representations.
   (a) 15   (b) 99
2. Match the following instruction mnemonic with its function.
   _____ DAA   (a) ASCII addition
   _____ DAS   (b) ASCII subtraction
   _____ AAS   (c) BCD subtraction
   _____ AAA   (d) BCD addition

SECTION 3.5: ROTATE INSTRUCTIONS

In many applications there is a need to perform a bitwise rotation of an operand. The rotation instructions ROR, ROL and RCR, RCL are designed specifically for that purpose. They allow a program to rotate an operand right or left. In this section we explore the rotate instructions, which frequently have highly specialized applications. In rotate instructions, the operand can be in a register or memory. If the number of times an operand is to be rotated is more than 1, this is indicated by CL. This is similar to the shift instructions. There are two type of rotations. One is a simple rotation of the bits of the operand, and the other is a rotation through the carry. Each is explained below.

Rotating the bits of an operand right and left

ROR rotate right

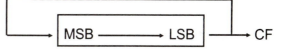

In rotate right, as bits are shifted from left to right they exit from the right end (LSB) and enter the left end (MSB). In addition, as each bit exits the LSB, a copy of it is given to the carry flag. In other words, in ROR the LSB is moved to the MSB and is also copied to CF, as shown in the diagram. If the operand is to be rotated once, the 1 is coded, but if it is to be rotated more than once, register CL is used to hold the number of times it is to be rotated.

```
        MOV   AL,36H       ;AL=0011 0110
        ROR   AL,1         ;AL=0001 1011  CF=0
        ROR   AL,1         ;AL=1000 1101  CF=1
        ROR   AL,1         ;AL=1100 0110  CF=1
or:
        MOV   AL,36H       ;AL=0011 0110
        MOV   CL,3         ;CL=3 number of times to rotate
        ROR   AL,CL        ;AL=1100 0110 CF=1
```

;the operand can be a word:

```
        MOV   BX,0C7E5H    ;BX=1100 0111 1110 0101
        MOV   CL,6         ;CL=6 number of times to rotate
        ROR   BX,CL        ;BX=1001 0111 0001 1111 CF=1
```

ROL rotate left

In rotate left, as bits are shifted from right to left they exit the left end (MSB) and enter the right end (LSB). In addition, every bit that leaves the MSB is copied to the carry flag. In other words, in ROL the MSB is moved to the LSB and is also copied to CF, as shown in the diagram. If the operand is to be rotated once, the 1 is coded. Otherwise, the number of times it is to be rotated is in CL.

```
        MOV   BH,72H      ;BH=0111 0010
        ROL   BH,1        ;BH=1110 0100 CF=0
        ROL   BH,1        ;BH=1100 1001 CF=1
        ROL   BH,1        ;BH=1001 0011 CF=1
        ROL   BH,1        ;BH=0010 0111 CF=1
or:
        MOV   BH,72H      ;BH=0111 0010
        MOV   CL,4        ;CL=4 number of times to rotate
        ROL   BH,CL       ;BH=0010 0111 CF=1
```

The operand can be a word:
```
        MOV   DX,672AH    ;DX=0110 0111 0010 1010
        MOV   CL,3        ;CL=3 number of times to rotate
        ROL   DX,CL       ;DX=0011 1001 0101 0011 CF=1
```

Write a program that finds the number of 1s in a byte.

From the data segment:
```
DATA1   DB    97H
COUNT   DB    ?
```
From the code segment:
```
        SUB   BL,BL       ;clear BL to keep the number of 1s
        MOV   DL,8        ;rotate total of 8 times
        MOV   AL,DATA1
AGAIN:  ROL   AL,1        ;rotate it once
        JNC   NEXT        ;check for 1
        INC   BL          ;if CF=1 then add one to count
NEXT:   DEC   DL          ;go through this 8 times
        JNZ   AGAIN       ;if not finished go back
        MOV   COUNT,BL    ;save the number of 1s
```

Program 3-8

Program 3-8 shows an application of the rotation instruction. The maximum count in Program 3-8 will be 8 since the program is counting the number of 1s in a byte of data. If the operand is a 16-bit word, the number of 1s can go as high as 16. Program 3-9 is Program 3-8, rewritten for a word-sized operand. It also provides the count in BCD format instead of hex. *Reminder:* AL is used to make a BCD counter because the DAA instruction works only on AL.

Write a program to count the number of 1s in a word. Provide the count in BCD.
```
DATAW1  DW    97F4H
COUNT2  DB    ?
        ...
        SUB   AL,AL       ;clear AL to keep the number of 1s in BCD
        MOV   DL,16       ;rotate total of 16 times
        MOV   BX,DATAW1   ;move the operand to BX
AGAIN:  ROL   BX,1        ;rotate it once
        JNC   NEXT        ;check for 1.  If CF=0 then jump
        ADD   AL,1        ;if CF=1 then add one to count
        DAA               ;adjust the count for BCD
NEXT:   DEC   DL          ;go through this 16 times
        JNZ   AGAIN       ;if not finished go back
        MOV   COUNT2,AL   ;save the number of 1s in COUNT2
```

Program 3-9

RCR rotate right through carry

In RCR, as bits are shifted from left to right, they exit the right end (LSB) to the carry flag, and the carry flag enters the left end

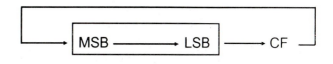

(MSB). In other words, in RCR the LSB is moved to CF and CF is moved to the MSB. In reality, CF acts as if it is part of the operand. This is shown in the diagram. If the operand is to be rotated once, the 1 is coded, but if it is to be rotated more than once, the register CL holds the number of times.

```
        CLC                 ;make CF=0
        MOV   AL,26H        ;AL=0010 0110
        RCR   AL,1          ;AL=0001 0011 CF=0
        RCR   AL,1          ;AL=0000 1001 CF=1
        RCR   AL,1          ;AL=1000 0100 CF=1
or:
        CLC                 ;make CF=0
        MOV   AL,26H        ;AL=0010 0110
        MOV   CL,3          ;CL=3 number of times to rotate
        RCR   AL,CL         ;AL=1000 0100 CF=1

;the operand can be a word
        STC                 ;make CF=1
        MOV   BX,37F1H      ;BX=0011 0111 1111 0001
        MOV   CL,5          ;CL=5 number of times to rotate
        RCR   BX,CL         ;BX=0001 1001 1011 1111 CF=0
```

RCL rotate left through carry

In RCL, as bits are shifted from right to left they exit the left end (MSB) and enter the carry flag, and the carry flag enters the right end

(LSB). In other words, in RCL the MSB is moved to CF and CF is moved to the LSB. In reality, CF acts as if it is part of the operand. This is shown in the diagram. If the operand is to be rotated once, the 1 is coded, but if it is to be rotated more than once, register CL holds the number of times.

```
        STC                 ;make CF=1
        MOV   BL,15H        ;BL=0001 0101
        RCL   BL,1          ;0010 1011 CF=0
        RCL   BL,1          ;0101 0110 CF=0
or:
        STC                 ;make CF=1
        MOV   BL,15H        ;BL=0001 0101
        MOV   CL,2          ;CL=2 number of times for rotation
        RCL   BL,CL         ;BL=0101 0110 CF=0

;the operand can be a word:

        CLC                 ;make CF=0
        MOV   AX,191CH      ;AX=0001 1001 0001 1100
        MOV   CL,5          ;CL=5 number of times to rotate
        RCL   AX,CL         ;AX=0010 0011 1000 0001 CF=1
```

SECTION 3.5: ROTATE INSTRUCTIONS 113

1. What is the value of BL after the following?
   ```
   MOV   BL,25H
   MOV   CL,4
   ROR   BL,CL
   ```
2. What are the values of DX and CF after the following?
   ```
   MOV   DX,3FA2H
   MOV   CL,7
   ROL   DX,CL
   ```
3. What is the value of BH after the following?
   ```
   SUB   BH,BH
   STC
   RCR   BH,1
   STC
   RCR   BH,1
   ```
4. What is the value of BX after the following?
   ```
   MOV   BX,FFFFH
   MOV   CL,5
   CLC
   RCL   BX,CL
   ```
5. Why does "ROR BX,4" give an error in the 8086? How would you change the code to make it work?

SECTION 3.6: BITWISE OPERATION IN THE C LANGUAGE

One of the most important and powerful features of the C language is its ability to perform bit manipulation. Due to the fact that many books on C do not cover this important topic, it is appropriate to discuss it in this section. This section describes the action of operators and provides examples.

Bitwise operators in C

While every C programmer is familiar with the logical operators AND (&&), OR (||), and NOT (!), many C programmers are less familiar with the bitwise operators AND (&), OR (|), EX-OR (^), inverter (~), Shift Right (>>), and Shift Left (<<). These bitwise operators are widely used in software engineering and control; consequently, their understanding and mastery are critical in system design and interfacing. See Tables 3-4 and 3-5. The following code shows Examples 3-5 through 3-7 using the C logical operators. Recall that "0x" in the C language indicates that the data is in hex format.

Table 3-4: Bitwise AND, OR, and EX-OR in C

A	B	A & B	A | B	A ^ B
0	0	0	0	0
0	1	0	1	1
1	0	0	1	1
1	1	1	1	0

Table 3-5: Bitwise Inverter in C

A	~A
0	1
1	0

```
0x35 & 0x0F = 0x05          /* ANDing: see Example 3-5 */
0x0504 | 0xDA68 = 0xDF6C    /* ORing:  see Example 3-6 */
0x54 ^ 0x78 = 0x2C          /* XORing: see Example 3-7 */
~0x37 = 0xC8                /* inverting 37H */
```

The last one is like the NOT instruction in x86 microprocessors:
```
MOV AL,37H    ;AL=37H
NOT AL        ;AFTER INVERTING 37, AL=C8H
```

Bitwise shift operators in C

There are two bitwise shift operators in C: Shift Right (> >) and Shift Left (<<). They perform exactly the same operation as SHR and SHL in Assembly language, as discussed in Section 3.3. Their format in C is as follows:

data >> number of bits to be shifted   /* shifting right */
data << number of bits to be shifted   /* shifting left */

The following shows Examples 3-9 through 3-11 using shift operators in C. Program 3-10 shows all of these examples with C syntax.

```
0x9A >> 3   = 0x13          /* shifting right 3 times: see Example 3-9 */
0x7777 >> 4   = 0x0777      /* shifting right 4 times: see Example 3-10 */
0x6 << 4      = 0x60        /* shifting left 4 times: see Example 3-11 */
```

```c
/* Program 3-10 Repeats Examples 3-5 through 3-11 in C  */

#include    <stdio.h>
main()

{

 // Notice the way data is defined in C for Hex format using 0x

unsigned char data_1 = 0x35;
unsigned int data_2 = 0x504;
unsigned int data_3 = 0xDA66;
unsigned char data_4= 0x54;
unsigned char data_5=0x78;
unsigned char data_6=0x37;
unsigned char data_7=0x09A;
unsigned char temp;
unsigned int temp_2;

temp=data_1&0x0F;           //ANDing
printf("\nMasking the upper four bits of %X (hex) we get %X (hex)\n",data_1,temp);

temp_2=data_2|data_3;       //ORing
printf("The result of %X hex ORed with %X hex is %X hex\n",data_2,data_3,temp_2);

temp= data_4^data_5;        //EX-ORing
printf("The results of %X hex EX-ORed with %X hex is %X hex\n",data_4,data_5,temp);

temp=~data_6;               //INVERTING
printf("The result of %X hex inverted is %X hex\n",data_6,temp);

temp=data_7>>3;             //SHIFTING Right
printf("When %X hex is shifted right three times we get %X hex\n",data_7,temp);

printf("When %X hex is shifted right four times we get %X hex\n",0x7777,0x7777>>4);

temp=(0x6<<4);              //SHIFTING Left
printf("When %X hex is shifted left %d times we get %X hex\n",0x6,4,temp);

}
```

Program 3-10

Program 3-10 demonstrates the syntax of bitwise operators in C. Next we show some real-world examples of their usage.

Packed BCD to ASCII conversion in C

Section 3.4 showed one way to convert a BCD number to ASCII. This conversion is widely used when dealing with a real-time clock chip. Many of the real-time clock chips provide very accurate time and date for up to ten years without the need for external power. There is a real-time clock in every x86 IBM PC or compatible computer. However, these chips provide the time and date in packed BCD. In order to display the data, it needs to be converted to ASCII. Program 3-11 is a C version of the packed BCD-to-ASCII conversion example discussed in Section 3.4. Program 3-11 converts a byte of packed BCD data into two ASCII characters and displays them using the C bitwise operators.

```
/* Program 3-11 shows packed BCD-to-ASCII conversion using logical bitwise operators in C */

#include       <stdio.h>
main()

{

unsigned char mybcd=0x29;        /* declare a BCD number in hex */
unsigned char asci_1;
unsigned char asci_2;
asci_1=mybcd&0x0f;               /* mask the upper four bits */
asci_1=asci_1|0x30;              /* make it an ASCII character */
asci_2=mybcd&0xf0;               /* mask the lower four bits */
asci_2=asci_2>>4;                /* shift it right 4 times */
asci_2=asci_2|0x30;              /* make it an ASCII character */
printf("BCD data %X is %c , %c in ASCII\n",mybcd,asci_1,asci_2);
printf("My BCD data is %c if not converted to ASCII\n",mybcd);

}
```

Program 3-11

Notice in Program 3-11 that if the packed BCD data is displayed without conversion to ASCII, we get the parenthesis ")". See Appendix F.

Testing bits in C

In many cases of system programming and hardware interfacing, it is necessary to test a given bit to see if it is high. For example, many devices send a high signal to state that they are ready for an action or to indicate that they have data. How can the bit (or bits) be tested? In such cases, the unused bits are masked and then the remaining data is tested. Program 3-12 asks the user for a byte and tests to see whether or not D0 of that byte is high.

```
/* Program 3-12 shows how to test bit D0 to see if it is high */
#include       <stdio.h>
main()

{
unsigned char status;
unsigned char temp;
printf("\nType in a Hex value\n");
scanf("%X",&status);             //get the data
temp=status&0x01;                //mask all bits except D0
if (temp==0x01)                  //is it high?
   printf("D0 is high");         //if yes, say so
else printf("D0 is low");        //if no, say no
}
```

Program 3-12

The assembly language version of Program 3-12 is as follows:

```
;assume AL=value (in hex)
        AND AL,01          ;MASK ALL BITS EXCEPT D0
        CMP AL,01          ;IS D0 HIGH
        JNE BELOW          ;MAKE A DECISION
        ....               ;YES D0 IS HIGH
BELOW:  ....               ;D0 IS LOW
```

Review Questions

1. What is the result of 0x2F&0x27 ?
2. What is the result of 0x2F|0x27 ?
3. What is the result of 0x2F^0x27 ?
4. What is the result of ~0x2F ?
5. What is the result of 0x2F >> 3 ?
6. What is the result of 0x27 << 4 ?
7. In Program 3-11 if mybcd=0x32, what is displayed if it is not converted to BCD?
8. Modify Program 3-12 to test D3.

SUMMARY

The 8- or 16-bit data items in 80x86 computers can be treated as either signed or unsigned data. Unsigned data uses the entire 8 or 16 bits for data representation. Signed data uses the MSB as a sign bit and the remaining bits for data representation. This chapter covered the arithmetic and logic instructions that are used for unsigned data. The instructions ADD and SUB perform addition and subtraction on unsigned data. Instructions ADC and SBB do the same, but also take the carry flag into consideration. Instructions MUL and DIV perform multiplication and division on unsigned data. Logic instructions AND, OR, XOR, and CMP perform logic operations on all the bits of their operands and were therefore included in this chapter. Shift and rotate instructions for unsigned data include SHR, SHL, ROR, ROL, RCL, and RCR. ASCII and BCD data operations for addition and subtraction were also covered. Finally, bitwise logic instructions were demonstrated using the C language.

PROBLEMS

1. Find CF, ZF, and AF for each of the following. Also indicate the result of the addition and where the result is saved.

   (a) MOV BH,3FH   (b) MOV DX,4599H   (c) MOV AX,255
       ADD BH,45H   MOV CX,3458H   STC
                                ADD CX,DX   ADC AX,00
   (d) MOV BX,0FF01H   (e) MOV CX,0FFFFH   (f) MOV AH,0FEH
       ADD BL,BH   STC   STC
                                ADC CX,00   ADC AH,00

2. Write, run, and analyze a program that calculates the total sum paid to a salesperson for eight months. The following are the monthly paychecks for those months: $2300, $4300, $1200, $3700, $1298, $4323, $5673, $986.
3. Rewrite Program 3-2 (in Section 3.1) using byte addition.
4. Write a program that subtracts two multibytes and saves the result. Subtraction should be done a byte at a time.   Use the data in Program 3-2.
5. State the three steps involved in a SUB and show the steps for the following data.

   (a) 23H-12H   (b) 43H-51H   (c) 99-99

6. Write, run, and analyze the result of a program that performs the following:

   (1)(a) byte1 × byte2   (b) byte1 × word1   (c) word1 × word2
   (2) (a) byte1 / byte2   (b) word1 / word2   (c) doubleword/byte1
   Assume byte1=230, byte2=100, word1=9998, word2=300 and doubleword =100000.

7. Assume that the following registers contain these HEX contents: AX = F000, BX = 3456, and DX = E390. Perform the following operations. Indicate the result and the register where it is stored. Give also ZF and CF in each case.
   Note: the operations are independent of each other.
   (a) AND DX,AX
   (b) OR DH,BL
   (c) XOR AL,76H
   (d) AND DX,DX
   (e) XOR AX,AX
   (f) OR BX,DX
   (g) AND AH,0FF
   (h) OR AX,9999H
   (i) XOR DX,0EEEEH
   (j) XOR BX,BX
   (k) MOV CL,04
      SHL AL,CL
   (l) SHR DX,1
   (m) MOV CL,3
      SHR DL,CL
   (n) MOV CL,5
      SHL BX,CL
   (o) MOV CL,6
      SHL DX,CL

8. Indicate the status of ZF and CF after CMP is executed in each of the following cases.
   (a) MOV BX,2500
      CMP BX,1400
   (b) MOV AL,0FFH
      CMP AL,6FH
   (c) MOV DL,34
      CMP DL,88
   (d) SUB AX,AX
      CMP AX,0000
   (e) XOR DX,DX
      CMP DX,0FFFFH
   (f) SUB CX,CX
      DEC CX
      CMP CX,0FFFFH
   (g) MOV BX,2378H
      MOV DX,4000H
      CMP DX,BX
   (h) MOV AL,0AAH
      AND AL,55H
      CMP AL,00

9. Indicate whether or not the jump happens in each case.
   (a) MOV CL,5
      SUB AL,AL
      SHL AL,CL
      JNC TARGET
   (b) MOV BH,65H
      MOV AL,48H
      OR AL,BH
      SHL AL,1
      JC TARGET
   (c) MOV AH,55H
      SUB DL,DL
      OR DL,AH
      MOV CL,AH
      AND CL,0FH
      SHR DL,CL
      JNC TARGET

10. Rewrite Program 3-3 to find the lowest grade in that class.
11. Rewrite Program 3-4 to convert all uppercase letters to lowercase.
12. In the IBM BIOS program for testing flags and registers, verify every jump (conditional and unconditional) address calculation. Reminder: As mentioned in Chapter 2, in forward jumps the target address is calculated by adding the displacement value to IP of the instruction after the jump and by subtracting in backward jumps.
13. In Program 3-6 rewrite BCD_ADD to do subtraction of the multibyte BCD.
14. Rewrite Program 3-7 to subtract DATA2 from DATA1. Use the following data.
    DATA1   DB   '0999999999'
    DATA2   DB   '0077777775'
15. Using the DT directive, write a program to add two 10-byte BCD numbers.
16. We would like to make a counter that counts up from 0 to 99 in BCD. What instruction would you place in the dotted area?
       SUB AL,AL
       ADD AL,1
       ....
17. Write Problem 16 to count down (from 99 to 0).
18. An instructor named Mr. Mo Allem has the following grading policy: "Curving of grades is achieved by adding to every grade the difference between 99 and the highest grade in the class." If the following are the grades of the class, write a program to calculate the grades after they have been curved: 81, 65, 77, 82, 73, 55, 88, 78, 51, 91, 86, 76. Your program should work for any set of grades.
19. If we try to divide 1,000,000 by 2:
    (a) What kind of problem is associated with this operation in 8086/286 CPUs?
    (b) How does the CPU let us know that there is a problem?

20. Which of the following groups of code perform the same operation as LOOP XXX?

(a) DEC CL   (b) DEC CH   (c) DEC BX   (d) DEC CX
     JNZ XXX   JNZ XXX   JNZ   JNZ XXX

21. Write a program that finds the number of zeros in a 16-bit word.

22. In Program 3-2, which demonstrated multiword addition, pointers were updated by two INC instructions instead of "ADD SI,2". Why?

23. Write a C program with the following components:

(a) have two hex values: data1=55H and data2=AAH; both defined as unsigned char,

(b) mask the upper 4 bits of data1 and display it in hex,

(c) perform AND, OR, and EX-OR operations between the two data items and then display each result,

(d) invert one and display it,

(e) shift left data1 four times and shift right data2 two times, then display each result.

24. Repeat the above problem with two values input from the user. Use the scanf("%X") function to get the hex data.

25. In the same way that the real-time clock chip provides data in BCD, it also expects data in BCD when it is being initialized. However, data coming from the keyboard is in ASCII. Write a C program to convert two ASCII bytes of data to packed BCD.

26. Write a C program in which the user is prompted for a hex value. Then the data is tested to see if the two least significant bits are high. If so, a message states "D0 and D1 are both high"; otherwise, it states which bit is not high.

27. Repeat the above problem for bits D0 and D7.

ANSWERS TO REVIEW QUESTIONS

SECTION 3.1: UNSIGNED ADDITION AND SUBTRACTION
1. destination
2. in 80x86 Assembly language, there are no memory to memory operations
3. MOV AX,DATA_2
    ADD DATA_1,AX
4. destination, source + destination + CF
5. in (a), the byte addition results in a carry to CF, in (b), the word addition results in a carry to the high byte BH
6. DEC CX
    JNZ ADD_LOOP
7.   43H   0100 0011   0100 0011
   - 05H   0000 0101 2's complement=+1111 1011
    3EH   0011 1110
   CF=0; therefore, the result is positive
8. AL = 95 - 4F - 1 = 45

SECTION 3.2: UNSIGNED MULTIPLICATION AND DIVISION

1. AX	2. DX and AX	3. AX	4. AL, AH
5. AX, DX	6. AL, AH	7. AX, DX	

SECTION 3.3: LOGIC INSTRUCTIONS AND SAMPLE PROGRAMS
1. (a) 4202   (b) CFFF   (c) 8DFD
2. the operand will remain unchanged; all zeros 3. all ones   4. all zeros
5. A0F2 =   1010 0000 1111 0010
   shift left: 0100 0001 1110 0100 CF =1
   shift again: 1000 0011 1100 1000 CF =0
   shift again: 0000 0111 1001 0000 CF =1
   A0F2 shifted left three times = 0790.
   A0F2 =   1010 0000 1111 0010
   shift right: 0101 0000 0111 1001 CF = 0
   shift again: 0010 1000 0011 1100 CF = 1
   shift again: 0001 0100 0001 1110 CF = 0
   A0F2 shifted right three times = 141E
6. SUB   7. false

SECTION 3.4: BCD AND ASCII OPERANDS AND INSTRUCTIONS
1.   (a) 15 = 0001 0101 packed BCD = 0000 0001 0000 0101 unpacked BCD
     (b) 99 = 1001 1001 packed BCD = 0000 1001 0000 1001 unpacked BCD
2.   DAA -- BCD addition;   DAS -- BCD subtraction;   AAS -- ASCII subtraction; AAA -- ASCII addition

SECTION 3.5: ROTATE INSTRUCTIONS
1.   BL = 52H, CF = 0   2. DX = D11FH, CF = 1
3.   BH = C0H   4. BX = FFEFH
5.   the source operand cannot be immediate; to fix it:
     MOV CL,4
     ROR BX,CL

SECTION 3.6: BITWISE OPERATION IN THE C LANGUAGE

1.   0x27   2.   0x2F   3.   0x08
4.   0xD0   5.   0x05   6.   0x70
7.   2
8.

```c
/* This program shows how to test Bit D3 to see if it is high */
#include    <stdio.h>
main()
{
unsigned char status;
unsigned char temp;
printf("\nType in a Hex value\n");
scanf("%X",&status);
temp=status&0x04;
if (temp==0x04)
  printf("D3 is high");
else printf("D3 is low");
}
```

CHAPTER 4

BIOS AND DOS PROGRAMMING IN ASSEMBLY AND C

OBJECTIVES

Upon completion of this chapter, you will be able to:

» Use INT 10H function calls to:
»   Clear the screen
»   Set the cursor position
»   Write characters to the screen in text mode
»   Draw lines on the screen in graphics mode
»   Change the video mode
» Use INT 21H function calls to:
»   Input characters from the keyboard
»   Output characters to the screen
»   Input or output strings
» Use the LABEL directive to set up structured data items
» Use INT 16H for keyboard access
» Use C function calls int86 and intdos to perform BIOS and DOS interrupts
» Use in-line assembly within C programs

There are some extremely useful subroutines within BIOS and DOS that are available to the user through the INT (interrupt) instruction. In this chapter, some of them are studied to see how they are used in the context of applications. First, a few words about the interrupt itself. The INT instruction is somewhat like a FAR call. When it is invoked, it saves CS:IP and the flags on the stack and goes to the subroutine associated with that interrupt. The INT instruction has the following format:

INT xx   ;the interrupt number xx can be 00 - FFH

Since interrupts are numbered 00 to FF, this gives a total of 256 interrupts in 80x86 microprocessors. Of these 256 interrupts, two of them are the most widely used: INT 10H and INT 21H. Each one can perform many functions. A list of these functions is provided in Appendices D and E. Before the service of INT 10H or INT 21H is requested, certain registers must have specific values in them, depending on the function being requested. Various functions of INT 21H and INT 10H are selected by the value put in the AH register, as shown in Appendices D and E. Interrupt instructions are discussed in detail in Appendix B.

SECTION 4.1: BIOS INT 10H PROGRAMMING

INT 10H subroutines are burned into the ROM BIOS of the 80x86-based IBM PC and compatibles and are used to communicate with the computer's screen video. Much of the manipulation of screen text or graphics is done through INT 10H. There are many functions associated with INT 10H. Among them are changing the color of characters or the background color, clearing the screen, and changing the location of the cursor. These options are chosen by putting a specific value in register AH. In this section we show how to use INT 10H to clear the screen, change the cursor position, change the screen color, and draw lines on the screen.

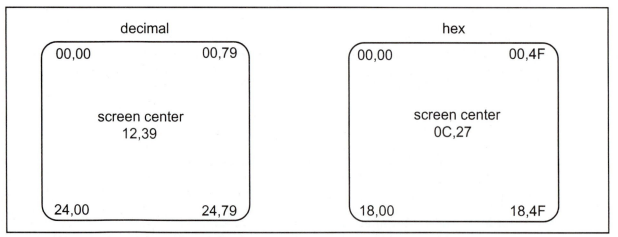

Figure 4-1. Cursor Locations (row,column)

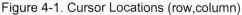

Monitor screen in text mode

The monitor screen in the IBM PC is divided into 80 columns and 25 rows in normal text mode (see Figure 4-1). This is the case for all monitors regardless of whether they are color or monochrome, as long as they are in text mode. When the computer is turned on, the monitor is set to the default text mode. The mode can be changed, depending on the type of the monitor. Monitor types include MDA, CGA, EGA, and VGA. The text screen is 80 characters wide by 25 characters long. Since both a row and a column number are associated with each location on the screen, one can move the cursor to any location on the screen simply by changing the row and column values. The 80 columns are numbered from 0 to 79 and the 25 rows are numbered 0 to 24. The top left corner has been assigned 00,00 (row = 00, column = 00). Therefore, the top right corner will be 00,79 (row = 00, column = 79). Similarly, the bottom left corner is 24,00 (row = 24, column = 00) and the bottom right corner of the monitor is 24,79 (row = 24, column = 79). Figure 4-1 shows each location of the screen in both decimal and hex.

Clearing the screen using INT 10H function 06H

It is often desirable to clear the screen before displaying data. To use INT 10H to clear the screen, the following registers must contain certain values before INT 10H is called: AH = 06, AL = 00, BH = 07, CX = 0000, DH = 24, and DL = 79. The code will look like this:

```
MOV    AH,06        ;AH=06 to select scroll function
MOV    AL,00        ;AL=00 the entire page
MOV    BH,07        ;BH=07 for normal attribute
MOV    CH,00        ;CH=00 row value of start point
MOV    CL,00        ;CL=00 column value of start point
MOV    DH,24        ;DH=24 row value of ending point
MOV    DL,79        ;DL=79 column value of ending point
INT    10H          ;invoke the interrupt
```

Remember that DEBUG assumes immediate operands to be in hex; therefore, DX would be entered as 184F. However, MASM assumes immediate operands to be in decimal. In that case DH = 24 and DL = 79.

In the program above, one of many options of INT 10H was chosen by putting 06 into AH. Option AH = 06, called the *scroll* function, will cause the screen to scroll upward. The CH and CL registers hold the starting row and column, respectively, and DH and DL hold the ending row and column. To clear the entire screen, one must use the top left cursor position of 00,00 for the start point and bottom right position of 24,79 for the end point.

Option AH = 06 of INT 10H is in reality the "scroll window up" function; therefore, one could use that to make a window of any size by choosing appropriate values for the start and end rows and columns. However, to clear the screen, the top left and bottom right values are used for start and stop points in order to scroll up the entire screen. It is much more efficient coding to clear the screen by combining some of the lines above as follows:

```
MOV    AX,0600H     ;scroll entire screen
MOV    BH,07        ;normal attribute
MOV    CX,0000      ;start at 00,00
MOV    DX,184FH     ;end at 24,79 (hex = 18,4F)
INT    10H          ;invoke the interrupt
```

INT 10H function 02: setting the cursor to a specific location

INT 10H function AH = 02 will change the position of the cursor to any location. The desired position of the cursor is identified by the row and column values in DX, where DH = row and DL = column. Video RAM can have more than one page of text, but only one of them can be viewed at a time. When AH = 02, to set the cursor position, page zero is chosen by making BH = 00.

It must be pointed out that after INT 10H (or INT 21H) has executed, the registers that have not been used by the interrupt remain unchanged. In other words, these registers have the same values after execution of the interrupt as before the interrupt was invoked. Examples 4-1 and 4-2 demonstrate setting the cursor to a specific location.

Example 4-1

Write the code to set the cursor position to row = 15 = 0FH and column = 25 = 19H.

Solution:

```
MOV    AH,02        ;set cursor option
MOV    BH,00        ;page 0
MOV    DL,25        ;column position
MOV    DH,15        ;row position
INT    10H          ;invoke interrupt 10H
```

Example 4-2

Write a program that (1) clears the screen and (2) sets the cursor at the center of the screen.

Solution:

The center of the screen is the point at which the middle row and middle column meet. Row 12 is at the middle of rows 0 to 24 and column 39 (or 40) is at the middle of columns 0 to 79. Therefore, by setting row = DH = 12 and column = DL = 39, the cursor is set to the screen center.

```
;clearing the screen
    MOV    AX,0600H      ;scroll the entire page
    MOV    BH,07         ;normal attribute
    MOV    CX,0000       ;row and column of top left
    MOV    DX,184FH      ;row and column of bottom right
    INT    10H           ;invoke the video BIOS service

;setting the cursor to the center of screen
    MOV    AH,02         ;set cursor option
    MOV    BH,00         ;page 0
    MOV    DL,39         ;center column position
    MOV    DH,12         ;center row position
    INT    10H           ;invoke interrupt 10H.
```

INT 10H function 03: get current cursor position

In text mode, one is able to determine where the cursor is located at any time by executing the following:

```
MOV    AH,03        ;option 03 of BIOS INT 10H
MOV    BH,00        ;page 00
INT    10H          ;interrupt 10H routine
```

After execution of the program above, registers DH and DL will have the current row and column positions, and CX provides information about the shape of the cursor. The reason that page 00 was chosen is that the video memory could contain more than one page of data, depending on the video board installed on the PC. In text mode, page 00 is chosen for the currently viewed page.

Changing the video mode

First it must be noted that regardless of what type of adapter is used (MDA, CGA, EGA, MCGA, or VGA), all are upwardly compatible. For example, the VGA emulates all the functions of MCGA, EGA, CGA, and MDA. Similarly, the EGA emulates the functions of CGA and MDA, and so on. Therefore, there must be a way to change the video mode to a desired mode. To do that, one can use INT 10H with AH = 00 and AL = video mode. A list of video modes is given in Appendix E, Table E-2.

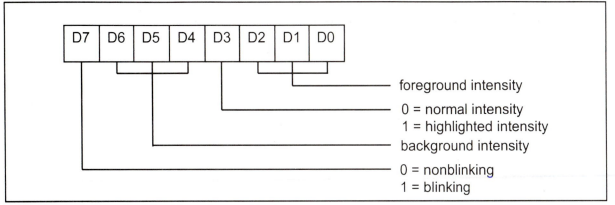

Figure 4-2. Attribute Byte for Monochrome Monitors

Attribute byte in monochrome monitors

There is an attribute associated with each character on the screen. The attribute provides information to the video circuitry, such as color and intensity of the character (foreground) and the background. The attribute byte for each character on the monochrome monitor is limited. Figure 4-2 shows bit definitions of the monochrome byte attribute.

Foreground refers to the actual character displayed. Normal, highlighted intensity and blinking are for the foreground only. The following are some possible variations of the attributes above.

Binary	Hex	Result
0000 0000	00	white on white (no display)
0000 0111	07	white on black normal
0000 1111	0F	white on black highlight
1000 0111	87	white on black blinking
0111 0111	77	black on black (no display)
0111 0000	70	black on white
1111 0000	F0	black on white blinking

For example, "00000111" would give the normal screen mode where the background is black and the foreground is normal intensity, nonblinking. "00001111" would give the same mode with the foreground highlighted. "01110000" would give a reverse video screen mode with the foreground black and the background normal intensity. See Example 4-3.

Attribute byte in CGA text mode

Since all color monitors and their video circuitry are upwardly compatible, in examples concerning color, in this chapter we use CGA mode, the common denominator for all color monitors. The bit definition of the attribute byte in CGA text mode is as shown in Figure 4-3. From the bit definition it can be seen that the background can take eight different colors by combining the prime colors red, blue, and green. The foreground can be any of 16 different colors by combining red, blue, green, and intensity. Example 4-4 shows the use of the

D7	D6	D5	D4	D3	D2	D1	D0
B	R	G	B	I	R	G	B
	background				foreground		

B = blinking
I = intensity
Both blinking and intensity are applied to foreground only.

Figure 4-3. CGA Attribute Byte

attribute byte in CGA mode. Table 4-1 lists the possible colors. As examples of some possible variations look at the following cases:

Binary	Hex	Color effect
0000 0000	00	Black on black
0000 0001	01	Blue on black
0001 0010	12	Green on blue
0001 0100	14	Red on blue
0001 1111	1F	High-intensity white on blue

Example 4-3

Write a program using INT 10H to:
(a) Change the video mode.
(b) Display the letter "D" in 200H locations with attributes black on white blinking (blinking letters "D" are black and the screen background is white).
(c) Then use DEBUG to run and verify the program.

Solution:

(a) INT 10H function AH = 00 is used with AL = video mode to change the video mode. Use AL = 07 for monochrome (MDA), EGA, or VGA; otherwise, use any of the 80x25 text modes, or use AL = 03 for CGA, which all color monitors emulate.

```
MOV     AH,00           ;SET MODE OPTION
MOV     AL,07           ;7 FOR MONOCHROME OR 03 FOR CGA TEXT
INT     10H             ;MODE OF 80X25 FOR ANY COLOR MONITOR
```

(b) With INT 10H function AH=09, one can display a character a certain number of times with specific attributes.

```
MOV     AH,09           ;DISPLAY OPTION
MOV     BH,00           ;PAGE 0
MOV     AL,44H          ;THE ASCII FOR LETTER "D"
MOV     CX,200H         ;REPEAT IT 200H TIMES
MOV     BL,0F0H         ;BLACK ON WHITE BLINKING
INT     10H
```

(c) Reminder: DEBUG assumes that all the numbers are in hex.

```
C>debug
-A
1131:0100 MOV AH,00
1131:0102 MOV AL,07           ;USE 03 IF MONITOR IS COLOR
1131:0104 INT 10
1131:0106 MOV AH,09
1131:0108 MOV BH,00
1131:010A MOV AL,44
1131:010C MOV CX,200
1131:010F MOV BL,F0
1131:0111 INT 10
1131:0113 INT 3
1131:0114
-
```

Now see the result by typing in the command -G. Make sure that IP = 100 before running it.

As an exercise, change the BL register to other attribute values given earlier. For example, BL = 07 white on black, or BL = 87H white on black blinking.

Example 4-4

Write a program that puts 20H (ASCII space) on the entire screen. Use high-intensity white on a blue background attribute for any characters to be displayed.

Solution:

```
        MOV   AH,00       ;SET MODE OPTION
        MOV   AL,03       ;CGA COLOR TEXT MODE OF 80X25
        INT   10H
        MOV   AH,09       ;DISPLAY OPTION
        MOV   BH,00       ;PAGE 0
        MOV   AL,20H      ;ASCII FOR SPACE
        MOV   CX,800H     ;REPEAT IT 800H TIMES
        MOV   BL,1FH      ;HIGH INTENSITY WHITE ON BLUE
        INT   10H
```

Graphics: pixel resolution and color

In the text mode, the screen is viewed as a matrix of rows and columns of characters. In graphics mode, the screen is viewed as a matrix of horizontal and vertical pixels. The number of pixels varies among monitors and depends on monitor resolution and the video board. In this section we show how to access and program pixels on the screen. Before embarking on pixel programming, the relationship between pixel resolution, the number of colors available, and the amount of video memory in a given video board must be clarified. There are two facts associated with every pixel on the screen, (1) the location of the pixel, and (2) its attributes: color and intensity. These two facts must be stored in the video RAM. Therefore, the higher the number of pixels and colors, the larger the amount of memory that is needed to store them. In other words, the memory requirement goes up as the resolution and the number of colors on the monitor go up. The CGA board can have a maximum of 16K bytes of video memory due to its inherent design structure. The 16K bytes of memory can be used in three different ways.

Table 4-1: The 16 Possible Colors

I	R	G	B	Color
0	0	0	0	black
0	0	0	1	blue
0	0	1	0	green
0	0	1	1	cyan
0	1	0	0	red
0	1	0	1	magenta
0	1	1	0	brown
0	1	1	1	white
1	0	0	0	gray
1	0	0	1	light blue
1	0	1	0	light green
1	0	1	1	light cyan
1	1	0	0	light red
1	1	0	1	light magenta
1	1	1	0	yellow
1	1	1	1	high intensity white

1. Text mode of 80 × 25 characters; This takes a total of 2K bytes (80 × 25 = 2000) for the characters plus 2K bytes of memory for their attributes, since each character has one attribute byte. That means that each screen (frame) takes 4K bytes, and that results in CGA supporting a total of four pages of data, where each page represents one full screen. In this mode, 16 colors are supported. To select this mode, use AL = 03 for mode selection in INT 10H option AH = 00.

2. Graphics mode of 320 × 200 (medium resolution); In this mode there are a total of 64,000 pixels (320 columns × 200 rows = 64,000). Dividing the total video RAM memory of 128K bits (16K × 8 bits = 128K bits) by the 64,000 pixels gives 2 bits

for the color of each pixel. These 2 bits give four possibilities. Therefore, the 320 × 200 resolution CGA can support no more than 4 colors. To select this mode, use AL = 04.

3. Graphics resolution of 640 × 200 (high resolution); In this mode there are a total of 128,000 pixels (200 × 640 = 128,000). Dividing the 16K bytes of memory by this gives 1 bit (128,000/128,000 = 1) for color. The bit can be on (white) or off (black). Therefore, the 640 × 200 high-resolution CGA can support only black and white. To select this mode, use AL = 06.

The 160 × 100 low-resolution mode used with color TV sets was bypassed in this discussion since no computer uses that anymore. From the discussion above one can conclude that with a fixed amount of video RAM, as the resolution increases the number of supported colors decreases. That is the reason that to create more colors in VGA boards, one must increase the memory on the video board since there must be a storage place to store the extra colors. Many VGA boards do provide the capacity to expand the video RAM up to 1 megabyte or more.

INT 10H and pixel programming

To address a single pixel on the screen, use INT 10H with AH = 0CH. The X and Y coordinates of the pixel must be known. The values for X (column) and Y (row) vary, depending on the resolution of the monitor. The registers holding these values are CX = the column point (the X coordinate) and DX = the row point (Y coordinate). If the display mode supports more than one page, BH = page number; otherwise, it is ignored. To turn the pixel on or off, AL = 1 or AL = 0 for black and white. The value of AL can be modified for various colors.

Drawing horizontal or vertical lines in graphics mode

To draw a horizontal line, choose values for the row and column to point to the beginning of the line and then continue to increment the column until it reaches the end of the line, as shown in Example 4-5.

Example 4-5

Write a program to:
  (a) Clear the screen.
  (b) Set the mode to CGA of 640 × 200 resolution.
  (c) Draw a horizontal line starting at column = 100, row = 50, and ending at column 200, row 50.

Solution:

```
        MOV    AX,0600H      ;SCROLL THE SCREEN
        MOV    BH,07         ;NORMAL ATTRIBUTE
        MOV    CX,0000       ;FROM ROW=00,COLUMN=00
        MOV    DX,184FH      ;TO ROW=18H,COLUMN=4FH
        INT    10H           ;INVOKE INTERRUPT TO CLEAR SCREEN
        MOV    AH,00         ;SET MODE
        MOV    AL,06         ;MODE = 06 (CGA HIGH RESOLUTION)
        INT    10H           ;INVOKE INTERRUPT TO CHANGE MODE
        MOV    CX,100        ;START LINE AT COLUMN =100 AND
        MOV    DX,50         ;ROW = 50
BACK:   MOV    AH,0CH        ;AH=0CH TO DRAW A LINE
        MOV    AL,01         ;PIXELS = WHITE
        INT    10H           ;INVOKE INTERRUPT TO DRAW LINE
        INC    CX            ;INCREMENT HORIZONTAL POSITION
        CMP    CX,200        ;DRAW LINE UNTIL COLUMN = 200
        JNZ    BACK
```

As an exercise, put INT 3 at the end of the program above and run it in DEBUG to get a feeling of the concept. To draw a vertical line, simply increment the vertical value held by the DX register and keep CX constant. The linear equation $y = mx + b$ can be used to draw any line.

Changing the background color

CGA graphics medium resolution provides 16 colors. Option AH = 0BH can be used to change the background color as shown in Example 4-6. Running that program in DEBUG causes the entire screen to change to blue, and when INT 3 is invoked, all the registers are displayed in very large letters. The cursor gets lost as well, since the screen is in graphics mode. To get out of that mode, simply type Q, followed by the return key to get out of DEBUG and back to DOS. Then use the DOS command "MODE CO80" (the letters CO followed by the number 80) to get back to the original screen mode.

Example 4-6

Write and run a program in DEBUG to:

(a) Change the video mode to 320 × 200 graphics with four colors (mode AL = 4)

(b) Make the entire screen blue

Solution:

```
;a
        MOV     AH,00           ;SET MODE
        MOV     AL,4            ;CGA STANDARD 300X 200 WITH 4 COLOR
        INT     10H
;b
        MOV     AH,0B           ;SET COLOR
        MOV     BH,0            ;SET BACKGROUND COLOR
        MOV     BL,1            ;BLUE (SEE TABLE 4-2)
        INT     10H
        INT     3               ;STOP
```

Review Questions

1. Interrupt 10H function calls perform what services?
2. The monitor in text mode has _____ columns and _____ rows. The top left position is (__,__) and the bottom right position is (__,__).
3. Fill in the blanks in the following program, which clears the screen. Write comments on each line stating the purpose of each line of code.
```
        MOV     AH,___
        MOV     AL,___
        MOV     BH,___
        MOV     CH,___
        MOV     CL,___
        MOV     DH,___
        MOV     DL,___
        INT     10H
```
4. INT 10 function AH = 03 was used. Afterward, DH = 05 and DL = 34. What does this indicate?
5. What is the purpose of the attribute byte for monochrome monitors?
6. In text mode, there is one attribute byte associated with each _____ on the screen.
7. Write the attribute byte to display background green, foreground white blinking.
8. State the purpose of the following program, which is for a monochrome monitor.

```
        MOV     AH,02
        MOV     BH,00
        MOV     DX,0000
        INT     10H
        MOV     AH,09
        MOV     BH,00
        MOV     AL,2AH
        MOV     CX,80
        MOV     BL,0F0H
        INT     10H
```

SECTION 4.2: DOS INTERRUPT 21H

INT 21H is provided by DOS in contrast to INT 10H, which is BIOS-ROM based. When MS-DOS (or its IBM version PC-DOS) is loaded into the computer, INT 21H can be invoked to perform some extremely useful functions. These functions are commonly referred to as DOS INT 21H function calls. A partial list of these options is provided in Appendix D. In this section we use only the options dealing with inputting information from the keyboard and displaying it on the screen. In previous chapters, a fixed set of data was defined in the data segment and the results were viewed in a memory dump. Starting with this chapter, data will come from the keyboard and after it is processed, the results will be displayed on the screen. This is a much more dynamic way of processing information and is the main reason for placing this chapter at this point of the book. Although data is input and output through the keyboard and monitor, there is still a need to dump memory to verify the data when troubleshooting programs.

INT 21H option 09: outputting a string of data to the monitor

INT 21H can be used to send a set of ASCII data to the monitor. To do that, the following registers must be set: AH = 09 and DX = the offset address of the ASCII data to be displayed. Then INT 21H is invoked. The address in the DX register is an offset address and DS is assumed to be the data segment. INT 21H option 09 will display the ASCII data string pointed at by DX until it encounters the dollar sign "$". In the absence of encountering a dollar sign, DOS function call 09 will continue to display any garbage that it can find in subsequent memory locations until it finds "$". For example, to display the message "The earth is but one country", the following is from the data segment and code segment.

DATA_ASC DB 'The earth is but one country','$'

```
        MOV     AH,09                   ;Option 09 to display string of data
        MOV     DX,OFFSET DATA_ASC      ;DX= offset address of data
        INT     21H                     ;invoke the interrupt
```

INT 21H option 02: outputting a single character to the monitor

There are occasions when it is necessary to output to the monitor only a single character. To do that, 02 is put in AH, DL is loaded with the character to be displayed, and then INT 21H is invoked. The following displays letter "J".

```
        MOV     AH,02           ;option 02 displays one character
        MOV     DL,'J'          ;DL holds the character to be displayed
        INT     21H             ;invoke the interrupt
```

This option can also be used to display '$' on the monitor since the string display option (option 09) will not display '$'.

INT 21H option 01: inputting a single character, with echo

This functions waits until a character is input from the keyboard, then echoes it to the monitor. After the interrupt, the input character will be in AL.

```
        MOV     AH,01           ;option 01 inputs one character
        INT     21H             ;after the interrupt, AL = input character (ASCII)
```

Program 4-1 combines INT 10H and INT 21H. The program does the following: (1) clears the screen, (2) sets the cursor to the center of the screen, and (3) starting at that point of the screen, displays the message "This is a test of the display routine".

```
TITLE      PROG4-1    SIMPLE DISPLAY PROGRAM
PAGE       60,132
           .MODEL  SMALL
           .STACK  64
;————————————
           .DATA
MESSAGE DB      'This is a test of the display routine','$'
;————————————
           .CODE
MAIN       PROC    FAR
           MOV     AX,@DATA
           MOV     DS,AX
           CALL    CLEAR              ;CLEAR THE SCREEN
           CALL    CURSOR             ;SET CURSOR POSITION
           CALL    DISPLAY            ;DISPLAY MESSAGE
           MOV     AH,4CH
           INT     21H                ;GO BACK TO DOS
MAIN       ENDP
;————————————
;THIS SUBROUTINE CLEARS THE SCREEN
CLEAR      PROC
           MOV     AX,0600H           ;SCROLL SCREEN FUNCTION
           MOV     BH,07              ;NORMAL ATTRIBUTE
           MOV     CX,0000            ;SCROLL FROM ROW=00,COL=00
           MOV     DX,184FH           ;TO ROW=18H,COL=4FH
           INT     10H                ;INVOKE INTERRUPT TO CLEAR SCREEN
           RET
CLEAR      ENDP
;————————————
;THIS SUBROUTINE SETS THE CURSOR AT THE CENTER OF THE SCREEN
CURSOR     PROC
           MOV     AH,02              ;SET CURSOR FUNCTION
           MOV     BH,00              ;PAGE 00
           MOV     DH,12              ;CENTER ROW
           MOV     DL,39              ;CENTER COLUMN
           INT     10H                ;INVOKE INTERRUPT TO SET CURSOR POSITION
           RET
CURSOR     ENDP
;————————————
;THIS SUBROUTINE DISPLAYS A STRING ON THE SCREEN
DISPLAY    PROC
           MOV     AH,09              ;DISPLAY FUNCTION
           MOV     DX,OFFSET MESSAGE  ;DX POINTS TO OUTPUT BUFFER
           INT     21H                ;INVOKE INTERRUPT TO DISPLAY STRING
           RET
DISPLAY    ENDP
           END     MAIN
```

Program 4-1

INT 21H option 0AH: inputting a string of data from the keyboard

Option 0AH of INT 21H provides a means by which one can get data from the keyboard and store it in a predefined area of memory in the data segment. To do that, registers are set as follows: AH = 0AH and DX = offset address at which the string of data is stored. This is commonly referred to as a *buffer* area. DOS requires that a buffer area be defined in the data segment and the first byte specifies the size of the buffer. DOS will put the number of characters that came in through the keyboard in the second byte and the keyed-in data is placed in the buffer starting at the third byte. For example, the following program will accept up to six characters from the keyboard, including the return (carriage return) key. Six locations were reserved for the buffer and filled with FFH. The following shows portions of the data segment and code segment.

```
                ORG      0010H
DATA1           DB       6,?,6 DUP (FF)              ;0010H=06, 0012H to 0017H = FF

                MOV      AH,0AH                       ;string input option of INT 21H
                MOV      DX,OFFSET DATA1              ;load the offset address of buffer
                INT      21H                          ;invoke interrupt 21H
```

The following shows the memory contents of offset 0010H:

0010	0011	0012	0013	0014	0015	0016	0017
06	00	FF	FF	FF	FF	FF	FF

When this program is executed, the computer waits for the information to come in from the keyboard. When the data comes in, the IBM PC will not exit the INT 21H routine until it encounters the return key. Assuming the data that was entered through the keyboard was "USA" <RETURN>, the contents of memory locations starting at offset 0010H would look like this:

0010	0011	0012	0013	0014	0015	0016	0017
06	03	55	53	41	0D	FF	FF
		U	S	A	CR		

The following is a step-by-step analysis:

0010H=06	DOS requires the size of the buffer in the first location
0011H=03	the keyboard was activated three times (excluding the RETURN key) to key in the letters U, S, and A
0012H=55H	this is ASCII hex value for letter U
0013H=53H	this is ASCII hex value for letter S
0014H=41H	this is ASCII hex value for letter A
0015H=0DH	this is ASCII hex value for CR (carriage return)

One might ask where the value 03 in 0011H came from. DOS puts that value there to indicate that three characters were entered. How can this character count byte be accessed? See the following:

```
                MOV      AH,0AH
                MOV      DX,OFFSET DATA1
                INT      21H

;After data has been keyed in, next fetch the count value
                MOV      BX,OFFSET DATA1
                SUB      CH,CH              ;CH=00
                MOV      CL,[BX]+1          ;move count to CL
```

To locate the CR value 0DH in the string and replace it, say with 00, simply code the following line next:

```
                MOV      SI,CX
                MOV      BYTE PTR[BX+SI]+2,00
```

The actual keyed-in data is located beginning at location [BX]+2.

Inputting more than the buffer size

Now what happens if more than six characters (five, the maximum length + the CR = 6) are keyed in? Entering a message like "USA a country in North America" <RETURN> will cause the computer to sound the speaker and the contents of the buffer will look like this:

0010	0011	0012	0013	0014	0015	0016	0017
06	05	55	53	41	20	61	0D
		U	S	A	SP	a	CR

Location 0015 has ASCII 20H for space and 0016 has ASCII 61H for "a" and finally, the 0D for RETURN key at 0017. The actual length is 05 at memory offset 0011H. Another question is: What happens if only the CR key is activated and no other character is entered? For example, in the following,

```
            ORG      20H
DATA4       DB       10,?,10 DUP (FF)
```

which puts 0AH in memory 0020H, the 0021H is for the count and the 0022H is the first location which will have the data that was entered. So if only the return key is activated, 0022H has 0DH, the hex code for CR.

0020	0021	0022	0023	0024	0025	0026	0027	0028	0029	002A	002B	002C	...
0A	00	0D	FF	FF	FF	FF	FF	FF	FF	FF	FF	FF	...

The actual number of characters entered is 0 at location 0021. Remember that CR is not included in the count. It must be noted that as data is entered it is displayed on the screen. This is called an *echo*. So the 0AH option of INT 21H accepts the string of data from the keyboard and echoes (displays) it on the screen as it is keyed in.

```
;Program 4-2 performs the following, (1) clears the screen, (2) sets the cursor at the beginning of the third line
;from the top of the screen,  (3) accepts the message "IBM perSonal COmputer" from the keyboard,
;(4) converts lowercase letters of the message to uppercase, (5) displays the converted results on the next line.

        TITLE   PROG4-2
        PAGE    60,132
                .MODEL SMALL
                .STACK 64
;------------------------
                .DATA
BUFFER  DB      22,?,22 DUP (?)          ;BUFFER FOR KEYED-IN DATA
        ORG     18H
DATAREA DB      CR,LF,22 DUP (?),'$'     ;AREA TO PLACE DATA AFTER CONVERSION
DTSEG   ENDS
CR      EQU     0DH
LF      EQU     0AH
;------------------------
                .CODE
MAIN    PROC    FAR
        MOV     AX,@DATA
        MOV     DS,AX
        CALL    CLEAR                    ;CLEAR THE SCREEN
        CALL    CURSOR                   ;SET CURSOR POSITION
        CALL    GETDATA                  ;INPUT A STRING INTO BUFFER
        CALL    CONVERT                  ;CONVERT STRING TO UPPERCASE
        CALL    DISPLAY                  ;DISPLAY STRING DATAREA
        MOV     AH,4CH
        INT     21H                      ;GO BACK TO DOS
MAIN    ENDP
;------------------------
;THIS SUBROUTINE CLEARS THE SCREEN
CLEAR   PROC
        MOV     AX,0600H                 ;SCROLL SCREEN FUNCTION
        MOV     BH,07                    ;NORMAL ATTRIBUTE
        MOV     CX,0000                  ;SCROLL FROM ROW=00,COL=00
        MOV     DX,184FH                 ;TO ROW=18H,4FH
        INT     10H                      ;INVOKE INTERRUPT TO CLEAR SCREEN
        RET
CLEAR   ENDP
;------------------------
```

Program 4-2 *(continued on next page)*

```
;THIS SUBROUTINE SETS THE CURSOR TO THE BEGINNING OF THE 3RD LINE
CURSOR  PROC
        MOV     AH,02                   ;SET CURSOR FUNCTION
        MOV     BH,00                   ;PAGE 0
        MOV     DL,01                   ;COLUMN 1
        MOV     DH,03                   ;ROW 3
        INT     10H                     ;INVOKE INTERRUPT TO SET CURSOR
        RET
CURSOR  ENDP
;_____
;THIS SUBROUTINE DISPLAYS A STRING ON THE SCREEN
DISPLAY  PROC
        MOV     AH,09                   ;DISPLAY STRING FUNCTION
        MOV     DX,OFFSET DATAREA       ;DX POINTS TO BUFFER
        INT     21H                     ;INVOKE INTERRUPT TO DISPLAY STRING
        RET
DISPLAY  ENDP
;_____
;THIS SUBROUTINE PUTS DATA FROM THE KEYBOARD INTO A BUFFER
GETDATA PROC
        MOV     AH,0AH                  ;INPUT STRING FUNCTION
        MOV     DX,OFFSET BUFFER        ;DX POINTS TO BUFFER
        INT     21H                     ;INVOKE INTERRUPT TO INPUT STRING
        RET
GETDATA ENDP
;_____
;THIS SUBROUTINE CONVERTS ANY SMALL LETTER TO ITS CAPITAL
CONVERT PROC
        MOV     BX,OFFSET BUFFER
        MOV     CL,[BX]+1               ;GET THE CHAR COUNT
        SUB     CH,CH                   ;CX = TOTAL CHARACTER COUNT
        MOV     DI,CX                   ;INDEXING INTO BUFFER
        MOV     BYTE PTR[BX+DI]+2,20H   ;REPLACE CR WITH SPACE
        MOV     SI,OFFSET DATAREA+2     ;STRING ADDRESS
AGAIN:  MOV     AL,[BX]+2               ;GET THE KEYED-IN DATA
        CMP     AL,61H                  ;CHECK FOR 'a'
        JB      NEXT                    ;IF BELOW, GO TO NEXT
        CMP     AL,7AH                  ;CHECK FOR 'z'
        JA      NEXT                    ;IF ABOVE GO TO NEXT
        AND     AL,11011111B            ;CONVERT TO CAPITAL
NEXT:   MOV     [SI],AL                 ;PLACE IN DATA AREA
        INC     SI                      ;INCREMENT POINTERS
        INC     BX
        LOOP    AGAIN                   ;LOOP IF COUNTER NOT ZERO
        RET
CONVERT ENDP
        END     MAIN
```

Program 4-2 *(continued from preceding page)*

Use of carriage return and line feed

In Program 4-2, the EQU statement was used to equate CR (carriage return) with its ASCII value of 0DH, and LF (line feed) with its ASCII value of 0AH. This makes the program much more readable. Since the result of the conversion was to be displayed in the next line, the string was preceded by CR and LF. In the absence of CR the string would be displayed wherever the cursor happened to be. In the case of CR and no LF, the string would be displayed on the same line after it had been returned to the beginning of the line by the CR and consequently, would write over some of the characters on that line.

Program 4-3 prompts the user to type in a name. The name can have a maximum of eight letters. After the name is typed in, the program gets the length of the name and prints it to the screen.

Program 4-4 demonstrates many of the functions described in this chapter.

```
TITLE        PROG4-3        READS IN LAST NAME AND DISPLAYS LENGTH
PAGE         60,132
             .MODEL  SMALL
             .STACK  64 (?)
;─────────────────
             .DATA
MESSAGE1  DB      'What is your last name?','$'
          ORG     20H
BUFFER1   DB      9,?,9 DUP (0)
          ORG     30H
MESSAGE2  DB      CR,LF,'The number of letters in your name is: ','$'
ROW       EQU     08
COLUMN    EQU     05
CR        EQU     0DH                      ;EQUATE CR WITH ASCII CODE FOR CARRIAGE RETURN
LF        EQU     0AH                      ;EQUATE LF WITH ASCII CODE FOR LINE FEED
;─────────────────
             .CODE
MAIN      PROC    FAR
          MOV     AX,@DATA
          MOV     DS,AX
          CALL    CLEAR
          CALL    CURSOR
          MOV     AH,09                    ;DISPLAY THE PROMPT
          MOV     DX,OFFSET MESSAGE1
          INT     21H
          MOV     AH,0AH                   ;GET LAST NAME FROM KEYBOARD
          MOV     DX,OFFSET BUFFER1
          INT     21H
          MOV     BX,OFFSET BUFFER1        ;FIND OUT NUMBER OF LETTERS IN NAME
          MOV     CL,[BX+1]                ;GET NUMBER OF LETTERS
          OR      CL,30H                   ;MAKE IT ASCII
          MOV     MESSAGE2+40,CL           ;PLACE AT END OF STRING
          MOV     AH,09                    ;DISPLAY SECOND MESSAGE
          MOV     DX,OFFSET MESSAGE2
          INT     21H
          MOV     AH,4CH
          INT     21H                      ;GO BACK TO DOS
MAIN      ENDP
;─────────────────
CLEAR     PROC                             ;CLEAR THE SCREEN
          MOV     AX,0600H
          MOV     BH,07
          MOV     CX,0000
          MOV     DX,184FH
          INT     10H
          RET
CLEAR     ENDP
;─────────────────
CURSOR PROC                                ;SET CURSOR POSITION
          MOV     AH,02
          MOV     BH,00
          MOV     DL,COLUMN
          MOV     DH,ROW
          INT     10H
          RET
CURSOR ENDP
          END     MAIN
```

Program 4-3

INT 21H option 07: keyboard input without echo

Option 07 of INT 21H requires the user to enter a single character but that character is not displayed (or echoed) on the screen. After execution of the interrupt, the PC waits until a single character is entered and provides the character in AL.

```
MOV     AH,07   ;keyboard input without echo
INT     21H
```

Using the LABEL directive to define a string buffer

A more systematic way of defining the buffer area for the string input is to use the LABEL directive. The LABEL directive can be used in the data segment to assign multiple names to data. When used in data segment it looks like this:

```
name        LABEL        attribute
```

The attribute can be either BYTE, WORD, DWORD, FWORD, QWORD, or TBYTE. Simply put, the LABEL directive is used to assign the same offset address to two names. For example, in the following,

```
JOE    LABEL BYTE
TOM    DB       20 DUP(0)
```

the offset address assigned to JOE is the same offset address for TOM since the LABEL directive does not occupy any memory space (see Appendix C for many examples of the use of the LABEL directive). Next we show how to use this directive to define a buffer area for the string keyboard input:

```
DATA_BUF      LABEL BYTE
MAX_SIZE      DB       10
BUF_COUNT     DB       ?
BUF_AREA      DB       10 DUP(20H)
```

Now in the code segment the data can be accessed by name as follows:

```
MOV    AH,0AH                    ;load string into buffer
MOV    DX,OFFSET DATA_BUF
INT    21H
MOV    CL,BUF_COUNT              ;load the actual length of string
MOV    SI,OFFSET BUF_AREA;SI=address of first byte of string
```

This is much more structured and easier to follow. By using this method, it is easy to refer to any parameter by its name. For example, using the LABEL directive, one can rewrite the CONVERT subroutine in Program 4-2 as follows:

```
In that data segment the BUFFER is redefined as
BUFFER                LABEL  BYTE
BUFSIZE               DB     22
BUFCOUNT DB           ?
REALDATA              DB     22 DUP(' ')
```

and in the code segment, in place of the CONVERT procedure:

```
CONVERT   PROC
          MOV    CL,BUFCOUNT        ;load the counter
          SUB    CH,CH              ;CX=counter
          MOV    DI,CX              ;index into data field
          MOV    BX,OFFSET REALDATA ;actual data address in buffer
          MOV    BYTE PTR[BX+DI],20H ;replace the CR with space
          MOV    SI,OFFSET DATAREA  ;SI=address of converted data
AGAIN:    MOV    AL,[BX]            ;move the char into AL
          CMP    AL,61H             ;check if is below 'a'
          JB     NEXT               ;if yes then go to next
          CMP    AL,7AH             ;check for above 'z'
          JA     NEXT               ;if yes then go to next
          AND    AL,11011111B       ;if not then mask it to capital
NEXT:     MOV    [SI],AL            ;move the character
          INC    SI                 ;increment the pointer
          INC    BX                 ;increment the pointer
          LOOP   AGAIN              ;repeat if CX not zero yet
          RET                       ;return to main procedure
CONVERT   ENDP
```

Write a program to perform the following:
(1) clear the screen
(2) set the cursor at row 5 and column 1 of the screen,
(3) prompt "There is a message for you from Mr. Jones. To read it enter Y ". If the user enters 'Y' or 'y' then the message "Hi! I must leave town tomorrow, therefore I will not be able to see you" will appear on the screen. If the user enters any other key, then the prompt "No more messages for you" should appear on the next line.

```
TITLE    PROGRAM 4-4
PAGE 60,132
              .MODEL  SMALL
              .STACK  64
;——————————————
              .DATA
PROMPT1  DB      'There is a message for you from Mr. Jones. '
         DB      'To read it enter Y','$'
MESSAGE  DB      CR,LF,'Hi!  I must leave town tomorrow, '
         DB      'therefore I will not be able to see you','$'
PROMPT2  DB      CR,LF,'No more messages for you','$'
DTSEG    ENDS
CR       EQU     0DH
LF       EQU     0AH
;——————————————
              .CODE
MAIN     PROC    FAR
         MOV     AX,@DATA
         MOV     DS,AX
         CALL    CLEAR           ;CLEAR THE SCREEN
         CALL    CURSOR          ;SET CURSOR POSITION
         MOV     AH,09           ;DISPLAY THE PROMPT
         MOV     DX,OFFSET PROMPT1
         INT     21H
         MOV     AH,07           ;GET ONE CHAR, NO ECHO
         INT     21H
         CMP     AL,'Y'          ;IF 'Y', CONTINUE
         JZ      OVER
         CMP     AL,'y'
         JZ      OVER
         MOV     AH,09           ;DISPLAY SECOND PROMPT IF NOT Y
         MOV     DX,OFFSET PROMPT2
         INT     21H
         JMP     EXIT
OVER:    MOV     AH,09           ;DISPLAY THE MESSAGE
         MOV     DX,OFFSET MESSAGE
         INT     21H
EXIT:    MOV     AH,4CH
         INT     21H             ;GO BACK TO DOS
MAIN     ENDP
;——————————————
CLEAR    PROC                    ;CLEARS THE SCREEN
         MOV     AX,0600H
         MOV     BH,07
         MOV     CX,0000
         MOV     DX,184FH
         INT     10H
         RET
CLEAR    ENDP
;——————————————
CURSOR   PROC                    ;SET CURSOR POSITION
         MOV     AH,02
         MOV     BH,00
         MOV     DL,05           ;COLUMN 5
         MOV     DH,08           ;ROW 8
         INT     10H
         RET
CURSOR   ENDP
         END     MAIN
```

Program 4-4

SECTION 4.2: DOS INTERRUPT 21H

This is the same as Program 4-4, rewritten with full segment definition.

```
TITLE     PROGRAM 4-4 REWRITTEN WITH FULL SEGMENT DEFINITION
PAGE 60,132
STSEG     SEGMENT
          DB 64 DUP (?)
STSEG     ENDS
;─────────────────
DTSEG     SEGMENT
PROMPT1   DB      'There is a message for you from Mr. Jones. '
          DB      'To read it enter Y','$'
MESSAGE   DB      CR,LF,'Hi!  I must leave town tomorrow, '
          DB      'therefore I will not be able to see you','$'
PROMPT2   DB      CR,LF,'No more messages for you','$'
DTSEG     ENDS
CR        EQU     0DH
LF        EQU     0AH
;─────────────────
CDSEG     SEGMENT
MAIN      PROC    FAR
          ASSUME CS:CDSEG,DS:DTSEG,SS:STSEG
          MOV     AX,DTSEG
          MOV     DS,AX
          CALL    CLEAR           ;CLEAR THE SCREEN
          CALL    CURSOR          ;SET CURSOR POSITION
          MOV     AH,09           ;DISPLAY THE PROMPT
          MOV     DX,OFFSET PROMPT1
          INT     21H
          MOV     AH,07           ;GET ONE CHAR, NO ECHO
          INT     21H
          CMP     AL,'Y'          ;IF 'Y', CONTINUE
          JZ      OVER
          CMP     AL,'y'
          JZ      OVER
          MOV     AH,09           ;DISPLAY SECOND PROMPT IF NOT Y
          MOV     DX,OFFSET PROMPT2
          INT     21H
          JMP     EXIT
OVER:     MOV     AH,09           ;DISPLAY THE MESSAGE
          MOV     DX,OFFSET MESSAGE
          INT     21H
EXIT:     MOV     AH,4CH
          INT     21H             ;GO BACK TO DOS
MAIN      ENDP
;─────────────────
CLEAR     PROC                    ;CLEARS THE SCREEN
          MOV     AX,0600H
          MOV     BH,07
          MOV     CX,0000
          MOV     DX,184FH
          INT     10H
          RET
CLEAR     ENDP
;─────────────────
CURSOR    PROC                    ;SET CURSOR POSITION
          MOV     AH,02
          MOV     BH,00
          MOV     DL,05           ;COLUMN 5
          MOV     DH,08           ;ROW 8
          INT     10H
          RET
CURSOR    ENDP
CDSEG     ENDS
          END     MAIN
```

Program 4-4, rewritten with full segment definition

Review Questions

1. INT ____ function calls reside in ROM BIOS, whereas INT ____ function calls are provided by DOS.
2. What is the difference between the following two programs?

   MOV AH,09   MOV AH,0AH
   MOV DX,OFFSET BUFFER   MOV DX,OFFSET BUFFER
   INT 21H   INT 21H
3. INT 21H function 09 will display a string of data beginning at the location specified in register DX. How does the system know where the end of the string is?
4. Fill in the blanks to display the following string using INT 21H.

   MESSAGE1   DB   'What is your last name?$'
       MOV   AH,_____
       MOV   DX,_____
       INT   21H
5. The following prompt needs to be displayed. What will happen if this string is output using INT 21H function 09?

   PROMPT1   DB   'Enter (round to nearest $) your annual salary'
6. Use the EQU directive to equate the name "BELL" with the ASCII code for sounding the bell.
7. Write a program to sound the bell.
8. Code the data definition directives for a buffer area where INT 21H Option 0AH will input a social security number.

SECTION 4.3: INT 16H KEYBOARD PROGRAMMING

The last section demonstrated the use of INT 21H function AH=07, which waits for the user to input a character. What if a program must run a certain task continuously while checking for a keypress? Such cases require the use of INT 16H, a BIOS interrupt used exclusively for the keyboard.

Checking a key press

To check a key press we use INT 16H function AH=01.

```
MOV AH,01        ;check for key press
INT 16H          ;using INT 16H
```

Upon return, ZF=0 if there is a key press; ZF=1 if there is no key press. Notice that this function does not wait for the user to press a key. It simply checks to see if there is a key press. The use of this function is best understood in the context of examples. Program 4-5 sends the ASCII bell character, 07 hex (see Appendix F), to the screen continuously. To stop the bell sound, the user must press any key.

Which key is pressed?

There are times when the program needs to know not only if a key has been pressed but also which key was pressed. To do that, INT 16H function AH=0 can be used immediately after the call to INT 16H function AH=01.

```
MOV AH,0         ;get key pressed
INT 16H          ;using INT 16H
```

Upon return, AL contains the ASCII character of the pressed key; its scan key is in AH. Notice that this function must be used immediately after calling INT 16H function AH=01. Program 4-6 demonstrates how it works. Keyboard scanning is discussed further in Chapter 18.

```
TITLE PROGRAM 4-5:KEYBOARD HIT USING INT 16H
;THIS PROGRAM SOUNDS THE BELL CONTINUOUSLY UNLESS ANY KEY IS PRESSED
                .MODEL SMALL
                .STACK
                .DATA
MESSAGE    DB  'TO STOP THE BELL SOUND PRESS ANY KEY$'
                .CODE
MAIN       PROC
           MOV   AX,@DATA
           MOV   DS,AX
           MOV   AH,09
           MOV   DX,OFFSET MESSAGE  ;DISPLAY THE MESSAGE
           INT   21H
AGAIN:     MOV   AH,02              ;SENDING TO MONITOR A SINGLE CHAR
           MOV   DL,07              ;SEND OUT THE BELL CHAR
           INT   21H
           MOV   AH,01              ;CHECK THE KEY PRESS
           INT   16H                ;USING INT 16H
           JZ    AGAIN              ;IF NO KEY PRESS STAY IN THE LOOP
           MOV   AH,4CH             ;IF ANY KEY PRESSED GO BACK TO DOS
           INT   21H
MAIN       ENDP
           END
```

Program 4-5

```
TITLE PROGRAM 4-6: MODIFIED VERSION OF PROGRAM 4-5
;THIS PROGRAM SOUNDS THE BELL CONTINUOUSLY UNTIL 'Q' OR 'q' IS PRESSED
                .MODEL SMALL
                .STACK
                .DATA
MESSAGE DB  'TO STOP THE BELL SOUND PRESS Q (or q) KEY$'

                .CODE
MAIN       PROC
           MOV   AX,@DATA
           MOV   DS,AX
           MOV   AH,09
           MOV   DX,OFFSET MESSAGE      ;DISPLAY THE MESSAGE
           INT   21H
AGAIN:     MOV   AH,02
           MOV   DL,07                  ;SOUND THE BELL BY SENDING OUT BELL CHAR
           INT   21H
           MOV   AH,01                  ;CHECK FOR KEY PRESS
           INT   16H                    ;USING INT 16H
           JZ    AGAIN                  ;IF NO KEY PRESS KEEP SOUNDING THE BELL
           MOV   AH,0                   ;TO GET THE CHARACTER
           INT   16H                    ;WE MUST USE INT 16H ONE MORE TIME
           CMP   AL,'Q'                 ;IS IT 'Q'?
           JE    EXIT                   ;IF YES EXIT
           CMP   AL,'q'                 ;IS IT 'q'
           JE    EXIT                   ;IF YES EXIT
           JMP   AGAIN                  ;NO. KEEP SOUNDING THE BELL
EXIT:      MOV   AH,4CH                 ;GO BACK TO DOS
           INT   21H
MAIN       ENDP
           END
```

Program 4-6

Review Questions

1. Which function of INT 16H is used for key press detection?
2. In the above question, how do you know if a key is pressed?
3. In the above question, how can the ASCII value for the pressed key be obtained?
4. Indicate the main difference between INT 21H function AH=07, and INT 16H function AH=01.
5. Write a simple program to sound the bell unless letter 'X' is pressed. If 'X' is pressed, the program should exit.

SECTION 4.4: INTERRUPT PROGRAMMING WITH C

Although C is a high-level language, it has strong bit manipulation capability. For this reason, some programmers refer to C as a "high-level assembly" language. For C/C++ programmers who do not have detailed knowledge of 80x86 Assembly language programming but want to write programs using DOS function calls INT 21H and BIOS interrupts, there is help from compilers in the form of *int86* and *intdos* functions. The int86 function is used for calling any of the PC's interrupts, while the intdos function is used only for the INT 21H DOS function calls.

Programming BIOS interrupts with C/C++

To use the int86 function, registers are first set to desired values, then int86 is called. Upon return from int86, the 80x86 registers can be accessed. In this regard, int86 is just like the "INT #" instruction in 80x86 Assembly language. To access the 80x86 registers, use the union of the REGS structure already defined by the C compiler. It has the following format, where regin and regout are variable names:

union REGS regin,regout;

The union of the REGS structure allows x86 registers to be accessed in either their 16- or 8-bit format. The 16-bit registers are referred to as x; 8-bit registers are referred to as h (for "halfword"). This is shown in Table 4-2.

Table 4-2: REGS Union Elements and Assembly Equivalent

16-bit		8-bit	
C language	**Assembly Language**	**C Language**	**Assembly Language**
regin.x.ax	AX	regin.h.al	AL
		regin.h.ah	AH
regin.x.bx	BX	regin.h.bl	BL
		regin.h.bh	BH
regin.x.cx	CX	regin.h.cl	CL
		regin.h.ch	CH
regin.x.dx	DX	regin.h.dl	DL
		regin.h.dh	DH
regin.x.si	SI		
regin.x.di	DI		
regin.x.cflag	CY		

The following code compares loading the registers and invoking the interrupt in C and Assembly language.

```
/*    C language                        Assembly language     */
union REGS regin,regout;
regin.h.ah=0x25;                    /* mov ah,25h  ;AH=25H    */
regin.x.dx=0x4567;                  /* mov dx,4567h ;DX=4567H */
regin.x.si=0x1290;                  /* mov si,1290h ;SI=1290H */
int86(interrupt#,&regin,&regout);   /* int #                  */
```

In the code above, interrupt # is a value from 00 to 255 (or 0x00 to 0xFF in hex, using the C syntax for hexadecimal numbers), and &regin and &regout are the addresses of the REGS variables. Upon returning from the int86 function, we can access the contents of registers just as in 80x86 Assembly language programs. This is shown as follows:

```
mydata=regout.h.ah;     /* mov mydata,ah  ;assign AH to mydata */
myvalu=regout.x.bx;     /* mov myvalu,bx  ;assign BX to myvalu */
```

Examples 4-7 and 4-8 demonstrate how int86 is used in C programming. Example 4-9 shows how to access registers upon returning from int86. Example 4-10 uses INT 21H, AH = 30H to display the DOS version.

Example 4-7

Use the int86 function to clear the screen. Show the equivalent INT 10 instruction.
Solution:

```
                    /* example 4-7A using 16-bit registers */
                    #include <dos.h>              /* int86 is part of this library */
                    main()
                    {
                    union REGS regin,regout;
                    regin.x.ax=0x0600;                    /* MOV AX,0600H        */
                    regin.h.bh=0x07;                      /* MOV BH,07H          */
                    regin.x.cx=0;                         /* MOV CX,0            */
                    regin.x.dx=0x184F;                    /* MOV DX,184FH        */
                    int86(0x10,&regin,&regout);           /* INT 10H             */
                    }
```

We can mix 8- and 16-bit registers as shown next:

```
                    /* example 4-7B using 8-bit registers */
                    #include <dos.h>              /* int86 is part of this library */
                    main()
                    {
                    union REGS regin,regout;
                    regin.h.ah=6;                         /* MOV AH,6            */
                    regin.h.al=0;                         /* MOV AL,0            */
                    regin.h.bh=07;                        /* MOV BH,7            */
                    regin.x.cx=0;                         /* MOV CX,0            */
                    regin.h.dl=0x4F;                      /* MOV DL=4FH          */
                    regin.h.dh=0x18;                      /* MOV DH=18H          */
                    int86(0x10,&regin,&regout);           /* INT 10H             */
                    }
```

Example 4-8

Use function int86 with INT 12H to find the size of conventional memory installed on a given PC.
Solution:
INT 12H provides the size of conventional memory in register AX.

```
                    #include <stdio.h>
                    #include <dos.h>
                    main()
                    {
                    unsigned int convmem;
                    union REGS regin,regout;
                    int86(0x12,&regin,&regout);
                    convmem=regout.x.ax;
                    printf("This PC has %dKB of Conventional memory\n" ,convmem);
                    }
```

Example 4-9

Use the int86 function to perform the following functions.

(a) save the current cursor position, (b) set the cursor to row 12, column 8, and
(c) display the message "Hello" using the printf function.

Solution:

```
#include <stdio.h>
#include <dos.h>
main()
{
unsigned char oldrow;
unsigned char oldcol;
union REGS regin,regout;
regin.h.ah=3;                      /* MOV AH,3 ;option 3 INT 10H   */
regin.h.bh=0;                      /* MOV BH,0 ;page 0                       */
int86(0x10,&regin,&regout);        /* INT 10H ;video INT                     */
oldrow=regout.h.dh;                /* MOV oldrow,DH ;save row                */
oldcol=regout.h.dl;                /* MOV oldcol,DL ;save col                */
printf("Cursor was at row=%d,column=%d \n",oldrow,oldcol);
regin.h.ah=2;                      /* MOV AH,2  ;option 2 of int 10H         */
regin.h.bh=0;                      /* MOV BH,0  ;Page zero                   */
regin.h.dl=8;                      /* MOV DL,8  ;col location                */
regin.h.dh=12;                     /* MOV DH,12 ;row location                */
int86(0x10,&regin,&regout);        /* INT 10H                                */
printf("Hello\n");
}
```

Example 4-10

Using function int86 with INT 21H option 30H, write a program to display the DOS version.

Solution:

```
#include <dos.h>
#include <stdio.h>
#include<conio.h>
main()
{
 union REGS regin,regout;
 unsigned char minor, major;
 clrscr();
 regin.h.ah=0x30;          //get DOS version using AH=30H of INT 21H
 int86(0x21,&regin,&regout);
 minor=regout.h.ah;
 major=regout.h.al;
 printf("The DOS version on this PC is %d.%d\n",major,minor);
}
```

Programming INT 21H DOS function calls with C/C++

Although we can use the int86 function for INT 21H DOS function calls, there is a specially designated function, intdos, that can be used for DOS function calls. The format of intdos is given below. Example 4-11 shows how to use intdos.

```
intdos(&regin,&regout);  /* to be used for INT 21H only */
```

SECTION 4.4: INTERRUPT PROGRAMMING WITH C

Example 4-11

Use INT 21H option 2AH to display the date in the form dd-mm-yy on the screen.
(a) Use intdos functions.   (b) Use the int86 function.

Solution:

Upon returning from the INT 21H function 2AH, DL contains the day, DH the month, CX the year.

(a) This program uses intdos.

```
#include <stdio.h>
#include <dos.h>
main()
{
unsigned int year;
unsigned char month;
unsigned char day;
union REGS regin,regout;
regin.h.ah=0x2A;
intdos(&regin,&regout);
day=regout.h.dl;
month=regout.h.dh;
year=regout.x.cx;
printf("Today's date is %d-%d-%d\n",month,day,year);
}
```

(b) In this program we can replace the intdos statement with

```
int86(0x21,&regin,&regout)
```

Accessing segment registers

Both int86 and intdos allow access to registers AX, BX, CX, DX, SI, and DI, but not segment registers CS, DS, SS, and ES. In some of the interrupt services, we need access to the segment registers as well. In such cases we must use int86x instead of int86, and intdosx instead of intdos. In using int86x and intdosx, we must also pass the argument SREGS. Functions int86x and intdosx have the following formats.

```
int86x(interrupt #,&regin,&regout,&regseg);
intdosx(&regin,&regout,&regseg);
struct SREGS regseg;                    //struct SREGS given below
```

Functions int86x and intdosx provide access only to registers ES and DS and not the segment registers CS and SS. The contents of SS and CS cannot be altered since their alteration will cause the program to crash. Fortunately, BIOS and DOS function calls that use segment registers do not request the alteration of CS and SS. Example 4-12 shows how to get the values of interrupt vector tables.

Accessing the carry flag in int86 and intdos functions

Upon return from many of the interrupt functions, we need to examine the carry flag. Functions int86, intdos, int86x and intdosx allow us to examine the carry flag bit only, and no other flag bits are available through these functions. To access the carry flag bit we write:

```
if(regout.x.cflag)
```

The structures of word, byte, and segment registers are shown below.

```
union REGS {
    struct WORDREGS {
        unsigned int ax;
        unsigned int bx;
        unsigned int cx;
        unsigned int dx;
        unsigned int si;
        unsigned int di;
        unsigned int cflag;
        } x;
    struct BYTEREGS {
        unsigned char al,ah;
        unsigned char bl,bh;
        unsigned char cl,ch;
        unsigned char dl,dh;
        } h;
    } *inregs;
union REGS *outregs;
    struct SREGS {
        unsigned int es;
        unsigned int cs;
        unsigned int ss;
        unsigned int ds;
        } *seregs;
```

Example 4-12

Using INT 21 option 35H, get the CS:IP in the interrupt vector table for INT 10H.

Solution:
From Appendix D, we have INT 21H, AH=35, and AL=interrupt number. Upon return, ES contains the code segment (CS) value and BX has the instruction pointer (IP) value from the vector table.

```
#include <stdio.h>
#include <dos.h>
main()
{
unsigned int ipvalu;
unsigned int csvalu;
union REGS regin,regout;
struct SREGS regseg;
regin.h.ah=0x35;               /* MOV AH,35H */
regin.h.al=0x10;               /* MOV AL,10H */
int86x(0x21,&regin,&regout,&regseg);
/* or we can use intdosx(&regin,&regout,&regseg) */
ipvalu=regout.x.bx;            /* MOV ipvalu,BX */
csvalu=regseg.es;              /* MOV csvalu,ES */
printf("The CS:IP of INT 10H is %X:%X \n " ,csvalu,ipvalu);
}
```

Mixing C with Assembly and checking ZF

The vast majority of interrupts in the PC use the carry flag to indicate special conditions such as errors. For example, functions AH=3CH and AH=3DH for INT 21H (see Appendix D) both deal with files and need the carry flag to indicate certain conditions. As discussed earlier, the REGS union provides only the CF (carry flag). If the zero flag needs to be checked to see if certain conditions are met, the only alternative is to mix C with assembly. This is called *in-line* Assembly, and is shown in Program 4-7. Any x86 valid instruction can be used in a C program if it is prefixed with "asm". In Program 4-7 a statement is displayed several times. Instead of using a C "for" loop, register CX is used as a counter while "JNZ" checks the zero flag.

```
// using in-line assembly to check the zero flag, this program displays a statement
// a certain number of times, using the cx register for a counter
// thanks to Mark Lessley for his input on this example
#include <stdio.h>
#include <conio.h>
main()
    {
    unsigned char row=10;
    unsigned char col=10;    //Byte size data
    unsigned int counter=5;
    clrscr();
                asm MOV CX,counter          //cx=counter
    AGAIN:      asm MOV AH,2                // AH=02 of INT 10H to set cursor
                asm MOV BH,0                //page 0
                asm MOV DH,row              // load the row value
                asm MOV DL,col              // load the column value
                asm INT 10H                 // call INT 10H to set cursor
                asm PUSH CX                 // save the counter
    printf("This is a test!");
    row++;
    col++;
                asm POP CX                  //restore the counter
                asm DEC CX;                 // decrement the counter
                asm JNZ AGAIN               // go back if ZF not high
    getch();
    }
```
Note: While C statements must end with a semicolon; it is optional for statements with the prefix asm.

Program 4-7

C function kbhit vs INT 16H keyboard input

The kbhit function is the C equivalent of INT 16H in Assembly language. Program 4-8 is a C version of Program 4-5.

```
/*keep sounding the bell unless any key is pressed */
#include <stdio.h>
#include  <conio.h>
 main()
 {
 clrscr();
 printf("To Stop the Bell Sound Press Any Key\n");
 while(!kbhit())          //continue as long as there is no keyboard hit
 printf("%c",0x07);    //send the Bell ASCII character to monitor
 }
```

Program 4-8

Program 4-9 checks for a specific key while checking the key press by using the getch() function in C.

```
/*Keep sounding the bell unless 'Q' or 'q' is pressed */
#include <stdio.h>
#include  <conio.h>
 main()
 {
 unsigned char data;
 clrscr();
 printf("To Stop the Bell Sound Press Q or q Key\n");
 do
   {
   while(!kbhit())  //keep sounding Bell unless a key is pressed
     printf("%c",0x07);
     data=getch();          //get the key press
   }
 while(!(data=='q'|| data=='Q'));   //continue as long as it is not Q or q
 }
```

Program 4-9

1. True or false. Function int86 can be used for any interrupt number.
2. True or false. Function intdos can be used for any interrupt number.
3. The int86 has_____ arguments, whereas intdos has _____.
4. True or false. Operand regin.h.al accesses the 16-bit register.
5. Is the following code correct?
   ```
   union REGS rin,rout;
   rin.x.ax=0x1250;
   ```
6. To access segment registers we use _____(int86x, int86, intdos).
7. The int86x function has _____ arguments and they are _____.
8. True or false. In the int86x and intdosx functions, only the ES and DS registers are accessible.

SUMMARY

INT 10H function calls provide the capability to manipulate text and graphics on the screen. These interrupts reside in ROM BIOS because speed is an important factor in these often-used routines. The function calls described in this chapter include calls to clear or scroll the screen, change the video mode, and write text or graphics to the screen. In text mode, the programmer works with a matrix of 80×25 characters. In pixel mode, the programmer works with a matrix of pixels, the number varying with the video mode used.

INT 21H function calls are provided by DOS to perform many useful functions. The function calls described in this chapter include calls to input or output characters or strings to the monitor. INT 16H function calls provide access to the keyboard through the BIOS programs.

Interrupt programming can also be performed in the C programming language through the use of structure REGS and C functions int86 and intdos. In addition, Assembly language instructions can be coded into C programs when the "asm" prefix is used. This is called *in-line* assembly.

PROBLEMS

1. Write a program that:
   (a) Clears the screen, and (b) sets the cursor position at row = 5 and column = 12.
2. What is the function of the following program?
   ```
   MOV  AH,02
   MOV  BH,00
   MOV  DL,20
   MOV  DH,10
   INT  10H
   ```
3. The following program is meant to set the cursor at position row = 14 and col = 20. Fix the error and run the program to verify your solution.
   ```
   MOV  AH,02
   MOV  BH,00
   MOV  DH,14H
   MOV  DL,20H
   INT  10H
   ```
4. Write a program that sets the cursor at row = 12, col = 15, then use the code below to get the current cursor position in register DX with DH = row and DL = col. Is the cursor position in DH and DL in decimal or hex? Verify your answer.
   ```
   MOV  AH,03
   MOV  BH,00
   INT  10H
   ```
5. In clearing the screen, does the sequence of code prior to INT 10H matter? In setting a cursor position? Verify by rearranging and executing the instructions.

SUMMARY 147

6. You want to clear the screen using the following program, but there are some errors. Fix the errors and run the program to verify it.

    ```
    MOV  AX,0600H
    MOV  BH,07
    MOV  CX,0000
    MOV  DX,184F
    INT  10H
    ```

7. Write a program that:
    (a) Clears the screen
    (b) Sets the cursor at row = 8 and column = 14
    (c) Displays the string "IBM Personal Computer"

8. Run the following program and dump the memory to verify the contents of memory locations 0220H to 022FH if "IBM PC with 8088 CPU" is keyed in.

    ```
                    ORG  220H
    BUFFER      DB   15, 16 DUP (0FFH)
    ```
    and for the code :
    ```
    MOV  AH,0AH
    MOV  DX,OFFSET BUFFER
    INT  21H
    ```

9. Write a program that:
    (a) Clears the screen.
    (b) Puts the cursor on position row = 15 and column = 20.
    (c) Displays the prompt "What is your name?"
    (d) Gets a response from the keyboard and displays it at row = 17 and column = 20.

10. Write a program that sets the mode to medium resolution, draws a vertical line in the middle of the screen, then draws a horizontal line across the middle of the screen.

11. Write a program to input a social security number in the form 123-45-6789 and transfer it to another area with the hyphens removed, as in 123456789. Use the following data definition.

    ```
    SS_AREA      LABEL BYTE
    SS_SIZE      DB    12
    SS_ACTUAL  DB    ?
    SS_DASHED  DB    12 DUP (?)
    SS_NUM       DB    9 DUP (?)
    ```

12. Write a program (use the simplified segment definition) to input two seven-digit numbers in response to the prompts "Enter the first number" and "Enter the second number". Add them together (using AAA, covered in Chapter 3) and display the sum with the message "The total sum is: ".

13. Show how to use the union REGS to set AX =9878H, BH=90H, and CL=F4H.

14. Write a C function to set the cursor using int86. Then use it to set the cursor to row=10, col=20 and display the message "HELLO".

15. Write a C program with the following objectives.
    (a) Clear the screen. Use int86.
    (b) Set the cursor to somewhere around the middle of the screen. Use int86.
    (c) Display the date and time continuously in the following format. Use intdos.

Time:	hr:min:sec
Date:	mon/day/yr

    (d) A prompt should ask for "Q" to quit. Use C functions.
    (e) When the user types in Q, it should quit displaying time and date and go back to DOS. Use C functions.

16. A programmer has declared the REGS union as follows. Would this work?

    ```
    union REGS inregs,outregs;
    ```

17. Write two versions of a C program that counts upward by increments of 1 and displays the count with a 1-second delay in between counting. When any key is pressed, it stops counting and goes back to DOS. The count should start from 0.
    (a) use the kbhit function   (b) use the INT 16 function 01

18. Repeat the (b) version of the above program, with the user entering the initial count instead of starting from zero.

ANSWERS TO REVIEW QUESTIONS

SECTION 4.1: BIOS INT 10H PROGRAMMING
1. perform screen i/o
2. 80, 25; 00,00 and 24,79
3.
```
   MOV AH,06   ;SELECT CLEAR SCREEN FUNCTION
   MOV AL,00   ;AH=0 TO SCROLL ENTIRE PAGE
   MOV BH,07   ;BH=07 FOR NORMAL ATTRIBUTE
   MOV CH,00   ;START AT ROW 00
   MOV CL,00   ;START AT COLUMN 00
   MOV DH,24   ;END AT ROW 24
   MOV DL,79   ;END AT ROW 79
   INT 10H     ;INVOKE THE INTERRUPT
```
4. indicates that the cursor is at row 5, column 34
5. it provides information about the foreground and background intensity, whether the foreground is blinking and/or highlighted
6. character
7. 10100111
8. the first time INT 10H is invoked, it sets the cursor to position 00,00; the second time it is invoked, it displays the character '*' 80 times with attributes of white on black, blinking.

SECTION 4.2: DOS INTERRUPT 21H
1. 10H, 21H
2. the leftmost code inputs a string from the keyboard into a buffer; the code on the right outputs a string from a buffer to the monitor
3. the end of the string is the dollar sign '$'
4. 0AH, OFFSET MESSAGE1
5. When the '$' within the string is encountered, the computer will stop displaying the string.
6. BELL   EQU   07H
7. Using the EQU in Answer 6, the code segment would include the following:
```
           MOV      AH,02
           MOV      DL,BELL
           INT 21H
```
8.
```
   SS_AREA    LABEL      BYTE
   SS_SIZE    DB  12
   SS_ACTUAL  DB  ?
   SS_NUM     DB  12 DUP (?)
```

SECTION 4.3: INT 16H KEYBOARD PROGRAMMING
1. INT 16H function AH=01
2. After return from INT 16H function AH=01, if ZF=1 there is no key press; if ZF=0 then a key has been pressed.
3. If ZF=0, then we use INT 16H function AH=0 to get the ASCII character for the pressed key.
4. INT 21H waits for the user to press the key; INT 16H scans the keyboard, allowing the program to continue executing other tasks while scanning for the key press.
5.
```
AGAIN:  MOV AH,02    ;USE FUNCTION AH=02 OF INT 21H
        MOV DL,07    ;SOUND THE BELL BY SENDING OUT BELL CHAR
        INT 21H
        MOV AH,01    ;CHECK FOR KEY PRESS
        INT 16H      ;USING INT 16H
        JZ  AGAIN    ;IF NO KEY PRESS KEEP SOUNDING THE BELL
        MOV AH,0     ;TO GET THE CHARACTER
        INT 16H      ;WE MUST USE INT 16H ONE MORE TIME
        CMP AL,'X'   ;IS IT 'X'?.
        JE  EXIT     ;IF YES EXIT
        JMP AGAIN    ;NO. KEEP SOUNDING THE BELL
EXIT:   MOV AH,4CH   ;GO BACK TO DOS
        INT 21H
```

SECTION 4.4: INTERRUPT PROGRAMMING WITH C
1. true
2. false; only for the INT 21H
3. 3, 2
4. false
5. Yes; we can use any name. Other commonly used names are inregs,outregs, and r1,r2.
6. int86x
7. four: INT #, &regin, &regout, &regseg
8. true

CHAPTER 5

MACROS AND THE MOUSE

<table>
<tr><td colspan="2">OBJECTIVES</td></tr>
<tr><td colspan="2">Upon completion of this chapter, you will be able to:</td></tr>
<tr><td>»</td><td>List the advantages of using macros</td></tr>
<tr><td>»</td><td>Define macros by coding macro definition directives MACRO and ENDM</td></tr>
<tr><td>»</td><td>Code Assembly language instructions to invoke macros</td></tr>
<tr><td>»</td><td>Explain how macros are expanded by the assembler</td></tr>
<tr><td>»</td><td>Use .XALL, .SALL, and .LALL directives to control macro expansion within the list file</td></tr>
<tr><td>»</td><td>Use the LOCAL directive to define local variables within macros</td></tr>
<tr><td>»</td><td>Use the INCLUDE directive to include macros from another file in a program</td></tr>
<tr><td>»</td><td>Use INT 33H to control mouse functions in text and graphics modes</td></tr>
<tr><td>»</td><td>Code Assembly language instructions to initialize the mouse</td></tr>
<tr><td>»</td><td>Code Assembly language instructions to set or get the mouse cursor position</td></tr>
<tr><td>»</td><td>Use INT 33H functions to retrieve mouse button press and release information</td></tr>
<tr><td>»</td><td>Limit mouse cursor positions by setting boundaries or defining exclusion areas</td></tr>
</table>

In this chapter we explore the concept of the macro and its use in Assembly language programming. The format and usage of macros are defined and many examples of their application are explored. In addition, this chapter demonstrates the use of INT 33H to control mouse functions in Assembly language programs.

SECTION 5.1: WHAT IS A MACRO AND HOW IS IT USED?

There are applications in Assembly language programming where a group of instructions performs a task that is used repeatedly. For example, INT 21H function 09H for displaying a string of data and function 0AH for keying in data are used repeatedly in the same program, as was seen in Chapter 4. So it does not make sense to rewrite them every time they are needed. Therefore, to reduce the time that it takes to write these codes and reduce the possibility of errors, the concept of macros was born. Macros allow the programmer to write the task (set of codes to perform a specific job) once only and to invoke it whenever it is needed wherever it is needed.

MACRO definition

Every macro definition must have three parts, as follows:

```
name        MACRO      dummy1,dummy2,...,dummyN
            ......
            ......
            ENDM
```

The MACRO directive indicates the beginning of the macro definition and the ENDM directive signals the end. What goes in between the MACRO and ENDM directives is called the body of the macro. The name must be unique and must follow Assembly language naming conventions. The dummies are names, or parameters, or even registers that are mentioned in the body of the macro. After the macro has been written, it can be invoked (or called) by its name, and appropriate values are substituted for dummy parameters. Displaying a string of data using function 09 of INT 21H is a widely used service. The following is a macro for that service:

```
STRING      MACRO      DATA1
            MOV        AH,09
            MOV        DX,OFFSET DATA1
            INT        21H
            ENDM
```

The above is the macro *definition*. Note that dummy argument DATA1 is mentioned in the body of macro. In the following example, assume that a prompt has already been defined in the data segment as shown below. In the code segment, the macro can be invoked by its name with the user's actual data:

```
MESSAGE1    DB         'What is your name?','$'
            ...
            STRING MESSAGE1
```

The instruction "STRING MESSAGE" *invokes* the macro. The assembler *expands* the macro by providing the following code in the .LST file:

```
1    MOV    AH,09
1    MOV    DX,OFFSET MESSAGE1
1    INT    21H
```

The (1) indicates that the code is from the macro. In earlier versions of MASM, a plus sign (+) indicated lines from macros.

Comments in a macro

Now the question is: Can macros contain comments? The answer is yes, but there is a way to suppress comments and make the assembler show only the lines that generate opcodes. There are basically two types of comments in the macro: listable and nonlistable. If comments are preceded by a single semicolon (;) as is done in Assembly language programming, they will show up in the ".lst" file, but if comments are preceded by a double semicolon (;;) they will not appear in the ".lst" file when the program is assembled. There are also three directives designed to make programs that use macros more readable, meaning that they only affect the ".lst" file and have no effect on the ".obj" or ".exe" files. They are as follows:

.LALL (List ALL) will list all the instructions and comments that are preceded by a single semicolon in the ".lst" file. The comments preceded by a double semicolon cannot be listed in the ".lst" file in any way.

.SALL (Suppress ALL) is used to make the list file shorter and easier to read. It suppresses the listing of the macro body and the comments. This is especially useful if the macro is invoked many times within the same program and there is no need to see it listed every time. It must be emphasized that the use of .SALL will not eliminate any opcode from the object file. It only affects the listing in the ".lst" file.

.XALL (eXecutable ALL), which is the default listing directive, is used to list only the part of the macro that generates opcodes.

Example 5-1 demonstrates the macro definition.

Example 5-1

Write macro definitions for setting the cursor position, displaying a string, and clearing the screen.

Solution:

```
CURSOR   MACRO   ROW,COLUMN
;THIS MACRO SETS THE CURSOR LOCATION TO ROW,COLUMN
;;USING BIOS INT 10H FUNCTION 02
         MOV   AH,02          ;SET CURSOR FUNCTION
         MOV   BH,00          ;PAGE 00
         MOV   DH,ROW         ;ROW POSITION
         MOV   DL,COLUMN      ;COLUMN POSITION
         INT   10H            ;INVOKE THE INTERRUPT
         ENDM

DISPLAY  MACRO   STRING
;THIS MACRO DISPLAYS A STRING OF DATA
;;DX = ADDRESS OF STRING. USES FUNCTION 09 INT 21H.
         MOV   AH,09              ;DISPLAY STRING FUNCTION
         MOV   DX,OFFSET STRING   ;DX = OFFSET ADDRESS OF DATA
         INT   21H                ;INVOKE THE INTERRUPT
         ENDM

CLEARSCR  MACRO
;THIS MACRO CLEARS THE SCREEN
;;USING OPTION 06 OF INT 10H
         MOV   AX,0600H       ;SCROLL SCREEN FUNCTION
         MOV   BH,07          ;NORMAL ATTRIBUTE
         MOV   CX,0           ;FROM ROW=00,COLUMN=00
         MOV   DX,184FH       ;TO ROW=18H,COLUMN=4FH
         INT   10H            ;INVOKE THE INTERRUPT
         ENDM
```

Remember that the comments marked with ";;" will not be listed in the list file as seen in the list file for Program 5-1.

Using the macro definition in Example 5-1, write a program that clears the screen and then at each of the following screen locations displays the indicated message:

at row 2 and column 4 "My name"   at row 12 and column 44 "what is"

at row 7 and column 24 "is Joe"   at row 19 and column 64 "your name?"

```
TITLE       PROG5-1
PAGE        60,132
;————————————————————
CLEARSCR    MACRO
;THIS MACRO CLEARS THE SCREEN
;;USING OPTION 06 OF INT 10H
            MOV     AX,0600H        ;SCROLL SCREEN FUNCTION
            MOV     BH,07           ;NORMAL ATTRIBUTE
            MOV     CX,0            ;FROM ROW=00,COLUMN=00
            MOV     DX,184FH        ;TO ROW=18H,COLUMN=4FH
            INT     10H             ;INVOKE THE INTERRUPT
            ENDM
;————————————————————
DISPLAY     MACRO  STRING
;THIS MACRO DISPLAYS A STRING OF DATA
;;DX = ADDRESS OF STRING.  USES FUNCTION 09 INT 21H.
            MOV     AH,09                 ;DISPLAY STRING FUNCTION
            MOV     DX,OFFSET STRING      ;DX = OFFSET ADDRESS OF DATA
            INT     21H                   ;INVOKE THE INTERRUPT
            ENDM
;————————————————————
CURSOR      MACRO  ROW,COLUMN
;THIS MACRO SETS THE CURSOR LOCATION TO ROW,COLUMN
;;USING BIOS INT 10H FUNCTION 02
            MOV     AH,02           ;SET CURSOR FUNCTION
            MOV     BH,00           ;PAGE 00
            MOV     DH,ROW          ;ROW POSITION
            MOV     DL,COLUMN       ;COLUMN POSITION
            INT     10H             ;INVOKE THE INTERRUPT
            ENDM
;————————————————————
            .MODEL SMALL
            .STACK 64
;————————————————————
            .DATA
MESSAGE1 DB       'My name ','$'
MESSAGE2 DB       'is Joe','$'
MESSAGE3 DB       'What is ','$'
MESSAGE4 DB       'your name?','$'
;————————————————————
            .CODE
MAIN        PROC    FAR
            MOV     AX,@DATA
            MOV     DS,AX
            .LALL                 ;LIST ALL
            CLEARSCR              ;INVOKE CLEAR SCREEN MACRO
            CURSOR 2,4            ;SET CURSOR TO ROW2,COL 2
            DISPLAY MESSAGE1      ;INVOKE DISPLAY MACRO
            .XALL                 ;LIST ALL EXECUTABLE
            CURSOR 7,24           ;SET CURSOR TO ROW 7,COL 24
            DISPLAY MESSAGE2      ;INVOKE DISPLAY MACRO
            .SALL                 ;SUPPRESS ALL
            CURSOR 12,44          ;SET CURSOR TO ROW 12,COL 44
            DISPLAY MESSAGE3      ;INVOKE DISPLAY MACRO
            CURSOR 19,64          ;SET CURSOR TO ROW 19,COL 64
            DISPLAY MESSAGE4      ;INVOKE DISPLAY MACRO
            MOV     AH,4CH
            INT     21H           ;GO BACK TO DOS
MAIN        ENDP
            END     MAIN
```

Program 5-1

Analysis of Program 5-1

Compare the ".asm" and ".lst" files to see the use of .LALL, .XALL, and .SALL. The .LALL directive was used for each macro and then .XALL was used for two of them. From then on, all were suppressed.

```
Microsoft (R) Macro Assembler  Version 5.10          1/13/92 00:17:15
PROG5-1                                              Page    1-1


                              TITLE   PROG5-1
                              PAGE    60,132
                          ;───────────────────────
                          CLEARSCR      MACRO
                          ;THIS MACRO CLEARS THE SCREEN
                          ;;USING OPTION 06 OF INT 10H
                                   MOV    AX,0600H       ;SCROLL SCREEN FUNCTION
                                   MOV    BH,07          ;NORMAL ATTRIBUTE
                                   MOV    CX,0           ;FROM ROW=00,COLUMN=00
                                   MOV    DX,184FH       ;TO ROW=18H,COLUMN=4FH
                                   INT    10H            ;INVOKE THE INTERRUPT
                                   ENDM
                          ;───────────────────────
                          DISPLAY       MACRO   STRING
                          ;THIS MACRO DISPLAYS A STRING OF DATA
                                   ;;DX = ADDRESS OF STRING.  USES FUNCTION 09 INT 21H.
                                   MOV    AH,09              ;DISPLAY STRING FUNCTION
                                   MOV    DX,OFFSET STRING   ;DX = OFFSET ADDRESS OF DATA
                                   INT    21H                ;INVOKE THE INTERRUPT
                                   ENDM
                          ;───────────────────────
                          CURSOR        MACRO   ROW,COLUMN
                          ;THIS MACRO SETS THE CURSOR LOCATION TO ROW,COLUMN
                          ;;USING BIOS INT 10H FUNCTION 02
                                   MOV    AH,02          ;SET CURSOR FUNCTION
                                   MOV    BH,00          ;PAGE 00
                                   MOV    DH,ROW         ;ROW POSITION
                                   MOV    DL,COLUMN      ;COLUMN POSITION
                                   INT    10H            ;INVOKE THE INTERRUPT
                                   ENDM
                          ;───────────────────────
0000                               .MODEL SMALL
0000                               .STACK 64
                          ;───────────────────────
0000                               .DATA
0000  4D 79 20 6E 61 6D   MESSAGE1 DB    'My name ','$'
      65 20 24
0009  69 73 20 4A 6F 65   MESSAGE2 DB    'is Joe','$'
0010  57 68 61 74 20 69   MESSAGE3 DB    'What is ','$'
      73 20  24
0019  79 6F 75 72 20 6E   MESSAGE4 DB    'your name?','$'
      61 6D 65 3F 24
                          ;───────────────────────
0024                               .CODE
0000                      MAIN  PROC    FAR
0000  B8 0000s                    MOV    AX,@DATA
0003  8E D8                       MOV    DS,AX
                                  .LALL                  ;LIST ALL
                                  CLEARSCR               ;INVOKE CLEAR SCREEN MACRO
```

List File for Program 5-1 *(continued on next page)*

```
                          1 ;THIS MACRO CLEARS THE SCREEN
                          1 ;
0005 B8 0600              1          MOV  AX,0600H        ;SCROLL SCREEN FUNCTION
0008 B7 07               1          MOV  BH,07           ;NORMAL ATTRIBUTE
000A B9 0000             1          MOV  CX,0            ;FROM ROW=00,COLUMN=00
000D BA 184F             1          MOV  DX,184FH        ;TO ROW=18H,COLUMN=4FH
0010 CD 10               1          INT   10H            ;INVOKE THE INTERRUPT
                                    CURSOR  2,4;CURSOR MACRO WILL SET CURSOR TO 2, 2
                          1 ;THIS MACRO SETS THE CURSOR LOCATION
                          1 ;
0012 B4 02               1          MOV  AH,02           ;SET CURSOR FUNCTION
0014 B7 00               1          MOV  BH,00           ;PAGE 00
0016 B6 02               1          MOV  DH,2            ;ROW POSITION
0018 B2 04               1          MOV  DL,4            ;COLUMN POSITION
001A CD 10               1          INT   10H            ;INVOKE THE INTERRUPT
                                    DISPLAY MESSAGE1  ;INVOKE DISPLAY MACRO
                          1 ;THIS MACRO DISPLAYS A STRING OF DATA
                          1 ;
001C B4 09               1          MOV  AH,09  ;DISPLAY STRING FUNCTION
001E BA 0000 R           1          MOV  DX,OFFSET MESSAGE1  ;DX = OFFSET ADDRESS OF DATA
0021 CD 21               1          INT   21H            ;INVOKE THE INTERRUPT
                                    .XALL                ;LIST ALL EXECUTABLE
                                    CURSOR  7,24         ;SET CURSOR TO ROW=7,COL= 24
0023 B4 02               1          MOV  AH,02           ;SET CURSOR FUNCTION
0025 B7 00               1          MOV  BH,00           ;PAGE 00
0027 B6 07               1          MOV  DH,7            ;ROW POSITION
0029 B2 18               1          MOV  DL,24           ;COLUMN POSITION
002B CD 10               1          INT   10H            ;INVOKE THE INTERRUPT
                                    DISPLAY MESSAGE2  ;INVOKE DISPLAY MACRO
002D B4 09               1          MOV  AH,09           ;DISPLAY STRING FUNCTION
002F BA 0009 R           1          MOV  DX,OFFSET MESSAGE2  ;DX = OFFSET ADDRESS OF DATA
0032 CD 21               1          INT   21H            ;INVOKE THE INTERRUPT
                                    .SALL                ;SUPPRESS ALL
                                    CURSOR  12,44        ;SET CURSOR TO ROW=12,COL=44
                                    DISPLAY MESSAGE3  ;INVOKE DISPLAY MACRO
                                    CURSOR  19,64        ;SET CURSOR TO ROW=19,COL=64
                                    DISPLAY MESSAGE4  ;INVOKE DISPLAY MACRO
0056 B4 4C                          MOV  AH,4CH
0058 CD 21                          INT   21H            ;GO BACK TO DOS
005A             MAIN               ENDP
                                    END    MAIN
```

List File for Program 5-1 *(continued from preceding page)*

LOCAL directive and its use in macros

In the discussion of macros so far, examples have been chosen that do not have a label or name in the body of the macro. This is because if a macro is expanded more than once in a program and there is a label in the label field of the body of the macro, these labels must be declared as LOCAL. Otherwise, an assembler error would be generated when the same label was encountered in two or more places. The following rules must be observed in the body of the macro:

1. All labels in the label field must be declared LOCAL.
2. The LOCAL directive must be right after the MACRO directive. In other words, it must be placed even before comments and the body of the macro; otherwise, the assembler gives an error.
3. The LOCAL directive can be used to declare all names and labels at once as follows:

```
LOCAL      name1,name2,name3
```

or one at a time as:

```
LOCAL      name1
LOCAL      name2
LOCAL      name3
```

To clarify these points, look at Example 5-2.

SECTION 5.1: WHAT IS A MACRO AND HOW IS IT USED?   **155**

Example 5-2

Write a macro that multiplies two words by repeated addition, then saves the result.

Solution:

The following macro can be expanded as often as desired in the same program since the label "back" has been declared as LOCAL.

```
MULTIPLY MACRO VALUE1, VALUE2, RESULT
     LOCAL  BACK
;   THIS MACRO COMPUTES RESULT = VALUE1 X VALUE2
;;  BY REPEATED ADDITION
;;VALUE1 AND VALUE2 ARE WORD OPERANDS; RESULT IS A DOUBLEWORD
             MOV    BX,VALUE1    ;BX=MULTIPLIER
             MOV    CX,VALUE2    ;CX=MULTIPLICAND
             SUB    AX,AX        ;CLEAR AX
             MOV    DX,AX        ;CLEAR DX
BACK:        ADD    AX,BX        ;ADD BX TO AX
             ADC    DX,00        ;ADD CARRIES IF THERE IS ONE
             LOOP   BACK         ;CONTINUE UNTIL CX=0
             MOV    RESULT,AX    ;SAVE THE LOW WORD
             MOV    RESULT+2,DX  ;SAVE THE HIGH WORD
             ENDM
```

Use the macro definition in Example 5-2 to write a program that multiplies the following:

    (1) 2000 x 500   (2) 2500 x 500   (3) 300 x 400

```
TITLE    PROG5-2
PAGE 60,132
;————————————————————
MULTIPLY  MACRO  VALUE1, VALUE2, RESULT
       LOCAL BACK
;THIS MACRO COMPUTES RESULT = VALUE1 X VALUE2
;;BY REPEATED ADDITION
;;VALUE1 AND VALUE2 ARE WORD OPERANDS; RESULT IS A DOUBLEWORD
             MOV    BX,VALUE1    ;BX=MULTIPLIER
             MOV    CX,VALUE2    ;CX=MULTIPLICAND
             SUB    AX,AX        ;CLEAR AX
             MOV    DX,AX        ;CLEAR DX
BACK:        ADD    AX,BX        ;ADD BX TO AX
             ADC    DX,00        ;ADD CARRIES IF THERE IS ONE
             LOOP   BACK         ;CONTINUE UNTIL CX=0
             MOV    RESULT,AX    ;SAVE THE LOW WORD
             MOV    RESULT+2,DX  ;SAVE THE HIGH WORD
             ENDM
;————————————————————
             .MODEL SMALL
             .STACK 64
;————————————————————
             .DATA
RESULT1  DW      2 DUP (0)
RESULT2  DW      2 DUP (0)
RESULT3  DW      2 DUP (0)
;————————————————————
             .CODE
MAIN         PROC   FAR
             MOV    AX,@DATA
             MOV    DS,AX
             MULTIPLY 2000,500,RESULT1
             MULTIPLY 2500,500,RESULT2
             MULTIPLY  300,400,RESULT3
             MOV    AH,4CH
             INT    21H                 ;GO BACK TO DOS
MAIN         ENDP
             END    MAIN
```

Program 5-2

Notice in Example 5-2 that the "BACK" label is defined as LOCAL right after the MACRO directive. Defining this anywhere else causes an error. The use of a LOCAL directive allows the assembler to define the labels separately each time it encounters them. The list file below shows that when the macro is expanded for the first time, the list file has "??0000". For the second time it is "??0001", and for the third time it is "??0002" in place of the "BACK" label, indicating that the label "BACK" is local. To clarify this concept, try Example 5-2 without the LOCAL directive to see how the assembler will give an error.

```
Microsoft (R) Macro Assembler Version 5.10        1/13/92  00:33:14
PROG5-2                                                  Page     1-1


                          TITLE    PROG5-2
                          PAGE 60,132
                          ;─────────────────────────────
                          MULTIPLY  MACRO  VALUE1, VALUE2, RESULT
                                  LOCAL BACK
                          ;THIS MACRO COMPUTES RESULT = VALUE1 X VALUE2
                          ;;BY REPEATED ADDITION
                          ;;VALUE1 AND VALUE2 ARE WORD OPERANDS; RESULT IS A DOUBLE
                                  MOV  BX,VALUE1      ;BX=MULTIPLIER
                                  MOV  CX,VALUE2      ;CX=MULTIPLICAND
                                  SUB  AX,AX          ;CLEAR AX
                                  MOV  DX,AX          ;CLEAR DX
                          BACK:   ADD  AX,BX          ;ADD BX TO AX
                                  ADC  DX,00          ;ADD CARRIES IF THERE IS ONE
                                  LOOP BACK           ;CONTINUE UNTIL CX=0
                                  MOV  RESULT,AX      ;SAVE THE LOW WORD
                                  MOV  RESULT+2,DX    ;SAVE THE HIGH WORD
                                  ENDM
                          ;─────────────────────────────
0000                              .MODEL SMALL
0000                              .STACK 64
0040              STSEG  ENDS
                          ;─────────────────────────────
0000                              .DATA
0000  0002 [      RESULT1  DW     2 DUP (0)
         0000
      ]

0004  0002 [      RESULT2  DW     2 DUP (0)
         0000
      ]

0008  0002 [      RESULT3  DW     2 DUP (0)
                          ;─────────────────────────────
000C                              .CODE
0000              MAIN     PROC   FAR
                                  ASSUME CS:CDSEG,DS:DTSEG,SS:STSEG
0000  B8 0000s            MOV    AX,@DATA
0003  8E D8               MOV    DS,AX
                          MULTIPLY 2000,500,RESULT1
0005  BB 07D0   1         MOV    BX,2000      ;BX=MULTIPLIER
0008  B9 01F4   1         MOV    CX,500       ;CX=MULTIPLICAND
000B  2B C0     1         SUB    AX,AX        ;CLEAR AX
000D  8B D0     1         MOV    DX,AX        ;CLEAR DX
000F  03 C3     1  ??0000: ADD   AX,BX        ;ADD BX TO AX
0011  83 D2 00  1         ADC    DX,00        ;ADD CARRIES IF THERE IS ONE
0014  E2 F9     1         LOOP   ??0000       ;CONTINUE UNTIL CX=0
0016  A3 0000 R 1         MOV    RESULT1,AX   ;SAVE THE LOW WORD
0019  89 16 0002 R 1      MOV    RESULT1+2,DX ;SAVE THE HIGH WORD
```

List File for Program 5-2 *(continued on next page)*

SECTION 5.1: WHAT IS A MACRO AND HOW IS IT USED?   157

```
                                    MULTIPLY 2500,500,RESULT2
001D  BB 09C4        1              MOV     BX,2500      ;BX=MULTIPLIER
0020  B9 01F4        1              MOV     CX,500       ;CX=MULTIPLICAND
0023  2B C0          1              SUB     AX,AX        ;CLEAR AX
0027  03 C3          1??0001:       ADD     AX,BX        ;ADD BX TO AX
0029  83 D2 00       1              ADC     DX,00        ;ADD CARRIES IF THERE IS ONE
002C  E2 F9          1              LOOP    ??0001       ;CONTINUE UNTIL CX=0
002E  A3 0004 R      1              MOV     RESULT2,AX   ;SAVE THE LOW WORD
0031  89 16 0006 R 1                MOV     RESULT2+2,DX ;SAVE THE HIGH WORD
                                    MULTIPLY 300,400,RESULT3
0035  BB 012C        1              MOV     BX,300       ;BX=MULTIPLIER
0038  B9 0190        1              MOV     CX,400       ;CS=MULTIPLICAND
003B  2B C0          1              SUB     AX,AX        ;CLEAR AX
003D  8B D0          1              MOV     DX,AX        ;CLEAR DX
003F  03 C3          1??0002:       ADD     AX,BX        ;ADD BX TO AX
0041  83 D2 00       1              ADC     DX,00        ;ADD CARRIES IF THERE IS ONE
0044  E2 F9          1              LOOP    ??0002       ;CONTINUE UNTIL CX=0
0046  A3 0008 R      1              MOV     RESULT3,AX   ;SAVE THE LOW WORD
0049  89 16 000A R 1                MOV     RESULT3+2,DX ;SAVE THE HIGH WORD
004D  B4 4C                         MOV     AH,4CH
004F  CD 21                         INT     21H          ;GO BACK TO DOS
0051                   MAIN  ENDP
                             END     MAIN
```

List File for Program 5-2   *(continued from preceding page)*

INCLUDE directive

Assume that there are several macros that are used in every program. Must they be rewritten every time? The answer is no if the concept of the INCLUDE directive is known. The INCLUDE directive allows a programmer to write macros and save them in a file, and later bring them into any file. For example, assume that the following widely used macros were written and then saved under the filename "MYMACRO1.MAC".

```
CLEARSCR    MACRO                   ;the clear screen macro
            MOV   AX,0600H
            MOV   BH,07
            MOV   CX,0000
            MOV   DX,184FH
            INT   10H
            ENDM

DISPLAY     MACRO  STRING           ;the string display macro
            MOV   AH,09
            MOV   DX,OFFSET STRING
            INT   21H
            ENDM

REGSAVE     MACRO                   ;this macro saves all the registers
            PUSH  AX
            PUSH  BX
            PUSH  CX
            PUSH  DX
            PUSH  DI
            PUSH  SI
            PUSH  BP
            PUSHF
            ENDM
```

```
REGRESTO    MACRO            ;this macro restores all the registers
            POPF
            POP   BP
            POP   SI
            POP   DI
            POP   DX
            POP   CX
            POP   BX
            POP   AX
            ENDM
```

Assuming that these macros are saved on a disk under the filename "MY-MACRO1.MAC", the INCLUDE directive can be used to bring this file into any ".asm" file and then the program can call upon any of the macros as many times as needed. When a file includes all macros, the macros are listed at the beginning of the ".lst" file and as they are expanded, they will be part of the program. To understand this, see the following program.

Program 5-3 includes macros to clear the screen, set the cursor, and display strings. These macros are all saved under the "MYMACRO2.MAC" filename. The ".asm" and ".lst" versions of the program that use the clear screen and display string macros only to display "This is a test of macro concepts" are shown on the following pages.

Notice that in the list file of Program 5-3, the letter "C" in front of the lines indicates that they are copied from another file and included in the present file.

```
TITLE      PROG5-3
PAGE       60,132
;
           INCLUDE  MYMACRO2.MAC
           .MODEL SMALL
           .STACK 64
;─────────────────────────────
           .DATA
MESSAGE1 DB       'This is a test of macro concepts','$'
           .CODE
MAIN       PROC   FAR
           MOV    AX,@DATA
           MOV    DS,AX
           CLEARSCR               ;INVOKE CLEAR SCREEN MACRO
           DISPLAY  MESSAGE1      ;INVOKE DISPLAY MACRO
           MOV    AH,4CH
           INT    21H             ;GO BACK TO DOS
MAIN       ENDP
           END    MAIN
```

Program 5-3

Review Questions

1. Discuss the benefits of macro programming.
2. List the three parts of a macro.
3. Explain and contrast the macro definition, invoking the macro, and expanding the macro.
4. True or false. A label defined within a macro is automatically understood by the assembler to be local.
5. True or false. In the list file for Program 5-3, the "C" at the beginning of a line indicates that it is a comment.

SECTION 5.1: WHAT IS A MACRO AND HOW IS IT USED?   159

```
                         TITLE    PROG5-3
                         PAGE 60,132
               ;
                         INCLUDE  MYMACRO2.MAC
C ;           MYMACRO2 (MAC)  FOR PROGRAM5-3
C ;————————————————————————————
C CURSOR  MACRO  ROW,COLUMN
C ;THIS MACRO SETS THE CURSOR LOCATION AT ROW,COLUMN
C ;;USING BIOS INT 10H FUNCTION 02
C         MOV  AH,02              ;SET CURSOR FUNCTION
C         MOV  BH,00              ;PAGE 00
C         MOV  DH,ROW             ;ROW POSITION
C         MOV  DL,COLUMN          ;COLUMN POSITION
C         INT  10H                ;INVOKE THE INTERRUPT
C         ENDM
C ;————————————————————————————
C DISPLAY  MACRO  STRING
C ;THIS MACRO DISPLAYS A STRING OF DATA
C ;;DX = ADDRESS OF STRING.  USES FUNCTION 09 INT 21H.
C         MOV  AH,09                 ;DISPLAY STRING FUNCTION
C         MOV  DX,OFFSET STRING      ;DX = OFFSET ADDRESS OF DATA
C         INT  21H                   ;INVOKE THE INTERRUPT
C         ENDM
C ;————————————————————————————
C CLEARSCR MACRO
C ;THIS MACRO CLEARS THE SCREEN
C ;;USING OPTION 06 OF INT 10H
C         MOV  AX,0600H  ;SCROLL SCREEN FUNCTION
C         MOV  BH,07              ;NORMAL ATTRIBUTE
C         MOV  CX,0               ;FROM ROW=00,COLUMN=00
C         MOV  DX,184FH            ;TO ROW=18H,COLUMN=4FH
C         INT  10H                ;INVOKE THE INTERRUPT
C         ENDM
C ;————————————————————————————
               ;————————————————————————————
0000                      .MODEL SMALL
0000                      .STACK 64
               ;————————————————————————————
0000                      .DATA
0000 54 68 69 73 20 69    MESSAGE1  DB     'This is a test of macro concepts','$'
     73 20 61 20 74 65
     73 74 20 6F 66 20
     6D 61 63 72 6F 20
     63 6F 6E 63 65 70
     74 73 24
               ;————————————————————————————
0000                      .CODE
0000                      MAIN    PROC FAR
0000 B8 0000s             MOV     AX,@DATA
0003 8E D8                MOV     DS,AX
                          CLEARSCR   ;INVOKE CLEAR SCREEN MACRO
0005 B8 0600              MOX     AX,0600H
```

List File for Program 5-3 *(continued on next page)*

```
0008 B7 07      1          MOV  BH,07             ;NORMAL ATTRIBUTE
000A B9 0000    1          MOV  CX,0              ;FROM ROW=00,COLUMN=00
000D BA 184F    1          MOV  DX,184FH          ;TO ROW=18H,COLUMN=4FH
0010 CD 10      1          INT  10H               ;INVOKE THE INTERRUPT
                           DISPLAY  MESSAGE1      ;INVOKE DISPLAY MACRO
0012 B4 09      1          MOV  AH,09             ;DISPLAY STRING FUNCTION
0014 BA 0000 R  1          MOV  DX,OFFSET MESSAGE1 ;DX =OFFSET ADDRESS OF DATA
0017 CD 21      1          INT  21H               ;INVOKE THE INTERRUPT
0019 B4 4C                 MOV  AH,4CH
001B CD 21                 INT  21H               ;GO BACK TO DOS
001D                  MAIN ENDP
                           END  MAIN
```

List File for Program 5-3   *(continued from preceding page)*

SECTION 5.2: MOUSE PROGRAMMING WITH INTERRUPT 33H

Next to the keyboard, the mouse is one of the most widely used input devices. This section describes how to use INT 33H to add mouse capabilities to programs.

INT 33H

The original IBM PC and DOS did not provide support for the mouse. For this reason, mouse interrupt INT 33H is not part of BIOS or DOS. This is in contrast to INT 21H and INT 10H, which are the DOS and BIOS interrupts, respectively. INT 33H is part of the mouse driver software that is installed when the PC is booted.

Detecting the presence of a mouse

While new PCs come with a mouse and driver already installed by the PC manufacturer, many older-generation PCs in use do not have a mouse. Therefore, the first task of any INT 33H program should be to verify the presence of a mouse and the number of buttons it supports. This is the purpose of INT 33H function AX=0. Upon return from INT 33H, if AX=0 then no mouse is supported. If AX=FFFFH, the mouse is supported and the number of mouse buttons will be contained in register BX. Although most mice have two buttons, right and left, there are some with middle buttons as well. See the following code.

```
     MOV   AX,0          ;mouse initialization option
     INT   33H
     CMP   AX,0          ;check AX contents upon return from INT 33H
     JE    EXIT          ;exit if AX=0 since no mouse available
     MOV   M_BUTTON,BX   ;mouse is there, save the number of buttons
     ...
EXIT:
```

Notice the following points about the way INT 33H is called.

1. In INT 21H and INT 10H, the AH register is used to select the functions. This is not the case in INT 33H. In INT 33H the register AL is used to select various functions and AH is set to 0. That is the reason behind the instruction "MOV AX,0".
2. Do not forget the "H" indicating hex in coding INT 33H. In the absence of the "H", the compiler assumes it is decimal and will execute DOS INT 21H since 33 decimal is equal to 21H.

Some mouse terminology

Before further discussion of INT 33H, some terminology concerning the mouse needs to be clarified. The *mouse pointer* (or *cursor*) is the pointer on the screen indicating where the mouse is pointing at a given time. In graphics mode, the mouse pointer (cursor) is an arrow; in text mode, the mouse pointer is a flashing block. In either mode, as the mouse is moved, the mouse cursor is also moved. While the movement of the mouse is measured in inches (or centimeters), the movement of the mouse cursor (arrowhead) on the screen is measured in units called *mickeys*. Mickey units indicate *mouse sensitivity*. For example, a mouse that can move the cursor 200 units for every inch of mouse movement has a sensitivity of 200 mickeys. In this case, one mickey represents 1/200 of an inch on the screen. Some mice have a sensitivity of 400 mickeys in contrast to the commonly used 200 mickeys. In that case, for every inch of mouse movement, the mouse cursor moves 400 mickeys.

Displaying and hiding the mouse cursor

The AX=01 function of INT 33H is used to display the mouse cursor.

```
MOV     AX,01
INT     33H
```

After executing the above code, the mouse pointer is displayed. If the video mode is graphics, the mouse arrow becomes visible. If the video mode is text, a rectangular block representing the mouse cursor becomes visible. In text mode, the color of the mouse cursor block is the opposite of the background color in order to be visible. It is best to hide the mouse cursor after making it visible by executing option AX=02 of INT 33H. This is shown in Example 5-3. Try Example 5-3 in DEBUG (remember to omit the "H" and place INT 3 as the last instruction when in DEBUG). Then try it with mode AH=03 for INT 10H to see the mouse cursor in text mode.

Example 5-3

(a) Use INT 10H option 0F to get the current video mode and save it in BL; (b) set the video mode to VGA graphics using option AH=10H of INT 10H; (c) initialize the mouse with AX=0, INT 33H; (d) make the mouse visible; (e) use INT 21H option AH=01 to wait for key press; (f) if any key is pressed restore the original video mode.

Solution:

```
MOV     AH,0FH      ;get the current video mode
INT     10H
MOV     BL,AL       ;and save it
MOV     AH,0        ;set the video mode
MOV     AL,10H      ;to VGA graphics
INT     10H
MOV     AX,0        ;initialize the mouse
INT     33H
MOV     AX,01       ;make the mouse cursor visible
INT     33H
MOV     AH,01       ;wait for key press
INT     21H
MOV     AX,2        ;when any key is pressed
INT     33H         ;make mouse invisible
MOV     AH,0
MOV     AL,BL       ;and restore original video mode
INT     10H
```

Video resolution vs. mouse resolution in text mode

As discussed in Chapter 4, the video screen is divided into 640x200 pixels in text mode. This means that in text mode of 80x25 characters, each character will use 8x8 pixels (80 x 8 = 640 and 25 x 8 = 200). When the video mode is set to text mode (AH=03 of INT 10H), the mouse will automatically adopt the same resolution of 640x200 for its horizontal and vertical coordinates. Therefore, in text mode when a program gets the mouse cursor position, the values are provided in pixels and must be divided by 8 to get the mouse cursor position in terms of character locations 0 to 79 (horizontal) and 0 to 24 (vertical) on the screen.

Video resolution vs. mouse resolution in graphics mode

In graphics, resolution is not only 640x200 but also 640x350 and 640x480. When the video resolution is changed to these video modes, the mouse also adopts the graphics resolutions. See Table 5-1.

Table 5-1: Video and Mouse Resolution for Some Video Modes

Video Mode	Video Resolution	Type	Mouse Resolution	Characters per Screen
AL=03	640x200	Text	640x200	80x25
AL=0EH	640x200	Graphics	640x200	80x25
AL=0FH	640x350	Graphics	640x350	80x44
AL=10H	640x350	Graphics	640x350	80x44
AL=11H	640x480	Graphics	640x480	80x60
AL=12H	640x480	Graphics	640x480	80x60

Getting the current mouse cursor position (AX=03)

Option AX=03 of INT 33H gets the current position of the mouse cursor. Upon return, the X and Y coordinates are in registers CX (horizontal) and DX (vertical). BX contains the button status as follows: D0= left button status, D1= right button status, D2= center button status. The status is 1 if down, 0 if up. Notice that the cursor position provided by this function is given in pixels. For example, the position returned will be in the range of 0 - 639 (horizontal) and 0 - 199 (vertical) for a 640x200 screen in most text and graphics video modes. However, the mouse cursor position is often needed in terms of character positions such as 80x25 and not in terms of pixels. To get the mouse cursor character position, divide both the horizontal and vertical values of CX and DX by 8. See Programs 5-4 and 5-5.

```
TITLE   PROGRAM 5-4: DISPLAYING MOUSE POSITION
;Performs the following tasks: (a) gets the current video mode and saves it, (b) sets the mode to a new video
;mode, (c) gets the mouse pointer position, converts it to character position and displays it continuously unless a
;key is pressed, (d) upon pressing any key, it restores the original video mode and exits to DOS.
PAGE    60,132
CURSOR     MACRO ROW,COLUMN
           MOV    AH,02H
           MOV    BH,00
           MOV    DH,ROW
           MOV    DL,COLUMN
           INT    10H
           ENDM
DISPLAY    MACRO STRING
           MOV    AH,09H
           MOV    DX,OFFSET STRING       ;load string address
           INT    21H
           ENDM
```

Program 5-4 *(continued on the next page)*

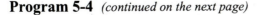

```
                .MODEL SMALL
                .STACK
                DATA
MESSAGE_1       DB 'PRESS ANY KEY TO GET OUT','$'
MESSAGE_2       DB 'THE MOUSE CURSOR IS LOCATED AT ','$'
POS_HO          DB ?,?, ' AND $'
POS_VE          DB ?,?,'$'
OLDVIDEO        DB ?                             ;current video mode
NEWVIDEO        DB 0EH                           ;new video mode
                .CODE
MAIN     PROC
         MOV     AX,@DATA
         MOV     DS,AX
         MOV     AH,0FH                           ;get current video mode
         INT     10H
         MOV     OLDVIDEO,AL                      ;save it
         MOV     AX,0600H                         ;clear screen
         MOV     BH,07
         MOV     CX,0
         MOV     DX,184FH
         INT     10H
         MOV     AH,00H                           ;set new video mode
         MOV     AL,NEWVIDEO
         INT     10H
         MOV     AX,0                             ; initialize mouse
         INT     33H
         MOV     AX,01                            ;show mouse cursor
         INT     33H
         CURSOR 20,20
         DISPLAY MESSAGE_1
AGAIN:   MOV     AX,03H                           ;get mouse location
         INT     33H
         MOV     AX,CX                            ;get the hor. pixel position
         CALL    CONVERT                          ;convert to displayable data
         MOV     POS_HO,AL                        ;save the LSD
         MOV     POS_HO+1,AH                      ;save  the MSD
         MOV     AX,DX                            ;get the vert. pixel position
         CALL    CONVERT                          ;convert
         MOV     POS_VE,AL                        ;save
         MOV     POS_VE+1,AH
         CURSOR 5,20                              ;
         DISPLAY MESSAGE_2 ;
         DISPLAY POS_HO
         DISPLAY POS_VE
         MOV     AH,01                            ;check for key press
         INT     16H
         JZ      AGAIN                            ;if no key press, keep monitoring mouse position
         MOV     AH,02                            ;hide mouse
         INT     33H
         MOV     AH,0                             ;restore original video mode
         MOV     AL,OLDVIDEO                      ;load original video mode
         INT     10H
         MOV     AH,4CH                           ;go back to DOS
         INT     21H
MAIN     ENDP
;-------------------------------------
;divide pixels position by 8 and convert to ASCII to make it displayable
;ax=pixels position (it is in hex)
;on return ax= two ASCII digits
CONVERT  PROC
         SHR     AX,1                             ;divide
         SHR     AX,1                             ;by 8
         SHR     AX,1                             ;to get screen position by character
         MOV     BL,10
         SUB     AH,AH
         DIV     BL                               ;divide by ten to convert from hex. to decimal
         OR      AX,3030H                         ;make it ASCII
         RET                                      ;return with AX=two ASCII digits
CONVERT  ENDP
         END     MAIN
```

Program 5-4 *(continued from preceding page)*

```
TITLE PROGRAM 5-5:MOUSE BOX PROGRAM
;Performs the following: (a) gets the current video mode and saves it, (b) sets the video mode to a new one and
;clears screen, (c) draws a colored box and gets the mouse position, (d) displays different messages depending
;on whether the mouse is clicked inside or outside the box  Pressing any key will return to DOS.
;Thanks to Travis Erck and Gary Hudson for their input on this program

CURSOR      MACRO  ROW,COLUMN
            MOV    AH,02H
            MOV    BH,00
            MOV    DH,ROW
            MOV    DL,COLUMN
            INT    10H
            ENDM
DISPLAY     MACRO  STRING
            MOV    AH,09H
            MOV    DX,OFFSET STRING      ;load string address
            INT    21H
            ENDM
FILL        MACRO  ROW_START,COL_START,ROW_END,COL_END,COLOR
            LOCAL  START,AGAIN
            MOV    DX,ROW_START
START:      MOV    CX,COL_START
AGAIN:      MOV    AH,0CH
            MOV    AL,COLOR
            INT    10H
            INC    CX
            CMP    CX,COL_END
            JNE    AGAIN
            INC    DX
            CMP    DX,ROW_END
            JNE    START
            ENDM

            .MODEL SMALL
            .STACK
            .DATA
MESSAGE_1   DB 'AN EXAMPLE OF HOW TO USE INTERRUPT 33H FOR MOUSE.','$'
MESSAGE_2   DB 'IT WORKS!','$'
MESSAGE_3   DB 'CLICK IN THE BOX TO SEE WHAT HAPPENS!','$'
MESSAGE_4   DB 'No, NO, NO I SAID IN THE BOX!','$'
MESSAGE_5   DB 'NOW PRESS ANY KEY TO GET BACK TO DOS. $'
OLDVIDEO    DB ?
NEWVIDEO    DB 12H

            .CODE
MAIN        PROC
            MOV    AX,@DATA
            MOV    DS,AX
            MOV    AH,0FH                ;get the current video mode
            INT    10H
            MOV    OLDVIDEO,AL    ;save it
            MOV    AX,0600H              ;clear screen
            MOV    BH,07
            MOV    CX,0
            MOV    DX,184FH
            INT    10H
            MOV    AH,00H                ;set new video mode
            MOV    AL,NEWVIDEO
            INT    10H
            CURSOR 0,0
            FILL 150,250,250,350,4       ;draw red box
            CURSOR 1,1
            DISPLAY MESSAGE_1;
            CURSOR 5,22
            DISPLAY MESSAGE_3;
            MOV    AX,0000H              ;initialize mouse
            INT    33H
            MOV    AX,01H
            INT 3  3H                    ;show mouse cursor
```

Program 5-5 *(continued on next page)*

```
BACK:        MOV     AX,03H          ;check for mouse button press
             INT     33H             ;now CX =COL and DX=ROW location
             CMP     BX,0001H        ;check to see if left button is pressed
             JNE     BACK            ;If not keep checking
             CMP     CX,250          ;see if on right side of box
             JB      NOT_INSIDE      ;if less it must be outside box
             CMP     CX,350          ;see if on left side of box
             JA      NOT_INSIDE      ;if not then it is outside the box
             CMP     DX,150          ;check for the top of the box
             JB      NOT_INSIDE      ;if not then outside the box
             CMP     DX,250          ;see if bottom of the box
             JA      NOT_INSIDE
             CURSOR 18,18            ;then it must be inside box
             DISPLAY MESSAGE_2       ;indicate mouse is inside the box
             JMP     EXIT            ;go prepare to exit to DOS
NOT_INSIDE:
             CURSOR 20,18            ;indicate mouse is not inside box
             DISPLAY MESSAGE_4
EXIT:        MOV     AH,02H          ;hide mouse before exiting to DOS
             INT     33H
             CURSOR 22,18
             DISPLAY MESSAGE_5
             MOV     AH,07           ;wait for a key press
             INT     21H
             MOV     AH,0            ;restore original video mode
             MOV     AL,OLDVIDEO
             INT     10H
             MOV     AH,4CH          ;exit
             INT     21H             ;to DOS
MAIN         ENDP
             END
```

Program 5-5 *(continued from preceding page)*

Setting the mouse pointer position (AX=04)

This function allows a program to set the mouse pointer to a new location anywhere on the screen. Before calling this function, the coordinates for the new location must be placed in registers CX for the horizontal (x coordinate) and DX for the vertical (y coordinate). These values must be in pixels in the range of 0 - 639 and 0 - 199 for 640x200 resolution. Coordinate (0,0) is the upper left corner of the screen. For example, to set the mouse cursor at location 9x5 (on a 80x25 screen), simply multiply both by 8 to get the pixel location. Therefore, a character coordinate of 9x5 becomes 72x40 in pixel coordinates.

Getting mouse button press information (AX=05)

This function is used to get information about specific button presses since the last call to this function. It is set up as follows.

        AX=05
        BX=0 for left button; 1 for right button; 2 for center button
Upon return:
        AX=button status where
            D0= Left button, if 1 it is down and if 0 it is up
            D1= Right button, if 1 it is down and if 0 it is up
            D2= Center button, if 1 it is down and if 0 it is up
        BX=button press count
        CX=x-coordinate at the last button press in pixels (horizontal)
        DX=y-coordinate at the last button press in pixels (vertical)

Notice in function AX=05 that upon returning from INT 33H, register AX has the button status (up or down), while register BX has the number of times the specific button is pressed since the last call to this function. Program 5-6 shows one way to use this function.

Monitoring and displaying the button press count program

Program 5-6 uses the AX=05 function to monitor the number of times the left button is pressed and then displays the count. It prompts the user to press the left button a number of times. When the user is ready to see how many times the button was pressed, any key can be pressed.

```
TITLE   PROGRAM 5-6: DISPLAY MOUSE PRESS COUNT
;THIS PROGRAM WAITS FOR THE MOUSE PRESS COUNT AND DISPLAYS IT WHEN
;ANY KEY IS PRESSED.
;PRESS ANY KEY TO GO BACK TO DOS
PAGE    60,132

CURSOR      MACRO ROW,COLUMN
            MOV     AH,02H
            MOV     BH,00
            MOV     DH,ROW
            MOV     DL,COLUMN
            INT     10H
            ENDM

DISPLAY     MACRO STRING
            MOV     AH,09H
            MOV     DX,OFFSET STRING ;LOAD STRING ADDRESS
            INT     21H
            ENDM

            .MODEL SMALL
            .STACK
            .DATA
MESSAGE_1       DB 'PRESS LEFT BUTTON A NUMBER OF TIMES:LESS THAN 99.','$'
MESSAGE_2       DB 'TO FIND OUT HOW MANY TIMES, PRESS ANY KEY','$'
MESSAGE_3       DB 'YOU PRESSED IT ','$'
P_COUNT         DB ?,?,' TIMES $'
MESSAGE_4       DB 'NOW PRESS ANY KEY TO GO BACK TO DOS','$'
OLDVIDEO        DB ?           ;current video mode
NEWVIDEO        DB 12H         ;new video mode
            .CODE
MAIN        PROC
            MOV     AX,@DATA
            MOV     DS,AX
            MOV     AH,0FH          ;get current video mode
            INT     10H
            MOV     OLDVIDEO,AL     ;save it
            MOV     AX,0600H        ;clear screen
            MOV     BH,07
            MOV     CX,0
            MOV     DX,184FH
            INT     10H
            MOV     AH,00H          ;set new video mode
            MOV     AL,NEWVIDEO
            INT     10H
            MOV     AX,0            ;initialize mouse
            INT     33H
            MOV     AX,01           ;show mouse cursor
            INT     33H
            CURSOR 2,1
            DISPLAY MESSAGE_1
            CURSOR 4,1
            DISPLAY MESSAGE_2
            MOV     AH,07           ;wait for key press
            INT     21H
            MOV     AX,05H          ;get mouse press count
            MOV     BX,0            ;check press count for left button
            INT     33H
```

Program 5-6 *(continued on next page)*

SECTION 5.2: MOUSE PROGRAMMING WITH INTERRUPT 33H   **167**

```
        MOV     AX,BX           ;BX=button press count
        MOV     BL,10
        DIV     BL
        OR      AX,3030H        ;convert it to ASCII
        MOV     P_COUNT,AL      ;save the number
        MOV     P_COUNT+1,AH
        CURSOR 10,2
        DISPLAY MESSAGE_3
        DISPLAY P_COUNT
        CURSOR 20,2
        DISPLAY MESSAGE_4
        MOV     AH,07           ;wait for a key press to get out
        INT     21H
        MOV     AH,02           ;hide mouse
        INT     33H
        MOV     AH,0            ;restore original video mode
        MOV     AL,OLDVIDEO     ;load original vide mode
        INT     10H
        MOV     AH,4CH          ;go back to DOS
        INT     21H
MAIN    ENDP
        END     MAIN
```

Program 5-6 *(continued from preceding page)*

Getting mouse button release information (AX=06)

This function is the same as AX=05 except that it provides information about the button release.

Call with:
        AX = 06
        BX =   0 for left button; 1 for right button; 2 for center button
Upon return:
        AX = button status where
            D0 = Left button, if 1 it is down and if 0 it is up
            D1 = Right button, if 1 it is down and if 0 it is up
            D2 = Center button, if 1 it is down and if 0 it is up
        BX = button release count
        CX = x coordinate at the last button release in pixels (horizontal)
        DX = y coordinate at the last button release in pixels (vertical)

Setting horizontal boundary for mouse pointer (AX=07)

When the mouse is initialized using the AX=0 function, the mouse cursor can move anywhere on the screen. This function and the following one set the x and y coordinate boundaries in which mouse movement is confined. Call as follows.

AX = 07
CX = minimum x coordinate in pixels (0 - 639)
DX = maximum x coordinate in pixels (0 - 639)

Notice that CX must be less than DX; otherwise, they are swapped.

Setting vertical boundary for mouse pointer (AX=08)

This and the previous function allow a program to set the x and y coordinates in which the mouse movement is confined.

AX = 08
CX = minimum y coordinate in pixels (0 - 199)
DX = maximum y coordinate in pixels (0 - 199)

Notice that CX must be less than DX; otherwise, they are swapped.

Setting an exclusion (off-limits) area for the mouse pointer (AX=10H)

This function is used to set an area that is off-limits for the mouse cursor. If the mouse cursor moves to the exclusion area, it disappears.

Call with:
```
AX = 10H
CX = upper x coordinate,       SI = lower x coordinate
DX = upper y coordinate,       DI = lower y coordinate
```

Getting mouse driver information (version) (AX=24H)

This function allows a program to determine which version of the mouse driver is installed. It also indicates the type of mouse, such as an internal bus mouse or an external mouse via a serial port type.

Call with:
```
AX = 24H
```
Upon return we have:
```
BH = major version       BL = minor version  (e.g., "7.20" 7 is major, 20 is minor)
CH = mouse type where
    1 = bus mouse, 2 = serial mouse, 3 = InPort mouse,
    4 = PS/2 mouse, 5 = HP mouse
CL= 0 if mouse is PS/2 type. If it is a serial mouse, the number is between 2 and
    7, indicating the IRQ number being used (such as IRQ2, IRQ3, etc.).
```

Program 5-7 gives an example of a C program using the mouse. Interrupt 33H is called using the int86 C function to see if there is a mouse present. If there is a mouse present, the program will print the number of buttons on the mouse.

```c
// This program uses INT 33H to see if mouse driver is installed and if so it displays
// the number of buttons on the mouse
#include <stdio.h>
#include <dos.h>
#include <conio.h>
main()
{
union REGS ri,ro;
clrscr();
ri.x.ax=0x0;          //check to see if there is mouse
int86(0x33,&ri,&ro);
if(ro.x.ax>0)
    printf("\n\nYou have %d button(s) on the mouse on this PC",ro.x.bx);
else
    printf("This PC has no mouse\n");
}
```

Program 5-7

Review Questions

1. Which function of INT 33H is used to detect the presence of a mouse in a PC? In which register do we expect to get that information? In which register do we find the number of buttons in the mouse?
2. The following code is an attempt to call INT 33H function 2. Is it correct? If not, correct it.
   ```
   MOV AH,02
   INT 33H
   ```
3. Why do we need to save the original video mode before changing it for the mouse?
4. In INT 33H function AX=03, how can a left button press be detected?
5. True or false. The mouse coordinate is the same as video mode.

SUMMARY

Macros are used by programmers to save time in coding and debugging. Whenever a set of instructions must be performed repeatedly, these become ideal candidates for a macro. Values can be passed to macros to be used by instructions within the macro. Programmers can place several often-used macros within a file that can be brought into one or more programs.

In recent years, the mouse has become a standard input device on the IBM PC. The original IBM PC BIOS and DOS did not provide for the mouse. INT 33H is part of the mouse driver software that is installed when the PC is booted. Function calls to INT 33H are used for mouse input and cursor control.

PROBLEMS

1. Every macro must start with directive _____ and end with directive _____.
2. Identify the name, body, and dummy argument in the following macro:
    ```
    WORK_HOUR MACRO   OVRTME_HR
                MOV     AL,40              ;WEEKLY HRS
                ADD     AL,OVRTME_HR    ;TOTAL HRS WORKED
                ENDM
    ```
3. Explain the difference between the .SALL, .LALL, and .XALL directives.
4. What is the total value in registers DX and AX after invoking the following macro?
    ```
    WAGES           MACRO           SALARY,OVERTIME,BONUSES
                    ;TOTAL WAGES=SALARY + OVERTIME + BONUSES
                    SUB  AX,AX                 ;CLEAR
                    MOV  DX,AX                 ;AX AND DX
                    ADD  AX,SALARY
                    ADD  AX,OVERTIME
                    ADC  DX,0                      ;TAKE CARE OF CARRY
                    ADD  AX,BONUSES
                    ADC  DX,0
                    ENDM
    ```
    The macro is invoked as
    ```
         WAGES  60000,25000,3000
    ```
5. In Problem 4, in the body of the macro, dummies were used as they are listed from left to right. Can they be used in any order? Rewrite the body (leave the dummies alone) by adding OVERTIME first.
6. In Problem 4, state the comments that are listed if the macro is expanded as:
    ```
         .LALL
         WAGES  X,Y,Z
    ```
7. Macros can use registers as dummies. Show the ".lst" file and explain what the macro in Problem 4 does if it is invoked as follows:
    ```
         WAGES           BX,CX,SI
    ```
8. Fill in the blanks for the following macro to add an array of bytes. Some blanks might not need to be filled.
    ```
    SUMMING         MACRO           COUNT,VALUES
                    LOCAL  ......
                    ;;this macro adds an array of byte size elements.
                    ;;ax will hold the total sum
                    MOV   CX,....           ;size of array
                    MOV   SI,OFFSET .... .;load offset address of array
                    SUB   AX,AX             ;clear ax
    AGAIN:          ADD   AL,[SI]
                    ADC   AH,0              ;add bytes and takes care of carries
                    INC   SI               ;point to next byte
                    LOOP  AGAIN            ;continue until finished
                    ENDM  .....
    ```

9. Invoke and run the macro above for the following data.
   In the data segment:

DATA1	DB	89,75,98,91,65
SUM1	DW	?
DATA2	DB	86,69,99,14,93,99,54,39
SUM2	DW	?
DATA3	DB	10 DUP (99)
SUM3	DW	?

   (*Hint*: For the format, see Problem 10.)

10. Insert the listing directives in Problem 9 as follows and analyze the ".lst" file.
    From the code segment:

.LALL		
SUMMING	5,DATA1	;adding and saving data1
.XALL		
SUMMING	....	;adding and saving data2
.....		
.SALL		;adding and saving data3
.....		

11. Rewrite Problem 8 to have a third dummy argument for SUM. Then rework Problem 9.

12. Rewrite Program 5-2 using the DD directive for RESULT1, RESULT2, and RESULT3.

13. Using INT 33H, write and test an Assembly language program to check the presence of a mouse in a PC. If a mouse driver is installed, it should state the number of buttons it supports. If no mouse driver is installed, it should state this.

14. Change the video mode of Program 5-4 to the following and verify the data given in Table 5-1.
    (a) 10H   (b) 12H

15. Change the size and color of the box to the coordinates and color of your choice in Program 5-5, then run and test it. Make the box large enough to display your first name followed by your last name inside it.

16. Modify Program 5-6 to indicate the number of times the right button is pressed.

17. Repeat the above problem to indicate the number of times the right button is released.

18. Write an Assembly language program using INT 33H to display the mouse driver version and number of buttons supported by the mouse in a given PC.
    Note: The following are additional problem assignments for the mouse in C.

19. Write a C language version of Program 5-4. Your program performs the following.
    (a) clears the screen
    (b) saves the video mode and changes it to mode AL=0EH (640x200)
    (c) gets the mouse pointer position continuously and displays it in the
        80x25 character screen coordinate
    (d) upon a pressing any key, it restores the original video mode and exits to DOS

20. Repeat the above problem. This time display the mouse pointer position in pixels. Test it for video modes AL=10H and AL=12H.

21. Write a C language program to state (display) which mouse button is being pressed at any given time. Upon pressing any key, the program goes back to DOS.

22. Write a C program version of Program 5-6 that performs the following:
    (a) save the video mode and change it to one of the graphics modes
    (b) ask the user to press the left button a number of times
    (c) display the number when a key is pressed
    (d) upon pressing any key, restore the original video mode and go back to DOS

23. Repeat the above problem for the right button.

24. Using INT 33H functions AX=0 and AX=24H, write a C program to display the following:
    (a) the number of mouse buttons
    (b) the mouse type (bus mouse, serial mouse, PS/2 mouse, etc.)
    (c) mouse driver version in a form such as "7.5"
    (d) the IRQ number, if used by the mouse

ANSWERS TO REVIEW QUESTIONS

SECTION 5.1: WHAT IS A MACRO AND HOW IS IT USED?

1. Macro programming can save the programmer time by allowing a set of frequently repeated instructions to be invoked within the program by a single line. This can also make the code easier to read.
2. The three parts of a macro are the MACRO directive, the body, and the ENDM directive.
3. The macro definition is the list of statements the macro will perform. It begins with the MACRO directive and ends with the ENDM directive. Invoking the macro is when the macro is called from within an assembly language program. Expanding the macro is when the assembly replaces the line invoking the macro with the assembly language code in the body of the macro.
4. False. A label that is to be local to a macro must be declared local with the LOCAL directive.
5. False. The "C" at the beginning of a line indicates that this line of code was brought in from another file by the INCLUDE directive.

SECTION 5.2: MOUSE PROGRAMMING WITH INTERRUPT 33H

1. AX=0; reg AX; number of buttons in reg. BX
2. It is wrong. Register AL=02 and AH=0.
   MOV AX,02
   INT 33H
3. In the absence of doing that, when we go back to DOS we lose our cursor if the mouse program has changed the video to graphics.
4. We check the contents of register BX for value 01.
5. True

CHAPTER 6

SIGNED NUMBERS, STRINGS, AND TABLES

OBJECTIVES

Upon completion of this chapter, you will be able to:

» **Represent 8- or 16-bit signed numbers as used in computers**

» **Convert a number to its 2's complement**

» **Code signed arithmetic instructions for addition, subtraction, multiplication, and division: ADD, SUB, IMUL, and IDIV**

» **Demonstrate how arithmetic instructions affect the sign flag**

» **Explain the difference between a carry and an overflow in signed arithmetic**

» **Prevent overflow errors by sign-extending data**

» **Code signed shift instructions SAL and SAR**

» **Code logic instruction CMP for signed numbers and demonstrate understanding of the effect on the flag register**

» **Code conditional jump instructions after CMP of signed data**

» **Explain the function of registers SI and DI in string operations**

» **Describe the operation of the direction flag in string instructions**

» **Code instructions CLD and STD to control the direction flag**

» **Describe the operation of the REP prefixes**

» **Code string instructions:**

»   **MOVSB and MOVSW for data transfer**

»   **STOS, LODS to store and load the contents of AX**

»   **CMPS to compare two strings of data**

»   **SCAS to scan a string for data matching that in AX**

»   **XLAT for table processing**

In the first section of this chapter we focus on the concept of signed numbers in software engineering. Signed number operations are explained along with examples. In the second section we discuss string operations and table processing.

SECTION 6.1: SIGNED NUMBER ARITHMETIC OPERATIONS

All data items used so far have been unsigned numbers, meaning that the entire 8-bit or 16-bit operand was used for the magnitude. Many applications require signed data. In this section the concept of signed numbers is discussed along with related instructions.

Concept of signed numbers in computers

In everyday life, numbers are used that could be positive or negative. For example, a temperature of 5 degrees below zero can be represented as -5, and 20 degrees above zero as $+20$. Computers must be able to accommodate such numbers. To do that, computer scientists have devised the following arrangement for the representation of signed positive and negative numbers: The most significant bit (MSB) is set aside for the sign (+ or $-$) and the rest of the bits are used for the magnitude. The sign is represented by 0 for positive (+) numbers and 1 for negative ($-$) numbers. Signed byte and word representations are discussed below.

Signed byte operands

D7	D6	D5	D4	D3	D2	D1	D0
sign				magnitude			

In signed byte operands, D7 (MSB) is the sign and D0 to D6 are set aside for the magnitude of the number. If D7 = 0, the operand is positive, and if D7 = 1, it is negative.

Positive numbers

The range of positive numbers that can be represented by the format above is 0 to +127.

```
   0      0000 0000
  +1      0000 0001
  +5      0000 0101
  ..      .... ....
  ..      .... ....
+127      0111 1111
```

If a positive number is larger than +127, a word-sized operand must be used. Word operands are discussed later.

Negative numbers

For negative numbers D7 is 1, but the magnitude is represented in 2's complement. Although the assembler does the conversion, it is still important to understand how the conversion works. To convert to negative number representation (2's complement), follow these steps:

1. Write the magnitude of the number in 8-bit binary (no sign).
2. Invert each bit.
3. Add 1 to it.

Examples 6-1, 6-2, and 6-3 demonstrate these three steps.

Example 6-1

Show how the computer would represent −5.

Solution:
1. `0000 0101`   5 in 8-bit binary
2. `1111 1010`   invert each bit
3. `1111 1011`   add 1   (hex = FBH)

This is the signed number representation in 2's complement for −5.

Example 6-2

Show −34H as it is represented internally.

Solution:
1. `0011 0100`
2. `1100 1011`
3. `1100 1100`   (which is CCH)

Example 6-3

Show the representation for −128$_{10}$.

Solution:
1. `1000 0000`
2. `0111 1111`
3. `1000 0000`   Notice that this is not negative zero (-0).

From the examples above it is clear that the range of byte-sized negative numbers is −1 to −128. The following lists byte-sized signed numbers ranges:

Decimal	Binary	Hex
−128	1000 0000	80
−127	1000 0001	81
−126	1000 0010	82
..	....	..
−2	1111 1110	FE
−1	1111 1111	FF
0	0000 0000	00
+1	0000 0001	01
+2	0000 0010	02
..	....	..
+127	0111 1111	7F

Word-sized signed numbers

D15	D14	D13	D12	D11	D10	D9	D8	D7	D6	D5	D4	D3	D2	D1	D0

⌐sign⌐   magnitude

In 80x86 computers a word is 16 bits in length. Setting aside the MSB (D15) for the sign leaves a total of 15 bits (D14 - D0) for the magnitude. This gives a range of −32768 to +32767. If a number is larger than this, it must be treated as a multiword operand and be processed chunk by chunk the same way as unsigned numbers (as discussed in Chapter 3). The following shows the range of signed word operands. To convert a negative to its word operand representation, the three steps discussed in negative byte operands are used.

SECTION 6.1: SIGNED NUMBER ARITHMETIC OPERATIONS   175

Decimal	Binary	Hex
−32 768	1000 0000 0000 0000	8000
−32 767	1000 0000 0000 0001	8001
−32 766	1000 0000 0000 0010	8002
. . . . . .	. . . .	. . . .
. . . . . .	. . . .	. . . .
−2	1111 1111 1111 1110	FFFE
−1	1111 1111 1111 1111	FFFF
0	0000 0000 0000 0000	0000
+1	0000 0000 0000 0001	0001
+2	0000 0000 0000 0010	0002
. . .	. . . .	. . . .
. . .	. . . .	. . . .
+32 766	0111 1111 1111 1110	7FFE
+32 767	0111 1111 1111 1111	7FFF

Overflow problem in signed number operations

When using signed numbers, a serious problem arises that must be dealt with. This is the overflow problem. The CPU indicates the existence of the problem by raising the OF (overflow) flag, but it is up to the programmer to take care of it. The CPU understands only 0s and 1s and ignores the human convention of positive and negative numbers. Now what is an overflow? If the result of an operation on signed numbers is too large for the register, an overflow has occurred and the programmer must be notified. Look at Example 6-4.

Example 6-4

Look at the following code and data segments:

```
DATA1   DB  +96
DATA2   DB  +70
        ...  ....
        MOV   AL,DATA1   ;AL=0110 0000 (AL=60H)
        MOV   BL,DATA2   ;BL=0100 0110 (BL=46H)
        ADD   AL,BL      ;AL=1010 0110 (AL=A6H= –90 invalid!)
```

```
+  96   0110 0000
+  70   0100 0110
+166   1010 0110  According to the CPU, this is –90, which is wrong. (OF = 1, SF = 1, CF = 0)
```

In the example above, +96 is added to +70 and the result according to the CPU was −90. Why? The reason is that the result was more than what AL could handle. Like all other 8-bit registers, AL could only contain up to +127. The designers of the CPU created the overflow flag specifically for the purpose of informing the programmer that the result of the signed number operation is erroneous.

When the overflow flag is set in 8-bit operations

In 8-bit signed number operations, OF is set to 1 if either of the following two conditions occurs:

1. There is a carry from D6 to D7 but no carry out of D7 (CF = 0).
2. There is a carry from D7 out (CF = 1) but no carry from D6 to D7.

In other words, the overflow flag is set to 1 if there is a carry from D6 to D7 or from D7 out, but not both. This means that if there is a carry both from D6 to D7 and from D7 out, OF = 0. In Example 6-4, since there is only a carry from D6 to D7 and no carry from D7 out, OF = 1. Examples 6-5, 6-6, and 6-7 give further illustrations of the overflow flag in signed arithmetic.

Example 6-5

Observe the results of the following:

```
MOV   DL,-128    ;DL=1000 0000 (DL=80H)
MOV   CH,-2      ;CH=1111 1110 (CH=FEH)
ADD   DL,CH      ;DL=0111 1110 (DL=7EH=+126 invalid!)
```

−128	1000 0000
+ − 2	1111 1110
−130	0111 1110 OF=1, SF=0 (positive), CF=1

According to the CPU, the result is +126, which is wrong. The error is indicted by the fact that OF=1

Example 6-6

Observe the results of the following:

```
MOV   AL,-2      ;AL=1111 1110 (AL=FEH)
MOV   CL,-5      ;CL=1111 1011 (CL=FBH)
ADD   CL,AL      ;CL=1111 1001 (CL=F9H=-7 which is correct)
```

−2	1111 1110
+ −5	1111 1011
−7	1111 1001 OF = 0, CF = 0 ,SF = 1 (negative); the result is correct since OF = 0.

Example 6-7

Observe the results of the following:

```
MOV   DH,+7      ;DH=0000 0111   (DH=07H)
MOV   BH,+18     ;BH=0001 0010   (BH=12H)
ADD   BH,DH      ;BH=0001 1001   (BH=19H=+25, correct)
```

+7	0000 0111
+ +18	0001 0010
+25	0001 1001 OF = 0, CF = 0 and SF = 0 (positive).

Overflow flag in 16-bit operations

In a 16-bit operation, OF is set to 1 in either of two cases:

1. There is a carry from D14 to D15 but no carry out of D15 (CF = 0).
2. There is a carry from D15 out (CF = 1) but no carry from D14 to D15.

Again the overflow flag is low (not set) if there is a carry from both D14 to D15 and from D15 out. The OF is set to 1 only when there is a carry from D14 to D15 or from D15 out but not from both. See Examples 6-8 and 6-9.

Example 6-8

Observe the results in the following:

```
MOV   AX,6E2FH   ;  28,207
MOV   CX,13D4H   ;+  5,076
ADD   AX,CX      ;= 33,283 is the expected answer
```

6E2F	0110 1110 0010 1111
+ 13D4	0001 0011 1101 0100
8203	1000 0010 0000 0011 = −32,253 incorrect! OF = 1, CF = 0, SF = 1

Example 6-9

Observe the results in the following:

```
MOV    DX,542FH    ;  21,551
MOV    BX,12E0H    ; + 4,832
ADD    DX,BX       ;=26,383
```

543F	0101 0100 0010 1111
+ 12E0	0001 0010 1110 0000
670F	0110 0111 0000 1111 = 26,383 (correct answer) OF = 0, CF = 0, SF = 0

Avoiding erroneous results in signed number operations

To avoid the problems associated with signed number operations, one can *sign-extend* the operand. Sign extension copies the sign bit (D7) of the lower byte of a register into the upper bits of the register, or copies the sign bit of a 16-bit register into another register. CBW (convert signed byte to signed word) and CWD (convert signed word to signed double word) are used to perform sign extension. They work as follows:

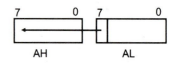

CBW will copy D7 (the sign flag) to all bits of AH. This is demonstrated below. Notice that the operand is assumed to be AL and the previous contents of AH are destroyed.

```
MOV    AL,+96    ;AL=0110 0000
CBW              ;now AH=0000 0000 and AL=0110 0000
```

or:

```
MOV    AL,-2     ;AL=1111 1110
CBW              ;AH=1111 1111 and AL=1111 1110
```

CWD sign extends AX. It copies D15 of AX to all bits of the DX register. This is used for signed word operands. This is illustrated below.

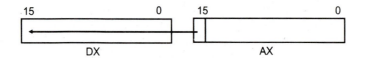

Look at the following example:
```
MOV    AX,+260    ;AX=0000 0001 0000 0100 or AX=0104H
CWD               ;DX=0000H and AX=0104H
```

Another example:
```
MOV    AX,-32766    ;AX=1000 0000 0000 0010B or AX=8002H
CWD                 ;DX=FFFF and AX=8002
```

As can be seen in the examples above, CWD does not alter AX. The sign of AX is copied to the DX register. How can these instructions help correct the overflow error? To answer that question, Example 6-10 shows Example 6-4 rewritten to correct the overflow problem.

In Example 6-10, if the overflow flag is not raised (OF = 0), the result of the signed number is correct and JNO (jump if no overflow) will jump to OVER. However, if OF = 1, which means that the result is erroneous, each operand must be sign extended and then added. That is the function of the code below the JNO instruction. The program in Example 6-10 works for addition of any two signed bytes.

Example 6-10

Rewrite Example 6-4 to provide for handling the overflow problem.

Solution:

```
DATA1    DB  +96
DATA2    DB  +70
RESULT   DW  ?
         ......
         SUB   AH,AH          ;AH=0
         MOV   AL,DATA1       ;GET OPERAND 1
         MOV   BL,DATA2       ;GET OPERAND 2
         ADD   AL,BL          ;ADD THEM
         JNO   OVER           ;IF OF=0 THEN GO TO OVER
         MOV   AL,DATA2       ;OTHERWISE GET OPERAND 2 TO
         CBW                  ;SIGN EXTEND IT
         MOV   BX,AX          ;SAVE IT IN BX
         MOV   AL,DATA1       ;GET BACK OPERAND 1 TO
         CBW                  ;SIGN EXTEND IT
         ADD   AX,BX          ;ADD THEM AND
OVER:    MOV   RESULT,AX      ;SAVE IT
```

The following is an analysis of the values in Example 6-10. Each is sign extended and then added as follows:

S	AH	AL	
0	000 0000	0110 0000	+96 after sign extension
0	000 0000	0100 0110	+70 after sign extension
0	000 0000	1010 0110	+166

As a rule, if the possibility of overflow exists, all byte-sized signed numbers should be sign extended into a word, and similarly, all word-sized signed operands should be sign extended before they are processed. This will be shown shortly in Program 6-1. Before discussing that, it is important to understand the division and multiplication of signed operands.

IDIV (Signed number division)

The Intel manual says that IDIV means "integer division"; it is used for signed number division. In actuality, all arithmetic instructions of the 8088/86 are for integer numbers regardless of whether the operands are signed or unsigned. To perform operations on real numbers, the 8087 coprocessor is used. Remember that real numbers are the ones with decimal points such as "3.56". Division of signed numbers is very similar to division of unsigned numbers discussed in Chapter 3. Table 6-1 summarizes signed number division. It is very similar to Table 3-2, which summarized unsigned number division.

Table 6-1: Signed Division Summary

Division	Numerator	Denominator	Quotient	Remainder
byte/byte	AL = byte CBW	register or memory	AL	AH
word/word	AX = word CWD	register or memory	AX	DX
word/byte	AX = word	register or memory	AL [1]	AH
doubleword/word	DXAX = doubleword	register or memory	AX [2]	DX

Notes:
1. Divide error interrupt if -127 > AL > +127.
2. Divide error interrupt if -32767 > AL > +32767.

Table 6-2: Signed Multiplication Summary

Multiplication	Operand 1	Operand 2	Result
byte × byte	AL	register or memory	AX [1]
word × word	AX	register or memory	DXAX [2]
word × byte	AL=byte CBW	register or memory	DXAX [2]

Notes:
1. CF = 1 and OF = 1 if AH has part of the result, but if the result is not large enough to need the AH, the sign bit is copied to the unused bits and the CPU makes CF = 0 and OF = 0 to indicate that.
2. CF = 1 and OF = 1 if DX has part of the result, but if the result is not large enough to need the DX, the sign bit is copied to the unused bits and the CPU makes CF = 0 and OF = 0 to indicate that.
   One can use the J condition to find out which of the conditions above has occurred. The rest of the flags are undefined.

IMUL (Signed number multiplication)

Signed number multiplication is similar in its operation to the unsigned multiplication described in Chapter 3. The only difference between them is that the operands in signed number operations can be positive or negative; therefore, the result must indicate the sign. Table 6-2 summarizes signed number multiplication; it is similar to Table 3-1.

```
TITLE       PROG 6-1            FIND THE AVERAGE TEMPERATURE
PAGE        60,132
;─────────────────────────────
            .MODEL STMALL
            .STACK 64
;─────────────────────────────
            .DATA
SIGN_DAT    DB      +13,-10,+19,+14,-18,-9,+12,-19,+16
            ORG     0010H
AVERAGE     DW      ?
REMAINDER   DW      ?
;─────────────────────────────
            .CODE
MAIN        PROC    FAR
            MOV     AX,@DATA
            MOV     DS,AX
            MOV     CX,9            ;LOAD COUNTER
            SUB     BX,BX           ;CLEAR BX, USED AS ACCUMULATOR
            MOV     SI,OFFSET SIGN_DAT  ;SET UP POINTER
BACK:       MOV     AL,[SI]         ;MOVE BYTE INTO AL
            CBW                     ;SIGN EXTEND INTO AX
            ADD     BX,AX           ;ADD TO BX
            INC     SI              ;INCREMENT POINTER
            LOOP    BACK            ;LOOP IF NOT FINISHED
            MOV     AL,9            ;MOVE COUNT TO AL
            CBW                     ;SIGN EXTEND INTO AX
            MOV     CX,AX           ;SAVE DENOMINATOR IN CX
            MOV     AX,BX           ;MOVE SUM TO AX
            CWD                     ;SIGN EXTEND THE SUM
            IDIV    CX              ;FIND THE AVERAGE
            MOV     AVERAGE,AX      ;STORE THE AVERAGE (QUOTIENT)
            MOV     REMAINDER,DX    ;STORE THE REMAINDER
            MOV     AH,4CH
            INT     21H             ;GO BACK TO DOS
MAIN        ENDP
            END     MAIN
```

Program 6-1

An application of signed number arithmetic is given in Program 6-1. It computes the average of the following Celsius temperatures: +13, –10, +19, +14, –18, –9, +12, –19, and +16.

The program is written in such a way as to handle any overflow that may occur. In Program 6-1, each byte of data was sign extended and added to BX, computing the total sum, which is a signed word. Then the sum and the count were sign extended, and by dividing the total sum by the count (number of bytes, which in this case is 9), the average was calculated.

The following is the ".lst" file of the program above. Notice the signed number format provided by the assembler.

```
Microsoft (R) Macro Assembler  Version 5.10          12/30/91  06:14:2
PROG6-1                                                     Page  1-1


                              TITLE    PROG6-1  FIND THE AVERAGE TEMPERATURE
                              PAGE 60,132
                              ;————————————————————————————
0000                                   .MODEL SMALL
0000                                   .STACK 64

0000                          ;————————————————————————————
                                       .DATA
0000  0D F6 13 0E EE F7   SIGN_DAT      DB      +13,-10,+19,+14,-18,-9,+12,-19,+16
      0C F7 10
0010                                   ORG     0010H
0010  0000                AVERAGE       DW      ?
0012  0000                REMAINDER     DW      ?

                          ;————————————————————————————
0014                                   .CODE
0000                      MAIN  PROC    FAR
0000  B8 0000s                   MOV     AX,@DATA
0003  8E D8                      MOV     DS,AX
0005  B9 0009                    MOV     CX,9            ;LOAD COUNTER
0008  2B DB                      SUB     BX,BX           ;CLEAR BX, USED AS ACCUMULATOR
000A  BE 0000 R                  MOV     SI,OFFSET SIGN_DAT ;SET UP POINTER
000D  8A 04           BACK:      MOV     AL,[SI]         ;MOVE BYTE INTO AL
000F  98                         CBW                     ;SIGN EXTEND INTO AX
0010  03 D8                      ADD     BX,AX           ;ADD TO BX
0012  46                         INC     SI              ;INCREMENT POINTER
0013  E2 F8                      LOOP    BACK            ;LOOP IF NOT FINISHED
0015  B0 09                      MOV     AL,9            ;MOVE COUNT TO AL
0017  98                         CBW                     ;SIGN EXTEND INTO AX
0018  8B C8                      MOV     CX,AX           ;SAVE DENOMINATOR IN CX
001A  8B C3                      MOV     AX,BX           ;MOVE SUM TO AX
001C  99                         CWD                     ;SIGN EXTEND THE SUM
001D  F7 F9                      IDIV    CX              ;FIND THE AVERAGE
001F  A3 0010 R                  MOV     AVERAGE,AX      ;STORE THE AVERAGE (QUOTIENT)
0022  89 16 0012 R               MOV     REMAINDER,DX    ;STORE THE REMAINDER
0026  B4 4C                      MOV     AH,4CH
0028  CD 21                      INT     21H             ;GO BACK TO DOS
002A                      MAIN  ENDP
                                 END     MAIN
```

List File for Program 6-1

Arithmetic shift

As discussed in Chapter 3, there are two types of shifts: logical and arithmetic. Logical shift, which is used for unsigned numbers, was discussed previously. The arithmetic shift is used for signed numbers. It is basically the same as the logical shift, except that the sign bit is copied to the shifted bits. SAR (shift arithmetic right) and SAL (shift arithmetic left) are two instructions for the arithmetic shift.

SAR (shift arithmetic right)

SAR   destination,count

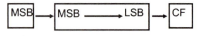

As the bits of the destination are shifted to the right into CF, the empty bits are filled with the sign bit. One can use the SAR instruction to divide a signed number by 2, as shown next:

```
MOV    AL,-10        ;AL=-10=F6H=1111 0110
SAR    AL,1          ;AL is shifted right arithmetic once
                     ;AL=1111 1011=FDH=-5
```

Example 6-11 demonstrates the use of the SAR instruction.

SAL (shift arithmetic left) and SHL (shift left)

These two instructions do exactly the same thing. It is basically the same instruction with two mnemonics. As far as signed numbers are concerned, there is no need for SAL. For a discussion of SHL (SAL), see Chapter 3.

Example 6-11

Using DEBUG, evaluate the results of the following:

```
MOV    AX,-9
MOV    BL,2
IDIV   BL        ;divide -9 by 2     results in FCH
MOV    AX,-9
SAR    AX,1      ;divide -9 by 2 with arithmetic shift    results in FBH
```

Solution:

The DEBUG trace demonstrates that an IDIV of −9 by 2 gives FCH (−4), whereas SAR −9 gives FBH (−5). This is because SAR rounds negative numbers down but IDIV rounds up.

Signed number comparison

CMP   dest,source

Although the CMP (compare) instruction is the same for both signed and unsigned numbers, the J condition instruction used to make a decision for the signed numbers is different from the unsigned numbers. While in unsigned number comparisons CF and ZF are checked for conditions of larger, equal, and smaller (see Chapter 3), in signed number comparison, OF, ZF, and SF are checked:

destination > source	OF=SF or ZF=0
destination = source	ZF=1
destination < source	OF=negation of SF

The mnemonics used to detect the conditions above are as follows:

JG	Jump Greater	jump if OF=SF or ZF=0
JGE	Jump Greater or Equal	jump if OF=SF
JL	Jump Less	jump if OF=inverse of SF
JLE	Jump Less or Equal	jump if OF=inverse of SF or ZF=1
JE	Jump if Equal	jump of ZF = 1

Example 6-12 should help clarify how the condition flags are affected by the compare instruction. Program 6-2 is an example of the application of the signed number comparison. It uses the data in Program 6-1 and finds the lowest temperature.

The concept of signed number arithmetic is so important and widely used that even the RISC processors in their attempt to streamline the instruction set could not eliminate these instructions.

Example 6-12

Show the DEBUG trace of the following instructions comparing several signed numbers.

```
        MOV   AL,-5
        CMP   AL,-9
        CMP   AL,-2
        CMP   AL,-5
        CMP   AL,+7
```

Solution:

```
C>debug
-a 100
103D:0100 mov al,-5
103D:0102 cmp al,-9
103D:0104 cmp al,-2
103D:0106 cmp al,-5
103D:0108 cmp al,7
103D:010A int 3
103D:010B
-t=100,5

AX=00FB BX=0000 CX=0000 DX=0000 SP=CFDE BP=0000 SI=0000 DI=0000
DS=103D ES=103D SS=103D CS=103D IP=0102  NV UP DI PL NZ NA PO NC
103D:0102 3CF7        CMP   AL,F7

AX=00FB BX=0000 CX=0000 DX=0000 SP=CFDE BP=0000 SI=0000 DI=0000
DS=103D ES=103D SS=103D CS=103D IP=0104  NV UP DI PL NZ NA PO NC
103D:0104 3CFE        CMP   AL,FE

AX=00FB BX=0000 CX=0000 DX=0000 SP=CFDE BP=0000 SI=0000 DI=0000
DS=103D ES=103D SS=103D CS=103D IP=0106  NV UP DI NG NZ AC PO CY
103D:0106 3CFB        CMP   AL,FB

AX=00FB BX=0000 CX=0000 DX=0000 SP=CFDE BP=0000 SI=0000 DI=0000
DS=103D ES=103D SS=103D CS=103D IP=0108  NV UP DI PL ZR NA PE NC
103D:0108 3C07        CMP   AL,07

AX=00FB BX=0000 CX=0000 DX=0000 SP=CFDE BP=0000 SI=0000 DI=0000
DS=103D ES=103D SS=103D CS=103D IP=010A  NV UP DI NG NZ NA PO NC
103D:010A CC          INT   3

-q
```

```
TITLE      PROG6-2        ;FIND THE LOWEST TEMPERATURE
PAGE       60,132
;
           .MODEL SMALL
           .STACK 64
;
           .DATA
SIGN_DAT   DB         +13,-10,+19,+14,-18,-9,+12,-19,+16
           ORG   0010H
LOWEST  DB    ?
;
           .CODE
MAIN       PROC   FAR
           MOV    AX,@DATA
           MOV    DS,AX
           MOV    CX,8              ;LOAD COUNTER (NUMBER ITEMS - 1)
           MOV    SI,OFFSET SIGN_DAT   ;SET UP POINTER
           MOV    AL,[SI]           ;AL HOLDS LOWEST VALUE FOUND SO FAR
BACK:      INC    SI                ;INCREMENT POINTER
           CMP    AL,[SI]           ;COMPARE NEXT BYTE TO LOWEST
           JLE    SEARCH            ;IF AL IS LOWEST, CONTINUE SEARCH
           MOV    AL,[SI]           ;OTHERWISE SAVE NEW LOWEST
SEARCH:  LOOP   BACK               ;LOOP IF NOT FINISHED
           MOV    LOWEST,AL         ;SAVE LOWEST TEMPERATURE
           MOV    AH,4CH
           INT    21H               ;GO BACK TO DOS
MAIN       ENDP
           END    MAIN
```

Program 6-2

Review Questions

1. In an 8-bit operand, bit _____ is used for the sign bit, whereas in a 16-bit operand, bit _____ is used for the sign bit.
2. Covert 16H to its 2's complement representation.
3. The range of byte-sized signed operands is −_____ to +_____. The range of word-sized signed operands is −_____ to +_____.
4. Explain the difference between a carry and an overflow.
5. Explain the purpose of the CBW and CWD instructions. Demonstrate the effect of CBW on AL = F6H. Demonstrate the effect of CWD on AX = 124CH.
6. The instruction for signed multiplication is _____. The instruction for signed division is _____.
7. Explain the difference between the SHR (discussed in Chapter 3) and SAR instructions.
8. For each of the following instructions, indicate the flag condition necessary for each jump to occur:
   (a) JLE   (b) JG

SECTION 6.2: STRING AND TABLE OPERATIONS

There are a group of instructions referred to as *string* instructions in the 80x86 family of microprocessors. They are capable of performing operations on a series of operands located in consecutive memory locations. For example, while the CMP instruction can compare only 2 bytes (or words) of data, the CMPS (compare string) instruction is capable of comparing two arrays of data located in memory locations pointed at by the SI and DI registers. These instructions are very powerful and can be used in many applications, as will be shown shortly.

Use of SI and DI, DS and ES in string instructions

For string operations to work, designers of the CPU must set aside certain registers for specific functions. These registers must permanently provide the source and destination operands. This is exactly what the 80x86 has done. In 8088/86 microprocessors, the SI and DI registers always point to the source and destination operands, respectively. Now the question is: Which segments are they combined with to generate the 20-bit physical address? To generate the physical address, the 8088/86 always uses SI as the offset of the DS (data segment) register and DI as the offset of ES (extra segment). This is the default mode. It must be noted that the ES register must be initialized for the string operation to work.

Byte and word operands in string instructions

In each of the string instructions, the operand can be a byte or a word. They are distinguished by the letters B (byte) and W (word) in the instruction mnemonic. Table 6-3 provides a summary of all the string instructions. Each one will be discussed separately in the context of examples.

Table 6-3: String Operation Summary

Instruction	Mnemonic	Destination	Source	Prefix
move string byte	MOVSB	ES:DI	DS:SI	REP
move string word	MOVSW	ES:DI	DS:SI	REP
store string byte	STOSB	ES:DI	AL	REP
store string word	STOSW	ES:DI	AX	REP
load string byte	LODSB	AL	DS:SI	none
load string word	LODSW	AX	DS:SI	none
compare string byte	CMPSB	ES:DI	DS:SI	REPE/REPNE
compare string word	CMPSW	ES:DI	DS:SI	REPE/REPNE
scan string byte	SCASB	ES:DI	AL	REPE/REPNE
scan string word	SCASW	ES:DI	AX	REPE/REPNE

DF, the direction flag

To process operands located in consecutive memory locations requires that the pointer be incremented or decremented. In string operations this is achieved by the direction flag. Of the 16 bits of the flag register (D0 - D15), bit 11 (D10) is set aside for the direction flag (DF). It is the job of the string instruction to increment or decrement the SI and DI pointers, but it is the job of the programmer to specify the choice of increment or decrement by setting the direction flag to high or low. The instructions CLD (clear direction flag) and STD (set direction flag) are specifically designed for that.

CLD (clear direction flag) will reset (put to zero) DF, indicating that the string instruction should increment the pointers automatically. This automatic incrementation sometimes is referred to as *autoincrement*.

STD (set the direction flag) performs the opposite function of the CLD instruction. It sets DF to 1, indicating to the string instruction that the pointers SI and DI should be decremented automatically.

REP prefix

The REP (repeat) prefix allows a string instruction to perform the operation repeatedly. Now the question is: How many times is it repeated? REP assumes that CX holds the number of times that the instruction should be repeated. In other words, the REP prefix tells the CPU to perform the string operation and then decrements the CX register automatically. This process is repeated until CX becomes zero. To understand some of the concepts discussed so far, look at Example 6-13.

Example 6-13

Using string instructions, write a program that transfers a block of 20 bytes of data.

Solution:

```
in the data segment:
DATA1 DB       'ABCDEFGHIJKLMNOPQRST'
       ORG     30H
DATA2 DB       20 DUP (?)

In the code segment:

    MOV    AX,@DATA
    MOV    DS,AX              ;INITIALIZE THE DATA SEGMENT
    MOV    ES,AX              ;INITIALIZE THE EXTRA SEGMENT
    CLD                       ;CLEAR DIRECTION FLAG FOR AUTOINCREMENT
    MOV    SI,OFFSET DATA1    ;LOAD THE SOURCE POINTER
    MOV    DI,OFFSET DATA2    ;LOAD THE DESTINATION POINTER
    MOV    CX,20              ;LOAD THE COUNTER
    REP    MOVSB              ;REPEAT UNTIL CX BECOMES ZERO
```

In Example 6-13, after the transfer of every byte by the MOVSB instruction, both the SI and DI registers are incremented automatically once only (notice CLD). The REP prefix causes the CX counter to be decremented and MOVSB is repeated until CX becomes zero. Notice in Example 6-13 that both DS and ES are set to the same value.

An alternative solution for Example 6-13 would change only two lines of code:

```
    MOV    CX,10
    REP    MOVSW
```

In this case the MOVSW will transfer a word (2 bytes) at a time and increment the SI and DI registers each twice. REP will repeat that process until CX becomes zero. Notice the CX has the value of 10 in it since 10 words is equal to 20 bytes.

STOS and LODS instructions

The STOSB instruction stores the byte in the AL register into memory locations pointed at by ES:DI and increments (if DF = 0) DI once. If DF = 1, then DI is decremented. The STOSW instruction stores the contents of AX in memory locations ES:DI and ES:DI+1 (AL into ES:DI and AH into ES:DI+1), then increments DI twice (if DF = 0). If DF = 1, DI is decremented twice.

The LODSB instruction loads the contents of memory locations pointed at by DS:SI into AL and increments (or decrements) SI once if DF = 0 (or DF = 1). LODSW loads the contents of memory locations pointed at by DS:SI into AL and DS:SI+1 into AH. The SI is incremented twice if DF = 0. Otherwise, it is decremented twice. LODS is never used with a REP prefix.

Testing memory using STOSB and LODSB

Example 6-14 uses string instructions STOSB and LODSB to test an area of RAM memory.

Example 6-14

Write a program that:
(1) Uses STOSB to store byte AAH into 100 memory locations.
(2) Uses LODS to test the contents of each location to see if AAH was there. If the test fails, the system should display the message "bad memory".

Solution:

Assuming that ES and DS have been assigned in the ASSUME directive, the following is from the code segment:

```
        ;PUT PATTERN AAAAH IN TO 50 WORD LOCATIONS
        MOV    AX,DTSEG           ;INITIALIZE
        MOV    DS,AX              ;DS REG
        MOV    ES,AX              ;AND ES REG
        CLD                       ;CLEAR DF FOR INCREMENT
        MOV    CX,50              ;LOAD THE COUNTER (50 WORDS)
        MOV    DI,OFFSET MEM_AREA ;LOAD THE POINTER FOR DESTINATION
        MOV    AX,0AAAAH          ;LOAD THE PATTERN
        REP    STOSW              ;REPEAT UNTIL CX=0
        ;BRING IN THE PATTERN AND TEST IT ONE BY ONE
        MOV    SI,OFFSET MEM_AREA ;LOAD THE POINTER FOR SOURCE
        MOV    CX,100             ;LOAD THE COUNT (COUNT 100 BYTES)
AGAIN:  LODSB                     ;LOAD INTO AL FROM DS:SI
        XOR    AL,AH              ;IS PATTERN THE SAME?
        JNZ    OVER               ;IF NOT THE SAME THEN EXIT
        LOOP   AGAIN              ;CONTINUE UNTIL CX=0
        JMP    EXIT               ;EXIT PROGRAM
OVER:   MOV    AH,09              ;{ DISPLAY
        MOV    DX, OFFSET MESSAGE ;{ THE MESSAGE
        INT    21H                ;{ ROUTINE
EXIT:   ...
```

In the program in Example 6-14, first AAH is written into 100 locations by using word-sized operand AAAAH and a count of 50. In the test part, LODS brings in the contents of memory locations into AL one by one, and each time it is eXclusive-ORed with AAH (the AH register has the hex value of AA). If they are the same, ZF = 1 and the process is continued. Otherwise, the pattern written there by the previous routine is not there and the program will exit. This, in concept, is somewhat similar to the routine used in the IBM PC's BIOS except that the BIOS routine is much more involved and uses several different patterns of data for the test and it can be used to test any part of RAM, either the main RAM or the video RAM.

The REPZ and REPNZ prefixes

These prefixes can be used with the CMPS and SCAS instructions for testing purposes. They are explained below.

REPZ (repeat zero), which is the same as REPE (repeat equal), will repeat the string operation as long as the source and destination operands are equal (ZF = 1) or until CX becomes zero.

REPNZ (repeat not zero), which is the same as REPNE (repeat not equal), will repeat the string operation as long as the source and destination operands are not equal (ZF = 0) or until CX become zero. These two prefixes will be used in the context of applications after the explanation of the CMPS and SCANS instructions.

SECTION 6.2: STRING AND TABLE OPERATIONS

187

CMPS (compare string) allows the comparison of two arrays of data pointed at by the SI and DI registers. One can test for the equality or inequality of data by use of the REPE or REPNE prefixes, respectively. The comparison can be performed a byte at a time or a word at time by using CMPSB or CMPSW.

For example, if comparing "Euorop" and "Europe" for equality, the comparison will continue using the REPE CMPS as long as the two arrays are the same.

```
;from the data segment:
DATA1  DB 'Europe'
DATA2  DB 'Euorop'
;from the code segment:
        CLD                             ;DF=0 for increment
        MOV     SI,OFFSET DATA1         ;SI=DATA1 offset
        MOV     DI,OFFSET DATA2         ;DI=DATA2 offset
        MOV     CX,06                   ;load the counter
        REPE    CMPSB                   ;repeat until not equal or CX=0
```

In the case above, the two arrays are to be compared letter by letter. The first characters pointed at by SI and DI are compared. In this case they are the same ('E'), so the zero flag is set to 1 and both SI and DI are incremented. Since ZF = 1, the REPE prefix repeats the comparison. This process is repeated until the third letter is reached. The third letters "o" and "r" are not the same; therefore, ZF is reset to zero and the comparison will stop. ZF can be used to make the decision as shown in Example 6-15.

Example 6-15

Assuming that there is a spelling of "Europe" in an electronic dictionary and a user types in "Euorope", write a program that compares these two and displays the following message, depending on the result:
1. If they are equal, display "The spelling is correct".
2. If they are not equal, display "Wrong spelling".

Solution:

```
DAT_DICT     DB 'Europe'
DAT_TYPED    DB 'Euorope'
MESSAGE1     DB 'The spelling is correct','$'
MESSAGE2     DB 'Wrong spelling','$'

;from the code segment:
          CLD                              ;DF=0 FOR INCREMENT
          MOV   SI,OFFSET DAT_DICT         ;SI=DATA1 OFFSET
          MOV   DI,OFFSET DAT_TYPED        ;DI=DATA2 OFFSET
          MOV   CX,06                      ;LOAD THE COUNTER
          REPE  CMPSB                      ;REPEAT AS LONG AS EQUAL OR UNTIL CX=0
          JE    OVER                       ;IF ZF=1 THEN DISPLAY MESSAGE1
          MOV   DX,OFFSET MESSAGE2         ;IF ZF=0 THEN DISPLAY MESSAGE2
          JMP   DISPLAY
OVER:     MOV   DX,OFFSET MESSAGE1
DISPLAY:  MOV   AH,09
          INT   21H
```

One could juggle the code in Example 6-15 to make it more efficient and use fewer jumps, but for the sake of clarity it is presented in this manner.

CMPS can be used to test inequality of two arrays using "REPNE CMPSB". For example, when comparing the following social security numbers, the comparison will continue to the last digit since no two digits in the same position are the same.

231-24-7659   564-77-1338

SCAS (scan string)

The SCASB string instruction compares each byte of the array pointed at by ES:DI with the contents of the AL register, and depending on which prefix of REPE or REPNE is used, a decision is made for equality or inequality. For example, in the array "Mr. Gones", one can scan for the letter "G" by loading the AL register with the character "G" and then using the "REPNE SCASB" operation to look for that letter.

```
in the data segment:
DATA1  DB 'Mr. Gones'

and in the code segment:
        CLD                          ;DF=0 FOR INCREMENT
        MOV    DI,OFFSET DATA1       ;DI=ARRAY OFFSET
        MOV    CX,09                 ;LENGTH OF ARRAY
        MOV    AL,'G'                ;SCANNING FOR THE LETTER 'G'
        REPNE SCASB                  ;REPEAT THE SCANNING IF NOT EQUAL OR
                                     ;UNTIL THE CX IS ZERO
```

In the example above, the letter "G" is compared with "M". Since they are not equal, DI is incremented and CX is decremented, and the scanning is repeated until the letter "G" is found or the CX register is zero. In this example, since "G" is found, ZF is set to 1 (ZF = 1), indicating that there is a letter "G" in the array.

Replacing the scanned character

SCASB can be used to search for a character in an array, and if it is found, it will be replaced with the desired character. See Example 6-16.

Example 6-16

Write a program that scans the name "Mr. Gones" and replaces the "G" with the letter "J", then displays the corrected name.

Solution:

```
in the data segment:
DATA1     DB  'Mr. Gones','$'

and in the code segment:
            MOV    AX,@DATA
            MOV    DS,AX
            MOV    ES,AX
            CLD                        ;DF=0 FOR INCREMENT
            MOV    DI,OFFSET DATA1     ;ES:DI=ARRAY OFFSET
            MOV    CX,09               ;LENGTH OF ARRAY
            MOV    AL,'G'              ;SCANNING FOR THE LETTER 'G'
            REPNE SCASB                ;REPEAT THE SCANNING IF NOT EQUAL OR
            JNE    OVER                ;UNTIL CX IS ZERO. JUMP IF Z=0
            DEC    DI                  ;DECREMENT TO POINT AT 'G'
            MOV    BYTE PTR [DI],'J'   ;REPLACE 'G' WITH 'J'
OVER:       MOV    AH,09               ;DISPLAY
            MOV    DX,OFFSET DATA1     ;THE
            INT    21H                 ;CORRECTED NAME
```

In string operations, after each execution, the pointer is incremented (that is, if DF = 0). Therefore, in the example above, DI must be decremented, causing the pointer to point to the scanned character and then replace it.

SECTION 6.2: STRING AND TABLE OPERATIONS   **189**

XLAT instruction and look-up tables

There is often a need in computer applications for a table that holds some important information. To access the elements of the table, 8088/86 microprocessors provide the XLAT (translate) instruction. To understand the XLAT instruction, one must first understand tables. The table is commonly referred to as a *look-up table*. Assume that one needs a table for the values of x^2, where x is between 0 and 9. First the table is generated and stored in memory:

```
SQUR_TABLE  DB      0,1,4,9,16,25,36,49,64,81
```

Now one can access the square of any number from 0 to 9 by the use of XLAT. To do that, the register BX must have the offset address of the look-up table, and the number whose square is sought must be in the AL register. Then after the execution of XLAT, the AL register will have the square of the number. The following shows how to get the square of 5 from the table:

```
MOV    BX,OFFSET SQUR_TABLE    ;load the offset address of table
MOV    AL,05                   ;AL=05 will retrieve 6th element
XLAT                           ;pull out of table the element
                               ;and put in AL
```

After execution of this program, the AL register will have 25 (19H), the square of 5. It must be noted that for XLAT to work, the entries of the look-up table must be in sequential order and must have a one-to-one relation with the element itself. This is because of the way XLAT works. In actuality, XLAT is one instruction, which is equivalent to the following code:

```
SUB    AH,AH       ;AH=0
MOV    SI,AX       ;SI=000X
MOV    AL,[BX+SI]  ;GET THE SIth ENTRY FROM BEGINNING
                   ;OF THE TABLE POINTED AT BY BX
```

In other words, if there was no XLAT instruction, the code above would do the same thing, and this is the way many RISC processors perform this operation. Now why would one want to use XLAT to get the square of a number from a look-up table when there is the MUL instruction? The answer is that MUL takes longer.

Code conversion using XLAT

In many microprocessor-based systems, the keyboard is not an ASCII type of keyboard. One can use XLAT to translate the hex keys of such keyboards to ASCII. Assuming that the keys are 0 - F, the following is the program to convert the hex digits of 0 - F to their ASCII equivalents.

```
data segment:
ASC_TABL DB '0','1','2','3','4','5','6','7','8'
         DB '9','A','B','C','D','E','F'
HEX_VALU DB ?
ASC_VALU DB ?
```

```
code segment:
        MOV    BX,OFFSET ASC_TABL  ;BX= TABLE OFFSET
        MOV    AL,HEX_VALU         ;AL=THE HEX DATA
        XLAT                       ;GET THE ASCII EQUIVALENT
        MOV    ASC_VALU,AL         ;MOVE IT TO MEMORY
```

Review Questions

1. In string operations, register _____ is used to point to the source operand and register _____ is used to point to the destination operand.
2. SI is used as an offset into the _____ segment, and DI is used as an offset into the _____ segment.
3. The _____ flag, bit _____ of the flag register, is used to tell the CPU whether to increment or decrement pointers in repeated string operations.
4. State the purpose of instructions CLD and STD.
5. If a string instruction is repeatedly executed because of a REP prefix, how does the CPU know when to stop repeating it?
6. In the following program segment, what condition will cause the REPNZ to fail?

```
MOV    SI, OFFSET DATA1
MOV    DI, OFFSET DATA2
MOV    CX,LENGTH
REPNZ  CMPSB
```

SUMMARY

Signed number representation in the 8086/88 is achieved by using the MSB (most significant bit) as a sign bit. In a byte operand, the sign bit is D7, and in a word operand, the sign bit is D15. A sign bit of zero indicates a positive number, and a sign bit of 1 indicates a negative number. Negative numbers are represented in 2's complement. Signed addition and subtraction instructions use the same instructions as unsigned addition and subtraction: ADD and SUB. However, signed multiplication and division use the instructions IMUL and IDIV instead of MUL and DIV. In signed number arithmetic, the programmer must check for the overflow problem. An overflow occurs when either there is a carry into the MSB, or there is a carry out and no carry into the MSB. The overflow problem can be avoided by use of the sign extension instructions CBW and CWD.

Arithmetic shift instructions work similarly to the logic shift instructions except that the arithmetic shift instructions must take the sign bit into account. Therefore, they copy the sign bit into the shifted bits. The compare (CMP) instruction works the same for signed numbers as it does for unsigned numbers, but different conditional jump instructions are used after the CMP in programs.

The 80x86 has many instructions that operate on strings of data. These instructions include STOS and LODS instructions to store and load data and the SCAS scanning instruction. String operations use registers DI and SI as pointers to data in the extra and data segments. These instructions can be repeated by using any of the various forms of the REP prefix. Whether the pointers DI and SI will be incremented or decremented with each repetition depends on the setting of the direction flag. There is also an instruction for table processing, the XLAT instruction.

PROBLEMS

1. Show how the 80x86 computer would represent the following numbers and verify each with DEBUG.
   (a) -23   (b) +12   (c) -28H   (d) +6FH
   (e) -128   (f) +127   (g) +365   (h) -32 767
2. Find the overflow flag for each case and verify the result using DEBUG.
   (a) (+15)+(-12)   (b) (-123)+(-127)   (c) (+25h)+ (+34)
   (d) (-127) + (+127)   (e) (+1000) + (-1000)
3. Sign-extend the following and write simple programs in DEBUG to verify them.
   (a) -122   (b) -999h   (c) +17h
   (d) +127   (e) -129

---

4. Modify Program 6-2 to find the highest temperature. Verify your program.
5. Which instructions are used to set and reset the direction flag? State the purpose of the direction flag.
6. The REP instruction can be used with which of the following instructions?
   (a) MOVSB (b) MOVSW (c) CMPSB
   (d) LODSB (e) STOSW (f) SCASW
7. In Problem 6, state the source and destination operand for each instruction.
8. Write and verify a program that transfers a block of 200 words of data.
9. Use instructions LODSx and STOSx to mask the 3 from a set of 50 ASCII digits and transfer the result to a different memory location. This involves converting from ASCII to unpacked BCD, then storing it at a different location; for example,

	source	destination
ASCII for '5'	0011 0101	0000 0101

10. Which prefix is used for the inequality case for CMPS and SCAS instructions?
11. Write a program that scans the initials "IbM" and replaces the lowercase "b" with uppercase "B".
12. Using the timing chart in Appendix B.2, compare the clock count of the instruction XLAT and its equivalent to see which is more efficient.
13. Write a program using a look-up table and XLAT to retrieve the y value in the equation $y = x^2 + 2x + 5$ for x values of 0 to 9.

ANSWERS TO REVIEW QUESTIONS

SECTION 6.1: SIGNED NUMBER ARITHMETIC OPERATIONS
1. d7, d15
2. 16H= 0001 0110$_2$ in 2's complement: 1110 1010$_2$
3. -128 to +127; -32,768 to +32,767 (decimal)
4. an overflow is a carry into the sign bit; a carry is a carry out of the register
5. the CBW instruction sign extends the sign bit of a byte into a word; the CWD instruction sign extends the sign bit of a word into a doubleword
   F6H sign extended into AX = FFF6H
   124C sign extended into DX AX would be DX = 0000 and AX = 124CH.
6. IMUL, IDIV
7. SHR shifts each bit right one position and fills the MSB with zero
   SAR shifts each bit right one position and fills the MSB with the sign bit
   in each; the LSB is shifted into the carry flag
8. (a) JLE will jump if OF is the inverse of SF, or if ZF = 1
   (b) JG will jump if OF equals SF, or if ZF = 0

SECTION 6.2: STRING AND TABLE OPERATIONS
1. SI, DI
2. data, extra
3. direction flag, bit 11 or D10
4. CLD clears DF to 0; STD sets DF to 1
5. when CX = 0
6. if CX = 0 or the point at which DATA1 and DATA2 are not equal

CHAPTER 7

MODULES; MODULAR AND C PROGRAMMING

<table>
<tr><td colspan="2">

OBJECTIVES

Upon completion of this chapter, you will be able to:

» Discuss the advantages of modular programming
» Break large programs into modules and code the modules and calling program
» Declare names that are defined in other modules by use of the EXTRN directive
» Declare names that are used in other modules by use of the PUBLIC directive
» Link subprograms together into one executable program
» Code segment directives to link data, code, or stack segments from different modules into one segment
» Code programs using the full segment definitions
» List the various methods of passing parameters to modules and discuss the advantages and disadvantages of each
» Code programs passing the parameters via registers, memory, or stack
» Code Assembly language within C programs by using in-line coding
» Code Assembly language modules for C programs
» Describe the C calling convention for parameter passing
» Link Assembly language modules with C programs

</td></tr>
</table>

In this chapter the concept of modules is presented along with rules for writing modules and linking them together. Some very useful modules will be given, along with the methods of passing parameters among various modules. In the final section we show how to combine Assembly language programs with C programs.

SECTION 7.1: WRITING AND LINKING MODULES

Why modules?

It is common practice in writing software packages to break down the project into small modules and distribute the task of writing those modules among several programmers. This not only makes the project more manageable but also has other advantages, such as:

1. Each module can be written, debugged, and tested individually.
2. The failure of one module does not stop the entire project.
3. The task of locating and isolating any problem is easier and less time consuming.
4. One can use the modules to link with high-level languages such as C, Pascal, or BASIC.
5. Parallel development shortens considerably the time required to complete a project.

In this section we explain how to write and link modules to create a single executable program.

Writing modules

In previous chapters, a main procedure was written that called many other subroutines. In those examples, if one subroutine did not work properly, the entire program would have to be rewritten and reassembled. A more efficient way to develop software is to treat each subroutine as a separate program (or module) with a separate filename. Then each one can be assembled and tested. After testing each program and making sure that each works, they can all be brought together (linked) to make a single program. To enable these modules to be linked together, certain Assembly language directives must be used. Among these directives, the two most widely used are EXTRN (external) and PUBLIC. Each is discussed below.

EXTRN directive

The EXTRN directive is used to notify the assembler and linker that certain names and variables which are not defined in the present module are defined externally somewhere else. In the absence of the EXTRN directive, the assembler would show an error since it cannot find where the names are defined. The EXTRN directive has the following format:

```
EXTRN name1:type              ;each name can be in a separate EXTRN
EXTRN name2:type
EXTRN name1:type,name2:type   ;or many can be listed in the same EXTRN
```

External procedure names can be NEAR, FAR, or PROC (which will be NEAR for small models or FAR for larger models). The following are the types for data names, with the number of bytes indicated in parentheses: BYTE (1), WORD (2), DWORD (4), FWORD (6), QWORD (8), or TBYTE (10).

PUBLIC directive

Those names or parameters defined as EXTRN (indicating that they are defined outside the present module) must be defined as PUBLIC in the module where they are defined. Defining a name as PUBLIC allows the assembler and linker to match it with its EXTRN counterpart(s). The following is the format for the PUBLIC directive:

```
PUBLIC name1          ;each name can be in a separate directive
PUBLIC name2

PUBLIC name1, name2   ;or many can be listed in the same PUBLIC
```

Example 7-1 should help to clarify these concepts. It demonstrates that for every EXTRN definition there is a PUBLIC directive defined in another module. In Example 7-1 the EXTRN and PUBLIC directives were related to the name of a FAR procedure.

END directive in modules

In Example 7-1, notice the entry and exit points of the program. The entry point is MAIN and the exit point is "END MAIN". Modules that are called by the main module have the END directive with no label or name after it. Notice that SUBPROG1 and SUBPROG2 each have the END directive with no labels after them.

Example 7-1

Assume there is a program that constitutes the main routine, and two smaller subroutines named SUBPROG1 and SUBPROG2. The subprograms are called from the main routine. The following shows the use of the EXTRN and PUBLIC directives:

Solution:

```
;—————————————————————————————————————
;one file will contain the main module:
        EXTRN   SUBPROG1:FAR
        EXTRN   SUBPROG2:FAR
        .MODEL  SMALL
        .CODE
MAIN    PROC    FAR
        ...
        CALL    SUBPROG1
        CALL    SUBPROG2
        ...
        MOV     AH,4CH
        INT     21H
MAIN    ENDP
        END     MAIN


;—————————————————————————————————————

;————— and in a separate file: —————————
        PUBLIC  SUBPROG1
        .MODEL SMALL
        .CODE
SUBPROG1   PROC FAR
        ...
        RET
SUBPROG1   ENDP
        END


;—————————————————————————————————————
;————— and in another file: —————————
        PUBLIC  SUBPROG2
        .MODEL SMALL
        .CODE
SUBPROG2   PROC FAR
        ...
        RET
SUBPROG2   ENDP
        END
```

Linking modules together into one executable unit

Assuming that each program module in Example 7-1 is assembled separately and saved under the filenames EXAMPLE1.OBJ, PROC1.OBJ, and PROC2.OBJ, the following shows how to link them together in MASM in order to generate a single executable file:

C> LINK EXAMPLE1.OBJ + PROC1.OBJ + PROC2.OBJ

Program 7-1 shows how the EXTRN and PUBLIC directives can also be applied to data variables. In Program 7-1, the main module contains a data segment and a stack segment, but the subroutine modules do not. Each module can have its own data and stack segment. While it is entirely permissible and possible that the modules have their own data segments if they need them, generally there is only one stack that is defined in the main program and it must be defined so that it is combined with the system stack. Later in this chapter we show how to combine many segments of different modules to generate one uniform segment for each segment of code, data, and stack.

Use the program shells in Example 7-1 to:
1. Add two words.
2. Multiply two words.
Each one should be performed by a separate module. The data is defined in the main module, and the add and multiply modules have no data segment of their own.

```
TITLE    PROG7-1MM    DEMONSTRATES MODULAR PROGRAMMING
PAGE 60,132
            EXTRN    SUBPROG1:FAR
            EXTRN    SUBPROG2:FAR
            PUBLIC   VALUE1, VALUE2, SUM, PRODUCT
            .MODEL   SMALL
;————————————————
            .STACK  64
;————————————————
            .DATA
VALUE1      DW      2050
VALUE2      DW      500
SUM         DW      2 DUP (?)
PRODUCT     DW      2 DUP (?)
;————————————————
            .CODE
MAIN        PROC    FAR
            MOV     AX,@DATA
            MOV     DS,AX
            CALL    SUBPROG1      ;CALL SUBPROG TO ADD VALUE1 + VALUE2
            CALL    SUBPROG2      ;CALL SUBPROG TO MUL VALUE1 * VALUE2
            MOV     AH,4CH
            INT     21H           ;GO BACK TO DOS
MAIN        ENDP
            END     MAIN
```

Program 7-1: Main Module

```
;THIS PROGRAM FINDS THE SUM OF TWO EXTERNALLY DEFINED WORDS
;AND STORES THE SUM IN A LOCATION DEFINED BY THE CALLING MODULE
;
TITLE       PROG7-1M2  PROGRAM TO ADD TWO WORDS
PAGE        60,132
            EXTRN    VALUE1:WORD
            EXTRN    VALUE2:WORD
            EXTRN    SUM:WORD
            PUBLIC   SUBPROG1
            .MODEL   SMALL
            .CODE
SUBPROG1 PROC    FAR
            SUB     BX,BX            ;INITIALIZE CARRY COUNT
            MOV     AX,VALUE1
            MOV     DX,VALUE2
            ADD     AX,DX            ;ADD VALUE1 + VALUE2
            ADC     BX,00            ;ACCUMULATE CARRY
            MOV     SUM,AX           ;STORE SUM
            MOV     SUM+2,BX         ;STORE CARRY
            RET
SUBPROG1 ENDP
            END
```

Program 7-1: Module 2

```
;THIS PROGRAM FINDS THE PRODUCT OF TWO EXTERNALLY DEFINED WORDS
;AND STORES THE PRODUCT IN A LOCATION DEFINED BY THE CALLING MODULE
;
TITLE       PROG7-1M3  PROGRAM TO MULTIPLY TWO WORDS
PAGE        60,132
            EXTRN    VALUE1:WORD
            EXTRN    VALUE2:WORD
            EXTRN    PRODUCT:WORD
            PUBLIC   SUBPROG2
            .MODEL   SMALL
            .CODE
SUBPROG2 PROC    FAR
            MOV     AX,VALUE1
            MOV     CX,VALUE2
            MUL     CX               ;MULTIPLY VALUE1 * VALUE2
            MOV     PRODUCT,AX       ;STORE PRODUCT
            MOV     PRODUCT+2,DX ;STORE PRODUCT HIGH WORD
            RET
SUBPROG2 ENDP
            END
```

Program 7-1: Module 3

Analysis of Program 7-1

Notice in the main module that each of the two subroutines was declared with the EXTRN directive, indicating that these procedures would be defined in another file. The external subroutines were defined as FAR in this case. In the files where each subroutine is defined, it is declared as PUBLIC, so that other programs can call it. In the main module, the names VALUE1, VALUE2, SUM, and PRODUCT were defined as PUBLIC, so that other programs could access these data items. In the subprograms, these data items were declared as EXTRN. These three programs would be linked together as follows:

C> LINK PROG7-1MM.OBJ + PROG7-1M2 + PROG7-1M3

The linker program resolves external references by matching PUBLIC and EXTRN names. The linker program will search through the files specified in the LINK command for the external subroutines. Notice that the filenames are unrelated to the procedure names. "MAIN" is contained in file "PROG7-1MM.OBJ".

Example 7-2 shows the shell of modular programs using the simplified segment definition. Modular programming with full segment definition is defined later in this section.

Example 7-2

Create a shell for modular programming using the simplified segment definition.

Solution:

Modular program shells for the simplified segment directives are as follows.

The main file will contain:

```
          .MODEL SMALL
          .STACK 64
          .DATA
          ....
          ....
          .CODE
          EXTRN  SUBPROG1:NEAR
          EXTRN  SUBPROG2:NEAR
MAIN:     MOV    AX,@DATA     ;this is the program entry point
          MOV    DS,AX
          CALL   SUBPROG1
          CALL   SUBPROG2
          MOV    AH,4CH
          INT    21H
          END    MAIN         ;this is the program exit point


;————— and in a separate file: ——————————————

          .MODEL  SMALL
          .CODE
          PUBLIC  SUBPROG1
SUBPROG1 PROC
          ...
          RET
SUBPROG1 ENDP
          END


;————— and in another file: ——————————————

          .MODEL  SMALL
          .CODE
          PUBLIC  SUBPROG2
SUBPROG2 PROC
          ...
          RET
SUBPROG2 ENDP
          END
```

Notice that in the main module of Example 7-2, the name MAIN has a colon after it and is used for the first executable instruction. This is the entry point of the program. The exit point of the program is indicated by the same label, which must be named in the END directive. No program can have more than one entry and one exit point. The label MAIN was chosen in this instance, but of course any name could have been chosen. Remember that the END directives in other modules do not have a label after the word "END". Program 7-2 is the same as Program 7-1, rewritten for the full segment definition. Compare the two programs to see the ease of the simplified segment definition. When using the simplified segment definition shown in Example 7-2, procedures will default to NEAR for small or compact models and to FAR for medium, large, or huge models.

Modular programming and full segment definition

Program 7-2 uses full segment definition to redefine all the segments of Program 7-1. An analysis of how the segments are combined as shown in the link map follows the program. The code segments were not made PUBLIC in this example. Notice that in order to combine various segments from different modules into one segment, the segment names must be the same.

```
        TITLE       PROG7-2MM       PROG7-1 REWRITTEN WITH FULL SEGMENT DEFINITION
        PAGE        60,132
                    EXTRN   SUBPROG1:FAR
                    EXTRN   SUBPROG2:FAR
                    PUBLIC  VALUE1, VALUE2, SUM, PRODUCT
;_____
STSEG       SEGMENT  PARA STACK 'STACK'
            DB       100 DUP(?)
STSEG       ENDS
;_____
DTSEG       SEGMENT  PARA 'DATA'
VALUE1      DW       2050
VALUE2      DW       500
SUM         DW       2 DUP (?)
PRODUCT     DW       2 DUP (?)
DTSEG       ENDS
;_____
CODSG_A     SEGMENT  PARA 'CODE'
MAIN        PROC     FAR
            ASSUME CS:CODSG_A,DS:DTSEG,SS:STSEG
            MOV     AX,DTSEG
            MOV     DS,AX
            CALL    SUBPROG1        ;CALL SUBPROG TO ADD VALUE1 + VALUE2
            CALL    SUBPROG2        ;CALL SUBPROG TO MUL VALUE1 * VALUE2
            MOV     AH,4CH
            INT     21H             ;GO BACK TO DOS
MAIN        ENDP
CODSG_A     ENDS
            END     MAIN
```

Program 7-2: Main Module

```
;THIS PROGRAM FINDS THE SUM OF TWO EXTERNALLY DEFINED WORDS
;AND STORES THE SUM IN A LOCATION DEFINED BY THE CALLING MODULE
TITLE       PROG7-2M2  PROGRAM TO ADD TWO WORDS
PAGE        60,132
            EXTRN    VALUE1:WORD
            EXTRN    VALUE2:WORD
            EXTRN    SUM:WORD
            PUBLIC   SUBPROG1
CODSG_B     SEGMENT  PARA 'CODE'
SUBPROG1 PROC    FAR
            ASSUME CS:CODSG_B
            SUB     BX,BX           ;INITIALIZE CARRY COUNT
            MOV     AX,VALUE1
            MOV     DX,VALUE2
            ADD     AX,DX           ;ADD VALUE1 + VALUE2
            ADC     BX,00           ;ACCUMULATE CARRY
            MOV     SUM,AX          ;STORE SUM
            MOV     SUM+2,BX        ;STORE CARRY
            RET
SUBPROG1 ENDP
CODSG_B     ENDS
            END
```

Program 7-2: Module 2

SECTION 7.1: WRITING AND LINKING MODULES   199

```
;THIS PROGRAM FINDS THE PRODUCT OF TWO EXTERNALLY DEFINED WORDS
;AND STORES THE PRODUCT IN A LOCATION DEFINED BY THE CALLING MODULE
TITLE        PROG7-2M3  PROGRAM TO MULTIPLY TWO WORDS
PAGE         60,132
             EXTRN    VALUE1:WORD
             EXTRN    VALUE2:WORD
             EXTRN    PRODUCT:WORD
             PUBLIC   SUBPROG2
CODSG_C   SEGMENT  PARA 'CODE'
SUBPROG2 PROC   FAR
             ASSUME CS:CODSG_C
             MOV    AX,VALUE1
             MOV    CX,VALUE2
             MUL    CX                ;MUL VALUE1 * VALUE2
             MOV    PRODUCT,AX    ;STORE PRODUCT
             MOV    PRODUCT+2,DX ;STORE PRODUCT HIGH WORD
             RET
SUBPROG2 ENDP
CODSG_C   ENDS
             END


     Start  Stop   Length Name            Class
     00000H 00063H 00064H STSEG           STACK
     00070H 0007BH 0000CH DTSEG           DATA
     00080H 00092H 00013H CODSG_A         CODE
     000A0H 000B5H 00016H CODSG_B         CODE
     000C0H 000D0H 00011H CODSG_C         CODE
```

Program 7-2: Module 3 and the Link Map

Analysis of Program 7-2 link map

The link map shows the start and end of each segment. Notice that each segment starts at a 16-byte boundary: 00070H, 00080H, etc. The code segment for the main module has the name "CODSG_A", starts at 00080H, and ends at 00092H, taking a total of 00013H bytes. It was classified as 'CODE'. The next code segment is defined under the name "CODSG_B". Notice that it starts at the 16-byte boundary 000A0H since it was defined as PARA. This means that from 00093H to 0009FH is unused. Similarly, the third module starts at 000C0H. Notice that each code segment is separate. They can all be merged together into one segment by using the PUBLIC option. This is shown in Example 7-3. To merge the code segments together, each code segment must have the same name and be declared PUBLIC.

Example 7-3

Show the link map for Program 7-2 rewritten to combine code segments (use PARA boundaries) using directive:
```
CDSEG    SEGMENT  PARA PUBLIC 'CODE'
```

Solution:
```
     Start    Stop     Length  Name        Class
     00000H 00063H 00064H STSEG            STACK
     00070H 0007BH 0000CH DTSEG            DATA
     00080H 000D0H 00051H CDSEG            CODE
```

The following are the SEGMENT directives using word boundaries:
```
STSEG    SEGMENT  WORD STACK 'STACK'
DTSEG    SEGMENT  WORD 'DATA'
CDSEG    SEGMENT  WORD PUBLIC 'CODE'
```

The following is the link map when the program used WORD boundaries:
```
     Start    Stop     Length  Name        Class
     00000H 00063H 00064H STSEG            STACK
     00064H 0006FH 0000CH DTSEG            DATA
     00070H 000AAH 0003BH CDSEG            CODE
```

In previous chapters, when a segment was defined using full segment definition, no other attributes were mentioned after it. It was simply written

```
name        SEGMENT
```

This kind of definition of segments was acceptable since there was only one of each segment of code, data, and stack. However, when there are many modules to be linked together, the segment definition must be adjusted. The complete segment definition used widely in modular programming is as follows:

```
name        SEGMENT alignment combine type class name
```

Appendix C (see SEGMENT) gives a complete description of the fields of the SEGMENT directive. A brief explanation of each field is given below.

The *alignment* field indicates whether a segment should start on a byte, word, paragraph, or page boundary. For example, if WORD is given in the alignment field, the segment will start at the next available word. When the WORD boundary is used, if a previous segment ended at offset 0048H, the next segment will start at 004AH. The default alignment is PARA, meaning that each segment will start on a paragraph boundary. A paragraph in DOS is defined as 16 bytes; therefore, each segment will start on a 16-byte boundary. When PARA is used, if the previous segment ended at 0048H, the next segment would begin at the next paragraph boundary, which is 0050H. Paragraph boundaries end in 0; they are evenly divisible by 16 (10H).

The *combine type* field indicates to the linker whether segments of the same type should be linked together. Typical options for combine type are STACK or PUBLIC. An example below shows how to use this field in the stack segment definition to combine the stack segment of a program with the system stack to eliminate the "Warning: no stack segment" message generated by the linker. If the combine type is PUBLIC, the linker will combine that segment with other segments of the same type in other modules. This can be used to combine code segments with various names under a single name.

The *class name* field has four options: 'CODE', 'STACK', 'DATA', and 'EXTRA'. It must be enclosed in single quotes. It is used in combining segments of the same type from various modules.

Complete stack segment definition

The following stack segment definition in the main module will eliminate the "Warning: no stack segment" message generated by the linker:

```
name        SEGMENT PARA    STACK 'STACK'
```

Complete data and code segment definitions

The following is a data segment definition that can be used if no other module has defined any data segment:

```
name        SEGMENT PARA    'DATA'
```

If any other module has defined a data segment then PUBLIC should be placed between PARA and 'DATA'. The following are the code and data segment definitions to combine segments from different modules:

```
name        SEGMENT PARA    PUBLIC 'CODE'
name        SEGMENT PARA    PUBLIC 'DATA'
```

Example 7-4 rewrites Example 7-2 to define segments using the complete segment definition.

Example 7-4

Create a shell for modular programming using the complete segment definition.

Solution:

The main file will contain:

```
TITLE       PROG   PROGRAM SHELL  WITH COMPLETE SEGMENT DEFINITION
PAGE        60,132
            EXTRN    SUBPROG1:FAR
            EXTRN    SUBPROG2:FAR
            PUBLIC   ...                       ;declare data here to be shared
;————————————————
STSEG       SEGMENT  PARA STACK 'STACK'
            DB       100 DUP(?)
STSEG       ENDS
;————————————————
DTSEG       SEGMENT  PARA 'DATA'
            ;define data here
DTSEG       ENDS
;————————————————
CODSG_A     SEGMENT  PARA 'CODE'
MAIN        PROC   FAR
            ASSUME CS:CODSG_A,DS:DTSEG,SS:STSEG
            MOV    AX,DTSEG
            MOV    DS,AX
            CALL   SUBPROG1      ;CALL SUBPROG
            CALL   SUBPROG2      ;CALL SUBPROG
            ...
            MOV    AH,4CH
            INT    21H               ;GO BACK TO DOS
MAIN        ENDP
CODSG_A     ENDS
            END    MAIN


;————————and in another file: ————————————————
TITLE       SUBPROG1  PROGRAM ...
PAGE        60,132
            EXTRN    ...               ;declare data that is defined externally
            PUBLIC   SUBPROG1          ;declare procedures that are called externally
CODSG_B     SEGMENT  PARA 'CODE'
SUBPROG1 PROC   FAR
            ASSUME CS:CODSG_B
            ; the instructions that perform the work of the subroutine go here
            RET
SUBPROG1 ENDP
CODSG_B     ENDS
            END

;————————and in another file: ————————————————
TITLE       SUBPROG2  PROGRAM TO ...
PAGE        60,132
            EXTRN    ...               ;declare data that is defined externally
            PUBLIC   SUBROG2           ;declare procedures that are called externally
CODSG_C     SEGMENT  PARA 'CODE'
SUBPROG2 PROC   FAR
            ASSUME CS:CODSG_C
            ; the instructions that perform the work of the subroutine go here
            RET
SUBPROG2 ENDP
CODSG_C     ENDS
            END
```

Review Questions

1. List three advantages of modular programming.
2. The _____ directive is used within a module to indicate that the named variable can be used by another module.
3. The _____ directive is used within a module to indicate that the named variable was defined in another module.
4. How does the system determine the entry and exit points of a program consisting of more than one module?
5. What is a paragraph?
6. Write the directive used in complete segment definition that will define the stack segment so that it will be combined with the system stack.
7. If a word-sized data item named TOTAL was defined in module1, code the directive to define TOTAL in module2.
8. If PARA were used for the alignment type of a code segment that ended at 56H, where would the next code segment begin?
9. Write the code segment directives for a calling program and a module so that they will be combined into one code segment.

SECTION 7.2: SOME VERY USEFUL MODULES

This section shows the development of two very useful programs that convert from hex to decimal, and vice versa. Then they are rewritten as modules that can be called from any program. Finally, the calling program is written.

Binary (hex)-to-ASCII (decimal) conversion

The result of arithmetic operations is, of course, in binary. To display the result in decimal, the number is first converted to decimal, and then each digit is tagged with 30H to put it in ASCII form so that it can be displayed or printed. The first step is to convert the binary number to decimal. Look at the following example, which converts 34DH to decimal.

```
34DH = (3 × 16²)   + (4 × 16¹)  + (D=13 × 160)
     = (3 × 256)   + (4 × 16)   + (13 × 1)
     = 768         + 64         + 13
     = 845
```

Another method to convert a hex number to decimal is to divide it repeatedly by 10 (0AH), storing each remainder, until the quotient is less than 10. The following steps would be performed:

34DH / A = 84 remainder 5
84H / A = 8 remainder 4
8 (< A, so the process stops)
Taking the remainders in reverse order gives: 845 decimal

Program 7-3 shows the conversion process for a word-sized (16-bit) number using the method of repeated division demonstrated above. Since a word-sized hex number is between 0 and FFFFH, the result in decimal can be as high as 65535. Therefore, a string length of 5 should be sufficient to hold the result. The binary number to be converted is in data item BINNUM. Notice in Program 7-3 that as each decimal digit (the remainder) is placed in DL, it is tagged with 30H to convert it to ASCII. It is then placed in a memory area called ASCNUM. The ASCII digits are placed in memory so that the lowest digit is in high memory, as is the convention of ASCII storage in DOS.

```
TITLE          PROG7-3    CONVERT BINARY TO ASCII
;USING SIMPLIFIED SEGMENT DEFINITION
PAGE           60,132
               .MODEL SMALL
;——————————————
               .STACK 64
               .DATA
BINNUM         DW      246DH
               ORG     10H
ASCNUM         DB      5 DUP ('0')
;——————————————
               .CODE
B2ASC_CON PROC FAR
               MOV     AX,@DATA
               MOV     DS,AX
               MOV     BX,10               ;BX=10 THE DIVISOR
               MOV     SI,OFFSET ASCNUM    ;SI = BEGINNING OF ASCII STRING
               ADD     SI,5                ;ADD LENGTH OF STRING
               DEC     SI                  ;SI POINTS TO LAST ASCII DIGIT
               MOV     AX,BINNUM           ;LOAD BINARY (HEX) NUMBER
BACK:          SUB     DX,DX               ;DX MUST BE 0 IN WORD DIVISION
               DIV     BX                  ;DIVIDE HEX NUMBER BY 10 (BX=10)
               OR      DL,30H              ;TAG '3' TO MAKE IT ASCII
               MOV     [SI],DL             ;MOVE THE ASCII DIGIT
               DEC     SI                  ;DECREMENT POINTER
               CMP     AX,0                ;CONTINUE LOOPING WHILE AX > 0
               JA      BACK
               MOV     AH,4CH
               INT     21H                 ;GO BACK TO DOS
B2ASC_CON ENDP
               END     B2ASC_CON
```

Program 7-3

ASCII (decimal)-to-binary (hex) conversion

When a user keys in digits 0 to 9, the keyboard provides the ASCII version of the digits to the computer. For example, when the key marked 9 is pressed, in reality the keyboard provides its ASCII version 00111001 (39H) to the system. In Chapter 3 we showed how in some cases, such as addition, the numbers can be processed in ASCII and there is no need to convert them to hex (binary). However, in the majority of cases the number needs to be converted to hex in order to be processed by the CPU. Look at the example of converting decimal 482 to hex. The following shows the steps to convert this number to hex:

$$482 / 16^2 = 482 / 256 = 1$$
$$482 - (1 \times 256) = 226 \quad 226 / 16^1 = 226 / 16 = 14 = E$$
$$226 - (14 \times 16) = 2$$
482 decimal = 1E2 hexadecimal

However, a computer would use a different method since it works in binary arithmetic, not decimal. First the 30H would be masked off each ASCII digit. Then each digit is multiplied by a weight (a power of 10) such as 1, 10, 100, or 1000 and they are then added together to get the final hex (binary) result. Converting decimal 482 to hex involves the following steps. First a user types in '482' through the PC ASCII keyboard, yielding 343832, the ASCII version of 482. Then the following steps are performed:

$$2 \times 1 \quad = \quad = \quad 2$$
$$8 \times 10 \quad = \quad 80 = \quad 50H$$
$$4 \times 100 = 400 = \quad \underline{190H}$$
$$\qquad\qquad\qquad 1E2 \text{ hexadecimal}$$

Program 7-4 converts an ASCII number to binary. It assumes the maximum size of the decimal number to be 65535. Therefore, the maximum hex result is FFFFH, a 16-bit word. It begins with the least significant digit, masks off the 3, and multiplies it by its weight factor. Register CX holds the weight, which is 1 for the least significant digit. For the next digit CX becomes 10 (0AH), for the next it becomes 100 (64H), and so on. The program assumes that the least significant ASCII digit is in the highest memory location of the data. This is consistent with the conventions of storing ASCII numbers with the most significant digit in the lower memory address and the least significant digit in the highest memory address. For example, placing '749' at memory offset 200 gives offset 200 = (37), 201 = (34), and 202 = (39). DOS 21H function call 0A also places ASCII numbers this way.

```
TITLE      PROG7-4   CONVERT ASCII TO BINARY
PAGE       60,132
           .MODEL SMALL
           .STACK 64
;—————————————
           .DATA
TEN        DW      10
ASCNUM     DB      '09325'
STRLEN     DB      5
           ORG     10H
BINNUM     DW      0
;—————————————
           .CODE
ASC2B_CON PROC FAR
           MOV     AX,@DATA
           MOV     DS,AX
           SUB     DI,DI              ;CLEAR DI FOR THE BINARY(HEX) RESULT
           MOV     SI,OFFSET ASCNUM   ;SI = BEGINNING OF ASCII STRING
           MOV     BL,STRLEN          ;BL = LENGTH OF ASCII STRING
           SUB     BH,BH              ;BH=0 USE BX IN BASED INDEX MODE
           DEC     BX                 ;BX IS OFFSET TO LAST DIGIT
           MOV     CX,1               ;CX = WEIGHT FACTOR
AGAIN:     MOV     AL,[SI+BX]         ;GET THE ASCII DIGIT
           AND     AL,0FH             ;STRIP OFF '3'
           SUB     AH,AH              ;CLEAR AH FOR WORD MULTIPLICATION
           MUL     CX                 ;MULTIPLY BY THE WEIGHT
           ADD     DI,AX              ;ADD IT TO BINARY (HEX )RESULT
           MOV     AX,CX              ;MULTIPLY THE WEIGHT FACTOR
           MUL     TEN                ;  BY TEN
           MOV     CX,AX              ;  FOR NEXT ITERATION
           DEC     BX                 ;DECREMENT DIGIT POINTER
           JNS     AGAIN              ;JUMP IF COUNTER >= 0
           MOV     BINNUM,DI          ;SAVE THE BINARY(HEX)RESULT
           MOV     AH,4CH
           INT     21H                ;GO BACK TO DOS
ASC2B_CON ENDP
           END     ASC2B_CON
```

Program 7-4

Programs 7-3 and 7-4 have been written and tested with sample data, and now can be changed from programs into modules that can be called by any program.

Binary-to-ASCII module

Program 7-5 is the modularized Program 7-3. The procedure is declared as public, so it can be called by another program. All values used are declared external since the data will be provided by the calling program. Therefore, this module does not need its own data segment. Notice the following points about the module:

1. Since this module will be called by another module, no entry point and exit point were given. Therefore, the END directive does not have the label B2ASC_CON.
2. The module must return to the caller and not DOS as was the case in Program 7-4.
3. This module does not need its own data or stack segments.

SECTION 7.2: SOME VERY USEFUL MODULES   205

```
TITLE        PROG7-5   BINARY TO DECIMAL CONVERSION MODULE
PAGE         60,132
;this module converts a binary (hex) number up to FFFFH to decimal
;  then makes it displayable (ASCII)
;CALLING PROGRAM SETS
;  AX = BINARY VALUE TO BE CONVERTED TO ASCII
;  SI = OFFSET ADDRESS WHERE ASCII VALUE IS TO BE STORED
             .MODEL SMALL
             PUBLIC  B2ASC_CON
             .CODE
B2ASC_CON PROC   FAR
             PUSHF                      ;STORE REGS CHANGED BY THIS MODULE
             PUSH   BX
             PUSH   DX
             MOV    BX,10               ;BX=10 THE DIVISOR
             ADD    SI,4                ;SI POINTS TO LAST ASCII DIGIT
B2A_LOOP:    SUB    DX,DX               ;DX MUST BE 0 IN WORD DIVISION
             DIV    BX                  ;DIVIDE HEX NUMBER BY 10 (BX=10)
             OR     DL,30H              ;TAG '3' TO REMAINDER TO MAKE IT ASCII
             MOV    [SI],DL             ;MOVE THE ASCII DIGIT
             DEC    SI                  ;DECREMENT POINTER
             CMP    AX,0                ;CONTINUE LOOPING WHILE AX > 0
             JA     B2A_LOOP
             POP    DX                  ;RESTORE REGISTERS
             POP    BX
             POPF
             RET
B2ASC_CON ENDP
             END
```

Program 7-5

```
TITLE        PROG7-6  ASCII TO BINARY CONVERSION MODULE
PAGE         60,132
;this module converts any ASCII number between 0 to 65535 to binary
;CALLING PROGRAM SETS
;  SI = OFFSET OF ASCII STRING
;  BX = STRING LENGTH - 1 (USED AS INDEX INTO ASCII NUMBER)
;THIS MODULE SETS
;  AX = BINARY NUMBER
;────────────────
             .MODEL  SMALL
             EXTRN   TEN:WORD
             PUBLIC  ASC2B_CON
             .CODE
ASC2B_CON PROC  FAR
             PUSHF                      ;STORE REGS CHANGED IN THIS MODULE
             PUSH   DI
             PUSH   CX
             SUB    DI,DI               ;CLEAR DI FOR THE BINARY (HEX) RESULT
             MOV    CX,1                ;CX = WEIGHT FACTOR
A2B_LOOP:    MOV    AL,[SI+BX]          ;GET THE ASCII DIGIT
             AND    AL,0FH              ;STRIP OFF '3'
             SUB    AH,AH               ;CLEAR AH FOR WORD MULTIPLICATION
             MUL    CX                  ;MULTIPLY BY THE WEIGHT
             ADD    DI,AX               ;ADD IT TO BINARY (HEX) RESULT
             MOV    AX,CX               ;MULTIPLY THE WEIGHT FACTOR
             MUL    TEN                 ;  BY TEN
             MOV    CX,AX               ;  FOR NEXT ITERATION
             DEC    BX                  ;DECREMENT DIGIT POINTER
             JNS    A2B_LOOP            ;JUMP IF OFFSET >= 0
             MOV    AX,DI               ;STORE BINARY NUMBER IN AX
             POP    CX                  ;RESTORE FLAGS
             POP    DI
             POPF
             RET
ASC2B_CON ENDP
             END
```

Program 7-6

ASCII-to-binary module

Program 7-6 is the modularized version of Program 7-4. Notice the following points about the module:

1. TEN is defined in the calling program.
2. This module must return to the caller and not DOS.

Calling module

Program 7-7 shows the calling program for the module that converts ASCII to binary. This program sets up the data segment, inputs the ASCII data from the keyboard, places it in memory, then calls the routine to convert the number to binary. Finally, the hex result is stored in memory.

```
TITLE        PROG7-7  CALLING PROGRAM TO CONVERT ASCII TO BINARY
PAGE         60,132
;
             PUBLIC TEN
             .MODEL SMALL
             .STACK 64
;───────────────────────────────────────────────────────────
             .DATA
ASC_AREA  LABEL    BYTE
MAX_LEN   DB       6
ACT_LEN   DB       ?
ASC_NUM   DB       6 DUP (?)
          ORG      10H
BINNUM    DW       0
PROMPT1   DB       'PLEASE ENTER A 5 DIGIT NUMBER','$'
TEN       DW       10
;───────────────────────────────────────────────────────────
             .CODE
             EXTRN ASC2B_CON:FAR
MAIN         PROC   FAR
             MOV    AX,@DATA
             MOV    DS,AX
             ;DISPLAY THE PROMPT
             MOV    AH,09
             MOV    DX,OFFSET PROMPT1
             INT    21H
             ;INPUT STRING
             MOV    AH,0AH
             MOV    DX,OFFSET ASC_AREA
             INT    21H
             MOV    SI,OFFSET ASC_NUM
             MOV    BH,00
             MOV    BL,ACT_LEN
             DEC    BX
             CALL   ASC2B_CON
             MOV    BINNUM,AX        ;SAVE THE BINARY (HEX) RESULT
             MOV    AH,4CH
             INT    21H              ;GO BACK TO DOS
MAIN         ENDP
             END    MAIN
```

Program 7-7

Review Questions

1. Show a step-by-step analysis of Program 7-3 with data F624H. Show the sequence of instructions and the data values.
2. Show a step-by-step analysis of Program 7-4 with data '1456'. Show the sequence of instructions and the data values.

SECTION 7.3: PASSING PARAMETERS AMONG MODULES

Occasionally, there is a need to pass parameters among different Assembly language modules or between Assembly language and BASIC, Pascal, or C language programs. The parameter could be fixed values, variables, arrays of data, or even pointers to memory. Parameters can be passed from one module to another through registers, memory, or the stack. In this section we explore passing parameters between Assembly language modules.

Passing parameters via registers

When there is a need to pass parameters among various modules, one could use the CPU's registers. For example, if a main routine is calling a subroutine, the values are placed in the registers in the main routine and then the subroutine is called upon to process the data. In such cases the programmer must clearly document the registers used for the incoming data and the registers that are expected to have the result after the execution of the subroutine. In Chapter 4 this concept was demonstrated with INT 21H and INT 10H. Program 7-7 demonstrated this method. In that program, registers BX and SI were set to point to certain data items before the module was called, and the called module placed its result in register AX prior to returning to the calling routine.

Passing parameters via memory

Although parameter passing via registers is widely used in many of the DOS and BIOS interrupt function calls, the limited number of registers inside the CPU is a major limitation associated with this method of parameter passing. This makes register management a cumbersome task. One alternative is to pass parameters via memory by defining an area of RAM and passing parameters to these RAM locations. DOS and IBM BIOS use this method frequently. The problem with passing parameters to a fixed area of memory is that there must be a universal agreement to the address of the memory area in order to make sure that modules can be run on the hardware and software of various companies. This kind of standardization is hard to come by. The only reason that BIOS and DOS use an area of memory for passing parameters is because IBM and Microsoft worked closely together to decide on the memory addresses. Another option, and indeed the most widely used method of passing parameters, is via the stack, as discussed next. Passing parameters via the stack makes the parameters both register and memory independent.

Passing parameters via the stack

The stack is a very critical part of every program and playing with it can be risky. When a module is called, it is the stack that holds the address where the program must return after execution. Therefore, if the contents of the stack are altered, the program can crash. This is the reason that working with the stack and passing parameters through it must be understood very thoroughly before one embarks on it.

Program 7-8, on the following page, demonstrates this method of parameter passing and is written with the following requirements. The main module gets three word-sized operands from the data segment, stores them on the stack, and then calls the subroutine. The subroutine gets the operands from the stack, adds them together, holds the result in a register, and then returns control to the main module. The main module stores the result of the addition. Following the program is a detailed stack contents analysis that will show how the parameters are stored on the stack by the main routine and retrieved from the stack by the called routine.

```
         TITLE      PROG7-8  PASSING PARAMETERS VIA THE STACK
         PAGE       60,132
         ;

                    .MODEL   SMALL
                    EXTRN    SUBPROG6:FAR
         ;——————————————
                    .STACK 64
         ;——————————————
                    .DATA
         VALUE1     DW       3F62H
         VALUE2     DW       1979H
         VALUE3     DW       ·25F1H
         RESULT     DW       2 DUP (?)
         ;——————————————
                    .CODE
         MAIN       PROC     FAR
                    MOV      AX,@DATA
                    MOV      DS,AX
                    PUSH     VALUE3         ;SAVE VALUE3 ON STACK
                    PUSH     VALUE2         ;SAVE VALUE2 ON STACK
                    PUSH     VALUE1         ;SAVE VALUE1 ON STACK
                    CALL     SUBPROG6       ;CALL THE ADD ROUTINE
                    MOV      RESULT,AX      ;STORE
                    MOV      RESULT+2,BX    ;  THE RESULT
                    MOV      AH,4CH
                    INT      21H
         MAIN       ENDP
                    END      MAIN
```

Program 7-8: Main Module

```
         ;—————————— in a separate file: ——————————————
         TITLE      SUBPROG6   MODULE TO ADD THREE WORDS BROUGHT IN FROM THE STACK
         PAGE       60,132
                    .MODEL   SMALL
                    PUBLIC   SUBPROG6
                    .CODE
         SUBPROG6   PROC     FAR
                    SUB      BX,BX          ;CLEAR BX FOR CARRIES
                    PUSH     BP             ;SAVE BP
                    MOV      BP,SP          ;SET BP FOR INDEXING
                    MOV      AX,[BP]+6      ;MOV VALUE1 TO AX
                    MOV      CX,[BP]+8      ;MOV VALUE2 TO CX
                    MOV      DX,[BP]+10     ;MOV VALUE3 TO DX
                    ADD      AX,CX          ;ADD VALUE2 TO VALUE1
                    ADC      BX,00          ;KEEP THE CARRY IN BX
                    ADD      AX,DX          ;ADD VALUE3
                    ADC      BX,00          ;KEEP THE CARRY IN BX
                    POP      BP             ;RESTORE BP BEFORE RETURNING
                    RET      6              ;RETURN AND ADD 6 TO SP TO BYPASS DATA
         SUBPROG6   ENDP
                    END
```

Program 7-8: Module 2

Stack contents analysis for Program 7-8

To clarify the concept of parameter passing through the stack, the following is a step-by-step analysis of the stack pointer and stack contents. Assume that the stack pointer has the value SP = 17FEH before the "PUSH VALUE3" instruction in the main module is executed.

1. VALUE3 = 25F1H is pushed and SP = 17FC (remember little endian: low byte to low address and high byte to high address).
2. VALUE2 = 1979H is pushed and then SP = 17FA.
3. VALUE1 = 3F62H is pushed and then SP = 17F8.

SECTION 7.3: PASSING PARAMETERS AMONG MODULES

4. CALL SUBPROG6 is a FAR call; therefore, both CS and IP are pushed onto the stack, making SP = 17F4. If it had been a near call, only IP would have been saved.

5. In the subprogram module, register BP is saved by PUSHing BP onto the stack, which makes SP = 17F2. In the subprogram, BP is used to access values in the stack. First SP is copied to BP since only BP can be used in indexing mode with the stack segment (SS) register. In other words, "MOV AX,[SP+4]" will cause an error. "MOV AX,[BP]+6" loads VALUE1 into AX. [BP]+6 = 17F2+6 = 17F8, which is exactly where VALUE1 is located. Similarly, BP+8 = 17F2+8 = 17FA is the place where VALUE2 is located, and BP+10 = 17F2H+10 = 17FCH is the location of VALUE3.

6. After all the parameters are brought into the CPU by the present module and are processed (in this case added), the module restores the original BP contents by POPping BP from stack. Then SP = 17F4.

7. RET 6: This is a new instruction. The RETurns shown previously did not have numbers right after them. The "RET n" instruction means first to POP CS:IP (IP only if the CALL was NEAR) off the top of the stack and then add n to the SP. As can be seen from the Program 7-8 diagram, after popping CS and IP off the stack, the stack pointer is incremented four times, making SP = 17F8. Then adding 6 to it to bypass the six locations of the stack where the parameters are stored makes the SP = 17FEH, its original value. Now what would happen if the program had a RET instruction instead of the "RET 6"? The problem is that every time this subprogram is executed it will cause the stack to lose six locations. If that had been done in the example above, when the same routine is called again the stack starts at 17F8 instead of 17FE. If this practice of losing some area of the stack continues, eventually the stack could be reduced to a point where the program would run out of stack and crash.

17F0		
17F1		
17F2	BP	
17F3		
17F4	IP	
17F5		
17F6	CS	
17F7		
17F8	62	VALUE1
17F9	3F	
17FA	79	VALUE2
17FB	19	
17FC	F1	VALUE3
17FD	25	
17FE		

Program 7-8: Stack Contents Diagram

Review Questions

1. List one advantage and one disadvantage of each method of parameter passing.
   (a) via register (b) via stack (c) via memory
2. Assume that we would like to access some parameters from the stack. Which of the following are correct ways of accessing the stack?
   (a) MOV AX,[BP]+20   (b) MOV AX,[SP]+20
   (c) MOV AX,[BP+DI]   (d) MOV AX,[SP+SI]

SECTION 7.4: COMBINING ASSEMBLY LANGUAGE AND C PROGRAMS

Although Assembly language is the fastest language available for a given CPU, it cannot be run on different CPUs. For example, Intel's 80x86 Assembly programs cannot be run on Motorola's 68000 series computers since the opcode, mnemonics, register names, and size are totally different. Therefore, a portable language is needed.

Why C?

Although the dream of a universal language among the peoples of the world is still unrealized, C language is becoming the universal language among all the various CPUs. Today, a large portion of programs written for computers, from PCs to supercomputers such as CRAY, are in the C language. C is such a universal programming language that it can be run on any CPU architecture with little or no modification. It is simply recompiled for that CPU. The fact that C is such a portable language is making it the dominant language of programmers. However, C is not as fast as Assembly language. Combining C and Assembly language takes advantage of C's portability and Assembly's speed. Today it is very common to see a software project written using 70 to 80% C and the rest Assembly language.

There are two ways to mix C and Assembly. One is simply to insert the Assembly code in C programs, which is commonly referred to as *in-line assembly*. The second method is to make the C language call an external Assembly language procedure. In this section we first discuss how to do in-line assembly coding and then show a C language program calling an Assembly procedure. Readers without a C programming background can bypass this section without loss of continuity. This section covers Borland's Turbo C++.

Inserting 80x86 assembly code into C programs

In this section we discuss in-lining with Borland's Turbo C++. For other C compilers, consult their C manual. The following code demonstrates how to change the cursor position to row = 10 and column = 20 in a C program. Assembly instructions are prefaced with "asm", which is a reserved word. Microsoft uses the keyword "_asm". Note that in Microsoft, not all interrupts may be supported in the latest versions of Visual C++. The following shows two variations of Borland's format for in-line assembly.

```
/* version 1: using keyword asm before each line of in-line code */
/* Microsoft uses keyword "_asm" */
main ()
{
asm     mov ah,2;        /* each line should end with semicolon or <CR> */
asm     mov bh,0;
asm     mov dl,20;       /* comments must be C style, not ";" assembly style */
asm     mov dh,10;
asm     int 10h;
}


/* version 2: using the keyword asm before a block of in-line code */
/* Microsoft uses keyword "_asm" */
main ()
{
asm     {
mov ah,2
mov bh,0
mov dl,20
mov dh,10
int 10h
        }
}
```

As shown above, each line of in-line code is prefaced by the keyword "asm", or a block of in-line code is prefaced by "asm". Each line must end in a semicolon or newline, and any comments must be in the correct form for C.

Example 7-5, on the following page, shows two programs that display a string of data. Solution A uses C language exclusively. Solution B uses Borland's Turbo C with in-line Assembly code. Notice that in mixing C with Assembly code, Assembly directives such as OFFSET in "MOV DX,OFFSET MESSAGE" are not recognizable by C.

Example 7-6 also shows in-line assembly. The in-line code sets the cursor at row = 10 and column = 20 and then displays a string of data using a combination of Borland C and Assembly.

SECTION 7.4: COMBINING ASSEMBLY LANGUAGE AND C PROGRAMS   **211**

Example 7-5

Solution A: A C language program

```
#include <stdio.h>
main ( )
{
        printf("The planet Earth. \n");
}
```

Solution B: Turbo C with in-line Assembly

```
char const *MESSAGE = "The planet Earth.\n$";
main ()
{
  asm   mov ah,9
  asm   mov dx, MESSAGE
  asm   int 21h
}
```

Example 7-6

Borland C and in-line Assembly code

```
main ()
  {
  int const row = 10;
  int const column = 20;
  char  const *MESSAGE = "The planet Earth. \n$";

  asm      {
  mov ah,2
  mov bh,0
  mov dl,column
  mov dh,row
  int 10h          /* set cursor position */
  mov ah,09
  mov dx,MESSAGE
  int 21h          /* display message */
          }
  }
```

C programs that call Assembly procedures

Although in-line assembly is fast and easy, in real-life applications it is common to write Assembly language subroutines and then make them available for C to call as if calling a C function. What is referred to in C language terminology as a *function* is called a *procedure (subroutine)* in Assembly language. Before embarking on writing Assembly routines to be used with C, one must first understand how parameters are passed from C to Assembly language. All high-level languages, such as C, BASIC, FORTRAN, and Pascal, pass parameters to subroutines (functions) that they are calling via the stack. Some of them pass the value itself (C, Pascal), while some others pass the address of the value (BASIC, FORTRAN). In BASIC, only the offset address is passed, while in FORTRAN both the segment and offset addresses are passed. Even the order in which they pass parameters differs among high-level languages. The terminology *calling convention* refers to the way that a given language passes parameters to the subroutines it calls. The following describes the C calling convention.

Example 7-7

```
extern cursor (int, int);
main ( )
{
cursor (15,12);
printf("This program sets the cursor");
}


;――――― in cursor.asm: ―――――――――――――
    .MODEL    SMALL
    .CODE
    PUBLIC    _CURSOR
;this procedure is written to be called by a C program
_CURSOR PROC
            PUSH   BP             ;save the BP (it is being altered)
            MOV    BP,SP          ;use BP as indexing into stack
            MOV    DH,[BP+4]      ;get the x (row) value from stack
            MOV    DL,[BP+6]      ;get the y (column) value from stack
            MOV    AH,02          ;set registers for INT call
            MOV    BH,00
            INT    10H
            POP    BP             ;restore BP
            RET
_CURSOR ENDP
```

C calling convention

How does C pass parameters to functions? It is extremely important to understand this since failure to do so can cause getting the wrong data from the stack when trying to access it through the Assembly subroutine. The following describes the C calling convention for mixing C with MASM Assembly language. An Assembly language procedure to be called by C must follow these rules:

1982		
1983		
1984	BP	SP=BP
1985		
1986	IP	BP +2 holds return address
1987		
1988	0F	BP+4 holds x value 15=FH
1989	00	
198A	0C	BP+6 holds y value 12=CH
198B	00	
198C		

Example 7-7 Stack Contents Diagram

1. The parameters are passed by value to the stack in reverse order of encountering them. For example, in the function prog (x,y,z), first z is passed, then y, and so on.
2. After parameters are passed in reverse order, C also saves the address (CS,IP). If C is compiled in the SMALL or COMPACT memory model (or if the procedure is NEAR) only the IP is saved. If C is compiled for MEDIUM, LARGE, or HUGE (or if the procedure is FAR), both CS and IP are passed to the stack (CS is passed first, then IP).
3. BP must be saved on the stack and then the parameters must be accessed by the BP register and displacement, since BP is the offset of the stack segment (SS) register.
4. The last instruction should be RET with no number after it, since it is the job of C to restore the stack to its original place when it takes back control.
5. Any name shared publicly with C must be prefaced with an underscore, and only the first eight characters of the name are recognized by C.
6. C passes the parameters by value except for arrays, which are passed by reference.
7. If C is compiled in the MEDIUM, HUGE, or LARGE model, use the FAR option for the Assembly language procedure. If C is compiled with the SMALL model, use the NEAR option for the Assembly language procedure.

To understand the concepts above, assume that there is a C function named cursor (x,y), where x and y are the column and row values, respectively. Example 7-7 shows how x and y are passed to the stack and then accessed by Assembly code. The step-by-step sequence of the stack contents is shown also.

When C calls the cursor function, it saves y first, then x, and then the return address IP, and finally, gives control to the Assembly code. The first instruction of the assembly procedure must be saving the BP register, "PUSH BP". The last instruction must be the RET instruction.

Notice that the first two instructions of the procedure must always be saving BP and moving SP to BP. Similarly, the last two instructions must be popping BP and RET. The body of Assembly code goes in between them. This way of accessing arguments in the stack is standard, and saving any other registers will have no effect on displacement calculation as long as the number of PUSH and POP instructions are equal.

Example 7-8 shows the cursor routine rewritten to save all registers altered by the routine. The stack contents analysis is shown in the diagram.

Example 7-8

```
        .MODEL SMALL
        .CODE
        PUBLIC  _CURSOR
        ;this procedure is written to be called by C language
_CURSOR PROC
        PUSH    BP              ;save the BP since contents are altered
        MOV     BP,SP
        PUSH    AX              ;push regs altered by this module
        PUSH    DX
        PUSH    BX
        MOV     DH,[BP+4]       ;get x the row value from stack
        MOV     DL,[BP+6]       ;get y the column value from stack
        MOV     AH,02           ;set up for INT call
        MOV     BH,00
        INT     10H
        POP     BX              ;restore registers
        POP     DX
        POP     AX
        POP     BP
        RET
_CURSOR ENDP
        END
```

How parameters are returned to C

In the preceding section we described how arguments are passed from C to the stack and from there to an Assembly procedure. What happens if a C function expects to receive an argument? When C expects an argument from an Assembly procedure, it expects to find the returned parameter in certain register(s), depending on the size of the parameters as shown in Table 7-1.

197E	BL	BX
197F	BH	
1980	DL	DX
1981	DH	
1982	AL	AX
1983	AH	
1984	BP	SP=BP
1985		
1986	IP	BP +2 holds return address
1987		
1988	0F	BP+4 holds x value 15=FH
1989	00	
198A	0C	BP+6 holds y value 12=CH
198B	00	

Example 7-8 Stack Contents Diagram

Table 7-1: Returned Values from Assembly Procedures

Register	Size	C Data Type
AL	1 byte	char, short
AX	2 bytes	int
DX:AX	4 bytes	long

In Table 7-1 the register indicates the register used for the return value. If the value returned is a pointer (address), AX will hold IP if it is NEAR and DX:AX will hold CS:IP if it is FAR. This is illustrated in Example 7-9, where the sum of x, y, and z is returned to C through DX and AX as expected by C. DX has the higher word and AX the lower word. In the stack frame illustration, first notice that since the procedure is FAR, both CS and IP are saved on the stack. Therefore, to access the C arguments, it is necessary to use BP+6, BP+8, and BP+10 displacements.

As a rule, if the Assembly procedure is NEAR, the last argument passed by C is accessed by the displacement of BP+4, and if it is FAR, it is accessed by BP+6 displacement. In order not to be bothered by these rules, new assemblers have become more user friendly, as shown in the next topic.

Example 7-9

Three values of int size are passed by a C function to an Assembly procedure. The assembly code adds them together and returns the total sum back to C, which displays the result.

```
extern unsigned long sum (int, int, int);
{
main()
printf("The sum  is equal to %u", sum (500,6500,200));
}
The following is sum.asm

            .MODEL MEDIUM
            .CODE
            PUBLIC _SUM
            ;this far procedure gets three words from the stack and adds
            ;them together.   At the end DX:AX has the total sum
_SUM        PROC    FAR
            PUSH    BP              ;save BP
            MOV     BP,SP           ;use it as SP
            SUB     AX,AX           ;clear AX
            MOV     DX,AX           ;and DX
            ADD     AX,[BP+6]       ;add the first
            ADC     DX,0            ;add the carry
            ADD     AX,[BP+8]       ;add the second
            ADC     DX,0            ;add the carry
            ADD     AX,[BP+10]      ;add the third
            ADC     DX,0            ;add the carry
            POP     BP              ;restore BP
            RET                     ;go back to C
_SUM        ENDP
            END
```

New assemblers and linking with C

In recent years some Assemblers have made linking with C much easier. Using MASM 5.1, or TASM 1.0 and higher, ends the need to worry about the displacement or about beginning the names common to C and Assembly with an underscore or about saving BP. The program in Example 7-9 is rewritten on the following page. Notice letter C in directive ".MODEL SMALL, C". This automatically makes the

17F0		
17F1		
17F2	BP	SP=BP
17F3		
17F4	IP	
17F5		
17F6	CS	
17F7		
17F8		x value pointed at by BP+6
17F9		
17FA		y value pointed at by BP+8
17FB		
17FC		z value pointed at by BP+10
17FD		

Example 7-9 Stack Contents Diagram

---

assembler calculate [BP+n] for all the parameters. Compare these two programs to see the convenience of the new assemblers. Example 7-9 in the new format follows.

```
              .MODEL MEDIUM, C
              .CODE
              PUBLIC    SUM
              ;this FAR procedure gets three words from the stack and adds
              ;them together.  At the end DX:AX has the total sum
SUM           PROC      FAR    DATA1:WORD, DATA2:WORD, DATA3:WORD
              SUB       AX,AX  ;CLEAR AX
              MOV       DX,AX  ;CLEAR DX
              ADD       AX,DATA1
              ADC       DX,0
              ADD       AX,DATA2
              ADC       DX,0
              ADD       AX,DATA3
              ADC       DX,0
              RET
SUM           ENDP
              END
```

Passing array addresses from C to the stack

The C language passes variables to the stack by value and arrays by a pointer. In other words, the offset address of the array is pushed onto the stack if the memory model is SMALL or MEDIUM; otherwise, both the segment and offset address of the array are pushed. Example 7-10 illustrates this point. It uses a C language array to define daily wages for a five-day week, using an unsigned int (0 to 255 range values) data definition. It then uses Assembly code to add them and return the total sum back to C to be displayed.

Example 7-10

```
int  wages [5] = {154, 169, 98, 129, 245};
extern unsigned short weekpay(int wages[]);
main()
{
printf("Weekly pay = %u", weekpay(wages));
}

weekpay.asm is as follows:
              .MODEL MEDIUM
              .CODE
              PUBLIC          _WEEKPAY
              ;this procedure adds five bytes together.
              ;At the end AX has the total sum
_WEEKPAY PROC FAR
              PUSH  BP              ;save BP
              MOV   BP,SP
              PUSH  SI              ;save SI
              SUB   AX,AX
              MOV   CX,5
              MOV   SI,[BP+6]
AGAIN:        ADD   AL,[SI]         ;add a day's wages
              ADC   AH,0
              INC   SI              ;increment pointer to next wage
              INC   SI
              LOOP  AGAIN
              POP   SI
              POP   BP
              RET
_WEEKPAY ENDP
              END
```

Linking Assembly language routines with C

The following steps describe how to link Borland C++ with MASM Assembly language routines.

1. Make sure that the Assembly language procedure declares the procedure as PUBLIC. The procedure name should begin with an underscore. For example, if the procedure is called "sum" in the C program, it should be "_sum" in the Assembly language routine. Make the Assembly language procedure NEAR for the small model and FAR for the medium model.
2. In the C program, declare the procedure as external.
3. Assemble the Assembly language program with MASM to produce the object file: for example, module1.obj.
4. Compile the C program to produce the object file: for example, prog1.obj.
5. Link them together to produce the executable file.

> C> link prog1.obj + module1.obj

In Borland C++, they can be linked together as follows:

> C> bcc prog1.c module1.asm

The "bcc" command will compile the C program. Use the TASM assembler to assemble the Assembly language program, and then link them together. Note that Borland C is case sensitive. If your procedure is called "_SUM" in the Assembly language program and "sum" in the C program, the linker will not be able to link them together. Make the procedure name lowercase in the Assembly language program. If you are using Borland C++, it is recommended not to use the "cpp" filename extension since this will cause the function name to be mangled, and therefore the linker will not be able to find the function. If you must use the "cpp" extension, you must compile the C program with the /S option to obtain an Assembly language listing, then see how the function name was mangled and use that name in your Assembly language program. For example, suppose that the function name "sum" was listed as "@sumq$iii". In that case, all references to "sum" in the Assembly language program will have to be changed to "@sumq$iii" in order to allow the program to be linked with the C++ program.

Review Questions

1. A C program can either call an Assembly language program or use _____ coding that inserts the Assembly language code into the C program.
2. Describe the C convention for passing parameters to functions.

SUMMARY

Modular programming involves breaking down a project into independent subprograms. Each subprogram accomplishes a specific set of tasks. Good programming practices dictate that the input and output variables to each subprogram be clearly documented. Variables within subprograms are defined by the EXTRN and PUBLIC directives. These provide the means by which the computer can locate variables. Various methods for passing parameters are used, including passing by register, by memory, or by the stack.

The C programming language has gained widespread popularity because of the ease with which code can be ported from one machine to another. However, it is often desirable to include Assembly language programs because of the increased speed that can be gained. Assembly language routines can be called by C programs, or the Assembly language code can be coded directly into the C program by a technique called in-line coding.

---

1. Fill in the blanks in the following program. The main program defines the data and calls another module to add 5 bytes of data, then saves the result. *Note:* Some blanks may not need anything.

```
                .MODEL SMALL
                .STACK 100H
                .DATA
                PUBLIC    _____,_____
DATA1           DB       25,12,34,56,98
RESULT          DW       ?
                .CODE
                EXTRN     _____:FAR
HERE:           MOV   AX,@DATA
                MOV   DS,AX
                CALL  SUM
                MOV   AH,4CH
                INT   21H
                END   _____
```

In another file there is the module for summing 5 bytes of data:

```
                .MODEL  SMALL
                _____DATA1:BYTE
                _____RESULT:WORD
COUNT           EQU     5
                .CODE
                _____SUM
SUM             PROC    _____
                MOV   BX,OFFSET DATA1
                SUB   AX,AX
                MOV   CX,COUNT
_____:         ADD   AL, BYTE PTR [BX]
                ADC   AH,0
                INC   BX
                LOOP  AGAIN
                MOV   RESULT,AX
                RET
_____          ENDP
                END   _____
```

2. If a label or parameter is not defined in a module, it must be declared as _____.
3. If a label or parameter is used by other modules, it must be declared as _____ in the present module.
4. List the options for the EXTRN directive when it is referring to a procedure.
5. List the options for the EXTRN directive when it is referring to a data item.
6. List the options for the PUBLIC directive when it is referring to a procedure.
7. List the options for the PUBLIC directive when it is referring to a data item.
8. Convert Program 4-1 to the modular format, making each of the INT subroutines a separate module. Each module should be NEAR. Assemble and test the program.
9. Write a program that accepts two unsigned numbers (each less than 999) from the keyboard, converts them to hex, takes the average, and displays the result on the monitor. Use the hex-to-decimal and ASCII-to-hex conversion modules in the text.
10. Write a program (similar to Program 7-1) with the following components.
    (a) In the main program, two values are defined: 1228 and 52400.
    (b) The main program calls two separate modules, passing the values by stack.
    (c) In the first module, the two numbers are multiplied and the result is passed back to the main module.

(d) The second module performs division of the two numbers (52400 /122) and passes both the quotient and remainder back to the main program to be stored.

(e) Analyze the stack and its contents for each module if SP = FFF8H immediately before the first CALL instruction in the main module.

11. Write an in-line assembly program to set the cursor to row 14, column 27, and then display the message "This is a test".

12. Modify Example 7-8 to add twelve monthly salaries. The total yearly salary cannot be higher than $65,535.

ANSWERS TO REVIEW QUESTIONS

SECTION 7.1: WRITING AND LINKING MODULES

1. (1) each module can be developed individually, allowing parallel development of modules, which shortens development time, (2) easier to locate source of bugs, (3) these modules can be linked with high-level languages such as C
2. PUBLIC   3. EXTRN
4. the module that is the entry and exit point will have a label after the END statement
5. a paragraph consists of 16 bytes and begins on an address ending in 0H
6. name   SEGMENT   PARA STACK 'STACK'   7. EXTRN   TOTAL:WORD
8. 60H   9. name   SEGMENT   PARA   PUBLIC 'CODE'

SECTION 7.2: SOME VERY USEFUL MODULES

1. 1st iteration: AX=F624   F624/A = 189D   remainder DL=2
   2nd iteration:AX=189D   189D/A=0276   remainder DL=1
   3rd iteration: AX=0276   0276/A=003F   remainder DL=0
   4th iteration:AX=003F   003F/A=6   remainder DL=3
   5th iteration:AX=0006   0006/A=0   remainder DL=6
   AX is now zero, so the conversion is complete: $F624H=63012_{10}$
2. 1st iteration: AL=36   06x1=6   DI=6
   2nd iteration:AL=35   05xA=32   DI=6+32=38
   3rd iteration:AL=34   04x64=190   DI=38+1C8
   4th iteration:AL=31   01x03E8=03E8   DI=1C8+03E8=05B0
   5th iteration:AL=30   0x2710=0   DI=05B0
   BX has been decremented from 4 to 0, is now -1, so the conversion is complete
   $01456_{10}=05B0H$

SECTION 7.3: PASSING PARAMETERS AMONG MODULES

1. (a) by register; one advantage is the execution speed of registers; one disadvantage is that there is a limited number of registers available so that not many values can be passed
   (b) by stack; one advantage is it does not use up available registers; one disadvantage is that errors in processing the stack can cause the system to crash
   (c) by memory; one advantage is a large area available to store data; one disadvantage is that the program would not be portable to other computers
2. (a) and (c) are correct, (b)and (d) are not correct because SP cannot be used in indexing mode with SS

SECTION 7.4: COMBINING ASSEMBLY LANGUAGE AND C PROGRAMS

1. in-line
2. parameters are passed by value except for arrays, which are passed by reference; parameters are passed in reverse order of argument list; after arguments are pushed onto the stack, C pushed CS:IP for FAR procedures or IP for NEAR procedures

CHAPTER 8

32-BIT PROGRAMMING FOR 386 AND 486 MACHINES

OBJECTIVES

Upon completion of this chapter, you will be able to:

» Discuss the major differences between the 8086/286 and the 80386/486

» List the registers of the 80386/486 machines

» Diagram the register sizes available in the 386/486

» Describe the difference between real and protected modes

» Explain the difference in register usage between the 8086/286 and 386/486

» Discuss how the increased register size of the 386/486 relates to an increased memory range

» Diagram how the "little endian" storage convention of 80x86 machines stores doubleword-sized operands

» Code programs for 386/486 machines using extended registers and new directives

» Code arithmetic statements using the extended registers of the 386/486

» Describe the factors resulting in the increased performance of 386/486 over previous generations of microprocessors

All programs discussed so far were intended for 16-bit machines such as 8088, 8086, and 80286 IBM and compatible computers. Although those programs will run on 80386- and 80486-based machines with much improved speed, the true power of 386 and 486 microprocessors shows up when they are switched to protected mode. What is protected mode? As mentioned in Chapter 1, the 386 and 486 can operate in two modes: real mode and protected mode. In real mode they function essentially the same as 8086/286 machines with the exception that 32-bit registers are available. They still have a capacity of addressing a maximum of 1 megabyte of memory. More important, they run all MS-DOS programs without any modification. In protected mode they can access up to 4 gigabytes of memory, but they also require a very complex operating system, one of whose tasks is to assign a privilege level of 0 to 3 (0 being the highest) to each program run on the CPU. At this time only the Unix operating system is taking advantage of protected mode in both the 386 and 486 microprocessors. In March 1992, IBM introduced OS/2 version 2.0, designed specifically for 386 systems. This is a 32-bit version of the OS/2 operating system written to take advantage of protected mode in 386 and higher microprocessors. This is in contrast to OS/2 version 1, which was designed for 16-bit 80286 protected mode. The combination of 32-bit processing power and an operating system with *multitasking* and *multiuser* capability makes the 386 and 486 computers comparable to the minicomputers of the 1970s at a fraction of the cost. The term *multiuser* refers to a system that can support more than one terminal/keyboard at a time. *Multitasking* refers to systems that can execute more than one program at a time.

In this chapter we discuss the characteristics of 386 and 486 microprocessors in real mode that affect programming. Then some program examples will be given that use the 32-bit capability of these machines. Finally, a timing comparison of several programs run across 80x86 machines will be given in order to appreciate the speed of 386/486-based computers. Protected mode and other capabilities of these processors are discussed further in Chapter 21.

SECTION 8.1: 80386/80486 MACHINES IN REAL MODE

In this section we concentrate on some of the most important differences between the 8086/286 and 386/486 in real mode. One major difference is the register size. While in the 8086/286 the maximum register size is 16 bits wide, in the 386/486 the maximum size of registers has been extended to 32 bits. All register names have been changed to reflect this extension. Therefore, AX has become EAX, BX is now EBX, and so on, as illustrated below and outlined in Table 8-1. For example, the 386/486 contains registers AL, AH, AX, and EAX with 8, 8, 16, and 32 bits, respectively. In the 86/286, register AX is accessible either as AL or AH or AX, while in the 386/486, register EAX can be accessed only as AL or AH or AX or EAX. In other words, the upper 16 bits of EAX are not accessible as a separate register. The same rule applies to EBX, ECX, and EDX. Registers DI, SI, BP, and SP have become EDI, ESI, EBP, and ESP, respectively. That means the 386/486 can access DI, EDI, SI, ESI, BP, EBP, SP, and ESP. All of these registers are accessible in real mode.

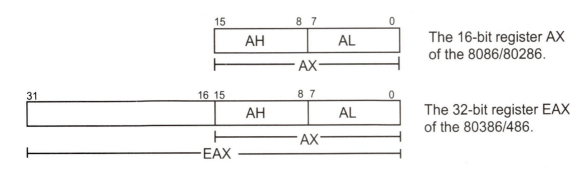

The 16-bit register AX of the 8086/80286.

The 32-bit register EAX of the 80386/486.

In addition to the CS, DS, SS, and ES segment registers, there are also two new segment registers which are accessible in real mode: FS and GS. With the addition of these new segment registers, there are a total of six segment registers, making it possible to access 384K bytes ($6 \times 64 = 384$), since each segment register can access up to 64K bytes of memory. Again, all these registers are accessible in real mode. Although both the flag register and IP are extended to 32 bits, only the lower 16 bits are available in real mode. To access all 32 bits of these registers, one must switch to protected mode. There are several control registers (CR0, CR1, CR2, and CR3) in protected mode but only bit 0 of CR0 is available in real mode. Bit 0 of CR0 is the protection-enable bit. When power is applied to the 386/486, it selects real mode automatically and PE (bit 0 of CR0) is low. To switch from real mode to protected mode, this bit must be set to 1. Again, only the Unix and OS/2 operating systems use protected mode at this time. MS-DOS, up to version 5, is not using the protected mode of the 386/486. This might change with future versions of MS-DOS.

Table 8-1: Registers of the 80386/486 by Category

Category	Bits	Register Names
General	32	EAX,EBX,ECX,EDX
	16	AX,BX,CX,DX
	8	AH,AL,BH,BL,CH,CL,DH,DL
Pointer	32	ESP (extended SP), EBP (extended BP)
	16	SP (stack pointer), BP (base pointer)
Index	32	ESI (extended SI), EDI (extended DI)
	16	SI (source index), DI (destination index)
Segment	16	CS (code segment), DS (data segment), SS (stack segment), ES (extra segment) FS (extra segment), GS (extra segment)
Instruction	32	EIP (extended instruction pointer)
Flag	32	EFR (extended flag register)
Control	32	CR0, CR1, CR2, CR3

Note: Only bit 0 of CR0 is available in real mode. All other control registers are available in protected mode only.

General registers are pointers in 386/486

Another major change from 86/286 to 386/486 is the ability of general registers such as EAX, ECX, and EDX to be used as pointers. As shown in previous chapters, AX, CX, and DX could not be used as pointers. For example, an instruction such as "MOV CL,[AX]" would have cause an error in the 86/286 CPU since only BX, SI, DI, and BP were allowed to be used as pointers to memory. This has changed starting with the 80386 microprocessor. In the 386/486 CPU, the following instructions are perfectly legal:

```
MOV     AX,[ECX]
ADD     SI,[EDX]
OR      EBX,[EAX]+20
```

It must be noted that when EAX, ECX, or EDX are used as offset addresses, DS is the default segment register. That means that SS is the default segment register for ESP and EBP, CS for EIP, and DS for all other registers. The segment override symbol (:) can be used to change the default segment register as shown next.

```
MOV        AX,FS:[ECX]    ;move contents of FS:ECX to AX
```

Calculation of physical addresses in real mode is the same as for the 86/286 as discussed in Chapter 1. In the example above, the physical addresses can be calculated by shifting left the segment register FS one hex digit and adding it to offset ECX. For example, if FS = 12E0 and ECX = 00000120, the physical address specified by FS:ECX would be 14000H (12E00 + 0120 = 14000H).

Table 8-2 summarizes addressing modes for the 386/486. There are additional addressing modes available for 386 and higher CPUs which will be covered in future volumes.

Table 8-2: Addressing Modes for the 80386/486

Addressing Mode	Operand	Default Segment
Register	register	none
Immediate	data	none
Direct	[OFFSET]	DS
Register indirect	[BX]	DS
	[SI]	DS
	[DI]	DS
	[EAX]	DS
	[EBX]	DS
	[ECX]	DS
	[EDX]	DS
	[ESI]	DS
	[EDI]	DS
Based relative	[BX]+disp	DS
	[BP]+disp	SS
	[EAX]+disp	DS
	[EBX]+disp	DS
	[ECX]+disp	DS
	[EDX]+disp	DS
	[EBP]+disp	SS
Indexed relative	[DI]+disp	DS
	[SI]+disp	DS
	[EDI]+disp	DS
	[ESI]+disp	DS
Based indexed relative	[R1][R2]+disp where R1 and R2 are any of the above	If BP is used, segment is SS; otherwise, DS is the segment

Note: In based indexed relative addressing, disp is optional.

386/486 maximum memory range in real mode: 1M

There is a dilemma in the 386/486 working under DOS in real mode. If in real mode the maximum range of memory is 1M (00000 to FFFFFH), what happens if a 32-bit register is used as an offset into the segment? The range of the 32-bit offset is 00000000H to FFFFFFFFH; therefore, using a 32-bit register as an offset will place the address range beyond 1M. This is a situation that the programmer must avoid in 386/486 computers working with DOS version 5 and lower. For example, to execute the instruction "MOV AX,[ESI]", the programmer must make the upper 16 bits of ESI all zeros. This means that the segment offset range in real mode 386/486 under DOS is 0000 to FFFFH. The following are some cases of legal and illegal codings for real mode of the 386/486 under DOS.

```
ADD     EAX,[BX]            ;LEGAL
ADD     ECX,[DX]            ;ILLEGAL! EDX CAN BE USED
                            ;AS POINTER BUT NOT DX
MOV     AX,WORD PTR [ECX]   ;LEGAL
ADD     EBX,[EDX]           ;LEGAL
```

Accessing 32-bit registers with commonly used assemblers

In Assembly language the directive ".386" is used to access the 32-bit registers of 386/486 computers in real mode under DOS and to employ the new instructions of the 386 microprocessor. Every new generation of 80x86 has some new instructions that do not execute on lower processors, meaning that they are upwardly compatible. In other words, using the ".386" directive in a program means that the program must be run only on 386 and higher (486 and 586) computers and cannot be run on 8088/86- and 286-based computers. In contrast, all programs in previous chapters were written to be run on any 80x86 computer. The following are additional assembler directives, which indicate the type of microprocessor supported by Microsoft's assembler (MASM) and Borland's Turbo assembler (TASM).

MASM	TASM	Meaning
.86	P8086	will run on any 80x86 computer (default)
.286	P286	will run on any 286 and higher computer; also allows use of new 286 instructions
.386	P386	will run on any 386 and higher computer; also allows use of new 386 instructions
.486	P486	will run on any 486 and higher computer; also allows use of new 486 instructions

Program 8-1 demonstrates the use of the ".386" directive and the 80386 32-bit instructions. The simplified segment definition was used. The program used the 32-bit register EAX to add and subtract values of various size to demonstrate 32-bit programming of the 386/486 under DOS. Now the question is how to run this program and see the register contents in 386/486 machines. Unfortunately, the DEBUG utility used in earlier chapters cannot be used since it shows only the 16-bit registers. In many assemblers, including MASM and TASM, there are advanced debugging tools that one can use to see the execution of the 386/486 programs. In the case of MASM, the CodeView utility, and for TASM, the Turbo Debugger, are tools that allow one to monitor the execution of the 386/486 in addition to 8086/88 and 286 programs.

Below is shown a trace of Program 8-1, using Microsoft's CodeView program. To examine the 32-bit registers in CodeView, press F2 to display registers; then select the Options menu from the top of the screen, and a drop-down menu appears. Select "386" from the drop-down menu to display the registers in 32-bit format.

Write a program using the 32-bit registers of the 386 to add the values 100000, 200000, and 40000. Then subtract from the total sum the values 80000, 35000, and 250. Place the result in memory locations allocated using the DD, doubleword directive, used for 32-bit numbers.

```
TITLE     ADD AND SUBTRACT USING 32-BIT REGISTERS IN 386 MACHINES
PAGE      60,132
          .MODEL SMALL
          .386
          .STACK  200H
          .DATA
RESULT    DD   ?
          .CODE
BEGIN:    MOV      AX,@DATA
          MOV      DS,AX
          SUB      EAX,EAX
          ADD      EAX,100000       ;EAX = 186A0H
          ADD      EAX,200000       ;EAX = 186A0H + 30D40 H =  493E4H
          ADD      EAX,40000        ;EAX = 493E4H + 9C40H = 53020H
          SUB      EAX,80000        ;EAX = 53020H - 13880H = 3F7A0H
          SUB      EAX,35000        ;EAX = 3F7A0H - 88B8H = 36EE8H
          SUB      EAX,250          ;EAX = 36338H - FAH = 36DEEH = 224750 decimal
          MOV      RESULT,EAX
          MOV      AH,4CH
          INT      21H
          END      BEGIN
```

Program 8-1

```
 File   View   Search   Run   Watch   Options   Language   Calls   Help | F8=Trace F5=Go

 4833:0000 B83648        MOV       AX,4836                       EAX=00036DEE
 4833:0003 8ED8          MOV       DS,AX                         EBX=00000000
 4833:0005 662BC0        SUB       EAX,EAX                       ECX=00000000
 4833:0008 6605A0860100  ADD       EAX,000186A0                  EDX=00000000
 4833:000E 6605400D0300  ADD       EAX,00030D40                  ESP=00000200
 4833:0014 6605409C0000  ADD       EAX,00009C40                  EBP=00000000
 4833:001A 662D80380100  SUB       EAX,00013880                  ESI=00000000
 4833:0020 662DB8880000  SUB       EAX,000088B8                  EDI=00000000
 4833:0026 662DFA000000  SUB       EAX,000000FA                   DS=....4836
 4833:002C 66A30400      MOV       DWord Ptr [0004],EAX           ES=....4823
 4833:0030 B44C          MOV       AH,4C                          FS=....0000
 4833:0032 CD21          INT       21                             GS=....0000
 4833:0034 0000          ADD       Byte Ptr [BX+SI],AL            SS=....4837
 4833:0036 0000          ADD       Byte Ptr [BX+SI],AL            CS=....4833
 4833:0038 0000          ADD       Byte Ptr [BX+SI],AL            IP=0000002C
 4833:003A 0000          ADD       Byte Ptr [BX+SI],AL
 4833:003C 0000          ADD       Byte Ptr [BX+SI],AL
 4833:003E 0000          ADD       Byte Ptr [BX+SI],AL            NV UP
                                                                 EI NG
                                                                 NZ AC
 >t                                                              PE NC
 >t
 >t                                                              DS:0004
 >                                                               00000000
```

Program 8-1: CodeView Screen of Program Execution

Little endian revisited

In analyzing how the 386/486 stores 32-bit data in memory or loads a 32-bit operand into a register, recall the little endian convention: The low byte goes to the low address and the high byte to the high address. For example, an instruction such as "MOV RESULT,EAX" in Program 8-1 will store the data in this way:

OFFSET	CONTENTS
RESULT	d0-d7
RESULT+1	d8-d15
RESULT+2	d16-d23
RESULT+3	d24-d31

Example 8-1

Assuming that SI=1298 and EAX = 41992F56H, show the contents of memory locations after the instruction "MOV [SI],EAX".

Solution: (in HEX)
  DS:1298 = (56)
  DS:1299 = (2F)
  DS:129A = (99)
  DS:129B = (41)

Review Questions

1. In the 80386/486, the bits of register EDX can be accessed either as DL, bits __ to __, or DH, bits __ to __, or DX, bits __ to __, or EAX, bits __ to __.
2. In the 386/486 segment size is ____ bytes.
3. What is real mode? What is protected mode?
4. True or false: The instruction "MOV DX,[AX]" is illegal in the 8086 but "MOV DX,[EAX]" is legal in the 386/486.
5. What is the default segment register when EAX is used as a pointer?
6. What is the purpose of the ".386" directive?
7. If DI = 148F and EBX = 6B2415F9, show the contents of memory after the instruction "MOV [DI],EBX" is executed.

SECTION 8.2: SOME SIMPLE 386/486 PROGRAMS

One way to increase the processing power of the microprocessor is to widen the register size. This allows processing large numbers as a whole rather than breaking them into smaller chunks to fit into small registers. The 32-bit registers have become standard in all recent microprocessors. Powerful supercomputers use 64-bit registers. In this section we show revisions of some earlier programs using the 32-bit capability of the 386/486 machines to see the impact of the wider registers in programming. By comparing 32-bit versions of these programs with the 16-bit versions, one can see the increased efficiency of 32-bit coding. The impact on speed is discussed in the final section.

Adding 16-bit words using 32-bit registers

Program 3-1b used 16-bit registers for adding several words of data. The sum was accumulated in one register and another register was used to add up the carries. This is not necessary when using 32-bit registers. First, refresh your memory by looking at Program 3-1b and then examine Program 8-2, a 32-bit version of the same program, written for 386/486 CPUs.

CHAPTER 8: 32-BIT PROGRAMMING FOR 386 AND 486 MACHINES

```
TITLE       REVISION OF PROGRAM 3-1B USING 32-BIT REGISTERS
PAGE        60,132
            .MODEL SMALL
            .386
            .STACK 200H
            .DATA
DATA1       DD  27345,28521,29533,30105,32375
SUM         DD  ?
COUNT       EQU    5
            .CODE
BEGIN:      MOV     AX,@DATA
            MOV     DS,AX
            MOV     CX,COUNT            ;CX is loop counter
            MOV     SI,OFFSET DATA1     ;SI is data pointer
            SUB     EAX,EAX             ;EAX will hold sum
BACK:       ADD     EAX,DWORD PTR[SI]   ;add next word to EAX
            ADD     SI,4                ;SI points to next dword
            DEC     CX                  ;decrement loop counter
            JNZ     BACK                ;continue adding
            MOV     SUM,EAX             ;store sum
            MOV     AH,4CH
            INT     21H
            END     BEGIN
```

Program 8-2

Rewrite Program 3-2 in Chapter 3 to add two 8-byte operands using 32-bit registers.

```
TITLE       ADD TWO 8-BYTE NUMBER USING 32-BIT REGISTERS IN THE 386
PAGE        60,132
            .MODEL SMALL
            .386
            .STACK  200H
            .DATA
DATA1       DQ  548FB9963CE7H
            ORG    0010H
DATA2       DQ  3FCD4FA23B8DH
            ORG    0020H
DATA3       DQ  ?
            .CODE
BEGIN:      MOV     AX,@DATA
            MOV     DS,AX
            CLC                         ;clear carry before first addition
            MOV     SI,OFFSET DATA1     ;SI is pointer for operand1
            MOV     DI,OFFSET DATA2     ;DI is pointer for operand2
            MOV     BX,OFFSET DATA3     ;BX is pointer for the sum
            MOV     CX,02               ;CX is the loop counter
BACK:       MOV     EAX,DWORD PTR [SI]  ;move the operand to EAX
            ADC     EAX,DWORD PTR [DI]  ;add the operand to EAX
            MOV     DWORD PTR [BX],EAX  ;store the sum
            INC     SI                  ;point to next dword of operand1
            INC     SI
            INC     SI
            INC     SI
            INC     DI                  ;point to next dword of operand2
            INC     DI
            INC     DI
            INC     DI
            INC     BX
            INC     BX                  ;point to next dword of sum
            INC     BX
            INC     BX
            LOOP    BACK                ;if not finished, continue adding
            MOV     AH,4CH
            INT     21H                 ;go back to DOS
            END     BEGIN
```

Program 8-3a

SECTION 8.2: SOME SIMPLE 386/486 PROGRAMS   227

Adding multiword data in 386/486 machines

In Program 3-2, two multiword numbers were added using 16-bit registers. Each number could be as large as 8 bytes wide. That program required a total of four iterations. Using the 32-bit registers of the 386/46 requires only two iterations, as shown in Program 8-3a on the previous page. This loop version of the multiword addition program is very long and inefficient. It can be made more efficient by saving the flag register that holds the carry bit of the first 32-bit addition on the stack and then adding four to each pointer instead of incrementing the pointers 4 times. The loop is shown below in Program 8-3b.

```
;this revision of Program 3-1a shows how to save the flags
;     before updating the pointers
    BACK:   MOV     EAX,DWORD PTR [SI]      ;move the operand to EAX
            ADC     EAX,DWORD PTR [DI]      ;add the  operand to EAX
            MOV     DWORD PTR [BX],EAX      ;store the sum
            PUSHF                           ;save the flags
            ADD     SI,4                    ;point to next dword of operand1
            ADD     DI,4                    ;point to next dword of operand2
            ADD     BX,4                    ;point to next dword of sum
            POPF                            ;restore the flags
            LOOP    BACK                    ;if not finished, continue adding
```

Program 8-3b

Due to the high penalty associated with branch instructions such as the LOOP and Jcondition instructions in the 386/486, it is better to use the nonloop version of this program, shown in Program 8-4.

First notice that the data is stored exactly the same way as in the loop version of the program. Data directive DQ is used to set up storage for the 8-byte numbers. First, the lower dword (4 bytes) of DATA1 is moved into EAX, and the lower dword of DATA2 is added to EAX. Then the upper dword of DATA1 is moved into EBX, and the upper dword of DATA2 is added to EBX, with any carry that may have been generated in the addition of the lower dwords. EAX now holds the lower 4 bytes of the result, and EBX holds the upper 4 bytes of the result.

Program 8-4 is much more efficient than using the loop concept. To see why and for a discussion of the impact of branching instructions on the performance of programs in the 80386/486, see Section 8.3.

```
    TITLE       ADD TWO 8-BYTE NUMBERS USING 32-BIT REGISTERS IN THE 386  (NO-LOOP VERSION)
    PAGE        60,132
                .MODEL SMALL
                .386
                .STACK  200H
                .DATA
    DATA1       DQ  548FB9963CE7H
                ORG     0010H
    DATA2       DQ      3FCD4FA23B8DH
                ORG     0020H
    DATA3       DQ ?
                .CODE
    BEGIN:  MOV     AX,@DATA
            MOV     DS,AX
            MOV     EAX,DWORD PTR DATA1     ;move lower dword of DATA1 into EAX
            ADD     EAX,DWORD PTR DATA2     ;add lower dword of DATA2 to EAX
            MOV     EBX,DWORD PTR DATA1+4   ;move upper dword of DATA1 into EBX
            ADC     EBX,DWORD PTR DATA2+4   ;add upper dword of DATA2 to EBX
            MOV     DWORD PTR DATA3,EAX     ;store lower dword of result
            MOV     DWORD PTR DATA3+4,EBX   ;store upper dword of result
            MOV     AH,4CH
            INT     21H
            END     BEGIN
```

Program 8-4

Multiplying a 32-bit operand by a 16-bit operand in the 386/486

As a final example of the power of 32-bit registers in 386/486 machines, in this section we look at the multiplication of a 32-bit operand by a 16-bit operand. See Figure 8-1. Comparing the 386/486 version with the 86/286 version of this program clearly reveals the coding efficiency of the 32-bit register of the 386/486 systems. First look at the 386/486 version of the multiplication of a 32-bit operand by a 16-bit operand, shown in Program 8-5. Notice that the 16-bit operand is placed in a 32-bit register in order to perform 32-bit arithmetic.

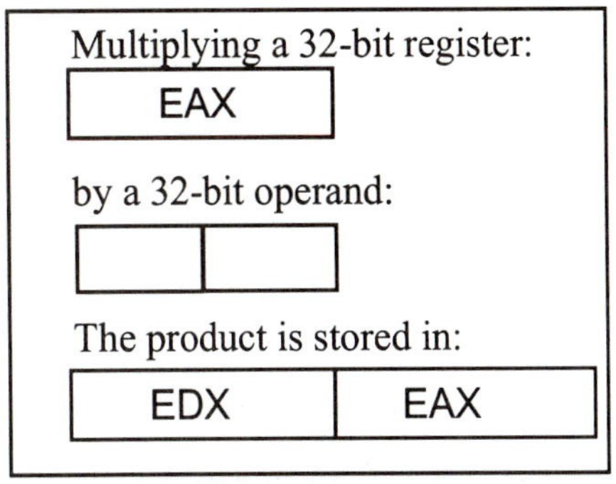

```
Multiplying a 32-bit register:
    EAX

by a 32-bit operand:


The product is stored in:
    EDX          EAX
```

Figure 8-1. 386/486 Multiplication

Write a program using the 386/486 to multiply a 32-bit operand by a 16-bit operand.

```
        TITLE       MULTIPLICATION OF DOUBLE WORD BY WORD USING 386/486
        PAGE        60,132
                    .MODEL SMALL
                    .386
                    .STACK   200H
                    .DATA
DATA1   DD  500000      ;MULTIPLICAND (UP TO 32-BIT SIZE DATA)
DATA2   DD  50000       ;MULTIPLIER (UP TO 16-BIT SIZE)
RESULT  DQ  ?                   ;PRODUCT (UP TO 48-BIT SIZE)
                    .CODE
MAIN:   MOV     AX,@DATA
        MOV     DS,AX
        MOV     EAX,DATA1                   ;32-BIT OPERAND
        MUL     DWORD PTR DATA2             ;TIMES 16-BIT OPERAND
        MOV     DWORD PTR RESULT,EAX        ;SAVE THE RESULT
        MOV     DWORD PTR RESULT+4,EDX
        MOV     AH,4CH
        INT     21H
        END     MAIN
```

Program 8-5

To appreciate the processing power of the 32-bit registers of the 386/486, the next topic shows a revision of Program 8-5, using the 16-bit registers of the 8086/286 processors.

32-bit by 16-bit multiplication using 8086/286 registers

For the sake of clarity in the following discussion, word size W1 and W2 will be used to represent the 32-bit multiplicand. To multiply that by the 16-bit operand W3, the following algorithm must be followed. Assume that all values are in hex.

```
        W2 W1              multiplicand
    x      W3              multiplier
    ---------------
        W3xW1              a 32-bit result
+   W3xW2                  a 32-bit result must be shifted left one hex
    ---------------        position, then added
        X3 X2 X1           a 48-bit result (X1, X2, X3 are word size)


W1 DW_____
W2 DW_____    ;W2:W1 IS A 32-BIT MULTIPLICAND
W3 DW_____    ;W3 IS A 16-BIT MULTIPLIER
```

```
            X1  DW ?
            X2  DW ?      ;X3X2X1 THE 48-BIT PRODUCT RESULT
            X3  DW ?
                ...
                MOV    AX,W1    ;GET THE LOW WORD
                MUL    W3       ;MULTIPLY
                MOV    X1,AX    ;SAVE THE LOW WORD OF THE PRODUCT
                MOV    X2,DX    ;SAVE THE HIGH WORD OF THE PRODUCT
                MOV    AX,W2    ;GET THE HIGH WORD
                MUL    W3       ;MULTIPLY
                ADD    X2,AX    ;ADD THE MIDDLE 16-BIT WORD
                ADC    DX,0     ;PROPAGATE THE CARRY TO DX
                MOV    X3,DX    ;SAVE THE HIGH WORD RESULT
```

Now after understanding the process above, Program 8-6 will show how it is actually coded. First, a DD directive is used to define a 32-bit data instead of using DW twice. Similarly, since there is no directive to define a 48-bit data, the DQ (define quad word) directive is used, which defines a 64-bit operand. The unused bits become zeros.

```
TITLE      MULTIPLICATION OF DOUBLEWORD BY WORD USING 86/286
PAGE       60,132
           .MODEL SMALL
           .STACK  200H
           .DATA
DATA1      DD  500000              ;MULTIPLICAND (UP TO 32-BIT SIZE DATA)
DATA2      DW 50000                ;MULTIPLIER (UP TO 16-BIT SIZE)
RESULT     DQ  ?                   ;PRODUCT (UP TO 48-BIT SIZE)
           .CODE
MAIN:      MOV    AX,@DATA
           MOV    DS,AX
           MOV    AX,WORD PTR DATA1       ;GET LOW WORD OF MULTIPLICAND
           MUL    WORD PTR DATA2          ;MULTIPLY THE MULTIPLIER
           MOV    WORD PTR RESULT,AX      ;SAVE LOW WORD OF THE PRODUCT
           MOV    WORD PTR RESULT + 2,DX  ;SAVE MIDDLE WORD OF PRODUCT
           MOV    AX,WORD PTR DATA1 + 2   ;GET THE HIGH WORD OF MULTIPLICAND
           MUL    WORD PTR DATA2          ;MULTIPLY THE MULTIPLIER
           ADD    WORD PTR RESULT + 2 ,AX ;ADD THE MIDDLE 16-BIT WORD
           ADC    DX,0                    ;PROPAGATE THE CARRY TO DX
           MOV    WORD PTR RESULT + 4,DX  ;SAVE THE HIGH WORD RESULT
           MOV    AH,4CH
           INT    21H
           END    MAIN
```

Program 8-6

Comparing these two programs, one can see why the 32-bit registers have become the standard for all new generations of microprocessors.

Review Questions

1. Compare the number of iterations for adding two 8-byte numbers for the following CPUs.
   (a) 8085 (a 8-bit) CPU
   (b) 8086/88/286
   (c) 386/486
   (d) Cray supercomputer (64-bit system)
2. What data directives are used to define 32-bit and 64-bit operands?
3. What directive is used in MASM to inform the assembler that the program is using 386 instructions? In TASM?

SECTION 8.3: 80x86 PERFORMANCE COMPARISON

The newer generations of the 80x86 family not only have powerful mainframe features such as protection capabilities, but they also execute the instructions of previous generations much faster. In the preceding section it was seen how efficient the 32-bit coding can be. In this section we compare the performance of the 80x86 family of processors. To do that, the number of clock cycles (ticks) that each instruction of a given program takes to execute for the 8086, 286, 386, and 486 CPUs will be examined. To fully understand the remaining material in this section, it is necessary to review the introduction to Appendix B, Section B.2. The clock cycles table in Appendix B does not show the total time taken for each processor to execute. This is because such a calculation in terms of microseconds or milliseconds depends on the hardware design of the system, primarily on the factors of working frequency and memory design. Further discussion of the hardware and its impact on the performance of the computer can be found in Chapter 21. Comparing the performance of the 80x86 family can take one of two approaches:

1. Taking a program written for the 8086 and calculating the number of clocks taken to execute it, unchanged, on each of Intel's 8086, 80286, 386, and 486 microprocessors.

2. Modifying the same program for 32-bit processing of the 386/486 and then calculating the total number of clocks to execute it for the 386 and 486 microprocessors.

Running an 8086 program across the 80x86 family

Intel has employed some very advanced techniques in pipelining to enhance the processing power of 386 and 486 microprocessors. For example, many instructions that took four or five clocks to execute on the 8086, take only one or two clocks on the 386 and 486 machines. This is shown next.

Problem 3-1b showed a program that calculated the total sum of five words of data. Since the loop is the most time-consuming part of this program, below is shown a comparison of the total number of clocks taken for one iteration across all of Intel's 80x86 processors. The number of clocks for each instruction is taken from Appendix B.

		8086	286	386	486
BACK: ADD	AX,[SI]	14	7	6	2
ADC	BX,0	4	3	2	1
INC	SI	2	2	2	1
INC	SI	2	2	2	1
DEC	CX	2	2	2	1
JNZ	BACK	16/4	7/3	7/3	3/1
total clocks per iteration		40	23	21	9

To calculate the total clocks for the five iterations, simply multiply the total number of clocks for one iteration by 5 and then adjust it for the last iteration, since the number of clocks for no jump in the last iteration is less than for jump in previous iterations. To adjust it, subtract the difference between the jump and no jump clock. For example, the total clocks for the five iterations in the 8086 column is $40 \times 5 = 200$, so adjusting for the last iteration involves subtracting 12 ($16 - 4 = 12$) from 200, which results in 188 clocks. The same procedure followed for the 80286 results in total clocks of $(23 \times 5) - 4 = 111$.

The data above shows clearly the power of the newer generation of the 80x86 family. The same program originally written for 8086 machines runs twice as fast on Intel's 386 and four times faster on the 486 microprocessor. This plus the fact that 386/486 microprocessors have 32-bit data buses transferring data in and out of the CPU makes the case for the 386/486 even stronger. Now the question is how much faster will it run if the program is rewritten to utilize the 32-bit processing power of the 386/486 CPU. This is shown next. Program 8-2 showed the 32-bit version of the same program. The iteration section of the program and the number of clocks for the 386 and 486 are as follows:

			386	486
BACK:	ADD	EAX,DWORD PTR [SI]	6	2
	ADD	SI,4	2	1
	DEC	CX	2	1
	JNZ	BACK	7/3	3/1
total clocks per iteration			17	7

Multiplying for the five iterations gives $(5 \times 17) - 4 = 81$ clocks for the 386 and $(5 \times 7) - 2 = 33$ clocks for the 486. By comparing the results above, one might conclude that rewriting all 16-bit programs for the 32-bit registers of the 386/486 is about 25% faster. That is true for some applications but not all. For example, examine the case of adding multibyte operands shown next. Compare the performance of Program 3-2 run across 80x86 CPUs. The following times assumed two clock cycles for m (memory fetch).

		8086	286	386	486
BACK:	MOV AX,[SI]	10	5	4	1
	ADC AX,[DI]	14	7	6	2
	MOV [BX],AX	10	3	2	1
	INC SI	2	2	2	1
	INC SI	2	2	2	1
	INC DI	2	2	2	1
	INC DI	2	2	2	1
	INC BX	2	2	2	1
	INC BX	2	2	2	1
	LOOP BACK	17/5	(10)/4	(13)/11	7/6
total per iteration		53	37	37	17
for four iterations:		200	142	146	67

Now compare this result with the 32-bit version of the same program. The modified version was discussed in the preceding section (Program 8-4) and the clock count is as follows:

	386	486
MOV EAX,DWORD PTR DATA1	4	1
ADD EAX,DWORD PTR DATA2	6	2
MOV EBX,DWORD PTR DATA1+4	4	1
ADC EBX,DWORD PTR DATA2+4	6	2
MOV DWORD PTR DATA3,EAX	2	1
MOV DWORD PTR DATA3+4,EBX	2	1
total clocks for the entire operation:	24	8

In the 386, the clock count is reduced from 146 to 24, and for the 486 it is reduced from 67 to 8. Using the same hardware but changing the software to take advantage of the 32-bit capability of the 386/486 sped up the processing power 6-fold (146 divided by 24) and 8.5-fold (67 divided by 8), respectively, for the 386 and 486.   The discussion above clearly indicates there is a very heavy penalty associated with branching in the advanced processors. It also shows that if one wants to spend the resources and rewrite the old 16-bit programs for the new 32-bit architecture, it can be well worth the effort.

1. Compare the execution clocks (refer to Appendix B) for "ADD BX,AX" (ADD reg,reg) for the 8086, 286, 386, and 486 CPUs.
2. Using the 8086 as a base, show the increase in speed as a percentage for each processor in Question 1.
3. For instruction JNZ, compare the branch penalty for the 8086, 286, and 386. Assume that m=2.
   (a) in clocks   (b) in percent (notake/take $\times$ 100)

SUMMARY

The 386/486 CPUs represent a major inprovement over the 8086/286 in several areas. Not only are they much faster in terms of execution speed, they also have much better processing power because they are 32-bit machines. They are also capable of accessing an address range of 4 gigabytes of memory. The 386/486 was designed in such a way that all programs written for the 8086/286 will run on it with no modification. Other changes in the 386/486 include two additional segment registers and the ability to use general registers as pointers.

PROBLEMS

1. In a 386/486 program, show the content of each register indicated in parentheses after execution of the instruction.
   (a) MOV EAX,9823F4B6H   (AL,AH,AX and EAX)
   (b) MOV EBX,985C2H   (BL,BH,BX,EBX)
   (c) MOV EDX,2000000   (DL,DH,DX,EDX)
   (d) MOV ESI,120000H   (SI,ESI)
2. Show the destination and its contents in each of the following cases.
   (a) MOV   EAX,299FF94H
       ADD   EAX,34FFFFH
   (b) MOV   EBX,500 000
       ADD   EBX,700 000
   (c) MOV   EDX,40 000 000
       SUB   EDX,1 500 000
   (d) MOV   EAX,39393834H
       AND   EAX,0F0F0F0FH
   (e) MOV   EBX,9FE35DH
       XOR   EBX,0F0F0F0H
3. Using the little endian convention show the contents of the destination in each case.
   (a) MOV [SI],EAX   ;ASSUME SI=2000H AND EAX=9823F456H
   (b) MOV [BX],ECX   ;ASSUME BX,348CH AND ECX=1F23491H
   (c) MOV EBX,[DI]   ;ASSUME DI=4044H WITH THE
                               ;FOLLOWING DATA.   ALL IN HEX.

   DS:4044=(92)
   DS:4045=(6D)
   DS:4046=(A2)
   DS:4047=(4C)
4. Compare the clock count for the 80x86 microprocessor in each case:
   (a) Write a program for the 8086/286 to transfer 50 words of data, one word
       (16 bits) at a time. Do not use string instructions.
   (b) Modify the program in part (a) to transfer 2 words (32 bits) at a time and
       calculate the clock count for the 386 and 486.
   In both parts (a) and (b), the clock count should be calculated for one iteration and
   for all iterations.

5. Instruction DAA, described in Chapter 3, works only on the AL register, regardless of which of Intel's 80x86 microprocessors is used. Write a program that adds two multibyte packed BCD numbers, each 10 bytes wide (use the DT directive) and compare the clock count for one iteration if it is run on Intel's 8086, 286, 386, and 486 CPUs.

ANSWERS TO REVIEW QUESTIONS

SECTION 8.1: 80386/80486 MACHINES IN REAL MODE
1. 0 to 7, 8 to 15, 0 to 15, 0 to 31
2. 64K bytes
3. real mode is similar to the operation of 8086 machines; protected mode assigns a priority to programs and has other advanced features that take advantage of the 386's power
4. true
5. DS
6. allows use of 386 instructions
7. DS:148F = F9,   DS:1490 = 15,   DS:1491 = 24,   DS:1492 = 6B

SECTION 8.2: SOME SIMPLE 386/486 PROGRAMS
1. (a) 8; (b) 4; (c) 2; (d) 1
2. DD, DQ
3. .386, P386

SECTION 8.3: 80X86 PERFORMANCE COMPARISON
1. 8086: 3   286: 2   386: 2   486: 1
2. 286 is a 33% improvement over 8086
   386 is a 0% improvement over 286
   486 is a 100% improvement over 386
3. (a) 8086: 8   286: 6   386: 6
   (b) 8086: 400   286: 300   386: 300

CHAPTER 9

8088, 80286 MICROPROCESSORS AND ISA BUS

OBJECTIVES

Upon completion of this chapter, you will be able to:

>> State the function of the pins of the 8088
>> List the functions of the 8088 data, address, and control buses
>> State the differences in the 8088 microprocessor in maximum mode versus minimum mode
>> Describe the function of the pins of the 8284 clock generator chip
>> Describe the function of the pins of the 8288 bus controller chip
>> Explain the role of the 8088, 8284A, and 8288 in the PC
>> Explain how bus arbitration between the CPU and DMA is accomplished
>> State the function of the pins of the 80286
>> Describe the differences between real and protected modes
>> Describe the operation of the 80286 data, address, and control buses
>> Describe the purpose of the expansion slots of the IBM PC AT (ISA) bus
>> Describe the ISA bus system

Since the original IBM PCs used 8088 and 80286 microprocessors, this chapter is a detailed hardware study of these two microprocessors, as well as the major signals of the ISA bus. In Section 9.1, a detailed look at the 8088 CPU, including pin descriptions, is provided. Two IC chips that support the 8088, the 8284 clock generator and the 8288 chip, are discussed in Section 9.2. Next, the IBM PC address, data, and control buses are covered in Section 9.3. In Section 9.4, the 80286 microprocessor is discussed, including pin descriptions. Finally, in Section 9.5, the PC's ISA buses are covered.

SECTION 9.1: 8088 MICROPROCESSOR

The first IBM PC used the 8088 microprocessor, and modern PCs still carry that legacy. In this section, the function of each pin of the 8088 CPU is described, as well as how the microprocessor chip is connected with some simple logic gates to create the address, data, and control signals. The 8088 is a 40-pin microprocessor chip that can work in two modes: minimum mode and maximum mode. Maximum mode is used when we need to connect the 8088 to an 8087 math coprocessor. If we do not need a math coprocessor, the 8088 is used in minimum mode. First we look at the 8088 in minimum mode since it is much simpler and easy to understand. Maximum mode and supporting chips are discussed in Section 9.2.

In 1978 Intel introduced the 16-bit microprocessor called the 8086. It was 16-bit both internally and externally. A year later Intel introduced the 8088 to allow the use of 8-bit peripheral chips and to make system boards cheaper. The 8088 is internally identical to the 8086, but has only an 8-bit external data bus. Since the original IBM PC introduced in 1981 used the 8088, we explore the 8088 instead of the 8086.

Microprocessor buses

Every microprocessor-based system must have three sets of separate buses: the address bus, the data bus, and the control bus. The address bus provides the path for the address to locate the targeted device, while the data bus is used to transfer data between the CPU and the targeted device. The control bus provides the signals to indicate the type of operation being executed, such as read or write. Next we discuss how these signals are provided by the 8088 microprocessor.

Data bus in 8088

Figure 9-1 shows the 8088/86 in minimum mode. Pins 9-16 (AD0 - AD7) are used for both data and addresses in the 8088. At the time of the design of this microprocessor in the late 1970s, due to IC chip packaging limitations, there was a great effort to use the minimum number of pins for external connections. Therefore, designers multiplexed the address and data buses, meaning that Intel used the same pins to carry two sets of information: address and data. Notice that the name of the pins reflects this dual function. In the 8088, the address/data bus pins are named AD0 - AD7, "AD" for "address/data." The ALE (address latch enable) pin signals whether the information on pins AD0 - AD7 is address or data. Every time the microprocessor sends out an address, it activates (sets high) the ALE to indicate that the information on pins AD0 - AD7 is the address (A0 - A7). This information must be latched, then pins AD0 - AD7 are used to carry data. When data is to be sent out or in, ALE is low, which indicates that AD0 - AD7 will be used as data buses (D0 - D7). This process of separating address and data from pins AD0 - AD7 is called *demultiplexing*.

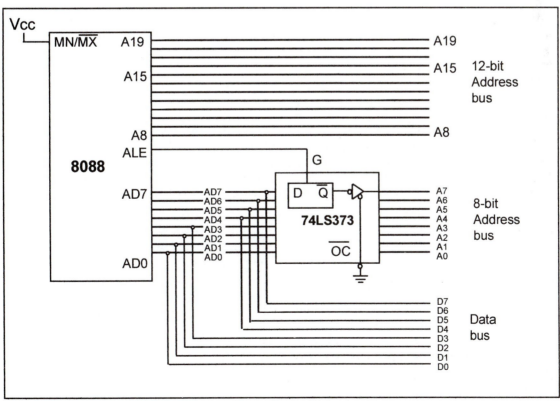

1 GND	Vcc 40
2 AD14	AD15 39
3 AD13	A16 38
4 AD12	A17 37
5 AD11	A18 36
6 AD10	A19 35
7 AD9	$\overline{BHE}$/S7 34
8 AD8	MN/$\overline{MX}$ 33
9 AD7	$\overline{RD}$ 32
10 AD6	HOLD 31
11 AD5	HLDA 30
12 AD4	$\overline{WR}$ 29
13 AD3	$\overline{IO}$/M 28
14 AD2	DT/$\overline{R}$ 27
15 AD1	$\overline{DEN}$ 26
16 AD0	ALE 25
17 NMI	$\overline{INTA}$ 24
18 INTR	$\overline{TEST}$ 23
19 CLK	READY 22
20 GND	RESET 21

8086

1 GND	Vcc 40
2 A14	A15 39
3 A13	A16 38
4 A12	A17 37
5 A11	A18 36
6 A10	A19 35
7 A9	$\overline{SS0}$ 34
8 A8	MN/$\overline{MX}$ 33
9 AD7	$\overline{RD}$ 32
10 AD6	HOLD 31
11 AD5	HLDA 30
12 AD4	$\overline{WR}$ 29
13 AD3	IO/$\overline{M}$ 28
14 AD2	DT/$\overline{R}$ 27
15 AD1	$\overline{DEN}$ 26
16 AD0	ALE 25
17 NMI	$\overline{INTA}$ 24
18 INTR	$\overline{TEST}$ 23
19 CLK	READY 22
20 GND	RESET 21

8088

Figure 9-1. The 8086 and 8088 in Minimum Mode
(Reprinted by permission of Intel Corporation, Copyright Intel, 1989)

Figure 9-2. Role of ALE in Address/Data Demultiplexing

SECTION 9.1: 8088 MICROPROCESSOR   237

Address bus in 8088

The 8088 has 20 address pins (A0 - A19), allowing it to address a maximum of one megabyte of memory ($2^{20} = 1M$). Pins AD0 - AD7 provide the A0 - A7 addresses with the assistance of a latch. To demultiplex the address signals from the address/data pins, a latch must be used to grab the addresses. The most widely used latch is the 74LS373 IC (see Figures 9-2 and 9-3). We can also use the 74LS573 chip since it is a variation of the 74LS373 chip. AD0 to AD7 of the 8088 go into the 74LS373 latch. ALE provides the signal for the latching action. For the 8088, the output of the 74LS373 provides the 8-bit address A0 - A7, while A8 - A15 come directly from the microprocessor (pins 2 - 8 and pin 39). The last 4 bits of the address come from A16 - A19, pin numbers 35 - 38. In any system, all addresses must be latched to provide a stable, high-drive-capability address bus for the system (see Figure 9-5).

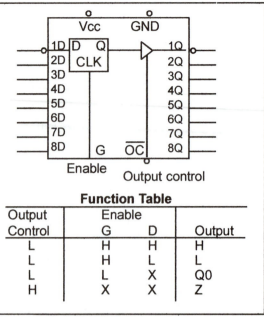

Function Table

Output Control	Enable		Output
	G	D	
L	H	H	H
L	H	L	L
L	L	X	Q0
H	X	X	Z

Figure 9-3. 74LS373 D Latch
(Reprinted by permission of Texas Instruments, Copyright Texas Instruments, 1988)

Table 9-1: Control Signal Generation

RD	WR	IO/M	Signal
0	1	0	$\overline{\text{MEMR}}$
1	0	0	$\overline{\text{MEMW}}$
0	1	1	$\overline{\text{IOR}}$
1	0	1	$\overline{\text{IOW}}$
0	0	x	Never happens

8088 control bus

There are many control signals associated with the 8088 CPU; however, for now we discuss those that deal with read and write operations. The 8088 can access both memory and I/O devices for read and write operations. This gives us four operations for which we need four control signals: MEMR (memory read), MEMW (memory write), IOR (I/O read), and IOW (I/O write).

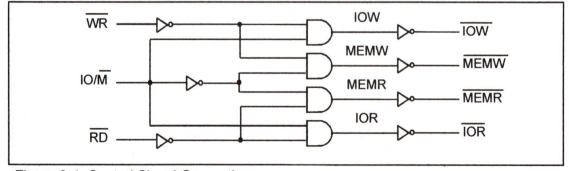

Figure 9-4. Control Signal Generation

The 8088 provides three pins for these control signals: RD, WR, and IO/M̄. The RD and WR pins are both active low. IO/M is low for memory and high for I/O devices. From these three pins, four control signals are generated: IOR, IOW, MEMR, and MEMW, as shown in Figure 9-4 and listed in Table 9-1. Notice that all of these signals must be active low since they go into the RD and WR inputs of memory and peripheral chips that are active low. Figure 9-5 shows the use of simple logic gates (inverters and ORs) to generate control signals. One can use CPLD (complex programmable logic devices) for that purpose and that is exactly what chipsets do in today's PCs.

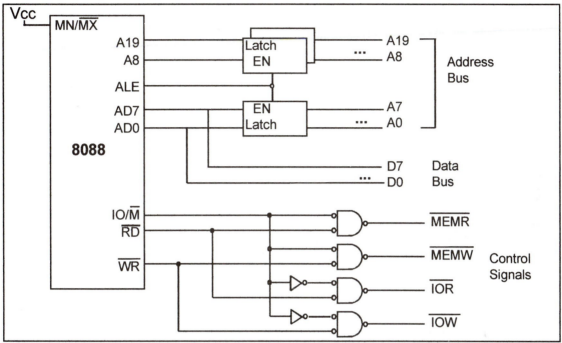

Figure 9-5. Address, Data, and Control Buses in 8088-based System

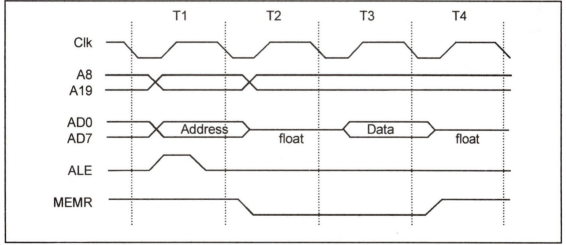

Figure 9-6. ALE Timing

Bus timing of the 8088

In Figure 9-6 the timing for ALE is shown. The 8088 uses 4 clocks for memory and I/O bus activities. For example, in the read timing, ALE latches the address in the first clock cycle. In the second and third clock cycles, the read signal is provided. Finally, by the end of the fourth clock cycle the data must be at the pins of the CPU to be fetched in. Notice that the entire read or write cycle time is only 4 clock cycles. If the task of reading or writing takes more than 4 clocks due to the slowness of memory or I/O devices, wait states (WS) can be requested from the CPU. This will be demonstrated in Chapter 10.

Other 8088 pins

Pins 24 - 32 of the 8088 have different functions depending on whether the 8088 is used in minimum mode or maximum mode. As stated earlier, maximum mode is used only when we want to connect the 8088 to an 8087 math coprocessor. In maximum mode, the 8088 needs supporting chips to generate the control signals, as described in the next section. Table 9-2 lists the functions of pins 24 - 32 of the 8088 in minimum mode.

Table 9-2: Pins 24 - 32 in Minimum Mode

Pin	Name and Function
24	INTA (interrupt acknowledge) Active-low output signal. Informs interrupt controller that an INTR has occurred and that the vector number is available on the lower 8 lines of the data bus.
25	ALE (address latch enable) Active-high output signal. Indicates that a valid address is available on the external address bus.
26	DEN (data enable) Active-low output signal. Enables the 74LS245. This allows isolation of the CPU from the system bus.
27	DT/R (data transmit/receive) Active-low output signal used to control the direction of data flow through the 74LS245 transceiver.
28	IO/M (input-output or memory) Indicates whether address bus is accessing memory or an I/O device. In the 8088, it is low when accessing memory and high when accessing I/O. This pin is used along with RD and WR pins to generate the four control signals MEMR, MEMW, IOR, and IOW.
29	WR (write) Active-low output signal. Indicates that the data on the data bus is being written to memory or an I/O device. Used along with signal IO/M (pin 28) to generate the MEMW and IOW control signals for write operations.
30	HLDA (hold acknowledge) Active-high output signal. After input on HOLD, the CPU responds with HLDA to signal that the DMA controller can use the buses.
31	HOLD (hold) Active-high input from the DMA controller that indicates that the device is requesting access to memory and I/O space and that the CPU should release control of the local buses.
32	RD (Read) Active-low output signal. Indicates that the data is being read (brought in) from memory or I/O to the CPU. Used along with signal IO/M (pin 28) to generate MEMR and IOR control signals for read operations.

Other pins of the 8088 are described below.

MN/MX (minimum/maximum)

Minimum mode is selected by connecting MN/MX (pin number 33) directly to +5V; maximum mode is selected by grounding this pin.

NMI (nonmaskable interrupt)

This is an edge-triggered (going from low to high) input signal to the processor that will make the microprocessor jump to the interrupt vector table after it finishes the current instruction. This interrupt cannot be masked by software, as we will see in Chapter 14.

INTR (interrupt request)

INTR is an active-high level-triggered input signal that is continuously monitored by the microprocessor for an external interrupt. This pin and INTA are connected to the 8259 interrupt controller chip, as we will see in Chapter 14.

CLOCK

Microprocessors require a very accurate clock for synchronization of events and driving the CPU. For this reason, Intel has designed the 8284 clock generator to be used with the 8088 processor. CLOCK is an input and is connected to the 8284 clock generator. It acts as the heartbeat of the CPU. Any irregularity causes the CPU to malfunction. The 8284 chip is used whether the 8088 is connected in minimum mode or in maximum mode. The details of the 8284 chip are covered in the next section.

READY

READY is an input signal used to insert a wait state for slower memories and I/O. It inserts wait states when it is low. The READY signal is needed to interface the CPU to low-speed memories and I/O devices.

TEST

In minimum mode this is not used. In maximum mode, however, this is an input from the 8087 math coprocessor to coordinate communications between the two processors.

RESET

To terminate the present activities of the microprocessor, a high is applied to the RESET input pin. A presence of high will force the microprocessor to stop all activity and set the major registers to the values shown in Table 9-3. The data in Table 9-3 has certain implications in the allocation of memory space to RAM and ROM that we will clarify next.

Table 9-3: IP and Segment Register Contents after Reset

Register	Contents
CS	FFFF
IP	0000
DS	0000
SS	0000
ES	0000

(Reprinted by permission of Intel Corporation, Copyright Intel Corp. 1983)

At what address does the 8088 wake up?

According to Table 9-3, when power is applied to the 8088, it wakes up at physical address FFFF0H, since a CS:IP address of FFFF:0000 leads to a physical address of FFFF0H. Therefore, we must have a non-volatile memory such as ROM at the FFFF0H address. This is discussed further in Chapter 10.

Review Questions

1. Describe the differences between the external data bus of the 8086 and 8088.
2. In the 8088, pins AD0 - AD7 are used for both data and addresses. How does the CPU indicate whether the information on these pins is data or an address?
3. The 8088 memory or I/O read cycles take _____ clock pulses to complete.
4. If we do not need an 8087 math coprocessor, the 8088 is connected in _____ mode.
5. Indicate whether each of the following pins are input pins, output pins, or both.
   (a) AD0 - AD7   (b) ALE   (c) A8 - A15
6. Give the status of the IO/M and RD pins when MEMR is active.
7. Give the status of the IO/M and WR pins when MEMW is active.

SECTION 9.2: 8284 AND 8288 SUPPORTING CHIPS

The original IBM PC introduced in 1981 used the 8088 in maximum mode with a socket for the 8087 math coprocessor. In maximum mode, the 8088 requires the use of the 8288 to generate some of the control signals. In this section we cover the 8088's supporting chips, the 8284 and 8288, and their use in maximum mode. Modern microprocessors such as the Pentium have all these chips incorporated into a single chip. Therefore, this section can be skipped unless you are interested in the design of the original PC.

Figure 9-7 shows the 8086/88 in maximum mode. Comparing Figure 9-7 with Figure 9-1, we see that pins 24 - 32 have different functions. To use the 8088 in maximum mode we must use the 8288 supporting chip. We describe the 8288 next and how it is used with the 8088 in maximum mode.

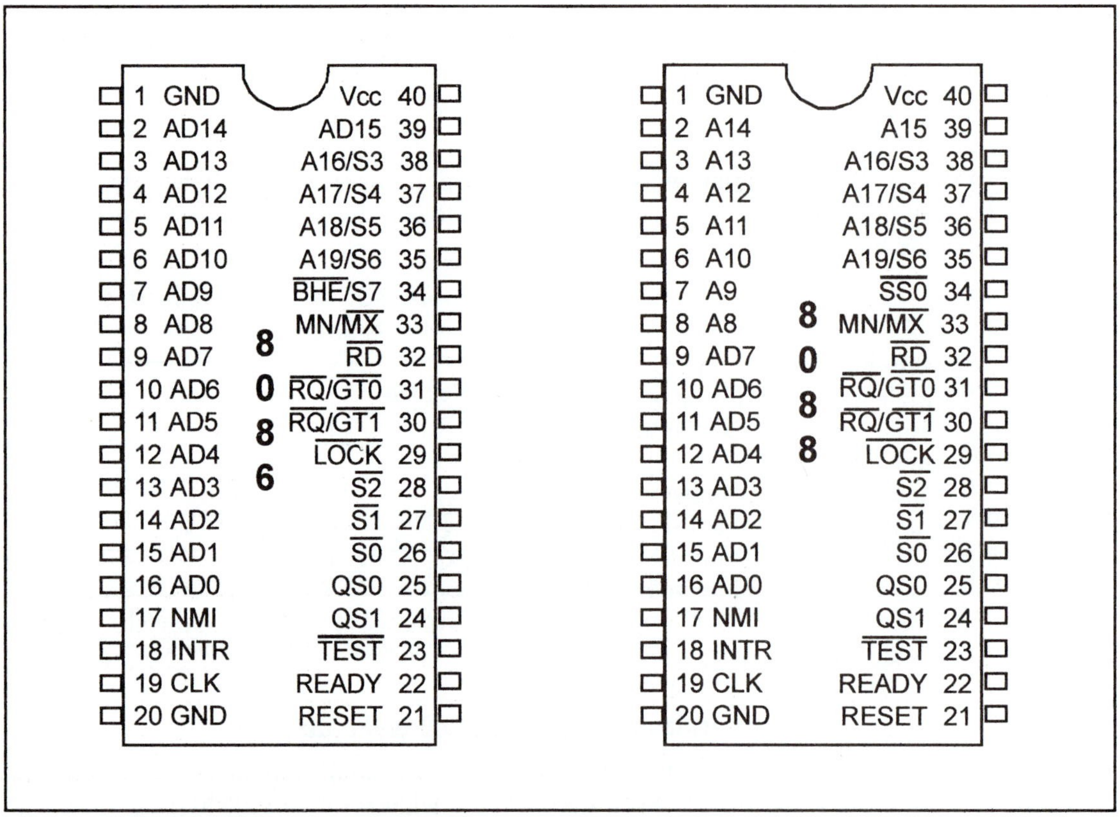

Figure 9-7. The 8086 and 8088 in Maximum Mode
(Reprinted by permission of Intel Corporation, Copyright Intel, 1989)

8288 bus controller

As shown in Figure 9-8, the 8288 is a 20-pin chip specially designed to provide all the control signals when the 8088 is in maximum mode. The input and output signals are described below.

Input signals

$\overline{S0}$, $\overline{S1}$, $\overline{S2}$ (status input)

Input to these pins comes from the 8088. Depending upon the input from the CPU, the 8288 will provide one of the commands or control signals shown in Table 9-4.

CLK (clock)

This is input from the 8284 clock generator, providing the clock pulse to the 8288 to synchronize all command and control signals with the CPU. The 8284 chip is discussed later in this section.

AEN (address enable)

AEN, an active-low signal, activates the 8288 command output at least 115 ns after its activation. In the IBM PC it is connected to the AEN generation circuitry.

CEN (command enable)

An active-high signal is used to activate/enable the command signals and DEN. In the IBM PC it is connected to the AEN generating circuitry.

IOB (input/output bus mode)

An active-high signal makes the 8288 operate in input/output bus mode rather than in system bus mode. Since the IBM PC is designed with system buses, it is connected to low.

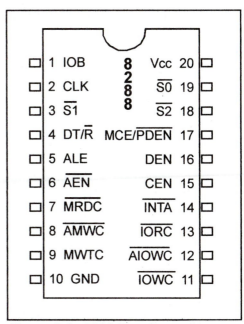

Figure 9-8. 8288 Bus Controller
(Reprinted by permission of Intel Corporation, Copyright Intel, 1983)

Table 9-4: Status Pins of the 8288 and Their Meaning

S2	S1	S0	Processor State	8288 Command
0	0	0	Interrupt acknowledge	INTA
0	0	1	Read input/output port	IORC
0	1	0	Write input/output port	IOWC, AIOWC
0	1	1	Halt	None
1	0	0	Code access	MRDC
1	0	1	Read memory	MRDC
1	1	0	Write memory	MWTC, AMWC
1	1	1	Passive	None

(Reprinted by permission of Intel Corporation, Copyright Intel Corp. 1989)

Output signals

The following are the output signals of the 8288 bus controller chip.

MRDC (memory read command)

This is active low and provides the MEMR (memory read) control signal. It activates the selected device or memory to release its data to the data bus.

MWTC (memory write command), AMWC (advanced memory write)

These two active-low signals are used to tell memory to record the data present on the data bus. These two are the same as the MEMW (memory write) signal, the only difference being that AMWC is activated slightly earlier in order to give extra time to slow devices.

$\overline{\text{IORC}}$ (I/O read command)

$\overline{\text{IORC}}$ is an active-low signal that tells the I/O device to release its data to the data bus. In the PC it is called the $\overline{\text{IOR}}$ (I/O read) control signal.

$\overline{\text{IOWC}}$ (I/O write command), $\overline{\text{AIOWC}}$ (advanced I/O write command)

Both are active-low signals used to tell the I/O device to pick up the data on the data bus. $\overline{\text{AIOWC}}$ is available a little bit early to give sufficient time to slow devices. It is unused in the IBM PC. In the PC, $\overline{\text{IOWC}}$ is labeled as $\overline{\text{IOW}}$.

$\overline{\text{INTA}}$ (interrupt acknowledge)

An active-low signal will inform the interrupting device that its interrupt has been acknowledged and will provide the vector address to the data bus. In the IBM PC this is connected to INTA of the 8259 interrupt controller chip.

DT/$\overline{\text{R}}$ (data transmit/receive)

DT/$\overline{\text{R}}$ is used to control the direction of data in and out of the 8088. In the IBM PC it is connected to DIR of the 74LS245. When the 8088 is writing data, this signal is high and will allow data to go from the A side to the B side of the 74LS245, so that data is released to the system bus. Conversely, when the CPU is reading data, this signal is low, which allows data to come in from the B to the A side of the 74LS245 data transceiver chip so that it can be received by the CPU.

DEN (data enable)

An active-high signal will make the data bus either a local data bus or the system data bus. In the IBM PC it is used along with a signal from the 8259 interrupt controller to activate G of the 74LS245 transceiver.

MCE/PDEN (master cascade enable/peripheral data enable)

This is used along with the 8259 interrupt controller in master configuration. In the IBM PC the 8259 is used as a slave; therefore, this pin is ignored.

ALE (address latch enable)

ALE is an active-high signal used to activate address latches. The 8088 multiplexes address and data through AD0 - AD7 in order to save pins. In the IBM PC, ALE is connected to the G input of the 74LS373, making demultiplexing of the addresses possible.

8284 CLOCK GENERATOR

The 8284 is used in both minimum and maximum modes since it provides the clock and timing for the 8088-based system. Figure 9-9 shows the 8284A, an 18-pin chip especially designed for use with the 8088/86 microprocessor. It provides not only the clock and synchronization for the microprocessor, but also the READY signal for the insertion of wait states into the CPU bus cycle. A description of each pin and how it is connected in the IBM PC follows.

Input pins

$\overline{\text{RES}}$ (reset in)

This is an input active-low signal to generate RESET. In the IBM PC, it is connected to the power-good signal from the power supply. When the power switch in the IBM PC is turned on, assuming that the power supply is good, a low signal is provided to this pin and the 8284 in turn will activate the RESET pin, forcing the 8088 to reset; then the microprocessor takes over. This is called a *cold boot*.

X1 and X2 (crystal in)

X1 and X2 are the pins to which a crystal is attached. The crystal frequency must be 3 times the desired frequency for the microprocessor. The maximum crystal for the 8284A is 24 MHz. The IBM PC is connected to a crystal of 14.31818 MHz.

F/$\overline{C}$ (frequency/clock)

This pin provides an option for the way the clock is generated. If connected to low, the clock is generated by the 8284 with the help of a crystal oscillator. If it is connected to high, it expects to receive clocks at the EFI pin. Since the IBM PC uses a crystal, this pin is connected to low.

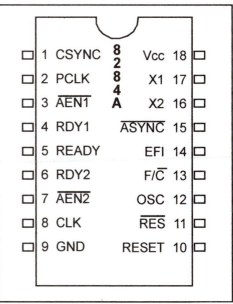

Figure 9-9. 8284A Chip
(Reprinted by permission of Intel Corporation, Copyright Intel, 1983)

EFI (external frequency in)

External frequency is connected to this pin if F/$\overline{C}$ has been connected to high. In the IBM PC this is not connected since a crystal is used instead of an external frequency generator.

CSYNC (clock synchronization)

This active-high signal is used to allow several 8284 chips to be connected together and synchronized. The IBM PC uses only one 8284; therefore, this pin is connected to low.

RDY1 and $\overline{AEN1}$

RDY1 is active high and $\overline{AEN1}$ (address enable) is active low. They are used together to provide a ready signal to the microprocessor, which will insert a WAIT state to the CPU read/write cycle. In the PC, RDY1 is connected to DMAWAIT, $\overline{AEN1}$ is connected to RDY/WAIT. This allows a wait state to be inserted by either the CPU or DMA.

RDY2 and $\overline{AEN2}$

These function exactly like RDY1 and $\overline{AEN1}$ but are designed to allow for a multiprocessing system. In the IBM PC, RDY2 is connected to low, $\overline{AEN2}$ is connected to high, which permanently disables this function since there is only one 8088 microprocessor in the system.

ASYNC

This is called ready synchronization select. An active low is used for devices that are not able to adhere to the very strict RDY setup time requirement. In the IBM PC this is connected to low, making the timing design of the system easier with slower logic gates.

Output signals

RESET

This is an active-high signal that provides a RESET signal to the 8088. It is activated by the RES input signal discussed earlier.

OSC (oscillator)

This provides a clock frequency equal to the crystal oscillator and is TTL compatible. Since the IBM crystal oscillator is 14.31818 MHz, OSC will provide this frequency to the expansion slot of the IBM PC.

CLK (clock)

This is an output clock frequency equal to one-third of the crystal oscillator, or EFI input frequency, with a duty cycle of 33%. This is connected to the clock input of the 8088 and all other devices that must be synchronized with the CPU. In the IBM PC it is connected to pin 19 of the 8088 microprocessor and other circuitry under the CLK88 label. This frequency, 4.772776 MHz (14.31818 divided by 3), is the processor frequency on which all of the timing calculations of the memory and I/O cycle are based.

PCLK (peripheral clock)

PCLK is one-half of CLK (or one-sixth of the crystal) with a duty cycle of 50% and is TTL compatible. In the IBM PC this 2.386383 MHz is provided to the 8253 timer to be used to generate speaker tones, and for other functions.

READY

This signal is connected to READY of the CPU. In the IBM PC it is used to signal the 8088 that the CPU needs to insert a wait state due to the slowness of the devices that the CPU is trying to contact.

Review Questions

1. Pin RESET is an _____ (input, output) for the 8284 and an _____ (input, output) for the 8088.
2. True or false. Regardless of whether the 8088 is in minimum or maximum mode, the 8284 clock generator is needed to provide a reliable clock.
3. True or false. The 8288 is used to provide control signals for the 8088 when it is in minimum mode.
4. The 8288 output pin _____ controls the direction data flows in and out of the CPU.

SECTION 9.3: 8-BIT SECTION OF ISA BUS

Previous sections have explained the 8088 CPU and supporting chips. This section will explain how they are all connected in the original IBM PC to produce the required buses to communicate with memory, input/output peripherals, and the 8-bit section of the ISA bus. The study of the 8-bit section of ISA is the main topic of this section.

A bit of bus history

The original IBM PC introduced in 1981 used an 8088 microprocessor, whose 8-bit data bus gave birth to the 8-bit section of the ISA bus. In 1984 when IBM introduced the IBM PC/AT using the 80286 microprocessor, the data bus was expanded to 16 bits. The 8-bit data bus can be seen as a subsection of the 16-bit ISA bus. Very often the 8-bit data bus was referred to as the IBM PC/XT (extended technology) bus in order to differentiate it from the IBM PC AT (advanced technology). Eventually the IBM PC AT bus became known as the ISA (Industry Standard Architecture) bus since the term "PC AT" was copyrighted by IBM. Throughout this book we use the terms PC/XT and PC interchangeably to refer to the 8-bit portion of the IBM PC AT (ISA) bus. The following is the description of the three main buses of IBM PC as generated by the 8088 and supporting chips.

Local bus vs. system bus

In the discussion of PC design we often see the terms *local bus* and *system bus*. The system bus not only provides necessary signals to all the chips (RAM, ROM, and peripheral chips) on the motherboard, but also goes to the expansion slot for any plug-in expansion card. In contrast, the local bus is connected directly to the CPU. Any communication with the CPU must go through the local bus. There is a bridge between the local bus and the system bus to make sure they are isolated from each other. Sometimes the system bus is referred to as a *global bus*. We use tri-state buffers to isolate the local bus and system bus. For example, 74LS245 is a widely used chip for the data bus buffer since it is bidirectional. See Figure 9-10. Figure 9-11 shows an example of local and system buses. Figure 9-11 gives an overview of the 8088 and its supporting chips as designed in the original PC. Notice the role of the 74LS245 and 74LS373 in isolating the local and system buses. Everything on the left of the 8288, 74LS373s, and 74LS245 represent the local bus and everything on the right side of those chips are the system bus. The 74LS245 and 74LS373s play the role of bridge to isolate the local and system buses. Now let's look at each of the buses.

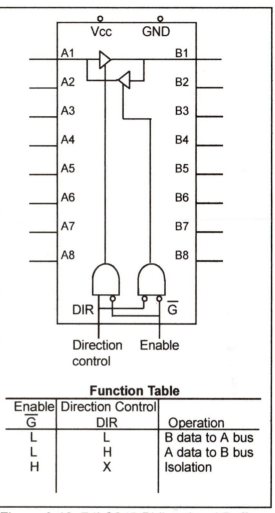

Function Table

Enable	Direction Control	
$\overline{G}$	DIR	Operation
L	L	B data to A bus
L	H	A data to B bus
H	X	Isolation

Figure 9-10. 74LS245 Bidirectional Buffer
(Reprinted by permission of Texas Instruments, Copyright Texas Instruments, 1988)

Address bus

Three 74LS373 chips in Figure 9-11 are used for two functions:
1. To latch the addresses from the 8088 and provide stable addresses to the entire computer. The address bus is a unidirectional bus. The 74LS373 chips are activated by control signals AEN and ALE. When AEN is low, the 8088 provides the address buses to the system. The 8288's ALE (connected to G) enables the 74LS373 to latch the addresses from the CPU, providing a 20-line stable address to memory, peripherals, and expansion slots. Demultiplexing addresses A0 - A7 is performed by the 74LS373 connected to pins AD0 - AD7 of the CPU. The CPU's A8 - A15 is connected to the second 74LS373, and A16 - A19 to the third one. Half of the third 74LS373 is unused.
2. To isolate the system address buses from local address buses. The system buses must be allowed to be used by the DMA or any other board through the expansion slot without disturbing the CPU. This is achieved by the 74LS373s through AEN. The AEN signal is described shortly.

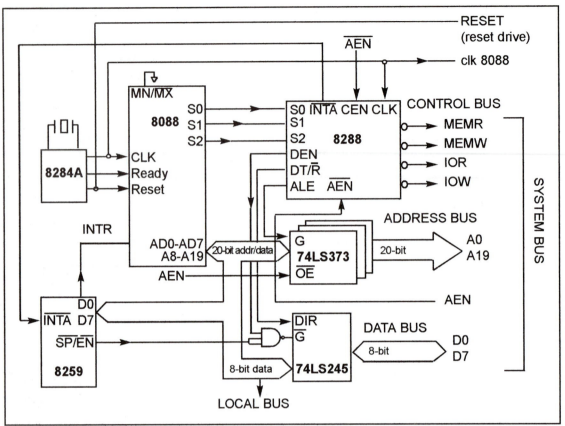

Figure 9-11. 8088 Connections and Buses in the PC/XT
(Reprinted by permission from "IBM Technical Reference" c. 1984 by International Business Machines Corporation)

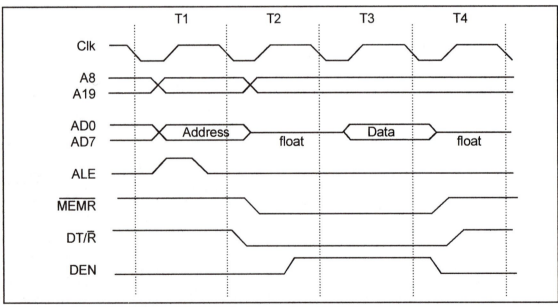

Figure 9-12. ALE, DEN, and DTR Timing for the 8088 System

Data bus

The bidirectional data bus goes through the 74LS245 transceiver (see Figures 9-10 and 9-11). DT/$\overline{R}$ and DEN are the two signals that activate the 74LS245. DT/$\overline{R}$ goes to DIR of the 74LS245 and makes the transceiver transmit information from the A side to the B side when DT/$\overline{R}$ is high. Conversely, when

CHAPTER 9: 8088, 80286 MICROPROCESSORS AND ISA BUS

DT/$\overline{\text{R}}$ makes DIR low, the transceiver transfers information from the B side to the A side, thereby receiving information from the system data bus and bringing it to the microprocessor. DEN (an active low signal) enables the 74LS245. This isolates the data buses to make them either a local bus or a system bus. When the 74LS245 is not active, the system data bus is isolated from the local data bus.

Control bus

The four most important control signals of the IBM PC are $\overline{\text{IOR}}$ (I/O read), $\overline{\text{IOW}}$ (I/O write), $\overline{\text{MEMR}}$ (memory read), and $\overline{\text{MEMW}}$ (memory write). They are provided by the 8288 chip as shown in Figure 9-11. The timing for the bus activity is shown in Figure 9-12.

One bus, two masters

While the 8088, the main processor, is designed for fetching and executing instructions, it is unacceptably slow for transferring large numbers of bytes of data such as in hard disk data transfers. Instead, the 8237 chip is used for data transfers of large numbers of bytes. The detailed function of this chip is explained in Chapter 15. All that is needed here is to know that the 8237's job is to transfer data and it must have access to all three buses to do that. Since no bus can serve two masters at the same time, there must be a way to allow either the 8088 processor or the 8237 DMA to gain control over the buses. This is called *bus arbitration* and is achieved by the AEN (address enable) generation circuitry.

AEN signal generation

When the system is turned on, the 8088 CPU is in control of all the buses. The CPU maintains control as long as it is fetching and executing instructions. As

Table 9-5: AEN Bus Arbitration

AEN	Bus Control
0	Buses controlled by CPU
1	Buses controlled by DMA

can be seen from Figure 9-13, AEN is the output signal of the D flip-flop. Since Q is either high or low, depending on the status of this signal, either the CPU or the DMA can access the buses. Table 9-5 shows the role of AEN in bus arbitration.

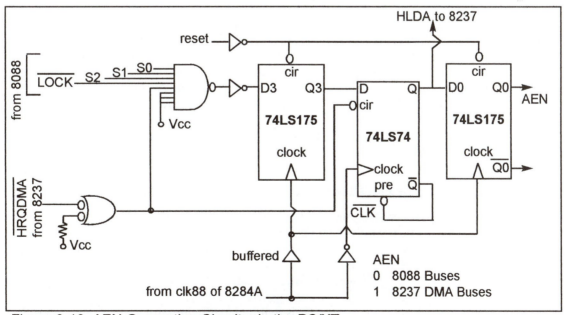

Figure 9-13. AEN Generation Circuitry in the PC/XT

(Reprinted by permission from "IBM Technical Reference" c. 1984 by International Business Machines Corporation)

Control of the bus by DMA

How does AEN become high, handing control of the system buses to DMA? The answer is that when DMA receives a request for service, it will notify the CPU that it needs to use the system buses by putting a LOW on HRQDMA (this is the same as the HOLD signal in minimum mode of the 8088). This in turn will provide a high on the D3 output of the 74LS175, assuming that the current memory cycle is finished and that LOCK is not activated. In the following clock cycle, HLDA (hold acknowledge) is provided to the DMA and AEN becomes high, giving control over the buses to the DMA.

Bus boosting

One more point that needs explaining is bus boosting of the control, data, and address buses to provide sufficiently strong signals to drive various IC chips. When a pulse leaves an IC chip it can lose some of its strength, depending on how far away the receiving IC chip is located. In addition, the more pins a signal is connected to, the stronger the signal must be to drive them all. Therefore, the signals must be amplified. Stated another way, every pin connected to a given signal has input capacitance, and the capacitances are in parallel; thus as far as that signal is concerned they are all added together, making one big capacitor load. This requires that the signal be strong enough to drive all the inputs (see Chapter 26 for more details on this topic). It is common to combine the functions of bus isolation and bus boosting into a single chip. For example, 74LS373 chips are used to boost the addresses provided by the 8088 microprocessor in addition to the bus isolation mentioned earlier. The signals provided by the CPU need boosting since the 8088 is a CMOS chip. CMOS has a much lower driving capability than TTL, of which 74LS373s are made. Likewise, the 74LS245 is used for both data bus booster and data bus isolation. Details of IC interfacing and how 74LS245 chips are used for signal amplification (boosting) are shown in Chapter 26.

8-bit section of the ISA bus

As stated earlier, the original IBM PC had an 8-bit data bus. Later with the introduction of the 80286, the 16-bit version of the bus became available. The 80286 bus became known as the ISA bus. The 8-bit bus is a subset of the 16-bit ISA bus and used in many peripheral boards. Figure 9-14 shows the 8-bit portion of the ISA bus expansion slot. From that figure notice that addresses A0 - A19 and data signals D0 - D7 are on the A side of the expansion slot. On the A side, also notice the AEN pin. On the B side are found control signals IOR, IOW, MEMR, and MEMW. The – sign on these and other control signals implies an active low signal. In Chapter 11 we use signals A0 - A19, D0 - D7, AEN, IOR, and IOW to design an I/O interfacing card. The rest of the signals in Figure 9- 14 will be covered in subsequent chapters. The signals asscociated with interrupts (IRQs) are covered in Chapter 14; signals associated with DMA (DREQs and DACKs) are covered in Chapter 15.

Review Questions

1. The system bus can be accessed either by the CPU or by _____.
2. The control signal that provides bus arbitration is _____.
3. True or false. After a cold boot, DMA is given control of the buses.
4. The bidirectional data bus goes through the 74LS245 transceiver. Signal _____ determines whether data is flowing from the A to B side or from the B to A side.
5. Bus _____ is required to provide strong signals to various IC chips in the IBM PC.

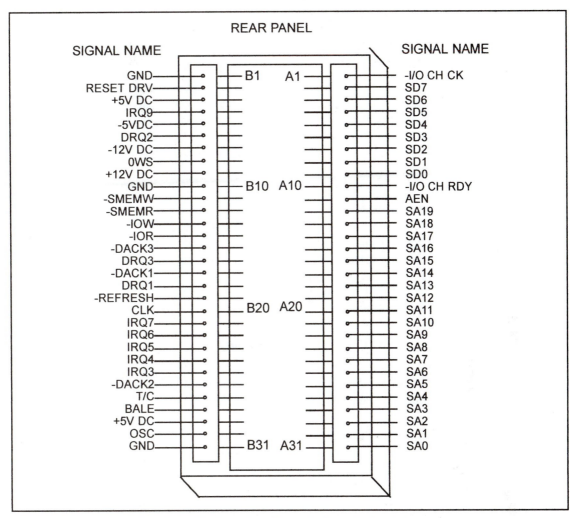

REAR PANEL

SIGNAL NAME			SIGNAL NAME
GND	B1	A1	-I/O CH CK
RESET DRV			SD7
+5V DC			SD6
IRQ9			SD5
-5VDC			SD4
DRQ2			SD3
-12V DC			SD2
0WS			SD1
+12V DC			SD0
GND	B10	A10	-I/O CH RDY
-SMEMW			AEN
-SMEMR			SA19
-IOW			SA18
-IOR			SA17
-DACK3			SA16
DRQ3			SA15
-DACK1			SA14
DRQ1			SA13
-REFRESH			SA12
CLK	B20	A20	SA11
IRQ7			SA10
IRQ6			SA9
IRQ5			SA8
IRQ4			SA7
IRQ3			SA6
-DACK2			SA5
T/C			SA4
BALE			SA3
+5V DC			SA2
OSC			SA1
GND	B31	A31	SA0

Figure 9-14. ISA (IBM PC AT) Bus Slot Signals Detail (8-bit Section)
(Reprinted by permission from "IBM Technical Reference" c. 1985 by International Business Machines Corporation)

SECTION 9.4: 80286 MICROPROCESSOR

The 80286 is a 68-pin microprocessor available in either of two packaging formats: LCC (leaded chip carrier) and PGA (pin grid array). This is in contrast to the 8088, which is a 40-pin DIP (dual in-line package). To package the 68-pin IC in DIP packaging would have made it a long IC physically and consequently more fragile. Such packaging would also necessitate a longer path for some signals and as a result make it unsuitable for use in high-frequency systems. Figure 9-15 shows the 80286 in LCC packaging.

The 80286 can work in one of two modes: real mode or protected mode. In real mode, the maximum memory it can access is 1M, 00000H to FFFFFH. To access the entire 16M bytes of memory, 000000H to FFFFFFH, it must work in protected mode. In real mode, the 80286 is a faster version of the 8086 with a few new instructions. When power is applied to the 80286, it starts up in real mode and can be switched to protected mode at any time through a software instruction. However, to use the 286 in protected mode requires an extremely complex memory management system. Since very few systems are using the 286 in protected mode, it is not discussed here (even in protected mode it is still a 16-bit computer, meaning that all registers are 16-bit, as opposed to 32-bit).

Pin descriptions

The following are pin descriptions of the 80286 microprocessor.

Pins A0 - A23 (address bus)

These output signals provide a 24-bit address to be used by the decoding circuitry to locate memory or I/O. When providing an address for memory, all 24 pins must be used (A0 - A23); therefore, it can access a maximum of 16M bytes of memory (2^{24} = 16M). To access an I/O address, only A0 - A15 are used. If the I/O address is a 16-bit address, A0 - A15 are used to provide the address, and pins A16 - A23 are low. If the I/O address is an 8-bit address, only A0 - A7 are used, and A8 - A23 are all low.

Pins D0 - D15 (data bus)

These pins provide the 16-bit path for data to be transferred in and out of the CPU. It must be noted that unlike the 8088, the data bus is not multiplexed. The use of separate pins for address and data results in higher pin counts, but saves time since it eliminates the need for a demultiplexer. This 2-byte data path to the CPU allows the transfer of data on both bytes or on either byte, depending on the operation. The 80286 coordinates the activity on the D0 - D15 data bus with the help of A0 and BHE.

Pin $\overline{BHE}$ (bus high enable)

This is an active-low output signal used to indicate that data is being transferred on D8 - D15. Table 9-6 shows how $\overline{BHE}$ and A0 are used to indicate whether the data transfer is on D0 - D7, D8 - D15, or the entire bus, D0 - D15.

Table 9-6: BHE, A0, and Byte Selection in the 80286

BHE	A0	Data Bus Status
0	0	Transferring 16-bit data on D0 - D15
0	1	Transferring a byte on the upper half of data bus D8 - D15
1	0	Transferring a byte on the lower half of data bus D0 - D7
1	1	Reserved (the data bus is idle)

(Reprinted by permission of Intel Corporation, Copyright Intel Corp. 1983)

Pin CLK (clock)

CLK is an input providing the working frequency for the 80286. The processor always works on half of this frequency. For example, if CLK = 16 MHz, the system is an 8-MHz system. In other words, for the 80286 computer to be an 8-MHz system, the CLK must be 16 MHz.

Pin M/$\overline{IO}$ (memory I/O select)

M/$\overline{IO}$ is an output signal used by the CPU to distinguish between I/O and memory access. When it is high, memory is being accessed, and when it is low, I/O is being addressed.

Pin COD/$\overline{INTA}$ (code/interrupt acknowledge)

This is an output signal used by the CPU to indicate whether it is performing memory read/write of data or an instruction fetch. It is also used to distinguish between the action of interrupt acknowledge and I/O cycle. This signal, along with the status signals and M/$\overline{IO}$, is used to define the bus cycle.

Pins $\overline{S1}$ and $\overline{S0}$ (status signals)

These status signals for the bus cycle are both output signals used by the CPU along with M/$\overline{IO}$ and COD/$\overline{INTA}$ to define the type of bus cycle.

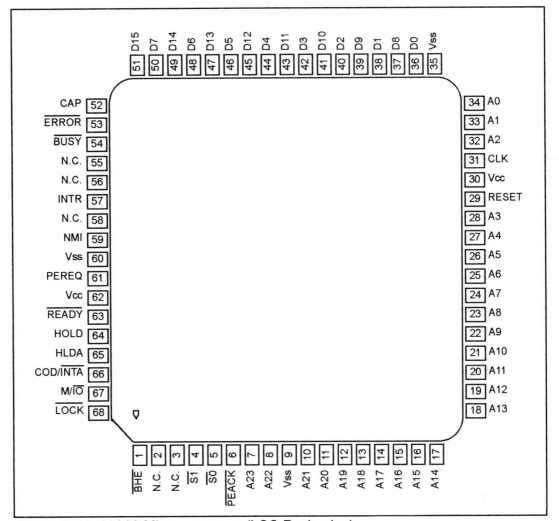

Figure 9-15. 80286 Microprocessor (LCC Packaging)
(Reprinted by permission of Intel Corporation, Copyright Intel Corp. 1983)

Pins HOLD and HLDA (hold and hold acknowledge)

HOLD and HLDA allow the CPU to control the buses. HOLD is an input signal to the 80286 and is active high. It is used by devices such as DMA to request permission to use the buses. In response, the CPU activates the output signal HLDA by putting a high on it to inform the requesting device that it has released the buses for the device's use. The DMA has control over the buses as long as HOLD is high, and in response the CPU keeps HLDA high. Whenever the DMA brings HOLD low, the CPU responds by making HLDA low, and regains control over the buses.

RESET pin

This is an input signal and is active high. When there is a low-to-high transition on RESET (and it stays high for at least 16 clocks), the 80286 initializes all registers to their predefined values and the output pins of the 80286 will have the status shown in Table 9-7. Of the above signals, the status of the following pins must be noted since they are used in the memory design of the IBM PC AT computers: A20 = 1, A21 = 1, A22 = 1, and A23 = 1.

As long as the RESET pin is high, no instruction or bus activity is allowed. The contents of the instruction pointer and segment registers of the 80286 after RESET are shown in Table 9-8.

It must be noted that when RESET of the 80286 is activated, it forces the 80286 to enter into real mode. In other words, the CPU wakes up in real mode. In real mode the 80286 (indeed, all the x86s from the 80286 to the Pentium4) processor can address only 1 megabyte since it uses only address lines A0 - A19. Since RESET also causes A20 - A23 to be high, the first instruction for the 286 must be at physical address FFFFF0H. This is due to the fact that at reset, CS = F000 and IP = FFF0, making the logical address of the first instruction F000:FFF0. This provides the physical address of FFFF0H on A0 - A19, and since A20 - A23 is high at reset, the physical address of the first instruction must be FFFFF0H. This is 16 bytes from the top of the 16M address range of the 80286. The

Table 9-7. Pin State (Bus Idle) During Reset

Pin Name	Signal Level during Reset
D0 - D15	High impedance
A0 - A23	High
W/R	Low
M/IO	High

(Reprinted by permission of Intel Corporation, Copyright Intel Corp. 1983)

Table 9-8: IP and Segment Registers After RESET

Register	Contents
CS	F000
IP	FFF0
DS	0000
SS	0000
ES	0000

(Reprinted by permission of Intel Corporation, Copyright Intel Corp. 1983)

80286 expects to have a far jump at location FFFFF0H and when the JMP is executed, the 286 puts 0s on pins A20 - A23, making it effectively a 1M range real mode system. Further implications of these facts are discussed in Chapter 10.

Pin INTR (interrupt request)

INTR is an input signal into the 80286 requesting suspension of the current program execution. It is used for external hardware interrupt expansion along with the 8259 interrupt controller chip. See Chapter 14 for more information.

Pin NMI (nonmaskable interrupt request)

NMI is an active-high input signal. When this pin is activated, the 80286 will automatically perform INT 2, meaning that there is no INTA response since INT 2 is assigned to it. See Chapter 14 for more details of this pin.

READY pin

READY is an active-low input signal used to insert a wait state and consequently prolong the read and write cycle for slow memory and I/O devices.

LOCK, PEREQ, BUSY, ERROR, and PEACK

These five signals coordinate bus activities with the math coprocessor.

Figure 9-16 shows the 80286 timing. Notice the 2-clock cycle time for read. This is discussed further in Chapter 10.

Review Questions

1. When power is applied to the 80286, which mode does it wake up in, real mode or protected mode?
2. The 286 can access _____ of memory in real mode and _____ in protected mode.
3. When RESET is set to high, what are the contents of the CS and IP registers?

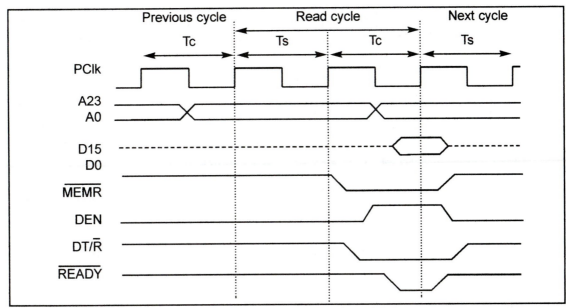

Figure 9-16. ALE, DEN, and DTR Timing for the 80286 CPU

SECTION 9.5: 16-BIT ISA BUS

The origin of technical specifications of many of today's x86 PCs is the 80286-based IBM PC/AT. Much of the PC/AT in turn is based on the original 8088-based IBM PC introduced in 1981. A major legacy of those original PCs is the ISA (Industry Standard Architecture) bus slot. Remember that ISA is another name for the PC/AT bus since PC/AT is a trademark copyrighed by IBM Corp. In this section we examine the address, data, and control buses of the ISA expansion bus and some of the issues related to them. Whether the microprocessor used in a PC is Intel's Pentium, 386, 486, or an equivalent AMD processor, if it has an ISA bus slot, the material in this section is relevant and needs to be understood if you want to design expansion cards for ISA slots.

Figure 9-17 shows the 80286 microprocessor, along with supporting chips used in the original PC/AT computers. The address, data, and control buses in this figure are used throughout the motherboard and are also provided to the ISA expansion slot. In today's PC the 80286 is replaced with Intel's Pentium or AMD's Athlon microprocessor, and all the control signals are provided by a chipset. A *chipset* is an IC chip containing all the circuitry needed to support the CPU in a given motherboard. For educational purposes throughout the book, we use simple logic gates from the original PC to discuss some design concepts, even though in the real world the chipset uses CPLDs (Complex Programmable Logic Devices) for design with all the circuitry details buried inside. Next, we examine the major signals of the ISA expansion slot.

Exploring ISA bus signals

In Section 9.3 we discussed the 8-bit section of the ISA bus. The 8-bit section uses a 62-pin connector to provide access to the system buses. In order to maintain compatibility with the original PC, the 16-bit ISA slot used the 8-bit section as a subset. A 36-pin connector was added to incorporate the new signals as shown in Figure 9-18. In designing a plug-in peripheral card for the ISA slot we need to understand the basic features of the ISA signals. The ISA bus has 24 address pins (A0 - A23), 16 data pins (D0 - D15), plus many control signals.

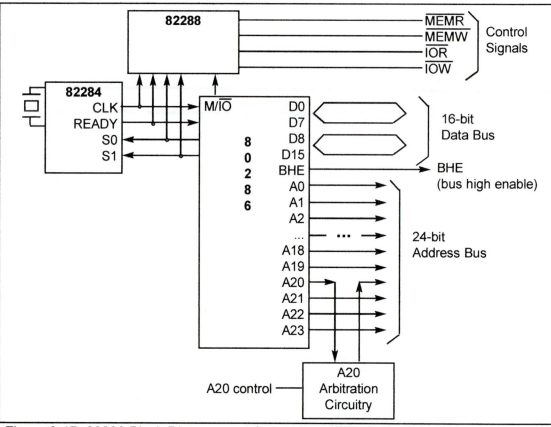

Figure 9-17. 80286 Block Diagram and Supporting Chips in the PC AT

Address bus

Addresses A0 - A19 are latched using ALE. These addresses are used throughout the motherboard and are also provided to the 62-pin part of the ISA slot as SA0 - SA19 (system address). See Figure 9-18. Notice that this is already latched and cannot be latched again by a plug-in card. The A20 - A23 part of the address is provided in the 36-pin section. In the 36-pin section of the ISA slot, A17 - A23 are also provided as LA17 - LA23 (latchable address). We need to use the ALE signal to latch these addresses in the design of plug-in cards. The ALE signal is provided as BALE (buffered ALE) and can be used to latch LA17 - LA23.

Data bus

The data bus is composed of pins D0 to D15. The data bus is buffered by a pair of 74ALS245 data bus transceivers that are used throughout the motherboard to access memory and ports. They are also provided at the expansion slot as SD0 - SD15 (system data). However, it must be noted that SD0 - SD7 are provided at the 62-pin part in order to make it compatible with the original 8088-based PC/XT, while SD8 - SD15 show up on the 36-pin part. This allows the 16-bit data bus to access any 16-bit peripheral. To select the upper byte or the lower byte of 16-bit data, we use BHE (bus high enable). BHE is latched and used on the system board and also provided at the expansion slot under SBHE (system bus high enable). We will see how to use this pin below.

Memory and I/O control signals

IOR and IOW are the two control signals used to access ports throughout the system. They show up on the 62-pin section of the ISA expansion slot. This

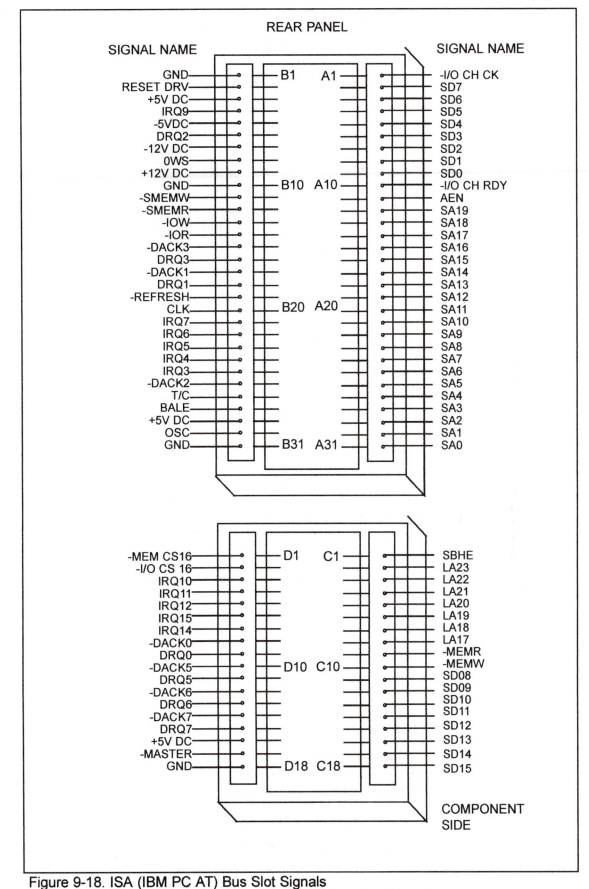

REAR PANEL

SIGNAL NAME				SIGNAL NAME
GND	B1	A1		-I/O CH CK
RESET DRV				SD7
+5V DC				SD6
IRQ9				SD5
-5VDC				SD4
DRQ2				SD3
-12V DC				SD2
0WS				SD1
+12V DC				SD0
GND	B10	A10		-I/O CH RDY
-SMEMW				AEN
-SMEMR				SA19
-IOW				SA18
-IOR				SA17
-DACK3				SA16
DRQ3				SA15
-DACK1				SA14
DRQ1				SA13
-REFRESH				SA12
CLK	B20	A20		SA11
IRQ7				SA10
IRQ6				SA9
IRQ5				SA8
IRQ4				SA7
IRQ3				SA6
-DACK2				SA5
T/C				SA4
BALE				SA3
+5V DC				SA2
OSC				SA1
GND	B31	A31		SA0

SIGNAL NAME				SIGNAL NAME
-MEM CS16	D1	C1		SBHE
-I/O CS 16				LA23
IRQ10				LA22
IRQ11				LA21
IRQ12				LA20
IRQ15				LA19
IRQ14				LA18
-DACK0				LA17
DRQ0				-MEMR
-DACK5	D10	C10		-MEMW
DRQ5				SD08
-DACK6				SD09
DRQ6				SD10
-DACK7				SD11
DRQ7				SD12
+5V DC				SD13
-MASTER				SD14
GND	D18	C18		SD15

COMPONENT
SIDE

Figure 9-18. ISA (IBM PC AT) Bus Slot Signals

(Reprinted by permission from "IBM Technical Reference" c. 1985 by International Business Machines Corporation)

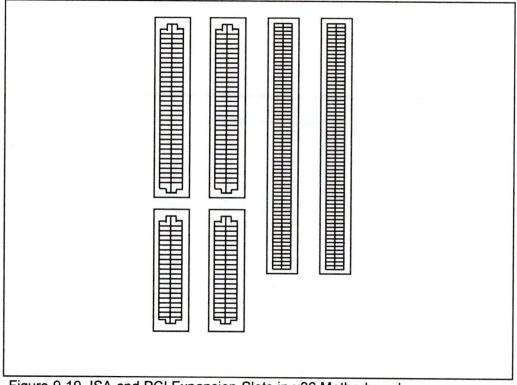

Figure 9-19. ISA and PCI Expansion Slots in x86 Motherboard

makes them 8088 PC/XT compatible. We will discuss how these signals are used in peripheral interfacing in Chapter 11.

Signals MEMR, MEMW, SMEMR, and SMEMW are used to access memory. There is a reason for duplicate memory read and write signals. To allow access to any memory within the range of 16 megabytes, read/write control signals are provided to the 36-pin section of the ISA expansion slot under the designations of MEMR and MEMW, respectively.

To maintain compatibility with the original 8088-based PC/XT, MEMR and MEMW are designated as SMEMR and SMEMW and are provided on the 62-pin part of ISA on the same strip as the XT bus systems. In other words, MEMR and MEMW can be used to access memory in any location, but to access memory within the 1 megabyte range, we must use SMEMR and SMEMW on the 62-pin part of the ISA bus. In this case they can be used only to address memory locations 0 - FFFFFH. Of course, to allow the same signals, MEMR and MEMW, from the support chip to show up in two distinctive places with two different names and functions requires some extra logic circuitry. Such details are buried inside the chipsets in today's PC.

Other control signals

Examining the ISA bus pins in Figure 9-18, we see numerous control signals that we have not seen before. The rest of the control signals in Figure 9-18 are related to the interrupt and DMA chips. IRQ and DMA signals are covered in subsequent chapters.

Figure 9-19 shows the expansion slots in today's x86 PC. In the vast majority of them we have both ISA and PCI buses. The PCI bus is a much faster and superior bus; it is discussed in Chapter 27. Some PC makers offer motherboards with PCI bus only and no ISA expansion slots.

ODD and EVEN bytes and BHE

In the 36-pin section of the ISA bus there is a pin called SBHE that we explain next. Pin C1 is the same as the BHE pin from the 80286 that we studied in the last section. The BHE pin has to do with the differences between the 8-bit and 16-bit data bus CPUs. Like all general-purpose microprocessors, the memory (and I/O) space of x86 microprocessors is byte addressable. That means that every address location can provide a maximum of one byte of data. If the CPU has an 8-bit data bus, like the 8088, then the addresses are designated as 0 to FFFFFH, as shown in Figure 9-20.

Notice in Figure 9-20 that the bus width for the data bus is only 8 bits. In other words, only 8 strips of wire connect the CPU's data bus to devices such as memory and I/O ports. Since the vast majority of memory and I/O devices also have an 8-bit data bus of D0 - D7, their interfacing to CPUs with an 8-bit data bus is simple and straightforward. The CPU's D0 - D7 data bus is connected directly to the D0 - D7 data bus of memory and I/O devices. This is a perfect match. If the CPU has a 16-bit data bus, like 8086/80286/80386SX microprocessors, then the address spaces are designated as odd and even bytes, as shown in Figure 9-21. In such cases, the D0 - D7 byte is designated as even and the D8 - D15 byte as odd. To distinguish between odd and even bytes, 8086/286/386SX CPUs provide an extra signal called BHE (bus high enable). BHE, in association with the A0 pin, is used to select the odd or even byte according to Table 9-9. In Figure 9-21, notice the odd and even banks. They are called odd and even banks since the memory chips have only an 8-bit data bus of D0 - D7 and two IC

Table 9-9: Distinguishing Between Odd and Even Bytes

BHE	A0		
0	0	Even word	D0 - D15
0	1	Odd byte	D8 - D15
1	0	Even byte	D0 - D7
1	1	None	

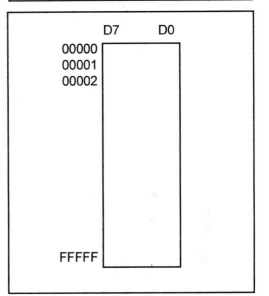

Figure 9-20. Memory Byte Addressing in 8088 (8-bit Data Bus)

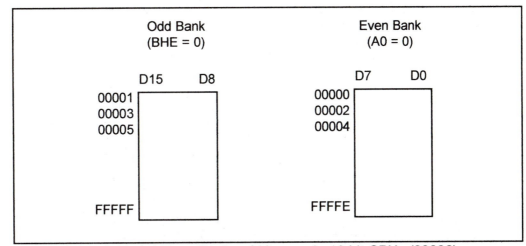

Figure 9-21. Odd and Even Banks of Memory in 16-bit CPUs (80286)

---

chips must be used, one for each byte. Although Figure 9-21 shows only 1 megabyte of memory space, the concept of odd and even bytes applies to the entire memory and I/O space of the x86 CPU. This is also the case for the 386 and 486 CPUs with a 32-bit data bus.

A20 gate and the case of high memory area (HMA)

A20 is an anomaly associated with 286 and higher microprocessors that needs to be discussed. In the 8088, when the segment register added to the offset is more than FFFFFH, it automatically wraps around and starts at 00000H. However, in 80286 and higher processors in real mode, such a wrap-around will not occur. Instead, the result will be 100000H, making A20 = 1. The problem is that A20 - A23 is supposed to be activated only when the CPU is in protected mode. To control activation of A20, IBM used a latch controlled by the keyboard in the original PC AT; however, with the introduction of PS computers, control of A20 can also be handled by port 92H. One can use this A20 gate (as it is commonly called) to create a high memory area (HMA). This concept is important for understanding HMA memory in x86 PCs and is discussed in Chapter 25. See Examples 9-1 and 9-2 for clarification on this issue. Notice that the process of enabling and disabling the A20 gate in Figure 9-17 is handled by a piece of software called the *A20 handler*, which is provided with MS-DOS and Windows operating systems.

Example 9-1

(a)   If the A20 gate is enabled, show the highest address that 286 (and higher processors) can access while still in real mode.
(b)   How far high above 1M is this address?

Solution:

(a)   To access the highest physical location in real mode, we must have CS = FFFFH and IP = FFFFH. We shift left the segment register CS and add the offset IP = FFFF:

CS shifted left one hex digit   FFFF0H
adding the offset IP   +   FFFFH
                                           10FFEFH

Therefore, the addresses FFFF0H - 10FFEFH are the range that the CPU can access while it is in real mode. This is a total of 64K bytes.

(b)   If the A20 gate is enabled, accessible memory locations above 1M are 100000 to 10FFEF. This is a total of 65,520 bytes, or 16 locations short of 1M + 64K.

Example 9-2

Assume that CS = FF25H. Find the lowest and highest physical addresses for
(a) the 8088   (b) the 80286
Specify the bit on A20.

Solution:

(a)   The lowest physical address is FF250H and the highest is 0F24FH (FF250H + FFFFH); since there are only lines A0 - A19 in the 8088, the 1 is dropped.
(b)   In the 286 the lowest address is the same as in the 8088, but the highest physical address is 10F24F; therefore, A20 = 1.

---

1. The first IBM PC AT used the _____ microprocessor.
2. What is the advantage of using the 74LS573 chip address latch over the 74LS373?
3. What is the purpose of the A20 handler circuitry?
4. What address area is called the high memory area in the 80286?
5. Since the control signals MEMR and MEMW are available on the 62-pin part of the expansion slot, why are they duplicated on the 36-pin part?

SUMMARY

This chapter began with a detailed examination of the 8088, including a description of all input and output pins and their functions. The IBM PC always uses the 8088 in maximum mode. Next, the 8284A clock generator was described, as well as its role in providing timing synchronization for the IBM PC. The 8288 bus controller was also described, which is essential for providing control signals for the 8088 CPU in maximum mode. Then, the system buses were examined in further detail in terms of their connections among the various IC chips and how control over them is achieved. In addition, an overview of the core of the IBM PC system, the CPU, the 8284A, and the 8288, was presented.

This chapter also explored the 80286 microprocessor. All the pins of the 80286 were described and an overview was given of the differences between real and protected mode. The address, data, and control buses of the 286 were described in terms of accessing the 16M memory range of the 80286. Finally, the ISA bus architecture of the PC AT and the expansion slot were examined. It was shown how these were designed so that devices could remain compatible with PC/XT expansion slots but also take advantage of unique features of the PC AT, such as extended memory.

PROBLEMS

SECTION 9.1: 8088 MICROPROCESSOR

1. State the main differences between the 8088 and 8086 pinouts. Are the two chips interchangeable?
2. ALE is an _____ (input, output) signal for the 8088.
3. What is the maximum number of bytes of memory addressable by the 8088, and why?
4. RESET is an _____ (input, output) signal for the 8088.
5. When the 8088 uses the pins for addresses, they are _____(input, output, both in and out), but when they are used for data, they will be _____ (input, output, both in and out).
6. To use a math coprocessor with the 8088, one must connect the 8088 in _____ (maximum, minimum) mode.
7. True or false. An address must be latched from pins AD0 - AD7 in the 8088.
8. Which of the following signals is provided by the 8088 CPU in minimum mode?
   (a) INTR (b) ALE (c) WR (d) IO/M
9. What is the advantage of demultiplexing address/data in the 8088 CPU?
10. What is the penalty (disadvantage) in terms of clocks in Problem 9?
11. ALE is activated in which T state?
12. Why are 8086-based systems more expensive compared to 8088-based systems?
13. To use the 8088 with the 8087 math coprocessor, is the minimum/maximum pin connected to low or high?

---

14. When the input signal RESET in the 8088 is activated, what are the contents of the IP and CS registers?

Use the following for the next 3 problems.

	MEMR*	MEMW*
(a)	0	0
(b)	0	1
(c)	1	0
(d)	1	1

* Active low.

15. Which of the above control signals is activated during the memory read cycle?
16. Which one is activated during memory write?
17. Which of the above absolutely cannot happen at the same time?

SECTION 9.2: 8284 AND 8288 SUPPORTING CHIPS

18. In maximum mode in an 8088-based system, which chip provides the ALE signal, the 8088 or the 8288?
19. Which of the following signals are provided by the 8288 chip?
    (a) IOR   (b) RESET   (c) IOW
    (d) NMI   (e) MEMR   (f) MEMW

Use the following for the next 3 problems.

	MEMR*	MEMW*
(a)	0	0
(b)	0	1
(c)	1	0
(d)	1	1

* Active low.

20. Which of the above control signals is activated during the memory read cycle?
21. Which one is activated during memory write?
22. Which of the above absolutely cannot happen at the same time?

SECTION 9.3: 8-BIT SECTION OF ISA BUS

23. When the computer is RESET, which master takes over, the 8088 or DMA?
24. To latch all the address bits of the 8088, how many 74LS373 chips are needed?
25. In the IBM PC, when AEN = 0 it indicates that the _____ (8088 CPU, DMA) is in charge of the buses. Which controls the buses when AEN = 1?
26. Which chip is used for the following?
    (a) bidirectional bus buffering
    (b) unidirectional bus buffering
27. In the IBM PC, the 74LS373 is used for which of the following?
    (a) address latch   (b) isolating the address bus
    (c) address bus boosting   (d) all of the above
28. Draw a block diagram for the 8088 minimum mode connection to the 74LS373 and 74LS245. (Modify Figure 9-5.)
29. To access the buses for interfacing with the CPU, AEN must be _____ (low, high).
30. In the 74LS245, to allow the transfer of data from side A to B, DIR = _____ and G = _____.
31. Answer Problem 30 if data is transferred from side B to A.
32. In the 74LS245, what happens if G = 1 and DIR = 0?
33. The 74LS245 chip is used for _____ (address, data) buses.
34. To allow the passage of data through the 74LS373, G = _____ and OE = _____. When is the data actually latched?

35. True or false. The 80286 is available in both LCC and PGA packages.
36. The salesclerk at the local computer store says that the 80286 has 24 bits for address and 16 bits for data; therefore, it has 2^{24} times 2 bytes = 33,554,432 bytes = 32M memory space. Is this person right? Give justification for your answer.
37. When A0 = 0 and BHE = 0, which section of the data bus (the high byte or the low byte or both) is transferring information?
38. When A0 = 0, it makes the address an _____ (odd, even) address.
39. True or false. If CLK is 20 MHz, the 80286-based system is a 10-MHz system.
40. True or false. The entire 16-megabyte memory space of the 80286 is accessible in real mode.
41. When power is applied to the 80286, it wakes up in _____ (real, protected) mode.
42. Indicate the contents of CS, IP, DS, SS, and ES when power is applied to the 286.
43. In what physical address does the 286 look for the first opcode?
44. Justify your answer in Problem 43.

SECTION 9.5: 16-BIT ISA BUS

45. The ISA expansion slot of the 80286 has two parts. How many pins does each part have? State also the number of pins for A, B, C. Which side is the component side?
46. In the ISA bus, which part of the expansion slot provides signals A20 - A23?
47. In the ISA bus, which part of the expansion slot provides signals D8 - D15?
48. Why is D0 - D7 provided on the 62-pin part of the expansion slot?
49. The BHE signal is provided on the _____ (62-pin, 36-pin) section of the ISA bus. Why?
50. If CS = FC48H and IP = 7652H, find the status of A20 for each member of the 80x86 family.
51. Which of the 8088 and 80286 microprocessors have the BHE pin?
52. To access memory anywhere in the 16M range, we must use _____ and _____ for the memory write and memory read control signals. Which part of the ISA bus provides them?
53. True or false. In the 8088, there is no A20 pin.
54. True or false. The 62-pin part of the ISA bus is almost the same as the 62-pin expansion slot of the original PC/XT.

ANSWERS TO REVIEW QUESTIONS

SECTION 9.1: 8088 MICROPROCESSOR
1. In the 8086, pins AD0 - AD15 are used for the data bus; the 8088 has an 8-bit external data bus, pins AD0 - AD7.
2. The ALE (address latch enable) pin signals whether the information is data or an address.
3. 4
4. minimum
5. (a) both (b) output (c) output
6. IO/M = 0 and RD = 0
7. IO/M = 0 and WR = 0

SECTION 9.2: 8284 AND 8288 SUPPORTING CHIPS
1. output, input
2. true
3. false
4. DTR

SECTION 9.3: 8-BIT SECTION OF ISA BUS
1. 8237 DMA   2. AEN
3. false   4. DT/R
5. buffering

SECTION 9.4: 80286 MICROPROCESSOR
1. real mode
2. 1 megabyte, 16 megabytes
3. CS = F000H and IP = FFF0

SECTION 9.5: 16-BIT ISA BUS
1. 80286
2. The advantage of the 573 is that all outputs are on one side and all inputs on the other, which reduces noise in high-frequency systems and makes the circuit board easier to design.
3. The A20 handler circuitry allows control of the A20 address bit by software, thereby solving the problem associated with the A20 pin in the 80286.
4. 100000H - FFFFFFH
5. They are provided on the 36-pin part to allow access to extended memory.

CHAPTER 10

MEMORY AND MEMORY INTERFACING

OBJECTIVES

Upon completion of this chapter, you will be able to:

>> Define the terms *capacity*, *organization*, and *speed* as used in semiconductor memories
>> Calculate the chip capacity and organization of semiconductor memory chips
>> Compare and contrast the variations of ROM: PROM, EPROM, EEPROM, Flash EPROM, and Mask ROM
>> Compare and contrast the variations of RAM: SRAM, DRAM, and NV-DRAM
>> Diagram methods of address decoding for memory chips
>> Diagram the memory map of the IBM PC in terms of RAM, VDR, and ROM allocation
>> Describe the checksum method of ensuring data integrity in ROM
>> Describe the parity bit method of ensuring data integrity in DRAM
>> Describe 16-bit memory design and related issues
>> Calculate memory cycle time and bus bandwidth for the ISA bus

This chapter explores memory and memory interfacing of the x86 PC. We first study the basics of semiconductor memory chips, then in Section 10.2 we present memory address decoding using simple logic gates. The memory map and memory space allocation of the PC are discussed in Section 10.3. Section 10.4 explores the issue of data integrity in RAM and ROM. Section 10.5 discusses the CPU's bus cycle time for memory and shows how to calculate bus bandwidth. The specifics of memory cycle time for the ISA bus are covered in Section 10.6.

SECTION 10.1: SEMICONDUCTOR MEMORY FUNDAMENTALS

In the design of all computers, semiconductor memories are used as primary storage for code and data. Semiconductor memories are connected directly to the CPU and they are the memory that the CPU first asks for information (code and data). For this reason, semiconductor memories are sometimes referred to as *primary memory*. The main requirement of primary memory is that it must be fast in responding to the CPU; only semiconductor memories can do that. Among the most widely used semiconductor memories are ROM and RAM. Before we discuss different types of RAM and ROM, we discuss terminology common to all semiconductor memories, such as capacity, organization, and speed.

Memory capacity

The number of bits that a semiconductor memory chip can store is called its *chip capacity*. It can be in units of Kbits (kilobits), Mbits (megabits), and so on. This must be distinguished from the storage capacity of computers. While the memory capacity of a memory IC chip is always given in bits, the memory capacity of a computer is given in bytes. For example, an article in a technical journal may state that the 64M chip has become popular. In that case, although it is not mentioned that 64M means 64 megabits, it is understood since the article is referring to an IC memory chip. However, if an advertisement states that a computer comes with 64M memory, since it is referring to a computer it is understood that 64M means 64 megabytes.

Memory organization

Memory chips are organized into a number of locations within the IC. Each location can hold 1 bit, 4 bits, 8 bits, or even 16 bits, depending on how it is designed internally. The number of bits that each location within the memory chip can hold is always equal to the number of data pins on the chip. How many locations exist inside a memory chip depends on the number of address pins. The number of locations within a memory IC always equals 2^x where x is the number of address pins. Therefore, the total number of bits that a memory chip can store is equal to the number of locations times the number of data bits per location. To summarize:

1. Each memory chip contains 2^x locations, where x is the number of address pins on the chip.
2. Each location contains y bits, where y is the number of data pins on the chip.
3. The entire chip will contain $2^x \times y$ bits, where x is the number of address pins and y is the number of data pins on the chip.
4. The $2^x \times y$ is referred to as the *organization* of the memory chip, where x is the number of address pins and y is the number of data pins on the chip.
5. For 2^x, use Table 10-1 to give the number of locations in K or M units.
6. $2^{10} = 1024 = 1$ K. Notice that in common speech, 1K is 1000 (as in discussing salaries or distance), but in computer terminology it is 1024.

Speed

Table 10-1: Powers of 2

One of the most important characteristics of a memory chip is the speed at which data can be accessed from it. To access the data, the address is presented to the address pins, and after a certain amount of time has elapsed, the data shows up at the data pins. The shorter this elapsed time, the better, and consequently, the more expensive the memory chip. The speed of the memory chip is commonly referred to as its *access time*. The access time of memory chips varies from a few nanoseconds to hundreds of nanoseconds, depending on the IC technology used in the design and fabrication.

The three important memory characteristics of capacity, organization, and access time will be used extensively in this chapter and throughout the book. Many of these topics will be explored in more detail in the context of applications in this and future chapters. Table 10-1 serves as a reference for the calculation of memory organization. See Examples 10-1 and 10-2 for clarification.

x	2^x
10	1K
11	2K
12	4K
13	8K
14	16K
15	32K
16	64K
17	128K
18	256K
19	512K
20	1M
21	2M
22	4M
23	8M
24	16M

Example 10-1

A given memory chip has 12 address pins and 8 data pins. Find:
(a) the organization   (b) the capacity

Solution:

(a)   This memory chip has 4096 locations ($2^{12} = 4096$), and each location can hold 8 bits of data. This gives an organization of 4096 × 8, often represented as 4Kx8.
(b)   The capacity is equal to 32K bits since there is a total of 4K locations and each location can hold 8 bits of data.

Example 10-2

A 512K memory chip has 8 pins for data. Find:
(a) the organization   (b) the number of address pins for this memory chip

Solution:

(a)   A memory chip with 8 data pins means that each location within the chip can hold 8 bits of data. To find the number of locations within this memory chip, divide the capacity by the number of data pins. 512K/8 = 64K; therefore, the organization for this memory chip is 64Kx8.
(b)   The chip has 16 address lines since $2^{16} = 64K$.

ROM (read-only memory)

ROM is a type of memory that does not lose its contents when the power is turned off. For this reason, ROM is also called *nonvolatile memory*. There are different types of read-only memory, such as PROM, EPROM, EEPROM, Flash ROM, and mask ROM. Each is explained next.

PROM (programmable ROM) or OTP ROM

PROM refers to the kind of ROM that the user can burn information into. In other words, PROM is a user-programmable memory. For every bit of the PROM, there exists a fuse. PROM is programmed by blowing the fuses. If the information burned into PROM is wrong, that PROM must be discarded since internal fuses are blown permanently. For this reason, PROM is also referred to as *OTP* (one-time programmable). The process of programming ROM is also called *burning* ROM and requires special equipment called a ROM burner or ROM programmer.

EPROM (erasable programmable ROM)

EPROM was invented to allow changes in the contents of PROM after it is burned. In EPROM, one can program the memory chip and erase it thousands of times. This is especially useful during development of the prototype of a microprocessor-based project. The only problem with EPROM is that erasing its contents can take up to 20 minutes. All EPROM chips have a window that is used to shine ultraviolet (UV) radiation to erase its contents. For this reason, EPROM is also referred to as UV-erasable EPROM or simply UV-EPROM. Figure 10-1 shows the pins for a 64Kbit UV-EPROM chip. Notice the A0 - A12 address pins and O0 - O7 (output) for D0 - D7 data pins. The OE (out enable) is for the read signal.

To program a UV-EPROM chip, the following steps must be taken:

1. Its contents must be erased. To erase a chip, it is removed from its socket on the system board and placed in EPROM erasure equipment to expose it to UV radiation for 15 - 20 minutes.
2. Program the chip. To program a UV-EPROM chip, place it in the ROM burner (programmer). To burn code and data into EPROM, the ROM burner uses 12.5 volts or higher, depending on the EPROM type. This voltage is referred to as VPP in the UV-EPROM data sheet.
3. Place the chip back into its socket on the system board.

As can be seen from the above steps, in the same way that there is an EPROM programmer (burner), there is also separate EPROM erasure equipment. The main problem, and indeed the major disadvantage of UV-EPROM, is that it cannot be programmed while in the system board (motherboard). To find a solution to this problem, EEPROM was invented.

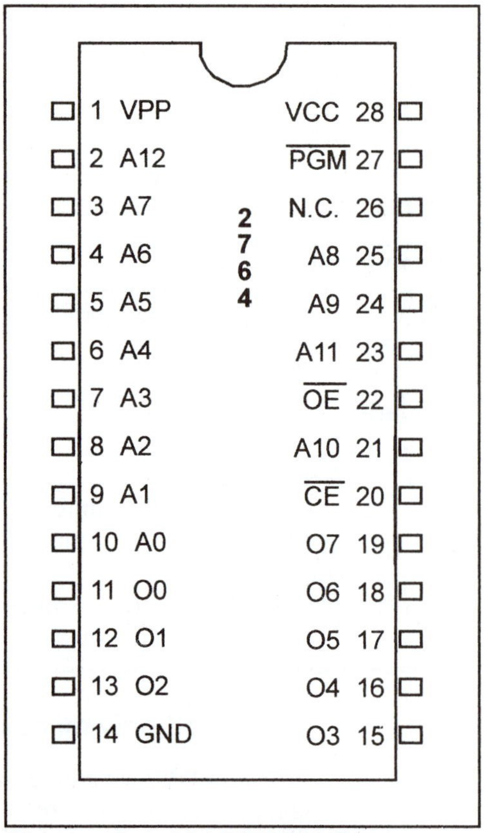

Figure 10-1. UV-EPROM Chip
(Reprinted by permission of Intel Corporation, Copyright Intel Corp., 1987)

EEPROM (electrically erasable programmable ROM)

EEPROM has several advantages over EPROM, such as the fact that its method of erasure is electrical and therefore instant, as opposed to the 20-minute erasure time required for UV-EPROM. In addition, in EEPROM, one can select which byte to be erased, in contrast to UV-EPROM, in which the entire contents of ROM are erased. However, the main advantage of EEPROM is the fact that one can program and erase its contents while it is still in the system board. It does not require physical removal of the memory chip from its socket. In other words, unlike UV-EPROM, EEPROM does not require an external erasure and programming device. To utilize EEPROM fully, the designer must incorporate into the system board the circuitry to program the EEPROM, using 12.5 V for VPP. EEPROM with VPP of 5 - 7 V is available, but it is more expensive. In general, the cost per bit for EEPROM is much higher than for UV-EPROM.

Table 10-2: Examples of ROM Memory Chips

Type	Part Number	Speed (ns)	Capacity	Organization	Pins	VPP
UV-EPROM	2716	450	16K	2Kx8	24	25
	2716-1	350	16K	2Kx8	24	25
	2716B	450	16K	2Kx8	24	12.5
	2732A-45	450	32K	4Kx8	24	21
	2732A-20	200	32K	4Kx8	24	21
	27C32	450	32K	4Kx8	24	25
	2764A-25	250	64K	8Kx8	28	12.5
	27C64-15	150	64K	8Kx8	28	12.5
	27128-20	200	128K	16Kx8	28	12.5
	27C128-25	250	128K	16Kx8	28	12.5
	27256-20	200	256K	32Kx8	28	12.5
	27C256-20	200	256K	32Kx8	28	12.5
	27512-25	250	512K	64Kx8	28	12.5
	27C512-25	250	512K	64Kx8	28	12.5
	27C010-12	120	1M	128Kx8	32	12.5
	27C201-12	120	2M	256Kx8	32	12.5
	27C401-12	120	4M	512Kx8	32	12.5
EEPROM	28C16A-25	250	16K	2Kx8	24	5
	2864A	250	64K	8Kx8	28	5
	28C256-15	150	256K	32Kx8	28	5
	28C256-25	250	256	32x8	28	5
Flash ROM	28F256-20	200	256K	32Kx8	32	12
	28F256-15	150	256K	32Kx8	32	12
	28F010-20	200	1M	128Kx8	32	12
	28F020-15	150	2M	256Kx8	32	12

Table 10-2 shows examples of some popular ROM chips and their characteristics. Notice the patterns of the IC numbers. For example, 27128-20 refers to UV-EPROM that has a capacity of 128K bits and access time of 200 nanoseconds. The capacity of the memory chip is indicated in the part number and the access time is given with a zero dropped. In part numbers, C refers to CMOS technology. While 27xx is for UV-E PROM, 28xx is for EEPROM.

Example 10-3

For ROM chip 27128, find the number of data and address pins, using Table 10-2.

Solution:

The 27128 has a capacity of 128K bits. Table 10-2 also shows that it has 16Kx8 organization, which indicates that there are 8 pins for data, and 14 pins for address ($2^{14} = 16$K).

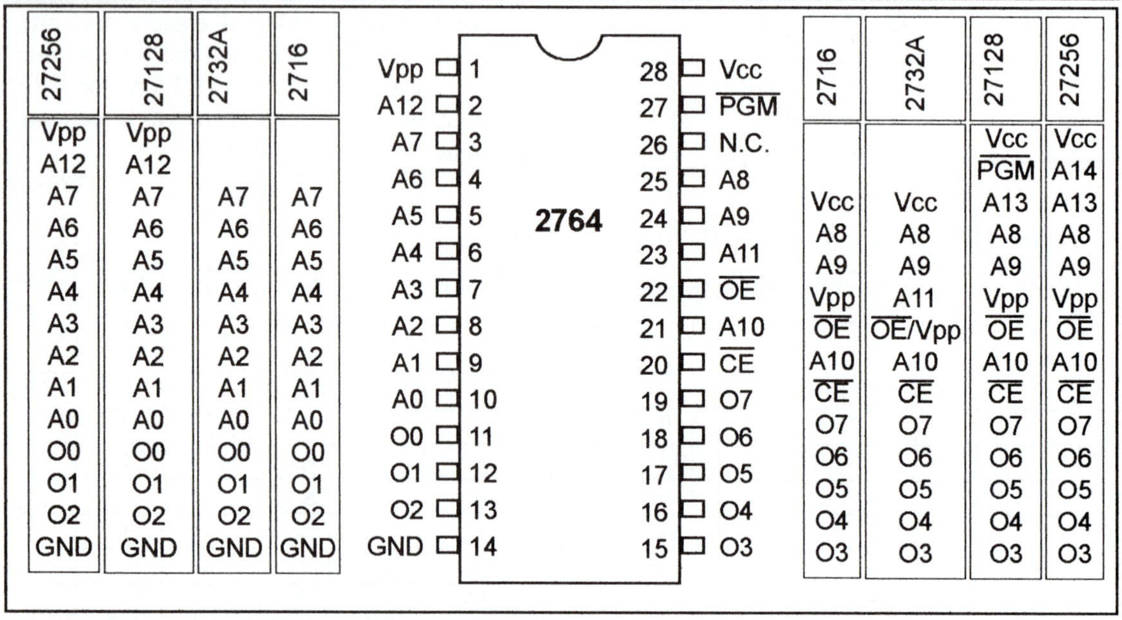

27256	27128	2732A	2716		2764			2716	2732A	27128	27256
Vpp	Vpp			Vpp □ 1	⌒	28 □ Vcc				Vcc	Vcc
A12	A12			A12 □ 2		27 □ $\overline{PGM}$				$\overline{PGM}$	A14
A7	A7	A7	A7	A7 □ 3		26 □ N.C.		Vcc	Vcc	A13	A13
A6	A6	A6	A6	A6 □ 4		25 □ A8		A8	A8	A8	A8
A5	A5	A5	A5	A5 □ 5	2764	24 □ A9		A9	A9	A9	A9
A4	A4	A4	A4	A4 □ 6		23 □ A11		Vpp	A11	Vpp	Vpp
A3	A3	A3	A3	A3 □ 7		22 □ $\overline{OE}$		$\overline{OE}$	$\overline{OE}$/Vpp	$\overline{OE}$	$\overline{OE}$
A2	A2	A2	A2	A2 □ 8		21 □ A10		A10	A10	A10	A10
A1	A1	A1	A1	A1 □ 9		20 □ $\overline{CE}$		$\overline{CE}$	$\overline{CE}$	$\overline{CE}$	$\overline{CE}$
A0	A0	A0	A0	A0 □ 10		19 □ O7		O7	O7	O7	O7
O0	O0	O0	O0	O0 □ 11		18 □ O6		O6	O6	O6	O6
O1	O1	O1	O1	O1 □ 12		17 □ O5		O5	O5	O5	O5
O2	O2	O2	O2	O2 □ 13		16 □ O4		O4	O4	O4	O4
GND	GND	GND	GND	GND □ 14		15 □ O3		O3	O3	O3	O3

Figure 10-2. Pin Configurations for 27xx ROM Family

Flash Memory

Since the early 1990s, Flash ROM has become a popular user-programmable memory chip, and for good reasons. First, the process of erasure of the entire contents takes only a few seconds, or one might say in a flash, hence its name: Flash memory. In addition, the erasure method is electrical and for this reason it is sometimes referred to as Flash EEPROM. To avoid confusion, it is commonly called Flash ROM. The major difference between EEPROM and Flash memory is the fact that when flash memory's contents are erased the entire device is erased, in contrast to EEPROM, where one can erase a desired section or byte. Although there are some flash memories recently made available in which the contents are divided into blocks and the erasure can be done block by block, unlike EEPROM, no byte erasure option is available. Because Flash ROM can be programmed while it is in its socket on the system board, it is widely used to upgrade the BIOS ROM of the PC or the operating system on Cisco routers. Some designers believe that flash memory will replace the hard disk as a mass storage medium. This would increase the performance of computers tremendously, since flash memory is semiconductor memory with access time in the range of 100 ns compared with disk access time in the range of tens of milliseconds. For this to happen, flash memory's program/erase cycles must become infinite, just like hard disks. *Program/erase cycle* refers to the number of times that a chip can be erased and programmed before it becomes unusable. At this time, the program/erase cycle is 500,000 for Flash and EEPROM, 2000 for UV-EPROM, and infinite for RAM and disks. In Table 10-2, notice that the part number for Flash ROM uses the 28Fxx designation, where F indicates the Flash type ROM.

Mask ROM

Mask ROM refers to a kind of ROM whose contents are programmed by the IC manufacturer. In other words, it is not a user-programmable ROM. The terminology *mask* is used in IC fabrication. Since the process is costly, mask ROM is used when the needed volume is high and it is absolutely certain that the contents will not change. It is common practice to use UV-EPROM or Flash for the development phase of a project, and only after the code/data have been finalized is mask ROM ordered. The main advantage of mask ROM is its cost, since it is significantly cheaper than other kinds of ROM, but if an error in the data is found, the entire batch must be thrown away.

RAM (random access memory)

RAM memory is called *volatile memory* since cutting off the power to the IC will mean the loss of data. Sometimes RAM is also referred to as RAWM (read and write memory), in contrast to ROM, which cannot be written to. There are three types of RAM: static RAM (SRAM), dynamic RAM (DRAM), and NV-RAM (nonvolatile RAM). Each is explained separately.

SRAM (static RAM)

Storage cells in static RAM memory are made of flip-flops and therefore do not require refreshing in order to keep their data. This is in contrast to DRAM, discussed below. The problem with the use of flip-flops for storage cells is that each cell requires at least 6 transistors to build, and the cell holds only 1 bit of data. In recent years, the cells have been made of 4 transistors, which is still too many. The use of 4-transistor cells plus the use of CMOS technology has given birth to a high-capacity SRAM, but the capacity of SRAM is far below DRAM. Table 10-3 shows some examples of SRAM. SRAMs are widely used for cache memory, which is discussed in Chapter 22. Figure 10-3 shows the pin diagram for the 6116 SRAM chip. The 6116 has an organization of 2Kx8, which gives a capacity of 16 Kbits, as indicated in the part number. The following is a description of the 6116 SRAM pins.

A0 - A11 are for address inputs, where 12 address lines gives $2^{16} = 2K$.

I/O0 - I/O7 are for data I/O, where 8-bit data lines gives an organization of 2Kx8.

WE (write enable) is for writing data into SRAM (active low).

OE (output enable) is for reading data out of SRAM (active low)

CS (chip select) is used to select the memory chip.

The functional diagram for the 6116 SRAM is given in Figure 10-4.

Figure 10-5 shows the following steps to write data into SRAM.

1. Provide the addresses to pins A0 - A11.
2. Activate the CS pin.
3. Make WE = 0 while RD = 1.
4. Provide the data to pins I/O0 - I/O7.
5. Make WE = 1 and data will be written into SRAM on the positive edge of the WE signal.

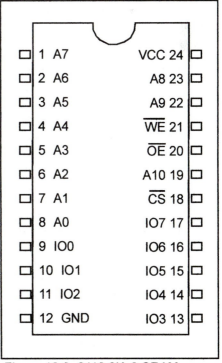

1 A7	VCC 24
2 A6	A8 23
3 A5	A9 22
4 A4	$\overline{WE}$ 21
5 A3	$\overline{OE}$ 20
6 A2	A10 19
7 A1	$\overline{CS}$ 18
8 A0	IO7 17
9 IO0	IO6 16
10 IO1	IO5 15
11 IO2	IO4 14
12 GND	IO3 13

Figure 10-3. 6116 2Kx8 SRAM

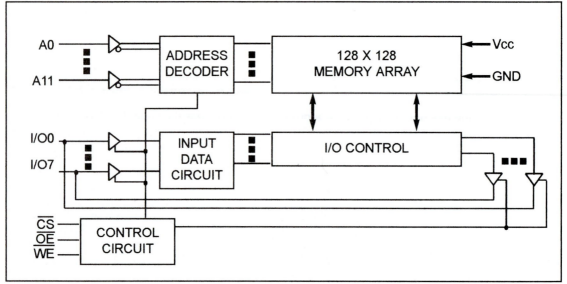

Figure 10-4. Functional Block Diagram for 6116 SRAM

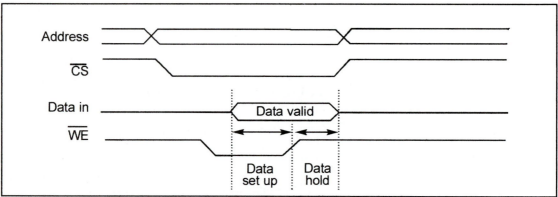

Figure 10-5. Memory Write Timing for SRAM

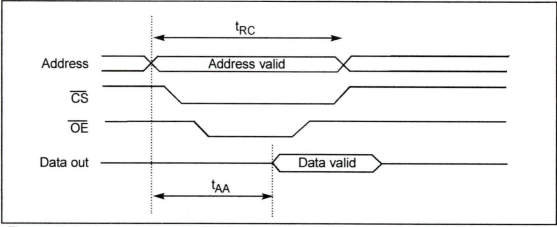

Figure 10-6. Memory Read Timing for SRAM

The following are steps to read data from SRAM. See Figure 10-6.
1. Provide the addresses to pins A0 - A11. This is the start of the access time (tAA).
2. Activate the CS pin.
3. While WE = 1, a high-to-low pulse on the OE pin will read the data out of the chip.

In the 6116 SRAM, the access time, tAA, is measured as the time elapsed from the moment the address is provided to the address pins to the moment that the data is available at the data pins. The speed for the 6116 chip can vary from 100 ns to 15 ns.

Examine the read cycle time for SRAM in Figure 10-6. The read cycle time (tRC) is defined as the minimum amount of time required to read one byte of data. That is, from the moment we apply the address of the byte to the moment we can begin the next read operation. In SRAM for which tAA =100 ns, tRC is also 100 ns. This implies that we can read the contents of consecutive address locations with each taking no more than 100 ns. Hence, in SRAM and ROM, tAA = tRC. They are not equal in DRAM, as we will find in Chapter 22.

DRAM (dynamic RAM)

Since the early days of the computer, the need for huge, inexpensive read/write memory was a major preoccupation of computer designers. In 1970, Intel Corporation introduced the first dynamic RAM (random access memory). Its density (capacity) was 1024 bits and it used a capacitor to store each bit. The use of a capacitor as a means to store data cuts down the number of transistors needed to build the cell; however, it requires constant refreshing due to leakage. This is in contrast to SRAM (static RAM), whose individual cells are made of flip-flops. Since each bit in SRAM uses a single flip-flop and each flip-flop requires 6 transistors, SRAM has much larger memory cells and consequently lower density. The use of capacitors as storage cells in DRAM results in much smaller net memory cell size.

The advantages and disadvantages of DRAM memory can be summarized as follows. The major advantages are high density (capacity), cheaper cost per bit, and lower power consumption per bit. The disadvantage is that it must be refreshed periodically, due to the fact that the capacitor cell loses its charge; furthermore, while it is being refreshed, the data cannot be accessed. This is in contrast to SRAM's flip-flops, which retain data as long as the power is on, which do not need to be refreshed, and whose contents can be accessed at any time. Since 1970, the capacity of DRAM has exploded. After the 1K-bit (1024) chip came the 4K-bit in 1973, and then the 16K chip in 1976. The 1980s saw the introduction of 64K, 256K, and finally 1M and 4M memory chips. The 1990s saw the 16M, 64M, and 256M DRAM chips. In 1980 when the IBM PC was being designed, 16K-bit chips were widely used, but currently motherboards use 256K, 1M, 4M, 16M, 64M and 256M chips. Keep in mind that when talking about IC memory chips, the capacity is always assumed to be in bits. Therefore, a 1M chip means a 1-megabit chip and a 256K chip means a 256K-bit chip. However, when talking about the memory of a computer system, it is always assumed to be in bytes. For example, if we say that a PC motherboard has 256M, it means 256 megabytes of memory.

Packaging issue in DRAM

In DRAM it is difficult to pack a large number of cells into a single chip with the normal number of pins assigned to addresses. For example, a 64K-bit chip (64Kx1) must have 16 address lines and 1 data line, requiring 16 pins to send in the address if the conventional method is used. This is in addition to VCC power, ground, and read/write control pins. Using the conventional method of data access, the large number of pins defeats the purpose of high density and small packaging, so dearly cherished by IC designers. Therefore, to reduce the number of pins needed for addresses, multiplexing/demultiplexing is used. The method used is to split the address into halves and send in each half of the address through the same pins, thereby requiring fewer address pins. Internally, the DRAM structure is divided into a square of rows and columns. The first half of the address is called the *row* and the second half is called the *column*. For example, in the case of DRAM of 64Kx1 organization, the first half of the address is sent in through the 8 pins A0 - A7, and by activating RAS (row address strobe), the internal latches inside DRAM grab the first half of the address. After that, the second half of the address is sent in through the same pins and by activating CAS (column address strobe), the internal latches inside DRAM latch this second half of the address. This results in using 8 pins for addresses plus RAS and CAS, for a total of 10 pins, instead of the 16 pins that would be required without multiplexing. To access a bit of data from DRAM,

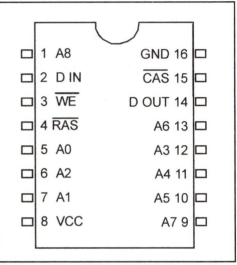

Figure 10-7. 256Kx1 DRAM

both row and column addresses must be provided. For this concept to work, there must be a 2 by 1 multiplexer outside the DRAM circuitry while the DRAM chip has its own internal demultiplexer. Due to the complexities associated with

Table 10-3: Examples of RAM Chips

Type	Part Number	Speed (ns)	Capacity	Organization	Pins
SRAM	6116-1	100	16K	2Kx8	24
	6116LP-70*	70	16K	2Kx8	24
	6264-10	100	64K	8Kx8	28
	62256LP-10*	100	256K	32Kx8	28
DRAM	4116-20	200	16K	16Kx1	16
	4116-15	150	16K	16Kx1	16
	4116-12	120	16K	16Kx1	16
	4416-12	120	64K	16Kx4	18
	4416-15	150	64K	16Kx4	18
	4164-15	150	64K	64Kx1	16
	41464-8	80	256K	64Kx4	18
	41256-15	150	256K	256Kx1	16
	41256-6	60	256K	256Kx1	16
	414256-10	100	1M	256Kx4	20
	511000P-8	80	1M	1Mx1	18
	514100-7	700	4M	4Mx1	20
NV-SRAM	DS1220	100	16K	2Kx8	24
	DS1225	150	64K	8Kx8	28
	DS1230	70	256K	32Kx8	28

* LP indicates low power.

DRAM interfacing (RAS, CAS, the need for external multiplexer and refreshing circuitry), many small microprocessor-based projects that do not require much RAM use SRAM instead of DRAM. Figure 10-7 shows the pins for a DRAM chip. Notice the RAS and CAS pins. Also notice the WE (write enable) pin for read and write actions. Table 10-3 provides some examples of DRAM chips.

DRAM, SRAM, and ROM organizations

Although the organizations for SRAMs and ROMs are always x8, DRAM can have x1, x4, x8, or even x16 organizations. In some memory chips (notably SRAM), the data pins are called I/O. In some DRAMs there are separate pins Din and Dout. The DRAMs with 1x organization are widely used for parity bit as we will soon see in this chapter. See Examples 10-4 and 10-5. As the density of the DRAM chips goes up, it makes sense to use higher-density chips to save space on the printed circuit board. For that reason, the memory configuration for various PCs is different depending on the date of manufacturing and the availability of the memory chip at the time of the design.

Example 10-4

Show possible organizations and number of address pins for the: (a) 256K DRAM chip, and (b) 1M DRAM chip.

Solution:

(a)   For 256K chips, possible organizations are 256Kx1 or 64Kx4. In the case of 256Kx1, there are 256K locations and each location inside DRAM provides 1 bit. The 256K locations are accessed through the 18-bit address A0 - A17 since 2^{18} = 256K. The chip has only A0 - A8 physical pins plus RAS and CAS and one pin for data in addition to VCC, ground, and the R/W pin that every DRAM chip must have. For 64Kx4, it requires 16 address bits to access each location (2^{16} = 64K), and each location inside the DRAM has 4 cells. That means that it must have 4 data pins, D0 - D3, 8 address pins, A0 - A7, plus RAS and CAS.

(b)   In the case of a 1M chip, there can be either 1Mx1 or 256Kx4 organizations. For 1Mx1, there are A0 - A9, 10 pins, to access 2^{20} = 1M locations with the help of RAS and CAS and one pin for data. The 256Kx4 has 9 (A0 - A8) and 4 (D0 - D3) pins, respectively, for address and data plus RAS and CAS pins.

Example 10-5

Discuss the number of pins set aside for addresses in each of the following memory chips.
(a) 16Kx4 DRAM   (b) 16Kx8 SRAM

Solution:
Since 2^{14} =16K:
(a)   For DRAM we have 7 pins (A0 - A6) for the address pins and 2 pins for RAS and CAS.
(b)   For SRAM we have 14 pins (A0 - A13) for address and no pins for RAS and CAS since they are associated only with DRAM.

NV-RAM (nonvolatile RAM)

While both DRAM and SRAM are volatile, there is a new type of RAM called NV-RAM, nonvolatile RAM. Like other RAMs, it allows the CPU to read and write to it; but when the power is turned off, the contents are not lost, just as for ROM. NV-RAM combines the best of RAM and ROM: the read and writability of RAM, plus the nonvolatility of ROM. To retain its contents, every NV-RAM chip internally is made of the following components:

1. It uses extremely power-efficient (extremely low power consumption) SRAM cells built out of CMOS.
2. It uses an internal lithium battery as a backup energy source.
3. It uses an intelligent control circuitry. The main job of this control circuitry is to monitor the VCC pin constantly to detect loss of the external power supply. If the power to the VCC pin falls below out-of-tolerance conditions, the control circuitry switches automatically to its internal power source, the lithium battery. In this way, the internal lithium power source is used to retain the NV-RAM contents only when the external power source is off.

It must be emphasized that all three of the components above are incorporated into a single IC chip, and for this reason nonvolatile RAM is much more expensive than SRAM as far as cost per bit is concerned. Offsetting the cost, however, is the fact that it can retain its contents up to ten years after the power has been turned off and allows one to read and write exactly the same as in SRAM. See Table 10-3 for NV-RAM parts made by Dallas Semiconductor. In the x86 PC, NV-RAM is used to save the system setup. This NV-RAM in PC is commonly referred to CMOS RAM.

Review Questions

1. The speed of semiconductor memory is in the range of _____.
2. Find the organization and chip capacity for each of the following with the indicated number of address and data pins.
   (a) 11 address, 8 data SRAM   (b) 13 address, 8 data ROM
   (c) 8 address, 4 data DRAM   (d) 9 address, 1 data DRAM
3. Find the capacity and number of pins set aside for address and data for memory chips with the following organizations.
   (a) 16Kx8 SRAM   (b) 32Kx8 EPROM   (c) 1Mx1 DRAM
   (d) 256Kx4 DRAM (e) 64Kx8 EEPROM   (f) 1Mx4 DRAM
4. Why is flash memory preferable to UV-EPROM in system development?
5. What kind of memory is used in the CMOS RAM of the x86 PC?

SECTION 10.2: MEMORY ADDRESS DECODING

Current system designs use CPLDs (complex programmable logic devices), in which memory and address decoding circuitry are integrated into one programmable chip. However, it is still important to understand how this task can be performed with common logic gates. In this section we show how to use simple logic gates to accomplish address decoding. The CPU provides the address of the data desired, but it is the job of the decoding circuitry to locate the memory chip where the desired data is stored. To explore the concept of decoding circuitry, we look at the use of NAND and 74LS138 chips as decoders. In this discussion we use SRAM or ROM for the sake of simplicity.

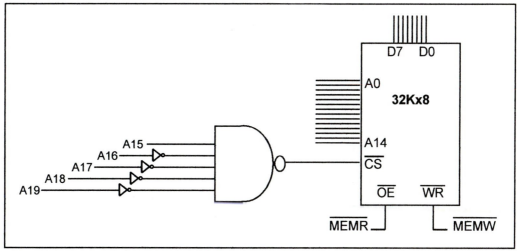

Figure 10-8. Using Simple Logic Gate as Decoder

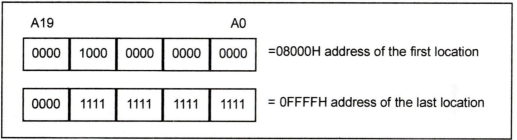

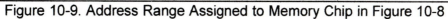

Figure 10-9. Address Range Assigned to Memory Chip in Figure 10-8

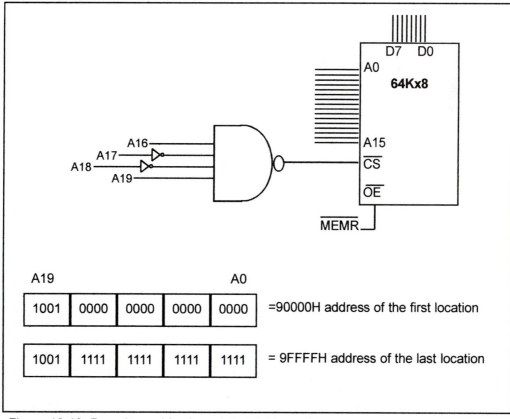

Figure 10-10. Decoder and Its Associated Address Range

Example 10-6

Referring to Figure 10-10 we see that the memory chip has 64K bytes of space. Show the calculation that verifies that address range 90000 to 9FFFFH is comprised of 64K bytes.

Solution:

To calculate the total number of bytes for a given memory address range, subtract the two addresses and add 1 to get the total bytes in hex. Then the hex number is converted to decimal and divided by 1024 to get K bytes.

$$
\begin{array}{ll}
\quad 9FFFF & \quad FFFF \\
\underline{-90000} & \underline{+\quad\ \ 1} \\
\quad 0FFFF & \quad 10000 \text{ hex} = 65,536 \text{ decimal} = 64K
\end{array}
$$

Simple logic gate as address decoder

As seen in the last section, memory chips have one or more CS (chip select) pins that must be activated for the memory's contents to be accessed. Sometimes the chip select is also referred to as chip enable (CE). In connecting a memory chip to the CPU, the data bus is connected directly to the data pins of the memory. Control signals MEMR and MEMW are connected to the OE and WR pins of the memory chip, respectively (see Figure 10-8). In the case of the address buses, while the lower bits of the address go directly to the memory chip address pins, the upper ones are used to activate the CS pin of the memory chip. It is the CS pin along with RD/WR that allows the flow of data in or out of the memory chip. In other words, no data can be written into or read from the memory chip unless CS is activated. The CS input is active low and can be activated using some simple logic gates, such as NAND and inverters. See Figures 10-8 and 10-10. Figure 10-9 shows the address range for the design in Figure 10-8.

In Figure 10-10, notice that the output of the NAND gate is active low and that the CS pin is also active low. That makes them a perfect match. Also notice that A19 - A16 must be = 1001 in order for CS to be activated. This results in the assignment of addresses 9000H to 9FFFFH to this memory block. Figures 10-8 and 10-10 show that for every block of memory, we need a NAND gate. The 74LS138 has 8 NAND gates in it; therefore, a single chip can control 8 blocks of memory. This was the method of memory addressing decoding used before the introduction of CPLD, and it is still the best method if you do not have access to CPLD.

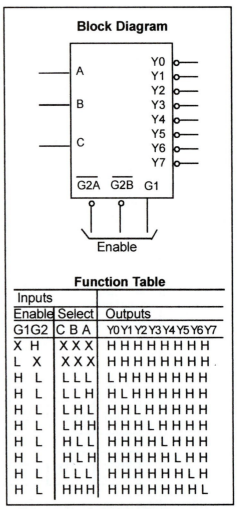

Block Diagram

Function Table

Inputs			Outputs
Enable	Select		
G1 G2	C B A	Y0 Y1 Y2 Y3 Y4 Y5 Y6 Y7
X H	X X X	H H H H H H H H
L X	X X X	H H H H H H H H .
H L	L L L	L H H H H H H H
H L	L L H	H L H H H H H H
H L	L H L	H H L H H H H H
H L	L H H	H H H L H H H H
H L	H L L	H H H H L H H H
H L	H L H	H H H H H L H H
H L	L L L	H H H H H H L H
H L	H H H	H H H H H H H L

Figure 10-11. 74LS138 Decoder
(Reprinted by permission of Texas Instruments, Copyright Texas Instruments, 1988)

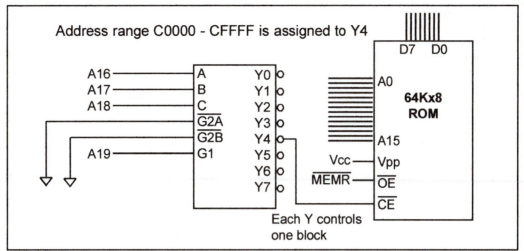

Address range C0000 - CFFFF is assigned to Y4

Figure 10-12. 74LS138 as Decoder

Using the 74LS138 as decoder

In the absence of CPLD or FPGA as address decoders, the 74LS138 chip is an excellent choice. The 3 inputs A, B, and C of the 74LS138 generate 8 active-low outputs Y0 - Y7, as shown in Figure 10-11. Each Y output is connected to the CS of a memory chip, allowing control of 8 memory blocks by a single 74LS138. This eliminates the need for using NAND and inverter gates. As shown in Figure 10-11, where A, B, and C select which output is activated, there are three additional inputs, G2A, G2B, and G1, that can be used for address or control signal selection. Notice that G2A and G2B are both active low, while G1 is active high. If any one of the inputs G1, G2A, or G2B is not connected, they must be activated permanently by either VCC or ground, depending on the activation level.

In Figure 10-12, we have A0 - A15 going from the CPU directly to A0 - A15 of the memory chip. A16 - A18 are used for the A, B, and C inputs of the 74LS138. A19 is controlling the G1 pin of the 74138. For the 74138 to be enabled, we need G2A = 0, G2B = 0, and G1 = 1. G2A and G2B are grounded. When G1 = 1, this 74138 is selected. Depending on the status of pins A, B, and C, one of the Ys is selected. To select Y4, we need CBA = 100 (in binary). That gives us the address range of C0000 to CFFFFH for the memory chip controlled by the Y4 output. For further clarification, see Example 10-7.

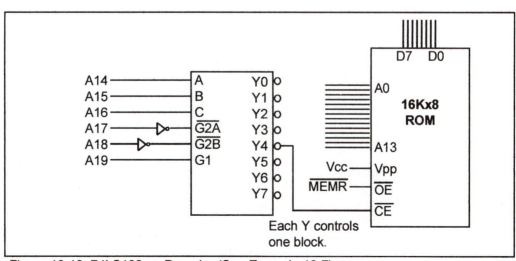

Figure 10-13. 74LS138 as Decoder (See Example 10-7)

Looking at the design in Figure 10-13, find the address range for (a) Y4, (b) Y2, and (c) Y7, and verify the block size controlled by each Y.

Solution:

(a) The address range for Y4 is calculated as follows.

```
A19 A18 A17 A16 A15 A14 A13 A12 A11 A10 A9 A8 A7 A6 A5 A4  A3 A2 A1  A0
1   1   1   1   0   0   0   0   0   0   0  0  0  0  0  0   0  0  0   0
1   1   1   1   0   0   1   1   1   1   1  1  1  1  1  1   1  1  1   1
```

The above shows that the range for Y4 is F0000H to F3FFFH. In Figure 10-13, notice that A19, A18, and A17 must be 1 for the decoder to be activated. Y4 will be selected when A16 A15 A14 = 100 (4 in binary). The remaining A13 - A0 will be 0 for the lowest address and 1 for the highest address.

(b) The address range for Y2 is E8000H to EBFFFH.

```
A19 A18 A17 A16 A15 A14 A13 A12 A11 A10 A9 A8 A7 A6 A5 A4  A3 A2 A1 A0
1   1   1   0   1   0   0   0   0   0   0  0  0  0  0  0   0  0  0  0
1   1   1   0   1   0   1   1   1   1   1  1  1  1  1  1   1  1  1  1
```

(c) The address range for Y7 is FC000H to FFFFFH. Notice that FFFFF – FC000H = 3FFFH, which is equal to 16,383 in decimal. Adding 1 to it because of the 0 location, we have 16,384. 16,384/1024 = 16K, the block (chip) size.

Review Questions

1. The MEMR signal from the CPU is connected to the _____ pin of the ROM chip.
2. The MEMW signal from the CPU is connected to the _____ pin of the RAM chip.
3. The CS pin of the memory chip is normally an _____ (active low, active high) signal.
4. The 74LS138 has total of _____ outputs.
5. The Y output of the 74138 is _____ (active low, active high).

SECTION 10.3: IBM PC MEMORY MAP

All x86 CPUs in real mode provide 20 address bits (A0 - A19). Therefore, the maximum amount of memory that they can access is one megabyte. How this 1M is allocated in the original PC is the main topic of this section. The 20 lines, A0 - A19, of the system address bus can take the lowest value of all 0s to the highest value of all 1s in binary. Converting these values to hexadecimal gives an address range of 00000H to FFFFFH. This is shown in Figure 10-14. Any address that is assigned to any memory block in the 8088-based original PC must fall between these two ranges. This includes all x86 microprocessors in real mode.

The 20-bit address of the 8088 provides a maximum of 1M (1024K bytes) of memory space. Of the 1024K bytes, the designers of the original IBM PC decided to set aside 640K for RAM, 128K for video display RAM (VDR), and 256K for ROM, as shown in Figure 10-15. In today's PC, 640K bytes is not that much, but the standard of the personal computer in 1980 was 64K bytes of memory. At that time, 640K seemed like more than anyone would ever need. Next we discuss the memory map of the PC.

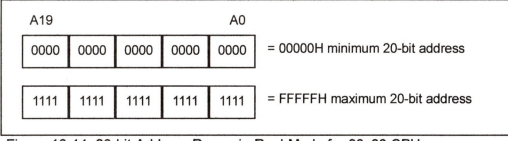

A19				A0	
0000	0000	0000	0000	0000	= 00000H minimum 20-bit address
1111	1111	1111	1111	1111	= FFFFFH maximum 20-bit address

Figure 10-14. 20-bit Address Range in Real-Mode for 80x86 CPUs

Conventional memory: 640K of RAM

In the x86 PC, the addresses from 00000 to 9FFFFH, including location 9FFFFH, are set aside for RAM. In early PCs, only 64K to 256K bytes of RAM were on the motherboard and the rest had to be expanded by adding a memory expansion plug-in card. In those early models, when a RAM memory board was installed, switches had to be set to inform BIOS and DOS of the added memory. In today's x86 PC this is done by the CMOS set-up process and this information is kept by the CMOS NV-RAM for the next cold boot. Of the 640K bytes of memory, some is used by the MS-DOS operating system (the amount depends on the version of DOS) and the rest of the available RAM is used by utilities and application programs. This 640K bytes of memory is commonly referred to as *conventional memory*. Notice that even though the vast majority of PCs use MS Windows for the operating system the above concepts are still valid since DOS is embedded into Windows to run legacy applications.

Of the total amount of RAM installed, the first 1K (00000 to 003FF = 1024 bytes) is set aside for the interrupt vector table (see Chapter 15). 00400 to 004FF is set aside for the BIOS temporary data area, as shown in Appendix H. Finally, a certain number of kilobytes is occupied by the operating system itself. The exact amount varies depending on the version of MS-DOS.

FFFFF	ROM 256K
C0000	
FFFFF A0000	VDR 128K
9FFFF	RAM 640K
00000	

Figure 10-15. Memory Map of the IBM PC

Example 10-8

Show the calculation that verifies that addresses 00000 to 9FFFFH comprise 640K bytes.

Solution:

To calculate the total number of bytes for a given memory address range, subtract the two addresses and add 1 to get the total bytes in hex. Then the hex number is converted to decimal and divided by 1024 to get K bytes.

$$
\begin{array}{cc}
9FFFF & 9FFFF \\
- 00000 & + 00001 \\
\hline
9FFFF & A0000 \text{ hex} = 655{,}360 \text{ decimal} = 640K
\end{array}
$$

BIOS data area

As mentioned earlier, the BIOS data area is a section of RAM memory used by BIOS to store some extremely important system information. A partial list of that information is given in Figure 10-16. The full list of the BIOS data area is in Appendix H. BIOS stores system information in the BIOS data area as it tests each section of the PC. The operating system navigates the system hardware with the help of information stored in the BIOS data area. For example, the BIOS data area tells the operating system how many serial and parallel ports are installed in the PC. We will examine this topic further in the serial and parallel port chapters.

```
Partial Listing of IBM PC RAM Memory Map for Interrupt, BIOS Data

     Memory Location          Bytes   Description
     0000:0000 to 0000:03FF   1024    interrupt table

     0000:0400 to 0000:0401   2       port address of com1
     0000:0402 to 0000:0403   2       port address of com2
     0000:0404 to 0000:0405   2       port address of com3
     0000:0406 to 0000:0407   2       port address of com4

     0000:0408 to 0000:0409   2       port address of lpt1
     0000:040A to 0000:040B   2       port address of lpt2
     0000:040C to 0000:040D   2       port address of lpt3
     0000:040E to 0000:040F   2       port address of lpt4

     0000:0410 to 0000:0411   2       list of installed hardware
     0000:0412 to 0000:0412   1       initialization flag

     0000:0413 to 0000:0414   2       memory size (K bytes)

     0000:0415 to 0000:0416   2       memory in I/O channel (if any)

     0000:0417 to 0000:0418   2       keyboard status flag
     0000:0419 to 0000:0419   1       alternate key entry storage
     0000:041A to 0000:041B   2       keyboard buffer pointer (head)
     0000:041C to 0000:041D   2       keyboard buffer pointer (tail)
     0000:041E to 0000:043D   32      keyboard buffer

     0000:043E to 0000:0448   11      diskette data area

     0000:0449 to 0000:0449   1       current video mode
```

Figure 10-16. The BIOS Data Area in PC (See Appendix H for full listing)

Video display RAM (VDR) map

To display information on the monitor of the PC, the CPU must first store that information in memory called *video display RAM (VDR)*. It is the job of the video controller to display the contents of VDR on the screen. Therefore, the address of the VDR must be within the CPU address range. In the x86 PC, from A0000 to BFFFFH, a total of 128K bytes of the CPU's addressable memory is allocated for video. Of that 128K, only a portion is used for VDR, the amount used depending on the mode in which the video system is being used (text or graphics), and the resolution. For example, the monochrome video mode uses only addresses starting at B0000 up to 4K bytes of RAM, color graphics mode uses addresses starting at B8000, and VGA has a starting address of A0000. See Table 10-4. For more details of each video mode and how many bytes of memory are used in text and graphics modes and their resolution, see Chapter 16. DOS/Windows can use the unused portion of the 128K-byte space allocated to video, as we will see in Chapter 25.

Table 10-4: Video Display RAM Memory Map

Adapters	Number of Bytes Used	Starting Address
CGA, EGA, VGA	16,384 (16 K)	B8000H
MDA, EGA, VGA	4096 (4K)	B0000H
EGA, VGA	65,536 (64K)	A0000H

ROM Address and Cold Boot in the PC

When power is applied to a CPU it must wake up at an address that belongs to ROM. Obviously, the first code executed by the CPU must be stored in nonvolatile memory. The IBM PC is no exception to this design rule. After RESET the 8088 has the values shown in Table 10-5. This means that upon RESET, the 8088 starts to fetch information from CS:IP of FFFF:0000, which gives the physical address FFFF0H. This is the reason that BIOS ROM is located at the upper address range of the memory map. As a result, when the PC is RESET, ROM BIOS is the memory block that is accessed first by the CPU. The ROM BIOS has, among other things, programs that do the testing of the CPU, ROM, and RAM. After those tests, it initializes all peripheral devices, sets up the system, and loads the operating system from hard disk into DRAM and hands over the control of the PC to the operating system. Since the microprocessor starts to fetch and execute instructions from physical location FFFF0H there must be an opcode sitting in that ROM location. In the x86 PC, the CPU finds the opcode for the FAR jump, EA, at location FFFF0H and the target address of the JUMP. You can verify that on your PC regardless of the microprocessor installed on the motherboard. Example 10-9 shows one such case using a simple Debug command. Notice in Example 10-9 that the date of ROM BIOS of a PC is stored in locations F000:FFF5 to F000:FFFD of BIOS ROM.

Table 10-5: 8088 After RESET

CPU	Contents
CS	FFFFH
DS	0000H
SS	0000H
ES	0000H
IP	0000H
Flags	Clear
Queue	Empty

(Reprinted by permission of Intel Corporation, Copyright Intel 1989)

Example 10-9

Using the DEBUG dump command, verify the JMP address for the cold boot and the BIOS date.

Solution:

From the directory containing DEBUG, enter the following:

```
C>DEBUG
-D FFFF:0 LF
FFFF:0000  EA 5B E0 00 F0 30 31 2F-31 35 2F 38 38 FF FC   j[`.p01/15/88.|
-Q
C>
```

The first 5 bytes showed the jump command "EA" and the destination "F0000:E05B". The next 8 bytes show the BIOS date, 01/15/88.

| Example 10-10 |

Suppose that you buy a software package and encounter a problem installing and running it on your computer. After contacting the technical support department of the manufacturer, you are told that the package is good for the BIOS ROM date of 10/8/88, but you are not told how to find the date. Use DEBUG to find the date for the ROM BIOS of a PC, PS, or compatible.

Solution:

In the DOS directory (or wherever you keep DEBUG), type the following at the DOS prompt:

```
C>DEBUG
-D F000:FFF5 FFFD
F000:FFF5 30 31 2F-31 35 2F 38 38 FF  01/15/88.
-Q
C>
```

The BIOS data is stored at F000:FFF5 through F000:FFFD. In the above case, the BIOS data is 1/15/88, which is earlier than the 10/8/88 date that you hoped to find.

Review Questions

1. What address range is called conventional memory? How many K bytes is that?
2. If the starting physical address of VDR is B0000H, what is the last address if it uses 16K bytes of RAM?
   (a) Show the beginning and ending physical addresses.
   (b) Give the corresponding logical addresses.
3. If the total ROM memory space used by BIOS and other expansion boards is 92K bytes, how many bytes are still unused?
4. What are the contents of CS and IP after the 8088 is reset (cold boot)?
5. What is the implication of Question 4?

SECTION 10.4: DATA INTEGRITY IN RAM AND ROM

When storing data, one major concern is maintaining data integrity. That is, ensuring that the data retrieved is the same as the data stored. The same principle applies when transferring data from one place to another. There are many ways to ensure data integrity depending on the type of storage. The checksum method is used for ROM, and the parity bit method is used for DRAM. For mass storage devices such as hard disks and for transferring data on the Internet, the CRC (cyclic redundancy check) method is employed. In this section we discuss the checksum and parity methods. The CRC method is discussed in Chapter 17.

Checksum byte

To ensure the integrity of the contents of ROM, every PC must perform a checksum calculation. The process of checksum will detect any corruption of the contents of ROM. One of the causes of ROM corruption is current surge, either when the PC is turned on or during operation. The checksum method uses a checksum byte. This checksum byte is an extra byte that is tagged to the end of a series of bytes of data. To calculate the checksum byte of a series of bytes of data, the following steps can be taken.

1. Add the bytes together and drop the carries.
2. Take the 2's complement of the total sum, and that is the checksum byte, which becomes the last byte of the stored information.

To perform the checksum operation, add all the bytes, including the checksum byte. The result must be zero. If it is not zero, one or more bytes of data have been changed (corrupted). To clarify these important concepts, see Examples 10-11 and 10-12.

Example 10-11

Assume that we have 4 bytes of hexadecimal data: 25H, 62H, 3FH, and 52H.
(a) Find the checksum byte.
(b) Perform the checksum operation to ensure data integrity.
(c) If the second byte 62H had been changed to 22H, show how checksum detects the error.

Solution:

(a)   The checksum is calculated by first adding the bytes.

```
  25H
+ 62H
+ 3FH
+ 52H
1 18H
```

The sum is 118H, and dropping the carry, we get 18H. The checksum byte is the 2's complement of 18H, which is E8H.

(b)   Adding the series of bytes including the checksum byte must result in zero. This indicates that all the bytes are unchanged and no byte is corrupted.

```
  25H
+ 62H
+ 3FH
+ 52H
+ E8H
2 00H  (dropping the carry)
```

(c)   Adding the series of bytes including the checksum byte shows that the result is not zero, which indicates that one or more bytes have been corrupted.

```
  25H
+ 22H
+ 3FH
+ 52H
+ E8H
1 C0H  dropping the carry, we get C0H.
```

Example 10-12

Assuming that the last byte of the following data is the checksum byte, show whether the data has been corrupted or not: 28H, C4H, BFH, 9EH, 87H, 65H, 83H, 50H, A7H, and 51H.

Solution:
The sum of the bytes plus the checksum byte must be zero; otherwise, the data is corrupted
28H + C4H + BFH + 9EH + 87H + 65H + 83H + 50H + A7H + 51H = 500H
By dropping the accumulated carries (the 5), we get 00. The data is not corrupted. See Figure 10-17 for a program that performs this verification.

```
                  2411    ;------------------------------------
                  2412    ;        ROS  CHECKSUM  SUBROUTINE         :
                  2413    ;------------------------------------
EC4C              2414    ROS_CHECKSUM PROC NEAR      ;NEXT_ROS_MODULE
EC4C  B90020      2415                 MOV   CX,8192  ;NUMBER OF BYTES TO ADD
EC4F              2416    ROS_CHECKSUM_CNT:   ;ENTRY PT. FOR OPTIONAL ROS TEST
EC4F  32C0        2417                 XOR   AL,AL
EC51              2418    C26:
EC51  0207        2419                 ADD   AL,DS:[BX]
EC53  43          2420                 INC   BX    ;POINT TO NEXT BYTE
EC54  E2FB        2421                 LOOP  C26   ;ADD ALL BYTES IN ROS MODULE
EC56  0AC0        2422                 OR    AL,AL ; SUM = 0?
EC58  C3          2423                 RET
                  2424    ROS_CHECKSUM ENDP
```

Figure 10-17. PC BIOS Checksum Routine

(Reprinted by permission from "IBM Technical Reference" c. 1984 by International Business Machines Corporation)

Checksum Program

When the PC is turned on, one of the first thing the BIOS does is to test the system ROM. The code for such a test is stored in the BIOS ROM. Figure 10-17 shows the program using the checksum method. Notice in the code how all the bytes are added together without keeping the track of carries. Then, the total sum is ORed with itself to see if it is zero. The zero flag is expected to be set to high upon return from this subroutine. If it is not, the ROM is corrupted.

Use of parity bit in DRAM error detection

System boards or memory modules are populated with DRAM chips of various organizations, depending on the time they were designed and the availability of a given chip at a reasonable cost. The memory technology is changing so fast that DRAM chips on the boards have a different look every year or two. While early PCs used 64K DRAMs, current PCs commonly use 256M chips. To understand the use of a parity bit in detecting data storage errors, we use some simple examples from the early PCs to clarify some very important design concepts. It must be noted that in today's PCs, these design concepts are still the same, even though the DRAMs have much higher density and CPLDs are used in place of TTL logic gates. You may wish to review DRAM organization and capacity, covered earlier in this chapter, before proceeding.

DRAM memory banks

The arrangement of DRAM chips on the system or memory module boards is often referred to as a *memory bank*. For example, the 64K bytes of DRAM can be arranged as one bank of 8 IC chips of 64Kx1 organization, or 4 banks of 16Kx1 organization. The first IBM PC introduced in 1981 used memory chips of 16Kx1 organization. Figure 10-18 shows the memory banks for 640K bytes of RAM using 256K and 1M DRAM chips. Notice the use of an extra bit for every byte of data to store the parity bit. With the extra parity bit, every bank requires an extra chip of x1 organization for parity check. Figure 10-19 shows DRAM design and parity bit circuitry for a bank of DRAM. First, note the use of the 74LS158 to multiplex the 16 address lines A0 - A15, changing them to the 8 address lines of MA0 - MA7 (multiplexed address) as required by the 64Kx1 DRAM chip. The resistors are for the serial bus line termination to prevent under-shooting and overshooting at the inputs of DRAM. They range from 20 to 50 ohms, depending on the speed of the CPU and the printed circuit board layout.

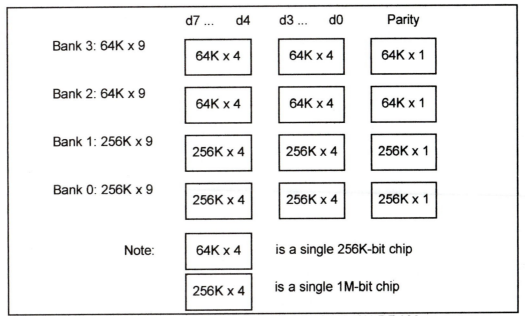

Figure 10-18. A Possible Memory Configuration for 640K DRAM

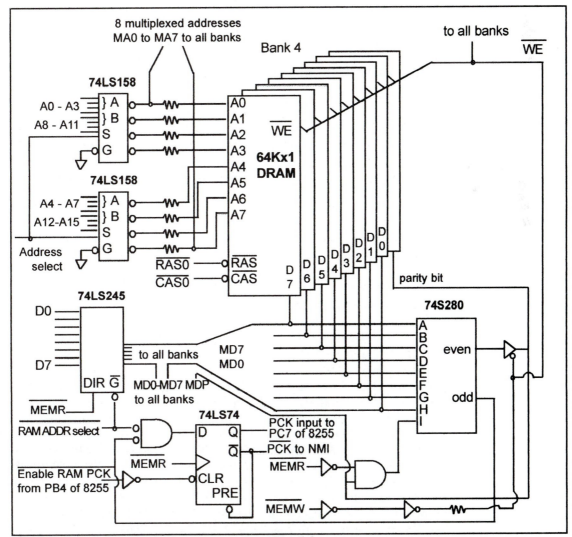

Figure 10-19. DRAM Connection in the IBM PC

(Reprinted by permission from "IBM Technical Reference" c. 1984 by International Business Machines Corporation)

SECTION 10.4: DATA INTEGRITY IN RAM AND ROM

287

A few additional observations about Figure 10-19 should be made. The output of multiplexer addresses MA0 - MA7 will go to all the banks. Likewise, memory data MD0 - MD7 and memory data parity MDP will go to all the banks. The 74LS245 not only buffers the data bus MD0 - MD7 but also boosts it to drive all DRAM inputs. Since the banks of the DRAMs are connected in parallel and the capacitance loading is additive, the data line must be capable of driving all the loads. Next we discuss how parity is used to detect RAM defects.

Parity bit generator/checker in the IBM PC

There are two types of errors that can occur in DRAM chips: soft error and hard error. In a *hard error*, some bits or an entire row of memory cells inside the memory chip get stuck to high or low permanently, thereafter always producing 1 or 0 regardless of what you write into the cell(s). In a *soft error*, a single bit is changed from 1 to 0 or from 0 to 1 due to current surge or certain kinds of particle radiation in the air. Parity is used to detect such errors. Including a parity bit to ensure *data integrity* in RAM is the most widely used method since it is the simplest and cheapest. This method can only indicate if there is a difference between the data that was written to memory and the data that was read. It cannot correct the error as is the case with some mainframes and supercomputers. In those computers and some of the x86-based servers, the EDC (error detection and correction) method is used to detect and correct the error bit. The early IBM PC and compatibles use the 74S280 parity bit generator and checker to implement the concept of the parity bit. The study of that chip should help us to understand the parity bit concept.

74S280 parity bit generator and checker

In order to understand the parity bit circuitry of Figure 10-19 it is necessary first to understand the 74LS280 parity bit generator and checker chip. This chip has 9 inputs and 2 outputs. Depending on whether an even or odd number of ones appears in the input, the even or odd output is activated according to Table 10-6.

As can be seen from Table 10-6, if all 9 inputs have an even number of 1 bits, the even output goes high, as in cases 1 and 4. If the 9 inputs have an odd number of high bits, the odd output goes high, as in cases 2 and 3. The way the IBM PC uses this chip is as follows. Notice that in

Table 10-6: 74280 Parity Check

Case	Inputs		Outputs	
	A - H	I	Even	Odd
1	Even	0	1	0
2	Even	1	0	1
3	Odd	0	0	1
4	Odd	1	1	0

Figure 10-19, inputs A - H are connected to the data bus, which is 8 bits, or one byte. The I input is used as a parity bit to check the correctness of the byte of data read from memory. When a byte of information is written to a given memory location in DRAM, the even-parity bit is generated and saved on the ninth DRAM chip as a parity bit with use of control signal $\overline{\text{MEMW}}$. This is done by activating the tri-state buffer using $\overline{\text{MEMW}}$. At this point, I of the 74S280 is equal to zero since $\overline{\text{MEMR}}$ is high. When a byte of data is read from the same location, the parity bit is gated into the I input of the 74S280 through $\overline{\text{MEMR}}$. This time the odd output is taken out and fed into a 74LS74. If there is a difference between the data written and the data read, the Q output (called PCK, parity bit check) of the 74LS74 is activated and Q activates NMI, indicating that there is a parity bit error, meaning that the data read is not the same as the data written. Consequently, it will display a parity bit error message. For example, if the byte of data written to a location has an even number of 1s, A to H has an even number of 1s, and I is zero, then the even-parity output of 74280 becomes 1 and is saved on parity bit DRAM. This is case 1 shown in Table 10-6. If the same byte of data is read and there is an even number of 1s (the byte is unchanged), 1 from the ninth bit DRAM, which is 1, is input to the 74S280, even becomes low, and odd becomes high, which is case 2 in Table 10-6. This high from the odd output will be inverted and fed to the 74LS74, making Q low. This means that $\overline{Q}$ is high, thereby indicating that the written byte is the same as the byte read and that no errors occurred. If the number of 1s in the byte has changed from even to odd and the 1 from the saved parity DRAM makes the number of inputs even (case 4 above), the odd output becomes low, which is inverted and passed to the 7474 D flip-flop. That makes Q = 1 and $\overline{Q}$ = 0, which signals the NMI to display a parity bit error message on the screen.

Review Questions

1. Find the checksum byte for the following bytes: 24H, 76H, F5H, 98H, 89H, 7AH, 61H, C2H.
2. Show a simple program in Assembly language to find the checksum byte of the 8 bytes of information (code or data) given in Question 1. Assume that SI equals the offset address of the data.
3. In a given PC we have only 512K of memory on the motherboard. Show possible configurations and the number of chips used to add memory up to the maximum allowed by the limits of conventional memory if we have each of the following. Include the parity bit in your configuration and count.
   (a) 64Kx1 (b) 64Kx4 and 64Kx1
4. To detect corruption of information stored in RAM and ROM memories, system designers use the _____ method for RAM and the _____ method for ROM.
5. Assume that due to slight current surge in the power supply, a byte of RAM has been corrupted while the computer is on. Can the system detect the corruption while the computer is on? Is this also the case for ROM?

SECTION 10.5: 16-BIT MEMORY INTERFACING

In the design of current x86 PCs, a single IC chip called a *chipset* has replaced 100 or so logic ICs connected together in the original PC. As a result, the details of CPU connection to memory and other peripherals are not visible for educational purposes. The 16-bit bus interfacing to memory chips is one of these details that is now buried within a chipset but still needs to be understood. In this section we explore memory interfacing for 16-bit CPUs. We use the 286 as an example but the concepts can apply to any 16-bit microprocessor. We also discuss the topics of memory cycle time and bus bandwidth.

ODD and EVEN banks

In a 16-bit CPU such as the 80286, memory locations 00000 - FFFFF are designated as odd and even bytes as shown in Figure 10-20. Although Figure 10-20 shows only 1 megabyte of memory, the concept of odd and even banks applies to the entire memory space of a given processor with a 16-bit data bus. To distinguish between odd and even bytes, the CPU provides a signal called BHE (bus high enable). BHE in association with A0 is used to select the odd or even byte according to Table 10-7.

Table 10-7: Distinguishing Between Odd and Even Bytes

BHE	A0		
0	0	Even word	D0 - D15
0	1	Odd byte	D8 - D15
1	0	Even byte	D0 - D7
1	1	None	

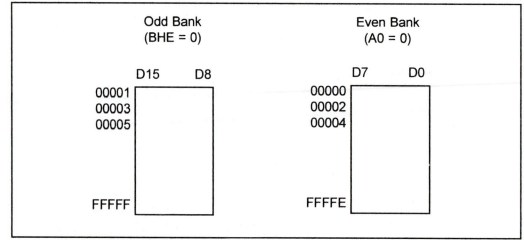

Figure 10-20. Odd and Even Banks of Memory

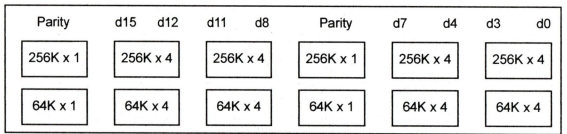

Figure 10-21. 640K Bytes of DRAM with odd and even banks designation

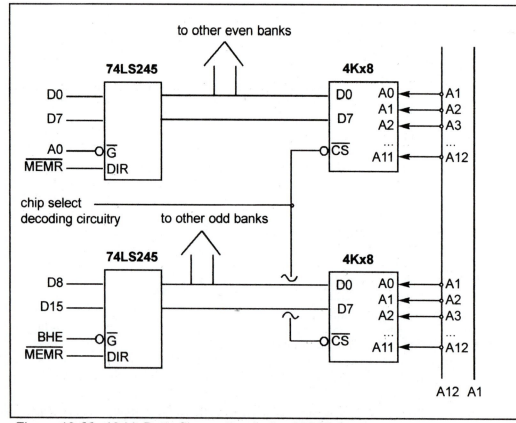

Figure 10-22. 16-bit Data Connection in the 80286 System

Examine Figure 10-20 to see how the odd and even addresses are designated for the 16-bit wide data buses. Figure 10-21 shows 640 KB of DRAM for 16-bit buses. Figure 10-22 shows the connection for the 16-bit data bus. In Figure 10-22, notice the use of A0 and BHE as bank selectors. Also notice the use of the 74LS245 chip as a data bus buffer.

Memory cycle time and inserting wait states

To access an external device such as memory or I/O, the CPU provides a fixed amount of time called a *bus cycle time*. During this bus cycle time, the read and write operation of memory or I/O must be completed. Here, we cover the memory bus cycle time. Bus cycle time for I/O devices is given in the next chapter. For the sake of clarity we will concentrate on reading memory, but the concepts apply to write operations as well. The bus cycle time used for accessing memory is often referred to as MC (memory cycle) time. From the time the CPU provides the addresses at its address pins to the time the data is expected at its data pins is called *memory read cycle time*. While in older processors such as the 8088 the memory cycle time takes 4 clocks, in the newer CPUs the memory cycle time is 2 clocks. In other words, in all x86 CPUs from the 286 to the Pentium, the memory cycle time is only 2 clocks. If memory is slow and its access time does not match the MC time of the CPU, extra time can be requested from the CPU to extend the read cycle time. This extra time is called a *wait state* (WS). In the 1980s, the clock speed for memory cycle time was the same as the CPU's clock speed. For example, in the 20 MHz 286/386/486 processors, the buses were working at the same speed of 20 MHz. This resulted in 2 × 50 ns = 100 ns for the memory cycle time (1/20 MHz = 50 ns). When the CPU's speed was under 100 MHz, the bus speed was comparable to the CPU speed. In the 1990s the CPU speed exploded to 1 GHz (gigahertz) while the bus speed maxed out at around 133 MHz. The gap between the CPU speed and bus speed is one of the biggest problem in the design of high-performance computers. To avoid the use of too many wait states in interfacing memory to CPU, cache memory and other high-speed DRAMs were invented. These are discussed in Chapter 22.

Example 10-13

Calculate the memory cycle time of a 20-MHz 8386 system with
(a) 0 WS,
(b) 1 WS, and
(c) 2 WS.
Assume that the bus speed is the same as the processor speed.

Solution:

1/20 MHz = 50 ns is the processor clock period. Since the 386 bus cycle time of zero wait states is 2 clocks, we have:

	80386 20 MHz
Memory cycle time with 0 WS	2 × 50 = 100 ns
Memory cycle time with 1 WS	100 + 50 = 150 ns
Memory cycle time with 2 WS	100 + 50 + 500 = 200 ns

It is preferred that all bus activities be completed with 0 WS. However, if the read and write operations cannot be completed with 0 WS, we request an extension of the bus cycle time. This extension is in the form of an integer number of WS. That is, we can have 1, 2, 3, and so on WS, but no 1.25 WS.

It must be noted that memory access time is not the only factor in slowing down the CPU, even though it is the largest one. The other factor is the delay associated with signals going through the data and address path. Delay associated with reading data stored in memory has the following two components:

1. The time taken for address signals to go from CPU pins to memory pins, going through decoders and buffers (e.g., 74LS245). This, plus the time it takes for the data to travel from memory to CPU, is referred to as a *path delay*.
2. The memory access time to get the data out of the memory chip. This is the largest of the two components.

The total sum of these two must equal the memory read cycle time provided by the CPU. Memory access time is the largest and takes about 80% of the read cycle time. See Example 10-14 for further clarification of these points. These concepts are critical in the design of microprocessor-based products.

Example 10-14

A 20-MHz 80386-based system is using ROM of 150 ns speed. Calculate the number of wait states needed if the path delay is 25 ns.

Solution:

If ROM access time is 150 ns and the path delay is 25 ns, every time the 80386 accesses ROM it must spend a total of 175 ns to get data into the CPU. A 20-MHz CPU with zero WS provides only 100 ns (2×50 ns = 100 ns) for the memory read cycle time. To match the CPU bus speed with this ROM we must insert 2 wait states. This makes the cycle time 200 ns (100 + 50 + 50 = 200 ns). Notice that we cannot ask for 1.5 WS since the number of WS must be an integer. That would be like going to the store and wanting to buy half an apple. You must get one complete WS or none at all.

Accessing EVEN and ODD words

As you recall from earlier chapters, Intel defines 16-bit data as a *word*. The address of a word can start at an even or an odd number. For example, in the instruction "MOV AX,[2000]" the address of the word being fetched into AX starts at an even address. In the case of "MOV AX,[2007]" the address starts at an odd address. In systems with a 16-bit data bus, accessing a word from an odd addressed location can be slower. This issue is important and applies to 32-bit and 64-bit systems with 386 and Pentium processors, as we will see in Chapter 22.

As shown in Figure 10-23, in the 8-bit system, accessing a word is treated like accessing two bytes regardless of whether the address is odd or even. Since accessing a byte takes one memory cycle, accessing any word will take 2 memory cycles. In the 16-bit system, accessing a word with an even address takes one memory cycle. That is because one byte is carried on D0 - D7 and the other on D8 - D15 in the same memory cycle. However, accessing a word with an odd address requires two memory cycles. For example, see how accessing the word in the instruction "MOV AX,[F617]" works as shown in Figure 10-24. Assuming that DS = F000H in this instruction, the contents of physical memory locations FF617H and FF618H are being moved into AX. In the first cycle, the 286 CPU accesses location FF617H and puts it in AL. In the second cycle, the contents of memory location FF618H are accessed and put into AH. The lesson to be learned from this is to try not to put any words on an odd address if the program is going to be run on a 16-bit system. Indeed this is so important that there is a pseudo-op specifically designed for this purpose. It is the EVEN directive and is used as follows:

```
                    EVEN

VALUE1      DW      ?
```

This ensures that VALUE1, a word-sized operand, is located in an even address location. Therefore, an instruction such as "MOV AX,VALUE1" or "MOV VALUE1,CX", will take only a single memory cycle.

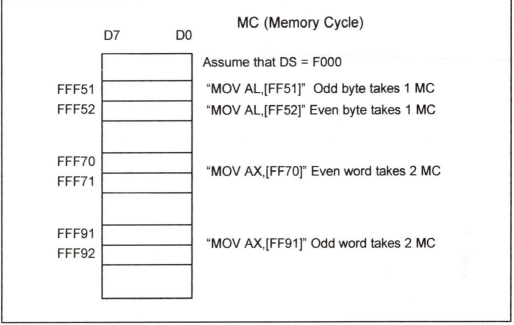

Figure 10-23. Accessing Even and Odd Words in the 8-bit CPU

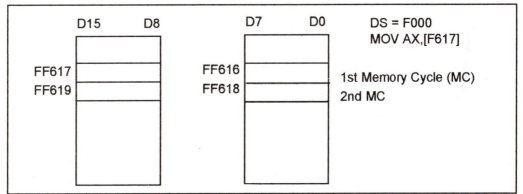

Figure 10-24. Accessing an Odd-Addressed Word in a 16-bit processor (80286)

Bus bandwidth

The main advantage of the 16-bit data bus is doubling the rate of transfer of information between the CPU and the outside world. The rate of data transfer is generally called *bus bandwidth*. In other words, bus bandwidth is a measure of how fast buses transfer information between the CPU and memory or peripherals. The wider the data bus, the higher the bus bandwidth. However, the advantage of the wider external data bus comes at the cost of increasing the size of the printed circuit board. Now you might ask why we should care how fast buses transfer information between the CPU and outside, as long as the CPU is working as fast as it can. The problem is that the CPU cannot process information that it does not have. In other words, the speed of the CPU must be matched with the higher bus bandwidth; otherwise, there is no use for a fast CPU. This is like driving a Porsche or Ferrari in first gear; it is a terrible underusage of CPU power. Bus bandwidth is measured in MB (megabytes) per second and is calculated as follows:

bus bandwidth = (1/bus cycle time) × bus width in bytes

In the above formula, bus cycle time can be either memory or I/O cycle time. The I/O cycle time is discussed in Chapter 11. Example 10-15 clarifies the concept of bus bandwidth. As can be seen from Example 10-15, there are two ways to increase the bus bandwidth. Either use a wider data bus, or shorten the bus cycle time, or do both. That is exactly what 386, 486, and Pentium processors have done. While the data bus width has increased from 16-bit in the 80286 to 64-bit in the Pentium, the bus cycle time is reaching a maximum of 133 MHz. Again, it must be noted that although the processor's speed can go to 1 GHz or higher, the bus speed is limited to around 133 MHz. The reason for this is that the signals become too noisy for the circuit board if they are above 100 MHz. This is even worse for the ISA expansion slot. The ISA bus speed is limited to around 8 MHz. This is because the ISA slot uses large and bulky connectors and they are too noisy for a speed of more than 8 MHz. In other words, in PCs with a 500 MHz Pentium, the CPU must slow down to 8MHz when accessing the ISA bus. The PCI bus was introduced to solve this limitation. It can go as high as 133 MHz and its data bus width is 64-bit. Chapter 27 provides further discussion of bus bandwidth.

Example 10-15

Calculate memory bus bandwidth for the following microprocessors if the bus speed is 20 MHz.

(a) 286 with 0 WS and 1 WS (16-bit data bus)
(b) 386 with 0 WS and 1 WS (32-bit data bus)

Solution:

The memory cycle time for both the 286 and 386 is 2 clocks, with zero wait states. With the 20 MHz bus speed we have a bus clock of 1/20 MHz = 50 ns.

  (a)   Bus bandwidth = (1/(2 × 50 ns)) × 2 bytes = 20 Mbytes/second (MB/S)
            With 1 wait state, the memory cycle becomes 3 clock cycles
            3 × 50 =150 ns and the memory bus bandwidth is = (1/150 ns) × 2 bytes = 13.3 MB/S

    (b)   Bus bandwidth = (1/(2 × 50 ns)) × 4 bytes = 40 MB/S
            With 1 wait state, the memory cycle becomes 3 clock cycles
            3 × 50 =150 ns and the memory bus bandwidth is = (1/150 ns) × 4 bytes = 26.6 MB/S

From the above it can be seen that the two factors influencing bus bandwidth are:

1. The read/write cycle time of the CPU
2. The width of the data bus

Notice in this example that the bus speed of the 286/386 was given as 20 MHz. That means that the CPU can access memory on the board at this speed. If this 286/386 is used on a PC board with an ISA expansion slot, it must slow down to 8 MHz when communicating with the ISA bus since the maximum bus speed for the ISA bus is 8 MHz. This is done by the chipset circuitry.

Review Questions

1. True or false. If A0 = 0 and BHE = 1, a byte is being transferred on the D0 - D7 data bus from an even-address location.
2. True or false. If we have A0 = 1 and BHE = 0, a byte is being transferred on the D0 - D7 data bus from an odd-address location.
3. True or false. If we have A0 = 1 and BHE = 0, a byte is being transferred on the D8 - D15 data bus.
4. True or false. If we have A0 = 0 and BHE = 0, a word is being transferred on the D0 - D15 data bus.
5. In the instruction "MOV AX,[2000]", the transferring of data into the accumulator takes _____ memory cycles for the 8088 and _____ for the 80286.
6. A 16-MHz 286 has a memory cycle time of _____ ns if it is used with a zero wait state.
7. To interface a 10-MHz 286 processor to a 350-ns access time ROM, how many wait states are needed?

SECTION 10.6: ISA BUS MEMORY INTERFACING

In Chapter 9 we covered the basics of ISA bus signals. In this section we provide more details of the ISA bus for memory interfacing including the memory read/write cycle time. In PCs with 386/486/Pentium microprocessors, the signals for the ISA expansion slots are provided by the chipset. The chipset makes sure that the signals for the ISA slot conform with the ISA bus standard regardless of the CPU's speed and data width. The ISA bus specifications and timing for memory are precise and must be understood if we want to design an ISA plug-in card with on-board memory. Next, we review once more the signals related to memory in the ISA expansion slot.

Address bus signals

SA0 - SA19 (system address)

The system address bus provides the address signals for the desired memory (or I/O) location. The chipset latches these signals and holds them valid throughout the bus cycle.

LA17 - LA23 (latchable address)

These signals, along with SA0 - SA19, allow access to 16M bytes of memory space from the ISA expansion slot. The chipset does not latch these signals. They must be latched by the board designed for the expansion slot.

SBHE (system byte high enable)

Because it is an active low signal, when it is low it indicates that data is being transferred on the upper byte (D8 - D15) of the data bus.

SD0 - SD15 (system data bus)

The system data bus (SD0 - SD15) is used to transfer data between the CPU, memory, and I/O devices.

Memory control signals

MEMW (memory write)

An active low control signal is used to write data into the memory chip. This signal is connected to the WE (write enable) pin of the memory chip. This signal can be used to access the entire 16M allowed by the ISA bus.

MEMR (memory read)

This active low control signal is used to read data from the memory chip. It is connected to the OE (output enable) pin of the memory chip. This signal can be used to access the entire 16M allowed by the ISA bus.

MEMW and MEMR are used for 16M memory. However, if the 1M memory 00000 - FFFFFH is chosen, the following control signals must be used.

SMEMW (system memory write)

An active low control signal used to write data into a memory chip. This signal is connected to the WE pin of the memory chip. This signal goes low when accessing addresses between 0 and FFFFFH (0 and 1M bytes).

SMEMW (system memory write)

An active low control signal used to read data from the memory chip. This signal is connected to the OE pin of the memory chip. This signal goes low when accessing addresses between 0 and FFFFFH (0 and 1M bytes).

Although the ISA bus has a 16-bit data bus (D0 - D15), either the 8-bit section (D0 - D7) or the entire 16 bits (D0 - D15) can be used. This is decided by the input pin MEMCS16, as explained next.

MEMCS16 (memory chip select 16)

This is an input signal and is active low. When not asserted, it indicates to the chipset that only the D0 - D7 portion of the data bus is being used. Notice that the 8-bit portion is the default mode and it is achieved by doing nothing to this pin. In contrast, when this signal is asserted low, both the low byte and high byte of the data bus (D0 - D15) will be used for data transfer. Therefore, to use the entire 16-bit data bus, this pin must be low.

The ISA bus allows the interfacing of slow memories by inserting wait states into the memory cycle time. This prolonging of memory cycle time is available for both 8-bit and 16-bit data transfers. The standard 8-bit data transfer has 4 WS in the memory read cycle time. As a result, the default memory cycle time is 6 clocks. The standard 16-bit data transfer uses 1 WS in the memory cycle time. That results in 3 clocks for the read/write cycle time. To shorten the memory cycle time we use the ZEROWS pin as explained next.

ZEROWS (zero wait state)

This is an input signal and is active low. The standard 16-bit ISA bus cycle time contains one WS unless ZEROWS is activated. By activating this pin (making it low), we are telling the CPU that the present memory cycle can be completed without a wait state. That results in performing the bus cycle time in 2 clocks. The standard 8-bit ISA bus cycle time contains 4 WS unless ZEROWS is activated. That means that when both MEMCS16 and ZEROWS are high (without asserting them low), the 8-bit data bus (D0 - D7) is being used and the data transfer is completed in 6 clocks. This default 8-bit read/write cycle time with its 4 WS is sufficient for interfacing even slow ROMs to the ISA bus. If MEMCS16 = 1 and ZEROWS = 0, the 8-bit memory cycle time has 1 WS instead of 4.

SYSCLK (system clock)

This is is an output clock providing the standard 8 MHz ISA bus clock. The 8 MHz clock results in 125 ns (1/8 MHz = 125 ns) period and all memory and I/O ISA bus timing is based on this. Therefore, a zero WS read cycle time for 16-bit bus takes 2×125 ns = 250 ns. The standard 8-bit bus with its 4 WS will be $(2 + 4) \times 125$ ns = 750 ns. The 8-bit bus with ZEROWS asserted low has 1 WS, making its memory cycle time $(2 + 1) \times 125$ ns = 375 ns. The SYSCLK pin is located on the B side of the 62-pin section of the ISA bus.

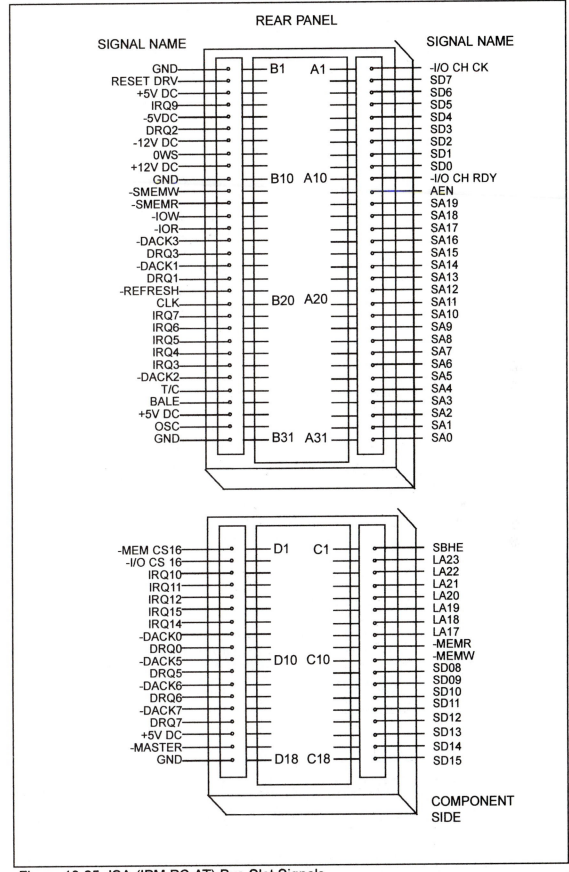

Figure 10-25. ISA (IBM PC AT) Bus Slot Signals

(Reprinted by permission from "IBM Technical Reference" c. 1985 by International Business Machines Corporation)

IOCHRDY(IO Channel Ready)

This is an input signal into the ISA bus and is active low. By driving it low, we are asking the system to extend the standard ISA bus cycle time. In response to asserting of this signal, the system will insert wait states into the memory (or I/O) cycle time until it is deasserted. This is rarely used for memory interfacing since the standard memory cycle time of the ISA bus provides plenty of time.

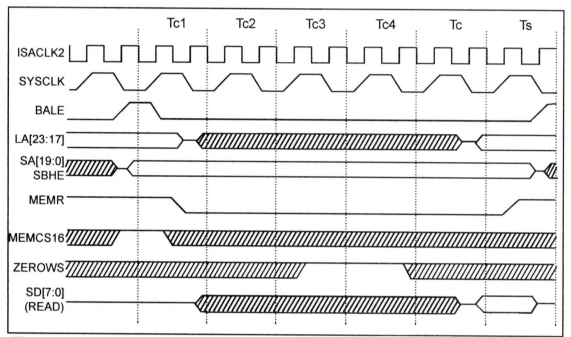

Figure 10-26. Standard 8-bit ISA Memory Read Cycle Time (4 WS)

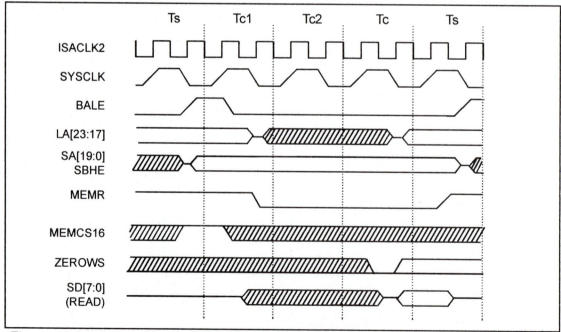

Figure 10-27. Zero WS 8-bit ISA Memory Read Cycle Time (1WS)

ISA bus timing for memory

Suppose that we are designing a data acquisition board for an ISA expansion slot, and the board requires ROM. What would be the best approach? We can use either the 8-bit or 16-bit data section of the ISA bus. There are some major differences between them that must be noted. Next we look at each separately.

8-bit memory timing for ISA bus

In the case of using the 8-bit data bus, we use D0 - D7 and A0 - A19 of the 62-pin section of the bus. More importantly, the ISA bus provides plenty of time for slow memories by inserting wait states into the read cycle time. First, we must remember that the memory cycle time is only 2 clocks (with 0 WS) and the maximum speed is 8 MHz. Since 1/8 MHz = 125 ns, the bus cycle time is 2×125 = 250 ns. In the case of the standard 8-bit read/write cycle time, the chipset inserts 4 WS clocks into the read cycle time as shown in Figure 10-26. In the case where ZEROWS is asserted low and MEMCS16 = 1, the read/write cycle time has 1 WS for the 8-bit bus as shown in Figure 10-27. Figure 10-28 shows the 8-bit memory interfacing for the ISA bus.

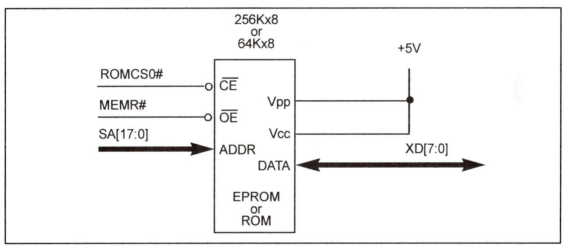

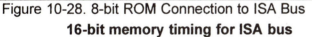

Figure 10-28. 8-bit ROM Connection to ISA Bus

16-bit memory timing for ISA bus

As mentioned earlier, the 8-bit data transfer is the default mode for the ISA bus expansion slots. In order to perform the 16-bit data transfer using lines D0 - D15, we must assert the MEMCS16 pin low. A 16-bit data bus transfer is twice as fast as an 8-bit data bus transfer. However, it also requires twice the board space in addition to having a higher power consumption. The 16-bit data read cycle time for the ISA bus with zero WS is shown in Figure 10-30. Notice that in all the 2-clock bus cycle CPUs and systems, the first clock is set aside for the addresses and the second clock is for data. In order to get a zeros wait state bus activity, the ZEROWS pin (active low) must be activated. This is because the standard ISA bus cycle contains one wait state. The standard 16-bit ISA bus cycle timing with 1 WS is shown in Figure 10-29, and Figure 10-30 shows the ISA cycle with zero WS. Combining the effect of MEMCS16 and ZEROWS pins gives the information in Table 10-8.

Another major issue in 16-bit ISA bus interfacing is the problem of odd and even banks. For example, assume that we are interfacing two ROM chips to D0 - D15 of the ISA bus, one connected to D0 - D7 and the other to D8 - D15. We must divide our information (code or data) into two parts and burn each part into one of the ROMs. In many ROM burners, there is an option for splitting the data into odd and even addressed bytes to support 16-bit data systems. The 16-bit data connection to two ROMs for the ISA slot is shown in Figure 10-31.

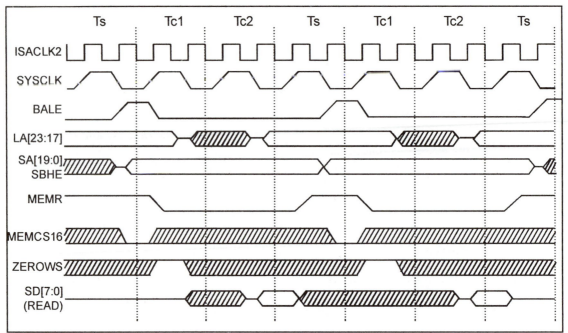

Figure 10-29. Standard 16-bit ISA Memory Read Cycle Time (1 WS)

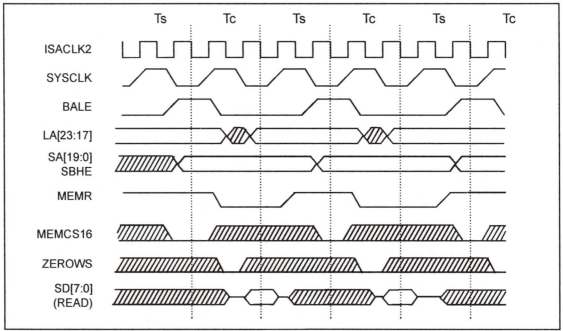

Figure 10-30. Zero WS 16-bit ISA Memory Read Cycle Time (0 WS)

Table 10-8: ISA Bus Memory Read/Write Cycle Time

MEMCS16	ZEROWS	Data Bus	Read Cycle Time
0	0	D0 - D15	250 ns
0	1	D0 - D15	375 ns
1	0	D0 - D7	375 ns
1	1	D0 - D7	750 ns

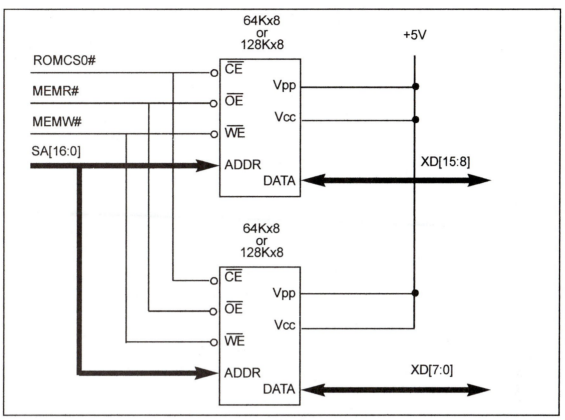

Figure 10-31. 16-bit ISA Bus Connection to ROM

Example 10-16

Calculate the bus bandwidth of the ISA bus for (a) 0 WS, and (b) 1 WS. Assume that all the data transfers are 16-bit (D0 - D15).

Solution:

Since the ISA bus speed is 8 MHz, we have 1/8 MHz =125 ns as a clock period. The bus cycle time for zero wait state is 2 clocks; therefore, we have:

(a) ISA bus cycle time with 0 WS is 2×125 ns = 250 ns.
    Bus bandwidth = 1/250 ns $\times$ 2 bytes = 8 megabytes/second.

(b) ISA bus cycle time with 1 WS is 250 ns + 125 ns = 375 ns.
    Bus bandwidth = 1/375 ns $\times$ 2 bytes = 5.33 megabytes/second.

ROM duplicate and x86 PC memory map

The memory map for the 1 megabyte memory range 00000 to FFFFFH is the same for all x86-based PCs. See Figure 10-32. As shown in Chapter 9, when 286 microprocessors are powered up, the CPU is in real mode and fetches the first opcode from physical memory location FFFFF0H. This is because CS = F000H, IP = FFF0H, and A20 - A23 are all high. This leads to a physical address of FFFFF0H, which is 16 bytes below the top of FFFFFFH, the 16 megabyte maximum memory range of the 286. After execution of the first opcode, address pins A20 - A23 all become 0s. Address pins A20 - A23 will not be activated again unless the 286 mode of operation is changed to protected mode. In other words, while the CPU wakes up in real mode at address FFFF0H, the first opcode is

---

fetched from FFFFF0H because A20 - A23 are all high. This is one reason that there is an exact duplicate of ROM at addresses 0F0000 - 0FFFFF and FF0000 - FFFFFF. This duplication allows access of BIOS ROM in both real and protected modes. This concept applies to the 386/486 and Pentium PCs and is shown in Figure 10-33. In these processors, the 32-bit address bus provides the memory space of 00000000 to FFFFFFFFH. We can verify this by using the system tools software that comes with Windows 9x and higher. You can experiment with this by going to *Accessories*, clicking on *System Tools*, and then clicking on *System Information*. Click on *Hardware Resources* and then click on *Memory*.

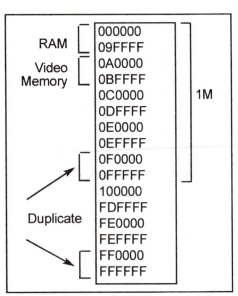

Figure 10-32. PC AT Memory Map

Shadow RAM

By using *System Tools* to explore the system memory of Pentium PCs, we can see the duplicates of ROM in the RAM memory space. The reason for that is the fact that the ROM access time is too slow for the 100 MHz bus speed. To speed up the ROM access time, its contents are copied into RAM and write protected. This is called *shadow RAM* and provides the ROM's contents to CPU at a much faster speed than ROM. From this point forward, every time the CPU needs to access the ROM's contents, it will get the information from RAM at a very high speed. As long as the PC is on, the DRAM containing the ROM information is write protected and will not be corrupted. It must be noted that the process of

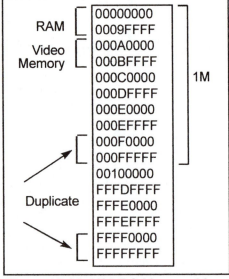

Figure 10-33. X86 PC Memory Map

creating shadow RAM is done when the system is booted, and when the PC is turned off, shadow RAM's contents are lost.

DIMM and SIMM memory modules

In the 1980s, PCs had sockets on the motherboard for DRAM chips. To expand the memory of a PC, you had to buy memory chips and plug them into the sockets. With the introduction of the 16-bit ISA bus, memory expansion boards became common. The problem with the memory expansion cards was that you were limited to the bus speed of the ISA expansion slot no matter how fast the memory chip. This led to the idea of a memory module as a way to expand memory for x86 PCs (386 and higher). The connectors for memory modules are much smaller in size and accommodate much faster memory than ISA connectors. It is also much easier to insert them into the motherboard. The only problem was the lack of a standard connector. This was solved with the introduction of SIP (single in-line package). Later, the SIMM (single in-line memory module) and DIMM (dual in-line memory module) were introduced. Currently, SIP is no longer in use, and SIMM and DIMM are the dominant memory modules. It is important to notice that the use of memory modules frees the motherboard designer from the agonizing choice of which organization and speed of DRAM to use. All that is required is to incorporate various organizations and speeds into the design of the motherboard and let the user select options via the CMOS set-up process.

Review Questions

1. The MEMCS16 pin of the ISA bus is an active _____ (low, high) signal.
2. If MEMCS16 = high, which portion of the data bus is used?
3. The ZEROWS pin of the ISA bus is an active _____ (low, high) signal.
4. If ZEROWS = high, the 8-bit memory cycle time takes _____ clocks.
5. The ISA bus has a maximum frequency of 8 MHz. Find its bus bandwidth for 0 WS.

SUMMARY

This chapter began with an introduction to semiconductor technology. Semiconductor memories were described in terms of their capacity, organization, and speed. The different types of ROM and examples of how each is used were examined. Address decoding for memory was discussed and the use of DRAM in the IBM PC was illustrated. This chapter explored the memory map of the IBM PC, how the 1M of memory is distributed among RAM, video display RAM, and ROM. The organization and addressing of the 640K bytes of RAM called conventional memory was discussed. In addition, error checking of RAM and ROM was explained. Also, 16-bit bus interfacing to memory and bus bandwidth issues for ISA buses were explored.

PROBLEMS

SECTION 10.1: SEMICONDUCTOR MEMORY FUNDAMENTALS

1. What is the difference between a 4M memory chip and 4M of computer memory as far as capacity is concerned?
2. True or false. The more address pins, the more memory locations are inside the chip.
3. True or false. The more data pins, the more each location inside the chip will hold.
4. True or false. The more data pins, the higher the capacity of the memory chip.
5. True or false. With a fixed number of address pins, the more data pins, the greater the capacity of the memory chip.
6. The speed of a memory chip is referred to as its _____.
7. True or false. The price of memory chips varies according to capacity and speed.
8. The main advantage of EEPROM over UV-EPROM is _____.
9. True or false. SRAM has a larger cell size than DRAM.
10. Which of the following, EPROM, DRAM, and SRAM, must be refreshed periodically?
11. Which memory is used for cache?
12. Which of the following, SRAM, UV-EPROM, NV-RAM, DRAM, and cache memory, is volatile memory?
13. RAS and CAS are associated with which memory?
    (a) EPROM (b) SRAM (c) DRAM (d) all of the above
14. Which memory needs an external multiplexer?
    (a) EPROM (b) SRAM (c) DRAM (d) all of the above

---

15. Find the organization and the capacity of memory chips with the following pins.
    (a) EEPROM A0 - A14, D0 - D7   (b) UV-EPROM A0 - A12, D0 - D7
    (c) SRAM A0 - A11, D0 - D7   (d) SRAM A0 - A12, D0 - D7
    (e) DRAM A0 - A10, D0   (f) SRAM A0 - A12, D0-7
    (g) EEPROM A0 - A11, D0 - D7   (h) UV-EPROM A0 - A10, D0 - D7
    (i) DRAM A0 - A8, D0 - D3   (j) DRAM A0 - A7, D0 - D7

16. Find the capacity, address, and data pins for the following memory organizations.
    (a) 16Kx8 ROM   (b) 32Kx8 ROM
    (c) 64Kx8 SRAM   (d) 256Kx4 DRAM
    (e) 64Kx8 ROM   (f) 64Kx4 DRAM
    (g) 1Mx4 DRAM   (h) 4Mx4 DRAM
    (i) 64Kx8 DRAM

SECTION 10.2: MEMORY ADDRESS DECODING

17. Find the address range of the following memory design.

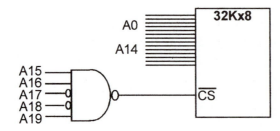

18. Using NAND gates and inverters, design decoding circuitry for the address range 0C0000H - 0C0FFFH.

19. Find the address range for Y0, Y3, and Y6 of the 74LS138 for the following design. This is the ROM interfacing with the 8088 CPU in the original PC.

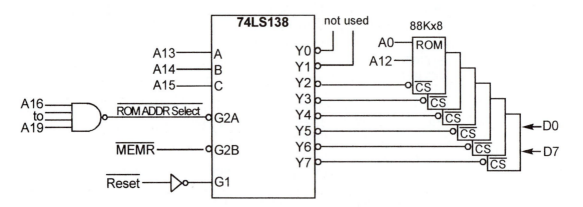

20. Using the 74138, design the memory decoding circuitry in which the memory block controlled by Y0 is in the range 00000H to 03FFFH. Indicate the size of the memory block controlled by each Y.

21. Find the address range for Y3, Y6, and Y7 in Problem 20.

22. Using the 74138 and OR gates, design memory decoding circuitry in which the memory block controlled by Y0 is in the 80000H to 807FFH space. Indicate the size of the memory block controlled by each Y.

23. Find the address range for Y1, Y4, and Y5 in Problem 22.

24. The CS pin of the memory chip is active _____ (low, high). What about the RD pin?
25. Which one can accommodate more inputs, the 74138 or CPLD?

SECTION 10.3: IBM PC MEMORY MAP

26. Indicate the address range and total kilobytes of memory allocated to the RAM, ROM, and video display RAM of the PC.
27. What address range is called conventional memory, and how many K bytes is it?
28. Can we increase the size of conventional memory? Explain your answer.
29. What are the contents of CS and IP in the 8088 upon RESET?
30. A user wants to add some EPROM to a PC. Can he/she use the address range 00000 - 9FFFFH? What happens if this range is used?
31. Give the logical and physical location where the BIOS ROM date is stored.
32. Suppose that the memory address range C0000H - C7FFFH is used in a certain plug-in adapter card. How many K bytes is that, and is this memory in the RAM or ROM allocated area?
33. If a video card uses only 4K bytes of VDR and the starting address is B0000H, what is the ending address of this VDR?
34. In a certain video card the starting address is B8000H and it uses only 16K bytes of memory. What is the ending address of this video card?
35. Why is ROM mapped where it is in the PC? Why can't we use addresses starting at 00000?
36. When the CPU is powered up, at what physical address does it expect to see the first opcode? In the PC, what opcode is there normally?

SECTION 10.4: DATA INTEGRITY IN RAM AND ROM

37. Find the checksum byte for the following bytes.
    34H, 54H, 7FH, 11H, E6H, 99H
38. For each of the following sets of data (the last byte is the checksum byte) verify if the data is corrupted.
    (a) 29H, 1CH, 16H, 38H, and 6DH (b) 29H, 1CH, 16H, 30H, and 6DH
39. To maintain data integrity, the checksum method is used for _____ type memory and the parity bit method for _____ memory.
40. True or false. ROM is tested for corruption during a cold boot-up, but data corruption in RAM can be detected any time the system is on.
41. A given PC needs only 320K bytes to reach the maximum allowed conventional memory. Show the memory configuration using 256Kx1 and 64Kx1 memory chips. How many chips are needed? (Include the parity bit.)
42. Repeat Problem 41 if we have 256Kx4 and 64Kx4 chips in addition to 64x1 and 256Kx1.
43. Why is it preferable to use higher density memory chips in memory design?
44. True or false. To access DRAM, the RAS address is provided first and then the CAS address.
45. True or false. The 74S280 is both a parity generator and a checker.
46. In the 74S280 we have 10010011 for A - H inputs and I = 0. What is the status of the even and odd output pins?
47. In the 74S280 we have 11101001 for A - H inputs and I = 0. What is the status of the even and odd output pins?
48. In the 74S280 we have 10001001 for A - H inputs and I = 1. What is the status of the even and odd output pins?

49. Odd and even banks are associated with which microprocessor, the 8088 or the 80286?
50. How many memory chips are needed if we use 256Kx8, 256Kx1, 64Kx8, and 64Kx1 memory chips for conventional memory of the 80286 PC/compatible?
51. State the status of A0 and BHE when accessing an odd-addressed byte.
52. State the status of A0 and BHE when accessing an even-addressed word.
53. State the status of A0 and BHE if we only want to access D0 - D7 of the data bus.
54. State the status of A0 and BHE if we only want to access D8 - D15 of the data bus.
55. What is the use of 74LS245 in memory interfacing?
56. What is the bus bandwidth unit?
57. Give the variables that affect the bus bandwidth.
58. True or false. One way to increase the bus bandwidth is to widen the data bus.
59. True or false. An increase in the number of address bus pins results in a higher bus bandwidth for the system.
60. Calculate the memory bus bandwidth for the following systems.
    (a) 80286 of 10 MHz and 0 WS
    (b) 80286 of 16 MHz and 0 WS

SECTION 10.6: ISA BUS MEMORY INTERFACING

61. Which of the control signals are used for ISA memory interfacing if the address of memory is in the range of F0000 - FFFFFH?
62. The ISA bus can access a maximum of _____ bytes. Why?
63. The MEMCS16 is an active _____ signal. Is this an input signal?
64. Explain the use of the MEMCS16 pin.
65. The ZEROWS is an active _____ signal. Is this an input signal?
66. Explain the use of the ZEROWS pin.
67. If the MEMCS16 pin is high, what portion of the data bus is being used?
68. If the MEMCS16 pin is low, what portion of the data bus is being used?
69. If the ZEROWS pin is high, give the memory cycle time for an 8-bit data transfer.
70. If the ZEROWS pin is high, give the memory cycle time for a 16-bit data transfer.
71. If the ZEROWS pin is asserted low, give the memory cycle time for a 16-bit data transfer.
72. If the ZEROWS pin is asserted low, give the memory cycle time for an 8-bit data transfer.
73. To achieve the best data transfer rate for ISA bus memory interfacing, what should be the status of the MEMCS16 and ZEROWS pins?
74. Fill the blanks for following cases.

MEMCS16	ZEROWS	Data bus used	Read Cycle time	Bus Bandwidth
0	0			
0	1			
1	0			
1	1			

75. Why do we use DIMM and SIMM sockets for memory expansion instead of the ISA bus slot?

ANSWERS TO REVIEW QUESTIONS

SECTION 10.1: SEMICONDUCTOR MEMORY FUNDAMENTALS
1. nanoseconds
2. (a) 2Kx8, 16K bits (b) 8Kx8, 64K (c) 64Kx4, 256K (d) 256Kx1, 256K
3. (a) 128K bits, 14 address, and 8 data (b) 256K, 15 address, and 8 data
   (c) 1M, 10 address, and 1 data (d) 1M, 9 address, and 4 data
   (e) 512K, 16, and 8 data (f) 4M, 10 address, and 4 data
4. It takes much less time to erase and does not need to be removed from the system board.
5. NV-RAM

SECTION 10.2: MEMORY ADDRESS DECODING
1. OE
2. WE
3. active low
4. 8
5. active low

SECTION 10.3: IBM PC MEMORY MAP
1. 00000 - 9FFFFH, 640K bytes
2. (a) B0000H-B0FFFH (b) B000:0000 - B000:0FFF
3. 256K - 92K = 164K
4. CS = FFFFH, IP = 0000
5. It indicates that the CPU fetches the first opcode at the physical address FFFF0H when the system is turned on. Therefore, no RAM can be mapped into the last segment of the 8088; the memory space must be occupied by a cold boot ROM.

SECTION 10.4: DATA INTEGRITY IN RAM AND ROM
1. adding the bytes: 24H + 76H + F5H + 98H + 89H + 7AH + 61H + C2H = 44DH. Dropping the carries, we get 4DH, and taking the 2's complement, we have B3H for the checksum byte.
2.
```
        MOV SI,OFFSET DATA   ;LOAD THE OFFSET ADDRESS
        MOV CX,08            ;LOAD THE COUNTER
        SUBAL,AL
BOO:    ADD AL,[SI]     ;ADD THE BYTE AND IGNORE THE CARRY
        INCSI           ;POINT TO NEXT BYTE
        LOOP BOO        ;CONTINUE UNTIL count IS ZERO
```
3. Since the maximum limit is 640K bytes, we need add only 128K bytes of RAM (640 - 512 = 128).
   (a) two banks each of 9 chips of 64Kx1, total = 18 chips
   (b) two banks each two 64Kx4 to contain data and 64Kx1 for parity, total = 6 chips
4. parity bit generation/checker, checksum
5. While the computer is on, any corruption in the contents of RAM is detected by the parity bit error checking circuitry when that data is accessed (read) again. However, the ROM corruption is not detected since the checksum detection is performed only when the system is booted.

SECTION 10.5: 16-BIT MEMORY INTERFACING
1. True
2. False
3. True
4. True
5. 2 and 1
6. 125 ns
7. 2WS

SECTION 10.6: ISA BUS MEMORY INTERFACING

1. low
2. D0 - D7
3. low
4. 6 clocks since 2 + 4 WS = 6
5. 8 megabytes /sec. See Example 10 -16.

CHAPTER 11

I/O AND THE 8255;
ISA BUS INTERFACING

OBJECTIVES

Upon completion of this chapter, you will be able to:

>> Code Assembly language instructions to read and write data to and from I/O ports
>> Diagram the design of peripheral I/O using the 74LS373 output latch and the 74LS244 input buffer
>> Describe the I/O address map of x86 PCs
>> List the differences in memory-mapped I/O versus peripheral I/O
>> Describe the purpose of the 8255 programmable peripheral interface chip
>> Code Assembly language instructions to perform I/O through the 8255
>> Describe how to interface the 8255 to the ISA bus
>> Understand the design of the PC Trainer
>> Code I/O Programming for Microsoft Visual C/C++
>> Code I/O Programming for Borland Turbo C/C++
>> Code I/O Programming for Linux C/C++
>> Code I/O Programming for Visual Basic

In addition to memory space, x86 microprocessors also have I/O space. This allows it to access ports. Ports are used either to bring data into the CPU from an external device such as the keyboard or to send data from the CPU to an external device such as a printer. In this chapter we study I/O instructions and I/O design for x86 PCs. In Section 11.1 we discuss I/O instructions and programming. In Section 11.2 we look at ways to design I/O ports for 8088-based systems. In Section 11.3, the I/O map of the x86 IBM PC is given. The 8255 chip and its programming are discussed in Section 11.4. The details of an 8255-based PC Trainer connected to the ISA bus will be given in Section 11.5. I/O programming using C/C++ and Visual Basic is covered in Section 11.6. Section 11.7 is dedicated to 16-bit data I/O ports for ISA buses.

SECTION 11.1: 8088 INPUT/OUTPUT INSTRUCTIONS

All x86 microprocessors, from the 8088 to the Pentium, can access external devices called *ports*. This is done using I/O instructions. The x86 CPU is one of the few processors that have I/O space in addition to memory space. While memory can contain both opcodes and data, I/O ports contain data only. There are two instructions for this purpose: "OUT" and "IN". These instructions can send data from the accumulator (AL or AX) to ports or bring data from ports into the accumulator. In accessing ports, we can use an 8-bit or 16-bit data port. Since 8-bit data ports in the 8088 are the most widely used, we will concentrate on them and introduce 16-bit data ports only in the last section of this chapter.

8-bit data ports

The 8-bit I/O operation of the 8088 is applicable to all x86 CPUs from the 8088 to the Pentium. The 8-bit port uses the D0 - D7 data bus to communicate with I/O devices. In 8-bit port programming, register AL is used as the source of data when using the OUT instruction and the destination for the IN instruction. This means that to input or output data from any other registers, the data must first be moved to the AL register. Instructions OUT and IN have the following formats:

```
         Inputting Data          Outputting Data
Format:   IN    dest,source       OUT   dest,source

(1)       IN    AL,port#          OUT   port#,AL


(2)       MOV   DX,port#          MOV   DX,port#
          IN    AL,DX             OUT   DX,AL
```

In format (1), port# is the address of the port and can be from 00 to FFH. This 8-bit address allows 256 input ports and 256 output ports. In this format, the 8-bit port address is carried on address bus A0 - A7. No segment register is involved in computing the address, in contrast to the way data is accessed from memory.

In format (2), the port# is also the address of the port, except that it can be from 0000 to FFFFH, allowing up to 65,536 input and 65,536 output ports. In this case, the 16-bit port address is carried on address bus A0 - A15, and no segment register (DS) is involved. This is the way Intel Corporation expanded the number of ports from 256 to 65,536 while maintaining compatibility with the earlier 8085 microprocessors. The use of a register as a pointer for the port address has an advantage in that the port address can be changed very easily, especially in cases of dynamic compilations where the port address can be passed to DX.

How to use I/O instructions

I/O instructions are widely used in programming peripheral devices such as printers, hard disks, and keyboards. The port address can be either 8-bit or 16-bit. For an 8-bit port address, we can use the immediate addressing mode. The following program sends a byte of data to a fixed port address of 43H.

```
MOV    AL,36H          ;AL=36H
OUT    43H,AL          ;send value 36H to port address 43H
```

The 8-bit address used in immediate addressing mode limits the number of ports to 256 for input plus 256 for output. To have a larger number of ports we must use the 16-bit port address instruction.

To use the 16-bit port address, register indirect addressing mode must be used. The register used for this purpose is DX. The following program sends values 55H and AAH to I/O port address 300H (a 16-bit port address). In other words, the program below toggles the bits of port address 300H continuously.

```
BACK: MOV    DX,300H         ;DX = port address 300H
      MOV    AL,55H
      OUT    DX,AL           ;toggle the bits
      MOV    AL,0AAH
      OUT    DX,AL           ;toggle the bits
      JMP    BACK
```

Notice that we can only use register DX for 16-bit I/O addresses; no other register can be used for this purpose. Also notice the use of register AL for 8-bit data. For example, the following code transfers the contents of register BL to port address 378H.

```
MOV DX,378H          ;DX=378 the port address
MOV AL,BL            ;load data into accumulator
OUT DX,AL            ;write contents of AL to port
                     ;whose address is in DX
```

To bring into the CPU a byte of data from an external device (external to the CPU) we use the IN instruction. Example 11-1 shows decision making based on the data that was input.

Example 11-1

In a given 8088-based system, port address 22H is an input port for monitoring the temperature. Write Assembly language instructions to monitor that port continuously for the temperature of 100 degrees. If it reaches 100, then BH should contain 'Y'.

Solution:

```
BACK:      IN    AL,22H   ;get the temperature from port # 22H
           CMP   AL,100   ;is temp = 100?
           JNZ   BACK     ;if not, keep monitoring
           MOV   BH,'Y    ;temp = 100, load 'Y' into BH
```

Just like the OUT instruction, the IN instruction uses the DX register to hold the address and AL to hold the arrived 8-bit data. In other words, DX holds the 16-bit port address while AL receives the 8-bit data brought in from an external port. The following program gets data from port address 300H and sends it to port address 302H.

```
MOV   DX,300H      ;load port address
IN    AL,DX        ;bring in data
MOV   DX,302H
OUT   DX,AL        ;send it out
```

Review Questions

1. In the x86 system, if we use only the 8-bit address bus A0 - A7 for port addresses, what is the maximum number of (a) input, and (b) output ports?
2. The x86 can have a maximum of how many I/O ports?
3. What does the instruction "OUT 24H,AL" do?
4. Write Assembly language instructions to accept data input from port 300H and send it out to port 304H.
5. Write Assembly language instructions to place the status of port 60H in CH.

SECTION 11.2: I/O ADDRESS DECODING AND DESIGN

In this section we show the design of simple I/O ports using TTL logic gates 74LS373 and 74LS244. For the purpose of clarity we use simple logic gates such as AND and inverter gates for decoders. It may be helpful to review the address decoding section for memory interfacing in the preceding chapter before you embark on this section. The concept of address bus decoding for I/O instructions is exactly the same as for memory. The following are the steps:
1. The control signals IOR and IOW are used along with the decoder.
2. For an 8-bit port address, A0 - A7 is decoded.
3. If the port address is 16-bit (using DX), A0 - A15 is decoded.

Using the 74LS373 in an output port design

In every computer, whenever data is sent out by the CPU via the data bus, the data must be latched by the receiving device. While memories have an internal latch to grab the data, a latching system must be designed for simple I/O ports. The 74LS373 can be used for this purpose. Notice in Figure 11-1 that in order to make the 74LS373 work as a latch, the OC pin must be grounded. For an output latch, it is common to AND the output of the address decoder with the control signal IOW to provide the latching action as shown in Figure 11-2.

IN port design using the 74LS244

Likewise, when data is coming in by way of a data bus, it must come in through a three-state buffer. This is referred to as *tristated*, which comes from the term tri-state buffer ("tri-state" is a registered trademark of National Semiconductor Corp.).

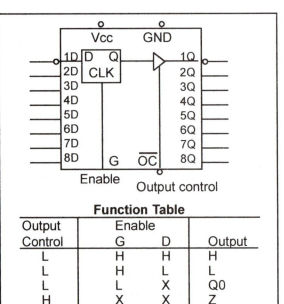

Enable
Output control

Function Table

Output Control	Enable G	D	Output
L	H	H	H
L	H	L	L
L	L	X	Q0
H	X	X	Z

Figure 11-1. 74LS373 D Latch
(Reprinted by permission of Texas Instruments, Copyright Texas Instruments, 1988)

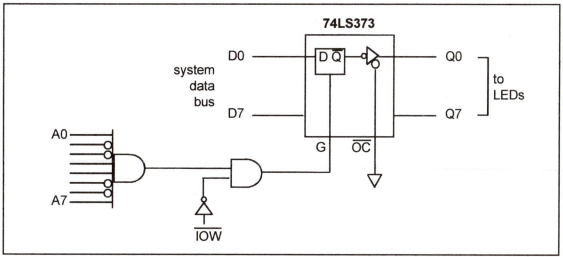

Figure 11-2. Design for "OUT 99H,AL"

Example 11-2

Show the design of an ouput port with an I/O address of 31FH using the 74LS373.

Solution:

31F9H is decoded, then ANDed with IOW to activate the G pin of the 74LS373 latch. This is shown in Figure 11-3.

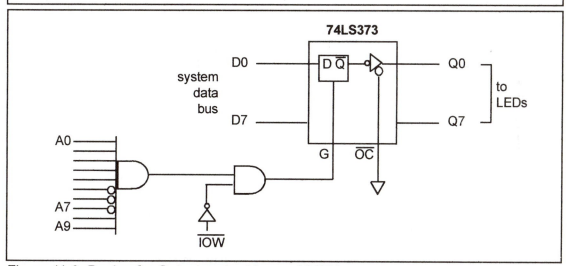

Figure 11-3. Design for Output port address of 31FH

As was the case for memory chips, such a tri-state buffer is internal and therefore invisible. For the simple input ports we use the 74LS244 chip. See Figure 11-4 for the internal circuitry of the 74LS244. In Figure 11-4, notice that since G1 and G2 each control only 4 bits of the 74LS244, they both must be activated for the 8-bit input. Examine Figure 11-5 to see the use of the 74LS244 as an entry port to the system data bus. Notice in Figures 11-5 and 11-6 how the address decoder and the IOR control signal together activate the tri-state input.

The 74LS244 not only plays the role of buffer, but also provides the incoming data with sufficient driving capability to travel all the way to the CPU. Indeed, the 74LS244 chip is widely used for buffering and providing high driving capability for unidirectional buses. The 74LS245 is used for bidirectional buses, as seen in Chapter 9.

SECTION 11.2: I/O ADDRESS DECODING AND DESIGN

Memory mapped I/O

Communicating with I/O devices using IN and OUT instructions is referred to as *peripheral I/O*. Some designers also refer to it as *isolated I/O*. However, there are many microprocessors, such as the new RISC processors, that do not have IN and OUT instructions. In such cases, these microprocessors use what is called *memory-mapped I/O*. In memory-mapped I/O, a memory location is assigned to be an input or output port. The following are the differences between peripheral I/O and memory-mapped I/O in the x86 PC.

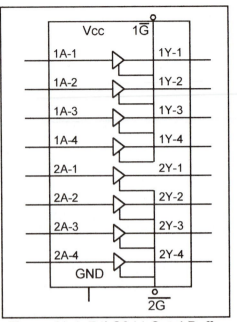

Figure 11-4. 74LS244 Octal Buffer
(Reprinted by permission of Texas Instruments, Copyright Texas Instruments, 1988)

1. In memory-mapped I/O, we must use instructions accessing memory locations to access the I/O ports instead of IN and OUT instructions. For example, an instruction such as "MOV AL,[2000]" will access an input port of memory address 2000 and "MOV [2010],AL" will access the output port.

2. In memory-mapped I/O, the entire 20-bit address, A0 - A19, must be decoded. This is in contrast to peripheral I/O, in which only A0 - A15 are decoded. Furthermore, since the 20-bit address involves both the segment and an offset, the DS register must be loaded before memory-mapped I/O is accessed. For example, if physical memory address 35000H is used for the input port, the following instructions can be used to access the port.

        MOV AX,3000H ;load the segment value
        MOV DS,AX
        MOV AL,[5000] ;bring in 1 byte from address 35000H

   Physical address 35000H is generated by shifting left DS one hex digit and adding it to offset address 5000 (30000 + 5000 = 35000H). Since all 20-bit addresses are decoded, the decoding circuitry for memory-mapped I/O is more expensive.

3. In memory-mapped I/O circuit interfacing, control signals MEMR and MEMW are used. This is in contrast to peripheral I/O, in which IOR and IOW are used.

4. In peripheral I/O we are limited to 65,536 input ports and 65,536 output ports, whereas in memory I/O the number of ports can be as high as 2^{20} (1,048,576). Of course, that many ports are never needed.

5. In memory-mapped I/O, one can perform arithmetic and logic operations on I/O data directly without first moving them into the accumulator. In memory-mapped I/O, data can be transferred into any register, rather than into the accumulator.

6. One major and severe disadvantage of memory-mapped I/O is that it uses memory address space, which could lead to memory space fragmentation.

Example 11-3

Show the design of "IN AL,9FH" using the 74LS244 as a tri-state buffer.

Solution:

9FH is decoded, then ANDed with IOR. To activate OC of the 74LS244, it must be inverted since OC is an active-low pin. This is shown in Figure 11-5.

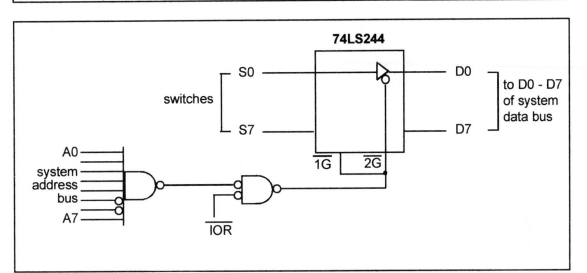

Figure 11-5. Design for "IN AL,9FH"

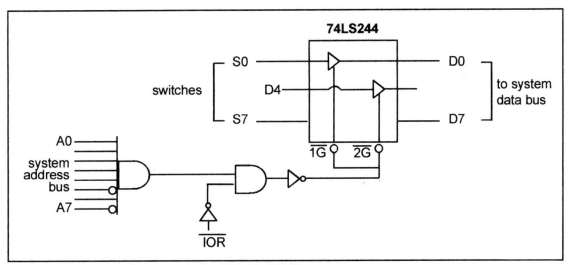

Figure 11-6. Input port design for "IN AL,5FH"

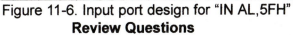

Review Questions

1. Designers use a _____ (latch, tri-state buffer) for output and a _____ (latch, tri-state buffer) for input.
2. Why do we use latches in I/O design?
3. Why is the 74LS373 called the transparent latch?
4. To use the 73LS373 as a latch, OC must be set to _____ permanently.
5. True or false. To access the maximum number of ports in the x86, we must decode addresses A0 - A15.
6. In memory-mapped I/O, which signal is used to select the (a) output, and (b) input devices?

SECTION 11.3: I/O ADDRESS MAP OF X86 PCS

Designers of the original IBM PC decided to make full use of I/O instructions. This led to assignment of different port addresses to various peripherals such as LPT and COM ports, and other chips and devices. The list of the designated I/O port addresses is referred to as the *I/O map*. Table 11-1 shows the I/O map for the x86 PC. A much more detailed I/O map of the x86 PC is given in Appendix G. Any system that needs to be compatible with the x86 IBM PC must follow the I/O map of Table 11-1. For example, the map shows that we can use I/O address 300 - 31F for a prototype card. This is shown below.

Absolute vs. linear select address decoding

In decoding addresses, either all of them or a selected number of them are decoded. If all the address lines are decoded, it is called *absolute decoding*. If only selected address pins are used for decoding, it is called *linear select decoding*. Linear select is cheaper, since the less input there is, the fewer the gates needed for decoding. The disadvantage is that it creates what are called *aliases*: the same port with multiple addresses. In cases where linear select is used, we must document port addresses in the I/O map thoroughly. In the first IBM PC, linear select decoding was used and that resulted in large numbers of address aliases, as we will see in future chapters. If you see a large gap in the I/O address map of the x86 PC, it is due to the address aliases of the original PC.

Prototype addresses 300 - 31FH in x86 PC

In the x86 PC, the address range 300H - 31FH is set aside for prototype cards to be plugged into the expansion slot. These prototype cards can be data acquisition boards used to monitor analog signals such as temperature, pressure, and so on. Interface cards using the prototype address space use the following signals on the 62-pin section of the ISA expansion slot:
1. IOR and IOW. Both are active low.
2. AEN signal: AEN = 0 when the CPU is using the bus.
3. A0 - A9 for address decoding.

Use of simple logic gates as address decoders

Figure 11-7 shows the circuit design for a 73LS373 latch connected to port address 300H of an x86 PC via an ISA expansion slot. Notice the use of signals A0 - A9 and AEN. AEN is low when the x86 microprocessor is in control of the buses. After all, it is the job of the CPU to control all the peripheral devices and not the DMA. In Figure 11-7, we are using simple logic gates such as NAND and inverter gates for the I/O address decoder. These can be replaced with the 74LS138 chip because the 74LS138 is a group of NAND gates in a single chip.

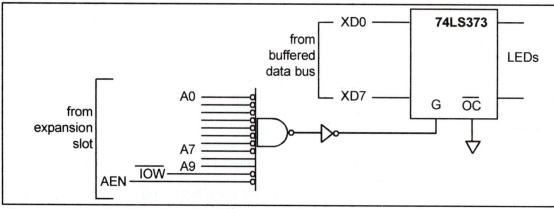

Figure 11-7. Using Simple Logic Gate for I/O Address Decoder (I/O Address 300H)

Hex Range	Device
000 - 01F	DMA controller 1, 8237A-5
020 - 03F	Interrupt controller 1, 8259A, Master
040 - 05F	Timer, 8254-2
060 - 06F	8042 (keyboard)
070 - 07F	Real-time clock, NMI mask
080 - 09F	DMA page register, 74LS612
0A0 - 0BF	Interrupt controller 2, 8237A-5
0C0 - 0DF	DMA controller 2, 8237A-5
0F0	Clear math coprocessor busy
0F1	Reset math coprocessor
0F8 - 0FF	Math coprocessor
1F0 - 1F8	Fixed disk
200 - 207	Game I/O
20C - 20D	Reserved
21F	Reserved
278 - 27F	Parallel printer port 2
2B0 - 2DF	Alternate enhanced graphics adapter
2E1	GPIB (adapter 0)
2E2 & 2E3	Data acquisition (adapter 0)
2F8 - 2FF	Serial port 2
300 - 31F	Prototype card
360 - 363	PC network (low address)
364 - 367	Reserved
368 - 36B	PC network (high address)
36C - 36F	Reserved
378 - 37F	Parallel printer port 1
380 - 38F	SDLC, bisynchronous 2
390 - 393	Cluster
3A0 - 3AF	Bisynchronous 1
3B0 - 3BF	Monochrome display and printer adapter
3C0 - 3CF	Enhanced graphics adapter
3D0 - 3DF	Color/graphics monitor adapter
3F0 - 3F7	Diskette controller
3F8 - 3FF	Serial port 1
6E2 & 6E3	Data acquisition (adapter 1)
790 - 793	Cluster (adapter 1)
AE2 & AE3	Data acquisition (adapter 2)
B90 - B93	Cluster (adapter 2)
EE2 & EE3	Data acquisition (adapter 3)
1390 - 1393	Cluster (adapter 3)
22E1	GPIB (adapter 1)
2390 - 2393	Cluster (adapter 4)
42E1	GPIB (adapter 2)
62E1	GPIB (adapter 3)
82E1	GPIB (adapter 4)
A2E1	GPIB (adapter 5)
C2E1	GPIB (adapter 6)
E2E1	GPIB (adapter 7)

Table 11-1: I/O Map for the x86 PC (See Appendix G for Further Details)

Use of 74LS138 as decoder

In current system board design, CPLD (complex programmable logic device) chips are used for supporting logics such as decoders. In the absence of CPLD, one could use NANDs, inverters, and 74LS138 chips for decoders as we saw in the preceding chapter for memory address decoding. The same principle applies to I/O address decoding. Figure 11-8 shows the 74LS138. As an example of the use of a 74LS138 for an I/O address decoder, examine Figure 11-9. Notice how each Y output can control a single device. Figure 11-9 shows the address decoding for an input port located at address 304H. The Y4 output, together with the IOR signal, controls the 74LS244 input buffer. Alternatively, Y0 along with the IOW control signal could be used to control a 74LS373 latch. In other words, each Y output controls a single I/O device. Contrast that with Figure 11-7. The 74LS138 is much more efficient than the use of simple logic gates as decoders.

IBM PC I/O address decoder

Figure 11-10 shows a 74LS138 chip used as an I/O address decoder in the original IBM PC. Notice that while A0 to A4 go to individual peripheral input addresses, A5, A6, and A7 are responsible for the output selection of outputs Y0 to Y7. In order to enable the 74LS138, pins A8, A9, and AEN all must be low. While A8 and A9 will directly affect the port address calculations, AEN is low only when the x86 is in control of the system bus. Since all the peripherals are programmed by the x86, AEN is low during CPU activity. See Table 11-2. We will discuss each Y output of 74LS138 in Figure 11-8 in subsequent chapters.

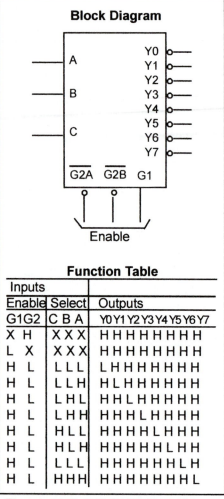

Block Diagram

Function Table

Inputs			
Enable	Select		Outputs
G1 G2	C B A		Y0 Y1 Y2 Y3 Y4 Y5 Y6 Y7
X H	X X X		H H H H H H H H
L X	X X X		H H H H H H H H
H L	L L L		L H H H H H H H
H L	L L H		H L H H H H H H
H L	L H L		H H L H H H H H
H L	L H H		H H H L H H H H
H L	H L L		H H H H L H H H
H L	H L H		H H H H H L H H
H L	L L L		H H H H H H L H
H L	H H H		H H H H H H H L

Figure 11-8. 74LS138 Decoder
(Reprinted by permission of Texas Instruments, Copyright Texas Instruments, 1988)

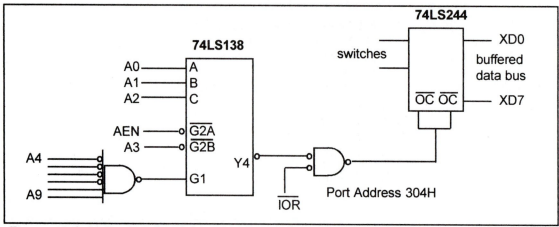

Figure 11-9. Using 74LS138 for I/O Address Decoder

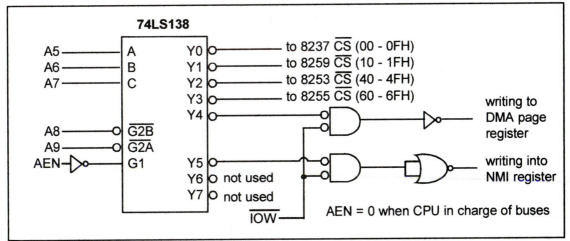

Figure 11-10. Port Address Decoding in the original IBM PC (PC/XT)

Table 11-2: Port Addresses decoding table on the Original PC

G1	G2A	G2B	C B A	
AEN	A9	A8	A7 A6 A5 A4 A3 A2 A1 A0	
0	0	0	0 0 0 0 0 0 0 0	00 Lowest port address
0	0	0	1 1 1 1 1 1 1 1	FF Highest port

Port 61H and time delay generation

Appendix G provides a detailed I/O map of the PC. In order to maintain compatibility with the IBM PC and run operating systems such as MS-DOS and Windows, the assignment of I/O port addresses must follow the standard set in Appendix G. Port 61H is a widely used port, the details of which are shown in Appendix G. We can use this port to generate a time delay, which will work in any PC with any type of processor from the 286 to the Pentium. I/O port 61H has eight bits (D0 - D7). Bit D4 is of particular interest to us. In all 286 and higher PCs, bit D4 of port 61H changes its state every 15.085 microseconds (μs). In other words, it stays low for 15.085 μs and then changes to high and stays high for the same amount of time before it goes low again. This toggling of D4 bit goes on indefinitely as long as the PC is on. Chapter 13 provides more details on this topic. The following program shows how to use port 61H to generate a delay of 1/2 second. In this program all the bits of port 310H are toggled with a 1/2 second delay in between.

```
;TOGGLING ALL BITS OF PORT 310H EVERY 0.5 SEC
        MOV    DX,310H
HERE:   MOV    AL,55H          ;toggle all bits
        OUT    DX,AL
        MOV    CX,33144        ;Delay=33144x15.085 us=0.5 sec
        CALL   TDELAY
        MOV    AL,0AAH
        OUT    DX,AL
        MOV    CX,33144
        CALL   TDELAY
        JMP    HERE
```

```
;CX=COUNT OF 15.085 MICROSEC
TDELAY      PROC    NEAR
            PUSH    AX              ;save AX
W1:         IN      AL,61H
            AND     AL,00010000B    ·
            CMP     AL,AH
            JE      W1              ;wait for 15.085 usec
            MOV     AH,AL
            LOOP    W1              ;another 15.085 usec
            POP     AX              ;restore AX
            RET
TDELAY      ENDP
```

In the above program, notice that when port 61H is read, all the bits are masked except D4. The program waits for D4 to change every 15.085 μs before it loops again.

Review Questions

1. What I/O address range is set aside for prototype cards?
2. In the x86 PC, give the status of the AEN signal when I/O ports are being addressed by the CPU.
3. In decoding addresses for I/O instructions, why do we need to include AEN, and what is the activation level?
4. Which bit of port 61H toggles every 15.085 μs?
5. Calculate the time delay if CX = 25,000 for Question 4.

SECTION 11.4: 8255 PPI CHIP

In this section we study the 8255 chip, one of the most widely used I/O chips. The 8255 is a 40-pin DIP chip (see Figure 11-11). It has three separately accessible ports, A, B, and C, which can be programmed, hence the name PPI (programmable peripheral interface). Notice that the individual ports of the 8255 can be programmed to be input or output. They can also be changed dynamically, in contrast to the 74LS244 and 74LS373, which are hard-wired.

Port A (PA0 - PA7)

This 8-bit port A can be programmed all as input or all as output.

Port B (PB0 - PB7)

This 8-bit port B can be programmed all as input or all as output.

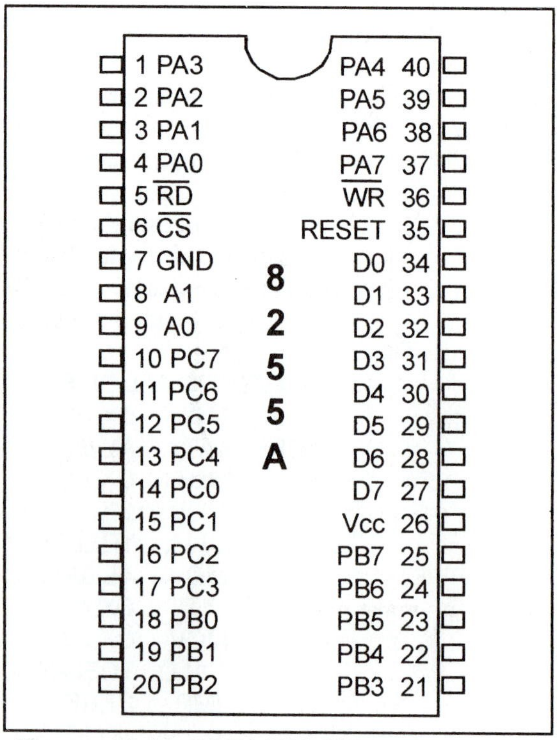

Figure 11-11. 8255 PPI Chip
(Reprinted by permission of Intel Corporation, Copyright Intel, 1983)

Port C (PC0 - PC7)

This 8-bit port C can be all input or all output. It can also be split into two parts, CU (upper bits PC4 - PC7) and CL (lower bits PC0 - PC3). Each can be used for input or output. Any of PC0 to PC7 can be programmed individually.

RD and WR

These two active-low control signals are inputs to the 8255. If the 8255 is using peripheral I/O design, IOR and IOW of the system bus are connected to these two pins. If the port uses memory-mapped I/O, MEMR and MEMW activate them.

RESET

This is an active-high signal input into the 8255 used to clear the control register. When RESET is activated, all ports are initialized as input ports. This pin must be connected to the RESET output of the system bus or ground, making it inactive. Like all IC input pins, it should not be left unconnected. This pin can be grounded.

A0, A1, and CS

While CS (chip select) selects the entire chip, address pins A0 and A1 select the specific port within the 8255. These three pins are used to access ports A, B, C, or the control register, as shown in Table 11-3.

Table 11-3: 8255 Port Selection

CS	A1	A0	Selects
0	0	0	Port A
0	0	1	Port B
0	1	0	Port C
0	1	1	Control register
1	x	x	8255 is not selected

(Reprinted by permission of Intel Corporation, Copyright Intel Corp. 1983)

Mode selection of the 8255A

While ports A, B, and C are used for I/O data, it is the control register that must be programmed to select the operation mode of the three ports A, B, and C. The ports of the 8255 can be programmed in any of the following modes.

1. Mode 0, simple I/O mode. In this mode, any of the ports A, B, CL, and CU can be programmed as input or output. In this mode, all bits are out or all are in. In other words, there is no control of individual bits. Mode 0 is the most widely used mode in current system I/O interfacing design. For the rest of this and the next chapter we use only this simple mode.
2. Mode 1. In this mode, ports A and B can be used as input or output ports with handshaking capabilities. This mode is not used due to timing incompatibility with devices such as the printer.
3. Mode 2. In this mode, port A can be used as a bidirectional I/O port with handshaking capabilities. This mode is rarely used.

Notice from Figure 11-12 that we must set $D7 = 1$ to get the above I/O modes of 0, 1, and 2. If $D7 = 0$, we have BSR mode. In BSR (bit set/reset) mode, the bits of port C are programmed individually. This mode is also rarely used.

The 8255 chip is programmed in any of the above modes by sending a byte (Intel calls it control word) to the control register of the 8255. For example, to make ports A, B, and C output ports, we make $D7 = 1$ according to Figure 11-12. To select simple I/O mode of 0, we need 1000 0000 as the control word. Similarly, to get PB as input, and PA and all of PC as output, we must have 1000 0010 or 82H for the control word. Examples 11-4, 11-5, and 11-6 demonstrate how the 8255 chip is programmed. Study these three examples in detail since they will be the basis of many lab assignments.

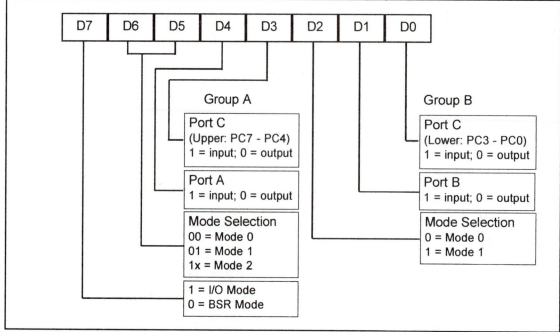

Figure 11-12. 8255 Control Word Format (I/O Mode)

(Reprinted by permission of Intel Corporation, Copyright Intel, 1983)

Example 11-4

(a) Find the control word if PA = out, PB = in, PC0 - PC3 = in, and PC4 - PC7 = out.
(b) Program the 8255 to get data from port A and send it to port B. In addition,
data from PCL is sent out to the PCU.
Use port addresses of 300H - 303H for the 8255 chip.
Solution:

(a) From Figure 11-12 we get the control word of 1000 0011 in binary or 83H.

(b) The code is as follows:

```
B8255C EQU   300H   ;Base address of 8255 chip
CNTL   EQU   83H    ;PA=out,PB=in,PCL=in,PCU=out

MOV   DX,B8255C+3  ;load control reg. address (300H+3=303H)
MOV   AL,CNTL      ;load control byte
OUT   DX,AL        ;send it to control register
MOV   DX,B8255C+1  ;load PB address
IN    AL,DX        ;get the data from PB
MOV   DX,B8255C    ;load PA address
OUT   DX,AL        ;send it to PA
MOV   DX,B8255C+2  ;load PC address
IN    AL,DX        ;get the bits from PCL
AND   AL,0FH       ;mask the upper bits
ROL   AL,1
ROL   AL,1         ;shift the bits
ROL   AL,1         ;to upper position
ROL   AL,1
OUT   DX,AL        ;send it to PCU
```

Example 11-5

The 8255 shown in Figure 11-13 is configured as follows: port A as input, B as output, and all the bits of port C as output.
(a) Find the port addresses assigned to A, B, C, and the control register.
(b) Find the control byte (word) for this configuration.
(c) Program the ports to input data from port A and send it to both ports B and C.

Solution:
(a) The port addresses are as follows:

$\overline{CS}$	A1	A0	Address	Port
11 0001 00	0	0	310H	Port A
11 0001 00	0	1	311H	Port B
11 0001 00	1	0	312H	Port C
11 0001 00	1	1	313H	Control register

(b) The control word is 90H, or 1001 0000.
(c) One version of the program is as follows:

```
        MOV   AL,90H      ;control byte PA=in, PB=out, PC=out
        MOV   DX,313H     ;load control reg address
        OUT   DX,AL       ;send it to control register
        MOV   DX,310H     ;load PA address
        IN    AL,DX       ;get the data from PA
        MOV   DX,311H     ;load PB address
        OUT   DX,AL       ;send it to PB
        MOV   DX,312H     ;load PC address
        OUT   DX,AL       ;and to PC
```

Using the EQU directive one can rewrite the above program as follows:

```
CNTLBYTE   EQU   90H      ;PA=in, PB=out, PC=out
PORTA      EQU   310H
PORTB      EQU   311H
PORTC      EQU   312H
CNTLREG    EQU   313H
           ....
           ...
           MOV   AL,CNTLBYTE
           MOV   DX,CNTLREG
           OUT   DX,AL
           MOV   DX,PORTA
           IN    AL,DX
           ;and so on.
```

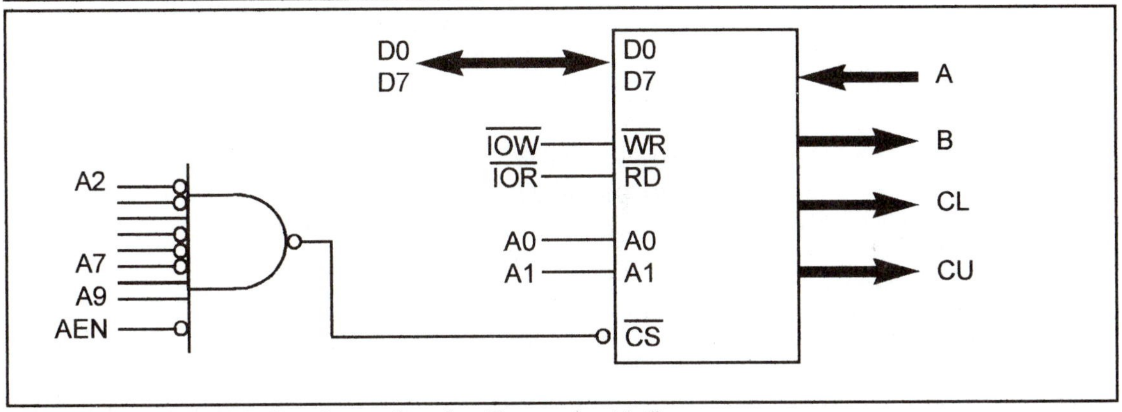

Figure 11-13. 8255 Configuration for Example 11-5

Example 11-6

Show the address decoding where port A of the 8255 has an I/O address of 300H, then write a program to toggle all bits of PA continuously with a 1/4 second delay. Use INT 16H to exit if there is a keypress.

Solution:

The address decoding for the 8255 is shown in Figure 11-14. The control word for all ports as output is 80H. The program below will toggle all bits of PA indefinitely with a delay in between. To prevent locking up the system, we press any key to exit to DOS.

```
                MOV    DX,303H      ;CONTROL REG ADDRESS
                MOV    AL,80H       ;ALL PORTS AS OUTPUT
                OUT    DX,AL
        AGAIN:  MOV    DX,300H
                MOV    AL,55H
                OUT    DX,AL
                CALL   QSDELAY      ;1/4 SEC DELAY
                MOV    AL,0AAH      ;TOGGLE BIT
                OUT    DX,AL
                CALL   QSDELAY
                MOV    AH,01
                INT    16H          ;CHECK KEYPRESS
                JZ     AGAIN        ;PRESS ANY KEY TO EXIT
                MOV    AH,4CH
                INT    21H          ;EXIT

QSDELAY         PROC   NEAR
                MOV    CX,16572     ;16,572x15.085 usec=1/4 sec
                PUSH   AX
W1:             IN     AL,61H
                AND    AL,00010000B
                CMP    AL,AH
                JE     W1
                MOV    AH,AL
                LOOP   W1
                POP    AX
                RET
QSDELAY         ENDP
```

Notice the use of INT 16H option AH=01 where the keypress is checked. If there is no keypress, it will continue. We must do that to avoid locking up the x86 PC.

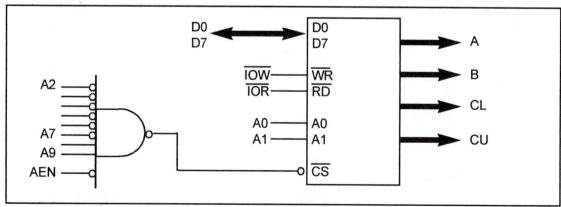

Figure 11-14. 8255 Configuration for Example 11-6

Intel calls mode 0 the *basic input output mode*. The more commonly used term is *simple I/O*. In this mode, any of ports A, B, or C can be programmed as input or output. It must be noted that in this mode a given port cannot be both an input and output port at the same time. One major characteristic of port C is that one can program CL (PC0 - PC3) and CU (PC4 - PC7) independently of each other, as shown in Example 11-4.

Review Questions

1. Find addresses for all 8255 ports if A7 - A2 = 111101 is used to activate CS.
2. Find the control word for an 8255 in mode 0 (simple I/O) if all the ports are configured as output ports.
3. Find the control word for an 8255 in mode 0 (simple I/O) if all the ports are configured as input ports.
4. Program an 8255 with the following specifications. All ports are output ports. Write 55H to the ports. After a delay, switch them all to AAH.
5. How are ports configured after the control register is loaded with 89H?

SECTION 11.5: PC INTERFACE TRAINER AND BUS EXTENDER

This section describes the PC Interface Trainer and the PC Bus Extender. These cards allow students to access the PC buses and can be used to interface real-world devices such as LCDs, stepper motors, and sensors to the x86 PC. These boards were designed by M. A. Mazidi to provide students with a relatively inexpensive learning tool. They are available from the supplier listed at the following Web site.

www.microdigitaled.com

The above site for Micro Digital Education is designed to support this and other textbooks by the authors.

PC I/O Bus Extender

The PC Bus Extender card is an ISA plug-in board that brings out some of the ISA signals through a 50-pin cable. This card allows one to interface an 8255 I/O board called the PC Interface Trainer to the x86 PC. The PC Bus Extender buffers only port addresses 300 - 31FH. It does not have a plug-and-play feature and must be used with an x86 PC in which I/O port addresses 300 - 31FH are free. Although some network cards use a portion of the address range 300 - 31FH, one can assign a different address space to them since these network cards have plug-and-play features. The 50-pin cable used by the PC Bus Extender can be as long as 5 feet.

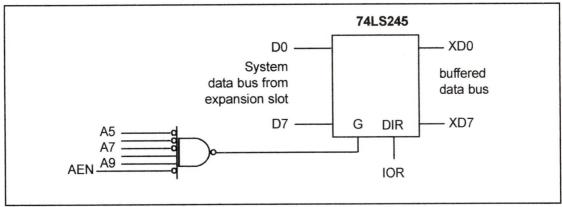

Figure 11-15. Buffering I/O Address Range 300 - 31FH

Buffering 300 - 31FH address range

When accessing the system bus via the expansion slot, we must make sure that the plug-in card does not interfere with the working of system buses on the motherboard. To do that we isolate (buffer) a range of I/O addresses using the 74LS245 chip. In buffering, the data bus is accessed only for a specific address range and access by any address beyond the range is blocked. Figure 11-15 shows how the I/O address range 300H - 31FH is buffered with the use of the 74LS245.

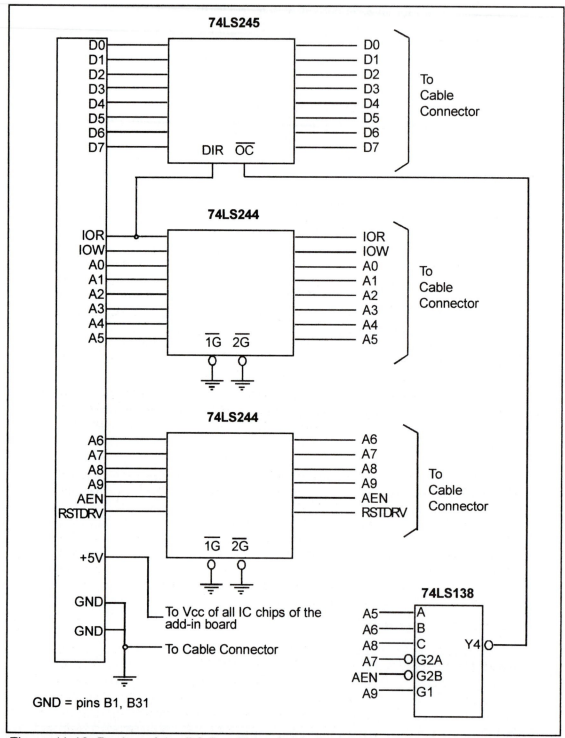

Figure 11-16. Design of the PC Bus Extender Card

CHAPTER 11: I/O AND THE 8255; ISA BUS INTERFACING

Figure 11-16 shows the circuit for buffering all the buses in the Bus Extender card. Notice the use of the 74LS244 to boost the address and control signals. This ensures the integrity of the signal transmitted via cables to the Trainer.

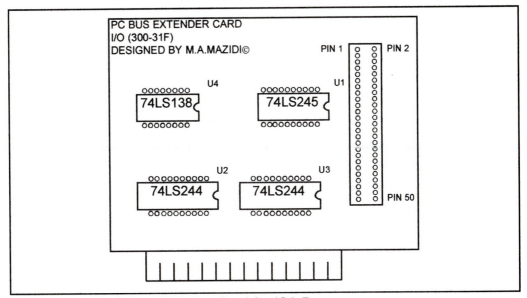

Figure 11-17. PC Bus Extender Card for ISA Bus

Installing the PC Bus Extender and booting the PC

The 50-pin cable with a female connector on each end brings the bus signals out of the PC. In this way, the PC case can be closed and secure. Although the cable is generally 2 or 3 feet long, it can be as long as 5 feet. The red (or some other color) stripe on your cable indicates pin 1. See Figure 11-17. When connecting the cable with connectors on both ends to H1, make sure that the red stripe and pin 1 of H1 match. Figure 11-18 shows the female connector. Figure 11-19 shows the connector cable with the stripe that indicates pin 1.

Install the card into an ISA expansion slot and turn on the PC power switch. If the PC boots up, you are ready to do PC interfacing with the PC Trainer. The PC Trainer is designed to be used specifically with the PC Bus Extender card.

Failure to boot

If the PC does not boot up properly, turn off the power to the PC and remove the bus extender card. First examine it to make sure that all the ICs on the Bus Extender card match the number and the direction as shown on the card. Assuming that everything is exactly the same as shown on the board, the failure of your PC to boot is due to I/O address conflict. This means that some other card(s) installed in your PC is using a portion of I/O addresses 300 - 31FH that are needed by the PC Bus Extender card. You need to free all the I/O addresses 300 - 31FH by changing switches on the plug-in board that caused the conflict before installing the PC Bus Extender card.

PC Interface Trainer

The PC Trainer (see Figure 11-21) is a trainer designed to be used with the PC Bus Extender discussed above. In version 1 of this Trainer, there were both 8255 and 8253 chips on board. Version 2 has only an 8255 chip on board in which PA, PB, and PC are accessible via a terminal block. With the 8255 chip installed on the module, one can perform real-world experiments such as interfacing an LCD, stepper motor, ADC, or sensors with the PC. Next we discuss the design of the PC Interface Trainer.

Design of the PC Trainer

Header 1 (H1) on the module brings in bus signals from the PC Bus Extender by a 50-pin cable. As discussed above, the addresses provided to this module are 300 - 31FH. Addresses 300 - 303 are used for the 8255 chip. The address decoding on the module is performed by the 74LS138. As shown in Figure 11-20, Y0 and Y1 are used on-board for the 8255 and 8253, respectively. Y1, Y2, Y3, and Y4 are provided through H2 for additional ports to be added off-board.

The role of H1 and H2

Note the role of H1 and H2 on the PC Trainer. H1 is connected to the PC Bus Extender cable, thereby bringing on-board the signals from the x86 PC ISA expansion slot. H2 is used only if additional ports are needed. In other words, H2 is used exclusively for future expansion of your I/O ports where the address decoding is already performed on the module by a 74LS138, while Y2, Y3, and Y4 provide the signal for the chip select. Notice that the H2 signals are identical to the signals of H1 except that pins 45, 47, and 49 provide the Y2, Y3, and Y4 outputs from the 74LS138. Table 11-4 shows the 74LS138 address assignment.

Connecting the Module Trainer to the PC and testing

The 50-pin cable coming from the PC Bus Extender goes to H1 of the PC Interface Trainer. Notice that the red stripe of your cable indicates pin 1 and must match pin 1 of H1. In some versions of the PC Trainer you need to provide an external +5 V DC source to power the board. In newer versions, it has an on-board voltage regulator that needs only an adaptor. The LED should indicate when the power is turned on. Now you are ready for testing the module.

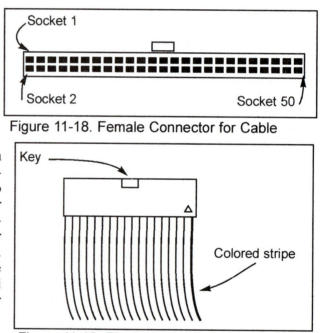

Figure 11-18. Female Connector for Cable

Figure 11-19. Pin 1 on Connector Cable

Table 11-4: Address Assignment of the 74LS138 for PC Trainer

Selector	Address	Assignment
Y0	300 - 303	Used by 8255 on the Module
Y1	304 - 307	Used by 8253 (not available in newer versions)
Y2	308 - 30B	Available via H2
Y3	30C - 30F	Available via H2
Y4	310 - 313	Available via H2

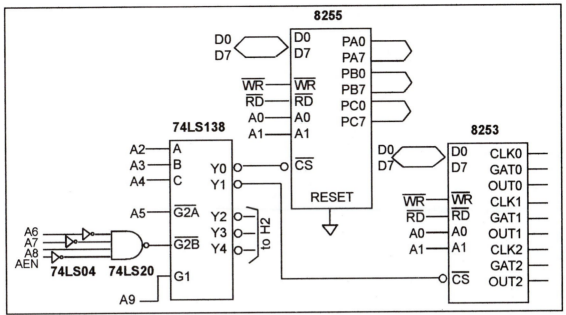

Figure 11-20. PC Interface Trainer Decoding Circuitry

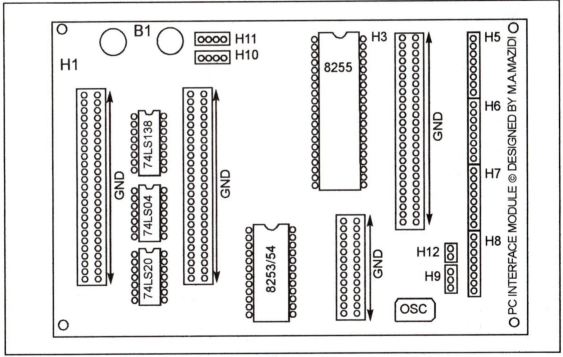

Figure 11-21. PC Interface Trainer Module (Version 1)

Testing the 8255 port

The headers on the board allow access to ports A, B, and C. Port addresses for the 8255 are shown in Table 11-4. Following is a series of simple programs to test ports A, B, and C of the 8255 using DEBUG. To perform the test, we write 55H (01010101 binary) and its complement AAH (10101010) to each port and check the result using a logic probe.

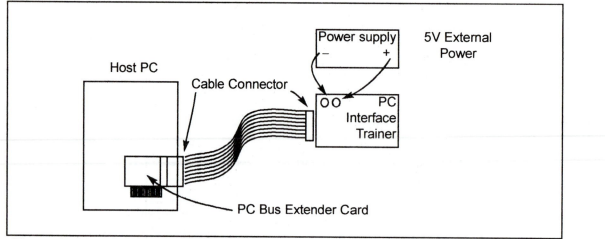

Figure 11-22. PC Interface Trainer Connection

Testing Port A

To test Port A, go to DEBUG in the DOS directory and write the following program:

```
C:\DOS>Debug
 -A 100
 MOV AL,80
 MOV DX,303
 OUT DX,AL
 MOV AL,55
 MOV DX,300
 OUT DX,AL
 INT 3

 -G=100 {RETURN}
```

After you run the above program, port A should have binary 01010101. Use a logic probe to verify the result. Repeat the above test by sending AAH to Port A. Replace the line "MOV AL,55" with "MOV AL,AA" as shown below:

```
 -A106
 MOV AL,AA
 -G=100
```

Using the logic probe, examine Port A to see if it contains binary 10101010. We can repeat the above tests for Ports B and C by simply changing the port to 301 and then to 302. Example 11-7 shows a test program to toggle the PA and PB bits. Notice that in order to avoid locking up the system, we use INT 16H to exit upon pressing any key. You can modify Example 11-7 to toggle all bits of PA, PB, and PC. Make sure to put a message on the PC screen to prompt the user to exit by pressing any key.

Notice that all the headers with the dual-row male connector are polarized. They use the even pins as ground in order to prevent crosstalk. Such use of even pins as ground allows the use of long cables to carry signals off the board while suppressing the noise.

Example 11-7

Write a program to toggle all bits of PA and PB of the 8255 chip on the PC Trainer. Put a 1/2 second delay in between "on" and "off" states. Use INT 16H to exit if there is a keypress.

Solution:

The program below toggles all bits of PA and PB indefinitely. Pressing any key exits the program.

```
            MOV    DX,303H        ;CONTROL REG ADDRESS
            MOV    AL,80H         ;ALL PORTS AS OUTPUT
            OUT    DX,AL
AGAIN:      MOV    DX,300H        ;PA ADDRESS
            MOV    AL,55H
            OUT    DX,AL
            INC    DX             ;PB ADDRESS
            OUT    DX,AL
            CALL   HSDELAY        ;1/2 SEC DELAY
            MOV    DX,300H        ;PA ADDRESS
            MOV    AL,0AAH
            OUT    DX,AL
            INC    DX             ;PB ADDRESS
            OUT    DX,AL
            CALL   HSDELAY        ;1/2 SEC DELAY
            MOV    AH,01
            INT    16H            ;CHECK KEYPRESS
            JZ     AGAIN          ;PRESS ANY KEY TO EXIT
            MOV    AH,4CH         ;
            INT    21H            ;EXIT

HSDELAY     PROC   NEAR
            MOV    CX,33144       ;33144x15.085 usec=1/2 sec
            PUSH   AX
W1:         IN     AL,61H
            AND    AL,00010000B
            CMP    AL,AH
            JE     W1
            MOV    AH,AL
            LOOP   W1
            POP    AX
            RET
HSDELAY     ENDP
```

Notice the use of INT 16H option AH=01 where the keypress is checked. If there is no keypress, it will continue.

How to get the PC Interface Trainer

You can the Bus Extender and PC Trainer from the vendors listed on the www.microdigitaled.com site. The PCI bus version of the Trainer is under development. See the above web site for more information.

Review Questions

1. What address range is used by the PC Bus Extender card?
2. True or false. The PC Bus Extender is a plug-and-play card.
3. What addresses are used by the 8255 chip on the PC Trainer board?
4. What addresses are decoded by the Y1 output of the 74LS138 chip?
5. In the program that tests the 8255, why do we use INT 16H?

SECTION 11.6: I/O PROGRAMMING WITH C/C++ AND VB

With the rise in popularity of C/C++ and Visual Basic in recent decades, it is fitting to explore how these languages are used in x86 I/O programming. In this section we discuss I/O operation in C/C++ and Visual Basic. We discuss I/O programming in Microsoft's Visual C/C++, Borland's Turbo C, Linux C/C++, and Visual Basic environments.

Visual C/C++ I/O programming

Microsoft Visual C++ programming is one of the most widely used programming languages on the Windows platform. Since Visual C++ is an object-oriented language, it comes with many classes and objects to make programming easier and more efficient. Unfortunately, there is no object or class for directly accessing I/O ports in the full Windows version of Visual C++. The reason for that is that Microsoft wants to make sure the x86 system programming is under full control of the operating system. This precludes any hacking into the system hardware. This applies also to Windows NT. In other words, none of the system INT instructions such as INT 21H and I/O operations that we have discussed in previous chapters are applicable in Windows NT and its subsequent versions. To access the I/O and other hardware features of the x86 PC in the NT environment you must use MS Developer's Software provided by Microsoft. The situation is different in the Windows 9x (95 and 98) environment. While INT 21H and other system interrupt instructions are blocked in Windows 9x, direct I/O addressing is available. To access I/O directly in Windows 9x, you must program Visual C++ in console mode. The instruction syntax for I/O operations is shown in Table 11-5. Notice the use of the undersign (_) in both the _outp and _inp instructions. It must also noted that while the x86 Assembly language makes a distinction between the 8-bit and 16-bit I/O addresses by using the DX register, there is no such distinction in C programming, as shown in Table 11-5. In other words, for the instruction "outp(port#,byte)" the port# can take any address value between 0000 - FFFFH.

Table 11-5. Input/Output Operations in Microsoft Visual C++

x86 Assembly	Visual C++
OUT port#,AL	_outp(port#,byte)
OUT DX,AL	_outp(port#,byte)
IN AL,port#	_inp(port#)
IN AL,DX	_inp(port#)

Visual C++ output example

Next we give some examples of I/O programming in Visual C++. Reexamine Example 11-7 in Assembly language. The Visual C++ version of that program is given in Example 11-8. In Example 11-8, we are toggling all the bits of PA and PB of the 8255 in the PC Trainer. Notice the following points.
1. The use of the _sleep function to create a delay.
2. The use of kbhit to exit upon any key press.
3. The use of 0x in _outp(0x300,0x80) to indicate that the values are in hex.

Visual C++ input example

As an example of inputting data in Visual C++, examine Example 11-9. We wrote the Assembly language version of this program in Example 11-5. In Example 11-9, we are getting a byte of data from port A and sending it to both PB and PC. Notice that when we bring a byte of data in, we save it using the variable

Example 11-8

Write a Visual C++ program to toggle all bits of PA and PB of the 8255 chip on the PC Trainer.
Use the kbhit function to exit if there is a keypress.
Solution:

```
//Tested by Dan Bent
#include<conio.h>
#include<stdio.h>
#include<iostream.h>
#include<iomanip.h>
#include<windows.h>
void main()
  {
  cout<<setiosflags(ios::unitbuf);           // clear screen buffer
  cout<<"This program toggles the bits for Port A and Port B.";
  _outp(0x303,0x80);       //MAKE PA,PB of 8255 ALL OUTPUT
  do
    {
    _outp(0x300,0x55);             //SEND 55H TO PORT A
    _outp(0x301,0x55);             //SEND 55H TO PORT B
    _sleep(500);                   //DELAY of 500 msec.
    _outp(0x300,0xAA);             //NOW SEND AAH TO PA, and PB
    _outp(0x301,0xAA);
    _sleep(500);
    }
    while(!kbhit());
  }
```

Example 11-9

Write a Visual C++ program to get a byte of data from PA and send it to both PB and PC of
8255 in PC Trainer.
Solution:

```
#include<conio.h>
#include<stdio.h>
#include<iostream.h>
#include<iomanip.h>
#include<windows.h>
#include<process.h>
//Tested by Dan Bent
void main()
  {
        unsigned char mybyte;
         cout<<setiosflags(ios::unitbuf);      // clear screen buffer
        system("CLS");
        _outp(0x303,0x90);      //PA=in, PB=out, PC=out
        _sleep(5);           //wait 5 milliseconds
        mybyte=_inp(0x300);      //get byte from PA
        _outp(0x301,mybyte);      //send to PB
        _sleep(5);
        _outp(0x302,mybyte);      //send to Port C
        _sleep(5);
        cout<<mybyte;            //send to PC screen also
        cout<<"\n\n";
  }
```

mybyte before we send it out. Make a habit of doing this every time you input data. Avoid combining a bunch of input and output operations together in a single line. That kind of dense code is very difficult for other programmers to read. Also, notice how the "unsigned char mybyte" line dictates the size of data as unsigned character. This allows the mybyte variable to be an 8-bit data, taking values of 00 - FFH.

I/O programming in Turbo C/C++

Borland is a major provider of software for the x86 PC. The company was founded in the early 1980s and became one of the early pioneers in the development of software for the x86 PC. They are also known by the name of Inprise. Check their web site www.borland.com. Their Turbo C/C++ is a widely used program compiler for x86 PCs. It supports I/O programming of ports as shown in Table 11-6. It must be noted that these I/O functions are no longer supported in Borland C++ Builder and you must write your own I/O functions using Assembly language. See the Micro Digital Education web site www.microdigitaled.com for more information on this and other topics.

Table 11-6: Input/Output Operation in Borland C++

x86 Assembly	Turbo C++
OUT port#,AL	outp(port#,byte)
OUT DX,AL	outp(port#,byte)
IN AL,port#	inp(port#)
IN AL,DX	inp(port#)

Example 11-10

Write a Borland (Inprise) Turbo C program to toggle all bits of PA and PB of the 8255 chip on the PC Trainer. Put a 500 ms (milliseconds) delay between the "on" and "off" states. Use the kbhit function to exit if there is a keypress.

Solution:

```
#include<conio.h>
#include<stdio.h>
void main()
  {
  printf("This program toggles the bits for Port A and Port B.");
  outp(0x303,0x80);       //MAKE PA,PB of 8255 ALL OUTPUT
  do
    {
    outp(0x300,0x55);          //SEND 55H TO PORT A
    outp(0x301,0x55);          //SEND 55H TO PORT B
    delay(500);                //DELAY of 500 msec.
    outp(0x300,0xAA);          //NOW SEND AAH TO PA, and PB
    outp(0x301,0xAA);
    delay(500);
    }
  while(!kbhit());
  }
```

Example 11-11

Write a Turbo C/C++ program to get a byte of data from PA and send it to both PB and PC of the 8255 in the PC Trainer.

Solution:

```c
#include<conio.h>
#include<stdio.h>
void main()
 {
        unsigned char mybyte;
        cls();
        printf("This program gets a byte from PA and sends it to PB,PC and screen\n.");
        outp(0x303,0x90);    //PA=in, PB=out, PC=out
        delay(5);           //wait 5 milliseconds
        mybyte=inp(0x300);     //get byte from PA
        outp(0x301,mybyte);    //send to PB
        dealy(5);
        outp(0x302,mybyte);    //send to Port C
        sleep(5);
        printf("The input from PA is equal to %X in hex \n",mybyte);
}
```

I/O programming in Linux C/C++

Linux is a popular operating system for the x86 PC. You can get a copy of the latest C/C++ compiler from http://gcc.gnu.org. Table 11-7 provides the C/C++ syntax for I/O programming in the Linux OS environment.

Table 11-7. Input/Output Operations in Linux

x86 Assembly	Linux C/C++
OUT port#,AL	outb(byte,port#)
OUT DX,AL	outb(byte,port#)
IN AL,port#	inb(port#)
IN AL,DX	inb(port#)

Compiling and running Linux C/C++ programs with I/O functions

To compile the I/O programs of Examples 11-12 and 11-13, the following points must be noted.
1. To compile with a keypress loop, you must link to library ncurses as follows:
   > gcc -lncurses toggle.c -o toggle
2. To run the program, you must either be root or root must change permissions on executable for hardware port access.
   Example: (as root or superuser)
   > chown root toggle
   > chmod 4750 toggle

   Now toggle can be executed by users other than root. More information on this topic can be found at www.microdigitaled.com.

Example 11-12

Write a C/C++ program for a PC with the Linux OS to toggle all bits of PA and PB of the 8255 chip on the PC Trainer. Put a 500 ms delay between the "on" and "off" states. Pressing any key should exit the program.

Solution:

```
//        This program demonstrates low level I/O
//        using C language on a Linux based system.
 //       Tested by Nathan Noel  //
#include <stdio.h>        // for printf()
#include <unistd.h>       // for usleep()
#include <sys/io.h>       // for outb() and inb()
#include <ncurses.h>      // for console i/o functions
int main ()
 {
 int n=0;                 // temp char variable
 int delay=5 e5;          // sleep delay variable

 ioperm(0x300,4,0x300);  // get port permission
 outb(0x80,0x303);        // send control word

 //----- begin ncurses setup ----------
 //--- (needed for console i/o) -------

 initscr();               // initialize screen for ncurses
 cbreak();                // do not wait for carriage return
 noecho();                // do not echo input character
 halfdelay(1);            // only wait for 1ms for input
                          // from keyboard
 //-----  end ncurses setup  ----------

 do                       // main toggle loop
   {
   printf("0x55 \n\r");   // display status to screen
   refresh();             // refresh() to update console
   outb(0x55,0x300);      // send 0x55 to PortA (01010101B)
   outb(0x55,0x301);      // send 0x55 to PortB (01010101B)
   usleep(delay);                // wait for 500ms (5 e5 microseconds)
   printf("0xAA \n\r");   // display status to screen
   refresh();             // refresh() to update console
   outb(0xaa,0x300);      // send 0xAA to PortA (10101010B)
   outb(0xaa,0x301);      // send 0xAA to PortB (10101010B)
   usleep(delay);                // wait for 500ms
                          // get input from keyboard
   n=getch();             // if no keypress in 1ms, n=0
                          // due to halfdelay()

   }
 while(n<=0);             // test for keypress
                          // if keypress, exit program
 endwin();                // close program console for ncurses
 return 0;                        // exit program
 }
```

Example 11-13

Write a C/C++ program for a PC with the Linux OS to get a byte of data from port A and send it to both port B and port C of the 8255 in the PC Trainer.

Solution:

```c
//          This program gets data from Port A and
//          sends a copy to both Port B and Port C.
//          Tested by: Nathan Noel -- 2/10/2002
//---------------------------------------------
#include <stdio.h>
#include <unistd.h>
#include <sys/io.h>
#include <ncurses.h>

int main ()
  {
  int n=0;                    // temp variable
  int i=0;                    // temp variable

  ioperm(0x300,4,0x300);// get permission to use ports
  outb(0x90,0x303);     // send control word for
                        // PortA=input, PortB=output, PortC=output

  initscr();            // initialize screen for ncurses
  cbreak();             // do not wait for carriage return
  noecho();             // do not echo input character
  halfdelay(1);         // only wait for 1ms for input

  do                    // main toggle loop
    {
    i=inb(0x300);       // get data from PortA
    usleep(1e5);  // sleep for 100ms

    outb(i,0x301);      // send data to PortB
    outb(i,0x302);      // send data to PortC

    n=getch();          // get input from keyboard
                        // if no keypress in 1ms, n=0
    }while(n<=0);       // test for keypress
                        // if keypress, exit program

  endwin();             // close program window
  return (0);           // exit program
  }
```

I/O programming in Visual Basic

Microsoft's Visual Basic is a widely used programming language for the x86 PC due to its quick development time. With all its capabilities, VB does not allow direct access of I/O ports. To solve this problem we have written a dll (dynamic link library) program in C/C++ for the Visual Basic I/O access. The VB source code for the I/O examples along with the dll files are available from the www.microdigitaled.com web site.

Review Questions

1. Show how to send 99H to port address 300H in Visual C/C++.
2. Show how to send 99H to port address 300H in Borland Turbo C/C++.
3. Show how to send 99H to port address 300H in Linux C/C++.
4. Show how to bring a byte from the address 302H in Visual C/C++.
5. Show how to bring a byte from the address 302H in Linux C/C++.

SECTION 11.7: 8-BIT AND 16-BIT I/O TIMING IN ISA BUS

As we have seen, interfacing 8-bit devices such as the 8255 to the x86 PC is a straightforward process because the 8-bit data pins match D0 - D7 of the ISA bus. Just as in 16-bit memory interfacing, there is a problem when we want to use the 16-bit (D0 - D15) data bus of the x86 for I/O operations. In this section we look at the timing and design of 16-bit data I/O and compare it with 8-bit I/O operations. First, we examine a few issues concerning the ISA bus.

8-bit and 16-bit I/O in ISA bus

The term *ISA computers* encompasses IBM PC AT, PS, and any x86 PCs with AT-type expansion slots, as explored in Chapter 9. These computers could use the 286, 386, 486, Intel Pentium, or any x86 microprocessor from AMD as their CPU, but they have ISA-type expansion slots. The following points must be noted about these types of computers:

1. In communications between the x86 CPU and I/O ports, typically I/O devices are slow and cannot respond to the CPU's normal speed. In such situations, wait states must be inserted into the I/O cycle. The 80286 and all higher microprocessors (386, 486, Pentium, etc.) have two clocks for the I/O cycle time when they are designed with 0 WS. In this regard, it is the same as memory cycle time for such processors. For example, a 100-MHz Pentium processor provides a total of 20 ns (2 × 10 ns since 1/200 MHz = 10 nanoseconds) for the I/O cycle time. While in recent years, memory speed has been increasing steadily, there has not been a corresponding increase in the speed of I/O components such as ADCs (analog-to-digital converters). In general, I/O devices are much slower than memory since the I/O is interfaced with nature whereas memory is a semiconductor device. For example, a temperature sensor converts temperatures to voltage levels and the voltages are converted to binary numbers using ADCs before they are provided to the CPU. The delays associated with each stage of conversion add to the I/O response time, causing a severe bottleneck. This is only one of the reasons why the ISA expansion slot speed is limited to 8 MHz, in spite of the fact that the CPU bus speed is 100 MHz in the Pentium. Therefore, to interface with slow I/O devices, one must insert wait states into the I/O cycle time to match the device speed.
2. When the CPU communicates with an ISA expansion slot, be it memory or a peripheral I/O port, it can only use an 8- or 16-bit data bus, even if the CPU is 32-bit, such as a 386, 486, or Pentium. In 386 and 486 PCs where the CPU

has a 32-bit data bus, memory (or even a peripheral) on the motherboard uses a 32-bit data bus, but when it goes to an ISA expansion slot it must use a 16-bit bus. When designing a plug-in card for the ISA expansion slot of a motherboard with a 32-or 64-bit CPU, we use the 8- or 16-bit data bus but not the 32/64 bit data buses. To access the x86's entire 32- or 64-bit data bus through the expansion slot, we must use a PCI slot. This is discussed in Chapter 27.

3. The ISA bus speed is limited to 8 MHz. It does not matter that the x86 CPU works on a frequency of 10 MHz or 100 MHz or even 1 GHz: When it communicates with devices (memory or I/O ports) through the ISA expansion slots it must slow down to 8 MHz. That means inserting many wait states in order to access the boards connected to ISA expansion slots. The good news is that the chipset on the motherboard will do all the above tasks.

These three limitations are commonly referred to as *I/O bottleneck* since they slow down the flow of information to/from the CPU when using I/O devices.

I/O signals of the ISA bus

Just as in memory, the ISA bus supports both 8- and 16-bit I/O operations. Furthermore, the problems associated with 8- and 16-bit data memory interfacing discussed in Chapter 10 also exist in I/O interfacing of the ISA bus. The following ISA slot pins are associated with I/O interfacing and must be understood.

SA0 - SA9 (system address)

The system address A0 - A9 bus provides the I/O addresses. This limits the I/O addresses supported by the ISA slot to 1024 ports.

SD0 - SD7 (system data bus)

The system D0 - D7 8-bit data bus or D0 - D15 16-bit data bus provides the data path between the CPU and the I/O device.

IOR and IOW

The IOR and IOW control signals are both active low and are connected to the read and write pins of I/O devices.

IOCS16 (I/O chip select 16)

This is an input into the ISA bus and is an active low signal. It informs the system that the I/O operation uses the entire 16-bit data bus D0 - D15. If this input is not driven low, the ISA bus uses the 8-bit data bus D0 - D7.

ZEROWS (zero wait state)

ZEROWS is an input pin into the ISA bus and is active low. If this signal is driven low, it tells the system that I/O and memory operations can be completed without any WS. We will see how this applies to I/O operation in this section.

IOCHRDY (I/O channel ready)

This is an input signal into the ISA bus and is active low. By driving it low, we are asking the system to extend the standard ISA bus cycle. In response to asserting this pin, the system will insert wait states into the I/O or memory cycle until it is deasserted. The function of this pin is the opposite of ZEROWS. In other words, by asserting this pin we are extending the I/O (or memory) read and write cycle time to allow the interfacing of slow devices to the ISA bus.

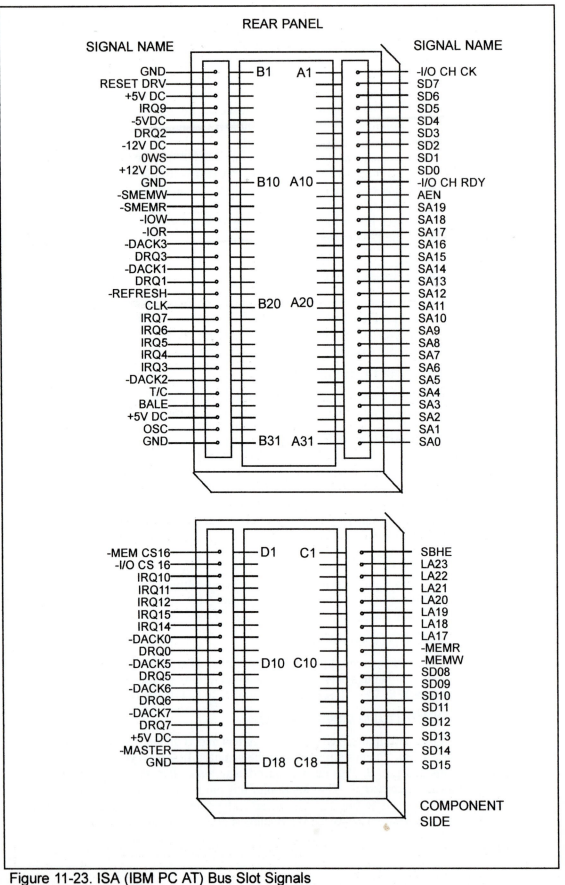

Figure 11-23. ISA (IBM PC AT) Bus Slot Signals

(Reprinted by permission from "IBM Technical Reference" c. 1985 by International Business Machines Corporation)

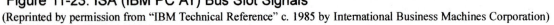

CHAPTER 11: I/O AND THE 8255; ISA BUS INTERFACING

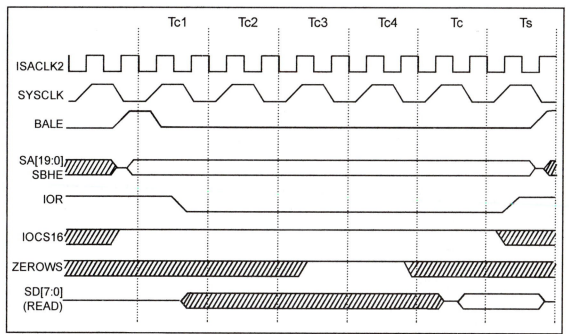

Figure 11-24. Standard 8-bit ISA I/O Read Cycle Time (4 WS)

8-bit I/O timing and operation in ISA bus

It would be helpful to review the discussion of 8-bit memory operation and interfacing in Chapter 10 since memory and I/O operations are very similar. The I/O operation of the ISA bus defaults to 8-bit and uses the D0 - D7 data bus to transfer data between the I/O device and the CPU. It completes the read (or write) cycle in 6 (2 + 4 WS) clocks if ZEROWS is not asserted low. In other words, the default I/O operation for ISA bus has 4 WS and uses the 8-bit data bus. This is an 8-bit standard I/O operation and is shown in Figure 11-24. With a maximum of 8 MHz for the ISA bus clock, I/O takes a total of 125 ns × 6 = 750 ns. Just like memory cycle time, we can shorten the I/O cycle time by asserting the ZEROWS pin low. This will cause the I/O operation to be completed in 3 (2 + 1

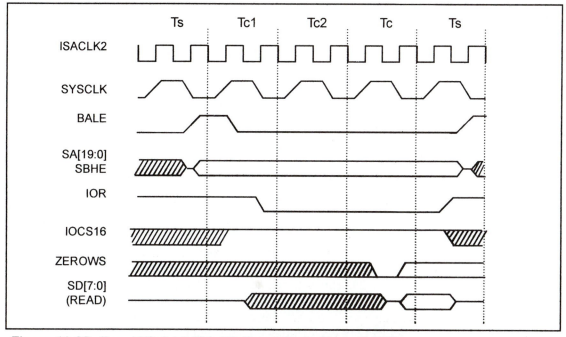

Figure 11-25. Zero WS 8-bit ISA I/O Read Cycle Time (1 WS)

WS) clocks instead of 6 (2 + 4 WS). Notice that the default is 4 WS unless ZEROWS is asserted. This is shown in Figure 11-25. Now if the default 4 WS I/O cycle time is not long enough, we can extend it by driving the IOCHRDY pin low. The extension happens as long as the IOCHRDY pin is low. The maximum extension is limited to 10 WS.

16-bit I/O operation and timing in ISA bus

Review the discussion of 16-bit memory operations and interfacing in Chapter 10 since memory and I/O operations are very similar. Just as in memory, the 16-bit I/O port uses data bus D0 - D15 to transfer data between the CPU and I/O devices. First let's look at 16-bit I/O instructions supported by the x86 family. Similar to memory, the x86 allows the uses of D0 - D15 for data transfer between the CPU and I/O devices. It must be noted that in the 16-bit I/O operation, the low byte uses D0 - D7 and the high byte uses the D8 - D15 data bus. The following is the 16-bit I/O instructions format.

16-bit data ports instruction

```
     Inputting Data        Outputting Data
(1)  IN    AX,port#        OUT   port#,AX

(2)  MOV   DX,port#        MOV   DX,port#
     IN    AX,DX           OUT   DX,AX
```

Notice that we must use AX instead of AL in the 16-bit I/O. To use 16-bit data I/O, we need two port addresses, one for each byte. Again, this is because I/O space is byte addressable, just like memory space. Look at the following:

```
MOV    AX,98F6H
OUT    40H,AX   ;send out AX to port 308H & 309H
```

In this case F6H, the content of AL, goes to port address 40H while 98H, the content of AH, is transferred to port address 41H. The low byte goes to the low port address and the high byte to the high port address. This is exactly like memory data transfers in that the low byte goes to the low address location and the high byte goes to the high address location (the little endian convention). This principle works the same for 16-bit port addresses as shown below:

```
MOV    DX,310H
MOV    AX,98F6H
OUT    DX,AX   ;send out AX to port 310H & 311H
```

Again the F6H is sent to port address 310H using data path D0 - D7 and 98H goes to port address 311H using the D8 - D15 data path. Next we will contrast 8-bit and 16-bit I/O via the ISA bus.

16-bit I/O timing and operation via ISA bus

As mentioned earlier, 8-bit data transfer is the default mode for the ISA bus expansion slots. To perform 16-bit data transfers using D0 - D15, we must assert the IOCS16 pin low. The 16-bit bus transfers data twice as fast as an 8-bit data bus. However, it requires twice the board space in addition to an increase in power consumption. The 16-bit data read cycle time for the ISA bus with one WS is shown in Figure 11-26. This is the standard read cycle time in the ISA bus. In other words, unlike memory, you cannot get zero wait state bus activity for 16-bit I/O operations. Therefore, ZEROWS has no effect on 16-bit I/O operations, and the standard 16-bit ISA bus cycle timing is completed in 3 clocks. While you cannot shorten the 16-bit I/O bus cycle time, you can extend it by asserting the IOCHRDY pin.

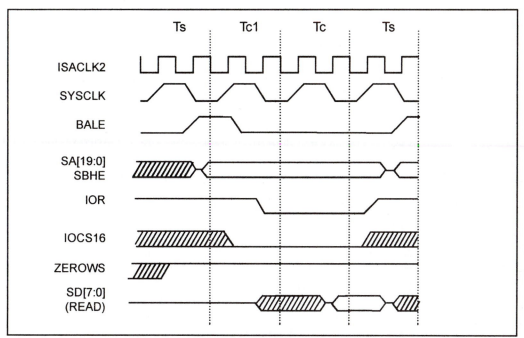

Figure 11-26. Standard 16-bit ISA I/O Read Cycle Time (1WS)

I/O bus bandwidth for ISA

The I/O bus bandwidth is the rate of data transfer between the CPU and I/O devices and is dictated by the bus speed and the data bus width used. Example 11-14 shows the calculation of bus bandwidth for the ISA bus.

Example 11-14

Find the ISA bus bandwidth for (a) 8-bit standard, (b) 8-bit with ZEROWS asserted, and (c) 16-bit standard.

Solution:

Since the ISA bus speed is 8 MHz, we have 1/8 MHz = 125 nanoseconds for the bus clock.

(a) The standard 8-bit I/O bus cycle for ISA uses 4 WS. Therefore, the bus cycle time is 6 clocks. Now cycle time = 6 × 125 ns = 750 and the bus bandwidth is 1/750 ns × 1 bytes = 1.33 megabytes/second.

(b) If ZEROWS is asserted in 8-bit I/O, we have 3 clocks for the I/O cycle time. Therefore, we have a cycle time of 3 × 125 = 375 ns and 1/375 ns × 1 byte = 2.66 megabytes/second for bus bandwidth.

(c) For 16-bit I/O data transfers we must assert the IOCS16 pin. The I/O cycle time is 3 clocks. Therefore, we have a bus bandwidth of 1 / (3 × 125) × 2 = 5.33 megabytes/second. For 16-bit I/O, we cannot assert ZEROWS to shorten the cycle time.

Interfacing 8-bit peripherals to a 16-bit data bus

As mentioned in the last chapter, microprocessors with a 16-bit data bus use odd and even byte spaces. This is done with the help of the A0 and BHE pins as shown below:

BHE	A0	
0	0	Even-addressed words (uses D0 - D15)
0	1	Odd-addressed byte (uses D8 - D15)
1	0	Even-addressed byte (uses D0 - D7)

In interfacing 16-bit I/O, the main issue is how to deal with odd and even address ports. The fact that data for even-address ports is carried on data bus D0 - D7 and data for odd address ports is carried on D8 - D15 makes port design a challenging issue. There are two solutions to this problem.

1. Simply use two separate PPI devices, such as the 8255. One is used for odd addresses and the other for even addresses. For example, in a design using this method, if port 74H is assigned to port A of the 8255, then port B has the address 76H, port C the address 78H, and so on. Another problem is out-putting the contents of register AX in an instruction such as "OUT 76H,AX". In this case, AL is carried to the 8255 with odd port addresses on D0 - D7 and AH is carried to the other 8255 with even port addresses on D8 - D15. This is extremely awkward and confusing for the programmer. Figure 11-27 shows the 8255 with odd and even port addresses.

2. The second solution is to connect all 8-bit peripheral ports to data bus D0 - D7. This is exactly what IBM PC/AT designers, and indeed all makers of the x86 ISA bus, have done. In such a design, one problem must be solved. What happens when instructions such as "OUT 75H,AL" are executed? This is the odd-address port and the data is provided by the CPU on D8 - D15, but the port is connected to D0 - D7. To solve this problem, one must use a latch to grab the data from bus D8 - D15 and provide it to D0 - D7, where the port is connected. The latch responsible for this is called the Hi/Lo byte copier in ISA bus literature. In order for the Hi/Lo byte copier to work properly, it needs some logic circuitry. It is the function of the bus control logic circuitry to detect the following cases and activate the Hi/Lo byte copier.

Case 1: Outputting a byte to odd-addressed ports

To write a byte to an odd-addressed port, the CPU provides the data on its upper data bus (D8 - D15) and makes A0 = 1 and BHE = 0 since the port address is an odd address. For example, in the instruction "OUT 41H,AL", the contents of AL are provided to D8 - D15 while BHE = 0 and A7 - A0 = 0100 0001. The bus control logic circuitry senses that the CPU is trying to send 8-bit data to an odd address through its D8 - D15 data bus. It activates the Hi/Lo byte copier, which copies the data from D8 - D15 to D0 - D7. The data is presented to the 8-bit peripheral device, which is connected to the lower data bus, D0 - D7.

Case 2: Inputting a byte from odd-addressed ports

To read a byte from an odd-addressed port, the CPU expects to receive the data on its upper data bus (D8 - D15) and makes A0 = 1 and BHE = 0. For example, in the instruction "IN AL,43", A7 - A0 = 0100 0011 and BHE = 0. The CPU expects the data to come in through D8 - D15. The input port device is connected to D0 - D7. The bus control logic circuitry senses that the CPU is trying to get 8 bits of data from a peripheral device through its D8 - D15 data pins. The port

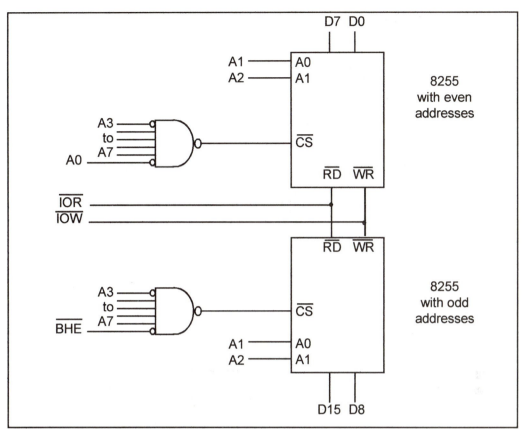

Figure 11-27. Odd and Even Ports with the 8255

is connected to D0 - D7. It activates the Hi/Lo byte copier and copies the data from D0 - D7 to D8 - D15 and the data is presented to the CPU. The details of bus control logic circuits are quite involved and in today's PCs are buried in the chipsets of x86 PCs.

Review Questions

1. In the ISA bus, we use address _____ to locate the I/O device.
2. What is the maximum number of I/O devices the ISA bus supports?
3. What is role of the ZEROWS pin for I/O cycle time?
4. The IOCS16 pin is an _____ (input, output) signal.
5. To use the 16-bit I/O capability of the ISA bus we must assert pin _____.
6. What is the minimum I/O cycle time for the 8-bit ISA bus?
7. What is the maximum bus bandwidth for the ISA bus?

SUMMARY

This chapter discussed the input/output ports of the x86 IBM PC and compatible computers. First, the Assembly language instructions used to read to or write from these ports were covered. This was followed by a look at the hardware design of I/O ports. The differences between memory-mapped and peripheral I/O were discussed. We also discussed the I/O map of the x86 PC. Then the 8255 programmable peripheral chip was examined in terms of pin layout, programming, mode selection, and port addressing. Then we described the PC Interface Trainer and PC Bus Extender, which can be attached to a PC in order to allow interfacing with real-world devices. Specific problems associated with interfacing I/O ports to the 16-bit data bus of the PC were addressed. It was also shown how to calculate the I/O cycle time and I/O bus bandwidth.

PROBLEMS

SECTION 11.1: 8088 INPUT/OUTPUT INSTRUCTIONS

1. True or false. While memory contains both code and data, ports contain data only.
2. In instruction "OUT 99H,AL", the port address is:
   (a) 8 bits   (b) 16 bits   (c) both (a) and (b)   (d) none of the above
3. In instruction "OUT DX,AL", the port address is:
   (a) 8 bits   (b) 16 bits   (c) either (a) or (b)   (d) none of the above
4. True or false. In instruction "IN AL,78H", register AL is the destination.
5. Explain what the instruction "IN AL,5FH" does.
6. In instruction "OUT DX,AL", assume that AL = 3BH and DX = 300H. Explain what the instruction does.

SECTION 11.2: I/O ADDRESS DECODING AND DESIGN

7. In the execution of an OUT instruction, which control signal is activated?
8. In the execution of an IN instruction, which control signal is activated?
9. True or false. Segment register DS is used to generate a port's physical address.
10. True or false. In "OUT 65H,AL", only address pins A0 - A7 are used by the 8088 to provide the address.
11. True or false. An input port is distinguished from an output port by the port address assigned to it.
12. True or false. An input port is distinguished from an output port by the IOR and IOW control signals.
13. _____ (Latch, Tri-state buffer) is used in the design of input ports.
14. _____ (Latch, Tri-state buffer) is used in the design of output ports.
15. _____ (IOR, IOW) is used in the design of input ports.
16. _____ (IOR, IOW) is used in the design of output ports.
17. Draw a logical design for "OUT 16H,AL" using AND and inverter gates in addition to a 74LS373.
18. Draw a logical design for "IN AL,81H" using AND and inverter gates in addition to a 74LS244.
19. Show one implementation of Problem 17 using NAND and inverter gates. Use as many as you need.
20. Show one implementation of Problem 18 using NAND and inverter gates. Use as many as you need.
21. True or false. Memory-mapped I/O uses control signals MEMR and MEMW.
22. True or false. In memory-mapped I/O, one can perform logical and arithmetic operations on the data without moving it into the accumulator first.
23. Show the logical design of "MOV [0100],AL" for memory-mapped I/O using AND and inverter gates and a 74LS373 latch. Assume that DS = B800H.
24. Why is memory-mapped I/O decoding more expensive?

SECTION 11.3: I/O ADDRESS MAP OF x86 PCs

25. Show the circuit connection to the PC bus for the following instructions. Use simple logic gates 74LS373 and 74LS244.
    (a) OUT 309H,AL   (b) IN AL,30CH
26. Repeat Problem 25 using a 74LS138 for the decoder.
27. Show the design of an 8255 connection to the PC bus using simple logic gates. Assume port address 304H as the base port address for the 8255.

28. Show the design of an 8255 connection to the PC bus using a 74LS138. Assume base address 31CH.
29. In the IBM PC, how many port addresses are available in the address space commonly referred to as prototype?
30. Which one is more economical, linear address select or absolute address decoding?
31. Explain address aliasing.
32. Which one creates aliases, the linear address select or absolute address decoding?
33. True or false. To design an IBM PC compatible system, one must follow the I/O map of the PC.
34. In accessing ports in the PC, why must the AEN = 0 signal be used in decoding?
35. What port address is used for a fixed delay in the x86 PC?
36. In x86 PC, the ___ bit of port address ___H toggles every ___ microseconds.
37. In Problem 36, to get a 1/4 second delay, we need to load the CX with what value?
38. In Problem 36, calculate the time delay if CX = 38,000.

SECTION 11.4: 8255 PPI CHIP

39. How many pins of the 8255 are used for ports, and how are they categorized?
40. What is the function of data pins D0 - D7 in the 8255?
41. What is the advantage of using the 8255 over the 74LS373 and 74LS244?
42. True or false. All three ports, A, B, and C, can be programmed for simple I/O.
43. True or false. In simple I/O programming of port A of the 8255, we can use PA0 - PA3 for output and PA4 - PA7 for an input port.
44. Show the decoding circuitry for the 8255 if we want port A to have address 68H. Use NAND and inverter gates.
45. Which of the following port addresses cannot be assigned to port A of the 8255, and why?
    (a) 32H (b) 45H (c) 89H (d) BAH
46. If 91H is the control word, indicate which port is input and which is output.
47. Find the control word if PA = input, PB = input, and PC0 - PC7 = output.
48. In the 8255, which mode is used if we want to simply send out data?
49. Write a program to monitor PA for a temperature of 100. If it is equal, it should be saved in register BL. Also, send AAH to port B and 55H to port C. Use the port address of your choice.
50. Write a program in Assembly language to get a byte of the data from PA, convert it to ASCII bytes, and store them in registers CL, AH, and AL. For example, an input of FFH will show as 255 (Note: FF in binary becomes 323535 in ASCII).

SECTION 11.5: PC INTERFACE TRAINER AND BUS EXTENDER

51. Give the I/O address space used by the PC Bus Extender. Indicate the reason for use of this space.
52. A PC Bus Extender is tested on (plugged into) two different PCs. One of the PCs does not boot up while the card is inserted but the same card works fine on the second PC (it boots up and works with the PC Interface Trainer). What do you think is the problem?
53. In the PC Interface Trainer, each Y output of the 74LS138 handles how many ports? List the I/O address space for Y0 through Y4.
54. Give the I/O address space used on the PC Trainer itself.
55. Give the I/O address space for the 8255 on the PC Trainer.

56. Give the I/O address space for Y1.
57. Give the I/O address space available for future additions on H2 of the PC Trainer.
58. True or false. The PC bus extender uses power from the PC power supply for its chips (74LS138, 74LS244, 74LS245, etc.).
59. True or false. The PC Interface Trainer needs an external power source.
60. Write a simple program in Assembly language (use DEBUG) that sends 99 hex to all ports of the 8255 on the PC Trainer.
61. Write the Assembly language code for the PC Trainer to get data from port A and send it to ports B and C.
62. Examine the program below.

```
MOV    AL,90H
MOV    DX,303H
OUT    DX,AL
MOV    DX,300H
IN     AL,DX
INC    DX
OUT    DX,AL
INC    DX
OUT    DX,AL
```

Explain what the program does.

SECTION 11.7: 8-BIT AND 16-BIT I/O TIMING IN ISA BUS

63. Explain the role of ZEROWS in I/O timing.
64. Explain the role of IOCS16 in 16-bit I/O.
65. In the ISA bus, the default mode for I/O operation is _____ (8-bit, 16-bit).
66. What is the clock speed for the ISA bus?
67. How many WS are used in the 8-bit standard I/O cycle?
68. In Problem 67, how much does it take to complete one I/O read cycle?
69. How many WS are used in the 8-bit I/O cycle when ZEROWS is asserted?
70. In Problem 68, how much does it take to complete one I/O read cycle?
71. How many WS are used in the 16-bit standard I/O cycle?
72. In Problem 71, how much does it take to complete one I/O read cycle?
73. The IOCS16 pin is an _____ (input, output) and an active _____ (low, high) signal.
74. The ZEROWS pin is an _____ (input, output) and an active _____ (low, high) signal.
75. Calculate the bus bandwidth for 8-bit standard I/O of the ISA bus.
76. Calculate the bus bandwidth for 16-bit standard I/O of the ISA bus.
77. What is the function of the CHANRDY pin?
78. Explain how we can extend the I/O cycle time of the ISA bus.

ANSWERS TO REVIEW QUESTIONS

SECTION 11.1: 8088 INPUT/OUTPUT INSTRUCTIONS
1. 256 input and 256 output ports
2. 65,536 input and 65,536 output ports
3. It sends the contents of 8-bit register AL to port address 24H.
4.
```
MOV  DX,300H        ;LOAD THE PORT ADDR
IN   AL,DX          ;GET THE DATA FROM PORT
MOV  DX,304H        ;LOAD THE PORT ADDR
OUT  DX,AL          ;SEND OUT THE DATA
```
5.
```
IN   AL,60H         ;GET DATA FROM PORT ADDRESS 60H
MOV  CH,AL          ;GIVE THE COPY TO CH REG
```

SECTION 11.2: I/O ADDRESS DECODING AND DESIGN
1. latch, tri-state buffer
2. The CPU provides the data on the data bus only for a short amount of time. Therefore, it must be latched before it is lost.
3. Assuming that OC = 0, the input data is transferred from D to Q when G goes from low to high, making it available right away; but it is actually latched when G goes from high to low. This reduces the time delay from D to Q.
4. low
5. true
6. MEMW* for output and MEMR* for input devices.

SECTION 11.3: I/O ADDRESS MAP OF x86 PC
1. 300 - 31FH
2. AEN = 0
3. The I/O devices are programmed by the CPU; therefore, with AEN = 0 it will make sure that the I/O device is accessed by the addresses provided by the CPU and not the DMA. AEN is active low when the CPU is using the buses.
4. D4 bit
5. 25,000 x 15.085 μs = 377.125 ms

SECTION 11.4: 8255 PPI CHIP
1. F4H, F5H, F6H, F7H for PA, PB, PC, and control register, respectively
2. 80H (see Figure 11-12)
3. 9BH (see Figure 11-12)
4.
```
MOV  AL,80H
OUT  CONTREG,AL
MOV  AL,55H
OUT  PORTA,AL
OUT  PORTB,AL
OUT  PORTC,AL
CALL DELAY
MOV  AL,0AAH
OUT  PORTA,AL
OUT  PORTB,AL
OUT  PORTC,AL
```
5. All are simple I/O. PA and PB are both out. PC0 - PC3 and PC4 - PC7 are both in (see Figure 11-12).

PROBLEMS

SECTION 11.5: PC INTERFACE TRAINER AND BUS EXTENDER
1. 300 - 31FH
2. false
3. 300 - 303H
4. 304 - 307H
5. To avoid locking up the PC.

SECTION 11.6: I/O PROGRAMMING WITH C/C++ AND VB
1. _outp(0x300,0x99);
2. outp(0x300,0x99);
3. outb(0x99,0x300);
4. mdata=_inp (0x302);
5. mdata=inb(0x302);

SECTION 11.7: 8-BIT AND 16-BIT I/O TIMING IN ISA BUS
1. A0-A9
2. With ten address lines, A0-A9, we get 1024 I/O devices.
3. By asserting it, we tell the system board to shorten the I/O bus cycle time.
4. input
5. IOCS16
6. 3 clock cycles (only 1WS) if we assert ZEROWS. That gives us 3 x 125 ns = 375 ns
7. The maximum bus bandwidth is achieved with the 16-bit data bus and it is 1/(3 x 125 ns) x 2 = 5.33 megabytes per second.

CHAPTER 11: I/O AND THE 8255; ISA BUS INTERFACING

CHAPTER 12

INTERFACING TO THE PC: LCD, MOTOR, ADC, AND SENSOR

OBJECTIVES

Upon completion of this chapter, you will be able to:

>> Diagram the interfacing of a PC to an LCD, and code the corresponding programs in Assembly and C/C++

>> Diagram the interfacing of a PC to a stepper motor, and code the corresponding programs in Assembly and C/C++

>> Diagram the interfacing of a PC to a DAC (digital-to-analog converter) device, and code the corresponding programs in Assembly and C/C++

>> Diagram the interfacing of a PC to an ADC (analog-to-digital converter) device, and code the corresponding programs in Assembly and C/C++

>> Show the interfacing of ADC devices to sensors

In this chapter we show PC interfacing to some real-world devices such as an LCD, stepper motor, ADC and DAC devices, and sensors. Section 12.1 describes interfacing and programming of an LCD. In Section 12.2, stepper motor interfacing is described. DAC (digital-to-analog converter) interfacing to PC is shown in Section 12.3 and ADC (analog-to-digital converter) interfacing to PC is shown in Section 12.4. Sensors, such as temperature sensors, and their interfacing are also described in Section 12.4.

SECTION 12.1: INTERFACING AN LCD TO THE PC

This section describes the operation modes of LCDs, then describes how to program and interface an LCD to a PC via the 8255 of the PC Interface Trainer.

LCD operation

In recent years the LCD is replacing LEDs (seven-segment LEDs or other multi-segment LEDs). This is due to the following reasons:

1. The declining prices of LCDs.
2. The ability to display numbers, characters, and graphics. This is in contrast to LEDs, which are limited to numbers and a few characters.
3. Incorporation of the refreshing controller into the LCD itself, thereby relieving the CPU of the task of refreshing the LCD. In the case of the LED, it must be refreshed by the CPU (or in some other way) to keep displaying the data.
4. Ease of programming for both characters and graphics.

Table 12-1: Pin Descriptions for LCD

Pin	Symbol	I/O	Description
1	V_{SS}	--	Ground
2	V_{CC}	--	+5V power supply
3	V_{EE}	--	Power supply to control contrast
4	RS	I	RS=0 to select command register, RS=1 to select data register
5	R/W	I	R/W=0 for write, R/W=1 for read
6	E	I/O	Enable
7	DB0	I/O	The 8-bit data bus
8	DB1	I/O	The 8-bit data bus
9	DB2	I/O	The 8-bit data bus
10	DB3	I/O	The 8-bit data bus
11	DB4	I/O	The 8-bit data bus
12	DB5	I/O	The 8-bit data bus
13	DB6	I/O	The 8-bit data bus
14	DB7	I/O	The 8-bit data bus

LCD pin descriptions

The LCD discussed in this section has 14 pins. The function of each pin is given in Table 12-1. Figure 12-1 shows the pin positions for various LCDs.

VCC, VSS, and VEE: While VCC and VSS provide +5V and ground, respectively, VEE is used for controlling the LCD contrast.

RS, register select: There are two registers inside the LCD and the RS pin is used for their selection as follows. If RS = 0, the instruction command code register is selected, allowing the user to send a command such as clear display, cursor at home, etc. If RS = 1, the data register is selected, allowing the user to send data to be displayed on the LCD (or data to be retrieved).

R/W, read/write: R/W input allows the user to write information into the LCD or read information from it. R/W = 1 when reading and R/W = 0 when writing.

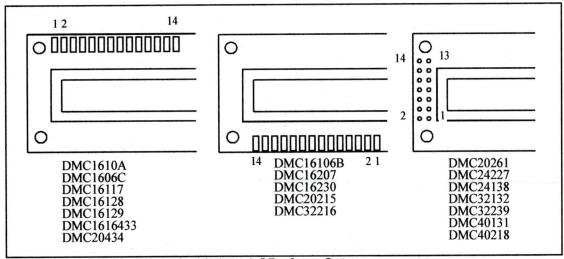

Figure 12-1. Pin Positions for Various LCDs from Optrex

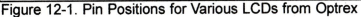

	DMC1610A	DMC16106B	DMC20261
	DMC1606C	DMC16207	DMC24227
	DMC16117	DMC16230	DMC24138
	DMC16128	DMC20215	DMC32132
	DMC16129	DMC32216	DMC32239
	DMC1616433		DMC40131
	DMC20434		DMC40218

E, enable: The enable pin is used by the LCD to latch information presented to its data pins. When data is supplied to data pins, a high-to-low pulse must be applied to this pin in order for the LCD to latch in the data present at the data pins. This pulse must be a minimum of 450 ns wide.

D0 - D7: The 8-bit data pins are used to send information to the LCD or read the contents of the LCD's internal registers.

To display letters and numbers, we send ASCII codes for the letters A - Z, a - z, and numbers 0 - 9 to these pins while making RS = 1.

There are also instruction command codes that can be sent to the LCD in order to clear the display or force the cursor to the home position or blink the cursor. Table 12-2 lists the instruction command codes.

Table 12-2: LCD Command Codes

Code (Hex)	Command to LCD Instruction Register
1	Clear display screen
2	Return home
4	Decrement cursor (shift cursor to left)
6	Increment cursor (shift cursor to right)
5	Shift display right
7	Shift display left
8	Display off, cursor off
A	Display off, cursor on
C	Display on, cursor off
E	Display on, cursor on
F	Display on, cursor blinking
10	Shift cursor position to left
14	Shift cursor position to right
18	Shift the entire display to the left
1C	Shift the entire display to the right
80	Force cursor to beginning of 1st line
C0	Force cursor to beginning of 2nd line
38	2 lines and 5x7 matrix

Note: This table is extracted from Table 12-4.

Sending commands to LCDs

To send any of the commands from Table 12-2 to the LCD, make pin RS = 0, and send a high-to-low pulse to the E pin to enable the internal latch of the LCD. The connection of an 8255 to an LCD is shown in Figure 12-2.

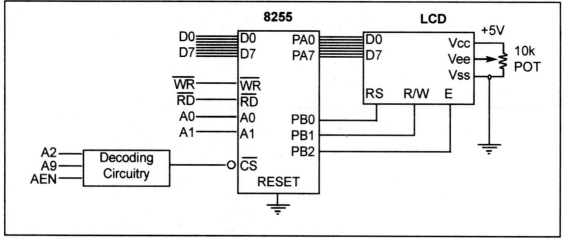

Figure 12-2. 8255 Connection to LCD

Notice the following for the connection in Figure 12-2:

1. The LCD's data pins are connected to Port A of the 8255.
2. The LCD's RS pin is connected to PB0 of Port B of the 8255.
3. The LCD's R/W pin is connected to PB1 of Port B of the 8255.
4. The LCD's E pin is connected to PB2 of Port B of the 8255.
5. Both Ports A and B are configured as output ports.

```
;The following sends all the necessary commands to the LCD
        MOV  AL,38H       ;initialize LCD for 2 lines & 5x7 matrix
        CALL COMNDWRT     ;write the command to LCD
        CALL DELAY        ;wait before issuing the next command
        CALL DELAY        ;this command needs lots of delay
        CALL DELAY
        MOV  AL,0EH       ;send command for LCD on,cursor on
        CALL COMNDWRT     ;write the command to LCD
        CALL DELAY        ;wait before issuing the next command
        MOV  AL,01        ;clear LCD
        CALL COMNDWRT
        CALL DELAY        ;wait
        MOV  AL,06        ;command for shifting cursor right
        CALL COMNDWRT
        CALL DELAY        ;wait

COMNDWRT PROC            ;this procedure writes commands to LCD
   PUSH DX               ;save DX
   MOV DX,PORTA
   OUT DX,AL             ;send the code to Port A
   MOV DX,PORTB          ;Port B address
   MOV AL,00000100B  ;RS=0,R/W=0,E=1 for H-TO-L pulse
   OUT DX,AL             ;to Port B
   NOP                   ;wait for high-to-low pulse to be
   NOP                   ;wide enough
   MOV AL,00000000B;RS=0,R/W=0,E=0 for H-TO-L pulse
   OUT DX,AL             ;
   POP DX                ;restore DX
   RET                   ;return to caller
COMNDWRT ENDP
```

In the above program, we must wait before issuing the next command; otherwise, it will jam the LCD. A delay of 20 ms should work fine. We can use the port 61H delay generation shown in Chapter 11. The code is shown below.

Sending data to the LCD

In order to send data to the LCD to be displayed, we must set pin RS = 1, and also send a high-to-low pulse to the E pin to enable the internal latch of the LCD. The following code sends characters to the LCD. Again, it places sufficient time delays between each data issue to ensure that the LCD is ready for new data.

```
        MOV   AL,'Y'       ;display 'Y' letter
        CALL  DATWRIT      ;issue it to LCD
        CALL  DELAY        ;wait before issuing the next character
        MOV   AL,'E'       ;display 'E' letter
        CALL  DATWRIT      ;issue it to LCD
        CALL  DELAY        ;wait before issuing the next character
        MOV   AL,'S'       ;display 'S' letter
        CALL  DATWRIT      ;issue it to LCD
        CALL  DELAY        ;wait

;data write to LCD without checking the busy flag
;AL=char sent to LCD

DATWRIT PROC
   PUSH DX   ;save DX
   MOV   DX,PORTA          ;DX=port A address
   OUT   DX,AL             ;issue the char to LCD
   MOV   AL,00000101B      ;RS=1,R/W=0, E=1 for H-to-L pulse
   MOV   DX,PORTB          ;port B address
   OUT   DX,AL             ;make enable high
   MOV   AL,00000001B   ;RS=1,R/W=0 AND E=0 for H-to-L pulse
   OUT   DX,AL
   POP   DX
   RET
DATWRIT ENDP

;delay generation using the PB4 bit of port 61H

DELAY PROC
        MOV CX,1325        ; 1,325x15.085 usec=20 msec
        PUSH  AX
W1:     IN    AL,61H
        AND   AL,00010000B
        CMP   AL,AH
        JE    W1
        MOV   AH,AL
        LOOP  W1
        POP AX
        RET
DELAY ENDP
```

Checking LCD busy flag

The above programs used a time delay before issuing the next data or command. This allows the LCD a sufficient amount of time to get ready to accept the next data. However, the LCD has a busy flag. We can monitor the busy flag and issue data when it is ready. This will speed up the process. To check the busy flag, we must read the command register (R/W = 1, RS = 0). The busy flag is the D7 bit of that register. Therefore, if R/W = 1, RS = 0. When D7 = 1 (busy flag = 1), the LCD is busy taking care of internal operations and will not accept any new information. When D7 = 0, the LCD is ready to receive new information. It is recommended by the LCD manufacturer's data sheet to monitor the busy flag before sending the data or command codes to the LCD. This ensures that the LCD is ready to receive data. See the code below.

```
                ;writing to LCD with checking the busy flag, AL=char
        MOV     AL,38H          ;initialize LCD for 2 lines & 5x7
        CALL    COMNDWRT        ;write the command to LCD
        MOV     AL,0EH          ;send command for LCD on,cursor on
        CALL    COMNDWRT        ;write the command to LCD
        MOV     AL,01           ;clear LCD
        CALL    COMNDWRT
        MOV     AL,06           ;command for shifting cursor right
        CALL    COMNDWRT
        MOV     AL,'Y'          ;display 'Y' letter
        CALL    DATWRT          ;issue it to LCD
        MOV     AL,'E'          ;display 'E' letter
        CALL    DATWRT          ;issue it to LCD
        MOV     AL,'S'          ;display 'S' letter
        CALL    DATWRT          ;issue it to LCD
        ....

;--------------

DATWRT PROC
        CALL    LCDREADY
        PUSH DX    ;save DX
        MOV     DX,PORTA            ;DX=port A address
        OUT     DX,AL               ;issue the char to LCD
        MOV     AL,00000101B ;RS=1,R/W=0, E=1 for H-to-L pulse
        MOV     DX,PORTB            ;port B address
        OUT     DX,AL               ;make enable high
        NOP
        NOP
        MOV     AL,00000001B  ;RS=1,R/W=0 AND E=0 for H-to-L
        OUT     DX,AL
        POP     DX
        RET
DATWRT ENDP

;------------
```

```
COMNDWRT PROC
       LCDREADY
       PUSH DX                  ;save DX
       MOV DX,PORTA
       OUT DX,AL                ;send the code to Port A
       MOV DX,PORTB             ;Port B address
       MOV AL,00000100B         ;RS=0,R/W=0,E=1 for H-TO-L pulse
       OUT DX,AL                ;to Port B
       NOP                      ;wait for high-to-low pulse to be
       NOP                      ;wide enough
       MOV AL,00000000B         ;RS=0,R/W=0,E=0 for H-TO-L pulse
       OUT DX,AL
       POP DX                   ;restore DX
       RET                      ;return to caller
COMNDWRT ENDP
  ;----------------
LCDREADY PROC
       PUSH AX
       PUSH DX
       MOV AL,90H   ;PA=input to read LCD status,PB=OUT
       MOV DX,CNTPORT           ;DX=control port address
       OUT DX,AL                ;issue to control reg
       MOV AL,00000110B         ;RS=0 busy flag is a
                                ;command R/W=1,E=1 (L-to-H for E)
       MOV DX,PORTB             ;port B address
       OUT DX,AL                ;issue it to port B
       MOV DX,PORTA             ;port A address
AGAIN:IN AL,DX                  ;read command reg busy flag is D7
       ROL AL,1                 ;send busy flag to carry flag
       JC AGAIN                 ;IF CF=1 LCD not ready try again
       MOV AL,80H               ;make PA=OUT to send character
       MOV DX,CONTPORT          ;DX=control port address
       OUT DX,AL                ;issue to 8255's control reg
       POP DX
       POP AX
       RET
LCDREADY ENDP
```

LCD cursor position

In the LCD, one can put data at any location. For the 20x2 LCD, the address for the first location of line 1 is 80H, and for line 2 it is C0H. The following shows address locations and how they are accessed.

RS	R/W	DB7	DB6	DB5	DB4	DB3	DB2	DB1	DB0
0	0	1	A	A	A	A	A	A	A

where AAAAAAA = 0000000 to 0100111 for line 1 and AAAAAAA = 1000000 to 1100111 for line 2. See Table 12-3. The upper address range can go as high as 0100111 for the 40-character-wide LCD while for the 20-character-wide LCD it goes up to 010011 (19 decimal = 10011 binary). Notice that the upper range 0100111 (binary) = 39 decimal, which corresponds to locations 0 to 39 for the LCDs of 40x2 size. From the above discussion we can get the addresses of cursor positions for various sizes of LCDs. See Figure 12-3. Note that all the addresses are in hex.

As an example of setting the cursor at the fourth location of line 1 we have the following:

```
MOV   AL,83H        ;LINE 1 POSITION 4
CALL  COMNDWRT
```

and for the 6th location of the 2nd line we have:

```
MOV   AL,0C5H
CALL  COMNDWRT
```

Notice that since the location addresses are in hex, 0 is the first location.

LCD programming in Visual C/C++

In Chapter 11 we showed how to program the x86 PC I/O port using C/C++ for MS Visual C/C++, Borland Turbo C/C++, and Linux C/C++. Example 12-1 shows LCD programming using Visual C/C++. For other environments, you can modify and test Example 12-1.

LCD timing and data sheet

Figures 12-4 and 12-5 show timing diagrams for LCD write and read timing, respectively. Notice that the write operation happens on the H-to-L pulse of the E pin while the read is activated on the L-to-H pulse of the E pin. Table 12-4 provides a more detailed list of LCD instructions.

Table 12-3: LCD Addressing

	DB7	DB6	DB5	DB4	DB3	DB2	DB1	DB0
Line 1 (min)	1	0	0	0	0	0	0	0
Line 1 (max)	1	0	1	0	0	1	1	1
Line 2 (min)	1	1	0	0	0	0	0	0
Line 2 (max)	1	1	1	0	0	1	1	1

16 x 2 LCD	80	81	82	83	84	85	86 through 8F
	C0	C1	C2	C3	C4	C5	C6 through CF
20 x 1 LCD	80	81	82	83	through 93		
20 x 2 LCD	80	81	82	83	through 93		
	C0	C1	C2	C3	through D3		
20 x 4 LCD	80	81	82	83	through 93		
	C0	C1	C2	C3	through D3		
	94	95	96	97	through A7		
	D4	D5	D6	D7	through E7		
40 x 2 LCD	80	81	82	83	through A7		
	C0	C1	C2	C3	through E7		

Note: All data is in hex.

Figure 12-3. Cursor Addresses for Some LCDs

Example 12-1

Write a Visual C/C++ program to display "Hello" on line 1 starting at the sixth position.

Solution:

```
#include<conio.h>
#include<stdio.h>
#include<iostream.h>
#include<iomanip.h>
#include<windows.h>
//tested by Dan Bent
void main()
  {
  unsigned int i;
  char message[5]="Hello";
  cout<<setiosflags(ios::unitbuf);

  _outp(0x303,0x80);     // control word for Port A, B, C
  _outp(0x300,0x38);     //init LCD for 2 lines & 5x7 matrix
  _outp(0x301,0x04);     //RS=0,R/W=0,E=1 for H-to-L pulse
  _outp(0x301,0x00);     //RS=0,R/W=0,E=0 for H-to-L pulse
  _sleep(500);           //delay 500 milliseconds
  _outp(0x300,0x0E);     //send command for LCD on, cursor on
  _outp(0x301,0x04);     //RS=0,R/W=0,E=1 for H-to-L pulse
  _outp(0x301,0x00);     //RS=0,R/W=0,E=0 for H-to-L pulse
  _sleep(250);           //delay 250 milliseconds
  _outp(0x300,0x01);     //clear LCD
  _outp(0x301,0x04);     //RS=0,R/W=0,E=1 for H-to-L pulse
  _outp(0x301,0x00);     //RS=0,R/W=0,E=0 for H-to-L pulse
  _sleep(250);
  _outp(0x300,0x06);     //shift cursor right
  _outp(0x301,0x04);     //RS=0,R/W=0,E=1 for H-to-L pulse
  _outp(0x301,0x00);     //RS=0,R/W=0,E=0 for H-to-L pulse
  _sleep(250);

  _outp(0x300,0x85);     //move cursor to beginning of line
  _outp(0x301,0x04);     //RS=0,R/W=0,E=1 for H-to-L pulse
  _outp(0x301,0x00);     //RS=0,R/W=0,E=0 for H-to-L pulse
  _sleep(250);

//write data to LCD
  for(i=0;i<strlen(message);i++)
    {
    _outp(0x300,(int)message[i]);
    _outp(0x301,0x05);         //RS=1,R/W=0,E=1 for H-to-L pulse
    _outp(0x301,0x01);         //RS=1,R/W=0,E=0 for H-to-L pulse
    _sleep(250);               //delay 250 milliseconds
    }
  }
```

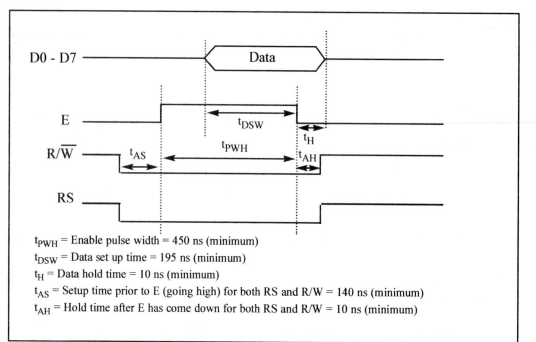

t_{PWH} = Enable pulse width = 450 ns (minimum)

t_{DSW} = Data set up time = 195 ns (minimum)

t_H = Data hold time = 10 ns (minimum)

t_{AS} = Setup time prior to E (going high) for both RS and R/W = 140 ns (minimum)

t_{AH} = Hold time after E has come down for both RS and R/W = 10 ns (minimum)

Figure 12-4. LCD Write Timing

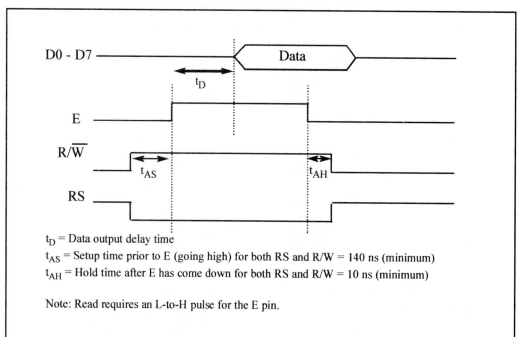

t_D = Data output delay time

t_{AS} = Setup time prior to E (going high) for both RS and R/W = 140 ns (minimum)

t_{AH} = Hold time after E has come down for both RS and R/W = 10 ns (minimum)

Note: Read requires an L-to-H pulse for the E pin.

Figure 12-5. LCD Read Timing

Review Questions

1. The RS pin is an _____ (input, output) pin for the LCD.
2. The E pin is an _____ (input, output) pin for the LCD.
3. The E pin requires an _____ (H-to-L, L-to-H) pulse to latch in information at the data pins of the LCD.
4. For the LCD to recognize information at the data pins as data, RS must be set to _____ (high, low).
5. Give the command codes for line 1, first character, and line 2, first character.

Table 12-4: List of LCD Instructions

Instruction	RS R/W DB7 DB6 DB5 DB4 DB3 DB2 DB1 DB0	Description	Execution Time (Max)
Clear Display	0 0 0 0 0 0 0 0 0 1	Clears entire display and sets DD RAM address 0 in address counter	1.64 ms
Return Home	0 0 0 0 0 0 0 0 1 –	Sets DD RAM address 0 as address counter. Also returns display being shifted to original position. DD RAM contents remain unchanged.	1.64 ms
Entry Mode Set	0 0 0 0 0 0 0 1 1/D S	Sets cursor move direction and specifies shift of display. These operations are performed during data write and read.	40 μs
Display On/ Off Control	0 0 0 0 0 0 1 D C B	Sets On/Off of entire display (D), cursor On/Off (C), and blink of cursor position character (B).	40 μs
Cursor or Display Shift	0 0 0 0 0 1 S/C R/L – –	Moves cursor and shifts display without changing DD RAM contents.	40 μs
Function Set	0 0 0 0 1 DL N F – –	Sets interface data length (DL), number of display lines (L), and character font (F).	40 μs
Set CG RAM Address	0 0 0 1 AGC	Sets CG RAM address. CG RAM data is sent and received after this setting.	40 μs
Set DD RAM Address	0 0 1 ADD	Sets DD RAM address. DD RAM data is sent and received after this setting.	40 μs
Read Busy Flag & Address	0 1 BF AC	Reads Busy flag (BF) indicating internal operation is being performed and reads address counter contents.	40 μs
Write Data CG or DD RAM	1 0 Write Data	Writes data into DD or CG RAM.	40 μs
Read Data CG or DD RAM	1 1 Read Data	Reads data from DD or CG RAM.	40 μs

Notes:
1. Execution times are maximum times when fcp or fosc is 250 kHz.
2. Execution time changes when frequency changes. Example: When fcp or fosc is 270 kHz: 40 μs × 250 / 270 = 37 μs.
3. Abbreviations:

DD RAM	Display data RAM
CG RAM	Character generator RAM
ACC	CG RAM address
ADD	DD RAM address, corresponds to cursor address
AC	Address counter used for both DD and CG RAM addresses.

1/D = 1	Increment	1/D = 0	Decrement
S = 1	Accompanies display shift		
S/C = 1	Display shift;	S/C = 0	Cursor move
R/L = 1	Shift to the right;	R/L = 0	Shift to the left
DL = 1	8 bits, DL = 0: 4 bits		
N = 1	1 line, N = 0 : 1 line		
F = 1	5 x 10 dots, F = 0 : 5 x 7 dots		
BF = 1	Internal operation;	BF = 0	Can accept instruction

SECTION 12.2: INTERFACING A STEPPER MOTOR TO THE PC

This section begins with an overview of the basic operation of stepper motors. Then we describe how to interface a stepper motor to the PC using the PC Trainer. Finally, we use Assembly language programs to demonstrate control of the angle and direction of stepper motor rotation.

Stepper motors

A stepper motor is a widely used device that translates electrical pulses into mechanical movement. In applications such as disk drives, dot matrix printers, and robotics, the stepper motor is used for position control. Every stepper motor has a permanent magnet rotor (also called the shaft) surrounded by a stator (see Figure 12-6). The most common stepper motors have four stator windings that are paired with a center-tapped common as shown in Figure 12-7. This type of stepper motor is commonly referred to as a four-phase stepper motor. The center tap allows the change of current direction in each of two coils when a winding is grounded, which results in a polarity change of the stator. Notice that while a conventional motor shaft runs freely, the stepper motor shaft moves in a fixed repeatable increment, which allows one to move it to a precise position. This repeatable fixed movement is possible as a result of the basic magnet theory where poles of the same polarity repel and opposite poles attract. The direction of the rotation is dictated by the stator poles. The stator poles are determined by the current sent through the wire coils. As the direction of current is changed, the polarity is also changed causing the reverse motion of the rotor. The stepper motor discussed

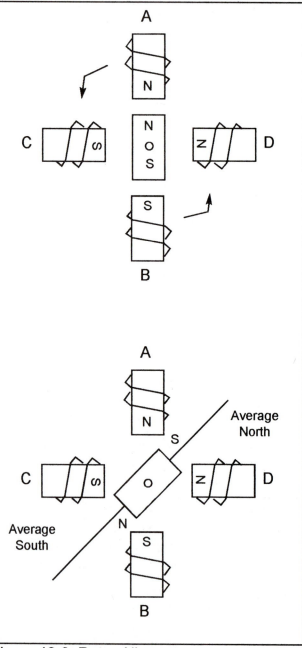

Figure 12-6. Rotor Alignment

here has a total of 6 leads: 4 leads representing the four stator windings and two commons for the center tapped leads. As the sequence of power is applied to each stator winding, the rotor will rotate. There are several widely used sequences where each has different degree of precision. Table 12-5 shows the normal 4-step sequence.

It must be noted that although we can start with any of the sequences in Table 12-5, once we start we must continue in the proper order. For example, if we start with step 3 (0110) we must continue in the sequence of steps 4, 1, 2, and so on.

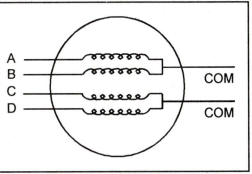

Figure 12-7. Stator Windings Configuration

Table 12-5: Normal 4-Step Sequence

Clockwise	Step #	Winding A	Winding B	Winding C	Winding D	Counter-clockwise
↓	1	1	0	0	1	↑
	2	1	1	0	0	
	3	0	1	1	0	
	4	0	0	1	1	

Step angle

How much movement is associated with a single step? This depends on the internal construction of the motor, in particular the number of teeth on the stator and the rotor. The step angle is the minimum degree of rotation associated with a single step. Various motors have different step angles. Table 12-6 shows some step angles for various motors. In Table 12-6, notice the term *steps per revolution*. This is the total number of steps needed to rotate one complete rotation or 360 degrees (e.g., 180 steps × 2 degrees = 360).

Table 12-6: Stepper Motor Step Angles

Step Angle	Steps per Revolution
0.72	500
1.8	200
2.0	180
2.5	144
5.0	72
7.5	48
15	24

It must be noted that perhaps contrary to one's initial impression, a stepper motor does not need to have more terminal leads for the stator to achieve smaller steps. All the stepper motors discussed in this section have 4 leads for the stator winding and 2 com wires for the center tap. Although some manufacturers have set aside only one lead for the common signal instead of two, they always have 4 leads for the stators.

With this background on stepper motors, next we see how we can interface them with the PC.

Stepper motor connection and programming

Example 12-2 shows the programming of the stepper motor as connected in Figure 12-8. Study this example very carefully since it contains some very important points on motor interfacing.

Example 12-2

Describe the 8255 connection to the stepper motor of Figure 12-8 and code a program to rotate it continuously.

Solution:

The following steps show the 8255 connection to the stepper motor and its programming.

1. Use an ohmmeter to measure the resistance of the leads. This should identify which COM leads are connected to which winding leads.
2. The common wire(s) are connected to the positive side of the motor's power supply. In many motors, +5 V is sufficient.
3. The four leads of the stator winding are controlled by the four bits of port A (PA0 - PA3). However, since the 8255 lacks sufficient current to drive the stepper motor windings, we must use a driver such as the ULN2003 to energize the stator. Instead of the ULN2003, we could use transistors as drivers. However, notice that if transistors are used as drivers, we must also use diodes to take care of inductive current generated when the coil is turned off. One reason that the ULN2003 is preferable to the use of transistors as drivers is that the ULN2003 has an internal diode to take care of back EMF.

```
            MOV    AL,80H
            MOV    DX,CNTRLPORT        ;LOAD CONTROL PORT ADDRESS
            OUT    DX,AL               ;PORT AS OUTPUT
            MOV    BL,33H
AGAIN:MOV   AH,01
            INT    16H                 ;CHECK KEY PRESS
            JNZ    EXIT                ;EXIT UPON KEY PRESS
            MOV    AL,BL
            MOV    DX,PORTA
            OUT    DX,AL
            MOV    CX,20000
HERE:  LOOP  HERE                      ;DELAY
            ROR    BL,1
            JMP    AGAIN
       EXIT:
```

In the above program we are sending the sequence 33H, 66H, CCH, and 99H to the stepper motor continuously. The motor keeps moving unless a key is pressed.

By changing the value of DELAY, we can change the speed of rotation. In your program use a fixed time delay. The fixed time delay generation was shown in Chapter 11.

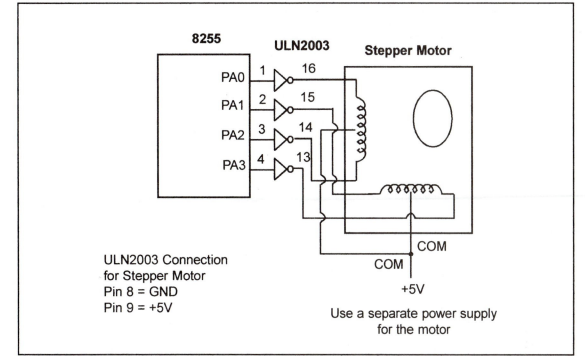

Figure 12-8. 8255 Connection to Stepper Motor

Steps per second and RPM relation

The relationship between the RPM (revolutions per minute), steps per revolution, and steps per second is intuitive and is as follows.

$$Steps\ per\ second = \frac{RPM \times Steps\ per\ revolution}{60}$$

The four-step sequence and number of teeth on rotor

The switching sequence shown above in Table 12-5 is called the 4-step switching sequence since after four steps the same two windings will be "ON". How much movement is associated with these four steps? After completing every four steps, the rotor moves only one tooth pitch. Therefore, in a stepper motor with 200 steps per revolution, its rotor has 50 teeth, since $4 \times 50 = 200$ steps are needed to complete one revolution. This leads to the conclusion that the minimum step angle is always a function of the number of teeth on the rotor. In other words, the smaller the step angle, the more teeth the rotor passes. See Example 12-3.

Looking at Example 12-3, one might wonder what happens if we want to move 45 degrees since the steps are 2 degrees each. To allow for finer resolutions, all stepper motors allow what is called an 8-step switching sequence. The 8-step sequence is also called half-stepping since in following the 8-step sequence each step is half of the normal step angle. For example, a motor with a 2-degree step angle can be used as a 1-degree step angle if the sequence of Table 12-7 is applied.

Example 12-3

Give the number of times the 4-step sequence in Table 12-5 must be applied to a stepper motor to make an 80-degree move if the motor has a 2-degree step angle.

Solution:

A motor with a 2-degree step angle has the following characteristics:

Step angle: 2 degrees   Steps per revolution: 180

Number of rotor teeth: 45   Movement per 4-step sequence: 8 degrees

To move the rotor 80 degrees, we need to send 10 four-step sequences right after each other, since 10×4 steps $\times 2$ degrees = 80 degrees.

Motor speed

The motor speed, measured in steps per second (steps/s), is a function of the switching rate. Notice in Example 12-2 that by changing the length of the time delay loop, we can achieve various rotation speeds.

Holding torque

The following is the definition of the holding torque: "With the motor shaft at standstill or zero RPM condition, the amount of torque, from an external source, required to break away the shaft from its holding position. This is measured with rated voltage and current applied to the motor." The unit is ounce-inch (or kg-cm).

Wave drive 4-step sequence

Table 12-7: Half-Step 8-Step Sequence

Clockwise	Step #	Winding A	Winding B	Winding C	Winding D	Counter-clockwise
↓	1	1	0	0	1	↑
	2	1	0	0	0	
	3	1	1	0	0	
	4	0	1	0	0	
	5	0	1	1	0	
	6	0	0	1	0	
	7	0	0	1	1	
	8	0	0	0	1	

Table 12-8: Wave Drive 4-Step Sequence

Clockwise	Step #	Winding A	Winding B	Winding C	Winding D	Counter-clockwise
↓	1	1	0	0	0	↑
	2	0	1	0	0	
	3	0	0	1	0	
	4	0	0	0	1	

In addition to the 8-step sequence and the 4-step sequence discussed earlier, there is another sequence called the wave drive 4-step sequence. It is shown in Table 12-8. Notice that the sequence of Table 12-8 is simply the combination of the wave drive 4-step and normal 4-step normal sequences shown in Tables 12-5 and 12-7, respectively. Experimenting with the wave drive 4-step is left to the reader. Example 12-4 shows the Turbo C++ version of the program to turn the stepper motor clockwise.

Example 12-4

Write a Turbo C++ program to turn the stepper motor clockwise continuously. Pressing any key should exit the program.

Solution:

```
//Turning the stepper motor clockwise continuously
#include <conio.h>
#include <stdio.h>
main()
{
outp(0x303,0x80);        //CONFIGURE 8255 AS OUT
printf("\n Turning the Stepper motor clockwise. Press any key to exit this program\n");
do
{
 outp(0x300,0x99);
 delay(500);                    //500 msec
outp(0x300,0xcc);
 delay(500);                    //500 msec
outp(0x300,0x66);
 delay(500);                    //500 msec
 outp(0x300,0x33);
 delay(500);                    //500 msec
 }
while(!kbhit());                 //PRESS ANY KEY TO STOP
return(0);
}
```

Notice that if a given motor requires more current than the ULN2003 can provide, we can use transistors, as shown in Figure 12-9.

Review Questions

1. Give the 4-step sequence of a stepper motor if we start with 0110.
2. A stepper motor with a step-angle of 5 degrees has _____ steps per revolution.
3. Why do we put a driver between the 8255 and the stepper motor?

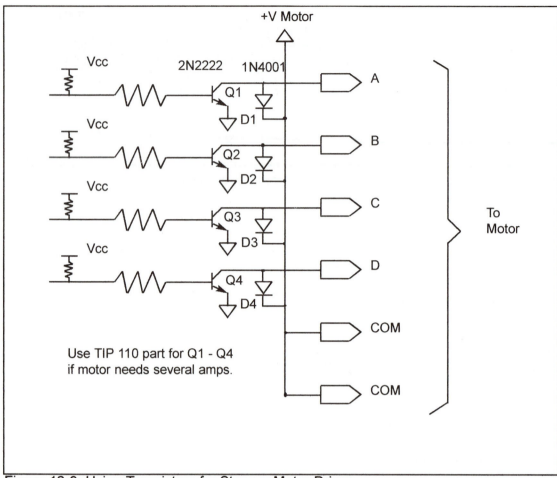

Figure 12-9: Using Transistors for Stepper Motor Driver

SECTION 12.3: INTERFACING DAC TO THE PC

This section will show how to interface a DAC (digital-to-analog converter) to a PC via the 8255 of the PC Interface Trainer. Then we demonstrate how to generate a sine wave on the scope using the DAC.

Digital-to-analog (DAC) converter

The digital-to-analog converter (DAC) is a device widely used to convert digital pulses to analog signals. In this section we discuss the basics of interfacing a DAC to a PC.

Recall from your digital electronics book the two methods of making a DAC: binary weighted and R/2R ladder. The vast majority of integrated circuit DACs, including the MC1408 used in this section, use the R/2R method since it can achieve a much higher degree of precision. The first criterion for judging a DAC is its resolution, which is a function of the number of binary inputs. The common ones are 8, 10, and 12 bits. The number of data bit inputs decides the resolution of the DAC since the number of analog output levels is equal to 2^n, where n is the number of data bit inputs. Therefore, the 8-input DAC such as the MC1408 provides 256 discrete voltage (or current) levels of output. Similarly, the 12-bit DAC provides 4096 discrete voltage levels. Although there are 16-bit DACs, they are expensive.

MC1408 DAC (or DAC 808)

In the MC1408 (DAC808), the digital inputs are converted to current (Iout). By connecting a resistor to the Iout pin, we convert the result to voltage. The total current provided by the Iout is a function of the binary numbers at the D0 - D7 inputs of the 1408 and the reference current (Iref), and is as follows.

$$I_{out} = I_{ref} \left(\frac{D7}{2} + \frac{D6}{4} + \frac{D5}{8} + \frac{D4}{16} + \frac{D3}{32} + \frac{D2}{64} + \frac{D1}{128} + \frac{D0}{256} \right)$$

where D0 is the LSB, D7 is the MSB for the inputs, and Iref is the input current that must be applied to pin 14. The Iref current is generally set to 2.0 mA. Figure 12-10 shows the generation of current reference (setting Iref = 2 mA) by using the standard 5-V power supply and 1K, 1.5K ohm standard resistors. Some also use the zener diode (LM336), which overcomes any fluctuation associated with the power supply voltage. Now assuming that Iref = 2 mA, if all the inputs to the DAC are high, the maximum output current is 1.99 mA (verify this for yourself).

Converting Iout to voltage in 1408 DAC

We connect the output pin Iout to a resistor, convert this current to voltage, and monitor the output on the scope. However, in real life this can cause inaccuracy since the input resistance of the load where it is connected will also affect the output voltage. For this reason, the Iref current output is isolated by connecting it to an op amp such as the 741 with Rf = 5K ohms for the feedback resistor. Assuming that R = 5K ohms, by changing the binary input, the output voltage changes as shown in Example 12-5.

Example 12-5

Assuming that R = 5K and I_{ref} = 2 mA, calculate V_{out} for the following binary inputs:

(a) 10011001 binary   (99H)   (b) 11001000 (C8H)

Solution:

(a) I_{out} = 2 mA (153/255) = 1.195 mA and V_{out} = 1.195 mA × 5K = 5.975 V

(b) I_{out} = 2 mA (200/256) = 1.562 mA and V_{out} = 1.562 mA × 5K = 7.8125 V

Generating a sine wave

To generate a sine wave, we first need a table whose values represent the magnitude of the sine of angles between 0 and 360 degrees. The values for the sine function vary from −1.0 to +1.0 for 0 to 360 degree angles. Therefore, the table values are integer numbers representing the voltage magnitude for the sine of theta. This method ensures that only integer numbers are output to the DAC by the x86 processor. Table 12-9 shows the angles, the sine values, the voltage magnitude, and the integer values representing the voltage magnitude for each angle with 30-degree increments. To generate Table 12-9, we assumed the full-scale voltage of 10V for the DAC output. Full-scale output of the DAC is achieved when all the data inputs of the DAC are high. Therefore, to achieve the full-scale 10V output, we use the following equation.

$$V_{out} = 5 \text{ V} + (5 \times \sin \theta)$$

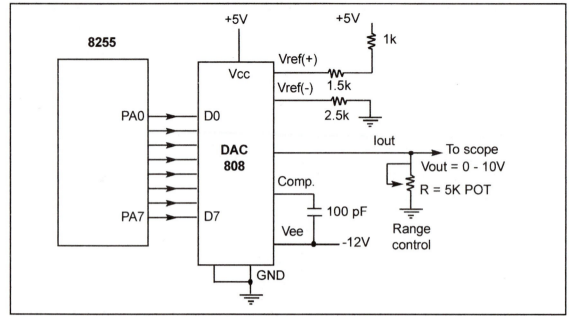

Figure 12-10. 8255 Connection to DAC808

To find the value sent to the DAC for various angles, we simply multiply the Vout voltage by 25.60 because there are 256 steps and full-scale Vout is 10 volts. Therefore, 256 steps / 10 V = 25.6 steps per volt. To further clarify this, look at Example 12-6.

Example 12-6

Verify the values of Table 12-9 for the following angles: (a) 30   (b) 60.

Solution:

(a) $V_{out} = 5\ V + (5\ V \times \sin\theta) = 5\ V + 5 \times \sin 30 = 5\ V + 5 \times 0.5 = 7.5\ V$

DAC input values = 7.5 V × 25.6 = 192 (decimal)

(b) $V_{out} = 5\ V + (5\ V \times \sin\theta) = 5\ V + 5 \times \sin 60 = 5\ V + 5 \times 0.866 = 9.33\ V$

DAC input values = 9.33 V × 25.6 = 238 (decimal)

The following program sends the values of Table 12-9 to the DAC.

```
      ;in data segment
TABLE DB 128,192,238,255,238,192,128,64,17,0,17,64,128
      ;in code segment
      ;PA is assumed to be output
A1:   MOV  CX,12        ;count
      MOV  BX,OFFSET TABLE
      MOV  DX,PORTA     ;port A address
NEXT: MOV  AL,[BX]
      OUT  DX,AL
      INC  BX
      CALL DELAY        ;let DAC recover
      LOOP NEXT
      JMP  A1           ;do it again
```

To produce a simple stair-step sine wave, we can use Example 12-7. Example 12-8 uses the Turbo C++ math functions to generate the look-up table values. You can use Visual C++ instead.

Example 12-7

In order to generate a stair-step ramp, set up the circuit in Figure 12-10 and connect the output to an oscilloscope. Then write a program to send data to the DAC to generate a stair-step ramp.

Solution:

```
              MOV   AL,80H                ;PA=OUT
              MOV   DX,303H
              OUT   DX,AL
              MOV   DX,300H
AGAIN:        MOV   AH,01
              INT   16H                   ;CHECK KEY PRESS
              JNZ   STOP                  ;EXIT UPON KEY PRESS
              SUB   AL,AL
BACK: OUT     DX,AL
              INC   AL
              CMP   AL,0
              JZ    AGAIN
              CALL  DELAY         ;LET DAC RECOVER
              JMP   BACK
STOP:                             ;EXIT
```

Table 12-9: Angle v. Voltage Magnitude for Sine Wave

Angle θ (degrees)	Sin θ	Vout (Voltage Magnitude) 5 V + (5 V × sin θ)	Values Sent to DAC (decimal) (Voltage Mag × 25.6)
0	0	5	128
30	0.5	7.5	192
60	0.866	9.33	238
90	1.0	10	255
120	0.866	9.33	238
150	0.5	7.5	192
180	0	5	128
210	−0.5	2.5	64
240	−0.866	0.669	17
270	−1.0	0	0
300	−0.866	0.669	17
330	−0.5	2.5	64
360	0	5	128

Review Questions

1. In a DAC, input is _____ (digital, analog) and output is _____ (digital, analog). Answer for ADC input and output as well.
2. DAC808 is a(n) ____-bit D-to-A converter.
3. The output of DAC808 is in _____ (current, voltage).

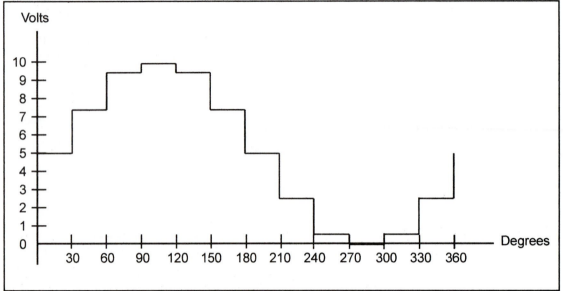

Figure 12-11. Angle v. Voltage Magnitude for Sine Wave

Example 12-8

Write a Turbo C++ to generate a sine wave on PA. Use the C++ math functions to generate the look-up table values. Pressing any key should exit the program.

Solution:

```
//GENERATING SINE WAVE VIA A DAC CONNECTED TO PORT A
#include <conio.h>
#include <stdio.h>
#include <math.h>
main()
{
outp(0x303,0x80);          //CONFIGURE 8255 AS OUT
unsigned char v1;          //v1 IS A BYTE SIZE DATA
float Vout,magnitude;
int a;
printf("\n Press any key to exit this program\n");
do
{
        for (a=0;a<360;a++)                   //FOR THE FULL 360 DEGREES
        {
        Vout=5.0+(5.0 * sin ((3.14*a)/180));  //VOLTAGE MAGNITUDE
        magnitude=Vout * 25.6;                //VALUE SENT TO DAC
        v1=(char)magnitude;                   //MAKE IT A BYTE SIZE
        delay(1);
        outp(0x300,v1);                       //OUTPUT IT TO PORT A
        }
}
while(!kbhit());                              //PRESS ANY KEY TO EXIT
return(0);
}
```

SECTION 12.4: INTERFACING ADC AND SENSORS TO THE PC

This section will explore interfacing ADC (analog-to-digital converter) chips and temperature sensors to a PC. After describing the ADC chips, we show how to interface them to the PC using the PC Interface Trainer. Then we examine the characteristics of the LM3/35 temperature sensor and show how to interface it with proper signal conditioning.

ADC devices

Analog-to-digital converters are among the most widely used devices for data acquisition. Digital computers use binary (discrete) values, but in the physical world everything is analog (continuous). Temperature, pressure (wind or liquid), humidity, and velocity are a few examples of physical quantities that we deal with every day. A physical quantity is converted to electrical (voltage, current) signals using a device called a *transducer*. Transducers are also referred to as *sensors*. There are sensors for temperature, velocity, pressure, light, and many other natural quantities, and they produce an output that is voltage (or current). Therefore, we need an analog-to-digital converter to translate the analog signals to digital numbers so that the PC can read them. Next we describe an ADC chip.

ADC0848 chip

The ADC0848 IC is an analog-to-digital converter in the family of the ADC0800 series from National Semiconductor Corp. Data sheets for this chip can be found at their web site, www.national.com. From there, go to Products > Analog-Data Acquisition > A-to-D Converter-General Purpose.

The ADC0848 has a resolution of 8 bits. It is an 8-channel ADC, thereby allowing it to monitor up to 8 different analog inputs. See Figures 12-12 and 12-13. The ADC0844 chip in the same family has 4 channels. The following is the discussion of the pins of the ADC0848.

CS: Chip select is an active low input used to activate the 848 chip. To access the 848, this pin must be low.

RD (read): This is an input signal and is active low. ADC converts the analog input to its binary equivalent and holds it in an internal register. RD is used to get the converted data out of the 848 chip. When CS = 0, if the RD pin is asserted low, the 8-bit digital output shows up at the D0 - D7 data pins. The RD pin is also referred to as output enable (OE).

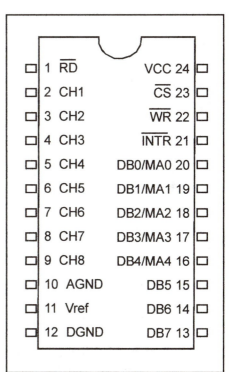

Figure 12-12. ADC0848 Chip

Vref is an input voltage used for the reference voltage. The voltage connected to this pin dictates the step size. For the ADC0848, the step size is Vref/256 since it is an 8-bit ADC and 2 to the power of 8 gives us 256 steps. See Table 12-10. For example, if the analog input range needs to be 0 to 4 volts, Vref is connected to 4 volts. That gives 4V/256 = 15.62 mV for step size. In another case, if we need the step size of 10 mV then Vref = 2.56 V, since 2.56 V/256 = 10 mV.

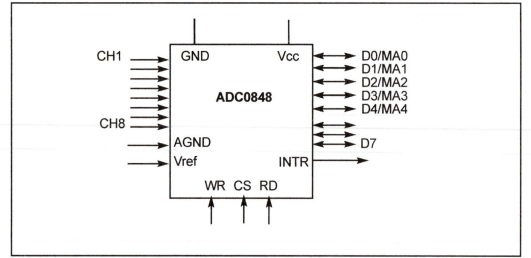

Figure 12-13. ADC0848 Block Diagram

DB0 - DB7 are the digital data output pins. With a D0 - D7 output, the 848 must be an 8-bit ADC. The step size, which is the smallest change, is dictated by the number of digital outputs and the Vref voltage. To calculate the output voltage, we use the following formula:

$$D_{out} = \frac{V_{in}}{step\ size}$$

where Dout = digital data output (in decimal), Vin = analog input voltage, and step size (resolution) is the smallest change, which is Vref/256 for an 8-bit ADC. See Example 12-9 for clarification. Notice that D0 - D7 are tri-state buffered and that the converted data is accessed only when CS = 0 and a low pulse is applied to the RD pin. Also, notice the dual role of pins D0 - D7. They are also used to send in the channel address. This is discussed next.

MA0 - MA4 (multiplexed address). The ADC0848 uses multiplexed address/data pins to select the channel. Notice in Figure 12-13 that a portion of the DB0 - DB7 pins are also designated as MA0 - MA4. The D0 - D7 pins are inputs when the channel's address is sent in. However, when the converted data is being read, D0 - D7 are outputs. While the use of multiplexed address/data saves some pins, it makes the I/O interfacing more difficult as we will soon see.

WR (write; a better name might be "start conversion"). This is an input into the ADC0848 chip and plays two important roles: (1) It latches the address of the selected channel present on the D0 - D7 pins, and (2) it informs the ADC0848 to start the conversion of analog input at that channel. If CS = 0 when WR makes a low-to-high transition, the ADC0848 latches in the address of the selected channel and starts converting the analog input value to an 8-bit digital number. The amount of time it takes to convert is a maximum of 40 microseconds for ADC0848. The conversion time is set by an internal clock.

Table 12-10: ADC0848 Vref vs. Step Size

Vref(V)	Step size (mV)
5	19.53
4	15.62
2.56	10
1.26	5
0.64	2.5

Note: Step size = Vref/256.

Example 12-9

For a given ADC0848, we have Vref = 2.56 V. Calculate the D0 - D7 output if the analog input is: (a)1.7 V, and (b) 2.1 V.
Solution:

Since the step size is 2.56/256 = 10 mV, we have the following.
(a)D_{out} = 1.7V/10 mV=170 in decimal, which gives us 10101011 in binary for D7-D0.

(b)D_{out} = 2.1V/10 mV=210 in decimal, which gives us 11010010 in binary for D7-D0.

CH1 - CH8 are 8 channels of the Vin analog inputs. In what is called single-ended mode, each of the 8 channels can be used for analog Vin where the AGND (analog ground) pin is used as a ground reference for all the channels. These 8 channels of input allow us to read 8 different analog signals, but not all at the same time since there is only a single D0 - D7 output. We select the input channel by using the MA0 - MA4 multiplexed address pins according to Table 12-11. In Table 12-11, notice that MA4 = low and MA3 = high for single-ended mode. The ADC0848 can also be used in differential mode. In differential mode, two channels, such as CH1 and CH2, are paired together for the Vin(+) and Vin(-) differential analog inputs. In that case Vin = CH1(+) – CH2(-) is the differential analog input. To use ADC0848 in differential mode, MA4 = don't care, and MA3 is set to low. For more on this, see the ADC0848 data sheet on the www.national.com web site.

Table 12-11: ADC0848 Analog Channel Selection (Single-Ended Mode)

Selected Analog Channel	MA4	MA3	MA2	MA1	MA0
CH1	0	1	0	0	0
CH2	0	1	0	0	1
CH3	0	1	0	1	0
CH4	0	1	0	1	1
CH5	0	1	1	0	0
CH6	0	1	1	0	1
CH7	0	1	1	1	0
CH8	0	1	1	1	1

Note: Channel is selected when CS = 0, RD =1, and an L-to-H pulse is applied to WR.

VCC is the +5 volt power supply.

AGND, DGND (analog ground and digital ground). Both are input pins providing the ground for both the analog signal and the digital signal. Analog ground is connected to the ground of the analog Vin while digital ground is connected to the ground of the VCC pin. The reason that we have two ground pins is to isolate the analog Vin signal from transient voltages caused by digital switching of the output D0 - D7. Such isolation contributes to the accuracy of the digital data output. Notice that in the single-ended mode the voltage at the channel is the analog input and AGND is the reference for the Vin. In our discussion, both the AGND and DGND are connected to the same ground; however, in the real world of data acquisition, the analog and digital grounds are handled separately.

INTR (interrupt; a better name might be "end of conversion"). This is an output pin and is active low. It is a normally high pin and when the conversion is

---

picked up. After INTR goes low, we make CS = 0 and apply a low pulse to the RD pin to get the binary data out of the ADC0848 chip. See Figure 12-14.

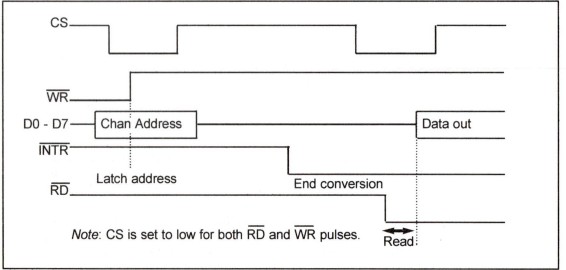

Figure 12-14. Selecting a Channel and Read Timing for ADC0848

Selecting an input channel

The following are the steps for data conversion by the ADC0848 chip.

1. While CS = 0 and RD = 1, provide the address of the selected channel (see Table 12-11) to the DB0 - DB7 pins and apply a low-to-high pulse to the WR pin to latch in the address and start the conversion. The channel's addresses are 08H for CH1, 09H for CH2, 0AH for CH3, and so on, as shown in Table 12-11. Notice that this process not only selects the channel, but also starts the conversion of the analog input at the selected channel.
2. While WR = 1 and RD = 1, keep monitoring the INTR pin. When INTR goes low, the conversion is finished and we can go to the next step. If INTR is high, keep polling until it goes low, signalling end-of-conversion.
3. After the INTR has become low, we must make CS = 0, WR = 1, and apply a low pulse to the RD pin to get the data out of the 848 IC chip.

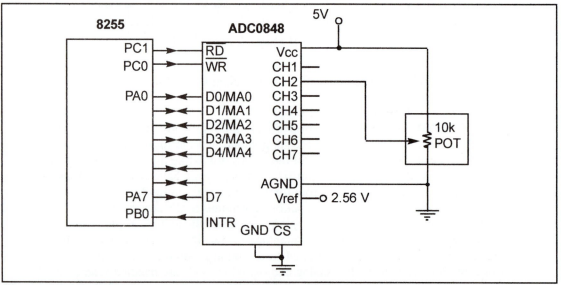

Figure 12-15. 8255 Connection to ADC0848 for CH2

CHAPTER 12: INTERFACING TO THE PC

ADC0848 connection to 8255

The following is a summary of the connection between the 8255 and the ADC0848 as shown in Figure 12-15.

PA0 - PA7 to D0 - D7 of ADC:	Channel selection (out), data read (in)
PB0 to INTR	Port B as input
PC0 to WR	Port C as output
PC1 to RD	Port C as output

Notice the following facts about the above connection.
1. Port A is an output when we select a channel, and it is an input when we read the converted data.
2. We must monitor the INTR pin of the ADC for end-of-conversion; therefore, we configure PB as input. Since both WR and RD are inputs into ADC, Port C is configured as an output port.

The following program is for Figure 12-15. It selects channel 2. After reading its data, the data is converted from binary (hex) to ASCII. In the program, CL = least significant digit (LSD) and AL = most significant digit (MSD).

```
        MOV   AL,82H ;PA=OUT,PB=IN, PC=OUT
        MOV   DX,CNT_PORT
        OUT   DX,AL
        MOV   AL,09           ;CHANNEL 2 ADDRESS (Table 12-11)
        MOV   DX,PORT_A
        OUT   DX,AL
        MOV   AL,02           ;WR=0,RD=1
        MOV   DX,PORT_C
        OUT   DX,AL
        CALL  DELAY           ;few usec
        MOV   AL,03           ;WR=1,RD=1
        OUT   DX,AL           ;TO LATCH CHANNEL ADDRESS
        CALL  DELAY           ;few usec
        MOV   AL,92H          ;PA=IN,PB=IN, AND PC=OUT
        MOV   DX,CNT_PORT
        OUT   DX,AL
        MOV   DX,PORT_B       ;GET READY TO MONITOR INTR
B1:     IN    AL,DX           ;MONITOR INTR
        AND   AL,01           ;MASK ALL BITS EXCEPT INTR
        CMP   AL,01           ;IS IT HIGH?
        JNE   B1              ;KEEP MONITORING FOR LOW
        MOV   AL,01           ;WR=1,RD=0 TO READ DATA
        MOV   DX,PORT_C
        OUT   DX,AL
        MOV   DX,PORT_A       ;PORT_A  TO GET  DATA
        IN    AL,DX           ;GET THE CONVERTED DATA
;converting 00-FFH hex value to decimal and then to ASCII.
; AL,AH,CL will have decimal values in ASCII
        MOV   BL,10
        SUB   AH,AH           ;CLEAR AH FOR WORD/BYTE DIV
        DIV   BL             ;AX/BL
        MOV   CL,AH          ;SAVE LSD IN CL REG
        SUB   AH,AH          ;
        DIV   BL             ;AX/BL FOR 2ND DIGIT
;make them all ASCII
        OR    AX,3030H
        OR    CL,30H
```

Notice the conversion of the above data to ASCII. In order to display ADC input on a screen or LCD, it must be converted to ASCII. However, to convert it to ASCII, it must be converted to decimal first. To convert a 00 - FF hex value to decimal we keep dividing it by 10 until the remainder is less than 10. Each time we divide it by 10 we keep the quotient as one of our decimal digits. In the case of an 8-bit data, dividing it by 10 twice will do the job. For example, if we have FFH it will become 255 in decimal. To convert from decimal to ASCII format, we OR each digit with 30H. Now all we have to do is to send the digits to the PC screen by using INT 21H or send them to the LCD as was shown in the first section of this chapter. One advantage of using C/C++ programs is that such a conversion is done by the compiler.

Interfacing a temperature sensor to PC

Transducers are used to convert physical quantities such as temperature, light intensity, flow, and speed to electrical signals. Depending on the transducer, the output produced is in the form of voltage, current, resistance, or capacitance. For example, temperature is converted to electrical signals using a transducer called a *thermistor*. The thermistor responds to temperature change by changing its resistance. However, its response is not linear, as shown in Table 12-12.

Table 12-12: Thermistor Resistance vs. Temperature

Temperature (C)	Tf (K ohms)
0	29.490
25	10.000
50	3.893
75	1.700
100	0.817

From William Kleitz, *Digital Electronics*

The complexity associated with writing software for such nonlinear devices has led many manufacturers to market the linear temperature sensor. Simple and widely used linear temperature sensors include the LM34 and LM35 series from National Semiconductor Corp. They are discussed next.

Table 12-13: LM34 Temperature Sensor Series Selection Guide

Part Scale	Temperature Range	Accuracy	Output
LM34A	−50 F to +300 F	+2.0 F	10 mV/F
LM34	−50 F to +300 F	+3.0 F	10 mV/F
LM34CA	−40 F to +230 F	+2.0 F	10 mV/F
LM34C	−40 F to +230 F	+3.0 F	10 mV/F
LM34D	−32 F to +212 F	+4.0 F	10 mV/F

Note: Temperature range is in degrees Fahrenheit.

LM34 and LM35 temperature sensors

The sensors of the LM34 series are precision integrated-circuit temperature sensors whose output voltage is linearly proportional to the Fahrenheit temperature. The LM34 requires no external calibration since it is inherently calibrated. It outputs 10 mV for each degree of Fahrenheit temperature. Table 12-13 is the selection guide for the LM34. The LM35 series sensors are also precision integrated-circuit temperature sensors whose output voltage is linearly proportional to the Celsius (centigrade) temperature. The LM35 requires no external calibration since it is inherently calibrated. It outputs 10 mV for each degree of centigrade temperature. Table 12-14 is the selection guide for the LM35.

Table 12-14: LM35 Temperature Sensor Series Selection Guide

Part	Temperature Range	Accuracy	Output Scale
LM35A	−55 C to +150 C	+1.0 C	10 mV/F
LM35	−55 C to +150 C	+1.5 C	10 mV/F
LM35CA	−40 C to +110 C	+1.0 C	10 mV/F
LM35C	−40 C to +110 C	+1.5 C	10 mV/F
LM35D	0 C to +100 C	+2.0 C	10 mV/F

Note: Temperature range is in degrees Celsius.

Signal conditioning and interfacing the LM35 to a PC

Signal conditioning is a widely used term in the world of data acquisition. The most common transducers produce an output in the form of voltage, current, charge, capacitance, and resistance. However, we need to convert these signals to voltage in order to send input to an A-to-D converter. This conversion (modification) is commonly called *signal conditioning*. Signal conditioning can be a current-to-voltage conversion or a signal amplification. For example, the thermistor changes resistance with temperature. The change of resistance must be translated into voltages in order to be of any use to an ADC. Look at the case of connecting an LM35 to an ADC0848. Since the ADC0848 has 8-bit resolution with a maximum of 256 steps and the LM35 (or LM34) produces 10 mV for every degree of temperature change, we can condition Vin of ADC0848 to produce a Vout of 2560 mV (2.56 V) for full-scale output. Therefore, in order to produce the full-scale Vout of 2.56 V for the ADC0848, we need to set Vref = 2.56. This makes Vout of the ADC0848 correspond directly to the temperature as monitored by the LM35. This is shown in Table 12-15.

Figure 12-17 shows connection of the temperature sensor to CH2 of the ADC0848. Notice that we use the LM336-2.5 zener diode to fix the voltage across the 10K POT at 2.5 volts. The use of the LM336-2.5 should overcome any fluctuations in the power supply. The LM336 has three leads. However, the third lead is unconnected.

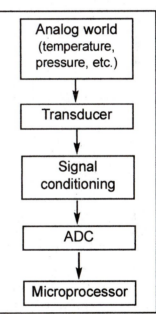

Figure 12-16. Getting Data to the CPU

Table 12-15: Temperature v. Vout of the ADC0848

Temp. (C)	Vin (mV)	Vout (D7 - D0)
0	0	0000 0000
1	10	0000 0001
2	20	0000 0010
3	30	0000 0011
10	100	0000 1010
30	300	0001 1110

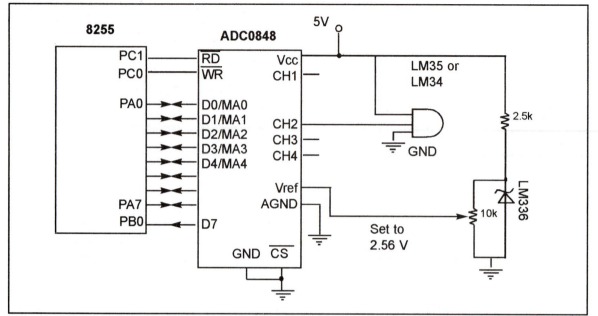

Figure 12-17. 8255 Connection to ADC0848 and Temperature Sensor

ADC808/809

Another popular ADC is the ADC808/809 chip. It has eight input channels allowing it to convert 8 different analog inputs. See Figure 12-18. It is an 8-bit ADC. The following is the pin description of the ADC808/809 chip.

OE (output enable): This is an input signal and is active high. ADC converts the analog input to its binary equivalent and holds it in an internal register. OE is used to get the converted data out of the ADC808 chip. If a low-to-high pulse is applied to the OE pin, the 8-bit digital output shows up at the D0 - D7 data pins. The OE pin is also referred to as RD (read).

SC (start conversion): This is an input pin and is used to inform the ADC808 to start the conversion process. If we apply a low-to-high pulse to this pin, the ADC808 starts converting the analog input value of Vin to an 8-bit digital number. The amount of time it takes to convert varies depending on the CLK value. When the data conversion is complete, the EOC (end of conversion) pin is forced low by the ADC808.

CLK is an input pin and is connected to an external clock source. The CLK speed dictates the conversion time. While the ADC0848 uses an internal clock, for the ADC808 the clock source is external. This way one can contol the conversion speed.

EOC (end of conversion): This is an output pin and is active low. It is a normally high pin and when the conversion is finished, it goes low to signal the CPU that the converted data is ready to be picked up. After EOC goes low, we send a low-to-high pulse to the OE pin to get the data out of the ADC808 chip.

Vref(+) and Vref(-) are both input voltages used for the reference voltage. The voltage connected to these pins dictates the step size. For the ADC808/809, the step size is [Vref(+) – Vref(–)] /256, since it is an 8-bit ADC and 2 to the power of 8 gives us 256 steps. For example, if the analog input range needs to be 0 to 4 volts, Vref(+) is connected to 4 volts and Vref(–) is grounded. That gives 4V/256 = 15.62 mV for the step size. In another case, if we need the step size of 10 mV, Vref = 2.56 V since 2.56 V/256 = 10 mV. Notice that if we connect the Vref(–) input to a voltage other than ground, the step size is calculated based on the differential value of the Vref(+) – Vref(–) inputs.

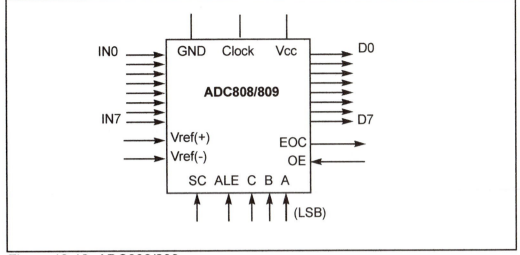

Figure 12-18. ADC808/809

D0 - D7 are the digital data output pins. These are tri-state buffered and the converted data is accessed only when OE is forced high. While the analog input voltage is in the range 0 to +5V, the output D0 - D7 is given in binary.

IN0 - IN7 are the 8 channels of the Vin analog inputs. These 8 channels of input allow us to read 8 different analog signals. However, they cannot all be read at the same time since there is only a single D0 - D7. We select the input channel by using the A, B, C address selector pins according to Table 12-16.

A, B, C, and ALE. The input channel is selected by using the A, B, C, and ALE pins. These are input signals into the ADC808/809 and the channel is selected according to Table 12-16. To select a channel, we provide the channel address to the A, B, and C pins according to Table 12-16 and then apply an L-to-H pulse to the ALE pin to latch in the address.

Table 12-16: ADC808/809 Analog Channel Selection

Selected Analog Channel	C	B	A
IN0	0	0	0
IN1	0	0	1
IN2	0	1	0
IN3	0	1	1
IN4	1	0	0
IN5	1	0	1
IN6	1	1	0
IN7	1	1	1

Note: Channel is selected when OE = 0, and an L-to-H pulse is applied to ALE.

How to read ADC808/809 data

Comparing the ADC808/809 with the ADC0848 shows that the ADC808/809 has a clock pin. This means that we must provide an external clock source. Therefore, the conversion speed varies according to the speed of the external clock source. Also, notice that the ADC808/809 has no CS pin.

The following are the steps to select a channel and read its data.

1. Provide the channel address (see Table 12-16) to pins A, B, and C.
2. Apply an L-to-H pulse to the ALE pin to latch in the channel address.
3. Apply an L-to-H pulse to the SC pin to start the conversion of analog input to digital data.
4. After the passage of 8 clocks, the EOC pin will go low to indicate that the data is converted and ready to be picked up. We can either use a small time delay and then read the data out, or monitor the EOC pin and read the data out after it goes low. Notice that if you use a time delay to wait before you read the data, the size of the delay varies depending on the speed of the clock connected to the clock pin of the ADC808/809.
5. Apply an L-to-H pulse to the OE pin and read the data.

Review Questions

1. In the ADC0848, the INTR signal is an _____ (input, output).
2. To begin conversion, send a(n)_____ pulse to _____.
3. Which pin of the ADC0848 indicates end-of-conversion?
4. The LM35 provides ____ mV for each degree of _____ (Fahrenheit, Celsius) temperature.
5. Both the ADC0848 and ADC808 are _____-bit converters.

SUMMARY

This chapter discussed PC interfacing to real-world devices such as LCDs, ADCs, DACs, and stepper motors. First, we described the operation modes of LCDs, then described how to program the LCD by sending data or commands to it via the 8255. Next, we showed stepper motor interfacing to the PC. DAC and ADC devices and their applications were also discussed. ADC and temperature sensors and their interfacing were explored. The issue of signal conditioning and its importance in sensor interfacing was discussed as well.

PROBLEMS

SECTION 12.1: INTERFACING AN LCD TO THE PC

1. The LCD discussed in this section has _____ (4, 8) data pins.
2. Describe the function of pins E, R/W, and RS in the LCD.
3. What is the difference between the VCC and VEE pins on the LCD?
4. Clear LCD is a _____ (command code, data item) and its value is ___ hex.
5. What is the hex value of the command code for display on, cursor on?
6. Give the state of RS, E, and R/W when sending a command code to the LCD.
7. Give the state of RS, E, and R/W when sending data character 'Z' to the LCD.
8. Which of the following is needed on the E pin in order for a command code (or data) to be latched in by the LCD?
   (a) H-to-L pulse (b) L-to-H pulse
9. True or false. For the above to work, the value of the command code (data) must be already at the D0 - D7 pins.

10. There are two methods of sending streams of characters to the LCD: (1) checking the busy flag, or (2) putting some time delay between each character without checking the busy flag. Explain the difference and advantage and disadvantage of each method. Also explain how we monitor the busy flag.

11. For a 16x2 LCD the location of the last character of line 1 is 8FH (its command code). Show how this value came about.

12. For a 16x2 LCD the location of the first character of line 2 is C0H (its command code). Show how this value came about.

13. For a 20x2 LCD the location of the last character of line 2 is 93H (its command code). Show how this value came about.

14. For a 20x2 LCD the location of the third character of line 2 is C2H (its command code). Show how this value came about.

15. For a 40x2 LCD the location of the last character of line 1 is A7H (its command code). Show how this value came about.

16. For a 40x2 LCD the location of the last character of line 2 is E7H (its command code). Show how this value came about.

17. Show the value (in hex) for the command code for the 10th location, line 1 on a 20x2 LCD. Show how you got your value.

18. Show the value (in hex) for the command code for the 20th location, line 2 on a 40x2 LCD. Show how you got your value.

19. Rewrite the COMNDWRT procedure (shown in Section 12.1) if port C is used for control signals. Assume that PC4=RS, PC5=R/W, PC6=E.

20. Repeat the above program for a data write procedure. Send the string "Hello" to the LCD without checking the busy flag.

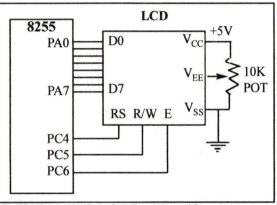

LCD Connection for Problem 19

SECTION 12.2: INTERFACING A STEPPER MOTOR TO THE PC

21. If a motor takes 90 steps to make one complete revolution, what is the step angle for this motor?

22. Calculate the number of steps per revolution for a step angle of 7.5 degrees.

23. Finish the normal 4-step sequence clockwise if the first step is 0011 (binary).

24. Finish the normal 4-step sequence clockwise if the first step is 1100 (binary).

25. Finish the normal 4-step sequence counterclockwise if the first step is 1001 (binary).

26. Finish the normal 4-step sequence counterclockwise if the first step is 0110 (binary).

27. What is the purpose of the ULN2003 placed between the 8255 and the stepper motor? Can we use that for 3A motors?

28. Which of the following cannot be a sequence in the normal 4-step sequence for a stepper motor?
    (a) CCH (b) DDH (c) 99H (d) 33H
29. What is the effect of a time delay between issuing each step?
30. In Question 29, how can we make a stepper motor go faster?

SECTION 12.3: INTERFACING DAC TO THE PC

31. True or false. DAC1408 is the same as DAC0808. Are they pin compatible?
32. Find the number of discrete voltages provided by the n-bit DAC for the following.
    (a) n=8 (b) n=10 (c) n=12
33. For DAC1408, if Iref = 2 mA show how to get the Iout of 1.99 when all inputs are high.
34. Find Iout for the following inputs. Assume Iref = 2 mA for DAC1408.
    (a) 10011001 (b) 11001100 (c) 11101110
    (d) 00100010 (e) 00001001 (f) 10001000
35. To get a smaller step, we need a DAC with _____ (more, fewer) digital inputs.
36. To get full-scale output what should be the inputs for DAC?

SECTION 12.4: INTERFACING ADC AND SENSORS TO THE PC

37. Give the status of CS and WR in order to start conversion for the ADC0848.
38. Give the status of CS and WR in order to get data from the ADC0848.
39. In the ADC0848 what happens to the converted analog data? How do we know that the ADC is ready to provide us the data?
40. In the ADC0848 what happens to the old data if we start conversion again before we pick up the last data?
41. In the ADC0848 INTR is an _____ (input, output) signal. What is its function in the ADC0848?
42. For an ADC0848 chip, find the step size for each of the following Vref.
    (a) Vref = 1.28V (b) Vref = 1V (c) Vref = 1.9V
43. In the ADC0848 what should be the Vref value if we want a step size of 20 mv?
44. In the ADC0848 what should be the Vref value if we want a step size of 5 mv?
45. In the ADC0848 how is the analog channel selected?
46. With a step size of 19.53 mV, what is the analog input voltage if all outputs are 1?
47. With Vref = 1.28V, find the Vin for the following outputs.
    (a) D7 - D0 = 11111111 (b) D7 - D0 = 10011001 (c) D7 - D0 = 1101100
48. What does it mean when it is said that a given sensor has a linear output?
49. The LM34 sensor produces _____ mv for each degree of temperature.
50. What is signal conditioning?

ANSWERS TO REVIEW QUESTIONS

SECTION 12.1: INTERFACING AN LCD TO THE PC

1. Input   2. Input   3. H-to-L
4. High   5. 80H and C0h

SECTION 12.2: INTERFACING A STEPPER MOTOR TO THE PC

1. 0110,0011,1001,1100 for clockwise, and 0110,1100,1001.0011 for counterclockwise
2. 72
3. Because the 8255 does not provide sufficient current to drive the stepper motor.

SECTION 12.3: INTERFACING DAC TO THE PC

1. Digital,analog. In ADC the input is analog, the output is digital.
2. 8
3. current

SECTION 12.4: INTERFACING ADC AND SENSORS TO THE PC

1. output
2. L-to-H WR pin
3. INTR
4. 10,both
5. 8

CHAPTER 13

8253/54 TIMER AND MUSIC

OBJECTIVES

Upon completion of this chapter, you will be able to:

» **Describe the function of each pin of the 8253/54 PIT (programmable interval timer)**

» **Program the three counters of the 8253/54 by use of the chip's control word**

» **Diagram how the 8253/54 timer is connected in the IBM PC**

» **Produce the "beep" sound in the IBM PC by programming the 8253 timer**

» **Write programs to play songs on the IBM PC**

» **Diagram the output pulses of the 8253/54 timer**

» **Describe the various modes of the OUT signal in the 8253/54**

In the PC there is a single clock used to synchronize activities of all peripheral chips connected to the CPU. That clock, which has the highest frequency in the system, belongs to the 80x86 CPU. There are functions within the PC that require a clock with a lower frequency. The 8253/54 PIT (programmable interval timer) is used to bring down the frequency to the desired level for various uses such as the beep sound in the PC. This chapter will first describe the 8253/54 timer and show the processes of initializing and programming it. Then interfacing and the use of the 8253/54 in the IBM PC are discussed. The third section will show how the 8253/54 can be used to generate various frequencies, including musical notes on the PC. Section 13.4 describes the various shapes of 8253/54 output pulses.

SECTION 13.1: 8253/54 TIMER DESCRIPTION AND INITIALIZATION

The 8253 chip was used in the IBM PC/XT, but starting with the IBM PC AT, the 8254 replaced the 8253. The 8254 and 8253 have exactly the same pinout. The 8254 is a superset of the 8253, meaning that all programs written for the 8253 will run on the 8254. The following are pin descriptions of the 8253/54.

A0, A1, $\overline{\text{CS}}$

Inside the 8253/54 timer, there are three counters. Each works independently and is programmed separately to divide the input frequency by a number from 1 to 65,536. Each counter is assigned an individual port address. The control register common to all three counters has its own port address. This means that a total of 4 ports are needed for a single 8253/54 timer. The ports are addressed by A0, A1, and $\overline{\text{CS}}$, as shown in Table 13-1. Each of the 3 counters has 3 pins associated with it, CLK (clock), GATE, and OUT, as shown in Figure 13-1.

Table 13-1: Addressing 8253/54

$\overline{\text{CS}}$	A1	A0	Port
0	0	0	Counter 0
0	0	1	Counter 1
0	1	0	Counter 2
0	1	1	Control register
1	x	x	8253/54 is not selected

(Reprinted by permission of Intel Corporation, Copyright Intel Corp. 1983)

CLK

CLK is the input clock frequency, which can range between 0 and 2 MHz for the 8253. For input frequencies higher than 2 MHz, the 8254 must be used; the 8254 can go as high as 8 MHz, and the 8254-2 can go as high as 10 MHz.

OUT

Although the input frequency is a square wave of 33% duty cycle, the shape of the output frequency coming from the OUT pin after being divided can be programmed. Among the options are square-wave, one-shot, and other square-shape waves of various duty cycles but no sine-wave or saw-tooth shapes.

GATE

This pin is used to enable or disable the counter. Putting HIGH (5 V) on GATE enables the counter, whereas LOW (0 V) disables it. In some modes a 0-to-1 pulse must be applied to GATE to enable the counter.

D0 - D7

The D0 - D7 data bus of the 8253/54 is a bidirectional bus connected to D0 - D7 of the system data bus. The data bus allows the CPU to access various registers inside the 8253/54 for both read and write operations. $\overline{\text{RD}}$ and $\overline{\text{WR}}$ (both active low) are connected to $\overline{\text{IOR}}$ and $\overline{\text{IOW}}$ control signals of the system bus.

Initialization of the 8253/54

Each of the three counters of the 8253/54 must be programmed separately. In order to program any of the three counters, the control byte must first be written into the control register, which among other things tells the 8253/54 what shape of output pulse is needed. In addition, the number that the input clock should be divided by must be written into that counter of the 8253/54. Since this number can be as high as FFFF (16-bit data) and the data bus for the 8253/54 timer is only 8 bits wide, the divisor must be sent in one byte at a time. The 8253/54 must be initialized before it is used.

Control word

Figure 13-2 shows the one-byte control word of the 8253/54. This byte, which is sent to the control register, has the following bits.

D0 chooses between a binary number divisor of 0000 to FFFFH or a BCD divisor of 0000 to 9999H. The lowest number that the input frequency can be divided by for both options is 0001. The highest number is 2^{16}

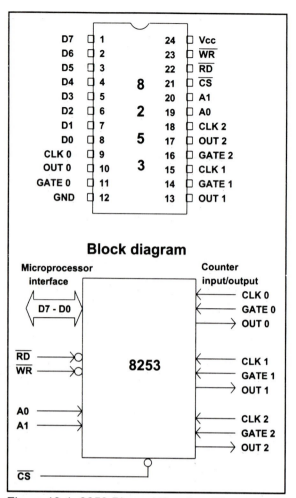

Block diagram

Figure 13-1. 8253 Pin and Function Diagram
(Reprinted by permission of Intel Corporation, Copyright Intel Corp. 1983)

for binary and 10^4 for BCD. To get the highest count (65,536 decimal and 10000 BCD), the counter is loaded with zeros.

D1, D2, and **D3** are for mode selection. There are six possible modes that determine the shape of the output signal.

Mode 0	Interrupt on terminal count
Mode 1	Programmable one-shot
Mode 2	Rate generator
Mode 3	Square wave rate generator
Mode 4	Software triggered strobe
Mode 5	Hardware triggered strobe

D4 and **D5** are for RL0 and RL1. The data bus of the 8253/54 is 8 bits (1 byte), but the number that the input frequency can be divided by (divisor) can be as high as FFFFH. Therefore, RL0 and RL1 are used to indicate the size of the divisor. RL0 and RL1 have three options: (1) read/write the most significant byte (MSB) only, (2) read/write the least significant byte (LSB) only, (3) read/write the LSB first followed immediately by the MSB.

The options for RL0 and RL1 show that programmers cannot only write the value of the divisor into the 8253/54 timer but read the contents of the counter at any given time, as well. Since all counters are down counters, and the count register is decremented, the count register's contents can be read at any time, thus using the 8253/53 as an event counter.

D6 and **D7** are used to select which of the three counters, counter 0, counter 1, or counter 2, is to be initialized by the control byte.

CHAPTER 13: 8253/54 TIMER AND MUSIC

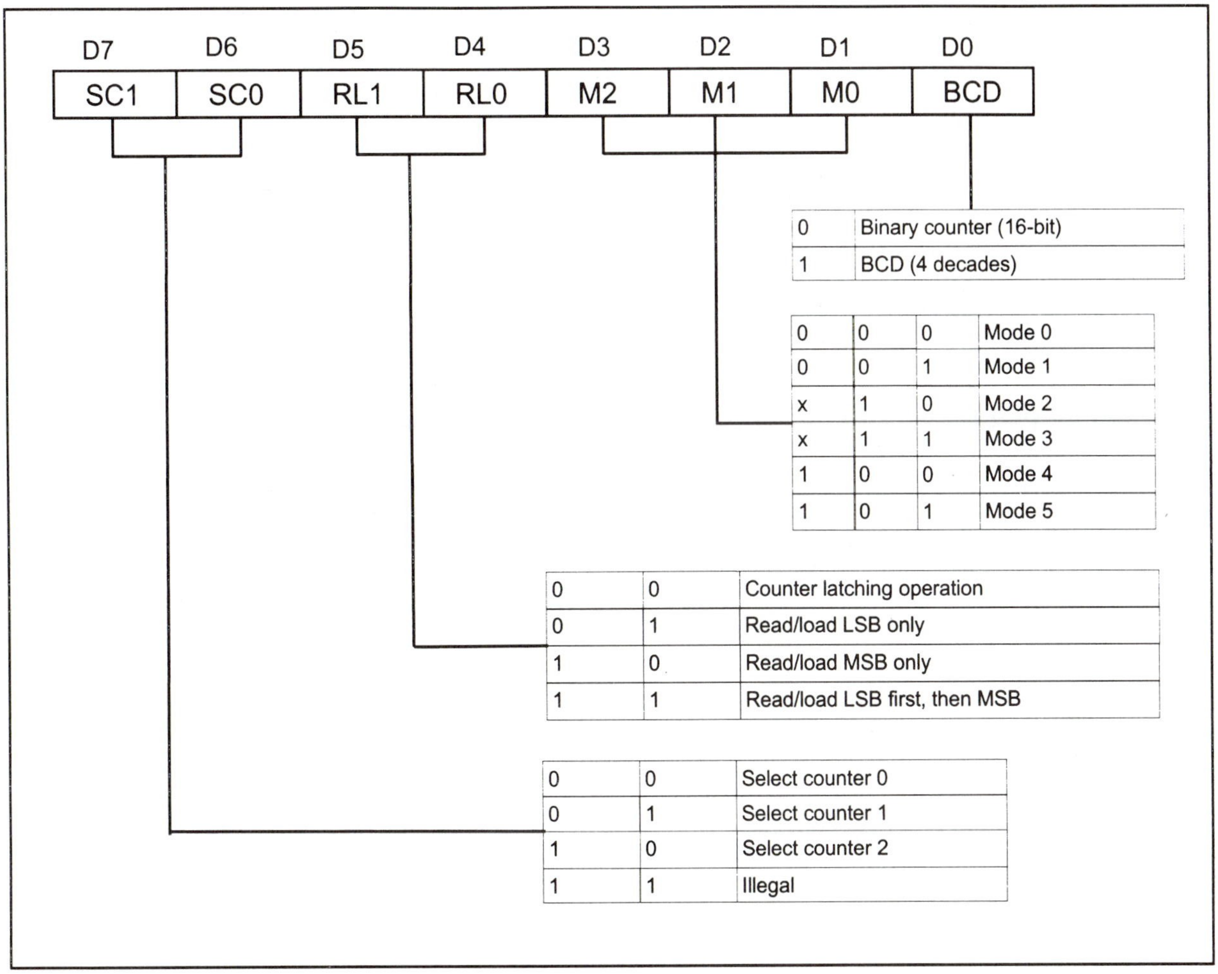

D7	D6	D5	D4	D3	D2	D1	D0
SC1	SC0	RL1	RL0	M2	M1	M0	BCD

0	Binary counter (16-bit)
1	BCD (4 decades)

0	0	0	Mode 0
0	0	1	Mode 1
x	1	0	Mode 2
x	1	1	Mode 3
1	0	0	Mode 4
1	0	1	Mode 5

0	0	Counter latching operation
0	1	Read/load LSB only
1	0	Read/load MSB only
1	1	Read/load LSB first, then MSB

0	0	Select counter 0
0	1	Select counter 1
1	0	Select counter 2
1	1	Illegal

Figure 13-2. 8253/54 Control Word Format
(Reprinted by permission of Intel Corporation, Copyright Intel Corp. 1983)

Example 13-1

Pin $\overline{CS}$ of a given 8253/54 is activated by binary address A7 - A2 =100101.
(a) Find the port addresses assigned to this 8253/54.
(b) Find the configuration for this 8253/54 if the control register is programmed as follows.

```
MOV    AL,00110110
OUT    97H,AL
```

Solution:

(a) From Table 13-1, we have the following:

CS		A1A0	Port	Port address (hex)
1001	01	00	Counter 0	94
1001	01	01	Counter 1	95
1001	01	10	Counter 2	96
1001	01	11	Control register	97

(b) Breaking down the control word 00110110 and comparing it with Table 13-1 indicates counter 0 since the SC bits are 00. The RL bits of 11 indicates that the low-byte read/write is followed by the high byte. The mode selection is mode 3 (square wave), and finally binary counting is selected since the D0 bit is 0.

To program a given counter of the 8253/4 to divide the CLK input frequency, one must send the divisor to that specific counter's register. In other words, although all three counters share the same control register, the divisor registers are separate for each counter. This is shown in Example 13-3.

Regarding the options bit D0 of the control byte, it must be noted that in BCD mode, if we program the counter for 9999, the input frequency is divided by that number. However, to divide the frequency by 10,000 we must send in 0 for both high and low bytes, as shown in Example 13-2.

We can program any of the counters for divisors of up to 65,536 if we use the binary option for D0. To program the counter for the divisor of 65,536, the counter must be loaded with 0 for the low byte and another 0 for the high byte of the divisor. In that case, D0=0 for the control byte.

Example 13-2

Use the port addresses in Example 13-1 to program:
(a) counter 0 for binary count of mode 3 (square wave) to divide CLK0 by number 4282 (BCD)
(b) counter 2 for binary count of mode 3 (square wave) to divide CLK2 by number C26A hex
(c) Find the frequency of OUT0 and OUT2 in (a) and (b) if CLK0 =1.2 MHz, CLK2 = 1.8 MHz.

Solution:

(a) To program counter 0 for mode 3, we have 00110111 for the control word. Therefore,

```
        MOV   AL,37H        ;counter 0, mode 3, BCD
        OUT   97H,AL        ;send it to control register
        MOV   AX,4282H      ;load the divisor (BCD needs H for hex)
        OUT   94H,AL        ;send the low byte
        MOV   AL,AH         ;to counter 0
        OUT   94H,AL        ;and then the high byte to counter 0
```

(b) By the same token:

```
        MOV   AL,B6H        ;counter2, mode 3, binary(hex)
        OUT   97H,AL        ;send it to control register
        MOV   AX,C26AH      ;load the divisor
        OUT   96H,AL        ;send the low byte
        MOV   AL,AH         ;to count 2
        OUT   96H,AL        ;send the high byte to counter 2
```

(c) The output frequency for OUT0 is 1.2MHz divided by 4282, which is 280 Hz. Notice that the program in part (a) used instruction "MOV AX,4282H" since BCD and hex numbers are represented in the same way, up to 9999. For OUT2, CLK2 of 1.8 MHz is divided by 49770 since C26AH = 49770 in decimal. Therefore, OUT2 frequency is a square wave of 36 Hz.

Example 13-3

Using the port addresses in Example 13-1, show the programming of counter 1 to divide CLK1 by 10,000, producing the mode 3 square wave. Use the BCD option in the control byte.

Solution:

```
        MOV   AL,77H        ;counter1, mode 3, BCD
        OUT   97H,AL        ;send it to control register
        SUB   AL,AL         ;AL =0 load the divisor for 10,000
        OUT   95H,AL        ;send the low byte
        OUT   95H,AL        ;and then the high byte to counter 1
```

Review Questions

1. True or false. Any code written for the 8253 will work on the 8254.
2. The 8253/54 can be used to _____ (divide, multiply) a square-wave digital frequency.
3. If $\overline{CS}$ of the 8253/54 is activated by A7 - A2 = 0110 00 binary, find the port address for this timer.
4. Find the control byte to program counter 2 for mode 1 (programmable one-shot), BCD count, low byte, followed by high-byte R/W.
5. True or false. To divide input frequency CLK1 by 5065, we must send the 5065 to the control register.
6. For Question 5, give the port address using the ports in Question 3.
7. To divide the CLK frequency by 52,900, which option for D0 of the control byte must be selected, and why?
8. If D0 =0 in the control byte, what is the highest number for the divisor?

SECTION 13.2: IBM PC 8253/54 TIMER CONNECTIONS AND PROGRAMMING

The IBM PC uses a 74LS138 to decode addresses for $\overline{CS}$ of the 8253 as shown in Figure 13-3. The port addresses are selected as indicated in Table 13-2, assuming zeros for x's. Chapter 12 contains a complete discussion of port selection.

Table 13-2: 8253/4 Port Address Calculation in the PC

Binary Address			Hex Address	Function
AEN A9 A8	**A7 A6 A5 A4**	**A3 A2 A1 A0**	**Hex Address**	**Function**
1 0 0	0 1 0 x	x x 0 0	40	Counter 0
1 0 0	0 1 0 x	x x 0 1	41	Counter 1
1 0 0	0 1 0 x	x x 1 0	42	Counter 2
1 0 0	0 1 0 x	x x 1 1	43	Control register

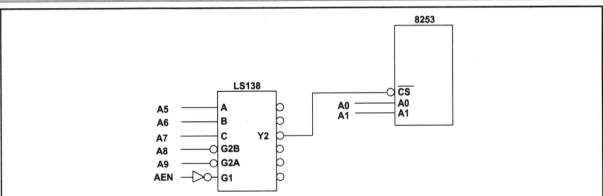

Figure 13-3. 8253 Port Selection in the PC/XT

The three clocks of the 8253, CLK0, CLK1, and CLK2, are all connected to a constant frequency of 1.1931817 MHz. This frequency is from PCLK of the 8284 chip after it has been divided by 2 with the use of D flip-flop 72LS175, as shown in Figure 13-4. PCLK of the 8284 (discussed in Chapter 9) is 2.3863633 MHz and must be divided by 2 since the maximum allowed input frequency of CLK of the 8253 is 2 MHz. GATE0 and GATE1, which enable counter 0 and counter 1, respectively, are connected to HIGH (5 V), thereby making those two counters enabled permanently. GATE2 of counter 2 can be enabled or disabled through PB0 of port B of the 8255. Now that the input frequency to each timer is known, programming and applications of each counter in the PC can be explained.

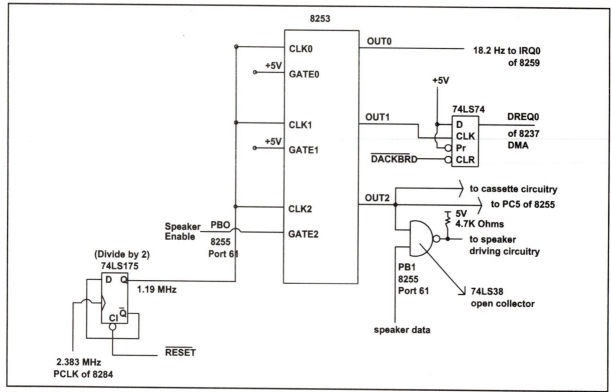

Figure 13-4. 8253 Chip Connections in the PC

Using counter 0

CLK0 of counter 0 is 1.193 MHz, and GATE0 is connected to high permanently. OUT0 of counter 0 is connected to IRQ0 (the highest-priority interrupt) of the 8259 interrupt controller to provide time-of-day (TOD) interrupt, among other services. The next question is: How often is IRQ0 activated, or in other words, what is the output frequency? IRQ0 is activated 18.2 times per second, or put another way, the OUT0 frequency is 18.2 Hz. If the frequency of CLK0 is 1.193 MHz and the output frequency should be 18.2 Hz, the counter must be programmed to divide 1.193 MHz by 65,536. The wave shape is a square wave (mode 3 of the 8253) in order to trigger IR0 on the positive edge of each pulse of the square wave so that a high pulse will not be mistaken for a multiple interrupt. Using the above information and Figure 13-2, the control word can be calculated in the following way:

D0 = 0 for the binary (or hex) value of the counter divisor. The timer is decremented after every input pulse until it reaches zero and then the original value is loaded again. Therefore, to divide the input frequency by 65,536, the timer is programmed with 0s for both high and low bytes.

D3 D2 D1 = 011, mode 3, for the square-wave output of 18.2 Hz frequency.

D4 D5 = 11 for reading/writing the LSB first, followed by the MSB.

D7 D6 = 00 for counter 0.

Summarizing the above gives the following control word:

```
D7  D6  D5  D4     D3  D2  D1  D0
 0   0   1   1      0   1   1   0  = 36H
```

The programming of counter 0 is as follows:

```
MOV    AL,36H        ;control word
OUT    43H,AL        ;to control register of 8253
MOV    AL,00         ;00 LSB and MSB of the divisor
OUT    40H,AL        ;LSB to timer 0
OUT    40H,AL        ;MSB  to timer 0
```

The IBM PC BIOS shows the same process as follows:

```
                  22        TIMER EQU 40H
          ..     ...        ...
E277  B036  695   MOV   AL,36H        ;SET TIM 0,LSB,MSB,MODE 3
E279  E643  696   OUT   TIMER+3,AL    ;WRITE TIMER MODE REG
E27B  B000  697   MOV   AL,0
E27D  E640  698   OUT   TIMER,AL      ;WRITE LSB TO TIMER 0 REG
          ..     ...        ...
E284  E640  704   OUT   TIMER,AL      ;WRITE MSB TO TIMER 0 REG
```

It must be noted that the function of IR0 is not only taking care of the time-of-day clock. BIOS also uses this interrupt to see if the motor on the floppy disk drive needs to be turned off. It also allows the user to use this interrupt for user-defined applications. At the rate of 18.2 Hz (or every 54.94 ms), BIOS will make this interrupt available by going to the vector table of INT 1CH. The user can define CS:IP of a service routine at the vector location belonging to INT 1CH and use it for any purpose, as will be seen in Chapter 14. If the user is not using this interrupt, control will automatically be returned to BIOS.

Using counter 1

In counter 1, CLK1 is connected to 1.193 MHz and GATE is high permanently. OUT1 generates a periodic pulse required to refresh DRAM memory of the computer. This refreshing must be done at least every 15 µs for each cell. As will be discussed in Chapter 15, in the IBM PC/XT the task of refreshing DRAM is performed by the 8237 DMA. It is up to the 8253's counter 1 to inform DMA periodically, lest the allowed time pass. To achieve this, OUT1 will provide DMA a pulse of approximately 15 µs duration or 66,278 Hz. This means that counter 1 must divide the input frequency 1.19318 MHz by 18 (1.19318 MHz divided by 18 = 66,278 Hz). Now why does the pulse have a duration of 15 µs? It is because there are 128 rows that must be refreshed in DRAMs of 64K- and 256K-bit capacity, and if they are refreshed every 15 µs, that makes a $15 \times 128 = 1.92$ ms refreshing period just below the required 2 ms. This will be explained in more detail in Chapter 15. Using Figure 13-2, the control byte can be figured out as follows:

$D0$ = 0 for binary option

$D3$ $D2$ $D1$ = 010 for mode 2 shape output. In this mode, OUT1 stays high for a total of 18 pulses and goes low for one pulse. This action is repeated continuously (see Section 13.4).

$D5$ $D4$ = 01 for the LSB only, since the byte is less than FF. CLK1 is divided by 18; therefore, 18 is the LSB and there is no need for the MSB.

$D7$ $D6$ = 01 for counter1

```
D7 ... D0
0101 0100 = 54H for the control word
```

The programming of the 8253 counter1 in the IBM BIOS is listed as follows, with slight modifications for the sake of clarity:

```
MOV   AL,54H     ; the control word
OUT   43H,AL     ; to control register
MOV   AL,18      ; 18 decimal, the divisor
OUT   41H,AL     ; to counter 1
```

Using counter 2

The output of counter 2 is connected to two different devices: the speaker and PC5 of the 8255. In early models of the IBM PC/XT, it was also connected to the cassette circuitry. That option has been eliminated in all the IBM PC and PS/2 computers built in recent years. Since counter 2 in the IBM PC is used to play music, it is important to understand counter 2 programming thoroughly.

Use of timer 2 by the speaker

In the IBM PC, CLK2 is connected to a frequency of 1.19318 MHz and GATE2 is programmed by PB0 of port 61H (port B). The IBM PC uses counter 2 to generate the beep sound. Although BIOS uses timer 2 for the beep sound, it can be changed to play any musical note, as will be shown in the next section. The beep sound has a frequency of 896 Hz of mode 3 (square wave). Dividing the input frequency of 1.19318 MHz by 896 Hz gives 1331 (0533 hex) for the value to be loaded to counter 2. This gives the following control word:

```
D7 ... D0
1011 0110     = B6H for binary option, mode 3 (square wave), LSB
```
first, then MSB, counter2. The program would be as follows:

```
MOV     AL,0B6H       ;control word
OUT     43H,AL
MOV     AL,33H        ;low byte
OUT     42H,AL
MOV     AL,05         ;high byte
OUT     42H,AL
```

or as IBM BIOS has written:

```
TIMER   EQU    40H

.........      ........
MOV     AL,10110110B   ;SET TIM 2,LSB,MSB,BINARY
OUT     TIMER+3,AL     ;WRITE THE TIMER MODE REG
MOV     AX,533H        ;DIVISOR FOR 1000 HZ (896 HZ)
OUT     TIMER+2,AL     ;WRITE TIMER 2 CNT - LSB
MOV     AL,AH
OUT     TIMER+2,AL     ;WRITE TIMER 2 CNT - MSB
```

Turning on the speaker via PB0 and PB1 of port 61H

The process of turning on the speaker is the same for all IBM PCs and compatibles from 8088-based to 80486 and Intel's Pentium-based systems. As can be seen from Figure 13-4, GATE2 must be high to provide the CLK to timer 2. This function is performed by PB0 of port 61H. Again from Figure 13-4, OUT2 of timer 2 is ANDed with PB1 of port 61H, then is input to the driving circuitry of the speaker. Therefore, to allow OUT2 to go to the speaker, PB1 of port 61H must be set to high as well. The following is the code to turn the speaker on, which is exactly the same as the IBM BIOS's code to sound the BEEP.

```
IN      AL,61H         ;GET THE CURRENT SETTING OF PORT B
MOV     AH,AL          ;SAVE IT
OR      AL,00000011B   ;MAKE PB0=1 AND PB1=1
OUT     61H,AL         ;TURN THE SPEAKER ON.
{HOW LONG THE BEEP SHOULD SOUND GOES HERE}
MOV     AL,AH          ;GET THE ORIGINAL SETTING OF PORT B
OUT     61H,AL         ;TURN OFF THE SPEAKER
```

The amount of time that a musical note is played is referred to as its time delay and is produced with the help of the main 80x86 processor in the PC.

Time delay for 80x86 PCs

CPU-generated time delays are often needed for various applications. Since creating delays is different in 8088/86 PCs versus 80286, 386, 486, and Pentium IBM-compatible PCs, they are described separately.

Creating time delays in 8088/86-based PC/XT, PS/2, and compatibles

In these PCs, the following routine can be used to generate a time delay:

```
             MOV      CX,N
AGAIN:       LOOP     AGAIN
```

Every LOOP instruction in the 8088/86 CPU takes a total of 17 clocks to be executed; therefore, the time delay is approximately $N \times T$ period $\times$ 17. For example, if CX = 28,000 and the system frequency is 4.7 MHz (the T = 210 ns for the original PC/XT), that gives an approximate time delay of 100 ms ($28,000 \times 210$ ns $\times$ 17). The reason that the delay is said to be approximate is that every so often the CPU is frozen to allow the DMA to refresh memory. In other words, the time for the above delay is really little more than 100 ms. In 8088-based IBM PC/XT BIOS, IBM designers used the above method to generate the delay for the BEEP sound as shown next:

```
             SUB      CX,CX
         G7: LOOP     G7
             DEC      BL
             JNZ      G7
```

With CX = 65,536, "LOOP G7" gives a delay of 250 ms (210 ns $\times$ 65,536 $\times$ 17 = 234 ms), taking into consideration the time for refreshing the system memory. BL contains the number of 250-ms delays. If the 8088/86 working frequency is 8 MHz (as in the IBM PS/2 model 25), then T =125 ns (1/8 MHz =125 ns) and the above time delay is much shorter. This means that the delay is not only frequency dependent but also CPU dependent, since in the 80286 the LOOP instruction takes 8 clocks instead of 17 clocks as in the 8088/86. The same instruction takes 11 and 7 clocks in 386 and 486 CPUs, respectively, as shown in Appendix B. This is the reason that starting with the PC AT and all 80286, 80386, 80486, and Intel Pentium computers, IBM provides a scheme to create a time delay using hardware that is not only frequency but also CPU independent.

Time delays in 80x86 IBM PC (for 286 and higher processors)

The following method of creating fixed hardware time delays was first implemented in the IBM PC AT and continued in all 286, 386, 486, and Pentium-based IBM and compatible computers. This makes sense since all these processors have working frequencies of 6 to 66 MHz in addition to the fact that the LOOP instruction timing varies among the processors. To create a processor independent delay, IBM made PB4 of port 61H toggle every 15.085 µs. That means that by monitoring PB4 of port 61H, a fixed time delay can be obtained, as shown next from IBM PC AT BIOS. Upon entering this subroutine called WAITF, register CX must hold the number of 15.085 µs time delays needed.

```
;(CX) = COUNT OF 15.085 MICROSECOND
WAITF        PROC NEAR
             PUSH AX
WAITF1:
             IN       AL,61H
             AND      AL,10H           ;CHECK PB4
             CMP      AL,AH            ;DID IT JUST CHANGE
             JE       WAITF1           ;WAIT FOR CHANGE
             MOV      AH,AL            ;SAVE THE NEW PB4 STATUS
             LOOP     WAITF1           ;CONTINUE UNTIL CX BECOMES 0
             POP      AX
             RET
WAITF        ENDP
```

SECTION 13.2: IBM PC 8253/54 TIMER CONNECTIONS AND PROGRAMMING   **395**

It must be noted that in 286, 386, 486, and Intel Pentium-based IBM and compatible PCs, port B (port 61H) is used both as input and output, in contrast with 8088/86-based IBM computers, in which port B is used only as output.

Now a time delay of any duration can be created regardless of the CPU frequency as long as it is a 286 and higher PC. For example, to create a half-second delay, set CX =33,144 (33,144 × 15.085 µs =1/2 second), and then call the above routine as done by IBM PC AT BIOS:

```
MOV   CX,33144        ;1/2-second delay
CALL  WAITF
```

Example 13-4

Using the BIOS WAITF routine, show how to create a 1.5-second time delay.

Solution:

Since the 1.5-second delay requires the counter to be set to 99,436 (1.5/15.085 µs = 99,436) and the maximum value of CX is 65,536, the following method is used to generate the 1.5-second delay.

```
          MOV   BL,03
BACK:     MOV   CX,33144        ;1/2-second delay
          CALL  WAITF
          DEC   BL
          JNZ   BACK
```

Review Questions

1. What port addresses are assigned to the 8253/54 timer on the PC motherboard?
2. Of the three counters of the 8253/54 timer on the PC motherboard, which one is used for the speaker, and what port address belongs to it?
3. True or false. In the PC, counters 0 and 1 are used for internal system use.
4. True or false. While the user can program counter 2, users cannot program counters 0 and 1 since they are for system use only.
5. True or false. In the PC, while GATE0 and GATE1 are high permanently, GATE2 can be controlled by the user.
6. In the PC, how is GATE2 controlled by the user?
7. What is the approximate time delay generated by the following sequence of instructions if the CPU is:
   (a) 8088 of 5 MHz   (b) 8086 of 8 MHZ
```
            MOV   AL,250
BACK:       SUB   CX,CX
AGAIN:      NOP
            LOOP  AGAIN
            DEC   AL
            JNZ   BACK
```
8. Find the time delay generated by the following code using the method of monitoring PB4 of port 61H in 286, 386, 486, and Pentium PCs and compatibles.
```
            MOV   DL,200
BACK:       MOV   CX,16572        ;delay=16572 x 15.085 microsec
WAIT:       IN    AL,61H
            AND   AL,10H          ;check PB4
            CMP   AL,AH           ;did it just change
            JE    WAIT            ;wait for change
            MOV   AH,AL           ;save the new PB4 status
            LOOP  WAIT            ;continue until CX becomes 0
            DEC   DL
            JNZ   BACK            ;try until DL is 0
```

SECTION 13.3: GENERATING MUSIC ON THE IBM PC

As mentioned earlier, counter 2 is connected to the speaker and it can be programmed to output any frequency that is desired. First, look at the list of piano notes and their frequencies given in Figure 13-5. Since the input frequency to counter 2 is fixed at 1.1931817 MHz for all 80x86-based IBM computers and compatibles, programs for playing music found in this section can run on any of them without modification. To play music, the input frequency of 1.1931817 MHz is divided by the desired output frequency to get the value that must be loaded into counter 2. This is shown in Example 13-5.

Example 13-5

Show the values to be loaded into counter 2 in order to have the output frequency for the notes (a) D3, (b) A3, and (c) A4.

Solution:

From Figure 13-5, notice that the frequency for note D3 is 147. The value that must be loaded into counter 2 is 1.1931 MHz divided by 147, which is 8116. Going through this procedure for each note gives the following:

Note	Frequency	Value Loaded into Counter 2	
		Decimal	Hex
D3	147 Hz	8116	1FB4
A3	220 Hz	5423	152F
A4	440 Hz	2711	0A97

Now that the values to be loaded into counter 2 are known, the program for getting the speaker to sound the notes for a certain duration is shown in Example 13-6.

For a delay of 250 ms in 80286 and higher IBM compatibles, the following routine can be used.

```
;this 250 ms delay can work only on 286,386,486, and Pentium PCs
;
DELAY    PROC     NEAR
         MOV      CX,16578     ;16578 x 15.08 microsec = 250 ms
         PUSH     AX
WAIT:
         IN       AL,61H
         AND      AL,10H       ;check PB4
         CMP      AL,AH        ;did it just change?
         JE       WAIT         ;wait for change
         MOV      AH,AL        ;save the new PB4 status
         LOOP     WAIT         ;decrement CX and continue
                               ;until CX becomes 0
         POP      AX
         RET
DELAY    ENDP
```

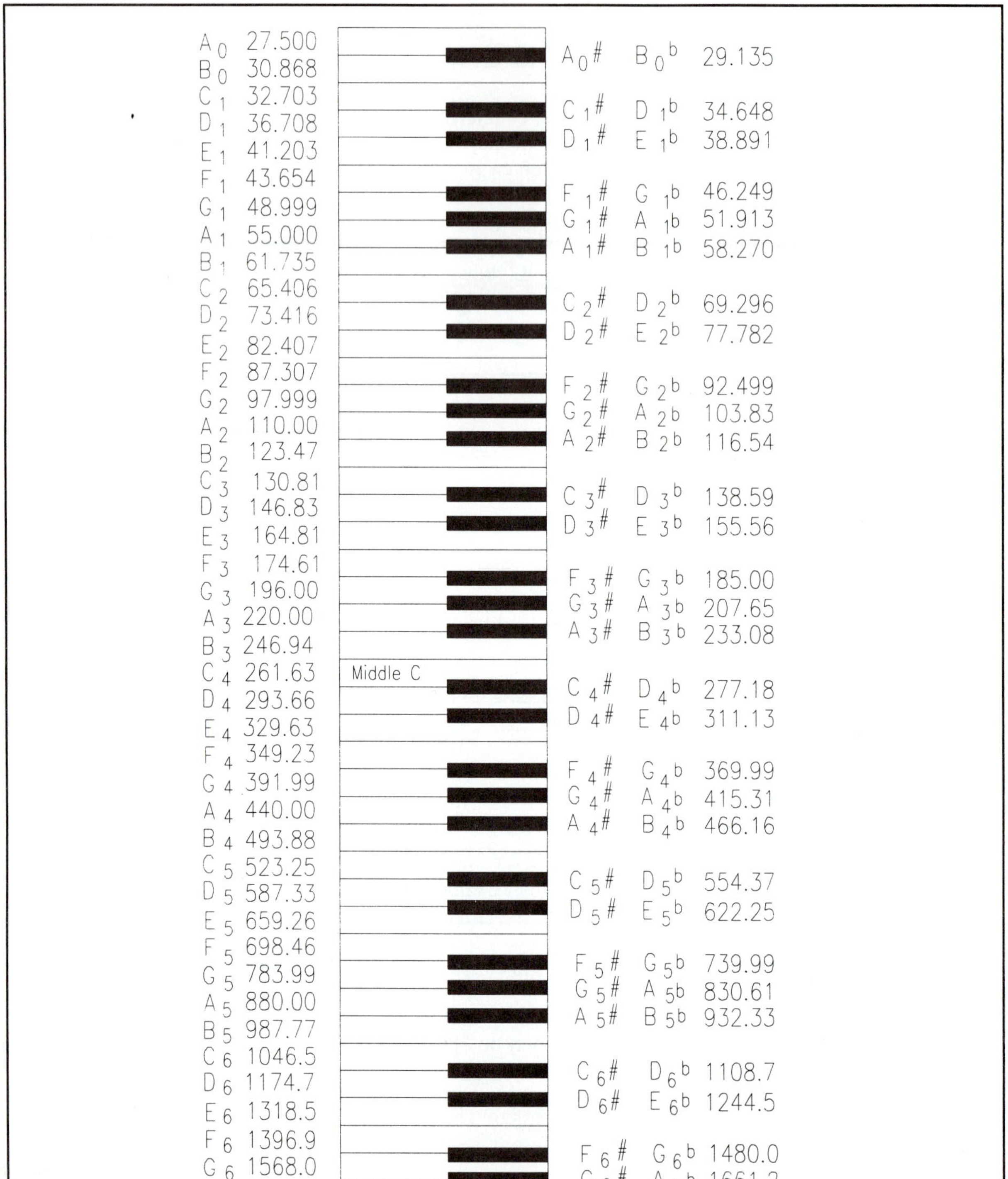

Figure 13-5. Piano Note Frequencies

White keys:

Note	Frequency
A_0	27.500
B_0	30.868
C_1	32.703
D_1	36.708
E_1	41.203
F_1	43.654
G_1	48.999
A_1	55.000
B_1	61.735
C_2	65.406
D_2	73.416
E_2	82.407
F_2	87.307
G_2	97.999
A_2	110.00
B_2	123.47
C_3	130.81
D_3	146.83
E_3	164.81
F_3	174.61
G_3	196.00
A_3	220.00
B_3	246.94
C_4 (Middle C)	261.63
D_4	293.66
E_4	329.63
F_4	349.23
G_4	391.99
A_4	440.00
B_4	493.88
C_5	523.25
D_5	587.33
E_5	659.26
F_5	698.46
G_5	783.99
A_5	880.00
B_5	987.77
C_6	1046.5
D_6	1174.7
E_6	1318.5
F_6	1396.9
G_6	1568.0
A_6	1760.0
B_6	1975.5
C_7	2093.0
D_7	2349.3
E_7	2637.0
F_7	2793.8
G_7	3136.0
A_7	3520.0
B_7	3951.1
C_8	4186.0

Black keys (sharps/flats):

Note	Frequency
$A_0\# \quad B_0^b$	29.135
$C_1\# \quad D_1^b$	34.648
$D_1\# \quad E_1^b$	38.891
$F_1\# \quad G_1^b$	46.249
$G_1\# \quad A_1^b$	51.913
$A_1\# \quad B_1^b$	58.270
$C_2\# \quad D_2^b$	69.296
$D_2\# \quad E_2^b$	77.782
$F_2\# \quad G_2^b$	92.499
$G_2\# \quad A_2^b$	103.83
$A_2\# \quad B_2^b$	116.54
$C_3\# \quad D_3^b$	138.59
$D_3\# \quad E_3^b$	155.56
$F_3\# \quad G_3^b$	185.00
$G_3\# \quad A_3^b$	207.65
$A_3\# \quad B_3^b$	233.08
$C_4\# \quad D_4^b$	277.18
$D_4\# \quad E_4^b$	311.13
$F_4\# \quad G_4^b$	369.99
$G_4\# \quad A_4^b$	415.31
$A_4\# \quad B_4^b$	466.16
$C_5\# \quad D_5^b$	554.37
$D_5\# \quad E_5^b$	622.25
$F_5\# \quad G_5^b$	739.99
$G_5\# \quad A_5^b$	830.61
$A_5\# \quad B_5^b$	932.33
$C_6\# \quad D_6^b$	1108.7
$D_6\# \quad E_6^b$	1244.5
$F_6\# \quad G_6^b$	1480.0
$G_6\# \quad A_6^b$	1661.2
$A_6\# \quad B_6^b$	1864.7
$C_7\# \quad D_7^b$	2217.5
$D_7\# \quad E_7^b$	2489.0
$F_7\# \quad G_7^b$	2960.0
$G_7\# \quad A_7^b$	3322.4
$A_7\# \quad B_7^b$	3729.3

A delay of 5 ms between notes can be achieved in the same way.

```
DELAY_OFF     PROC  NEAR
              MOV   CX,331         ;331 x 15.08 micro sec = 5 ms
              PUSH  AX
WAIT:         IN    AL,61H
              AND   AL,10H         ;check PB4
              CMP   AL,AH          ;did it just change?
              JE    WAIT           ;wait for change
              MOV   AH,AL          ;save the new PB4 status
              LOOP  WAIT           ;continue until CX becomes 0
              POP   AX
              RET
DELAY_OFF     ENDP
```

The following creates a delay for the 8088-based PC/XT of 4.7 MHz.

```
DELAY         PROC  NEAR
              SUB   CX,CX
G7:           LOOP  G7
              RET
ELAY          ENDP
```

Another way to get an approximate DELAY_OFF of 5 ms for the 8088 PC/XT of 4.7-MHz computers is

```
DELAY_OFF     PROC  NEAR
              MOV   CX,1400
G1:           LOOP  G1             ;1400 x 210 ns x 17 =5 ms
              RET
DELAY_OFF     ENDP
```

Playing "Happy Birthday" on the PC

This background should be sufficient to develop a program to play any song. The tune for the song "Happy Birthday" is given below.

Lyrics	Notes	Freq. (Hz)	Duration
hap	C4	262	1/2
py	C4	262	1/2
birth	D4	294	1
day	C4	262	1
to	F4	349	1
you	E4	330	2
hap	C4	262	1/2
py	C4	262	1/2
birth	D4	294	1
day	C4	262	1
to	G4	392	1
you	F4	349	2
hap	C4	262	1/2
py	C4	262	1/2
birth	C5	523	1
day	A4	440	1
dear	F4	349	1
so	E4	330	1
so	D4	294	3
hap	B4b	466	1/2
py	B4b	466	1/2
birth	A4	440	1
day	F4	349	1
to	G4	392	1
you	F4	349	2

Example 13-6

Program counter 2 to play the following notes: D3, A3, A4, for durations of 250, 500, and 500 ms, respectively. Place a 5-ms silence between each note.

Solution: This program uses the values calculated in Example 13-5.

```
                MOV   AL,0B6H              ;control byte:counter2,lsb,msb,binary
                OUT   43H,AL               ;send the control byte to control reg
;load the counter2 value for D3 and play it for 250 ms
                MOV   AX,1FB4H             ;for D3 note
                OUT   42H,AL               ;the low byte
                MOV   AL,AH
                OUT   42H,AL               ;the high byte
;turn the speaker on
                IN    AL,61H               ;get the current setting of port b
                MOV   AH,AL                ;save it
                OR    AL,00000011B         ;make pb0 =1 and pb1 =1
                OUT   61H,AL               ;turn the speaker on
                CALL  DELAY                ;play this note for 250 ms
                MOV   AL,AH                ;get the original setting of port b
                OUT   61H,AL               ;turn off the speaker
                CALL  DELAY_OFF            ;speaker off for this duration
;load the counter2 value for A3 and play it for 500 ms
                MOV   AX,152FH             ;for A3 note
                OUT   42H,AL               ;the low byte
                MOV   AL,AH
                OUT   42H,AL               ;the high byte
;turn the speaker on
                IN    AL,61H               ;get the current setting of port b
                MOV   AH,AL                ;save it
                OR    AL,00000011B         ;make PB0 =1 and PB1 =1
                OUT   61H,AL               ;turn the speaker on
                CALL  DELAY                ;play for 250 ms
                CALL  DELAY                ;play for another 250 ms
                MOV   AL,AH                ;get the original setting of port b
                OUT   61H,AL               ;turn off the speaker
                CALL  DELAY_OFF            ;speaker off for this duration
;load the counter2 value for A4 and play it for 500 ms
                MOV   AX,0A97H             ;for A4 note
                OUT   42H,AL               ;the low byte
                MOV   AL,AH
                OUT   42H,AL               ;the high byte
;turn the speaker on
                IN    AL,61H               ;get the current setting of port b
                MOV   AH,AL                ;save it
                OR    AL,00000011B         ;make PB0 =1 and PB1 =1
                OUT   61H,AL               ;turn the speaker on
                CALL  DELAY                ;play for 250 ms
                MOV   AL,AH                ;get the original setting of port b
                OUT   61H,AL               ;turn off the speaker
                CALL  DELAY_OFF            ;speaker off for this duration
```

In previous examples concerning counter 2, values loaded into that counter were calculated by dividing 1.1931817, the input to CLK2, by the desired OUT2 frequency. The 80x86 can do that calculation as well, by loading 1.1931817 MHz into registers DX:AX and then dividing it by the desired output frequency using the DIV instruction. As a result, the AX register will have the values needed to be loaded into the counter.

Review Questions

1. Find the frequency and the value that must be loaded into the register for counter 2 to play the following notes.
   (a) C4   (b) D3   (c) E4   (d) F4
2. Write pseudocode to program counter 2 to play a note.
3. Of the steps in Question 2, which must the 80x86 be involved in, and why?

SECTION 13.4: SHAPE OF 8253/54 OUTPUTS

This section begins with a discussion of the shape of output pulses of the 8253/54 for all three counters of the IBM PC. Then the various modes of the 8253/54 are discussed.

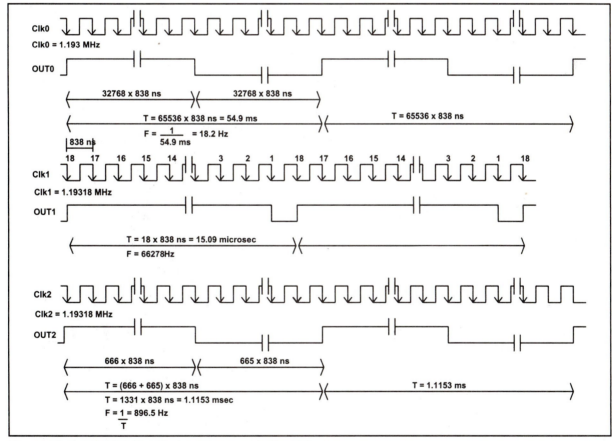

Figure 13-6. 8253/54 Out Timing Diagrams in the PC

OUT0 pulse shape in IBM BIOS

As was seen in Section 13.2, IBM BIOS programmed counter 0 to create mode 3, which is a square-wave shape. Since counter 0 is loaded with the number 65,536 and the clock period of input frequency 838 ns (1/1.193 MHz = 838 ns), the period of the OUT0 pulse is equal to $65{,}536 \times 838$ ns $= 54.9$ ms (18.2 Hz). Now if the number N loaded into the counter is even, both the high pulse and low pulse are the same length ($N/2 \times 838$ ns). If the number N is odd, the high pulse is $(N + 1)/2$

$\times$ 838 ns and the low pulse is $(N-1)/2 \times 838$ ns wide. In other words, if the number loaded into counter is an odd number, the high portion of the square-wave output pulse is slightly wider than the low portion. In the case of counter 0, BIOS loads it with the value 65,536, which is an even number; therefore, the high portion and the low portion of each output pulse is equal to $32,768 \times 838$ ns. Another important note is that OUT0 continuously sends out square-wave pulses. This is due to the fact that when the PC is turned on, BIOS programs load the counter once and let it go. That means that there is no need to reprogram it every time when the count reaches zero since 8253/54 automatically reloads the value 65,536 when the counter counts down to 0. This automatic reloading is done internally without the help of the 80x86 CPU. Note that GATE0 is permanently set to high, making counter 0 enabled permanently.

OUT1 pulse shape in the IBM BIOS

IBM BIOS programmed counter 1 in mode 2, rate generator, with the value 18 loaded into the counter. With a CLK1 period of 838 ns, OUT1 will be high for a total of 17×838 ns and will go low for one pulse of 838 ns, which makes the period T of OUT1 equal to 18×838 ns. At the end of the eighteenth pulse, counter 1 internally, without the help of the 80x86 CPU, reloads the original value of 18 and the process continues as long as the power is on. Note that GATE1 is set high permanently, making counter 1 enabled permanently.

OUT2 pulse shape in the IBM BIOS

Although the mode 3 square wave was discussed under counter 0, there is one difference between the square waves for counter 0 and counter 2. As mentioned earlier, if the number N loaded into the counter is even, both the high pulse and low pulse are the same length ($N/2 \times 838$ ns). However, if the number N is odd, the high pulse is $(N+1)/2 \times 838$ ns and the low pulse is $(N-1)/2 \times 838$ ns. IBM BIOS loads the value 1331 into counter 2. Since 1331 is an odd number, the OUT2 pulse is high for a total of $(1331 + 1)/2 = 666 \times 838$ ns and is low for a total of $(1331 - 1)/2 = 665 \times 838$ ns. This is shown in Figure 13-6. In other words, the duty cycle of the system is slightly more than 50%, but the period, T, is still 1331×838 ns $((666 + 665) \times 838$ ns). Next we discuss all 6 modes of the 8253/54.

8253/54 modes of operation

The 8253/54 has a total of six modes of operation, modes 0 through 5. These six modes of operation are available to each of the three counters inside the 8253/54. Next we describe each mode with examples. It must be noted that these examples are given only to clarify the concepts behind these modes and you must not program your PC's 8253/54 counters for these values, although you can connect an 8253/54 to the PC bus and test these examples. The 6 modes of the 8253/54 fall into two categories as far as activation is concerned. In the first category, after the counter is programmed, OUT will have the desired output only if GATE = 1. In the second category, after the counter is programmed, OUT will have the desired OUT only if a 0-to-1 pulse is applied to the GATE. This second category is called *hardware triggerable*. It is also called *programmable*. Next we describe each mode.

Mode 0: interrupt on terminal count

The output in this mode is initially low, and will remain low for the duration of the count if GATE = 1. The width of the low ouput is as follows.

Width of low pulse = N x T

where N is the clock count loaded into counter, and T is the clock period of the CLK input. When the terminal count is reached, the output will go high and remain high until a new control word or new count number is loaded.

Example 13-7

Assume that GATE1 = 1 and CLK1 = 1 MHz, and the clock count N =1000. Show the output of OUT1 if it is programmed in mode 0.

Solution:

The clock period of CLK1 is 1 μs; therefore, OUT1 is low for 1000 x 1 μs = 1 ms, before it goes high, as shown in the following diagram.

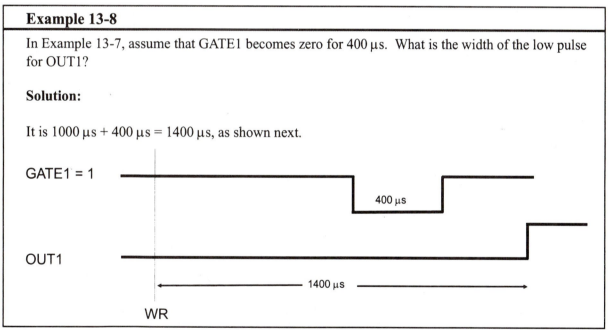

In this mode, if the GATE input becomes low at the middle of the count, the count will stop and the output will be low. The count resumes when the gate becomes high again. This in effect adds to the total time the output is low. The amount added is the time that the GATE input was kept low.

Example 13-8

In Example 13-7, assume that GATE1 becomes zero for 400 μs. What is the width of the low pulse for OUT1?

Solution:

It is 1000 μs + 400 μs = 1400 μs, as shown next.

Mode 1: programmable one-shot

This mode is also called *hardware triggerable one-shot*. The triggering must be done through the GATE input by sending a 0-to-1 pulse to it. In 8253/54 modes that are programmable (triggerable) such as mode 1, the following two steps must be performed for the counter to work.

1. Load the count registers.
2. A 0-to-1 pulse must be sent to the GATE input to trigger the counter.

Contrast this with mode 0, in which the counter produces the output immediately after the counter is loaded as long as GATE = 1. In mode 1 after sending the 0-to-1 pulse to GATE, OUT becomes low and stays low for a duration of $N \times T$, then becomes high and stays high until the GATE is triggered again. This is like any one-shot, in that when it is triggered the output is activated and stays active for a period of time, then returns to the inactive state. In the case of the 8253/54, we can program the period in which the one-shot is active. The width of the low pulse at the ouput is $N \times T$ where T is the clock period of CLK and N is the count number. If, at the middle of the count, GATE is triggered again, the counter is reloaded with N and the count starts all over again. See Example 13-9.

Example 13-9

(a) If CLK1 = 1 MHz and N = 500, show the output for OUT1 if it is programed for mode 1.
(b) Assume that after 150 clock pulses, GATE1 is retriggered. Show the output for OUT1.

Solution:

(a) Notice that OUT1 becomes low only when GATE1 goes from 0 to 1.

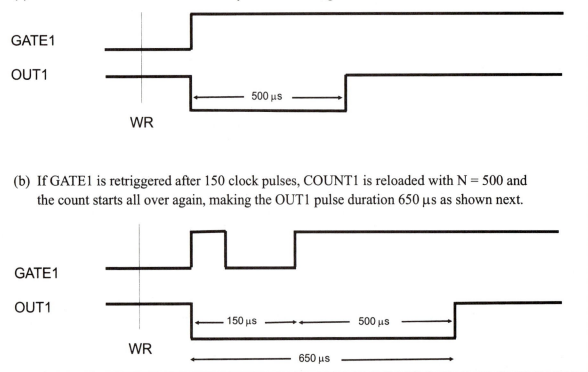

(b) If GATE1 is retriggered after 150 clock pulses, COUNT1 is reloaded with N = 500 and the count starts all over again, making the OUT1 pulse duration 650 μs as shown next.

Notice in this mode that the count starts only when a 0-to-1 pulse is applied to the GATE input. This is the reason it is called programmable or hardware triggerable. This is unlike many other modes where the counter starts upon loading the count. In other words, in the hardware triggerable after loading the count, we must also send a 0-to-1 pulse to the GATE input to trigger the count.

Mode 2: rate generator

Mode 2 is also called *divide-by-N counter*. In this mode, if GATE = 1, OUT will be high for the $N \times T$ clock period, goes low only for one clock pulse, then the count is reloaded automatically, and the process continues indefinitely. This mode in effect produces a divide-by-N counter. In this mode, the period of OUT is equal to $(N + 1) \times T$ where for $N \times T$, OUT is high and for 1 clock pulse, OUT is low. See Example 13-10.

Example 13-10

If CLK2 =1 MHz, GATE2 =1, and N =750, show OUT2 if COUNT2 is programmed for mode 2.

Solution:

Notice that the count is reloaded automatically and the counter countinues to produce OUT2.

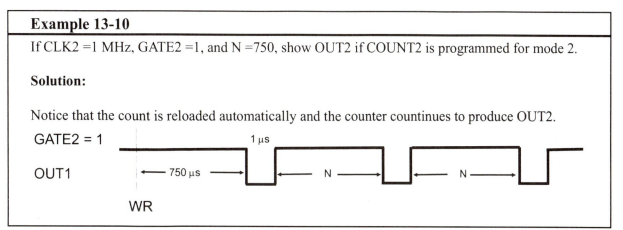

Mode 3: square wave rate generator

In this mode if GATE=1, OUT is a square wave where the high pulse is equal to the low pulse if N is an even number. In that case, the high part and low part of the pulse have the same duration and are equal to $(N/2)$ x T where N is the clock count and T is the CLK period. In this mode, the count is reloaded automatically when the terminal count is reached, thereby producing a continuous square wave with frequency of $1/N$ of the CLK frequency. Mode 3 is similar to mode 2, except that OUT in mode 3 is a square wave of 50% duty cycle. If N is an odd number, the high pulse is one clock pulse longer, as discussed in the case of OUT2 of the PC.

Mode 3 in reality divides the CLK input by N, producing a square wave just like the input, except that the frequency is divided by N. This mode is widely used as a frequency divider and audio-tone generator, as we saw in the case of the IBM PC. In this mode, if the GATE input becomes low, the count down will stop and the

Example 13-11

If CLK2 =1 MHz, GATE1 =1, N =1000, show OUT1 if COUNT1 is programmed for mode 3.

Solution:

Since the clock period is 1 µs, OUT1 is high for 500 µs and low for 500 µs, producing the square wave of 1 ms period continuously, as shown next.

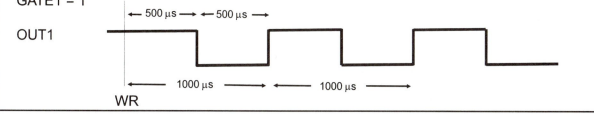

count will resume only after GATE = 1. See Example 13-11.

Mode 4: software triggered strobe

In this mode if GATE = 1, the output will go high upon loading the count. It will stay high for the duration of N x T, where N is the count and T is the clock period. After the count reaches zero (terminal count), it becomes low for one clock pulse, then goes high again and stays high until a new command word or new count is loaded. To repeat the strobe, the count must be reloaded again. In other words, this mode does not automatically reload the count upon reaching the terminal count.

Mode 4 is similar to mode 2, except that the counter is not reloaded automatically. In this mode, if the GATE input becomes low, the count will stop and the output will be high. The count resumes only when the gate becomes high again.

Example 13-12

If CLK0 = 1 MHz, GATE0 = 1, and N = 600, show the shape of OUT0 where counter 0 is programmed for mode 4.

Solution:

Since the CLK0 period is 1 μs, after the count is loaded OUT0 will be high for 600 μs and will go low for 1 μs. Then it will go high again and stay high until the counter is reprogrammed, as shown below.

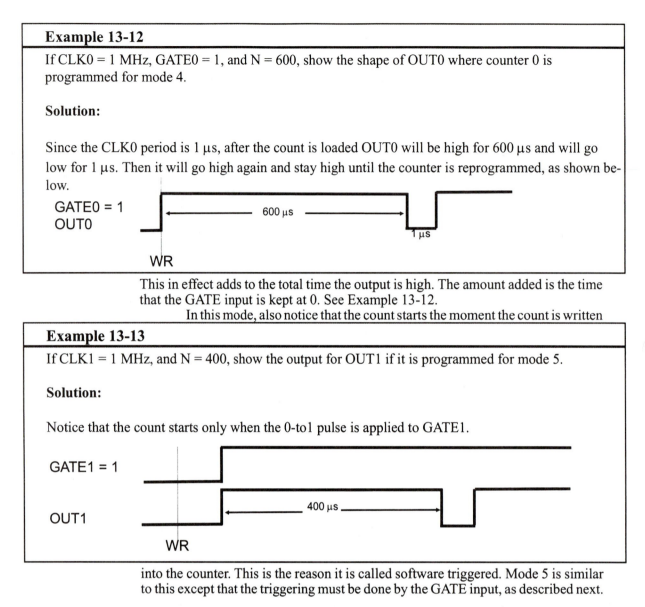

This in effect adds to the total time the output is high. The amount added is the time that the GATE input is kept at 0. See Example 13-12.

In this mode, also notice that the count starts the moment the count is written

Example 13-13

If CLK1 = 1 MHz, and N = 400, show the output for OUT1 if it is programmed for mode 5.

Solution:

Notice that the count starts only when the 0-to1 pulse is applied to GATE1.

into the counter. This is the reason it is called software triggered. Mode 5 is similar to this except that the triggering must be done by the GATE input, as described next.

Mode 5: hardware triggered strobe

This mode is similar to mode 4 except that the triggering must be done with the GATE input. In this mode, the count begins only when a 0-to-1 pulse is sent to the GATE input. This is unlike mode 4 where the counter started upon loading the

Example 13-14

In Example 13-13, assume that GATE1 is retriggered after 150 pulses. Show the output for OUT1.

Solution:

If GATE1 is retriggered after 150 clock pulses into the countdown, COUNT1 is reloaded with N = 400 and the counts begins again, making the OUT1 pulse duration 550 μs, as shown next.

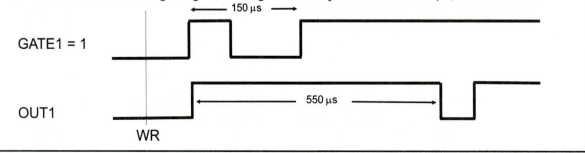

count, as long as GATE = 1. In other words, in this mode after the count is loaded, we must also send a low-to-high pulse to the GATE to start the counter.

Testing the 8253/54 timer of the PC Interface Trainer

In Section 12.7 we discussed the PC Interface Trainer. The Trainer has an 8253/54 timer. Table 13-3 shows its port addresses. To test the 8343/54 timer, we program counter 0 to divide the TTL-output crystal frequency by 5000, as follows:
1. Connect one of the SIP pins of the OSC in H9 to the SIP pin of CLK0 in H8.
2. Connect GATE0 in H8 to VCC on H11
3. Now go to DEBUG and run this program:

```
C:\>DOS\DEBUG
-A100
MOV AL,37
MOV DX,0307
OUT DX,AL
MOV AX,5000
MOV DX,304
OUT DX,AL
MOV AL,AH
OUT DX,AL
INT 3

-G=100 {RETURN}
```

Table 13-3: 8253/54 Port Addresses

Counter	Address
0	304H
1	305H
2	306H
Control register	307H

Using the scope, examine the frequency at OSC and OUT0 pins. The OSC pin has the frequency of the crystal oscillator. Some modules have a 1-MHz and some have a 2-MHz TTL-output crystal oscillator. Check yours to see which is installed. This program divides the OSC frequency (a square wave) by 5000 using the mode 3 (square wave) option of the 8253/54 timer. Use the OSCOPE to verify.

Review Questions

1. In the PC, why is there no need to reprogram the counter 0 register when the counter reaches 0?
2. True or false. In the mode 3 square-wave-shape output of the 8253/54, the duty cycle is not exactly 50% if the divisor is an odd number.
3. Assume that we have connected an 8253/54 to a PC and programmed counter 0 for a divisor of 200 in mode 2 (rate generator). If CLK0 = 3 MHz, find the time that OUT0 is high and low. Assume that GATE0 = 5 V.
4. Answer Question 3 if it is programmed for mode 3 (square wave).

SUMMARY

This chapter began with a look at the pin layout of the 8253 programmable interval timer chip. We showed how to program the three counters of the 8253 by use of the control word. Next we explained how the 8253/54 timer is connected in the IBM PC and gave examples of how it was programmed in BIOS. We showed how to turn on the speaker and produce the beep sound for any desired time interval. Once these concepts were mastered, we showed how to program the 8253/54 to play a song on the IBM PC. Finally, timing considerations were covered in terms of the shape of 8253/54 output waves.

PROBLEMS

SECTION 13.1: 8253/54 TIMER DESCRIPTION AND INITIALIZATION
Note: Problems 1 - 10 are not necessarily IBM PC compatible.

1. True or false. Each of the 8253/54 counters must be programmed independently.
2. CLK of the 8253/54 is an _____ (input, output) _____ (square, sine) wave.
3. Design the decoder for the 8253/54, where A7 - A2 =0010 11 is used to activate CS. Use NAND and inverters only. Give the port address for each port of this design.
4. Which of the following addresses cannot be assigned to counter 0 of the 8253/54, and why? 23H, 54H, 97H, 51H, FCH, 59H
5. Give the highest number by which a single counter of the 8253/54 can divide the input frequency, and what value is loaded into the counter. Give your answer for both binary and BCD options.
6. True or false. If the divisor is larger than 255, we must send the low byte first, then the high byte to the counter.
7. Find the control word to program counter 1 for mode 3, binary count, low byte first, followed by high byte.
8. Write a program for Problem 7 if CLK1 =1.6 MHz and OUT1 =1200 Hz. Use the port addresses in Problem 3.
9. Repeat Problem 8 for OUT1 = 250 Hz.
10. In Problem 8, what would be the OUT1 frequency if it is programmed for the maximum divisor? What if the maximum divisor BCD option were used?

SECTION 13.2: IBM PC 8253/54 TIMER CONNECTIONS, PROGRAMMING

11. State the CLK frequency of all three counters of the 8253/54 in the IBM PC.
12. State the source of CLK in Problem 11.
13. What port addresses are assigned to the 8253/54 in the PC? Can they be changed?
14. State the function of each counter in the 8253/54 of the PC.
15. True or false. A PC user can program counter 2 only, and should not program counters 0 and 1.
16. State the status of the GATE input for each of the counters of 8253/54 in the PC.
17. Why is a time delay based on the microprocessor's instruction clock count not widely used?
18. Find the time delay generated by the following sequence of instructions, assuming that it is for an 8-MHz 8088.

```
            SUB  CX,CX
AGAIN:      NOP
            NOP
            NOP
            LOOP AGAIN
```

19. True or false. A fixed hardware time delay is available only in PCs with 80286, 386, 486, or Pentium microprocessors.
20. Write a program to generate a 10-second delay using a fixed hardware delay.

SECTION 13.3: GENERATING MUSIC ON THE IBM PC

21. To generate the following notes, state the value programmed into the divisor of counter 2 in the PC. A3, G5, B6
22. Write a program to play the song "Mary Had a Little Lamb," shown at the right.

SECTION 13.4: SHAPE OF 8253/54 OUTPUTS

23. In which mode is counter 0 in the PC programmed?
24. State the duty cycle and duration of the high and low parts of OUT0 in the PC.

Lyrics	Note	Freq (Hz)	Length
Mar	E4	330	1
y	D4	294	1
had	C4	262	1
a	D4	294	1
lit	E4	330	1
tle	E4	330	1
lamb	E4	330	2
lit	D4	294	1
tle	D4	294	1
lamb	D4	294	2
lit	E4	330	1
tle	G4	392	1
lamb	G4	392	2
Mar	E4	330	1
y	D4	294	1
had	C4	262	1
a	D4	294	1
lit	E4	330	1
tle	E4	330	1
lamb	E4	330	1
whose	E4	330	1
fleece	D4	294	1
was	D4	294	1
white	E4	330	1
as	D4	294	1
snow.	C4	262	4

"Mary Had a Little Lamb"

25. In which mode is counter 1 in the PC programmed?
26. State the duty cycle and duration of the high and low parts of OUT1 in the PC.
27. In which mode is counter 2 in the PC normally programmed?
28. State the duty cycle and duration of the high and low parts of OUT2 in the IBM PC for each of the following divisors.
    (a) 1200   (b) 1825
29. Assume that CLK1 = 1.5 MHz. Show OUT1 for mode 1 if N = 1200.
30. Assume that CLK2 = 1.8 MHz. Show OUT2 for mode 5 if N = 1450.

ANSWERS TO REVIEW QUESTIONS

SECTION 13.1: 8253/54 TIMER DESCRIPTION AND INITIALIZATION

1. true
2. divide
3. 60H is the base address and 63H is the address for the control register
4. B3H
5. False; it must be sent to the counter 1 register.
6. the port address of 61H
7. D0 =0 since the maximum BCD number is 10,000 but the binary (hex) option goes as high as 65,536
8. 65,536

SECTION 13.2: IBM PC 8253/54 TIMER CONNECTIONS AND PROGRAMMING

1. 40H, 41H, 42H, and 43H
2. counter 2 at port 42H
3. true
4. true
5. true
6. using PB0 of port address 61H
7. The inner loop is 17 +3 =20 clocks since NOP takes 3 T state clocks. Therefore the inner loop is 65,536 x 20 =1,310,720 T states.
          delay = N x (65,536 x 20 x T)
          N = number of outer loop
    (a) delay = 250 x 65,536 x 20 x 200 ns = 65.536 seconds since 1/5 MHz = 200 ns
    (b) delay = 250 x 65,536 x 20 x 125 ns = 40.96 seconds since 1/8 MHz = 125 ns
8. Monitoring of PB4 of port address 61 provides us 16,572 x 15.085 μs = 0.25 sec. 200 x 0.25 sec. = 50 secs.

SECTION 13.3: GENERATING MUSIC ON THE IBM PC

1. Since CLK2 =1.193187 MHz, we must divide this input frequency by the desired OUT2 frequency of each note to get the value to be loaded into counter 2. Therefore, we have:
    (a) 262, 4554   (b) 147, 8116   (c) 330, 3616   (d) 349, 3419
2. The sequence is as follows
    (a) Load the control byte for the 8253/54.
    (b) Load the divisor into port 42H.
    (c) Get the status of port 61 and save it.
    (d) Turn the speaker on by setting high both PB0 and PB1.
    (e) Let the 8253/54 play the note.
    (f) Use the 80x86 to generate a time delay for the duration of the note.
    (g) Turn off the speaker by restoring the original status of port 61.
3. all of them except step (e) since playing of the notes is performed by the 8253/54, independent of the 80x86

SECTION 13.4: SHAPE OF 8253/54 OUTPUTS

1. If GATE is high after the counter counts to zero, the 8253/54 automatically reloads the original divisor and continues.
2. true
3. The total number of pulses is 201. This means that it stays high 200 and goes low for one pulse. Since T =333 ns, we have 200 x 333 ns = 66,600 ns for high pulse and 333 ns for low pulse duration.
4. Since the divisor is an even number we have a square wave of 50% duty with 100 x 333 ns = 33300 ns for each of the high and low parts of the square pulse.

CHAPTER 14

INTERRUPTS AND THE 8259 CHIP

OBJECTIVES

Upon completion of this chapter, you will be able to:

» **Explain how the IBM PC executes interrupts by using the interrupt vector table and interrupt service routines**

» **List the differences between interrupts and CALL instructions**

» **Describe the differences between hardware and software interrupts**

» **Examine the ISR for any interrupt, given its interrupt number**

» **Describe the function of each pin of the 8259 programmable interrupt controller (PIC) chip**

» **Explain the purpose of each of the 4 control words of the 8259 and demonstrate how they are programmed**

» **Diagram how the 8259 is interfaced in the IBM PC/XT**

» **Diagram how the 8259 is interfaced in IBM PC AT machines**

This chapter first discusses the concept of interrupts in the 8088/86 CPU, then in Section 14.2 we look at the interrupt assignment of the IBM PC and MS DOS. The third section examines the 8259 interrupt controller chip in detail, and use of the 8259 chip in the IBM PC/XT is discussed in Section 14.4. In Section 14.5, hardware interrupts and use of the 8259 chip in 80x86-based PC AT computers are discussed. We also discuss sources of hardware interrupts in the PC, followed by a discussion on methods of writing software to take advantage of interrupts.

SECTION 14.1: 8088/86 INTERRUPTS

An interrupt is an external event which informs the CPU that a device needs its service. In the 8086/88 there are a total of 256 interrupts: INT 00, INT 01, ... , INT FF (sometimes called TYPEs). When an interrupt is executed, the microprocessor automatically saves the flag register (FR), the instruction pointer (IP), and the code segment register (CS) on the stack, and goes to a fixed memory location. In 80x86 PCs, the memory location to which an interrupt goes is always four times the value of the interrupt number. For example, INT 03 will go to address 0000CH ($4 \times 3 = 12 = 0CH$). Table 14-1 is a partial list of the *interrupt vector table*.

Table 14-1: Interrupt Vector

INT Number	Physical Address	Logical Address
INT 00	00000	0000:0000
INT 01	00004	0000:0004
INT 02	00008	0000:0008
INT 03	0000C	0000:000C
INT 04	00010	0000:0010
INT 05	00014	0000:0014
...	...	...
INT FF	003FC	0000:03FC

Interrupt service routine (ISR)

For every interrupt there must be a program associated with it. When an interrupt is invoked it is asked to run a program to perform a certain service. This program is commonly referred to as an *interrupt service routine* (ISR). The interrupt service routine is also called the *interrupt handler*. When an interrupt is invoked, the CPU runs the interrupt service routine. Now the question is, where is the address of the interrupt service routine? As can be seen from Table 14-1, for every interrupt there are allocated four bytes of memory in the interrupt vector table. Two bytes are for the IP and the other two are for the CS of the ISR. These four memory locations provide addresses of the interrupt service routine for which the interrupt was invoked. Thus the lowest 1024 bytes ($256 \times 4 = 1024$) of memory space are set aside for the interrupt vector table and must not be used for any other function. Figure 14-1 provides a list of interrupts and their designated functions as defined by Intel Corporation.

Example 14-1

Find the physical and logical addresses in the interrupt vector table associated with:
(a) INT 12H   (b) INT 8

Solution:

(a) The physical addresses for INT 12H are 00048H - 0004BH since ($4 \times 12H = 48H$). That means that the physical memory locations 48H, 49H, 4AH, and 4BH are set aside for the CS and IP of the ISR belonging to INT 12H. The logical address is 0000:0048H - 0000:004BH.

(b) For INT 8, we have $8 \times 4 = 32 = 20H$; therefore, memory addresses 00020H, 00021H, 00022H, and 00023H in the interrupt vector table hold the CS:IP of the INT 8 ISR. The logical address is 0000:0020H - 0000:0023H.

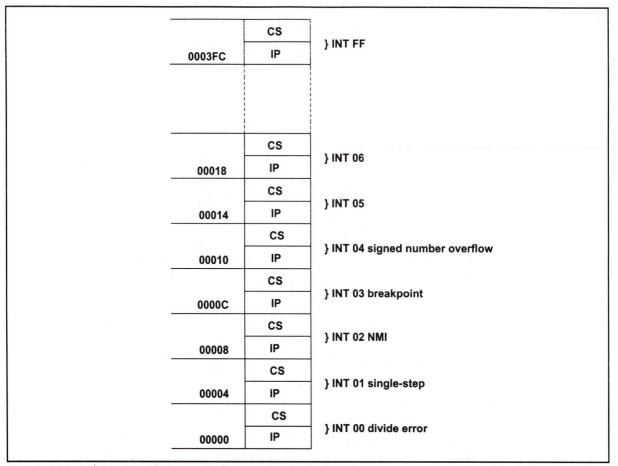

Figure 14-1: Intel's List of Designated Interrupts for the 8086/88
(Reprinted by permission of Intel Corporation, Copyright Intel Corp. 1989)

Difference between INT and CALL instructions

If the INT instruction saves the CS:IP of the following instruction and jumps indirectly to the subroutine associated with the interrupt, what is the difference between that and a CALL FAR instruction, which also saves the CS:IP and jumps to the desired subroutine (procedure)? The differences can be summarized as follows:

1. A "CALL FAR" instruction can jump to any location within the 1 megabyte address range of the 8088/86 CPU, but "INT nn" goes to a fixed memory location in the interrupt vector table to get the address of the interrupt service routine.
2. A "CALL FAR" instruction is used by the programmer in the sequence of instructions in the program but an externally activated hardware interrupt can come in at any time, requesting the attention of the CPU.
3. A "CALL FAR" instruction cannot be masked (disabled), but "INT nn" belonging to externally activated hardware interrupts can be masked. This is discussed in a later section.
4. A "CALL FAR" instruction automatically saves only CS:IP of the next instruction on the stack, while "INT nn" saves FR (flag register) in addition to CS:IP of the next instruction.
5. At the end of the subroutine that has been called by the "CALL FAR" instruction, the RETF (return FAR) is the last instruction, whereas the last instruction in the interrupt service routine (ISR) for "INT nn" is the instruction IRET (interrupt return). The difference is that RETF pops CS, IP off the stack but the IRET pops off the FR (flag register) in addition to CS and IP.

Categories of interrupts

"INT nn" is a 2-byte instruction where the first byte is for the opcode and the second byte is the interrupt number. This means that we can have a maximum of 256 (INT 00 - INT FFH) interrupts. Of these 256 interrupts, some are used for software interrupts and some are for hardware interrupts.

Hardware interrupts

As we saw in Chapters 9 and 10, there are three pins in the 80x86 that are associated with hardware interrupts. They are INTR (interrupt request), NMI (nonmaskable interrupt), and INTA (interrupt acknowledge). The use of INTA will be discussed in Section 14.3. INTR is an input signal into the CPU which can be masked (ignored) and unmasked through the use of instructions CLI and STI. However, NMI, which is also an input signal into the CPU, cannot be masked and unmasked using instructions CLI and STI, and for this reason it is called a nonmaskable interrupt. INTR and NMI are activated externally by putting 5 V on the pins of NMI and INTR of the 80x86 microprocessor. When either of these interrupts is activated, the 80x86 finishes the instruction which it is executing, pushes FR and the CS:IP of the next instruction onto the stack, then jumps to a fixed location in the interrupt vector table and fetches the CS:IP for the interrupt service routine (ISR) associated with that interrupt. At the end of the ISR, the IRET instruction causes the CPU to get (pop) back its original FR and CS:IP from the stack, thereby forcing the CPU to continue at the instruction where it left off when the interrupt came in.

Intel has embedded "INT 02" into the 80x86 microprocessor to be used only for NMI. Whenever the NMI pin is activated, the CPU will go to memory location 00008 to get the address (CS:IP) of the interrupt service routine (ISR) associated with NMI. Memory locations 00008, 00009, 0000A, and 0000B contain the 4 bytes of CS:IP of the ISR belonging to NMI. In contrast, this is not the case for the other hardware pin, INTR. There is no specific location in the vector table assigned to INTR. The reason is that INTR is used to expand the number of hardware interrupts and should be allowed to use any "INT nn" which has not been previously assigned. The 8259 programmable interrupt controller (PIC) chip can be connected to INTR to expand the number of hardware interrupts up to 64. In the case of the IBM PC, one Intel 8259 PIC chip is used to add a total of 8 hardware interrupts to the microprocessor. IBM PC AT, PS/2 80286, 80386, 80486, and Intel Pentium computers use two 8259 chips to allow up to 16 hardware interrupts. The design of hardware interrupts and the use of the 8259 in the IBM PC/XT are covered in Sections 14.3 and 14.4, while 80x86-based AT-type PC interrupts are covered in Section 14.5.

Software interrupts

If an ISR is called upon as result of the execution of an 80x86 instruction such as "INT nn", it is referred to as a *software interrupt* since it was invoked from software, not from external hardware. Examples of such interrupts are DOS "INT 21H" function calls and video interrupts "INT 10H", which were covered in Chapter 4. These interrupts can be invoked in the sequence of code just like a CALL or any other 80x86 instruction. Many of the interrupts in this category are used by the MS DOS operating system and IBM BIOS to perform essential tasks that every computer must provide to the system and the user. Within this group of interrupts there are also some predefined functions associated with some of the interrupts. They are "INT 00" (divide error), "INT 01" (single step), "INT 03" (breakpoint), and "INT 04" (signed number overflow). Each is described below. These interrupts are shown in Figure 14-1. Aside from "INT 00" to "INT 04", which have predefined functions, the rest of the interrupts from "INT 05" to "INT FF" can be used to implement either software or hardware interrupts.

Interrupts and the flag register

Among bits D0 to D15 of the flag register, there are two bits that are associated with the interrupt: D9, or IF (interrupt enable flag), and D8, or TF (trap or single-step flag). In addition, OF (overflow flag) can be used by the interrupt. See Figure 14-2.

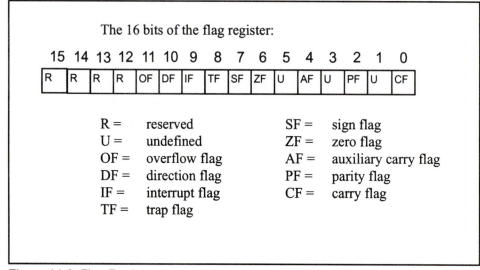

The 16 bits of the flag register:

15	14	13	12	11	10	9	8	7	6	5	4	3	2	1	0
R	R	R	R	OF	DF	IF	TF	SF	ZF	U	AF	U	PF	U	CF

R =   reserved   SF =   sign flag
U =   undefined   ZF =   zero flag
OF =   overflow flag   AF =   auxiliary carry flag
DF =   direction flag   PF =   parity flag
IF =   interrupt flag   CF =   carry flag
TF =   trap flag

Figure 14-2. Flag Register
(Reprinted by permission of Intel Corporation, Copyright Intel Corp. 1989)

The interrupt flag is used to mask (ignore) any hardware interrupt that may come in from the INTR pin. If IF = 0, all hardware interrupt requests through INTR are ignored. This has no effect on interrupts coming from the NMI pin or "INT nn" instructions. The instruction CLI (clear interrupt flag) will make IF = 0. To allow interrupt requests through the INTR pin, this flag must be set to one (IF = 1). The STI (set interrupt flag) instruction can be used to set IF to 1. Section 14.3 will show how to use STI and CLI to mask or allow interrupts through the INTR pin. The trap flag (TF) is explained below when "INT 01", the single-step interrupt, is discussed.

Processing interrupts

When the 8088/86 processes any interrupt (software or hardware), it goes through the following steps:

1. The flag register (FR) is pushed onto the stack and SP is decremented by two, since FR is a 2-byte register.
2. IF (interrupt enable flag) and TF (trap flag) are both cleared (IF = 0 and TF = 0). This masks (causes the system to ignore) interrupt requests from the INTR pin and disables single-stepping while the CPU is executing the interrupt service routine. Depending on the nature of the interrupt procedure, a programmer can unmask the INTR pin by the STI instruction.
3. The current CS is pushed onto the stack and SP is decremented by 2.
4. The current IP is pushed onto the stack and SP is decremented by 2.
5. The INT number (type) is multiplied by 4 to get the physical address of the location within the vector table to fetch the CS and IP of the interrupt service routine.
6. From the new CS:IP, the CPU starts to fetch and execute instructions belonging to the ISR program.
7. The last instruction of the interrupt service routine must be IRET, to get IP, CS, and FR back from the stack and make the CPU run the code where it left off.

Functions associated with INT 00 to INT 04

As mentioned earlier, interrupts INT 00 to INT 04 have predefined tasks (functions) and cannot be used in any other way. The function of each is described next.

INT 00 (divide error)

This interrupt belongs to the category of interrupts referred to as *conditional* or *exception interrupts*. Internally, they are invoked by the microprocessor whenever there are conditions (exceptions) that the CPU is unable to handle. One such situation is an attempt to divide a number by zero. Since the result of dividing a number by zero is undefined, and the CPU has no way of handling such a result, it automatically invokes the divide error exception interrupt. In the 8088/86 microprocessor, out of 256 interrupts, Intel has set aside only INT 0 for the exception interrupt. There are many more exception-handling interrupts in 80286, 80386, 80486, and Pentium CPUs, which are discussed in Section 14.5. INT 00 is invoked by the microprocessor whenever there is an attempt to divide a number by zero. In the IBM PC and compatibles, the service subroutine for this interrupt is responsible for displaying the message "DIVIDE ERROR" on the screen if a program such as the following is executed:

```
MOV     AL,92          ;AL=92
SUB     CL,CL          ;CL=0
DIV     CL             ;92/0=undefined result
```

INT 0 is also invoked if the quotient is too large to fit into the assigned register when executing a DIV instruction. Look at the following case:

```
MOV     AX,0FFFFH      ;AX=FFFFH
MOV     BL,2           ;BL=2
DIV     BL             ;65535/2 =32767 larger than 255
                       ;maximum capacity of AL
```

Put INT 3 at the end of the above two programs in DEBUG and see the reaction of the PC. For further discussion of divide error interrupts due to an oversized quotient, see Chapter 3.

INT 01 (single step)

In executing a sequence of instructions, there is often a need to examine the contents of the CPU's registers and system memory. This is often done by executing the program one instruction at a time and then inspecting registers and memory. This is commonly referred to as *single-stepping*, or performing a *trace*. Intel has designated INT 01 specifically for implementation of single-stepping. To single-step, the trap flag (TF), D8 of the flag register, must be set to 1. Then after execution of each instruction, the 8086/88 automatically jumps to physical location 00004 to fetch the 4 bytes for CS:IP of the interrupt service routine, whose job is, among other things, to dump the registers onto the screen. Now the question is, how is the trap flag set or reset? Although Intel has not provided any specific instruction for this purpose (unlike IF, which uses STI and CLI instructions to set or reset), one can write a simple program to do that. The following shows two methods of making TF= 0. The first is:

```
PUSHF
POP     AX
AND     AX,1111111011111111B
PUSH    AX
POPF
```

---

and the second method is:

```
PUSHF
MOV      BP,SP
AND      [BP] + 0,1111111011111111B
POPF
```

Recall that TF is D8 of the flag register. The analysis of the above two programs is left to the reader. To make TF =1, one simply uses the OR instruction in place of the AND instruction above.

INT 02 (nonmaskable interrupt)

All Intel 80x86 microprocessors have a pin designated NMI. It is an active-high input. Intel has set aside INT 2 for the NMI interrupt. Whenever the NMI pin of the 80x86 is activated by a high (5 V) signal, the CPU jumps to physical memory location 00008 to fetch the CS:IP of the interrupt service routine associated with NMI. Section 14.4 contains a detailed discussion of its purpose and application.

INT 03 (breakpoint)

To allow implementation of breakpoints in software engineering, Intel has set aside INT 03 solely for that purpose. Whereas in single-step mode, one can inspect the CPU and system memory after the execution of each instruction, a breakpoint is used to examine the CPU and memory after the execution of a group of instructions. In recent years, some very powerful software debuggers have been written using INT 1 and INT 3. Even in writing simple programs on the IBM PC, the use of single-step (trace) and breakpoints (INT 3) is indispensable. One interesting point about INT 3 is the fact that it is a 1-byte instruction. This is in contrast to all other interrupt instructions of the form "INT nn", which are 2-byte instructions.

INT 04 (signed number overflow)

This interrupt is invoked by a signed number overflow condition. There is an instruction associated with this, INTO (interrupt on overflow). For a detailed discussion of signed number overflow, see Chapter 6. If the instruction INTO is placed after a signed number arithmetic or logic operation such as IMUL or ADD, the CPU will activate INT 04 if OF = 1. In cases where OF = 0, the INTO instruction is not executed but is bypassed and acts as a NOP (no operation) instruction. To understand that, look at the following example.

```
MOV      AL,DATA1
MOV      BL,DATA2
ADD      AL,BL          ;add BL to AL
INTO
```

Suppose in the above program that DATA1 =+64 = 0100 0000 and DATA2 = +64 = 0100 0000. The INTO instruction will be executed and the 8086/88 will jump to physical location 00010H, the memory location associated with INT 04. The carry from D6 to D7 causes the overflow flag to become 1.

	+ 64	0100 0000	
+	+ 64	0100 0000	
	+128	1000 0000	OF=1 and the result is not +128

The above incorrect result causes OF to be set to 1. INTO causes the CPU to perform "INT 4" and jump to physical location 00010H of the vector table to get the CS:IP of the service routine. Suppose that the data in the above program was DATA1 =+64 and DATA2 =+17. In that case, OF would become 0; the INTO is not executed and acts simply as a NOP (no operation) instruction.

Review Questions

1. True or false. When any interrupt (software or hardware) is activated, the CPU jumps to a fixed and unique address.
2. There are _____ bytes of memory in the interrupt vector table for each "INT nn".
3. How many K bytes of memory are used by the interrupt vector table, and what are the beginning and ending addresses of the table?
4. The program associated with an interrupt is referred to as _____.
5. What is the function of the interrupt vector table?
6. What physical memory locations in the interrupt vector table hold the CS:IP of INT 10H?
7. The 8086/88 has assigned INT 2 to NMI. Can that be changed?
8. Which interrupt is assigned to divide error exception handling?

SECTION 14.2: IBM PC AND MS DOS ASSIGNMENT OF INTERRUPTS

Of the 256 possible interrupts in the 80x86, some are used by the PC peripheral hardware (BIOS), some are used by the Microsoft Disk Operating System, and the rest are available for programmers of software applications. Table 14-2 lists many of the PC/XT interrupts, the logical and physical addresses of their service subroutines, and their purpose. Some interrupts are explained throughout this book. It must be mentioned that depending on the computer and the DOS version, some logical addresses could be different from those shown in Table 14-2.

Examining the interrupt vector table of your PC

Example 14-3 shows how to use DEBUG's dump command to examine the interrupt vector table of a PC, regardless of which 80x86 CPU it contains.

Example 14-2

For a given ISR, the logical address is F000:FF53. Verify that the physical address is FFF53H.

Solution:
Since the logical address is F000:FF53, this means that CS = F000H and IP = FF53H. Shifting left the segment register one hex digit and adding it to the offset gives the physical address FFF53H.

Example 14-3

(a) Use the DEBUG dump command to dump the contents of memory locations 00000 - 0002FH.
(b) Find the CS:IP of divide error, NMI, and INT 8.

Solution:
(a) It is very possible that the data you get on your PC will be different from the following dump, depending on the DOS version, the BIOS chip date of your PC, and activation of shadow memory.

```
C>debug
-D 0000:0000-002F
0000:0000 E8 56 2B 02 56 07 70 00-C3 E2 00 F0 56 07 70 00 .V+.V.p.....V.p.
0000:0010 56 07 70 00 54 FF 00 F0-47 FF 00 F0 47 FF 00 F0 V.p.T...G...G...
0000:0020 A5 FE 00 F0 87 E9 00 F0-DD E6 00 F0 DD E6 00 F0 ................
```

(b) For the divide error interrupt (INT 0), CS:IP is located at addresses 0, 1, 2, 3. Remember that because of the little endian convention, the low address has the low value; therefore, IP =56E8H and CS =022BH. By the same token, NMI's INT 2 is located at the vector table addresses of 8, 9, A, and B. Therefore, we have IP =E2C3H and CS =F000H. The CS:IP of INT 8 ISR is located in addresses starting at 00020 since 8 × 4 =32 =20H, IP = FEA5, CS = F000.

Table 14-2: IBM PC/XT Interrupt System

Interrupt	Logical Addr.	Physical Addr.	Purpose
0	00E3:3072	03EA2	Divide error
1	0600:08ED	068ED	Single-step (trace command in DEBUG)
2	F000:E2C3	FE2C3	Nonmaskable interrupt
3	0600:08E6	068E6	Breakpoint
4	0700:0147	07147	Signed number arithmetic overflow
5	F000:FF54	FFF54	Print screen (BIOS)
6, 7			reserved
8	F000:FEA5	FFEA5	IRQ0 of 8259 (BIOS timer interrupt)
9	F000:E987	FE987	IRQ1 of 8259 (BIOS keyboard interrupt)
A			IRQ2 of 8259 (reserved)
B			IRQ3 of 8259 (reserved for serial com2)
C			IRQ4 of 8259 (reserved for serial com1)
D			IRQ5 of 8259 (reserved for hard disk XT)
E	F000:EF57	FEF57	IRQ6 of 8259 (floppy diskette)
F	0070:0147	00847	IRQ7 of 8259 (parallel printer LPT1)
10	F000:F065	FF065	Video I/O (BIOS)
11	F000:F84D	FF84D	Equipment configuration check (BIOS)
12	F000:F841	FF841	Memory size check (BIOS)
13	F000:EC59	FEC59	Disk I/O (BIOS)
14	F000:E739	FE739	RS-232 I/O (BIOS)
15	F000:F859	FF859	Cassette I/O (BIOS)
16	F000:E82E	FE82E	Keyboard I/O (BIOS)
17	F000:EFD2	FEFD2	Parallel printer I/O (BIOS)
18	F600:0000	F6000	Load ROM BASIC
19	F000:E6F2	FE6F2	Load boot-strap (BIOS)
1A	F000:FE6E	FFE6E	Time-of-day (BIOS)
1B	0070:0140	00840	Ctrl-Brk control (BIOS)
1C	F000:FF53	FFF53	Timer control
1D	F000:F0A4	FF0A4	Video parameters table
1E	0000:0522	00522	Floppy disk parameters table
1F	00E3:0B07	01937	Graphics character table (DOS 3.0 and up)
20	PSP:0000	-----	DOS program terminate
21	Relocatable	-----	DOS function calls
22	PSP:000A	-----	DOS terminate address
23	PSP:000E	-----	DOS Ctrl-Brk exit address
24	PSP:0012	-----	DOS critical error-handling vector
25	Relocatable	-----	DOS absolute disk read
26	Relocatable	-----	DOS absolute disk write
27	Relocatable	-----	DOS terminate but stay resident (TSR)
28 - 2E		-----	Reserved for DOS
2F	Relocatable	-----	Multiplex interrupt
30 - 3F		-----	Reserved for DOS
40		FEC59	Diskette I/O (XT)
41		FE401	Fixed (hard) disk parameters (XT)
42 - 5F		-----	Reserved for DOS
60 - 66		-----	User defined
67		-----	Expanded memory manager
68 - 7F		-----	Not used
80 - 85		-----	Reserved for BASIC
86 - F0		-----	BASIC interpreter
F1 - FF		-----	Not used

(Reprinted by permission from "IBM BIOS Technical Reference" c. 1987 by International Business Machines Corporation)

CHAPTER 14: INTERRUPTS AND THE 8259 CHIP

From the CS:IP address of the ISR, it is possible to determine who provides the service: DOS or BIOS. This is shown in Example 14-4.

Example 14-4

Examine the answers for Example 14-3(b) to determine whether DOS or BIOS provides the ISR for the divide error and NMI.

Solution:

In the case of INT 0 (divide error), the logical address for CS:IP is 022B:56E8. This results in a physical address of 07998H for the divide error interrupt service routine. This area of memory belongs to MS DOS, as discussed in Chapter 11. For NMI interrupt 2, we have a logical address of CS:IP =F000:E2C3, which corresponds to physical address FE2C3H. This is the BIOS ROM area.

Analyzing an IBM BIOS interrupt service routine

To understand the structure of an ISR, we examine the interrupt service routine of INT 12H from IBM BIOS. The interrupt 12H service is available on any PC with an 80x86 microprocessor.

INT 12H: checking the size of RAM on the IBM PC

IBM PC BIOS uses INT 12H to provide the amount of installed conventional (0 to 640K bytes) RAM memory on the system. By *system* is meant both the motherboard and expansion boards. One of the functions of the BIOS POST (power-on self-test) is to test and count the total K bytes of conventional RAM memory installed on the system and write it in memory locations 00413H and 00414H, which have been set aside for this purpose in the BIOS data area. In Chapter 11, we showed the data area used by BIOS. The job of INT 12H is to copy that value from memory locations 00413H and 00414H into AX and return. In other words, after executing INT 12H, AX will contain the total K bytes of conventional RAM memory on the system. This value is in hex and must be converted to decimal to get values of 1 to 640K bytes. The interrupt service routine for INT 12H looks as follows in the IBM PC Technical Reference.

```
MEM_SIZE     PROC  FAR
             STI                    ;interrupt back on
             PUSH  DS               ;save segment
             SUB   AX,AX            ;set DS = 0
             MOV   DS,AX            ;for BIOS data area
             MOV   AX,[0413]        ;conv mem size in 413,414
             POP   DS               ;recover segment
             IRET                   ;return to caller
MEM_SIZE     ENDP
```

As mentioned in Chapter 11, addresses 400H to 4FFH are used by the IBM BIOS data area, and addresses from 500H to 5FFH are set aside for DOS and the BASIC language parameters. When IBM developed the PS/2 computers and needed more data area for BIOS, they used a few K bytes from the top of installed RAM in the 640K conventional memory area. This is the reason that in the IBM PS and compatibles, the total RAM for the 640 conventional memory given in AX is a few (1 to 2) K bytes less than the actual installed RAM size since these few K bytes are used by the extended BIOS data area.

It must be noted that although BIOS finds the size of conventional RAM memory in the system when the computer is turned on, INT 12H can be run at any time to retrieve that information. This is shown in Example 14-5.

Example 14-5

Execute INT 12H followed by INT 3 (breakpoint) in DEBUG. Verify the memory size.

Solution:

```
C>DEBUG
-a
1131:0100 INT 12
1131:0102 INT 3
1131:0103
-G
AX=0280 BX=0000 CX=0000 DX=0000 SP=FFEE BP=0000 SI=0000 DI=0000
DS=1132 ES=1132 SS=1132 CS=1132 IP=0102 NV UP EI PL NZ NA PO NC
1131:0102 CC            INT    3
-Q
```

AX=0280 which is in hex format. Converting it to decimal gives a size of 640K bytes of memory installed on this computer.

Example 14-6

Use the DEBUG D (dump) command to dump memory locations 0040:0000 to 0040:001FH and inspect the contents of locations 0040:13 and 0040:14. Does this match the result of Example 14-5?

Solution:

```
C>DEBUG
-D 0040:0000 1F
0040:0000 F8 03 F8 02 00 00 00 00-78 03 78 02 00 00 00 00  ........x.x.....
0040:0010 6F 94 00 80 02 40 02 00-00 00 22 00 22 00 66 21  o...@....".".f!
-Q
```

Location 13 contains 80 and 14 contains 02; 0280 matches the result of Example 14-5.

Review Questions

1. Find the logical address, CS, and IP values for the ISR if it is located at the ROM physical address of FF065H.
2. In Question 1, if the ISR belongs to INT 10H, find the exact contents of memory locations in the interrupt vector table.
3. Assume that after using the dump command of DEBUG to dump a section of the interrupt vector table, we have the following:
   0000:0030 ....-57 EF 00 FF 00 00 00 00
   To what interrupt number does the above dump belong?
4. For Question 3, show the logical address and exact values for CS, IP of the ISR.
5. In a given IBM PS/2 model 25, the documentation states that the motherboard comes with 512K bytes of RAM, but running the INT 12H shows AX = 01FE. Explain the difference.
6. Which one is the last instruction in the ISR of INT 12H, RET or IRET?

SECTION 14.3: 8259 PROGRAMMABLE INTERRUPT CONTROLLER

The 80x86 has only pins INTR and INTA for interrupts, but one can use these two pins to expand the number of interrupts. Intel Corporation has provided an IC chip called the 8259 programmable interrupt controller (PIC) to make the job of expanding the number of hardware interrupts much easier. See Figures 14-3 and 14-4. This section covers the 8259 IC chip pins and programming options. Note that this section is about the 8259 chip. The ports and programs covered in this section are not related to the PC.

CAS0 - CAS2

CAS0, CAS1, and CAS2 can be used to set up several 8259 chips to expand the number of hardware interrupts of the 8088/86 up to 64 by cascading the 8259 chips in a master-slave configuration. This section will focus on slave mode. Section 14.5 discusses both the master and slave configurations as used in PC/AT-type computers. To use the 8259 in slave mode, the chip must be programmed and CAS0 to CAS2 are ignored.

SP/EN

SP/EN (slave programming/enable) in buffered mode is an output signal from the 8259 to activate the transceiver (EN). In nonbuffered mode it is an input signal into the 8259, SP=1 for the master and SP=0 for the slave.

INT

INT is an output that is connected to INTR of the 80x86.

INTA

INTA is input to the 8259 from INTA of the 80x86.

IR0 - IR7

Inputs IR0 to IR7 (interrupt request) are used as hardware interrupts. When a HIGH is put on any of IR0 to IR7, the 8088/86 will jump to a vector location. For each IR there exists a physical memory location in the interrupt vector table. The 80x86 has 256 hardware or software interrupts (INT 00 - INT FF).

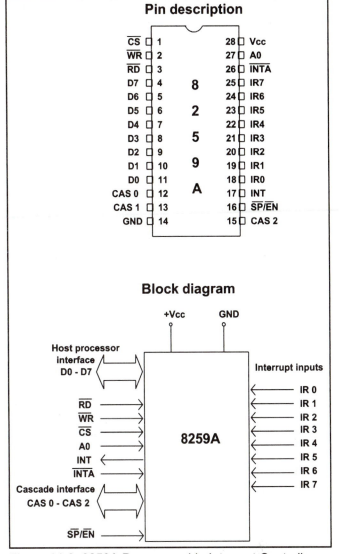

Figure 14-3. 8259A Programmable Interrupt Controller
(Reprinted by permission of Intel Corporation, Copyright Intel Corp. 1983)

8259 control words and ports

The four control words associated with the 8259 are ICW1 (initialization command word), ICW2, ICW3, and ICW4. ICW3 is used in master mode only and is discussed in Section 14.5. As can be seen from the pins of the 8259, there is only one address line A0 to communicate with the chip. Table 14-3 shows the values that A0 and CS must take to initialize the 8259.

Table 14-3: 8259 Initialization

CS	A0	Initialization
0	0	ICW1
0	1	ICW2, ICW3, ICW4
1	x	8259 is not addressed

(Reprinted by permission of Intel Corporation, Copyright Intel Corp. 1983)

ICW1 (initialization command word 1)

In looking at Table 14-3, the question might arise: How can the 8259 make a distinction between ICW2, ICW3, and ICW4 when they are sent to the same address? This is one of the functions of ICW1. D0, the LSB of ICW1, will tell the 8259 if it should look for ICW4 or not. In a similar manner, if D1 is high it knows that the system is configured in slave mode and it should not expect any ICW3 in the initialization sequence. The initialization sequence must always start with ICW1, followed by ICW2, and finally the last one, if needed. There is no jumping ahead. D2 is always set low (= 0) for the 80x86. D3 chooses between level triggering or edge triggering of the input signals IR0 - IR7. In edge triggering, a low-to-high input is recognized as an interrupt request. In level triggering, a high on the IR is recognized as an interrupt request. D4 must always be high. D5, D6, and D7 are all low for the 80x86 microprocessors (they are used only for the 8080/85). See Figure 14-5.

ICW2 (initialization command word 2)

It is the function of ICW2 to assign interrupt numbers to IR0 - IR7. While the lower three bits, D0, D1, and D2, vary from 000 to 111, they, along with D3 - D7 (T3 through T7), form the 8-bit INT type number assigned to the corresponding IR0 through IR7. That means that D3 - D7 can only be programmed according to the assignment of the INT type, with the lower bits being provided by the 8259, depending on which of IR0 to IR7 is activated. See Figure 14-5.

ICW3 (initialization command word 3)

ICW3 is used only when 2 or more 8259s are cascaded. In this mode, a single 8259 can be connected to 8 slave 8259s, thereby providing up to 64 hardware interrupts. In cascade mode, there are separate ICW3 words for the master and the slave. For the master, it indicates which IR has a slave connected to it, and a separate ICW3 informs the slave which IR of the master it is connected to. See Figure 14-6.

ICW4 (initialization command word 4)

D0 indicates the processor mode (PM), the choice of microprocessor. D0 equals 1 for the 8088/86 and 0 for the 8080/8085. When D1, which is AEOI (automatic end of interrupt), is high it eliminates the need for an EOI instruction to be present before the IRET (interrupt return) instruction in the interrupt service routine. When D1 is zero, the EOI must be issued using the OCW (operation command word) to the 8259. In other words, if D1 =0, the last three instructions of the interrupt service routine for IR0 - IR7 must be issuing the EOI followed by IRET. The significance of this is discussed shortly when the OCW is discussed. D2 and D3 are for systems where data buses are buffered with the use of bidirectional transceivers. The 8259 can work either in buffered or nonbuffered mode. D4 is for SFNM (special fully nested mode). This mode must be used when the 8259 is in master mode and then D4 =1; otherwise, it is 0. D5 - D7 must be zero, as required by the 8259. See Figure 14-6.

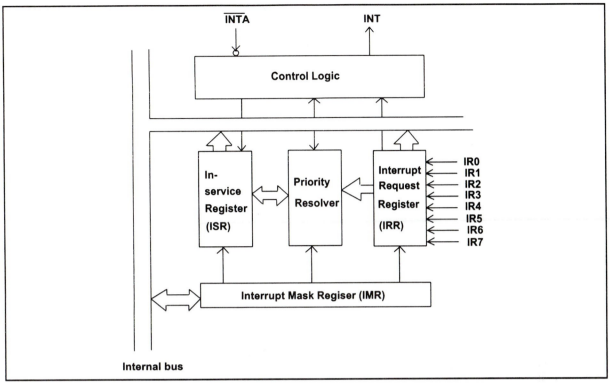

Figure 14-4. Partial Block Diagram of the 8259A
(Reprinted by permission of Intel Corporation, Copyright Intel Corp. 1983)

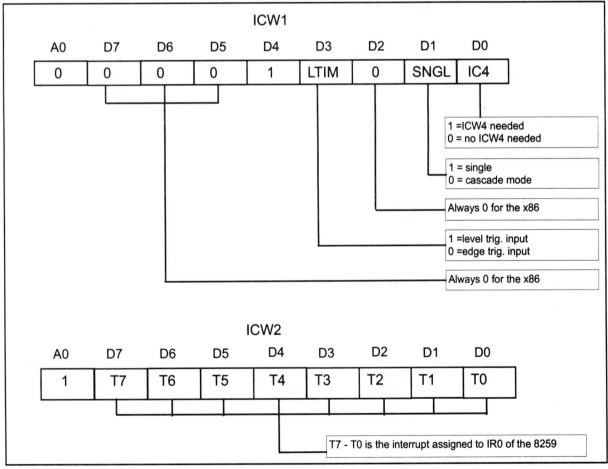

Figure 14-5. ICW Formats (ICW1 and ICW2) for the 8259
(Reprinted by permission of Intel Corporation, Copyright Intel Corp. 1983)

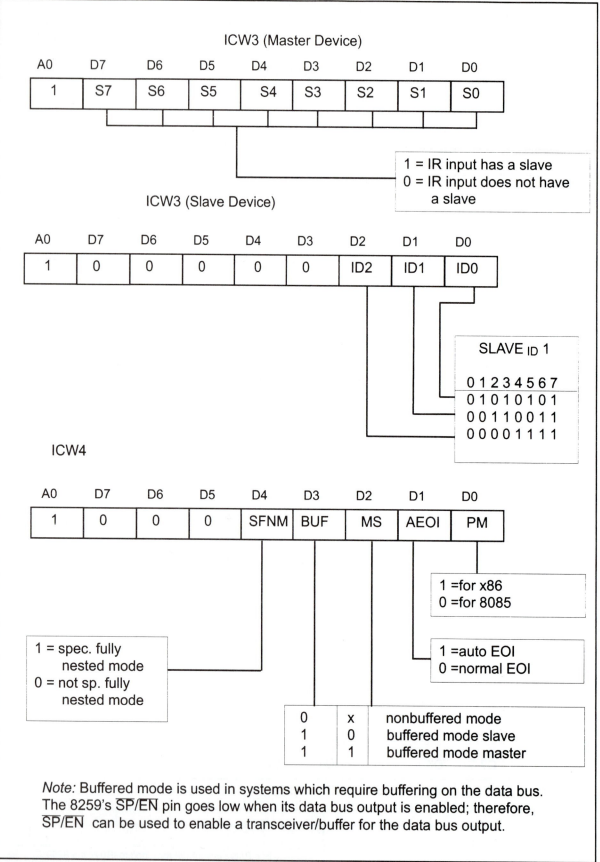

Figure 14-6. ICW Formats (ICW3 and ICW4) for the 8259
(Reprinted by permission of Intel Corporation, Copyright Intel Corp. 1983)

Example 14-8

(a) Find the ICWs of the 8259 if it is used with an 8088/86 CPU, single, level triggering IRs, and IR0 is assigned "INT 50H". The 8259 is in slave buffered mode with normal EOI.

(b) Show the program to initialize the 8259 using the port addresses in Example 14-7.

(c) Find the addresses associated with IR0, IR1, and IR2 in the interrupt vector table.
Note: This example is not PC-compatible and is given only for an exercise.

Solution:

(a) From Figure 14-5, we get the following for each of the ICWs:

ICW1

D0 =1	ICW4 needed
D1 =1	single
D2 =0	this is always zero for 80x86 CPUs
D3 =1	level triggering
D4 =1	required by the ICW1 itself
D5 =D6 =D7 =0	this is always zero for 80x86 CPUs

This gives ICW1 = 00011011 = 1BH. To get ICW2, look at Table 14-4. Always equate ICW2 to the INT # assigned to IR0: ICW2 = 01010000 = 50H. Notice that "INT nn", assigned to IR0, can decide only bits D7 - D3 (T7 - T3 in Figure 14-5) of ICW2. This means that the "INT nn" assigned to IR0 must have the lower three bits = 000; therefore, it can take either values of X0H or X8H, where X is a hex number. For example "INT 45H" cannot be assigned to IR0.

No ICW3 is needed since it is single and not cascaded.

ICW4

D0 =1	8088/86
D1 =0	normal (we must issue EOI before IRET instruction)
D2 =0;D3=1	slave buffered mode
D4 =0	not nested
D5 =D6 =D7 =0	required by the ICW4

We get ICW4 =00001001 =09H.

(b) The program is as follows:

```
MOV   AL,1BH      ;ICW1
OUT   26H,AL      ;TO PORT 26H
MOV   AL,50H      ;ICW2
OUT   27H,AL      ;TO PORT 27H
MOV   AL,09       ;ICW4
OUT   27H,AL      ;TO PORT 27H
```

(c) If "INT 50H" is assigned to IR0, then IR1 and IR2 have "INT 51H" and "INT 52", respectively, and so on. The vector memory locations associated with the IRs are as follows:

		Vector Location	
IRQ (Pin of 8259)	INT	Logical Address	Physical Address
IR0	50H	0000:0140H-0143	00140H-00143
IR1	51H	0000:0144H-0147	00144H-00147
IR2	52H	0000:0148H-014B	00148H-0014B

SECTION 14.3: 8259 PROGRAMMABLE INTERRUPT CONTROLLER   425

Table 14-4: INT Numbers for Hardware Interrupts in Example 14-8

Binary Data for ICW2		8259 Interrupt	
D7 D6 D5 D4	D3 D2 D1 D0	Input	INT Type
0 1 0 1	0 0 0 0	IR0	INT 50
0 1 0 1	0 0 0 1	IR1	INT 51
0 1 0 1	0 0 1 0	IR2	INT 52
0 1 0 1	0 0 1 1	IR3	INT 53
0 1 0 1	0 1 0 0	IR4	INT 54
0 1 0 1	0 1 0 1	IR5	INT 55
0 1 0 1	0 1 1 0	IR6	INT 56
0 1 0 1	0 1 1 1	IR7	INT 57

Masking and prioritization of IR0 - IR7 interrupts

One might ask what happens if more than one of IR0 - IR7 is activated at the same time? Can we mask any of the interrupts? What about responding to another interrupt while an interrupt is being serviced? To answer all these questions, the function of the OCW (operation command word) must be understood. This is discussed next.

OCW (operation command word)

After ICW1, ICW2, and ICW4 have been issued in sequence to the 8259 chip in order to initialize it, the 8088/86 is ready to receive hardware interrupts through the 8259's IR0 - IR7 pins. After the process of initialization, the OCW (operation command word) can be sent to mask any of IR0 - IR7, or change the priority assigned to each IR. There are three operation command words: OCW1,

Table 14-5: Addresses for 8259 OCWs

$\overline{CS}$	A0	Operation Command Word
0	0	OCW2, OCW3
0	1	OCW1
1	x	8259 is not addressed

OCW2, and OCW3. See Table 14-5. With the help of OCWs, a programmer can dynamically change the priority associated with each of IR0 - IR7, or mask any of them. Example 14-9 shows how the OCWs are sent to the 8259.

Example 14-9

Find the port addresses for the OCWs of the 8259 in Example 14-7.

Solution:
From Table 14-5 and Example 14-7:

A7	A6	A5	A4	A3	A2	A1	A0		
0	0	1	1	0	1	1	0	= 26H	Port for OCW2 and OCW3
0	0	1	1	0	1	1	1	= 27H	Port for OCW1

Figure 14-7 shows the OCWs for the 8259. Below is a discussion of each. Before discussing the OCWs, the existence of three registers inside the 8259 must be noted. They are the ISR (in-service register), IRR (interrupt request register), and IMR (interrupt mask register). See Figure 14-4.

CHAPTER 14: INTERRUPTS AND THE 8259 CHIP

OCW1 (operation command word 1)

OCW1 is used to mask any of IR0 - IR7. Logic 1 is for masking (disabling) and 0 is for unmasking (enabling). For example, 11111000 is the OCW1 to enable (unmask) IR0, IR1, IR2, and disable (mask) the rest (IR3 - IR7). When this byte is written to the 8259 (by making A0 =1 and $\overline{CS}$ =low), it goes into the internal register called IMR (interrupt mask register).

Example 14-10

Write the code to unmask (enable) IR0 - IR7. Use the ports in Example 14-9.

Solution:

To enable IR0 - IR7, use 0 for M0 - M7 in OCW1 of Figure 14-7.

OCW1 =0000 0000 =00H

```
            MOV    AL,00       ;OCW1 to unmask IR0 - IR7
            OUT    27,AL       ;issue OCW1 to IMR
```

There are occasions when one needs to know which IR is disabled and which one is enabled. In that case, simply read OCW1, which is the contents of IMR. For example, to read OCW1 using the port addresses in Example 14-10, code "IN AL,27H". By examining the contents of AL, one can find out which IRs are enabled and which ones are disabled.

OCW2 (operation command word 2)

This command word is used to assign a specific priority to the IRs. Three methods for assigning priority to IR0 - IR7 are discussed below.

Fully nested mode

This assigns the highest priority to IR0 and the lowest to IR7. In this case, if IR3 and IR5 are activated at the same time, first IR3 is served and then IR5. What happens if IR3 is being served when both IR2 and IR4 request service? In that case, IR3 is put on hold, then IR2 is served. After IR2 is served, IR3 is completed and finally IR4 is served. This is the default mode when the 8259 is initialized. The 8259 can be programmed to change the default mode to assign the highest priority to any IR. For example, the following shows OCW2 if IR6 has been assigned the highest priority, then IR7 has the next priority, and so on.

```
IR0  IR1  IR2  IR3  IR4  IR5  IR6  IR7     interrupt pin on 8259

2    3    4    5    6    7    0    1       priority 0=highest and 7=lowest
```

Automatic rotation mode

In this scheme, when an IR has been served it will take the lowest priority and will not be served until every other request has had a chance. This prevents interrupt starvation, where one device monopolizes the interrupt service.

Specific rotation mode

In this scheme the 8259 can be programmed to make the rotation follow a specific sequence rather than IR0 to IR7, which is the case for the automatic rotation mode. In this mode, the IR served will be stamped as the lowest priority, meaning that it will not be served until every other request has had a chance. The only difference between this mode and automatic rotation is the sequence of rotation.

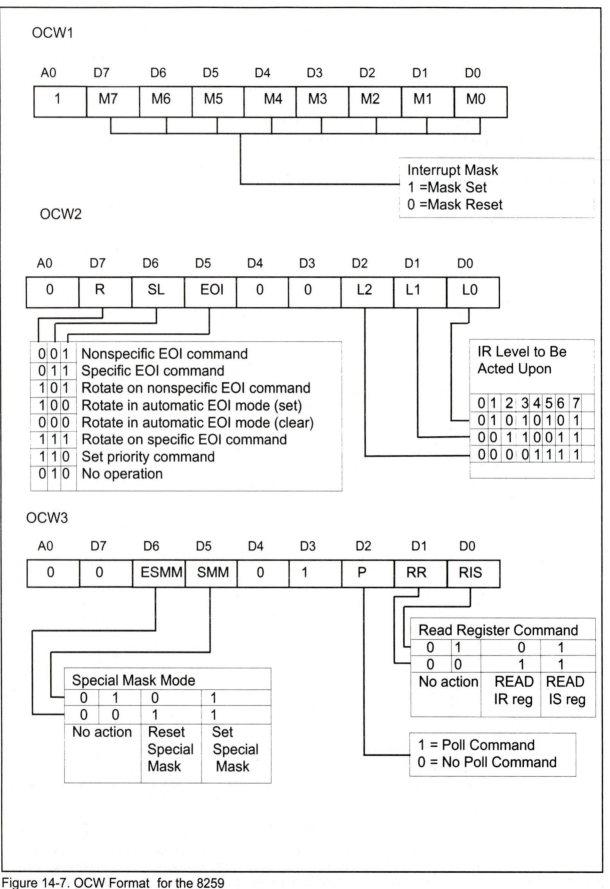

Figure 14-7. OCW Format for the 8259
(Reprinted by permission of Intel Corporation, Copyright Intel Corp. 1983)

Having concluded this brief description of priority schemes, the following will discuss the OCW2 bits. D2 - D0 are used to assign a new priority scheme to an IR other than the default. For example, to assign the highest priority to IR6, use D2 - D0 = 110. D4 - D3 must always be 0 for OCW2. D5, EOI (end of interrupt), is used to issue an end-of-interrupt command to the 8259. This is a very important function and is widely used in IBM BIOS. The following paragraph discusses this at length. D6 (SL, select) and D7 (R, rotation) bits are used to program the 8259 for the various priority schemes discussed earlier. A frequently used bit combination for OCW2 in the IBM PC/XT, AT, PS/2, and compatibles is 0010000 = 20H.

Importance of the EOI (end of interrupt) command

Why is it necessary to issue an EOI command to the 8259? To understand the answer to this question, consider the following case. Assume that an 8259 has been initialized and is in the default fully nested mode (where IR0 has the highest priority and IR7 the lowest). Now assume that IR3 is activated and the CPU acknowledges the interrupt by sending back a signal through INTA. Then the CPU goes to the vector table and gets CS:IP of the interrupt service routine and starts to execute the routine. When the CPU acknowledges IR3, the 8259 marks (sets to 1) the bit associated with IR3 in its ISR (in-service register) to indicate that this is being serviced now. Issuing EOI to the 8259 indicates that servicing IR3 is now complete and the bit associated with IR3 in register ISR can be reset to zero, thereby allowing IR3 to come in again. Of course, the EOI must be issued at the end of the service routine; otherwise, IR3 might keep interrupting itself again and again. If it is not issued and the CPU goes back to the main program after it finishes servicing IR3, it will not be able to be serviced again since in the ISR register the bit indicates that IR3 is being serviced. The important point of this is that the last three instructions of any interrupt service routine for IR0 - IR7 must be issuing the EOI, followed by IRET (see Example 14-11). It should be noted that while IR3 is being serviced, all of IR2, IR1, and IR0 are allowed to come in and interrupt it since they have higher priority, but no lower-priority interrupts of IR4 - IR7 are responded to. For example, if IR3 is being serviced and IR1 is activated, only IR0 can interrupt, and IR2, IR4 - IR7 will not be responded to. If the programmer has failed to issue the EOI at the end of IR1 and IR3, he has simply put these two IRs out of circulation in addition to IR2, IR4, IR5, IR6, and IR7. Only IR0 will be responded to by the 8259, since the ISR has marked IR3 and consequently all the lower-priority interrupts IR4 - IR7. Then IR1 puts the lower-priority IR2 out of circulation.

Example 14-11

Show the last 3 instructions of the interrupt service routine for IR1 of the 8259 in Example 14-8.

Solution:

Before the interrupt service routine returns control to the main program, it issues OCW2 to the 8259.

```
INT_SERV      PROC  FAR          ;routine for IR1
              ...   ...
              MOV   AL,20H        ;the EOI byte for OCW2
              OUT   26H,AL        ;to port designated for OCW2
              IRET                ;return from ISR to the main program
```

OCW3 (operation command word 3)

OCW3 is used, among other functions, to read the 8259's internal registers IRR (interrupt request register) and ISR (in-service register). D0 and D1 allow the program to read these registers in order to see which of IR0 - IR7 is pending for service and which one is being served (as mentioned earlier, OCW1 is used to peek into the IMR). The rest of the bits are for changing the masking mode and other advanced functions of the 8259. Interested readers should refer to Intel manuals.

Review Questions

1. A single 8259 can add up to _____ hardware interrupts to an 80x86 CPU.
2. INTR is an _____ (input, output) signal for the 8259 but it is an _____ (input, output) signal for the 80x86.
3. True or false. CAS0, CAS1, and CAS2 are used for master/slave mode only.
4. Indicate the logic level (high or low) on input pins A0 and $\overline{CS}$ in order to send ICW1 to the 8259.
5. True or false. ICWs can be sent to the 8259 in random order.
6. The 8259 can receive ICWs in the sequence ICW1, ICW2, ICW3, ICW4 or ICW1, ICW2, ICW4. How does it know which option is being programmed?
7. True or false. When the ISR (interrupt service routine) of IR5 is being executed, the 8259 prevents requests from the same interrupt by marking bit IR5 in its in-service register.
8. True or false. Fully nested mode is the default mode.
9. In fully nested mode, which IR has the highest priority?
10. Assume that an 8259 is configured in fully nested mode and the CPU is executing the interrupt service routine for IR5. During the execution of IR5, which interrupts can come in and which ones are blocked?

SECTION 14.4: USE OF THE 8259 CHIP IN THE IBM PC/XT

The 8088-based PC/XT uses only one 8259 chip to extend the number of hardware interrupts to eight, but starting with the 286 and higher processor computers, two 8259 chips are used to extend the hardware interrupts up to 15. Since the PC/XT is a subset of the PC AT, we discuss the hardware interrupts of the AT computers in a later section.

Interfacing the 8259 to the 8088 in IBM PC/XT computers

To interface the 8259 to the 8088, there must be two port addresses assigned to the 8259. One is for ICW1 and the second one is for ICW2 and ICW4. Figure 14-8 shows the address decoding for the 8259 in the IBM PC/XT. Since the chip select is activated by Y1 and all the x's for don't care must be zero, the addresses can be calculated in the manner indicated as shown in Table 14-6. These are the port addresses that were given in Chapter 12.

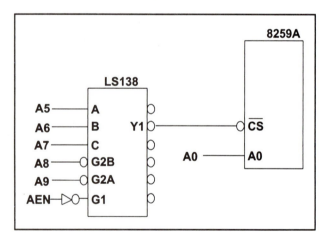

Figure 14-8. Chip Select Decoder of the 8259A

Table 14-6: Port Addresses of ICWs and OCWs

AEN A9 A8	A7 A6 A5 A4	A3 A2 A1 A0	Hex Address	Function
1 0 0	0 0 1 x	x x x 0	20	Port address ICW1
1 0 0	0 0 1 x	x x x 1	21	Port addresses ICW2, ICW3, ICW4
1 0 0	0 0 1 x	x x x 0	20	Port addresses OCW2, OCW3
1 0 0	0 0 1 x	x x x 1	21	Port address OCW1

Initialization words of the 8259 in the IBM PC/XT

Next the IBM PC/XT initialization words for the 8259 will be explained. From earlier discussions and Figure 14-5, the following configuration for the control words ICW1, ICW2, and ICW4 can be calculated:

ICW1	IBM PC/XT Configuration
D0=1	ICW4 needed
D1=1	single
D2=0	for 80x86 this must be zero
D3=0	edge triggering
D4=1	required by ICW1 itself
D5=D6=D7=0	0s for 8088/86-based systems

ICW1 = 00010011 = 13H

As was explained in Section 14.2, out of the 256 interrupts of the 8088/86, IBM PC/XT designers assigned INT 08 - INT 0F for expansion of hardware interrupts. These interrupts are used by IR0 - IR7 of the 8259 and commonly referred to as IRQ0 - IRQ7. INT 08 is for IRQ0, INT 09 is for IRQ1, and so on. It is the function of ICW2 to inform the 8259 which interrupt numbers are assigned to IRQ0 - IRQ7. This is done by equating ICW2 of the 8259 to the interrupt number assigned to IRQ0. In other words, ICW2 is the interrupt number for IR0, which in the case of the IBM PC/XT and compatibles is INT 08. The 8259 is only programmed only for the value of IRQ0, so the 8259 generates the INT numbers for IR1 through IR7. These are listed in Table 14-7. Summarizing the above discussion gives ICW2 = 00001000 = 08H.

Table 14-7: IBM PC/XT Hardware Interrupts

Binary Data ICW2		8259 Interrupt	
D7 D6 D5 D4	D3 D2 D1 D0	Input	INT Type
0 0 0 0	1 0 0 0	IR0	INT 08
0 0 0 0	1 0 0 1	IR1	INT 09
0 0 0 0	1 0 1 0	IR2	INT 0A
0 0 0 0	1 0 1 1	IR3	INT 0B
0 0 0 0	1 1 0 0	IR4	INT 0C
0 0 0 0	1 1 0 1	IR5	INT 0D
0 0 0 0	1 1 1 0	IR6	INT 0E
0 0 0 0	1 1 1 1	IR7	INT 0F

ICW3 is used only when multiple 8259 chips are connected in master/slave mode, which is the case in 80286, 386, 486, and Pentium PCs. This discussion is presented in Section 14.5. Next the ICW4 configuration will be examined.

ICW4	IBM PC Configuration
D0=1	8088/86
D1=0	normal (issue EOI before IRET)
D2=0,D3=1	slave buffered mode
D4=0	not nested
D5=D6=D7=0	required by ICW4

ICW4 = 00001001 = 09H

This gives the following code for 8259 initialization.

```
MOV     AL,13H          ;the ICW1
OUT     20H,AL
MOV     AL,8            ;the ICW2
OUT     21H,AL
MOV     AL,9            ;the ICW4
OUT     21H,AL
```

The IBM PC/XT BIOS version of the above program is as follows:

```
LOC OBJ         LINE        SOURCE
0020            19 INTA00 EQU    20H              ;8259 PORT
0021            20 INTA01 EQU    21H              ;8259 PORT
....            ...     ......    ...      ...       .........

                553 ;      INITIALIZE THE 8259 INTERRUPT CONTROLLER CHIP
                554 ;-------------------------------------------------
E1B4            555 C21:
E1B4  B013      556           MOV    AL,13H        ; ICW1 - EDGE, SNGL, ICW4
E1B6  E620      557           OUT    INTA00,AL
E1B8  B008      558           MOV    AL,8          ;SETUP ICW2 - INT TYPE 8 (8-F)
E1BA  E621      559           OUT    INTA01,AL
E1BC  B009      560           MOV    AL,09         ;SETUP ICW3 - BUFFERED,8086 MODE
E1BE  E621      561           OUT    INTA01,AL
........         ...      ...      .........
```

(Reprinted by permission from "IBM BIOS Technical Reference" c. 1984 by International Business Machines Corporation)

Now that the 8259 is initialized, it is ready to accept an interrupt on any of the inputs IRQ0 - IRQ7, thereby expanding the number of hardware interrupts for the 8088/86. What if the 8259 IC is defective? The 8259 is tested by a program in BIOS. There is a collection of programs in BIOS that is responsible for testing and initialization of the CPU and peripheral chips. This is commonly referred to as the POST (power-on self-test).

Since it is possible that one of the bits of the IMR (interrupt mask register) has become stuck to zero or one during the fabrication of the chip and escaped detection, the following program from IBM BIOS tests the IMR of the 8259 chip by writing 0s and 1s to it and reading them back. Reminder: To access the IMR, use OCW1, which has the port address 21H in the IBM PC (see Table 14-6). In the following program, if the test fails, the system will beep.

```
                620 ;TEST THE IMR REGISTER
                621
E217  BA2100    622           MOV    DX,0021      ;POINT INTR. CHIP ADDR 21
E21A  B000      623           MOV    AL,0         ;SET IMR TO ZERO
E21C  EE        624           OUT    DX,AL
E21D  EC        625           IN     AL,DX        ;READ IMR
E21E  0AC0      626           OR     AL,AL        ;IMR = 0?
E220  7515      627           JNZ    D6           ;GO TO ERR ROUTINE IF NOT 0
E222  B0FF      628           MOV    AL,0FF       ;DISABLE DEVICE INTERRUPTS
E224  EE        629           OUT    DX,AL        ;WRITE TO IMR
E225  EC        630           IN     AL,DX        ;READ IMR
E226  0401      631           ADD    AL,1         ;ALL IMR BIT ON?
E228  750D      632           JNZ    D6           ;NO - GO TO ERR ROUTINE
....  ....       ...      ....     ..            ....................
```

(Reprinted by permission from "IBM BIOS Technical Reference" c. 1984 by International Business Machines Corporation)

Sequences of hardware interrupts with the 8259

When a high is put on any of IR0 - IR7, how does the microprocessor become involved? As mentioned earlier, $\overline{INTA}$ of the 8088/86 is connected to $\overline{INTA}$ of the 8288 bus controller, and INTR of the 8259 is connected to INTR of the 8088/86. The following is the sequence of events after an IR of the 8259 is activated.

1. After an IR is activated, the 8259 will respond by putting a high on INTR, thereby signaling the CPU for an interrupt request.
2. The 8088/86 puts the appropriate signals on S0, S1, and S2 (S0 = 0, S1 = 0, and S2 = 0), indicating to the 8288 that an interrupt has been requested.
3. The 8288 issues the first $\overline{\text{INTA}}$ to the 8259.
4. The 8259 receives the first $\overline{\text{INTA}}$ and does internal housekeeping, which includes resolution of priority (if more than one IR has been activated) and resolution of cascading.
5. The 8288 issues the second $\overline{\text{INTA}}$ to the 8259.
6. The second $\overline{\text{INTA}}$ pulse makes the 8259 to put a single interrupt vector byte on the data bus which the 8088/86 will latch in. The value of the single byte depends on ICW2 and which IR has been activated, as discussed earlier.
7. The 8088/86 uses this byte to calculate the vector location, which is four times the value of the INT type.
8. The 8088/86 pushes the flag register onto the stack, clears IF (Interrupt Flag) and TF (Trap Flag), thereby disabling further external interrupt requests and disabling single-step mode, and finally pushes the present CS:IP registers onto the stack.
9. The 8088/86 reads CS:IP of the interrupt service routine from the vector table and begins execution of the interrupt routine.

Next we see which devices in the IBM PC/XT use the 8 hardware interrupts, IRQ0 to IRQ7, of the 8259.

Sources of hardware interrupts in the IBM PC/XT

With the use of the 8259, the IBM PC/XT has 8 interrupts, IR0 to IR7, plus NMI of the 8088/86. First the assignment of IR0 to IR7 will be discussed, then NMI and its use in the IBM PC/XT.

Of the 8 interrupts for the 8259, IBM has used two of them, IR0 and IR1, for internal use of the system. The other six, IR2 through IR7, are available through the expansion slots. Of those used internally, IR0 is for channel 0 of the 8253 timer to update the time-of-day (TOD) clock, and IR1 is dedicated to the keyboard. IR1 is activated whenever the serial-in-parallel-out shift register of the keyboard has a byte of data. IR2 to IR7 are generally used with the following assignments. The following two are used on the motherboard:

| INT | 08 | IRQ0 | channel 0 of 8253 timer to update TOD |
| INT | 09 | IRQ1 | keyboard input data |

The following are available through the expansion slot bus and used widely in industry, as indicated. Figure 14-9 summarizes the hardware interrupt assignment in the IBM PC/XT.

INT	0AH	IRQ2	reserved
INT	0BH	IRQ3	serial COM2
INT	0CH	IRQ4	serial COM1
INT	0DH	IRQ5	alternative printer
INT	0EH	IRQ6	floppy disk
INT	0FH	IRQ7	parallel printer LPT1

Sources of NMI in the IBM PC

The last hardware interrupt to be discussed for the 8088/86 computer is the NMI (nonmaskable interrupt). This interrupt is actually one of the pins of the CPU (similar to the TRAP pin in the 8080/85), and unlike INTR there is no need for the $\overline{\text{INTA}}$ pin to acknowledge it. Furthermore, it cannot be masked (disabled) by software as is the case for INTR, which can be masked at any time through use of the instruction CLI (clear interrupt flag). It is for this reason that the IBM PC has used the NMI for parity bit checking of DRAM to make sure that all read/write memory is working properly. In the absence of RAM memory, the operating system would not be loaded and the computer could not function.

SECTION 14.4: USE OF THE 8259 CHIP IN THE IBM PC/XT

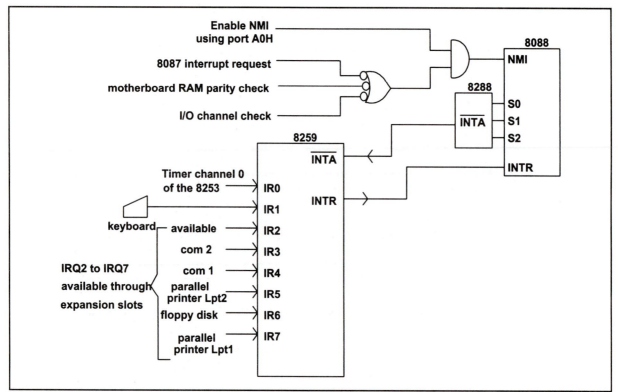

Figure 14-9. PC/XT Sources of Hardware Interrupts
(Reprinted by permission from "IBM BIOS Technical Reference" c. 1984 by International Business Machines Corporation)

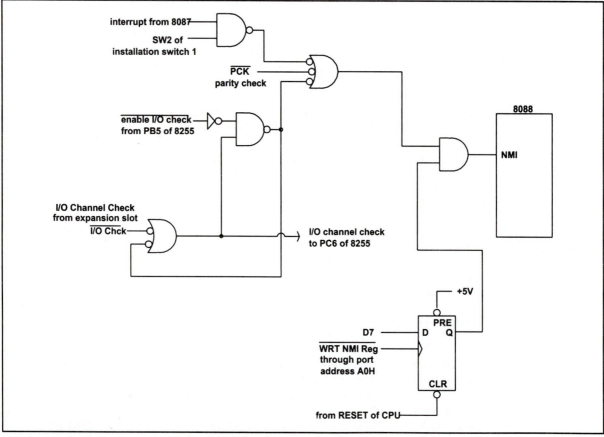

Figure 14-10. Sources of NMI in the PC/XT
(Reprinted by permission from "IBM BIOS Technical Reference" c. 1984 by International Business Machines Corporation)

If the NMI is so important to the system, which devices can activate it, and can they be masked at all? First, as can be seen from Figure 14-10, there are three sources of activation of the NMI:

1. NPIRQ (numerical processor interrupt request)
2. Read/write $\overline{PCK}$ (parity check)
3. IOCHK (input/output channel check)

Since three different sources can activate NMI, how does the system know which one is requesting interrupt service at any given time? The IBM PC system recognizes which of these interrupt requests has been activated by checking input port C of the 8255. It looks at PC6 of the 8255 to see if it is IOCHK and at PC7 to see if it is $\overline{PCK}$. The NMI service routine software must check PC6 and PC7 and determine which one has requested service. If neither of these two is requesting service, the request must have come from the 8087 coprocessor on the motherboard (in IBM terminology, planer). IBM BIOS checks the source of each and as it finds them, displays an appropriate messages on the video screen. The BIOS code is shown next.

```
E2C3        746              ORG   0E2C3H
E2C3        747 NMI_INT      PROC  NEAR
E2C3 50     748              PUSH  AX              ;SAVE ORIG CONTENTS OF AX
E2C4 E462   749              IN    AL,PORT_C
E2C6 A8C0   750              TEST  AL,0C0H         ;PARITY CHECK?
E2C8 7415   751              JZ    D14             ;NO, EXIT FROM ROUTINE

..........  752              ...                   ;ADDR OF ERROR MSG
E2CE A840   753              TEST  AL,40H          ;I/O PARITY CHECK
E2D0 7504   754              JNZ   D13             ;DISPLAY ERROR MEG
....  ....  755              ...   ...             ;MUST BE PLANER
E2D6        756 D13:
...   ....  757              ...   ...     } sends the message to
                                           } video and halts the system.
...   ....                   ...
E2DF        762 D14:
E2DF 58     763              POP   AX              ;RESTORE ORIGINAL AX
E2E0 CF     764              IRET
            765 NMI_INTENDP
```

(Reprinted by permission from "IBM BIOS Technical Reference" c. 1984 by International Business Machines Corporation)

Is there any way that NMI can be masked? The answer is yes. As can be seen from Figure 14-10, NMI is masked by a RESET signal from the CPU with CLR of the D flip-flop when the computer is first turned on. It can also be unmasked or masked through port A0H by setting D7 of the data bus to 1 (unmask) or 0 (mask). Again from the IBM PC/XT BIOS we see the following:

```
            1261         ;    ENABLE NMI INTERRUPTS
            1262
E5BC B080   1263              MOV   AL,80H         ;ENABLE NMI INTERRUPTS
E5BE E6A0   1264              OUT   0A0H,AL
....  ....  ....             ...   .......    ....................
```

(Reprinted by permission from "IBM BIOS Technical Reference" c. 1984 by International Business Machines Corporation)

Review Questions

1. True or false. The IBM PC/XT uses only one 8259.
2. What ports are assigned to ICWs in the PC/XT?
3. In the PC/XT, the IRQ are _____ (edge-, level-triggered).
4. Of the 256 possible interrupts of the 8088, which ones are assigned to IRQ0 - IRQ7 of the 8259?
5. True or false. IRQ0 and IRQ1 can be used by the system but not by the user.
6. Which IRQ of the 8259 is available on the expansion slot?
7. True or false. The 80x86 can mask and unmask the NMI by using the STI and CLI instructions.
8. True or false. If there is a problem with the memory of the PC, NMI is activated.

SECTION 14.5: INTERRUPTS ON 80286 AND HIGHER 80x86 PCs

When the first PC was introduced, only the six hardware interrupts, IRQ2 - IRQ7, were available through the PC/XT expansion slot. The other two, IRQ0 and IRQ1, were used by the motherboard. With the introduction of the 80286-based PC AT, another eight interrupts, IRQ8 - IRQ15, were added. IBM implemented the additional hardware interrupts with the use of a second 8259 programmable interrupt controller. To make their computers IBM compatible, all clone makers followed IBM's lead. Subsequent 386, 486, and Intel Pentium-based PCs and PS computers have remain faithful to the original IBM PC AT computer. In this section we study the hardware interrupts and hardware interrupt assignment for 286 and later PCs.

IBM PC AT hardware interrupts

In the design of the 80286-based IBM PC AT, IBM designers had to make sure that it was compatible with the 8088-based PC/XT. This lead to the use of IRQ0 and IRQ1 for the system timer and keyboard, respectively, as was the case in the PC/XT. IBM made the first 8259 a master, and added the second 8259 in slave mode. To do that, it connected the INT pin of the slave 8259 to IRQ2 of the master 8259. The master and slave 8259s communicate with each other through pins IRQ2, INT, CAS0, CAS1, and CAS2. See Figure 14-11. A detailed discussion of 8259 design in master and slave modes in IBM PC AT and compatibles follows.

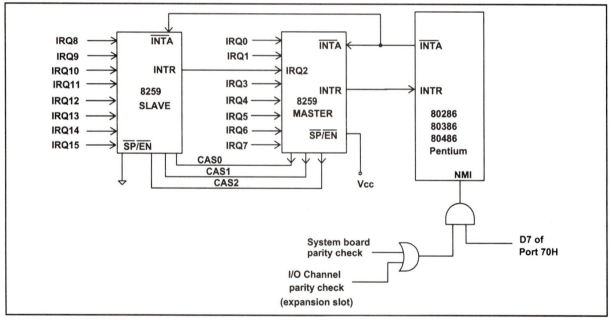

Figure 14-11. 8259 Chips in Master/Slave Relation for 286 and Higher PCs

8259 in master mode

To implement the 8259 in master mode, the following steps must be taken:
1. The 8259 $\overline{\text{SP/EN}}$ pin must be connected to 5 V, V_{CC}, to make it the master in nonbuffered mode. See Figure 14-11.
2. CAS0, CAS1, and CAS2 of the master 8259 are connected to the same pins of the slave 8259. CAS0, CAS1, and CAS2 are outputs for the master 8259.
3. INT and $\overline{\text{INTA}}$ pins of the master are connected to INTR and $\overline{\text{INTA}}$ of the CPU.
4. IRQ2 of the master is connected to the INT pin of the slave 8259.
5. The master 8259 must be programmed specifically to operate in master mode. This is done through the ICWs, as will be demonstrated below.

The port addresses assigned to the master are the same as the port addresses in the PC/XT, 20H and 21H. The following is a program to initialize the 8259 master from the IBM PC/AT BIOS Technical Reference, with slight modifications for the sake of clarity.

```
;INITIALIZE THE 8259 INTERRUPT # 1 CONTROLLER CHIP
;-----------------------------------------------------------------------
MOV      AL,11H             ;ICW1 - EDGE, MASTER, ICW4
OUT      20H,AL
MOV      AL,8               ;SETUP ICW2 - INTERRUPT TYPE 8 (8-F)
OUT      21H,AL
MOV      AL,04H             ;SETUP ICW3 - MASTER LEVEL 2
OUT      21H,AL
MOV      AL,01H             ;SETUP ICW4 - MASTER x86 MODE
OUT      21H,AL
```

ICW1, ICW2, and ICW4 for the 8259 master are basically the same as in the PC/XT 8259, with some minor differences. In the ICW1 initialization, D1 of the ICW1 is 0 since it is used in cascade mode. Cascade mode is chosen whenever there is more than one 8259. ICW2 is the same in the PC/XT and the PC AT. ICW4 in the PC AT is in nonbuffered mode. In nonbuffered mode, master/slave is determined by the SP pin (see the $\overline{SP/EN}$ pin description). The most important difference is ICW3. In cascade mode, ICW3 must be coded into both master and slave 8259s. In the case of the master, it must be informed which of the IRQs has a slave 8259 connected to it. Since IRQ2 of the master is connected to INTR of the slave 8259, we have 00000100 = 04H for ICW3. See Figure 14-6.

8259 in slave mode

To implement the 8259 in slave mode, the following steps must be taken:
1. $\overline{SP/EN}$ must be grounded (SP/EN = 0) for the 8259 in slave mode for nonbuffered mode. See Figure 14-11.
2. CAS0, CAS1, and CAS2 of the slave 8259 are connected to the same pins of the master 8259. These signals are inputs for the slave 8259.
3. $\overline{INTA}$ of the slave is connected to $\overline{INTA}$ of the 80x86.
4. The INT pin of the slave 8259 is connected to IRQ2 of the slave.
5. The slave 8259 must be programmed specifically through its ICWs to operate in slave mode.

Since the second 8259 must be accessed independent of the first 8259, the port addresses assigned to it are different and must be unique. These addresses in IBM PC AT and compatibles are shown in Table 14-8. The following is a program to initialize the slave 8259 from the IBM PC/AT BIOS Technical Reference, with slight modification for the sake of clarity.

Table 14-8: Port Addresses for the Slave 8259 in ISA PCs

Port	ICWs	OCWs
A0H	ICW1	OCW2, OCW3
A1H	ICW2, ICW3, ICW4	OCW1

```
;INITIALIZE THE 8259 INTERRUPT # 2 CONTROLLER CHIP
;-----------------------------------------------------------
MOV      AL,11H   ;ICW1 - EDGE, SLAVE, ICW4
OUT      A0H,AL
MOV      AL,70H   ;SETUP ICW2 - INTERRUPT TYPE 70H (70-77)
OUT      A1H,AL
MOV      AL,02H   ;SETUP ICW3 - SLAVE LEVEL 2
OUT      A1H,AL
MOV      AL,01H   ;SETUP ICW4 - x86 MODE SLAVE
OUT      A1H,AL
```

In the above codes notice ICW2 and ICW3. They are different from the master 8259 in the following way. In the IBM PC AT, IR0 - IR7 of the second 8259 are designated as IRQ8 - IRQ15 and were assigned the interrupt type numbers INT

70H to INT 77H. This is reflected in the value for ICW2. The 8259 used in slave mode must be informed of the IRQ number of the master that is connected to its INT out pin. This is the function of ICW3. Since IRQ2 of the master is connected to INT of the slave, ICW3 =00000010. See Figure 14-6. The following is the INT number assigned to IRQ8 - IRQ15 of the second 8259.

IRQ8	INT 70H
IRQ9	INT 71H
IRQ10	INT 72H
IRQ11	INT 73H
IRQ12	INT 74H
IRQ13	INT 75H
IRQ14	INT 76H
IRQ15	INT 77H

AT-type computers interrupt assignment

In the PC/XT, IRQ2 is available through the expansion slot. However, in AT-type computers, since IRQ2 is used as an input from the second 8259, IBM had to replace it with another IRQ. IRQ9 (i.e., IR1 of the second 8259) is used in place of IRQ2. Looking at the ISA expansion slot 62-pin section (see Chapter 10), we see IRQ9 in place of IRQ2 on exactly the same pin. The replacement of IRQ2 with IRQ9 makes the AT computers PC/XT compatible. For this scheme to work, IRQ9 must be redirected to IRQ2 internally by the software, and that is exactly what IBM has done. The following is the code from the IBM AT BIOS showing the process of redirecting IRQ9 to IRQ2.

```
;--HARDWARE INT 71H --(IRQ LEVEL 9 ) TO INT 0AH (IRQ 2)
;REDIRECT SLAVE INTERRUPT 9 TO INTERRUPT LEVEL 2
    RE_DIRECT      PROC    NEAR
                   PUSH    AX               ;SAVE (AX)
                   MOV     AL,20H           ;EOI
                   OUT     0A0H,AL          ;TO SLAVE INTERRUPT CONTROLLER
                   POP     AX               ;RESTORE (AX)
                   INT     0AH              ;GIVE CONTROL TO HARDWARE LEVEL 2
                   IRET                     ;RETURN
    RE_DIRECT      ENDP
```

In the above code notice the issuing of EOI (end of interrupt) to the slave 8259. This redirection process shown above makes any changes in the design of the hardware and software of plug-in boards using the IRQ2 unnecessary. In such plug-in boards, hardware interrupts come in through the same pin on the 62-pin expansion slot, are captured by IRQ9, and redirected to IRQ2. This is as if the IRQ2 had responded directly to the interrupt.

Case of missing IRQs on the AT expansion slot

Examining the AT bus we see that of the 8 interrupts, IRQ8 to IRQ15, two of them, IRQ8 and IRQ13, are missing. In IBM PC/AT 286 and higher computers, IRQ8 is used on the motherboard by the Motorola MC146818 real-time clock CMOS RAM chip. Therefore, it is not available on the expansion slot pin. As was discussed in Section 14.3, the IBM PC/XT used NMI for parity bit checking of RAM both on the motherboard and memory expansion RAM cards through the slot, in addition to interrupts from the 8087 math coprocessor. Starting with the AT-type 286 and higher PCs, IRQ13 (INT type 75H) is used for the interrupt associated with the math coprocessors. Although IRQ13 captures the 80287 math coprocessor interrupt, it is still handled by NMI for the sake of compatibility with the PC/XT. The following code (again with slight modification for clarity) from BIOS of the IBM PC AT documentation shows this process.

```
;---HARDWARE INT 75H--(IRQ LEVEL 13   )-------
; SERVICE X287 INTERRUPTS
;    THIS ROUTINE FIELDS X287 INTERRUPTS AND CONTROL
;    IS PASSED TO NMI INTERRUPT HANDLER FOR
;    COMPATIBILITY
;------------------------------------------
INT_287    PROC    NEAR
           PUSH    AX              ;SAVE (AX)
           SUB     AL,AL
           OUT     0F0H,AL         ;REMOVE THE INTERRUPT REQUEST
           MOV     AL,20H          ;ISSUE EOI
           OUT     0A0H,AL         ;TO SLAVE
           OUT     20H,AL          ;TO MASTER
           POP     AX              ;RESTORE (AX)
           INT     02              ;GIVE CONTROL TO NMI(INT TYPE 2)
           IRET                    ;RETURN
INT_287    ENDP
```

Again notice in the above code the issuing of an end-of-interrupt (EOI) command to both 8259s. For the discussion of how and why the 80x87 math coprocessors get the attention of the CPU in case it encounters a problem, see Chapter 20.

Notice on the ISA expansion slot that IRQ10, IRQ11, IRQ12, IRQ14, and IRQ15 are on the 32-pin section and IRQ9, IRQ3, IRQ4, IRQ5, IRQ6, and IRQ7 are on the 62-pin section. Table 14-9 shows the hardware interrupt assignment in 286, 386, 486, and Pentium IBM and compatible computers.

Table 14-9: Hardware Interrupt Assignment for ISA PCs

IRQ	INT Number	Use
IRQ0	INT 08	8254 TIMER OUT0
IRQ1	INT 09H	Keyboard
IRQ2	INT 0AH	Interrupt from 8259 #2
IRQ3	INT 0BH	Serial port COM2 (and COM4)
IRQ4	INT 0CH	Serial port COM1 (and COM3)
IRQ5	INT 0DH	Parallel port 2: LPT2
IRQ6	INT 0EH	Floppy disk Controller
IRQ7	INT 0FH	Parallel port 1: LPT1
IRQ8	INT 70H	CMOS real-time clock
IRQ9	INT 71H	Software redirected to INT 0AH
IRQ10	INT 72H	Available
IRQ11	INT 73H	Available
IRQ12	INT 74H	PS/2 mouse
IRQ13	INT 75H	Math coprocessor
IRQ14	INT 76H	Hard disk
IRQ15	INT 77H	Available

80x86 microprocessor generated interrupts (exceptions)

As mentioned in Section 14.1, when the CPU encounters an unusual situation such as dividing a number by zero, it generates an exception. The 8088/86 had only one exception, divide error or INT 0. In the 8088/86, Intel Corporation left the first 32 interrupts (INT 00 to INT 1FH) reserved for future microprocessors.

However, designers of the first IBM PC ignored this and assigned many of these interrupts to hardware and software interrupts on the system. By not adhering to Intel's specifications, IBM has created a massive headache for software designers of protected mode 386 and later systems. This is due to the fact that Intel continued to assign the processor exception cases generated by the 80x86 CPU to INT 5 and higher with each new member of the x86 family. This is shown in Table 14-10.

Many of the interrupts in Table 14-10 are used by the x86 in protected mode. Since the IBM OS/2, Microsoft Windows, and NT operating systems use the x86 in protected mode, they have to map all these interrupts to new interrupts to avoid interrupt conflict with the IRQs of the PC. Note that Pentium exceptions are the same as the 486, with the addition of INT 12H for machine check.

Table 14-10: 80x86 Microprocessor Interrupt Assignment

Interrupt	8086/88	286	386	486
00	Divide error	Divide error	Divide error	Divide error
01	Single step	Single step	Single step debugging exceptions	Single step debugging exceptions
02	Nonmaskable interrupt	Nonmaskable interrupt	Nonmaskable interrupt	Nonmaskable interrupt
03	Breakpoint	Breakpoint	Breakpoint	Breakpoint
04	INTO detected overflow	INTO detected overflow	INTO detected overflow	INTO detected overflow
05		Bound range exceeded	Bound range exceeded	Bound range exceeded
06		Invalid instruction	Invalid instruction	Invalid instruction
07		Coprocessor not available	Coprocessor not available	Coprocessor not available
08		Double exception detected	Double exception detected	Double exception detected
09		Coprocessor	Coprocessor protection error	(Reserved) protection error
0A		Invalid task state segment	Invalid task state segment	Invalid task state segment
0B		Segment not present	Segment not present	Segment not present
0C		Stack fault	Stack fault	Stack fault
0D		Protection fault	Protection fault	Protection fault
0E			Page fault	Page fault
10		Coprocessor error	Coprocessor error	Coprocessor error
11				Alignment check

Note: Pentium has assigned a new exception called machine check with INT 12H.
(Reprinted by permission of Intel Corporation, Copyright Intel Corp. 1990)

Interrupt priority

The last topic in this section is the concept of priority for INT 00 to INT FFH. What happens if two interrupts want the attention of the CPU at the same time? Which has priority? As far as the 80x86 is concerned, the INTR pin is considered a single interrupt. Therefore, the resolution of priority among the IRQs is up to the 8259. Assume that the INT instruction (such as INT 21H) and INTR both want to be processed. The INT instruction has a higher priority than either INTR or NMI. If both NMI and INTR are activated at the same time, NMI is responded to first since NMI has a higher priority than INTR. Table 14-11 shows the interrupt processing order for the 80286 microprocessor from Intel's manual (1 is the highest priority).

Table 14-11: 80286 Interrupt Priority

Order	Interrupt
1	INT instruction or exception
2	Single step
3	NMI
4	Processor extension segment overrun
5	INTR

For the IRQs coming through INTR, the 8259 resolves priority depending on the way the 8259 is programmed. In the 80x86 IBM PC, PS, and compatibles, IRQ0 has the highest priority and IRQ7 is assigned the lowest priority. It must be noted that since IRQ8 - IRQ15 of the slave 8259 are connected to IRQ2 of the master 8259, they have higher priority than IRQ3 to IRQ7 of the master 8259. Figure 14-12 shows the IRQ0 - IRQ15 priority.

```
IRQ0    HIGHEST PRIORITY
IRQ1
IRQ8
IRQ9
IRQ10
IRQ11
IRQ12
IRQ13
IRQ14
IRQ15
IRQ3
IRQ4
IRQ5
IRQ6
IRQ7    LOWEST PRIORITY
```

Figure 14-12. IRQ Priority in the x86 PC

More about edge- and level-triggered interrupts

As discussed previously, in the 8259 there are two ways to activate the interrupt input IRQ, depending upon how it is programmed. One is level-triggered mode and the other is edge-triggered mode.

Level-triggered mode

In level-triggered mode, the 8259 will recognize a high on the IRQ input as an interrupt request. The request on the IRQ line must remain high until the first $\overline{INTA}$ is acknowledge from the 8259. It is only then that the high must be removed from the IRQ input immediately. If the IRQ input remains high after the end-of-interrupt (EOI) command has been issued, the 8259 will generate another interrupt on the same IRQ input. Therefore, to avoid multiple interrupt generation, the IRQ input must be brought low before the EOI is issued

Edge-triggered mode

In edge-triggered mode, the 8259 will recognize an interrupt request only when a low-to-high pulse is applied to an IRQ input. This means that after the low-to-high transition on the IRQ input, the 8259 will acknowledge the interrupt

---

request by activating $\overline{\text{INTA}}$ and the high-level input will not generate further interrupts even after the EOI is issued. Therefore, the designer does not need to worry about quickly removing the high to avoid generating multiple interrupts, as is the case for level-triggered mode. This is due to the fact that in edge-triggered mode, before another interrupt can be requested, the IRQ input must be brought back to low. Notice in both edge- and level-triggered modes that the IRQ must stay high until after the falling edge of the first $\overline{\text{INTA}}$ pulse in order to acknowledge the interrupt request. It is interesting to note the role of the IRR (interrupt request register) of the 8259. In level-triggered mode, the IRR latch is always ready to recognize a high on the IRQ as a request for interrupt. But in edge-triggered mode, the IRR latch is disabled after the request is acknowledged and will not latch another interrupt until that IRQ input goes back to low. The disadvantage of edge-triggered mode is the problem of false interrupt caused by a good-sized spike as a result of noise on the IRQ line, especially in high-speed systems. This, plus the concept of interrupt sharing, is discussed next.

Interrupt sharing in the x86 PC

In the design of the IBM PC/AT, each IRQ is assigned to a single device only and no two devices can use the same IRQ line to get the attention of the CPU. This limitation has caused some headaches in recent years due to the fact that we are running out of IRQs available to new peripheral devices. Another source of problems is the edge-triggered IRQ. Since the PC AT IRQs are positive edge-triggered, they are susceptible to false interrupts as a result of noise (spike) on the IRQ input. This problem has led the designers of new bus systems such as EISA and Micro Channel to change the IRQ from edge triggering to level triggering. In both EISA and Micro Channel buses, the interrupts are level triggered and each IRQ can be shared among two or more devices. The interrupt sharing in these buses in concept is similar to NMI sharing discussed in Section 14.4. As we saw in Section 14.4, the NMI can be activated by several different sources, but it is the job of BIOS to detect who the source is and respond to that source only. In the same way, BIOS of EISA and Micro Channel PCs are equipped to detect the source of IRQ activation if an IRQ is shared among several devices. For further discussion of these buses, see Chapter 27.

Review Questions

1. True or false. Cascading 8259s refers to a scheme of connecting multiple slave 8259s to a single master 8259.
2. True or false. In cascading the 8259, we must program all 8259s for ICW3.
3. True or false. There is no difference between ICW1 of the master and slave 8259 of the PC AT.
4. What port addresses are assigned to the slave 8259 in the PC AT?
5. Indicate the port addresses used by the ICWs of the slave 8259 in the PC AT.
6. What port addresses are used by the OCWs of the slave 8259 in the PC AT?
7. If ICW2, ICW3, ICW4 all go into the same port address, how does the 8259 avoid getting confused?
8. Since ICW1 and OCW2 share the same port address on the 8259, how does it distinguish between them?
9. If OCW2 and OCW3 share the same port address, how does the 8259 distinguish between them?
10. If a given design has used IRQ2 for some ISR, does this make it incompatible with the AT system?
11. True or false. IRQ13 is used for math processor error detection in AT machines.
12. If any of IRQ8 - IRQ15 is activated, an EOI must be issued to both 8259s. Why?
13. Which has the higher priority, IRQ10 or IRQ7?
14. What port addresses are used to issue the EOI to the slave 8259?
15. With 286 and later processors, an invalid instruction causes an exception. Which interrupt is assigned to it by Intel?

　　　　　　　　　　CHAPTER 14: INTERRUPTS AND THE 8259 CHIP

SUMMARY

This chapter began with a look at the interrupt vector table, which is used to store the CS:IP of the interrupt service routine that services the interrupt. The differences between INT and CALL instructions were outlined and the two major categories of interrupts, hardware and software, were explored. We also described how the IBM PC executes interrupts. Next we looked at the assignment of interrupts by BIOS and DOS. Then the 8259 programmable interrupt controller chip was examined in terms of its pin layout, control word programming, and priority handling. How the 8259 is interfaced into the IBM PC was covered, as well as addressing and programming specific control words. The various sources of hardware interrupts, both maskable and nonmaskable, were discussed. Finally, use of 8259 chips in 80286 and higher microprocessors was discussed.

PROBLEMS

SECTION 14.1: 8088/86 INTERRUPTS

1. Assume that the 8088/86 is executing an instruction with 17 clock counts. Meanwhile, the INTR pin is activated. Does the CPU finish the current instruction before it responds to INTR? How does the CPU resume from where it left off?
2. Give the logical and physical addresses in the interrupt vector table associated with each of the following interrupts.
    (a) INT 5   (b) INT 21H
3. What does ISR stand for, and what is it? Give another name for ISR.
4. Where is the address of each ISR kept?
5. Compare the number of bytes of stack memory used by each of the following.
    (a) CALL FAR (b) interrupt activation
6. Vector table addresses 003F8H - 003FB belong to which interrupt?
7. Give the logical and physical addresses used by the interrupt vector table.
8. How many bytes are used by the interrupt vector table, and why?
9. Why should we not use the first 1K of address space in 8088/86-based systems?
10. Indicate the interrupt(s) set aside for exception handling in the 8088/86.
11. Give the interrupt number (type) assigned to each of the following.
    (a) divide error (b) single step   (c) NMI
12. True or false. When an interrupt through INTR is executed, IF = 0 and TF = 0.
13. True or false. CLI blocks both INTR and NMI.
14. Show how to set TF to high.
15. Show how to set IF to each of the following.
    (a) low   (b) high
16. True or false. Instruction INTO is executed only if the overflow flag is high (OF =1).
17. The last instruction in the ISR is _____ , whereas the last instruction in a FAR subroutine is _____.
18. What is the difference between RETF and IRET in terms of stack activity?
19. Show the stack frame where CS, IP, and FR are stored for both an interrupt and a CALL FAR routine. Assume that SP = FFE0H.
20. In which of the following sequences are the stack contents popped off by IRET?
    (a) IP, FR, CS   (b) FR, IP, CS   (c) FR, CS, IP   (d) none of the above

SECTION 14.2: IBM PC AND MS DOS ASSIGNMENT OF INTERRUPTS

21. Answer the following questions, assuming that vector table locations 0000:001C to 0000:001F have the contents indicated below.
    0000:0010 ....-............. 47 FF 00 F0
    (a) Which interrupt does this belong to?
    (b) What is the logical address and physical address of the ISR?

22. In Problem 21, does BIOS or DOS provide the service?
23. INT 12H provides the size of which of the following memories?
    (a) high memory area   (b) extended memory
    (c) conventional memory   (d) expansion memory installed in the expansion slot
24. If the start of an ISR is located in BIOS ROM at FFE6EH, what are the values of CS and IP in the vector table?
25. In Problem 24, if the ISR belongs to INT 1CH, show the exact contents of the vector table.
26. In what BIOS data area location is the size of conventional memory stored?

SECTION 14.3: 8259 PROGRAMMABLE INTERRUPT CONTROLLER

Note: These problems are not necessarily IBM PC compatible.

27. True or false. In the 8259, to program ICW1, we must have A0 =1 and $\overline{CS}$ =0.
28. For the 8259, indicate which of the following is input and which is output.
    (a) IR0 - IR7   (b) INT   (c) $\overline{INTA}$
    (d) A0   (e) $\overline{CS}$ (f) $\overline{RD}$
29. Find the addresses for each ICW of the 8259 if CS is activated by A7 - A1 =1001 010.
30. Find the ICW1 and ICW2 if the 8259 is used with an 8088/86, single, edge triggering, no ICW4, and IR0 is assigned INT 88H.
31. Show the programming of ICW0 and ICW2 in Problem 30. Use the port addresses of Problem 29.
32. Which of the following interrupts cannot be assigned to IR0 of the 8259, and why?
    (a) 99H   (b) 98H   (c) CCH
    (d) 22H   (e) 10H   (f) F8H
33. Find the INT type number assigned to IR0 and IR7 if IR3 is assigned INT 1BH.
34. Find the INT number assigned to IR0, IR4, and IR6 if IR2 is assigned INT 32H.
35. Which of the OCWs is used to mask a given IR of the 8259?
36. EOI is issued by which of the OCWs?
37. What is the default mode for the prioritization of IR0 to IR7?
38. Find the port addresses assigned to each of the OCWs in Problem 29.
39. Show the program to enable IR2 and IR4 and mask the rest of the IRs. Use the port addresses in Problem 38.
40. OCW2, OCW3, and ICW1 go to the same port address in the 8259 when A0 =0 and CS = 0. How does the 8259 distinguish between them? How does it distinguish between OCW2 and OCW3 since both go to the same port address?

SECTION 14.4: USE OF THE 8259 CHIP IN THE IBM PC/XT

41. Why is signal AEN used in accessing the 8259 in the PC?
42. The PC/XT uses the 8259 in _____ (single, cascade) mode.
43. Indicate the IRQs level of triggering in the IBM PC (edge-, level-triggered).
44. What interrupt numbers are assigned to the 8259 in the PC/XT?
45. What port addresses are assigned to the 8259 in the PC/XT?
46. True or false. IRQ0 and IRQ1 are used by the system board and are not available.
47. Which of the IRQs of the 8259 are available on the expansion slot?
48. Indicate on which side, A or B, of the expansion slot IRQs are located.
49. What is the binary and hex value for the EOI, and to which port is it issued in the IBM PC?
50. Which IRQ has the highest priority, and why?
51. True or false. The 8288 chip issues two $\overline{INTA}$s to the 8259 when INTR of the 8088 is activated.
52. True or false. In the PC/XT, there is more than one source of NMI activation.
53. In the PC/XT, can NMI be blocked? If yes, how?
54. True or false. In the PC/XT, the parity bit error from both memory of the system board and the memory board of the expansion slot can activate NMI.

55. True or false. In the 80x86 PC, INTR of the 80x86 comes from the primary (master) 8259 chip.____

56. True or false. $\overline{\text{INTA}}$ from the 80x86 goes to both 8259 chips in the 80x86 PC.

57. In the 80x86, what port addresses are assigned to the 8259s?

58. In the 80x86 PC, what interrupt numbers are assigned to the second 8259 chip?

59. What IRQs are available on the AT expansion slot?

60. Why is there no IRQ2 on the AT expansion slot? Is there any replacement for it?

61. Discuss the issuing of EOI for the IRQs of the primary and secondary 8259 with specific answers for the following.
    (a) Give the port addresses it is issued to.
    (b) Explain in what cases it must be issued to both 8259s.

62. True or false. With every generation of the 80x86, more exception interrupts are added but they are downward compatible.

63. True or false. The INT instruction and the exception interrupt have a lower priority than NMI.

64. True or false. The NMI has a higher priority than INTR.

65. If NMI, IRQ10, and IRQ6 are all activated at the same time, explain the sequence when the system responds and executes them.

66. If IRQ3, IRQ7, and IRQ15 are all activated at the same time, in what order are they serviced? What impact does issuing EOI have on servicing them?

67. The failure to issue EOI at the end of the ISR blocks the servicing of IRQs with _____ (lower, higher) priority.

68. Explain the implications if the STI instruction is the first instruction of an ISR.

69. Explain the implications if there is no STI instruction in an ISR.

70. Which chip takes care of the priority of the various IRQs, the 80x86 or the 8259?

71. True or false. STI and CLI have no impact on the INT instruction, exception, and NMI.

72. What is the activation level for the IRQs in the IBM PC as set by BIOS?

73. Of IRQ10 and IRQ4, which has the higher priority, and why?

74. Give the advantages of level-triggered interrupts.

75. Discuss interrupt sharing and state which PC buses are equipped with that.

ANSWERS TO REVIEW QUESTIONS

SECTION 14.1: 8088/86 INTERRUPTS
1. true
2. 4
3. 1K bytes beginning at 00000 and ending at 003FFH
4. interrupt service routine (ISR) or interrupt handler
5. to hold the CS:IP of each ISR
6. 00040H, 41H, 42H, and 43H
7. No; it is internally embedded into the CPU.
8. INT 0

SECTION 14.2: IBM PC AND MS DOS ASSIGNMENT OF INTERRUPTS
1. F000:F065, CS = F000 and IP = F065
2. INT 10H is assigned memory locations 00040H, 41H, 42H, and 43H in the interrupt vector table. That means that we have 00040 = (65), 00041 = (F0), 00042 = (00), and 00043 = (F0).
3. The dash (-) tells us this is the 8th boundary; therefore, the 0000:0038 address in interrupt vector table belongs to INT 14 (0E hex).
4. The logical address is F000:EF57; therefore, CS = F000 and IP = EF57.
5. The calculation shows that it is 510K bytes. This means that the other 2K bytes are used by the extended BIOS data area and are not available to the user's application programs.
6. IRET

SECTION 14.3: 8259 PROGRAMMABLE INTERRUPT CONTROLLER
1. 8
2. output, input
3. true
4. $\overline{CS}$ = 0 and A0 = 0
5. false
6. The bit D1 in ICW1 indicates if it is for single or cascade. If cascade, it expects to receive all ICWs, from 1 to 4, but if it is single, it does not expect ICW3.
7. true
8. true
9. IR0
10. Higher-priority interrupts IR0, IR1, IR2, IR3, IR4 can come in, but IR6 and IR7 are blocked since they have lower priority.

SECTION 14.4: USE OF THE 8259 CHIP IN THE IBM PC/XT
1. true
2. 20H and 21H
3. edge
4. INT 08 to INT 0FH
5. true
6. IRQ2 through IRQ7
7. false
8. true

SECTION 14.5: INTERRUPTS ON 80286 AND HIGHER 80x86 PCs
1. true
2. true
3. true
4. A0H and A1H
5. ICW1 uses A0H while ICW2, ICW3, and ICW4 use A1H.
6. OCW1 uses A1H while OCW2 and OCW3 use A0H.
7. ICW1 sets the stage for the sequence of ICW2, ICW3, and ICW4. After the ICW2 is sent in, ICW3 must be sent in (if cascaded) and then ICW4. If single, ICW3 is bypassed, and if no ICW4 is needed in ICW1, that leaves only ICW2 using the port.
8. Notice that D4 in ICW1 is 1, while it is 0 in OCW2.
9. Notice that D3 and D4 differ in OCW2 and OCW3.
10. No, since in place of IRQ2, IRQ9 is activated and then it is redirected to IRQ2 by BIOS.
11. true
12. The EOI must be sent to the slave 8259 since it originates the interrupt request but also to the master 8259 since the slave sends the request through IRQ2 of the master.
13. IRQ10
14. The EOI is an OCW2; therefore, it must go to port address A0H.
15. INT 06

CHAPTER 15

DIRECT MEMORY ACCESSING; THE 8237 DMA CHIP

OBJECTIVES

Upon completion of this chapter, you will be able to:

» Describe the concept of DMA, direct memory accessing
» List the pins of the 8237A DMA chip and describe their functions
» Explain how bus arbitration is achieved between DMA and the CPU
» Explain how the 4 channels of the 8237 are used in the PC
» Program the 4 channels of the 8237 for data transfer
» Describe the purpose of the 8 control registers of the 8237
» Demonstrate how to program the control registers of the 8237
» Diagram how the 8237 is interfaced in the PC/XT
» Describe how DRAM refreshing is achieved via channel 0 of the 8237
» Diagram how the 8237 is interfaced in the PC AT and 80x86 compatibles

For a computer to work efficiently, there must be a way to transfer a large amount of data in a short amount of time. In the IBM PC, this is accomplished with the help of what is called *direct memory access* (DMA), and that is the subject of this chapter. In Section 15.1, the concept of DMA is explained. In Section 15.2, the Intel 8237 DMA chip is described. The third section studies the 8237's connection to the 8088/86 CPU in the IBM PC/XT. Section 15.4 shows an example of the use of DMA in refreshing DRAM memory. Finally in Section 15.5, the DMA for 80x86-based PC AT-type computers is discussed.

SECTION 15.1: CONCEPT OF DMA

In computers there is often a need to transfer a large number of bytes of data between memory and peripherals such as disk drives. In such cases, using the microprocessor to transfer the data is too slow since the data first must be fetched into the CPU and then sent to its destination. In addition, the process of fetching and decoding the instructions themselves adds to the overhead. For this reason, Intel created the 8237 DMAC (direct memory access controller) chip, whose function is to bypass the CPU and provide a direct connection between peripherals and memory, thus transferring the data as fast as possible. While the 8237 can transfer a byte of data between an I/O peripheral and memory in only 4 clocks, the 8088 would take 39 clocks:

			Number of Clocks
BACK:	MOV	AL,[SI]	10
	OUT	PORT,AL	10
	INC	SI	2
	LOOP	BACK	17
	;total clocks		39

One problem with using DMA is that there is only one set of buses (one set of each bus: data bus, address bus, control bus) in a given computer and no bus can serve two masters at the same time. The buses can be used either by the main CPU 80x86 or the 8237 DMA. Since the 80x86 has primary control over the buses, it must give permission to DMA to use them. How is this done? The answer is that any time the DMA needs to use the buses to transfer data, it sends a signal called HOLD to the CPU and the CPU will respond by sending back the signal HLDA (hold acknowledge) to indicate to the DMA that it can go ahead and use the buses. While the DMA is using the buses to transfer data, the CPU is sitting idle, and conversely, when the CPU is using the bus, the DMA is sitting idle. After DMA finishes its job it will make HOLD go low and then the CPU will regain control over the buses. See Figure 15-1.

For example, if the DMA is to transfer a block of data from memory to an I/O device such as a disk, it must know the address of the beginning of the block (address of first byte of data) and the number of bytes (count) it needs to transfer. Then it will go through the following steps.

1. The peripheral device (such as the disk controller) will request the service of DMA by pulling DREQ (DMA request) high.
2. The DMA will put a high on its HRQ (hold request), signaling the CPU through its HOLD pin that it needs to use the buses.
3. The CPU will finish the present bus cycle and respond to the DMA request by putting high on its HLDA (hold acknowledge), thus telling the 8237 DMA that it can go ahead and use the buses to perform its task. HOLD must remain active high as long as DMA is performing its task.
4. DMA will activate DACK (DMA acknowledge) which tells the peripheral device that it will start to transfer the data.

5. DMA starts to transfer the data from memory to peripheral by putting the address of the first byte of the block on the address bus and activating MEMR, thereby reading the byte from memory into the data bus; it then activates IOW to write it to the peripheral. Then DMA decrements the counter and increments the address pointer and repeats this process until the count reaches zero and the task is finished.
6. After the DMA has finished its job it will deactivate HRQ, signaling the CPU that it can regain control over its buses.

This above discussion indicates that DMA can only transfer information; unlike the CPU, it cannot decode and execute instructions. Notice also that when the CPU receives a HOLD request from DMA, it finishes the present bus cycle (but not necessarily the present instruction) before it hands over control of the buses to the DMA. This is in contrast to a hardware interrupt, in which the CPU finishes the present instruction before it responds with INTA. One could look at the DMA as a kind of CPU without the instruction decoder/executer logic circuitry. For the DMA to be able to transfer data it is equipped with the address bus, data bus, and control bus signals IOR, IOW, MEMR, and MEMW.

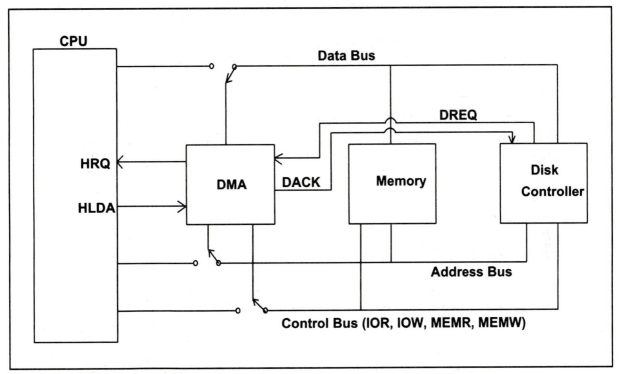

Figure 15-1. DMA Usage of System Bus

Review Questions

1. True or false. When the DMA is working, the CPU is sitting idle.
2. True or false. When the CPU is working, the DMA is sitting idle.
3. True or false. No bus can serve two masters at the same time.
4. True or false. The main CPU (80x86) has control over all the system buses.
5. To get control over the system bus the _____ (INTR, HOLD) pin of the 80x86 is activated.
6. The 80x86 CPU informs the peripheral that it relinquishes control over the system bus through its _____ pin.
7. The HOLD is an _____ (input, output) for the 80x86 CPU.
8. The HLDA is an _____ (input, output) for the 80x86 CPU.

SECTION 15.2: 8237 DMA CHIP PROGRAMMING

The Intel 8237 DMA controller is a 40-pin chip. It has four channels for transferring data, and each must be used for one device. For example, one is used for the floppy disk, one for the hard disk, and so on. Of course, only one device can use the DMA to transfer data at a given time. With every channel there are 2 associated signals, DREQ (DMA request) and DACK (DMA acknowledge). DREQ is an input to DMA coming from the peripheral device (such as the hard disk controller) and DACK is an output signal from the 8237 going to the peripheral device. From the 8237 DMA, there is only one HOLD and one HLDA that are connected to HOLD and HLDA of the 80x86. This means that four channels from four different devices can request use of the system buses, but DMA decides who gets control based on the way its priority register has been programmed. Every channel of the 8237 DMA must be initialized separately for the address of the data block and the count (the size of the block) before it can be used. This initialization involves writing into each channel:

1. The address of the first byte of the block of data that must be transferred (called the *base address*).
2. The number of bytes to be transferred (called the *word count*).

After initialization, each channel can be enabled and controlled with the use of a control word. There are many modes of operation and these various modes and options must be programmed into the 8237's internal registers. To access these registers, the 8237 provides 4 address pins, A0 - A3, along with the CS (chip select) pin. Since each channel needs separate addresses for the base address and the word count, a total of 8 ports is set aside for those alone. Table 15-1a shows the internal addresses of the 8237 registers for each channel. Example 15-1 shows how these addresses are generated.

Example 15-1

Find the port addresses for the base address and word count of each channel of the 8237 for Figure 15-2 ($\overline{CS}$ is activated by A7 - A4 =1001 binary).

Solution:

From Table 15-1a, one can get the addresses found in Table 15-2.

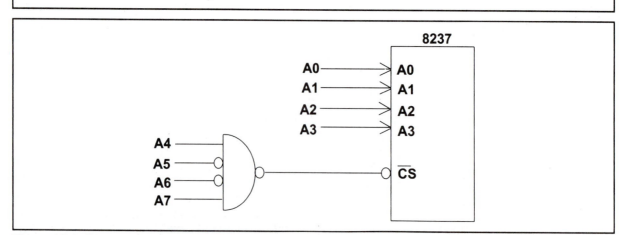

Figure 15-2. Diagram for Example 15-1

CHAPTER 15: DIRECT MEMORY ACCESSING; THE 8237 DMA CHIP

Table 15-1a: 8237 Internal Addresses for Writing Transfer Addresses and Counts

CH	Register	R/W	$\overline{CS}$	$\overline{IOR}$	$\overline{IOW}$	A3	A2	A1	A0
0	Base and current address	Write	0	1	0	0	0	0	0
	Current address	Read	0	0	1	0	0	0	0
	Base and current word count	Write	0	1	0	0	0	0	1
	Current word count	Read	0	0	1	0	0	0	1
1	Base and current address	Write	0	1	0	0	0	1	0
	Current address	Read	0	0	1	0	0	1	0
	Base and current word count	Write	0	1	0	0	0	1	1
	Current word count	Read	0	0	1	0	0	1	1
2	Base and current address	Write	0	1	0	0	1	0	0
	Current address	Read	0	0	1	0	1	0	0
	Base and current word count	Write	0	1	0	0	1	0	1
	Current word count	Read	0	0	1	0	1	0	1
3	Base and current address	Write	0	1	0	0	1	1	0
	Current address	Read	0	0	1	0	1	1	0
	Base and current word count	Write	0	1	0	0	1	1	1
	Current word count	Read	0	0	1	0	1	1	1

(Reprinted by permission of Intel Corporation, Copyright Intel Corp. 1983)

Table 15-1b: 8237 Internal Addresses for Commands/Status

A3	A2	A1	A0	$\overline{IOR}$	$\overline{IOW}$	Operation
1	0	0	0	0	1	Read status register
1	0	0	0	1	0	Write command register
1	0	0	1	0	1	Illegal
1	0	0	1	1	0	Write request register
1	0	1	0	0	1	Illegal
1	0	1	0	1	0	Write single mask register bit
1	0	1	1	0	1	Illegal
1	0	1	1	1	0	Write mode register
1	1	0	0	0	1	Illegal
1	1	0	0	1	0	Clear byte pointer flip-flop
1	1	0	1	0	1	Read temporary register
1	1	0	1	1	0	Master clear
1	1	1	0	0	1	Illegal
1	1	1	0	1	0	Clear mask register
1	1	1	1	0	1	Illegal
1	1	1	1	1	0	Write all mask register bits

(Reprinted by permission of Intel Corporation, Copyright Intel Corp. 1983)

Table 15-2: 8237A Address Selection for Example 15-1

Binary Address		Hex		Read/
A7 A6 A5 A4	A3 A2 A1 A0	Address	Function	Write
1 0 0 1	0 0 0 0	90	CHAN0 memory address register	R/W
1 0 0 1	0 0 0 1	91	CHAN0 count register	R/W
1 0 0 1	0 0 1 0	92	CHAN1 memory address register	R/W
1 0 0 1	0 0 1 1	93	CHAN1 count register	R/W
1 0 0 1	0 1 0 0	94	CHAN2 memory address register	R/W
1 0 0 1	0 1 0 1	95	CHAN2 count register	R/W
1 0 0 1	0 1 1 0	96	CHAN3 memory address register	R/W
1 0 0 1	0 1 1 1	97	CHAN3 count register	R/W

The two sets of information needed in order to program a channel of the 8237 DMA to transfer data are (1) the address of the first byte of data to be transferred, and (2) how many bytes of data are to be transferred.

For set 1, the channel's memory address register must be programmed. Since the memory address register of the 8237 is 16 bits and the data bus of the 8237 is 8 bits, one byte at a time, right after each other, is sent in to the same port address. For set 2, the channel's count register is programmed. The count can go as high as FFFFH. Since the count register is 16 bits and the data bus of the DMA is only 8 bits, it takes two consecutive writes to program that register. This is shown in Example 15-2.

Example 15-2

Assume that channel 2 of the DMA in Example 15-1 is to transfer a 2K (2048) byte block of data from memory locations starting at 53400H. Use the port addresses of Example 15-1 for the DMA to program the memory address register and count register of channel 2.

Solution:

The port address for the channel 2 memory address register and count register in Example 15-1 are 94H and 95H, respectively. The initialization will look as follows:

```
        MOV   AX,3400H      ;load lower 4 digits of start address
        OUT   94H,AL        ;send out the low byte of the address
        MOV   AL,AH         ;
        OUT   94H,AL        ;send out the high byte of the address
        MOV   AX,2048       ;load block size into AX
        OUT   95H,AL        ;send out the low byte of the count
        MOV   AL,AH         ;
        OUT   95H,AL        ;send out the high byte of the count
```

The contents of the memory address and count registers can be read in the same manner (low byte first, then high byte) to monitor these registers at any time. From looking at the above program one might ask, since the system address bus is 20 bits and the memory address is 53400H, why does this program use 16-bit addresses? This is a limitation of the 8237 DMA. In the 8237, not only is the register holding the address of the block 16 bits, but in addition there are only 16 address pins that carry the addresses. The IBM PC solves this problem by using an external 4-bit register to hold the upper bits of the address in 1M memory range. How to do that is discussed in Section 15.3.

8237's internal control registers

Although the 8237 has four channels and each channel must be programmed separately for the base address and count, there is only one set of control/command registers used by all channels. These registers are shown in Table 15-1a. To understand how to access those registers, look at Example 15-3.

Of these 8 registers, only the most essential ones will be explained in detail here. The reader can refer to Intel manuals for information concerning others. The functions of 8237 pins are described in Section 15.4 in the context of some real-life designs, such as the IBM PC/XT.

Example 15-3

Use the circuit in Example 15-1 to find the address of the 8237 DMA control registers.

Solution:

Using Table 15-1b and substituting for A7 - A3 gives the information in Table 15-3.

Table 15-3: Address Selection for Example 15-3

Binary Address		Hex	Register Name	Read/
A7 A6 A5 A4	A3 A2 A1 A0			Write
1 0 0 1	1 0 0 0	98	Status/command register	R/W
1 0 0 1	1 0 0 1	99	Request register	W
1 0 0 1	1 0 1 0	9A	Single mask register bit	W
1 0 0 1	1 0 1 1	9B	Mode register	W
1 0 0 1	1 1 0 0	9C	Clear byte pointer	W
1 0 0 1	1 1 0 1	9D	Master clear/temporary register	R/W
1 0 0 1	1 1 1 0	9E	Clear mask register	W
1 0 0 1	1 1 1 1	9F	Mask register bits	W

Command register

This is an 8-bit register used for controlling the operation of the 8237 (see Figure 15-3). It must be programmed (written into) by the CPU. It is cleared by the RESET signal from the CPU or the master clear instruction of the DMA. The function of each bit is described below.

The 8237 is capable of transferring data (1) from a peripheral device to memory (reading from disk), (2) from memory to a peripheral device (writing the file into disk), or (3) from memory to memory. One example of the use of the memory-to-memory option is what is called *shadow RAM*. In computers such as 386- and 486-based systems, the access time of ROM is too long. However, the system can copy the ROM into a portion of RAM and allow the CPU to access it from RAM, which has a much shorter access time than ROM.

D0 gives the option to use only channels 0 and 1 for transferring a block of data from memory to memory. Why the need for two channels? Channel 0 must be used for the source and channel 1 for the destination. Channel 0 reads the byte into a temporary register inside the 8237, and then channel 1 will write it to the destination. This is in contrast to I/O-to-memory or memory-to-I/O transfers, in which the data is read into the data bus and transferred to the destination, all without being saved anywhere temporarily.

D1 is used only when the memory-to-memory option is enabled and can be used to disable the memory incrementation/decrementation of channel 0 in order to write a fixed value into a block of memory.

D2 is used to enable or disable DMA.

D3 gives the option to choose between the normal memory cycle of 4 clock pulses and compressed timing of 2 clock pulses per memory cycle. There are 4 clock pulses per byte after the initial delay, assuming that the high byte address is already latched. If every byte of transfer requires both high-byte and low-byte addresses, an extra clock pulse for the address latch is required, which makes the bus cycle 5 clock pulses. The same is the case for the compressed option, making it 3 clock ticks per bus cycle.

D4 gives the option of using the four channels on fixed priority or rotating priority. If fixed priority is chosen, DREQ0 has the highest priority and DREQ3 has the lowest priority. If more than one DREQ is activated at the same time, it will always respond to the one with the highest priority. In rotating mode, DREQ0 again has the highest priority and DREQ3 the lowest, but the system rotates through DREQ0, DREQ1, DREQ2, and DREQ3 in that order, servicing one request from each if present. In other words, when DREQ0 is served it will not be given a chance until the rest of the DREQs are given a chance. This prevents monopolization by the DREQ with the highest priority.

D5 allows time for the write signal to be extended for slow devices.

D6 gives the option of programming the activation level of DREQ. It can be an active-high or active-low signal.

D7 gives the option of programming the activation level of DACK. It can be an active-high or active-low signal.

The command byte is issued to this register through port address X8H, where X is the combination provided to activate $\overline{\text{CS}}$, as shown in Example 15-4.

Example 15-4

Program the command register of the 8237 in Example 15-3 for the following options: no memory-to-memory transfer, normal timing, fixed priority, late write, DREQ and DACK both active high.

Solution:
From Figure 15-3, the command byte would be 1000 0000 =80H and the program is

```
MOV   AL,80H        ;load the command byte into AL
OUT   98H,AL        ;issue the command byte to port 98H
```

Example 15-5

Assume that the CPU is doing some very critical processing and that the 8237 DMA should be disabled. Use the ports in Example 15-3 to show the program.

Solution:
To disable the 8237, send 0000 0100 =04H to the command register as follows:

```
MOV   AL,04H
OUT   98H,AL
```

Status register

This is an 8-bit register that can only be read by the CPU through the same port address as the command register. This register is often referred to as RO (read only) in PC documentation. As mentioned above, the port is X8 hex, where X is for $\overline{\text{CS}}$. It contains various information about the operating state of the four channels. The lower four bits, **D0 - D3**, are used to indicate if channels 0 - 3 have reached their TC (terminal count). TC is set high when the count register has been decremented to zero. This gives the option to monitor the count register by software. This monitoring also can be done by hardware through the $\overline{EOP}$ pin of the 8237, as we will see in the next section. The upper four bits, **D4 - D7,** of the status register keep count of pending DMA requests. This information can be used by the CPU to see which channel has a pending DMA request. See Figure 15-4.

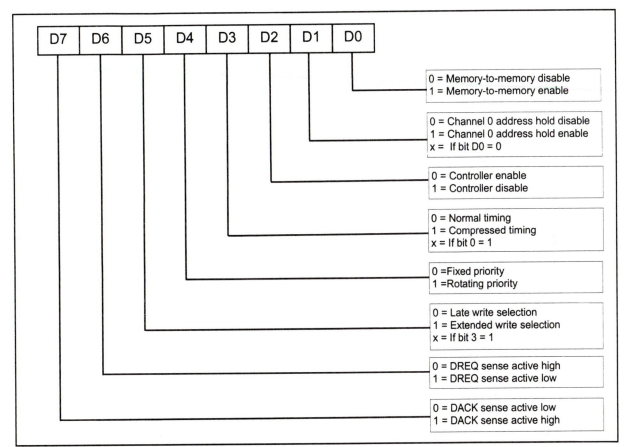

Figure 15-3. 8237 Command Register Format
(Reprinted by permission of Intel Corporation, Copyright Intel Corp. 1983)

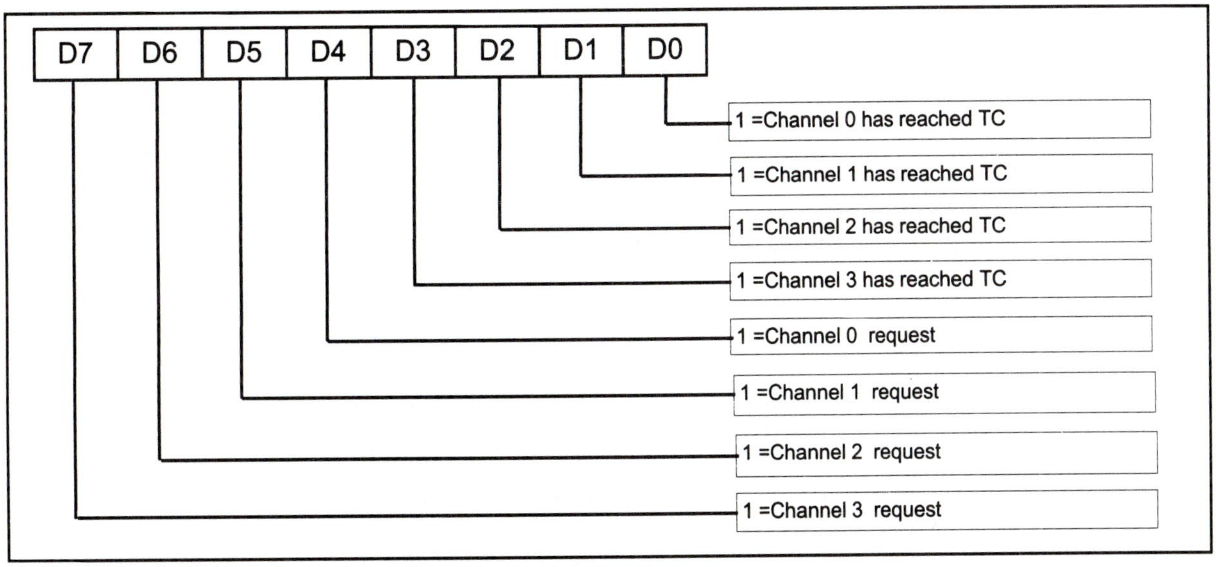

Figure 15-4. 8237 Status Register Format
(Reprinted by permission of Intel Corporation, Copyright Intel Corp. 1983)

Mode register

This register can only be written to by the CPU through port address XBH, where X is the address combination for $\overline{CS}$ activation. Of the 8 bits of the mode register, the lower two, **D0** and **D1**, are used for channel selection. The other 6 bits are used to select various operation modes to be used for the channel selected by bits D0 and D1. **D2** and **D3** specify data transfer mode. In the write transfer option DMA transfers from an I/O device (such as a disk) to memory by activating IOR and MEMW. Reading from memory to an I/O is a read transfer and is achieved by activating MEMR and IOW. The verify transfer is called pseudo and is like a read or write except that it does not generate any control signals, such as IOR, MEMR, etc. **D4** is used for autoinitialization. If enabled, the memory address register and the count register are reloaded with their original values at the end of a DMA data transfer (when the count register becomes zero). In this way those registers are programmed only once and the original values are saved internally. **D5** gives the option to increment or decrement the address. **D6** and **D7** determine the way the 8237 is used. The options are:

1. Demand mode, where the transfer of data continues until DREQ is deactivated or the terminal count has been reached. This ensures that the DMA can finish the job without interruption even though it means monopolization of the system buses by DMA for the duration of the transfer of the entire block of data.

2. Block mode, which is the same as demand mode except that DREQ can be deactivated after the DMA cycle starts and the process of data transfer will go on until the TC (terminal count) state has been reached. In other words, there is no need to keep the DREQ high for the duration of the data transfer.

3. Single mode, where if DREQ is held active, the DMA transfers one byte of data, then allows the 80x86 to gain control of the system bus by deactivating its HRQ for one bus cycle. This process goes on alternating access to the system bus between the CPU and DMA until the TC has been reached, and then autoinitialization will happen if that choice has been made in the control word. This is the option used in all PC, PS, and compatibles since the DMA and CPU alternately share the system buses, allowing both to do their job without either monopolizing the buses.

4. Cascade mode, in which several DMAs can be cascaded to expand the number of DREQs to more than 4. This option is used in IBM PC AT and higher computers, as we will see in Section 15.5. The 8088 PC/XT uses only one 8237. Example 15-6 shows the programming of the mode register.

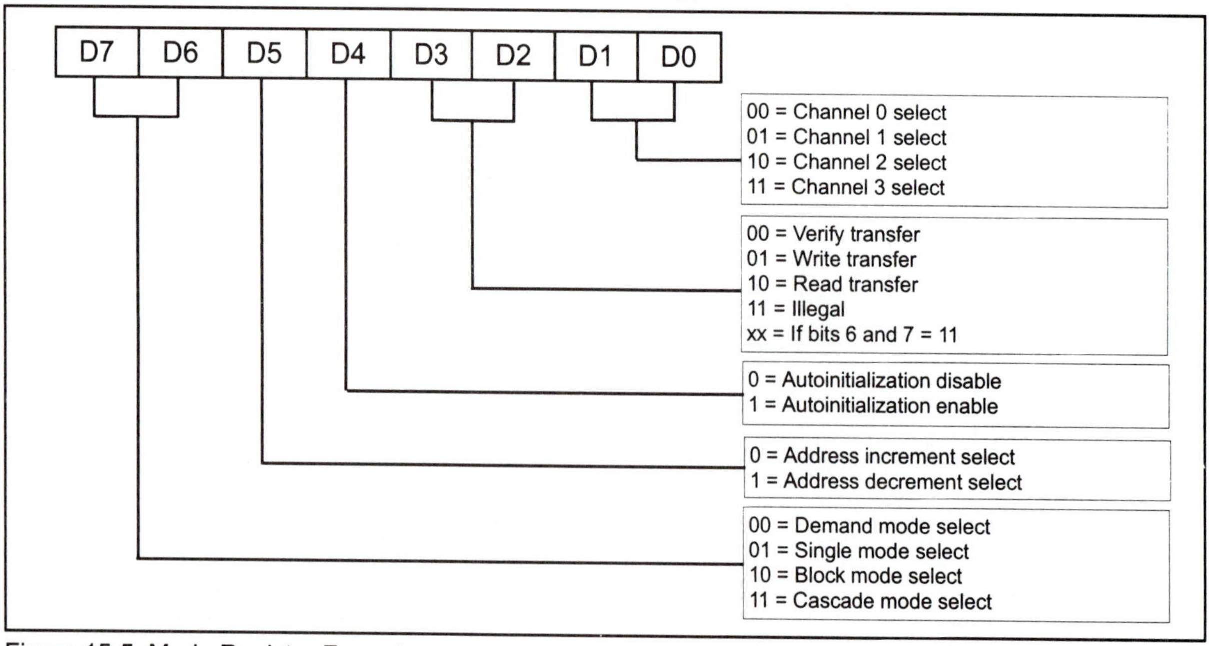

Figure 15-5. Mode Register Format
(Reprinted by permission of Intel Corporation, Copyright Intel Corp. 1983)

Single mask register

This register can only be written to by the CPU through port address XA hex, where X is for $\overline{CS}$. Of the 8 bits of this register, only three are used. **D0** and **D1** select the channel. **D2** clears or sets the mask bit for that channel. It is through this register that the DREQ input of a specific channel can be masked (disabled) or unmasked (enabled). For example, if the value 00000101 is written to this register, it will mask (block) DREQ1 and the DMA will not respond to DREQ of channel 1 when DREQ1 is activated. While the command register can be used to disable the whole DMA chip, this register allows the programmer to disable or enable a specific channel. The only problem is that only one channel can be masked or unmasked at a time. To mask or unmask more than one channel, the all mask register is used. Figure 15-6 shows the single mask register format.

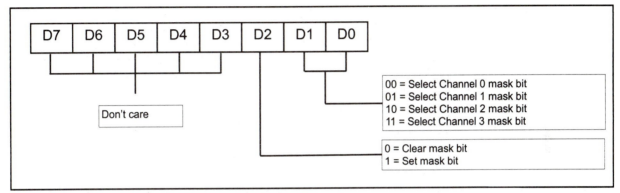

Figure 15-6. 8237 Single Mask Register Format
(Reprinted by permission of Intel Corporation, Copyright Intel Corp. 1983)

All mask register

In function, this register is similar to the single mask register except that all 4 channels can be masked or unmasked with one write operation. For example, if 00000010 is written to this register, it will mask the OUT of channel 1 and unmask (enable) the other channels. See Figure 15-7. Again this register can only be written to by the CPU through the port address XFH, where X is for $\overline{CS}$ activation.

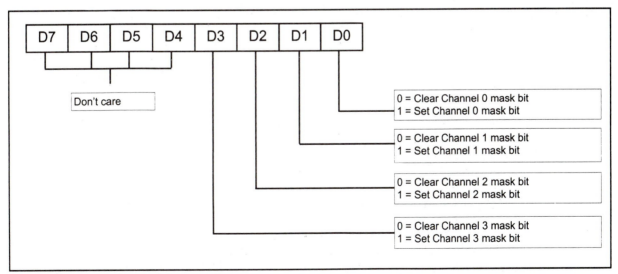

Figure 15-7. 8237 All Mask Register Format
(Reprinted by permission of Intel Corporation, Copyright Intel Corp. 1983)

Example 15-6

Program the 8237's mode register of Example 15-3 to select channel 2 to transfer from memory to I/O using autoinitialization, address increment, single-byte transfer.

Solution:
From Figure 15-5, with these options, the mode register must have 01011010 =5AH. The port address for the command register is 9BH, which results in

```
MOV    AL,5AH
OUT    9BH,AL
```

Example 15-7

Program the 8237 of Example 15-3 to enable channel 2.

Solution:
From Figure 15-6, the value for the single mask register to enable (unmask) channel 2 is 0000 00010 =02H and is sent to port 9AH as follows:

```
MOV    AL,02
OUT    9AH,AL
```

Master clear/temporary register

This register must only be written to by the CPU through port address XDH, where X is for $\overline{CS}$ activation. The byte sent to this register does not matter since it simply clears the status, command, request, and mask registers and forces the DMA to the idle cycle. This is the same as activating the hardware RESET of the 8237. If an attempt is made to read from this register, DMA will provide the last byte of data that was transferred during the memory-to-memory transfer. Note that when the DMA is doing an I/O-to-memory or memory-to-I/O transfer, it transfers the data directly between these two sections of the computer without bringing it into the DMA, but in memory-to-memory transfers it must bring each byte into the DMA before it sends it to the destination, since it has to switch the contents of the address bus for the source and destination. This is similar to string instruction MOVSB, except that it is performed by the DMA instead of the CPU.

Clear mask register

This register can be written to by the CPU only through port address XEH, where X is for $\overline{CS}$. The bit patterns written to it do not matter. Its function is to clear the mask bits of all 4 channels, thereby enabling them to accept the DMA request through the DREQs.

Review Questions

1. How many address bits are used to select a register inside the 8237?
2. For an 8237, why are addresses X0H to XFH used to access its internal registers?
3. True or false. To use a channel to transfer data, both the memory address register and count register for that channel must be programmed.
4. State the functions of the memory address register and the count register.
5. Show instructions to program channel 0 memory address and count registers to transfer 4K starting from offset 1440H. Use port addresses from Example 15-1.
6. In the fixed-priority scheme, which channel has the highest priority?
7. True or false. Programming some control registers is optional (depending on how the 8237 is used), but the command register must always be programmed.
8. True or false. The command register is accessed by $\overline{CS}$ =0 and A3 - A0 =1000.
9. True or false. The mode register is accessed by $\overline{CS}$ =0 and A3 - A0 =1011.
10. True or false. The level of activation, high or low, for DREQ and DACK of each channel can be programmed.

SECTION 15.3: 8237 DMA INTERFACING IN THE IBM PC/XT

As shown in Figure 15-8, the 8237 DMA has 8 addresses, A0 - A7. Four of these, A0 - A3, form a bidirectional address bus, sending addresses into the 8237 to select one of the 16 possible registers, assuming that chip select is activated. In the IBM PC, chip select is activated by Y0 of the 74LS138 as shown in Figure 15-9. The address selection of the registers inside the 8237 is summarized as shown in Table 15-4, assuming zero for each x. The conditions for A6, A7, A8, A9, and AEN were discussed in Chapter 12 and will not be repeated here. From Table 15-4 it can be seen that port addresses 0 to 7 are assigned to the 4 channels, and 08 - 0F are assigned to the control registers commonly used by all the channels.

IOR	1		40	A7
IOW	2		39	A6
MEMR	3		38	A5
MEMW	4		37	A4
+5V	5		36	EOP
READY	6	8	35	A3
HLDA	7		34	A2
ADSTB	8	2	33	A1
AEN	9		32	A0
HRQ	10	3	31	Vcc
CS	11		30	DB0
CLK	12	7	29	DB1
RESET	13	A	28	DB2
DACK2	14		27	DB3
DACK3	15		26	DB4
DREQ3	16		25	DACK0
DREQ2	17		24	DACK1
DREQ1	18		23	DB5
DREQ0	19		22	DB6
GND	20		21	DB7

Figure 15-8: 8237A DMA Pin Layout
(Reprinted by permission of Intel Corporation, Copyright Intel Corp. 1983)

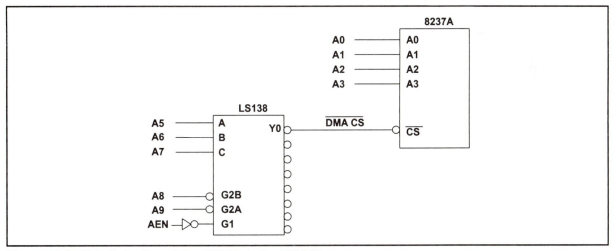

Figure 15-9. Chip Selection of the 8237A in the PC/XT

8237 and 8088 connections in the IBM PC

Since the DMA must be capable of transferring data between I/O and memory without any interference from the CPU, it must have all the required control, data, and address buses. Looking at Figure 15-10, one can see that the 8237 has its own data bus, D0 - D7. This is a bidirectional bus connected to the system bus D0 - D7. It also has all four control buses, $\overline{IOR}$, $\overline{IOW}$, $\overline{MEMR}$, and $\overline{MEMW}$. However, its address bus, A0 - A7, is only 8 bits. If the 8237 can transfer up to 64K bytes of data between I/O and memory, it must have 16 address lines, A0 - A15. Where are the other 8 address pins, A8 - A15? The answer is that the high byte of the 16-bit address changes only once, while D0 - D7 are used by the 8237 to send out the upper part of the address whenever A0 - A7 rolls over from FF to 00. There must be a device to latch and hold the A8 - A15 part of the address from the D0 - D7 data bus. This is the function of the 74LS373. The function of the ADSTB (address strobe) is to activate the latch whenever the 8237 provides the upper 8-bit address through the data bus. Similar to ALE, the ADSTB goes high only when D0 - D7 are used to provide the upper address, meaning that as long as ADSTB stays low, D0 - D7 are a normal data bus. Figure 15-11 diagrams the 8237

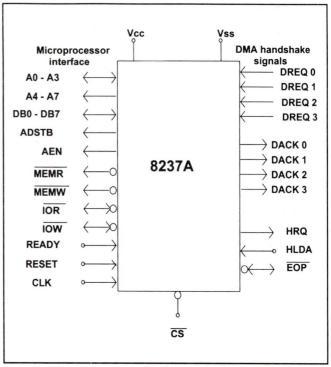

Figure 15-10. Block Diagram of the 8237A DMA
(Reprinted by permission of Intel Corporation, Copyright Intel Corp. 1983)

Table 15-4: PC 8237 Internal Register Port Addresses

Binary Address		Hex		Read/
A7 A6 A5 A4	A3 A2 A1 A0	Address	Function	Write
0 0 0 x	0 0 0 0	00	CHAN0 memory address register	R/W
0 0 0 x	0 0 0 1	01	CHAN0 count register	R/W
0 0 0 x	0 0 1 0	02	CHAN1 memory address register	R/W
0 0 0 x	0 0 1 1	03	CHAN1 count register	R/W
0 0 0 x	0 1 0 0	04	CHAN2 memory address register	R/W
0 0 0 x	0 1 0 1	05	CHAN2 count register	R/W
0 0 0 x	0 1 1 0	06	CHAN3 memory address register	R/W
0 0 0 x	0 1 1 1	07	CHAN3 count register	R/W
0 0 0 x	1 0 0 0	08	Status/command register	R/W
0 0 0 x	1 0 0 1	09	Request register	W
0 0 0 x	1 0 1 0	0A	Single mask register bit	W
0 0 0 x	1 0 1 1	0B	Mode register	W
0 0 0 x	1 1 0 0	0C	Clear byte pointer	W
0 0 0 x	1 1 0 1	0D	Master clear/temporary register	R/W
0 0 0 x	1 1 1 0	0E	Clear mask register	W
0 0 0 x	1 1 1 1	0F	Mask register bits	W

circuit connection.

It should be noted that A0 - A3 and all control buses are bidirectional, so that when programming the 8237 they can be used to communicate with the internal registers. As long as the CPU is idle and the DMA is in control transferring data, A0 - A7 and all the control buses are unidirectional.

One last point about A0 - A15 from the 8237 is that the system bus can be used by the CPU only when the 8237 is not functioning. This is ensured through the AEN signal. This signal was discussed in Chapter 9 and is summarized here.

AEN
0   80x86 is in control of the system bus
1   8237 DMA is in control of the system bus

Example 15-8

If the 8237 DMA is programmed to transfer data from memory locations starting at offset 3450H to an I/O device, explain the role of A0 - A7, D0 - D7, and ADSTB pins of 8237 in producing the 16-bit addresses A0 - A15.

Solution:

The upper byte of the address, A8 - A15, is provided through the data bus, D0 - D7, and ADSTB goes high, making the 74LS373 latch the 34H. The lower part of the address 50 is provided through A0 - A7. As the addresses are incremented producing 50, 51, 52, and so on, the A0 - A7 are changed in every memory cycle until DMA reaches address 34FFH. At this point, A0 - A7 is rolled over and D0 - D7 will provide 35H to the 74LS373 latch. The process continues until the count reaches zero.

The rest of the pins in Figure 15-10 are described below.

RESET is the input coming from the RESET of 8284.

$\overline{\text{CS}}$ is from the 74LS138 decoder, as shown earlier.

READY input is from the RDYDMA of the wait-state generation circuitry. The purpose of this is to extend the memory cycle of the DMA.

CLK is from the CLK of the 8284 and is equal to the working frequency, which is 4.7 MHz (210 ns clock cycle) in the PC/XT.

HOLD and **HLDA** are connected to the pins with the same name on the 80x86 CPU.

EOP (end of process) is inverted and becomes TC (terminal count). This signal is activated whenever the count register of any of the four channels is decremented to zero. This signal could be used with the DACK of a specific channel to prevent multiple DMA requests from that channel at the same time or could be used to inform the requesting device that the DMA has finished the job and it should deactivate its DREQ. In other words, EOP is a hardware pin indicating that the counter has reached zero. Using software one can monitor the count register of each channel by reading the status register, as was shown in Section 15.2.

DREQ0 and **DACK0** are the signals for channel 0 and are used for refreshing DRAM as explained in Section 15.4. While the DREQ is active high, the DACK0 is programmed to be active low by BIOS, as shown in Section 15.4.

DREQ1 - DREQ3 and **DACK1 - DACK3** are the signals for channel 1 to channel 3, and are available through the expansion slot. The assignment of these channels is discussed next.

SECTION 15.3: 8237 DMA INTERFACING IN THE IBM PC/XT   461

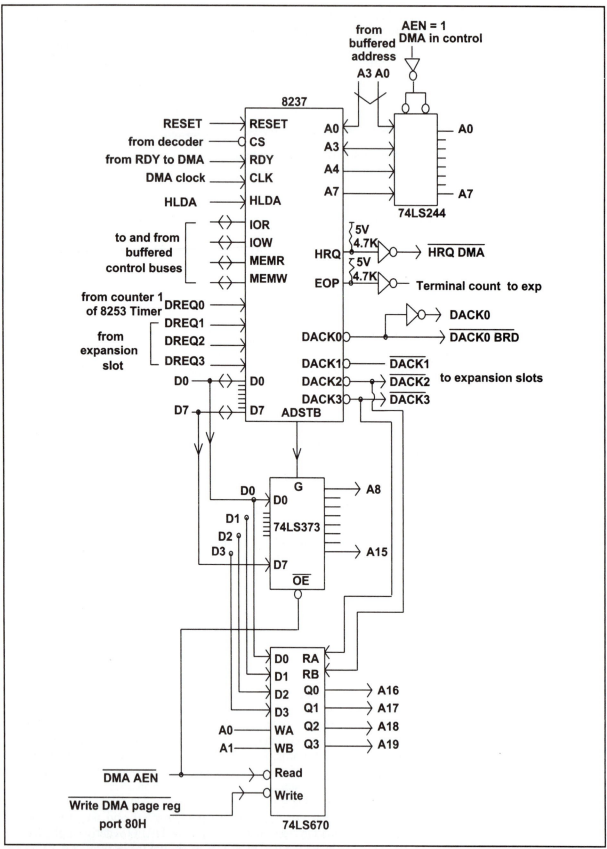

Figure 15-11. The 8237 DMA Circuit Connection in the PC/XT
(Reprinted by permission from "IBM Technical Reference" c. 1984 by International Business Machines Corporation)

Channel assignment of the 8237 in the IBM PC/XT

In the IBM PC/XT, each of the four channels of the 8237 is assigned in the following fashion.
1. Channel 0 for refreshing DRAM. In the PC/AT and compatibles this practice was abandoned.
2. Channel 1 is unused, but in many implementations it is used for networks.
3. Channel 2 usually is used for the floppy disk controller.
4. Channel 3 normally is used for the hard disk controller.

Inspecting IBM BIOS shows that 8237 channels 1, 2, and 3 have been initialized by programming the mode register. The mode register, which must be sent to port address 0BH, is as follows for channel 1 (from Figure 15-5):

D1,D0 = 01   for channel 1
D3,D2 = 00   for verify transfer
D4 = 0 autoinitialization disable
D5 = 0 for address increment
D7,D6 = 01   for single byte mode

D7   D0
0100 0001 = 41H mode register for channel 1

For channels 2 and 3, the value for the mode register is the same except that D0 and D1 are changed to 10 and 11, respectively. Therefore, channels 2 and 3 have mode register values of 42 and 43. The program could look like the following code.

```
MOV     DX,000BH      ;load the mode register address
MOV     AL,41H        ;chan1 mode reg value.
OUT     DX,AL
MOV     AL,42H        ;chan2 mode reg value
OUT     DX,AL
MOV     AL,43H        ;chan3 mode reg value
OUT     DX,AL
```

The way IBM BIOS does the initialization is slightly more compact:

```
E136   B20B   474            MOV   DL,0BH        ;DX=000B
....   ....   ...            ...........
E142   B103   481            MOV   CL,3
E144   B041   482            MOV   AL,41H        ;SET MODE FOR CHANNEL 1
E146          483 C18A:
E146   EE     484            OUT   DX,AL
E147   FEC0   485            INC   AL            ;POINT TO NEXT CHANNEL
E149   E2FB   486            LOOP  C18A
```

(Reprinted by permission from "IBM BIOS Technical Reference" c. 1984 by International Business Machines Corp.)

These channels are programmed by the device that uses them when the device is installed. For example, the hard disk controller ROM programs channel 3 according to its specifications.

DMA page register

Since the 8237 can provide only A0 - A15, the 74LS670 4x4 file register is used to provide the rest of the A16 - A19 physical address bits for each of the 4 channels. See Figure 15-13. These registers can only be written to by the 8086/88 CPU through the D0 - D3 data bus using the OUT instruction. Since the 74LS670 is a 4x4 register, there are four port locations assigned, one for each of them. As discussed in Chapter 12 and shown in Figure 15-12, the 74LS138 is used to decode the addresses. The address calculation of those ports is as follows assuming that x's are all zero:

AEN	A9	A8	A7	A6	A5	A4	A3	A2	A1	A0		Hex	
0	0	0	1	0	0	x	x	x	0	0	=	80	4-bit reg (not used PC/XT)
0	0	0	1	0	0	x	x	x	0	1	=	81	4-bit reg for channel 2
0	0	0	1	0	0	x	x	x	1	0	=	82	4-bit reg for channel 3
0	0	0	1	0	0	x	x	x	1	1	=	83	4-bit reg for channel 1

Of these four registers (each 4-bit), only three are used since channel 0 is used for refreshing DRAMs and does not require the upper 4-bit address (as will be seen in the next section). To read from these registers, 8237's DACK must be programmed to be active low using the command mode register. Depending on which DREQ is active, it can read the contents of its designated register as follows:

DACK2 RB	DACK3 RA	
1	1	channel 1
1	0	channel 3
0	1	channel 2
0	0	never happens since DACK2 and DACK3 can never be active at the same time

Writing into the DMA page register is done only by the 8088 CPU since addresses A0 and A1 from the CPU are connected to WA and WB, respectively, as shown in Figure 15-12. The reading of the page register can happen only when the DMA is in control since RA and RB are connected to DACK3 and DACK2 and DACK is not activated until the related DREQ is activated.

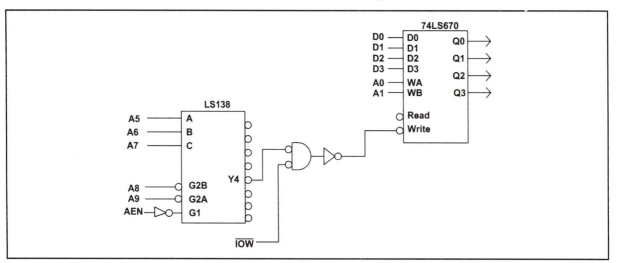

Figure 15-12. DMA File Register Address Decoding (PC/XT)

DMA data transfer rate of the PC/XT

It takes many more clocks to transfer one byte of data between an I/O device and memory using the 8088 CPU than using the 8237 DMA chip. What is the data transfer rate of DMA channels 1, 2, and 3? The wait-state circuitry inserts one clock into the memory cycle of channels 1 - 3 through the RDYDMA signal. Since the DMA on its own requires 5 clock pulses, that gives a total of 6 clock pulses for DMA channels 1 - 3. In between each DMA cycle there is one CPU memory cycle, which is 4 clock pulses. This gives a total of 10 clock pulses for the transfer of every byte of data by channels 1 - 3. Since every clock pulse is 210 ns (1/4.7 MHz) in the IBM PC/XT, it takes 10×210 ns = 2100 ns for the transfer of one byte of data by the DMA. Thus the transfer rate of the 8237 in the IBM PC/XT is 476,190 bytes per second (1/2100 ns = 476,190). This assumes that there is no other device inserting a wait state into the memory cycle. Both DMA and the 8088/86 CPU work on the same 4.7 MHz frequency, unlike other 80x86 PCs.

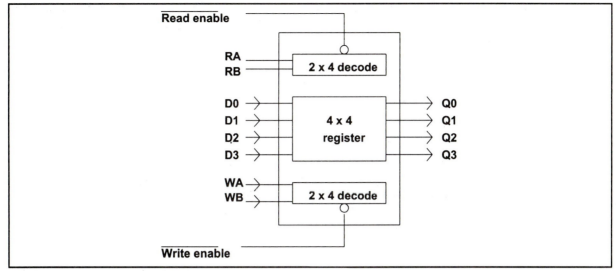

Figure 15-13. Inside the 74LS670
(Reprinted by permission of Texas Instruments, Copyright Texas Instruments Corporation, 1988)

Review Questions

1. What port addresses are assigned to the 8237 in the PC/XT?
2. What port address is assigned to the command register of the 8237 in the PC/XT?
3. If the 8237 has only A0 - A7, how is the 16-bit address A0 - A15 provided?
4. What is the function of ADSTB? Is it an out or in signal pin for the 8237?
5. The 8237 in the IBM PC/XT is programmed to have channel ___ as the highest priority and channel ___ as the lowest priority.
6. Which IC chip provides the 4-bit address A16 - A19 of the memory location accessed by the DMA?
7. What port addresses are assigned to the DMA file register?
8. In the IBM PC/XT, which DMA channels are used internally by the motherboard for refreshing DRAM, and which are available through the expansion slot?
9. In the IBM PC/XT for DMA 8237 channels 1, 2, and 3, what is the DMA data-transfer bus cycle? How much time is that if each clock is 210 ns?
10. Rework Question 9 for channel 0.

SECTION 15.4: REFRESHING DRAM USING CHANNEL 0 OF THE 8237

Since DREQ0 has the highest priority in the 8237, the PC designers at IBM assigned to it the task of refreshing DRAM. In other words, if the floppy disk and the request for the refresh come at the same time, DMA will take care of the refresh request before answering the disk request for service. Otherwise, DRAM would lose its data. At power up, BIOS initializes channel 0 of the 8237 to prepare it for the DRAM refreshing task. The following shows the values for the various registers and then the programming of the 8237 for initialization.

1. The memory address register is 0000, which does not need to be programmed and can use the default value of 0000.
2. The count register is FFFF for all the 64K locations written to port address 01.
3. The mode register that must be sent to port address 0BH is as follows:

D1,D0 = 00   for channel 0
D3,D2 = 10   for read transfer
D4 = 1 autoinitialization
D5 = 0 for address increment
D7,D6 = 01   for single byte mode

D7 ____ D0
0101 1000 = 58H

---

4. The command register that is written to port address 08H is as follows:

D0 = 0 memory-to-memory disable
D1 = x
D2 = 0 DMA controller enable
D3 = 0 normal timing
D4 = 0 fixed priority
D5 = 0 late write
D6 = 0 DREQ active high
D7 = 0 DACK active low

D7   D0
0000 0000 = 00H

5. The single mask register bit that goes to port address 0AH is as follows:

D1,D0 = 00 for channel 0
D2 = 0 for clear mask bit to allow DREQ0 to come in
D7,D6,D5,D4,D3 = xxxxx

D7   D0
0000 0000 = 00H

The initialization process is as follows:

```
MOV    AL,0FFH    ;count register value
OUT    01,AL      ; to count register of chan0 (Low Byte)
OUT    01,AL      ; to count register of chan0 (High byte)
MOV    AL,58H     ;mode register value
OUT    0BH,AL     ;write to mode register
MOV    AL,00      ;command register value
OUT    08,AL      ;write to command register
MOV    AL,00      ;single mask register value
OUT    0AH,AL     ;write to single mask register
```

The IBM BIOS version of the same program is:

```
.... ....   ...      ...............................
0008        ...   26   DMA08 EQU 08;DMA STATUS REG PORT ADDR
0000              27   DMA   EQU 00 ;DMA CHAN 0 ADDR REG PORT ADDR
.... ....   ...      ...      ...    .....................
                  464   ;INITIALIZE AND START DMA FOR MEMORY REFRESH
... ...     ...      ...............................
E12F  B0FF  ...   470        MOV    AL,0FFH;SET CNT OF 64K FOR RAM REF
E131  E601        471        OUT    DMA+1,AL
E133  50          472        PUSH   AX
E134  E601        473        OUT    DMA+1,AL
E136  B202        474        MOV    DL,0BH        ;DX=000B
E138  B058        475        MOV    AL,058H       ;SET DMA  MODE,CH0,READ
                                                  ;AUTO INIT
E13A  EE          476        OUT    DX,AL         ;WRITE DMA MODE REG
E13B  B000        477        MOV    AL,00         ;ENABLE DMA CONTROLLER
E13D  E608        478        OUT    DMA+8,AL      ;SETUP DMA COMMAND REG
E13F  50          479        PUSH   AX
E140  E60A        480        OUT    DMA+10,AL     ;ENABLE CHAN 0 FOR REFRESH
.... ....   ...      ....      ...................................
```

Notice in lines 472 and 479 the "PUSH AX" instruction in between two OUT instructions. These PUSH instruction have no software significance except to prevent any hardware timing problems associated with two consecutive OUT instructions issued to the same peripheral chip.

Refreshing DRAM with the 8237

DRAM cells must be refreshed at least once every 15.6 microseconds. This rule applies to all DRAMs regardless of their capacity. Inside DRAMs, the storage cells are arranged in a matrix of rows and columns (always 4 arrays of n by n cells). It would be impossible to refresh all the cells one at a time; therefore, refreshing is done one row at time. When a row is refreshed, all the columns in that row are refreshed at the same time, so there is no need to refresh each column in every row individually. Now one might ask, what happens when the density of DRAM goes up with every new generation of memory chips? Does it mean that all rows in 1M- and 4M-bit memory chip have to be refreshed? The answer is yes. If the answer is yes, doesn't this take up too much time? To avoid taking too much time for refreshing, DRAM designers arranged the matrices of rows and columns in such a way that for every 128 rows refreshing needs to be done only once every 2 ms, or to put it another way, every cell regardless of the size of the DRAM chip must be refreshed once every 15.6 µs (2 ms divided by 128). For example, in 64K-bit chips refreshing the 128 rows need only be done every 2 ms. In the 256K-bit chips there are 256 rows, all of which must be refreshed within 4 ms (4 ms divided by 256 = 15.6 µs). The case is the same for 1M bit chips; the 512 rows must be refreshed within 8 ms. Again, dividing 8 ms by 512 rows gives 15.6 µs per row. This is the standard rule in the industry and will remain so for the foreseeable future even though the capacity of DRAMs is approaching 64M bits.

DRAM Capacity	Number of Cell Matrices	Number of Rows	Minimum for All Rows	Refresh Time for One Row
64Kb	4 -- 128x128	128	2 ms	15.6 µs
256Kb	4 -- 256x256	256	4 ms	15.6 µs
1Mb	4 -- 512x512	512	8 ms	15.6 µs
4Mb	4 -- 1024x1024	1024	16 ms	15.6 µs

Refreshing in the IBM PC/XT

In the IBM PC/XT, DREQ0 is activated by a special pulse every 15.06 microseconds to perform the task of refreshing the DRAMs. In Chapter 13 we discussed how this pulse is generated by the 8253 timer chip. As seen above, DMA's channel 0 is initialized by BIOS for the count of 65,536 starting at memory location 00, incrementing addresses with options of read transfer, autoinitialization, increment address, and single-byte mode. The DMA continuously does the read transfer cycle by activating $\overline{MEMR}$, providing the addresses to the inputs of DRAMs and activating RAS, thereby sequentially cycling through all 128 rows as required by the DRAMs. In the course of the READ transfer, DMA provides the $\overline{MEMR}$ and $\overline{IOW}$ control signals, but only $\overline{MEMR}$ is used to perform refreshing. $\overline{IOW}$ control is unused since no data is available at the DOUT of the DRAMs. Some call this reading the data and writing them to a *dummy* port. It must be made clear that when a row is being refreshed by activation of the RAS, every row of every bank of memory on the system is being refreshed at the same time.

DMA cycle of channel 0

Since channel 0 is used to refresh DRAM, the wait-state circuitry will not insert any wait state into its cycle time. This is in order to shorten refreshing time. Therefore the 8237's cycle time is its default of 4 clocks for channel 0. This gives a total of 128 rows × 4 clock cycles × 210 ns = 0.10752 ms for the time to refresh DRAMs every 2 ms. Dividing this number by 2 ms (1.93 ms since 128 × 15.09 = 1.93 ms) gives 5%. In other words, only 5% of the time, the CPU cannot access memory since DRAM cannot be accessed during refreshing. The method of refreshing the DRAMs just discussed is commonly referred to as RAS-only refresh and is shown in Figure 15-14, along with the partial refresh circuitry of the PC.

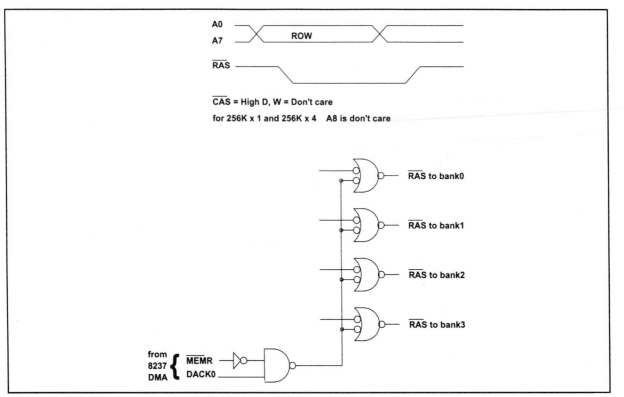

CAS = High D, W = Don't care

for 256K x 1 and 256K x 4   A8 is don't care

RAS to bank0

RAS to bank1

RAS to bank2

RAS to bank3

from { MEMR
8237
DMA { DACK0

Figure 15-14. RAS* Only Refresh Timing in the IBM PC/XT

Review Questions

1. Give the port addresses used by the base address and count registers of channel 0.
2. Who activates the DREQ of channel 0, and how often?
3. Why is the 8237 programmed to make channel 0 the highest-priority channel?
4. Each row of the DRAM, regardless of its size, must be refreshed every _____.
5. True or false. While the DRAM is being refreshed, the CPU cannot access it.
6. What is cycle time of channel 0 of the DMA in the PC/XT?

SECTION 15.5: DMA IN 80x86-BASED PC AT-TYPE COMPUTERS

In Section 15.4, DMA in 8088-based PC/XT computers was discussed. Starting with the PC AT, IBM added an extra 8237 DMA, and clone makers followed IBM. In this section we discuss DMA in the 286, 386, 486, and Pentium PC, PS, and compatibles. The 8088-based PC/XT had only three DMA channels available through the expansion slot. All these channels were designed for 8-bit data transfer. To expand the capability of the PC, designers of the 80286 IBM PC AT added the second 8237 and made it a 16-bit data transfer DMA.

8237 DMA #1

To maintain compatibility with the PC/XT, DREQ1, DREQ2, and DREQ3 of DMA #1 are available through the expansion slot and are for 8-bit data transfer capable of transferring data between 8-bit I/O and the 16-MB memory range of the PC AT. The ports assigned to DMA#1 are exactly the same as in the PC/XT. IBM abandoned the idea of refreshing DRAM using DMA channel 0 and instead replaced it with DRAM refresher circuitry. This made channel 0 available through the AT expansion slot. The signal associated with channel 0 is DREQ0 and DACK0 and is accessed though the 36-edge of the PC AT bus. The following points must be noted regarding channels 0, 1, 2, and 3 of DMA #1 in 80x86 PC AT-type computers.

1. Channels 0, 1, 2, and 3 can be used only for data transfer between 8-bit I/O and system memory. The system memory address can be on an odd-byte or even-byte boundary.
2. Since the count register is a 16-bit register, each of channels 0, 1, 2, and 3 can transfer up to a 64K byte block of data.
3. Each channel 0, 1, 2, or 3 can transfer data in 64K byte blocks throughout the 16M system memory address space. The memory addresses are provided as follows:

A23	A16	A15	A0
Provided by the DMA page register		Provided by 8237 DMA#1	

In the design of the PC AT and consequently all ISA bus computers, different port addresses are assigned to the DMA page registers of channels 0 - 4 than in the PC/XT. This is shown next:

Page Register	I/O Port Address (Hex)
DMA channel 0	0087
DMA channel 1	0083
DMA channel 2	0081
DMA channel 3	0082

8237 DMA #2

The second 8237 DMA is connected as master (level 1) and its channel 0 is used for cascading of DMA#1 as shown in Figure 15-15. The other three channels of this DMA are available through the expansion slot (36-edge) under DREQ5 and DACK5, DREQ6 and DACK6, and DREQ7 and DACK7 designations. These three channels must be used for 16-bit data transfer. The port addresses are assigned to the 8237 DMA#2, as shown in Table 15-5. In this address assignment, even addresses are assigned to the registers of the 8237 since it is connected to D0 - D7 of the system bus. The Hi/Lo copier discussed in Chapter 12 is not applicable in this instance since it works only when the 80x86 CPU is accessing the ports and in AT-type computers a master other than the 80x86 CPU motherboard can take over the system bus. This is done by the AT expansion pin master and is available to channels 5, 6, and 7 only.

Table 15-5: 80x86 Second DMA Ports

Address (Hex)	Register Function
0C0	CH0 memory address register
0C2	CH0 word count register
0C4	CH1 memory address register
0C6	CH1 word count register
0C8	CH2 memory address register
0CA	CH2 word count register
0CC	CH3 memory address register
0CE	CH3 word count register
0D0	Command/status register
0D2	Request register
0D4	Mask register
0D6	Mode register
0D8	Clear byte pointer
0DA	Master clear register
0DC	Clear mask register
0DE	All mask register

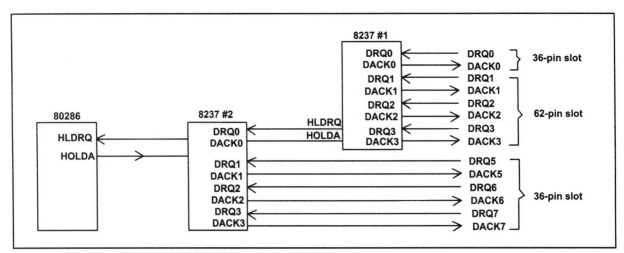

Figure 15-15. 80286 (and Higher) PC/AT DMA

Points to be noted regarding 16-bit DMA channels

Channels 5, 6, and 7 of DMA #2 are used exclusively for 16-bit data transfer between the 16MB memory address space and I/O peripherals. The following points must be noted regarding their use.

1. Channels 5, 6, and 7 must be used for 16-bit data transfers between 16-bit system memory and 16-bit I/O adapters. Notice that the I/O must support 16-bit data.
2. The number of 16-bit (2-byte) words to be transferred is programmed into the count register of channels 5, 6, and 7. Since the count register is a 16-bit register, each channel can transfer up to 65,536 words or 128K bytes between I/O and memory.
3. The memory address for a DMA memory transfer must be on an even-byte address boundary.
4. Channels 5, 6, and 7 transfer data in blocks that have a maximum size of 128K bytes throughout the 16M system memory.
5. Since channels 5, 6, and 7 cannot transfer data on an odd-byte boundary, A0 and BHE are both forced to 0. The rest of the addresses are provided as follows:

A23	A17 A16	A1
Provided by the DMA page register	Provided by 8237 DMA#2	

6. DMA #2 can be accessed (programmed) by another master from the expansion slot using the MASTER input signal on the 36-pin part of the ISA bus.

The following are the port addresses assigned to the DMA page registers of both the DMA#1 and DMA#2. Notice that channels 0 - 3 page register addresses are different than in the PC/XT.

PC/AT ISA-type Computers Page Register	DMA Page Reg Ports I/O Port Address (Hex)
DMA channel 0	0087
DMA channel 1	0083
DMA channel 2	0081
DMA channel 3	0082
DMA channel 5	008B
DMA channel 6	0089
DMA channel 7	008A
Refresh	008F

It must be noted that in the case of channels 5, 6, and 7, only the 7 bits (A23 - A17) of the desired memory address are provided by the DMA page register. These 7 bits can be programmed into the page register with data bits D7 through D1, and as a result, D0 is ignored.

DMA channel priority

The BIOS of the PC AT programs both DMAs to have channel 0 as the highest priority. This means that of the 7 DMA channels available through the expansion slot of the 80x86 ISA bus PC, channel 0 has the highest priority and channel 7 the lowest priority. This is due to the fact that the master DMA (8237 #2) has channel 0 as the highest priority and since the slave 8237 #1 is connected to it, channels 0 through 3 have higher priority than channels 5, 6, and 7. Therefore, we have the following:

	80x86 ISA DMA Channels Priority
channel 0	Highest priority
channel 1	
channel 2	
channel 3	
channel 5	
channel 6	
channel 7	Lowest priority

I/O cycle recovery time

In the IBM PC/AT and indeed all the ISA bus systems, two back-to-back I/O instructions to the same chip are not allowed. The following code is from the PC AT technical reference. It shows that instruction "JMP SHORT $+2" must be inserted before the second OUT command to the same chip because "MOV AL,AH" will not allow enough recovery time for the chip.

```
OUT    IO_ADD,AL        ;first I/O instruction
JMP    SHORT $+2        ;insert this to allow enough time
MOV    AL,AH            ;
OUT    IO_ADD,AL        ;second I/O instruction
```

The JMP is a 2-byte instruction that will empty the 80x86 CPU queue and thus will provide enough recovery time for the DMA chip. Now after the above explanation, we can examine the PC AT BIOS for the initialization of both 8237s.

```
DMA08   EQU    008H            ;DMA STATUS REGISTER PORT ADDRESS
DMA     EQU    000H            ;DMA CH 0 ADDRESS REG PORT ADDRESS
DMA18   EQU    0D0H            ;2ND DMA STATUS PORT ADDRESS
DMA1    EQU    0C0H            ;2ND DMA CH 0 ADDRESS REG ADDRESS
....    ...  ....            ...............................
....    ...  ....            ...............................
        MOV    AL,40H          ;SET MODE FOR CHANNEL 0
        OUT    DMA+0BH,AL
        MOV    AL,0C0H         ;SET CASCADE MODE ON CHANNEL 4
        OUT    DMA18+06H,AL
        JMP    $+2             ;I/O DELAY
        MOV    AL,41H          ;SET MODE FOR CHANNEL 1
        OUT    DMA+0BH,AL
        OUT    DMA18+06H,AL    ;SET MODE FOR CHANNEL 5
        JMP    $+2             ;I/O DELAY
        MOV    AL,42H          ;SET MODE FOR CHANNEL 2
        OUT    DMA+0BH,AL
        OUT    DMA18+06H,AL    ;SET MODE FOR CHANNEL 6
        JMP    $+2             ;I/O DELAY
        MOV    AL,43H          ;SET MODE FOR CHANNEL 3
        OUT    DMA+0BH,AL
        OUT    DMA18+06H,AL    ;SET MODE FOR CHANNEL 7
```

Example 15-9

In the PC AT BIOS initialization, verify the following
(a) port address for the mode register of DMA#1, (b) port address for the mode register of DMA#2
(c) options used for the mode register of channel 0, (d) options used for the mode register of channel 4

Solution:

(a) In the instruction "OUT DMA+0BH,AL", since DMA is equated with address 00, the port address is 0BH, which is compatible with the PC/XT.
(b) DMA18 =0D0H; therefore, DMA18 +06H =0D6H, which is compatible with Table 15-5.
(c) AL=40H is sent to the mode register of channel 0, which gives the following from Figure 15-4: enable verify operation, disable autoinitialization, enable address increment, enable single transfer mode.
(d) For channel 4, since it is used for the cascading of the second 8237, all the options are the same as channel 0 (see part c of this example) except that D7 D6=11, making the control word for the mode register C0H, as shown in the instruction "MOV AL,0C0H".

DMA transfer rate

In the 80286 and higher PCs, both the DMA and the CPU clock speeds vary, depending on the system. For example, the original PC/AT used 6 MHz for the 80286 and 3 MHz for the 8237. To calculate the DMA data transfer rate of any 80x86 ISA system, we must know the clock speed and the cycle time for both the CPU and DMA. Example 15-10 shows such a calculation.

Example 15-10

In the 286 PC AT, the 80286 is at 6 MHz and 3MHz for the DMA. The CPU bus cycle uses one wait state and the bus cycle for the DMA is 5 clocks. Calculate the DMA data transfer rate if the system bus is used alternately by the CPU and DMA.

Solution:

For the 80286 with 0 wait states, it uses 2 clocks as was shown in Chapter 11. Now for one WS we have 3×167 nsec =500 nsec memory bus cycle for the CPU since 1/6 MHz =167 nsec. For the DMA of 3 MHz we have 5×333 =1665 nsec for the cycle time. Since every CPU cycle is alternated by the DMA cycle, we have the bus data transfer rate of 1/(500 +1665) ns =451 Kbytes/sec. In this example we assumed the CPU cycle is the memory cycle and not an I/O instruction cycle, which is normally longer.

Table 15-6 provides the CPU and DMA speed and cycle times for some of the IBM PS/2 models. In this table it must be noted that IBM does not use the 8237 chip since the maximum speed of the 8237 is 5 MHz. Instead, an ASIC (application-specific IC) that is completely compatible with the 8237 is used.

Table 15-6: CPU and DMA Speed and Cycle Time for PS/2

Model	CPU Speed (MHz)	WS	Cycle Time (ns)	DMA Speed (MHz)	DMA Cycle Time (ns)
50	10 MHz	1	300 ns	10 MHz	300 ns
60	16 MHz	0	125 ns	10 MHz	300 ns
60/E	20 MHz	0	100 ns	10 MHz	300 ns

Example 15-11

Calculate the data transfer rate of channels 5, 6, and 7 for the PS/2 model 60, assuming that the DMA cycle is alternated with CPU memory access time.

Solution:

From Table 15-6, adding the CPU and DMA cycle times we have 125 +300 =425 ns. Data transfer rate is 2 × (1/425 ns) = 4.7 Mbytes/second. Remember that channels 5, 6, and 7 are 16-bit channels.

In Figure 15-16, notice how the DMA first reads the data from memory by the $\overline{\text{MEMR}}$ signal and immediately writes the same data into an I/O device such as a disk controller, by activating $\overline{\text{IOW}}$.

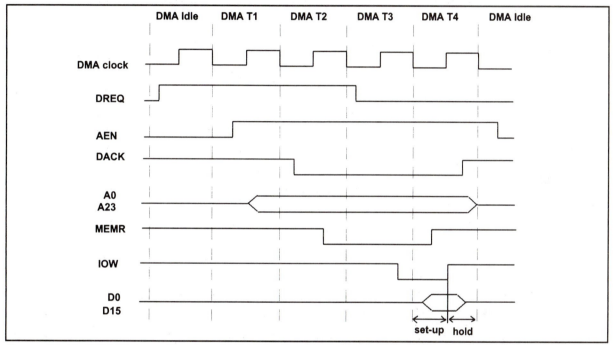

Figure 15-16. DMA Memory Read and I/O Write Bus Cycle for Many 286 and Later PCs

It must be noted that since channels 5, 6, and 7 require that the data be on an even-boundary address and MS DOS does not conform to this standard, in the PC AT the hard disk does not use a 16-bit DMA channel. Instead, 80x86 string instructions are used to transfer between I/O and memory.

Review Questions

1. True or false. In the AT-type 80x86 PCs, all channels of the DMA #1 are available on the expansion slot.
2. How many channels of the DMA are available on the AT bus?
3. Indicate on what part of the AT (62- or 36-pin) bus the channels of the DMA are accessible.
4. True or false. The port addresses assigned for DMA#1 on the PC AT are the same as in the PC/XT.
5. True or false. All the port addresses for DMA#2 are even port addresses.
6. True or false. Channels 0 - 4 can access only 1M address space of the PC AT, not the entire 16M.
7. The DMA page register for channels 0 - 4 must provide ____ bits of the address.
8. True or false. In the PC AT, the DMA page register addresses for DMA#1 is not compatible with the PC/XT.
9. True or false. Channels 5, 6, and 7 can be used for 8- or 16-bit data transfers.
10. Why is bus control in the IBM PC, PS, and compatibles alternated between DMA and the CPU?

SUMMARY

DMA, direct memory access, is a system that accesses memory directly without going through the CPU. This chapter began with a look at how bus arbitration is achieved between the CPU and DMA by use of control signals DREQ, DACK, HRQ, HLDA, and HOLD. DMA is implemented in the PC by use of the 8237A DMA chip, which has 4 channels for transferring data, each of which is used for a separate device such as a floppy or hard disk. Each channel is programmed separately by sending it the base address and word count for the data transfer. The 8237 contains 8 control registers. In the PC/XT channel 0 is used for refreshing DRAM (highest priority), channel 1 is often used for networks, and channel 2 is for the floppy disk controller. 286 AT and later PCs use two 8237s. The second allows 16-bit transfers and is accessible through the expansion slot.

PROBLEMS

SECTION 15.1: CONCEPT OF DMA

1. Compare the rate of data transfer between the 8088 CPU and DMA. For DMA, assume that it takes 4 clocks to transfer a byte. How many times faster is DMA?
2. Calculate the time needed to transfer 512 bytes by the 8088 and by DMA in Problem 1. Assume 200 ns for each clock period.
3. Explain the difference between the CPU's response to signals INTR and HOLDR.
4. For the CPU, HOLDR is an _____ (input, output) signal and HOLDA is an _____ (input, output) signal.
5. In response to activation of HOLDR, the CPU finishes the current _____ before handing the buses to DMA.
   (a) instruction   (b) bus cycle   (c) subroutine
6. The DMA cannot take over the buses until signal _____ is activated by the CPU.
7. Why it is much less expensive to design a DMA chip than a CPU chip?
8. At what point does the CPU regain control over the buses?

SECTION 15.2: 8237 DMA CHIP PROGRAMMING

9. There are total of _____ port addresses assigned to an 8237.
10. Which of the following port address cannot be assigned to the 8237 DMA, and why? (This question is not PC compatible.)
    (a) 88H   (b) 80H   (c) 92H   (d) F0H
11. For a DMA channel to transfer data it must have two sets of information. State these. How many port addresses are assigned to each set?
12. Explain why a total of 8 port addresses are set aside for the 8237 channels.
13. If $\overline{CS}$ is activated by A7 - A4 =0101, give the port addresses assigned to the four channels of the 8237.
14. In Problem 13, what are the port addresses of the 8237 internal control registers?
15. Which register inside the 8237 is used to program the activation level (low or high) of the DREQ and DACK pins? In Problem 14, what port address is that?
16. In fixed priority, which channel has the highest priority? Which has the lowest? How is this different from rotating priority?
17. Assume that the 8237 is programmed for fixed priority. If DREQ2 and DREQ4 are activated at the same time, who gets serviced first?
18. The _____ (80x86 CPU, 8237 DMA) resolves channel priority.
19. State the function of the status register TC bits in the 8237. Can we write into it?
20. Program the mode register in Problem 14 for I/O-to-memory transfer, autoinitialization, address decrement, and block mode for channel 2.
21. Program the single mask register in Problem 14 to enable channel 3.
22. Show the programming of the memory address and count registers of channel 3 to transfer 8K bytes of data from I/O to memory starting at address 1500H.

SECTION 15.3: 8237 DMA INTERFACING IN THE IBM PC/XT

23. To access a memory block, explain how many address bits are provided by the 8237, and how they are provided.
24. The 8237 can access 64K-byte blocks of memory. Explain how the 8237 in the IBM PC can access the entire 1M address range.
25. True or false. For every DREQ there is a DACK in the 8237.
26. True or false. For every DREQ there is a HOLD in the 8237.
27. State if each of the following pins is input, output, or both for the 8237.
    (a) HOLD   (b) HOLDA   (c) DREQs   (d) DACKs   (e) A3 - A0
    (f) ADDSTB   (g) IOR   (h) IOW   (i) MEMR   (j) MEMW

SECTION 15.4: REFRESHING DRAM USING CHANNEL 0 OF THE 8237

28. Why was channel 0 used in the PC/XT for DRAM refreshing?
29. Each cell of DRAM must be refreshed every _____ seconds.
30. If a given DRAM has 2048 rows, each row must be refreshed within _____ ms before it loses the data.
31. Why are DRAMs refreshed one row at time rather than one cell at a time?
32. In the PC/XT, what does the CPU do while the 8237 is refreshing DRAM?

SECTION 15.5: DMA IN 80x86-BASED PC AT-TYPE COMPUTERS

33. True or false. In the 80x86 PC AT-type computers, the use of channel 0 for DRAM refreshing was abandoned.
34. In PC AT-type computers, how many channels are available through the expansion slot? Indicate on what part (62-pin or 36-pin) they are available.
35. How wide is the page register for DMA channels in PC AT computers, and why?
36. True or false. The port addresses for 8237 #2 are the same in the XT and AT.
37. True or false. In the PC AT, the port addresses assigned to channels 0 - 4 of the 8237 #1 are the same as in the PC/XT.
38. Why are the port addresses assigned to DMA#2 of the PC AT all even addresses?
39. State which channels of the PC AT must be used for an 8-bit data transfer and which for a 16-bit data transfer.
40. True or false. Channels 0 - 3 of the PC AT must be used to transfer data in the 1M address range.
41. The starting address for the memory address registers of channels 0 - 3 must be an _____ boundary address.
    (a) even   (b) odd   (c) does not matter
42. The starting address for the memory address registers of channels 5 - 7 must be an _____ boundary address.
    (a) even   (b) odd   (c) does not matter
43. If channels 2 and 6 count registers are both programmed for value 16,384 calculate how many bytes are transferred by each channel. Show your calculation.
44. If DREQ2 and DREQ6 are activated at the same time, who gets serviced first?
45. If a DMA cycle is alternated with the CPU cycle, calculate the rate of data transfer by DMA for the following cases for 16-bit channels. The first column gives the CPU memory cycle time and CPU speed. For the DMA the first number is the cycle time and the second number is the DMA speed.

	CPU	DMA
(a)	3 clocks at 25 MHz	4 clocks at 8 MHz
(b)	2 clocks at 33 MHz	3 clocks at 10 MHz
(c)	4 clocks at 50 MHz	3 clocks at 16 MHz

46. In early PCs, the DMA clock speed was the same or half the CPU speed. Do you think it is the same for the 80486 or Pentium PCs? Explain your answer.
47. Why do you think there are two separate performance benchmarks for memory-intensive and disk-intensive applications?
48. Although 386 and higher PCs have 32-bit address buses, the DMA uses only 24-bit addresses in ISA-type PCs. Explain why and state the implication.

ANSWERS TO REVIEW QUESTIONS

SECTION 15.1: CONCEPT OF DMA
1. true
2. true
3. true
4. true
5. HOLD
6. HLDA
7. input
8. output

SECTION 15.2: 8237 DMA CHIP PROGRAMMING
1. A0, A1, A2, A3, and $\overline{CS}$
2. because A0 - A3 gives rise to only 16 possibilities, 0 - F hex
3. true
4. The memory address of the first byte of the block of data to be transferred is loaded in the memory address register and the number of bytes to be transferred is loaded into the count byte register.
5.
```
        MOV AX,1440H   ;LOAD LOWER 4 DIGITS OF START ADDRESS
        OUT 90H,AL     ;SEND OUT THE LOW BYTE OF THE ADDRESS
        MOV AL,AH
        OUT 90H,AL     ;SEND OUT THE HIGH BYTE OF THE ADDRESS
        MOV AX,4048    ;LOAD BLOCK SIZE INTO AX
        OUT 91H,AL     ;SEND OUT THE LOW BYTE OF THE COUNT
        MOV AL,AH
        OUT 91H,AL     ;SEND OUT THE HIGH BYTE OF THE COUNT
```
6. channel 0
7. true
8. true
9. true
10. true

SECTION 15.3: 8237 DMA INTERFACING IN THE IBM PC/XT
1. 00 - 0F hex
2. 0B hex
3. It is provided through the D0 - D7 data pins of the 8237 to the 74LS373 latch only whenever A0 - A7 rolls over from FF to 00.
4. The 8237 informs the 74LS373 to latch A8 - A15 using the ADSTB pin. The ADSTB is an out-signal pin for the 8237.
5. 0, 3
6. the 74LS670 (a file register), which is a simple 16-bit register organized in 4x4 fashion
7. 80H - 83H, one for each of four channels
8. Channel 0 and channels 1, 2, and 3 are available through the expansion slot.
9. It is 5 clocks or 1.05 μs
10. 4 clocks or 840 ns

SECTION 15.4: REFRESHING DRAM USING CHANNEL 0 OF THE 8237
1. 00 port address for the address register and 01 for count register
2. The 8253/54 timer channel, every 15 μs
3. This is due to the fact that memory refreshing must be done within a fixed period of time; otherwise, the DRAM contents are lost. To do that, channel 0 must be assigned the highest priority.
4. 2 ms
5. true
6. 4 x 210 =840 ns

SECTION 15.5: DMA IN 80X86-BASED PC AT-TYPE COMPUTERS
1. true   2. only 7
3. channels 1 - 3 on the 62-pin and channels 0, 5, 6, and 7 on the 36-pin
4. true   5. true
6. false   7. A16 - A23
8. true   9. False; they can be used only for 16-bit data transfers.
10. This is because in the mode register initialization, the DMA is programmed for the single mode. This means that for every DMA cycle there is a CPU cycle in between, making the DMA bus bandwidth much lower than if the DMA had control over the buses for the entire duration of the data transfer. The real reason is that the buses must be released by DMA in order to allow refreshing of DRAM before it loses the data.

CHAPTER 16

VIDEO AND VIDEO ADAPTERS

OBJECTIVES

Upon completion of this chapter, you will be able to:

» Determine the quality of a monitor by technical features such as resolution, dot rate, horizontal and vertical frequency, and dot pitch

» Describe how images are produced on the screen by the method called raster scanning

» Explain the function of the video adapter board and its two components: video display RAM and the video controller

» Describe the differences in text and graphics modes

» Contrast and compare the video modes MDA, CGA, EGA, VGA, and MCGA

» State the purpose of the attribute byte and how it affects storage space in video display RAM

» Write Assembly language programs to manipulate text data on the screen using INT 10H

» Describe the relation between the number of colors available for a monitor and the amount of video memory needed

» Write an Assembly language program to program pixels on the screen

» Write an Assembly language program to change the cursor shape

Although the quality of video monitors has improved dramatically since the introduction of the first IBM PC in 1981, the principles behind them have remained the same. This chapter will look at the video system of the IBM PC and 80x86 compatibles. In Section 16.1 we describe the fundamental concepts of video monitors, then in Section 16.2 we look at all the various popular IBM PC video monitor adapters, such as MDA, CGA, EGA, and VGA. In Section 16.3 we show text mode programming of these adapters using INT 10H. Section 16.4 will look at graphics mode and graphics programming using INT 10H.

SECTION 16.1: PRINCIPLES OF MONITORS AND VIDEO ADAPTERS

Video monitors use a method called *raster scanning* to display images on the monitor screen. This method uses a beam of electrons to illumine phosphorus dots, called *pixels*, on the screen. This electron gun rasters from the top left corner of the screen to the bottom right, one line at a time. As the gun turns on and off, it moves from left to right toward the end of the line, at which time it is turned off to move back to the beginning of the next line. This moving back while the gun is off is called *horizontal retrace*. When it reaches the bottom right of the screen, the gun is turned off and moves to the top left of the screen. This turning off and moving back to the top is called *vertical retrace*. Figure 16-1 shows two methods of scanning. One is noninterlaced (normal) scanning and the other is interlaced scanning. The concept for both is the same, but in *interlaced* scanning (which is the same method used in television sets), each frame is scanned twice. First the odd lines are scanned and then the gun comes back to scan the even lines. This method can create flicker, but allows better vertical resolution at a cheaper cost. Noninterlaced monitors provide much better flicker-free images than interlaced monitors and for this reason are widely used as monitors of most PCs.

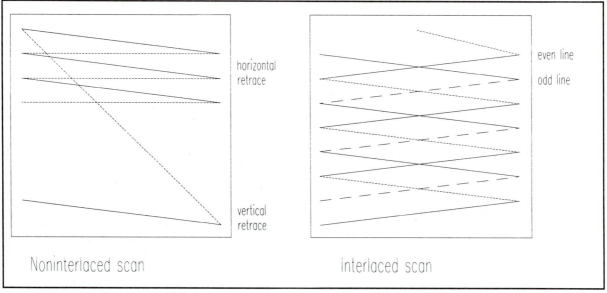

Figure 16-1. CRT Scanning Methods

How to judge a monitor

The resolution of the screen depends upon the following factors:
1. The number of pixels (dots) per scanned line
2. The speed at which the gun can turn on and off the phosphorus coating on the surface of the tube
3. The speed at which it can scan and retrace a horizontal line
4. The number of scan lines per screen (frame)
5. The speed at which it finishes one frame and performs the vertical retrace.

While in a television set, horizontal scanning is done at the rate of 15,750 times per second (15.75 KHz) and vertical scanning at 60 times per second (60 Hz), on the IBM PC monochrome monitor using the monochrome display adapter (MDA) the frequencies are 18.432 KHz and 50 Hz, respectively. Knowing these two frequencies enables one to calculate the maximum number of scan lines per screen by dividing the horizontal frequency by the vertical one as follows:

$$\text{number of scanned lines per screen (not all visible)} = \frac{\text{horizontal frequency (HF)}}{\text{vertical frequency (VF)}}$$

Example 16-1

In a IBM PC monochrome monitor with HF =18.432 KHz and VF =50 Hz, calculate the number of scanned lines per screen.

Solution:

The number of scanned lines = HF/VF; therefore, 18,432 divided by 50 = 368 lines per screen.

Not all 368 horizontal lines in Example 16-1 are visible on the screen since some lines are for overscan and some are used for vertical retrace time. *Overscan* refers to the lines above or below the visible portion of the screen; these lines ensure clear edges at the top and bottom of the screen. In the IBM PC MDA, only 350 lines are visible on screen and of the remaining 18 (368 − 350), some (about 3 or 4) are used for overscanning. The time that would have been taken for scanning the rest (approximately 14) is used for the vertical retrace. Now that the number of scan lines is known, the next question is, how many pixels (dots) are there per line? This is calculated by dividing the video frequency (sometimes called dot frequency) by the horizontal frequency:

$$\text{number of pixels per scan line (not all visible)} = \frac{\text{dot frequency}}{\text{horizontal frequency}}$$

In the IBM monochrome display adapter (MDA), dot frequency is 16.257 MHz, which when divided by HF =18.43 KHz, gives 882 pixels per line. Again, all 882 are not visible. With the IBM PC monochrome adapter, only 720 pixels are visible for each scan line. The time set aside for the remaining 162 is used for the time taken by horizontal retrace and overscanning on the left and right sides of the screen. Again, this overscanning allows sharp edges on the right and left sides of the screen. From the above discussion it can be seen that the three most critical factors in a monitor are:

1. The video frequency (also referred to as dot rate, pixel rate, or video bandwidth)
2. The horizontal frequency
3. The vertical frequency

From these three parameters, the number of pixels per line and the number of lines per screen can be calculated, keeping in mind that not all pixels and lines are visible on the screen, due to overscanning and retrace times. The number of visible pixels is given by the manufacturer of the adapters or monitors. Looking at Table 16-1 for the IBM monochrome adapter, the number of pixels is 720 x 350, which means that there are 720 pixels per line and 350 lines per screen, giving a total of 252,000 pixels. The total number of pixels (dots) per screen is a major factor in assessing a monitor's *resolution*, which is one of its most critical characteristics. The total number of pixels per screen is determined by the size of the pixel and how far apart pixels are spaced. For this reason one must look at what is called the dot pitch in monitor specifications.

Dot pitch

Dot pitch is the distance in between adjacent pixels (dots) and is given in millimeters. For example, a dot pitch of 0.31 means that the distance between pixels is 0.31 mm. Consequently, the smaller the size of the pixel itself and the smaller the space in between them, the higher the total number of pixels and the better the resolution. Dot pitch varies from 0.6 inch in some low-resolution monitors to 0.2 inch in higher-resolution monitors. In some video monitor specifications, it is given in terms of the number of dots per square inch, which is the same way it is given for laser printers, for example, 300 DPI (dots per inch).

Dot pitch and monitor size

Monitors, like televisions, are advertised according to their diagonal size. For example, a 14-inch monitor means that its diagonal measurement is 14 inches. There is a relation between the number of horizontal and vertical pixels, the dot pitch, and the diagonal size of the image on the screen. The diagonal size of the image must always be less than the monitor diagonal size. The following simple equation can be used to relate approximately these three factors to the diagonal measurement. It is derived from the Pythagorean theorem:

$$(\text{image diagonal size})^2 = (\text{number of horizontal pixels} \times \text{dot pitch})^2 + (\text{number of vertical pixels} \times \text{dot pitch})^2$$

Since the dot pitch is in millimeters, the above size would be in mm, so it must be multiplied by 0.039 to get the size of the monitor in inches.

As can be seen from the above discussion, one can use a lower vertical frequency to get a higher number of vertical pixels. However, this can result in flickering, as happens in 1024 x 768 interlaced monitors. The interlaced method is a cheap way of increasing the vertical pixels by halving the vertical frequency and making the frame be scanned in two successive sweeps of odd and even fields.

Example 16-2

A manufacturer has advertised a 14-inch monitor of 1024 x 768 resolution with a dot pitch of 0.28. Calculate the diagonal size of the image on the screen. It must be less than 14 inches.

Solution:

The calculation is as follows:
$$(\text{diagonal size})^2 = (1024 \times 0.28 \text{ mm})^2 + (768 \times 0.28 \text{ mm})^2$$
$$\text{diagonal size (inches)} = 358 \text{ mm} \times 0.039 \text{ inch per mm} = 13.99 \text{ inches}$$

Phosphorous materials

Another factor that determines the quality of the monitor is the phosphorus material used, since the brightness of the pixels depends on two factors:

1. The intensity of the electron beam, which decides how bright or dark each pixel should be. This can be controlled by software, as will be seen in Section 16.4.
2. The phosphorous material used. After the pixel has been illuminated, some phosphorous materials retain their brightness for longer periods of time than others. This characteristic, called *persistence*, is fixed in the monitor and cannot be changed. In the early IBM PC monochrome monitors, a high-persistence type of phosphorus was used to create a better look on the monitor in case there was a need for frequency compensation. In situations where the frequency cannot be adjusted, such as the IBM monochrome monitors, a phosphorus with longer persistence is used to compensate.

Color monitors

In color monitors the principles are the same, except that every phosphorus dot is made of three colors: red, green, and blue, hence the name RGB (red/green/blue) monitors. Color monitors require three different wires to carry three electronic beams, one for each color, unless the monitor is of the composite brand. In composite monitors there is one single wire that carries all three colors. The process of combining, then separating, the three colors diminishes the quality of the image compared to true RGB monitors. In older RGB monitors, for every color wire there was a separate gun, but in newer models a single gun carries three wires for the three beams. Another difference between color and monochrome monitors is the presence of the shadow mask on color monitors. The *shadow mask* is a metal plate with many holes that is placed just before the phosphorus-coated screen in order to coordinate the shooting of the electron beam of each gun through a single hole. This ensures that the red gun illumines the red dot only, the blue gun illumines the blue dot only, and the green gun illumines the green dot only.

Since each pixel has three dots of color: red, green, and blue, how is the dot pitch measured? The answer is that the dot pitch on color monitors is the distance between two dots of the same color, or as some manufacturers advertise, the distance between two consecutive holes of the shadow mask. While in monochrome monitors, it is by changing the intensity of the electron gun that the shades of gray-black-white are generated, in color monitors it is out of the combination of the three primary colors that all other colors are generated. In other words, by changing the intensity of the red, green, and blue triad, one can create all the colors (see the next section for some examples).

Analog and digital monitors

Another monitor characteristic to be explained is *digital* versus *analog*. In digital monitors such as the MDA- and CGA-based monitors, one uses a number of bits to specify variations of color and intensity. To increase these variations one must employ large numbers of bits. Analog monitors, because they can accommodate many more variations, have much better quality pictures. To understand the difference between digital and analog systems, imagine that we have defined a temperature of 20 degrees as cold and 100 degrees as hot. These would be represented in digital as 0 and 1 for cold and hot, but an analog system could accommodate temperature variations of 20 to 100 in increments of one, allowing many more variations. Another example is the state of a light bulb. In digital it is represented as 0 and 1 for on and off. In analog, one can accommodate many more variations, similar to the concept of using a dimmer switch.

Video display RAM and video controller

Communication between the system board (motherboard) and the video display monitor is through the video adapter board. Among the components every video board must have is a video controller and video display RAM. The information displayed on the monitor (either text or graphics) is stored in memory called video display RAM (VDR), also called video buffer. In order for the information to be displayed, it must be written first into video RAM by the CPU; then it is the job of the video adapter's controller (processor) to read the information from video RAM and convert it to the appropriate signals to be displayed on the screen. In other words, there is a separate controller, often called a CRT controller or video processor, apart from the main 80x86 CPU. The video controller's sole job is to take care of the video section of the computer. Since the CRT controller is built specifically for that purpose, it can perform the tasks associated with video much more efficiently than can a CPU such as the 80x86. That also means that video RAM must be accessible to both the main processor and the video processor. In the IBM PC and compatibles, of 1 megabyte of addressable memory, from address A0000H to BFFFFH is set aside for video display RAM. Of this 128K bytes of memory, only some is used; the amount depends on the resolution of the video adapter and the selected mode: text mode or graphics. For example, when displaying text the IBM monochrome adapter uses only 4K bytes of memory, starting at memory address B0000H. Of this 4K bytes, 2K bytes are for the full screen of characters (80 characters per line and 25 lines per screen = 2000 bytes) and another 2000 bytes are for the attributes. The *attribute* byte provides various information, such as color, intensity, and blinking, to the video circuitry. As will be seen later, every time a byte of character data is accessed, its attribute is automatically fetched as well. The memory requirement of each video board and the number of colors that it can handle will be given in the next section.

If the same video display RAM is accessed by both the microprocessor and the video controller, how can two masters access the same RAM at the same time? There are several solutions to this dilemma.

1. The CPU can access the video RAM only during the time when the video controller is doing the retrace.
2. To use a more expensive, specially designed kind of RAM called VRAM (video RAM). This kind of RAM allows the transfer of data by the video controller at a much higher rate than is allowed by normal DRAM.
3. Another approach used in some high-performance graphics system is to use dual-port RAMs. This kind of RAM has two sets of data pins, allowing both the CPU and video controller to access the video RAM with much less conflict, since it eliminates the time wasted by a multiplexer.

It must be noted that if the CPU tries to access the VDR while the video controller is accessing it, the CPU is blocked since the screen must be refreshed by the video controller before it is lost. In other words, the video controller has a higher priority than the main CPU in accessing the VDR. If by software manipulation one blocks the video controller's access to the VDR, it will result in snow on the screen.

Video systems have improved dramatically in recent years, due to the fact that the speed of the CPU has reached 50 MHz and can therefore transfer data from (or to) the VDR at a much faster rate during the retrace time.

Character box

Video boards can be programmed in two modes: text and graphics. While in graphics mode the individual pixels are accessed and manipulated, in text mode characters, which are a group of pixels, are accessed. In text mode, horizontal and vertical pixels are grouped into what are called character boxes. Each character box can display a single character. The size of the character box matrix varies from adapter to adapter. For example, in IBM's MDA (monochrome display adapter) there is a 9 pixel by 14 pixel character box. Since every character is 9 pixels wide and 14 pixels high, one can calculate the number of character columns per screen by dividing the number of pixels on a horizontal line by 9, and can calculate the number of rows per screen by dividing the number of pixels on a vertical line by 14. Conversely, one can calculate the horizontal and vertical pixels by using the size of the character box, number of rows, and the number of columns per screen:

pixels per scan line = number of character columns × pixel width of char. box
raster lines = number of rows per screen × pixel height of char. box

---

Example 16-3

If the MDA character box is 9 x 14 (9 pixels wide and 14 pixels high) and the resolution of MDA is 720 x 350, verify the fact that MDA in text mode can display 80 x 25 characters per screen.

Solution:

720 horizontal scan lines divided by 9, the width of character box, gives 80 columns of characters. Dividing 350 vertical pixels by 14, the height of the character box, results in 25 rows of characters.

Example 16-4

In a given adapter, the character box is 8 x 14 and the adapter in text mode displays 80 x 25 characters. Calculate the pixel resolution.

Solution:

The total number of horizontal pixels is 640 (8 x 80) and the vertical number is 350 (14 x 25). Therefore, it has 640 x 350 resolution.

To get better-looking characters, the character box size must be increased, which translates to more pixels horizontally and vertically. From the above discussion, one can conclude that for a fixed-size monitor to display a fixed number of rows and columns of characters, the number of horizontal and vertical pixels is the most important factor. Since the number of horizontal and vertical pixels is directly proportional to the horizontal and vertical dot frequencies, in judging CRT monitors one must look for higher HF (horizontal frequency), VF (vertical frequency), and DF (dot frequency). Figure 16-2 illustrates the character box.

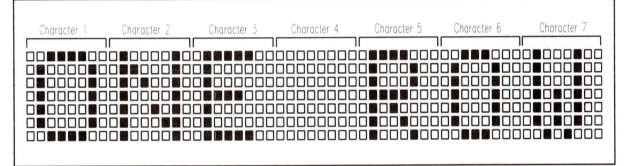

Figure 16-2. Character Boxes
(Reprinted by permission of Intel Corporation, Copyright Intel Corp. 1983)

While in early PCs, monitors had fixed frequencies, the advent of multisync monitors in recent years has allowed variation of these three frequencies to be input from various video boards. This means changing the video adapter board but staying with the same multisync monitor as the technology and the need changes.

The next section will describe various video adapters for the IBM PC and their characteristics in detail, such as number of pixels and color capabilities.

Review Questions

1. The way images are displayed on the monitor screen is referred to as _____.
2. If the dot frequency (DF) is increased but HF and VF remain constant, it will increase the number of _____ (horizontal, vertical) scan lines.
3. True or false. The smaller the dot pitch, the better the monitor.
4. Of the three frequencies DF, HF, and VF, state which has the:
   (a) highest frequency   (b) lowest frequency
5. True or false. In the design of monitors, for a fixed-size monitor one must increase the dot pitch to get more pixels.
6. True or false. For any information to be displayed it must be stored in the VDR.
7. True or false. The VDR memory address range (memory space) must be accessible to both the main CPU and the video processor.
8. True or false. To display crisper characters, one must design more pixels into the character box.

SECTION 16.2: VIDEO ADAPTERS AND TEXT MODE PROGRAMMING

When the IBM PC was introduced in 1981, it had two video monitor options: MDA (monochrome display adapter) and CGA (color graphics adapter). While CGA allowed both graphics and text mode options, MDA allowed only text mode. Although CGA had color for both graphics and text, the text was not very crisp. On the other hand, MDA had excellent text but did not support graphics. Not until 1985 was EGA (enhanced graphics adapter) introduced in order to provide both graphics and text on the same monitor. In 1987, when IBM introduced the new PS/2 line of products, they also introduced new video standards called VGA (video graphics adapter) and MCGA (multicolor graphics array). In MDA and CGA, IBM used the Motorola 6845 CRT controller to design the adapter board and in EGA used a set of proprietary LSI chips, but in all these three cases the adapter board had to be plugged into one of the expansion slots. Starting with PS models, IBM put the video circuitry on the motherboard. PS/2 models 25 and 30 use MCGA, and models 50, 60, and 80 use VGA. Currently, many manufacturers make VGA-compatible adapter plug-in boards. This section discusses each of these boards separately, then text mode programming is explained.

Table 16-1: Adapter Characteristics

Adapter	Year Intro.	Dot Rate MHz	Horizontal Rate KHz	Vertical Rate Hz	Monitor Type	Pixels
CGA	1981	14.318	15.75	60	Comp. dig. RGB	640 x 200
MDA	1981	16.257	18.43	50	Digital mono	720 x 350
EGA	1984	14.318	15.75	60	Digital RGB	640 x 200
		16.257	18.43	50	Digital mono	720 x 350
		16.257	21.85	60	Digital RGB	640 x 350
VGA	1987	25.175	31.5	70	Analog	640 x 400
		28.175	31.5	70	Analog	720 x 400
		25.175	31.5	70	Analog	640 x 450
		25.175	31.5	60	Analog	640 x 350

CGA (color graphics adapter)

CGA was one of two display adapters offered with the IBM PC when it was introduced to the market in 1981. Unlike MDA, which could only support text applications, CGA had the capability of providing both text and graphics on the screen. In addition, it supported color. Since its character box is 8 x 8, text resolution was not as good as MDA, which forced users to choose between the good text quality of MDA and the color and graphics capability of CGA. CGA displays a maximum of 80 characters per line and 25 lines per screen with a resolution of only 640 x 200. It used the Motorola 6845 CRT controller. Programming modes are selected using INT 10H video subroutines contained in the ROM BIOS. Table 16-2 shows video modes for CGA.

Video RAM in CGA

The video display RAM of the CGA starts at B8000H and goes up to 16K bytes. However, to implement the entire 16K bytes of video RAM using static RAM would be too expensive and for that reason DRAM is used. For the CGA starting at memory location B800:0000, even addresses hold the characters to be displayed and odd addresses the attributes of characters as shown next:

```
           Address
        Logical      Physical    Contents
        B800:0000    B8000       row 1, column 1 character
        B800:0001    B8001       attribute for row 1, column 1 character
        B800:0002    B8002       row 1 column 2 character
        B800:0003    B8003       attribute for row 1, column 2 character

        ...          ...
        B800:07CE    B87CE       row 25, column 80 character
        B800:07CF    B87CF       attribute for row 25, column 80 character
```

If the full screen text of 80 x 25 takes 4K bytes (2K bytes for the characters and 2K bytes for the attributes), then 16K bytes of memory can hold 4 pages of text at any given time. Only one page can be viewed at a time, and one can switch to one of the other 3 pages at any time without delay. The page that is being displayed at any given time is commonly referred to as the *active page*. Since the 40 x 25 text option requires only 2000 bytes for both characters and attributes, the 16K bytes of VDR can hold a maximum of 8 pages of text.

Table 16-2: Video Modes for CGA

AL	Pixels	Chars.	Char. Box	Text/Graphics	Colors	Buffer Pages	Start
00H	320 x 200	40 x 25	8 x 8	Text	16*	8	B8000h
01H	320 x 200	40 x 25	8 x 8	Text	16	8	B8000h
02H	640 x 200	80 x 25	8 x 8	Text	16*	4	B8000h
03H	640 x 200	80 x 25	8 x 8	Text	16	4	B8000h
04H	320 x 200	40 x 25	8 x 8	Graphics	4	1	B8000h
05H	320 x 200	40 x 25	8 x 8	Graphics	4*	1	B8000h
06H	640 x 200	80 x 25	8 x 8	Graphics	2	1	B8000h

* Color burst off.

Attribute byte in CGA text mode

As mentioned earlier, even addresses are for the characters to be displayed and odd addresses are for the attributes. Graphics mode is discussed separately in the next chapter. The bit definition of the attribute byte in text mode of CGA is as shown in Figure 16-3.

From the above bit definition it can be seen that the background and foreground can take 8 and 16 different colors, respectively, by combining the primary colors red, blue, and green. Table 16-3 lists those possible colors. Example 16-5 demonstrates how the attribute byte is used.

D7	D6	D5	D4	D3	D2	D1	D0
B	R	G	B	I	R	G	B
	Background			Foreground			

B = blinking, I = intensity
Both blinking and intensity are applied to foreground only.

Figure 16-3. CGA Text Mode Attribute Byte

Example 16-5

Find the attribute byte (in binary and hex) for the following color options:
(a) blue on black   (b) green on blue   (c) high-intensity white on blue.

Solution:

	Binary	Hex	Color Effect
(a)	0000 0001	01	Blue on black
(b)	0001 0010	12	Green on blue
(c)	0001 1111	1F	High-intensity white on blue

MDA (monochrome display adapter)

MDA was one of two display adapters offered with the IBM PC when it was introduced to the market in 1981. It supported a monochrome monitor which displayed 80 alphanumeric characters per line and 25 lines per screen and used Motorola's 6845 CRT controller. The resolution of MDA proved very useful for text applications, but users who needed to use graphics had to use the other adapter offered with the IBM PC at that time, CGA. MDA characteristics are shown in Table 16-4.

Video RAM in MDA

MDA uses video RAM addresses starting at B0000H. This is in contrast to CGA, with memory starting at B8000H. As shown below, the starting physical address of the video RAM in the monochrome is B0000H or B000:0000 using segment:offset. Starting at B0000, the contents of even locations are the characters to be displayed and the odd locations are the attributes of each character.

Table 16-3: The 16 Possible Colors

I	R	G	B	Color
0	0	0	0	Black
0	0	0	1	Blue
0	0	1	0	Green
0	0	1	1	Cyan
0	1	0	0	Red
0	1	0	1	Magenta
0	1	1	0	Brown
0	1	1	1	White
1	0	0	0	Gray
1	0	0	1	Light blue
1	0	1	0	Light green
1	0	1	1	Light cyan
1	1	0	0	Light red
1	1	0	1	Light magenta
1	1	1	0	Yellow
1	1	1	1	High-intensity white

```
                 Address
Logical      Physical     Contents
B000:0000    B0000        row 1, column 1 character
B000:0001    B0001        attribute for row 1, column 1 character
B000:0002    B0002        row 1, column 2 character
B000:0003    B0003        attribute  for row 1,column 2 character
...          ...
B000:07CE    B07CE        row 25, column 80 character
B000:07CF    B07CF        attribute for row 25,column 80 character
```

Table 16-4: Video Modes for MDA

AL	Pixels	Characters	Character Box	Text/ Graphics	Colors	Buffer Pages	Start
07H	720 x 350	80 x 25	9 x 14	Text	Mono	8	B0000h

Attribute byte in IBM MDA

The attribute byte for the characters of MDA uses the same background-foreground scheme as CGA; however, since it supports only black and white, not all possible combinations can be used. Figure 16-4 shows bit definitions of the monochrome byte attribute.

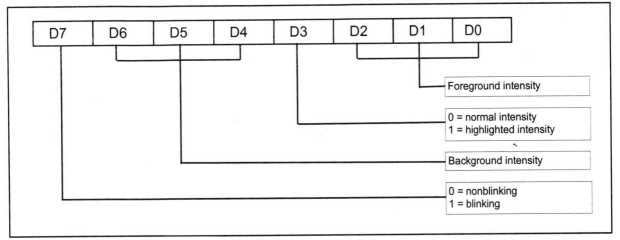

Figure 16-4. Attribute Byte in MDA

Example 16-6

Find the attributes associated with following attribute bytes in MDA.
(a) 07H   (b) 0FH   (c) 70H

Solution:

(a) 07H = 00000111 gives background black, foreground normal intensity, nonblinking.
(b) 0FH = 00001111 gives the same as (a) except with foreground highlighted.
(c) 70H = 01110000 gives black on white, a reverse video screen mode in which the foreground is black and the background is white, nonblinking.

EGA (enhanced graphics adapter)

The EGA adapter, introduced in 1985, gave the PC user the best characteristics of both MDA and CGA, since it could be configured to emulate either. It had graphics and 16 color capabilities, like the CGA but with much improved resolution. Although its resolution of 640 x 350 was still not as good as MDA, it was a significant improvement over the resolution of CGA. IBM used a set of proprietary LSI chips as a CRT controller instead of the Motorola 6845 chip. As shown in Table 16-5, all previous modes of CGA and MDA (modes 0 - 7) are supported in addition to new modes 0DH - 10H.

EGA video memory and attribute

Although the EGA video board can accommodate a maximum of 256K bytes of memory, it can use only 128K bytes of memory space from A0000H to BFFFFH. When in graphics mode, the starting address is at A0000H, but in text mode the address location varies depending on which mode it is emulating, as shown in Table 16-5. EGA can emulate both CGA and MDA. When EGA is programmed to emulate CGA text, the address for the video is B8000H, but for the MDA emulation the address is B0000H. EGA in graphics mode can display 16 colors out of 64 possible colors. Section 16.4 will discuss relating the number of colors supported to memory installed on the video boards.

SECTION 16.2: VIDEO ADAPTERS AND TEXT MODE PROGRAMMING   **487**

Table 16-5: Video Modes for EGA

AL	Pixels	Characters	Character Box	Text/ Graphics	Colors	Buffer Pages	Start
00H	320 x 350	40 x 25	8 x 14	Text	16*	8	B8000h
01H	320 x 350	40 x 25	8 x 14	Text	16	8	B8000h
02H	640 x 350	80 x 25	8 x 14	Text	16*	8	B8000h
03H	640 x 350	80 x 25	8 x 14	Text	16	8	B8000h
04H	320 x 200	40 x 25	8 x 8	Graphics	4	1	B8000h
05H	320 x 200	40 x 25	8 x 8	Graphics	4*	1	B8000h
06H	640 x 200	80 x 25	8 x 8	Graphics	2	1	B8000h
07H	720 x 350	80 x 25	9 x 14	Text	Mono	4	B0000h
08H - 0CH not used for PC, PS							
0DH	320 x 200	40 x 25	8 x 8	Graphics	16	2/4	A0000h
0EH	640 x 200	80 x 25	8 x 8	Graphics	16	1/2	A0000h
0FH	640 x 350	80 x 25	9 x 14	Graphics	Mono	1	A0000h
10H	640 x 350	80 x 25	8 x 14	Graphics	16	2	A0000h

* Color burst off.

Note: Modes 08, 09, and 0AH are used only by the IBM PC Jr.; 0BH and 0CH are used by the EGA video BIOS and are not available.

MCGA (multicolor graphics array)

The MCGA video system is used with analog monochrome or color monitors on PS/2 models 25 and 30. It has improved resolution and color selection over CGA. The advantage of analog systems is their improved color and shading capabilities. In MCGA there are up to 64 brightness levels in monochrome monitors and in color monitors there are up to 262,144 colors of which 256 at a time can be selected for a palette. Table 16-6 gives video modes for MCGA. Notice that it emulates all of CGA modes 0 - 6 in addition to two new modes, 11H and 13H.

Table 16-6: Video Modes for MCGA

AL	Pixels	Characters	Character Box	Text/ Graphics	Colors	Buffer Pages	Start
00H	320 x 400	40 x 25	8 x 16	Text	16*	8	B8000h
01H	320 x 400	40 x 25	8 x 16	Text	16	8	B8000h
02H	640 x 400	80 x 25	8 x 16	Text	16*	8	B8000h
03H	640 x 400	80 x 25	8 x 16	Text	16	8	B8000h
04H	320 x 200	40 x 25	8 x 8	Graphics	4	1	B8000h
05H	320 x 200	40 x 25	8 x 8	Graphics	4*	1	B8000h
06H	640 x 200	80 x 25	8 x 8	Graphics	2	1	B8000h
11H	640 x 480	80 x 30	8 x 16	Graphics	2	1	A0000h
13H	320 x 200	40 x 25	8 x 8	Graphics	256	1	A0000h

* Color burst off.

VGA (video graphics array)

VGA is a single-chip video controller designed by IBM which performs many tasks previously done by several chips in EGA. However, many use the term *VGA* to refer to the entire adapter. VGA is used with analog monochrome or color monitors on PS/2 models 50, 60, 80, and 90. It has excellent resolution of up to 720 x 400 for text modes and 640 x 480 for graphics modes. In many PC and compatible computers, VGA is already on the motherboard, but it can also be purchased as an adapter board to be plugged into one of the expansion slots. Table 16-7 shows video modes for VGA. Notice that it emulates all the modes of CGA, MDA, EGA, plus new modes 11H, 12H, and 13H, which are not available with the earlier adapters.

Table 16-7: Video Modes for VGA

AL	Pixels	Characters	Character Box	Text/ Graphics	Colors	Buffer Pages	Start
00H	360 x 400	40 x 25	9 x 16	Text	16*	8	B8000h
01H	360 x 400	40 x 25	9 x 16	Text	16	8	B8000h
02H	720 x 400	80 x 25	9 x 16	Text	16*	8	B8000h
03H	720 x 400	80 x 25	9 x 16	Text	16	8	B8000h
04H	320 x 200	40 x 25	8 x 8	Graphics	4	1	B8000h
05H	320 x 200	40 x 25	8 x 8	Graphics	4*	1	B8000h
06H	640 x 200	80 x 25	8 x 8	Graphics	2	1	B8000h
07H	720 x 400	80 x 25	9 x 16	Text	Mono	8	B0000h
08H - 0CH not used							
0DH	320 x 200	40 x 25	8 x 8	Graphics	16	8	A0000h
0EH	640 x 200	80 x 25	8 x 8	Graphics	16	4	A0000h
0FH	640 x 350	80 x 25	8 x 14	Graphics	Mono	2	A0000h
10H	640 x 350	80 x 25	8 x 14	Graphics	16	2	A0000h
11H	640 x 480	80 x 30	8 x 16	Graphics	2	1	A0000h
12H	640 x 480	80 x 30	8 x 16	Graphics	16	1	A0000h
13H	320 x 200	40 x 25	8 x 8	Graphics	256	1	A0000h

* Color burst off.

Video memory and attributes in VGA

Up to 1 megabyte of DRAM can be installed on VGA boards. This extra memory is used to store pixels and their attributes. Since in graphics mode, VGA can display up to 256 colors out of 262,144 possible colors at once, it requires 1 megabyte of DRAM to store them. How the 1M of memory is mapped into the 128K-byte address space A00000 - BFFFFH is discussed in Section 16.3, which covers text mode programming.

When VGA is programmed to emulate CGA text, the address for the video is B8000H, but for MDA emulation the address is B0000H. This is in order to be compatible with previous adapters. When VGA is in text mode it uses mode 3, as we will see in the next section.

Table 16-8: Video Modes and Their Definitions

AL	Pixels	Chars	Box	T/G	Colors	Mode	Buf	Start
00H	320 x 200	40 x 25	8 x 8	Text	16*	CGA	8	B8000h
	320 x 350	40 x 25	8 x 14	Text	16*	EGA	8	B8000h
	360 x 400	40 x 25	9 x 16	Text	16*	VGA	8	B8000h
	320 x 400	40 x 25	8 x 16	Text	16*	MCG	8	B8000h
01H	320 x 200	40 x 25	8 x 8	Text	16	CGA	8	B8000h
	320 x 350	40 x 25	8 x 14	Text	16	EGA	8	B8000h
	360 x 400	40 x 25	9 x 16	Text	16	VGA	8	B8000h
	320 x 400	40 x 25	8 x 16	Text	16	MCGA	8	B8000h
02H	640 x 200	80 x 25	8 x 8	Text	16*	CGA	8	B8000h
	640 x 350	80 x 25	8 x 14	Text	16*	EGA	8	B8000h
	720 x 400	80 x 25	9 x 16	Text	16*	VGA	8	B8000h
	640 x 400	80 x 25	8 x 16	Text	16*	MCGA	8	B8000h
03H	640 x 200	80 x 25	8 x 8	Text	16	CGA	8	B8000h
	640 x 350	80 x 25	8 x 14	Text	16	EGA	8	B8000h
	720 x 400	80 x 25	9 x 16	Text	16	VGA	8	B8000h
	640 x 400	80 x 25	8 x 16	Text	16	MCGA	8	B8000h
04H	320 x 200	40 x 25	8 x 8	Graphics	4	CGA,EGA,VGA,MCGA	1	B8000h
05H	320 x 200	40 x 25	8 x 8	Graphics	4*	CGA,EGA,VGA,MCGA	1	B8000h
06H	640 x 200	80 x 25	8 x 8	Graphics	2	CGA,EGA,VGA,MCGA	1	B8000h
07H	720 x 350	80 x 25	9 x 14	Text	Mono	MDA	8	B0000h
	720 x 350	80 x 25	9 x 14	Text	Mono	EGA	4	B0000h
	720 x 400	80 x 25	9 x 16	Text	Mono	VGA	8	B0000h
08H - 0CH not used								
0DH	320 x 200	40 x 25	8 x 8	Graphics	16	EGA	2/4	A0000h
	320 x 200	40 x 25	8 x 8	Graphics	16	VGA	8	A0000h
0EH	640 x 200	80 x 25	8 x 8	Graphics	16	EGA	1/2	A0000h
	640 x 200	80 x 25	8 x 8	Graphics	16	VGA	4	A0000h
0FH	640 x 350	80 x 25	9 x 14	Graphics	Mono	EGA	1	A0000h
	640 x 350	80 x 25	8 x 14	Graphics	Mono	VGA	2	A0000h
10H	640 x 350	80 x 25	8 x 14	Graphics	4	EGA	1/2	A0000h
	640 x 350	80 x 25	8 x 14	Graphics	16	VGA	2	A0000h
11H	640 x 480	80 x 30	8 x 16	Graphics	2	VGA,MCGA	1	A0000h
12H	640 x 480	80 x 30	8 x 16	Graphics	16	VGA	1	A0000h
13H	320 x 200	40 x 25	8 x 8	Graphics	256	VGA,MCGA	1	A0000h

* Color burst off.

Super VGA (SVGA) and other video adapters

Aside from video standards mentioned so far, there are some other adapters that are widely used. Among them are HGC, 8514/A, XGA, and SVGA. The Hercules Graphics Card (HGC) introduced in 1982 by Hercules Computer Technology adds the graphics option to monochrome monitors. This allows both text and graphics pixel programming on monochrome monitors. While 8514/A and XGA are standards set by IBM for high-performance PCs and workstations, SVGA is the commonly used standard. Both SVGA and XGA support up to 1024 x 768 resolution. The 8514/A video standard was not pushed by IBM; therefore, it never became a widely used standard. However, XGA is popular with some developers. This is due to the fact that IBM is providing the documentation to video board makers.

Review Questions

1. True or false. CGA supports both text and graphics modes.
2. What colors are defined by the following attribute bytes in CGA text mode?
   (a) 0   (b) 14H
3. The character displayed is the _____ (foreground, background).
4. True or false. Blinking is available to both background and foreground.
5. True or false. MDA supports both text and graphics modes.
6. The MDA video address starts at _____.
7. True or false. EGA supports both text and graphics mode programming.
8. True or false. Any program written for MDA or CGA will run on EGA.
9. True or false. VGA emulates MDA, CGA, and EGA.
10. True or false. A program written for VGA native mode will run on EGA.

SECTION 16.3: TEXT MODE PROGRAMMING USING INT 10H

In the IBM PC and 80x86 compatibles, programming the video screen is handled by a set of Assembly language programs burned into the BIOS ROM chip. These programs can be accessed through the use of the INT 10H instruction. Certain registers must be set to fixed values prior to invoking the INT 10H instruction. Functions are selected through the value put in register AH. In this section we examine some of these options, with emphasis on text mode programming. At the end of this section, character generator ROM is discussed. Graphic mode options are discussed in Section 16.4.

Finding the current video mode

To find the current video mode set AH = 0F and use INT 10H as follows:

```
MOV     AH,0FH       ;AH=0F
INT     10H
```

Reminder: DEBUG assumes that numbers are in hex. If you are assembling the above program in DEBUG, make sure to remove the H, and also put INT 3 as the last instruction.

The current video mode is also stored in the BIOS data area in location 00449H. You can verify that by dumping that memory location using DEBUG.

Changing the video mode

To change the video mode, use INT 10H with AH = 00 and AL = video mode. A list of video modes is given in Table 16-8. Regardless of what mode is selected, all of modes MDA, CGA, EGA, and MCGA are supported by the VGA monitor. In the same manner, EGA emulates the functions of CGA and MDA since every new board is downwardly compatible. Example 16-7 shows the use of INT 10H functions.

Example 16-7

Write the following program on an EGA or VGA monitor using INT 10H to:
 (a) change the video mode to 03
 (b) display letter "A" in 80 locations with the attributes of red on blue
 (c) use the DEBUG utility to run and verify the program
 (d) use DEBUG to dump the video RAM contents

Solution:

(a)
```
        MOV  AH,00          ;set mode option
        MOV  AL,03          ;CGA text mode of 80x25
        INT  10H            ;monitor is monochrome
```

(b) Using INT 10H function AH = 09, one can display a character a certain number of times with specific attributes.

```
        MOV  AH,09          ;display option
        MOV  BH,00          ;page 0
        MOV  AL,41H         ;ASCII for letter "A"
        MOV  CX,50H         ;repeat it 80 (50h) times
        MOV  BL,14H         ;Red on blue
        INT  10H
```

(c) Reminder: DEBUG assumes that all the numbers are in HEX.

```
C>debug
-A
1131:0100 MOV AH,00
1131:0102 MOV AL,03
1131:0104 INT 10
1131:0106 MOV AH,09
1131:0108 MOV BH,00
1131:010A MOV AL,41
1131:010C MOV CX,50
1131:010F MOV BL,14
1131:0111 INT 10
1131:0113 INT 3
1131:0114
-
```

Now see the result by typing in command -G. Make sure that IP =100 before you run it.

(d) When EGA and VGA monitors emulate CGA in text mode, the video memory address starts at B8000H. Dumping memory immediately after running the above program gives the following. Notice the character and the attribute byte stored in even and odd addresses.

```
-D B800:0 4F
B800:0000 41 14 41 14 41 14 41 14-41 14 41 14 41 14 41 14 A.A.A.A.A.A.A.A.
B800:0010 41 14 41 14 41 14 41 14-41 14 41 14 41 14 41 14 A.A.A.A.A.A.A.A.
B800:0020 41 14 41 14 41 14 41 14-41 14 41 14 41 14 41 14 A.A.A.A.A.A.A.A.
B800:0030 41 14 41 14 41 14 41 14-41 14 41 14 41 14 41 14 A.A.A.A.A.A.A.A.
B800:0040 41 14 41 14 41 14 41 14-41 14 41 14 41 14 41 14 A.A.A.A.A.A.A.A.
```

Example 16-8

Modify Example 16-7 using mode 7 to emulate MDA. Use attribute 87H (white on black blinking).

Solution:
The following shows the code as it would be entered in DEBUG.

```
1131:0100 MOV AH,00
1131:0102 MOV AL,07
1131:0104 INT 10
1131:0106 MOV AH,09
1131:0108 MOV BH,00
1131:010A MOV AL,41
1131:010C MOV CX,50
1131:010F MOV BL,87
1131:0111 INT 10
1131:0113 INT 3
1131:0114
-
```

After running the above program, we dump the MDA video buffer starting at memory address B000:0 and see the data and attribute bytes.

```
-D B000:0 F
B000:0000 41 87 41 87 41 87 41 87-41 87 41 87 41 87 41 87 A.A.A.A.A.A.A.A.
-
```

Setting the cursor position (AH = 02)

The cursor position is set by making AH = 2, DH = row, and DL = column. In, addition, BH = 0 for page 0, the active page being viewed. The following sets the cursor to row 12, column 28.

```
MOV    AH,02          ;set the cursor option
MOV    BH,0           ;page 0
MOV    DH,12          ;row 12
MOV    DL,28          ;col 28
INT    10H            ;invoke interrupt
```

Getting the current cursor position (AH = 03)

To get the current cursor position, AH = 03 of INT 10H must be used:

```
MOV    AH,03          ;get cursor position
MOV    BH,0           ;page 0
INT    10H
```

After running the above code, registers DH and DL have the row and column positions of the cursor (in hex), respectively. Register CX provides information about the cursor shape. See Examples 16-16, 16-17, and 16-18.

Scrolling the window up to clear the screen (AH = 06)

The options 06 and 07 are called scroll functions. They are used to scroll a part or all of the screen up or down. One of the most widely used applications of option AH = 06 is to clear the screen, as shown next.

```
                MOV     AH,06           ;scroll up option
                MOV     AL,0            ;the entire screen
                MOV     BH,07           ;normal attribute
                MOV     CL,0            ;col 0 (top left col)
                MOV     CH,0            ;row 0 (top left row)
                MOV     DL,79           ;col 79 (bottom right col)
                MOV     DH,24           ;row 24 (bottom right row)
                INT     10H
```

A more efficient version of the above code is

```
                MOV     AX,0600H
                MOV     BH,07
                MOV     CX,0
                MOV     DX,184FH        ;18H=24 AND 4FH=79
                INT     10H
```

Writing a character in teletype mode (AH = 0E)

Although INT 21H option AH = 02 can be used to write a single ASCII character to the screen, there are occasions such as in TSR (terminate and stay resident) routines in which DOS functions (we will show why in Chapter 24) should not be used. In these cases, option AH = 0EH of INT 10H of BIOS ROM is used, and this is commonly done in TSR programming. Function AH = 0EH writes an ASCII character in register AL to the monitor at the current cursor position and moves the cursor to the next position. If the ASCII number is 07H (bell), 08H (backspace), 0DH (carriage return), or 0AH (line feed), it performs the appropriate action.

Example 16-9

ASCII value 07 is for the bell. It is a nonprinting character; the speaker beeps if it is sent to the monitor. Write code to sound the bell twice with some time delay in between.

Solution:

```
                MOV     AH,0EH          ;write character option
                MOV     AL,07           ;sound the bell
                MOV     BH,0            ;page 0
                INT     10H
                SUB     CX,CX           ;CX=0
        WAIT:   LOOP    WAIT            ;waste some time
                INT     10H             ;sound the bell again
```

AH = 0EH can be used to display a string of characters as shown next:

```
        MYDATA      DB      'HELLO'

                    MOV     CX,5            ;CX=number of bytes
                    MOV     SI,OFFSET MYDATA   ;DS:SI string address
        AGAIN:      LODSB                   ;load character into AL
                    MOV     AH,0EH          ;option OEH of INT 10H
                    INT     10H
                    LOOP    AGAIN           ;display the next char
```

The above method of displaying a string without using INT 21H was the method used in TSR programs until option AH =13H of INT 10H was introduced. Function AH =13H is discussed next.

Writing a string in teletype mode (AH =13H)

This is the string version of option 0E. The early PC/XT BIOS did not support this function. It is available on PC/XT of BIOS date 1/10/1986 and all later versions. It is used widely in TSR programs to display a string of ASCII data. The following are the register values prior to the interrupt call. Example 16-10 demonstrates the use of this option.

AH =13H
AL =0   BL=has string's attribute. No update for cursor position.
AL =1   BL=has string's attribute. Cursor position is updated.
AL =2   string contains both character and attributes. Each character is followed by attribute. No update for cursor position.
AL=3   string contains both character and attributes. Each character is followed by attribute. Cursor position is updated.
BL=attribute
BH=page number
CX=string length
DH=row position
DL=column position
ES:BP=starting address of string

Example 16-10

Write a program to display 'HELLO' at screen location row 3, column 25 using BIOS INT 10H.

Solution:

```
MYDATA        DB      'HELLO'

              MOV     AX,1300H          ;AH =13H option AL =0
              MOV     BL,07             ;normal attribute
              MOV     BH,0              ;video page 0
              MOV     CX,5              ;display 5 characters
              MOV     DH,3              ;at row 3
              MOV     DL,25             ;and column 25
              MOV     BP,OFFSET MYDATA  ;offset of string
              MOV     SI, SEG MYDATA    ;get the segment address
              MOV     ES,SI             ;ES =segment of string
              INT     10H
```

Character generator ROM

To display characters on the screen, every video board must have access to the pixel patterns of the characters. In CGA, the patterns are burned into BIOS ROM starting at F000:FA60H. To decipher the patterns, first remember that the CGA character box is 8x8. Therefore, for every ASCII character there must be 64 (8 x 8 =64) bits for each pattern. This means that every 8 bytes of the ROM provides the pattern for one character. Now let's see the patterns by using DEBUG.

```
A>DEBUG
-D F000:FA6E L4F
F000:FA60    FF FF FF FF FF FF FF FF-FF FF FF FF FF FF 00 00
F000:FA70    00 00 00 00 00 00 7E 81-A5 81 BD 99 81 7E 7E FF
F000:FA80    DB FF C3 E7 FF 7E 6C FE-FE FE 7C 38 10 00 10 38
F000:FA90    7C FE 7C 38 10 00 38 7C-38 FE FE 7C 38 7C 10 38
F000:FAA0    38 7C FE 7C 38 7C 00 00-18 3C 3C 18 00 00 FF FF
```

Starting with FA6E, the first 8 bytes are for blank (null), resulting in all 00s. The second 8 bytes are for the patterns for happy face, and so on. Inspecting the contents of ROM BIOS reveals the following character definition table.

Address	Patterns in Hex	ASCII	Hex	Dec
F000:FA6E	00,00,00,00,00,00,00,00	NULL	00	00
F000:FA76	7E,81,A5,81,BD,99,81,7E	HAPPY FACE	01	01
F000:FA7E	7E,FF,DB,FF,C3,E7,FF,7E	...	..	..
....	....	...	..	..
F000:FBEE	7C,C6,CE,DE,F6,E6,7C,00	0	30	48
F000:FBF6	30,70,30,30,30,30,FC,00	1	31	49
....	....	...	..	..
F000:FC36	78,CC,CC,7C,0C,18,70,00	9	39	57
....	....	...	..	..
F000:FC76	30,78,CC,CC,FC,CC,CC,00	A	41	65
F000:FC7E	FC,66,66,7C,66,66,FC,00	B	42	66
....	....	...	..	..
F000:FD3E	FE,C6,8C,18,32,66,FE,00	Z	5A	90
....	....	...	..	..
F000:FD76	00,00,78,0C,76,CC,76,00	a	61	97
F000:FD7E	E0,60,60,7C,66,66,DC,00	b	62	98
....	....	...	..	..
F000:FE36	00,00,CC,CC,CC,7C,0C,F8	y	79	121
F000:FE3E	00,00,FC,98,30,64,FC,00	z	7A	122
....	....	...	..	..
F000:FE66	00,10,38,6C,C6,C6,FE,00	DELTA	7F	127

For example, for the happy face character we have 7E, 81, A5, 81, BD, 99, 81, and 7E for the hex patterns of the bits that form the character of the screen. Example 16-11 demonstrates this.

Example 16-11

Draw the patterns for the happy face (02H) and letter "A" (41H) on an 8x8 box.

Solution:
See Figure 16-5.

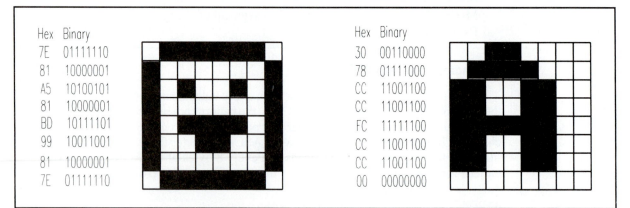

Figure 16-5. Diagram for Example 16-11

BIOS ROM contains only the patterns for the first 128 ASCII characters (00 - 7FH). For the extended character (80 - FFH) patterns, the GRAFTABL utility in MS-DOS can be used.

How characters are displayed in text mode

To see how characters are displayed on the screen, analyze the MDA block diagram shown in Figure 16-6. For CGA, the process of generating the signals is exactly the same as in MDA except that the video circuitry generates R (red), B (blue), G (green), as well as vertical, horizontal, intensity, and dot pattern signals.

The function of the multiplexer in Figure 16-6 is to allow access to the video RAM by both the CPU and 6845 CRT controller. The character generator ROM has patterns for all the ASCII characters. We just examined the content of this ROM holding the CGA patterns. First, the 6845 CRT controller is initialized by the CPU. Then to display characters, the 80x86 CPU writes the characters and their attributes into the video RAM. The job of the CRT controller is to fetch the characters and send them to the character generator ROM to get the patterns of every character on each row for the scan lines. As shown in Figure 16-6, the RA0 - RA4 (row address) output pins from the 6845 fetch the specific row of the character and send it to a parallel-in-serial-out register (often referred to as a serializer). The job of this register is to provide the patterns to the video circuitry, one at a time, to be output serially to the monitor with the attribute.

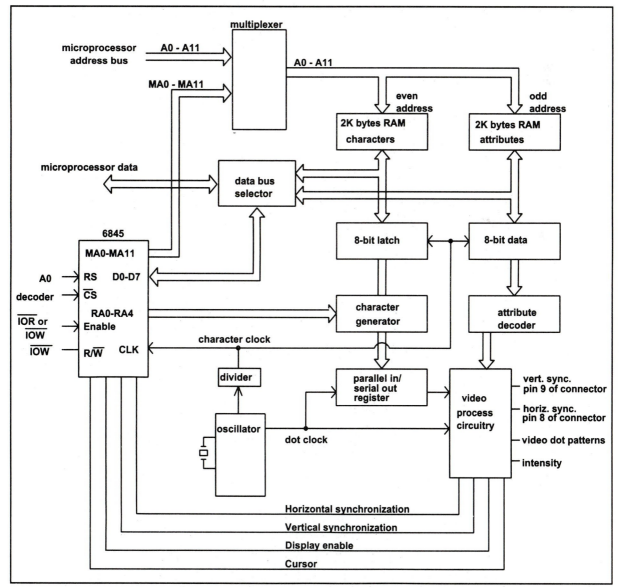

Figure 16-6. IBM PC Monochrome Block Diagram
(Reprinted by permission from "IBM Technical Reference Options and Adapters" c. 1981 by International Business Machines Corporation)

SECTION 16.3: TEXT MODE PROGRAMMING USING INT 10H

Character definition table in VGA

In the discussion of CGA, the character definition table was examined by inspecting the contents of character generator BIOS ROM. The character box in CGA is 8 x 8 and as a result, the text is not very sharp. In VGA, the character box is 8 x 16 and the patterns for all the character are stored in ROM memory. The address for that memory varies from computer to computer. To get the address of the character definition table, use INT 10H with AH =11H, AL =30H, and BH =06. On return from INT 10H, ES:BP has the address. The number of bytes used to form patterns for each character is given in CX, and DL has the row number minus one. This is shown in Example 16-13. Example 16-12 diagrams two characters for VGA.

Example 16-12

Draw the patterns for VGA characters of happy face (02H) and letter "A" (41H) on a 9x16 box. Contrast this with CGA in Example 16-11.

Solution:

The patterns for these characters are as follows. Figure 16-7 shows the diagram.
See Example 16-13 for how to get the patterns.

```
00,00,7E,81,A5,81,81,BD,99,81,81,7E,00,00,00,00      happy face
00,00,10,38,6C,C6,C6,FE,C6,C6,C6,C6,00,00,00,00         A
```

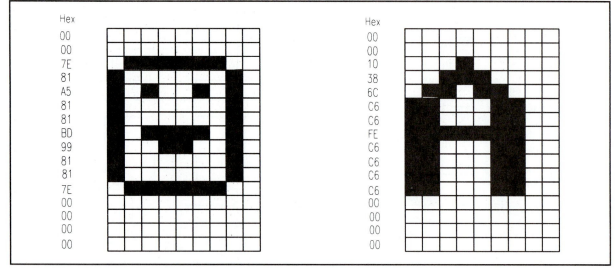

Figure 16-7. Diagram for Example 16-12

Changing the cursor shape using INT 10H

Using BIOS INT 10H options one can examine or even change the cursor shape. Option AH = 03 of INT 10H will not only provide the current position of the cursor but also the beginning and ending lines of the cursor as shown in Example 16-14. Notice the beginning and ending values are given in hex. Example 16-15 shows how to change the beginning and ending lines of the cursor using option AH= 01 of INT 10H. Example 16-16 shows further examples of changing the shape of the cursor.

Example 16-13

Use DEBUG to find the address of the character definition table of a given VGA board.

Solution:

```
C>\DOS\DEBUG
-A
17D9:0100 MOV AH,11
17D9:0102 MOV AL,30
17D9:0104 MOV BH,06
17D9:0106 INT 10
17D9:0108 INT 3
17D9:109
-G


AX=XXXX BX=XXXX CX=0010 DX=0018 SP=XXXX BP=61E7 SI=XXXX DI=XXXX
DS=17D9 ES=E000 SS=XXXX ..................................
17D9:0108 CC          INT 3
-
```

CX =10H gives 16 bytes used for each character. DL =18H =24, which indicates that there are a total of 25 rows of characters per screen. Dumping the contents of memory location at ES:BP. which is E000:61E7, shows the following pattern:

```
-D E000:61E7
E000:61E7                   00-00 00 00 00 00 00 00 00   .......
E000:61F0  00 00 00 00 00 00 00 00-00 7E 81 A5 81 81 BD 99   ......
E000:6200  81 81 7E 00 00 00 00 00-00 7E FF DB FF FE C3 E7   .....
E000:6210  FF FF FE 00 00 00 00 00-00 00 00 6C FE FE FE FE   .....
..........................................................
```

and so on

Going through these memory locations reveals:

BYTE PATTERNS	ASCII	HEX
00,00,00,00,00,00,00,00,00,00,00,00,00,00,00,00	null	00
00,00,7E,81,A5,81,81,BD,99,81,81,7E,00,00,00,00	happy face	01
....	....	..
00,00,38,6C,C6,C6,D6,D6,D6,C6,C6,6C,38,00,00,00	0	30
00,00,18,38,78,18,18,18,18,18,7E,00,00,00,00	1	31
00,00,7C,C6,06,0C,18,30,60,C0,C6,FE,00,00,00,00	2	32
....	...	..
....	...	..
00,00,10,38,6C,C6,C6,FE,C6,C6,C6,C6,00,00,00,00	A	41
00,00,FC,66,66,66,7C,66,66,66,66,FC,00,00,00,00	B	42
....	..	..
....	..	..

and so on

Example 16-14

Using DEBUG, find the beginning and ending lines of the cursor on your PC.

Solution:

```
        MOV   AH,03
        INT   10H
        INT   3
AX=0300  BX=0000  CX=0D0E  DX=1600  SP=CFDE  BP=0000  SI=0000  DI=0000
DS=15B0  ES=15B0  SS=15B0  CS=15B0  IP=0104  NV UP DI PL NZ NA PO NC
15B0:0104 CC    INT 3
```

Registers CH provides the beginning line (0D, or 13) and CL provides the ending line (0E, or 14), for the cursor, while registers DH and DL provide the current row and column position of the cursor.

Example 16-15

On a VGA monitor, (a) get the current cursor shape and save it, (b) change the cursor line to 2,12 (c) wait for a character (using INT21H) to be typed, and (d) restore the original cursor shape.

Solution:
Option AH =01 of INT 10H allows one to set the top and bottom lines of the cursors. Registers CH and CL must be programmed for the beginning and ending cursor lines, respectively.

(a) MOV AH,03   (c)   MOV AH,01
    INT 10H   INT 21H
    PUSH CX
(b) MOV AH,01   (d)   POP CX
    MOV CX,0C02   MOV AH,01
    INT 10H   INT 10H

Example 16-16

If we set the beginning and ending lines for the cursor more than the height of the character box, the cursor will disappear. Repeat Example 16-15 to make the cursor disappear.

Solution:
(a) MOV AH,03   (c)   MOV AH,01
    INT 10H   INT 21H
    PUSH CX
(b) MOV AH,01   (d)   POP CX
    MOV CX,2000H   MOV AH,01
    INT 10H   INT 10H

Review Questions

1. True or false. Functions AH = 0 to AH = 6 of INT 10H are native modes in CGA.
2. For MDA, AH = _____ is the native mode.
3. When VGA emulates CGA, it uses what addresses for the video buffer?
4. When VGA emulates MDA, it uses what addresses for the video buffer?
5. What is the size of the character box in CGA?
6. True or false. The patterns for CGA characters are provided in BIOS ROM.
7. How many bytes of memory does it take to store the pattern for one CGA character?
8. True or false. The patterns for VGA characters are provided in BIOS ROM.

SECTION 16.4: GRAPHICS AND GRAPHICS PROGRAMMING

In all the video programming examples given so far, characters have been used as units to be addressed and a character was treated as a group of pixels. In this section, programming individual pixels will be discussed. In graphics mode, pixel accessing is also referred to as bit-mapped graphics. IBM refers to it as APA (all points addressable) versus AN (alphanumerical) for text mode. First, the relationship between pixel resolution, the number of colors supported, and the amount of video memory in a given video board is clarified.

Graphics: pixel resolution, color, and video memory

There are two facts associated with every pixel on the screen:
1. The location of the pixel
2. Its attributes: color and intensity

These two facts must be stored in the video RAM. The higher the number of pixels and colors options, the larger the amount of memory that is needed to store them. In other words, the memory requirement goes up as the resolution and the number of colors supported go up. The number of colors displayed at one time is always 2^n where n is the number of bits set aside for the color. For example when 4 bits are assigned for the color of the pixel, this allows 16 combinations of colors to be displayed at one time because $2^4 = 16$. The relation between the video memory, resolution, and color for each video adapter is discussed separately.

Example 16-17

In certain video graphics, a maximum of 256 colors can be displayed at one time. How many bits are set aside for the color of the pixels?

Solution:
To display 256 colors at once, we must have 8 bits set for color since $2^8 = 256$.

The case of CGA

The CGA board can have a maximum of 16K bytes of video memory since the 6845 has only 14 address pins ($2^{14} = 16K$). We showed in Section 16.2 how this 16K bytes of memory can hold up to 4 pages of data, where each page represents one full screen of 80 x 25 characters. In graphics mode, the number of colors supported varies depending on the resolution, as shown next.

320 x 200 (Medium resolution)

In this mode there are a total of 64,000 pixels (320 columns × 200 rows = 64,000). Dividing the total video RAM memory of 128K bits (16K × 8 bits = 128K bits) by the 64,000 pixels gives 2 bits for the color of each pixel. These 2 bits give rise to 4 colors since $2^2 = 4$. Therefore, the 320 x 200 resolution CGA can support only up to 4 different colors at a time. See Figure 16-8. These 4 colors can be selected from a palette of 16 possible colors. To select this mode, use set mode option AH= 0 of INT 10H with AL = 04 for mode. After setting the video mode to AL = 04, we must use option 0BH of INT 10H to select the color of the pixel displayed on the screen.

640 x 200 (High resolution)

In this mode there are a total of 128,000 pixels (200 × 640 =128,000). Dividing the 16K bytes of memory by this gives 1 bit (128,000/128,000 =1) for color. The bit can be on (white) or off (black). Therefore, the 640 x 200 high-resolution CGA can support only two colors: black and white. To select this mode, use set mode option AH = 0 of INT 10H with AL = 06 for mode.

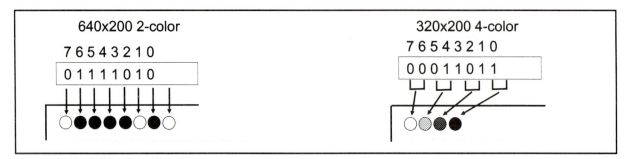

Figure 16-8. CGA Pixel Mapping

Note that for a fixed amount of video RAM, as the resolution increases, the number of supported colors decreases. This discussion bypasses 16-color 160 x 100 low resolution, which is used with color TV sets since no PC supports it.

The case of EGA

In the EGA board, the memory buffer was increased to a maximum of 256K bytes. This allowed both the number of colors and the number of pixels supported in graphics mode to increase. Although EGA can have up to 64 colors, only 16 of them can be displayed on the screen at a time. This is in contrast to CGA, which displayed only 4 colors of a 16-color palette. EGA graphics memory starts at A0000H and goes to a maximum of AFFFFH, using only 64K bytes of the PC's memory space. How is this 256K bytes of memory accessed through a 64K-byte address window? To solve this problem, IBM designers used 4 parallel planes, each 64K bytes, to access the entire 256K bytes of video RAM. In this scheme, each plane holds one bit of the 4-bit color. The assignment of 4 bits for color allows a maximum of 16 colors to be displayed at any given time. In the EGA card, IBM introduced what are called *palette registers*. There are a total of 16 palette registers in the EGA, each holding 8 bits. EGA uses only 6 bits out of the 8 bits of the palette register, giving rise to a maximum of 64 hues.

Video memory size and color relation for EGA

In EGA, to support 640 x 350 pixels with 16 colors requires a minimum of $640 \times 350 \times 4 = 896,000$ bits of memory, but because of the concept of the plane and the 64K-byte address space of A0000H - AFFFFH, the memory must be 256K bytes, although some portions of video memory are unused.

In EGA, one can use 64K bytes for the video RAM, making only 16K bytes available for each color plane, but this results in reducing the number of colors supported. EGA is downwardly compatible with CGA in graphics mode, the same as in text mode. To program the palette registers of the EGA, use option AL =0 of INT 10H. INT 10H has many options for pixel programming of EGA and VGA.

The case of VGA

In VGA, the number of pixels was increased to 640 x 480 with support for 256 colors displayed at one time. The color palette was increased to $2^{18} = 262,144$ hues. The number of palette registers was also increased to 256. Each palette register holds 18 bits, 6 bits for each of the red, green, and blue colors. VGA was the first analog monitor introduced by IBM. All previous monitors were digital. In analog VGA, the analog colors of red, green, and blue replace the digital red, green, and blue of the digital display, allowing substantial increases on the number of colors supported. This gives rise to the use of what is called a video DAC (digital-to-analog converter). Each color of red, green, and blue has a 6-bit D/A converter, allowing 64 combinations for each color, making a total of 18 bits used for the palette which gives rise to total of 262,144 (2^{18}) hues. If the video DAC size is expanded from 6 to 8, the number of combinations for the three signals will be $256 \times 256 \times 256 = 2^{24} = 16,777,216$ hues for the color palette, which is referred to as 16.7 million colors in many advertisements.

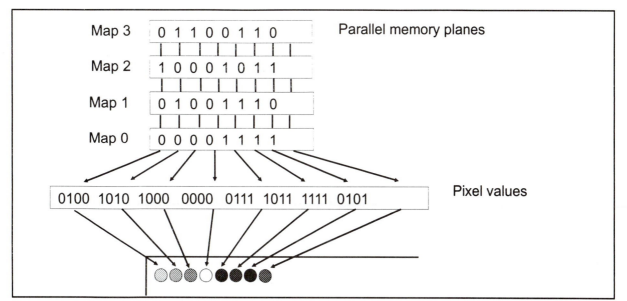

Figure 16-9. 16-Color EGA and VGA Mode Pixel Mapping

Video memory size and color relation for VGA

In VGA, 640 x 480 resolution with support for 256 colors displayed at one time will require a minimum of $640 \times 480 \times 8 = 2,457,600$ bits of memory, but due to the architectural design of VGA, there must be 256K bytes of memory available on the video board. Using the concept of planes means that each plane has 64K bytes. See Figure 16-9. VGA is downward compatible with both CGA and EGA in graphics mode. To access one of the 256 palette registers of VGA, set AH =10H and use option AL =10H of INT 10H. As mentioned earlier, for the AH=10H mode there are many options available for pixel programming of both EGA and VGA. These options are selected through register AL.

The case of SVGA

In SVGA and IBM's XGA, all the resolutions of 800 x 600, 1024 x 768, and 1024 x 1024 are supported. The memory requirement for these boards can reach millions of bytes, depending on the number of colors supported. For example, SVGA of 800 x 600 pixels with 256 colors displayed at the same time requires a minimum of $800 \times 600 \times 8 = 3,840,000$ bits of memory, or a total of 480,000 bytes. Due to the use of bit planes, a total of 512K bytes of DRAM is needed. See Table 16-9. Another example is the total memory required by 800 x 600 resolution with 16 million colors. In this case we need $800 \times 600 \times 24 = 11,520,000$ bits or 1,440,000 bytes, or 1406K bytes. Due to the use of bit planes, it uses 1.5M bytes of DRAM (see Table 16-9).

Table 16-9: Video Memory Requirements by Resolution

Resolution	16 Colors (4 bits)	256 Colors (8 bits)	65,536 Colors (16 bits)	16,777,216 Colors (24 bits)
640 x 480	256K	512K	1M	1M
800 x 600	256K	512K	1M	1.5M
1024 x 768	512K	1M	1.5M	2.5M
1280 x 1024	1M	1.5M	2.5M	4M
1600 x 1200	1M	2M	4M	6M

INT 10H and pixel programming

To address a single pixel on the screen, use INT 10H with AH = 0CH. To do that, X and Y coordinates of the pixel must be known. The values for X (column) and Y (row) vary depending on the resolution of the monitor. CX holds the column point (the X coordinate) and DX the row point (Y coordinate). If the display mode supports more than one page, then BH=page number; otherwise, it is ignored. To turn the pixel on or off, AL = 1 or AL = 0 for black and white. The value of AL can be modified for various colors.

Drawing horizontal or vertical lines in graphics mode

To draw a horizontal line, choose values for the row and column points at the beginning of the line and then continue to increment the column until it reaches the end of line as shown in Example 16-18.

Example 16-18

Using INT 10H, write a program to:
    (a) clear the screen
    (b) set the mode to CGA of 640 x 200 resolution
    (c) draw a horizontal line starting at column 50 and row 50 and ending at column 200, row 50

Solution:

```
        ;clear the screen
                MOV     AX,0600H
                MOV     BH,07
                MOV     CX,0
                MOV     DX,184F
                INT     10H
        ;set the mode to 06 (CGA high resolution)
                MOV     AH,00
                MOV     AL,06
                INT     10H
        ;draw the horizontal line from (50,50) to (200,50)
                MOV     CX,50           ;col pixel=50
                MOV     DX,50           ;row pixel=50
BACK:           MOV     AH,0C           ;0CH option to write a pixel
                MOV     AL,01           ;turn on the pixel
                INT     10H
                INC     CX              ;increment horizontal position
                CMP     CX,200          ;check for the last position
                JNZ     BACK            ;if not, continue
```

Review Questions

1. What is the maximum amount of memory that can be installed on the CGA card?
2. The 320 x 200 resolution CGA can support _____ colors.
3. True or false. In 640 x 200 resolution, the pixel color can be black or white.
4. As the number of pixels goes up, there is _____ (more, less) video memory for storage of color bits.
5. If a total of 24 bits is set aside for color, how many colors are available?
6. Calculate the total video memory needed for 1024 x 768 resolution with 16 colors displayed at the same time.

SUMMARY

The first section of this chapter focused on the technology of video monitors in terms of how images are displayed on the screen by the method called raster scanning. The features by which monitors are judged were given, including resolution, dot pitch, dot rate, and horizontal and vertical frequency. The video adapter board handles communication between the motherboard and the monitor. The information to be displayed on the monitor is stored in video display RAM (VDR). The CRT controller reads the data in VDR and converts it into signals to be sent to the monitor. Video monitors work in either text or graphics mode. In text mode, the screen is programmed in units called character boxes, whereas in graphics mode the screen is programmed pixel by pixel.

Next, the available video adapters were discussed: MDA, CGA, EGA, VGA, and MCGA. In text modes, each character requires 2 bytes of VDR, one for the byte of data to be displayed and one for the attribute byte, which gives information such as foreground and background color, blinking, and intensity. BIOS INT 10H can be used to set the video mode, cursor position and shape, clear the screen, write characters to the screen and draw pixels on the screen.

PROBLEMS

SECTION 16.1: PRINCIPLES OF MONITORS AND VIDEO ADAPTERS

1. Calculate the number of scan lines and dots per lines for each of the following.
   (a) 640 x 200 EGA, DF=14.318 MHz, HF=15.75 KHz, and VF=60 Hz
   (b) 640 x 350 EGA, DF=16.257 MHz, HF=21.85 KHz, and VF=60 Hz
   (c) 640 x 480 VGA, DF=25.175MHz, HF=31.5 KHz, and VF=60 Hz
2. In Problem 1, find the number of scan lines and pixels used for overscan and retrace.
3. The following table is from *PC Magazine*, July 1993, showing the recommended dot pitch for various resolutions and monitor size. Calculate the diagonal size used by the image on screen for each case.

Monitor size	640 x 480	800 x 600	1,024 x 768	1,280 x 1,024
14"	0.35	0.28	0.22	0.18
15"	0.38	0.30	0.24	0.19
17"	0.43	0.34	0.27	0.22
20"	0.50	0.40	0.31	0.25

4. A person wants to use a 14-inch monitor of 0.5 mm dot pitch for 640 x 480 VGA color resolution applications. Show by mathematics why he (she) cannot do that.
5. Dot pitch refers to the size of the _____ (pixel, distance in between pixels).
6. True or false. In color monitors each pixel has its own unique color.
7. For 640 x 200 resolution used for the 80 x 25 characters per screen, find the size of the character box.
8. A 320 x 200 resolution used for the 8 x 8 character box allows only _____ characters per screen.

SECTION 16.2: VIDEO ADAPTERS AND TEXT MODE PROGRAMMING

9. Give the total memory space and address range set aside for the video memory in the PC.
10. Give the video RAM starting address for each: MDA, CGA, EGA, and VGA.
11. The _____ (even, odd) location is used for the character while _____ (even, odd) location is used for the attribute.
12. Find the attribute byte for each of the following.
    (a) red on black   (b) red on blue
    (c) yellow on blue

SUMMARY

13. Give the resolution and the character box size for MDA.
14. For MDA, what attributes are associated with F0H?
15. Which video modes are supported by all three: CGA, EGA, and VGA?
16. How would you prove that CGA can hold a maximum of 8 pages of 40 x 25 characters?

SECTION 16.3: TEXT MODE PROGRAMMING USING INT 10H

17. True or false. VGA supports all previous video standards.
18. Give the starting memory address used by VGA emulating CGA text mode.
19. Give the starting memory address used by VGA emulating MDA.
20. The native mode for CGA text of 80 x 25 characters per page is selected by option AH = ____.
21. The following is the video memory dump using DEBUG.
    B800:0000   45 12 45 12 45 12 45 12.....
    (a) Which mode is this CGA or MDA?
    (b) Give the character and its attribute being displayed.
22. Repeat Problem 21 for the following dump.
    B0000:0000   64 F0 64 F0 64 F0 64 F0
23. Write a program that produces the memory dump in Problem 21. Use DEBUG.
24. Write a program that produces the memory dump in Problem 22. Use DEBUG.
25. Draw the character boxes for the letter B and digits 2 and 9, both in CGA and VGA.

SECTION 16.4: GRAPHICS AND GRAPHICS PROGRAMMING

26. True or false. The more color a given video board supports, the more DRAM it needs.
27. True or false. The more pixels a given video board supports, the more DRAM it needs.
28. The 16-bit color depth can support how many color hues?
29. Why can 620 x 200 CGA support only black and white?
30. In a given VGA board with 256K bytes of memory, how can it fit into space A0000 - AFFFFH?
31. To support 16,777,216 colors, the number of bits set aside for color depth must be ____.
32. Verify the memory requirements of the video board resolution and color depth of Table 16-9 for the following cases.
    (a) 640 x 480 of 16,256, 65,536 and 16 million colors
    (b) 1024 x 768 of 16,256, 65,536 and 16 million colors
    (c) 1600 x 1200 of 256 and 65,536 colors
33. Write a program to change the video mode to 620 x 200 CGA high-resolution graphics and then draw two vertical lines splitting the screen into three equal sections.
34. True or false. Both EGA and VGA use the same starting address for graphics.
35. Write a program using INT 10H to change the cursor shape to beginning and ending lines of 5,9.

ANSWERS TO REVIEW QUESTIONS

SECTION 16.1: PRINCIPLES OF MONITORS AND VIDEO ADAPTERS

1. raster scan   2. horizontal
3. true
4. (a) D-F   (b) V-F
5. false   6. true
7. true   8. true

SECTION 16.2: VIDEO ADAPTERS AND TEXT MODE PROGRAMMING

1. true
2. (a) 00H =0000 0000B is black on black.   (b) 14H =0001 0100 is red on blue.
3. foreground
4. false; to foreground only   5. false; only text mode
6. B0000H
7. true   8. true
9. true   10. false

SECTION 16.3: TEXT MODE PROGRAMMING USING INT 10H

1. true   2. 97
3. B8000H   4. B0000H
5. 8 x 8   6. true
7. 8 bytes   8. false

SECTION 16.4: GRAPHICS AND GRAPHICS PROGRAMMING

1. 16K bytes   2. 4
3. true   4. less
5. 16.7 million
6. 1024 x 768 x 4=3,145,728 bits = 384K bytes, but it uses 512KB due to bit planes.

CHAPTER 17

SERIAL DATA COMMUNICATION AND THE 16450/8250/51 CHIPS

OBJECTIVES

Upon completion of this chapter, you will be able to:

» **List the advantages of serial communication over parallel communication**
» **Explain the difference between synchronous and asynchronous communication**
» **Define the terms** *simplex*, *half duplex*, **and** *full duplex* **and diagram their implementation in serial communication**
» **Describe how start and stop bits frame data for serial communication**
» **Contrast and compare the measures** *baud rate* **and** *bps* **(bits per second)**
» **Describe the RS232 standard**
» **Contrast and compare DTE (data terminal) versus DCE (data communication) equipment**
» **Describe the purpose of handshaking signals such as DTR, RTS, and CTS**
» **Show how the DOS MODE command can be used to initialize COM ports**
» **Code Assembly language instructions to perform serial communication using BIOS INT 14H**
» **Describe the function of UART and USART chips**
» **Describe the BISYNC and SDLC synchronous standard protocols**
» **Describe the purpose and use of the 8250 and 8251 chips**

Computers transfer data in two ways: parallel and serial. In parallel data transfers, often 8 or more lines (wire conductors) are used to transfer data to a device that is only a few feet away. Examples of parallel transfers are printers and hard disks using cables with many wire strips. Although in such cases a lot of data can be transferred in a short amount of time by using many wires in parallel, the distance cannot be great. To transfer to a device located many meters away, the serial method is used. In serial communication, the data is sent one bit at a time, in contrast to parallel communication, in which the data is sent a byte or more at a time. Serial communication and the study of associated chips are the topics of this chapter. In Section 17.1 we study the basics of serial data communications. In the second section we look at serial communication in the IBM PC and BIOS INT 14H. In Section 17.3 we study National Semiconductor's 8250 UART chip (and its variations) since it is the most widely used chip for the PC and compatibles' COM port. In Section 17.4 we examine Intel's 8251 USART chip.

SECTION 17.1: BASICS OF SERIAL COMMUNICATION

When a microprocessor communicates with the outside world it provides the data in byte-sized chunks. In some cases, such as printers, the information is simply grabbed from the 8-bit data bus and presented to the 8-bit data bus of the printer. This can work if the cable is not too long since long cables diminish and even distort signals. In addition, an 8-bit data path is expensive. For these reasons, serial communication is used for transferring data between two systems located at distances of hundreds of feet to millions of miles apart. Figure 17-1 diagrams serial versus parallel data transfers.

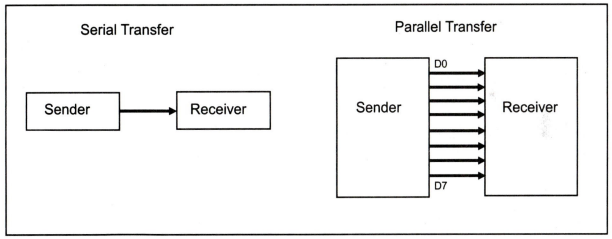

Figure 17-1. Serial vs. Parallel Data Transfer

The fact that in serial communication a single data line is used instead of the 8-bit data line of parallel communication makes it not only much cheaper but also makes it possible for two computers located in two different cities to communicate over the telephone.

For serial data communication to work, the byte of data must be grabbed from the 8-bit data bus of the microprocessor and converted to serial bits using a parallel-in-serial-out shift register; then it can be transmitted over a single data line. This also means that at the receiving end there must be a serial-in-parallel-out shift register to receive the serial data and pack them into a byte and present it to the system at the receiving end. Of course, if data is to be transferred on the telephone line, it must be converted from 0s and 1s to audio tones, which are sinosoidal-shaped signals. This conversion is performed by a peripheral device called a *modem*, which stands for "modulator/demodulator."

When the distance is short, the digital signal can be transferred as it is on a simple wire and requires no modulation. This is how IBM PC keyboards transfer data between the keyboard and the motherboard. However, for long-distance data transfers using communication lines such as a telephone, serial data communication requires a modem to modulate (convert from 0s and 1s to audio tones) and demodulate (converting from audio tones to 0s and 1s).

Serial data communication uses two methods, asynchronous and *synchronous*. The synchronous method transfers a block of data (characters) at a time while the asynchronous transfers a single byte at a time.

It is possible to write software to use either of these methods, but the programs can be tedious and long. For this reason, there are special IC chips made by many manufacturers for serial data communications. These chips are commonly referred to as UART (universal asynchronous receiver-transmitter) and USART (universal synchronous-asynchronous receiver-transmitter). The COM port in the IBM PC uses the 8250 UART, which is discussed in Section 17.3. The synchronous method and the Intel USART 8251 are discussed in Section 17.4.

Half- and full-duplex transmission

In data transmission if the data can be transmitted and received, it is a *duplex* transmission. This is in contrast to *simplex* transmissions such as printers, in which the computer only sends data. Duplex transmissions can be half or full duplex, depending on whether or not the data transfer can be simultaneous. If data is transmitted one way at a time, it is referred to as *half duplex*. If the data can go both ways at the same time, it is *full duplex*. Of course, full duplex requires two wire conductors for the data lines (in addition to ground), one for transmission and one for reception, in order to transfer and receive data simultaneously. See Figure 17-2.

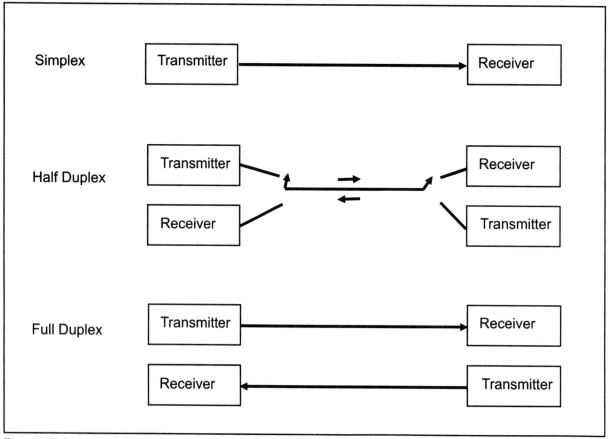

Figure 17-2. Simplex, Half- and Full-Duplex Transfers

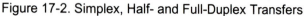

Asynchronous serial communication and data framing

The data coming in at the receiving end of the data line in a serial data transfer is all 0s and 1s; it is difficult to make sense of the data unless the sender and receiver agree on a set of rules, a *protocol*, on how the data is packed, how many bits constitute a character, and when the data begins and ends.

Start and stop bits

Asynchronous serial data communication is widely used for character-oriented transmissions, and block-oriented data transfers use the sychochrounous method. In the asynchronous method, each character is put in between start and stop bits. This is called *framing*. In data framing for asynchronous communications, the data, such as ASCII characters, are packed in between a start bit and a stop bit. The start bit is always one bit but the stop bit can be one or two bits. The start bit is always a 0 (low) and the stop bit(s) is 1 (high). For example, look at Figure 17-3 where the ASCII character "A", binary 0100 0001, is framed in between the start bit and 2 stop bits. Notice that the LSB is sent out first.

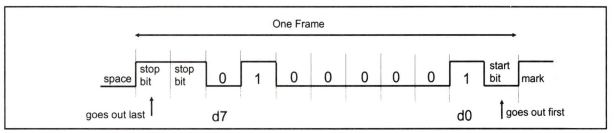

Figure 17-3. Framing ASCII "A" (41H)

In Figure 17-3, when there is no transfer the signal is 1 (high), which is referred to as *mark*. The 0 (low) is referred to as *space*. Notice that the transmission begins with a start bit followed by D0, the LSB, then the rest of the bits until the MSB (D7), and finally, the 2 stop bits indicating the end of the character "A".

In asynchronous serial communications, peripheral chips and modems can be programmed for data that is 5, 6, 7, or 8 bits wide. This in addition to the number of stop bits, 1 or 2. While in older systems ASCII characters were 7-bit, due to extended ASCII characters, 8 bits are required for each character. Small non-ASCII keyboards use 5- and 6-bit characters. In some older systems, due to the slowness of the receiving mechanical device, 2 stop bits were used to give the device sufficient time to organize itself before transmission of the next byte. However, in modern PCs the use of 1 stop bit is common. Assuming that we are transferring a text file of ASCII characters using 2 stop bits, we have a total of 11 bits for each character since 8 bits are for the ASCII code, and 1 and 2 bits are for start and stop bits, respectively. Therefore, for each 8-bit character there are an extra 3 bits, or more than 30% overhead.

Example 17-1

Calculate the total number of bits used in transferring 5 pages, each with 80x25 characters. Assume 8 bits per character and 1 stop bit.

Solution:

For each character, a total of 10 bits is used, 8 bits for each character, 1 stop bit, and 1 start bit. Therefore, the total number of bits is $80 \times 25 \times 10 = 20,000$ bits per page. For 5 pages, 100,000 bits will be transferred.

In some systems in order to maintain data integrity, the parity bit of the character byte is included in the data frame. This means that for each character (7- or 8-bit, depending on the system) we have a single parity bit in addition to start and stop bits. The parity bit is odd or even. In the case of an odd-parity bit the number of data bits, including the parity bit, has an odd number of 1s. Similarly, in an even-parity bit the total number of bits, including the parity bit, is even. For example, the ASCII character "A", binary 0100 0001, has 0 for the even-parity bit. UART chips allow programming of the parity bit for odd-, even-, and no-parity options, as we will see in the next section. If a system requires the parity, the parity bit is transmitted after the MSB, and is followed by the stop bit.

Data transfer rate

The rate of data transfer in serial data communication is stated in *bps* (bits per second). Another widely used terminology for bps is *baud rate*. However, the baud and bps rates are not necessarily equal. This is due to the fact that baud rate is the modem terminology and is defined as number of signal changes per second. In modems, there are occasions when a single change of signal transfers several bits of data. As far as the conductor wire is concerned, the baud rate and bps are the same, and for this reason in this book we use the terms *bps* and *baud* interchangeably.

The data transfer rate of a given computer system depends on communication ports incorporated into that system. For example, the early IBM PC/XT could transfer data at the rate of 100 to 9600 bps. However in recent years, PCs, PS, and 80x86 compatibles transfer data at rates as high as 19,200 bps. It must be noted that in asynchronous serial data communication, the baud rate is generally limited to 100,000 bps.

Example 17-2

Calculate the time it takes to transfer the entire 5 pages of data in Example 17-1 using:
(a) 2400 bps   (b) 9600 bps

Solution:

(a) 100,000/2400 = 41.67 seconds
(b) 100,000/9600 = 10.4 seconds

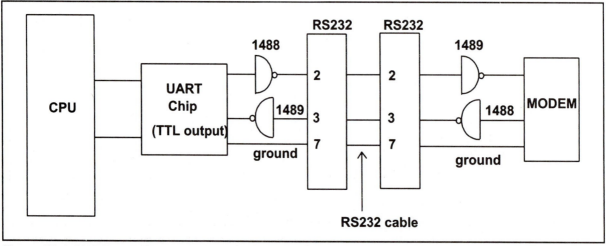

Figure 17-4. UART-to-RS232 Connections using MC1488 and MC1489 Chips

---

To allow compatibility among data communication equipment made by various manufacturers, an interfacing standard called RS232 was set by the Electronics Industries Association (EIA) in 1960. In 1963 it was modified and called RS232A. RS232B and RS232C were issued in 1965 and 1969, respectively. In this book we refer to it simply as RS232. Today, RS232 is the most widely used serial I/O interfacing standard. However, since the standard was set long before the advent of the TTL logic family, the input and output voltage levels are not TTL compatible. In the RS232 a 1 is represented by –3 to –25 V, while the 0 bit is +3 to +25 V, making –3 to +3 undefined. For this reason, to connect any RS232 to a microprocessor-based system we must use voltage converters such as MC1488 and MC1489 to convert the TTL logic levels to the RS232 voltage level, and vice versa. MC1488 is used to translate from the TTL to RS232 voltage level and MC1489 to convert from the RS232 to the TTL level. MC1488 and MC1489 IC chips are commonly referred to as *line drivers* and *line receivers*. This is shown in Figure 17-4.

RS232 pins

Table 17-1 provides the pins and their labels for the RS232 cable, commonly referred to as the DB-25 connector. In labeling, DB-25P refers to the plug connector (male) and DB-25S is for the socket connector (female). See Figure 17-5.

Due to the fact that not all the pins are used in modern microcomputers, IBM introduced the DB-9 version of the serial I/O standard, which uses only 9 pins, as shown in Table 17-2. The IBM PC 9-pin serial port is shown in Figure 17-6.

Table 17-1: RS232 Pins

Pin	Description
1	Protective ground
2	Transmitted data (TxD)
3	Received data (RxD)
4	Request to send ($\overline{\text{RTS}}$)
5	Clear to send ($\overline{\text{CTS}}$)
6	Data set ready ($\overline{\text{DSR}}$)
7	Signal ground (GND)
8	Data carrier detect ($\overline{\text{DCD}}$)
9/10	Reserved for data set testing
11	Unassigned
12	Secondary data carrier detect
13	Secondary clear to send
14	Secondary transmitted data
15	Transmit signal element timing
16	Secondary received data
17	Receive signal element timing
18	Unassigned
19	Secondary request to send
20	Data terminal ready ($\overline{\text{DTR}}$)
21	Signal quality detector
22	Ring indicator
23	Data signal rate select
24	Transmit signal element timing
25	Unassigned

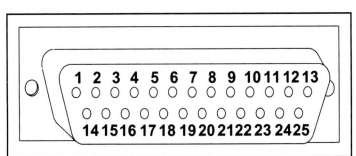

Figure 17-5. RS232 DB-25P (Male) Connector

Other serial I/O interface standards

In the RS232 cable, as the length of the cable is increased, the signals encounter more capacitance, making high-speed data transfer unreliable. We can use RS232 to transfer data at the rate of 100,000 bps and higher if we are willing to cut the cable length to 5 feet or less. To increase the rate of the data transfer and cable length, the electrical specifications of the RS232 had to be redefined. This led to new cable standards called RS422A and RS423. Table 17-3 shows the comparison of these two standards with the RS232.

Data communication classification

Current terminology classifies data communication equipment as DTE (data terminal equipment) or DCE (data communication equipment). *DTE* refers to terminals and computers that send and receive data, while *DCE* refers to communication equipment, such as modems, that are responsible for transferring the data. Notice that all the RS232 pin function definitions of Tables 17-1 and 17-2 are from the DTE point of view.

The simplest connection between DTE and DCE requires a minimum of three pins, TxD, RxD, and ground, as shown in Figure 17-7. However, the minimum connection between two DTE devices, such as two PCs, requires pins 2 and 3 to be interchanged as shown in Figure 17-7. In looking at Figure 17-7, keep in mind that the RS232 signal definitions are from the point of view of DTE.

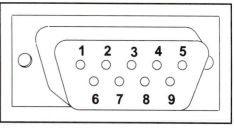

Figure 17-6. IBM PC 9-Pin Male Connector

Table 17-2: IBM PC 9-Pin Signals

Pin	Description
1	Data carrier detect ($\overline{\text{DCD}}$)
2	Received data (RxD)
3	Transmitted data (TxD)
4	Data terminal ready (DTR)
5	Signal ground (GND)
6	Data set ready ($\overline{\text{DSR}}$)
7	Request to send ($\overline{\text{RTS}}$)
8	Clear to send ($\overline{\text{CTS}}$)
9	Ring indicator (RI)

Table 17-3: RS232 Comparison with RS422 and RS423

	RS232	RS422	RS423
Max. cable length (ft)	50	4000	4000
Maximum speed (baud)	20K	10M/40 ft	100K/30 ft
		1M/400 ft	10K/300 ft
		100K/4000 ft	1K/4000 ft
Logic 1 voltage level	−3 to −25	A > B	−4 to −6
Logic 0 voltage level	+3 to +25	B > A	+4 to +6

Examining the RS232 handshaking signals

To ensure fast and reliable data transmission between two devices, the data transfer must be coordinated. In Chapter 12 we showed an example of handshaking signals in the 8255 chip and its connection to a printer. In the same way, there are handshaking signals in serial data communication. Just as in the case of the printer, due to the fact that in serial data communication the receiving device may have no room for the data there must be a way to inform the sender to stop sending data.

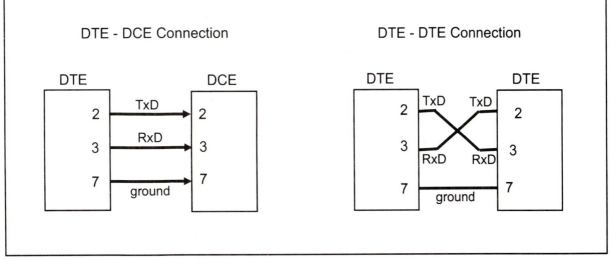

Figure 17-7. DTE-DCE and DTE-DTE Connections

Some of the pins of the RS-232 are used for handshaking signals. They are described below. They are so widely used that understanding them is essential in the study of any UART chip.

1. DTR (data terminal ready). When the terminal (or a PC COM port) is turned on, after going through a self-test, it sends out signal DTR to indicate that it is ready for communication. If there is something wrong with the COM port, this signal will not be activated. This is an active-low signal and can be used to inform the modem that the computer is alive and kicking. This is an output pin from DTE (PC COM port) and an input to the modem.

2. DSR (data set ready). When a DCE (modem) is turned on and has gone through the self-test, it asserts DSR to indicate that it is ready to communicate. Therefore, it is an output from the modem (DCE) and input to the PC (DTE). This is an active-low signal. If for any reason the modem cannot make a connection to the telephone, this signal remains inactive, indicating to the PC (or terminal) that it cannot accept or send data.

3. RTS (request to send). When the DTE device (such as a PC) has a byte to transmit, it asserts RTS to signal the modem that it has a byte of data to transmit. RTS is an active-low output from the DTE and an input to the modem.

4. CTS (clear to send). In response to RTS, when the modem has room for storing the data it is to receive, it sends out signal CTS to the DTE (PC) to indicate that it can receive the data now. This input signal to DTE is used by DTE to start transmission.

5. CD (carrier detect, or DCD, data carrier detect). The modem asserts signal CDC to inform the DTE (PC) that a valid carrier has been detected and that contact between it and the other modem is established. Therefore, CDC is an output from the modem and an input to the PC (DTE).

6. RI (ring indicator). An output from the modem (DCE) and an input to a PC (DTE) indicates that the telephone is ringing. It goes on and off in synchronization with the ringing sound. Of the 6 handshake signals, this is the least often used, due to the fact that modems take care of answering the phone. However, if in a given system the PC is in charge of answering the phone, this signal can be used.

From the above description, PC and modem communication can be summarized as follows: While signals DTR and DSR are used by the PC and modem, respectively, to indicate that they are alive and well, it is RTS and CTS that actually control the flow of data. When the PC wants to send data it asserts RTS, and in response, if the modem is ready (has room) to accept the data, it sends back CTS. If, for lack of room, the modem does not activate CTS, the PC will deassert DTR and try again. RTS and CTS are also referred to as hardware control flow signals.

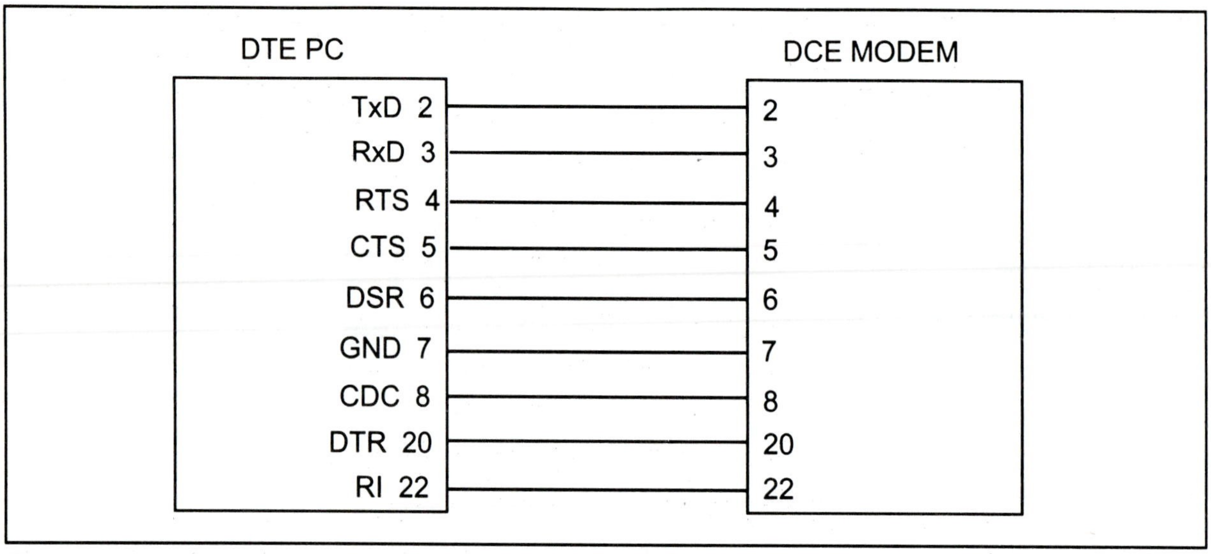

DTE PC   DCE MODEM

TxD 2 ──────────────── 2
RxD 3 ──────────────── 3
RTS 4 ──────────────── 4
CTS 5 ──────────────── 5
DSR 6 ──────────────── 6
GND 7 ──────────────── 7
CDC 8 ──────────────── 8
DTR 20 ──────────────── 20
RI 22 ──────────────── 22

Figure 17-8. DTE and DCE Connections with Handshaking

This concludes the description of the 9 most important pins of the RS232 handshake signals plus TxD, RxD, and ground. Ground is also referred to as SG (signal ground). In the next section we will see DOS and BIOS serial communication programming in the IBM PC.

Review Questions

1. The transfer of data using parallel lines is _____ (faster, slower) but _____ (more expensive, less expensive).
2. In communications between two PCs in New York and Dallas, we use _____ (serial, parallel) data communication.
3. In serial data communication, which method fits block-oriented data?
4. True or false. Sending data to a printer is duplex.
5. True or false. In duplex we must have two data lines.
6. The start and stop bits are used in the _____ (synch, asynch) method.
7. Assuming that we are transmitting letter "D", binary 100 0100, with odd-parity bit and 2 stop bits, show the sequence of bits transferred.
8. In Question 7, find the overhead due to framing.
9. Calculate the time it takes to transfer 400 characters as in Question 7 if we use 1200 bps. What percentage of time is wasted due to overhead?
10. True or false. RS232 is not TTL-compatible.
11. What voltage levels are used for binary 0 in RS232?
12. If in a given PC the COM port is defective, which handshake signal indicates this to the modem?

SECTION 17.2: ACCESSING IBM PC COM PORTS USING DOS AND BIOS

To relieve users and programmers from the tedious details of the 8250 UART chip, both DOS and BIOS provide means of accessing the IBM PC serial COM ports. IBM BIOS INT 14H can be used for serial data communication. In DOS, we can use the MODE command to initialize the COM ports for data size, baud rate, and so on. First some introductory comments on the number of ports in the PC will be given.

IBM PC COM ports

In the IBM PC and 80x86 compatibles, as many as 4 COM ports can be installed. They are numbered 1, 2, 3, and 4 (BIOS numbers them as 0, 1, 2, and 3). When the PC is turned on, it is the job of the POST (power-on self-test) to test the UART chip for each of the 4 COM ports. If they are installed, their I/O port addresses are written to memory locations 0040:0000 - 0040:0007. Since the I/O address assigned to each UART is a 16-bit address, it takes 2 bytes for each installed UART. The BIOS data area memory locations 0040:0000 and 0040:00001 will have the I/O port address for COM 0, and 0040:0002, 0040:0003 locations have the I/O port address for COM 1, and so on. If no COM hardware is connected for any of COM 0 - COM 3, these address have 0s in them. See Example 17-3.

Example 17-3

A nationally known computer columnist is asked by a reader how he/she can find the number of COM ports installed in a PC and which one is installed. What do you think the answer should be?

Solution:

Dumping memory locations 0040:0000 - 0040:0007 in DEBUG on the computer, as shown below, showed that there is only one COM port with starting address 03F8H.

```
C>DEBUG
-d 0040:0000 L08
0040:0000 F8 03 00 00 00 00 00 00
```

Using the DOS MODE command

In MS-DOS (or IBM PC-DOS), the MODE command is used to set the serial communication parameters, such as data size, baud rate, parity bit, and number of stop bits. The baud rates supported by the PC/XT, PC/AT BIOS are 110, 150, 300, 600, 2400, 4800, and 9600. In the IBM PS models and compatibles, 19,200 baud is also supported by BIOS. For example, look at the following command:

MODE COM1:24,O,8,1

This command sets parameters for COM1, 2400 baud rate, odd parity, 8 data bits, and 1 stop bit. It must be noted that COM ports are assigned numbers 0 - 3 by BIOS, but DOS uses 1 - 4. The general format of the MODE command is

MODE COMm:b,p,d,s

where m signifies the COM port number with options of 1 - 4, and b is the baud rate. Only the initial two numbers are used, with the following options:

11	110
15	150
30	300
60	600
12	1200
24	2400
48	4800
96	9600
19	19,200 (for IBM PS and compatibles)

where p is for parity [options include N (none), O (odd), E (even)], d is for the size of the character (5, 6, 7, or 8 bits; the default value is 7), s is for the number of stop bits (options are 1, 1.5, or 2). If the baud rate is 110, 2 stop bits must be used; otherwise, it is 1. Note that we can go beyond 19,200 baud if we bypass BIOS and program the 8250 UART directly. The next topic will show how to use BIOS INT 14H to perform serial communication.

AH   **INT 14H Function**

00   **Initialize COM Port**

Additional Call Registers	Result Registers
AL = parameter (see below)	AH = port status (see below)
DX = port number (0 if COM1, 1 if COM2, etc.)	AL = modem status (see below)

Note 1: The parameter byte in AL is defined as follows

7 6 5 4 3 2 1 0	Indicates
x x x	Baud rate (000=110, 001=150, 010=300, 011=600, 100=1200, 101=2400, 110=4800, 111=9600)
x x	Parity (01=odd, 11=even, x0=none)
x	Stop bits (0 = 1, 1 = 2)
x x	Word length (10=7 bits, 11=8 bits)

Note 2: The port status returned in AH is defined as follows

7 6 5 4 3 2 1 0	Indicates
1	Timed-out
1	Transmit shift register empty
1	Transmit holding register empty
1	Break detected
1	Framing error detected
1	Parity error detected
1	Overrun error detected
1	Received data ready

Note 3: The modem status returned in AL is defined as follows

7 6 5 4 3 2 1 0	Indicates
1	Received line signal detect
1	Ring indicator
1	DSR (data set ready)
1	CTS (clear to send)
1	Change in receive line signal detect
1	Trailing edge ring indicator
1	Change in DSR status
1	Change in CTS status

Figure 17-9. BIOS INT 14H Functions *(continued on following page)*

01   Write Character to COM Port

Additional Call Registers	Result Registers
AL = character	AH bit 7 =0 if successful, 1 if not
DX = port number (0 if COM1, 1 if COM2, etc.)	AH bits 0 - 6 = status if successful
	AL = character

Note: The status byte in AH, bits 0 - 6, after the call is as follows:

6 5 4 3 2 1 0	Indicates
1	Transmit shift register empty
1	Transmit holding register empty
1	Break detected
1	Framing error detected
1	Parity error detected
1	Overrun error detected
1	Receive data ready

02   Read Character from COM Port

Additional Call Registers	Result Registers
DX = port number (0 if COM1, 1 if COM2, etc.)	AH bit 7 =0 if successful, 1 if not
	AH bits 0 - 6 = status if successful
	AL = character read

Note: The status byte in AH, bits 1 - 4, after the call is as follows:

4 3 2 1	Indicates
1	Break detected
1	Framing error detected
1	Parity error detected
1	Overrun error detected

03   Read COM Port Status

Additional Call Registers	Result Registers
DX = port number (0 if COM1, 1 if COM2, etc.)	AH =port status
	AL =modem status

Note: The port status and modem status returned in AH and AL are the same format as in INT 14H function 00H, described above.

Figure 17-9. BIOS INT 14H Functions *(continued from preceding page)*

Data COM programming using BIOS INT 14H

The serial communication ports of the PC can be accessed using the BIOS-based INT 14H. Various options of INT 14H are chosen with the AH registers as shown in Figure 17-9. Starting with the PS/2 and all subsequent PS computers, function AH = 04 allows the programmer to set the baud rate to 19,200. In addition, the data bit size can be set to 5, 6, 7, and 8, not just 7, 8 as was the case with AH = 0. The options supported for AH = 4 and AH = 5 are shown in Appendix E.

Using BIOS INT 14H we can send and receive characters with another PC via a COM port. The process is as follows.

1. To send a character we use INT 14H, AH=1, AL=character.
2. To receive a character we use INT 14H, AH=3 to get the COM port's status in register AH. Notice that this is the status of the COM port and not the status of the MODEM, which is given in AL. Then check D0 of the status port, which is called *received data ready.* If it is high, a character has been received via the COM port and is sitting inside the 8250 UART. To read the received character we use INT 14H, AH=2 where AL holds the character upon return.

Example 17-4

This example involves serial data communication between two PCs via COM port 2 (COM 1 is already used by the mouse).
(a) Show the minimum signal (wire) connection between the two PCs,
(b) Write an assembly language program where any key press from one PC is transfered to the other. Pressing ESC should exit the program. In the program, initialize the COM port for 4800 baud, 8-bit data, no parity, 1 stop bit.

Solution:

(a) This drawing shows the minimum connection needed. One can use what is called a break-out box to connect two COM ports or use a null modem cable.

(b) The following steps need to be coded in the program:
(1) Check for key press and if a key has been pressed, get it and write it to the COM port to be transfered. Also check for ESC to exit.
(2) If there is no key pressed, go check the status of the COM port. If a character has been received, read it and display it on the screen.
(3) Go to step (1).

To test this example connect two PCs and run the program on the following page on both of them.

PC to PC Connection

Review Questions

1. The maximum number of COM ports allowed in the PC is _____.
2. Give the maximum and minimum baud rates supported by the PC, XT, AT, and compatibles if we use BIOS.
3. Give the maximum and minimum baud rates supported by the IBM PS and compatibles if we use BIOS.
4. Using the MODE command, show how to set a data format of 8 bits per character, no parity, 1 stop bit, and 1200 bps for COM port 2.
5. Repeat Question 4 using INT 14H.

```
TITLE       SERIAL DATA COMMUNICATION BETWEEN TWO PCS
            .MODEL SMALL
            .STACK
            .DATA
MESSAGE DB  'Serial communication via COM2, 4800 ,No P,1 Stop,8-BIT DATA.',0AH,0DH
            DB ' ANY KEY PRESS IS SENT TO OTHER PC.',0AH,0DH
            DB ' PRESS ESC TO EXIT','$'
            .CODE
MAIN        PROC
            MOV AX,@DATA
            MOV DS,AX
            MOV AH,09
            MOV DX,OFFSET MESSAGE
            INT 21H
            ;initializing COM 2
            MOV AH,0             ;initialize COM port
            MOV DX,1             ;COM 2
            MOV AL,0C3H          ;4800 ,NO P,1 STOP,8-BIT DATA
            INT 14H
            ;checking key press and sending key to COM2 to be transfered
AGAIN:      MOV AH,01            ;check for key press using INT 16H ,AH=01
            INT 16H              ;if ZF=1, there is no key press
            JZ NEXT              ;If no key go check COM port
            MOV AH,0             ;yes, there is a key press, get it
            INT 16H              ;notice we must use INT 16H twice,2nd time
            ;with AH=0 to get the char itself. AL=ASCII char pressed
            CMP AL,1BH           ;is it esc key?
            JE EXIT              ;yes EXIT
            MOV AH,1             ;no. send the char to COM 2 port
            MOV DX,01
            INT 14H
            ;check COM2 port to see there is char. if so get it and display it
NEXT:       MOV AH,03            ;get COM 2 status
            MOV DX,01
            INT 14H
            AND AH,01            ;AH has COM port status, mask all bits except D0
            CMP AH,01            ;check D0 to see if there is a char
            JNE AGAIN            ;no data, go to monitor keyboard
            MOV AH,02            ;yes, COM2 has data: get it
            MOV DX,01
            INT 14H              ;get it
            MOV DL,AL            ;and display it using INT 21H
            MOV AH,02            ;DL has char to be displayed
            INT 21H
            JMP AGAIN            ;keep monitoring keyboard
EXIT:       MOV AH,4CH           ;exit to DOS
            INT 21H
MAIN        ENDP
            END
```

SECTION 17.3: INTERFACING THE NS8250/16450 UART IN THE IBM PC

The National Semiconductor 8250 and its variations are the most widely used UART in the PC. Due to the fact that the 8250 had a minor bug, the 8250A replaced it. Later, National Semiconductor made an improved version of the 8250A and called it 16450. All the programs written for the 8250/8250A will run on the 16450. There is also a CMOS version of the 16450 available called 16C450. In this section we provide an overview of the 8250/16450 chip and its interfacing in the IBM PC, emphasizing some internal registers used in the communications software. The 8250A and 16450 chips are pin compatible.

Table 17-4. 8250A Register Addresses

DLAB	A2	A1	A0	Description
0	0	0	0	Receive buffer register for read, transmitter holding register for write
0	0	0	1	Interrupt enable register
x	0	1	0	Interrupt identification register (read only)
x	0	1	1	Line control register (data format register)
x	1	0	0	MODEM control register
x	1	0	1	Line status register
x	1	1	0	MODEM status register
x	1	1	1	Scratch register
1	0	0	0	Divisor latch register (LSB)
1	0	0	1	Divisor latch register (MSB)

(Reprinted by permission of National Semiconductor, Copyright National Semiconductor 1990)

8250 pin descriptions

The 8250 is a 40-pin IC with an 8-bit data bus. It receives a single character from the CPU, frames it, then transmits it serially. In the same way, it can receive serial data, strip away the start and stop bits, make a character out of it, and present it to the CPU. It also generates all the necessary modem handshake signals. The following describes the pins and how the chip is used in the IBM PC.

A0, A1, A2

These pins are used to access the internal registers of the 8250 according to Table 17-4. In Table 17-4, notice that in order to transmit a character, it is written into the transmitter hold register when A0 = 0, A1 = 0, A2 = 0 and the DLAB bit, which is D7 of the line control (data format) register, is low. In the same way, the character is received by reading the same register when DLAB = 0. To program the 8250 for the baud rate, the DLAB bit of the line control (data format) register is set to high and then the divisor bytes are sent to the registers with addresses A2 = 0, A1 = 0, A0 = 0, and A0 = 1. An example of this will be shown shortly. In the IBM PC, A0, A1, and A2 of the 8250 are connected to the signals with the same name on the system bus.

CS0, CS1, $\overline{CS2}$ (chip selects)

These are used to activate the chip. Notice that CS0 and CS1 are active high and CS2 is active low. In the IBM PC an active-low logic decoder activates $\overline{CS2}$, and CS0 and CS1 are connected to high permanently.

D0 - D7

This is the bidirectional data bus connected to the data bus of the CPU. Notice in Figure 17-10 that for the IBM PC COM port the 74LS245 data bus transceiver is activated only for the I/O address range of 3F8H - 3FFH.

S$_{in}$ and S$_{out}$

These are the serial data pins, which become RxD and TxD of the RS232 after conversion from TTL to RS232 voltage levels.

RTS, CTS, DTR, DSR, DCD, and RI

These are the modem signals discussed in the preceding section.

X$_{in}$ and X$_{out}$

X$_{in}$ (external crystal input) and X$_{out}$ are used for connection to the crystal oscillator. When the frequency is generated off-chip, it is connected to X$_{in}$, and X$_{out}$ is ignored. In the IBM PC, X$_{in}$ is connected to a frequency of 1.8432 MHz.

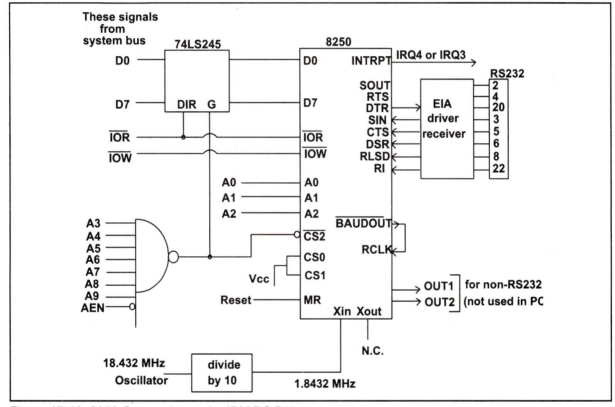

Figure 17-10. 8250 Connection to the IBM PC Buses

Example 17-5

(a) Find the I/O port address range set aside for COM0 of the IBM PC in Figure 17-10.
(b) Give the port address for each of the 8250 registers.

Solution:

(a) $\overline{CS2}$ is activated by A9 - A3. Therefore, we have

A9 A8	A7 A6 A5 A4 A3 A2 A1 A0	Hex
1 1	1 1 1 1 1 0 0 0	3F8
1 1	1 1 1 1 1 1 1 1	3FF

(b) The I/O addresses of the registers are as shown in Table 17-5.

MR

Master reset is an active-high input used to clear all the registers except the receiver buffer, transmitter hold, and divisor latches.

RCLK

The receiver clock is an input frequency clock equal to 16 times the baud rate clock of the receiver part.

BAUDOUT

This is is an output pin with clock frequency of sixteen times the baud rate of the transmitter part of the chip. Normally this is connected to the RCLK input pin making the transmitter and receiver of the chip work on the same baud rate. In such cases, the transmit and receive baud rate is always equal to the divisor word register value times sixteen.

The 8250 registers

There are total of 8 user-accessible registers inside the 8250. We describe the function of each with some examples of usage.

Table 17-5. IBM PC COM0 Registers

Hex Address	Description
3F8	Holds received character and the character to be transmitted. When DLAB = 0.
3F9	Interrupt enable register. When DLAB = 0.
3F8	Divisor latch register (LSB). When DLAB = 1.
3F9	Divisor latch register (MSB). When DLAB = 1.
3FA	Interrupt identification register (read only)
3FB	Line control (data format) register
3FC	MODEM control register
3FD	Line status register
3FE	MODEM status register
3FF	Scratch register

Transmitter holding register (A2 A1 A0 =000, and DLAB =0)

To transfer a byte serially, the CPU must write it to this register. In this case, the DLAB bit of the line control (data format) register must be 0. After a byte is written into this register, the 8250 frames it with proper start and stop bits and transfers it serially through the S_{out} pin.

Receiver buffer register (A2 A1 A0 =000, and DLAB =0)

When the 8250 receives the data through the S_{in} pin, it strips away the framing bits, makes it a byte, and holds it in this register for the CPU to read it. For the CPU to read this register the DLAB bit of the line control (data format) register must be equal to 0.

Interrupt enable register (A2 A1 A0 =001, and DLAB =0)

Bits D7 - D4 of this register are always 0, and the rest are used for the hardware-based interrupt to notify the CPU of certain conditions. While there is only one INTR pin on the 8250, there are four sources that can activate it. The interrupt enable register is used to mask or unmask any of these sources, as shown in Figure 17-11.

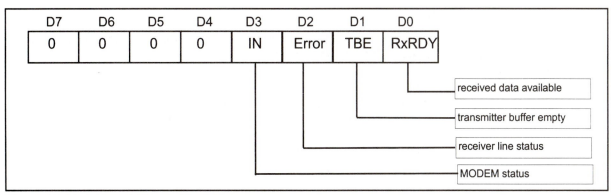

Figure 17-11. Interrupt Enable Register
(Reprinted by permission of National Semiconductor, Copyright National Semiconductor 1990)

D0 (received data available) If D0 =1, when a byte of data is received through the Sin pin, the INTR pin is activated to notify the CPU that a byte has been received.

D1 (transmit holding register empty) The 8250 moves the byte from the transmitter holding register into an internal parallel-in-serial-out register in order to transmit it; then it has room for a new byte. As soon as this happens, if D1 =1, INTR is activated to notify the CPU that it has room for another byte.

D2 (receiver line status) Whenever an error is detected in the course of receiving data, the INTR pin is activated if D2 = 1. The error could be due to a framing error or parity error or overrun error or break condition. To see which one is the source of error, the line status register is tested, assuming that D2 = 1.

D3 (MODEM status) When D3 =1, INTR is activated if any of the RS232 status lines changes during the reception or the transmission.

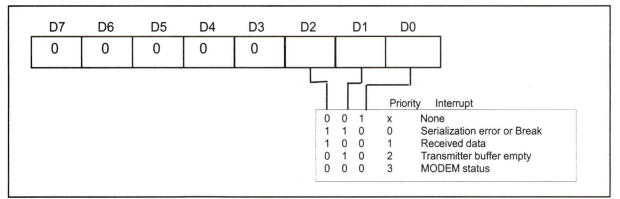

Figure 17-12. Interrupt Identification Register
(Reprinted by permission of National Semiconductor, Copyright National Semiconductor 1990)

Table 17-6: Interrupt Identification Register Priority

D2	D1	D0	Type of INTR	Source of INTR	Priority
1	1	0	Receiver line status	Error (overrun, parity, framing) or break interrupt	Highest
1	0	0	Receiver data available	Receiver data available	Second
0	1	0	Transmit holding register empty	Transmit holding register empty	Third
0	0	0	MODEM status	CTS, DSR, RI, or DCD	Fourth
x	x	1	None	None	--

(Reprinted by permission of National Semiconductor, Copyright National Semiconductor 1990)

Interrupt identification register (A2 A1 A0 =010)

From the above discussion, one might wonder how the CPU can know which is the source of INTR activation if all these conditions can activate the single INTR pin. This is exactly the function of the interrupt identification register. See Figure 17-12. This is a read-only register used to poll to determine the source of activation of the INTR pin. While D0 indicates if an interrupt is pending, D1 and D2 provide the source of the interrupt with the highest priority. They are prioritized as shown in Table 17-6. Another major use of this register is its ability to implement the polling method in monitoring INTR. The polling method first checks D0 to see if there is an interrupt pending and if there is one, D1 and D2 are tested to see which one. D3 to D7 of this register is always 0.

Line control (data format) register (A2 A1 A0 =011)

Framing information is sent to this register as shown in Figure 17-13. Example 17-6 shows how this register is programmed.

Example 17-6

Use the I/O port addresses in Example 17-5 to program the data format register for the following data format: 7 bits character, 1 stop bit, odd parity, and break control set to off.

Solution:

From Figure 17-13, we have 0000 1010 =0AH for the data format register, and the program is as follows:

```
MOV   DX,3FBH      ;the data format reg port address
MOV   AL,0AH       ;DLAB =0, no brk contl, odd, 1 stop, 7 bits
OUT   DX,AL        ;issue it
```

Notice in the above example that DLAB can be 0 or 1 for the data format register, but to access the buffer/hold register it must be 0, and for the divisor latch it must be 1, as we will see shortly.

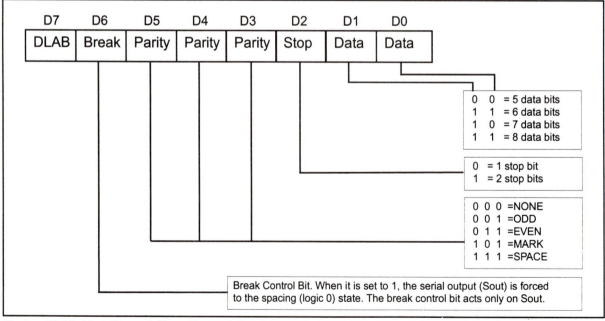

Figure 17-13. Line Control (Data Format) Register
(Reprinted by permission of National Semiconductor, Copyright National Semiconductor 1990)

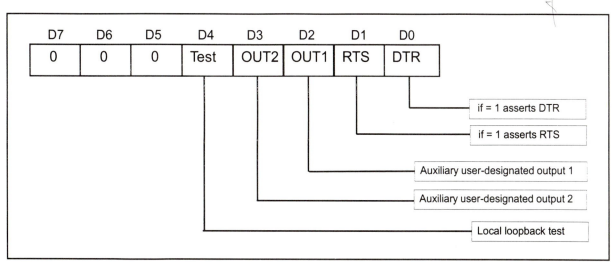

Figure 17-14.. MODEM Control Register
(Reprinted by permission of National Semiconductor, Copyright National Semiconductor 1990)

D7	D6	D5	D4	D3	D2	D1	D0
DCD	RI	DSR	CTS	Delta DCD	Delta RI	Delta DSR	Delta CTS

Figure 17-15. MODEM Status Register (MSR)
(Reprinted by permission of National Semiconductor, Copyright National Semiconductor 1990)

Modem control register (A2 A1 A0 =100)

This register, also referred to as the output control register, is used to assert the DTR and RTS pins of the 8250. See Figure 17-14. When D0 =1, it activates (active low) DTR, and when D1 =1, RTS is activated (set to 0). These can be used to test the RS232 pins. D2 and D3 are used for the OUT pins of the 8250 and are used in systems which are not RS232 compatible.

D4 allows testing of the 8250 using a method called local loopback. When D4 =1 we have S_{out} =1 (marking), S_{in} is disconnected, the PISO (parallel-in-serial-out) register is connected to SIPO (serial-in-parallel-out) register, and the modem control inputs (DSR, CTS, RI, and DCD) are all disconnected externally and connected internally to the modem control output signals.

D5, D6, and D7 are always 0.

Modem status register (A2 A1 A0 =110)

Referred to also as the RS232 input status register. Provides the current status of the modem control line as shown in Figure 17-15. The lower four bits indicate a change in status of the input to the chip since the last time it was read by the CPU. The term *delta* means change. Bits 0, 1, 2, and 3 are set to 1 when there is a change in signals CTS, DSR, RI, and DCD, respectively. Bit 2 is set to 1 when the RI signal has changed from low to high. Note that if bit 4 (loop test) of the modem control register is set to 1, bits 4, 5, 6, and 7 of the modem status register are equivalent to RTS, DTR, OUT1, and OUT2, respectively, of the modem control register. Notice that this register is the same as the modem status byte upon return from BIOS INT 14H.

Line status register (A2 A1 A0 =101)

When D2 of the interrupt enable register is set to high, one can monitor the line status register to see which of the errors (parity error, framing error, and so on) has occurred. The line status register bits are shown in Figure 17-16, and the bit definitions are as follows:

D0. When the 8250 receives the serial data and strips away the framing bits, it creates a byte to be given to the CPU. D0 indicates that a byte has been send to the receiver buffer to be picked up by the CPU.

D1. If the 8250 cannot keep up with the stream of incoming serial bits, the old byte is overwritten by the new data. This can happen if the CPU is slow and does not pick up the data previously received. D1 =1 indicates that the previous byte has been overrun by the new data.

D2 is set to 1 if the the parity bit of the data received does not match the data format register setup.

D3 indicates (when it equals 1) if the stop bit of the incoming data does not match the data format register.

D4. This is set to 1 if the S_{in} pin is low (space) for a period of one byte transfer (start bit + data bits + parity + stop bit). This is referred to as *break*.

D5. Transmitter holding register. If D5 =1, it indicates that the 8250 has room for a new byte to be transmitted. In other words when the byte is transferred from the transmit hold register into the serial shift register, D5 is set to 1. When the CPU writes a byte to transmit hold register, it becomes 0.

D6 =1 when both the transmit hold register and serial shift register are both empty.

D7 always equals 0.

Notice that this register is the same as the port status upon return from INT 14H, except that D7 is set to 1 for time-out by BIOS.

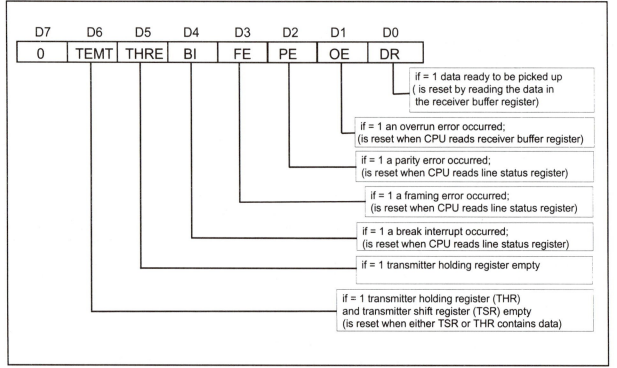

Figure 17-16. Line Status Register
(Reprinted by permission of National Semiconductor, Copyright National Semiconductor 1990)

Example 17-7

Find the divisor latch value for a baud rate of (a) 110, (b) 300, and (c) 2400. Assume that X_{in} = 1.8432 MHz.

Solution:
Dividing 1.8432 MHz by 16, we get 115,200 (1.8432 MHz/16 =115,200). To get the value for the divisor latch of 110 baud rate, we must divide 115,200 by 110 to get 1047. Similarly, 115,200/300 =384 and 115,200/2400 =48.

Table 17-7: Baud Rates and Divisors for 1.8432 MHz

Baud Rate	Divisor (Decimal)	Divisor (Hex)
110	1047	0417
300	384	0180
600	192	00C0
1200	96	0060
2400	48	0030
4800	24	0018
9600	12	000C

(Reprinted by permission of National Semiconductor, Copyright National Semiconductor 1990)

Example 17-8

Program the divisor latch for 300 baud, assuming that X_{in} =1.8432 MHz. Use the I/O port addresses of Example 17-5.

Solution:

As was shown in Example 17-7, the divisor value is 384. Therefore, we have
;set D7 of the line control register to 1 for accessing the DLAB

```
        MOV   AL,80H          ;10000000 (binary) to access DLAB
        MOV   DX,3FBH         ;the address of line control reg
        OUT   DX,AL           ;make D7 =1 for DLAB
;now send the divisor value
        MOV   AX,384          ;300 baud rate
        MOV   DX,3F8H         ;divisor latch address (LSB)
        OUT   DX,AL           ;issue the low byte
        MOV   AL,AH           ;
        INC   DX             ;the divisor latch address (MSB)
        OUT   DX,AL           ;issue the high byte
```

Example 17-9

Program the divisor latch for 2400 baud, assuming that X_{in} =1.8432 MHz. Use the I/O port addresses of Example 17-5.

Solution:

As was shown in Example 17-7, the divisor value is 48. Therefore, we have
;set D7 of the line control register to 1 for accessing the DLAB

```
        MOV   AL,80H          ;10000000 (bin) to access DLAB
        MOV   DX,3FBH         ;the address of line control reg
        OUT   DX,AL           ;make D7=1 for the DLAB
;now send the divisor value
        MOV   AX,48           ;2400 baud rate
        MOV   DX,3F8H         ;divisor latch address (LSB)
        OUT   DX,AL           ;issue the low byte
        MOV   AL,AH           ;
        INC   DX             ;the divisor latch address (MSB)
        OUT   DX,AL           ;issue the high byte
```

SECTION 17.3: INTERFACING THE NS8250/16450 UART IN THE IBM PC 529

Scratch pad register (A2 A1 A0 =111)

This is not used by the 8250 and is available to the CPU as an extra register for the purpose of a scratch pad.

Divisor latch LSB (A2 A1 A0 =000 and DLAB =1) and divisor latch MSB (A2 A1 A0 =001 and DLAB =1)

The baud rate of the 8250 is programmed through these two registers. The input frequency of X_{in} is the master clock. This clock is divided by the 16-bit integer contents of the divisor latches and again by the number 16 to get the desired baud rate, as shown in equation (17-1).

$$Divisor\ value = \frac{X_{in}\ clock\ frequency}{baud\ rate \times 16} \qquad (17-1)$$

Table 17-7 shows some of the divisor latch values in both decimal and hex for the 1.8432-MHz crystal frequency.

To program the divisor latches we set the DLAB bit of the line control (data format) register to 1 before issuing the divisor values, as shown in Examples 17-8 and 17-9. There are cases where the divisor is byte size instead of word size, but both upper and lower bytes must still be issued.

Limitation of the 8250/16450 UART and the 16550 replacement

A major limitation of the 8250/16450 is that it keeps interrupting the CPU for every single byte of data that it receives or is to be transmitted. In early PCs that was not a problem since everything in the PC was slow, but in today's world of high-performance PCs and workstations, such a limitation can be a source of severe bottleneck, especially in multitasked systems. Therefore, National Semiconductor introduced the 16550AF, which has an internal buffer of 16 bytes (instead of only 1 byte in 16450) to store data for transmission and reception. The 16550AF is fully 8250/16450 compatible. In many IBM PS models, this chip or an ASIC version of it is used to relieve the CPU from constant interruption. In the 16550AF the CPU can write a 16-byte block of data into its transmission buffer and let it transfer. When the buffer becomes empty it notifies the CPU for another block of data. In the same way, the 16-byte receive hold buffer keeps all the data received, and when the buffer becomes full, it interrupts the CPU to pick them up. This is much more efficient than interrupting the CPU for every byte of data. Although in such cases the CPU provides a block of data to the 16550 chip, the data is transmitted or received serially one byte at a time with proper framing.

Review Questions

1. State the number of user-accessible registers inside the 8250/16450.
2. True or false. The 8250/16450 can handle both asynchronous and synchronous serial data communication.
3. Which of the following addresses cannot be the 8250/16450 base address?
   (a) 2590H   (b) 2582H   (c) 2580H
4. True or false. To get the byte received by the 8250/16450, we must have A2 A1 A0 =000 and DLAB =1.
5. To which register does the DLAB bit belong, and what is its PC I/O port address?
6. Verify the calculation of divisor latch value for 4800 baud rate.
7. Using the IBM PC I/O port addresses, show the programming of Question 6.
8. True or false. The 16550 is compatible with the 8250/16450 chip.

SECTION 17.4: INTEL 8251 USART AND SYNCHRONOUS COMMUNICATION

This section provides programming examples of Intel's 8251A USART chip, and a brief discussion of the synchronous serial data communication.

Intel's 8251 USART chip

This is a 28-pin chip capable of doing both asynchronous and synchronous serial data communication. It supports all the control handshaking of the modem. It has several internal registers accessible, as shown in Table 17-8.

Table 17-8: 8251 Register Selection

$\overline{\text{CS}}$	$\overline{\text{C/D}}$	Description
0	0	Data register
0	1	Mode, command, and status registers
1	x	The 8251 is not selected

Example 17-10

Find the I/O port addresses assigned to the 8251 if $\overline{\text{CS}}$ is activated by A7 - A1 = "1001100" and A0 is connected to $\overline{\text{C/D}}$.

Solution:

1001 1000=98H for the data register;   1001 1001=99H for the status register

---

To allow the receiver and transmitter to work on the same baud rate, TxC (transmission) and RxC (receiver clock) are connected to the same frequency. The baud rate selections are x1, x16, and x64. This means that the baud rate selection times 1, 16, or 64 must be equal to the RxC and TxC clock frequency.

Example 17-11

Using the data in Figure 17-18, find the baud rate if each of the following options is selected. Assume RxC = TxC = 19,200 Hz.   (a) x16 (b) x64

Solution:
(a) 19,200/16 =1200   (b) 19,200/64 =300

The 8251 has the mode register and command register both with the same I/O address port as shown in Table 17-8 ($\overline{\text{CS}}$ = 0 and $\overline{\text{C/D}}$ = 1). To distinguish between them one must always program the mode register first to select the data format and baud rate; then writing to the same port is considered accessing the command register. To access the mode register again one must either reset the system by the RESET hardware pin or D6 of the command register as shown in Figure 17-19. The status register is a read-only register. Figure 17-17 shows the bits.

As an example of how to program the status register, look at Example 17-13. One can use polling to monitor the TxRDY bit of the status register (Figure 17-17) to transfer a byte of the data to 8251 to be transmitted serially. See Example 17-13.

Synchronous serial data communication

One of the problems with the asynchronous method discussed so far is the overhead. In some instances the overhead can be as high as 50%. The overhead can be even higher if characters are 5 or 6 bits wide. To solve this problem, the synchronous method was developed. In synchronous serial data communication, start and stop bits are replaced with special bytes of code, but instead of transferring only one byte, hundreds of bytes are transferred. In this method a block of bytes is framed in between the control codes and then transferred. The control codes are different for various protocols. There are two widely synchronous standard protocols: BISYNC and SDLC. In BISYNC protocol, there are no start and stop bits; instead, for each block of data there are one or more bytes called synch characters. After the synch character there is the character called STX (start of text), followed by the block of data bytes. The size of the data block can be 100 or more bytes of information. The data block is followed by the control byte called ETX (end of text). The last byte of the frame is the BCC (block check character) byte, which is used for error detection. In Chapter 11 we saw the checksum method used to maintain data integrity in block of data. There are other methods of error checking, such as

---

Example 17-12

Program the 8251 of Example 17-10 for asynchronous mode with a data format of 8 bits, 1 stop bit, even parity, and 1200 baud rate.

Solution:

From Figure 17-18 we have 01111110 for the mode register therefore:

```
MOV   AL,01000000B        ;reset the 8251 by d6 of command reg
OUT   99H,AL              ;to ensure mode reg is being accessed next
MOV   AL,01111110B        ;1 stop, even, 8 bits, 16x
OUT   99H,AL              ;write into mode register
```

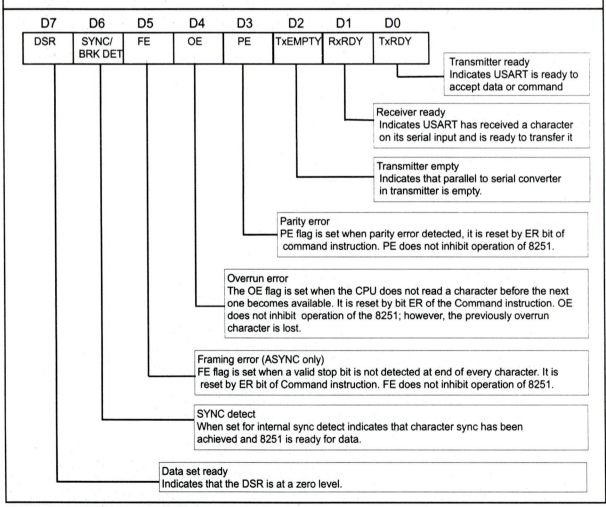

Figure 17-17. 8251 Status Register
(Reprinted by permission of Intel Corporation, Copyright Intel Corp. 1983)

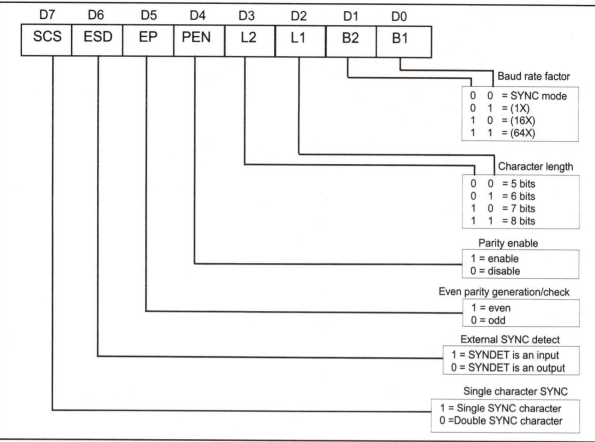

Figure 17-18. 8251 Mode Register
(Reprinted by permission of Intel Corporation, Copyright Intel Corp. 1983)

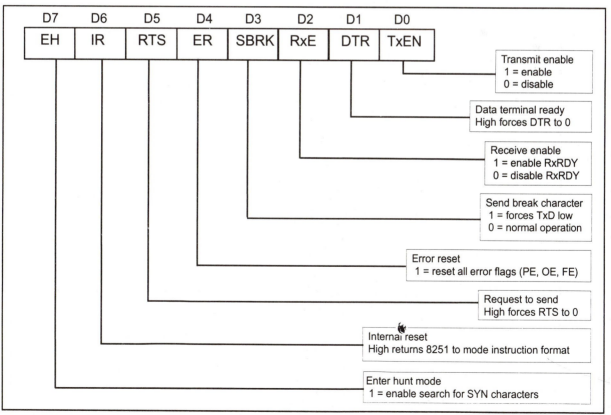

Figure 17-19. 8251 Command Register
(Reprinted by permission of Intel Corporation, Copyright Intel Corp. 1983)

CRCs (cyclic redundancy checks) and Hamming code. Shortly we will describe the CRC method and leave the Hamming code to be pursued by interested readers on their own. It must be noted that the checksum and CRC methods detect the error, but do not correct it. Hamming code not only detects the error, but also corrects it.

Table 17-9 shows some of the ASCII codes used for the BISYNC protocol. In some cases an SOH (start of header) header is inserted between the SYN and STX bytes. The header contains information such as destination and type of data block. This is shown in Figure 17-20.

Table 17-9: Some Codes Used in BISYNC Protocol

Char	Meaning	ASCII
SOH	Start of header	01H
STX	Start of text	02H
ETX	End of text	03H
EOT	End of transmission	04H
ENQ	Inquiry	05H
ETB	End-of-transmission block	0FH
DLE	Data link escape	10H
SYNC	Sync	16H
ETB	End-of-transmission block	17H
PAD	End-of-frame block	FFH

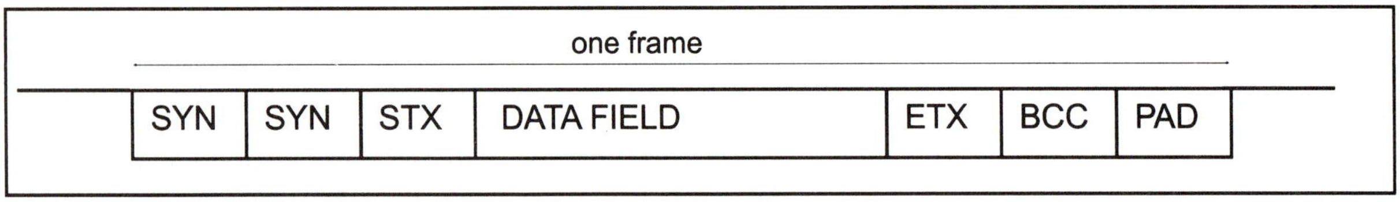

one frame

| SYN | SYN | STX | DATA FIELD | ETX | BCC | PAD |

Figure 17-20. Data Framing in BISYNC

Example 17-13

Write a program that transfers the message "The planet Earth","$" into the 8251. The "$" indicates the end of the message: Use the data format and the baud rate of Example 17-12.

Solution:

```
DATA      DB       "The planet Earth","$"
;from the code segment:
          MOV  AL,00            ;issuing dummy byte 00 to mode reg
          OUT  99H,AL           ;is recommended.  do it three times
          OUT  99H,AL
          OUT  99H,AL
          MOV  AL,01000000B     ;reset the 8251 by D6 of command reg
          OUT  99H,AL           ;to ensure mode reg is being accessed next
          MOV  AL,01111110B     ;1 stop, even, 8 bits, 16x
          OUT  99H,AL           ;write into mode register
          MOV  SI,OFFSET DATA   ;SI=offset address of data
B1:       IN   AL,99H           ;monitor the status reg
          AND  AL,00000001B     ;test the d0 for txrdy and
          JZ   B1               ;keep checking
          MOV  AL,[SI]          ;get the byte
          CMP  AL,"$"           ;is it the end?
          JE   B2               ;if yes then stop
          OUT  98H,AL           ;if not send it to data reg
          JMP  B1               ;keep sending
B2:       RET
```

SDLC (Serial data link control)

SDLC is the synchronous serial communication standard developed by IBM and is widely used in computer networking, such as in the IBM token ring. It uses 8-bit patterns of 01111110 in place of a SYNC byte. This byte is referred to in IBM literature as the flag byte. After the flag byte comes the byte for the address where the data is to be sent on the network of computers. Next is the control byte, containing information about the sequence of data, among other things. Then comes the data. The data in SDLC is in bits, thousands of bits. This is in contrast to BISYNC, which handles thousands of bytes. The data fields can go as high as 200,000 bits per frame. Following the data fields is a 16-bit field for error checking, and finally, the flag byte, indicating the end of the frame. See Figure 17-21.

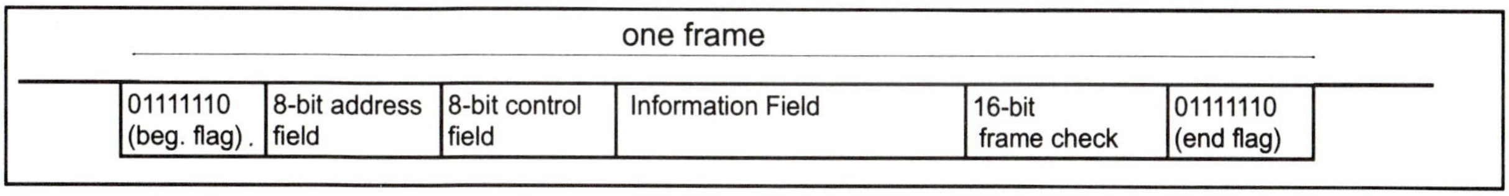

one frame						
01111110 (beg. flag) .	8-bit address field	8-bit control field	Information Field	16-bit frame check	01111110 (end flag)	

Figure 17-21. Data Framing in SDLC

Cyclic redundancy checks

One of the most widely used methods of error checking in synchronous serial data communications is that of cyclic redundancy checks, or CRCs. This method is also widely used in error checking of disk storage. CRC is used for error checking a stream of bits, in contrast to checksum, which is byte-oriented, but both are used for error checking blocks of data. In the CRC method, two bytes called the CRC bytes are appended to the stream of data and transmitted with the data. At the destination, by hardware or software, the data and the CRC bytes are tested for data integrity. The CRC bytes are calculated using the formula

$$\frac{M(X) \times X^n}{G(X)}$$

The polynomial M(X) represents the bit stream. This is multiplied by X^n where n is the number of bits in the stream. The product of these two terms is divided by a polynomial called the generator polynomial. This division will result in a quotient, Q(X), and a remainder, R(X). It is the remainder that forms the CRC bytes.

Example 17-14 shows the CRC method used for a 16-bit data stream, 4D92H. In binary this is 0100 1101 1001 0010. To get M(X), first the bits are reversed: 0100 1001 1011 0010. This series of bits is interpreted as a series of coefficients of a polynomial, as shown below.

M(0100 1001 1011 0010) =

$0x^{15} + 1x^{14} + 0x^{13} + 0x^{12} + 1x^{11} + 0x^{10} + 0x^9 + 1x^8 +$
$1x^7 + 0x^6 + 1x^5 + 1x^4 + 0x^3 + 0x^2 + 1x^1 + 0x^0$

Removing the terms with 0 coefficients yields

$M(X) = x^{14} + x^{11} + x^8 + x^7 + x^5 + x^4 + x^1$

Now M(X) is multiplied by x^{16} since the data is a 16-bit stream.

$M(X) \times X^n = (x^{14} + x^{11} + x^8 + x^7 + x^5 + x^4 + x^1) \times x^{16}$

$= x^{30} + x^{27} + x^{24} + x^{23} + x^{21} + x^{20} + x^{17}$

The dividend of our equation will be $x^{30} + x^{27} + x^{24} + x^{23} + x^{21} + x^{20} + x^{17}$. This will be divided by the generator polynomial, G(X). We will use G(X) = $x^{16} + x^{15} + x^2 + 1$. This is called the bisync protocol generator. The SDLC protocol uses generator polynomial G(X) = $x^{16} + x^{12} + x^5 + 1$. Notice that both begin with x^{16}. This ensures that the remainder will be less than x^{16} and can therefore be represented by 16 bits, or 2 bytes. The polynomial division shown in Example 17-14 results in remainder R(X) = $x^{15} + x^{13} + x^{12} + x^{10} + x^8 + x^7 + x^5 + x^3 + x$. Representing this as a series of coefficients gives the following.

$$R(X) = x^{15} + x^{13} + x^{12} + x^{10} + x^8 + x^7 + x^5 + x^3 + x =$$
$$1x^{15} + 0x^{14} + 1x^{13} + 1x^{12} + 0x^{11} + 1x^{10} + 0x^9 + 1x^8 +$$
$$1x^7 + 0x^6 + 1x^5 + 0x^4 + 1x^3 + 0x^2 + 1x^1 + 0x^0$$

or simply,

1011 0101 1010 1010.

Reversing the bits gives 0101 0101 1010 1101, which is 55ADH, the CRC bytes. This example used a small stream of bits, 16, for the sake of simplicity. The CRC bytes take as much room as the bit stream itself. Normally, this method is applied to a large block of data consisting of hundreds of bytes and generating the 2-byte CRC code. This 2-byte code is appended to the end of the data stream. When the receiver reads in the bit stream plus the CRC bytes and performs division by G(X), the remainder will be 0 if there were no errors in transmission.

Review Questions

1. True or false. The 8251 can perform both asynchronous and synchronous communication.
2. If $\overline{CS}$ of the 8251 is activated by A7 - A1 =1111 010, find the I/O port addresses assigned to the 8251. Assume that A0 is connected to C/$\overline{D}$.
3. If x16 is selected and we need a baud rate of 4800, what frequency must be connected to RcX and TxC?
4. Show a program to set the data format of 7 bits, odd, 2 stops, and x64. Use Question 2 for port addresses.
5. Find the data format information if the mode byte is CAH.
6. State the ways that the 8251 can be reset.
7. True or false. The BISYNC frame consists of thousands of bytes while the SDLC consists of thousands of bits.
8. True or false. The first byte in the synchronous frame is always a special byte or code.

SUMMARY

This chapter began with an introduction to the fundamentals of serial communication. Serial communication, in which data is sent one bit a time, is used in situations where data is sent over significant distances since in parallel communication, where data is sent a byte or more a time, great distances can cause distortion of the data. Serial communication has the additional advantage of allowing transmission over phone lines. Serial communication uses two methods: synchronous and asynchronous. In synchronous communication, data is sent in blocks of bytes, in asynchronous, data is sent in bytes. Data communication can be simplex (can send but cannot receive), half-duplex (can send and receive, but not at the same time), or full-duplex (can send and receive at the same time). RS232 is a standard for serial communication connectors. Two terms commonly used to classify communications equipment are DTE (data terminal equipment), which refers to devices such as computers which send or receive data, and DCE (data communications equipment), which refers to devices such as modems which handle the data communications. Handshaking protocol in RS232 communications is done with the aid of signals such as DTR, DSR, RTS, CTS, and others.

This chapter also showed how to use the DOS MODE command to initialize the 4 COM ports of the IBM PC and how to code Assembly language instructions to perform serial communication using BIOS INT 14H.

The National Semiconductor 8250/16450 UART chips were covered. They are used in the IBM PC for serial communications. They can receive a character from the CPU, frame it, then transmit it serially, or they can receive serial data, strip away the framing bits, make a character out of it, and send it to the CPU.

The final topic covered was two protocols used in synchronous communications: BISYNC and SDLC. In BISYNC, no start and stop bits are used; instead, control codes called sync characters begin a block and the block is ended with ETX (end of text) and BLC (block check character). SDLC is the standard developed by IBM and used in their token ring networks. In this protocol, blocks are prefaced by a flag byte, and the address where they are going, then a control byte, followed by the block, and the error field and flag bytes.

Example 17-14

Find the CRC byte for the data stream 4D92H using divisor $x^{16} + x^{15} + x^2 + 1$.

Solution:

Data stream 4D92H = 0100 1101 1001 0010. Reversing the bits gives 0100 1001 1011 0010. $M(X)$ is calculated by using these bits as coefficients of the polynomial. $M(X)$ for this bit stream is $x^{14} + x^{11} + x^8 + x^7 + x^5 + x^4 + x^1$. This is multiplied by x^{16} because there are 16 bits in the data stream (Recall that in multiplying powers of x we add the exponents). Multiplying $M(X)$ by x^{16} gives $x^{30} + x^{27} + x^{24} + x^{23} + x^{21} + x^{20} + x^{17}$, the dividend. The polynomial division is shown below.

$$
\begin{array}{r}
x^{14} + x^{13} + x^{12} + x^8 + x^5 + x
\end{array}
$$

$x^{16} + x^{15} + x^2 + 1$ ⟌

$x^{30} + x^{27} + x^{24} + x^{23} + x^{21} + x^{20} + x^{17}$

$x^{30} + x^{29} + x^{16} + x^{14}$

$x^{29} + x^{27} + x^{24} + x^{23} + x^{21} + x^{20} + x^{17} + x^{16} + x^{14}$

$x^{29} + x^{28} + x^{15} + x^{13}$

$x^{28} + x^{27} + x^{24} + x^{23} + x^{21} + x^{20} + x^{17} + x^{16} + x^{15} + x^{14} + x^{13}$

$x^{28} + x^{27} + x^{14} + x^{12}$

$x^{24} + x^{23} + x^{21} + x^{20} + x^{17} + x^{16} + x^{15} + x^{13} + x^{12}$

$x^{24} + x^{23} + x^{10} + x^8$

$x^{21} + x^{20} + x^{17} + x^{16} + x^{15} + x^{13} + x^{12} + x^{10} + x^8$

$x^{21} + x^{20} + x^7 + x^5$

$x^{17} + x^{16} + x^{15} + x^{13} + x^{12} + x^{10} + x^8 + x^7 + x^5$

$x^{17} + x^{16} + x^3 + x$

$x^{15} + x^{13} + x^{12} + x^{10} + x^8 + x^7 + x^5 + x^3 + x$

The remainder is $x^{15} + x^{13} + x^{12} + x^{10} + x^8 + x^7 + x^5 + x^3 + x$. Representing this as a series of coefficients gives 1011 0101 1010 1010. Reversing the bits gives 0101 0101 1010 1101, which is 55ADH, our CRC bytes.

PROBLEMS

SECTION 17.1: BASICS OF SERIAL COMMUNICATION

1. Which is more expensive, parallel or serial data transfer?
2. True or false. 0- and 5-V digital pulses can be transferred on the telephone without being converted (modulated).
3. Show the framing of the letter ASCII "Z" (0101 1010), even parity, 1 stop bit.
4. If there is no data transfer and the line is high, it is called _____ (mark, space).
5. What is space?
6. Calculate the overhead percentage if the data size is 6, 2 stop bits, even parity.
7. True or false. RS232 voltage specification is TTL compatible.
8. What is the function of the MC1488 and MC1489 chips?
9. True or false. RS232P refers to a male connector.
10. How many pins of the RS232 are used by the IBM serial cable, and why?
11. True or false. The longer the cable, the higher the data transfer baud rate.
12. The function definition of the RS232 pins is stated from the point of view of _____ (DTE, DCE).
13. If two PCs are connected through the RS232 without the modem, they are both configured as a _____ (DTE, DCE) -to- _____ (DTE, DCE) connection.
14. State the 9 most important signals of the RS232.
15. Calculate the total number of bits transferred if 200 pages of ASCII data are sent using asynchronous serial data transfer. Assume a data size of 8 bits, 1 stop bit, no parity.

SECTION 17.2: ACCESSING IBM PC COM PORTS USING DOS AND BIOS

16. In the IBM PC, what is the maximum number of COM ports that can be installed? Use DEBUG to find which COM port is installed on your PC.
17. What are the lowest and highest baud rates supported by the IBM PC or PS COM ports using BIOS programming?
18. Show the COM2 port setting of 9600 baud rate, 8 data bits, 1 stop bit, no parity bit, using each of the following.
    (a) DOS command   (b) BIOS INT 14H
19. Modify the program for Example 17-4 to display the key pressed not only on the receiving PC's monitor but also on the monitor of the PC that sent it.
20. Compare the options available with INT 14H when AH = 0 and the more recent one, AH = 04. See Appendix E.

SECTION 17.3: INTERFACING THE NS8250/16450 UART IN THE IBM PC

21. True or false. The 8250 is a USART chip.
22. True or false. There are a total of 8 user-accessible registers inside the 8250.
23. Which register has the DLAB bit, and what is its function?
24. What I/O port addresses are assigned to the first COM port in the PC?
25. When an IBM PC receives a byte of data through the serial data line, the 80x86 can read it from which I/O port address? Use the I/O addresses in Problem 24.
26. Why are D0 - D7 of the 8250 bidirectional?
27. The INTR pin is an _____ (input, output) for the 8250 chip.
28. If there is only one INTR pin and many sources can activate it, how does the CPU know which one is the source of activation?

The next three problems are not IBM PC/PS compatible. They are given only for exercise.

29. Assuming that Xin of 8250 is connected to 3.072 MHz-frequency, fill in the following table for all the desired baud rates.

Desired Baud Rate	Divisor to Generate x16 Clock (decimal)
110	
300	
600	
1200	
2400	
3600	
4800	
9600	

30. Assuming that $\overline{CS}$ of the 8250 is activated by address A7 - A3 =10010, determine the following.
(a) The I/O address range assigned to this 8250.
(b) The I/O address assigned to each register.
31. Program the divisor latch of the 8250 in Problems 29 and 30 for 1200 baud rate.
32. What is data overrun?
33. Why are RCLK and BAUDOUT connected to the same frequency?
34. True or false. Before the divisor latches are accessed, the DLAB bit must be set to high.

SECTION 17.4: INTEL 8251 USART AND SYNCHRONOUS COMMUNICATION

35. True or false. The 8251 is a UART chip.
36. Find the I/O port addresses assigned to the 8251 if A7 - A1=0110001 activates the $\overline{CS}$ pin. Assume that A0 is connected to C/$\overline{D}$.
37. Assuming that RxC =TxC =153,600 Hz, program the 8251 to transfer "One small step for man, one giant step for mankind" for 7-bit data size, odd parity, 1 stop bit, and X64. Use the I/O port addresses of Problem 36.
38. Explain the difference between the asynchronous and synchronous methods. Which has the lower overhead?
39. How many bytes are set aside for the CRC byte?
40. The data format _____ (BISYNC, SDLC) is byte-oriented.

ANSWERS TO REVIEW QUESTIONS

SECTION 17.1: BASICS OF SERIAL COMMUNICATION

1. faster, more expensive
2. serial
3. synchronous
4. False; it is simplex.
5. true
6. asynch
7. With 100 0100 binary we have 1 as the odd-parity bit. The bits as transmitted in the sequence:
   (a) 0 (start bit)   (b) 0   (c) 0   (d) 1   (e) 0   (f) 0   (g) 0   (h) 1   (i) 1 (parity)
   (j) 1 (first stop bit)   (k) 1 (second stop bit)
8. 4 bits
9. 400 x 11 = 4400 bits total bits transmitted. 4400/1200 = 3.667 seconds, 4/7 = 58%.
10. true
11. +3 to +25 V
12. DTR

SECTION 17.2: ACCESSING IBM PC COM PORTS USING DOS AND BIOS

1. 4
2. 110 to 9600 (can be higher if we bypass BIOS)
3. 110 to 19,200 (can be higher if we bypass BIOS)
4. C>mode com2:12,n,8,1
5. MOV AH,O
    MOV AL,83H
    MOV DX,1
    INT 14H

SECTION 17.3: INTERFACING THE NS8250/16450 UART IN THE IBM PC

1. 8
2. false; asynchronous only
3. (b)
4. False; DLAB must be 0.
5. It is the D7 bit of the line control (data format) register with I/O port address of 3FBH.
6. 1.8432 MHz/16 = 115,200.   115,200/4800 =24 =18 hex.
7. ;SET D7 OF THE LINE CONTROL REGISTER TO 1 FOR ACCESSING THE DLAB
```
        MOV     AL,80H              ;10000000 (BIN) TO ACCESS DLAB
        MOV     DX,3FBH             ;THE ADDRESS OF LINE CONTROL REG
        OUT     DX,AL               ;MAKE D7=1 FOR THE DLAB
;NOW SEND THE DIVISOR VALUE
        MOV     AX,24               ;4800 BAUD RATE
        MOV     DX,3F8H             ;DIVISOR LATCH ADDRESS (LSB)
        OUT     DX,AL               ;ISSUE THE LOW BYTE
        MOV     AL,AH               ;
        INC     DX                  ;THE DIVISOR LATCH ADDRESS (MSB)
        OUT     DX,AL               ;ISSUE THE HIGH BYTE
```
8. True

SECTION 17.4: INTEL 8251 USART AND SYNCHRONOUS COMMUNICATION

1. true
2. F4H and F5H
3. 4800 x 16 =76,800 Hz
4.
```
    MOV     AL,01000000B        ;RESET THE 8251 BY D6 OF COMMAND REG
    OUT     0F5H,AL             ;TO ENSURE MODE REG IS BEING ACCESSED NEXT
    MOV     AL,11011011B        ;2 STOP BITS, ODD, 7 BITS, X64
    OUT     0F5,AL              ;WRITE INTO MODE REGISTER
```
5. 2 stop bits, no parity, 7bits, X16
6. One way is by RESET hardware pin and another one is by outputting the 40H to command register.
7. true
8. true

CHAPTER 18

KEYBOARD AND PRINTER INTERFACING

OBJECTIVES

Upon completion of this chapter, you will be able to:

» **Diagram how a keyboard matrix is connected to the I/O ports of a PC**
» **Describe the processes of key press detection and key identification performed by a microprocessor with pseudocode or flowchart**
» **Describe the respective functions of the keyboard microcontroller, INT 09, and the motherboard in keyboard input**
» **Code Assembly language instructions using INT 16H to get and check the keyboard input buffer and status bytes**
» **Diagram how data from the keyboard is stored in the keyboard buffer ring**
» **State the differences between hard contact and capacitance keyboards**
» **List the 36-pin assignments of the Centronics printer interface**
» **Describe the BIOS programming for the 4 parallel printer ports LPT1 - LPT4**
» **Diagram the I/O port assignment for printer data, status, and control**
» **Code Assembly language instructions using INT 17H to check printer status, initialize printers, and send data to the printer**
» **Describe the printer time-out problem and how it can be alleviated**
» **Discuss the evolution of the PC's parallel port**
» **Contrast and compare parallel port types, including SPP, PS/2, EPP and ECP**
» **Explain interfacing of LPT ports to devices such as LCDs and stepper motors**

Along with video monitors, keyboards and printers are the most widely used input/output devices of the PC, and a basic understanding of them is essential. In this chapter, we first discuss keyboard fundamentals, along with key press and key detection mechanisms. In Section 18.2 we study hardware interfacing and BIOS programming of the keyboard in the IBM PC. The interfacing and programming of parallel port printers in the IBM PC are discussed in Section 18.3. Section 18.4 gives an overview of the bidirectional data bus in parallel ports. Specifically, parallel port types of SPP, PS/2, EPP and ECP are described, and techniques for interfacing the LPT port to devices such as LCDs and stepper motors are detailed.

SECTION 18.1: INTERFACING THE KEYBOARD TO THE CPU

At the lowest level, keyboards are organized in a matrix of rows and columns. The CPU accesses both rows and columns through ports; therefore, with two 8-bit ports, an 8 x 8 matrix of keys can be connected to a microprocessor. When a key is pressed, a row and a column make a contact; otherwise, there is no connection between rows and columns. In IBM PC keyboards, a single microcontroller (consisting of a microprocessor, RAM and EPROM, and several ports all on a single chip) takes care of hardware and software interfacing of the keyboard. In such systems, it is the function of programs stored in the EPROM of the microcontroller to scan the keys continuously, identify which one has been activated, and present it to the main CPU on the motherboard. More details of the IBM PC keyboard design are presented in Section 18.2. In this section we look at the mechanism by which the microprocessor scans and identifies the key. For clarity we use 8086/88 Assembly language instructions in examples.

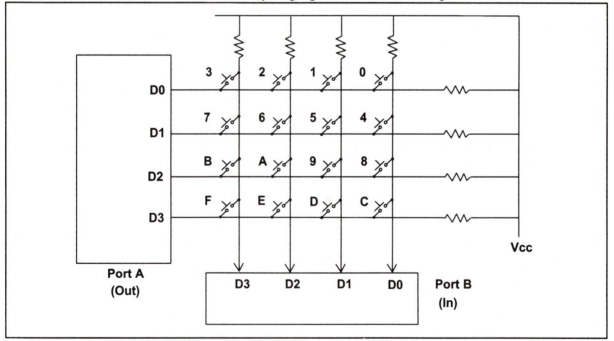

Figure 18-1. Matrix Keyboard Connection to Ports

Scanning and identifying the key

Figure 18-1 shows a 4 x 4 matrix connected to two ports. The rows are connected to an output port and the columns are connected to an input port. If no key has been pressed, reading the input port will yield 1s for all columns since they are all connected to high (V_{CC}). If all the rows are grounded and a key is pressed, one of the columns will have 0 since the key pressed provides the path to ground. It is the function of the microprocessor to scan the keyboard continuously to detect and identify the key pressed. How it is done is explained next.

CHAPTER 18: KEYBOARD AND PRINTER INTERFACING

Grounding rows and reading the columns

To detect the key pressed, the microprocessor grounds all rows by providing 0 to the output latch, then it reads the columns. If the data read from the columns is D3 - D0 =1111, no key has been pressed and the process continues until a key press is detected. However, if one of the column bits has a zero, this means that a key press has occurred. For example, if D3 - D0 =1101, this means that a key in the D1 column has been pressed. After a key press is detected, the microprocessor will go through the process of identifying the key. Starting with the top row, the microprocessor grounds it by providing a low to row D0 only; then it reads the columns. If the data read is all 1s, no key in that row is activated and the process is moved to the next row. It grounds the next row, reads the columns, and checks for any zero. This process continues until the row is identified. After identification of the row in which the key has been pressed, the next task is to find out which column the pressed key belongs to. This should be easy since the CPU knows at any time which row and column are being accessed. Look at Example 18-1.

Example 18-1

From Figure 18-1, identify the row and column of the pressed key for each of the following.
(a) D3 - D0 = 1110 for the row, D3 - D0 = 1011 for the column
(b) D3 - D0 = 1101 for the row, D3 - D0 = 0111 for the column

Solution:

From Figure 18-1 the row and column can be used to identify the key.
(a) The row belongs to D0 and the column belongs to D2; therefore, the key number 2 was pressed.
(b) The row belongs to D1 and the column belongs to D3; therefore, the key number 7 was pressed.

Program 18-1 is the Assembly language program for the detection and identification of the key activation. In this program, it is assumed that PORT_A and PORT_B are initialized as output and input, respectively. Program 18-1 goes through the following four major stages:

1. To make sure that the preceding key has been released, 0s are output to all rows at once, and the columns are read and checked repeatedly until all the columns are high. When all columns are found to be high, the program waits for a short amount of time before it goes to the next stage of waiting for a key to be pressed.
2. To see if any key is pressed, the columns are scanned over and over in an infinite loop until one of them has a 0 on it. Remember that the output latches connected to rows still have their initial zeros (provided in stage 1), making them grounded. After the key press detection, it waits 20 ms for the bounce and then scans the columns again. This serves two functions: (a) it ensures that the first key press detection was not an erroneous one due to a spike noise, and (b) the 20-ms delay prevents the same key press from being interpreted as a multiple key press. If after the 20-ms delay the key is still pressed, it goes to the next stage to detect which row it belongs to; otherwise, it goes back into the loop to detect a real key press.
3. To detect which row the key press belongs to, it grounds one row at a time, reading the columns each time. If it finds that all columns are high, this means that the key press cannot belong to that row; therefore, it grounds the next row and continues until it finds the row the key press belongs to. Upon finding the row that the key press belongs to, it sets up the starting address for the look-up table holding the scan codes for that row and goes to the next stage to identify the key.
4. To identify the key press, it rotates the column bits, one bit at a time, into the carry flag and checks to see if it is low. Upon finding the zero, it pulls out the scan code for that key from the look-up table; otherwise, it increments the pointer to point to the next element of the look-up table.

While the key press detection is standard for all keyboards, the process for determining which key is pressed varies. The look-up table method shown in Program 18-1 can be modified to work with any matrix up to 8 x 8. Figure 18-2 provides the flowchart for Program 18-1 for scanning and identifying the pressed key.

There are IC chips such as National Semiconductor's MM74C923 that incorporate keyboard scanning and decoding all in one chip. Such chips use combinations of counters and logic gates (no microprocessor) to implement the underlying concepts presented in Program 18-1.

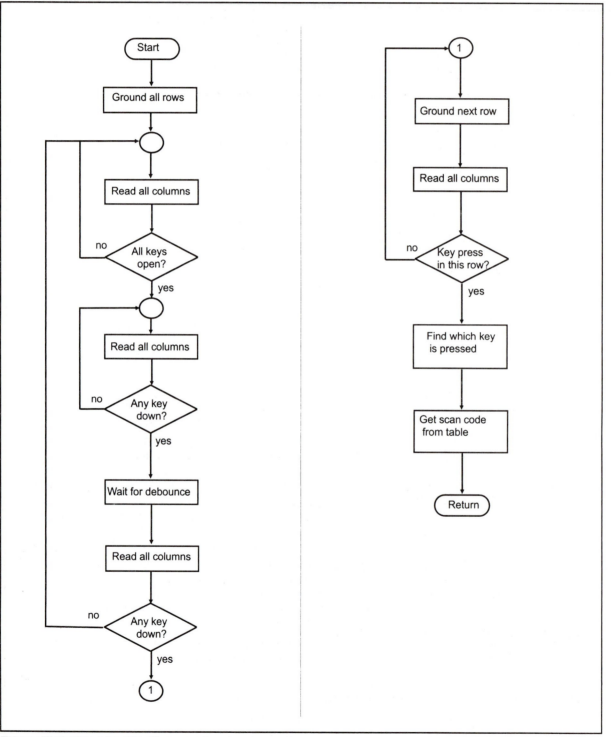

Figure 18-2. Flowchart for Program 1

```
;the following look-up scan codes are in the data segment
KCOD_0      DB   0,1,2,3            ;key codes for row zero
KCOD_1      DB   4,5,6,7            ;key codes for row one
KCOD_2      DB   8,9,0AH,0BH        ;key codes for row two
KCOD_3      DB   0CH,0DH,0EH,0FH    ;key codes for row three

;the following is from the code segment
            PUSH   BX                           ;save BX
            SUB    AL,AL                         ;AL=0 to ground all rows rows at once
            OUT    PORT_A,AL                     ;to ensure all keys are open (no contact)
K1:         IN     AL,PORT_B                     ;read the columns
            AND    AL,00001111B                  ;mask the unused bits (D7-D4)
            CMP    AL,00001111B                  ;are all keys released
            JNE    K1                            ;keep checking for all keys released
            CALL   DELAY                         ;wait for 20 ms
K2:         IN     AL,PORT_B                     ;read columns
            AND    AL,00001111B                  ;mask D7-D4
            CMP    AL,00001111B                  ;see if any key pressed?
            JE     K2                            ;if none keep checking
            CALL   DELAY                         ;wait 20 ms for debounce
;after the debounce see if still pressed
            IN     AL,PORT_B                     ;read columns
            AND    AL,00001111B                  ;mask D7-D4
            CMP    AL,00001111B                  ;see if any key closed?
            JE     K2                            ;if none keep polling
;now ground one row at a time and read columns to find the key
            MOV    AL,11111110B                  ;ground row 0 (D0=0)
            OUT    PORT_A,AL
            IN     AL,PORT_B                     ;read all columns
            AND    AL,00001111B                  ;mask unused bits (D7-D4)
            CMP    AL,00001111B                  ;see which column
            JE     RO_1                          ;if none go to grounding row 1
            MOV    BX,OFFSET KCOD_0              ;set BX=start of table for column 0 keys
            JMP    FIND_IT                       ;identify the key
RO_1:       MOV    AL,11111101B                  ;ground row 1 (D1=0)
            OUT    PORT_A,AL
            IN     AL,PORT_B                     ;read all columns
            AND    AL,00001111B                  ;mask unused bits (D7-D4)
            CMP    AL,00001111B                  ;see which column
            JE     RO_2                          ;if none go to grounding row 2
            MOV    BX,OFFSET KCOD_1              ;set BX=Start of table for column 1 keys
            JMP    FIND_IT                       ;identify the key
RO_2:       MOV    AL,11111011B                  ;ground row 2 (D2=0)
            OUT    PORT_A,AL
            IN     AL,PORT_B                     ;read all columns
            AND    AL,00001111B                  ;mask unused bits (D7-D4)
            CMP    AL,00001111B                  ;see which column
            JE     RO_3                          ;if none go to grounding row 3
            MOV    BX,OFFSET KCOD_2              ;set BX=start of table for column 2 keys
            JMP    FIND_IT                       ;identify the key
RO_3:       MOV    AL,11110111B                  ;ground row 3 (D3=0)
            OUT    PORT_A,AL
            IN     AL,PORT_B                     ;read all columns
            AND    AL,00001111B                  ;mask unused bits (D7-D4)
            CMP    AL,00001111B                  ;see which column
            JE     K2                            ;if none then false input repeat the process
            MOV    BX,OFFSET KCOD_3              ;set BX=start of table for column 3 keys
;A key press has been detected and the row identified.  Now find which key.
FIND_IT:    RCR    AL,1                          ;rotate the column input to search for 0
            JNC    MATCH                         ;if zero, go get the code
            INC    BX                            ;if not point at the next code
            JMP    FIND_IT                       ;and keep searching
;GET THE CODE FOR THE KEY PRESSED AND RETURN
MATCH:      MOV    AL,[BX]                       ;get the code pointed by BX
            POP    BX                            ;return with AL=code for pressed key
            RET

;FOR THE DELAY GENERATION SEE CHAPTER 13
```

Program 18-1. Key Press Detection and Identification

SECTION 18.1: INTERFACING THE KEYBOARD TO THE CPU   545

Review Questions

1. True or false. To see if any key is pressed, all rows are grounded.
2. If D3 - D0 = 0111 is the data read from the columns, which column does the key pressed belong to?
3. True or false. Key press detection and key identification require two different processes.
4. In Figure 18-1, if the row has D3 - D0 =1110 and the columns are D3 - D0 =1110, which key is pressed?
5. True or false. To identify the key pressed, one row at a time is grounded.

SECTION 18.2: PC KEYBOARD INTERFACING AND PROGRAMMING

In the IBM PC and compatibles, a microcontroller is used for both detection and identification of keys. This microcontroller has a microprocessor in addition to a few hundred bytes of RAM, a few Kbytes of EPROM, and a few I/O ports, all on one chip. The microcontroller used widely in the IBM PC and compatibles is Intel's 8042 (or some variation). The 8042 is programmed to detect and identify the key press. A scan code is assigned to each key and the microcontroller provides the scan code for the pressed key to the motherboard. To allow the keyboard to be detachable from the system board, the keyboard is connected to the system board through a cable. Such an arrangement necessitates the use of serial data communication to transfer the scan code to the main CPU (serial data transfer was covered in Chapter 17). IBM PC AT keyboards use the following data frame when sending the scan code serially to the motherboard. For each scan code, a total of 11 bits are transferred from the keyboard to the motherboard.

> one start bit (always 0)
> 8 bits for scan code
> odd-parity bit
> one stop bit (always 1)

In the PC/XT motherboard, a serial-in-parallel-out shift register, 74LS322, is used to receive the serial data coming in through the keyboard cable. The 74LS322 strips away the framing portion, makes an 8-bit scan code, and presents it to port A of the 8255 with I/O port address of 60H. On the IBM PC, AT, PS and compatibles, the 74LS322 and supporting logic were replaced by another 8042. This allows the option of programming the keyboard itself. Therefore, two 8042 microcontrollers, one on the keyboard and one on the motherboard, are responsible for keyboard bidirectional communication in the IBM PC, AT, PS, and compatible systems.

Make and break

In the IBM PC, the key press and release are represented by two different scan codes. The key press is referred to as a *make,* and the release of the same key is called a *break.* When a key is pressed (a make), the keyboard sends one scan code, and when it is released (a break), it sends another scan code. The scan code for the break is always 127 decimal (80H) larger than the make scan code. For example, if a given key produces a scan code of 06 on make, the scan code for the break is 86H (06 +80H =86H).

IBM PC scan codes

The IBM PC/XT keyboard has 83 keys, arranged in three major groupings:

1. The standard typewriter keys
2. Ten function keys, F1 to F10
3. 15-key keypad

These 83 keys are shown in Table 18-1. With the introduction of the PC AT, IBM added one more key, "Sys Rq", to make a total of 84 keys. The locking shift keys were made more noticeable by providing LED indicators for them. Later, IBM introduced what is called the *advanced keyboard*, known more commonly as the *enhanced keyboard*. The number of keys was increased to 101 for the U.S. market. Tables 18-1, 18-2, and 18-3 provide the scan codes for both the PC/XT and enhanced keyboards.

In Table 18-1, notice that the same scan code is used for a given lowercase letter and its capital. The same is true for all the keys with dual labels. If the scan code is the same for both of them, how does the system distinguish between them? This is taken care of by the keyboard shift status byte. The BIOS data area location 0040:0017H holds the shift status byte. The meaning of each bit is given in Figure 18-3.

The BIOS data area location 0040:0018H holds the second keyboard status byte. The meaning of each bit is given in Figure 18-4. Notice that some of the bits are used for the 101 enhanced keyboards.

Table 18-1: PC Scan Codes for 83 PC/XT Keys

Hex	Key	Hex	Key	Hex	Key	Hex	Key
01	Esc	15	Y and y	29	~ and ‘	3D	F3
02	! and 1	16	U and u	2A	LeftShift	3E	F4
03	@ and 2	17	I and i	2B	| and \	3F	F5
04	# and 3	18	O and o	2C	Z and z	40	F6
05	$ and 4	19	P and p	2D	X and x	41	F7
06	% and 5	1A	{ and [	2E	C and c	42	F8
07	^ and 6	1B	} and]	2F	V and v	43	F9
08	& and 7	1C	enter	30	B and b	44	F10
09	* and 8	1D	ctrl	31	N and n	45	NumLock
0A	(and 9	1E	A and a	32	M and m	46	ScrollLock
0B	) and 0	1F	S and s	33	< and ,	47	7 and Home
0C	_ and -	20	D and d	34	> and .	48	8 and UpArrow
0D	+ and =	21	F and f	35	? and /	49	9 and PgUp
0E	backspace	22	G and g	36	RightShift	4A	- (gray)
0F	tab	23	H and h	37	PrtSc and *	4B	4 and LeftArrow
10	Q and q	24	J and j	38	Alt	4C	5 (keypad)
11	W and w	25	K and k	39	Spacebar	4D	6 and RightArrow
12	E and e	26	L and l	3A	CapsLock	4E	+ (gray)
13	R and r	27	: and ;	3B	F1	4F	1 and End
14	T and t	28	" and ’	3C	F2	50	2 and DownArrow
						51	3 and PgDn
						52	0 and Ins
						53	. and Del

(Reprinted by permission from "IBM BIOS Technical Reference" c. 1987 by International Business Machines Corporation)

Table 18-2: Combination Key Scan Codes

Hex	Keys	Hex	Keys	Hex	Keys	Hex	Keys
54	Shift F1	60	Ctrl F3	6C	Alt F5	78	Alt 1
55	Shift F2	61	Ctrl F4	6D	Alt F6	79	Alt 2
56	Shift F3	62	Ctrl F5	6E	Alt F7	7A	Alt 3
57	Shift F4	63	Ctrl F6	6F	Alt F8	7B	Alt 4
58	Shift F5	64	Ctrl F7	70	Alt F9	7C	Alt 5
59	Shift F6	65	Ctrl F8	71	Alt F10	7D	Alt 6
5A	Shift F7	66	Ctrl F9	72	Ctrl PrtSc	7E	Alt 7
5B	Shift F8	67	Ctrl F10	73	Ctrl LeftArrow	7F	Alt 8
5C	Shift F9	68	Alt F1	74	Ctrl RightArrow	80	Alt 9
5D	Shift F10	69	Alt F2	75	Ctrl End	81	Alt 10
5E	Ctrl F1	6A	Alt F3	76	Ctrl PgDn		
5F	Ctrl F2	6B	Alt F4	77	Ctrl Home		

(Reprinted by permission from "IBM BIOS Technical Reference" c. 1987 by International Business Machines Corporation)

Table 18-3: Extended Keyboard Scan Codes

Hex	Keys	Hex	Keys	Hex	Keys	Hex	Keys
85	F11	8E	Ctrl -	97	Alt Home	A0	Alt DownArrow
86	F12	8F	Ctrl 5	98	Alt UpArrow	A1	Alt PgDn
87	Shift F11	90	Ctrl +	99	Alt PgUp	A2	Alt Insert
88	Shift F12	91	Ctrl DownArrow	9A		A3	Alt Delete
89	Ctrl F11	92	Ctrl Insert	9B	Alt LeftArrow	A4	Alt /
8A	Ctrl F12	93	Ctrl Delete	9C		A5	Alt Tab
8B	Alt F11	94	Ctrl Tab	9D	Alt RightArrow	A6	Alt Enter
8C	Alt F12	95	Ctrl /	9E			
8D	Ctrl UpArrow	96	Ctrl *	9F	Alt End		

(Reprinted by permission from "IBM BIOS Technical Reference" c. 1987 by International Business Machines Corporation)

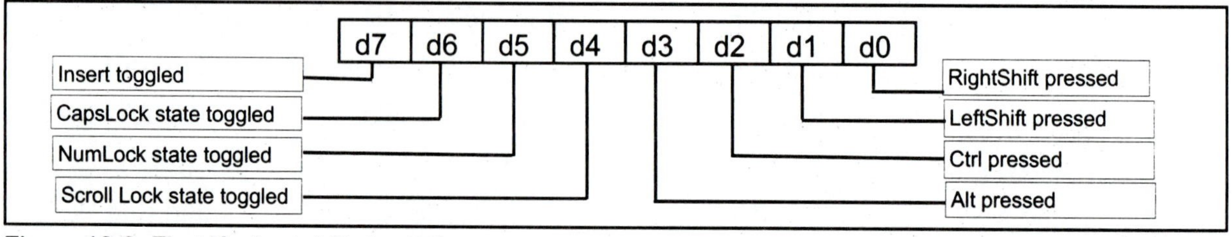

Figure 18-3. First Keyboard Status Byte

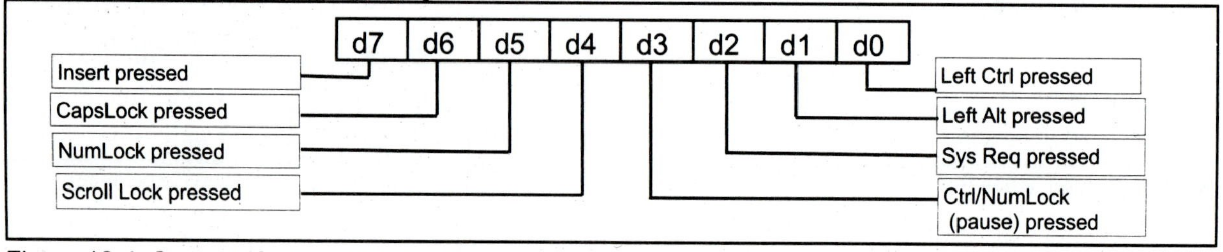

Figure 18-4. Second Keyboard Status Byte

When a key is pressed, the interrupt service routine of INT 9 receives the scan code and stores it in memory locations called a *keyboard buffer*, located in the BIOS data area. However, to relieve programmers from the details of keyboard interaction with the motherboard, IBM has provided INT 16H. We first look at the services provided by the BIOS INT 16H and then we study the details of how the keyboard interacts with the motherboard through hardware INT 09.

BIOS INT 16H keyboard programming

INT 16H, AH =0 (read a character)

This option checks the keyboard buffer for a character. If a character is available, it returns its scan code in AH and its ASCII code in AL. If no character is available in the buffer, it waits for a key press and returns it. For characters such as F1 - F10 for which there is no ASCII code, it simply provides the scan code in AH and AL = 0. Therefore, if AL = 0, a special function key was pressed. This option simply provides the code for the character and does not display it.

INT 16H, AH =01 (find if a character is available)

This option, which is similar to the AH = 0 option, checks the keyboard buffer for a character. If a character is available, it returns its scan code in AH and its ASCII code in AL and sets ZF = 0. If no character is available in the buffer, it does not wait for a key press and simply makes ZF =1 to indicate that.

INT 16H, AH =02 (return the current keyboard status byte)

This option provides the keyboard status byte in the AL register. The keyboard status byte (also referred to as the keyboard flag byte) is located in the BIOS data area memory location 0040:0017H. For the meaning of each bit of the shift status byte, see Figure 18-3.

Example 18-2

Run the following program in DEBUG. Interpret the result after typing each of the following. Run the program for each separately.   (a) Z   (b) F1   (c) ALT

```
    MOV   AH,0
    INT   16H
    INT   3
```

Solution:

(a) AX = 2C7A AH = 2C, the scan code, and AL = 7A, ASCII for 'Z'
(b) AX = 3B00 AH = 3B, the scan code for F1, and AL = 00, because F1 is not an ASCII key
(c) Nothing happens because there is no scan code for the Alt key. The status of keys such as Alt is found in the keyboard status byte stored in BIOS at 40:17 and 40:18.

Example 18-3

Run the following program in DEBUG while the right shift key is held down and the CapsLock key light is on. Verify it also by dumping the 0040:0017 location using DEBUG.

```
    MOV   AH,02
    INT   16H
    INT   3
```

Solution:

Running the program while the RightShift and CapsLock keys are activated gives
AH =41H =0100 0001 in binary, which can be checked against Figure 18-3. In DEBUG,
-d 0:417 418
will provide the keyboard status byte 41-00.

Due to additional keys on the IBM extended keyboard, BIOS added the following additional services to INT 16H.

INT 16H, AH=10H (read a character)

This is the same as AH = 0 except that it also accepts the additional keys on the IBM extended (enhanced) keyboard.

INT 16H, AH =11H (find if a character is available)

This is the same as AH = 1 except that it also accepts the additional keys on the IBM extended (enhanced) keyboard.

INT 16H, AH =12H (return the current status byte)

This is the same as AH = 2 except that it also provides the shift status byte of the IBM extended (enhanced) keyboard to AH. See Figure 18-5.

Example 18-4

Run the following program in DEBUG. Interpret the result after typing each of the following. Run the program for each separately. (a) F11   (b) ALT F11   (c) ALT TAB

```
MOV   AH,10H
INT   16H
INT   3
```

Solution:

After running the program above in DEBUG for each case, we have the following.

(a) AX = 8500, where 85H is the scan code for F11
(b) AX = 8B00, where 8BH is the scan code for Alt-F11
(c) AX = A500, where A5H is the scan code for Alt-Tab

All of the cases above have AL = 00 since there is no ASCII code for these keys.

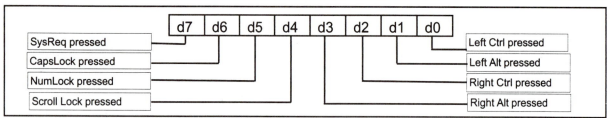

Figure 18-5. Enhanced Keyboard Shift Status Byte

Example 18-5

Write and test a program in DEBUG that increments counter CX whenever Shift-F7 is activated; otherwise, it should exit.

Solution:

The unassembled program in DEBUG follows.

```
-U 100 113
16B7:0100 B402        MOV    AH,02
16B7:0102 CD16        INT    16
16B7:0104 F6C403      TEST   AH,03
16B7:0107 740A        JZ     0113
16B7:0109 B400        MOV    AH,00
16B7:010B CD16        INT    16
16B7:010D 80FC5A      CMP    AH,5A
16B7:0110 7501        JNZ    0113
16B7:0112 41          INC    CX
16B7:0113 CC          INT    3
-
```

Hardware INT 09 role in the IBM PC keyboard

To understand fully the principles underlying the IBM PC keyboard, it is necessary to know how INT 09 works. The IBM PC keyboard communicates with the motherboard through hardware interrupt IRQ1 of the 8259. As mentioned in Chapter 14, IRQ1 (INT 09) of the 8259 is used by the keyboard. The way the INT 09 interrupt service works is as follows.

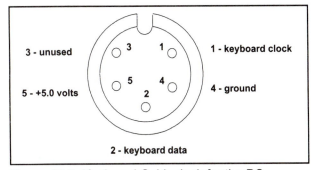

Figure 18-6. Keyboard Cable Jack for the PC
(Reprinted by permission from "IBM Technical Reference" c. 1984 by International Business Machines Corporation)

1. The keyboard microcontroller scans the keyboard matrix continuously. When a key is pressed (a make), it is identified and its scan code is sent serially to the motherboard through the keyboard cable (see Figure 18-6). The circuitry on the motherboard receives the serial bits, gets rid of the frame bits, and makes one byte (scan code) with the help of its serial-in-parallel-out shift register, then presents this 8-bit scan code to port A of the 8255 at I/O address of 60H, and finally activates IRQ1.
2. Since IRQ1 is set to INT 09, its interrupt service routine (ISR) residing in BIOS ROM is invoked.
3. The ISR of INT 09 reads the scan code from port 60H.
4. The ISR of INT 09 tests the scan code to see if it belongs to one of the shift keys (RightShift and LeftShift), Alt, Ctrl keys, and so on. If it is, the appropriate bit of the keyboard status bytes in BIOS memory locations 0040:0017H and 0018H are set. However, it will not write the scan code to the keyboard buffer. If the scan code belongs to any key other than a special key (Shift, Alt, Ctrl, and so on), INT 09 checks to see if there is an ASCII code for the key. If there is one, it will write both the ASCII and scan codes into the keyboard buffer. If there is no ASCII code for the key, it puts 00 in place of ASCII code and the scan code in the keyboard buffer.
5. Before returning from INT 09, the ISR will issue EOI to unmask IRQ1, followed by the IRET instruction. This allows IRQ1 activation to be responded to again.
6. When the key is released (a break), the keyboard generates the second scan code by adding 80H to it and sends it to the motherboard.
7. The ISR of INT 09 checks the scan code to see if there is 80H difference between the last code and this one. This is easy since all it has to do is to test D7 (80H =10000000 binary). If D7 is high, this is interpreted as meaning that the key has been released and the system ignores the second scan code. However, if the key is held down more than 0.5 seconds, it is interpreted as a new key and INT 09 will write it into the keyboard buffer next to the preceding one. Holding down the keyboard for more than 0.5 seconds is commonly referred to as *typematic* in IBM literature, which means repeating the same key.

From the above steps the following points must be emphasized:
1. The keyboard sends two separate scan codes for make and break to the motherboard.
2. It is the function of the ISR of INT 09 to read the scan code sent by the keyboard and convert it to ASCII (if any), then save both the scan code and ASCII code in the keyboard buffer of the motherboard.
3. If any of the special keys, such as Shift, Alt, or Ctrl, is pressed, INT 09 sets the appropriate bits to 1 in the BIOS data area of 0040:0017H and 0018H, but it will not deliver the scan code to the keyboard buffer.
4. If any undefined combination of keys are pressed, INT 09 is activated but it will ignore them since there is no associated scan code. If such key combinations are used by a given program, it is the job of the programmer to intercept them by hooking into INT 09.

Keyboard overrun

On the keyboard side, the 8042 circuitry must serialize the scan code and send it through the cable to the motherboard. On the motherboard side, there is circuitry responsible for getting the serial data and making a single byte of scan code out of the streams of bits, and holding it for the CPU to read. What happens if the CPU falls behind and cannot keep up with the number of keystrokes? Such a situation is called *keyboard overrun*. The motherboard beeps the speaker when an overrun occurs. The beeping process works as follows. The circuitry on the keyboard has a buffer of its own to store a maximum of 20 key strokes. When this buffer becomes full, it stops receiving keystrokes and sends a special byte called an *overrun byte* (which is FFH in the PC/XT) to the motherboard. After getting the scan code, INT 09 first checks to see if the scan code received is the overrun byte, FFH. If it is, it will sound the speaker; otherwise, it tests for the shift keys and so on, as explained earlier. In other words, the BIOS ROM on the motherboard is responsible for beeping the speaker in the event of keyboard overrun. The following program shows this process. It provides the beginning and ending codes for the INT 09 interrupt service routine, taken from the IBM PC/XT BIOS with some modification for the sake of clarity.

```
;KEYBOARD INT 09 INTERRUPT ROUTINE for PC/XT
KB_INT    PROC  FAR
          STI                     ;ALLOW FURTHER INTERRUPT
          PUSH  AX                ;SAVE ALL THESE
          PUSH  BX                ;REGISTERS
          PUSH  CX
          PUSH  DX
          PUSH  SI
          PUSH  DI
          PUSH  DS
          PUSH  ES
          ...
          ...
          IN    AL,60H            ;READ IN THE CHARACTER FROM PORT# 60H
          ...
          ...
          CMP   AL,0FFH           ;OVERRUN CHARACTER? (PC/AT USES 00)
          JNZ   K16               ;NO. TEST FOR SHIFT KEY
          JMP   K62               ;SOUND THE BEEPER FOR BUFFER FULL
          ...
          ...
          CLI                     ;TURN OFF INTERRUPTS
          MOV   AL,20H            ;ISSUE EOI (END-OF-INTERRUPT) TO 8259
          OUT   20H,AL            ;AT PORT ADDRESS 20H
          POP   ES
          POP   DS
          POP   DI
          POP   SI
          POP   DX
          POP   CX                ;RESTORE ALL THE
          POP   BX                ;REGISTERS
          POP   AX
          IRET                    ;RETURN FROM INTERRUPT
KB_INT    ENDP
```

Keyboard buffer in BIOS data area

As mentioned above, the INT 09 interrupt routine gets the scan code from the keyboard and stores it in some memory locations in the BIOS data area. These memory locations are referred to as the *keyboard buffer*. This keyboard buffer in the BIOS data area should not be confused with the buffer inside the keyboard itself, whose overrun causes the speaker to beep.

If there is an ASCII code, INT 09 also stores the ASCII code for the key in the keyboard buffer; otherwise, it puts 0 there instead. Where this keyboard buffer is located and how it is used by INT 09 are discussed next.

BIOS keyboard buffer

A total of 32 bytes (16 words) of memory in the BIOS data area is set aside for the keyboard buffer. It starts at memory address 40:001EH and goes to 40:003DH, which corresponds to physical addresses 0041EH and 0043DH. Each two consecutive locations are used for a single character, one for the scan code and the other one for the ASCII code (if any) of the character. How does INT 9 know in which word of this 16-word buffer it should put the next character, and how does INT 16H know which of the characters in the keyboard buffer to extract? To answer these questions, we must explain the role of keyboard buffer pointers. There are two keyboard buffer pointers: the *head pointer* and the *tail pointer*. See Table 18-4.

Table 18-4: BIOS Data Area Used by Keyboard Buffer

Address of Head Pointer	Address of Tail Pointer	Keyboard Buffer
41A and 41B	41C and 41D	41E to 43D

Tail pointer

Memory locations 0040:001CH and 0040:001DH (physical addresses 0041CH and 0041DH) hold the address for the tail. This means that at any given time, memory locations 0041CH and 0041D hold the address where INT 09 should store the next character. It is the job of INT 09 to put the character in the keyboard buffer and advance the tail by incrementing the word contents of memory location 0041C, where the tail pointer is held.

Head pointer

INT 16H gets the address of where to extract the next character from memory locations 41AH and 41BH, the head pointer. As INT 16H reads each character from the keyboard buffer, it advances the head pointer held by memory locations 41AH and 41BH.

The above discussion can be summarized as follows. As INT 09 inserts the character into the keyboard buffer it advances the tail, and as INT 16H reads the character from the keyboard buffer it advances the head. When they come to the end of the keyboard buffer they both wrap around, creating a ring of 16 words where the head is continuously chasing the tail. This is shown in Figure 18-7.

Notice in Figure 18-7 that if the keyboard buffer is empty, the head address is equal to the tail address. As INT 09 inserts characters into the buffer, the tail is moved. If the buffer is not read by INT 16H, it becomes full, which causes the tail to be right behind the head. Look at Example 18-6.

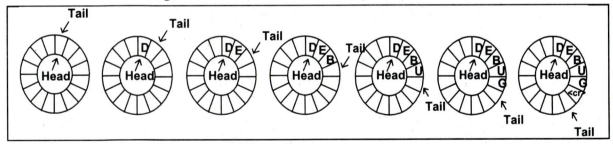

Figure 18-7. Keyboard Buffer Head and Tail

PC keyboard technology

The kind of keyboard shown in Figure 18-1 is referred to as a *hard contact* keyboard. When a key is pressed, a physical contact between the row and column causes the column to be pulled to ground. Although these kinds of keyboards are cheaper to make, they have the disadvantage of deteriorating at the contact points of rows and columns, eventually becoming too stiff to use, due to metal oxidation

---

of the contact points. The alternative to hard contact keyboards are *capacitive* keyboards. In such keyboards there is no physical contact between the rows and columns; instead, there is a capacitor for each point of the matrix. In capacitive keyboards when a key is pressed, the change in capacitance is detected by a sense amplifier and produces the logic level, indicating that a key has been pressed. Capacitive keyboards last much longer than do hard contact keyboards. All IBM PC and PS machines use capacitive keyboards except the now abandoned PC Jr. models, which used hard contact keyboards.

Example 18-6

Using DEBUG, dump the location where the head and tail pointer are held. Compare them. Is the buffer full or empty?

Solution:
```
C>DEBUG
-D 0:410 41F
0000:0410 63 44 F0 80 02 00 01 40-00 00 3C 00 3C 00 20 39 cD.....@..<.<. 9
-
```

In this case, the head and tail pointers point to the same location; therefore, the buffer is empty.

Review Questions

1. Show the bits transferred from the keyboard to the motherboard when "J" is pressed.
2. How does the PC recognize the difference between the key press and key release?
3. Find the make and break scan codes for the letter "X" in both hex and binary.
4. True or false. The CPU is notified through INT 09 only for key press, not for key release.
5. Does the Alt key have a scan code? If yes, what is it?
6. True or false. The CPU stores the scan code for the right SHIFT in the buffer.
7. Find the contents of the keyboard status byte if CapsLock and Alt are pressed.
8. True or false. INT 09 is responsible for finding the ASCII code for a given key if there is one.
9. True or false. The keyboard buffer holds both the scan code and the ASCII code for a given key.
10. True or false. The beep sound indicates that the BIOS keyboard buffer is full.
11. What does it mean when the head and tail pointer have the same values?
12. As INT 09 puts the scan code into keyboard buffer it advances the _____ (tail, head) pointer.

SECTION 18.3: PRINTER AND PRINTER INTERFACING IN THE IBM PC

In this section we describe the standard printer interface, called the *Centronics printer interface*. Then we study the IBM PC printer interfacing and provide some examples of printer programming using BIOS INT 17H.

Centronics printer interface pins

The Centronics-type parallel printer interface is the printer interface standard in the 80x86 PC. It is also referred to as Epson FX-100 standard. It is a 36-pin interface connector where the pins are labeled as 1 to 36. Many of the 36 pins are used for ground, allowing many signals to have their own ground return lines, which reduces electrical noise. The 36 pins can be grouped as follows.

1. The data lines, which carry the data sent by the PC to the printer.
2. Printer status signals, which indicate the status of the printer at any given time.
3. Printer control signals, which are used to tell the printer what to do.
4. Ground signals, which provide an individual ground return line for each data line and for certain control and status lines.

Table 18-5: Centronics Printer Specifications

Serial #	Return #	Signal	Direction	Description
1	19	$\overline{\text{STROBE}}$	IN	STROBE pulse to read data in. Pulse width must be more than 0.5 μs at receiving terminal. The signal level is normally "high"; read-in of data is performed at the "low" level of this signal.
2	20	DATA 1	IN	These signals represent information of the 1st to 8th bits of parallel data, respectively. Each signal is at "high" level when data is logical "1", and "low" when logical "0".
3	21	DATA 2	IN	
4	22	DATA 3	IN	
5	23	DATA 4	IN	
6	24	DATA 5	IN	
7	25	DATA 6	IN	
8	26	DATA 7	IN	
9	27	DATA 8	IN	
10	28	$\overline{\text{ACKNLG}}$	OUT	Approximately 0.5 μs pulse; "low" indicates data has been received and printer is ready for data.
11	29	BUSY	OUT	A "high" signal indicates that the printer cannot receive data. The signal becomes "high" in the following cases: (1) during data entry, (2) during printing operation, (3) in "off-line" status, (4) during printer error status.
12	30	PE	OUT	A "high" signal indicates that printer is out of paper.
13	--	SLCT	OUT	Indicates that the printer is in the state selected.
14	--	$\overline{\text{AUTOFEEDXT}}$	IN	With this signal being at "low" level, the paper is fed automatically one line after printing. (The signal level can be fixed to "low" with DIP SW pin 2-3 provided on the control circuit board.)
15	--	NC		Not used.
16	--	0V		Logic GND level.
17	--	CHASISGND	--	Printer chassis GND. In the printer, chassis GND and the logic GND are isolated from each other.
18	--	NC	--	Not used.
19 - 30	--	GND	--	"Twisted-pair return" signal'; GND level.
31	--	$\overline{\text{INIT}}$	IN	When this signal becomes "low" the printer controller is reset to its initial state and the print buffer is cleared. Normally at "high" level; its pulse width must be more than 50 μs at receiving terminal.
32	--	$\overline{\text{ERROR}}$	OUT	The level of this signal becomes "low" when printer is in "paper end", "off-line" and "error" state.
33	--	GND	--	Same as with pin numbers 19 to 30.
34	--	NC	--	Not used.
35	--			Pulled up to +5 V dc through 4.7 K ohms resistance.
36	--	$\overline{\text{SLCT IN}}$	IN	Data entry to the printer is possible only when the level of this signal is "low". (Internal fixing can be carried out with DIP SW 1 - 8. The condition at the time of shipment is set "low" for this signal.)

(Reprinted by permission from "IBM Technical Reference Options and Adapters" c. 1981 by International Business Machines) Corporation)

Data lines and grounds

Input pins DATA 1 to DATA 8 provide a parallel pathway for 8-bit data sent by the PC to the printer. Notice in Table 18-5 that pins 20 to 28 are used for individual ground return lines, one for each data pin. Table 18-6 describes the DB-25 printer pins. Figure 18-8 shows the connector.

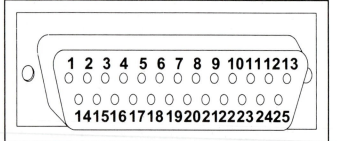

Figure 18-8. DB-25P (Male) Printer Connector
(Reprinted by permission from "IBM Technical Reference" c. 1988 by International Business Machines Corporation)

Printer status signals

These are all output pins from the printer to the PC used by the printer to indicate its own status. They are as follows.

PE (pin 12) is used by the printer to indicate that it is out of paper.

BUSY (pin 11) is high if the printer is not ready to accept a new character. This pin is high when the printer is off-line or when it is printing and cannot accept any data. The PC monitors this pin continuously and as long as this pin is high, it will not transfer data to the printer.

ERROR (pin 32) is normally a high output and is activated (goes low) when there are conditions such as out-of-paper, off-line state, or jammed printhead in which the printer cannot print.

SLCT (pin 13) is active high and goes from the printer to the PC when the printer is turned on and online, indicating that the printer is being selected.

ACKNLG (pin 10) is used by the printer to acknowledge receipt of data and that it can accept a new character.

Printer control signals

STROBE (pin 1) and ACKNLG are the most widely used signals among control and status pins. When the PC presents a character to the data pins of the printer, it activates the STROBE pin of the printer, telling it that there is a byte sitting at the data pins. When the printer picks up the data and is ready for another byte, it sends back the ACKNLG signal. While the STROBE is used by the CPU to tell the printer that there is a byte of data, it is the printer that must acknowledge the data receipt and its readiness for accepting another byte through the ACKNLG line. The ACKNLG signal can be used by the CPU to go and get another byte of data to be presented to the printer. See Chapter 12 for more about handshaking.

INIT (pin 31) is an input into the printer and is normally high. When it is activated (active low) it resets the printer. Upon receiving this signal, the printer goes through a sequence of internal initialization, including clearing its own internal buffer.

Table 18-6: DB-25 Printer Pins

Pin	Description
1	Strobe
2	Data bit 0
3	Data bit 1
4	Data bit 2
5	Data bit 3
6	Data bit 4
7	Data bit 5
8	Data bit 6
9	Data bit 7
10	Acknowledge
11	Busy
12	Out of paper
13	Select
14	Auto feed
15	Error
16	Initialize printer
17	Select input
18	Ground
19	Ground
20	Ground
21	Ground
22	Ground
23	Ground
24	Ground
25	Ground

(Reprinted by permission from "IBM Technical Reference" c. 1988 by International Business Machines Corporation)

There are two other control signals in the printer: AUTO FEED XT and SLCT IN. See Table 18-5 for their descriptions. The following are the steps in computer and printer communication.

1. The computer checks to see if a BUSY signal from the printer indicates that the printer is ready (not busy).
2. The computer puts 8-bit data on the data line connected to the printer data pins.
3. The computer activates the $\overline{\text{STROBE}}$ pin by making it low. Prior to asserting the printer input $\overline{\text{STROBE}}$ pin, the data must be at the printer's data pins at least for 0.5 μs . This is data setup time.
4. The $\overline{\text{STROBE}}$ must stay low for at least 0.5 μs before the computer brings it back to high. The data must stay at the printer's data pins at least 0.5 μs after the $\overline{\text{STROBE}}$ pin is deasserted (brought back to high).
5. The activation of $\overline{\text{STROBE}}$ causes the printer to assert its BUSY output pin high, indicating to the computer to wait until it finishes taking care of the last byte.
6. When the printer is ready to accept another byte, it sends the $\overline{\text{ACKNLG}}$ signal back to the computer by making it low. The printer keeps the $\overline{\text{ACKNLG}}$ signal low only for 5 μs. At the rising edge of $\overline{\text{ACKNLG}}$, the printer makes the BUSY (not BUSY =ready) pin low to indicate that it is ready to accept the next byte.

The CPU can use either the $\overline{\text{ACKNLDG}}$ or BUSY signals from the printer to initiate the process of sending another byte to printer. Some systems use BUSY and some use ACKNLG.

IBM PC printer interfacing

In the IBM PC, the POST (power-on self-test) portion of BIOS is programmed to check for printers connected to parallel ports. As they are identified, the base I/O port address of each is written into the BIOS data area 0040:0008 to 0040:000FH just like the COM port discussed in Chapter 9. A total of 8 bytes of memory in the BIOS data area can store the base I/O address of 4 printers, each taking 2 bytes.

Table 18-7: BIOS I/O Base Addresses for LPT

I/O Base Address	LPT
0040:0008 - 0040:0009	LPT1
0040:000A - 0040:000B	LPT2
0040:000C - 0040:000D	LPT3
0040:000E - 0040:000F	LPT4

Memory locations 00408 to 0040FH can be checked to see which LPT (line printer) port is available. Memory locations 0040:0008H and 0040:0009H (physical locations 00408H and 00409H) hold the base I/O address of LPT1, and so on, as shown in Table 18-7. If no printer port is available, 0s are found.

It must be emphasized that the base I/O port addresses assigned to LPTs can vary from system to system. This is due to the fact that the POST (power-on self-test) portion of BIOS will check for the existence of a printer port first at I/O address 03BCH, then at 0378H, and finally, at 0278H. Whichever is found first will be written into BIOS data area 408H, where the base I/O address for LPT1 is expected; the second one found is written to 40AH address for LPT2; and so on.

Printer interfacing circuitry uses only 3 I/O ports starting at the base address: one I/O port for the LPT's data lines, one for the LPT's status lines, and one for the LPT's control lines. For example, if the base I/O port address for LPT1 is 378H, the I/O port address 378H is used for the data, 379H for the status, and 37AH for the control signals. See Table 18-8 for the assignments.

Table 18-8: IBM PC Printer Ports and Their Functions

Line Printer	Data Port (R/W)	Status Port (Read Only)	Control Port (R/W)
LPT 1	03BCH	03BDH	03BEH
LPT 2	0378H	0379H	037AH
LPT 3	0278H	0279H	027AH

Example 18-7

Using DEBUG, determine which printer port(s) are available.

Solution:

C>DEBUG
-D 40:08 L8
0040:0008 78 03 00 00 00 00 00 00   x.......

This shows that the base I/O address of LPT1 is 0378H. No other printers are connected to the parallel ports.

Example 18-8

Two extremely reliable technical reference documents state that the base LPT1 I/O address is:
in document A, the base I/O address is 3BCH, and in document B, it is 378H. The two documents belong to different manufacturers. Which document is correct?

Solution:

They are both right: manufacturer A used port address 3BCH for address decoding and manufacturer B used 378H. To verify that both are right using DEBUG, simply dump the BIOS area 0040:0008 on both PCs and examine the contents of memory locations 0040:0008 and 0040:0009. They should match the documentation.

Figure 18-9 shows the printer's data, status, and control ports.

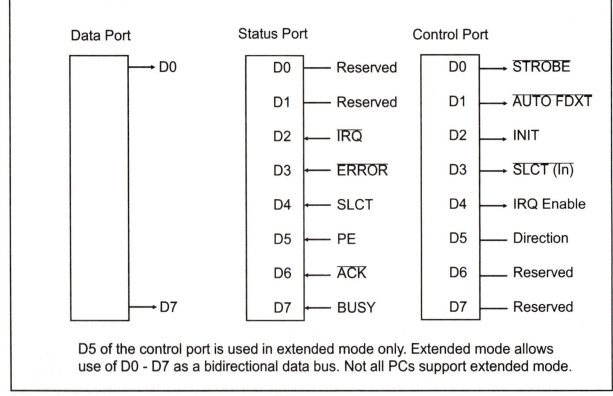

D5 of the control port is used in extended mode only. Extended mode allows use of D0 - D7 as a bidirectional data bus. Not all PCs support extended mode.

Figure 18-9. Printer's Data, Status, and Control Ports

Programming the IBM PC printer with BIOS INT 17H

BIOS INT 17H provides three services: printing a character, initializing the printer port, and getting the printer status port. These options are selected according to the value set in the AH register. This is described as follows.

INT 17H, AH=0 (print a character)

If this option is selected, INT 17H expects to have the LPT number in register DX (0 for LPT1, 1 for LPT2, and 2 for LPT3) and the ASCII character to be printed in the AL register. Upon return, INT 17H provides the status of the selected printer port as follows.

Bit No.	Function
7	1 = Not BUSY(ready), 0=BUSY
6	1 = Acknowledge
5	1 = Out of paper
4	1 = Printer selected
3	1 = I/O error
2,1	unused
0	1 = Printer time-out

INT 17H, AH =01 (initialize the printer port)

This option initializes the printer by setting the printer to the top-of-page position in spite of the fact that most printers do that automatically when they are turned on. Before this function is called, AH is set to 1 and DX contains the printer number (0=LPT1, 1=LPT2, and 2=LPT3). After calling AH = status, the situation is as shown under option 0.

Example 18-9

Using the INT 17H, show how to print the character "A" on the LPT1.

Solution:

```
MOV   AH,0        ;print character option
MOV   DX,0        ;select LPT1
MOV   AL,41H      ;ASCII code for letter "A"
INT   17H         ;call BIOS
```

If for any reason it cannot print the character, it sets AH = 01 which is 0000 00001, meaning that it tried for certain period of time and could not print. To examine the inner working of BIOS INT 17H that is responsible for printing characters, see the end of this section.

INT 17H, AH =02 (get the printer port status)

This option allows a programmer to check the status of the printer. Before calling the function, AH is set to 2 and DX holds the printer number (0=LPT1, 1=LPT2, and 2=LPT3). After calling, AH = status, the situation is the same as shown under option 0.

Example 18-10

Run the following program in DEBUG to check the LPT1 printer states. Run it once with the printer off-line, then run it again with the printer on-line. Interpret the AH register upon return.

```
          MOV   AH,2
          MOV   DX,0
          INT   17H
```

Solution:
```
C:>DEBUG
-A
16B7:0100 MOV AH,2
16B7:0102 MOV DX,0
16B7:0105 INT 17
16B7:0107 INT 3
16B7:0108
-G

AX=0800  BX=0000  CX=0000  DX=0000  SP=CFDE  BP=0000  SI=0000  DI=0000
DS=16B7  ES=16B7  SS=16B7  CS=16B7  IP=0107    NV UP DI PL NZ NA PO NC
16B7:0107 CC              INT     3
-G=100

AX=9000  BX=0000  CX=0000  DX=0000  SP=CFDE  BP=0000  SI=0000  DI=0000
DS=16B7  ES=16B7  SS=16B7  CS=16B7  IP=0107    NV UP DI PL NZ NA PO NC
16B7:0107 CC              INT     3
-Q
```

The first execution of the program occurred when the printer was off-line. It returned AH = 08. This indicates I/O error. The program was run again with the printer on-line. This time it returned AH = 90, which indicates "not busy."

What is printer time-out?

Occasionally, the printer time-out message will appear on the screen. This means that the printer port is installed but the printer is not ready to print. This could be due to the fact that the printer is turned off, the printer is not on-line, or some other condition in which the printer is connected to the PC but not ready to print. Upon detecting that the printer port is installed, BIOS tries repeatedly for a period of 20 seconds to see if it is ready to accept data. If the printer is not ready, the PC gives up (time-out) and displays a message to indicate that. Can the PC be forced not to give up so soon and try a little bit longer? The answer is yes. The amount of time that BIOS tries to get a response from the printer is stored in BIOS data area 0040:0078 to 0040:007B. Location 0040:0078 holds the time-out time for LPT1, 0040:0079 the time for LPT2, and so on. At boot time, these locations are initialized to 20 seconds.

ASCII control characters

Certain characters in ASCII are used to control the printer. Table 18-9 shows the most commonly used printer control characters in ASCII.

Table 18-9: ASCII Printer Control Characters

ASCII Symbol	Hex Code	Function
BS	08	Backspace
HT	09	Horizontal tab
LF	0A	Line feed (advances one line)
VT	0B	Vertical tab
FF	0C	Form feed (advances to next page)
CR	0D	Carriage return (return to left margin)

Inner working of BIOS INT 17H for printing a character

Below is a listing of a portion of BIOS INT 17H with some modification for the sake of clarity. It shows how a character is printed by monitoring the BUSY signal from the printer and issuing a STROBE. Figure 18-10 diagrams printer timing.

```
;This skeleton of BIOS INT 17H showing how a character is issued to printer is taken from
;the IBM PC/XT Technical Reference. The instructions not shown here include:
;(a) Loading DX with base  I/O address of printer port from BIOS data area 0040:0008-000F
;Reminder: The base I/O address is the address of the printer data bus.
;(b) Loading of time-out value into BL reg from BIOS data area 0040:0078H-007B
;Therefore we have the following upon going into this portion of BIOS,
;BL=time-out value
;DX=has the base I/O address, which is LPT's data port
;AL=character to be printed

            OUT     DX,AL           ;OUTPUT CHARACTER TO BE PRINTED
            INC     DX              ;POINT TO STATUS PORT
B3:         SUB     CX,CX           ;TIMER VALUE FOR BUSY
B3_1:       IN      AL,DX           ;GET STATUS
            MOV     AH,AL           ;SAVE IT IN AH
            TEST    AL,80H          ;IS BUSY LINE HIGH? (SEE FIG. 10-9, D7 = BUSY)
            JNZ     B4              ;IF READY THEN OUTPUT THE STROBE
            LOOP    B3_1            ;TRY AGAIN
            DEC     BL              ;DROP LOOP COUNT
            JNZ     B3              ;GO UNTIL TIME OUT ENDS
            OR      AH,01           ;SET ERROR FLAG
            AND     AH,0F9H         ;TURN OFF OTHER BITS
            .....
            IRET                    ;RETURN WITH ERROR FLAG BIT SET
            ....
B4:         MOV     AL,0DH          ;SET STROBE HIGH
            INC     DX              ;DX=I/O PRINTER CNTL REG
            OUT     DX,AL           ;STROBE IS BIT 0 OF CNTR REG
            MOV     AL,0CH          ;SET STROBE LOW
            OUT     DX,AL           ;AND SEND IT TO PRINTER CONTROL PORT
```

Notice the steps taken to print a character in the above listing.
1. Send the character to the D7 - D0 latch connected to data pins of the printer.
2. Test to see if BUSY is low (NOT BUSY). If ready (NOT BUSY), issue the STROBE to ask the printer to get the data by making STROBE =high and then STROBE =low.
3. If the printer is BUSY, try again until the time-out is finished. The time-out forces the CPU to check the printer for a period of time before it gives up.
4. After trying repeatedly, if for whatever reason the printer does not respond, go back and set the time-out bit to indicate that.

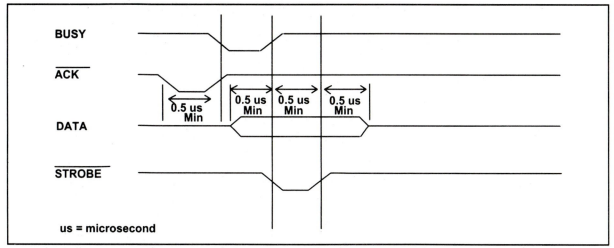

Figure 18-10. Printer Timing
(Reprinted by permission from "IBM Technical Reference" c. 1988 by International Business Machines Corporation)

Review Questions

1. The Centronics printer standard uses _____ (serial, parallel) data transfer.
2. Give one reason why there are 8 bits for the data lines in the Centronics standard.
3. The status signals of the printer are _____ (in, out) for the printer and _____ (in, out) for the computer.
4. The control signals of the printer are _____ (in, out) for the printer and _____ (in, out) for the computer.
5. STROBE is an _____ (in, out) signal for the printer and _____ (in, out) for the computer.
6. ACKNLG is an _____ (in, out) signal for the printer and _____ (in, out) for the computer.
7. D1 - D8 are _____ (in, out) signals for the printer and _____ (in, out) for the computer.
8. BUSY is an _____ (in, out) signal for the printer and _____ (in, out) for the computer.
9. Out-of-paper is an _____ (in, out) signal for the printer and _____ (in, out) for the computer.
10. How does the computer know if the printer got the last byte sent and is ready for the next one?
11. State the role and level of activation for the STROBE signal.
12. If the base I/O address of a given LPT is 3BCH, give the I/O address for each of the following lines of the printer.
    (a) control   (b) status   (c) data
13. Assuming that the I/O base address of LPT1 is 378H, show a simple Assembly language program that monitors the BUSY line of the printer.
14. What is *time-out* in IBM PC terminology?
15. Give the ASCII codes for the carriage return and line feed in hex.

SECTION 18.4: BIDIRECTIONAL DATA BUS IN PARALLEL PORTS

Since the introduction of the first IBM PC in 1981, the PC's parallel port has gone through various changes. In this section we give an overview of SPP, PS/2, EPP and ECP parallel port types and provide some parallel port interfacing tips. First, we discuss the characteristics of each parallel port type.

SPP

SPP stands for *standard parallel port*. This refers to the parallel port of the first IBM PC introduced in 1981. The data bus in SPP is unidirectional and is designed to send data from the PC to the printer. At that time, designers never thought that someone might want to use the LPT's data bus for input. In SPP, the internal logic circuitry is set for data output only and any attempt to use the data bus for input can damage the LPT port. For this reason you should never try to modify the LPT port unless you know what you are doing. Some designers use the status and control port of the SPP to send data in. In such cases, the pull-up resistors are used to prevent damage to the LPT's parallel port. For further information, refer to web page http://www.lvr.com.

PS/2

The first change in the data bus portion of the LPT port occurred in 1987 with the introduction of PS/2 models. By then, designers had seen the potential use of parallel ports for fast data acquisition. Therefore, internal circuitry of the data section of the LPT port in the PS/2 was changed to make it bidirectional. However, upon boot-up, BIOS configured the LPT port as SPP, meaning that it was to be used only for data output. At the same time, the C5 bit of the control port (base +2) was modified to allow the user to change the data port direction. At boot-up, C5 is low

(C5=0) meaning that the data port is for output. By making C5=1, we can make the data port an input port. Recall from the last section that control port C0 - C4 was already used by the SPP. Therefore, in the PS/2 LPT port, C5 of the control port is used for data port direction, while C6 and C7 are reserved.

Example 18-11

Assume that the I/O base address = 278H for LPT2 in a PS/2-compatible PC. Show how to change the control bit C5 to make the data port an input port.

Solution:

The I/O base address = 278H is for the data port. This means that we have 279H for the status port and 27AH for the control port.

```
MOV   DX,27AH          ;DX=control port address
IN    AL,DX            ;get the current information
OR    AL,00100000      ;make C5=1 without changing anything else
OUT   DX,AL            ;now data port is an input port
```

Examine the I/O addresses 278H, 279H, 27AH in Appendix G. It says that the data port is RW (read/write), the status port is RO (read only), and the control port is RW.

How to detect a PS/2-type bidirectional data bus

The following are steps in detecting if your LPT data port is bidirectional.
1. Put the data port in bidirectional mode by writing 1 to C5 of the control port (C5=1).
2. Write a known value (such as 55H, AAH, or 99H) to the data port.
3. Read back the value from the data port.
4. If the read value matches the value written to the data port, the data port is not bidirectional.

By making the data port bidirectional in the PS/2, IBM set a new standard, allowing many devices such as tape backup, scanners, and data acquisition instruments to use the LPT port instead of a PC expansion slot. However, there is one problem with PS/2-type LPT ports: They are too slow. This slowness led Intel and Xircom, along with other companies, to set a new LPT standard called EPP.

EPP

EPP stands for *enhanced parallel port*. It is the same as the PS/2, but much faster. Recall from Section 18.3 that handshaking signals such as the strobe signal are generated by software. In EPP, a higher speed was achieved by delegating the handshaking signals to the hardware circuitry on the LPT port itself. The EPP standard also added new registers to the I/O port address space beyond base address +2. In EPP, the I/O space goes from base to base+7. For example, if the base address is 278H, 279H and 27AH are the same as SPP. However, I/O addresses 27BH through 27FH are also used or reserved.

ECP

ECP stands for *extended capability port*. The need for an even faster LPT port led to ECP. The ECP has all the features of EPP plus DMA (direct memory address) capability, allowing it to transfer data via the DMA channel. It also has data compression capability. The DMA and data compression capabilities make ECP an ideal port for high-speed peripherals such as laser printers and scanners. This is the reason that Hewlett-Packard joined with Microsoft in developing the ECP standard. While the ECP-type LPT port is supposed to support SPP, PS/2 and EPP, not all of them can emulate EPP.

To see if a given ECP supports EPP, examine the PC technical documentation or check the CMOS setup on your PC.

In order to unify these various types of LPT ports, a committee of the IEEE has put together specification IEEE 1284. Refer to web page http://www.ieee.org.

Using an LPT port for output

Due to the fact that not all PCs have a single LPT standard, any detailed discussion of data input via the data port is avoided. Many people have damaged their LPT ports by not knowing what type of port they had. Regardless of the type of LPT on your PC, we know that it works for sending data out of the PC. Therefore, we can use the LPT port to send data to devices such as LCDs and stepper motors. In connecting the LPT ports to any device, first make sure that they are buffered using the 74LS244 chip. This is shown in Figure 18-11.

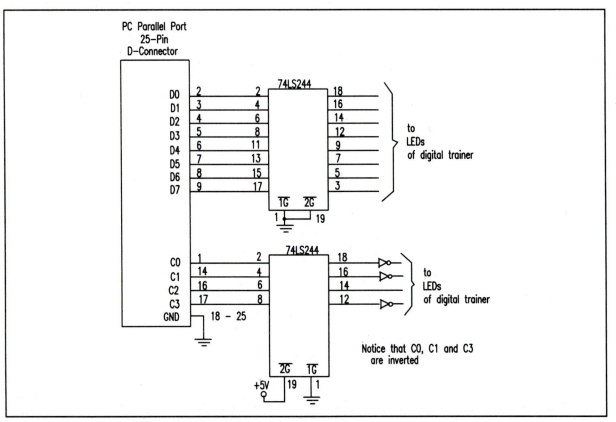

Figure 18-11. Buffering Data and Control Ports

LCD connection to the parallel port

In Chapter 12 we showed the LCD interfacing. Figure 18-12 shows the LCD connection to the parallel port. Notice in writing programs for the LCD, that in Figure 18-12 you cannot check the LCD's busy flag. The LCD command code and data must be sent to the data port one at a time with a time delay in between each. Refer to Chapter 12 for more discussion of this topic.

Stepper motor connection to the parallel port

In Chapter 12 we showed the interfacing of a stepper motor to a PC via the expansion slot. Figure 18-13 shows parallel port connection to a stepper motor. Make sure you understand the material in Chapter 12 before embarking on setting up and writing programs for such a circuit.

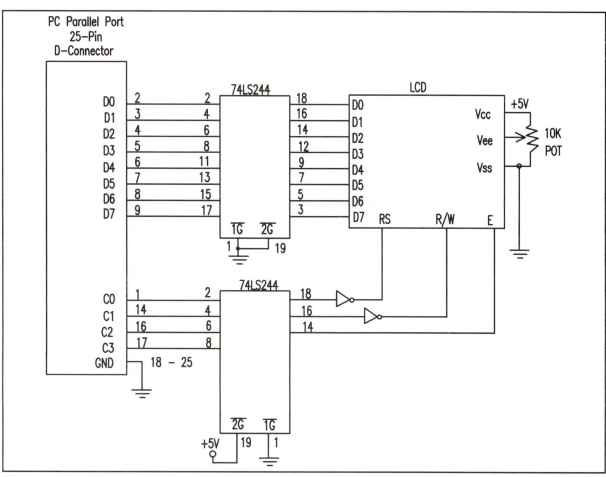

Figure 18-12. LCD Connection to Parallel Port

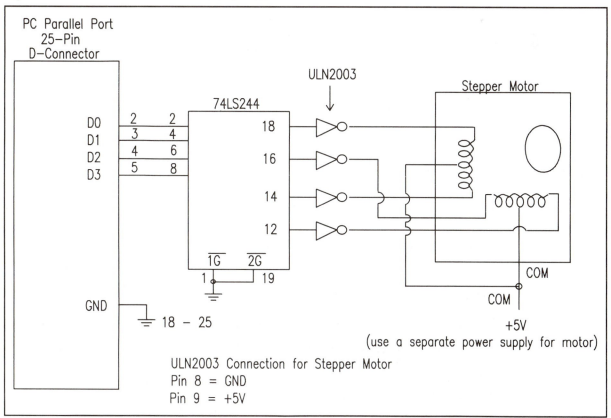

Figure 18-13. Stepper Motor Connection to Parallel Port

SECTION 18.4: BIDIRECTIONAL DATA BUS IN PARALLEL PORTS   565

Data input buffering

Assuming that the data port of LPT supports a PS/2-type bidirectional bus, one can use the circuit in Figure 18-14 to buffer it. Notice the use of 10K-ohms pull-up resistors. This is needed to prevent damage to the data port.

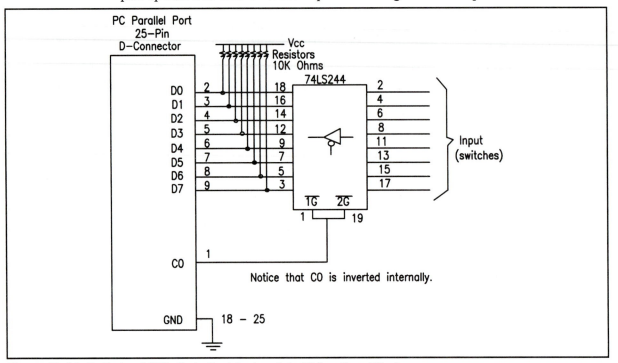

Figure 18-14. Buffering LPT's Data Port for Input in Bidirectional Ports

BIOS data area and LPT I/O address

When accessing the PC's parallel port for data acquisition, your program should get the base I/O address from the BIOS data area. This makes the program dynamic and able to run on any PC. In Assembly language, use the following code:

```
PUSH   DS              ;save DS
PUSH   AX              ;save AX
SUB    AX,AX           ;AX=0
MOV    DS,AX           ;DS=0 for BIOS data area
MOV    DX,[408]        ;get the LPT1 I/O base address
POP    AX              ;restore AX
POP    DS              ;restore DS
;now DX has the I/O base address of LPT1
```

In C we can use the following code:

```
main()
{
        ...
        unsigned int far *xptr;
        xptr = (unsigned int far *) 0x00000408;
        outp(*xptr,mybyte);       /* send mybyte to LPT's data port */
        ...
}
```

Example 18-12 demonstrates how to detect the presence of LPT1 and determine its port I/O address.

Example 18-12

Write a C program to detect the installation of LPT1 and report the I/O port address assigned to it.

Solution:

BIOS detects all the LPTs installed on the PC and reports the I/O port addresses to BIOS memory locations 00408H - 40FH, where 408H and 409H hold the I/O port address for LPT1, 40AH and 40BH for LPT2, and so on. If no LPT is installed, zeros are found in these memory locations.

```
/* this program detects the installation of LPT1 and reports the I/O port address
assigned to it. */
        #include <stdio.h>
        #include <dos.h>
        main()
        {
        unsigned int far *xptr;            /* a far pointer */
        xptr=(unsigned int far *) 0x00000408;  /* assign address */
        if(*xptr >0)
        printf("I/O base address assigned to LPT1 is %X \n",*xptr);
        else printf("LPT1 = None found");
        }
```

SUMMARY

This chapter looks at two of the most commonly used PC peripherals: the keyboard and the printer. Section 18.1 looks at the interfacing of the keyboard to the PC. Internally, keyboards are composed of a matrix of rows and columns of keys. The rows and columns are accessed through input/output ports. A microcontroller scans the keys continuously to determine if a key was pressed, and then identifies the key pressed and presents it to the main CPU.

The second section focuses on keyboard interfacing and programming in the IBM PC. IBM PCs use 8042 chips for keyboard detection and identification. INT 16H can be used by programmers to check the keyboard input buffer, read the buffer, check the status byte, and other functions. However, the system uses INT 9 to interface the keyboard to the motherboard. The keyboard microcontroller scans the keyboard continuously for a key press (a make), then INT 09 converts the scan code to ASCII, sends it to the buffer, and updates the status bytes. Separate codes are sent to the motherboard for a key release (a break) and a key press (a make). Hard contact keyboards have a hardware connection at the row and column intersections. The more expensive but longer-lasting capacitive keyboards use a capacitor at the intersection of row and column.

The third section described the standard printer interface called the *Centronics printer interface*. It describes the standard assignment of the 36 pins of the connector between the printer and the motherboard. The BIOS of the IBM PC allows for up to four parallel printers: LPT1 - LPT4. BIOS INT 17H allows the programmer to initialize printer ports, write characters to the printer, check the printer status, and perform other functions.

The fourth section gave an overview of SPP, PS/2, EPP and ECP parallel port types and provided some parallel port interfacing tips. In addition, this section gave an overview of using the LPT port to send data to devices such as LCDs and stepper motors.

PROBLEMS

SECTION 18.1: INTERFACING THE KEYBOARD TO THE CPU

1. In reading the columns of a keyboard matrix, if no key is pressed we should get all _____ (1s, 0s).
2. In Figure 18-1, to detect the key press, which of the following is grounded?
   (a) all rows   (b) one row at time   (c) both (a) and (b)
3. In Figure 18-1, to identify the key pressed, which of the following is grounded?
   (a) all rows   (b) one row at time   (c) both (a) and (b)
4. For Figure 18-1, indicate the column and row for each of the following.
   (a) D3 - D0 = 0111   (b) D3 - D0 = 1110
5. Indicate the steps to detect the key press.
6. Indicate the steps to identify the key pressed.
7. Modify Program 18-1 and Figure 18-1 for a 4 x 5 keyboard (4 rows, 5 columns).
8. Modify Program 18-1 and Figure 18-1 for a 6 x 6 keyboard.
9. Indicate an advantage and a disadvantage of using an IC chip for keyboard scanning and decoding instead of using a microprocessor.
10. What is the best compromise for the answer to Problem 9?

SECTION 18.2: PC KEYBOARD INTERFACING AND PROGRAMMING

11. In the IBM PC for each key press (make), _____ bits are transferred to the main CPU. What are these bits?
12. Find the break code for the following make codes.
    (a) 34H   (b) 1AH   (c) 5FH
13. Identify make and break among the following codes.
    (a) 9BH   (b) 89H   (c) 17H   (d) C2H   (e) 79H
14. Since keys "5" and "%" have the same scan code, how are they distinguished?
15. Find the scan code for the following.
    (a) ALT F2   (b) SHIFT F4   (c) &
    (d) V   (e) Pg Up   (f) F6
16. Which option of INT 16H is used to get the status bytes of enhanced keyboards?
17. Write a program to display a prompt such as "I will play for you the 'Happy Birthday' music if you guess the key I am thinking of. The key is one of the ALT Fs." Using INT 16H to monitor the scan codes continuously, if Alt F9 is activated the PC should play "Happy Birthday" and exit to DOS (for "Happy Birthday" music, see Chapter 13). If any other key is pressed it should display a message such as "Try again" and continue. The Esc key should exit to DOS.
18. INT 09 is assigned to which IRQ of the 8259?
19. True or false. INT 09 is activated for both the make and break scan codes.
20. True or false. If CapsLock is pressed, INT 09 saves it in the keyboard buffer.
21. What value does the BIOS keyboard subroutine save in the keyboard buffer if there is no ASCII code for a given key?
22. When there is keyboard overrun, which generates the sound beep, the circuitry inside the keyboard or the motherboard?
23. The keyboard shift status byte indicates the status of which keys?
24. Which of the following keys are non-ASCII keys?
    (a) HOME   (b) !   (c) Arrow   (d) *
25. Give the content (in hex and binary) of the keyboard shift status byte if only the NumLock and CapsLock are on.
26. If the content of the first keyboard shift byte is 10000001, what does it mean?
27. Give the physical address of memory locations in the BIOS data area set aside for each of the following keyboard components.
    (a) shift status byte (the first one)   (b) buffer
    (c) buffer's tail address   (d) buffer's head address
28. When the address of the head and tail are the same, what does it mean?

29. The key buffer is _____ (empty, full) if the address of the tail is one number higher than the address of the head.
30. What keyboard technology is used in the IBM enhanced keyboards?

SECTION 18.3: PRINTER AND PRINTER INTERFACING IN THE IBM PC

31. State the four categories of Centronics printer pins.
32. Of the following pins, which belongs to the printer's status signal and which belongs to the printer's control signal categories? Indicate which is input and which is output from the point of view of the PC.
    (a) BUSY   (b) $\overline{\text{STROBE}}$ (c) $\overline{\text{ACKNLDG}}$
    (d) SLCT   (e) INIT   (f) PE
33. Which pin is used by the printer to indicate that it is out of paper?
34. True or false. Each data line has its own ground return line.
35. What is the function of the BUSY signal? _____
36. In response to STROBE, the printer makes $\overline{\text{ACKNLDG}}$ _____ (low, high).
37. When does the BUSY signal go low?
38. True or false. The base I/O address for the printer port can be 378H or 3BCH.
39. True or false. Sending a form feed sets the printer to the top-of-page position.
40. Upon return from the INT 17H option AH=01, which register holds the error code and what does error code 90H mean?
41. In Problem 40, what is the error code for "out of paper"?
42. The PC is on but the printer is off. What message do you get on the PC screen?
43. Explain the "time out" message on the PC screen.
44. What are the ASCII codes for the line feed and carriage return?
45. Using DEBUG, check to see the time-out value for the LPT1 on your computer. Is this in seconds or milliseconds?

ANSWERS TO REVIEW QUESTIONS

SECTION 18.1: INTERFACING THE KEYBOARD TO THE CPU
1. true   2. column 3   3. true   4. 0   5. true

SECTION 18.2: PC KEYBOARD INTERFACING AND PROGRAMMING
1. The scan code is 24H. This has the odd-parity bit of 1; therefore, the following bits are transferred from the keyboard to the motherboard: 1 1 00100100 0.
2. The scan code for break is always 80H larger than the scan code for make.
3. 2DH=00101101 and ADH=10101101
4. false, for both   5. Yes, it is 38h.   6. false.
7. 48H   8. true   9. true
10. False, the buffer inside keyboard is full.
11. The keyboard buffer on the motherboard is empty.   12. tail

SECTION 18.3: PRINTER AND PRINTER INTERFACING
1. parallel   2. since the characters are 8-bit ASCII code
3. out, in   4. in, out   5. in, out
6. out, in   7. in, out   8. out, in
9. out, in   10. Through the $\overline{\text{ACKNLG}}$ or BUSY signals; either one can be used.
11. It must be high normally. When the computer has a byte of data for printer it makes it go low to inform the printer.
12. (a) 3BEH for control   (b) 3BDH for status   (c) 3BCH for data
13.
```
        MOV    DX,379H      ;LPT1 STATUS PORT ADDRESS
A1:     IN     AL,DX        ;GET THE LPT1 STATUS
        TEST   AL,80H       ;IS D7=1 (BUSY SIGNAL)
        JNZ    ...          ;NOT BUSY
        JMP    A1           ;TRY AGAIN
```
14. When the PC tests the printer status port and cannot get any response from the printer, it tries again for a certain time period, then if it does not get any response, sets the time-out bit.
15. 0DH and 0AH

CHAPTER 19

FLOPPY DISKS, HARD DISKS, AND FILES

OBJECTIVES

Upon completion of this chapter, you will be able to:

>> **Contrast and compare the terms** *primary storage* **and** *secondary storage*
>> **Analyze the storage capacity of floppy disks in terms of number of sectors, sectors per cluster, sectors per track, and total capacity of the disk**
>> **Discuss floppy disk organization in terms of the boot record, FAT, and the directory**
>> **List the information given in the boot record of a disk**
>> **List the information given in directory entries of a disk**
>> **Describe the contents of the FAT and how the operating system uses it to locate, update, and delete files**
>> **Analyze the capacity of hard disks in terms of sectors, tracks, clusters, cylinders, and platters**
>> **Define hard disk terminology: partitioning, interleaving, low-level and high-level formatting, parking the head, and MTBF**
>> **Define the components of hard disk access time: seek time, settling time, and latency time**
>> **Diagram encoding techniques FM, MFM, and RLL**
>> **Explain the interfacing standards for hard disks: ST506, ESDI, IDE, and SCSI**
>> **Code Assembly language programs to access files on disks**

This chapter will examine the characteristics of secondary storage devices such as floppy and hard disks, their file organization, and how they are used in the IBM PC and compatibles. The chapter includes a discussion of data encoding techniques, interfacing standards, and definitions of hard disk terminology. In Section 19.1 we look at floppy disk organization. Hard disk organization, characteristics, and terminology are covered in Section 19.2. Disk files and file programming are covered in Section 19.3.

SECTION 19.1: FLOPPY DISK ORGANIZATION

In the early days of the personal computer, cassette tape was used to store information. Due to its long access time, it was abandoned as a secondary storage medium. The term *secondary storage* refers to memory other than RAM. RAM is called *primary storage* since the CPU asks for the information that it needs from RAM first. The floppy disks initially used in the PC were one-sided, meaning that only one side of the disk could be accessed by the disk drive. To access the other side, the disk had to be flipped over. Today, all floppy disks are double-sided and the disk drive is equipped with two heads, one for each side, allowing the disk to be read on both sides without flipping it over. This section concentrates on double-sided disks only since they are the universal standard.

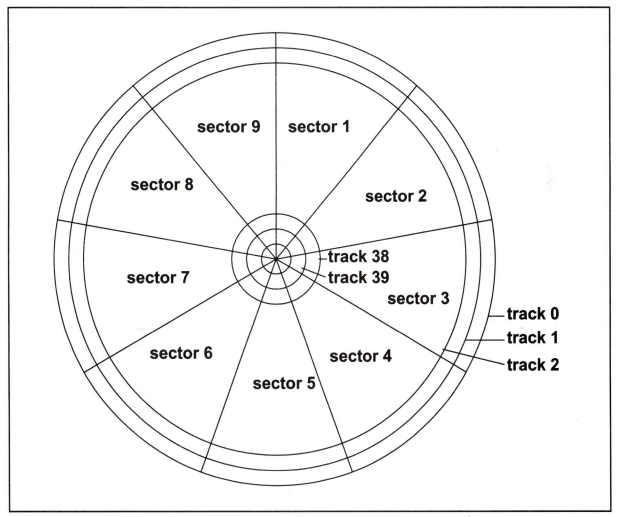

Figure 19-1. Sectors and Tracks of a 5 1/4" Diskette

Capacity of the floppy disk

In order to store data on the disk, both sides are coated with magnetic materials. The principles behind the process of reading and writing (storing) digital data 0 and 1 on disks is the same as is used in any magnetic-based medium. Each side of the disk is organized into tracks and sectors as shown in Figure 19-1. Tracks are organized as concentric circles and their number per disk varies from disk to disk, depending on the size and technology. Each track is divided into a number of sectors, and again the number of sectors per track varies, depending on the density of the disk and the version of the DOS operating system. Each sector stores 256 or 512 bytes of information, depending on the sector density. In what is commonly referred to as a *double-density* disk, the storage capacity of a single sector is 512 bytes. This format is supported by DOS 2.0 and higher. In addition to the number of tracks and sectors in the floppy disk, the physical size of the disk varies. Among the available sizes are 5 1/4 and 3 1/2 inches. The number of tracks and sectors, and the total capacity of double-sided floppy disks commonly used for the IBM PC and compatibles, are shown in Table 19-1 along with the supporting DOS version. The total density of the various disk types can be verified by using 512 bytes for the sector density. For example, the 5 1/4" with 40 tracks and 9 sectors per track will be as follows:

40 tracks x 9 sectors per track = 360 sectors per disk side
360 sectors x 512 bytes per sector = 184,320 bytes per side
184,320 x 2 sides = 368,640 bytes per disk, or 360K bytes per disk

The sectors of a disk are grouped into clusters. Cluster size varies among formats, but a common size for floppy disks is 2 sectors per cluster. The file allocation table, or FAT, which will be discussed later, keeps track of what clusters are used to store which files.

Table 19-1: Sectors and Tracks for Double-Sided Disks

Diskette Type	Tracks per Side	Sectors per Track	Bytes	MS DOS Version
5 1/4"	40	8	320K	1 and above
5 1/4"	40	9	360K	2 and above
5 1/4" high density	80	15	1.2M	3 and above
3 1/2"	80	9	720K	3.2 and above
3 1/2" high density	80	18	1.44M	3.3 and above

Formatting disks

Once a floppy disk is formatted, the computer can read from or write to that disk. *Formatting* organizes the sectors and tracks in a way that makes it possible for the disk controller to access the information on the disk. When a disk is formatted, a number of sectors are set aside for various functions and the remaining sectors are used to store the user's files. The formatting process sets aside a specific number of sectors for the boot record, directory, and FAT (file allocation table), each of which is explained in detail below. It also copies some system files onto the disk if it was formatted with the "/s" option, which makes it a bootable disk. The difference between bootable and nonbootable disks will be explained later.

Floppy disk organization

Regardless of the type of disk that is used, the first sector of the disk (side 0, track 0, sector 0) is always assigned to hold the boot record; then some sectors are used for storage of the FAT (file allocation table) copies 1 and 2. The number of sectors set aside for FAT depends upon the disk density. After the FAT, the directory is stored in consecutive sectors. Again, the number of sectors used by the directory depends on disk density. In assigning sectors for the FAT and the directory, DOS uses all the sectors of track 0 side 0, then goes to side 1 and uses all the sectors of track 0 of side 1, then comes back to side 0 and uses track 1, then goes to side 1 track 1, and so on. Floppy disk organization is shown in Figure 19-2, which shows the layout of 9-sector, 40-track, double-sided 5 1/4" floppy disks.

Since floppy disks with sectors of 9 and higher are the industry standard, the remainder of this discussion will focus on them. Before moving to the next topic it should be noted that the number of sectors assigned for the boot record, FAT, and directory are fixed for a given kind of disk and operating system version and it is only after assigning these sectors to those essential functions that DOS uses the remaining sectors to store files. The number of sectors set aside for each of the above can be calculated from the information in the boot record. The boot record, FAT, and directory are explained in detail next.

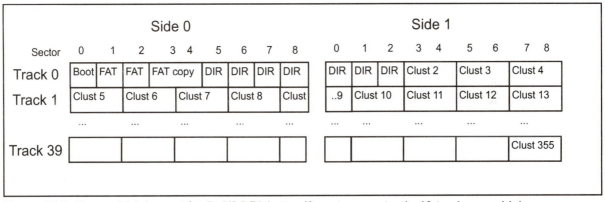

Figure 19-2. Floppy Disk Layout for 5 1/2 " Diskette (9 sectors per track, 40 tracks per side)

Looking into the boot record

When a disk is formatted, the first sector is used for the boot record. It is from the boot record that the computer will know the disk type, sector density, total number of sectors in the disk, and other essential information needed by BIOS and the operating system. Table 19-2 describes each byte of the boot record.

In order to understand the boot record's function, the boot record of several different disks will be analyzed. To access the boot record, the DEBUG program can be used to dump the information into memory and analyze it, byte by byte. First, a few reminders about DEBUG. The -L command can be used to load the specific sectors of a given disk into a specific area of RAM memory. Figure 19-1 showed sectors numbered from 1 to 9 for a given track; however, sectors are actually numbered in hex starting at 0. For example, a 5 1/4" 360K diskette has sectors numbered logically from 0 to 2D0H. The -L command is followed by the address that data from the disk should be loaded into, followed by the drive number, then the starting sector number, and finally, the number of sectors to be loaded. The drive number has the following options: 0 for drive A, 1 for drive B, 2 for drive C, and so on. All numbers in DEBUG commands are given in hex. Example 19-1 first loads into memory beginning at cs:100 from drive B, starting at sector 0, and loads one sector only. After the load, the -d command is used to dump memory onto the screen. One more reminder: In the IBM PC and compatibles, the least significant byte of data is always stored in the lower memory location. For example, to store 35F6H in memory locations 1300 and 1301, location 1300 will contain F6 and 1301 will contain 35.

SECTION 19.1: FLOPPY DISK ORGANIZATION 573

Table 19-2: Boot Record Layout

Offset	Bytes	Contents	Notes
00 - 02	3	E9 XX XX or EB XX 90	The first byte is always the opcode for the JMP command, either E9, the opcode for intrasegment JMP, or EB, the opcode for short JMP. If the JMP is intrasegment, the displacement is 2 bytes. If the JMP is short, the displacement is 1 byte. In the latter case, the opcode for the second byte is 90, NOP.
03 - 0A	8	Manufacturer name and version	This is the manufacturer's name and the version of DOS under which the disk has been formatted.
0B - 0C	2	Bytes per sector	This gives the density of the sectors: that is, the total number of bytes that can be stored on one sector.
0D	1	Sectors per allocation unit	This byte gives the number of sectors that are grouped into a cluster. A cluster is two or more sectors grouped together as a unit of storage. Every file uses a minimum of one cluster and there is no limit for the maximum as long as there are free clusters available.
0E - 0F	2	Reserved sectors	This represents the number of sectors reserved, starting at sector 0.
10	1	Number of FATs	Normally, there are 2 copies of the FAT.
11 - 12	2	Number of root-directory entries	
13 - 14	2	Total sectors in a disk	This gives the total number of sectors in the entire disk. For the hard disk of more than 32M capacity, use bytes 20 - 23.
15	1	Medium type	This code defines the type of disk. Refer to Table 19-3 for a list of codes and media types.
16 - 17	2	Number of sectors per FAT	This indicates how many sectors are used for each FAT.
18 - 19	2	Sectors per track	This gives the total number of sectors for each track.
1A - 1B	2	Number of heads	While this is always 2 for double-sided floppy disks, it is much more for hard disks as will be seen later in the hard disk section of this chapter.
1C - 1F	4	Number of hidden sectors	
20 - 23	4	Total sectors In logical volume	This is used only for hard disks of volume size greater than 32MB.
24	1	Physical drive number	
25	1	Reserved	
26	1	Extended boot signature record	
27 - 2A	4	32-bit binary volume ID	
2B - 35	10	Volume label	
36 - 3D	8	Reserved	
3E - ??		Bootstrap	

Example 19-1

Load the contents of drive B, sector 0 into memory starting at CS:100. Dump it to the screen and analyze it.

Solution:

The following DEBUG command loads drive 1 (B), starting at sector 0, for only one sector.

```
C>DEBUG
-L CS:100 1 0 1
-D CS:100 17F
142B:0100   EB 34 90 49 42 4D 20 20-33 2E 33 00 02 02 01 00
142B:0110   02 70 00 A0 05 F9 03 00-09 00 02 00 00 00 00 00
142B:0120   00 00 00 00 00 00 00 00-00 00 00 00 00 00 00 12
142B:0130   00 00 00 00 01 00 FA 33-C0 8E D0 BC 00 7C 16 07
142B:0140   BB 78 00 36 C5 37 1E 56-16 53 BF 2B 7C B9 0B 00
142B:0150   FC AC 26 80 3D 00 74 03-26 8A 05 AA 8A C4 E3 F1
142B:0160   06 15 89 47 02 C7 07 2B-7C FB CD 13 72 67 A0 10
142B:0170   7C 98 F7 26 16 7C 03 06-1C 7C 03 06 0E 7C A3 3F
```

Using Table 19-2, look at the first 3 bytes. The first byte in this example is EB, the jump opcode, followed by 34, the displacement memory location where the the first byte of the opcode for the boot-strap subroutine in sector 0 is located. The third byte is 90, NOP. The next 8 bytes give the manufacturer: 49 42 4D 20 20 represents "IBM " in ASCII, and 33 2E 33 represents 3.3 for the DOS version. The next 2 bytes give the number of bytes per sector, 00 02; remembering to reverse the order of the bytes gives 0200H, which is 512 bytes per sector. The next byte gives the number of sectors per cluster; in this case, 2 sectors are grouped together into a cluster. The next two bytes, 0001, represent the number of reserved sectors. The next byte gives the number of FAT tables, in this case there are 2 tables, or 2 copies of the FAT, the main one and a backup. The number of directory entries is given in the next two bytes: 0070H. The number of sectors contained on this disk is given next: 05A0H, which is 1440 in decimal. F9, the next byte, gives the type of disk. Table 19-3 shows that this is a 3 1/2" double-sided diskette with 9 sectors per track. The next two bytes give the number of sectors per FAT. In the above example, 0003 indicates there are 3 sectors used for each FAT. The next 2 bytes give the number of sectors per track, 0009, or 9 sectors/track. The next two bytes give the number of heads, 0002. The next two bytes represent the number of hidden sectors, which is zero in this case. The remaining bytes contain the bootstrap, since this disk was formatted in DOS 3.3. If a higher version of DOS had been used, other information would be given before the bootstrap.

From the above information, one can easily calculate the total capacity of the disk. There are 1440 sectors on the disk, and each sector contains 512 bytes. Using this information yields

$$1440 \times 512 = 737{,}280 / 1024 = 720K \text{ bytes for this disk}$$

One can also determine the number of tracks by dividing the total sectors by number of sectors per track:

1440 / 9 =160 tracks for both sides of the disk, which gives 80 tracks per side.

This result is confirmed by Tables 19-1 and 19-3, which show this to be a 3 1/2" disk with 80 sectors and 9 tracks per side. The number of sectors set aside for the storage of FATs and the directory can also be seen from the boot record. How this is done will be discussed later.

SECTION 19.1: FLOPPY DISK ORGANIZATION   575

Example 19-2

Load into memory, starting at cs:100 from disk A, sectors 0 to 1 (2 is the number of sectors to be loaded) and then dump them in order to examine them.

Solution:

```
C>DEBUG
-L CS:100 0 0 2
-D CS:100 14F
142B:0100   EB 34 90 4D 53 44 4F 53-33 2E 33 00 02 01 01 00
142B:0110   02 E0 00 60 09 F9 07 00-0F 00 02 00 00 00 00 00
142B:0120   00 00 00 00 00 00 00 00-00 00 00 00 00 00 00 12
142B:0130   00 00 00 00 01 00 FA 33-C0 8E D0 BC 00 7C 16 07
142B:0140   BB 78 00 36 C5 37 1E 56-16 53 BF 2B 7C B9 0B 00
etc.
```

The following is a brief analysis of the above boot record:

Bytes 0 - 2	contain JMP to the boot code and NOP
Bytes 3 - A	indicate MS DOS 3.3 is the operating environment
Bytes B - C	give the number of bytes per sector: 0200H or 512
Byte D	gives the number of sectors per cluster: 01
Bytes E - F	give the number of reserved sectors: 0001
Byte 10	gives the number of FAT tables: 02
Bytes 11-12	give the number of directory entries: 00E0H or 224
Bytes 13-14	give the number of sectors: 0960H or 2400
Byte 15	is the media descriptor byte: F9 (see Table 19-3)
Bytes 16-17	give the number of FAT sectors: 0007
Bytes 18-19	give the number of sectors per track: 000F (15)
Bytes 1A-1B	give the number of heads: 0002
Bytes 1C-1D	give the number of hidden sectors: 0000

The remaining bytes contain the boot code.

Table 19-3: First Byte of FAT and Storage Media

FAT First Byte	Storage Media
F0	Double-sided 3 1/2" diskette; 18 sectors/track
F8	Hard disk
F9	Double-sided 5 1/4" diskette; 15 sectors/track
	Double-sided 3 1/2" diskette; 9 sectors/track
FC	Single-sided 5 1/4" diskette; 9 sectors/track
FD	Double-sided 5 1/4" diskette; 9 sectors/track
FE	Single-sided 5 1/4" diskette; 8 sectors/track
FF	Double-sided 5 1/4" diskette; 8 sectors/track

Directory

After DOS allocates one sector for the boot record and several sectors for the FAT, it allocates some sectors for the directory. Which and how many sectors are allocated for the directory vary among the different disk types. This section examines the structure of the directory of the floppy disk before delving into the concept of FAT since understanding the directory is prerequisite to the study of FAT. According to Figure 19-2, the directory for a 9-sector 40-track 5 1/4" diskette is located at side 0, track 0, sector 5. The following example used DEBUG to load sector 5 of floppy disk A into memory, and dump a portion of it to the screen.

Example 19-3

Load, dump, and analyze the sector containing the directory of a 360K floppy in drive A.

Solution:

```
C>DEBUG
-L CS:100 0 5 1
-D CS:100 L400
1131:0100 45 44 20 20 20 20 20 20-45 58 45 20 00 00 00 00    ED      EXE ...
1131:0110 00 00 00 00 00 00 E0 96-98 0C 02 00 9B 57 01 00    ..............W.
1131:0120 45 44 20 20 20 20 20 20-48 4C 50 20 00 00 00 00    ED      HLP ...
1131:0130 00 00 00 00 00 00 1C 91-81 0C 58 00 44 B1 00 00    ..........X.D..
```

Byte (H)	Display Value	Interpretation	
0 - 7	45 44 20 20 20 20 20 20	"ED "	
8 - A	45 58 45	"EXE"	
B	20	Low-order byte = 0, no attribute indicated	
C - 15	00 00 00 00 00 00 00 00 00 00	Reserved	
16-17	E0 96 (use 96 E0)	1001 0110 1110 0000	
		00000 = 0 2-second incr.	
		110111 = 55 minutes	
		10010 = hour 18 (6 p.m.)	
18-19	98 0C (use 0C 98)	0000 1100 1001 1000	
		11000 = day 24	0100 = month 4
		0000110 = year 6	1980 + 6 = 1986
1A-1B	02 00	Starting cluster 0002	
1C-1F	9B 57 01 00	File size in bytes	00 01 57 9B = 87963

Analyzing with help from Figure 19-3, the first 8 bytes represent the name of the file "ED". The file name can be up to 8 bytes long and if it is shorter, is padded with blanks (ASCII code 20). Bytes 8 to A are the extension: "EXE". Byte B is the attribute byte; in this case the file is an archive file. From byte C to byte 15, a total of 8 bytes, is reserved for future use by Microsoft, and is filled with zeros. Bytes 16 and 17 hold the time of day that the file was created, 96E0, or 1001 0110 1110 0000. The first 5 bits, 00000, represent seconds in 2-second intervals. The next 6 bits, 110111, represent the minutes. Converting 110111 from binary into decimal gives 55 minutes. The last 5 bits, 10010, represent the hour, which in decimal is 18, or 6:00 p.m. Bytes 18 and 19, 0C98, represent the date the file was created or last modified. These bytes in binary are: 0000 1100 1001 1000. The day is given in the first 5 bits: 11000, which is 24 in decimal. The month is given in the next 4 bits: 0100, which is 4. The year is given in the last 7 bits: 0000110, which is year 6 in decimal. The year is stored as the number of years since 1980. Adding 6 to 1980 gives 1986, the year this file was last modified. Bytes 1A and 1B give the starting cluster number: 0200. Finally, bytes 1C to 1F give the file size in bytes, 0001579BH, or decimal 87963, the number of bytes in this file.

Example 19-4

Load, dump, and analyze the directory of a 360K diskette in drive A.

Solution:

```
C>DEBUG
-L CS:100 0 5 1
-D CS:100 3FF
1131:0100  49 42 4D 42 49 4F 20 20-43 4F 4D 27 00 00 00 00   IBMBIO  COM'....
1131:0110  00 00 00 00 00 00 00 60-9E 0B 02 00 F1 3F 00 00   .......`.....?..
1131:0120  49 42 4D 44 4F 53 20 20-43 4F 4D 27 00 00 00 00   IBMDOS  COM'....
1131:0130  00 00 00 00 00 00 00 60-9E 0B 12 00 3D 6F 00 00   .......`....=o..
```

Byte (H)	Display Value	Interpretation
0 - 7	49 42 4D 42 49 4F 20 20	"IBMBIO "
8 - A	43 4F 4D	"COM"
B	27	Attribute: 00100111
		Read-only, hidden, system, archive
C - 15	00 00 00 00 00 00 00 00 00 00	Reserved
16-17	00 60 (use 60 00)	0110 0000 0000 0000
		00000 = 0 2-second incr.
		000000 = 0 minutes
		01100 = hour 12
18-19	9E 0B (use 0B 9E)	0000 1011 1001 1110
		11110 = day 30
		1100 = month 12
		0000101 = year 51980 + 5 = 1985
1A-1B	02 00	Starting cluster 0002
1C-1F	F1 3F 00 00	File size in bytes 00 00 3F F1 = 16,369

Second directory entry:

Byte (H)	Display Value	Interpretation
0 - 7	49 42 4D 44 4F 53 20 20	"IBMDOS "
8 - A	43 4F 4D	"COM"
B	27	Same as above
C - 15	00 00 00 00 00 00 00 00 00 00	Reserved
16-17	00 60	0110 0000 0000 0000
		00000 = 0 2-second incr.
		000000 = 0 minutes
		01100 = hour 12
18-19	9E 0B (use 0B 9E)	0000 1011 1001 1110
		11110 = day 30
		1100 = month 12
		0000101 = year 51980 + 5 = 1985
1A-1B	1200	Starting cluster: 0012
1C-1F	3D 6F 00 00	File size in bytes 00 00 6F 3D = 28,477

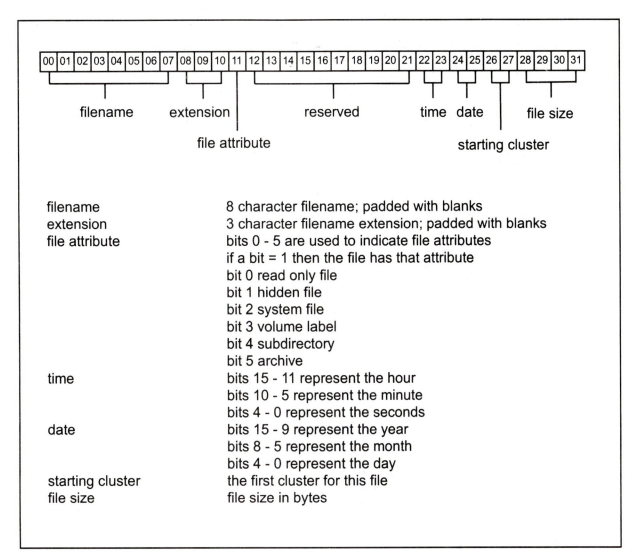

| 00 | 01 | 02 | 03 | 04 | 05 | 06 | 07 | 08 | 09 | 10 | 11 | 12 | 13 | 14 | 15 | 16 | 17 | 18 | 19 | 20 | 21 | 22 | 23 | 24 | 25 | 26 | 27 | 28 | 29 | 30 | 31 |

filename   extension   reserved   time   date   file size

file attribute   starting cluster

filename	8 character filename; padded with blanks
extension	3 character filename extension; padded with blanks
file attribute	bits 0 - 5 are used to indicate file attributes
	if a bit = 1 then the file has that attribute
	bit 0 read only file
	bit 1 hidden file
	bit 2 system file
	bit 3 volume label
	bit 4 subdirectory
	bit 5 archive
time	bits 15 - 11 represent the hour
	bits 10 - 5 represent the minute
	bits 4 - 0 represent the seconds
date	bits 15 - 9 represent the year
	bits 8 - 5 represent the month
	bits 4 - 0 represent the day
starting cluster	the first cluster for this file
file size	file size in bytes

Figure 19-3. Directory Entry Layout

Bootable and nonbootable disks

Examples 19-3 and 19-4 showed the first directory entries of two different disks. Example 19-4 shows a bootable disk; Example 19-3 shows one that is not bootable. If the disk is formatted as a system disk (bootable), the first two files are IBMBIO.COM and IBMDOS.COM, which are followed by COMMAND.COM. The first two are hidden files; therefore, they will not be listed when DIR is used. However, if the disk is formatted as a nonbootable disk, it will not have those three files on it after it is formatted. The job of IBMBIO.COM is to provide low-level (hardware) communication (interface) between BIOS and DOS. The high-level (software) interface is provided by the IBMDOS.COM file. This is the section of DOS that contains INT 21H, among other things. Among the functions of COM-MAND.COM is to provide the DOS prompt ">", read, interpret, and execute commands typed in by the user. The first two files have been given different names by MS DOS from Microsoft: IO.SYS and MSDOS.SYS instead of IBMBIO.COM and IBMDOS.COM. Beginning with DOS 4.0, these three files no longer have to be the first directory entries and they can be located anywhere in the directory. The SYS command can be used to copy these files to a nonbootable disk to make it bootable.

FAT (file allocation table)

If the boot record tells BIOS and the operating system the kind of disk, and the directory provides the lists of all the files contained on the disk, how does DOS locate a given file? Does it check every one of the hundreds of sectors to see if the file is there? This would obviously take an inordinate amount of time. It is the function of the FAT to provide a road map for the operating system to find where each file is located. In fact, the FAT is so critical to the operating system's ability to locate files that two copies of the FAT are kept on the disk, one for use and another one for backup in case something happens to the first one. If both are damaged, the operating system cannot find any file on that disk. The FAT is always located in the sectors following the boot record sector. The number of sectors used by the FAT varies depending on the size and density of the disk. Next we will describe the contents of the FAT and how the operating system uses it to locate, update, and delete files.

The first two entries in a FAT contain the media descriptor byte, followed by F's to pad the remaining space. For the remaining entries in a FAT, there is a one-to-one correspondence between each FAT entry and each cluster on the disk. In other words, if there are 355 clusters in the diskette available for data storage, there will be 355 FAT entries. This is the case for the diskette shown in Figure 19-2. This figure shows that the clusters are numbered starting with number 2. The reason is to make sure that there is one-to-one correspondence between the cluster number and the FAT entry since the entry 1 and 0 is F9FFFF. Each FAT entry indicates the status of that cluster: if it is free, unused, reserved, bad, or part of a file. The starting cluster of a file is stored in the directory entry for that file. When DOS reads a file, it first reads this cluster, then checks the FAT entry for that cluster to see if there is a pointer to another cluster or if there is a code indicating that this was the last cluster in that file. If there is a number of another cluster, it will read that cluster, then check that cluster's FAT entry, and repeat the process until an end-of-file code is found. In other words, DOS finds all the clusters of a file by following the links in the FAT entries.

The FAT entries examined in this section will be 12-bit entries, but MS DOS also supports FAT entries of 16 bits for drives of more than 6M capacity. Table 19-4 lists the special codes that a FAT entry may contain and their meaning.

Table 19-4: FAT Entry Codes and Their Meanings (12-bit FAT)

Code	Meaning
000H	Unused cluster: cluster has never been used
001H	Free cluster: cluster was used previously but is now free
002-FEFH	Cluster is used by a file
FF0-FF6H	Reserved
FF7H	Bad cluster, cannot be used
FF8-FFFH	Last cluster of a file

Any value other than those shown in Table 19-4 will be a pointer to the next cluster of a file. The following diagram and explanation should help clarify the above points.

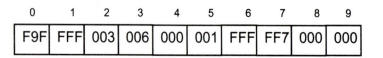

CHAPTER 19: FLOPPY DISKS, HARD DISKS, AND FILES

The diagram above is a conceptual picture of the first 10 entries of a 12-bit FAT table. The first two entries contain the media descriptor byte F9 (see Table 19-3), followed by FFFF. The remaining entries indicate the status of the corresponding clusters in memory. The first file stored on a formatted disk will normally begin in cluster 2, the first cluster available for storage. Suppose that DOS wants to read in this file. First it checks its directory entry and notes that the first cluster for this file is cluster 2. It reads in cluster 2, then checks FAT entry 2. FAT entry 2 indicates that there is more information for this file and that it can be located in cluster 3. DOS then reads cluster 3 and checks FAT entry 3. FAT entry 3 indicates that there is more information for this file in cluster 6. DOS then reads cluster 6 and checks FAT entry 6. FAT entry 6 indicates that this was the end of the file. The pointers in FAT entries 2, 3, and 6 above are each one link in the chain of that file. The above was the conceptual picture of a FAT. The following is how those 10 FAT entries would look in memory:

```
Byte:    0  1  2  3  4  5  6  7  8  9  A  B  C  D  E  ...

Data:    F9 FF FF 03 60 00 00 10 00 FF 7F FF 00 00 00 ...
```

For a 12-bit FAT table, as shown above, each FAT entry requires 1.5 bytes. In order to access the FAT entry for a given cluster, DOS multiplies the cluster number by 1.5 and uses the integer portion of the result as an offset into the FAT. If the FAT entry was an even number, DOS discards the high-order hex digit. If it was an odd entry, DOS discards the low-order hex digit. For example, to get the FAT entry for cluster 2, DOS multiplies 2 times 1.5, giving 3. Using 3 as an offset into the FAT will retrieve 0360; reversing the bytes: 6003. Since 2 (for cluster 2) is an even number, the high-order hex digit is discarded, leaving 003. When DOS has finished reading cluster 3, it will check FAT entry 3. To access this, DOS multiplies 3 times 1.5, which is 4.5, the integer portion of which is 4. Using 4 as an offset into the FAT gives 60 00; reversing gives 0060. Since 3 (FAT entry) is an odd number, the low-order hex digit is discarding, leaving 006. This process continues until FFF is found, indicating that the cluster was the last cluster for that file.

Additional copies of the FAT are updated whenever a change is made to a file. DOS primarily uses the first copy and compares both copies when a disk is first used to make sure that they are the same. Many of the errors that can occur in a FAT can be corrected with the DOS CHKDSK utility.

Table 19-5: Summary of Various Disk Data

Disk Type	360K	720K	1.2M	1.44M	10M HD	20M HD
Bytes per sector	200	200	200	200	200	200
FAT tables	2	2	2	2	2	2
Directory entries	70	70	E0	E0	200	200
Total sectors	2D0	5A0	960	B40	50F3	A307
FAT sectors	2	3	7	9	8	29
Sectors per track	9	9	F	12	11	11
Total bytes	368,640	737,280	1,228,800	1,474,560	10,610,176	21,368,320

Note: All data are in hex except the total bytes.

How to calculate sector locations of the FAT and the directory

Table 19-5 shows the summary of various data for different disks (including 10M and 20M hard disks) extracted from their boot records. This data can be used to calculate the sector location of FAT and the directory. Example 19-5 demonstrates this. Table 19-6 provides the sector map for different disks.

Table 19-6: Sector Map for Various Hard and Floppy Disks

Disk Type	360K	720K	1.2M	1.44M	10M HD	20M HD
Boot record	0	0	0	0	0	0
FAT 1	1-2	1-3	1-7	1-9	1-8	1-20
FAT 2	3-4	4-6	8-E	A-12	9-10	2A-52
Directory	5-B	7-D	F-1C	13-20	11-30	53-72
Data	C-2D0	E-5A0	1D-960	21-B40	31-50F3	73-A307

Note: All data is in hex.

Example 19-5

Use the data in Table 19-5 to find the sector map of a 1.2M capacity diskette.

Solution:

According to Table 19-5, there are two FATs, each taking 7 sectors, and there is a total of E0H entries for the directory. Sector 0 is used for the boot record, so FAT copy 1 will occupy sectors 1 through 7 and FAT copy 2 will occupy sectors 8 - E. The directory starts at sector F. How many sectors will it need? This is calculated by multiplying the number of entries, E0, by the size of each entry, 32 bytes.

E0H = 224 decimal
224 entries × 32 bytes = 7168 bytes
7168 bytes divided by 512 bytes per sector = 14 sectors used for the directory

Therefore, the directory will occupy sectors F to 1C. The first file on this 1.2M disk can start at sector 1D. See Table 19-6. This can be verified by looking at the disk through DEBUG. Remember that the first byte of the FAT is the media descriptor byte.

Review Questions

1. True or false. All sectors have the same capacity (total number of bytes that can be stored).
2. True or false. The 3 1/2" disks can be 1.2M or 1.44M capacity.
3. The very first sector always belongs to the _____ (FAT, boot record).
4. The sectors belonging to the _____ (FAT, directory) are located next to the boot sector.
5. Why are there two copies for each FAT?
6. True or false. The number of sectors set aside for the FAT varies among the various-sized disks.
7. The boot record provides the total _____ (byte capacity, number of sectors) per disk.
8. How does the operating system know how many sectors each track is divided into?
9. Sectors set aside for directories are always _____ (before, after) the FAT sector.
10. If 3 1/2" disks can have two different capacities, how does the operating system make a distinction between them?

SECTION 19.2: HARD DISKS

This section will look at the characteristics of hard disks and their organization with emphasis on performance factors such as access time and finally, interfacing standards. The *hard disk*, referred to sometimes as *fixed disk*, or *winchester disk* in IBM literature, is judged according to three major criteria: capacity, access time (speed of accessing data), and interfacing standard. Before delving into each category, an explanation should be given for the use of different names such as fixed disk, winchester disk, and hard disk to refer to the same device. The term *hard disk* comes from the fact that it uses hard solid metal platters to store information instead of plastic as is the case in floppy disks. It is also called *fixed disk* because it is mounted (fixed) at a place on the computer and is not portable like the floppy disk (although some manufacturers make removable hard disks). Why is it also called the *winchester disk*? When IBM made the first hard disk for mainframes it was capable of storing 30 megabytes on each side and therefore was called a 30/30 disk. The 30/30 began to be called the winchester 30/30, after the rifle, and soon it came to be known simply as the winchester disk.

Hard disk capacity and organization

One of the most important factors in judging a hard disk is its *capacity*, the number of bytes it can store. Capacity of hard disks ranges from 5 megabytes to many gigabytes (a gigabyte is 1024 megabytes). The 5 megabyte disk was used in the early days of the PC and is no longer made. At this time when the capacity of hard disks is increasing to the gigabyte level, 100 to 500 megabyte capacity disks are in common use for the 286, 386, 486, and Pentium computers. Regardless of the capacity of the hard disk, they all use hard metal platters to store data. In general, the higher the number of platters, the higher the capacity of the disk. Just as in the floppy disk, both sides of each platter in the disk are coated with magnetic material. Likewise, it uses a storage scheme that divides the area into sectors and tracks just as the floppy disk does. There is one read/write head for each side of every platter, and these heads all move together. For example, a hard disk with 4 plates might have 8 read/write heads, one for each side, and they all move from the outer tracks into inner tracks by the same arm. Hard disks give rise to more complex organization and hence a new term: the *cylinder*, which consists of all the tracks of the same radius on each platter. Since all the read/write heads move together from track to track it is logical to talk about cylinders in addition to tracks in the hard disk. Why do all the heads move together? The answer is that it is too difficult and expensive to design a hard disk controller that controls the movement of so many different heads. In addition, it would prolong the access time since it must stop one head and then activate a different head continuously until it reaches the end of the file. Using the concept of the cylinder, all the tracks of the same radius are accessed at the same time, and if the end of the file is not reached, all the heads move together to the next track. The number of read/write heads varies from one hard disk to another. The number is usually twice the number of platters but is sometimes 1 less than this number, as seen in Table 19-8. In some disks, one side of one platter is set aside for internal use and is not available for data storage by the user. Knowing the concepts of read/write heads and cylinders makes it possible to calculate the total number of tracks and the total capacity of the hard disk. The total capacity of a disk is calculated in the same way as it is for floppy disks:

number of tracks = number of cylinders x number of heads
HD capacity = number of tracks x number of sectors x sector density

Depending on the hard disk, often there are 17 to 36 sectors per track and 512 bytes for each sector.

Example 19-6

Verify the capacity of the Seagate 225 hard disk using the data in Table 19-8.

Solution:

As shown in Table 19-8, the ST225 has 4 heads, 615 cylinders, 17 sectors per track, and 512 bytes per sector:

Total sectors = 4 heads × 615 cylinders × 17 sectors per track = 41,820

The total capacity of the hard disk is calculated by multiplying the total sectors by the capacity of each sector:

capacity = 41820 × 512 = 21,411,840 bytes = 21.4 M

Notice in the above calculation that meg is 1 million and not 2^{20} as is the case of RAM and ROM. Sometimes "meg" is used to mean 1,000,000 and sometimes it is used to mean 1,048,576 (2^{20}).

Example 19-7

Load sector 0 of the hard disk, which is disk 2. Dump it to the screen and analyze the boot record.

Solution:

```
C>DEBUG
-L CS:100 2 0 1
-D CS:100 13F
142B:0100   EB 3C 90 4D 53 44 4F 53-35 2E 30 00 02 10 01 00
142B:0110   02 00 02 00 00 F8 CB 00-3F 00 10 00 3F 00 00 00
142B:0120   31 B0 0C 00 80 00 29 EC-16 1B 26 4D 41 5A 49 44
```

The following is a brief analysis of the above boot record, analyzed by using Table 19-2.

Bytes 0 - 2	EB, a JMP instruction to the boot code
Bytes 3 - A	operating system: MS DOS 5.0
Bytes B - C	number of bytes per sector: 0200H or 512
Byte D	number of sectors per cluster: 10H=16 (always a power of 2)
Bytes E - F	number of reserved sectors: 0001
Byte 10	number of FAT tables: 02
Bytes 11-12	number of directory entries: 0200H or 512
Bytes 13-14	number of sectors on disk (for hard disk see bytes 20 - 23)
Bytes 15	media descriptor byte: F8 (see Table 19-3)
Bytes 16-17	number of FAT sectors: 00CB=203
Bytes 18-19	number of sectors per track: 003FH or 63
Bytes 1A-1B	number of heads: 0010H=16 heads
Bytes 20-23	number of sectors on hard disk: 0CB031H=831,537

the remaining bytes contain the boot code

The disk capacity is calculated by multiplying the number of sectors by the capacity of each sector:
capacity = 831,537 × 512 = 425,746,944 bytes. Dividing it by 1,048,576 gives 406.023 megabytes.

Partitioning

Partitioning the disk is the process of dividing the hard disk into many smaller disks. This is done more frequently on disks larger than 32 megabytes. For example, a given hard disk of 80 megabytes capacity can be partitioned into three smaller logical disks with the DOS program FDISK. They are called *logical disks* since it is the same physical disk, but as far as DOS is concerned, they will be labeled disks C, D, and E. In the above case of the 80 megabyte disk, disk C will have 32M, disk D 32M, and the remaining 16M are for disk E if the default partitioning mode is used. A hard disk can be divided into many logical disks of variable sizes with the names C, D, E, F, G, ..., Z with no disk larger than 32 megabytes. It must be stated that any disk of 32M and lower can be partitioned into many logical disks but the disks of more than 32M must be partitioned for DOS versions up to 3.3. It is only under DOS 4 that the 32M limit has been removed.

After the hard disk has been partitioned, high-level formatting should be performed next. High-level formatting in the hard disk achieves exactly the same function as formatting a floppy disk. The C drive must be formatted with the system option (FORMAT C: /S) so that the system can boot from drive C. The remaining disks D through Z are in nonbootable format. The reason is that DOS always checks drive A first, then drive C for the system boot. In the absence of a bootable disk it goes automatically to the next drive until it finds one, and if it does not find any, it will display an appropriate message.

Hard disk layout

Although theoretically there is no difference in the layout of hard and floppy disks in terms of the boot record, FAT, and directory, because the capacity of the hard disk is much larger, more sectors are assigned to each of the above functions. Below is a description of the hard disk layout.

Hard disk boot record

As mentioned before, the first sector of the disk is set aside for the boot record regardless of the type of disk. Example 19-7 shows the analysis of the boot record of a hard disk.

Hard disk FAT

As in the floppy disk, the sectors immediately after the boot sector (sector 0) in the hard disk are used by the FAT. In order to make sure that too many sectors are not taken by the FAT, as the size of the disk increases, DOS increases the size of a cluster, explained below.

Clusters

In the 80x86 IBM PC, the sector size is always 512 bytes but the size of the cluster varies among disks of various sizes. The cluster size is always a power of 2: 1, 2, 4, 8, and so on. Example 19-7 showed a disk with a cluster size of 16 sectors. The fact that a file of 1-byte size takes a minimum of 1 cluster is important and must be emphasized. This means that a number of small files on a disk with a large number of sectors per cluster will result in wasted space on the hard disk. Let's look at an example. In a hard disk with a cluster size of 16 sectors (16 x 512 = 8192 bytes), storing a file of 26,000 bytes requires 4 clusters. The result is a waste of 6768 bytes since 4 x 8192 = 32,768 bytes, and 32,768 − 26,000 bytes = 6768.

Hard disk directory

The number of entries in the root directory in hard disks is a maximum of 512. Therefore it is essential to organize a hard disk into subdirectories. Otherwise, the hard disk may have available space, but no available directory entries and you will not be able to store any more files.

Speed of the hard disk

One of the most important and widely cited hard disk performance factors is its speed, or how fast the requested data is available to the user. The hard disk access time is in the range of 10 - 80 ms and is still dropping. This access time is much longer than the speed of primary DRAM memory, which is in the range of 50 - 250 ns and lower, which is the reason that disk caching is used, as will be discussed later in this chapter. The access time of the hard disk given by manufacturers is broken down into several smaller times indicating the speed of different sections of the hard disk. The components of access time are seek time, settling time, and latency time. *Seek time* is the amount of time that the read/write head takes to find the desired cylinder or track. The outer tracks (the outer cylinder) obviously take less time to find since the head is parked on the outermost track and moves into inner tracks. Manufacturers always give the average seek time in the data sheet. To reduce seek time, mainframe computers use hard disks with several heads parked at different cylinders (tracks), which reduces seek time drastically but also increases the cost substantially since the read/write heads and associated circuitry are one of the most expensive components in manufacturing the hard disk. For microprocessor-based personal computers in recent years, higher speeds for seek time have been achieved by replacing the stepper motor with a voice coil as a means to step the head from one cylinder (track) to the next.

Settling time, the second factor in access time, is the time it takes the head to stop vibrating before it can begin reading the data. Some manufacturers include settling time when they give the average seek time.

Rotational latency is the time it takes for the head to locate on the specific sector. In other words, after the head is settled, the platter is rotating at a certain RPM (revolutions per minute) rate. Rotational latency time depends on the distance between the head and the desired sector, but in no case is it more than the time for one revolution. This means that rotational latency is directly proportional to the RPM of the hard disk. The RPM for various disks varies between 2400 and 3600 (and as high as 7200 in some recent ones). Again, the average rotational latency must be taken into consideration. For example, if a given disk has 3600 RPM, which is 60 rotations per second, this is 16.6 ms for each full revolution. Since the desired sector could be directly beneath the head or at the end of the track, the average latency due to rotation is 8.3 ms [(16.6 + 0) / 2 = 8.3].

Data encoding techniques in the hard disk

Binary data is recorded on the magnetically coated platter by various encoding techniques that have different data transfer rates. Current techniques include FM, MFM, and RLL. This section will describe each one and discuss its advantages and disadvantages.

FM (frequency modulation)

FM takes its name from the fact that encoding 1 or 0 results in different frequencies, as can be seen in Figure 19-4. In FM encoding, there is a minimum of one pulse for each digit, regardless of whether it is 0 or 1. In this method, at the beginning of the clock there is always a pulse. If the data to be encoded happens to be 0, there will not be any more pulses, but if the digit is 1, there is a pulse for that as well. As can be seen in Figure 19-4, there is a minimum of 8 pulses if all the data is 0s and 16 pulses if the data is all 1s. Figure 19-4 shows the pulses for encoding "11000101" with flux transitions using the FM technique.

MFM (modified frequency modulation)

MFM is a much more efficient encoding method, achieved with the modification of the FM technique by eliminating the automatic pulse for each digit. In MFM, the pulse for 1 is still there. For encoding 0, there is a pulse at the beginning of the period unless it was preceded by a 1, in which case there is no pulse. Figure 19-4 shows the encoding of the same data "11000101" and the flux transitions. The head uses the flux transition (change of north and south) to read (decode) or write (encode) the binary data 0 or 1. A comparison of the flux transition of FM and MFM techniques in Figure 19-4 shows that MFM will encode data 11000101 with 6 transitions, while FM requires a total of 12 flux transitions for the same data. Trying any combination of binary numbers leads to the conclusion that MFM encoding is always twice as efficient as FM. This is the reason that FM encoding, also known as *single density*, was replaced by MFM, which is also known as *double density*.

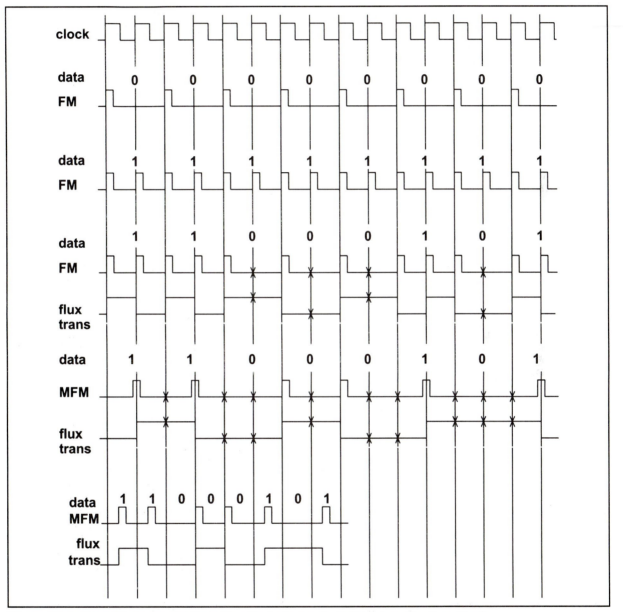

Figure 19-4. Time Diagrams for Encoding Techniques (x indicates no flux transition)

RLL (run length limit)

In the search for even more efficient encoding techniques, an elaborate scheme was developed called RLL (run length limit). In RLL encoding, the minimum and maximum number run lengths are even higher than in the MFM method. What is run length? The *run length* is the number of no-flux transitions between two consecutive transitions. Since it is easier to understand pulses and no pulses instead of flux transitions and no flux transitions, the former will be used to explain RLL encoding. First, a look at the minimum and maximum run length limits for FM and MFM is called for. To calculate the minimum and maximum run length limit for FM encoding involves simply counting the number of no pulses (0s) between two consecutive pulses (1s), as illustrated in Figure 19-4. These are marked by x. As can be seen, there is a minimum of zero no-pulses and a maximum of 1 no-pulses between two consecutive pulses. Similarly, it can be shown that the minimum and maximum number of no-pulses between consecutive pulses for MFM is 1 and 3. Trying any byte of data will show that the RLL factor for MFM is 1,3. As shown in Figure 19-4, the no-pulse is the same as for no-flux transition. The fewer the number of flux transitions, the better for the read/write head. In other words, if there were a scheme to increase the number of no-pulses (no flux transition) between the pulses, much better results would be achieved in the encoding. This is exactly what the RLL 2,7 has done. In RLL 2,7 encoding, there can be a minimum of 2 and a maximum of 7 no-pulses in between the consecutive pulses, no matter what byte of data is encoded. The method that is used is that every 2 to 4 bits of data is replaced by a 4- to 8-bit code according to Table 19-7. For example, using the above code results in the following coding for data 00111010:

Table 19-7: RLL Encoding

Data	RLL 2,7 Code
000	000100
10	0100
010	100100
0010	00100100
11	1000
011	001000
0011	00001000

```
data to be encoded      0011      10    10
RLL 2,7 code            00001000 0100  0100
```

Counting the 0s, there is a minimum of 3 and a maximum of 4 0s in between two consecutive 1s in the RLL 2,7 code for data "00111010". No matter what byte of data one encodes according to RLL 2,7, the result will never have less than 2 and more than 7 0s between two consecutive 1s. For example, the run length limit of the data "11001111", shown below, is 3,7.

```
data to be encoded      11    0011       11
RLL 2,7 code            1000  00001000  1000
```

In recent years there has been another RLL encoding scheme, called RLL 3,9, which is commonly referred to as ERLL (enhanced RLL) or ARLL (advanced RLL), which achieves even higher density data encoding than RLL 2,7.

Interfacing standards in the hard disk

To ensure that hard disks made by different manufacturers are compatible, common standards for interfacing the hard disk and personal computers have been devised. These standards are ST506, ESDI, and SCSI, which are explained below.

ST506 and ST506-412

ST506 is one of the oldest and most widely used interfacing standards for the IBM PC. It was developed by disk manufacturer Seagate Technology. The ST506 standard uses the disk controller to read data serially from the disk, pack it into an 8-bit data item, and deliver it to the buses of the PC. In other words, although it provides the data in parallel form through the 20-pin cable to the computer expansion slot, it still transfers data from the hard disk surface to the buffer of the controller serially. This is called *internal transfer rate*. For this reason there are various ratings for the data transfer, depending on the encoding used. The maximum data transfer rate for the ST506 is 5 megabits per second with MFM encoding. FM encoding is rarely used today. If the ST506 uses RLL, the rate of data transfer increases to 7.5 Mbit per second. An upgrade of the ST506, referred to as ST506-412 or simply ST412, provides an enhancement in the way that the controller makes the head move from track to track. In ST506, the head steps through the tracks one track at a time in order to reach the desired track, but in the ST506-412 with a feature called *seek buffering*, the controller moves the head between the nonadjacent tracks with a single and continuous move rather than one move at a time. This results in a shorter access time. From now on instead of ST506-412, this book will use ST412. The external transfer rate assumes packing of the serial bits into 8-bit data packages and is calculated by dividing the internal rate by 8. See Table 19-8.

Table 19-8: Hard Disk Specifications

Model	ST225	ST1111E	ST3600A	ST3491A	ST3600N /ND	ST3610 N/ND/NC
Drive capacity						
Unformatted megabytes	25.6	111.9	600	na	600	635
Formatted megabytes (512 bytes/sector)	21.4	98.9	528	428	525	535
Interface	ST412	ESDI	IDE	Fast IDE	Fast SCSI-2	Fast SCSI-2
Performance						
Internal transfer rate (Mbits/sec)	5.0	10	19 - 36	32	19 - 35	25 - 41
External transfer rate (Mbytes/sec)	0.625	1.25	8	up to 13.3	10 Sync	10 Sync
Track-to-track (ms)	20	4	1.5/2	5	1.5/2	1.5/2
Average seek (ms)	65	15	10.5/12	14	10.5/12	10.5/12
Maximum seek (ms)	150	35	26	34	26	26
Spindle speed (RPM)	3,600	3,600	4,500	3,811	4,500	5.400
Average latency (ms)	8.3	8.33	6.67	7.87	6.67	5.56
Drive configuration/data organization						
Discs	2	3	4	na	4	4
Read/write heads	4	5	7	16	7	7
Sectors per track	17	36	na	51	na	na
Sectors per drive	41,820	192,960	na	835,584	na	na
Bytes per sector	512	512	512	512	512	512
Cylinders	615	1,072	1,874	1,024	1,872	1,872
Recording method	MFM	RLL (2,7)	RLL (1,7)	RLL (1,7)	RLL (1,7)	RLL (1,7)
Reliability						
MTBF (POH)	100,000	150,000	200,000	300,000	200,000	200,000
Auto-park	No	Yes	Yes	Yes	Yes	Yes
Cache (Kbytes)	None	None	256	120	256	256

(Reprinted by permission of Seagate Corporation, Copyright Seagate Corp. 1991, 1994)

ESDI (enhanced small device interface)

The ESDI standard was developed by a group of disk drive manufacturers in 1983. There are some differences between ESDI and ST412:
1. ESDI can achieve a data transfer rate of up to 20 Mbits per second in contrast to the 7.5 Mbits/second of the ST412.
2. With the same RPM as ST412, it can have more sectors per track. The number of sectors for ESDI can vary between the 20s and the 50s.
3. While in ST412 the defect information must be provided manually during low-level formatting, for ESDI the defect map is already stored on the drive.
4. In the ST412 standard, the number of cylinders, heads, and sectors is stored either in CMOS RAM of the system or in the ROM of the hard disk controller, in contrast to ESDI, where the configuration information is already provided and there is no need to store it externally.

IDE (integrated device electronics)

IDE is the standard for current PCs. In IDE, the controller is part of the hard disk. In other words, there is no longer a need to buy a hard disk and a separate controller as is often the case for ST412. One of the reasons that the IDE drives have a better data transfer rate is the integration of many of the controller's functions into the drive itself with the use of VLSI chips. For example, in the ST412 standard the hard disk read/write heads would read the data and transfer it to the controller through the cable, and then the data is separated from the clock pulses by what is called data separator circuitry. By eliminating cable degradation, IDE and SCSI (discussed next) reach a much higher external data transfer rate.

The major limitation of IDE is its capacity. It is limited to 504 MB, due to a PC BIOS limitation. Since IDE uses PC BIOS, which supports only 1024 cyclinders, 16 heads, and 63 sectors per track, the hard disk capacity is limited to 504 MB ($1024 \times 16 \times 63 \times 512$ bytes/sector = 504 MB). This is not the case with SCSI drives since they bypass PC BIOS and their capacity can go as high as 8G.

SCSI (small computer system interface)

SCSI (pronounced "scuzzy") is one of the most widely used interface standards not only for high-performance IBMs and compatibles but also for non-80x86 computers by other manufacturers, such as Apple and Sun Micro. The main reason is that unlike IDE, SCSI is the standard for all kinds of peripheral devices, not just hard disks. One can daisy chain up to 7 devices, such as CD-ROM, optical disk, tape drive, floppy disk drive, networks, and other I/O devices, using the SCSI standard. See Figure 19-5.

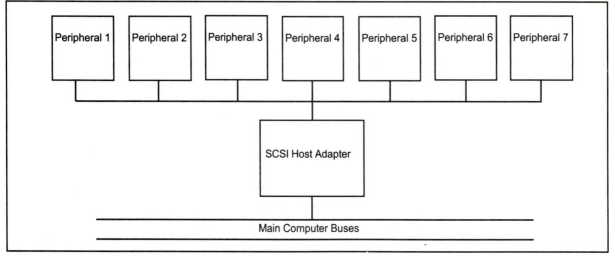

Figure 19-5. Peripheral Devices in SCSI "Daisy Chain"

All the characteristics discussed for ESDI and IDE apply equally to SCSI. In addition, SCSI can have an internal data transfer rate of up to 80 Mbits/second. It must be noted that SCSI hard drives always have the controllers embedded into them and there is no need for a separate controller. The only thing needed is an adapter to convert the SCSI signals to signals compatible with the bus expansion slot of the host computers.

Interleaving

As the read/write head moves along the track, it must read each sector and pass it to the controller. The controller in turn will deliver this data to the host computer through the buses. If the head and the controller cannot keep up with the stream of data passing under the head, there are two choices: either the rotation should be slower or interleaving should be used. Using a slower rotation, for example 600 RPM instead of 3600, will give an unacceptably long access time. That brings us to *interleaving*. While common sense tell us that the sectors should be numbered on each track sequentially, since the head and controller cannot process the data in sector 1 in time to be ready for sector 2 by the time it is under the head, it would have to wait for the next rotation to read sector 2. Likewise, while sector 2 is being processed, sector 3 has already passed under the head and to read that it must wait for the next revolution. This means to read all 17 sectors of each track it will take 17 revolutions. This is 1:1 interleaving and is as bad as slowing down the RPM. In 2:1 interleaving the sectors are numbered and accessed alternately. If the controller is not fast enough, 3:1 can be used. In 3:1 interleaving, every third sector is numbered and accessed. It will take 2 complete revolutions to access all the sectors in 2:1, and 3 revolutions for 3:1 interleaving. This is much better than all other choices discussed above. In 3:1 interleaving, used widely in PC/XT computers, the computer accesses sector 1, and by the time it finishes processing it, sector 2 is under the head. The two sectors in between sector numbers 1 and 2 give the controller time to get ready for accessing the next sector. Note that in today's high-performance computers using IDE and SCSI controllers, 1:1 interleaving can read the entire track with one revolution, due to their fast controllers and wide data buses. Figure 19-6 shows the concept of interleaving.

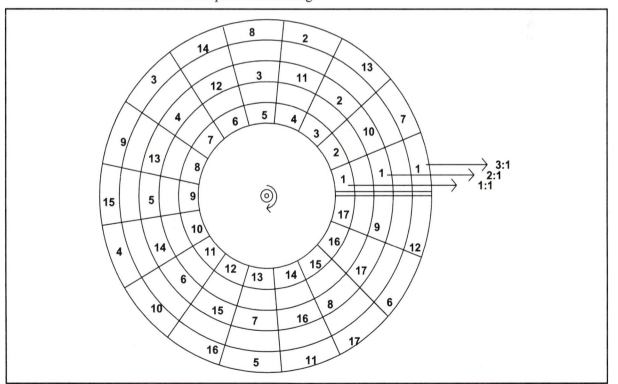

Figure 19-6. Hard Disk Interleaving

Low- and high-level formatting

During low-level formatting, every sector of the disk is examined and bad sectors are marked. Every sector will be given what is called an ID header. Among the information included in the ID header are sector number, cylinder number, and head. This information is never changed unless low-level formatting is done again. Low-level formatting is performed by the hard disk manufacturer. High-level formatting sets aside sectors for the FAT and the root directory. High-level formatting is done by the FORMAT command of DOS.

Parking the head

When the computer is moved, there is the possibility that the magnetic materials beneath the read/write head can be damaged. For this reason the head should be moved (parked) to that disk's landing zone before the system is moved. Even turning the computer on and off can cause incremental damage by electric shocks, which can accumulate and eventually make a disk dysfunctional. Parking the head is strongly recommended before the computer is turned off. Many controllers park the head automatically (see Table 19-8), but users of the others may use a utility program provided by the disk manufacturer to park the head.

Disk caching

Due to the long access time of the hard disk, disk caching is used to speed up the disk access time. There are two types of disk caching. In one type, the disk manufacturer puts some (64K - 1M) fast memory on the disk. This is called hardware disk cache (see Table 19-8). In the other type, a section of memory on the PC motherboard is set aside for disk caching. Obviously, the larger the size of this memory, the more files can be stored there and accessed by the CPU, assuming that there is extra memory to spare. Using a section of motherboard DRAM for disk caching is done by SMARTDRV.SYS, which comes with MS DOS. This kind of disk caching is called software disk cache.

Disk reliability

MTBF (mean time between failures) is a measure of reliability and durability of the disk when the power is on. This factor is given in hours. For example, the ST225 has a MTBF of 100,000 hours. Dividing it by 24 hours gives an MTBF value of 4166.6 days or 11.4 years (4116.6/365). Of course, manufacturers will not power on the disk for that long and then test it since they would be out of business by then. They use extremely reliable statistical analysis to figure out the MTBF. See the MTBF of various disks in Table 19-8.

Review Questions

1. Why in hard disks is the number of heads more than two?
2. True or false. Early DOS versions did not allow the hard disk to be partitioned to sizes bigger than 32 megabytes.
3. In hard disks, which sector is set aside for the boot record?
4. True or false. FM recording density is twice that of MFM.
5. True or false. SCSI interfacing standard is also used for devices other than hard disks, such as CD ROM.
6. How many rotations does it take to read all the sectors of a given track in a 3:1 interleaved hard disk?
7. True or false. Each file begins on a new cluster even if the previous cluster has some empty space.
8. How many bytes are used for each directory entry?
9. How many sectors are in a cluster?
10. What does "MTBF" stand for, and what does it measure?

SECTION 19.3: DISK FILE PROGRAMMING

In this section we first explain the concept of the file handle and then show how to perform file processing using INT 21H.

File handle and error code

For DOS version 1.0, in order to access a file programmers had to use what is called the FCB (file control block). However, starting with DOS version 2, the file handle was introduced. Since the use of file handle is the standard and recommended way of accessing files in recent applications, we will consider this method only.

When a file is created, DOS assigns it a 16-bit number called a *file handle*. From then on, any time that file is accessed it is with the help of the file handle.

The following two steps must be taken in order to create and access a file.

1. Use an ASCIIZ string for drive, path, and the file name. This string must end with 00, which is called ASCII zero.
2. Use INT 21H option 3CH to create the file. If the file creation is successful, DOS provides the file handle for that file and sets the carry flag to 0. This file handle must be saved and used in all subsequent accesses of this file. If the file creation is not successful, DOS provides an error code and sets the carry flag to 1. The following describes DOS 21H functions to create a file, write to and read from a file, and close a file. The remaining services provided by INT 21H are shown in Appendix D. Table D-1 provides a list of error codes.

INT 21H, AH =3CH (creating a file)

Prior to this function, AH = 3CH, DX points to the ASCIIZ string containing the filename, and CX contains the attribute byte. See Table 19-9. Note that if you use the option with an existing filename, it will write over the old information in the file, essentially deleting it. If the function was able to successfully create a file, the carry flag is set to zero, the file handle is placed in AX, and a directory entry is created for the file. All subsequent references to the file are done through the file handle. If any errors occurred, CF will be set to 1 and the error code will be in AX. See Table D-1 for a list of error codes.

Table 19-9: File Attribute Byte

Attribute	Meaning
00H	Normal file
01H	Read-only file
02H	Hidden file
04H	DOS system file
08H	Volume label (11 bytes)
10H	Subdirectory
20H	Archive file

```
MOV    AH,3CH              ;create file option
MOV    CX,0                ;normal file
MOV    DX,OFFSET FNAME_A   ;DX points to ASCIIZ string
INT    21H                 ;invoke interrupt
JC     ERROR               ;go display error message
MOV    HANDLE_A,AX         ;save file handle
```

INT 21H, AH =40H (writing to a file)

Prior to invoking this function, AH = 40H, CX = the number of bytes to write, BX contains the file handle, DX points to the beginning of the offset address of the data to be written. If the operation was successful, CF = 0 and AX = the number of bytes written, and if not successful CF = 1 and AX = the error code.

```
OUTBUF    DB          'Hello'
          MOV         AH, 40H              ;write to file option
          MOV         CX,5                 ;write 5 bytes
          MOV         BX,HANDLE_A          ;BX contains handle
          MOV         DX,OFFSET OUTBUF     ;DX points to data to write
          INT         21H                  ;invoke interrupt
```

INT 21H, AH =3EH (closing a file)

Before your program terminates, your should close all files. This will ensure that any records you have written to the file that may still be in the memory buffer will be added to the file. Closing the file will also update the directory entry and FAT entries for that file. Before the call, BX contains the file handle and AH = 3EH.

```
          MOV         AH,3EH               ;close file option
          MOV         BX,HANDLE_A          ;set up handle
          INT         21H                  ;invoke interrupt
```

INT 21H, AH =3FH (reading a file)

Prior to this function call, AH = 3FH, BX contains the file handle, CX = the number of bytes to read, and DX points to the input buffer (offset address). If the operation was successful, CF = 0 and AX = the number of bytes read. If an error occurred, CF = 1 and AX holds the error code.

```
          MOV         AH,3FH               ;read file option
          MOV         CX,80                ;read 80 bytes
          MOV         BX,HANDLE_A          ;BX holds the handle
          MOV         DX,OFFSET INBUF      ;DX points to input buffer
          INT         21H                  ;invoke interrupt
```

INT 21H, AH =3DH (open a file)

Prior to this function call, AL =mode (see list below) and DX points to the ASCIIZ string. If the operation was successful, CF = 0 and AX = the file handle. If an error occurred, CF = 1 and AX holds the error code.

```
AL mode:   76543210 (bits)    Result
                  000          open for read
                  001          open for write
                  010          open for read/write
                0              reserved
             000               give others compatible access
             001               read/write access denied to others
             010               write access denied to others
             011               read access denied to others
             100               give full access to others
           0                   file inherited by child process
           1                   file private to current process


          MOV         AH,3DH               ;open file option
          MOV         AL,0                 ;open for read
          MOV         DX,OFFSET FNAME_A    ;DX points to ASCIIZ string
          INT         21H                  ;invoke interrupt
```

```
TITLE       FILE I/O
PAGE        60,132
            .MODEL   SMALL
            .CODE
;--------------- MAIN procedure -----------------------------
;this program clears screen, reads keyboard input into buffer, writes buffer to file, closes file
;then opens file, reads file to buffer, writes buffer to monitor, closes file, exits when user is ready
MAIN        PROC     FAR
            MOV      AX,@DATA
            MOV      DS,AX
            CALL     CLR_SCR            ;clear the screen
            MOV      AH,02             ;set cursor position
            MOV      BH,00
            MOV      DL,10             ;column position
            MOV      DH,10             ;row position
            INT      10H
            CALL     CREATE_F          ;create an output file
            MOV      AH,0AH            ;read keyboard input
            MOV      DX,OFFSET BUFFER_1
            INT      21H
            CALL     WRITE_F           ;write buffer to file
            CALL     CLOSE_F           ;close file
            MOV      AH,02             ;set cursor position
            MOV      BH,00
            MOV      DL,10             ;column position
            MOV      DH,20             ;row position
            INT      10H
            CALL     OPEN_F            ;open file for read
            CALL     READ_F            ;read file to buffer
            MOV      AH,09             ;copy buffer to screen
            MOV      DX,OFFSET BUFFER_2
            INT      21H
            CALL     CLOSE_F           ;close file
            MOV      AH,02             ;set cursor position
            MOV      BH,00
            MOV      DL,10             ;column position
            MOV      DH,24             ;row position
            INT      10H
            MOV      DX,OFFSET EXIT_MSG  ;print "ANY KEY TO EXIT" message
            CALL     ERR_MSG
            MOV      AH,07             ;exit when user hits any key
            INT      21H
            MOV      AH,4CH            ;set up to return
            INT      21H               ;return to DOS
MAIN        ENDP
;--------------- this procedure creates an output file -----------------------
CREATE_F    PROC     NEAR
            MOV      AH,3CH            ;create file function
            MOV      CX,0              ;normal file
            MOV      DX,OFFSET FILE_1  ;DX points to ASCIIZ
            INT      21H               ;invoke interrupt
            JC       C_ERR
            MOV      HANDLE_F1,AX      ;save handle if OK
            JMP      C_EXIT
C_ERR:      MOV      DX,OFFSET ERR_CREATE  ;display error message
            CALL     ERR_MSG
C_EXIT:     RET
CREATE_F    ENDP
;--------------- this procedure reads file to buffer -----------------------
READ_F      PROC     NEAR
            MOV      AH,3FH            ;read from file
            MOV      BX,HANDLE_F1      ;use handle
            MOV      CX,25             ;number of bytes to read
            MOV      DX,OFFSET BUFFER_2 ;DX points to buffer
            INT      21H               ;invoke interrupt
            JNC      R_EXIT
            MOV      DX,OFFSET ERR_READ  ;display error message
            CALL     ERR_MSG
R_EXIT:     RET
READ_F      ENDP
```

Program 19-1. Using INT 21H File Handling *(continued on next page)*

```
;-------------- this procedure writes buffer to file -----------------------
WRITE_F    PROC    NEAR
           MOV     AH,40H                              ;write to file
           MOV     BX,HANDLE_F1
           MOV     CX,25                               ;number of bytes to write
           MOV     DX,OFFSET BUFFER_1+2                ;DX points to buffer
           INT     21H                                 ;invoke interrupt
           JNC     W_EXIT
           MOV     DX,OFFSET ERR_WRITE                 ;display error message
           CALL    ERR_MSG
W_EXIT:    RET
WRITE_F    ENDP
;-------------- this procedure closes a  file -----------------------
CLOSE_F    PROC    NEAR
           MOV     AH,3EH                              ;close file function
           MOV     BX,HANDLE_F1                        ;use handle
           INT     21H                                 ;invoke interrupt
           JNC     CL_EXIT
           MOV     DX,OFFSET ERR_CLOSE                 ;display error message
           CALL    ERR_MSG
CL_EXIT:   RET
CLOSE_F    ENDP
;-------------- this procedure opens a file -----------------------
OPEN_F     PROC    NEAR
           MOV     AH,3DH                              ;open file function
           MOV     AL,0                                ;read only
           MOV     DX,OFFSET FILE_1                    ;DX points to ASCIIZ
           INT     21H                                 ;invoke interrupt
           JC      O_ERR
           MOV     HANDLE_F1,AX                        ;save handle if OK
           JMP     O_EXIT
O_ERR:     MOV     DX,OFFSET ERR_OPEN                  ;display error message
           CALL    ERR_MSG
O_EXIT:    RET
OPEN_F     ENDP
;-------------- this procedure displays a message ---------------
ERR_MSG    PROC    NEAR                                ;DX points to msg before call
           MOV     AH,09H                              ;output to monitor
           INT     21H                                 ;invoke interrupt
           RET
ERR_MSG    ENDP
;-------------- this procedure clears the screen --------------------
CLR_SCR    PROC    NEAR
           MOV     AX,0600H                            ;scroll screen
           MOV     BH,07
           MOV     CX,0000                             ;scroll the
           MOV     DX,184FH                            ;  entire screen
           INT     10H
           RET
CLR_SCR    ENDP
;-------------- data area  ------------------------------------------
                   .DATA
HANDLE_F1      DW 0                          ;file1 handle
FILE_1         DB  'C:\FILE1.ASC',0          ;file1 ASCIIZ string
BUFFER_1       DB  25,?,25 DUP (' ')
BUFFER_2       DB  25 DUP (' '),'$'
ERR_CREATE     DB  0DH,0AH,'** Error creating file **$'
ERR_OPEN       DB  0DH,0AH,'** Error opening file **$'
ERR_READ       DB  0DH,0AH,'** Error reading file **$'
ERR_WRITE      DB  0DH,0AH,'** Error writing to file **$'
ERR_CLOSE      DB  0DH,0AH,'** Error closing file **$'
EXIT_MSG       DB  0DH,0AH,'PRESS ANY KEY TO EXIT$'

               .STACK   32
               END  MAIN
```

Program 19-1 *(continued from preceding page)*

1. True or false. FCB and file handles can be used interchangeably with the same DOS 21H function calls.
2. What happens if you use function 3CH to create a file, but the file already exists?
3. What might happen if your program terminates before closing output files?
4. What is wrong with the following ASCIIZ string?
   PATH_NAME   DB   'C:\PROGRAMS\TEST.DAT'
5. What is wrong with the following file handle?
   FILE_HNDL   DB   ?

SUMMARY

This chapter began with a look at floppy disk organization, a secondary storage medium. *Primary storage* refers to RAM and *secondary storage* refers to storage media other than RAM. Information is stored on floppy disks in tracks. There are many tracks on each side of a diskette. Each track consists of many clusters, which consists of one or more sectors, which consists of many bytes of information. Important information about the disk format is kept in the boot record of the disk. The directory lists the files on a disk and keeps track of important information about the files. The FAT (file allocation table) is like a road map telling the computer which files are stored in which sectors.

The second section looked at hard disks. Hard or fixed disks are judged according to three major criteria: capacity, access time (speed of accessing data), and interfacing standard. The capacity of a hard disk is found by multiplying the number of tracks times the number of sectors times the number of bytes per sector. In hard disk organization, the term *cylinder* refers to all the tracks of the same radius on each platter. Hard disks can be partitioned into many smaller logical disks. The access time of the disk has several components: *seek time*, the amount of time it takes the head to find the cylinder; *settling time,* the amount of time it takes the head to stop vibrating so it can read the data; and *rotational latency*, the time it takes for the head to locate the specific sector. Encoding techniques are the means by which logical 0 and 1 are translated into pulses and stored on disk. Common encoding techniques are FM, MFM, and RLL. The two major interfacing standards for interfacing hard disks and the PC are IDE and SCSI. Interleaving is used to access consecutive sectors faster. Low-level formatting, usually performed by the manufacturer, checks each sector and marks bad sectors. High-level formatting, done by the user using the DOS FORMAT command, sets aside sectors for the FAT and boot record.

The third section of this chapter showed how to access files on disk with Assembly language instructions. INT 21H function calls are used to open and close, read and write, create and delete files, among other functions.

PROBLEMS

SECTION 19.1: FLOPPY DISK ORGANIZATION

1. Why is the disk called secondary storage?
2. Verify the disk capacity of each of the following.
   (a) 3 1/2"   (b) 3 1/2" high density
3. The first sector of every floppy disk is set aside for _____.
4. True or false. In accessing sectors, the disk controller goes to side 0, track 0, sectors 0 to n, then side 2, track 0, sector 0 to sector n, and so on.
5. In the floppy disk of the PC, the cluster size is _____ sectors.

6. After the boot sector, sectors are assigned to the _____(FAT, directory).
7. Select a floppy disk of your choice, then using DEBUG, dump the boot sector and analyze the contents. Specifically, show the sector density, number of sectors per track, total number of sectors in the disk, disk capacity, and maximum number of directories per disk.
8. How many bytes are set aside for each directory entry?
9. If the sector size is 512 bytes, how many directories can each sector hold?
10. Using DEBUG, dump a sector containing directory entries of a 3 1/2" floppy disk and analyze the entries for 2 of the files.
11. In the date section of a directory entry, the year is stored as the number of years since _____.
12. For a disk to be bootable, which file must it contain? Show for both MS DOS and PC DOS (IBM version).
13. Why doesn't the DIR command list MSDOS.SYS and IO.SYS on the screen?
14. In PC DOS and MS DOS, how many bits are set aside for each FAT entry?
15. Indicate the meaning of the following FAT entries.
    (a) 001   (b) 000   (c) FF8
16. Use Table 19-5 to verify the sector map of a 1.2M disk. Check your result against Table 19-6.
17. Repeat Problem 16 for a 1.44M disk.
18. What is maximum number of files that a 720K disk can have?
19. Repeat Problem 18 for a 1.44M disk.
20. A given floppy disk has thousands of bytes of free space, but the user cannot open a new file. What do you think is the problem?

SECTION 19.2: HARD DISKS

21. True or false. The terms *hard disk* and *fixed disk* refer to the same thing.
22. What is a cylinder, and how is used in the hard disk?
23. The total number of tracks in a given hard disk is equal to _____.
24. Calculate the total number of sectors and the capacity of the following hard disks. Assume 512 bytes per sector for both cases.
    (a) 6 heads, 820 cylinders, and 26 sectors per track
    (b) 5 heads, 1072 cylinders, and 36 sectors per track
25. Using DEBUG, dump the boot sector of your hard disk and analyze it, part by part. Specifically, show the sector density, number of sectors per track, total number of sectors in the disk, disk capacity, maximum number of directories per disk, and the number of sectors per cluster.
26. True or false. The number of sectors per cluster is always a power of 2.
27. For Problem 25, show the hard disk layout. Identify the sectors, by hex number, that are used for the FAT, directory, and data storage.
28. With DOS 4 and higher, a logical disk can have _____ bytes of storage.
29. Discuss seek time, settling time, latency time, and how they relate to disk access time.
30. Which of the following has the shortest access time? Which has the longest access time?
    (a) 2400 RPM   (b) 3600 RPM   (c) 4800 RPM
31. True or false. The SCSI interfacing standard is used only for hard disks.
32. State the number of peripheral devices that can be daisy chained if SCSI is used.
33. To read all the sectors of a given track, how many times must the track rotate under the head for each of the following interleaving factors?
    (a) 1:1   (b) 1:3   (c) 1:5
34. What does MTBF stand for, and what is its use?

SECTION 19.3: DISK FILE PROGRAMMING

35. What is ASCIIZ?
36. If there are tens of files open at a given time, how does DOS distinguish among them?

37. Indicate the type of file associated with each of the following attributes.
    (a) 00   (b) 02   (c) 04
38. When INT 21H DOS function call AH =3CH is used to create a file, how does DOS indicate that a file has been created successfully?
39. True or false. In using DOS function call INT 21H to access an already created file, one must have the file handle.
40. True or false. When using INT 21H DOS function call to write into a file, one must define a memory data area in the data segment.
41. At what point are we certain that the data is written into file and the directory is updated?
    (a) when the file is created   (b) when the file is written to
    (c) when the file is closed
42. Using Assembly language and INT 21H, create a file, name it "MYFILE", and write to it the following: "In the world of microprocessors and personal computers, the mother of all battles looms ahead when the 80x86 clone makers and IBM/Motorola Power PC RISC battle Intel." After the file is closed, use the TYPE command to dump it on screen. Verify the size of the file by using the DIR command.

ANSWERS TO REVIEW QUESTIONS

SECTION 19.1: FLOPPY DISK ORGANIZATION

1. true
2. true
3. boot record
4. FAT
5. To make it possible to use the second for backup in case something happened to the first one, since the FAT is the road map for finding where data is located on the disk.
6. true
7. number of sectors
8. This information is provide in the boot sector.
9. after
10. by the media byte in the boot sector

SECTION 19.2: HARD DISKS

1. because there is more than one platter in hard disks and each platter has two heads
2. true
3. sector 0
4. false
5. true
6. 3 rotations
7. true
8. 32 bytes
9. 1, 2, 4, or 8; it is always a power of 2.
10. Mean time between failures; it is a measure of disk reliability.

SECTION 19.3: DISK FILE PROGRAMMING

1. false
2. The old information in the file is effectively deleted.
3. Closing a file writes any information from the buffer to the file, so if you terminate before closing, the information remaining in the buffer may be lost.
4. It is not an ASCIIZ string because it does not end in ASCII zero (null).
5. A file handle should be 16 bits (use DW directive).

CHAPTER 20

THE 80x87 MATH COPROCESSOR

OBJECTIVES

Upon completion of this chapter, you will be able to:

» **Diagram the bit assignment of floating-point IEEE standards for single- and double-precision data**
» **Convert data from real numbers to IEEE floating-point format**
» **Diagram the bit assignment for 8087 data types: word integer, short integer, long integer, packed decimal, short real, long real, and temporary real**
» **Code Assembly language data directives for 8087 data types**
» **List the registers of the 80x87**
» **Contrast and compare the use of registers in 8087 versus 8086 programming**
» **Write Assembly language programs using 8087 instructions**
» **Describe how the 8087 coprocessor works with the main CPU**

This chapter will examine the 80x87 math coprocessor. The math coprocessors found in 486 and Pentium processors have their origin in the 8087 coprocessor. As far as data types and instructions are concerned, there have been few changes since the introduction of the 8087, aside from the fact that 387, 487, and Pentium processors run 8087 instructions much faster. In the first section of this chapter, we study the IEEE standard for floating-point numbers and the Intel 8087 math coprocessor's data format. In Section 20.2, the 8087 instructions are discussed along with some sample programs run on a PC with a math processor. Section 20.3 covers interfacing of the 8087 to 8088/86-based PCs, and in Section 20.4 we provide an overview of 80x87 instructions and their clock counts.

SECTION 20.1: MATH COPROCESSOR AND IEEE FLOATING-POINT

Using a general-purpose microprocessor such as the 8088/86 to perform mathematical functions such as log, sine, and others is very time consuming, not only for the CPU but also for programmers writing such programs. In the absence of a math coprocessor, programmers must write subroutines using 8088/86 instructions for mathematical functions. Although some of these subroutines are already written and can be purchased at a small cost, no matter how good the subroutine, its CPU run time (8088/86, 286, 386) will still be quite long. Table 20-1 provides a comparison of the number of clocks used by the 8087 and 8086 to perform some mathematical functions. One can appreciate the advantage of having a coprocessor by comparing the run time of some programs, such as SPICE (a package for circuit analysis which uses floating-point operations extensively) on a computer with a coprocessor and one without a coprocessor. In some cases the difference is hours.

Table 20-1: Comparison of 8087 and 8086 Clock Times

Instruction	Approximate Execution Time (μs) (5-MHz clock)	
	8087	**8086 Emulation**
Multiply (single precision)	19	1,600
Multiply (double precision)	27	2,100
Add	17	1,600
Divide (single precision)	39	3,200
Compare	9	1,300
Load (single precision)	9	1,700
Store (single precision)	18	1,200
Square root	36	19,600
Tangent	90	13,000
Exponentiation	100	17,100

(Reprinted by permission of Intel Corporation, Copyright Intel Corp. 1989)

IEEE floating-point standard

Up to the late 1970s, real numbers (numbers with decimal points) were represented differently in binary form by different computer manufacturers. This made many programs incompatible for different machines. In 1980, an IEEE committee standardized the floating-point data representation of real numbers. This standard, much of which was contributed by Intel, based on the 8087 math coprocessor, recognized the need for different degrees of precision by different applications; therefore, it established *single precision* and *double precision*. Since almost all software and hardware companies, including IBM, Intel, and Microsoft, now abide by these standards, each one is explained thoroughly. RISC processors also use IEEE floating-point standards.

IEEE single-precision floating-point numbers

IEEE single-precision floating-point numbers use only 32 bits of data to represent any real number in the range 2^{128} to 2^{-126}, for both positive and negative numbers. This translates approximately to a range of 1.2×10^{-38} to $3.4 \times 10^{+38}$ in decimal numbers, again for both positive and negative values. In Intel coprocessor terminology, these single-precision 32-bit floating-point numbers are referred to as *short real*. Assignment of the 32 bits in the single-precision format is

Bit	Assignment
31	Sign bit: 0 for positive (+) and 1 for negative (-)
23 - 30	Biased exponent
22 - 0	The fraction, also called significand

To make the hardware design of the math processors much easier and less transistor consuming, the exponent part is added to a constant of 7FH (127 decimal). This is referred to as a *biased exponent*. Conversion from real to floating point involves the following steps.

1. The real number is converted to its binary form.
2. The binary number is represented in scientific form: 1.xxxx E yyyy
3. Bit 31 is either 0 for positive or 1 for negative.
4. The exponent portion, yyyy, is added to 7F to get the biased exponent, which is placed in bits 23 to 30.
5. The significand, xxxx, is placed in bits 22 to 0.

Examples 20-1, 20-2, and 20-3 demonstrate this process. In Section 20.2 we will verify all the above examples using an assembler.

IEEE double-precision floating-point numbers

Double-precision FP (called *long real* by Intel) can represent numbers in the range 2.3×10^{-308} to 1.7×10^{308}, both positive and negative. A total of 53 bits (bits 0 to 52) are for the significand, 11 bits (bit 53 to 62) are for the exponent, and finally, bit 63 is for the sign. The conversion process is the same as for single precision in that the real number must first be represented as 1.xxxxxxx E YYYY, then YYYY is added to 3FF to get the biased exponent.

Example 20-1

Convert 9.75_{10} to single-precision (short real) floating point.

Solution:
decimal 9.75 = binary 1001.11 = scientific binary 1.00111 E 3
sign bit 31 is 0 for positive
exponent bits 30 to 23 are 1000 0010 (3 + 7F = 82H) after biasing
significand bits 22 to 0 are 00111000000000000000 ...00
Putting it all together gives the following binary form, under which is written the hex form:

```
0100 0001 0001 1100 0000 0000 0000 0000
   4    1    1    C    0    0    0    0
```

This can be verified by using an assembler, such as MASM, as will be seen later in this chapter.

Example 20-2

Convert 0.078125_{10} to short real FP (single precision).

Solution:

decimal 0.078125 = binary 0.000101 = scientific binary 1.01 E -4
sign bit 31 is 0 for positive
exponent bits 30 - 23 are 0111 1011 (-4 + 7F = 7B) after biasing
significand bits 22 - 0 are 01000000....000
This number will be represented in binary and hex as

```
0011 1101 1010 0000 0000 0000 0000 0000
  3    D    A    0    0    0    0    0
```

Example 20-3

Convert -96.27_{10} to single-precision FP format.

Solution:

decimal 96.27 = binary 1100000.01000101000111101 =
scientific binary 1.10000001000101000111101 E 6
sign bit 31 is 1 for negative
exponent bits 30 - 23 are 1000 0101 (6 + 7F = 85H) after biasing
fraction bits 22 - 0 are 10000001000101000111101
The final form in binary and hex is

```
1100 0010 1100 0000 1000 1010 0011 1101
  C    2    C    0    8    A    3    D
```

It must be noted that conversion of the decimal portion 0.27 to binary can be continued beyond the point shown above, but because the fraction part of the single precision is limited only to 23 bits, this was all that was shown. For that reason, double-precision FP numbers are used in some applications to achieve a higher degree of accuracy.

Example 20-4

Convert 152.1875_{10} to double-precision FP.

Solution:

decimal 152.1875 = binary 10011000.0011 =
scientific binary 1.00110000011 E 7
bit 63 is 0 for positive
exponent bits 62 - 53 are 10000000110 (7 + 3FF = 406) after biasing
fraction bits 52 - 0 are 00110000011000.....000

```
0100 0000 0110 0011 0000 0110 0000 0000 0000 ...  ... 0000
  4    0    6    3    0    6    0    0    0  ...  ...    0
```

This example will be verified by an assembler in the next section.

Other data formats of the 8087

In addition to short real (single precision) and long real (double precision) representations for real numbers, the 8087 also supports 16-, 32-, and 64-bit integers. They are referred to as *word integers*, *short integers*, and *long integers*, respectively, and are shown in Figure 20-1. These forms are sometimes referred to as *signed integer numbers*. No decimal points are allowed in integers, in contrast to real numbers, in which decimal points are allowed. There are also two 80-bit data formats in the 8087 coprocessor, packed decimal and temporary real. The packed decimal format has 18 packed BCD numbers, which require a total of 72 bits ($18 \times 4 = 72$). Bits 71 to 0 are used for the numbers, bits 73 to 78 are always 0, and bit 79 is for the sign. The temporary real format is used internally by the 8087 and is shown in Figure 20-1. In the temporary real format, the conversion goes through the same process as shown above, except that the biased exponent is calculated by adding the constant 3FFFH.

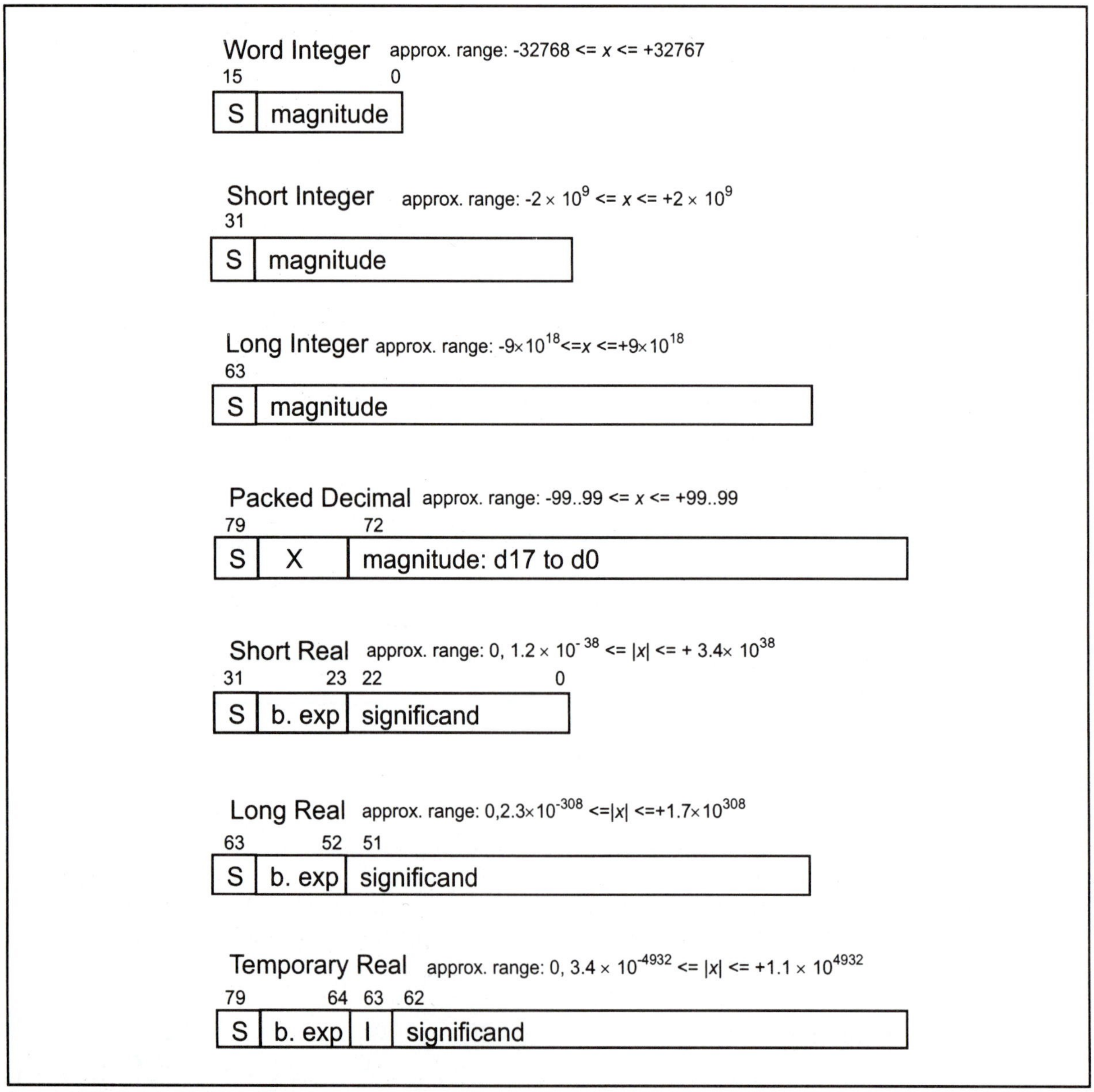

Word Integer   approx. range: $-32768 \le x \le +32767$

15	0
S	magnitude

Short Integer   approx. range: $-2 \times 10^9 \le x \le +2 \times 10^9$

31	
S	magnitude

Long Integer approx. range: $-9 \times 10^{18} \le x \le +9 \times 10^{18}$

63	
S	magnitude

Packed Decimal approx. range: $-99..99 \le x \le +99..99$

79	72	
S	X	magnitude: d17 to d0

Short Real   approx. range: $0, 1.2 \times 10^{-38} \le |x| \le + 3.4 \times 10^{38}$

31	23	22	0
S	b. exp	significand	

Long Real   approx. range: $0, 2.3 \times 10^{-308} \le |x| \le +1.7 \times 10^{308}$

63	52	51	
S	b. exp	significand	

Temporary Real   approx. range: $0, 3.4 \times 10^{-4932} \le |x| \le +1.1 \times 10^{4932}$

79	64	63	62	
S	b. exp	I	significand	

Figure 20-1. 80x87 Data Formats
(Reprinted by permission of Intel Corporation, Copyright Intel Corp. 1992)

Review Questions

1. True or false. In the absence of a math processor, the general-purpose processor must perform all math calculations.
2. True or false. The 80x87 follows the IEEE floating-point standard.
3. Single-precision IEEE FP standard uses _____ bits to represent data.
4. Double-precision IEEE FP standard uses _____ bits to represent data.
5. To get the biased exponent portion of IEEE single-precision floating-point data we add _____.
6. To get the biased exponent portion of IEEE double-precision floating-point data we add _____.

SECTION 20.2: 80x87 INSTRUCTIONS AND PROGRAMMING

This section shows the 80x87 registers, plus some 8087 instructions and their use in sample programs. A full and comprehensive discussion of each instruction and its programming use is beyond the scope of this book. However, the examples in this section provide an introduction to 8087 programming.

Assembling and running 80x87 programs on the IBM PC

To run any program with 8087 instructions, first the programmer must make sure that the PC has one of the following coprocessors on the motherboard: 8087, 80287, 80387, or a compatible coprocessor. In the case of the 80486 and Pentium, the coprocessor is not a separate chip but is integrated with the main processor on a single chip. The 80486SX requires a math coprocessor, the 487SX. This section shows how to assemble several 8087 programs using the Microsoft Assembler, MASM, run them on a PC, and analyze the result. These programs can also be run on Turbo Assembler, TASM, from Borland. First, the assembler directives for data types of the 8087 are explained. In MASM and compatible assemblers, there are different directives to define the different data types of the coprocessor. They are as follows:

DD	(Define double word) for short real (single precision)
DQ	(Define quad word) for long real (double precision)
DD	(Define double word) for short integer
DQ	(Define quad word) for long integer
DT	(Define ten bytes) for packed decimal
DT	(Define ten bytes) for temporary real

Recall that the word size in the 80x86 family is 16 bits. Therefore, when using DD to define a double word, the result is 32 bits. This is different from some other processors, notably RISC processors, in which a word is defined as 32 bits. It is worth repeating a point made in Chapter 0: that although a byte is defined as 8 bits universally, a word is defined differently by different companies. For example, the Cray computer defines a word as 64 bits.

Verifying the Solution for Examples 20-1 to 20-4

Program 20-1 is a portion of the .LST file produced when a program is assembled. It verifies the conversion from decimal to the internal machine representation given in Examples 20-1 through 20-4.

```
                                .8087
                                PAGE 60,132
0000                            .MODEL  SMALL
0000            20 [           .STACK 32
          ????
                 ]
0040
                         ;————————————
0000                            .DATA
0000                            ORG 00H
0000 00 00 1C 41    EX1         DD 9.75        ;example 1
0010                            ORG 10H
0010 00 00 A0 3D    EX2         DD 0.078125    ;example 2
0020                            ORG 20H
0020 3D 8A C0 C2    EX3         DD -96.27      ;example 3
0030                            ORG 30H
0030 00 00 00 00 00 06 EX4      DQ 152.1875    ;example 4
     63 40

                         ;-----------------------------
```

```
Example 20-1 data:
    hex:          41 1C 00 00
    binary:       0100 0001 0001 1100 0000 0000 0000 0000
    sign:         0 for positive
    biased exp:   1000 0010  normalize: 82 - 7F = 3
    significand:  0011100..00
    scientific binary:  1.00111000..00 E3 = 1001.11000...00
    decimal:      9.75


Example 20-2 data:
    hex:          3D A0 00 00
    binary:       0011 1101 1010 0000 0000 0000 0000 0000
    sign:         0 for positive
    biased exp:   0111 1011  normalize: 7B - 7F = -4
    significand:  01000.00
    scientific binary: 1.01 E-4 = .000101
    decimal:      0.078125


Example 20-3 data:
    hex:          C2 C0 8A 3D
    binary:       1100 0010 1100 0000 1000 1010 0011 1101
    sign:         1 for negative
    biased exp:   1000 0101  normalize: 85 - 7F = 6
    significand:  100000010001010000111101
    scientific binary: 1.100000010001010000111101 E6 = 1100000.0010000101000111101
    decimal:      -96.2700078


Example 20-4 data:
    hex:          40 63 06 00 00 00 00 00
    binary:       0100 0000 0110 0011 0000 0110 000..00
    sign:         0 for positive
    biased exp:   10000000110  normalize: 406 - 3FF = 7
    significand:  00110000011
    scientific binary: 1.00110000011 E7 =   10011000.0011
    decimal:      152.1875
```

Program 20-1

80x87 registers

There are only 8 general-purpose registers in the 80x87. Rather than having different-size registers for different-size operands, all the registers of the 8087 are 80 bits wide. Every time the 8087 loads an operand, it automatically converts it to this 80-bit format. This gives uniformity to the registers and makes programming, as well as 8087 hardware design, much easier. Although these 8 registers have been numbered from 0 to 7, they are accessed like a stack, meaning that a last-in-first-out policy is used. At any given time, the top of the stack is referred to as ST(0), or simply ST, and all other registers, regardless of their number, are referred to according to their positions compared to the top of the stack, ST. The programming examples below will demonstrate the use of registers in the 8087. Example 20-5 will show a complete Assembly language program using the 8087 coprocessor. First, a few points should be noted:

1. All 80x87 mnemonics start with the letter "f" to distinguish them from 80x86 instructions.
2. The 80x87 must be initialized to make sure that the top of the stack will be register number 7.
3. Whenever a register is not identified specifically, ST [which is ST(0)] is assumed automatically.
4. ST(0) is the top of the stack, ST(1) is one register below that, and ST(2) is two registers below ST(0), and so on. In other words, for register ST(m), the number in parentheses, m, has nothing to do with the register number. There is a way to find out which register number, 0 - 7, is ST(0), the top of the stack.
5. In the following programming examples, all values of X, Y, and Z have been defined in the data segment and allocated memory locations. The same is true for variables such as SUM, for storing the result.

Example 20-5

Write an 8087 program that loads three values for X, Y, and Z, adds them, and stores the result.

Solution:

```
finit                ;initialize the 8087 to start at the top of stack
fld     X            ;load X into ST(0). now ST(0)=X
fld     Y            ;load Y into ST(0). now ST(0)=Y and ST(1)=X
fld     Z            ;load Z into ST(0). now ST(0)=Z,ST(1)=Y,ST(2)=X
fadd    ST(1)        ;add Y to Z and save the result in ST(0)
fadd    ST(2)        ;add X to (Y+Z) and save it in ST(0)
fst     sum          ;store ST(0) in memory location called sum.
```

Now the same program can be written as follows:

```
finit
fld     X            ;load x, now ST(0) = x
fld     Y            ;load y, now ST(0)= y, ST(1) = x
fld     Z            ;load z, now ST(0)=z, ST(1)=y, ST(2)=x
fadd                 ;adds y to z
fadd    ST(2)        ;adds x to (y + z)
fst     sum
```

Program 20-2 shows the actual MASM code and execution. Figure 20-2 shows the registers.

```
;Program for Example 20-5 to load 3 numbers and compute their sum
        .8087
PAGE        60,132
.MODEL      SMALL
.STACK      32
;——————————
            .DATA
            ORG     00H
X           DD      9.75
            ORG     10H
Y           DD      13.09375
            ORG     20H
Z           DD      29.0390625
            ORG     30H
SUM         DD      ?
;——————————
            .CODE
START       PROC    FAR
            MOV     AX,@DATA
            MOV     DS,AX
            CALL    CSUM
            MOV     AH,4CH
            INT     21H
START       ENDP
;——————————
CSUM        PROC    NEAR
            FINIT               ;initialize 8087 stack
            FLD     X           ;load X into ST(0)
            FLD     Y           ;load Y into ST(0)
            FLD     Z           ;load Z into ST(0)
            FADD    ST(0),ST(1) ;ST(0) = Y + Z
            FADD    ST(0),ST(2) ;ST(0) = X + (Y + Z)
            FST     SUM         ;store ST(0) in sum
            RET
CSUM        ENDP
            END     START
```

```
C>DEBUG EX5.EXE
-U CS:0 C
16EF:0000 B8B617         MOV     AX,17B6
16EF:0003 8ED8           MOV     DS,AX
16EF:0005 E80400         CALL    000C
16EF:0008 B44C           MOV     AH,4C
16EF:000A CD21           INT     21
16EF:000C 9B             WAIT
-G

Program terminated normally
-D 17B6:0 3F
17B6:0000 00 00 1C 41 00 00 00 00-00 00 00 00 00 00 00 00   ...A............
17B6:0010 00 80 51 41 00 00 00 00-00 00 00 00 00 00 00 00   ..QA............
17B6:0020 00 50 E8 41 00 00 00 00-00 00 00 00 00 00 00 00   .PhA............
17B6:0030 00 88 4F 42 00 00 00 00-00 00 00 00 00 00 00 00   ..@.............
-Q
C>
```

Program 20-2

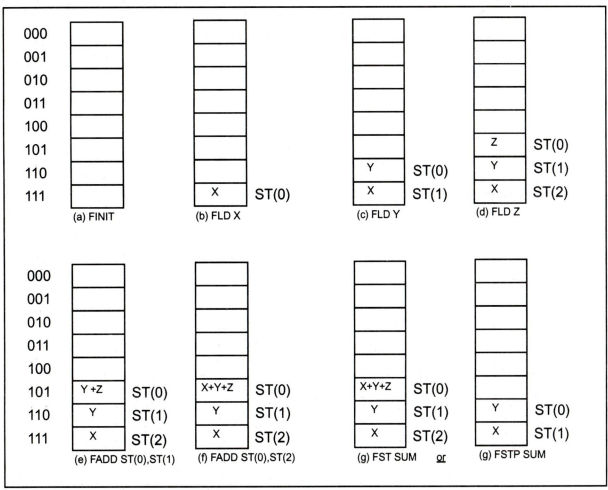

Figure 20-2. Stack Diagram for Example 20-5

Example 20-5 Data Analysis

X: hex: 41 1C 00 00 binary: 0100 0001 0001 1100 0000 0000 0000 0000
    sign: 0   for positive   biased exp: 1000 0010 normalize: 82 - 7F = 3
    significand: 0011 1000 00..00
    sci. binary: 1.00111000..00 E3 = 1001.11000.00
    decimal: 9.75

Y: hex: 41 51 80 00 binary: 0100 0001 0101 0001 1000 0000 0000 0000
    sign: 0 for positive biased exp: 1000 0010 normalize: 82 - 7F = 3
    significand: 1010001100..00
    sci. binary: 1.1010001100..00 E3 = 1101.00011
    decimal: 13.09375

Z: hex: 41 E8 50 00   binary: 0100 0001 1110 1000 0101 0000 0000 0000
    sign: 0 for positive biased exp: 1000 0011 normalize: 83 - 7F = 4
    significand: 1101000010100..00
    sci. binary: 1.1101000010100.00 E4 = 11101.0000101
    decimal: 20.0390625

SUM:hex: 42 4F 88 00   binary: 0100 0010 0100 1111 1000 1000 0000 0000
    sign: 0 for positive biased exp: 1000 0100 normalize: 84 - 7F = 5
    significand: 10011111000100..00
    sci. binary: 1.100111110001 E5 = 110011.1110001
    decimal: 51.8828125

Often, an application requires the use of both real and integer numbers. Real numbers can be rounded into integers by using the 80x87 instruction FRNDINT, as shown in Program 20-3, which includes a procedure to round real numbers. The sample data used is the real number 5.5. In addition, the data is analyzed to see how the number was rounded. FRNDINT rounds real numbers to integers by rounding up, rounding down, truncating, or rounding to the nearest integer. How real numbers are rounded is determined by the RC (round control) bits in the control word (see Figure 20-5).

```
        .8087
        PAGE 60,132
                .MODEL SMALL
        ; PROGRAM TO ROUND A REAL NUMBER TO AN INTEGER
                .STACK 32
        ;————————————
                .DATA
                ORG 00H
        REALNUM   DD        5.5
                ORG 10H
        INTNUM    DD        ?
        ;————————————
                .CODE
        START     PROC FAR
                  MOV    AX,@DATA
                  MOV    DS,AX
                  CALL   RND_NUM
                  MOV    AH,4CH
                  INT    21H
        START     ENDP
        ;————————————
        ;PROCEDURE TO ROUND A REAL NUMBER TO AN INTEGER
        RND_NUM   PROC    NEAR
                  FINIT                     ;initialize 8087
                  FLD      REALNUM          ;load real
                  FRNDINT                   ;round to integer
                  FIST     INTNUM           ;store integer
                  RET
        RND_NUM   ENDP
        ;————————————
                  END      START



        C>debug ex6.exe
        -g
        Program terminated normally
        ...
        ...
        -d 1065:0 1f
        1065:0000 00 00 B0 40 00 00 00 00-00 00 00 00 00 00 00 00   ..0@............
        1065:0010 06 00 00 00 00 00 00 00-00 00 00 00 00 00 00 00   ................
        -q
```

The verification of the above dump is left to the reader as an exercise. It must be noted that the control word can be accessed in order to see which rounding method was used.

Program 20-3

Program 20-4 calculates the area of a circle. The 8087 has instructions that load the top of the stack, ST(0), with a constant. For example, FLDPI loads PI into ST. To calculate the square of a number, the register is multiplied by itself with FMUL. FMUL can have two operands, such as "FMUL ST(2),ST(4)", where ST(4) is multiplied by ST(2) and the result is placed in ST(2). If no operands are given, the operation is assumed to be "FMUL ST(0),ST(1)", so that the first two stack registers are multiplied together and the result is stored in ST(0).

```
        .8087
        PAGE 60,132
                    .MODEL SMALL
        ;PROGRAM TO CALCULATE AREA OF A CIRCLE (radius 91.67)
                    .STACK 32
        ;——————————————
                    .DATA
                    ORG     00H
        R           DD      91.67
                    ORG     10H
        AREA        DD      ?
        ;——————————————
                    .CODE
        START       PROC    FAR
                    MOV     AX,@DATA
                    MOV     DS,AX
                    CALL    CIRC_AREA
                    MOV     AH,4CH
                    INT     21H
        START       ENDP
        ;——————————————
        ;PROCEDURE TO CALCULATE THE AREA OF A CIRCLE
        CIRC_AREA PROC   NEAR
                    FINIT                   ;initialize 8087
                    FLD     R               ;load radius
                    FMUL    ST(0),ST(0)     ;square R
                    FLDPI                   ;load PI
                    FMUL    ST(0),ST(1)     ;multiply PI by R squared
                    FSTP    AREA            ;store AREA
                    RET
        CIRC_AREA ENDP
        ;——————————————
                    END     START
```

The data dumped in DEBUG looked as follows:

```
-d 1065:0 1f
1065:0000  0A 57 B7 42 00 00 00 00-00 00 00 00 00 00 00 00   .W7B............
1065:0010  0C 40 CE 46 00 00 00 00-00 00 00 00 00 00 00 00   .@NF............
```

r = 42 B7 57 0A Hex
    binary = 0100 0010 1011 0111 0101 0111 0000 1010
    sign: 0 for positive
    biased exp: 1000 0101 normalize: 85 - 7F = 6
    significand: 01101110101011100001010
    sci. binary: 1.0110111010101110000 1010 E6 = 1011011.10101011100001010
    decimal: 91.6702481689454

area = 46 CE 40 0C
    binary =   0100 0110 1100 1110 0100 0000 0000 1100
    sign: 0 for positive
    biased exp: 1000 1101 normalize: 8D - 7F = 14
    significand: 10011100100000000001100
    sci. binary: 1.10011100100000000001100 E 14 = 110011100100000.000001100
    decimal: 26400.0232375

Program 20-4

SECTION 20.2: 80x87 INSTRUCTIONS AND PROGRAMMING

Trig functions

Example 20-6 uses trig functions. The instruction FPTAN (partial tangent) calculates Y/X = TAN Z, where Z is the angle in radians and must be 0 < Z < PI/4. Z is stored in ST(0) prior to execution of FPTAN. After the execution ST(0) = X and ST(1) = Y. Then X and Y are used to calculate the hypotenuse R. After that, it is easy to calculate the sine, cosine, tangent, and cotangent. This process is shown in Program 20-5.

Example 20-6

Write, run, and analyze an 8087 program to calculate sin, cosine, tan, cotan of a 30-degree angle.

Solution:

First the 30-degree angle must be converted to radians: (PI/180) × 30 = 0.523598776 radian.
The program is Program 20-5. The data dump in DEBUG, after Program 20-5 is run, is as follows.

```
-d  1065:0 8F
1065:0000  92 0A 06 3F 00 00 00 00-00 00 00 00 00 00 00 00   ...?...........
1065:0010  1B 06 BA 3F 00 00 00 00-00 00 00 00 00 00 00 00   ..:?...........
1065:0020  45 CD 56 3F 00 00 00 00-00 00 00 00 00 00 00 00   EMV?...........
1065:0030  45 CD D6 3F 00 00 00 00-00 00 00 00 00 00 00 00   EMV?...........
1065:0040  00 00 00 3F 00 00 00 00-00 00 00 00 00 00 00 00   ...?...........
1065:0050  D8 B3 5D 3F 00 00 00 00-00 00 00 00 00 00 00 00   X3]?...........
1065:0060  3A CD 13 3F 00 00 00 00-00 00 00 00 00 00 00 00   :M.?...........
1065:0070  D8 B3 DD 3F 00 00 00 00-00 00 00 00 00 00 00 00   X3]?...........
1065:0080  1E 2B C0 50 B8 65 10 8E-D8 E8 10 00 E8 23 00 E8   .+@P8e..Xh..h#.h
```

Data analysis:

```
sin = 3F 00 00 00
      binary:       0011 1111 0000 0000 0000 0000 0000 0000
      sign:         0      for positive
      biased exp:   0111 1110  normalize: 7E - 7F = -1
      significand:  000000..00
      sci. binary:  1.00..00 E-1 = .1
      decimal:      .5

cos = 3F 5D B3 D8
      binary:       0011 1111 0101 1101 1011 0011 1101 1000
      sign:         0      for positive
      biased exp:   0111 1110 normalize: 7E - 7F = -1
      significand:  10111011011001111011000
      sci. binary:  1.10111011011001111011000 E-1   = .110111011011001111011000
      decimal:      .8660254476

tan = 3F13CD3A
      binary:       0011 1111 0001 0011 1100 1101 0011 1010
      sign:         0      for positive
      biased exp:   0111 1110 normalize: 7E = 7F = -1
      significand:  00100111100110100111010 E-1
      sci. binary:  1.00100111100110100111010 E-1   = .100100111100110100111010
      decimal:      .577350259

cot = 3F DD B3 D8
      binary:       0011 1111 1101 1101 1011 0011 1101 1000
      sign:         0      for positive
      biased exp:   0111 1111  normalize: 7F - 7F = 0
      significand:  10111011011001111011000 E0
      sci. binary:  1.10111011011001111011000
      decimal:      1.732050896
```

```
        .8087
        PAGE 60,132
                        .MODEL SMALL
        ;program to calculate SIN, COS, TAN, and COT of a 30-degree angle
                        .STACK 32
        ;————————————
                        .DATA
                        ORG     00H
        ANGLE           DD      0.523598776    ;angle in radians for 30 degrees
                        ORG     10H
        X               DD      0
                        ORG     20H
        Y               DD      0
                        ORG     30H
        R               DD      0
                        ORG     40H
        SIN             DD      0
                        ORG     50H
        COS             DD      0
                        ORG     60H
        TAN             DD      0
                        ORG     70H
        COT             DD      0
        ;————————————
                        .CODE
        START           PROC    FAR
                        MOV     AX,@DATA
                        MOV     DS,AX
                        CALL    CALC_X_Y
                        CALL    CALC_R
                        CALL    CALC_SIN
                        CALL    CALC_COS
                        CALL    CALC_TAN
                        CALL    CALC_COT
                        MOV     AH,4CH
                        INT     21H
        START           ENDP
        ;————————————
        ;procedure to calculate X and Y given an angle
        CALC_X_Y  PROC  NEAR
                        FINIT                   ;initialize 8087
                        FLD     ANGLE           ;load ANGLE onto stack
                        FPTAN                   ;calculate X and Y
                        FSTP    X               ;store X and POP
                        FSTP    Y               ;store Y and POP
                        RET
        CALC_X_Y  ENDP
        ;————————————
        ;procedure to calculate hypotenuse given X and Y
        CALC_R    PROC NEAR
                        FINIT                   ;initialize 8087
                        FLD     X               ;load X onto stack
                        FMUL    ST(0),ST(0)     ;square X
                        FLD     Y               ;load Y onto stack
                        FMUL    ST(0),ST(0)     ;square Y
                        FADD    ST(0),ST(1)     ;calculate X**2 + Y**2
                        FSQRT                   ;take square root
                        FST     R               ;store R
                        RET
        CALC_R    ENDP
        ;————————————
```

Program 20-5 *(continued on following page)*

SECTION 20.2: 80x87 INSTRUCTIONS AND PROGRAMMING   613

```
        ;procedure to calculate SIN, given R and X
        CALC_SIN  PROC   NEAR
                  FINIT                      ;initialize 8087
                  FLD    R                   ;load R onto stack
                  FLD    Y                   ;load Y onto stack
                  FDIV   ST(0),ST(1)         ;SIN = Y/R
                  FST    SIN                 ;store SIN
                  RET
        CALC_SIN  ENDP
        ;————————————————
        ;procedure to calculate COS, given R and X
        CALC_COS  PROC   NEAR
                  FINIT                      ;initialize 8087
                  FLD    R                   ;load R onto stack
                  FLD    X                   ;load X onto stack
                  FDIV   ST(0),ST(1)         ;COS = X/R
                  FST    COS                 ;store COS
                  RET
        CALC_COS  ENDP
        ;————————————————
        ;procedure to calculate TAN, given X and Y
        CALC_TAN  PROC   NEAR
                  FINIT                      ;initialize 8087
                  FLD    X                   ;load X onto stack
                  FLD    Y                   ;load Y onto stack
                  FDIV   ST(0),ST(1)         ;TAN = Y/X
                  FST    TAN                 ;store TAN
                  RET
        CALC_TAN  ENDP
        ;————————————————
        ;procedure to calculate COT, given X and Y
        CALC_COT  PROC   NEAR
                  FINIT                      ;initialize 8087
                  FLD    Y                   ;load Y onto stack
                  FLD    X                   ;load X onto stack
                  FDIV   ST(0),ST(1)         ;COT = X/Y
                  FST    COT                 ;store COT
                  RET
        CALC_COT  ENDP
        ;————————————————
                  END    START
```

Program 20-5 *(continued from preceding page)*

Note in Example 20-6 that in order to calculate sine and cosine we had to use the tangent; however, starting with the 80387 coprocessor there are specific instructions such as FSIN (sine) and FCOS (cosine) for these purposes. To invoke these instructions one must use the .387 directive for the assembler.

Integer numbers

Although performance of real numbers in the 80x87 is very impressive, integer operations should not be overlooked. One way to appreciate this performance is to compare the addition of two multibyte numbers, each 64 bits, on the 8088/86/286 and on the 8087. Since in the 8086/88/286, AX is only 16 bits wide, it will take a loop of 4 iterations to add a 64-bit number plus the overhead of moving the four 16-bit words in for each number and moving the result in and out of the CPU. The same addition can be performed by the 80x87 with only 4 instructions. In the 80x87, integer number instructions are distinguished from real number instructions by the letter "I". For example, the instruction FILD loads an integer number into ST(0) while the FLD would do the same thing for real numbers. One important point about differences between real and integer negative numbers is that integer negative numbers are stored in 2's complement. In real negative numbers, the only difference between a number and its negative is the sign bit. These numbers are not stored in 2's complement. Program 20-6 will show the assembler representation of a negative integer.

```
;This program adds two positive integer numbers and stores the result.
.8087
PAGE 60,132
                .MODEL SMALL
                .STACK  32
;——————————
                .DATA
                ORG     00H
INT1            DD        -50000
                ORG     10H
INT2            DD         25000000
                ORG     20H
SUM             DD          ?
;——————————
                .CODE
START           PROC    FAR
                MOV     AX,@DATA
                MOV     DS,AX
                CALL    ADD_INT
                MOV     AH,4CH
                INT     21H
START           ENDP
;——————————
;procedure to add two integers
ADD_INT         PROC    NEAR
                FINIT                        ;initialize 8087
                FILD    INT1                 ;load integer 1 onto stack
                FIADD   INT2                 ;add the second integer
                FIST    SUM                  ;store result in SUM
                RET
ADD_INT         ENDP
;——————————
                END     START
```

The data dumped in DEBUG is as follows:

```
-d 1065:0 2f
1065:0000  B0 3C FF FF 00 00 00 00-00 00 00 00 00 00 00 00   0...........
1065:0010  40 78 7D 01 00 00 00 00-00 00 00 00 00 00 00 00   @x}..........
1065:0020  F0 B4 7C 01 00 00 00 00-00 00 00 00 00 00 00 00   p4|..........
```

Data Analysis

INT1 = FF FF 3C B0
    binary: 1111 1111 1111 1111 0011 1100 1011 0000
    sign: 1   for negative
    binary: 11111111111111110011110010110000
    reverse bits: 00000000000000001100001101001111
    add 1:   00000000000000001100001101010000
    decimal: -50000

INT2 = 01 7D 78 40
    binary: 0000 0001 0111 1101 0111 1000 0100 0000
    sign: 0   for positive
    integer: 000 0001 0111 1101 0111 1000 0100 0000
    decimal: 25,000,000

SUM = 01 7C B4 F0
    binary: 0000 0001 0111 1100 1011 0100 1111 0000
    sign: 0 for positive
    integer: 000 0001 0111 1100 1011 0100 1111 0000
    decimal: 24,950,000

Program 20-6

Review Questions

1. In Assembly language programming, which data directive is used for single-precison data?
2. In Assembly language programming, which data directive is used for double-precison data?
3. State the number of general-purpose registers in the 80x87.
4. 80x87 registers are accessed acording to _____ (LIFO, FIFO).
5. True or false. While in the 8086, "AX" always refers to the same physical register, in the 80x87 "ST(2)" could be assigned to different physical registers at different times.
6. The top of stack is referred to as ST(_____).
7. ST(1) is the register _____ (above, below) ST(0).
8. What does "FADD ST(4)" do? What are the operands, and where is the result kept?
9. What is the purpose of instruction "FINIT"?
10. Instructions using integer data have letter_____ as part of their mnemonics.
11. What is the difference between the instructions ADD and FADD?
12. True or false. The 8087 has an instruction named "FSIN" to calculate sine.

SECTION 20.3: 8087 HARDWARE CONNECTIONS IN THE IBM PC/XT

Every 8088/86-based IBM PC and compatible comes with a general-purpose processor such as an 8088 or 8086 on the motherboard and a socket for the 8087 coprocessor. This section describes the pin connections of the 8088 and 8087 found in the IBM PC/XT. In addition, it will also discuss the way these two processors communicate with each other and with the system bus.

8087 and 8088 connection in the IBM PC/XT

Figure 20-3 shows the pin layout of the 8087. The 8087 connection to the 8088 in the IBM PC and compatibles is an excellent example of what is called *tightly coupled multiprocessors*. They both access the same address, data, and control buses. In addition, they communicate with each other through a local bus. Figure 20-4 shows this connection. The following is description of the signal connection.

1. The 8088 and 8087 receive the same signals, CLK, READY, and RESET, from the 8284. This ensures that they are synchronized.
2. S0, S1, and S2 are going from the 8088 or 8087 to the 8288, which allows either of these two processors to provide the status signal to the 8288.

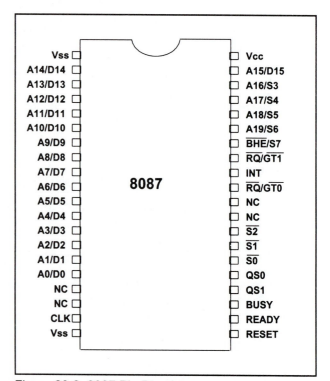

Figure 20-3. 8087 Pin Diagrams
(Reprinted by permission of Intel Corporation, Copyright Intel Corp. 1992)

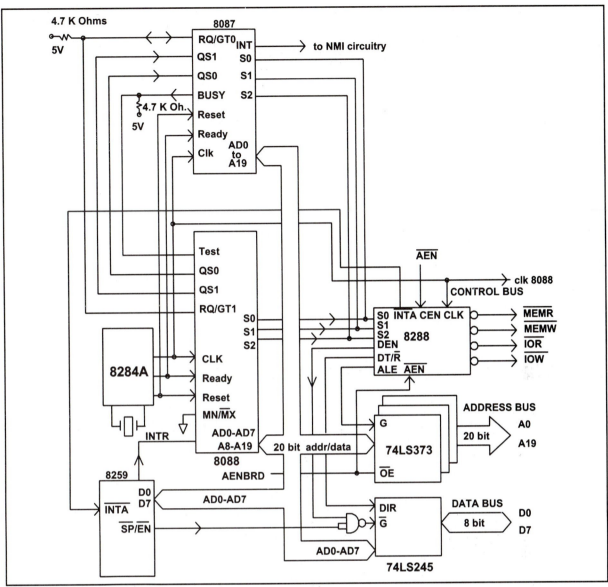

Figure 20-4. 8088-to-8087 Connections (PC/XT)
(Reprinted by permission from "IBM Technical Reference" c. 1984 by International Business Machines Corporation)

3. The Queue Status, QS1 and QS2, from the 8088 go to the 8087, allowing it to know the status of the queue of the 8088 at any given time.

4. The $\overline{\text{TEST}}$ signal to the 8088 comes from BUSY of the 8087. By activating (going low) the BUSY signal, the 8087 informs the 8088 that it finished execution of the instruction which it has been WAITing for.

5. $\overline{\text{RQ/GT1}}$ (request/grant) of the 8088 is connected to $\overline{\text{RQ/GT0}}$ of the 8087, allowing them to arbitrate mastery over the buses. There are two sets of RQ/GT: RQ/GT1 and RQ/GT0. RQ/GT1 of the 8087 is not used and is connected to V_{CC} permanently. This extra RQ/GT is provided in case there is a third microprocessor connected to the local bus.

6. Both the 8088 and 8087 share buses AD0 - AD7 and A8 - A19, allowing either one to access memory. Since the 8087 is designed for both the 8088 and 8086, signal BHE is provided for the 8086 processor. It is connected to V_{CC} if the 8087 is used with the 8088. If the microprocessor used was an 8086, BHE from the 8086 is connected to BHE of the 8087.

7. INT of the 8087 is an output signal indicating error conditions, also called exceptions, such as divide by zero. Error conditions are given in the status word. Assuming the bit for that error is not masked and an interrupt is enabled, whenever any of these

errors occurs, the 8087 automatically activates the INT pin by putting high on it. In the IBM PC and compatibles, this signal is connected to the NMI circuitry as discussed in Chapter 14. Since there is only one INT for all error conditions (exceptions) of the 8087, it is up to the programmer to write a program to check the status word to see which has caused the error.

8. The 8088, often called the *host processor*, must be connected in maximum mode to be able to accommodate a coprocessor such as the 8087.

How the 8088 and 8087 work together in the IBM PC/XT

These two processors work together doing what each can do best. For this to happen, they must work in such a fashion that they don't get in each other's way. This is achieved by each getting a copy of the instructions as they are fetched from memory. Since all the instructions of the 8087 have 9BH in the most significant byte of the opcode, the 8088/86 ignores these instructions. In reality, 9BH is the opcode for the 8088/86 ESCAPE instruction. Likewise, the 8087 ignores any opcode that lacks 9BH. It must be made clear that although both receive a copy of each fetched opcode, only the 8088/86 can fetch opcodes since it is the only device that has the instruction pointer. Now one might ask how the 8088/86 makes sure it is not flooding the 8087 by fetching instructions for the coprocessor faster than the 8087 can process them. The first rule of working together is that the 8088/86 cannot fetch another 8087 instruction until the 8087 has finished execution of the present instruction. This is especially critical since the 8088 can fetch memory in 4 clocks, while some 8087 instructions take over 100 clocks to execute. It is the job of the assembler to put a WAIT instruction between two consecutive 8087 instructions. Verify this by looking at the EXE file of any previous example. In addition, when the 8087 is executing an instruction, it activates the BUSY pin automatically by putting high on it. This pin is connected to the $\overline{\text{TEST}}$ pin of the 8088/86 as mentioned earlier. Next, the 8088/86 fetches the next instruction, which is a WAIT instruction that has been inserted by the assembler, and executes it, thereby going into an internal loop while continuously monitoring the $\overline{\text{TEST}}$ input pin to see when this pin goes low. When the 8087 finishes execution of the present instruction, it pulls down (low) the READY pin, indicating through the $\overline{\text{TEST}}$ pin to the 8088/86 that it can now send the next instruction to the 8087. Although the 8088/86 fetches opcodes for the 8087, it cannot fetch operands for the 8087 because the 8088/86 has no knowledge of the nature of 8087 operands since it does not execute 8087 instructions. So the question is: How are 8087 operands accessed from memory? For the sake of clarity, read and write cases will be looked at separately. Assume that the 8087 needs to read an operand. When the 8088/86 initiates the operand read cycle, the 8087 grabs the 20-bit address and saves it internally. If the operand is a single word (like a word integer), the read cycle has been initiated and the word will come into both processors. Only the 8087 will use the data; the 8088/86 will ignore it. However, if the operand is 32 bits or longer, the 8087 will take over the buses by sending a low pulse on its RQ/GT0 to the RQ/GT1 of the 8088/86 (see Figure 20-4). The 8088/86 in turn will send back a low pulse through the same pin, thereby allowing the 8087 to take over the buses. Remember, RQ/GT is a bidirectional bus. When the 8087 takes over the buses, it will use them until it brings in the last byte of the operand. It is only then that by activating RQ/GT (making it low), control of the buses is given back to the 8088/86.

For example, in the case of a DT operand, the 8087 has control over the buses for the time needed to fetch all 10 bytes and then it gives back the buses. In the case of writing an operand by the coprocessor (e.g., FST data), the 8088/86 initiates the write cycle, but the 8087 ignores it since the 8086 does not have the operand. This is called a *dummy cycle*. All the 8087 does during the dummy cycle is grab the address of the first memory location where the operand is to be stored and keep it until the data is ready, and then it requests the use of the buses by activating the RQ/GT pin. From then on, the process is the same as the read cycle, meaning that it will use the buses until it writes the last byte of the operand.

All the cases discussed so far have been taken care of by either the assembler or the hardware and there was no need for the programmer to be worried. However,

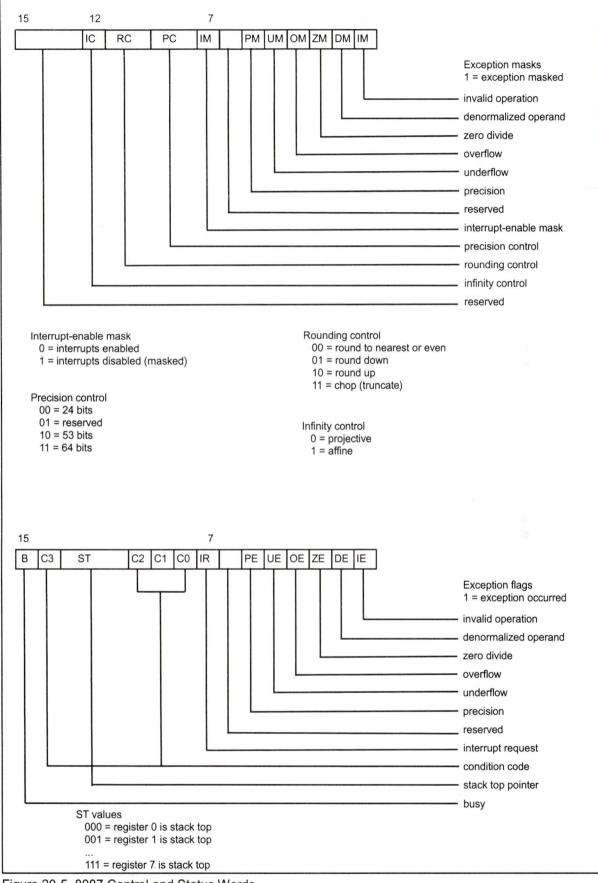

Figure 20-5. 8087 Control and Status Words
(Reprinted by permission of Intel Corporation, Copyright Intel Corp. 1992)

one case where the programmer must get involved is when the result of an operation performed by the 8087 needs to be used by the 8088/86. The programmer must ensure that this result is available before the 8088/86 uses it. This is done by putting an FWAIT instruction in between the 8087 instruction and the subsequent 8088/86 instruction, which needs to use the result. Look at the following case. Two operands have been added and the result stored using the 8087. Then the 8088/86 uses that result. X and Y are single-precision operands. The result of their addition will be a 32-bit real number used by the AX and DX registers, which are 16-bit registers.

```
FLD     X
FLD     Y
FADD    ST(0),ST(1)
FST     SUM
FWAIT
MOV     AX,SUM
MOV     DX,SUM+2
```

Notice that the FWAIT instruction has been put between "FST SUM" and "MOV AX,SUM" by the programmer and not the assembler. In the absence of that FWAIT, the 8088/86 will move erroneous data to AX. Of all the cases of 8087 and 8088/86 cooperation that have been discussed, the last one is the most important as far as programmers are concerned.

In the next section a brief summary of 80x87 instructions and their clock counts are provided. Figure 20-5 shows the 8087 control and status words.

Review Questions

1. True or false. The 8087 coprocessor has its own address and data buses.
2. The 8087 coprocessor produces which of the following signals on its own?
   (a) MEMR   (b) MEMW   (c) INTR
3. Indicate the direction of the following signals as far as the 8087 is concerned.
   (a) S0, S1, S2   (b) QS0, QS1   (c) RESET
   (d) READY   (e) BUSY   (f) RQ/GT0
4. Why is CLK for both the 8088 and the 8087 connected to the same frequency?
5. True or false. The 8087 fetches the opcode in addition to operands.
6. Explain your answer to Question 5.

SECTION 20.4: 80x87 INSTRUCTIONS AND TIMING

There are few changes as far as instructions and registers are concerned from the 8087 to the math processor inside the Pentium, except for a few new instructions and much lower clock counts for instruction execution. The new instructions introduced in the 80387 are FSIN (sine), FCOS (cosine), FSINCOS (sine and cosine), FPREM1 (partial remainder), and FUCOM and its variations. This section provides an overview of the 80x87 instructions and their clock counts.

Real transfers

FLD src   ;pushes source operand onto ST(0)
                 ;source may be ST(i) or memory
                 ;Tip: FLD ST(0) duplicates stack top

FST dest   ;copies ST(0) to destination
                 ;dest may be ST(i) or short or long real variable

FSTP dest   ;copies ST(0) to dest then pops ST(0)
                    ;dest may be ST(i) or short or long or temporary real memory
                    ;Tip: "FSTP ST(0)" is equivalent to popping the stack with no data transfer

FXCH dest   ;swaps contents of ST(0) and destination
                    ;FXCH with no operands swaps ST(0) and ST(1)
                    ;Tip: frequently used to move a register to the top before
                    ;using an instruction which assumes ST(0)

Integer transfers

FILD src   ;converts source to temporary real and pushes onto ST(0)

FIST dest   ;rounds ST(0) to integer and copies to destination
                    ;dest may be a word or short integer

FISTP dest   ;functions the same as FIST but then pops ST(0)
                    ;dest may be any binary integer data type

Packed decimal transfers

FBLD src   ;converts source contents to temporary real then pushes onto ST(0)

FBSTP dest   ;converts ST(0) to BCD and stores at destination, then pops stack

Addition

FADD dest,src ;adds src to dest, storing result in dest
                    ;If no operands are given, ST(0) becomes ST(0) + ST(1)
                    ;If one operand is given, destination is ST(0), source is operand
                    ;source may be ST(i) or real data variable
                    ;dest may be ST(i)

FADDP dest,src real   ; adds src to dest then pops ST(0)
                    ;dest may be ST(i), src is ST(0)

FIADD src   ;adds src to ST(0)

Subtraction

FSUB dest,src ;subtracts src from dest, stores result in dest
                    ;If no operands are given, ST(1) becomes ST(1) - ST(0)
                    ;If one operand is given, it will be src with ST(0) as the destination

FSUBP dest,src ;subtract src from dest and store in dest
                    ;src is ST(0) dest is ST(i)

FISUB src   ;subtract source from ST(0) and store in ST(0)

Reversed subtraction

FSUBR dest,src ;functions the same as FSUB but subtracts dest from src
                    ;and stores the result in dest (R is for reverse)
FSUBRP dst,src ;functions the same as FSUBP but subtracts dest from src
                    ;instead of src from destination

FISUBR src   ;operates the same as FISUB but subtracts ST(0) from source
                    ;and stores in ST(0)

Multiplication

FMUL dest,src ; multiplies dest by src and stores result in dest
                    ;If no operands are given, ST(0) becomes ST(0) x ST(1)
                    ;If one operand is given, destination is ST(0), source is operand
                    ;source may be ST(i) or real data variable
                    ;dest may be ST(i)

FMULP dest,src ;multiplies dest by src, stores result in dest and pops src
                    ;src is ST(0), dest is ST(i)

FIMUL src   ;multiplies ST(0) by src and stores result in ST(0)

Division

FDIV dest,src   ;divides dest by src and stores result in dest
                    ;If no operands are given, ST(0) becomes ST(0) / ST(1)
                    ;If one operand is given, ST(0) is divided by the src operand
                    ;source may be ST(i) or real data variable
                    ;dest is ST(i)

FDIVP dest,src ;divides dest by src and stores result in dest, then pops src
                    ;src is ST(0),dest is ST(i)

FIDIV src   ;divides ST(0) by src and stores result in ST(0)

Reversed division

FDIVR dest,src ;functions identical to FDIV except src is divided by dest

FDIVRP dst,src ;functions the same as FDIVP except src is divided by dest

FIDIVR src   ;functions the same as FIDIV except src is divided by dest

Other arithmetic instructions

FSQRT   ;replaces ST(0) with its square root

FSCALE   ;replaces ST(0) with ST(0) x 2n, where n is the integer in ST(1)
                    ;this provides fast method of multiplying by integral powers of 2

FPREM   ;ST(1) is repeatedly subtracted from ST(0) until ST(0) < ST(1)
                    ;same as ST(0) mod ST(1)

FRNDINT   ;rounds ST(0) to an integer
                    ;rounds according to RC (round control) bits in the control word

FXTRACT   ;extracts the exponent from ST(0) and   places in ST(1)
                  ;extracts the significand from ST(0) and places in ST(0)

FABS   ;replaces ST(0) with its absolute value

FCHS   ;reverses sign bit in ST(0)

Compare instructions

The following instructions compare ST(0) with the source operand and set condition code bits C3, C2, and C0 of the status word as follows:

C3	C2	C0	
0	0	0	ST(0) >source
0	0	1	ST(0) <source
1	0	0	ST(0) = source
1	1	1	numbers cannot be compared

The source operand may be ST(i) or a real number. If no source operand is given, ST(0) is assumed.

FCOM src   ;compares ST(0) with source operand

FCOMP src   ;compares ST(0) with source and pops ST(0)

FCOMPP   ;compares ST(0) and ST(1) and pops both

FICOM src   ;compares ST(0) to source, which may be a long or short integer

FICOMP src   ;same as FICOM but pops ST(0)

FTST   ;compares ST(0) with zero

FXAM   ;tests ST(0) to see if it is zero, infinity, unnormalized, or empty and sets condition codes as follows:

C3	C2	C1	C0	Meaning
0	0	0	0	+unnormal
0	0	0	1	+NAN (not a number)
0	0	1	0	-unnormal
0	0	1	1	-NAN
0	1	0	0	+normal
0	1	0	1	+infinite
0	1	1	0	-normal
0	1	1	1	-infinite
1	0	0	0	+0
1	0	0	1	empty
1	0	1	0	-0
1	0	1	1	empty
1	1	0	0	+denormal
1	1	0	1	empty
1	1	1	0	-denormal
1	1	1	1	empty

Transcendental instructions

FPTAN   ;computes tangent of theta = y/x
                  ;theta is in ST(0) and must be between 0 and pi/4
                  ;after the ratio is computed, y replaces theta in ST(0)
                  ;and x is pushed onto the stack, becoming the new top of stack

FPATAN	;computes theta = arctan (y/x)
	;x is in ST(0) and y in ST(1)
	;ST(0) is popped, theta is written over y in ST(1), the new stack top
F2XM1	;computes $y = 2^x - 1$
	;x is taken from ST(0) and must be in the range: -1 to +1
	;y replaces x in ST(0)
FYL2X	;computes $z = y$ times $\log_2 x$.
	;x is taken from ST(0) and y from ST(1)
	;x must be greater than 0
	;z replaces y, which becomes the new stack top as x is popped off
FYL2XP1	;computes $z = y$ times $\log_2 (x+1)$
	;x is from ST(0) and is in the range: $0 < |x| < (1 - x^{1/2} / 2)$
	;y is from ST(1)
	;z replaces y, which becomes the new stack top as x is popped off

The following instructions are available only in 387 and later coprocessors.

FSIN	;computes sin of ST(0) but provides x = ST(0) and y = ST(1)
	;to get the sin of ST(0), perform y/x using FDIV or
	;to get the cos of ST(0), perform x/y using FDIVR
FCOS	;same as FSIN, takes cos of ST(0) and places x = ST(0) and y = ST(1)
	;to get cos of ST(0), perform y/x using FDIV or
	;to get sin of ST(0), perform x/y using FDIVR
FSINCOS	;computes both sin and cos of ST(0)
	;places sin in ST(0) and cos in ST(1)

Constant instructions

FLDZ	;pushes +0.0 onto the stack
FLD1	;pushes +1.0 onto the stack
FLDPI	;pushes pi onto the stack
FLD2T	;pushes $\log_2 10$ onto the stack
FLDL2E	;pushes $\log_2 e$ onto the stack
FLDLG2	;pushes $\log_{10} 2$ onto the stack
FLDLN2	;pushes $\log_e 2$ onto the stack

Many mathematical equations can be implemented using constant and transcendental functions. For example, $x^y = 2^{y \log x}$. If $z = y \log_2 x$, FYL2X can be used to calculate z. Then F2XM1 can be used to calculate $2^z - 1$. Then 1 can be added to this to get 2^z, which is equal to x^y.

The instruction sequence would be

```
FLD     Y
FLD     X
FYL2X
F2XM1
FLD1
FADD
FST     SUM
```

Other frequently used functions can likewise be calculated, for example e^y and 10^y, substituting e or 10 for x in the above equations. In addition, a little creativity will allow a programmer to use the constant and transcendental functions frequently. For example, if the calculation $\log_2 x$ is needed, the FYL2X ($y \log_2 x$) function can be used by making $y = 1$. Likewise, if 2^x is needed, the F2XM1 function can be used, after which 1 can be added to the result.

Processor control instructions

Many of the following instructions have two mnemonics; the second one has an extra N. This N instructs the CPU not to prefix the instruction with a wait state. The no-wait forms should be used when CPU interrupts are disabled and the 8087 might generate an interrupt, which would create an endless wait. Wait forms are used when the CPU interrupts are enabled.

FINIT or FNINIT	;resets the processor
FDISI or FNDISI	;sets the interrupt enable mask in the control word ;thereby disabling interrupts in the 8087
FENI or FNENI	;clears the interrupt enable mask in the control word ;thereby enabling interrupts in the 8087
FLDCW src	;replaces the control word with the contents of src
FSTCW dest or FNSTCW	;writes the control word to dest
FSTSW dest or FNSTSW	;writes status word to dest
FCLEX or FNCLEX	;clears exception flags, busy flag ;and interrupt request flag in the status word
FSAVE dest or FNSAVE	;writes to dest the 94-byte save area, which ;includes the environment and the stack
FRSTOR src	;restores the 94-byte save area from src
FSTENV dest or FNSTENV	;stores environment (control, status ;and tag words and exception pointers) to dest
FLDENV src	;restores environment previously saved with FSTENV instruction
FINCSTP	;increments status word's stack pointer
FDECSTP	;decrements status word's stack pointer
FFREE dest	;marks dest as an empty register
FNOP	;stores ST(0) to ST(0), therefore performs no operation
FWAIT	;same as CPU's wait instruction ;used to synchronize CPU and 8087

CODE	DESCRIPTION	8087	80287	80387	80487
F2XM1	Calculates y = 2^x - 1	310-630	310-630	211-476	140-279
FABS	Converts ST to abs val	10-17	10-17	22	3
FADD	Add reg,reg	70-100	70-100	t=23-31, f=26-34	8-20
FADD	Add memreal	(s=90-120, s=95-125)+EA	s=90-120, l=95-125	s=24-32, l=29-37	s=8-20, l=8-20,
FADDP	Add and pop	75-105	75-105	23-31	8-20
FIADD	Integer add	(w=102-137, d=108-143)+EA	w=102-137, d=57-72	w=71-85, d=57-72	w=20-35, d=19-32
FBLD	See FLD				
FBSTP	See FST				
FCHS	Reserves sign ST	10-17	10-17	24-25	6
FCLEX/	Clear exceptions				
FNCLEX	No wait	2-8	2-8	11	7
FCOM	Compare	40-50	40-50	24	4
	memreal	(s=60-70, l=65-75)+EA	s=60-70, l=65-75	s=26 l=31	s=4, l=4
FCOMP	Compare, pop	42-52	42-52	26	4
	memreal	(s=63-73, l=67-77)+EA	s=63-73, l=67-77	s=26 l=31	s=4, l=4
FCOMPP	Compare, pop, pop	45-55	45-55	26	5
FICOM	Compare int	(w=72-86, d=78-91)+EA	w=72-86, d=78-91	w=71-75, d=56-63	w=16-20, d=15-17
FICOMP	Compare int, pop	(w=74-88, d=80-93)+EA	w=74-88, d=80-93	w=71-75, d=56-63	w=16-20, d=15-17
FCOS	Cosine	--	--	123-772	257-354
FDECSTP	Dec stack pointer	6-12	6-12	22	3
FDISI/	Disable interrupt				
FNDISI	No wait	2-8	2	2	
FDIV	Divide	193-203	193-203	t=88,f=91	73
	memreal	(s=215-225, l=220-230)+EA	s=215-225, l=220-230	s=89 l=94	s=73, l=73
FDIVP	Divide, pop	197-207	197-207	91	73
FIDIV	Int divide	(w=224-238, d=230-243)+EA	w=224-238, d=230-243	w=136-140, d=120-127	w=85-89, d=84-86
FDIVR	Divide reversed	194-204	194-204	t=88,f=91	73
FDIVR	memreal	(s=216-226, l=221-231)+EA	s=216-226, l=221-231	s=89 l=94	s=73 l=73
FDIVRP	Div rev, pop	198-208	198-208	91	73
FIDIVR	Int div reversed	(w=225-239, d=231-245)+EA	w=225-239, d=231-245	w=135-141, d=121-128	w=85-89, d=84-86
FENI/	Enable interrupt				
FNENI	No wait	2-8	2	2	
FFREE	Free register	9-16	9-16	18	3
FIADD	See FADD				
FISUB	See FSUB				
FISUBR	See FSUBR				
FIMUL	See FMUL				
FIDIV	See FDIV				
FIDIVR	See FDIVR				
FICOM/					
FICOMP	See FCOM				
FILD	See FLD				
FINCSTP	Incr stack ptr	6-12	6-12	21	3
FINIT/	Init coprocessor				
FNINIT	No wait	2-8	2-8	33	17
FIST/					
FISTP	See FST				
FLD	Load	17-22	17-22	14	4
	memreal	(s=38-56, l=40-60, t=53-65)+EA	s=38-56, l=40-60, t=53-65	s=20, l=25, t=44	s=3 l=3 t=6
FILD	Int load	(w=46-54, d=52-60, q=60-68)+EA	w=46-54, d=52-60, q=60-68	w=61-65, d=45-52, q=56-67	w=13-16, d=9-12 q=10-18

CODE	DESCRIPTION	8087	80287	80387	80487
FBLD	BCD	(290-310)+EA	290-310	266-275	70-103
FLD1	Load 1	15-21	15-21	24	4
FLDZ	Load zero	11-17	11-17	20	4
FLDPI	Load pi	16-22	16-22	40	8
FLDL2E	Load log2(e)	15-21	15-21	40	8
FLDL2T	Load log2(10)	16-22	16-22	40	8
FLDLG2	Load log10(2)	18-24	18-24	41	8
FLDLN2	Load loge(2)	17-23	17-23	41	8
FLDCW	Load control word	(7-14)+EA	7-14	19	4
FLDENV	Load env state	(35-45)+EA	35-45	71	
	real/virt 16-bit	--	--	--	44
	real/virt 32-bit	--	--	--	44
	pm 16-bit	--	--	--	34
	pm 32-bit	--	--	--	34
FMUL	Multiply	130-145 (90-105)	130-145 (90-105)	t=46-54 (49) f=29-57 (52)	16
FMUL	memreal	(s=110-125, l=154-168)+EA	s=110-125, l=154-168	s=27-35, l=32-57	s=11, l=14
FMULP	Multiply, pop	134-148 (94-108)	134-148 (94-108) (52)	29-57	16
FIMUL	Int multiply	(w=124-138, d=130-144)+EA	w=124-138, d=130-144	w=76-87, d=61-82	w=23-27, d=22-24
FNOP	No op	10-16	10-16	12	3
FPATAN	Partial arctangent	250-800	250-800	314-487	218-303
FPREM	Partial remainder	15-190	15-190	74-155	70-138
FPREM1	Parital rem (IEEE)	--	--	95-185	72-167
FPTAN	Partial tangent	30-540	30-540	181-497	200-273
FRNDINT	Round to int	16-50	16-50	66-80	21-30
FRSTOR	Restore saved state	(197-207)+EA *	308		
	real/virt 16-bit	--	--	--	131
	real/virt 32-bit	--	--	--	131
	pm 16-bit	--	--	--	120
	pm 32-bit	--	--	--	120
FSAVE/	Save coprocessor state				
FNSAVE	No wait	(197-207)+EA *	375-376		
	real/virt 16-bit	--	--	--	154
	real/virt 32-bit	--	--	--	154
	pm 16-bit	--	--	--	143
	pm 32-bit	--	--	--	143
FSCALE	Scale	32-38	32-38	67-86	30-32
FSETPM	Set protected mode	--	2-8	12	
FSIN	Sine	--	--	122-771	257-354
FSINCOS	Sine and cosine	--	--	194-809	292-365
FSQRT	Square root	180-186	180-186	122-129	83-87
FST	Store	15-22	15-22	11	3
	memreal	(s=84-90, l=96-104)+EA	s=84-90, l=96-104	s=44, l=45	s=7, l=8,
FSTP	Store, pop	17-24	17-24	12	3
	memreal	(s=86-92, l=98-106, t=52-58)+EA	s=86-92, l=98-106, t=52-58	s=44, l=45, t=53	s=7, l=8, t=6
FIST	Int store	(w=80-90, d=82-92)+EA	w=80-90, d=82-92	w=82-95, d=79-93	w=29-34, d=28-34
FISTP	Int store, pop	(w=82-92, d=84-94, q=94-105)+EA	w=82-92, d=84-94, q=94-105	w=82-95, d=79-93, q=80-97	w=29-34, d=29-34, q=29-34
FBSTP	Store BCD, pop	(520-540)+EA	520-540	512-534	172-176
FSTCW/	Store control word				
FNSTCW	No wait	12-18	12-18	15	
FSTENV/	Store env state				
FNSTENV	No wait	(40-50)+EA	40-50	103-104	
	real/virt 16-bit	--	--	--	67
	real/virt 32-bit	--	--	--	67
	pm 16-bit	--	--	--	56
	pm 32-bit	--	--	--	56
FSTSW/	Store status word				

SECTION 20.4: 80x87 INSTRUCTIONS AND TIMING   627

CODE	DESCRIPTION	8087	80287	80387	80487
FNSTSW	No wait	12-18	12-18	15	3
	Store in AX	--	10-16	13	3
FSUB	Subtract	70-100	70-100	t=29-37, f=26-34	8-20
	memreal	(s=90-120, l=95-125)+EA	s=90-120, l=95-125	s=24-32, l=28-36	s=8-20, l=8-20
FSUBP	Subtract, pop	75-105	75-105	26-34	8-20
FISUB	Int subtract	(w=102-137, d=108-143)+EA	w=102-137, d=108-143	w=71-83, d=57-82	w=20-35, d=19-32
FSUBR	Subtract reversed	70-100	70-100	t=29-37, f=26-34	8-20
	memreal	(s=90-120, l=95-125)+EA	s=90-120, l=95-125	s=25-33, l=29-37	s=8-20, l=8-20
FSUBRP	Subtract rev, pop	75-105	75-105	26-34	8-20
FISUBR	Int subtract rev	(w=103-139, d=109-144)+EA	w=103-139, d=109-144	w=72-84, d=58-83	w=20-35, d=19-32
FTST	Test for zero	38-48	38-48	28	
FUCOM	Unordered compare	--	--	24	4
FUCOMP	Unordered comp, pop	--	--	26	4
FUCOMPP	Unord comp, pop, pop	--	--	26	5
FWAIT	Wait	4	3	6	
FXAM	Examine	12-23	12-23	30-38	8
FXCH	Exchange registers	10-15	10-15	18	
FXTRACT	Extract exp & sig	27-55	27-55	70-76	16-20
FYL2X	Y log2(x)	900-1100	900-1100	120-538	196-329
FYL2XP1	Y log2(x+1)	700-1000	700-1000	257-547	171-326

SUMMARY

The first section of this chapter examined data types used in 80x87 programming. First, the IEEE floating-point standards were examined for 32-bit single precision (short real in Intel terminology) and 64-bit double precision (long real in Intel terminology). The conversion from real numbers to IEEE floating point was demonstrated. The 80x87 supports data types: word integer, short integer, long integer, packed decimal, short real, long real, and temporary real, ranging in size from 16 bits to 80 bits. The second section of this chapter showed how to code several Assembly language programs using 80x87 instructions. The third section explained how the 80x87 and the 8088/86 CPU are interfaced in the IBM PC/XT. The fourth section listed 80x87 instructions and their clock counts.

PROBLEMS

SECTION 20.1: MATH COPROCESSOR, IEEE FLOATING-POINT STANDARDS

1. What is the disadvantage of using a general-purpose processor to perform math operations?
2. The IEEE single-precision standard uses _____ bytes to represent a real number.
3. The IEEE double-precision standard uses _____ bytes to represent a real number.
4. Show the bit assignment of the IEEE single-precision standard.
5. Convert (by hand calculation) each of the following real numbers to IEEE single-precision standard.
    (a) 15.575   (b) 89.125   (c) −1022.543   (d) −0.00075
6. Use the last 4 digits of your ID number and put the decimal point in the middle. Convert it to single-precision IEEE standard (e.g., 9823 is 98.23).
7. What data types are called short real and long real in Intel's literature?

8. Show the bit assignment of the IEEE double-precision standard.
9. In single precision FP (floating point), the biased exponent is calculated by adding _____ to the _____ portion of a scientific binary number.
10. In double-precision FP, the biased exponent is calculated by adding _____ to the _____ portion of a scientific binary number.
11. Convert the following to double-precision FP.
   (a) 12.9823   (b) 98.76123
12. How many bits are set aside for the magnitude portion of Intel's long integer?
13. Which bits of packed decimal are used for the sign?
14. Packed decimal uses only _____ bits of an 80-bit operand.
15. In Intel's temporary real, the data type is _____ bytes wide.

SECTION 20:2: 80x87 INSTRUCTIONS AND PROGRAMMING

16. Indicate the data directive used for the following data types.
   (a) single-precision FP   (b) double-precision FP
   (c) packed decimal
17. Using the assembler of your choice, verify your calculation of Problems 5 and 11.
18. Write and run an 80x87 program to calculate $z = (x^2 + y^3)^{1/2}$, where $x = 3.12$ and $y = 5.43$.
19. Write and run an 80x87 program to calculate $y = 2x^2 + 5x + 12.34$, where $x = 1.25$.
20. Write and run an 80x87 program to calculate the area of a circle if $r = 25.5$.
21. Write and run an 80x87 program to calculate $3(\pi r^3)/4$ if $r = 25.5$.
22. Write and run an 8087 program to calculate sine of a 45-degree angle.

SECTION 20.3: 8087 HARDWARE CONNECTIONS IN THE IBM PC/XT

23. Why is an 80x87 interfaced with an 80x86 called tightly coupled multiprocessors?
24. True or false. The 8087 generates the IOR and IOW signals.
25. True or false. The 8087 generates the MEMR and MEMW signals.
26. Which is the host processor, the 80x86 or the 80x87?
27. Indicate the direction of READY, BUSY, and RESET as far as the 8087 is concerned.
28. What puts the FWAIT instruction between the 8087 and an 8086 instruction?
29. True or false. One cannot move an operand directly between the 8087 and 8086, or vice versa. Data must be exchanged via a temporaray memory location.
30. Explain your answer to Problem 29.

SECTION 20.4: 80x87 INSTRUCTIONS AND TIMING

31. Compare the clock counts for the FADD instruction for all 80x87 processors. Assume that the data is single-precision FP.
32. Compare the clock counts for the FDIV instruction for all 80x87 processors. Assume that the data is single-precision FP.
33. Compare the clock counts for the FSQRT instruction for all 80x87 processors. Assume that the data is single-precision FP.
34. Give the list of new instructions introduced in the 80387.
35. Which of the following processors have an on-chip coprocessor?
   (a) 80386   (b) 80486SX   (c) 80486
   (d) Pentium   (e) 80486DX2   (f) 80386SX
36. What kind of processor is the 80487SX, and where is it used?
37. For the 80x87 to calculate a trig function, the angle must be in _____ (degrees, radians).
38. Write the 80387 program to calculate SIN and COS of a 30-degree angle. Use a PC with 387 or higher to verify and analyze your program.

PROBLEMS

ANSWERS TO REVIEW QUESTIONS

SECTION 20.1: MATH COPROCESSOR AND IEEE FLOATING-POINT STANDARDS

1. true
2. true
3. 32
4. 64
5. 7FH
6. 3FFH

SECTION 20.2: 80x87 INSTRUCTIONS AND PROGRAMMING

1. DD (define double word)
2. DQ (define quad word)
3. 8
4. LIFO
5. true
6. ST(0)
7. below
8. It means ST(4)+ST(0) and the result is placed in ST(0).
9. to initialize the registers to top of the stack
10. I
11. The assembler generates the opcode to be used by the 80x86 for the ADD instruction while it produces the opcode for the FADD instruction to be used by the 80x87.
12. False. 387 and latter coprocessors have the FSIN instruction.

SECTION 20.3: 8087 HARDWARE CONNECTIONS IN THE IBM PC/XT

1. true
2. only INTR, which it calls INT
3. (a) and (e) are out; (b), (c), and (d) are in; (f) is bidirectional
4. This ensures the synchronization of fetching the opcodes and operands.
5. false
6. Only the 8088/86 fetches opcodes since it has an instruction pointer (IP) register. While the 8088/86 fetches all opcodes, both the 8088/86 and 8087 receive a copy of opcodes, but the 8088/86 ignores any opcode that has 9BH in front of it.

CHAPTER 21

386 MICROPROCESSOR: REAL vs. PROTECTED MODE

OBJECTIVES

Upon completion of this chapter, you will be able to:

» **List the additional features implemented on the 80186, 80286, and 80386**

» **State the purpose of designing two modes, real and protected, into the 80386**

» **Code Assembly language instructions using the new scaled index addressing mode of the 386**

» **Code Assembly language instructions using the new instructions of the 386**

» **State the purpose of each pin of the 80386 microprocessor**

» **Describe the data misalignment problem with 386 programs and discuss how to resolve the problem**

» **Describe how protection of user and system programs is accomplished in the 386 in protected mode**

» **Contrast and compare the two methods of virtual memory implementation: paging and segmentation**

» **Describe the methods of converting from logical to physical addresses in 386 protected mode**

This chapter emphasizes unique features of the 80386 microprocessor, from both hardware and software perspectives. In Section 21.1 we look at the 386 in real mode. The hardware of the 386 is examined in Section 21.2. Section 21.3 provides an introduction to protected mode of the 386.

SECTION 21.1: 80386 IN REAL MODE

In this section first we look at Intel's 80186 microprocessor and then unique features of the 286 and 386 from the perspective of real mode programming.

What happened to the 80186/188?

Intel has a very successful product called the 80186 (and 80188). This chip is alive and doing very well in the embedded controller market, where it is used to replace multiple devices with a single component. The 80186/88 was never used by IBM in their family of PC products. Some clone makers, notably Tandy Corp., used it in their PCs. Prior to the introduction of the 80186/88, Intel did a survey and found that many are using the 8086/88 along with other peripheral chips, such as the 8237 DMA controller, the 8254 timer, and the 8259 interrupt controller. This led to putting a portion of these chips along with the 8086/88 microprocessor on a single chip and calling it the 80186/88. Internally, the 80186 and 80188 are identical, but externally the 80186 has a 16-bit data bus and the 80188 has an 8-bit external data bus. In this regard they are similar to the 8086 and 8088. The address bus is still 20-bit, making a 1 megabyte memory system. The data bus is multiplexed with the address bus. The 80186/88 is a 68-pin chip that includes the following on-chip functions: (1) clock generator, (2) two 20-bit DMA channels, (3) three 16-bit programmable counters, (4) interrupt controller, (5) programmable wait-state generator, and (6) programmable chip select decoder unit. Although very few 80186/88 microprocessors are used in PCs, millions of them are found in embedded systems such as pocket translators, digital cellular phones, and so on.

80186/88 New instructions

The 80186/88 microprocessor supports all 8086/88 instructions in addition to some new ones. The new instructions of the 80186/88 are as follows:

```
BOUND       dest,source
ENTER       disp,level
LEAVE
IMUL        result,source,immediate data
INS         dest,port
OUTS        port,dest
SAR         dest,immediate count
SHR
SAL
RCR
ROR
RCL
ROL
PUSH        immediate data
PUSHA
POPA
```

Some of the above instructions, such as ENTER and LEAVE, are intended for implementation in high-level languages but many others can be used in everyday Assembly language programs. For example, look at the shift and rotate instructions. In the 8086/88 to shift or rotate an operand more than once required putting the count in CL; immediate operands could not be used. Starting with the 80186/88, immediate counts are allowed. Look at Example 21-1.

Example 21-1

Show Assembly language code to shift right operand 26H by 5 bits in each of the following systems.
(a) 8086/88 (b) 80186/88

Solution:

(a) 8086/88   (b) 80186/88

```
MOV     AL,26H              MOV     AL,26H
MOV     CL,5                SHR     AL,5
SHR     AL,CL
```

Other useful new instructions of the 80186/88 are PUSHA (push all) and POPA (pop all). Very often in writing a procedure (subroutine) all the registers need to be saved on the stack. In the 8086/88 one must code PUSH and POP for each 16-bit register separately; however, in the 80186/88 the use of PUSHA and POPA can save a lot of coding. See Example 21-2.

Example 21-2

Show a sequence of 8086/88 instructions equivalent to 80186/88 PUSHA and POPA.

Solution:

```
            8086/88                  80186/88

            PUSH  AX                 PUSHA
            PUSH  CX
            PUSH  DX
            PUSH  BX
            PUSH  SP
            PUSH  BP
            PUSH  SI
            PUSH  DI
            ...                      ...
            POP   DI
            POP   SI
            POP   BP
            POP   SP
            POP   BX
            POP   DX
            POP   CX
            POP   AX                 POPA
            RET                      RET
```

Note that POPA restores all registers except SP, which is ignored. In other words, it does not disturb the present stack frame.

One can test these new instructions on 286 PCs and later machines. However, using these instructions means that the program will not run on 8088/86 XT machines. There are two dominant trends in software for 80x86-based systems.
1. Software that runs on any 80x86 machine, including 8086/88-based systems.
2. Software that is 32-bit 386 based and must be run on 386 and higher machines.
   Also note that the DEBUG utility does not support the new 80186/88 instructions since it is intended to run on any 80x86 PC, including the 8088/86.

SECTION 21.1: 80386 IN REAL MODE   633

80286 Microprocessor

The demand for a more powerful CPU led Intel Corporation to use more than 100,000 transistors to design a new microprocessor called the 80286. This processor is downwardly compatible with the 80186/88 processor. The 80286 microprocessor was a major improvement over the core 8086 in the following ways.

1. There are separate pins for the address and data buses and thus no need for demultiplexing the buses as was the case in the 8088/86 microprocessor. This increased number of pins required abandoning DIP (dual in-line packaging). Instead, PGA (pin grid array) packaging was chosen.

2. The memory cycle time was reduced to 2 clocks from 4 clocks in the 8088/86. This made memory interfacing quite a challenge, especially for frequencies of 20 MHz and beyond. Memory design of high-performance computers is discussed in Chapter 22.

3. Introduction of virtual memory in the 80286 was the most drastic change over the 8088/86. The 80286 works in two different modes: real mode and protected mode. In real mode, the 80286 is simply a faster 8086, capable of handling only 1M byte of memory. It executes all the instructions of the 8086 with fewer clock cycles, as shown in Appendix E. In order to use the entire 16 megabytes of memory space, the 80286 must work in protected mode. When the 80286 microprocessor is turned on, it automatically starts from real mode and can be switched to protected mode. When in protected mode, all the address buses A0 - A23 can be used, thereby giving a total of 16M bytes of addressable physical memory (RAM and ROM). It is in this mode that most of the changes over the 8086 have been introduced into the 80286. Due to the declining price of the 80386, very few systems use the 286 in protected mode; therefore, we bypass any discussion of the 80286 in protected mode. However since protected mode of the 80286 is a subset of the 80386, many of the concepts of 386 protected mode apply to the 286.

Major changes in the 80386

The 80386 microprocessor started a new trend in the 80x86 family. Although it is downwardly compatible with the 8088/86 and 80286, there are some major changes in its architecture. The following are some of the major changes that have been introduced in the 80386.

1. The data bus was increased from 16 bits to 32 bits, both internally and externally.

2. All the registers were extended to 32 bits, thereby making the 80386 a 32-bit microprocessor.

3. The address bus was increased to 32 bits, thus providing 4 gigabytes (2^{32}) of physical memory addressing capability.

4. The paging virtual memory mechanism was introduced, making the 80386 capable of using both segmentation and paging. More about paging and segmentation is provided in Section 21.3.

5. A new addressing mode called *scaled index* was added.

6. Many new bit-manipulation instructions were added. These instructions work in both real mode and protected mode.

7. The 386 can be switched from protected to real mode by software. This is a major improvement over the 286, which had to be reset to switch back to real mode.

To reduce the cost of board design, Intel made the 80386SX microprocessor available with a 16-bit external data bus, but internally it remained a 32-bit processor, 100% compatible with the 80386. In terms of memory bandwidth, it is slower than the 80386 since it takes two memory cycles (each 2 clock cycles) to address a 32-bit word instead of only 1 memory cycle, as is the case in the 80386. Intel also made the 80386SX with only 24-bit address buses, the same as the 80286. In other words, the 80386SX is the same as the 80286 externally, but internally it is a 32-bit processor fully compatible with all 386 computers.

In real mode, the 80386 can access a maximum of 1 megabytes using address pins A19 - A0. However, in protected mode the 80386 can accesses 4 gigabytes of memory through using the 32-bit address bus.

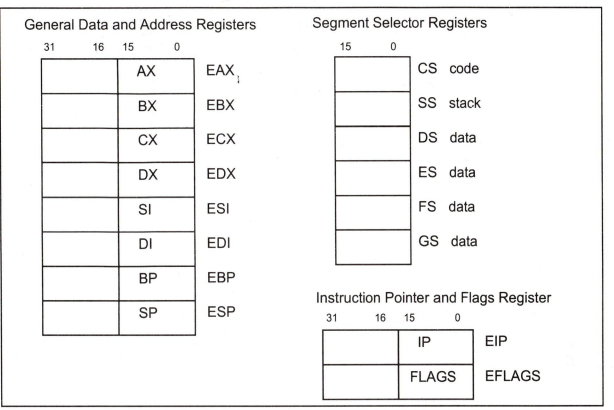

General Data and Address Registers

AX	EAX
BX	EBX
CX	ECX
DX	EDX
SI	ESI
DI	EDI
BP	EBP
SP	ESP

Segment Selector Registers

	CS code
	SS stack
	DS data
	ES data
	FS data
	GS data

Instruction Pointer and Flags Register

IP	EIP
FLAGS	EFLAGS

Figure 21-1. Selected Intel 386 Registers

80386 Real mode programming

In the design of the 80386, Intel made such massive design changes that it is radically different from the 80286, yet it is still capable of running all the code written for the 286 and 8086/88. Next we describe some of these new features that are available in both real and protected modes.

32-Bit registers

In the 80386, register sizes were extended from 16 bits to 32 bits and register names were changed to reflect this. For example, EAX is the extended AX, EBX is the extended BX, and so on. See Figure 21-1. In order to access 32-bit registers, the letter E must be included in the coding. The four general-purpose registers AX, BX, CX, and DX are still accessible in their 8086 formats in addition to the extended format. For example, register EAX is accessible as AL, AH, AX, and EAX. Notice that the upper 16-bit part is not accessible as a separate register. One way to access it is to shift EAX right. See Example 21-3.

Example 21-3

Load EAX with 7698E35FH and move it among the 8-, 16-, and 32-bit registers of the 386.

Solution:

```
        MOV   EAX,7698E35FH      ;EAX=7698E35F (AX=E35F,AH=E3,AL=5F)
        MOV   EDX,EAX            ;EDX=EAX=7698E35F
        MOV   CH,AL              ;CH=AL=5F
        MOV   DI,AX              ;DI=AX=E35F
        MOV   ESI,EDX            ;ESI=EDX=7698E35F
        ROR   EAX,16             ;rotate right EAX 16 times (EAX=E35F7698H)
        MOV   BX,AX              ;BX=AX=7698H
        MOV   CL,AL              ;CL=AL=98H
```

Which end goes first?

In storing data, the 386 followed the tradition of the 8086/286 in placing the least significant byte (little end of the data) in the low address. As discussed previously, this is referred to as *little endian*. See Example 21-4.

Example 21-4

Show how data is placed after execution of the following code.

```
                MOV     EAX,7698E39FH        ;EAX=7698E39F
                MOV     [4524],AX
                MOV     [8000],EAX
```

Solution:

For "MOV [4524],AX" we have

DS:4524 =(9F)
DS:4525 =(E3)

and for "MOV [8000],EAX" we have

DS:8000 = (9F)
DS:8001 = (E3)
DS:8002 = (98)
DS:8003 = (76)

In Example 21-4, notice how the least significant byte (the little end of the data) 9FH goes to the low address 8000, and the most significant byte of the data 76H goes to the high address 8003. This means that the little end of the data goes in first, hence the name little endian. In the Motorola 68000 family, data is stored the opposite way: the big end (most significant byte) goes into the low address first, and for this reason it is called big endian. Some recent RISC processors, such as Power PC (developed jointly by IBM and Motorola), allow selection of mode, big endian or little endian. The software overhead of converting from one camp to the other led Intel to introduce a new instruction called SWAP in the 80486, specifically to take care of this problem (see Chapter 23).

General registers as pointers

Another major change introduced in the 80386 is the use of general registers such as EAX, ECX, and EDX as pointers. As you might recall from Chapter 1, the 8088/86/286 can use only BX, SI, and DI as pointers into the data segments. But starting with the 386, all 32-bit general-purpose registers can be used for pointers into data segments. Look at the following cases for valid and invalid instructions.

```
MOV    BX,WORD PTR [EAX]      ;move into BX word pointed to by EAX
MOV    BX,WORD PTR [AX]       ;invalid AX can't be used as pointer
MOV    EAX,DWORD PTR [ECX]    ;move into EAX DWORD pointed to by ECX
MOV    AL,BYTE PTR [EDX]      ;move into AL BYTE pointed to by EDX
MOV    EBX,WORD PTR [CX]      ;invalid CX can't be used as pointer
MOV    EAX,DWORD PTR [EDI]    ;move into EAX DWORD pointed to by EDI
```

The 386 also allows the use of displacement for 32-bit register pointers. Therefore, instructions such as "MOV AL,[ECX+100]" are perfectly valid. Of course, the 386 supports all the addressing modes of the 8086/286 discussed in Chapter 1. Table 21-1 shows some of the addressing modes supported by the 386.

Table 21-1: Addressing Modes for the 80386

Addressing Mode	Operand	Default Segment
Register	Register	None
Immediate	Data	None
Direct	[OFFSET]	DS
Register indirect	[BX]	DS
	[SI]	DS
	[DI]	DS
	[EAX]	DS
	[EBX]	DS
	[ECX]	DS
	[EDX]	DS
	[ESI]	DS
	[EDI]	DS
Based relative	[BX]+disp	DS
	[BP]+disp	SS
	[EAX]+disp	DS
	[EBX]+disp	DS
	[ECX]+disp	DS
	[EDX]+disp	DS
	[EBP]+disp	SS
Indexed relative	[DI]+disp	DS
	[SI]+disp	DS
	[EDI]+disp	DS
	[ESI]+disp	DS
Based indexed relative	[R1][R2]+disp	If BP is used, segment is SS;
	R1 and R2 are any of the above	otherwise, DS is the segment

Note: In based indexed relative addressing, disp is optional.

Scaled index addressing mode

One of the most powerful addressing modes introduced in the 386 is scaled index addressing mode. It allows access of multidimensional arrays with ease. In scaled index addressing mode, any of the 32-bit registers, except ESP, can be used as a pointer that is multiplied by a scaled factor of 1, 2, 4, or 8. The scaling (multiplication) factors 1, 2, 4, and 8 correspond to byte, word, doubleword, and quadword operands, respectively. Look at Example 21-5 to see how the effective address is calculated in cases where the scaled index addressing mode is used. Only the 32-bit register pointers can be used for this mode. They are shown in Table 21-2.

Table 21-2: 386 Scaled Index Addressing Mode

Scaled Index	Default Segment
[EAX]	DS
[EBX]	DS
[ECX]	DS
[EDX]	DS
[ESI]	DS
[EDI]	DS
[EBP]	SS
[ESP]	SS

Example 21-6 shows how the scaled index addressing mode is used. It must be noted that we cannot use a 16-bit register as a scaled index. In other words, the instruction "MOV AL,[ESI+BX*4]" is invalid.

Example 21-5

Find the effective address in each of the following cases. Assume that ESI = 200H, ECX = 100H, EBX = 50H, and EDI = 100H.

(a) MOV AX,[2000+ESI*4]   (b) MOV AX,[5000+ECX*2]
(c) MOV ECX,[2400+EBX*4]   (d) MOV DX,[100+EDI*8]

Solution:

(a) EA (effective address) is 2000H + 200H × 4 = 2000 + 800H = 2800H. Therefore, the logical address of the operand moved into AX is DS:2800H.

(b) By the same token we have EA = 5000H + 100H × 2 = 5000H + 200 = 5200H.

(c) EA = 2400H + 4 × 50H = 2400H + 140H = 2540H.

(d) 100H + 8 × 100H = 100H + 800H = 900H.

Example 21-6

Using the scaled index addressing mode, write an Assembly language program to add 5 operands of 32-bit size and save the result.

Solution:

```
              .MODEL SMALL
              .386
              .STACK 300H
              .DATA
MYDATA        DD      234556H,0F983F5H,6754AE2H,0C5231239H,0AF34ACB4H
RESULT        DQ      ?
              .CODE
              MOV     AX,@DATA
              MOV     DS,AX
              SUB     EBX,EBX              ;EBX=0
              MOV     EDX,EBX              ;clear EDX
              MOV     EAX,EBX              ;clear EAX
              MOV     CX,5                 ;set the counter to 5
BACK:         ADD     EAX,[MYDATA+EBX*4]   ;add the 32-bit operand
              ADC     EDX,0                ;save the carry
              INC     EBX                  ;point to next 32-bit data
              DEC     CX                   ;decrement the counter
              JNZ     BACK                 ;repeat until counter is zero
              MOV     DWORD PTR RESULT,EAX ;save the lower 32 bits
              MOV     DWORD PTR RESULT+4,EDX ;save the upper 32 bits
              ;place code here to return to DOS
```

In this program, we first define the 32-bit data using the DD directive, and the RESULT is defined as 64-bit using the DQ directive. Notice that EBX is initially zero; therefore, the instruction "ADD EAX,[MYDATA +EBX*4]" adds the first 32-bit operand to EAX since the effective address is MYDATA. "INC EBX" makes EBX =1; therefore, in the next iteration the effective address is [MYDATA+1 * 4], and likewise in the next iteration the effective address is [MYDATA+2 * 4], which is MYDATA+8, and so on. For example, if the offset address for MYDATA is 2000H, the effective address is 2000H for the first iteration, 2004H for the second iteration, 2008H for third iteration, 200CH for the fourth iteration, and so on.

It must be noted that for an Assembly language program to be run under DOS, the effective address should not exceed FFFFH. In other words, if EBX is used as a pointer, you must make sure that the upper 16 bits of the EBX register are all zero, since DOS works in real mode.

Some new 386 instructions

There are many new instructions in the 386 which work in both real and protected modes. A detailed look at each new instruction and how it is used is beyond the scope of this volume. Here are some of the new instructions with examples.

MOVSX and MOVZX instructions

As we discussed in Chapter 6, the 8086 has sign-extend instructions such as CBW (D7 of AL is copied into all the AH bits) and CWD (D15 of the AX is copied into all bits of the DX). In the 386, there is a new instruction CDQ (convert doubleword to quadword) in which the sign bit of EAX, D31, is copied to all the bits of EDX. Notice that in all sign-extend instructions, the accumulator sign is extended. To overcome this limitation, Intel introduced the MOVSX and MOVZX instructions. In the MOVSX, the sign bit of any register (or even a memory location) can be extended (copied) into any register. Similarly, MOVZX zero-extends the contents of a register or memory location. The MOVSX instruction is used to sign-extend the operand in signed number arithmetic to prevent overflow problems. The MOVZX instruction is used in unsigned arithmetic. Look at Example 21-7.

Example 21-7

Find the contents of destination registers after execution of the following code.

```
(a) MOV    BL,-5              (b) MOV    DL,+9
    MOVSX  CX,BL                  MOVSX  EBX,DL
(c) MOV    AL,95H             (d) MOV    BH,83H
    MOVZX  ECX,AL                 MOVZX  AX,BH
```

Solution:

MOVSX copies the source register into the lower bits of the destination register and copies the sign bit into all upper bits of the destination register. Therefore, we have the following.

```
(a) MOV    BL,-5       ;BL=1111 1011B =FBH (2's complement)
    MOVSX  CX,BL       ;CL=FBH,CH=FF since BL is copied into
                       ;CL and the sign bit (D7) is copied into
                       ;all CH bits. BL is unchanged
```

(b) DL =0000 1001B =09H. Then BL=09 and D8 - D31 of EBX are all zero, the sign bit of DL. Therefore, EBX =00000009.

```
(c) MOV    AL,95H      ;AL =1001 0101B =95H
    MOVZX  ECX,AL      ;AL =CL =95H and D8 - D31 of ECX are all zeros
                       ;therefore, ECX =00000095H
```

(d) BH=1000 0011B = 83H. Then AL =BH =83H and D8 - D15 of AX are all zeros. Therefore, AX = 0083H.

Bit scan instructions

The 386 has new instructions allowing a program to scan an operand from LSB to MSB or from MSB to LSB, to find the first high bit (=1). If the scanning is done from the least significant bit (D0) toward higher bits, the BSF (bit scan forward) instruction is used. If the scanning is done from the most significant bit (D31) toward the lower bits, the BSR (bit scan reverse) instruction is used. In these instructions whenever the first high is found, the scanning is stopped and the position of the bit is written into the destination register. The bit position is numbered from D0 (LSB) to D31(MSB), regardless of the direction of scanning. See Example 21-8.

Example 21-8

Find the register contents after the execution of the following code.
(a) MOV   BX,4578H
     BSF   DX,BX   ;scan BX and put the position of the first high into DX
(b) MOV   ECX,3A9H
     BSR   EAX,ECX   ;scan ECX from D31 down and put position of first high into EAX

Solution:

(a) DX=03 since in scanning 4578H =0100 0101 0111 1000B from right to left yields 1 in D3
(b) EAX=9 since in scanning 000003A9H = 0000 0000 0000 0000 0000 0011 1010 1001
    from D31 toward D0 yields the first high in D9; therefore, EAX =9.

Review Questions

1. The 80188/86 is a(n) _____ (8-, 16-bit) processor.
2. What is the size of the external data bus on the 80186?
3. In which 80x86 was the concept of virtual memory introduced?
4. In which 80x86 was the protected mode concept introduced?
5. The 80286 works in which of the following?
   (a) real mode   (b) protected mode
   (c) both (a) and (b)   (d) 8086 virtual mode
6. True or false. The 32-bit registers of the 386 can be accessed only in protected mode.
7. Find the contents of BL, BH, BX, and EBX after execution of instruction "MOV EBX,99FF77AAH".
8. The 80386 uses the _____ (little endian, big endian) convention.
9. List all the 32-bit registers that can be used as pointers into the data segment.
10. In the instruction "MOV EBX,[EAX+ESI*8]", find the effective address if EAX =2000 and ESI =100 (both in hex).
11. Scaled index addressing mode can be used with which of the following registers?
   (a) SI   (b) EDI   (c) EAX
   (d) DX   (e) ECX   (f) CX
12. Find the contents of EDX after execution of the following code.
       MOV   DL,-9
       MOVSX EDX,DL
13. Find the contents of ECX after execution of the following code.
       MOV   DL,-5
       MOVZX ECX,DL
14. Find the contents of DX and AX after execution of the following code.
       MOV   BX,1998H
       BSF   DX,BX
       BSR   AX,BX

SECTION 21.2: 80386: A HARDWARE VIEW

We present a hardware view of the 386 in this section. To avoid confusion, Intel calls the 80386 with a 32-bit external data bus the 80386DX, and 80386SX refers to the 386 with a 16-bit external data bus. In this book we use the 80386 to refer to the 80386DX. Figures 21-2 and 21-3 provide a block diagram and pin layout of the 80386, respectively. Signal functions are provided in Table 21-3.

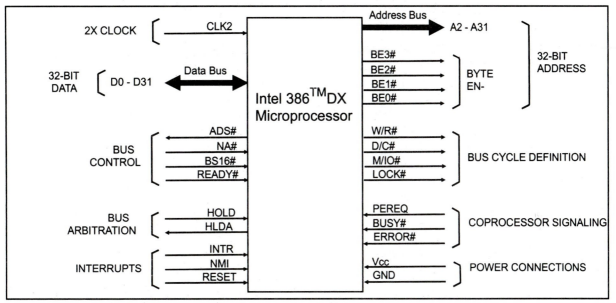

Figure 21-2. 80386 Block Diagram (# indicates active low)
(Reprinted by permission of Intel Corporation, Copyright Intel Corp. 1992)

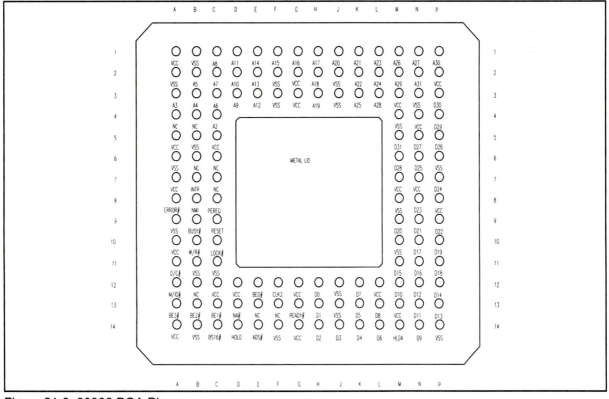

Figure 21-3. 80386 PGA Pins
(Reprinted by permission of Intel Corporation, Copyright Intel Corp. 1992)

Table 21-3: Intel 386 DX PGA Pinout

Signal/Pin		Signal/Pin		Signal/Pin		Signal/Pin		Signal/Pin		Signal/Pin	
A2	C4	A24	L2	D6	L14	D28	M6	V_{CC}	C12	V_{SS}	F2
A3	A3	A25	K3	D7	K12	D29	P4	V_{CC}	D12	V_{SS}	F3
A4	B3	A26	M1	D8	L13	D30	P3	V_{CC}	G2	V_{SS}	F14
A5	B2	A27	N1	D9	N14	D31	M5	V_{CC}	G3	V_{SS}	J2
A6	C3	A28	L3	D10	M12	D/C#	A11	V_{CC}	G12	V_{SS}	J3
A7	C2	A29	M2	D11	N13	ERROR#	A8	V_{CC}	G14	V_{SS}	J12
A8	C1	A30	P1	D12	N12	HLDA	M14	V_{CC}	L12	V_{SS}	J13
A9	D3	A31	N2	D13	P13	HOLD	D14	V_{CC}	M3	V_{SS}	M4
A10	D2	ADS#	E14	D14	P12	INTR	B7	V_{CC}	M7	V_{SS}	M8
A11	D1	BE0#	E12	D15	M11	LOCK#	C10	V_{CC}	M13	V_{SS}	M10
A12	E3	BE1#	C13	D16	N11	M/IO#	A12	V_{CC}	N4	V_{SS}	N3
A13	E2	BE2#	B13	D17	N10	NA#	D13	V_{CC}	N7	V_{SS}	P6
A14	E2	BE3#	A13	D18	P11	NMI	B8	V_{CC}	P2	V_{SS}	P14
A15	F1	BS16#	C14	D19	P10	PEREQ	C8	V_{CC}	P8	W/R#	B10
A16	G1	BUSY#	B9	D20	M9	READY#	G13	V_{SS}	A2	N.C.	A4
A17	H1	CLK2	F12	D21	N9	RESET	C9	V_{SS}	A6	N.C.	B4
A18	H2	D0	H12	D22	P9	V_{CC}	A1	V_{SS}	A9	N.C.	B6
A19	H3	D1	H13	D23	N8	V_{CC}	A5	VSS	B1	N.C.	B12
A20	J1	D2	H14	D24	P7	V_{CC}	A7	VSS	B5	N.C.	C6
A21	K1	D3	J14	D25	N6	V_{CC}	A10	VSS	B11	N.C.	C7
A22	K2	D4	K14	D26	P5	V_{CC}	A14	VSS	B14	N.C.	E13
A23	L1	D5	K13	D27	N5	V_{CC}	C5	VSS	C11	N.C.	F13

(Reprinted by permission of Intel Corporation, Copyright Intel Corp. 1992)

Overview of pin functions of the 80386

D31 - D0 (data bus)

These provide the 32-bit data path to the system board. They are grouped into 8-bit data chunks, D0 - D7, D8 - D15, D16 - D23, and D24 - D31. Each 8-bit data bus is accessed by a separate byte enable pin (BE).

A31 - A2 and BE0, BE1, BE2, BE3

These provide the 32-bit address path to the system board. Notice the absence of A0, A1, or BHE seen in earlier generations of the 80x86. Since the 80386 supports data types of byte (8 bits), word (16 bits), and double word (32 bits), the external buses must be able to access any of the 4 banks of memory connected to the 32-bit data bus. BE0 - BE3 are used to access each bank independently. BE, which stands for byte enable, is active low and used for bank selection. According to Table 21-4, to select D7 - D0, BE0 is used, BE1 is for D15 - D8, etc. See Figure 21-4.

Table 21-4: Data Bus Selection and BE

Data Bus	Byte Enable
D7 - D0	BE0
D15 - D8	BE1
D23 - D16	BE2
D31 - D24	BE3

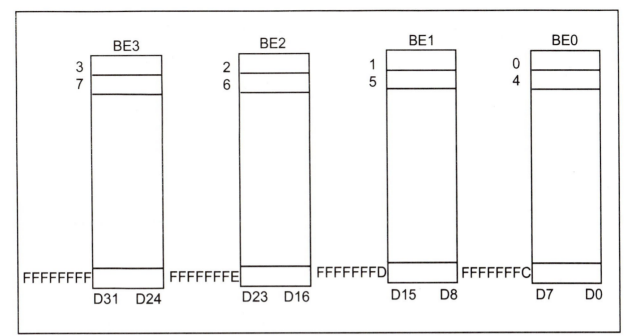

Figure 21-4. 80386 Banks

Example 21-9

Indicate which part of the data bus is selected for the following BEs.
(a) BE3 BE2 BE1 BE0 = 0000   (b) BE3 BE2 BE1 BE0 = 0011
(c) BE3 BE2 BE1 BE0 = 1100   (d) BE3 BE2 BE1 BE0 = 1101

Solution:

(a) D31 - D0, the entire 32-bit data bus
(b) D31 - D16, the upper 16-bit data bus
(c) D15 - D0, the lower 16-bit data bus
(d) the 8 bits of D15 - D8

Figure 21-5 shows the above data selection graphically. Note that BE is active low.

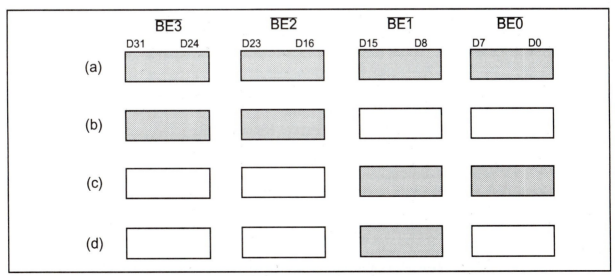

Figure 21-5. Graphical Representation of Example 21-9 (Selected Byte Is Shaded)

W/R, D/C, and M/IO

These signals provide the bus cycle definitions and the type of the bus cycle according to Table 21-5.

CLK2

This provides the timing for the 386. The frequency connected to CLK2 is always twice the system frequency. For example, a 16-MHz 386 system requires CLK2 to be 32 MHz.

Table 21-5: 80386 Bus Cycle Definition

M/$\overline{\text{IO}}$	D/$\overline{\text{C}}$	W/$\overline{\text{R}}$	Bus Cycle Type
0	0	0	Interrupt acknowledge
0	0	1	Does not occur
0	1	0	Data read (I/O)
0	1	1	Date write (I/O)
1	0	0	Memory code read
1	0	1	Halt (shutdown)
1	1	0	Memory data read
1	1	1	Memory data write

(Reprinted by permission of Intel Corporation, Copyright Intel Corp. 1992)

Example 21-10

A 80386 system is advertised as 33 MHz. What frequency is connected to CLK2?

Solution:

CLK2 = 66 MHz because the frequency connected to CLK2 is always twice the system frequency

---

ADS, BS16, NA, and READY

$\overline{\text{ADS}}$ (address status), $\overline{\text{BS16}}$ (bus size), NA (next address request), and $\overline{\text{READY}}$ are bus control signals. These signals allow the implementation of an efficient bus control circuitry. For example, using $\overline{\text{BS16}}$ allows the 80386 to be connected to the 16-bit data bus instead of 32-bit. The use of $\overline{\text{NA}}$ (next address) provides the option of address pipelining, where the address of the next memory cycle is provided in the last clock cycle of the present memory cycle.

RESET

This is a level-sensitive input signal into the 80386. When a low-to-high signal is applied to RESET, the 80386 will suspend all operations and the registers are initialized to fixed values. The RESET state of EIP and CS must be noted, along with the state of A31 - A2 and $\overline{\text{BE0}}$ - $\overline{\text{BE3}}$, because this has some major implications as far as where the boot ROM should be located (see Table 21-6). This means that the microprocessor will fetch the first opcode from memory location FFFFFFF0. This is 16 bytes from the 4 gigabyte maximum address range of FFFFFFFFH. At this location, there is either a JMP FAR or CALL FAR instruction. Upon executing the JMP or CALL instruction, the 386 makes A31 - A20 all zero, thereby forcing it to stay within the 1 megabyte address range of real mode. This is the case for all 386, 486, and Intel Pentium chips. All these processors wake up in real mode but the address where the first opcode must be found is located in the extended memory space and not in the first megabyte address space of real mode. This means that for 386 and higher PC systems, there are duplicate ROMs in both the 4 gigabyte and 1 megabyte address spaces, as shown in Figure 21-6.

The remaining signals of the 386 are similar to the 80286, and readers can refer to Chapter 10 for their meanings.

Table 21-6: RESET State

Item	Contents
CS	F000
EIP	0000FFF0
A31 - A2	All high
BE3 - BE0	All low

(Reprinted by permission of Intel Corporation, Copyright Intel Corp. 1992)

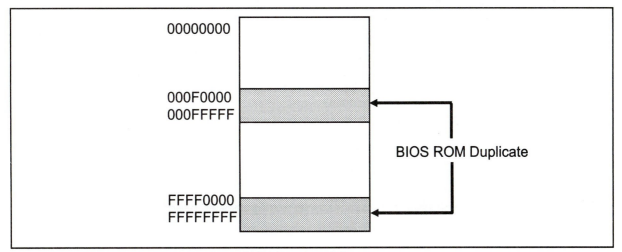

Figure 21-6. BIOS ROM Duplicate for 386/486/Pentium PC

Bus bandwidth in the 386

With zero wait states, it takes the 386 two clocks to perform the read or write cycle. A 2-clock bus cycle is standard in all high-performance microprocessors, including RISC processors. This leads to a very high bus bandwidth. The two clocks of the 386 memory (or I/O) bus cycle time for zero wait states are shown in Figure 21-7. In the case of pipelined read/write cycle time, the next address is provided in the last T clock of the present cycle, thus providing some extra time for the decoder logic circuitry, path delay, and memory access time. Although in pipelined mode, the next address is provided in the last stage of the present cycle, the read and write cycle time still consists of 2 clocks for the zero-wait-state system.

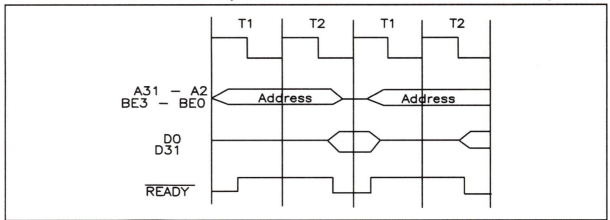

Figure 21-7. 386 Bus Cycle Time (Nonpipelined)
(Reprinted by permission of Intel Corporation, Copyright Intel Corp. 1992)

Example 21-11

Calculate the 386 bus bandwidth of a 33-MHz system with each of the following.
(a) 0 WS   (b) 1 WS

Solution:

With the T state of 30 ns (1/33 MHz=30 ns), we have memory cycle time of 60 ns and 90 ns for (a) and (b), respectively.
(a) The bus bandwidth is (1/60 ns) × 4 = 66.66 megabytes/second.
(b) The bus bandwidth is (1/90 ns) × 4 = 44.44 megabytes/second.

Data misalignment in the 386

The case of misaligned data has a major effect on the 386 bus performance. If the data is aligned, for every memory read cycle the 80386 brings in 4 bytes of data using the D31 - D0 data bus. Such data alignment is referred to as *doubleword alignment*. To make data doubleword aligned, the least significant digits of the hex addresses must be 0, 4, 8, or C (hex). Look at Example 21-12.

Example 21-12

Show the data transfer of the following cases and indicate the memory cycle time if the system frequency is 25 MHz. Assume that EAX = 4598F31EH and the system is in real mode.

(a) MOV [2950],EAX
(b) MOV [299A],EAX

Solution:

The system frequency of 25 MHz makes the cycle time 80 ns (1/25 MHz =40 ns and each memory cycle is 2 clocks, giving 80 ns).

(a) In this instruction, the 4-byte content of EAX is moved to memory location with starting offset address of 2950H on the 32-bit data bus of D31 - D0. This address is doubleword aligned since the least significant digit is 0. Therefore, it takes only one memory cycle or 80 ns to transfer the data.

(b) In the first memory cycle, locations with addresses of 2998H, 2999H, 299AH, and 299BH are accessed, but only 299AH and 299BH are used for storing AL and AH. In the second memory cycle, the address offsets of 299CH, 299DH, 299EH, and 299FH are accessed where only 299CH and 299D are used to store the upper 16 bits of EAX. This means that we have a total of 160 ns. If possible, this must be avoided since nonaligned data slows the data access.

I/O address space in the 386

The 80386 can access a maximum of 65,536 input ports and 65,536 output ports using the IN and OUT instructions. In this regard, the 386 is exactly like 8086/88/286 microprocessors.

Review Questions

1. BHE and A0 are associated with processor _____ (80386SX, 80386DX).
2. The 80386SX is _____ (16, 32) bits externally.
3. Exactly how many pins are set aside for the address in the 386?
4. The BE2 pin is associated with which part of the data bus?
5. An 80386 of 20 MHz requires a crystal frequency of _____.
6. Give the first physical address location where the 80386 looks for an opcode upon RESET.
7. With the same frequency, the 80386SX has bus bandwidth _____ (twice, half) that of the 80386.
8. Find the memory cycle time for an 80386 of 20 MHz.

SECTION 21.3: 80386 PROTECTED MODE

The 80386 protected mode discussion applies equally to 486 and Pentium chips. Due to the complexity associated with 80386 protected mode, many long chapters are needed for this subject, and for this reason, here we simply provide an overview of the 386 in protected mode.

Protection mechanism in the 386

As discussed in Chapter 1, physical addresses in the 8086 are calculated by shifting left the segment register and adding it to the offset. This is also the case for the 80286 and subsequent 80x86 processors in real mode. However, in protected mode, the physical address of blocks of data or code is held by a look-up table and the segment register is no longer shifted left to calculate the physical address. Instead, it is used as an index into a look-up table in which the physical address of the operand or code is held.

Another important change introduced in the 80386 is the protection mechanism. The lack of protection of the operating system or users' programs is one of the weaknesses of 8088/86-based MS DOS. This weakness is due to the inability of the 8088/86 to block general instructions from accessing the core (kernel) of the operating system. In the 8088/86, since any program can go from any code segment to any code segment, it is easy to crash the system. In contrast, the 80386 provides resources to the operating system that prevent the user from either accidentally or maliciously taking over the core (kernel) of the operating system and forcing the system to crash. Of course, this idea of protection is nothing new; it is commonly used in mainframes and minicomputers, where it is often referred to as user and supervisor mode. The 386 provides protection by allowing any data or code to be assigned a privilege level. The four privilege levels are 0, 1, 2, and 3, where the privilege level of 0 is the highest and level 3 is the lowest. While operating systems are always assigned the highest privilege level (level 0), the user and applications such as word processors are assigned the lowest privilege level (level 3). Since the user is assigned the lowest privilege level, any attempt by the user to take over the operating system is blocked. Higher privilege levels can access lower levels but not the other way around. Again, it must be emphasized that the protection mechanism can be used only when the 80386 is switched to protected mode.

Virtual memory

Another major feature of the 80386 is the ability to access *virtual memory*. A CPU with virtual memory is fooled into thinking that it has access to an unlimited amount of physical (DRAM) memory. DRAM primary memory is also called *main memory*. In this scheme, every time the CPU looks for certain information, the operating system will first search for it in main DRAM memory and if it is not there, it will bring it into RAM from secondary memory (hard disk). What happens if there is no room in RAM? It is the job of the operating system to swap data out of RAM and make room for new data. Which data will be swapped out depends on how the operating system is designed. Some operating systems use the LRU (least recently used) algorithm to swap data in and out of primary memory (DRAM). In the LRU method, the operating system keeps account of which data has been used the least number of times in a certain period, and when there is need for room it will swap out the least recently used data to hard disk to make room for the new data. The total amount of RAM on the computer could be only 16M with a hard disk capacity of 500M bytes, but the CPU is fooled into thinking that it has access to all 500M of memory. Among the operating systems, IBM OS/2, Microsoft Windows NT, and all the variations of Unix, such as Nextstep and Sun Micro's Solaris, use the capability of the 80386's virtual memory. Since MS DOS was written for the 8088/86 microprocessors, it does not have virtual memory.

To implement virtual memory, two methods are used: segmentation and paging. In segmentation, the size of the data swapped in and out can vary from 1 byte to few megabytes (in 80386, 80486, and Pentium, the upper limit can be as high as 4 gigabytes). In paging, the size is a multiple of one page of 4096 (4K) bytes. Paging is used widely since it prevents memory fragmentation, where available memory becomes fragmented into small sections of varied sizes. When this happens, the operating system must continuously move files around to make room for the new files, which could be any size. Paging makes the job of the operating system much easier since all the files will be a multiple of 4K bytes. If the size of a file is not a multiple of 4K bytes (which is the case most of the time), the operating system will leave the unused portion empty and the next file will be placed on a 4K boundary. This is similar to the cluster in floppy and hard disks. As shown in Chapter 19, the disk allocates memory to each file in clusters. For example, if 4 sectors are used for each cluster, each cluster can store 2048 (4 × 512) bytes per sector. If a given file is 12,249 bytes, the operating system will assign a total of 7 clusters or 14,168 (7 × 2024 =14,168) bytes. All bytes between 14,168 and 12,249 are unused. This results in wasting some memory space on the disk but at the same time makes the design of the disk controller and operating system much easier. This concept applies as well to the paging method of virtual memory as far as the allocation of main memory (DRAM) to data and code is concerned. One can briefly define the segmentation and paging virtual memory mechanisms in the following statement. While in segmentation virtual memory, the file can be any byte size, located anywhere it can fit into main memory. In paging virtual memory, the file is always a multiple of 4096 bytes and located on a 4K-byte boundary in main memory.

All high-performance RISC microprocessors use paging virtual memory only and none use the segmentation method. The reason that 386, 486, and Pentium processors support segmentation (in addition to paging) is due to the fact that they had to stay compatible with the 8086's 64K-byte segment size.

Segmentation and descriptor table

In segmentation virtual memory, the segment registers are used as selectors into the descriptor table, where all the information about a given piece of data and code is kept. The descriptor table uses 8 bytes of space to provide the following information about a given piece of code or data.

1. 4 bytes for the A0 - A31 address, where the code (or data) is located in main memory. This allows the 386 to access any memory location within its 4 gigabyte address space. Notice in Figure 21-8 that A23 - A0 is provided by bytes 2, 3, and 4, but A31 - A24 is provided by byte 7.
2. L0 - L19: This 20-bit limit is used for checking the segment size and is limited to 1 megabyte. Notice that bytes 0 and 1 provide L0 - L15, and D0 - D3 of byte 6 is set aside for L16 - L19. This provides the scheme whereby the 1 megabyte limit imposed on data or code is checked. Since the limit for the segment-oriented 8086/286 is 64K bytes (2^{16} = 64K), the upper 4 bits must all be zeros. However, in the 386, the segment limit can be raised to 4 gigabytes. To do that the G (granularity) bit is set to high. If G = 0, L0 - L19 is used as a number of bytes for the limit, but if G =1, L0 - L19 is used as a multiple of 4K for the segment limit. This gives $2^{20} \times 2^{12} = 2^{32} = 4$ gigabytes address range, making it possible for the 386 to have segments as large as 4 gigabytes. This is quite a relief for software writers of database and other application packages since the size of data (e.g., a big array) can go as high as 4 gigabytes and is no longer limited to 64K. In the case of the 286 when the size of the data section of the program was larger than 64K, they had to do lots of software manipulation to overcome this limitation. This also explains the origins of the memory models of SMALL, MEDIUM, LARGE, and so on, widely used in Assembly and C programs.
3. The access byte allows protection of a given piece of data or code by assigning the privilege levels of 0, 1, 2, and 3 to it, where 0 indicates the highest privilege level and 3 is the lowest privilege level. D0 - D7 of the access byte are described next.

```
  31                                                              0   BYTE
                                                                      ADDR.
 ┌──────────────────────────────────┬──────────────────────────────┬─────┐
 │ SEGMENT BASE 15......0            │ SEGMENT LIMIT 15......0       │  0  │
 ├──────────┬───┬─┬─────┬───────────┼─┬─────┬─┬──────┬─┬───────────┼─────┤
 │ BASE 31..24│G│D│0│ AVL│ LIMIT 19..16│P│ DPL │S│ TYPE │A│ BASE 23..16│ +4 │
 └──────────┴───┴─┴─────┴───────────┴─┴─────┴─┴──────┴─┴───────────┴─────┘
```

BASE	Base Address of the segment
LIMIT	The length of the segment
P	Present Bit 1 = Present 0 = Not Present
DPL	Descriptor Privilege Level 0 - 3
S	Segment Descriptor 0 = System Descriptor 1 = Code or Data Segment Descriptor
TYPE	Type of Segment (3 bits: X, E, R/W)
A	Accessed Bit
G	Granularity Bit 1 = Segment length is page granular 0 = Segment length is byte granular
D	Default Operation Size (code segment descriptors only) 1 = 32-bit segment 0 = 16-bit segment
0	Bit must be zero for compatibility with future processors
AVL	Available field for user or OS
Note:	In a maximum-size segment (i.e., a segment with G=1 and segment limit 19...0 = FFFFFH), the lowest 12 bits of the segment base should be zero (i.e., segment base 11...000 = 000H).

Figure 21-8. Descriptor Table Entry

A (accessed) bit

If the data or code is accessed (used), A =1; otherwise, A =0. This allows the operating system to monitor the A bit periodically to see if the CPU is using this piece of code or data. If a piece of code or data has not been used recently, the next time the operating system needs to make a room in main memory for new pieces of code (or data), it can move this code (or data) back to the hard disk. The A bit also allows the operating system to decide if a given piece of information (code or data) needs to be saved. For example, if a piece of data has not been accessed, the operating system can trash it and does not need to waste time saving it on the hard disk. On the other hand, if the data was accessed and it was written into, the operating system must save a copy of it on the hard disk before it abandons it to create room in main memory for some other data or code.

R/W (read/write) bit

This bit allows code or data to be read protected or write protected. For example, the core of the operating system can be write protected, which prevents the user from writing into it and crashing the system. In the case of DOS, any program can use the DEBUG utility and alter the core of the operating system residing in main memory (DRAM), and crash the PC.

X bit

This has a different meaning for the data segment and code segment. In the case of data, it indicates if the segment should expand downward as the stack segment grows, or upward as the data segment grows. In the case of the code segment, it is used to enforce certain rules of privilege level access.

E bit

This indicates if the information is executable (E = 1), such as code, or nonexecutable (E = 0), such as data and stack. This bit also affects the way the X and R/W bits are interpreted.

S bit

This indicates if the descriptor belongs to the code and data segment (S =1) or if it is a system segment descriptor (S =0).

SECTION 21.3: 80386 PROTECTED MODE   **649**

DPL (descriptor privilege level) bits

This allows one of the combinations, 00, 01, 10, or 11, to be assigned to the code or data, indicating the privilege level.

P (present) bit

This indicates if the piece of code or data is present in main memory (DRAM). If it is present (P =1), the CPU will process it. If it is not present (P =0), the CPU causes an exception and the exception handler of the operating system will bring the desired piece of code or data into main memory from the hard disk. When the operating system does so, it sets P =1 to indicate that the information is now present in main memory.

Example 21-13

From Figure 21-8 we have the following access byte for code and data.

P   DPL   1   1   ...   ...   A   (access byte for code segment)

P   DPL   1   0   ...   ...   A   (access byte for data segment)

Discuss the following access bytes.
(a) 10011011   (b) 10010111   (c) 11110001

Solution:

(a) This is an access byte for code segment, present, accessed, and privilege level of 00 (highest).
(b) This is an access byte for data segment, present, accessed, privilege level of 00 (highest), and both read and write accessible.
(c) This is an access byte for data segment, present, accessed, privilege level of 11 (lowest), and write protected.

The descriptor table is built by the operating system for every piece of code and data. The descriptor table register (DTR) inside the 386 holds the physical address of where the table is located in the 4 gigabyte address space, which means that the descriptor table register (DTR) is a 32-bit register. When the CPU changes the contents of a segment register (CS, DS, and so on), it uses the segment value as an index into the descriptor table and pulls into the CPU from the descriptor table all 8 bytes belonging to this segment. These 8 bytes are saved in the invisible part of the segment register inside the 386, which means that every segment register inside the 386 has an 8-byte extension which is not visible to the programmer. The pulling of an 8-byte table into the CPU for every change of segment register is time consuming but afterward, the CPU has all the information it needs to access a piece of code or data. The addition of two new segment registers, FS and GS, in the 386, plus the presence of CS, DS, SS, and ES, helps the CPU always to have a total of 6 descriptor table entries available inside the CPU. If code or data is not held by one of these 6 descriptor table entries, the CPU must go through the long process (it takes 22 clock cycles) of pulling them into the CPU. As we will show later in this section, this problem is solved in the paging method.

Looking at the 8 bytes of the descriptor table, one might ask why Intel did not assign 32-bit physical addresses of desired code or data in consecutive bytes, instead of using bytes 2, 3, 4, and then byte 7. The reason is the 80286 CPU. In the 286 protected mode, bytes 2, 3, and 4 are used for the 24-bit address (A0 - A23), and bytes 6 and 7 had to be zero. This led Intel to use byte 7 for the A31 - A24 part of the physical address of the 386. Byte 6 is used for raising the limit and the G bit, among other things. See Figure 21-8.

Local and global descriptor tables

There are two types of descriptor tables for the 386: the local descriptor table (LDT) and the global descriptor table (GDT). The GDT is used for the system, and individual tasks can have their own LDT. How do the segment registers know which one they are accessing? The third bit (TI) of the segment register (referred to as the selector) always indicates which table should be used. See Figure 21-9.

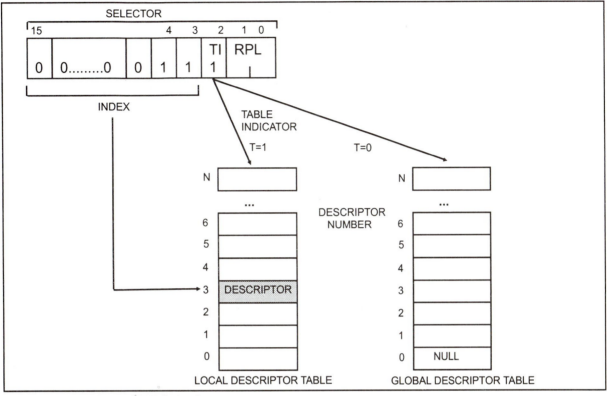

Figure 21-9. LDT and GDT Selection
(Reprinted by permission of Intel Corporation, Copyright Intel Corp. 1992)

64 Terabytes of virtual memory

As seen in Figure 21-9, the 14 bits of the selector (segment) register can have 16,384 (2^{14}) possible combinations. Each possible value can access a descriptor that can hold addresses of memory chunks as large as 4 gigabytes. Therefore, we have $2^{14} \times 2^{32} = 64$ terabytes of virtual memory for the 386 (recall that *tera* is defined as 2^{40}). To put it another way: The 386 can access 64 terabytes of hard disk (virtual memory) as long as the virtual memory is broken down into 4 gigabyte pieces, since it has only 32 address pins. While the segment limit in the 8086/286 is 64K bytes, the segment limit in the 386 was raised to 4G. One of the drawbacks of 386 segmentation is its variable segment size, which leads to memory fragmentation. Another is the absence of what is called a *dirty bit* in the access byte of the descriptor table. Assume that there is some memory that can be written into. The accessed (A) bit indicates if the data has been accessed but does not indicate if any new data was written into it. Why should the operating system care if the memory is altered (written into)? If the data is altered, it is the job of the operating system to save it on the disk to make sure that the hard disk always has the latest data. If the dirty bit is zero (D=0), it means that the data has not been altered and the operating system can abandon it when it needs room for new data (or code) since the original copy is on the hard disk. This will save time for the operating system. If the dirty bit is one (D=1), the operating system must save the data before it is lost or abandoned. Both problems of variable segment size and lack of a dirty bit in segmentation are fixed in the paging method of virtual memory.

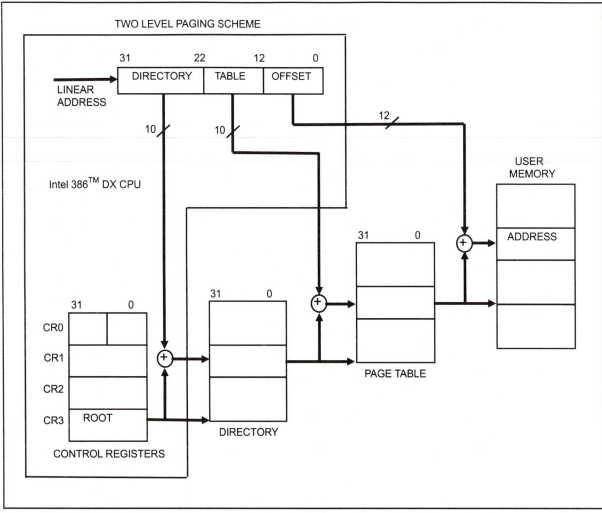

TWO LEVEL PAGING SCHEME

Figure 21-10. Paging Mechanism
(Reprinted by permission of Intel Corporation, Copyright Intel Corp. 1992)

Paging

Paging of virtual memory was a new addition to the 386, but the segmentation method was left over from the 80286. All RISC and Motorola 680x0 processors support paging virtual memory. In paging virtual memory, main memory is divided into fixed 4K-byte chunks instead of variable sizes of 1 byte to 4 gigabytes, as in segmentation. If a given piece of code or data is not present in main memory, the operating system brings it into main memory from the hard disk, 4K at a time. This is a much more manageable size of memory to transfer than, for example, a 64K-byte segment. Since the size of memory is reduced to 4K bytes, the 386 keeps a table for the 32 most recently used pages present in main memory to prevent the CPU from swapping data in and out of main memory unnecessarily. This table is called the *translation lookaside buffer* (TLB) and is kept inside the 386. To understand the importance of the TLB, let's look at the the way paging works. First, the term *linear address* in the 386 must be clarified. The 32-bit address of the operand is called the linear address. This linear address can be a direct value such as in the instruction "MOV EAX,[50000000]" or may be pointed to by any of registers EDI, ESI, EBX, EDX, and so on, as in the instruction, "MOV EAX,[EBX]". This linear address must be translated into a physical address to be put on the A31 - A0 address pins and sent out for the address decoder to find the location in RAM or ROM. In other words, the address 50000000H in instruction "MOV EAX,[50000000]" does not refer to an actual RAM and ROM address 50000000H. See Figure 21-10.

Going from a linear address to a physical address

In paging, the linear address is divided into three parts. The upper 10 bits (A31 - A22) are used for an entry into what is called a *page directory*. There is a 32-bit register, CR3, inside the 386 that holds the physical base address of the page directory. Since the upper 10 bits of the linear address points to the entry in the page table directory, there can be 1024 page directories (2^{10} =1024). Each entry in the page directory is 4 bytes of page table descriptor. Of the 4 bytes of each page table descriptor, the upper 20 bits are used to point to another table, where the physical address of the 4K page frame is held. How is the correct entry in the table located? A21 - A11 (10 bits total) of the linear address are used to point to one of the page table entries. Again, each entry in this second table has 4 bytes. The upper 20 bits are for A31 - A12 of the physical address of where data is located. The lower 12 bits of the physical address are the lower 12 bits of the linear address. See Figures 21-11 and 21-12. In other words, only the lower 12 bits of the linear address match the lower 12 bits of the physical location in RAM or ROM where data is located, and the upper 20 bits of the linear address must go through two levels of translation tables to get the actual physical address of the beginning page where the data is held. This seems like a very long and inefficient process, and it is. This is the reason for the TLB (translation lookaside buffer). The TLB inside the 386 holds the list of the most recently (commonly) used physical addresses of the page frames. When the CPU wants to access a piece of information (data or code) by providing the linear address, it first compares the 20-bit upper address with the TLB to see if the table entry for the desired page is already inside the CPU. This results in two possibilities: (1) If it matches, it picks the 20-bit physical address of the page and combines it with the lower 12 bits of the linear address to make a 32-bit physical address to put on the 32 address pins to fetch the data (or code); (2) if it does not match, the CPU must fetch into TLB the page table entry from memory.

Each entry in the page table has 4 bytes. Of these 4 bytes, 20 bits are used to hold the A31 - A11 physical address of the page frame. The rest are used for the P (present) bit, D (dirty) bit, R/W (read/write) bit, A (accessed) bit, and finally, U/S (user/supervisor) bit, which indicates the privilege level of given data or code. In the segmentation method there were 2 bits for privilege level, giving rise to 4 levels of protection of 0, 1, 2, and 3, where level 3 was assigned to the lowest level and level 0 to the highest level. However, in the paging method, there is only 1 bit for privilege level, which is called U/S (user/supervisor). If U/S =0, it is user privilege level and is equivalent to level 3 in segmentation. If U/S=1, it is supervisor level, belonging to the operating system and system kernel (BIOS). The supervisor privilege level is equivalent to level 0, 1, 2 in the segmentation method.

31 12	11 10 9	8	7	6	5	4	3	2	1	0
PAGE TABLE ADDRESS 31..12	OS RESERVED	0	0	D	A	0	0	U -- S	R -- W	P

Figure 21-11. Page Directory Entry   (Points to Page Table)
(Reprinted by permission of Intel Corporation, Copyright Intel Corp. 1992)

31 12	11 10 9	8	7	6	5	4	3	2	1	0
PAGE FRAME ADDRESS 31..12	OS RESERVED	0	0	D	A	0	0	U -- S	R -- W	P

Figure 21-12. Page Table Entry   (Points to Page)
(Reprinted by permission of Intel Corporation, Copyright Intel Corp. 1992)

The bigger the TLB, the better

Since the TLB in the 386 keeps the list of addresses for the 32 most recently used pages, it allows the CPU to have access to 128K bytes ($32 \times 4 = 128$) of code and data at any time without going through the time-consuming process of converting the linear address to a physical address (two-stage table translation). See Figure 21-13. Therefore, one way to enhance the processor is to increase the number of pages held by the TLB. This is what the Pentium has done, as we will see in Chapter 23. Table 21-7 compares paging and segmentation.

Table 21-7: Paging and Segmentation Comparison

Feature	Paging	Segmentation
Size	4K bytes	Any size
Levels of privilege	2	4
Base address	4K-byte aligned	Any address
Dirty bit	Yes	No
Access bit	Yes	Yes
Present bit	Yes	Yes
Read/write protection	Yes	Yes

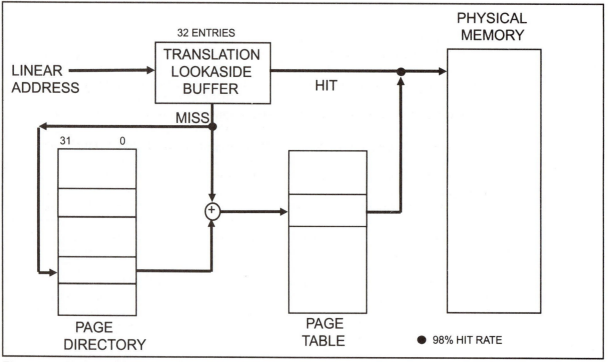

Figure 21-13. Translation Lookaside Buffer
(Reprinted by permission of Intel Corporation, Copyright Intel Corp. 1992)

Virtual 8086 mode

A major dilemma for designers of the Intel 386 was how to enhance the 386 and still run 8088/86 software based on MS DOS in protected mode. They solved this dilemma by adding the virtual 8086 mode to the 386. In virtual 8086 mode, the 386 partitions memory into 1 megabyte sections, each assigned to one task. It also runs each task as if it is an 8086 program, not concerned with privilege levels. In other words, the 8086 virtual mode of the 386 microprocessor allows any program written for DOS to be run unchanged under one task, where each task can have its own 1 megabyte of memory. This means that in virtual 8086 mode, the 386 uses the SEG:OFFSET concept used in the 8088/86 microprocessor. Both Microsoft Windows 3.x and the IBM OS/2 2.0 use the virtual 8086 mode of the 80386 microprocessor. These operating systems use the 386's virtual 8086 mode to run multiples of programs written for the 8088/86. The difference is that in MS Windows, only one task can be active at a time and all other tasks are sitting idle (dormant) while one task is being run, but in OS/2 2.0 each task is given a slice of the CPU's time, and many tasks can be active concurrently. For example, a word processor can be used while the modem/FAX is receiving and sending data and a spreadsheet program such as Lotus 123 is doing some calculations and a disk is being formatted. Of course, since there is only one microprocessor taking care of all these tasks, it is the job of the OS/2 2.0 operating system to slice the CPU time and assign each task time on a circular rotational basis. If there are too many tasks and all are active, they all seem to be slow since each task gets less time (attention) from the CPU. Of course, one way to solve this slowness is to use high-performance CPUs with 60 - 100 MHz speed, such as the Pentium. Note that while OS/2 2.0 uses preemptive multitasking, Windows 3.x uses cooperative multitasking. In cooperative multitasking, two or more applications cooperate with each other in taking turns to use the CPU alternately. If one application misbehaves, it can cause the whole system to be unstable and crash. In preemptive multitasking, a task can be interrupted preemptively at any point by another program. If a task is interrupted by another task, its present state will be saved by the operating system and it will be serviced after the new task is given a chance to use the CPU.

Review Questions

1. True or false. In protected mode, the 386 physical address is calculated by shifting the segment register value and adding the offset.
2. Virtual memory refers to _____ (main DRAM, hard disk) memory.
3. How does the operating system decide which code (or data) should be abandoned to make room for new code?
4. In protected mode (segmentation), where is the physical address of the desired code or data located?
5. Of the 8 bytes of the descriptor table entry, which one(s) are used for the physical address? Assume that they are numbered from 0 to 7.
6. When a piece of code is run, which bit of the access byte is modified?
7. In 386 segmentation, level 3 is assigned the _____ (lowest, highest) privilege.
8. In 386 segmentation, level 0 is assigned the _____ (lowest, highest) privilege.
9. How many privilege levels are there in 386 paging?
10. True or false. In 386 paging, the linear and physical addresses are the same.
11. To get the physical address in 386 paging the linear address must go through ___ (1, 2) stage(s) of translation.
12. The virtual 8086 mode was introduced in the _____ (80286, 80386).
13. True or false. In MS Windows 3.0 and 3.1, only one task can be active at a time.
14. Why is OS/2 2.0 but not Windows 3.x a true multitasking operating system?

SUMMARY

This chapter began with an overview of the additional features and new instructions included in some of the microprocessors introduced after the 8086, namely, the 80186, 80286, and 80386. The 386 runs previous-generation software in real mode, and has a new mode called protected mode, which allows more sophisticated software engineering.

The second section looked at the hardware of the 80386DX and 80386SX. Each pin of the microprocessor was explained, as well as issues of bus bandwidth and data misalignment.

The third section of this chapter gave an introduction to protected mode, introduced to provide a protection system based on user's priority. Another new feature of protected mode is virtual memory, which is implemented by one of two methods: segmentation and paging. While in segmentation virtual memory, the file could be any byte size, located anywhere it can fit into main memory; in paging virtual memory, the file is always a multiple of 4096 bytes and located on a 4K-byte boundary in main memory. In segmentation virtual memory, the segment registers are used as selectors into the descriptor table, where all the information about a given piece of data and code is kept, whereas in paging virtual memory, the physical address is retrieved from two layers of look-up tables. Conversion from linear to physical addresses is another variation in protected mode. Physical addresses in protected mode are retrieved from a look-up table rather than calculated by shifting left the segment register and adding the offset as is done in real mode.

PROBLEMS

SECTION 21.1: 80386 IN REAL MODE

1. Which microprocessors support the instructions PUSHA and POPA?
2. Explain the function of PUSHA. It is equivalent to what set of instructions?
3. Explain the function of POPA. It is equivalent to what set of instructions?
4. Which microprocessors support "SHL dest,immediate"?
5. Find the contents of the destination register for each of the following.
   (a) MOV   AX,43H
        SHL   AX,4
   (b) MOV   BX,8000H
        SHR   BX,16
   (c) MOV   CX,0AAAH
        ROL   CX,8
   (d) MOV   CX,0AAAH
        ROL   CX,12
6. True or false. The 80286 was the first 80x86 to abandon multiplexing of the address and data buses.
7. In which 80x86 microprocessor was the concept of virtual memory introduced?
8. In which of the 80x86 microprocessors was the 2-clock memory cycle introduced?
9. Which of the following instructions will cause an error in the 386?
   (a) MOV EBX,AX   (b) MOV ECX,BX
   (c) ADD ECX,EDX   (d) ADD EDX,AL
   (e) MOV EBX,SI   (f) ADD SI,DI
10. Show how data is stored in "MOV [3500],EBX". Assume that EBX =9834F543H.
11. Show how data is stored in "MOV ES:[1000],ECX" (ECX =07B324H).
12. Which registers can be used for the scaled index addressing mode?
13. Write a 386 program to add a factor of 100 to an array of 10 DWORD data. Use the scaled index addressing mode.
14. Write a 386 program to add two multibyte data items of 8-byte size and store the result. Use the scaled index addressing mode.
15. Indicate all the registers that can be used for pointers in the 386. Also give their default segments.
16. Find the destination register contents after execution of each of the following.
    (a)   MOV   BX,-12
           MOVSX EBX,BX
    (b)   MOV   CL,-8
           MOVSX EDX,CL
    (c)   MOV   AH,7
           MOVZX ECX,AH
    (d)   MOV   AX,99H
           MOVZX EBX,AX

17. Find the contents of EAX and EBX after execution of the following.

        MOV   ECX,307F455H
        BSF   EAX,ECX
        BSR   EBX,ECX

18. Find the contents of AX and DX after execution of the following.

        MOV   BX,98H
        BSF   AX,BX
        BSR   DX,BX

19. What is the purpose of instructions MOVSX and MOVZX?
20. True or false. In the instruction "MOVSX REG,REG", the source and destination registers must match in size.

SECTION 21.2: 80386: A HARDWARE VIEW

21. BE0 - BE3 are active _____ (low, high).
22. True or false. The address and data bus in the 386 are multiplexed.
23. Which part of the data bus is activated if BE0 = 0 and BE1 = 0 (at the same time)?
24. Which part of the data bus is activated if BE2 = 0 and BE3 = 0 (at the same time)?
25. Which part of the data bus is activated if BE0 = 0 and BE3 = 0 (at the same time)?
26. Which part of the data bus is activated if BE1 = 1 and BE2 = 0 (at the same time)?
27. A 25-MHz 386 is connected to CLK2 of _____ MHz.
28. Show the status of CS, IP, A31 - A2, and BE3 - BE0 in the 386 upon RESET.
29. What are the implications of your answer to Problem 28?
30. For what addresses in the 386 PC is BIOS ROM duplicated, and why?
31. Draw the bus cycle for the nonpipelined 386. Show the address, data, and READY signals.
32. Find the total bus cycle necessary to transfer the operand in the instruction "MOV [2002],ECX".
33. For aligned data, the addresses for DWORD type data in the 386 must have _____ as the lower hex digit.
34. Find the memory cycle time for a 33-MHz 386.

SECTION 21.3: 80386 PROTECTED MODE

35. What is virtual memory?
36. True or false. The CPU requests data from virtual memory before it requests data from main memory.
37. While main memory is made of _____ (DRAM, hard disk), virtual memory is _____ (DRAM, hard disk).
38. What is the difference between the real and protected modes of the 386 in terms of memory space?
39. To access the entire 4 gigabytes of the 386, the CPU must be in _____ mode.
40. True or false. The 286 supports both segmentation and paging virtual memory.
41. True or false. The 386 supports both segmentation and paging virtual memory.
42. True or false. In the 286, the segment size can be 1 byte to 16 megabytes.
43. True or false. In the 386, the segment size can be 1 byte to 4 gigabytes.
44. For the 386, what is the page size in paging virtual memory?
45. How many bytes does each entry in the descriptor table use?
46. State the difference between real mode and protected mode as far as the physical address of the operand is concerned.
47. How many bits are set aside for the addresses in the descriptor table, and where are they located in the descriptor table?
48. To make the descriptor table of a 386 286-compatible, we must make bytes 7 and 8 all _____ (0s,1s).
49. How many bits are set aside for the segment limits in the descriptor table, and where are they located in the descriptor table?
50. True or false. Every piece of data or code accessed by the 386 in protected mode must have an access byte.

PROBLEMS

51. 00 is the _____(lowest, highest) privilege level and 11 is the _____ (lowest, highest) one.
52. What is the function of bit A in the access byte of a 386 descriptor table entry?
53. What is the function of bit P in the access byte of a 386 descriptor table entry?
54. State the characteristic of each of the following access bytes. State for each if it is for code or data.
    (a) 10010001   (b) 11110001   (c) 11110011
    (d) 11111011   (e) 10011011   (f) 11111011
55. What does TLB stand for, and what is it used for?
56. In the 386, state the difference between the linear and the physical address.
57. In paging to get the address of code or data, the 386 converts from _____ (linear address, physical address) to _____ (linear address, physical address).
58. In the 386, before the address of the data or code is fetched it is checked against the values held by the_____.
59. What is the number of entries in the 386 TLB?
60. True or false. In virtual 8086, the addresses are calculated by shifting the segment register left and adding it to the offset.
61. State the differences between paging and segmentation virtual memory.
62. How many privilege levels are there in paging?

ANSWERS TO REVIEW QUESTIONS

SECTION 21.1: 80386 IN REAL MODE
1. 16
2. 16
3. 80286
4. 80286
5. c
6. false
7. BL=AA, BH=77, BX=77AA, EBX=99FF77AA
8. little endian
9. EAX, EBX, ECX, EDX, ESI, and EDI
10. EA=2000+8 x100=2800H
11. b, c, e only
12. EDX=FFFFFFF7H
13. ECX=000000FBH
14. DX=3, AX=000C

SECTION 21.2: 80386: A HARDWARE VIEW
1. 80386SX   2. 16
3. 34 pins since A31 - A2 is 30 and 4 pins for the BE0, BE1, BE2, and BE3
4. D23-D16   5. 40 MHz
6. FFFFFFF0H   7. half
8. 1/20 MHz=50 ns; therefore, it is 100 ns.

SECTION 21.3: 80386 PROTECTED MODE
1. false
2. hard disk
3. according to rule of least recently used
4. the descriptor table
5. bytes 2, 3, 4, and 7
6. A (access) bit
7. lowest
8. highest
9. two: user and supervisor
10. false   11. 2 stages
12. 80386   13. true
14. because in OS/2 2.0 more than one task can be active, but in Windows 3.x only one task is active for a given period

CHAPTER 22

HIGH-SPEED MEMORY INTERFACING AND CACHE

OBJECTIVES

Upon completion of this chapter, you will be able to:

» Explain how the introduction of wait states is implemented in the IBM PC to coordinate the memory cycle times of 80x86 CPUs and high-speed memory

» Define terms used in memory design, such as *memory cycle time* and *memory access time*

» Describe the various types of DRAM: standard mode, page mode, static column mode, and nibble mode

» Describe how the interleaving method is implemented to solve the problem of back-to-back DRAM access and the required precharge time

» Discuss the advantages of using DRAM for main memory and SRAM for cache

» Diagram the three types of cache organization: fully associative, direct mapped, and set associative

» Explain the write-back and write-through methods of updating main memory as cache data is altered

» Describe the cache replacement policies LRU and FIFO

» Contrast and compare EDO and FPM DRAM

» Describe the operation and purpose of SDRAM

» Explain the components and function of Rambus technology

The potential power of high-performance microprocessors can be exploited only if memory is fast enough to respond to the microprocessor's need to fetch code and data. There is no use in choosing a fast processor and then interfacing it with slow memory. In this chapter we deal with issues of high-speed memory design. In Section 22.1 we look at read and write cycle times of the 80x86 family. In Section 22.2 we discuss various types of DRAMs, such as page mode, static column mode, and nibble mode, and the method of interleaved memory design. In Section 22.3, the cache memory option is discussed, and the way 386, 486, and Pentium processors use cache memory to increase system throughput is examined. Section 22.4 examines the newer and faster DRAMS of EDO, SDRAM, and Rambus technologies.

SECTION 22.1: MEMORY CYCLE TIME OF THE 80X86

When interfacing a microprocessor to memory, the first issue is how much time is provided by the CPU for one complete read or write cycle. In other words, what is the memory cycle time of the CPU? In the 8088/86 microprocessor, the memory cycle time consists of 4 clocks, which leaves plenty of time to access memory. The slowest 8088/86, with a working frequency of 5 MHz, has an 800-ns memory cycle (4×200 ns = 800, T =1/5 MHz = 200 ns) and the fastest 8088/86, with 10-MHz speed, will have a 400-ns memory cycle. A memory cycle of 400 ns means that the CPU can access memory every 400 ns, and not faster. This is enough time to access even the slow and inexpensive DRAMs. However, for the 286, 386, 486, and Pentium, memory cycle time consists of only two T clocks. This makes memory design a challenging task, especially when the speed of the CPU goes beyond 20 MHz. Table 22-1 shows the memory cycle times for various speeds of 80x86 microprocessors. From Table 22-1 it can be seen that as the frequency of the CPU is increased, the maximum amount of time allowed to access memory is decreased, forcing the designer either to use fast and expensive memory or to introduce wait states into the memory cycle.

Introducing wait states into the memory cycle

When the memory timing requirement of the CPU cannot be met, one option that designers have is to introduce wait states. All 80x86 microprocessors have the READY pin. When the microprocessor initiates the memory cycle, meaning that it puts the addresses on the address bus, the time at which it must have the data at the pins of the data bus is fixed and is shown in Table 22-1. This fixed amount of time can be extended by activating the READY pin. Every time that the READY pin is activated, the CPU adds one extra clock to the memory cycle. For example, the 25-MHz 80386 has a memory cycle of 80 ns (2×40 =80) with zero wait states. If READY is activated only once during the memory cycle, it adds one clock of 40 ns to the memory cycle, thereby giving the memory and decoding circuitry a total of 120 ns to get the information to the data pins of the CPU. This 120 ns is spent on the following parameters: (1) memory decoding logic circuitry and address bus buffers (boosters) such as 74xx244, (2) access time of memory, and (3) the time it takes for signals to travel from memory data pins to the data pins of the CPU, going through any logic gates on the pathway, such as 74xx245 transceivers. Of these three parameters, memory access time is normally the longest, assuming the use of fast logic gates such as 74FXXX or 74ALSXXX. For more about logic families, see Chapter 26. If the allocated memory cycle time is not enough, more wait states are needed, making the memory cycle time longer.

In a 25-MHz 386 with 1 wait state, there is a 120-ns memory cycle time, meaning that the CPU can perform read and write operations no faster than every 120 ns. What happens if 140 ns is needed? Since the wait state is an integer multiple of the clock cycle (1, 2, 3, and so on), there is no other choice but to have 2 wait states. In other words, there is no such thing as 1.5 wait states. Wait states degrade computer performance, as shown in Example 22-1. It does not make sense to buy a high-frequency CPU, then interface it with slow memory. The next section will look at possible solutions to this problem.

Table 22-1: Memory Cycle Times for 80x86

Mem. Cycle	CPU	Clock Rate (MHz)	Clock Cycle (ns)	Memory Cycle (ns)
4 clocks	8088/86	5	200	800
	8088/86	8	125	500
	8086	10	100	400
2 clocks	80286	6	166.6	333.3
	80286	8	125	250
	80286	10	100	200
	80286	16	62.5	125
	80286	20	50	100
2 clocks	80386DX	16	62.5	125
	80386DX	20	50	100
	80386DX	25	40	80
	80386DX	33	30	60
2 clocks	80386SX	16	62.5	125
	80386SX	20	50	100
	80386SX	25	40	80
2 clocks*	80486DX	25	40	80
	80486DX	33	30	60
	80486DX	40	25	50
	80486DX	50	20	40
2 clocks*	80486SX	16	62.5	125
	80486SX	33	30	60
2 clocks*	Pentium	60	16.6	33
	Pentium	66	15	30
	Pentium	150 MHz (66 MHz bus frequency)	15	30
2 clocks*	Pentium Pro	200 MHz (66 MHz bus frequency)	15	30

* From external DRAM or the secondary cache.

Note: All memory cycle times are with zero wait states.
Note: In Pentium and Pentium Pro of over 100 MHz, the bus frequency is less than 100 MHz (often 66 - 80 MHz).

Example 22-1

Find the effective memory performance of a 25-MHz 386 CPU with one wait state.

Solution:

Since the 0 WS memory cycle is 80 ns (1/25 MHz =40 and 2 × 40 =80 ns), for 1 WS we have a memory cycle time of 120 ns. That means that the memory performance is the same as that of a 16.6-MHz 80386 (120 ns/2 =60 ns, then 1/60 ns =16.66 MHz) as far as memory accessing is concerned. This is 67% performance of the 80386 with zero wait states.

Review Questions

1. Find the read/write cycle time of the following systems
   (a) 40-MHz 386 with 0 WS   (b) 50-MHz 486 with 1 WS
   (c) 66-MHz Pentium with 1 WS
2. A given CPU has a read/write cycle time of 50 ns. What does this mean?
3. Find the effective working frequency for memory access in each of the following.
   (a) 40-MHz 386 with 1 WS   (b) 50-MHz 486 with 1WS
4. If a given CPU has a read cycle time of 60 ns and 10 ns is used for the decoder and address/data path delay, how much is for memory access time?
5. If a given system is designed with 1 WS and has a 90-ns memory cycle time, find the CPU's frequency if the read/write cycle time of this CPU is 2 clocks.

SECTION 22.2: PAGE, STATIC COLUMN, AND NIBBLE MODE DRAMS

To understand interfacing memory to high-performance computers, the different types of available RAM must first be understood. Although SRAMs are fast, they are expensive and consume a lot of power due to the use of flip-flops in the design of the memory cell, as we discussed in Chapter 11. At the opposite end of the spectrum is DRAM, which is cheaper but is slow (compared to CPU speed) and needs to be refreshed periodically. The refreshing overhead together with the long access time of DRAM is a major problem in the design of high-performance computers. The problem of the time taken for refreshing DRAM is minimal since it uses only a small percentage of bus time, but the solution to the slowness of DRAM is very involved. One common solution is using a combination of a small amount of SRAM, called *cache* (pronounced *cash*), along with a large amount of DRAM, thereby achieving the goal of near zero wait states. Before we discuss such solutions, we must understand what resources are available to high-performance system designers. To this end, the different types of available DRAM will be discussed, and cache memory is discussed in Section 22.3. First we clarify some widely used terminology such as memory cycle time and memory access time.

Memory access time vs. memory cycle time

Memory access time is defined as the time interval in between the moment the addresses are applied to the memory chip address pins and the time the data is available at the memory's data pins. The memory data sheets refer to it as t_{AA} (address access time). Another commonly used time interval is t_{CA} (access time from CS), which is measured from the time the chip select pin of memory is activated to the time the data is available. In some cases, notably EEPROM, t_{OE} is the time interval between the moment OE (READ) is activated to the time the data is available. However, memory access time t_{AA} is the one most often advertised.

Memory cycle time is the time interval between two consecutive accesses to the memory chip. For example, a memory chip of 100 ns cycle time can be accessed no faster than 100 ns, which means that two back-to-back reads can be performed no faster than 200 ns, and 3 back-to-back reads will take 300 ns, and so on. It must be noted that while in SRAM the memory cycle time is equal to memory access time, this is not so in DRAM memory, as discussed next.

Types of DRAM

There are different types of DRAM, which are categorized according to their mode of data access. These modes include standard mode, page mode, static column mode, and nibble mode. Although each mode is discussed separately below, often two of the above modes exist on the same DRAM chip. For example, page mode DRAM has standard mode as well.

DRAM (standard mode)

Standard mode (also called random access) DRAM, which has the longest memory cycle time, requires the row address to be provided first and then the column address for each cell. Each group is latched in by the activation of RAS (row address select) and CAS (column address select) inputs, respectively. The access time is from the time that the row address is provided to the time that the data is available at the output data pin of the DRAM chip. This is the access time that is commonly advertised and is called t_{RAC} (RAS access time, the access time from the moment RAS is provided). This is acceptable if we are accessing a random cell within DRAM. However, since most of the time data and code processed by the CPU are in consecutive memory locations and the CPU does not jump around to random locations (unless there is a JMP or CALL instruction), the DRAM will be accessed with back-to-back read operations. Unfortunately, DRAM cannot provide the code (or data) in the amount of time called t_{RAC} if there is a back-to-back read from the same DRAM chip because DRAM needs a precharge time (t_{RP}) after each RAS has been deactivated to get ready for the next access. This leads us back to the concept of memory cycle time for DRAM memory chips. The memory cycle time for memory chips is the minimum time interval between two back-to-back read/write operations. In SRAM and ROM, the access time and memory cycle time are always equal, but that is not the case for DRAMs. In DRAM, due to the fact that after RAS makes the transition to the inactive state (going from low to high) it must stay high for a minimum of t_{RP} (RAS precharge) to precharge the internal device circuitry for the next active cycle. Therefore, in DRAM we have the following approximate relationship between the memory access time and memory cycle time.

$t_{RC} = t_{RAC} + t_{RP}$ (This is for standard mode)
read cycle time = RAS access time + RAS precharge time

For example, if DRAM has an access time of 100 ns, the memory cycle time is really about 190 ns (100 ns access time plus 90 ns precharge time). To access a single location in such a DRAM, 100 ns is enough, but to access more than one successively, 190 ns is required for each access due to the precharge time that is needed internally by DRAM to get ready to access the next capacitor cell.

The read cycle time not being equal to the access time is one of the major differences between SRAM and DRAM. Although in SRAM the read cycle time is equal to the access time, in DRAM of standard mode the read cycle time is about twice the access time normally advertised (t_{ACC}). This could make a difference in the total time spent by the CPU to access memory. Look at Example 22-2. From the above discussion and Example 22-2 we can conclude that for successive accesses of random locations inside the DRAM the CPU must spent a minimum of t_{RC} time on each access. Tables 22-2 and 22-3 show DRAM and SRAM memory cycle times, respectively. See Figure 22-1 for DRAM and SRAM timing.

DRAM interfacing using the interleaving method

One of the methods used to overcome the problem of precharge time in DRAMs is the *interleaving method* of DRAM interfacing. In this method, two sets of banks are placed next to each other and the CPU accesses each set of banks alternately. In this way the precharge time of one set of banks is hidden behind the access time of the other one. This means that while the CPU is accessing one set of banks, the other set is being precharged. Look at Figure 22-2. Assume that the 80386SX is working on 20 MHz frequency; therefore, the CPU has a memory cycle time of 100 ns. Using DRAM with access time of 70 ns and the precharge of 65 ns gives a DRAM cycle time of 135 ns (70 + 65 =135). This is much longer than the 100 ns provided by the CPU. Using interleaved memory design can solve this problem. In this case when the 386SX accesses bank set A, it goes on to access bank set B while set A takes care of its precharge time. Similarly, when the CPU accesses set A, the B set banks will have time to precharge.

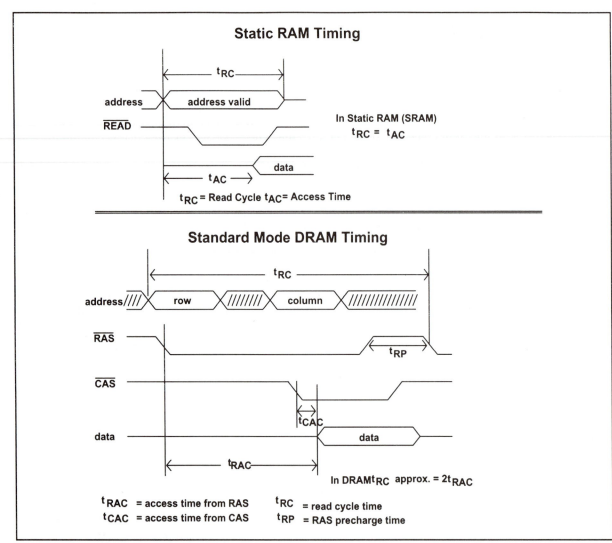

Static RAM Timing

In Static RAM (SRAM)
$t_{RC} = t_{AC}$

t_{RC} = Read Cycle t_{AC} = Access Time

Standard Mode DRAM Timing

In DRAM t_{RC} approx. = $2t_{RAC}$

t_{RAC} = access time from RAS t_{RC} = read cycle time
t_{CAC} = access time from CAS t_{RP} = RAS precharge time

Figure 22-1. DRAM vs. SRAM Timing

Example 22-2

Compare the minimum CPU time needed to read 150 random memory locations of a given bank in each of the following.
(a) DRAM with T_{ACC} =100 ns and T_{RC} =190
(b) SRAM of T_{ACC} =100

Solution:

(a) DRAM requires 190 ns to access each location. Therefore, a total of 150×190 =28,500 ns would be spent by the CPU to access all those 150 memory locations.

(b) In the case of SRAM, the CPU spends only 150×100 ns =15,000. This would have been needed since T access = T read cycle (t_{ACC} =t_{RC}).

Table 22-2: DRAM Access Time vs. Cycle Time (4M x 1)

DRAM	RAS Access (t_{RAC}) (ns)	Read Cycle (t_{RC}) (ns)	RAS Precharge (t_{RP}) (ns)
MCM44100-60	60	110	45
MCM44100-70	70	130	50
MCM44100-80	80	150	60

(Reprinted by permission of Motorola Corporation, Copyright Motorola Corp. 1993)

Table 22-3: SRAM Access Time vs Cycle Time

SRAM (IDT Product)	Address Access (t_{AA}) (ns)	Read Cycle (t_{RC}) (ns)
IDT71258S25	25	25
IDT71258S35	35	35
IDT71258S45	45	45
IDT71258S25	70	70

(Reprinted by permission of Integrated Device Technology, Copyright IDT, 1993)

Example 22-3

Calculate the time to access 1024 random bits of a 1Mx1 chip if t_{RC}=85 ns and t_{RAC}=165 ns.

Solution:

For standard mode (also called random) we have the following for reading 1024 bits:

time to read 1024 random bits =1024 × t_{RC} =1024 × 165 ns =168,960 ns

---

Example 22-4

Show the time needed to access all 1024 memory locations of Example 22-3 if the interleaved method of memory interfacing is used.

Solution:

In the interleaved method, since the precharge time of one bank is hidden behind the access time of the other bank, each memory location is accessed in t_{RAC} as far as the CPU is concerned; therefore, 1024 × 85 =87,040 ns is the total amount of time spent by the CPU to access 1024 locations.

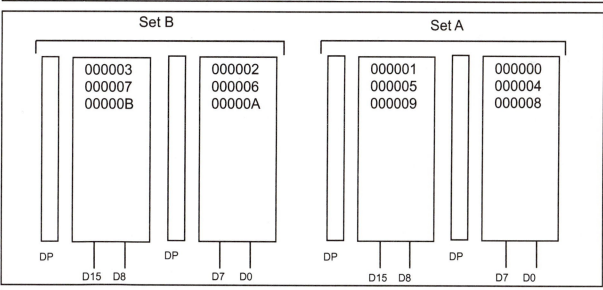

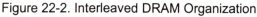

Figure 22-2. Interleaved DRAM Organization

Interleaved drawback

The major drawback of interleaved memory is memory expansion. In expanding the memory based on the interleaved method, a minimum of two sets of banks must be added every time additional memory is required. Look at Example 22-5. Many inexpensive personal computers based on 386SX, 386DX, and 486SX of 16 - 25 MHz frequency use the interleaved memory design method to avoid using expensive cache memory without sacrificing performance.

SECTION 22.2: PAGE, STATIC COLUMN, AND NIBBLE MODE DRAMS   **665**

Example 22-5

Assume that we are using 1Mx1 DRAM organization in Figure 22-2. If each set is 4 megabytes, find the following.
(a) the chip count   (b) the minimum memory addition and the chip count

Solution:

(a) Assuming 1Mx9 for each bank where each bank takes care of 8 bits of data, there are 9 chips for every byte. That means a total of 36 DRAM chips for each set, or a total of 72 1Mx1 chips for the first 8 megabytes of interleaved memory.

(b) From then on, any memory addition must be in multiples of 4 megabytes since each set needs 2M; therefore, we need another 36 of 1Mx1 DRAM chips to raise the total memory of the system to 12M.

Example 22-6

A 386SX PC has 1M of DRAM installed using the interleaved design method. Show the memory organization and DRAM chip count assuming that only 256Kx1 and 256Kx4 DRAM chips are used.

Solution:

Since the 386SX has a 16-bit data bus, it uses 512K bytes for each set of A and B, or four banks of 256Kx9, where each set consists of two banks of 256Kx9. Therefore, the total chip count is 12 since each bank uses 3 chips (two of 256Kx4 and one of 265Kx1 for parity bit). This is shown as follows.

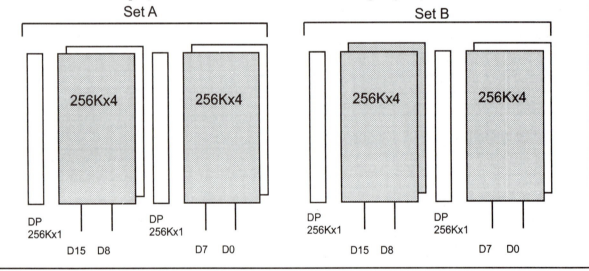

Example 22-7

Show the minimum memory addition and the chip count for Example 22-6. Assume that the available DRAM chips are 256Kx1 and 256Kx4.

Solution:

The minimum memory addition is 1M. Since we have two banks for each set of interleaved memory, we have 2 of 256Kx4 and 1 of 256Kx1 for parity, which means 3 chips for each bank. Therefore, the minimum memory addition requires 12 chips, 8 of which are 256Kx4 and 4 are 256Kx1 for parity bits, resulting in 1megabyte.

Page mode DRAM

The storage cells inside DRAM are organized in a matrix of N rows and N columns. In reading a given cell, the address for the row (A1 - An) is provided first and RAS is activated; then the address for the column (A1 - An) is provided and CAS is activated. In DRAM literature the term *page* refers to a number of column cells in a given row. See Examples 22-8 and 22-9.

The idea behind page mode is that since in most situations memory locations are accessed consecutively, there is no need to provide both the row and column address for each location, as was the case in DRAM with standard timing. Instead, in page mode, first the row address is provided, RAS latches in the row address, and then the column addresses are provided and CAS toggles back and forth, latching in the column addresses until the last column of a given page is accessed. Then the address of the next row (page) is provided and the process is repeated. While the access time of the first cell is the standard access time using both row and column (t_{RAC}), in accessing the second cell on to the last cell of the same page (row), the access time is much shorter. This access time is often referred to as t_{CAC} (T of column access). In page mode DRAM when we are in a given page, each successive cell can be accessed no faster than t_{PC} (page cycle time). See Figure 22-3. Table 22-4 gives page mode timing parameters. In DRAM of page mode both the standard mode and page mode are supported.

Example 22-8

Show how memory storage cells are organized in each of the following DRAM chips.
(a) 256Kx1   (b) 1Mx1   (c) 4Mx1

Solution:

(a) As discussed in Chapter 11, the 256Kx1 has 9 address pins (A0 - A8); therefore, cells are organized in a matrix of $2^9 \times 2^9 = 512 \times 512$, giving 512 rows, each consisting of 512 columns of cells.
(b) 1024×1024 (c) 2048×2048

Example 22-9

Assuming that the DRAMs in Example 22-8 are of page mode, show how each chip is organized into pages. Find the number of columns per page for (a), (b), and (c).

Solution:

(a) For 1Mx1 we have 512 pages, where each page has 512 columns of cells.
(b) 1024 pages, where each page has 1024 bits (columns).
(c) 2048 pages each of 2048 bits

Example 22-10

Calculate the total time spent by the CPU to access an entire page of memory if the memory banks are page mode DRAM of 1Mx1 with t_{RC} =165 ns, t_{RAC} =85 ns, and t_{PC} =50 ns.

Solution:

For page mode we have the following for reading 1024 bits:
Time to read 1024 bits of the same page $= t_{RAC} + 1023 \times t_{PC}$
$$= 85 \text{ ns} + 1023 \times 50 \text{ ns} = 51{,}235 \text{ ns}$$

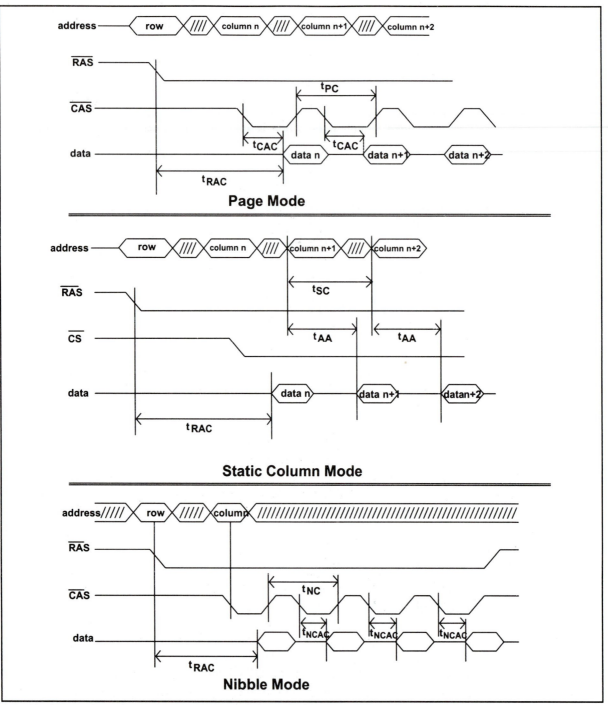

Figure 22-3. DRAM Page, Static Column, and Nibble Modes

Table 22-4: Page Mode DRAM Timing Parameters (4Mx1)

Page Mode DRAM	Access Time from RAS, t_{RAC} (ns)	Read Cycle Time, t_{RC} (ns)	Access Time from CAS, t_{CAC} (ns)	Page Cycle Time, t_{PC} (ns)
MCM44100-60	60	110	15	40
MCM44100-70	70	130	20	45
MCM44100-80	80	150	20	50

(Reprinted by permission of Motorola Corporation, Copyright Motorola Corp. 1993)

Static column mode

Static column mode makes accessing all the columns of a given row much simpler by eliminating the need for CAS. In this mode, the first location is accessed with a standard read cycle where the row address is latched by RAS followed by the column address and what is called CS (chip select) clock. From then on, CS is incremented internally. As long as RAS and CS remain low, the contents of successive cells appear at the data output pin of DRAM until the last column of a given row is accessed and then the process is moved to the next row. This means that the initial access time of the first cell is the standard access time (t_{RAC}), but each subsequent column in that row is accessed in a time called t_{AA} (access time from column address). Due to the fact that there is no setup and hold time for column address select (CAS), the use of static-column-mode DRAM lends itself to memory design of high-frequency systems. A large percentage of 80386 and higher processor computers use static column DRAM for main memory.

In static column mode where the initial standard access time is t_{RAC}, when we are in given a page, any cell can be accessed with the access time of t_{AA}, but all the successive bits can be accessed no faster than t_{SC} (static column cycle time). See Figure 22-3. Table 22-5 gives static-column-mode timing parameters.

Table 22-5: Static Column DRAM Timing Parameters (4Mx1)

Static Column DRAM	T RAS Access, t_{RAC} (ns)	T Read Cycle, t_{RC} (ns)	T Column Access, t_{AA} (ns)	Cycle Time, t_{SC} (ns)
MCM54102A-60	60	110	30	35
MCM54102A-70	70	130	35	40
35MCM54102A-80	80	150	40	45

(Reprinted by permission of Motorola Corporation, Copyright Motorola Corp., 1993)

Example 22-11

Calculate the total time spent by the CPU to access the entire page of memory if the memory banks are static-column-mode DRAMs of 1Mx1 with t_{RC} =165 ns, t_{RAC} =85 ns, and t_{SC} =50 ns.

Solution:

For static column mode we have the following for reading 1024 bits:

time to read 1024 bits of the same page

$$= t_{RAC} + 1023 \times t_{SC}$$
$$= 85 \text{ ns} + 1023 \times 50 \text{ ns} = 51{,}235 \text{ ns}$$

Comparing Examples 22-10 and 22-11, if for both the page mode and static column mode the time spent by the CPU is the same, what is the advantage of static column mode? The answer is that static-column-mode DRAM design is simpler since there is no circuit or timing requirement for the CAS pin. Notice in Figure 22-3 that we need to keep both RAS and CS (chip select) low in order to access successive cells. Here is what Motorola Application Note AN986 says about the superiority of the static-column-mode DRAM: "This mode is useful in applications that require less noise than page mode. Output buffers are always on when the device is in this mode and the CS clock is not cycled, resulting in fewer transients and simpler operation.... Static column consists of changing column addresses while holding the RAS and CS clocks active."

Nibble mode

In nibble mode, 4 bits (a nibble) can be accessed by providing RAS first and CAS second, then holding RAS active while the CAS is being toggled. In this regard, nibble mode is just like page mode except that only 4 bits are accessed instead of all the columns of a given row. After the initial standard read, row and column address counters are incremented internally, accessing the three subsequent bits

---

without the need for the column address. Notice from Figure 22-3 in nibble mode that unlike the page and static column modes, there is no need to provide the external column address.

In nibble mode, the first bit is accessed in the standard manner with the standard access time (t_{RAC}) but all read operations following the initial access are performed at the rate of t_{NCAC}, where t_{NCAC} refers to nibble mode access time.

The major differences between this and page mode are that (a) there is no need to provide the column address after the initial access, and (b) only 4 bits are accessed rather than all the columns of a given row as in page mode. This can be used to implement the burst mode of many 486, Pentium, and RISC processors, as we will see in Chapter 23. In nibble mode, after the initial access each bit can be accessed no faster than t_{NC} (nibble mode cycle time). See Table 22-6.

To avoid the interchip delay associated with using many logic gates (often called glue logic since they are soldered together), there are many DRAM controllers which support the various DRAM operation modes. The operation modes are compared in Example 22-13.

Table 22-6: Nibble Mode DRAM Timing Parameters (4Mx1)

Nibble Mode DRAM	T RAS Access, t_{RAC} (ns)	T Read Cycle, t_{RC} (ns)	T Nibble Mode Access, t_{NCAC} (ns)	Nibble Cycle, t_{NC} (ns)
MCM54101A-60	60	110	20	40
MCM54101A-70	70	130	20	40
MCM54101A-80	80	150	20	40

(Reprinted by permission of Motorola Corporation, Copyright Motorola Corp. 1993)

Example 22-12

Calculate the total time spent by the CPU to access each of the following.
(a) 4 bits of nibble mode DRAM
(b) all 1024 bits of nibble mode DRAM if t_{RAC} =85 ns, t_{NC} =40 ns, and t_{RP} =70 ns

Solution:

(a) To access the 4 bits of nibble mode we have:
time to read all 4 bits of nibble mode $= t_{RAC} + 3t_{NC}$
$= 85 + 3 \times 40 = 205$ ns

(b) To access all 1024 bits using 256 access of 4-bit nibble mode we have:
time to read all 1024 bits of nibble mode $= 256 (t_{RAC} + 3t_{NC} + t_{RP})$
$= 256 (85 + 3 \times 40 + 70) = 70,400$ ns

Example 22-13

Calculate the time spent to access 4 bits for each of the following. Use the data in Table 22-7.
(a) standard mode   (b) page mode
(c) static column mode   (d) Compare these with nibble mode DRAM.

Solution:

(a) Time to read 4 bits $= 4t_{RC} = 4 \times 165 = 660$ ns.
(b) Time to read 4 bits $= t_{RAC} + 3t_{PC} = 85 + 3 \times 50 = 235$ ns.
(c) Time to read 4 bits $= t_{RAC} + 3t_{SC} = 85 + 3 \times 50 = 235$ ns.
(d) In Example 22-12 we have $= 205$ ns.

Timing comparison of DRAM modes

A summary of DRAM timing is given in Tables 22-7 and 22-8. Much of this material is taken from Motorola Application Note AN986.

Table 22-7: Timing for 1Mx1 85 ns DRAM Chip

Access Time (ns)	Standard	Page	Static Column	Nibble
Access time from row, t_{RAC}	85	85	85	85
Access time from column, t_{CAC}		25		
Access time from column, t_{AA}			45	
Access time from column, t_{NCAC}				20
Cycle time				
Read cycle time, t_{RC}	165			
Page mode cycle time, t_{PC}		50		
Static column time, t_{SC}			50	
Nibble mode cycle time, t_{NC}				40

(Reprinted by permission of Motorola Corporation, Copyright Motorola Corp. 1993)

Table 22-8: Timing Comparison for Various DRAM Modes

NS	Standard (Random)	Page	Static Column	Nibble
Time to read 4 bits	660	235	235	205
Time to read 1024 bits	168,960	51,235	51,235	70,400

This concludes the discussion of DRAM operation modes. It must be noted that in many systems one of the above modes is implemented in order to eliminate the need for the wait state to access every bit of DRAM. As seen from the above discussion, even the best of any of the above modes still cannot eliminate the need for the wait state entirely unless SRAM is used for the entire memory, which is prohibitively expensive. The best solution is to use a combination of SRAM and DRAM, which is discussed next.

Review Questions

1. In which memory is the read cycle time equal to the memory access time?
2. A given DRAM is advertised to have an access time of 50 ns. What is the approximate memory cycle time for this DRAM?
3. A given DRAM has a 120-ns memory read cycle time. What is its access time (t_{RAC})?
4. In DRAM, a read cycle consists of _____ and _____.
5. Assume an 80386 of interleaved memory with 2M bytes initial DRAM for each of the following.
   (a) Show how the banks are organized.
   (b) What is the minimum memory addition?
6. True or false. In page mode, the initial read takes t_{RAC}.
7. For page mode DRAM, while we are in a given page, we can access successive memory locations no faster than _____ .
8. Calculate the time the CPU must spend to access 100 locations all within the same page if t_{RAC} =60 ns and t_{PC} =30 ns.
9. Calculate the access time for 4 bits in nibble mode if t_{RAC} =50 ns, t_{NC} =20 ns.
10. The higher the system frequency, the less noise can be tolerated in the system. Which is preferable in a 20-MHz system, static column or page mode DRAM?

SECTION 22.3: CACHE MEMORY

The most widely used memory design for high-performance CPUs implements DRAMs for main memory along with a small amount (compared to the size of main memory) of SRAM for cache memory. This takes advantage of the speed of SRAM and the high density and cheapness of DRAM. As mentioned earlier, to implement the entire memory of the computer with SRAM is too expensive and to use all DRAM degrades performance. Cache memory is placed between the CPU and main memory. See Figure 22-4.

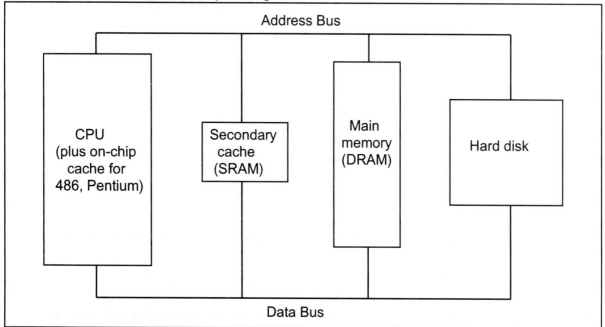

Figure 22-4. CPU and Its Relation to Various Memories

When the CPU initiates a memory access, it first asks cache for the information (data or code). If the requested data is there, it is provided to the CPU with zero wait states, but if the data is not in cache, the memory controller circuitry will transfer the data from main memory to the CPU while giving a copy of it to cache memory. In other words, at any given time the cache controller has knowledge of which information (code or data) is kept in cache; therefore, upon request for a given piece of code or data by the CPU the address issued by the CPU is compared with the addresses of data kept by the cache controller. If they match (hit) they are presented to the CPU with zero WS, but if the needed information is not in cache (miss) the cache controller along with the memory controller will fetch the data and present it to the CPU in addition to keeping a copy of it in cache for future reference. The reason a copy of data (or code) fetched from main memory is kept in the cache is to allow any subsequent request for the same information to result in a hit and provide it to the CPU with zero wait states. If the requested data is available in cache memory, it is called a *hit*; otherwise, if the data must be brought in from main memory, it is a *miss*.

In most computers with cache, the hit rate is 85% and higher. By combining SRAM and DRAM, cache memory's access time matches the memory cycle of the CPU. In the 80386/486 microprocessor with a frequency of 33 MHz and above, the use of cache is absolutely essential. For example, in the 33-MHz 80386-based computer with only a 60-ns read cycle time, only static RAM with an access time (cycle time) of 45 ns can provide the needed information to the CPU without inserting wait states. We have assumed that 15 ns (60 − 45 =15) is used for the delay associated with the address and data path. To implement the entire 16M of main memory of a 33-MHz 386/486 system with 45 ns SRAM is not only too expensive but the power dissipation associated with such a large amount of SRAM would require a complex cooling system used only for expensive mini- and mainframe computers. The problem gets worse if we use a 486 of 50 MHz or a Pentium of 60 MHz.

It must be noted that when the CPU accesses memory, it is most likely to access the information in the vicinity of the same addresses, at least for a time. This is called the principle of *locality of reference*. In other words, even for a short program of 50 bytes, the CPU is accessing those 50 memory locations from cache with zero wait states. If it were not for this principle of locality and that the CPU accesses memory randomly, the idea of cache would not work. This implies that JMP and CALL instructions are bad for the performance of cache-based systems. The *hit rate*, the number of hits divided by the total number of tries, depends on the size of the cache, how it is organized (cache organization), and the nature of the program.

Cache organization

There are three types of cache organization:
1. fully associative
2. direct mapped
3. set associative

The following is a discussion of each organization with their advantages and disadvantages. For the sake of clarity and simplicity, an 8-bit data bus and 16-bit address bus are assumed.

Fully associative cache

In fully associative cache, only a limited number of bytes from main memory are held by cache along with their addresses. The SRAMs holding data are called *data cache* and the SRAMs holding addresses of the data are called *tag cache*. This discussion assumes that the microprocessor is sending a 16-bit address to access a memory location that has 8 bits of data and that the cache is holding 128 of the possible 65,536 (2^{16}) locations. This means that the width of the tag is 16 bits since it must hold the address, and that the depth is 128. When the CPU sends out the 16-bit address, it is compared with all 128 addresses kept by the tag. If the address of the requested data matches one of the addresses held by the tags, the data is read and is provided to the CPU (a hit). If it is not in the cache (a miss), the requested data must be brought in from main memory to the CPU while a copy of it is given to cache. When the information is brought into cache, the contents of the memory locations and their associated addresses are saved in the cache (tag cache holds the address and data cache holds the data).

In fully associative cache, the more data that is kept, the higher the hit rate. An analogy is that the more books you have on a table, the better the chance of finding the book you want on the table before you look for it on the book shelf. The problem with fully associative is that if the depth is increased to raise the hit rate, the number of comparisons is too time consuming and inefficient. For example, 1024 depth fully associative requires 1024 comparisons, and that is too time consuming even for fast comparators. On the other hand, with a depth of 16 the CPU ends up waiting for data too often because the operating system is swapping information in and out of cache since its size is too small and it must save the present data in the cache before it can bring in new data. This replacement policy is discussed later. In the above example of 128 depth, the amount of SRAM for tag is 128×16 bits and 128×8 for data, that is, 256 bytes for tag and 128 bytes for data cache for a total of 384 bytes. Although the above example used a total of 384 bytes of SRAM, it is said that the system has 128 bytes of cache. In other words, the data cache size is what is advertised. The SRAM inside the cache controller provides the space for storing the tag bits. Tag bits are not included in cache size. In Figure 22-5, DRAM location F992 contains data 85H. The left portion of the figure shows when the data is moved from DRAM to cache.

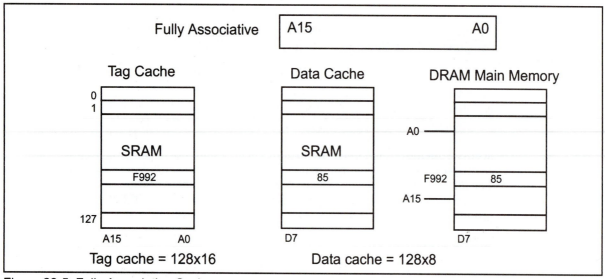

Figure 22-5. Fully Associative Cache

Direct-mapped cache

Direct-mapped cache is the opposite extreme of fully associative. It requires only one comparison. In this cache organization, the address is divided into two parts: the index and the tag. The index is the lower bits of the address, which is directly mapped into SRAM, while the upper part of the address is held by the tag SRAM. From the above example, A0 to A10 are the index and A11 to A15 are the tag. Assuming that CPU addresses location F7A9H, the 7A9 goes to the index but the data is not read until the contents of tag location 7A9 of tag is compared with 11110B. If it matches (its content is 11110), the data is read to the CPU; otherwise, the microprocessor must wait until the contents of location F7A9 are brought from main memory DRAM into the CPU while a copy of it is issued to cache for future reference. There is only one unique location with index address of 7A9, but 32 possible tags (2^5 =32). Any of these possibilities, such as C7A9, 27A9, or 57A9, could be in tag cache. In such a case, when the tag of a requested address does not match the tag cache, a cache miss occurs. Although the number of comparisons has been reduced to one, the problem of accessing information from locations with the same index but different tag, such as F7A9 and 27A9, is a drawback. The SRAM requirement for this cache is shown below. While the data cache is 2K bytes, the tag requirement is 2K × 5 =10K bits or about 1.25K bytes. See Figure 22-6.

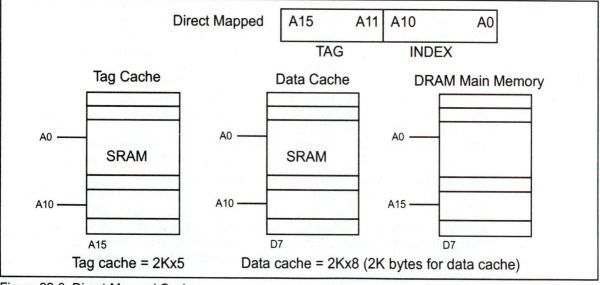

Figure 22-6. Direct-Mapped Cache

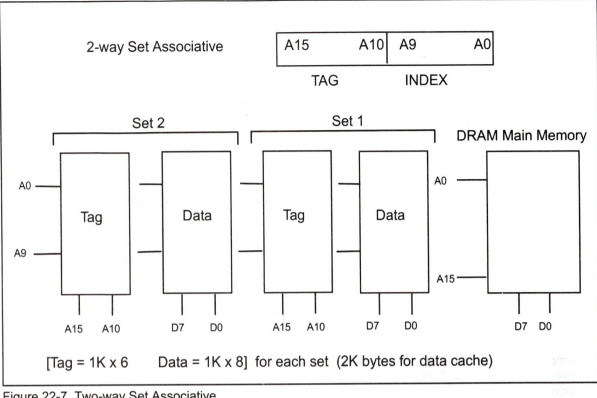

Figure 22-7. Two-way Set Associative

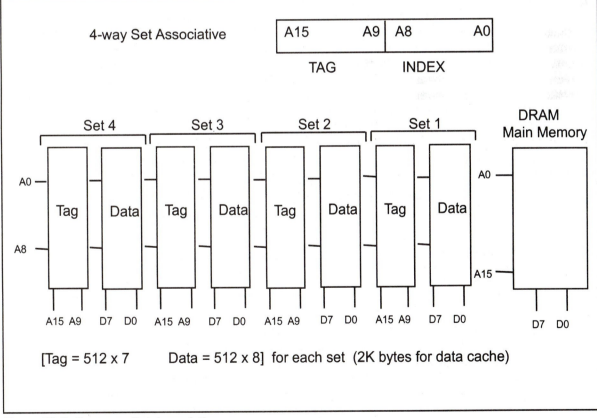

Figure 22-8. Four-way Set Associative

Set associative

This cache organization is in between the extremes of fully associative and direct mapped. While in direct mapped, there is only one tag for each index; in set associative, the number of tags for each index is increased, thereby increasing the hit rate. In 2-way set associative, there are two tags for each index, and in 4-way there are 4 tags for each index. See Figures 22-7 and 22-8. Comparing direct-mapped and 2-way set associative, one can see that with only a small amount of extra SRAM, a better hit rate can be achieved. In this organization, if the microprocessor is requesting the contents of memory location 41E6H, there are 2 possible tags that could hold it, since cache circuitry will access index 1E6H and compare the contents of both tags with "0100 00". If any of them matches it, the data of index location 1E6 is read to the CPU, and if none of the tags matches "0100 00", the miss will force the cache controller to bring the data from DRAM to cache, while a copy of it is provided to the CPU at the same time. In 4-way set associative, the search for the block of data starting at 41E6 is initiated by comparing the 4 tags with "0100 000", which will increase the chance of having the data in the cache by 50%, compared with 2-way set associative. As seen in the above example, the number of comparisons in set associative depends on the degree of associativity. It is 2 for 2-way set associative, 4 for 4-way set associative, 8 for 8-way, n for n-way set associative, and in the thousands for fully set associative. The higher the set, the better the performance, but the amount of SRAM required for tag cache is also increased, making the 8-way and 16-way associate increased costs unjustifiable compared to the small increase in hit rate. The increase in the set also increases the number of tag comparisons. Most cache systems that use this organization are implemented in 4-way set associative (e.g., 80486 on-chip cache).

From a comparison of these two cache organizations, the difference between them in organization and SRAM requirements can be seen. In 2-way, the tag of 1Kx6 and data of 1Kx8 for each set gives a total of 14K bits [2 × (1K × 6 + 1K × 8) = 28K bits]. In 4-way, there is 512 × 7 for the tag and 512 × 8 for data, giving a total of 32K bits [(512 × 7 + 512 × 8) × 4 = 32K bits] of SRAM requirement. Only with an extra 4K bits will the hit rate improve substantially. As the degree of associativity is increased, the size of the index is reduced and added to the tag and this increases the tag cache SRAM requirement, but the size of data cache remains the same for all cases of direct map, 2-way, and 4-way associative. These concepts are clarified further in Examples 22-14, 22-15, and 22-16.

Example 22-14

This example shows direct-mapped cache for 16M main memory.

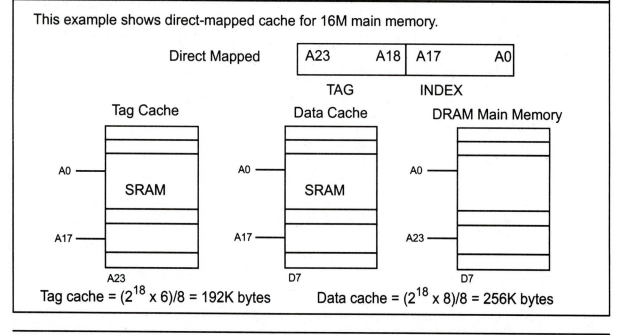

Tag cache = $(2^{18}$ x 6)/8 = 192K bytes   Data cache = $(2^{18}$ x 8)/8 = 256K bytes

---

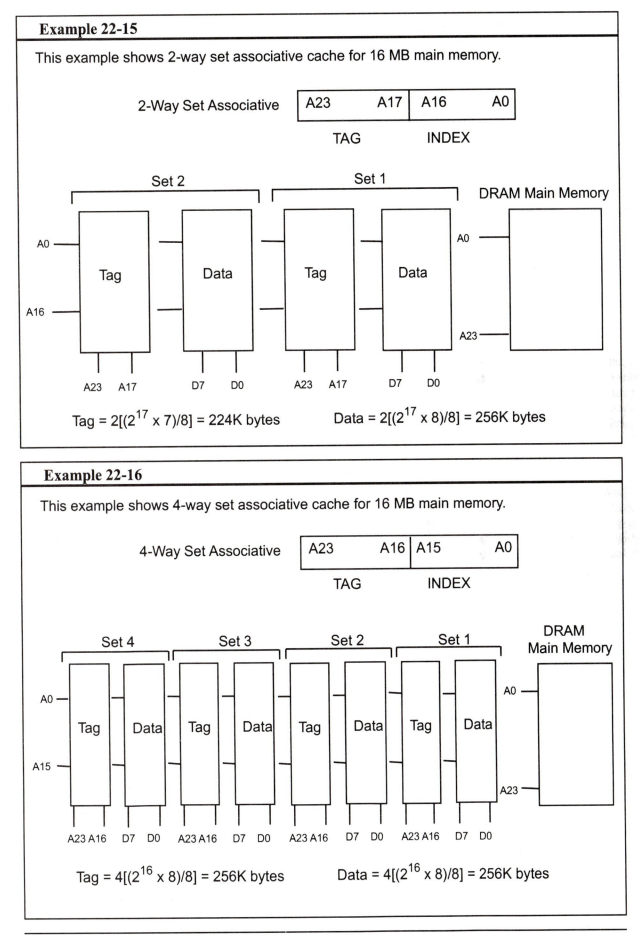

Example 22-15

This example shows 2-way set associative cache for 16 MB main memory.

2-Way Set Associative

A23	A17	A16	A0
TAG		INDEX	

Set 2   Set 1   DRAM Main Memory

A0

A16

Tag   Data   Tag   Data

A0

A23

A23 A17   D7 D0   A23 A17   D7 D0

Tag = $2[(2^{17} \times 7)/8]$ = 224K bytes   Data = $2[(2^{17} \times 8)/8]$ = 256K bytes

Example 22-16

This example shows 4-way set associative cache for 16 MB main memory.

4-Way Set Associative

A23	A16	A15	A0
TAG		INDEX	

Set 4   Set 3   Set 2   Set 1   DRAM Main Memory

A0

A15

Tag Data   Tag Data   Tag Data   Tag Data

A0

A23

A23 A16 D7 D0   A23 A16 D7 D0   A23 A16 D7 D0   A23 A16 D7 D0

Tag = $4[(2^{16} \times 8)/8]$ = 256K bytes   Data = $4[(2^{16} \times 8)/8]$ = 256K bytes

Updating main memory

In systems with cache memory, there must be a way to make sure that no data is lost and that no stale data is used by the CPU, since there could be copies of data in two places associated with the same address, one in main memory and one in cache. A sound policy on how to update main memory will ensure that a copy of any new data written into cache will also be written to main memory before it is lost since the cache memory is nothing but a temporary buffer located between the CPU and main memory. To prevent data inconsistency between cache and main memory, there are two major methods of updating the main memory: (1) write-through and (2) write-back. The difference has to do with main memory traffic.

Write-through

In write-through, the data will be written to cache and to main memory at the same time. Therefore, at any given time, main memory has a copy of valid data contained in cache. At the cost of increasing bus traffic to main memory, this policy will make sure that main memory always has valid data, and if the cache is overwritten, the copy of the latest valid data can be accessed from main memory. See Figure 22-9.

Write-back (copy-back)

In the write-back (sometimes called copy-back) policy, a copy of the data is written to cache by the processor and not to main memory. The data will be written to main memory by the cache controller only if cache's copy is about to be altered. The cache has an extra bit called the *dirty bit* (also called the *altered bit*). If data is written to cache, the dirty bit is set to 1 to indicate that the cache data is new data which exists only in cache and not in main memory. At a later time, the cache data is written to main memory and the dirty bit is cleared. In other words, when the dirty bit is high it means that the data in cache has changed and is different from the corresponding data in main memory; therefore, the cache controller will make sure that before erasing the new data in cache, a copy of it is given to main memory. Getting rid of information in cache is often referred to as *cache flushing*. This updating of the main memory at a convenient time can reduce the traffic to main memory so that main memory buses are used only if cache has been altered. If the cache data has not been altered and is the same as main memory, there is no need to write it again and thereby increase the bus traffic as is the case in the write-through policy. See Figure 22-9.

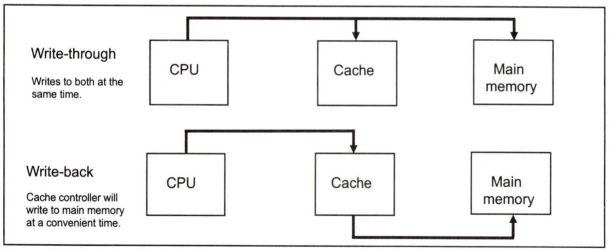

Figure 22-9. Methods of Updating Main Memory

Before concluding this section, two more cache terminologies will be described which are commonly used in the technical literature: cache coherency and cache replacement policy. Table 22-9 gives an overview of cache performance.

Table 22-9: Cache System Performance

Cache Configuration			Cache Performance
Size	Associativity	Line Size (bytes)	Hit Rate (%)
1K	Direct	4	41
8K	Direct	4	73
16K	Direct	4	81
32K	Direct	4	86
32K	2-Way	4	87
32K	Direct	8	91
64K	Direct	4	88
64K	2-Way	4	89
64K	4-Way	4	89
64K	Direct	8	92
64K	2-Way	8	93
128K	Direct	4	89
128K	2-Way	4	89
128K	Direct	8	93

(Reprinted by permission of Intel Corporation, Copyright Intel Corp. 1986)

Cache coherency

In systems in which main memory is accessed by more than one processor (DMA or multiprocessors), it must be ensured that cache always has the most recent data and is not in possession of old (or stale) data. In other words, if the data in main memory has been changed by one processor, the cache of that processor will have the copy of the latest data and the stale data in the cache memory is marked as dirty (stale) before the processor uses it. In this way, when the processor tries to use the stale data, it is informed of the situation. In cases where there is more than one processor and all share a common set of data in main memory, there must be a way to ensure that no processor uses stale data. This is called *cache coherency*.

Cache replacement policy

What happens if there is no room for the new data in cache memory and the cache controller needs to make room before it brings data in from main memory? This depends on the *cache replacement policy* adopted. In the LRU (least recently used) algorithm, the cache controller keeps account of which block of cache has been accessed (used) the least number of times, and when it needs room for the new data, this block will be swapped out to main memory or flushed if a copy of it already exists in main memory. This is similar to the relation between virtual memory and main memory. The other replacement policies are to overwrite the blocks of data in cache sequentially or randomly, or use the FIFO (first in, first out) policy. Depending on the computer's design objective and its intended use, any of these replacement policies can be adopted.

Cache fill block size

If the information asked for by the CPU is not in cache and the cache controller must bring it in from main memory, how many bytes of data are brought in whenever there is a miss? If the block size is too large (let's say 500 bytes), it will be too slow since the main memory is accessed normally with 1 or 2 WS. At the other extreme, if the block is too small, there will be too many cache misses. There must be a middle-of-the-road approach. The block size transfer from the main memory to CPU (and simultaneous copy to cache) varies in different computers, anywhere between 4 and 32 bytes. In cache controllers used with 386/486 machines, the block size is 32 bytes. This is called the *8-line cache refill policy*, where each line is 4 bytes of the 32-bit data bus. Advances in IC fabrication have allowed putting some cache on the CPU chip. This on-chip cache is called L1 (level 1) cache whereas cache outside the CPU is called L2 (level 2)

Review Questions

1. Cache is made of _____ (DRAM, SRAM).
2. From which does the CPU asks for data first, cache or main memory?
3. Rank the following from fastest to slowest as far as the CPU is concerned.
   (a) main memory   (b) register   (c) cache memory
4. In fully associative cache of 512 depth, there will be ____ comparisons for each data request.
5. Which cache organization requires the least number of comparisons?
6. A 4-way set associative organization requires _____ comparisons.
7. What does *write-through* refer to?
8. Which one increases the bus traffic, write-through or write-back?
9. What does LRU stand for, and how is it used?
10. A cache refill policy of 4 lines refers to _____.

SECTION 22.4: EDO, SDRAM, AND RAMBUS MEMORIES

In recent years the need for faster memory has led to the introduction of some very high-speed DRAMs. In this section we look at three of them: EDO (extended data-out), SDRAM (synchronous DRAM), and RDRAM (Rambus DRAM). In the mid-1990s, the speed of x86 processors went over 100 MHz and subsequently Intel and Digital Equipment began talking about 300 - 400 MHz CPUs. However, a major problem for these high-speed CPUs is the speed of DRAM. After all, cache has to be filled with information residing in main memory DRAM. Before we discuss some high-speed DRAMs, it needs to be noted that "300 MHz" CPU does not mean that its bus speed is also 300 MHz. For microprocessors over 100 MHz, the bus speed is often a fraction of the CPU speed. This is due to the expense and difficulty (e.g., crosstalk, electromagnetic interference) associated with the design of high-speed motherboards and the slowness of memory and logic gates. For example in many 150-MHz Pentium systems, the bus speed is only 66 MHz.

Table 22-10: 70 ns 4M DRAM Timing

	FPM	EDO
Speed (ns)	70	70
t_{RAC} (ns)	70	70
t_{RC} (ns)	130	130
t_{PC} (ns)	40	30

Note: 256Kx16 DRAM
From Micron Technology

EDO DRAM: origin and operation

Earlier in this chapter we discussed page mode DRAM. It needs to be noted that page mode DRAM has been modified and now is referred to as *fast page mode* DRAM. Note that DRAM data books of the mid-1990s refer only to fast page DRAM (FPM DRAM) and not page mode. The following describes the operation and limitations of fast page DRAM and how it led to EDO DRAM.

Table 22-11: 60, 50 ns 4M DRAM Timing

	FPM	EDO	EDO
Speed (ns)	60	60	50
t_{RAC} (ns)	60	60	50
t_{RC} (ns)	110	110	100
t_{PC} (ns)	35	25	20

Note: 256Kx16 DRAM
From Micron Technology

1. The row address is provided and latched in when RAS falls. This opens the page.
2. The column address is latched in when CAS falls and data shows up after t_{CAC} has elapsed. However, the next column of the same row (page) cannot be accessed faster than t_{PC} (page cycle time). This means that accessing consecutive columns of opened pages is limited by the t_{PC}. The t_{PC} timing itself is influenced by how long CAS has to stay low before it goes up. Why don't DRAM designers pull up the CAS faster in order to shorten the t_{PC}? This seems like a very logical suggestion. However, there is a problem with this approach in fast page mode: When the CAS goes high, the data output is turned off. So if CAS is pulled high too fast (to shorten the t_{PC}), the CPU is deprived of the data. One solution is to change the internal circuitry of fast page DRAM to allow the data to be available longer (even if CAS goes high). This is exactly what happened. As a result of this change, the name EDO (extended data-out) was given to avoid confusion with fast page mode DRAM. This is the reason that EDO is sometimes called *hyper-page* since it is the hyper version of fast page DRAM. Tables 22-10 and 22-11 show a comparison of FPM and EDO DRAM timing. Notice in both cases that all the parameters are the same except t_{PC}. For the EDO version of page mode, the t_{PC} is 10 ns less than fast page mode.

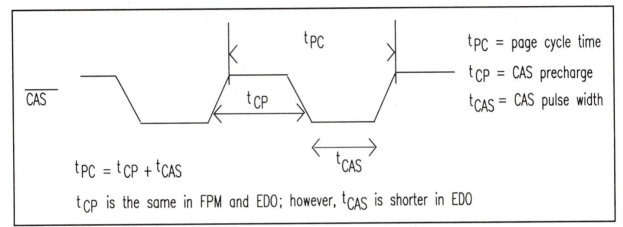

Figure 22-10. T$_{PC}$ Timing in Page Mode DRAM

In examining t_{PC} timing in Figure 22-10, notice that t_{PC} (page cycle time) consists of two portions: t_{CP} (CAS precharge time) and t_{CAS} (CAS pulse width). The t_{CP} is similar across 70 ns, 60 ns, and 50 ns DRAMs of FPM and EDO (about 10 ns). It is t_{CAS} that varies among these DRAMs. In EDO this portion is made as small as possible. Figure 22-11 compares FPM and EDO timing.

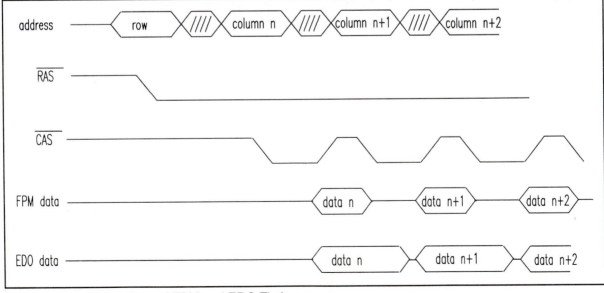

Figure 22-11. Comparison of FPM and EDO Timing

SECTION 22.4: EDO, SDRAM, AND RAMBUS MEMORIES

SDRAM (synchronous DRAM)

When the CPU bus speed goes beyond 75 MHz, even EDO is not fast enough. SDRAM is a memory for such systems. First, let us see why it is called synchronous DRAM. In all the traditional DRAMs (page mode, fast page and EDO), CPU timing is not synchronized with DRAM timing, meaning that there is no common clock between the CPU and DRAM for reference. In those systems it is said that the DRAM is *asynchronous* with the microprocessor since the CPU presents the address to DRAM and memory provides the data in the master/slave fashion. If data cannot be provided on time, the CPU is notified with the NOT READY signal. In response to NOT READY, the CPU inserts a wait state into its bus timing and waits until the DRAM is ready. In other words, the CPU bus timing is dependent upon the DRAM speed. This is not the case in synchronous DRAM. In systems with SDRAM, there is a common clock (called the *system clock*) that runs between the microprocessor and SDRAM. All bus activities (address, data, control) between the CPU and DRAM are synchronized with this common clock. That is, the common clock is the point of reference for both the CPU and SDRAM and there is no deviation from it and hence no waiting by the CPU. See Figure 22-12 for SDRAM timing. As shown in Figure 22-12, the system clock is the common clock that the address, data, and control signals are synchronized with. As you examine the timing figures in EDO and page mode, you will not find such a clock.

Synchronous DRAM and burst mode

The presence of the common system clock between the CPU and SDRAM lends itself to what is called *burst I/O*. Although burst I/O will do both read and write, we will discuss the read operation for the sake of simplicity. In burst read, the address of the first location is provided as normal. RAS is first, followed by CAS. However, since in the cache fill we read several consecutive locations (depending on whether the cache has 4, 8, 16, or 32 lines), there is no need to provide the full address of each line and pay the timing penalty for address setup and hold time. Why not simply program the burst SDRAM to let it know how many consecutive locations are needed according to the cache design? That is exactly the idea behind many SDRAMs. They are capable of being programmed to output up to 256 consecutive locations inside DRAM. In other words, the number of burst reads can be 1, 2, 4, 8, 16, or 256, and burst SDRAM can be programmed in advance for any number of these reads. The number of burst reads is referred to as *burst length*. In many recent SDRAMs, the burst length can be as high as a whole page. Burst read shortens memory access time substantially. For example if burst length is programmed for 8, for the first location we need the full address of RAS followed by

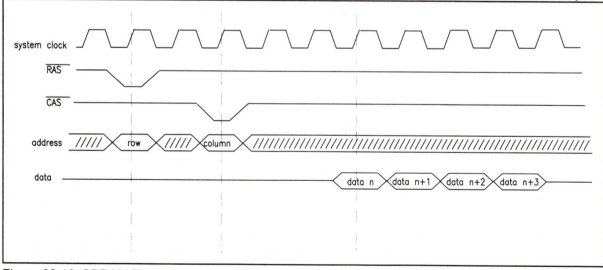

Figure 22-12. SDRAM Timing

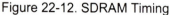

CAS. However, for the second, third, ..., eighth, we can get the data out of the SDRAM with a minimum delay, limited only by the internal circuitry of DRAM. Starting with the x486, processors use the concept of burst read in their bus timing.

SDRAM and interleaving

In order to increase performance, SDRAMs use the concept of interleaving discussed in Section 22.2. In traditional interleaved design, the board designer must arrange the DRAM in an interleaved fashion in order to hide the precharge time of one bank behind the access time of the other one. In SDRAM, this interleaving is done internally. In other words inside the SDRAM itself, DRAM cells are organized in such a way that while one bank is being refreshed the other one is being accessed. By incorporating both the burst mode and interleaving concepts into SDRAM, it is predicted that SDRAM memory can be used for a bus frequency as high as 125 MHz but not beyond that.

Figure 22-12 shows SDRAM timing. How many clocks after CAS will the data appear at the data pins? This can be programmed. It is called *read latency* and can be 1, 2, or 3 clocks. In Figure 22-12, the read latency is 3 since the data appears at the data buses 3 clocks after CAS.

It should be noted that SDRAM and EDO standards are both set by industry and every DRAM maker supports them. SDRAM and EDO are not proprietary technologies.

Example 22-17

Assume a bus frequency of 100MHz. Discuss bus timing for (a) EDO of 50 ns speed where $t_{CP}=20$, (b) SDRAM of $t_{CK}=10$ns.

Solution:

1/100MHz = 10ns is the system clock.

(a) In EDO when the page is opened, the fastest it can provide data is t_{PC}, which is 20ns. Therefore, we need at least one wait state.

(b) In SDRAM of $t_{CK}=10$ ns, the first address is strobed into the DRAM and subsequent data bursts are provided at 10 ns intervals. Therefore, no wait state is needed. Of course, for both of the above cases any bus overhead was ignored.

Rambus DRAM

In contrast to EDO and SDRAM, Rambus is proprietary DRAM architecture. DRAM manufacturers license this technology from Rambus Inc. in exchange for royalty payments. DRAMs with Rambus technology are referred to as RDRAM in technical literature.

Overview of Rambus technology

The heart of Rambus technology is a proprietary interface for chip-to-chip bus connection. This high-speed bus technology is composed of three sections: (1) a Rambus interface, (2) a Rambus channel, and (3) Rambus DRAM. The Rambus interface standard must be incorporated into both DRAM and the CPU. While many DRAM makers are introducing DRAM with a Rambus interface (called RDRAM), not every microprocessor is equipped with a Rambus interface. However, Intel has indicated that it will equip future generations of the x86 with a Rambus interface. However, if a given microprocessor is not equipped with a Rambus interface, one can design a memory controller with the Rambus interface and place it between the CPU and RDRAM. Such a controller is referred to as a *Rambus channel master* and the RDRAM is called a *Rambus channel slave*. See Figures 22-13 and 22-14.

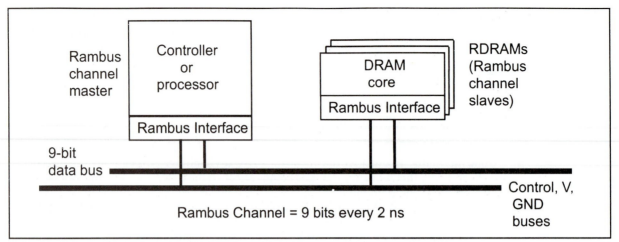

Figure 22-13. A Rambus-Based System

(Courtesy of Rambus, Inc.)

In Rambus technology, only the master can generate a request since it contains intelligence. Slave devices such as RDRAM respond to requests by the master. This eliminates any need for addition of intelligence circuitry to the RDRAM, thereby increasing its die size. This also means that data transfers can happen only between master and slave and there is never any direct data transfer between slaves. However, master capability can be added to devices other than the CPU such as peripheral devices, graphic processors, and memory controllers. The following describes additional features of Rambus channel technology.

1. The Rambus channel has only a 9-bit data bus.
2. There are only two DRAM organizations available for RDRAMs: x8 or x9. For example, a 16M-bit RDRAM is organized as 2Mx8. This is in contrast to DIMM memory modules where there are 72 pins for data alone. Such a large number of pins without a sufficient number of ground pins limits bus speed for these memory designs to less than 100 MHz. To reduce crosstalk and EMI (electromagnetic interference), we can add ground pins but that in turn makes the DRAM memory module too large (see Chapter 26 for the role of ground in reducing crosstalk).
3. Since the data bus is limited to 9 pins in a Rambus channel, by adding a sufficient number of ground pins one can push the speed of the bus to 500 MHz. To counter the impact of a limited bus size on bus bandwidth, the Rambus employs the method of block transfer. This is explained next.

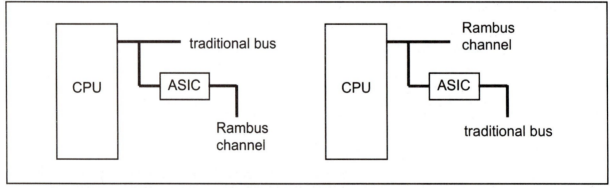

Figure 22-14. CPUs With and Without Rambus Channel (Courtesy of Rambus, Inc.)

Rambus protocol for block transfer

In Rambus the data is transferred in blocks. Such a block-oriented data transfer requires a set of protocols in which the packet types are defined very strictly. There are three types of packets in the Rambus protocol: (a) request, (b) acknowledge, and (c) data. The following steps show how the read operation works according to Rambus protocol. See Figure 22-15.

1. The master issues a request packet specifying the initial starting address of the needed data, plus the number of bytes needed to be transferred (the maximum for byte count is 256 bytes). This is considered one transaction.
2. RDRAM receives the request packet and decodes the addresses and byte count. If it has the requested data, an acknowledge packet is sent back to the master.
3. The acknowledge packet has three possibilities:
   (a) the addressed data does not exist,
   (b) the addressed data does exist but it is too busy to transfer the data. Try again later. This is called *nack*.
   (c) the addressed data does exist and it is ready to transfer them. This is called *okay*.
4. If the acknowledge packet has an okay in it, the RDRAM starts to transfer the data packet immediately.

An interesting aspect of this protocol is that the delay associated with receiving the acknowledge and sending the data packets can be programmed into configuration registers of both master and slave during BIOS system initialization.

At the time of this writing, the popular Nintendo 64 video game is among the many users of Rambus technology.

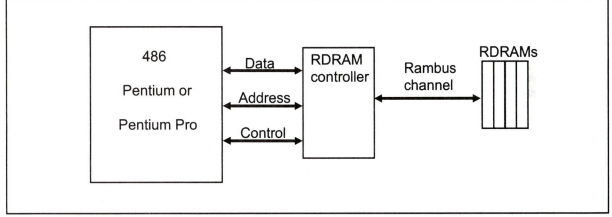

Figure 22-15. x86 System Using Rambus DRAM

(Courtesy of Rambus, Inc.)

Review Questions

1. A 200-MHz Pentium has a bus frequency of _____.
2. A 100-MHz Pentium has a bus frequency 2/3 of the CPU. What is the read cycle time for this processor?
3. When a page is opened, what limits us in accessing consecutive columns?
4. True or false. In EDO, when CAS goes up the data output is turned off.
5. Which of the following DRAMs has a common synchronous clock with the CPU?
   (a) FPM   (b) EDO   (c) SDRAM   (d) all of the above
6. True or false. SDRAM incorporates interleaved memory internally.
7. Can anyone incorporate the Rambus interface in their device?
8. Who issues the request in a Rambus system?
9. Who issues the acknowledge in a Rambus system?
10. Can normal EDO or FPM DRAMs be used for the Rambus channel?

SUMMARY

This chapter examined high-speed memory and its use in high-performance systems. The first section looked at memory cycle times of various 80x86 microprocessors and the introduction of wait states. One or more wait states must be introduced whenever the CPU clock cycle is faster than the memory clock cycle.

The second section defined some terminology commonly used in memory design. *Memory access time* is the time interval in between the moment the addresses are applied to the memory chip address pins and the time the data is available at the memory's data pins. *Memory cycle time* is the time interval in between two consecutive accesses to the memory chip. In addition, Section 22.2 explored various types of DRAM. These types are categorized by the mode of data access. Standard mode (or random access) DRAM has the longest access time. It requires the row address to be provided first and then the column address for each cell. In page mode DRAM, cells are organized in an N x N matrix, such that it is easy to access the cells of a given row consecutively, or in other words, to access all the columns of a page. Static column mode makes accessing all the columns of a given row much simpler by eliminating the need for CAS. A large percentage of 80386 and higher computers use static column DRAM for main memory. Nibble mode is similar to page mode except that it allows rapid access to 4 contiguous bits (a nibble), rather than all the bits of a page. Section 22.2 also discussed a problem that arises in back-to-back accesses to DRAM. After one access, the memory needs a precharge time to get ready for the next access. The interleaving method solves this problem by using two sets of banks, which are accessed alternately.

Section 22.3 covered cache memory. The most widely used memory design for high-performance CPUs implements DRAMs for main memory and a small amount of SRAM for cache memory. When the CPU needs data, it first checks cache. If it is there (a hit), it can be brought into the CPU much more quickly than from main memory. If the data is not found in cache (a miss), the data must be brought into cache and the CPU. Three types of cache organization were discussed: (1) fully associative, (2) direct mapped, and (3) set associative. Two methods, write-through and write-back, have been developed to ensure that the data in main memory and cache are consistent. As cache becomes full, data is swapped to main memory to make room for new data. How to select which data to swap out is called a *cache replacement policy*. Commonly used schemes are LRU and FIFO.

Section 22.4 covered newer and faster memory technologies, including EDO, SRAM, and Rambus. EDO (extended data-out) DRAM evolved from fast-page DRAM, and differs from it in that the data is made available longer to the CPU. In synchronous DRAM (SDRAM), SDRAM and the CPU are both synchronized to the system clock so that the CPU does not have to wait for the data. SDRAM is improved further by interleaving the DRAM cells so that one bank can be refreshed while another is being accessed. Rambus is a proprietary technology that uses a specialized interface, channel, and DRAM to create a high-speed DRAM. In Rambus DRAM, or RDRAM, data is transferred in blocks.

PROBLEMS

SECTION 22.1: MEMORY CYCLE TIME OF THE 80X86

1. Calculate the memory cycle time for each of the following systems.
   (a) 386 of 33 MHz, 1 WS   (b) 486 of 50 MHz, 1 WS
   (c) 386 of 25 MHz, 2 WS   (d) Pentium of 60 MHz, 1 WS
2. If the memory cycle time is 90 ns, a 386 system of 33 MHz needs ___ WS.
3. If the memory cycle time is 80 ns, a 486 system of 50 MHz needs ___ WS.
4. If the memory cycle time is 45 ns, a Pentium system of 66 MHz needs ___ WS.
5. If the memory cycle time is 200 ns, a 386SX system of 20 MHz needs ___ WS.
6. Find the effective memory performance of a 486 system of 50 MHz with 2 WS. Compare the performance degradation with a 0-WS system.
7. Find the effective memory performance of a 386 system of 33 MHz with 1 WS. Compare the performance degradation with a 0-WS system.
8. Find the effective memory performance of a Pentium system of 60 MHz with 1 WS. Compare the performance degradation with a 0-WS system.

9. If a given system with a 2-clock memory cycle time has a memory cycle time of 60 ns and is designed with 1 WS, find the CPU frequency.
10. In a 33-MHz 486 0 WS system, a minimum of 20 ns is used for data and address path delay and address decoding. What is the maximum memory cycle time?

SECTION 22.2: PAGE, STATIC COLUMN, AND NIBBLE MODE DRAMS

11. In which memory are the cycle time and access time equal?
12. What is the difference between the t_{AA} and t_{CA} SRAM data sheet?
13. Define the memory cycle time for a memory chip.
14. Define the memory cycle time for the CPU.
15. What is t_{RC} and t_{RAC} in DRAM? State the difference between them.
16. Show the relation (approximate) between t_{RC}, t_{RAC}, and t_{RP}.
17. A given DRAM has t_{RAC} =60 ns. What is the t_{RC} (approximate)?
18. A given DRAM has t_{RAC} =85 ns. What is the t_{RC} (approximate)?
19. A given DRAM has t_{RC} =110 ns. What is the t_{RAC} (approximate)?
20. A given DRAM has t_{RC} =90 ns. What is the t_{RAC} (approximate)?
21. Calculate the time needed to access 2048 bits of 60-ns 4Mx1. Use Table 22-2.
22. Calculate the time needed to access 2048 bits of 4Mx1 of 70 ns. Use Table 22-2.
23. Draw a timing diagram for standard mode SRAM and DRAM memory cycle.
24. What is the minimum memory addition to a 386 system with interleaved memory design if each bank of 8 bits is set for 1Mx8? The parity bit is not included.
25. Calculate the chip count for Problem 24 if 1Mx4 chips are used.
26. What is the minimum memory addition to a 486 system with interleaved memory design if each bank of 8 bits is set for 4Mx8? The parity bit is not included.
27. Calculate the chip count for Problem 26 if 4Mx4 chips are used.
28. Show the hex address for 386/486 interleaved memory banks.
29. Calculate the time needed to access 2048 bits of one page for page mode DRAM of 4Mx1 of 70 ns. Use Table 22-4.
30. Calculate the time needed to access 2048 bits of one page for page mode DRAM of 4Mx1 of 60 ns. Use Table 22-4.
31. Calculate the time needed to access 2048 bits of one page for static column mode DRAM of the 4Mx1 of 70 ns. Use Table 22-5.
32. Calculate the time needed to access 2048 bits of one page for static column mode DRAM of 4Mx1 of 60 ns. Use Table 22-5.
33. Calculate the time needed to access each of the following. Use Table 22-6.
   (a) 4 bits (b) 2048 bits of nibble mode DRAM of 60 ns
34. Calculate the time needed to access each of the following. Use Table 22-6 and assume that t_{RP} equals one-half of t_{RC}.
   (a) 4 bits   (b) 2048 bits of nibble mode DRAM of 70 ns

SECTION 22.3: CACHE MEMORY

35. List the three different cache organizations.
36. What is the principle of locality of reference?
37. What does LRU stand for, and to what does it refer in cache memory?
38. What do *write-through* and *write-back* refer to? Define each one and state an advantage and a disadvantage for each.
39. What does a line size of 16 bytes mean?
40. Calculate the tag and data cache sizes needed for each of the following cases if the memory requesting address to main memory is 20 bits (A19 - A0). Assume a data bus of 8 bits. Draw a block diagram for each case.
   (a) fully associative of 1024 depth
   (b) direct mapped where A15 - A0 is for the index
   (c) 2-way set associative where A14 - A0 is for the index
   (d) 4-way set associative   (e) 8-way set associative
41. In Problem 40, compare the size of data cache and tag cache parts (b), (c), (d), and (e). What is your conclusion?

PROBLEMS

42. Calculate the tag and data cache sizes needed for each of the following cases if the memory requesting address to main memory is 24 bits (A23 - A0). Assume a data bus of 8 bits. Draw a block diagram for each case.
(a) fully associative of 1024 depth
(b) direct mapped where A19 - A0 is for the index
(c) 2-way set associative where A18 - A0 is for the index
(d) 4-way set associative   (e) 8-way set associative
43. In Problem 42, compare the size of data cache and tag cache for (b), (c), (d), and (e).What is your conclusion based on this comparison?
44. Give 3 factors affecting the cache hit.
45. What does the law of diminishing returns mean when applied to cache?
46. The 486 cache is organized as _____.
47. Pentium cache is organized as _____.

SECTION 22.4: EDO, SDRAM, RAMBUS MEMORIES

48. The CPU speed (in Pentium and higher processors) is often a _____ (multiple, fraction) of the bus speed.
49. Calculate the memory read cycle time of a CPU with a bus speed of 300 MHz. Assume a 2-clock read cycle time.
50. In the above question, discuss the difficulties associated with the design of such a high-speed bus.
51. In Pentium processors with speeds of 100 MHz and higher, the bus speed is _____ the CPU speed.
(a) the same as (b) a fraction of (c) a multiple of
52. In DRAM technology, EDO stands for _____ and FPM stands for _____.
53. True or false. Both EDO and FPM are page mode DRAMs.
54. What does "opening a page" mean in page mode DRAMs? What is the role of signals RAS and CAS in opening a page?
55. In _____ (EDO, FPM) DRAM, the data is turned off when CAS goes high.
56. In FPM DRAM, what happens if CAS goes high too soon and what is the consequence?
57. When a page is opened, reading consecutive columns is limited by the speed of _____.
58. In the design of DRAM, why is it desired to pull CAS high as soon as possible?
59. What is the t_{PC} for a 50-ns DRAM?
60. In comparison of EDO and FPM DRAM of 60-ns and 70-ns speed, indicate which timing parameters are the same and which are different.
61. For EDO DRAM, t_{PC} is normally 10 ns _____ (less than, greater than) the t_{PC} of FPM DRAM.
62. The t_{PC} timing is made of two parts. They are _____ and _____. One of them is constant across all DRAMs of 70-ns, 60-ns, 50-ns speeds. Which one is that?
63. What does SDRAM stand for?
64. What is the most important difference between SDRAM and traditional DRAMs of FPM and EDO?
65. The SDRAM of 75 MHz can provide data every _____ ns after a page has been opened.
66. The SDRAM of 120 MHz can provide data every _____ ns after a page has been opened.
67. It is predicted that SDRAM can be used for bus speeds of as high as _____ MHz.
68. What is burst mode memory? Define burst length.
69. In SDRAM, what is the size of burst length?
70. True or false. The x86 processors starting with the 486 support burst mode read.
71. What is the difference between interleaved memory design on board and interleaved in SDRAM?
72. True or false. EDO and SDRAM memories are proprietary technology requiring licenses.
73. What does "RDRAM" stand for, and what is the difference between that and SDRAM and EDO DRAMs?

74. Rambus technology consists of three parts. Name them.
75. True or false. Both master and slave sections of Rambus technology must have a Rambus Interface.
76. A Rambus channel has a(n) _____-bit data bus.
77. RDRAM is a _____ (master, slave) in Rambus technology.
78. True or false. In Rambus technology, the data transfer happens between master and slave only.
79. True or false. In Rambus technology the data transfer never happens between slaves.
80. Name and describe the three types of packets for communication protocol in Rambus technology.
81. In the above question, "okay" and "nack" are part of which packet?
82. Explain the role of nack and okay for data transfer in Rambus technology.
83. The bus speed in Rambus can go as high as _____ MHz.
84. What happens if a CPU does not have a Rambus interface?
85. True or false. EDO and SDRAM can be used in place of RDRAM in a Rambus channel.

ANSWERS TO REVIEW QUESTIONS

SECTION 22.1: MEMORY CYCLE TIME OF THE 80X86
1. (a) 1/40 MHz =25 ns; therefore, 2 x 25 = 50 ns; (b) 2 x 20 (for 0 WS) + 20 (1 WS)=60 ns;
   (c) 2 x 15 ns + 15 = 45 ns
2. It means that the CPU cannot access memory faster than every 50 ns.
3. (a) The read cycle time is 75 ns; therefore, the effective working frequency is the same as 26.6 MHz of 0 WS (1/37.5 ns =26.6 MHz).
   (b) The read cycle time is 60 ns; therefore, the effective working frequency is the same as 33 MHz (1/30 ns =33 MHz).
4. A total of 50 ns is left for the memory access time.
5. Since 2+1 WS =3 clocks for each read cycle time, 30 ns (90/3=30) for the CPU clock duration; therefore, the CPU frequency is 33 MHz (1/30 ns = 33 MHz).

SECTION 22.2: PAGE, STATIC COLUMN AND NIBBLE MODE DRAMS
1. SRAM and ROM   2. 100 ns   3. 60 ns
4. t_{RAC} (RAS access time), t_{RP} (RAS precharge time)
5. (a) There are two sets of 1Mbytes; therefore, each set consists of 4 banks of 256Kx9 memory where each bank belongs to 1 byte of the D31 - D0 data bus. (b) 2M
6. true
7. t_{PC}
8. total time = t_{RAC}+99 x t_{PC} = 60+99 x 30 =3030 ns
9. total time = t_{RAC}+3 x t_{NC} =50 ns +3 x 20 ns =110 ns
10. static column

SECTION 22.3: CACHE MEMORY
1. SRAM   2. cache   3. register, cache, and main memory
4. 512   5. direct map   6. 4
7. The CPU writes to cache and main memory at the same time when updating main memory.
8. write-through
9. LRU (least recently used) is a cache replacement policy. When there is a need for room in the cache memory the cache controller flushes the LRU data to make room for new data.
10. When the cache is filled with new data, it is done a minimum of 4 lines (4 x 4=16 bytes) at a time.

SECTION 22.4: EDO, SDRAM, AND RAMBUS MEMORIES

1. often less than 100M Hz; many times it is only 66 MHz.
2. 2/3 x 100 MHz = 66 MHz. Now 1/66 MHz = 15 ns. 2 x 15 ns = 30 ns read cycle time.
3. the t_{PC} (page cycle time)
4. false
5. SDRAM
6. true
7. Yes, as long as you get a license from Rambus Inc.
8. Master (Rambus controller)
9. RDRAM slave
10. No. It must be RDRAM.

CHAPTER 23

486, PENTIUM, PENTIUM PRO AND MMX

OBJECTIVES

Upon completion of this chapter, you will be able to:

» List the design enhancements of the 80486 over previous-generation 80x86 microprocessors

» Discuss the advantages of the 5-stage 486 pipeline over the previous 2-stage 8086 pipeline

» Explain how the burst cycle is used to increase memory cycle times for read and write operations

» Compare execution speeds of the 386 and 486 for Assembly language programs

» List three ways that designers can increase the processing power of a CPU

» List design enhancements of the Pentium over previous-generation 80x86 microprocessors

» Describe the impact on performance of the 64-bit data bus of the Pentium

» Describe superscalar architecture and Harvard architecture and their use in the Pentium

» List the unique features of RISC architecture compared to CISC and describe the impact on processing speed and program development

The 8086/88 microprocessor is the product of technology of the 1970s. Advances made in integrated circuit technology in the 1980s made ICs with 1 million transistors possible. This led to the design of some very powerful microprocessors. This chapter will look at Intel's 486 and Pentium microprocessors and will examine the merit of RISC processors and their potential power. In Section 23.1 the 80486 microprocessor is studied. Intel's Pentium is discussed in Section 23.2. Section 23.3 explores RISC processors and their performance is compared with that of 80x86 CISC processors.

SECTION 23.1: THE 80486 MICROPROCESSOR

The 80486 is the first 1-million transistor microprocessor (actually, 1.2 million) packaged in 168-pin PGA packaging. It is not only compatible with all previous Intel 80x86 microprocessors, but is also much faster than the 80386. When Intel went from the 286 to the 386, register widths were increased from 16 bits to 32 bits. In addition, the external data bus size was increased from 16 bits to 32 bits, and the address bus became 32 bits instead of 24 bits as in the 80286. However, the 32-bit core of the 386 microprocessor is preserved in the 486 microprocessor. This is due to the fact that many studies have shown that 32-bit registers can take care of more than 95% of the operands in high-level languages. Like the 386, the 486 has a 32-bit address bus and a 32-bit data bus. The data bus is D0 - D31 and the address bus is A2 - A31 in addition to BE0 - BE3, just as in the 386. In the design of the 486, Intel uses four times as many transistors as used in the 386 to enhance its processing power while keeping a 32-bit microprocessor.

Enhancements of the 486

The following are the ways the 486 is enhanced in comparison to the 386.

Enhancement 1

By heavily pipelining the fetching and execution of instructions, many 486 instructions are executed in only 1 clock cycle instead of in 3 clocks as in the 386. By using a large number of transistors, the fetching and execution of each instruction is split into many stages, all working in parallel. This allows the processing of up to five instructions to be overlapped. Pipelining in the 486 will be discussed further at the end of this section.

Enhancement 2

By putting 8K bytes of cache with the core of the CPU all on a single chip, the 486 eliminates the interchip delay of external cache. In other words, while in the 386 the cache is external, the 486 has 8K bytes of on-chip cache to store both code and data. Although the 486 has 8K bytes of on-chip cache, 128K to 256K bytes of off-chip cache are also present in many systems. Off-chip cache (level two) is commonly referred to as *secondary cache,* while on-chip cache is called *first-level cache*. The 8K on-chip cache of the 486 has 2-way set associative organization and is used for storing both data and code. It uses the write-through policy for updating main memory.

Enhancement 3

Intel used some of 1.2 million transistors to incorporate a math coprocessor on the same chip as the CPU. While in all previous 80x86 microprocessors the math coprocessor was a separate chip, in the 80486 the math coprocessor is part of a single IC along with the CPU. This reduces the interchip delay associated with a multichip system such as the 386 and 387 but at the same time made the cost of a 80486 high compared to a 386 since the 80486 is in reality two chips in one: the main CPU and math coprocessor. For many people who did not need a math coprocessor this extra price was not justified. Therefore, Intel introduced the 80486SX, which is the main CPU, and a separate math coprocessor named 80487SX.

Enhancement 4

Another major addition to the 486 is the use of 4 pins for data parity (DP), which allows implementation of parity error checking on the system board. The four pins DP0, DP1, DP2, and DP3 are bidirectional, and each is used for 1 byte of the D31 - D0 data bus. When the 486 writes data it also provides the even-parity bit for each byte through the DP0 - DP3 pins. When it reads the data it expects to receive the even parity bit for each byte on the DP0 - DP3 pins. After comparing them internally, if there is a difference between the data written and the data read, it activates the pin PCHK (parity check) to indicate the error. This means that PCHK is an output pin while DP0 - DP3 are bidirectional I/O. It must be noted that inconsistency between data written and data read has no effect on the execution of code by the CPU. It is the responsibility of the system designer to incorporate error detection by using the PCHK pin in their design. In the above discussion of parity, the word *data* is meant to refer to both code and data. Figure 23-1 shows the memory organization of the 486.

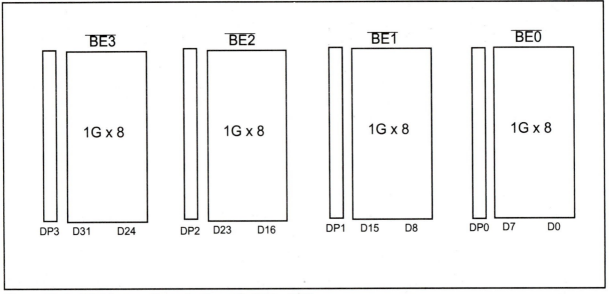

Figure 23-1. 486 Memory Organization with DP0 - DP3

Enhancement 5

Another enhancement of the 486 involves the burst cycle. The memory cycle time of the 486 with the normal zero wait states is 2 clocks. In other words, it takes a minimum of 2 clocks to read from or write to external memory or I/O. In this regard, the 486 is like the 386. To increase the bus performance of the 486, Intel provides an additional option of implementing what is called a *burst cycle*. The 486 has two types of memory cycles, nonburst (which is the same as the 386) and burst mode. In the burst cycle, the 486 can perform 4 memory cycles in just 5 clocks. The way the 80486 performs the burst cycle read is as follows. The initial read is performed in a normal 2-clock memory cycle time, but the next three reads are performed each with only one clock. Therefore, four reads are performed in only 5 clocks. This is commonly referred to as 2-1-1-1 read, which means 2 clocks for the first read and 1 clock for each of the following three reads. This is in contrast to 386, which is 2-2-2-2 for reading 4 doublewords of aligned data. Of course, burst cycle reading is most efficient if the data and codes are in 4 doubleword (32-bit) consecutive locations. In other words, the burst cycle can be used to fetch a maximum of 16 bytes of information into the CPU in only 5 clocks, provided that they are aligned on doubleword boundaries. There are two pins, BRDY (burst ready) and BLAST (burst last), used specifically to implement the burst cycle. BRDY is an input into the 486 and BLAST is an output from the 486. See Figure 23-2.

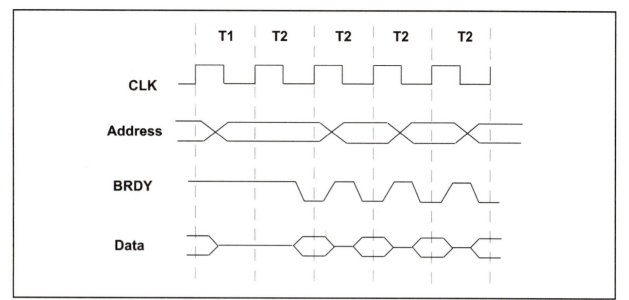

Figure 23-2. Burst Cycle Read in the 486

Example 23-1

Calculate and compare the bus bandwidth of the following systems. Assume that both are working with 33 MHz and that the 386 is 0 WS. Also assume that the data is aligned and is in 4 consecutive doubleword memory locations.
(a) 386   (b) burst mode of the 486

Solution:

(a) In the 386, since each memory cycle time takes 2 clocks we have
memory cycle time = 2 (1/33 MHz) = 2 × 30 ns = 60 ns
bus bandwidth = (1/60 ns) × 4 bytes = 66 megabytes/second

(b) In burst mode, the 486 performs 4 memory cycles in only 5 clocks; therefore, the average memory cycle time in burst mode is 1.25(5/4 = 1.25) clocks for each 32-bit (doubleword) of data fetched as long as they are aligned and located in consecutive memory locations. This results in bus bandwidth = [1/(1.25 × 30 ns)] 4 bytes = 106.66 megabytes/second

---

Enhancement 6

The 486 supports all 386 instructions in addition to 6 new ones. They are shown in Table 23-1. Three of the new instructions, INVD, INVLPG, and WBINVD, are added specifically for dealing with the on-chip cache and the TLB entries. The XADD instruction first loads the destination operand into the source and then loads the total sum of both the destination and the original source into the

Table 23-1: New 486 Instructions

Instruction	Meaning
BSWAP	Byte swap
CMPXCHG	Compare and exchange
INVD	Invalidate cache
INVLPG	Invalidate TLB entry
WBINVD	Write back and invalidate cache
XADD	Exchange and add

---

destination. The CMPXCHG instruction compares the accumulator, AL, AX, or EAX, with the destination operand, which could be a register or memory. If they are equal, the ZF = 1 and the source is copied into the destination. If they are not equal, ZF = 0 and the destination is copied into the accumulator. For example, the instruction "CMPXCHG BX,CX" copies CX into BX only if CX =AX; otherwise, it copies BX into accumulator AX.

As mentioned previously, some systems use little endian while others use the big endian convention of storing data. To allow the implementation of either, BSWAP is provided. The BSWAP instruction converts the contents of a 32-bit register from the little endian to big endian, or vice versa.

Example 23-2

Find the contents of memory location ES:4000 after running the following program.

```
        MOV    EAX,[2000]        ;load EAX from memory DS:2000
        BSWAP EAX                ;change little endian to big endian
        MOV    ES:[4000],EAX     ;save the result at ES:4000
```

Assume that memory locations DS:2000 - DS:2003 have the following contents.
    DS:2000=(87)
    DS:2001=(54)
    DS:2002=(F2)
    DS:2003=(99)

Solution:

The first instruction brings in the data in the little endian format where the least significant byte is fetched into the least significant byte of EAX, which is the AL register. BSWAP makes the 87H the most significant byte and puts 99H into the AL register. Therefore, after the execution of the last instruction we have the following:

    ES:4000=(99)
    ES:4001=(F2)
    ES:4002=(54)
    ES:4003=(87)

The addition of the BSWAP instruction makes the job of operating system software writers much easier in converting their software from little endian to big endian, or vice versa.

CLK in the 80486

Another difference is the clock frequency provided to the 486. As mentioned in Chapter 22, the CLK input frequency, which provides the fundamental timing for the internal working of the CPU, is twice the system frequency for 386 microprocessors. In the case of the 486, the CLK is the same as the system frequency.

Example 23-3

A 50-MHz clock is connected to CLK in each of the following systems. Find the system frequency.
(a) 386   (b) 486

Solution:

For the 80386, the system frequency is 25 MHz and for the 486 the system frequency is 50 MHz. In other words, these systems are advertised as 386 of 25 MHz and 486 of 50 MHz.

High memory area (HMA) and the 80486

Pin A20M (A20 mask) is an active-low input and was added to mask the A20 address bus. The problem of address wraparound for the 8086 was discussed in Chapter 9 and the 286/386 treatment of A20 was discussed in Chapter 10. Starting with the 486, Intel provided an input pin called A20M to allow masking of the A20 address. By asserting A20M, which makes A20M = 0, the 486 wraps around the address if it goes beyond the 1 megabyte address range. In this way the 486 acts like the 8086. However, if A20M = 1, the A20 address is provided to the external address bus just like the 286 and 386. It must be noted that the A20M pin is used only when the 486 is in real mode. In protected mode, the A20M pin's has no effect on the A20 address bit.

Example 23-4

Assuming that the 486 is working in real mode, find the status of the A20 address bit for (a) and (b).
(a) A20M =0
(b) A20M =1 if CS =FFFFFH and IP =FFFFH
(c) To use the high memory area, what must the A20M pin be set to?

Solution:

The physical address is

	FFFF0	segment shifted left
+	FFFF	add the offset
	10FFEF	

(a) If input pin A20M =0, the 1 is dropped and address bit A20 =0, just like the 8086/88.
(b) If input pin A20M =1, the 1 is passed to the address bit A20 and is provided to the system, which means that A20 =1, just like the 286 and 386.
(c) As shown in Chapter 10, A20 provides an extra 65,520 bytes of memory just above the 1M range while the CPU is still in real mode. This extra memory is called HMA (high memory area). Therefore, to access the HMA in the 486-based PC, the A20M input pin to 486 must be set to high (A20M=1), which tells the 486 not to mask address bit A20.

386, 486 Performance comparison

As stated earlier, most of the instructions in the 486 are executed with only one clock. This leads to a much lower clock count for a given program run on the 486 as opposed to the 386. We examine this concept in Example 23-5.

More about pipelining

In the 8085 there was no pipelining. At any given moment, it either fetched or it executed. It could not do both at the same time. In the 8085, while the buses were fetching the instructions (opcodes) and data, the CPU was sitting idle, and in the same way, when the CPU was executing instructions, buses were sitting idle. However, in the 8086/88 the fetch and execute were performed in parallel by two sections inside the CPU called the BIU (bus interface unit) and EU (execution unit). The 8086 has an internal queue where it keeps the opcodes that are prefetched and waiting for the execution unit to process them. In the sequence of instructions, if there is a jump (JMP, JNZ, JNC, and so on) or CALL, the prefetched buffer (queue) is flushed and the bus interface unit of the CPU brings in instructions from the target location while the the execution unit waits for the new instruction. Since the introduction of the 8086 in 1978, microprocessor designers have come to rely more

---

and more on the concept of pipelining to increase the processing power of the CPU. The next development was to expand the concept of a pipeline to the three stages of fetch, decode, and execute. In the 486, the pipeline stage is broken down even further, to 5 stages as follows:

1. fetch (prefetch)
2. decode 1
3. decode 2
4. execute
5. register write-back

Due to such a large number of addressing modes in the 80x86, a two-stage decoder is used for the calculation and protection check of operand addresses. The register write-back is the stage where the operand is finally delivered to the register. For example, in the instruction "ADD EAX,[EBX+ECX*8+200]", after it is fetched, the two decoding stages are responsible for calculating the physical address of the source operand, checking for a valid address, and getting it into the CPU. There it is added together with EAX during the execution stage, and finally, the addition result is written into EAX, the destination register. Figure 23-3 shows the 486 pipeline.

This concludes the discussion of the 80486 microprocessor. For the performance comparison of the 8086, 286, 386, and 486, see Chapter 8. The performance comparison of the 386 and 486 and Pentium is shown in the next section.

Example 23-5

Compare the clock count for the loop part of the following program run on the 386 and 486. This program transfers a block of DWORD data. Assume that the block size is 10.

```
        MOV   CX,10                  ;count=10
        MOV   SI,OFFSET ARRAY1       ;load address of source
        MOV   DI,OFFSET  RESULT      ;load address of destination
AGAIN:  MOV   EAX,DWORD PTR [SI]     ;get the element
        MOV   [DI],EAX               ;store it
        ADD   SI,4                   ;point to next element
        ADD   DI,4                   ;point to next element of result
        DEC   CX                     ;decrement the counter
        JNZ   AGAIN                  ;and go back if not zero
```

Solution:

			386	486
AGAIN:	MOV	EAX,DWORD PTR [SI]	4	1
	MOV	[DI],EAX	2	2
	ADD	SI,4	2	1
	ADD	DI,4	2	1
	DEC	CX	2	1
	JNZ	AGAIN	7/3	3/1

Total for one iteration	19	9

Notice the branch penalty for the JNZ instructions. If it goes back, it takes 7 clocks for the 386 and 3 for the 486. If it falls through, it takes only 3 and 1 for the 386 and 486, respectively. Also notice that "MOV [DI],EAX" takes 2 clocks since EAX must be provided first by the previous instruction. This is called a *data dependency.* In the next section we compare 386 and 486 performance with that of the Pentium processor.

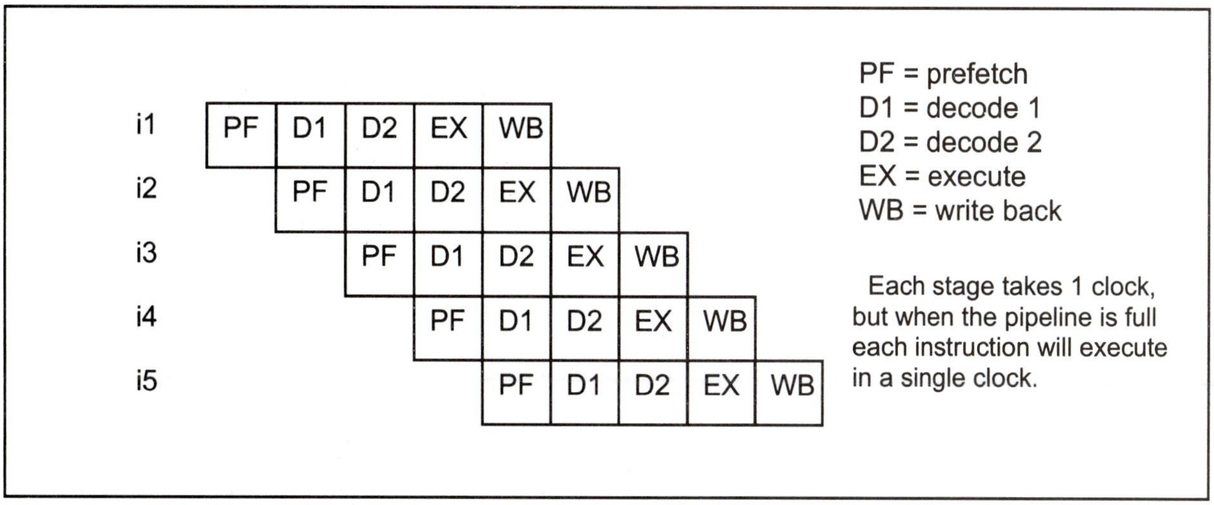

Figure 23-3. 486 Pipeline Stages

PF = prefetch
D1 = decode 1
D2 = decode 2
EX = execute
WB = write back

Each stage takes 1 clock, but when the pipeline is full each instruction will execute in a single clock.

Review Questions

1. How many pins does the 80486 have, and what kind of packaging is used for it?
2. True or false. The 486 is a 32-bit microprocessor.
3. The 80486 has a(n) _____-bit external and a(n) _____-bit internal data bus.
4. State the difference between the 80486 and the 80486SX.
5. On-chip cache is referred to as _____, while off-chip cache is called _____.
6. State the size of the on-chip cache for the 486 and the cache organization.
7. Calculate the bus bandwidth of a 486 burst read for a 50-MHz system.
8. If the 486 is advertised as 33 MHz, the clock frequency connected to the CLK pin is _____.
9. Pin A20M is an _____ (input, output) signal.
10. A20 (the twentieth address bit) is an _____ (input, output) signal for the 486.

SECTION 23.2: INTEL'S PENTIUM

Intel put 3.1 million transistors on a single piece of silicon using a 273-pin PGA package to design the next generation of 80x86. It is called Pentium instead of 80586. The name Pentium was chosen to distinguish it from clones because it is hard to copyright a number such as 80586. There are 3 ways available to microprocessor designers to increase the processing power of the CPU.

1. Increase the clock frequency of the chip. One drawback of this method is that the higher the frequency, the more the power dissipation and the more difficult and expensive the design of the microprocessor and motherboard.
2. Increase the number of data buses to bring more information (code and data) into the CPU to be processed. While in the case of DIP packaging this option was very expensive and unrealistic, in today's PGA packaging this is no longer a problem.
3. Change the internal architecture of the CPU to overlap the execution of more instructions. This requires a lot of transistors. There are two trends for this option, superpipeline and superscalar. In *superpipelining*, the process of fetching and executing instructions is split into many small steps and all are done in parallel. In this way the execution of many instructions is overlapped. The number of instructions being processed at a given time depends on the number of pipeline stages, commonly termed the *pipeline depth*. Some designers use as many as 8 stages of pipelining. One limitation of superpipelining is that the speed of the execution is limited to the the slowest stage of the pipeline. Compare this to making pizza. You

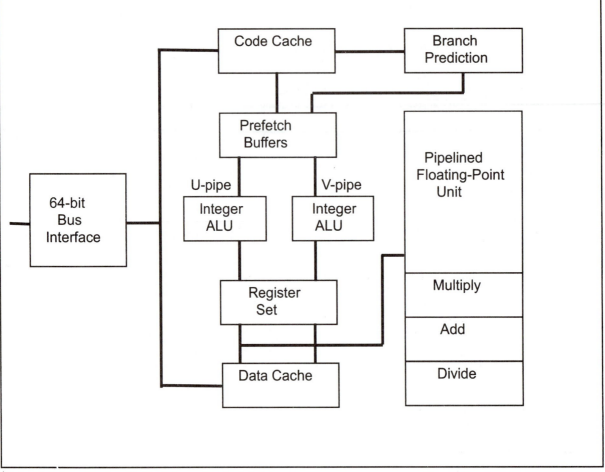

Figure 23-4. Inside the Pentium
(Reprinted by permission of Intel Corporation, Copyright Intel Corp. 1993)

can split the process of making pizza into many stages, such as flattening the dough, putting on the toppings, and baking, but the process is limited to the slowest stage, baking, no matter how fast the rest of the stages are performed. What happens if we use two or three ovens for baking pizzas to speed up the process? This may work for making pizza but not for executing programs, since in the execution of instructions we must make sure that the sequence of instructions is kept intact and that there is no out-of-step execution. The difficulties associated with a stalled pipeline (a slowdown in one stage of the pipeline, which prevents the remaining stages from advancing) has made CPU designers abandon superpipelining in favor of superscaling. In *superscaling*, the entire execution unit has been doubled and each unit has 5 pipeline stages. Therefore, in superscalar, there is more than one execution unit and each has many stages, rather than one execution unit with 8 stages as in the case of a superpipelined processor. In some superscalar processors, there are two execution units each with 4 pipeline stages instead of having a single execution unit with 8 pipeline stages as superpipelining proponents would have it. In other words, in superscaling we have two (or even three) execution units and as the instructions are fetched they are issued to the various execution units. Using the analogy of pizza, superscalar is like doubling or tripling the entire crew flattening the dough, putting toppings on, and baking. Of course, you will need a lot more people involved in the process and you have to have more ovens, but at the same time you are doubling or tripling the pizza output. In cases of recent microprocessor architecture, a vast majority of designers have chosen superscaling over superpipelining. This requires numerous transistors to duplicate several execution units, just like needing more people in our pizza-making analogy. Fortunately, advances in IC design have allowed designers access to a couple of million transistors to throw around for the

CHAPTER 23: 486, PENTIUM, PENTIUM PRO AND MMX

implementation of powerful superscaling. There are some problems with superscaling, such as data dependency issues, which can be solved by the compiler, as we will discuss below.

Intel used all three methods to increase the processing power of the Pentium. Currently, Intel is shipping the 60- and 66-MHz Pentium and planning a 100-MHz model, as well. The Pentium has a 64-bit external data bus and is a superscalar processor with two execution units to process integer data. This is in addition to a separate execution unit for floating-point data.

Features of the Pentium

The following are some of the major features of the Pentium processor.

Feature 1

In the Pentium, the external data buses are 64-bit, which will bring twice as much code and data into the CPU as the 486. However, just like the 386 and 486, Pentium registers are 32-bit. Bringing in twice as much as information can work only if there are two execution units inside the processor, and this is exactly what Intel has done. The Pentium uses 64 pins, D0 - D63, to access external memory banks, which are 64 bits wide. D0 - D7 is the least significant byte, and D56 - D63 is the most significant byte. Accessing 8 bytes of external data bus requires 8 BE (byte enable) pins, BE0 - BE7, where BE0 is for D0 - D7, BE1 for D8 - D15, and so on. This is shown in Figure 23-5 and Table 23-2.

Table 23-2: Pentium Byte Enable Signals

Byte Enable Signal	Associated Data Bus Signals
BE0#	D0-D7 (byte 0, the least significant)
BE1#	D8-D15 (byte 1)
BE2#	D16-D23 (byte 2)
BE3#	D24-D31 (byte 3)
BE4#	D32-D39 (byte 4)
BE5#	D40-D47 (byte 5)
BE6#	D48-D55 (byte 6)
BE7#	D56-D63 (byte 7, the most significant)

(Reprinted by permission of Intel Corporation, Copyright Intel Corp. 1993)

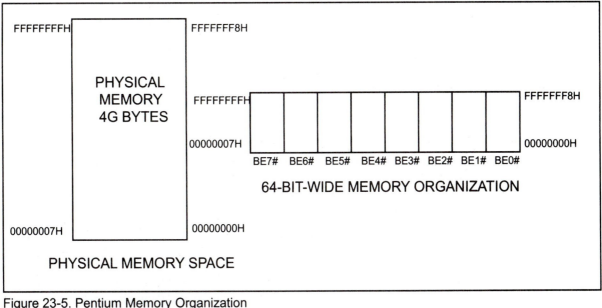

Figure 23-5. Pentium Memory Organization
(Reprinted by permission of Intel Corporation, Copyright Intel Corp. 1993)

While in the 486 there were four DP (data parity) pins, one for each of the 4 bytes of the data bus, in the Pentium there are 8 DP pins to handle the 8 bytes of data pins D0 - D63. The Pentium has A31 to A3 for the address buses. This is shown in Figure 23-6. Just like the 486, the Pentium also has the A20M (A20 Mask) input pin for the implementation of HMA (high memory area).

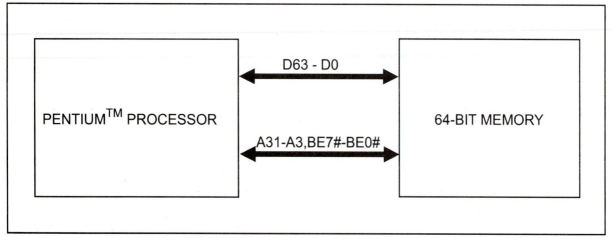

Figure 23-6. Pentium Address Buses
(Reprinted by permission of Intel Corporation, Copyright Intel Corp. 1993)

Feature 2

The Pentium has a total of 16K bytes of on-chip cache: 8K is for code and the other 8K is for data. In the 486 there is only 8K of on-chip cache for both code and data. The data cache can be configured as write-back or write-through, but to prevent any accidental writing into code cache, the 8K of code cache is write protected. In other words, while the CPU can read or write into the data cache, the code cache is write protected to prevent any inadvertent corruption. Of course, when there is a cache miss for code cache, the CPU brings code from external memory and stores (writes) it in the cache code, but no instruction executing in the CPU can write anything into the code cache. The replacement policy for both data and code caches is LRU (least recently used).

Both the on-chip data and code caches are accessed internally by the CPU core simultaneously. However, since there is only one set of address buses, the external cache containing both data and code must be accessed one at a time and not simultaneously. Some CPUs, notably RISC processors, use a separate set of address and data pins (buses) for the data and another set of address and data buses for the code section of the program. This is called *Harvard architecture* and will be discussed in the next section. The Pentium accesses the on-chip code and data caches simultaneously using Harvard architecture, but not the secondary (external) off-chip cache and data. The Pentium's cache organization for both the data and code caches is 2-way set associative. Each 8K is organized into 128 sets of 64 bytes, which means $2^7 \times 2^6 = 2^{13} = 8192 = 8K$ bytes. Each set consists of 2 lines of cache, and each line is 32 bytes wide.

Feature 3

The on-chip math coprocessor of the Pentium is many times faster than the one on the 486. It has been redesigned to perform many of the instructions, such as add and multiply, ten times faster than the 486 math coprocessor. In microprocessor terminology, the on-chip math coprocessor is commonly referred to as a *floating point unit* (FPU) while the section responsible for the execution of integer-type data is called the *integer unit* (IU). The FPU section of the Pentium uses an 8-stage pipeline to process instructions, in contrast to the 5-stage pipeline in the integer unit. See Figure 23-7.

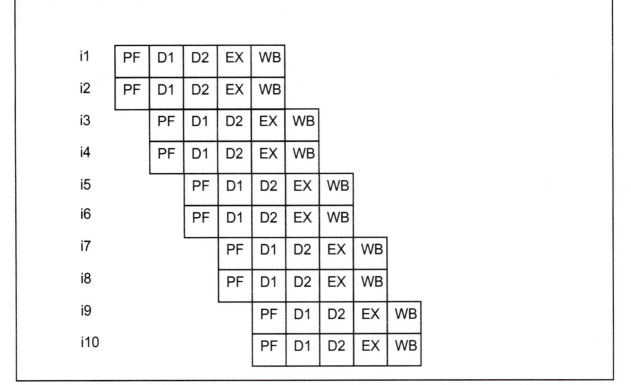

i1	PF	D1	D2	EX	WB				
i2	PF	D1	D2	EX	WB				
i3		PF	D1	D2	EX	WB			
i4		PF	D1	D2	EX	WB			
i5			PF	D1	D2	EX	WB		
i6			PF	D1	D2	EX	WB		
i7				PF	D1	D2	EX	WB	
i8				PF	D1	D2	EX	WB	
i9					PF	D1	D2	EX	WB
i10					PF	D1	D2	EX	WB

Figure 23-7. Pentium Pipeline

Feature 4

Another unique feature of the Pentium is its superscalar architecture. A large number of transistors were used to put two execution units inside the Pentium. As the instructions are fetched, they are issued to these two execution units. However, issuing two instructions at the same time to different execution units can work only if the execution of one does not depend on the other one, in other words, if there is no *data dependency*. As an example, look at the following instructions.

```
ADD    EAX,EBX         ;add EBX to EAX
NOT    EAX             ;take 1's complement EAX
INC    DI              ;increment the pointer
MOV    [DI],EBX        ;move out EBX
```

In the above code, the ADD and NOT instructions cannot be issued to two execution units since EAX, the destination of the first instruction, is used immediately by the second instruction. This is called *read-after-write dependency* since the NOT instruction wants to read the EAX contents, but it must wait until after the ADD is finished writing it into EAX. The problem is that ADD will not write into EAX until the last stage of the pipeline, and by then it is too late for the pipeline of the NOT instruction. This prevents the NOT instruction from advancing in the pipeline, therefore causing pipeline to be stalled until the ADD finishes writing and then the NOT instruction can advance through the pipeline. This kind of register dependency raises the clock count from one to two for the NOT instruction. What if the instructions are rescheduled, as follows?

```
ADD    EAX,EBX         ;add EBX to EAX
INC    DI              ;increment the pointer
NOT    EAX             ;take 1's complement of EAX
MOV    [DI],EBX        ;move out EBX
```

If they are rescheduled as shown above, each can be issued to separate execution units, allowing parallel execution of both instructions by two different units of the CPU. Since the clock count for each instruction is one, just like the 486, having two execution units leads to executing two instructions by pairing them together, thereby using only one clock count for two instructions. In the case of the above program, if it is run on the Pentium it will take only 2 clocks instead of 4 as is the case of 486 microprocessor, assuming that two instructions are paired together. This reordering of instructions to take advantage of the two internal execution units of the Pentium is the job of the compiler and is called *instruction scheduling*. Currently, compilers are being equipped to do instruction scheduling to remove dependencies. The role of the compiler to reschedule instructions in order to take advantage of the superscalar capability of the Pentium must be emphasized. The process of issuing two instructions to the two execution units is commonly referred to as *instruction pairing*. The two integer execution units of the Pentium are called "U" and "V" pipes. Each has 5 pipeline stages. While the U pipe can execute any of the instructions in the 80x86 family, the V pipe executes only simple instructions such as INC, DEC, ADD, SUB, MUL, DIV, NOT, AND, OR, EXOR, and NEG. These simple instructions are executed in one clock as long as the operands are "REG,REG" or "REG,IMM" and have no register dependency. For example, instructions such as "ADD EAX,EBX", "SUB ECX,2000", and "MOV EDX,1500" are simple instructions requiring one clock, but not "ADD DWORD PTR [EBX+EDI+500],EAX", which needs 3 clocks.

Feature 5

Branch prediction is another new feature of the Pentium. In Chapter 8, we discussed the branch penalty associated with jump and CALL instructions. The penalty for jumping is very high for a high-performance pipelined microprocessor such as the Pentium. For example, in the case of the JNZ instruction, if it jumps, the pipeline must be flushed and refilled with instructions from the target location. This takes time. In contrast, the instruction immediately below the JNZ is already in the pipeline and is advancing without delay. The Pentium processor has the capability to predict and prefetch code from both possible locations and have them advanced through the pipeline without waiting (installing) for the outcome of the zero flag. The ability to predict branches and avoid the branch penalty combined with the instruction pairing can result in a substantial reduction in the clock count for a given program. See Example 23-6.

Feature 6

As discussed in Chapter 21, the 386/486 has a page size of 4K for page virtual memory. The Pentium provides the option of 4K or 4M for the page size. The 4K page option makes it 386 and 486 compatible, while the 4M page size option allows mapping of a large program without any fragmentation. The 4M page size in the Pentium reduces the frequency of a page miss in virtual memory.

Feature 7

As discussed in Chapter 21, the 386 (and 486) has only 32 entries for the TLB (translation lookaside buffer), which means that the CPU has instant knowledge of the whereabouts of only 128K of code and data. If the desired code or data is not referenced in the TLB, the CPU must go through the long process of converting the linear address to a physical address. The Pentium has two sets of TLB, one for code and one for data. For data, the TLB has 64 entries for 4K pages. This means that the CPU has quick access to 256K ($64 \times 4K = 256K$) of data. The TLB for the code is 32 entries of 4K page size. Therefore, the CPU has quick access to 128K of code at any give time. Combining the TLBs for the code and data, the Pentium has quick access to 384K (128 + 256) of code and data before it resorts to updating the TLB for the page miss. Contrast this to 128K for the 486. If the page size of 4M is chosen, the TLB for the data has 8 entries while the TLB for the code has 32 entries.

Example 23-6

Compare the clock count for the program in Example 23-5, run on a 486 and a Pentium. Assume that the compiler has done the code scheduling to allow instruction pairing for the Pentium.

Solution:

First compare the following rearranged code with the code in Example 23-5. Here we have rescheduled instructions "ADD SI,4" and "ADD BX,4" to avoid register dependency. This allows pairing of two instruction and issuing them to execution units of the Pentium. For example, the instructions "MOV EAX,DWORD PTR [SI]" and "ADD SI,4" are issued simultaneously, one to each execution unit. This results in the execution of both instructions in only one clock.

```
AGAIN:  MOV   EAX,DWORD PTR [SI]  ⟩  1 clock
        ADD   SI,4
        MOV   [DI],EAX             ⟩  1 clock
        ADD   DI,4
        DEC   CX                   ⟩  1 clock
        JNZ   AGAIN
```

Total clock count for one iteration: 3

In the 486, the execution of "JNZ AGAIN" takes 3 clocks every time it jumps to AGAIN, but for the Pentium it takes only 1 clock since the CPU has predicted the branch, fetched it, and the instructions at label AGAIN are in the pipeline advancing. This way, regardless of the outcome of the JNZ, both the instruction below JNZ and the first instruction at label AGAIN are in two separate pipelines, advancing. If ZF =0, the other pipeline is trashed, and if ZF=1(the end of loop), the instruction below the JNZ is executed and the branch prediction pipeline is abandoned. In the above program each iteration takes only 3 clocks on the Pentium compared with 9 clocks in the 486. While branch prediction is performed by the internal hardware of the Pentium, instruction scheduling must be done by the compiler.

Feature 8

The Pentium has both burst read and burst write cycles. This is in contrast to the 486, which has only the burst read. This means that in the 486 any write to consecutive doubleword locations must be performed with the normal 2 clock cycles. This is not the case in the Pentium.

The Pentium has features that lend themselves to implementation of multiple microprocessors (multiprocessors) working together. It also has features called *error detection* and *functional redundancy* to preserve and ensure data and code integrity.

Intel's overdrive technology

To increase both the internal and external clock frequency of the CPU requires faster DRAM, high-speed motherboard design, high-speed peripherals, and efficient power management due to a high level of power dissipation. As a result, the system is much more expensive. To solve this problem, Intel came up with what is called *overdrive technology*, also referred to as *clock doubler* and *tripler*. The idea of a clock doubler or tripler is to increase the internal frequency of the CPU while the external frequency remains the same. In this way, the CPU processes code and data internally faster while the motherboard costs remain the same. For example, the 486DX2-50 uses the internal frequency of 50 MHz but the external frequency by which the CPU communicates with memory and peripherals is only 25 MHz. This allows the instructions stored in the queue of 486 to be executed at twice the speed of fetching them from the system buses. With the advent of the 32- and 64-bit external buses, on-chip cache, and the burst cycle reading (reading 16 bytes in only 5 clocks), the amount of code and data fetched into the queue of the CPU is sufficient to keep the execution unit of CPU busy even if it is working with twice or three times the speed of external buses. This is the reason that Intel is designing processors with clock triplers. In that case, if the CPU's external buses are working at the speed of 33 MHz, the CPU works at 99 MHz speed. The design of a system board of 33 MHz costs much less than that of a 100-MHz system board. With slower memory and peripherals one can get instruction throughput of three times the bus throughput. As designers move to wider data buses, such as 128-bit-wide buses, the use of clock doublers and triplers is one way of keeping the system board cost down without sacrificing system throughput. The Intel 486DX4 is an example of a clock-tripler CPU. Note that "X4" does not mean that the external frequency is 4 times the internal frequency.

Review Questions

1. The Pentium chip has _____ pins.
2. The Pentium has _____ data pins.
3. True or false. The Pentium is a 32-bit processor.
4. What is the total cache on the Pentium? How much is for data, and how much is for code?
5. Which is write protected, data or code cache?
6. True or false. The on-chip data and code cache are accessed simultaneously.
7. True or false. The branch prediction task is performed by circuitry inside the Pentium.
8. Why is the Pentium called a superscalar processor?
9. True or false. Instruction scheduling is done by circuitry inside the Pentium.
10. True or false. The general-purpose registers of the Pentium are the same as those in the 386 and 486.

SECTION 23.3: RISC ARCHITECTURE

In the early 1980s a controversy broke out in the computer design community, but unlike most controversies, it did not go away. Since the 1960s, in all mainframe and minicomputers, designers put as many instructions as they could think of into the microinstructions of the CPU. Some of these instructions performed complex tasks. An example is adjusting the result of decimal addition to get BCD nibble-type data. Naturally, microprocessor designers followed the lead of minicomputer and mainframe designers. Since these microprocessors used such a large number of instructions and many of them performed highly complex activities, they came to be known as CISC (complex instruction set computer). According to several studies in the 1970s, many of these complex instructions etched into the brain of the CPU were never used by programmers and compilers. The huge cost of implementing a large number of instructions (some of them complex) into the microprocessor, plus the fact that more than 60% of the transistors on the chip are used by the instruction decoder, made some designers think of simplifying and reducing the number of instructions. As this was developed, it came to be known as RISC (reduced instruction set computer).

Features of RISC

The following are some of the features of RISC. It must be noted that recently CISC processors such as the Pentium have used some of the following features in their design.

Feature 1

RISC processors have a fixed instruction size. In a CISC microprocessor such as the 80x86, instructions can be 1, 2, or even 6 bytes. For example, look at the following instructions.

CHAPTER 23: 486, PENTIUM, PENTIUM PRO AND MMX

```
CLC                    ;a 1-byte instruction
SUB    DX,DX           ;a 2-byte instruction
ADD    EAX,[SI+8]      ;a 5-byte instruction
JMP    FAR             ;a 5-byte instruction
```

This variable instruction size makes the task of the instruction decoder very difficult since the size of the incoming instruction is never known. In a RISC microprocessor, the size of all instructions is fixed at 4 bytes (32 bits). In cases where instructions do not require all 32 bits, they are filled with zeros. Therefore, the CPU can decode the instructions quickly. This is like a bricklayer working with bricks of the same size as opposed to using bricks of variable sizes. Of course, it is much more efficient using the same-size bricks.

Feature 2

RISC uses load/store architecture. In CISC microprocessors, data can be manipulated while it is still in memory. For example, in 80x86 instructions such as "ADD [BX],AL", the microprocessor must bring the contents of the memory location pointed at by BX into the CPU, add it to AL, then move the result back to the memory location pointed at by BX. In RISC, designers did away with this kind of instruction. In RISC, instructions can only load from memory into registers or store registers into memory locations. There is no direct way of doing arithmetic and logic instructions between registers and contents of memory locations. All these instructions must be performed by first bringing both operands into the registers inside the CPU, then performing the arithmetic or logic operation, and then sending the result back to memory. This idea was first implemented by the CRAY 1 supercomputer in 1976 and is commonly referred to as *load/store architecture*.

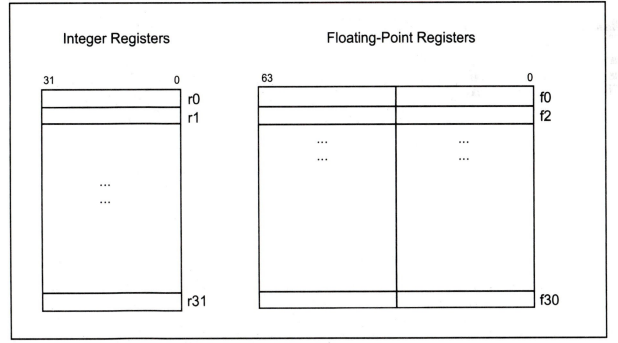

Figure 23-8. RISC Integer and Floating-Point Registers for Intel I860

Feature 3

One of the major characteristics of RISC architecture is a large number of registers. All RISC microprocessors have 32 registers, r0 - r31, each 32 bits wide. See Figure 23-8. Of these 32 registers, only a few of them are assigned to a dedicated function. For example, r0 is automatically assigned the value zero and no other value can be written to it. One advantage of a large number of registers is that it avoids

the use of the stack to store parameters. Although a stack can be implemented on a RISC processor, it is not as essential as in CISC since there are so many registers available. It must be noted that RISC processors, in addition to 32 general-purpose registers, also have another 32 registers for floating-point operations. The floating-point register can be configured as 64-bit in order to handle double-precision operands.

Feature 4

RISC processors have a small instruction set. RISC processors have only the basic instructions such as ADD, SUB, MUL, DIV, LOAD, STORE, AND, OR, EXOR, SHR, SHL, CALL, and JMP. For example, there are no such instructions as INC, DEC, NOT, NEG, DAA, DAS, and so on. Since RISC has very few instructions, it is the job of the programmer (compiler) to implement those instructions by using available RISC instructions. One example is an immediate load instruction such as "MOV AX,25 " which does not exist in Intel's 860 RISC. Instead, some other instructions, such as the OR instruction, can be used to implement an immediate move as shown in the following example for the Intel 860 RISC processor.

```
or   25,r0,r8      ;OR 25 with r0 and put result in r8
```

In Intel 860 RISC syntax, the destination register is the last register, r8 in the above example. Since r0 is always zero, ORing any number with it will result in that number. The above example will place 25 in r8. Another example is that there is no INC (increment) command. The ADD instruction is used instead, as in the following example:

```
add 1,r15,r15     ;add 1 to r15 and place result in r15
```

The limited number of instructions is one of the criticisms leveled at the RISC processor since it makes the job of Assembly language programmers much more tedious and difficult compared to CISC Assembly language programming. This is one reason that RISC is used more commonly in high-level language environments such as C rather than Assembly language environments. It is interesting to note that some defenders of CISC have called it "*complete* instruction set computer" instead of "complex instruction set computer" since it has a complete set of every kind of instruction. How many of them are used is another matter. The limited number of instructions in RISC leads to programs that are large, as Example 23-7 shows. Although this can lead to using more memory, since DRAM memory is so cheap this is not a problem. However, before the advent of semiconductor memory in the 1960s, CISC designers had to pack as much action as possible into a single instruction.

Feature 5

At this point, one might ask, with all the difficulties associated with RISC programming, what is the gain? The most important characteristic of the RISC processor is the fact that more than 95% of instructions are executed with only one clock, in contrast to CISC instructions. Although in the 80486 microprocessor some instructions are executed with one clock, with the the use of RISC concepts in designing it, it is still a CISC processor for the reasons discussed above. Even the other 5% of the RISC instructions that are executed with 2 clocks can be executed with one clock cycle by juggling instructions around (code scheduling). Code scheduling is the job of the compiler. What did designers do with all those transistors saved using the RISC implementation? In the case of Intel 860 RISC processors, these extra transistors are used to implement the math coprocessor, powerful cache and cache controller, and a very powerful graphics processor all on a single chip. In many computers, such as 386-based systems, all these functions are performed by separate chips.

Example 23-7

In the 80x86, the NOT instruction performs the 1's complement operation, but RISC does not have such an instruction. How is the 1's complement operation performed in RISC? Show code to take the 1's complement of 25H on both I860 RISC and 80x86.

Solution:

RISC has an EXOR instruction. If we EXOR the operand with all 1s, the operand is inverted.
In I860 RISC we have

```
or      25H,r0,r8       ;OR 25H with r0 and put result in r8(r8=25H)
or      FFH,r0,r5       ;OR FFH with r0 and put result in r5(r5=FFH)
xor     r8,r5,r9        ;XOR r8 with r5 and put result in r9 (r9=DAH)
```

Since each instruction is 4 bytes (32-bit), the three instructions take a total of 12 bytes of memory. In the 80x86, a CISC-type processor, we have the following, which takes only 4 bytes of memory (to see, use DEBUG to assemble):

```
MOV   AL,25H
NOT   AL
```

Feature 6

RISC processors have separate buses for data and code. In all 80x86 processors, like all other CISC computers, there is one set of buses for the address (e.g., A0 - A24 in the 80286) and another set of buses for data (e.g., D0 - D15 in the 80286) carrying opcodes and operands in and out of the CPU. To access any section of memory, regardless of whether it contains opcode or data operands, the same address bus and data bus are used. In RISC processors, there are 4 sets of buses: (1) a set of data buses for carrying data (operands) in and out of the CPU, (2) a set of address buses for accessing data operands, (3) a set of buses to carry the opcodes, and (4) a set of address buses to access the opcodes. The use of separate buses for code and data operands is commonly referred to as *Harvard architecture*.

Feature 7

Since CISC has such a large number of instructions, each with so many different addressing modes, microinstructions (microcode) are used to implement them. The implementation of microinstructions inside the CPU takes more than 60% of transistors in many CISC processors. However, in case of RISC, due to their small set of instructions, they are implemented using the hardwire method. Hardwiring of RISC instructions takes no more than 10% of the transistors. It is interesting to note that in the Pentium, a CISC processor, the V-pipe executes only simple instructions and it is hardwired while the U-pipe executes any of 80x86 instructions and uses microinstructions.

Comparison of sample program for RISC and CISC

Since RISC has established itself as the architecture of the 1990s and beyond, an example will be given of a program written for Intel's 80x86 CISC and Intel's I860 RISC. Then they will be compared. The next program example will compare total clocks for a program that transfers a block of 32-bit-size words from some memory location to another memory location. First, several points about the I860 RISC must be discussed.

SECTION 23.3: RISC ARCHITECTURE 707

1. In instructions such as "add r3,r5,r2", r3 is added to r5 and placed in r2. This is in contrast to the 80x86, in which the destination register is the first register.
2. r0 is always equal to zero, regardless of the operation performed on it.
3. The load instruction cannot be followed by the store instruction, which tries to use the value that is just being loaded: in other words, no read after write (RAW).
4. Some instructions, such as branch instructions, are delayed, which means that the next instruction after the branch will be executed since the pipeline already has fetched it before the branch is taken; therefore, if we cannot put a useful instruction after the branch, a NOP should be used.
5. There are several ways of encoding the NOP instruction. One is "add r0,r0,r0", which adds r0 to r0 and places the result in r0; since r0 is always zero, the instruction does nothing but waste time. Another would be to use shift left, such as "shl r0,r0,r0".
6. Some other instruction must be used to accomplish a MOV.
7. The logic instruction is used to perform the compare job.

Now look at the following Intel 860 RISC program, first written with total disregard to code scheduling and then written with code scheduling. The number of clocks for one round of loop is calculated in each case.

```
;this is a program for Intel 860 RISC processor to transfer a
;block of 20 dwords (each 32-bit) from memory locations starting at
;the address pointed at by r3 to memory locations pointed at by r4.
;r2 is the counter;   There is no code scheduling in the following example.
```

			clocks
	or 20,r0,r2	;load the r2 with 20(count=20)	
bak:	ld.l 0(r3),r5	;load r5 from content of mem loc 0+r3	1
	add r0,r0,r0	;NOP since r5 cannot be used by store	1
	st.l r5,0(r4)	;store r5 into mem loc of 0+r4	1
	add 4,r3,r3	;point at the next dword source data	1
	add 4,r4,r4	;point at the next dword destin data	1
	add -1.r2,r2	;decrement counter: r2=r2-1	1
	or r0,r2,r2	;set condition code to high if r2=0	1
	bnc.t bak	;go to bak if CC=0. execute next instru too	1
	add r0,r0,r0	;NOP for delayed branch	1
		total clocks for one loop iteration	9

In the above program, ld.1 and st.1 are for 32-bit operands. For byte and 16-bit bit operands, they would be ld.b and st.b, ld.s and st.s, respectively. Next the same program is juggled around and the NOPs are removed for better performance. As mentioned in the last section, this juggling is called code scheduling.

		number of clocks	
	or 20,r0,r2		
	;loop starts next		
bak:	ld.l 0(r3),r5	;	1
	add 4,r3,r3	;	1
	st.l r5,0(r4)	;	1
	add -1,r2,r2	;	1
	or r0,r2,r2	;	1
	bnc.t bak	;	1
	add 4,r4,r4	;	1
		total clock count for one iteration	7

It is assumed that the above program is run on a RISC processor with no superscalar capability. Using a superscalar RISC will cut the clock count to half, or 4 clocks for the same program. This is much better than the 386 and 486 microprocessors shown in Example 23-5. It is comparable with the Pentium as shown in Example 23-6. While the Pentium uses 3.1 million transistors to achieve such an impressive performance, a RISC processor with the same performance level can be designed using less than 1 million transistors. There is only one problem. It will not run the massive number software packages written for the 80x86 MS DOS PC.

IBM/Motorola RISC

IBM and Motorola together have a new RISC processor called the Power PC 601. It uses only 2.8 million transistors, with a power consumption of 8.5 watts versus 16 watts in the Pentium. Apple has chosen the Power PC RISC for the next generation of Macintosh computers. Many other PC makers are also planning to use the Power PC RISC processors, but for these computers to run MS DOS and Windows-based software, they must emulate them instead of running native. In other words, while MS DOS runs native on the 80x86, for the RISC processor or any non-80x86, there is no choice but to emulate. Assuming that a software emulator capable of running 100% of 80x86-based MS DOS application software on the RISC machine could be developed, the performance gain could be much higher than that of the Pentium in spite of the emulation overhead. In addition to the Power PC 601 RISC processor, there are some other notable RISC processors vying for a share of the desktop PC market. Among them are Digital Equipment Alpha, MIPS 4000, Hewlett-Packard PA-RISC, and SunMicro SPARC. The Power PC is a major force challenging 80x86 dominance since both Apple and IBM are using it in their products. Alpha, R4000, PA-RISC, and SPARC will probably compete for third place. It is unlikely that any RISC processor will take the lead over the 80x86 in the immediate future.

Among RISC processors, the IBM/Motorola Power PC 601 is the one most likely to challenge the dominance of the Intel 80x86. A comparison of the major characteristics of the Pentium and the Power PC 601 is provided in Table 23-3. To take full advantage of the power of RISC, software developers must write applications specifically for RISC, rather than using emulation.

Before concluding this discussion of RISC processors, it is interesting to note that RISC technology was explored by the scientists in IBM in the mid-1970s, but it was David Patterson of the University of California at Berkeley who in 1980 brought the merits of RISC concepts to the attention of computer scientists.

Table 23-3: Pentium vs. Power PC 601

Feature	Pentium	Power PC 601
Number of transistors (million)	3.1	2.8
Power dissipation at 66 MHz (watts)	16	8.5
Die size (mm^2)	262	120
Technology	BICMOS 0.8 μm	CMOS 0.6 μm
Number of general-purpose registers	8	32
Number of floating-point registers	8 (stack base)	32
Cache (KBytes)	16 (code 8K, data 8K)	32K (unified)
Register size	32-bit	32-bit
Running MS DOS software	Native	Emulation
Number of instructions issued per clock cycle	2	3 (one is FP)
Architecture	Superscalar	Superscalar

Review Questions

1. What do *RISC* and *CISC* stand for?
2. True or false. The 386 executes the vast majority of its instructions in 3 clock cycles, while RISC executes them in one clock.
3. RISC processors normally have ___ general-purpose registers, each ___-bits.
4. True or false. Instructions such as "ADD AX,[DI]" do not exist in RISC.
5. What is the size of instructions in RISC?
6. True or false. While CISC instructions are variable sizes, RISC instructions are all the same size.
7. Which of the following operations do not exist for the ADD instruction in RISC?
   (a) register to register (b) immediate to register
   (c) memory to register
8. How many floating-point registers do we have in RISC and the 80x87?
9. Why can floating-point registers in RISC be configured as 64-bit?
10. True or false. Harvard architecture uses the same address and data buses to fetch both opcode and data.

SECTION 23.4: PENTIUM PRO PROCESSOR

In this section we discuss the main features of Intel's Pentium Pro processor. Intel's Pentium Pro is the sixth generation of the x86 family of microprocessors. For this reason, early literature about this chip referred to it as P6. Intel officially calls this chip *Pentium Pro* to emphasize its superiority over the Pentium generation. Intel used 5.5 million transistors to make the Pentium Pro. The first Pentium Pro introduced in 1995 had a speed of 150 Mhz and consumed 23 watts of power at that speed. Since then, Intel has introduced Pentium Pro chips with higher speeds and various power consumption ratings.

There are no major surprises in the Pentium Pro in the sense that it runs all the software written for the 8086/88, 286, 386, 486, and Pentium microprocessors and its 32-bit registers are exactly the same as the 386. In other words, the register size was not increased to 64 bits as has been done by some RISC processors such as Digital Equipment's Alpha chip.

For the first time, Intel also attached level 2 (L2) cache to the Pentium Pro all on a single package but with two separate dies. This packaging is called *dual cavity* by Intel. The integration of a 256K-bytes L2 cache with the processor into a single package reduces interchip delay between the L2 cache and the CPU. While such an integration cuts memory access delay, it also made many SRAM makers mad since they lost another chunk of PC business to Intel. Notice that the Pentium Pro CPU has only 16K bytes of L1 cache on the same die just like the Pentium processor while 256KB (or 512KB) L2 cache is on the separate die. In addition to the 5.5 million transistors used for the Pentium Pro CPU and its 16KB L1 cache, the L2 cache uses over 10 millions transistors depending on the size of L2 cache. There is a possibility that Intel will introduce a Pentium Pro without L2 cache. See Table 23-4 for further comparison of Pentium and Pentium Pro processors.

Pentium Pro: internal architecture

Intel finally yielded to the rise of RISC concepts in the design of the Pentium Pro. In the Pentium Pro, all x86 instructions brought into the CPU are broken down into one or more small and easy to execute instructions. These easily executable instructions are called *micro-operations* (uops) by Intel. This is similar to the concept in RISC except that in RISC architecture the instruction set is very simple and easy to execute, and the instructions stored in memory are exactly the same as the ones inside the CPU. In contrast, Intel had to maintain code compatibility for the Pentium Pro with all previous x86 processors, all the way back to 8086. Therefore, Intel had no choice but to convert the x86 instructions produced by the compiler/assembler into micro-operations internally inside the CPU. An interesting aspect of converting x86 instructions into micro-ops internally is that it uses what

Table 23-4: Comparison of Pentium and Pentium Pro

Feature	Pentium	Pentium Pro
Year introduced	1993	1995
Number of transistors	3.3 million	5.5 million
Number of pins	273	387
External data bus	64 bits	64 bits
Address bus	32 bits	36 bits
Physical memory (maximum)	4 GB	64 GB
Virtual memory	64 TB	64 TB
Data types (register sizes)	8, 16, 32 bits	8, 16, 32 bits
Cache (L1)	16K bytes (data 8K, code 8K)	16K bytes (data 8K, code 8K)
Cache (L2)	External	256KB/512KB
Superscalar	2-Way	3-Way
Number of execution units	3	5
Branch prediction	yes	yes
Out-of-order execution	no	yes

is called *triadic* instruction formats. In triadic instruction format, there are two source registers and one destination register. An example of triadic format is "ADD R1,R5,R8" in which registers R1 and R5 are added together and placed in R8. The contents of source registers R1 and R5 are not altered. Contrast this with "ADD AX,BX" in which there is only one source register (BX). For more examples of the triadic instruction format, see Section 23.3 on the RISC processors.

The use of a triadic instruction set in the Pentium Pro architecture means that there are a large number of registers inside the Pentium Pro that are not accessible or visible to the programmer. In other words, as far as the programmer (or compiler) is concerned, only the traditional register set EAX, EBX, ECX, etc. is available and visible to the programmers of the Pentium Pro. This ensures that compatibility with previous generations of the x86 is maintained.

Pentium Pro is both superpipelined and superscalar

As mentioned above, in the Pentium Pro all x86 instructions are converted into micro-ops with triadic formats before they are processed. This conversion allows an increase in the pipeline stages with little difficulty. Intel uses a 12-stage pipeline for the Pentium Pro. In contrast to the 5-pipestage of the Pentium, although each pipestage of the 12-pipestage Pentium Pro performs less work, there are more stages. This means that in the Pentium Pro, more instructions can be worked on and finished at a time. The Pentium Pro with its 12-stage pipeline is referred to as *superpipelined*. Since it also has multiple execution units capable of working in parallel, it is also *superscalar*. Another advantage of the 12-pipestage is that it can achieve a higher clock rate (frequency) with the given transistor technology. This is one reason that the earliest Pentium chips had a frequency of only 60 MHz while the earliest Pentium Pro has a frequency of 150 MHz. Intel also used what is called *out-of-order execution* to increase the performance of the Pentium Pro. This is explained next.

What is out-of-order execution?

In Pentium architecture, when one of the pipeline stages is stalled, the prior stages of fetch and decode are also stalled. In other words, the fetch stage stops fetching instructions if the execution stage is stalled, due for example to a delay in

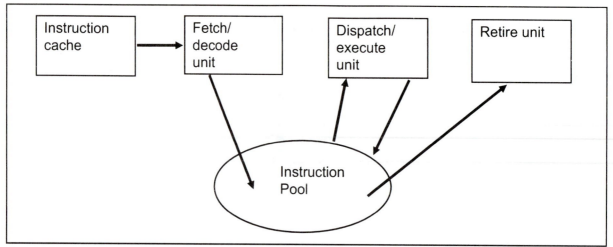

Figure 23-9. Pentium Pro Instruction Execution

memory access. This dependency of fetch and execution has to be resolved in order to increase CPU performance. That is exactly what Intel has done with the Pentium Pro and is called *decoupling* the fetch and execution phases of the instructions. In the Pentium Pro, as x86 instructions are fetched from memory they are decoded (converted) into a series of micro-ops, or RISC-type instructions, and placed into a pool called the instruction pool. See Figure 23-9. This fetch/decode of the instructions is done in the same order as the program was coded by the programmer (or compiler). However, when the micro-ops are placed in the instruction pool they can be executed in any order as long as the data needed is available. In other words, if there is no dependency, the instructions are executed out of order, not in the same order as the programmer coded them. In the case of the Pentium Pro, the dispatch/execute unit schedules the execution of micro-ops from the instruction pool subject to the availability of needed resources and stores the results temporarily. Such a speculative execution can go 20 - 30 instructions deep into the program. It is the job of the *retire unit* to provide the results to the programmer's (visible) registers (e.g., EAX, EBX) according to the order the instructions were coded. Again, it is important to note that the instructions are fetched in the same order that they were coded, but executed out of order if there is no dependency, but ultimately retired in the same order as they were coded. This out-of-order execution can boost performance in many cases. Look at Example 23-8.

Example 23-8

For the following code, indicate the instructions that can be executed out of order in the Pentium Pro.

```
i1)  LOAD (R2), R4      ;LOAD R4 FROM MEMORY POINTED AT BY R2
i2)  ADD  R3,R4,R7      ;R3+R4--->R7
i3)  ADD  R6,R8,R10     ;R6+R8--->R10
i4)  SUB  R5,R1,R9      ;R5-R1--->R9
i5)  ADD  R6,1,R12      ;R6+1---->R12
```

Solution:

Instruction i2 cannot be executed until the data is brought in from memory (either cache or main memory DRAM). Therefore, i2 is dependent on i1 and must wait until the R4 register has the data. However, instructions i3, i4 and i5 can be executed out of order and in parallel with each other since there is no dependency among them. After the execution of i2, all the instructions i2, i3, i4, and i5 can be retired instantly since they all have been executed already. This would not be the case if these instructions were executed in the Pentium since its pipeline would be stalled due to the memory access for R4. In that case, instructions i3, i4, and i5 could not even be fetched let alone decoded and executed.

Example 23-9

The following x86 code (a) sets the pointer for three different arrays, and the counter value, (b) gets each element of ARRAY_1, adds a fixed value of 100 to it, and stores the result in ARRAY_2, (c) complements the element and stores it in ARRAY_3. Analyze the execution of the code in light of the out-of-order execution and branch prediction capabilities of the Pentium Pro.

```
i1)               MOV EBX,ARRAY_1  ;LOAD POINTER
i2)               MOV ESI,ARRAY_2  ;LOAD POINTER
i3)               MOV EDI,ARRAY_3  ;LOAD POINTER
i4)               MOV ECX,COUNT    ;LOAD COUNTER
i5)  AGAIN:       MOV EAX,[EBX]    ;LOAD THE ELEMENT
i6)               ADD EAX,100      ;ADD THE FIX VALUE
i7)               ADD EBX,4        ;UPDATE THE POINTER
i8)               MOV [ESI],EAX    ;STORE THE RESULT
i9)               ADD ESI,4        ;UPDATE THE POINTER
i10)              NOT EAX          ;COMPLEMENT THE RESULT
i11)              MOV [EDI],EAX    ;AND STORE IT
i12)              ADD EDI,4        ;UPDATE THE POINTER
i13)              LOOP AGAIN       ;STAY IN THE LOOP
i14)              MOV AX,4C00H     ;EXIT
i15)              INT 21H
```

Solution:

The fetch/decode unit fetches and converts instructions into micro-ops. Since there is no dependency for instructions i1 through i5, they are dispatched, executed and retired except for i5. Notice that the pointer values are immediate values; therefore, they are embedded into the instruction when the fetch/decode unit gets them. Now i5 is a memory fetch which can take many clocks, depending on whether the needed data is located in cache or main memory. Meanwhile i6, i8, i10, and i11 must wait until the data is available. However i7, i9, i12 can be executed out of order knowing that the updated values of pointers EBX, EDI, ESI are kept internally until the time comes when they will be committed to the visible registers by the retire unit. More importantly, the LOOP instruction is predicted to go to the target address of AGAIN and i5, i6, ... are dispatched once more for the next iteration. This time the memory fetch will take very few clocks since in the previous data fetch, the CPU read at least 32 bytes of data using the Pentium Pro 64-bit (8 bytes) data bus and the burst read mode, transferring into the CPU 4 sets of 8-byte data. This process will go on until the last round of the LOOP instruction where ECX becomes zero and falls through. At this time due to misprediction, all the micro-instructions belonging to instructions i5, i6, i7, ... (start of the loop) are removed and the whole pipeline restarts with instructions belonging to i14, i15, and so on.

Due to the fact that memory fetches (due to cache misses) can take many clock cycles and result in underutilization of the CPU, out-of-order execution is a way of finding something to do for the CPU. Simply put, the idea of out-of-order execution is to look deep into the stream of instructions and find the ones that can be executed ahead of others, providing that resources are available. Again, it is important to note that the Pentium Pro will not immediately provide the results of out-of-order executions to programmer-visible registers such as EAX, EBX, etc., since it must maintain the original order of the code. Instead, the results of out-of-order executions are stored in the pool and wait to be retired in the same order as they were coded. Therefore, programmer-visible registers are updated in the same sequence as expected by the programmer.

Branch prediction

The Pentium Pro, like the Pentium before it, has branch prediction, but with greater capability. When the Pentium Pro encounters branch instructions (such as JNZ), it creates a list of them in what is called the *branch target buffer* (BTB). The BTB predicts the target of the branch and starts executing from there. When the branch is executed, the result is compared with what the prediction section of the CPU said it would do. If they match, the branch is retired. If not, all instructions behind the branch are removed from the pool and the correct branch target address is provided to the BTB. From there the BTB refills the pipeline with instructions from the new target address. See Example 23-9.

Note the following points concerning the reordering of store instructions from Intel documentation, "Stores are never performed speculatively since there is no transparent way to undo them. Stores are also never re-ordered among themselves. A store is dispatched only when both the address and the data are available and there are no older stores awaiting dispatch."

Bus frequency vs. internal frequency in Pentium Pro

Frequently you may see an advertisement for a 150-MHz or 200-MHz Pentium or Pentium Pro PC. It is important to note that the stated frequency is the internal frequency of the CPU and not the bus frequency. This is due to the fact that designing a 150-MHz motherboard is very difficult and expensive. Such a design requires a very fast logic family and memory in addition to a massive simulation to avoid crosstalk and signal radiation. The bus frequency for such systems is currently less than 100 MHz. Example 23-10 examines the speed of the logic family and memory.

Example 23-10

Find the memory read cycle time of a 100-MHz bus with zero wait states (0 WS).

Solution:

The bus clock is 1/100 MHz = 10 ns. Each memory cycle takes 2 bus clocks. Therefore, the total time budgeted for the address/data pathway delay and memory access time is only 20 ns.

Programming the Pentium Pro is examined in Section 23.6. Next, we discuss MMX technology.

Review Questions

1. *Pentium Pro* is the official name designated by Intel. What was it called before such a designation?
2. True or false. Both the Pentium and Pentium Pro have 16KB L1 cache.
3. True or false. Both the Pentium and Pentium Pro have L2 cache on the same package.
4. True or false. The x86 instruction set is in triadic form.
5. Which of the x86 processors has out-of-order execution?
6. Which unit inside the Pentium Pro commits the final results of operations to registers EAX, EBX, etc.?
7. True or false. The Pentium Pro is a superpipelined processor.

SECTION 23.5: MMX TECHNOLOGY

In this section, we discuss the MMX (MultiMedia extension) technology used in some of the Intel processors.

DSP and multimedia

To run high-quality multimedia applications with sound and graphics requires very fast and sophisticated mathematical operations. Such complex operations are normally performed by a highly specialized chip called DSP (digital signal processing). DSP chips are the main engines performing tasks such as 2- and 3-D graphics, video and audio compression, fax/modem, PC-based telephoning with live pictures, and image processing.

There are three approaches to equip the PC with DSP capability.

1. Use a full-fledged DSP chip on the board along with the main CPU. This is the best and ideal approach since there are some very powerful DSP chips out there. However, the problem is that there is no industry-wide standard to be followed by the PC designers and the lack of such a standard can lead to incompatibility both in hardware and software.
2. Use the x86 and x87 FP (floating-point) instructions to emulate the function of DSP. This is slow and performance is unacceptable.
3. The third approach is to incorporate some DSP functions into the x86 microprocessor. This approach leaves everyone at the mercy of Intel, yet it brings compatibility and a unified approach to the issue. Although the performance is not as good as the first approach, it is much better than the second approach.

The third approach is exactly what happened. In early 1997 Intel introduced a series of Pentium and Pentium Pro chips with somewhat limited DSP capability called *MMX technology*. In the case of Intel's MMX technology, software compatibility, both on the BIOS and operating system levels, was the most important goal. It needs to be noted that although MMX does not have a rich set of instructions normally associated with DSP chips such as Texas Instruments' 320xxx, it still performs many of the DSP functions reasonably well.

Register aliasing by MMX

As stated earlier, one of the main goals of MMX technology was to maintain compatibility with other x86 processors with no MMX capability. To assure that, Intel uses the FP (floating-point) register set of the x87 math coprocessor as the working register for MMX instructions instead of introducing a whole new set of registers. This is called *register aliasing*, meaning that the same physical register has different names. While the x87 FP registers are 80 bits wide, the MMX uses only 64 bits of it. The x87 floating-point registers are called ST(0), ST(1), ..., ST(7) when they are used by the x87 instruction set but the same registers are called MM0, MM1, ..., MM7 when used by the MMX portion of the CPU. See Figure 23-10. Register aliasing by MMX has some major implications:

63		0
	MM7	
	MM6	
	MM5	
	MM4	
	MM3	
	MM2	
	MM1	
	MM0	

Figure 23-10. MMX Register Set

1. We must not use the registers to store MMX data and FP (floating point) data at the same time since it is the same physical register.
2. We must not mix MMX instructions with FP instructions. Mixing MMX and FP instructions slows down the application since it takes many clock cycles to switch between MMX and x87 instructions. The best method is to have separate program modules for x87 instructions and MMX instructions with no intermixing.

3. When leaving an MMX program module, make sure that all the MMX registers are cleared before issuing any x87 instructions. The same is true if switching from x87 to MMX. All FP registers must be popped to leave them empty.
4. As shown in Chapter 20, FP registers are accessed by the x87 instructions in the stack format. However, when these same registers are accessed by the MMX instruction set, each one is accessed directly by its name, MM0 - MM7. These MMX registers cannot be used to address memory and must be used only to perform calculations on data.

Data types in MMX

As mentioned earlier, the MMX uses only 64 bits of the 80-bit wide FP registers. Therefore, the largest MMX data size is 64-bit. However, the 64-bit register can be used for four different data types. See Figure 23-11. They are as follows.

1. Quadword (one 64-bit)
2. Packed doubleword (two 32-bit)
3. Packed word (four 16-bit)
4. Packed byte (eight 8-bit)

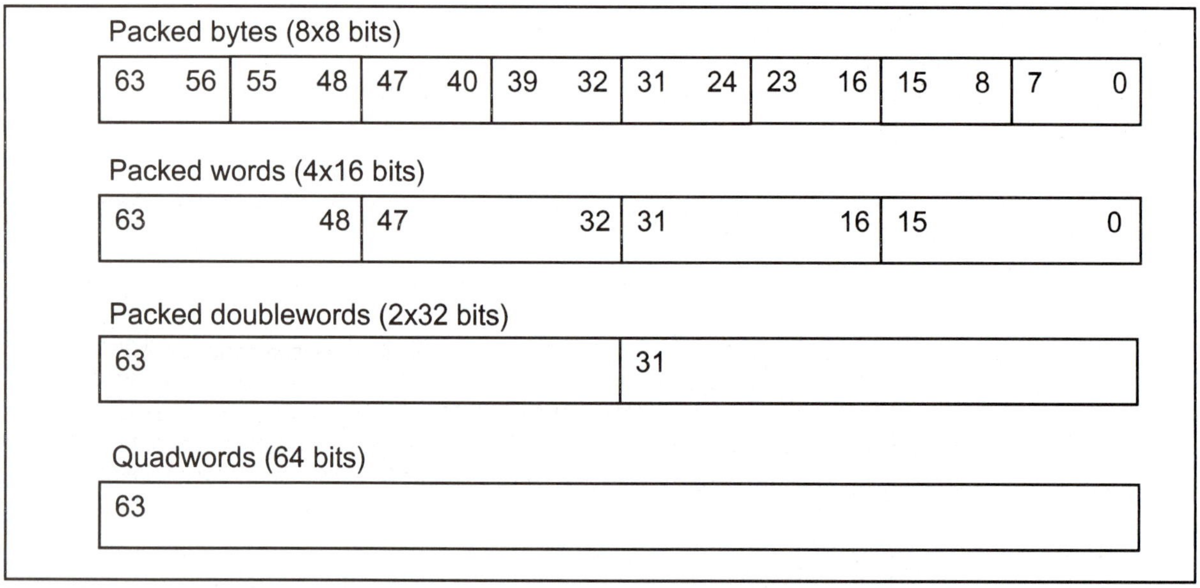

Packed bytes (8x8 bits)

63	56	55	48	47	40	39	32	31	24	23	16	15	8	7	0

Packed words (4x16 bits)

63	48	47	32	31	16	15	0

Packed doublewords (2x32 bits)

63	31

Quadwords (64 bits)

63

Figure 23-11. MMX Data Types

All four data types of the MMX are integers and are referred to as *packed data*. It must be noted that the contents of the MMX registers can be treated as any of the four different types of eight bytes, four words, two doublewords, or one quadword. It is the job of the MMX instruction to specify the data type. For example, the instruction Packed Add has three different formats depending on the data type. They are as follows:

PADDB (Add Packed Byte) adds two groups of 8 packed bytes
PADDW (Add Packed Word) adds two groups of 4 packed words
PADDD (Add Packed double) adds two groups of 2 packed doublewords

It is interesting to note that Intel has introduced to the x86 instruction set a total of 57 new instructions just for the MMX. Since currently there is no assembler or compiler equipped with the MMX instruction set, we will not give any MMX programming example. To find out if a given Pentium is an MMX chip, we must use a Pentium instruction called CPUID. This is discussed in the next section.

Review Questions

1. Why do we not use the FP x87 instructions for DSP multimedia?
2. MMX is available for which of the x86 processors?
   (a) 486   (b) Pentium   (c) Pentium Pro
3. True or false. MMX aliases the x86 registers.
4. What are the names of the MMX registers?
5. True or false. MMX instructions access registers in stack format.

SECTION 23.6: PROCESSOR IDENTIFICATION IN INTEL X86

In this section we discuss how to identify the x86 CPU on the PC using Assembly language programs. We will also discuss how to use the CPUID instruction of the Pentium to identify x86 processors with MMX technology.

Before embarking on the task of writing a CPU identification program, it must be noted that MS DOS 5 (or higher) comes with a utility called MSD (Microsoft Diagnostics). MSD provides useful information about hardware features of the PC including the type of 80x86 CPU installed on the motherboard. If you go to the DOS directory and enter "MSD <return>", the MSD utility will interrogate the PC and then inform you about hardware features of your PC such as the kind of microprocessor installed, RAM size, BIOS date, and so on. The utility is self-explanatory; just type in the highlighted letter to see the desired features.

Program to identify the CPU

Many software packages require a certain processor in order to run. For example, IBM OS/2 2.x and Microsoft Windows 95 must be run on machines with 386 or higher processors (486, Pentium, and so on). CPU identification is so important to the new generation of operating systems and software packages that starting with the Pentium, Intel has introduced a new instruction to do just that. The problem is how to identify microprocessors prior to the Pentium. According to Intel, for identifying the microprocessor by way of software one must examine the bits of the flag register. Notice that in the 8088/86/286 the flag register is a 16-bit register but in the 386/486/Pentium it is a 32-bit register. Table 23-5 shows the status of the flag bits used in identifying the processor type. These bits can be examined at any time and not just at boot-up.

Table 23-5: Flag Bits for CPU Identification

CPU	Flag Bits
8088/86	Bits 12 through 15 are always 1
80286	Bits 12 through 15 are always 0 (in real mode)
80386	Bit 18 is always 0 (in real and protected mode)
80846	Bit 21 cannot be changed, therefore it is 486; if bit 21 can be changed to 1 and 0, then it must be a Pentium.
Pentium and Pentium Pro	Starting with the Pentium, one can use a new instruction, CPUID, to get information such as family and model of the processor. However, it is the ability to set or reset bit 21 of the flag which indicates whether the CPUID instruction is supported or not. The CPUID instruction can be executed any time in protected mode or real mode. For Intel's Pentium and higher microprocessors, prior to execution of the CPUID instruction we must set EAX = 1. After the execution of CPUID, bits D8 - D11 of EAX have the family number. The family number is 5 for the Pentium and is 6 for the Pentium Pro.

It must also be noted that in the 80x86 family there is no instruction that can exchange the contents of the flag register and a general-purpose register directly. Therefore, to examine the contents of the flag register in an 80x86, we must use the stack as outlined in the following steps:

1. Push the flag register onto the stack.
2. Get (Pop) it into a register such as AX, BX.
3. Manipulate bits d15 - d12 (or any other bits).
4. Push it back onto the stack.
5. Pop it back into the flag register from the stack.
6. Push the flag back onto the stack again.
7. Get (Pop) the new flag bits back into a register.
8. Examine bits d12 - d15 to see if the changes in step 3 took effect.

The following code shows the above steps.

```
Step 1)    PUSHF                    ;push the flag into stack
Step 2)    POP      BX              ;and get it into BX
Step 3)    AND      BX,0FFFH        ;mask bits d15-d12
Step 4)    PUSH     BX              ;send it back into stack
Step 5)    POPF                     ;bring it back into flag reg
Step 6)    PUSHF                    ;store the flag on stack
Step 7)    POP      BX              ;get it into BX again to examine
Step 8a)   AND      BX,0F000H       ;mask all bits except d12-d15
Step 8b)   CMP      BX,....
```

Notice in the above code that instructions "PUSHF" and "POPF" are used for pushing and popping the 16-bit flag register. However, to access the 32 bits of the flag register in the 386/486/Pentium, instructions PUSHFD and POPFD must be used. These steps are coded in Program 23-1. That code does not make a distinction between the 8088 and 8086. The fact that the queue size for the 8086 is 6 bytes and for the 8088 is 4 bytes can be used to distinguish them.

CPUID instruction and MMX technology

Not all Pentium and Pentium Pro microprocessors come with MMX technology. To find out if a microprocessor is equipped with MMX technology, we can use Pentium instruction CPUID. According to Intel, upon return from instruction CPUID, if D23 of EDX is high, it has MMX technology. MMX identification is performed as follows.

```
          ;after making sure CPUID instruction is supported
          MOV      EAX,1               ;REQUEST FOR FEATURE FLAG
          CPUID                        ;CPUID INSTRUCTION
          TEST     EDX,00800000H       ;BIT 23 OF EDX INDICATES MMX
          JNZ      MMX_YES
          ...                          ;NO MMX
MMX_YES: ...
```

Review Questions

1. How can a program determine if the processor is 386?
2. To identify the 486 processor, toggle _____.
3. CPUID instruction was first introduced with the _____ processor
4. To identify the Pentium, use CPUID with EAX=_____.
5. Can the CPUID instruction help determine if the chip supports MMX technology? If so, how?

Note: To assemble the code below you need MASM 6.11 (or TASM 4.0) which support the Pentium instructions such as CPUID (the directive .586 is for that purpose). If you are using MASM 5.x then remove both the .586 directive and CPUID instruction and replace them with opcode for CPUID which is (0FA2H) in the following manner

```
            DW  0FA2H                ;opcode for CPUID instruction

;this routine identifies the PC'S 80X86 microprocessor
;upon return from this subroutine, AX contains microprocessor code
;where 0=8088/86,1=286,3=386,4=486, 5=Pentium, 6=Pentium Pro

GET_CPUID PROC
;
;see if it is 8086/88 by checking bits d12-d15 of flag reg
            PUSHF                    ;push the flag into stack
            POP   BX                 ;and get it into BX
            AND   BX,0FFFH           ;mask bits d15-d12
            PUSH  BX                 ;send it back into stack
            POPF                     ;bring it back into flag reg
            PUSHF                    ;store the flag back on stack again
            POP   BX                 ;and get it back into BX
            AND   BX,0F000H          ;mask all bits except d12-d15
            CMP   BX,0F000H          ;are the d12-d15 all zeros?
            MOV   AX,0               ;make AX=0 code for 8088/86
            JE    OVER               ;if yes then AX=0 code for 8086/88
;see if it is 80286 by checking bits of d12-d15 of flag reg
            OR    BX,0F000H          ;if not try setting d12-d15 to high
            PUSH  BX                 ;push it into stack
            POPF                     ;make d12-d15 of flag reg all 1s
            PUSHF                    ;get the flag back into stack
            POP   BX                 ;get it back to examine the bits
            AND   BX,0F000H          ;mask all bits except d12-d15
            CMP   BX,0F000H          ;are d12-d15 all 1s
            MOV   AX,1               ;make AX=1 code for 286
            JE    OVER               ;if yes set AX=1 code for 286
;see if it is 386 by checking bit 18 of flag bit
            .386
            PUSHFD                   ;if not it is 386 or higher. push flag
            POP   EBX                ;and get it into EBX
            MOV   EDX,EBX            ;save it
            XOR   EBX,40000H         ;flip bit 18
            PUSH  EBX                ;sent it into stack
            POPFD                    ;get it into flag
            PUSHFD                   ;get it back into stack
            POP   EBX                ;get the new flag back into EBX
            MOV   AX,3               ;make AX=3 code for 386
            XOR   EBX,EDX            ;see if bit 18 is toggled
            JE  OVER                 ;if yes then AX=3 CODE for the 386
;see if it is 486 or higher. try changing bit 21 of flag reg
            MOV   AX,4               ;if not it is 486 or higher (AX=4 for 486)
            PUSHFD                   ;see if bit 21(ID bit) can be altered
            POP   EBX                ;in order to use the CPUID instruction
            MOV   EDX,EBX            ;save original flag bit in EDX
            XOR   EBX,200000H        ;flip bit 21
            PUSH  EBX                ;save it on the stack
            POPFD                    ;get it into flag reg
            PUSHFD                   ;get flag back into stack
            POP   EBX                ;get it into EBX to examine bit 21
            XOR   EBX,EDX            ;see if bit 21 changes
            JE  OVER                 ;if yes AX=04 code for 486
;see which pentium (586,or 686) by using CPUID instruction
            MOV   EAX,1              ;set EAX=1 before executing CPUID
            .586                     ;use Pentium instruction
            CPUID                    ;after execution of CPUID, bits D8-D11 of EAX have family number
            .386                     ;back to 386 instructions
            AND   EAX,0F00           ;mask all bits except the family bit
            SHR   EAX,8              ;move d8-d11 to lower nibble then AX=5 for Pentium, 6 for Pentium Pro
            .8086
OVER:       RET                      ;return with AX=processor number
GET_CPUID ENDP
```

Program 23-1

SUMMARY

This chapter began with a look at Intel's 80486 microprocessor. The 486 is completely compatible with all previous 80x86 microprocessors and was designed to include several new enhancements, including (1) pipelining with 5 stages in the fetch/execute cycle which allows many instructions to execute in 1 clock cycle, (2) 8K of on-chip cache, (3) math coprocessor on the same chip as the CPU, (4) parity checking on the CPU, (5) read burst cycles, and (6) several new instructions. The second section explored Intel's latest and most powerful microprocessor, the Pentium. Unique features of the Pentium include (1) a 64-bit data bus combined with 2 execution units to increase execution speed; (2) 16K on-chip cache designed with Harvard architecture, which means that there are separate address and data buses to access code and data cache at the same time; (3) enhanced on-chip math processor; (4) superscalar architecture, two execution units within one CPU; (5) branch prediction, (6) increased virtual memory page size, which decreases the chances of a miss; (7) expanded TLB (translation lookaside buffer), which increases the amount of data and code that the CPU can access quickly; and (8) burst cycles for both read and write operations.

The third section gave an overview of the RISC versus CISC controversy that is affecting microprocessor design. RISC architecture contains several unique features, including (1) fixed instruction size; (2) load/store architecture, meaning that operations are performed on registers only (memory is loaded into registers prior to the operation); (3) a large number of registers; (4) a small instruction set; (5) most instructions execute in 1 clock cycle; and (6) Harvard architecture, separate buses for code and data. Furthermore, the execution speeds of programs written in RISC versus CISC systems were compared to show the advantage of RISC, increased processing speed, and the drawback, the increased program size and tedium of Assembly programming.

The fourth section discussed the main features of the Pentium Pro microprocessor, the sixth generation of the x86 family. The Pentium Pro features both superscalar and superpipelined architecture. It achieves improvements in speed by use of out-of-order execution and branch prediction. Section 5 gave an overview of MMX technology, used for multimedia processing. Finally, in Section 6 a method for processor identification in x86 machines was described.

PROBLEMS

SECTION 23.1: THE 80486 MICROPROCESSOR

1. The 486 chip uses _____ pins.
2. The 486 is a(n) _____-bit microprocessor.
3. What is the size of on-chip cache in the 486?
4. Off-chip cache is referred to as _____ cache.
5. True or false. On-chip cache in the 486 is used to hold both data and code.
6. State the differences between the 486 and 486SX microprocessors.
7. The 486 has a(n) _____-bit external data bus.
8. How many data parity pins does the 486 have?
9. The 486 can access _____ bytes of memory using the _____ address pins.
10. How many BE pins does the 486 have?
11. Nonburst read and write cycles take _____ clocks.
12. What does "2-2-2-2" cycle mean?
13. What does "2-1-1-1" cycle mean?
14. The 486 fetches _____ bytes of code and data into CPU using a burst cycle. How many clocks does the burst cycle take?
15. Calculate the bus bandwidth of a 486 with 25 MHz in each of the following.
    (a) nonburst cycle   (b) burst cycle

16. Calculate the bus bandwidth of a 486 with 33 MHz in each of the following.
    (a) nonburst cycle   (b) burst cycle
17. Which 486 instruction converts from the little endian to big endian, or vice versa?
18. Show how the data is placed in memory for the following program before and and after the execution of BSWAP.

        MOV   EAX,23F46512H
        MOV   [4000],EAX
        BSWAP EAX
        MOV   [6000],EAX

19. If a 486 is advertised as 25 MHz, what clock frequency is connected to CLK?
20. What is the purpose of the A20M pin in the 486?
21. Assume that CS =FFFFH and IP =76A0H. Calculate the physical address of the instruction in each of the following states.
    (a) A20M =0   (b) A20M=1
22. What is HMA, and how does it relate to the A20M pin in the 486?
23. The 486 uses a pipeline of _____ stages.
24. Give the names of the pipeline stages in the 486.

SECTION 23.2: INTEL'S PENTIUM

25. The number of pipeline stages in a superpipeline system is _____ (less, more) than in a superscalar system.
26. Which has one or more execution units, superpipeline or superscalar?
27. The Pentium uses _____ transistors and has _____ pins.
28. The Pentium has a(n) _____-bit external data bus whose pins are named _____.
29. The Pentium is a(n) _____-bit microprocessor.
30. State how many BE pins the Pentium has and their purpose.
31. BE pins are active _____ (low, high).
32. If BE7 - BE0 =11110000, which part of the data buses is activated?
33. If BE7 - BE0 =00000000, which part of the data buses is activated?
34. How many DP (data parity) bits does the Pentium have?
35. Find and compare the Pentium bus bandwidth for the following 60-MHz systems.
    (a) nonburst mode   (b) burst cycle mode
36. What is the size of on-chip cache in the Pentium?
37. In Problem 36, how much cache is for data and how much for code?
38. Which part of on-chip cache in the Pentium is write protected, data or code?
39. True or false. The Pentium has an on-chip math coprocessor.
40 What does instruction pairing mean in the Pentium?
41. The Pentium uses _____ (superscalar, superpipeline) architecture.
42. What is instruction pairing, and when can it happen?
43. What is data dependency, and how is it avoided?
44. Write a program for the 386/486/Pentium to calculate the total sum of 10 double-word operands. Use looping.
45. Compare the clock count of the loop in Problem 44 for each of the following. Use branch prediction for (c) and (d). Note that if two instructions are paired and one takes 2 clocks and the other takes only 1 clock, the clock count is 2.
    (a) 386   (b) 486   (c) a Pentium with instruction pairing but no code scheduling
    (d) Pentium with the instruction pairing and code scheduling
46. Calculate the bus bandwidth for a 486DX2-50 and 486DX4-100. Note that the 486DX2-50 is 25 MHz and the 486DX4-100 is 33 MHz.
47. Draw the pipeline stages for the pairing of instructions in the Pentium.
48. True or false. The Pentium has the A20M pin.

SECTION 23.3: RISC ARCHITECTURE

49. Why is RISC called load/store architecture?
50. In RISC, all instructions are _____-byte.
51. Which of the following instructions do not exist in RISC?
    (a) ADD reg,reg (b) MOV r,immediate (c) OR reg,mem

52. State the steps in a RISC program to add a register to a memory location.
53. What is the advantage of having all the instructions the same size?
54. Why are RISC programs larger than CISC programs?
55. The vast majority of RISC instructions are executed in _____ (1, 2, 3) clocks.
56. What is Harvard architecture? Is it unique to RISC? Can a CISC system use Harvard architecture?
57. Code a RISC program to add 10 operands of 4-byte (doubleword) size and save the result. Do not be concerned with carries.
58. Show the code scheduling and clock count for Problem 57.
59. What is a delayed branch?
60. MS DOS runs native on which of the following processors?
    (a) IBM/Motorola Power PC RISC   (b) Intel 80x86
    (c) Digital Equipment Alpha RISC
61. Generally, RISC processors have _____ registers, each 32 bits wide.
62. Which register in RISC always has value zero in it, no matter what operation is performed on it?
63. Discuss the terms porting, emulating, and running native.

SECTION 23.4: PENTIUM PRO PROCESSOR

64. The Pentium Pro is ____-bit internally and ____-bit externally.
65. The Pentium Pro has _____ address bits.
66. The Pentium Pro capable of addressing _____ bytes of memory.
67. The Pentium Pro has _____ pins for data bus.
68. A Pentium Pro is advertised as 200 MHz . Is this an internal CPU frequency or a bus frequency?
69. What is the difference between the L2 cache of Pentium and Pentium Pro systems?
70. Do the 5.5 million transistors used for the Pentium Pro include the L2 cache transistor count?
71. Which of the x86 processors has out-of-order execution?
72. Are the triadic registers of the Pentium Pro visible to the programmer?
73. True or false. Instructions are fetched according to the order in which they were written.
74. True or false. Instructions are executed according to the order in which they were written.
75. True or false. Instructions are retired according to the order in which they were written.
76. The visible registers EAX, EBX, etc., are updated by which unit of the CPU?
77. True or false. Among the instructions, STORES are never executed out of order.
78. Which of the x86 processors have branch prediction capability?

SECTION 23.5: MMX TECHNOLOGY

79. True or false. The MMX uses x86 and x87 instructions to emulate DSP functions.
80. A given system has both a general-purpose CPU (such as the x86) and a DSP chip. Discuss the role of each chip.
81. Does the 486 system have MMX technology?
82. Explain the concept of register aliasing.
83. Which registers are aliased in MMX technology?
84. Explain the difference between the way x87 registers are accessed and the way MMX accesses the same registers.
85. Indicate which group of instructions can be intermixed.
    (a) x86, x87   (b) x86, MMX   (c) x87, MMX
86. When leaving MMX, what is the last thing that a programmer should do?
87. When leaving x87, what is the last thing that a programmer should do?
88. How many bits of the x87 register are used by MMX?
89. A quadword has _____ bits.
90. Give other data formats than quadword which can be viewed by MMX instructions.

---

91. True or false. Every x86 supports the CPUID instruction.
92. Explain how to identify the 386 microprocessor.
93. Explain how to identify the 486 microprocessor.
94. Explain how to identify the Pentium microprocessor.
95. Explain how to identify the Pentium Pro microprocessor.
96. Explain how to identify the Pentium Pro with MMX technology.
97. How can the CPUID instruction be coded into a program if the assembler does not support it?

ANSWERS TO REVIEW QUESTIONS

SECTION 23.1: THE 80486 MICROPROCESSOR
1. 168 pins in PGA
2. true
3. 32, 32
4. The 80486 has the math coprocessor on-chip; the 80486SX has the math coprocessor 80487SX (a separate chip).
5. primary cache, secondary cache or L1 and L2 cache
6. 8K bytes, 2-way set associative
7. 1/50 MHz =20 ns is clock cycle. In burst cycle 5 clocks of 20 ns can access 16 bytes of memory (4 memory cycle x 4 bytes of 32-bit data bus); therefore, (1/100 ns) x16 bytes = 160 megabytes/second
   Another way would be average clock per 32-bit access is (5 x 20)/4=25 ns and (1/25 ns) x 4 bytes =160 megabytes/second
8. 33 MHz   9. input
10. output, like all other address bits in 80x86 processors

SECTION 23.2: INTEL'S PENTIUM
1. 273   2. 64
3. true   4. 16K bytes: 8K for code and 8K for data
5. code cache   6. true
7. true
8. since it has two execution units (pipelines) capable of executing two instructions with one clock
9. false; by the compiler   10. true

SECTION 23.3: RISC ARCHITECTURE
1. reduced instruction set computer, complex instruction set computer
2. true
3. 32, 32   4. true
5. They are all 4 bytes.   6. true
7. c   8. 32, 8 stack based
9. to take care of double-precision floating-point operands and single-precision operands
10. False, it uses separate buses for data and code.

SECTION 23.4: PENTIUM PRO PROCESSOR
1. P6   2. true
3. false   4. false
5. so far, the Pentium Pro   6. retire unit
7. true

SECTION 23.5: MMX TECHNOLOGY
1. The x87 instruction set is mainly for math functions such as sine, cosine, log, and so on and does not lend itself to DSP-type operations.
2. (b) & (c)
3. false
4. MA0 - MA7
5. false

SECTION 23.6: PROCESSOR IDENTIFICATION IN INTEL X86
1. bit 18 of flag register is always high.
2. bit 21
3. Pentium
4. EAX=1
5. Yes. After return from CPUID we test the bit 23 of the EDX register. If it is high, the chip has MMX technology.

CHAPTER 24

MS DOS STRUCTURE, TSR, AND DEVICE DRIVERS

OBJECTIVES

Upon completion of this chapter, you will be able to:

» Outline the steps performed by DOS in the cold boot process
» Describe the purpose and use of the CONFIG.SYS file
» Describe the purpose and use of the AUTOEXEC.BAT file
» Discuss the role of the COMMAND.COM file in the operating system
» Describe advantages and disadvantages of TSR programs
» Write Assembly language instructions to create a program that stays resident in conventional memory after termination
» Write Assembly language instructions to hook a TSR program into hardware or software interrupt
» State the purpose of device drivers
» Contrast and compare character-type and block-type device drivers
» Describe how the PC is informed of hardware added to the system

Chapters 9 through 15 covered hardware design of the IBM PC, PC AT, PS, and 80x86 compatibles. In this chapter we examine the DOS operating system and how application software interacts with DOS. In Section 24.1 we look at the structure of DOS and the process of loading the operating system. Section 24.2 is dedicated to a discussion of TSR (terminate and stay resident) programs and how to write them. In Section 24.2 we also look at the characteristics of device drivers.

SECTION 24.1: MS DOS STRUCTURE

MS DOS (and its IBM version PC DOS) is the most widely used operating system for desktop and laptop computers. Prior to the advent of MS DOS, CP/M was widely used for personal computers, while mainframes and minicomputers used either Unix or proprietary operating systems such as VMS and IBM's OS. Due to the massive base of MS DOS application software, even some recent variations of Unix have a DOS box. In such systems, the screen is split into many simultaneously visible windows, one or more of which is assigned to DOS applications. This allows DOS to be run under the Unix operating system.

DOS genealogy

Although DOS has come a long way, it is still a baby compared to Unix or OS/2. MS DOS has gone through many changes since its introduction in 1981. Table 24-1 lists the major changes introduced in various DOS versions throughout the years. Each new version also included new commands.

Table 24-1: MS DOS Genealogy

MS DOS Version	Year Introduced	Additional Features
1.0	1981	The original DOS
1.25	1982	Double-sided 320K diskettes
2.0	1983	Hierarchical file structure
2.11	1983	Fixed bugs in 2.0
3.0	1984	1.2M diskettes and large hard drives
3.1	1985	Network features
3.2	1985	3.5" 720K diskettes
3.3	1987	3.5" 1.44M diskettes
4.0	1988	Hard drive partitions greater than 32M, plus EMS
4.1	1988	Fixed bugs in 4.0
5.0	1991	Takes up less RAM
6.0	1993	Disk compression, antivirus features

From cold boot to DOS prompt

A cold boot occurs when the power switch of the PC is turned on. A warm boot is when the computer is reset using the CTRL-ALT-DEL keys. What happens from the moment the PC is turned on (cold boot) to the time the DOS prompt ">" appears on the screen? The following is the sequence of steps performed between boot-up and the appearance of the DOS prompt. Figure 24-1 shows the major steps.
1. Upon cold boot, the RESET pin of the CPU is activated, resulting in CS = F000 and IP = FFF0 for 286 and higher (on the 8086/88, CS = FFFF and IP = 0000).
2. The CPU fetches the first opcode from BIOS ROM, at physical address FFFF0.

3. The assembly programs contained in BIOS ROM test the CPU, test and initialize peripheral chips, test RAM memory, perform checksum on BIOS ROM, initialize peripheral devices LPT, COM, and floppy and hard disks, and call the bootstrap interrupt service routine. This process is often referred to as the *power-on self-test* (POST). If, for example, the keyboard is unplugged and the PC is powered on, it is the POST that will sound a beep and write an error code to the monitor.

4. The bootstrap subroutine in BIOS ROM is executed. It reads the boot sector (sector 0 of the disk) into the memory of the PC, then transfers control to that program.

5. The boot sector contains an Assembly language program called the disk bootstrap subroutine. The disk bootstrap checks the first sector of the root directory to see if files IO.SYS and MSDOS.SYS are present on the disk. If they are present, they are read into memory and control is transferred to IO.SYS. If they are not present, the user is notified and asked to insert a system disk (one formatted with option "/s") and press any key when ready. The memory locations where disk bootstrap, IO.SYS, and MSDOS.SYS files are stored at this point is some arbitrary location.

6. The SYSINIT subroutine, one of the subroutines of IO.SYS, tests and keeps track of up to 640K of contiguous conventional memory. After the amount and location of contiguous memory are known, SYSINIT copies both IO.SYS and MSDOS.SYS from the arbitrary memory locations to fixed and final memory locations.

7. SYSINIT calls MSDOS.SYS to perform some internal housekeeping chores. Afterward, SYSINIT takes over again.

8. Next, SYSINIT asks DOS to check for the existence of a file called CONFIG.SYS in the root directory. If the optional CONFIG.SYS file is found, it is loaded into memory and executed. CONFIG.SYS allows the user to customize MS DOS, as well as to add nonstandard hardware devices such as a mouse or CD ROM to the PC. Programs responsible for communication between these new hardware devices, the PC hardware, and the operating system are called *device drivers*. More about device drivers will be given Section 24.2. As CONFIG.SYS is executed, each device driver is loaded into conventional memory and assigned a fixed memory location. The more device drivers in CONFIG.SYS, the more of the 640K of conventional memory is used and the less is left for application software such as word processors and spreadsheets. Although device drivers take up conventional memory, they also allow user-installed input and output devices to override the devices supported by BIOS.

9. The last thing SYSINIT does is to ask DOS to load the command interpreter from drive A. The default command interpreter is called COMMAND.COM, but it can be replaced with a shell command interpreter through use of the DOS SHELL command in CONFIG.SYS. If drive A is not available or does not have COMMAND.COM, drive C is searched for the COMMAND.COM file. COMMAND.COM is responsible for interpretation of DOS commands such as DIR, COPY, and so on. If the COMMAND.COM is on drive A, it is loaded into memory. Then DOS looks for another file on drive A called AUTOEXEC.BAT, loads and executes it if it is present, and finally, DOS asks for the system DATE and TIME, then displays the prompt, ">". If drive A does not contain a system disk, COMMAND.COM and AUTOEXEC.BAT are loaded from the C drive. In AT and higher machines, DOS requests for the time and date from the user are bypassed, and instead this information is received internally from a battery-operated clock-calendar, which is kept by an RTC (real-time clock) CMOS RAM chip.

10. At this point, the command interpreter COMMAND.COM has taken over the system. It displays the prompt and waits for the user's response. SYSINIT is terminated and the memory space used by it is freed.

   The *kernel*, the nucleus of an operating system, contains the most frequently used programs, which primarily control processes. The kernel of DOS, including COMMAND.COM, will stay in conventional memory until the PC is turned off. The size of memory taken up by COMMAND.COM and the DOS kernel varies from DOS version to DOS version. In Chapter 25 we will show how DOS can be moved (loaded) into the high memory area to free conventional memory for application software. We will also show how device drivers and TSRs can be moved to make more of conventional memory available for application software.

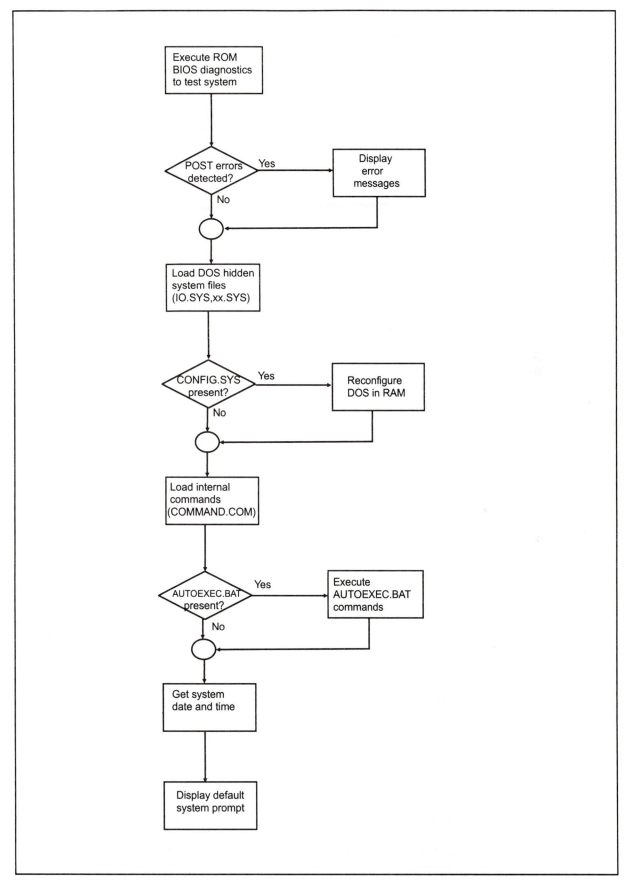

Figure 24-1. Boot Process

DOS standard device names

DOS has reserved some names to be used for standard devices connected to the PC. Since they have special meaning for DOS, they should not be used for user-defined filenames or device names. They are listed in Table 24-2.

Table 24-2: DOS Names for Standard Devices

Device Name	Meaning
AUX	Auxiliary device (often another name referring to COM1)
CLOCK$	System clock (date and time)
COM1	Serial communication port #1
COM2	Serial communication port #2
COM3	Serial communication port #3
COM4	Serial communication port #4
CON	Console for system input (keyboard) and output (video)
LPT1	Line printer #1
LPT2	Line printer #2
LPT3	Line printer #3
LPT4	Line printer #4
NUL	NULL (an input and output that does not exist)
PRN	Printer attached to LPT1

More about CONFIG.SYS and how it is used

Assume that you want to design and attach some kind of device such as a card reader to your PC. How is the PC notified of the presence of a card reader? CONFIG.SYS allows the user to incorporate new technology into the PC. CONFIG.SYS is an optional file which the PC reads and executes only during system boot. This is done right before the system loads COMMAND.COM. This optional file not only allows users to add a new nonstandard device such as a mouse and CD-ROM but also allows customizing DOS for a given system. Depending on the intended use of the PC, CONFIG.SYS can be modified to improve system performance dramatically. This is especially true for the way DOS allocates and uses the 640K of conventional memory. The setting of the parameters in the CONFIG.SYS file can have a dramatic impact on how much memory DOS leaves for application software, as we will see in Chapter 25. First, we describe some of the commands that can be used in the CONFIG.SYS file. Additional commands are covered when we discuss DOS memory management in Chapter 25. Note that the CONFIG.SYS file must be located in the root directory of the system disk.

BREAK

In DOS, Ctrl-break (or Ctrl-c) can be used to stop execution of a program and return to the DOS prompt. DOS checks for Ctrl-break whenever it is inputting data from the keyboard or outputting to the monitor or printer. It does not check during other operations, such as disk read/write operations. This means that if DOS is reading disk memory, it will essentially ignore Ctrl-break until the operation has completed. The BREAK command can be used to make DOS check for Ctrl-break during any call to the operating system's service functions. The disadvantage of this, however, is that it can slow overall system performance somewhat since DOS has to check continually if the user has entered Ctrl-break. Placing command "break on" in CONFIG.SYS will set up DOS for extended control-break checking. Command "break off" will tell DOS to check for Ctrl-break only during keyboard, monitor, and printer I/O; "break off" is the default setting.

BUFFERS

The simplest form of the BUFFERS command is "buffers=n", where n is a number from 1 to 99. This tells DOS to set aside n disk buffers in RAM. When DOS reads to or writes from disk, the data is stored temporarily in buffers that hold 512 bytes each. Word processors often require the system to have 20 or more buffers. The advantage of having more buffers is that it will speed up program execution since the system will require fewer disk accesses. The disadvantage to adding more buffers is that they take up conventional memory, leaving less for applications. The default setting for buffers varies from 2 to 15, depending on the amount of RAM on the system and the type of disk drive. If the operating system is loaded into the high memory area (by placing "DOS=HIGH" in CONFIG.SYS), the buffers will be there as well. This saves conventional memory.

An additional parameter can be added to the command, "buffers=n,m" where m is a number from 1 to 8 that specifies the number of sectors in the *read-ahead buffer*. Each sector is 512 bytes. The read-ahead buffer is recommended for systems performing many sequential disk reads. This allows the system to read additional sectors into the buffer (read-ahead) each time the disk is accessed. The default for look-ahead sectors is 0, meaning that no read-ahead buffer is set up. This parameter is available for DOS version 5 and higher.

FILES

This command determines how many file handles can be active at a time. For example, "files=20" means that there can be a maximum of 20 files open at a time. The number of files can be from 8 to 255, with 8 being the default value.

DEVICE

The device command allows the user to add additional device drivers to the system, such as for a mouse or a scanner. For example, "device=c:\acme\mouse.sys" could be used to load the indicated device driver. There are several standard device drivers provided by DOS: ANSI.SYS, DISPLAY.SYS, DRIVER.SYS, EGA.SYS, PRINTER.SYS, RAMDRIVE.SYS, EMM386.EXE, HIMEM.SYS, and SMARTDRV.SYS. Device drivers are covered in Section 24.2.

SHELL

The shell command can be included in the CONFIG.SYS file if you want to use a command interpreter other than COMMAND.COM or if you want to use a version of COMMAND.COM not found in your root directory. For example, command "shell=b:\olddos\command.com" will cause the system to use the file specified by that path for the command interpreter.

What is AUTOEXEC.BAT and how is it used?

AUTOEXEC.BAT is an optional file that contains a group (a batch) of DOS commands that are executed automatically when the PC is turned on. This saves the user from having to type in the commands each time the system is booted. The DOS commands contained in the batch file are executed in sequence without any external intervention from the user. After the PC is booted, CONFIG.SYS is loaded and executed, COMMAND.COM is loaded, and finally DOS searches for the AUTO-EXEC.BAT file in the root directory of the system disk. If the file is present, it is loaded into memory and each command in the file is executed, one after the other. An example of a batch command is "prompt pg", which sets up the DOS prompt to display the current directory, as well as the ">". Another example is the command "win", which is placed in the AUTOEXEC.BAT file when the user wants the system to run Windows upon boot.

Types of DOS commands

All DOS commands can be categorized according to when and where they can be invoked. The categories are internal, external, CONFIG.SYS, batch, and internal/external (internal or external). Each category is described next.

Internal DOS commands

After COMMAND.COM is loaded into the 640K conventional memory, it stays there, taking up precious memory. To keep the size of COMMAND.COM small, Microsoft includes only the most important DOS commands in COM-MAND.COM. DOS commands included in COMMAND.COM are referred to as *internal* DOS commands. Commands such as COPY, DEL, TYPE, MD, and CD are internal commands. Remember that DOS is not case sensitive.

COPY PROG1.ASM PROG2.ASM	copies first file to second
DEL B:PROG1.ASM	deletes prog1.asm on drive b
TYPE PROG1.ASM	lists file to screen
MD PROGRAMS	creates directory named programs
CD \TOM\PROGRAMS	changes current directory to \tom\programs

External DOS commands

There are many DOS commands that are not part of COMMAND.COM and are provided as separate files on the DOS disk. These are called *external* DOS commands (external from the point of view of COMMAND.COM or one might say not residing in conventional memory). When an external DOS command is executed, it is loaded from disk into memory and executed. After that, it is released from memory. To use it again, it must be loaded from the disk again. When external DOS commands are used, the path must be specified; otherwise, it is assumed to be in the current directory. Commands such as FORMAT, DISKCOPY, BACKUP, EDLIN, and UNDELETE are examples of external commands.

FORMAT B:	format diskette in drive b
DISKCOPY A: B:	copy diskette in drive a to disk in drive b
BACKUP C:\SUE\PROGRAMS*.* A:	back up all files in that directory to disk a
EDLIN PROG1.ASM	edit file with DOS line editor program
UNDELETE PROG1.ASM	recover deleted file prog1.asm

External/internal DOS commands

There are DOS commands that are external the first time they are used, but from then on they become internal in that they become resident in the 640K memory. When the computer is turned off, they are lost and become external until they are invoked again. DOSKEY is an example of this kind.

DOSKEY   start program to recall and edit DOS commands and macros

CONFIG.SYS DOS commands

CONFIG.SYS commands are used only in the CONFIG.SYS file. There are some that can be used in both CONFIG.SYS and as internal commands. DEVICE, FILES, and SHELL are examples of CONFIG.SYS commands. DEVICE is described in Section 24.2. FILES and SHELL are described above.

Batch file DOS commands

These commands are used primarily in batch files. They are also used in Shell programs. ECHO and REM are examples of BATCH commands.

ECHO OFF	DOS will not echo commands or prompts to the monitor
REM	used to enter comments in batch files

Review Questions

1. What is the difference between a cold boot and a warm boot?
2. The POST is performed upon _____ (cold boot, warm boot).
3. The program performing the POST is held by _____ (DOS, BIOS).
4. A message is displayed by _____ (POST, DOS) if the keyboard is not connected.
5. What is the function of the bootstrap on BIOS ROM?
6. True or false. File IODOS.SYS is loaded before CONFIG.SYS is loaded.
7. True or false. The AUTOEXEC.BAT file is loaded before the CONFIG.SYS file.
8. What is the difference between internal and external DOS commands?
9. What does it mean if a command is external/internal?
10. Indicate the category of each of the following DOS commands.
    (a) DEL   (b) TYPE   (c) DEVICE   (d) FORMAT

SECTION 24.2: TSR AND DEVICE DRIVERS

This section first examines the concept of TSR (terminate and stay resident) programs and how they are written. Device drivers are also covered in this section. To see the reason that TSR programs were developed, we must first understand how DOS runs programs. DOS uses what is called a *memory control block* (MCB) to keep track of the 640K conventional memory of the PC. One of the functions of DOS is to allocate memory for program execution. DOS uses a portion of the 640K bytes of conventional memory for itself and the rest is available to application programs. The amount of 640K used by DOS varies from version to version. When an application program contained in the disk is executed, DOS allocates a section of the 640K to it, and loads it from the disk into that section. After the program is executed, it is abandoned and its allocated memory space is returned to the pool of available memory. In other words, after the program is terminated, the memory space allocated to it is reallocated and a new program can be loaded into that space. If DOS did not abandon a terminated program and free its memory, after a while all the 640K memory would be filled and nothing would be left for other programs.

Executing but not abandoning the program

The idea behind a TSR program is to execute the program but not abandon it after it has executed. In other words, DOS keeps it resident in the 640K after it has finished executing. The advantage of a TSR program is that it is always resident in main memory and does not need to be brought in from the slow disk in order to execute. The disadvantage is that each TSR program takes some memory away from the 640K, and consequently, leaves less for other application programs. When DOS is loaded, only the core (kernel) of the DOS stays in the 640K memory and the rest of the 640K is available to application and utility software. As we make more and more of application and utility software TSR, it leaves less and less of the 640K for application programs.

As we saw in the preceding section, many DOS commands, such as COPY, DIR, and DEL, are internal and are part of COMMAND.COM, which stays resident in the 640K memory. There are also external commands that are on the disk and must be brought in from disk to be executed. TSR allows a programmer to make software internal in the sense that it is always in the 640K memory and there is no need to bring it in from disk.

How to make a program resident

To make a program resident, INT 21H option AH =31H is used to exit to DOS instead of using AH =4CH. Prior to function call 31H, the DX register must hold the size of the resident program in multiples of paragraphs (16 bytes). DOS will keep resident the number of paragraphs of code indicated by the value in DX and will abandon the rest. The following instructions will make 12 paragraphs (12 × 16 = 192 bytes) of code resident. This code is placed at the end of the program where function call AH = 4CH is normally placed.

```
MOV     AH,31H        ;option 31H OF INT 21H
MOV     DX,12         ;make resident 12 paragraphs (192 bytes)
INT     21H           ;invoke DOS function call
```

Invoking the TSR

Since TSR programs are resident, there is no need to get them from disk, but how are they invoked? Most often, TSR programs are activated by means of a hardware interrupt such as INT 09 (keyboard interrupt) or INT 08 (timer clock interrupt), but TSR can also be activated by exception interrupts such as zero divide error interrupt 00. TSR programs can be awakened by software or hardware interrupts. To awaken the TSR, it must be hooked into the interrupt. Next we show how to hook into an interrupt to activate a TSR.

Hooking into hardware interrupts

In Chapter 14 we discussed hardware interrupts INT 08 and 09. As discussed in Chapter 18, every time a key is pressed, INT 09 is activated and executes the interrupt service routine (ISR) associated with it. In Chapter 18 we also discussed the notion of a hot key. Many commonly used TSR programs are awakened by hot keys. When a TSR is awakened by a hot key, we must monitor the keyboard input to the motherboard and check for the desired hot key. If the hot key is detected, the TSR is activated; otherwise, we should let it go on to perform its service and not interfere with normal operation of the keyboard. Failure to let INT 09 perform its normal function results in keyboard lock-up, where no key press is detected by the main motherboard. To detect a specific hot key, we must replace CS:IP of INT 09 (the keyboard interrupt) in the interrupt vector table with CS:IP of our own keyboard interrupt handler. In doing so, we must save CS:IP of INT 09 and pass control to it before or after we check for the desired hot key. In this way, the keyboard can go about doing its business while we also get a chance to see what key was activated. This is called *hooking into the interrupt*. It is also referred to as *interrupt chaining*.

Replacing the CS:IP values in the interrupt vector table

INT 21H DOS function calls provide two services for manipulating CS:IP in the interrupt vector table. Option AH =35H gets the CS and IP values for the current interrupt, and option AH =25H allows placing new values for CS:IP in the interrupt vector table. Each is discussed next.

INT 21H option AH =35H gets the current values of CS:IP in the interrupt vector table for a given interrupt number so that we can examine it or save it. Before calling function AH=35, we must set AL =interrupt number. Upon return from INT 21H, the ES register has the code segment (CS) value and BX has the offset address (IP) value of the ISR belonging to the interrupt number. This is shown in Example 24-1.

INT 21H option AH=25H allows one to replace the current CS:IP in the vector table with a new value of CS:IP. Prior to calling INT 21H with option AH =25H, AL =interrupt number, DS =segment, and DX =offset address of the interrupt service routine (interrupt handler). This is shown in Example 24-3. Note in this example that SEG is referring to the segment address of procedure MY_INT and is loaded indirectly into the DS register.

Writing a simple TSR

To demonstrate how to create TSR programs, we will write two simple TSR programs, one that is activated by the INT 09 keyboard interrupt and the other by the INT 08 clock. Both will make the beep sound.

Example 24-1

(a) Execute INT 21H option AH=35 for the INT 0 (divide error).

(b) Verify the result with the DEBUG dump command.

Solution:

(a)
```
C>DEBUG
-A 100
16B7:0100 MOV AH,35    ;OPTION 35 OF INT 21H
16B7:0102 MOV AL,0     ;GET CS:IP OF INT 00
16B7:0104 INT 21
16B7:0106 INT 3
16B7:0107
-G

AX=3500  BX=108A  CX=0000  DX=0000  SP=CFDE  BP=0000  SI=0000  DI=0000
DS=16B7  ES=0116  SS=16B7  CS=16B7  IP=0106   NV UP DI PL NZ NA PO NC
16B7:0106 CC             INT    3
```

(b) Now using the Dump command, we can verify this as follows:

```
-D 0:0 LF
0000:0000 8A 10 16 01 ED 08 C6 13-16 00 F7 07 E6 08 C6   ...m.F...w.f.F
-Q
```
Notice that the value returned in BX matches the first 2 bytes dumped, the interrupt vector location for interrupt 0, and that the value in ES matches the second 2 bytes.

Example 24-2

Using INT 21H, show how CS:IP of INT F8H is retrieved from the interrupt vector table and saved.

Solution:

```
        OLD_IP_INT_F8     DW    ?
        OLD_CS_INT_F8     DW    ?

        MOV  AH,35H
        MOV  AL,0F8H                 ;get CS:IP of INT F8H
        INT  21H
        MOV  OLD_IP_INT_F8,BX  ;save IP
        MOV  OLD_CS_INT_F8,ES  ;save CS
```

Example 24-3

Assume that we have an interrupt handler (interrupt service routine) procedure named MY_INT for INT F8H. Show how to set the interrupt vector table for it.

Solution:

```
        MOV  AH,25H               ;option 25 to set interrupt vector
        MOV  AL,0F8H              ;set it for INT F8H
        MOV  DX,OFFSET MY_INT ;load the offset part
        MOV  BX,SEG MY_INT       ;load the segment part
        MOV  DS,BX               ;seg address of interrupt handler
        INT  21H
```

First, a few words about the beep sound created by INT 10H. INT 10H option 0EH outputs one character at a time to the video. There are some characters that are nondisplayable, such as ASCII 0AH (line feed), 0DH (carriage return), and 07 (beep). In such cases, INT 10H simply performs the action. For example, by sending 07, the ASCII code for beep, to the monitor using INT 10H service 0EH, the speaker will beep. Execute the following code in DEBUG to see (hear) what happens.

```
MOV     AH,0E
MOV     AL,07
INT     10      ;DEBUG assumes that all numbers are in hex
INT     3
```

TSR with hot keys

To hook into INT 09 of the keyboard, we first must make sure that CS:IP of the INT 09 belonging to the keyboard interrupt service routine of BIOS is saved. Then we replace it with CS:IP of our interrupt. In the following TSR program, every time hot keys ALT F10 are pressed, the speaker will beep. Notice how in the LOAD procedure, CS:IP of INT 09 is saved and then set to the new INT 09 procedure

```
;upon activation of ALT F10 the system will beep
;this program needs to be converted to a COM file using the EXE2BIN before it is run
;See Vol. 1 for the process of COM file creation.

CODESG   SEGMENT
         ASSUME CS:CODESG
         ORG    100H
MAIN:    JMP    SHORT LOAD
OLDINT9  DD  ?                              ;32-bit area to save the CS:IP of INT 09
;-------------- this portion remains resident
NEWINT9  PROC
         PUSH   AX
         MOV    AH,2                         ;get the keyboard status byte
         INT    16H                          ;using  BIOS INT 16
         TEST   AL,00001000B                 ;check for ALT
         JZ     OVER                         ;if no ALT key then exit
         IN     AL,60H                       ;get the scan code
         CMP    AL,44H                        ;see if it is F10 key
         JNE    OVER                         ;if no then exit
         MOV    AH,0EH                        ;if yes beep the speaker
         MOV    AL,07                         ;using BIOS
         INT    10H                          ;INT 10H
OVER:    POP    AX                            ;restore the reg
         JMP    CS:OLDINT9                    ;and perform INT 09
NEWINT9  ENDP
;-------------- this portion is run once only during the initialization
         ASSUME CS:CODESG,DS:CODESG
LOAD     PROC   NEAR
         MOV    AH,35H                        ;get the vector values
         MOV    AL,09H                        ;for INT 09
         INT    21H
         MOV    WORD PTR OLDINT9,BX              ;save
         MOV    WORD PTR OLDINT9+2,ES           ;them
         MOV    AH,25H                        ;set the vector for
         MOV    AL,09H                        ;the new INT 09
         MOV    DX,OFFSET NEWINT9             ;DX=IP, DS=CS set by COM
         INT    21H
         MOV    DX,(OFFSET LOAD - OFFSET CODESG)  ;find how how many bytes  resident
         ADD    DX,15                            ;make it
         MOV    CL,4                             ;multiple of 16 bytes
         SHR    DX,CL
         MOV    AH,31H                           ;and make it
         INT    21H                              ;resident
LOAD     ENDP
CODESG   ENDS
         END    MAIN
```

Program 24-1. TSR Hooked into INT 09

address. In the body of the new ISR (interrupt service routine), we first check for the ALT key using INT 16H option 2. If ALT is pressed, we check the contents of port 60H to examine the scan code. Only if the scan code belongs to F10 will the speaker beep. Notice that regardless of which route NEWINT9 takes, it must end up taking care of the original INT 09 residing in BIOS. Again it must be emphasized that failure to do so will result in locking up the keyboard. To get out of a locked keyboard, we must reboot the system.

Notice in Program 24-1 that after checking for the ALT key, we checked for the F10 scan code directly from port 60H. This is due to the fact that there is no scan code for every key combination. If a given key combination has a scan code, we can monitor it using option AH =0 of INT 16H, as was shown in Example 18-5. Also it must be noted that port 60H holds the scan code of the previous key indefinitely until it is replaced by the next keystroke.

TSR programs must be converted to COM files before they are run. DOS program EXE2BIN converts an EXE file to a COM file, as shown in Chapter 2.

```
CODESG      SEGMENT
            ASSUME CS:CODESG
            ORG 100H
MAIN:       JMP SHORT LOAD
OLDINT8     DD ?
COUNT       DW    275              ;275 x 54.94 ms =15 seconds
;---------------- this portion remains resident
NEWINT8     PROC
            DEC    CS:COUNT         ;is the time up?
            JNZ    EXIT
            MOV    CS:COUNT,275     ;if yes initialize the count
            MOV    AH,0EH           ;and
            MOV    AL,7             ;beep the speaker
            INT    10H              ;
            MOV    AH,0EH           ;do it again
            MOV    AL,7
            INT    10H
EXIT:       JMP    CS:OLDINT8       ;take care of  INT 08
NEWINT8     ENDP
;---------------- this portion is run once only
            ASSUME CS:CODESG,DS:CODESG
LOAD        PROC   NEAR
            MOV    AH,35H                     ;get the current CS:IP
            MOV    AL,08H                     ;for INT 08
            INT    21H
            MOV    WORD PTR OLDINT8,BX  ;save them
            MOV    WORD PTR OLDINT8+2,ES
            MOV    AH,25H                     ;set CS:IP
            MOV    AL,08H                     ;for the new INT 08
            MOV    DX,OFFSET NEWINT8     ;DX=offset IP, DS=CS set by COM
            INT    21H
            MOV    DX,(OFFSET LOAD-OFFSET CODESG)  ;find how many bytes resident
            ADD    DX,15                     ;round it
            MOV    CL,4                      ;to paragraph
            SHR    DX,CL                     ;size and
            MOV    AH,31H                    ;make it
            INT    21H                       ;resident
LOAD        ENDP
CODESG      ENDS
            END    MAIN
```

Program 24-2. TSR Hooked into INT 08

Hooking into timer clock INT 08

Another way to awaken a TSR is to use INT 08. As discussed in Chapters 13 and 14, hardware INT 08 is activated every 54.94 ms. In Program 24-2, the TSR is hooked into INT 08 and the speaker beeps every 15 seconds.

In many TSR programs, the original interrupt is serviced by simulating the interrupt instruction instead of "JMP CS:OLDINTERRUPT". The following shows how this is done.

```
NEWINT    PROC  FAR                          ;notice this is a far procedure
          ...
          PUSHF                              ;simulate
          CALL    CS:OLDINTERRUPT            ;the hardware interrupt
          IRET                               ;return from interrupt
NEWINT    ENDP
```

DOS is not reentrant

In the resident section of both of the above TSR programs, we used BIOS to do the work of the TSR and there is no DOS function call. The reason is that DOS is not reentrant. If we interrupt DOS while it is executing some DOS function calls, DOS will crash since in resuming where it left off all the parameters needed to finish the function call have been changed. This happens even if we save all registers of the CPU before we interrupt DOS. The fact that we cannot interrupt DOS is what we mean when we say that DOS is nonreentrant. For example, while DOS is reading the disk and a serial communication TSR program interrupts the process, the disk data read will be lost since DOS is not reentrant.

One way to solve this problem is to check the DOS busy flag to ensure that DOS is not doing anything before we interrupt it. The DOS busy flag is provided by AH =34H of INT 21H. Therefore, to use DOS function calls inside the resident portion of a TSR program, we must check the DOS busy flag and wait until DOS is not busy before we make a DOS function call. This is a good rule in writing any kind of TSR or resident device driver, regardless of whether we are using DOS function calls or BIOS services in the resident portions of the program. To check the DOS busy flag, we use the AH =34H option of INT 21H. Upon return, ES:BX is pointing to the DOS busy flag. If the flag is zero, it is is safe to use DOS function calls; otherwise, we must wait until it becomes zero. This is shown next.

```
          PUSH    ES                ;save  ES
          MOV     AH,34H            ;get DOS busy flag
          INT     21H
          MOV     AL,ES:[BX]        ;move the DOS busy flag into CPU
          POP     ES                ;restore ES
          CMP     AL,0              ;is it safe?
          JNZ     NOT_SAFE          ;no. bypass the following
          ....                      ;yes it is safe go ahead
NOT_SAFE:
```

Before we conclude the discussion about TSR programs, recall that one can use option AH =13H of INT 10H to display a string. See Example 14-10.

Device drivers

When the IBM PC was introduced in 1981, the keyboard was the only input device. For a computer and operating system to adapt to the changes and chances of the world, they must be able to integrate new devices, such as a mouse, CD-ROM, and scanners into the system. A program that allows such devices to interact with the PC system is called a *device driver*. The DOS operating system comes with a set of device drivers for standard devices such as the keyboard and line printers, referred to as default device drivers. We examined many of these device drivers in previous chapters. Since it is impossible for DOS to include device drivers for every possible device (now and future), it allows users to add their own device drivers, called *installable device drivers*. For example, assume that we have a plotter to interface with the PC. To do that, we go through the following steps.

1. Write and test a program that interfaces the plotter with the PC. This program will be called a plotter device driver which is installable.
2. Install the plotter device driver through the use of the DEVICE command in the CONFIG.SYS file. This will inform DOS about the existence of our plotter device driver. Upon booting the PC, the IO.SYS section of DOS first loads its own device drivers. Installable (user) device drivers, such as the one belonging to the plotter, are installed later by CONFIG.SYS.

Our plotter device driver is installed at boot time only if it has followed certain guidelines. Device drivers are used not only for hardware devices such as plotters, mouse, and CD-ROM, but device drivers can also belong to a utility such as memory management or disk management programs. Microsoft's RAM-DRIVE.SYS and Quarterdeck's QEMM386.SYS are such device drivers.

Device driver categories

Device drivers fall into two categories, character-type and block-type device drivers. Each is explained below.

Character-type device driver

In a character-type device driver the data is sent and received one byte at a time. Examples of character-type devices are keyboard and serial communication ports. Each character-type device connected to the PC is assigned a name referred to as its device name by MS DOS. Certain names, such as LPT1 and PRN, are reserved and must not be used. The reserved names of character-type devices were given in Table 24-2. If a reserved name is used for an installable device driver, DOS will let the user's device supersede the default device. For example, if we use PRN as the name of the device for a special kind of printer, DOS will load our device driver instead of the default DOS's PRN device driver. In DOS, the device name can be between 1 and 8 characters long.

Binary vs. ASCII data

For character-type devices, the data can be cooked or raw. If the data is packaged in a ASCII format, it is called *cooked*, but if the data is in binary format without any regard to ASCII standard, it is called *raw*. The major difference is that for ASCII characters, as each character is received (or sent) it is checked to see if it is Control-c. Upon detection of Control-c input, the operation is aborted by DOS, and consequently, DOS will transfer control to INT 23H. However, for non-ASCII character-type data (also called binary format), there is no check for Control-c between each character received or sent.

Block-type device driver

This refers to types of devices in which the data is sent and received a block at a time. The size of the block is generally 512 bytes. In this case, the driver always reads or writes the same number of bytes of data (the same-size block). Examples of block-type devices are floppy disks, hard disks, magnetic tape, and other mass storage devices. Unlike character-type devices, block-type devices are designated and referred to by a single letter, such as A, B, C, and so on. There is another difference between character-type and block-type data as far as the name is concerned. The device name in the character-type device driver always refers to only one physical peripheral device, but a single device driver of the block type can be divided into sections, and each separate section can be referred to by a different name. For example, there is only one keyboard (character-type device) in the system that is referred to as CON, but there can be a single physical hard disk that is partitioned into logical disk drives, designated as C, D, E, and so on.

Every device driver in MS DOS consists of three parts: the device header, device strategy routine (referred to as strat), and device interrupt routine (referred to as intr). Readers are referred to the lab book for a look inside device driver specifications and examples of how to write them.

Review Questions

1. Why does DOS abandon a program after it is executed?
2. In what part of the PC's memory do TSR programs take residence?
3. What is the danger of too many TSRs?
4. Explain the role of DOS INT 21H function 25H.
5. In the case of INT 08 and INT 09, why must the CS:IP of an interrupt be saved before we set new values in the interrupt vector table?
6. List the categories of device drivers.
7. How do we inform DOS about the existence of our device driver when the system is booted?
8. Can a device driver also be TSR?
9. Can a TSR program belong to a device driver?
10. State the absolute minimum function that we must include in a TSR program hooked into INT 09 and INT 08.

SUMMARY

This chapter examined the DOS operating system that provides the interface between the system hardware and the user. When a PC is turned on (a cold boot), MS DOS performs BIOS diagnostics on memory and peripherals, loads a bootstrap program, loads optional files CONFIG.SYS and AUTOEXEC.BAT, which allow the user to customize his or her system, and finally, displays the DOS prompt for user interaction. COMMAND.COM provides the interface between the user and the system. DOS commands can be categorized as internal, external, CONFIG.SYS, or batch.

The second section began with an examination of TSR (terminate and stay resident) programs. The advantage of these programs is that they stay resident after execution and do not have to be retrieved from disk every time they are run. The disadvantage is that they take up valuable conventional memory. TSR programs are activated by software or hardware interrupts. Hooking the TSR into an interrupt involves saving the old CS:IP of that interrupt and replacing it with the CS:IP of your own interrupt service routine, which invokes the TSR, then returns control to the original interrupt service routine. Device drivers are programs that allow a device to interact with the PC system. Users inform the system of the addition of new hardware by placing a DEVICE command in the CONFIG.SYS file. There are two categories of devices: character and block. Character devices process data one character at a time, whereas block devices process data in blocks, which are usually 512 bytes in size.

PROBLEMS

SECTION 24.1: MS DOS STRUCTURE

1. Turning on the PC is a _____ (warm, cold) boot.
2. The memory and peripheral chips are checked during _____ (warm, cold) boot.
3. The programs to test the memory and peripheral chips are kept by _____ (BIOS, DOS).
4. The beep sound indicating an error or defect in peripheral devices is generated by _____ (DOS, BIOS).
5. Which sector on disk holds the bootstrap subroutine?
6. For a disk to be a system disk, it must contain at least files _____ and _____.
7. What are the IBM versions of files for Problem 6?
8. True or false. The CONFIG.SYS file is loaded before COMMAND.COM.
9. Which is loaded first, the AUTOEXEC.BAT file or CONFIG.SYS?
10. Does the CONFIG.SYS file take any memory from conventional memory?
11. Gives some examples of DOS standard device names.

In this chapter we discuss how MS DOS manages the memory of the IBM PC and 80x86 compatibles. The concepts of conventional memory, upper memory block (UMB), high memory area (HMA), and expanded and extended memories are discussed in detail. We will also look at the DOS memory command (MEM) to examine the memory of the PC. In addition, techniques such as loading high are examined. In Section 25.1 we discuss all terminology and concepts related to the 80x86 IBM PC, PS, and DOS memory management. Section 25.2 covers loading high DOS, device drivers, and TSRs, in addition to the benefits of loading high.

SECTION 25.1: 80x86 PC MEMORY TERMINOLOGY AND CONCEPTS

In exploring the memory of the 80x86-based PC, there are a number of terms and concepts that must be understood thoroughly. Among them are the terms conventional memory, upper memory block, high memory area, expanded memory, extended memory, and shadow RAM. We will examine the origin and definition of each of these terms one by one, since understanding of these concepts is critical to understanding the rest of this chapter.

Conventional memory

The 8088/86 with its 20-pin address bus is capable of accessing only 1 megabyte of memory since 2^{20} =1,048,576 =1 megabyte, or 1024K bytes. This results in a memory address space of 00000 - FFFFFH. The designers of the original IBM PC set aside only the first 640K of the 1024K address space for RAM. The first 640K bytes of memory are located at the contiguous address range of 00000 to 9FFFFH and is called *conventional memory* (see Figure 25-1). Some refer to this as *lower memory*. Upon cold boot, BIOS will test this 640K address space for RAM and store the amount of memory installed in BIOS data area 0040:0017. While in early PCs only 64K to 256K of conventional memory was installed, in today's PCs the entire 640K is installed. DOS gets the amount of installed conventional memory from BIOS data area 40:0017 and allocates it as the need arises. Of the total conventional memory, the first 1K, from address 00000 to 003FFH, is set aside for the interrupt vector table, as was shown in Chapter 14. The area from 00400 to 004FFH, a total of 256 bytes, is set aside for the BIOS data area. Memory from 00500 to 005FFH, 256 bytes, is set aside to be used for DOS parameters (the DOS data area). Locations 00700H - 9FFFFH are available to DOS to allocate according to its own needs and the system configuration, and the rest is available for application software. Every IBM PC and compatible based on the 80x86 has conventional memory. This includes 8088- and 8086-based PCs.

Upper memory area

The designers of the first IBM PC set aside 128K, from A0000H to BFFFFH, for video RAM. The amount of this 128K space used depends on the video card and video mode used by the PC. From address C0000H to FFFFFH, a total of 256K, was set aside for ROM. Any kind of ROM, such as a hard disk controller, network card, video board, and BIOS ROM, must be located in the address range of C0000 to FFFFFH. Generally in the PC, memory from F0000 to FFFFF is used by BIOS ROM, and some of this address space is used by adapter cards such as hard disk controllers and network cards. In many IBM PS models, in addition to F0000 - FFFFF, addresses E0000 - EFFFF are also used for BIOS. In some technical literature, the 1M address space is divided into 16 segments of 64K bytes each and numbered in hex 0, 1, 2, ..., A, B, C, D, E, F. Such literature states that segments 0 - 9 are used for conventional memory, segments A and B are set aside for video RAM, and segments C, D, E, and F are set aside for ROM.

In PC literature, the term *upper memory area* refers to the address range A0000 to FFFFF, a total of 384K. Depending on the options and adapter cards installed, not all of the 384K of the upper memory area is used. The ROM address space for users' plug-in adapter cards is the address space C0000 to EFFFFH, since F0000 to FFFFF is used by BIOS. See Figure 25-1.

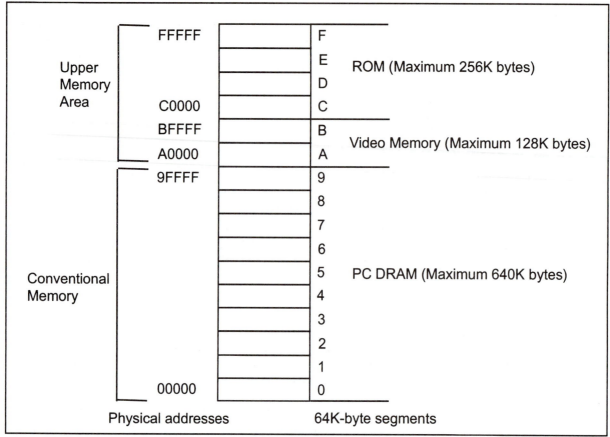

Figure 25-1. Memory Addressing of the 80x86 1 Megabyte

IBM standard using ROM space in the upper memory area

Assume that we are designing an adapter card for data acquisition to be plugged into a PC. If this adapter card has ROM memory on it, and the ROM memory is being accessed by the 80x86 CPU, we must follow two guidelines.

1. We must know what portion of the PC's ROM address space of C0000 - EFFFFH is unused and map ROM only into the unused space.
2. In deciding on the contents of ROM, we must follow IBM's recommended specification. According to IBM, the first byte of ROM must be 55H, the second byte must be AAH, and the third byte must indicate the total number of 512-byte blocks in the ROM (length/512). Finally, the fourth byte must contain executable code (an opcode).

Since DOS 5 and many memory management packages such as QEMM (from Quarterdeck) put into use the unused address space of the upper memory area, it is important to follow these specifications. Following this specification makes the job of DOS and other memory management packages such as QEMM much easier in locating the unused memory addresses of the upper memory area. The above discussion is summarized as shown in Table 25-1. ROM must be located in the unused address space of C0000 - FFFFFH.

Table 25-1: IBM Specifications for Adapter Card ROM

Byte	Contents
Byte 0	55H (this is the first byte of ROM).
Byte 1	AAH (this is the second byte of ROM).
Byte 2	Length indicator, represents the number (H) of 512-byte blocks in ROM (length/512).
Byte 3	The first opcode of the procedure contained in the ROM.

```
┌─────────────────────────────────────────────────────────────────────────────┐
│ Example 25-1                                                                   │
├─────────────────────────────────────────────────────────────────────────────┤
│ Using the dump command of DEBUG, dump the contents of the segment C000 to     │
│ examine the VGA adapter ROM contents. In many 386/486/Pentium systems, the    │
│ address is C000:0000H. If the area you dump contains all FFs, this means that │
│ this portion of ROM address space is unused.                                  │
│                                                                                │
│ Solution:                                                                      │
│                                                                                │
│ C>DEBUG                                                                        │
│ -D C000:0                                                                      │
│ C000:0000  55 AA 40 EB 7B 37 34 30-30 93 FF FF 83 00 98 24  U*@k{7400......$  │
│ C000:0010  CE 01 00 00 00 00 00 00-00 00 00 00 00 00 49 42  N.............IB  │
│ C000:0020  4D 00 00 00 00 00 00 00-00 00 00 00 00 00 00 00  M...............  │
│ C000:0030  20 37 36 31 32 39 35 35-32 30 00 00 00 00 00 00   761295520......  │
│ C000:0040  33 31 11 34 0C 20 B0 00-00 00 31 00 01 04 00 00  31.4. 0...1.....  │
│ C000:0050  31 39 39 31 2F 31 32 2F-31 33 20 31 36 3A 35 36  1991/12/13 16:56  │
│ C000:0060  00 00 00 00 E9 9A 70 00-E9 9A 70 00 E9 9A 70 00  ....i.p.i.p.i.p.  │
│ C000:0070  00 00 00 00 00 00 00 00-00 00 00 00 00 00 00 00  ................  │
│ -                                                                              │
│                                                                                │
│ Notice that the data corresponds to the specifications in Table 25-1. The     │
│ first byte is 55, the second is AA, and the third byte is 40H, the number of  │
│ 512-byte blocks in ROM. The fourth byte is an opcode for jump instructions.   │
└─────────────────────────────────────────────────────────────────────────────┘
```

Every IBM PC and compatible based on the 80x86, including the 8088/86 PC, uses some portion of the upper memory area space. The amount depends on the video board installed, the video mode the PC is working under, and the installed adapter cards.

The EMM386.EXE driver included in DOS 5 and 6 searches, finds, and keeps track of unused sections of the upper memory area. Since the unused address space of the upper memory area is not contiguous, they are categorized as blocks referred to as *upper memory blocks* (UMBs). Starting with DOS 5, UMBs are put to use. For example, TSR and device drivers could be moved from conventional memory to the upper memory using EMM386.EXE, thereby making more conventional memory available to the user. Again, it must be noted that as EMM386.EXE finds unused memory space in the memory address range of A0000 - FFFFF, it calls them upper memory blocks since they are not contiguous. See Figure 25-2. In Section 25.2, we discuss in more detail the UMB concept and the way DOS 5 and later versions put it to use.

Expanded memory

As discussed earlier, IBM designers set aside only 640K for RAM, and the rest was set aside for video memory and ROM. In the early 1980s, many application programs, such as spreadsheets, wanted much more RAM. This led to the development of a memory concept called *expanded memory*. In 1985, Lotus Corporation, the maker of the Lotus 1-2-3 spreadsheet, joined forces with Intel to solve the problem of memory limitation due to the 640K barrier. They proposed a set of specifications called EMS (expanded memory specifications) to standardize different attempts being made by different companies to use more than 640K of RAM. Later, Microsoft joined the group and the specifications came to be known as LIM (Lotus/Intel/Microsoft) EMS. The support of Microsoft was essential since its cooperation in implementing EMS specifications in future versions of DOS would make the success of expanded memory a certainty. In expanded memory, a method called *bank switching* is employed to allow the microprocessor to access information beyond the 640K barrier.

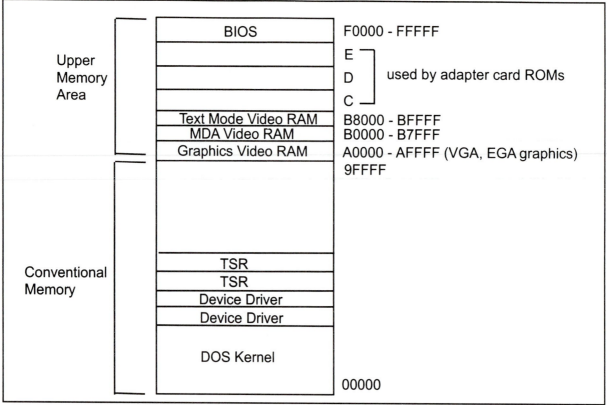

Figure 25-2. 1M Memory Map

The LIM expanded memory specification (EMS version 3.0) proposed using 64K of contiguous memory in the upper memory area as a window to 8 megabytes of expanded memory. This 64K contiguous memory space was divided into four sections called *page frames*, where each page frame accessed only 16K. The 16K address for the first page must start at a 16K boundary such as 0000, 4000, 8000, C000, and so on. However, since this must be 64K of contiguous memory, many expanded memory boards used addresses D0000 - DFFFF for the four page frames of expanded memory. This 64K bytes plays the role of a window to a larger memory. The size of this larger memory was first decided to be a maximum of 8 megabytes and was used for storing data only. No code was allowed to be stored in the expanded memory under EMS version 3.0. The 8M could be installed with four expansion boards, each handling 2M (each board consists of 128 pages, each page was 16K bytes). Starting with EMS version 4.0, which included many suggestions and improvements from AST Corporation and others, the maximum expanded memory was increased from 8M to 32M. The 16K pages may be located anywhere in the unused area above 640K, so there was no longer a need for 64K bytes of contiguous unused space to implement expanded memory. This made the job of adapter card designers much easier since the installation of these boards resulted in the fragmentation of the memory space between A0000 and EFFFFH. With EMS version 4, any time there is 16K of free space anywhere between A0000 and FFFFFH, it will use it as a window to expanded memory to as high as 32 megabytes. The following is a summary of the major features of EMS version 4.

1. EMS V4 allowed expanded memory to contain both code and data. This is in contrast to V3, in which expanded memory could store data only. In V3 code was kept in conventional memory and data was stored in expanded memory. This limitation was removed in EMS V4.

2. In EMS version 4, the limitation of 4 pages of 16K was removed; therefore, it can handle as many pages as it can find space for in the upper memory area as long as they fall on a 16K boundary.

3. In EMS version 4, expanded memory can be as high as 32 megabytes.

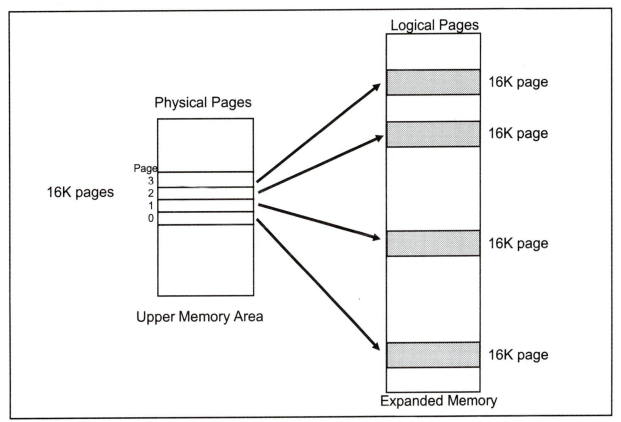

Figure 25-3. Expanded Memory

Regarding the use of expanded memory, two points must be emphasized. First, for an application program to use expanded memory, it must be equipped specifically to do so. This kind of software is called *EMS aware*. EMS aware application software requests to use expanded memory are handled by a piece of software called an *expanded memory manager* (EMM). When you buy the expanded memory board it comes with the EMM manager. In that case, you must install the expanded memory manager (EMM) software that comes with that expanded memory board and not some other EMM manager by some other expanded memory board maker. However, many of today's 386/486/Pentium-based PCs, which come with many megabytes of memory already installed on the motherboard, use EMM386.EXE, which is provided with DOS to convert installed memory into expanded memory. In such cases there is no need to buy expanded memory boards. Note that if the application software is not equipped to use expanded memory, expanded memory cannot do anything for that package even though you have expanded memory in your PC and the EMM (expanded memory manager) software is installed.

The second point to be emphasized about using expanded memory is that the bank switching scheme is totally different from DMA (discussed in Chapter 15). While in DMA the information is physically copied to RAM, bank switching is like a window to the information stored in EMS memory boards and there is no physical transfer of data from the expanded memory to the memory pages situated in the upper memory area (A0000 to F0000 addresses). This scheme of bank switching uses the memory locations accessible by the addresses of the microprocessor as a window to memory locations beyond the reach of CPU's physical addresses. This scheme was common in many computers based on the 8085, Z80, and 6502, which could only access 64K bytes due to their 16 address pins. In PC expanded memory terminology, the page frame addresses on 16K boundaries located in the address space of A0000H through FFFFFH are called *physical pages* and the 16K page frames located on the expanded memory board are referred to as *logical pages*. Every IBM PC and compatible based on the 80x86 is capable of having expanded memory. This includes the 8088- and 8086-based PCs. See Figure 25-3.

SECTION 25.1: 80x86 PC MEMORY TERMINOLOGY AND CONCEPTS

Example 25-2

A given LIM V4 expanded memory board has a total of 8M of memory. How many pages is that?

Solution:

In expanded memory, each page is 16K. Dividing the total 8M by 16K gives 512 pages.
8M/16K = 512

Extended memory

With the introduction of the 80286-based IBM PC AT, two more terms were added to the PC memory vocabulary: *high memory area* and *extended memory*. The 80286 microprocessor has 24 address pins that allow access of up to 16M of memory, from 000000 to FFFFFFH. The 32 address pins of the 386/486/Pentium allow access of a maximum of 4 gigabytes of memory, from 00000000 to FFFFFFFFH. In all these processors, the microprocessor can access a maximum of 1M when the CPU is in real mode. To access memory beyond 100000H, the microprocessor must be switched to protected mode and that requires a highly complex operating system such as OS/2, Windows NT, or Unix. In PC memory terminology, memory space beyond 100000H (beyond 1M) is called *extended memory*. To standardize accessing memory beyond 1M, the *extended memory specification* (XMS) standard was developed in 1988 by Lotus, Intel, Microsoft, and AST. Microsoft's HIMEM.SYS included in DOS 5 (and later) is one example of memory manager software that conforms to the XMS standard. It is common in today's 386/486/Pentium PC to have 4M or more of RAM installed. In such cases, while the first 1M is used as conventional memory and the UMB, the rest can be configured either as extended memory or expanded memory or both, as we will see in the next section.

High memory area (HMA)

When 286, 386, 486, and Pentium processors are switched to protected mode, the entire address space supported by the CPU can be accessed. The 286 in protected mode can access 16M, from 000000 to FFFFFFH. For the 386/486/Pentium it is 4 gigabytes, from 00000000 to FFFFFFFFH. However, in real mode, all these processors provide only 1M of memory space. The exception to this rule is that there are 65,520 bytes of memory from 100000H to 10FFEFH which are accessible without the use of an operating system with protected mode capability. In other words, the CPU in real mode can access another 65,520 bytes of memory beyond the 1M address space. This memory address space is called the *high memory area* (HMA). As discussed in Chapter 10, the existence of the high memory area is due to an anomaly in the A20 address pin of the 286 processor. For the 8086/88, the physical address is generated by shifting left the CS register and adding IP. If there is a carry, it is dropped since there is no A20 pin in the 8086/88. In the 286/386/486/Pentium, the carry is forwarded to the A20 pin. This is in spite of the fact that the 286/386/486/Pentium is in real mode and the real mode uses only the A0 - A19 address pins to access the 1M from 00000 to FFFFFH. As we discussed in Chapter 10, A20 to A23 of the 286 is used only when the processor is in protected mode and the address spaces of 100000H to FFFFFFH are accessible when the CPU is working in protected mode. Accessing a portion of this address space while the CPU is in real mode was a dilemma for designers of the IBM PC AT. They decided to use a logic circuitry to capture the A20 bit. The logic circuitry is commonly referred to as the A20 gate and the software to control it is called the A20 driver. To maintain compatibility with the 286, all subsequent Intel processors, such as the 386, 486, and Pentium, continued to provide the A20 bit even when the processor is in real mode. The high memory area belongs to 80286-, 386-, 486- and Pentium-based PCs only, and no 8086/88 has HMA. See Figure 25-4.

Example 25-3

Which type of PC supports conventional memory, expanded memory, HMA, and extended memory?

Solution:

Any 80x86-based PC (8088/86, 286, 386, 486, Pentium) supports conventional and expanded memory; however, HMA and extended memory are supported by 286, 386, 486, and Pentium PCs only.

Example 25-4

What is the highest physical address accessible in real mode for the IBM PC and compatibles with the following processors?   (a) 8088/86   (b) 286, 386, 486, and Pentium

Solution:

To get the highest physical address, we must have CS =FFFF and IP =FFFFH. For (a) we have the highest address of FFFFF since it wraps around when FFFF0 is added to FFFF and the carry is dropped. For the processors in (b), the highest address in real mode is 10FFEFH:

$$
\begin{array}{r}
\text{FFFF0} \\
+ \quad \text{FFFF} \\
\hline
\text{10FFEF}
\end{array}
$$

since the carry is passed to the A20 address pin. Now assuming that the A20 gate is turned on, these 65,520 bytes beyond the 1M address of FFFFFH are available in real mode.

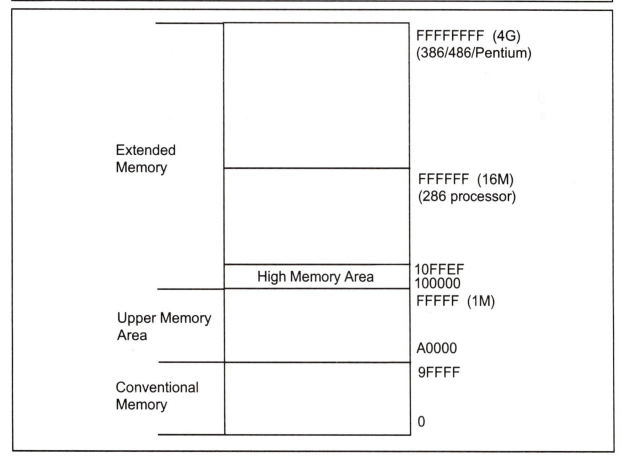

Figure 25-4. Extended Memory and High Memory Area

Shadow RAM

In high-speed 386, 486, and Pentium systems, for the CPU to access ROM requires two or three wait states since ROM has very long access time. Due to the fact that DRAM has a much shorter access time than ROM, the contents of ROMs in the upper memory area are mapped into DRAM locations of the same addresses as ROM using special mapping hardware and write-protected to prevent any data corruption. This DRAM holding the write-protected copy of the ROM is then called *shadow RAM*. From then on, any time the CPU asks for ROM BIOS service, or for that fact any service provided by ROM, the information is provided by the shadow RAM at much faster speed. In this way the CPU accesses the DRAM memory instead of the slow ROM for the same service. Although shadow RAM can be implemented in 8088/86/286-based PCs, the gain in performance is minimal. However, in 386-, 486-, and Pentium-based PCs in which applications are dominated by INT 10H graphics, the use of shadow RAM can boost performance dramatically, sometimes up to 40 - 50%. In some chip sets, shadow RAM is provided by the chip and the system's DRAM is not used.

DOS MEM command: mem [/program/debug/classify]

To examine the memory of the PC, DOS 5 provides the MEM command. The letters /p/d/c can be used for the switches program, debug, and classify. Only one switch can be used at a time. In other words, "mem /d/p" is not allowed. Switch "/p" displays the status of currently loaded programs. To see the display, type "mem /p | more".

In DOS 6, command "mem /p" displays the currently programs one page at a time and there is no need for the "| more". Notice that the display includes the physical address in hex where the programs are located in conventional memory, their size (in hex), and the name and type of program. At the end of the display, it also lists the total conventional memory installed, how much is available to application programs, (the largest executable program size), the total extended memory installed, and how much of it is allocated.

The status of currently loaded programs and device drivers is displayed by "mem /d". The status of currently loaded programs in conventional memory and upper memory area is displayed by "mem /c". DOS 6 has two new switches: "mem /m", which displays any memory space belonging to a given program, and "mem /f", which displays all free space in conventional and upper memory.

Example 25-5

Use the DOS MEM command to determine if your system has expanded and/or extended memory.

Solution: This shows that this computer has 1M of expanded memory and 7M of extended memory.

```
C:>mem

   655360 bytes total conventional memory
   655360 bytes available to MS-DOS
   574384 largest executable program size

  1048576 bytes total EMS memory
  1048576 bytes free EMS memory

  7340032 bytes total contiguous extended memory
        0 bytes available contiguous extended memory
  1048576 bytes available XMS memory
          MS-DOS resident in High Memory Area
```

Review Questions

1. True or false. All 80x86 PCs, including the 8088/86, have conventional memory.
2. Why is conventional memory limited to 640K?
3. What address space belongs to conventional memory?
4. For a PC with 512K of DRAM memory installed, the conventional memory is _____ (640K, 512K, 128K).
5. In Question 4, at what address range is memory located?
6. What address space belongs to video memory?
7. Any adapter card ROM must use the unused portion of the address space _____ to _____.
8. What address ranges are referred to as the upper memory area?
9. What address ranges are set aside for ROM and how many kilobytes is that?
10. True or false. All 80x86 PCs, including the 8088/86, have upper memory area.
11. Which of these physical addresses are located on a 16K boundary?
    (a) E3400H   (b) DC000H   (c) E9000H   (d) D0000H
12. Can LIM version 4 of EMS use addresses A0000H - AFFFFH for the page frame?
13. True or false. Any 80x86 PC, including an 8088/86, can have expanded memory.
14. True or false. Any 80x86 PC, including an 8088/86, can have extended memory.
15. The high memory area is a portion of _____ (extended, expanded) memory.
16. Shadow RAM speeds up the CPU's access of _____.
17. What is the UMB, and why is it called that?
18. What is the last address of the HMA?

SECTION 25.2: DOS MEMORY MANAGEMENT AND LOADING HIGH

For MS DOS-based software applications, every byte of conventional memory is precious. The dilemma is that the 640K conventional memory must be shared among DOS, TSRs, device drivers, and software applications. The fact that DOS first allocates memory for itself before it gives memory to other programs may seem selfish, but given the fact that no application software can run without DOS, this must be tolerated. However, starting with DOS 5, the operating system allows a good portion of itself to be moved to the high memory area (HMA), which releases more conventional memory for application software. It also allows the system to move TSR programs and device drivers to the upper memory area, releasing even more conventional memory. In this section, the process of loading high is discussed and we will show how DOS makes more conventional memory available by moving DOS, TSRs, and device drivers to the HMA or UMB.

Loading high into HMA

The resident portion of DOS normally takes 65K of conventional memory, depending on the DOS version. The load high option, introduced first with DOS 5, reduces the amount of conventional memory used by DOS to less than 20K by moving a majority of the resident portion of DOS into the high memory area. To do that, the HIMEM.SYS driver that comes with DOS 5 (and 6) must first be installed. Assuming that HIMEM.SYS is in the DOS directory, the following statement in the CONFIG.SYS file will install the HIMEM.SYS driver:

DEVICE=C:\DOS\HIMEM.SYS

To move DOS to the high memory area and free the conventional memory used by DOS, the following statement is placed in the CONFIG.SYS file:

DOS=HIGH

The existence of the above two statements in CONFIG.SYS increases the largest executable program size substantially, as shown in Example 25-6.

Example 25-6

Using the "MEM /D" command, show the difference in memory when DOS is loaded normally versus when it is loaded high with the DOS=HIGH command in CONFIG.SYS.

Solution:

For our system, when DOS was not high, "MEM /D" produced the following:

```
655360 bytes total conventional memory
655360 bytes available to MS-DOS
582320 largest executable program size

7340032 bytes total contiguous extended memory
7340032 bytes available contiguous extended memory
```

When the "DOS=HIGH" statement was added to CONFIG.SYS, "MEM /D" produced

```
655360 bytes total conventional memory
655360 bytes available to MS-DOS
634000 largest executable program size

7340032 bytes total contiguous extended memory
      0 bytes available contiguous extended memory
7274496 bytes available XMS memory
        MS-DOS resident in High Memory Area
```

Notice that moving DOS high makes the largest executable program size larger because portions of the system have been moved out of the 640K conventional memory.

While DOS 5 and subsequent versions allow a good portion of DOS to be moved into high memory, it still keeps almost a fourth of it in conventional memory. In other words, a portion of DOS, however small, will stay in conventional memory, no matter what we do. As a result of moving DOS into the HMA, the size of the largest executable program will increase by about 40K.

Finding holes in the upper memory area

Since the size of the HMA is small and loading DOS high will use most of it, DOS allows moving TSR programs and device drivers into the upper memory area instead. To do that, it must be determined which part of the UMB is available. To this end, the upper memory area is searched for unused memory space, and each unused section of the upper memory area is marked as an upper memory block (UMB). It is the job of the EMM386.EXE driver that comes with DOS 5 and later versions to find and keep track of UMBs. More important, EMM386.EXE allows a user to convert extended memory into expanded memory, as we will show below. The major feature of EMM386.EXE is that it also allows a user to move TSRs and device drivers into the upper memory area. The following two points about EMM386.EXE must be emphasized.

1. EMM386.EXE works only on 386-, 486-, and Pentium-based PCs, not on 8088/86-, or 286-based PCs. To find and utilize the UMB for the 8088/86/286 PC, third-party memory managers such as QRAM from Quarterdeck Corp, and QEMM-386 can be used.
2. HIMEM.SYS must be installed first in order for EMM386.EXE to be able to find the UMBs. This is due to the fact that the HIMEM.SYS belongs to Microsoft's XMS driver library and sets the standard for accessing memory beyond 1M, but it is also used in accessing memory beyond 640K.

From the above discussion we conclude that in order to use the UMB for 386/486/Pentium-based PCs, both HIMEM.SYS and EMM386.EXE must be present in the CONFIG.SYS, as shown next.

DEVICE=C:\DOS\HIMEM.SYS
DEVICE=C:\DOS\EMM386.EXE

The above commands in CONFIG.SYS will find and mark the UMB to be used for moving TSRs and device drivers into them. Before we discuss how to load TSRs and device drivers into the UMB, we must examine the options and switches available to EMM386.EXE since these switches affect loading high device drivers and TSRs. In addition, the size and the amount of UMB available can vary depending on how the switches for the EMM386.EXE arc set. See Figure 25-5.

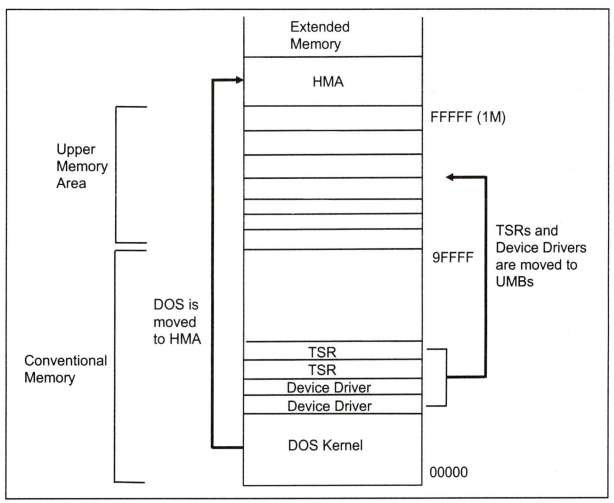

Figure 25-5. Moving Software High

EMM386.EXE options and switches

The EMM386.EXE command has several options and switches for configuring memory. The following are some of the options and switches for EMM386.EXE. A complete list is found in the MS DOS 5 (or 6) user's guide and reference book that comes with DOS.

DEVICE=[DRIVE:][PATH]EMM386.EXE [MEMORY][Mx or FRAME=address] [Pn=address][X=mmmm-nnnn][I=mmmm-nnnn] [RAM or NOEM][L=minXMS]

[MEMORY] option of EMM386.EXE

The memory parameter is used to specify the amount of expanded memory. The number given for the amount of desired expanded memory is in kilobytes. The following command tells EMM386.EXE to provide 3M (3072K = 3M) of expanded memory.

DEVICE=C:\DOS\EMM386.EXE 3072

Of course, the system must have at least 3M of extended memory in order for EMM386.EXE to simulate expanded memory out of extended memory. With DOS, if no number is specified for the memory option, only 256K of expanded memory is provided (for DOS 6, it is much more than that).

Example 25-7

In a given 486 with 8M of memory on the motherboard, we are using a software package that requires the availability of 4M of expanded memory. Show the EMM386.EXE statement in CONFIG.SYS for such a configuration.

Solution:

DEVICE=C:\DOS\EMM386.EXE 4096

This is exactly like having a 4M expanded memory board installed on the system. In reality, it is emulating expanded memory using extended memory.

[Mx or FRAME=address or Pmmmm] option of EMM386.EXE

Prior to EMS version 4, specific addresses, from D0000 to DFFFFH, had to be used for the expanded memory page frame. EMS V4 removed that limitation. EMM386.EXE of DOS 5 incorporates this feature by allowing the address to be set to other locations, using the Mx or FRAME parameter. The three options for specifying the address of the page frame for expanded memory are explained next.

[Mx], where m is a number between 1 and 14. The segment address of the page frame is specified according to the following table:

1=C000H	8=DC00H
2=C400H	9=E000H
3=C800H	10=8000H
4=CC00H	11=8400H
5=D000H	12=8800H
6=D400H	13=8C00H
7=D800H	14=9000H

Option 10 to 14 uses addresses 80000H - 9FFFFH for expanded memory. However, this area belongs to conventional memory. How can EMM386.EXE convert it to expanded memory? Assume that you have only 512K of memory installed on an older 386, and you buy an 8M expanded memory card. Using this option, you can specify that addresses 80000H to 9FFFFH are to be used for expanded memory. In such cases, DOS assumes that you wish to use conventional memory. This can work only if your expanded memory board and EMS driver are V4.

The "Frame=address" option allows the user to specify the base address of the expanded memory page frame directly. Although by default, EMM386.EXE uses D0000 to DFFFFH for the expanded memory frame, it can be forced to change it to some other address. For example, if D0000 - DFFFFH is used by an adapter card such as a data acquisition board and we know that E0000 - EFFFFH is available, we can assigned it as the expanded memory page frame. This is shown in the following command:

DEVICE=C:\DOS\EMM386.EXE 4096 FRAME=E000

Notice in the command above that when providing the frame address it must be a segment address and not a physical address. In cases such as IBM PS models where segment E000 is used for BIOS, if this statement is placed in CONFIG.SYS, EMM386.EXE will warn you of the conflict. The valid values for the address of the FRAME switch are 8000H - 9000H and C000H - E000H in increments of 400H. Notice that 400H is a segment address, which by shifting left one hex digit becomes 4000H, a 16K boundary as required by the EMS standard. Note that of the two options Mx or FRAME, only one of them can be used in a given CONFIG.SYS file.

[Pn=address] option of EMM386.EXE

Due to fragmentation of the upper memory area, sometimes it is difficult to find 64K contiguous bytes, such as D0000 - DFFFF or E0000 - EFFFF for the page frame. The "Pn=address" switch allows a user to include as many segment addresses as desired if they are located on a 16K boundary. In this switch, n is the page number, which can take values from 0 to 256, and the segment addresses are values that can range from 8000 to 9C00 and C000 to EC00. For example, the following two statements have the same effect; one uses the FRAME switch and the other one uses the Pn=address switch.

DEVICE=C:\DOS\EMM386.EXE 4096 FRAME=E000
DEVICE=C:\DOS\EMM386.EXE 4096 P0=E000 P1=E400 P2=E800 P3=EC00

These commands work with EMS V 3.2, since page frames for expanded memory must be contiguous; however, in EMS V 4.0, they need not be contiguous.

[I=mmmm-nnnn] option of EMM386.EXE

Due to the fact that much of the memory space in the upper memory area is not utilized, one can use the I switch to specify which address ranges should be included when EMM386.EXE determines the UMB. This is especially true for video adapter boards and applications. For example, in the case where the CGA or MDA video boards are installed, address space A0000 - AFFFFH is unused. Why not inform EMM386.EXE to utilize A0000 - AFFFFF as the upper memory block, since by default EMM386.EXE avoids A0000 - BFFFFH, assuming that it belongs to the video boards? EMM386.EXE, by default, includes only segments C000 and D000 in its search for UMBs and avoids the rest of the address space, assuming it is used. For example, many IBM PS and compatible systems use the address range E0000 - EFFFFH for the extended BIOS. If a given system is not using the E000 segment, we can inform EMM386.EXE to include it in its pool of UMBs. Valid addresses for the I switch are A000H to FFFFH and are rounded down to the nearest 4K boundary.

Example 25-8

An early model of a 386-based PC clone is connected to a CGA video board. Show the EMM386.EXE setting to include segments E000, A000, and B000 in the UMB. Notice that the lower part of B000 is not used by CGA either.

Solution:

DEVICE=C:\DOS\EMM386.EXE I=E000-EFFF I=A000-AFFF I=B000-B7FF

This command will sets aside blocks starting at E000, A000, and B000 -B7FF for the UMB.

[X=mmmm-nnnn] option of EMM386.EXE

This option is used to inform EMM386.EXE to exclude a section of the upper memory area when searching for UMBs. The valid addresses for the X switch are A000H - FFFFH and are rounded down to the nearest 4 K boundary. In cases where the address ranges overlap for I and X, X takes precedence over I. Look at the following command:

DEVICE=C:\DOS\EMM386.EXE I=E000-EFFF I=A000-BFFF X=B800-BFFF

Notice in the above command that although the 128K video RAM in the address range A000 - BFFF is included in UMB, X takes precedence by excluding the B800 - BFFF section since that is the area used by CGA. Note that EMM386.EXE rounds down to the nearest 4K boundary for both the I and X switches. Look at the following command:

DEVICE=C:\DOS\EMM386.EXE X=E480-EFFF

In the above case, it is rounded down to E400, the nearest 4K boundary. To get the segment address of the 4K boundary, simply drop the lower two digits. For example, segment C938 becomes C900 and DA93 becomes DA00. This implies that the smallest block that we can include or exclude is 4K bytes.

[RAM] option of EMM386.EXE

This allows EMM386.EXE to provide both expanded memory and upper memory.

[NOEM] option of EMM386.EXE

This allows EMM386.EXE to provide upper memory only and no expanded memory. If neither RAM nor NOEM is specified, EMM386.EXE provides the expanded memory only and prevents any use of the upper memory block. In other words, to have the upper memory area (where one can move the TSR and device drivers) and also to have expanded memory, the RAM option must be used. In the case where the NOEM option is selected, the upper memory area can be used for moving high TSRs and device drivers, but no expanded memory can be used.

Loading high TSR and device driver into upper memory area

The fact that about 48K of the resident portion of DOS is moved to the high memory area leaves only a small portion of the HMA unused. As we discussed earlier, after loading DOS high, 48K of the 64K (almost 64K since it is 65,520 bytes) of the HMA is used, leaving only about 19K left unused. This is not enough for many TSRs and device drivers. In addition, depending on the way the BUFFER and FILES parameters are set, the entire HMA could be used by DOS, leaving nothing for any TSR or device driver. This fact led to the idea of moving TSRs and device drivers to the upper memory area. To be able to load TSRs and device drivers into the UMB, we must have the statement "DOS=UMB" in the CONFIG.SYS file. The following are the steps in locating and using the UMB in 386/486/Pentium-based PCs only.

1. Install HIMEM.SYS using the DEVICE command in CONFIG.SYS.
   DEVICE=C:\DOS\HIMEM.SYS
2. Install EMM386.EXE using the DEVICE command in CONFIG.SYS.
   DEVICE=C:\DOS\EMM386.EXE NOEMS
3. Place DOS=UMB in CONFIG.SYS.
   DOS=UMB
If you are also moving DOS to the HMA, step 3 would be as follows:
   DOS=HIGH, UMB

After the above three steps the CONFIG.SYS looks as follows:
```
DEVICE=C:\DOS\HIMEM.SYS
DEVICE=C:\DOS\EMM386.EXE NOEM
DOS=HIGH, UMB
```

Now we are ready to move TSRs and device drivers to the UMB using the LOADHIGH and DEVICEHIGH commands. This is done in steps 4 and 5.

4. Place the LOADHIGH command in the CONFIG.SYS file to move any TSR to the UMB. For example, to move the DOSKEY TSR into the upper memory area, we have the following:

```
LOADHIGH C:\DOS\DOSKEY.COM
```

Therefore, CONFIG.SYS will look like this:

```
DEVICE=C:\DOS\HIMEM.SYS
DEVICE=C:\DOS\EMM386.EXE NOEM
DOS=HIGH, UMB
LOADHIGH C:\DOS\DOSKEY.COM
```

DOSKEY is a TSR provided with DOS 5 that, among other things, allows a user to view, edit, and reuse the commands that were entered previously. If we do not move the TSR DOSKEY to the upper memory area, it takes about 4K of conventional memory.

5. Use the DEVICEHIGH command in the CONFIG.SYS file to move any device driver to the UMB. Notice that this is DEVICEHIGH and not DEVICE. For example, to move the device driver ANSI.SYS into the upper memory area, we have the following statement in CONFIG.SYS:

```
DEVICEHIGH=C:\DOS\ANSI.SYS
```

Therefore, the above five steps of our CONFIG.SYS will look like this:

```
DEVICE=C:\DOS\HIMEM.SYS
DEVICE=C:\DOS\EMM386.EXE NOEM
DOS=HIGH, UMB
LOADHIGH C:\DOS\DOSKEY.COM
DEVICEHIGH=C:\DOS\ANSI.SYS
```

It must be noted that steps 4 or 5 must be placed after steps 1 - 3. "LH" is short for the LOADHIGH command.

According to the MS DOS reference manual, the ANSI.SYS device driver contains functions that control graphics, cursor movement, and key reassignment. In many software packages, the presence of ANSI.SYS in CONFIG.SYS is required. If we do not move ANSI.SYS into the upper memory area. it takes about 4K of conventional memory.

In all the examples so far, we used the NOEM option, which provided the UMB to move the device drivers and TSRs into but did not provide any expanded memory. Next we see how to allocate expanded memory and also have the UMB.

Emulating expanded memory and using UMB in 386/486/Pentium PC

In many of today's 386/486/Pentium PCs, the motherboard already has 8M or more of memory installed. This memory is set up for extended memory, which is to be used by advanced operating systems such as OS/2 and Windows NT. This extended memory is also used by much DOS-based application software that is equipped to use the extended memory. However, some application software requires expanded memory. Does this mean that we must buy an EMS board and install it even though we already have 8M of memory installed on the motherboard? The answer is no. In the 386/486/Pentium, extended memory can be converted to expanded memory. This is often referred to as *emulating* or *simulating expanded memory*. In DOS 5 we can set the EMM386.EXE switches in such a way that we can emulate (simulate) expanded memory and set up UMBs to move the device drivers and TSRs into. The following CONFIG.SYS emulates (simulates) 4M of expanded memory out of the extended memory while it also provides the UMB to move the TSR and device driver into.

```
DEVICE=C:\DOS\HIMEM.SYS
DEVICE=C:\DOS\EMM386.EXE RAM 4096
DOS=HIGH, UMB
LOADHIGH C:\DOS\DOSKEY.COM
DEVICEHIGH=C:\DOS\ANSI.SYS
```

Notice that this is similar to the previous CONFIG.SYS except that "NOEM" in the "DEVICE=C:\DOS\EMM386.EXE....." statement has been replaced with "RAM 4096" to provide both expanded memory and UMB.

How expanded memory is accesssed

Assume that we have an EMS aware software package and we are simulating 5M of expanded memory on a given PC. If we have 8 physical frames in the upper memory area, how is the 5M of expanded memory seen by the 8 page frames? Although in this system, the 5M of expanded memory provides a total of 320 (5M/16K = 320) logical page frames, only 8 of these are visible at a time. As the EMS aware application software requests expanded memory, the EMS driver allocates to it 1 page (16K) at a time. This process continues until all 8 pages are allocated. The request for the ninth page is responded to by the EMS driver by moving one of the previously assigned physical page frames to this ninth one. As requests for more memory come in, a new logical page frame becomes visible to one of the 8 page frames. In other words, software may use 30 pages (30 × 16K = 480K) of expanded memory, but only 8 of these are visible to the CPU at a time, since only 8 physical pages in the upper memory area are used for the page frames. This is one reason that EMS is slightly slower than conventional memory; that is, it must move the page frame windows.

Review Questions

1. If a given PC has only 640K of conventional RAM, can we load high?
2. True or false. When we load DOS high, the entire DOS is moved into the HMA.
3. Which piece of software is responsible for moving DOS to the HMA?
4. True or false. To load high, CONFIG.SYS must contain DOS = HIGH.
5. Which piece of software is responsible to find and allocate the UMB?
6. In Question 5, to run the software it must be based on what microprocessor?
7. The statement "DEVICE=C:\DOS\EMM386.EXE 2048" allocates how much expanded memory?
8. In Question 7, does the statement permit use of the UMB?
9. Show how C8000H - CFFFFH is excluded from the UMB.
10. By default, EMM386.EXE uses address range _____ to _____ for the expanded memory page frame.
11. If an address range is marked by both the I and X switches of EMM386.EXE, which one takes effect?
12. True or false. The NOEM switch in the EMM386.EXE command allows the use of UMB only.
13. True or false. The RAM switch in EMM386.EXE allows the use of UMB only.
14. What command is used to move a device driver into the UMB?
15. What command is used to move a TSR into the UMB?

SUMMARY

This chapter outlined the memory allocation of the IBM PC. The 1M of memory is composed of 640K (00000 to 9FFFF) of conventional DRAM memory used for running DOS and application software, and 384K of upper memory area, 256K of which is used for ROM (C0000 to FFFFF) and 128K of which is used for video memory (A0000 to BFFFF). The upper memory area is organized into UMBs (upper memory blocks). Expanded memory was developed to expand RAM beyond the 640K barrier that exists in the PC. Expanded memory is implemented by using 64K page frames in upper memory as windows to expanded memory boards. Modern PCs can contain up to 32M of expanded memory. Extended memory is memory space beyond 1M, and was made possible with the development of the 286. Most extended memory can be accessed only when the microprocessor is in protected mode. A small portion of extended memory, called the high memory area (HMA), can be accessed in real mode. Loading DOS into the HMA frees conventional memory. Shadow RAM is used to improve system performance by copying frequently used portions of ROM into faster RAM of the same address.

The DOS HIMEM.SYS driver helps the user manage extended memory. The "DOS=HIGH" command can be placed in CONFIG.SYS to load DOS into the HMA in order to free conventional memory. The EMM386.EXE program can be used to simulate expanded memory with extended memory and to access UMBs for storing DOS, device drivers, and TSRs.

PROBLEMS

SECTION 25.1: 80x86 PC MEMORY TERMINOLOGY AND CONCEPTS

1. What address ranges are called conventional memory?
2. The PC's 1M memory space consists of _____ segments of 64K size. Number them in hex.
3. In Problem 2, indicate the segments belonging to conventional memory.
4. Give the total number of bytes and physical addresses used for the interrupt vector table, BIOS data area, and DOS data area.
5. Why is conventional memory limited to 640K?
6. True or false. Every 80x86 PC including the 8088/86 has conventional memory.
7. What address ranges are called the upper memory area?
8. How many 64K segments does the upper memory area contain? Give the segment numbers in hex (see Problem 2).
9. In adding an adapter card, we use the address space of _____ to _____ for ROM.
10. Of the upper memory area, what addresses are assigned to video memory, and what addresses are assigned to ROM?
11. What segment addresses are set aside for video RAM?
12. In the PC, what addresses are assigned to BIOS? In an IBM PS?
13. Give the IBM specifications for the first 3 bytes of an adapter ROM.
14. Why is it important to follow the IBM specifications for Problem 13?
15. Why are spaces in the upper memory area called upper memory blocks?
16. What is LIM EMS?
17. What is the maximum expanded memory in LIM version 3? LIM version 4?
18. What is the size of a page frame in LIM versions 3 and 4?
19. What addresses are used for the page frame in LIM versions 3 and 4?
20. For an address to be used for the page frame it must be on a 16K boundary. Which of the following can be used for a page frame?
    (a) D6000H (b) DC000H (c) DE000H (d) E0000H
21. What does it mean when it is said that application software is EMS aware?
22. PCs based on which microprocessors can have extended memory? Answer the same question for the high memory area.
23. How many bytes is the HMA, and what are its addresses?

24. True or false. For software to access HMA, it must use protected mode.
25. If CS = FF56H and IP = 34C6, find the physical address in each of Intel's CPUs from the 8086 to the Pentium.
26. What is the A20 gate, and what is the A20 gate handler?
27. Run command "MEM /P" and analyze the result.
28. What is shadow RAM?

SECTION 25.2: DOS MEMORY MANAGEMENT AND LOADING HIGH

29. True or false. To load high in DOS, we must have the HIMEM.SYS driver.
30. Show the statements in the CONFIG.SYS file to move DOS high.
31. In moving DOS high, a portion of DOS is moved into the _____ (HMA, UMB).
32. Which software is responsible for finding the UMBs, and which processor does this software run on?
33. Show the statements in the CONFIG.SYS file to find the UMBs.
34. For a 386 PC with 6M on the motherboard, what is the maximum extended memory?
35. What does the term *simulating expanded memory* mean?
36. In Problem 34, show the EMM386.EXE switches that simulate 5M of expanded memory and UMB.
37. In a given 486 PC that has 8M of memory on the motherboard, do we need to buy EMS memory cards to have expanded memory? Why or why not?
38. Show the EMM386.EXE frame switch to set the page frame to E000H.
39. Show the EMM386.EXE Pn switch to set the page frame to C800H.
40. Show the I and X switches to include the B000 segment, but exclude the portion belonging to the MDA card.
41. In statement DEVICE=C\DOS\EMM386, where I = C000 - CFFF and X = C800 - CFFF, which section will be used for the UMB?
42. What is the difference between the RAM and NOEM switches for the UMB?
43. Show the EMM386 command to have both EMS and UMB.
44. What is the difference between loading high DOS and a device driver?
45. Show the CONFIG.SYS command to load high SMARTDRV.SYS with no expanded memory.
46. Repeat Problem 45 with a switch to provide 2M of EMS, and in addition, DOSKEY should be loaded into upper memory and DOS should be loaded high.

ANSWERS TO REVIEW QUESTIONS

SECTION 25.1: 80x86 PC MEMORY TERMINOLOGY AND CONCEPTS
1. true
2. because the original PC memory map set aside only 640K for RAM, with the rest of 1M set aside for video RAM and ROM

3. 00000 - 9FFFFH	4. 512K	5. 00000 - 7FFFFH
6. A0000 - BFFFFH	7. C0000H - EFFFFH	8. A0000H - FFFFFH
9. C0000H - FFFFFH, 256K	10. true	11. b and d
12. yes	13. true	14. false; only 80286 and higher
15. extended	16. ROM	

17. Upper memory block; it is unused memory space in the address space of A0000H - FFFFFH. It is called block since they are not contiguous.
18. 10FFEFH

SECTION 25.2: DOS MEMORY MANAGEMENT AND LOADING HIGH
1. No, it must have some extended memory as well.

2. false; only about 2/3 of it	3. HIMEM.SYS	4. true
5. EMM386.EXE	6. 386 or higher processors	7. 2M
8. no, only the expanded memory	9. X=C800 - CFFF	10. D0000 - DFFFFH
11. X	12. true	

13. False; it allows the expanded memory and UMB both.

14. DEVICEHIGH	15. LOADHIGH

CHAPTER 26

IC TECHNOLOGY AND SYSTEM DESIGN CONSIDERATIONS

OBJECTIVES

Upon completion of this chapter, you will be able to:

» **Contrast and compare MOS and bipolar transistors**

» **Evaluate logic families according to speed, power dissipation, noise immunity, input/output interface compatibility, and cost**

» **Trace the evolution of Intel 80x86 microprocessors in terms of IC technology**

» **Define** *IC fan-out* **and describe why connecting an output to too many inputs can cause false logic**

» **Describe** *capacitance derating* **and its effect on system design and the use of buffers to decrease its effect**

» **Discuss power in system design, including static and dynamic currents, power dissipation, and sleep mode for peripherals**

» **Define** *ground bounce* **and** *VCC bounce* **and describe methods that designers use to avoid false signal generation that they may cause**

» **Define** *crosstalk* **and describe ways to avoid crosstalk in system design**

» **Define** *transmission line ringing* **and discuss ways to reduce its effect**

» **Define** *hard* **and** *soft errors* **in DRAM**

» **Define** *measures of DRAM reliability*: **FIT and MTBF**

» **Contrast and compare parity checking with EDC**

The invention of the transistor and the subsequent advent of integrated circuit (IC) technology is believed by many to be the start of the second industrial revolution. In this chapter we provide an overview of IC technology and interfacing. In addition, we look at the computer system as a whole and examine some general considerations in system design. In Section 26.1 we provide an overview of IC technology. IC interfacing and system design considerations are examined in Section 26.2. In Section 26.3 we discuss how high-performance systems maintain data integrity with error detection and correction (EDC) circuitry.

SECTION 26.1: OVERVIEW OF IC TECHNOLOGY

In this section we examine IC technology and discuss recent developments in advanced logic families. Since this is an overview, it is assumed that you already have had an introduction to logic families on a level presented in many basic digital books, such as *Digital Systems* by R. Tocci.

The transistor was invented in 1947 by three scientists at Bell Laboratory. In the 1950s, transistors replaced vacuum tubes in many electronics systems, including computers. It was not until in 1959 that the first integrated circuit was successfully fabricated and tested by Jack Kilby of Texas Instruments. Prior to the invention of the IC, the use of transistors, along with other discrete components such as capacitors and resistors, was common in computer design. Early transistors were made of germanium, which was later abandoned in favor of silicon. This was due to the fact that the slightest rise in temperature resulted in massive current flows in germanium-based transistors. In semiconductor terms, it is because the band gap of germanium is much smaller than silicon, resulting in a massive flow of electrons from the valence band to the conduction band when the temperature rises even slightly. By the late 1960s and early 1970s, the use of the silicon-based IC was widespread in mainframe and minicomputers. Transistors and ICs were based on P-type materials. Due to the fact that the speed of electrons is much higher (about two and a half times) than the speed of the hole, N-type devices replaced P-type devices. By the mid-1970s, NPN and NMOS transistors had replaced the slower PNP and PMOS transistors in every sector of the electronics industry, including in the design of microprocessors and computers. Since the early 1980s, CMOS (complementary MOS) has become the dominant method of IC design. Next we provide an overview of differences between MOS and bipolar transistors.

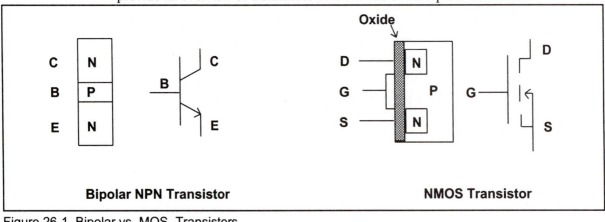

Figure 26-1. Bipolar vs. MOS Transistors

MOS vs. bipolar transistors

There are two type of transistors: bipolar and MOS (metal-oxide semiconductor). Both have three leads. In bipolar transistors, the three leads are referred to as the emitter, base, and collector, while in MOS transistors they are named source, gate, and drain. In bipolar, the carrier flows from the emitter to the collector and the base is used as a flow controller. In MOS, the carrier flows from the source to the drain and the gate is used as a flow controller. In NPN-type bipolar transistors, the electron carrier leaving the emitter must overcome two voltage barriers before it reaches the collector (see Figure 26-1). One is the N-P junction of the emitter-base and the other is the P-N junction of the base-collector. The voltage barrier of the base-collector is the most difficult one for the electron to overcome (since it is reversed biased) and it causes the most power dissipation. This led to the design of the unipolar type transistor called MOS. In N-channel MOS transistors, the electrons leave the source reaching the drain without going through any voltage barrier. The absence of any voltage barrier in the path of the carrier is one reason why MOS dissipates much less power than bipolar transistors. The low power dissipation of MOS allows putting millions of transistors on a single IC chip. In today's million-transistor microprocessors and DRAM memory chips, the use of MOS technology is indispensable. Without the MOS transistor, the advent of desktop personal computers would not have been possible, at least not so soon. The use of bipolar transistors in both the mainframe and minicomputer of the 1960s and 1970s required expensive cooling systems and large rooms due to their bulkiness. MOS transistors do have one major drawback: They are slower than bipolar transistors. This is due partly to the gate capacitance of the MOS transistor. For MOS to be turned on, the input capacitor of the gate takes time to charge up to the turn-on (threshold) voltage, leading to a longer propagation delay.

Overview of logic families

Logic families are judged according to (1) speed, (2) power dissipation, (3) noise immunity, (4) input/output interface compatibility, and (5) cost. Desirable qualities are high speed, low power dissipation, and high noise immunity (since it prevents the occurrence of false logic signals during switching transition). In interfacing logic families, the more inputs that can be driven by a single output, the better. This means that high-driving-capability outputs are desired. This plus the fact that the input and output voltage levels of MOS and bipolar transistors are not compatible mean that one must be concerned with the ability of one logic family in driving the other one. In terms of the cost of a given logic family, it is high during the early years of its introduction and prices decline as production and use rise.

The case of inverters

As an example of logic gates, we look at a simple inverter. In a one-transistor inverter, while the transistor plays the role of a switch, R is the pull-up resistor. See Figure 26-2. However, for this inverter to work effectively in digital circuits, the R value must be high when the transistor is "on" to limit the current flow from V_{CC} to ground in order to have low power dissipation ($P = VI$, where V = 5 V). In other words, the lower the I, the lower the power dissipation. On the other hand, when the transistor is "off", R must be a small value to limit the voltage drop across R, thereby making sure that V_{OUT} is close to V_{CC}. This is a contradictory demand on R. This is one reason that logic gate designers use active components (transistors) instead of passive components (resistors) to implement the pull-up resistor R.

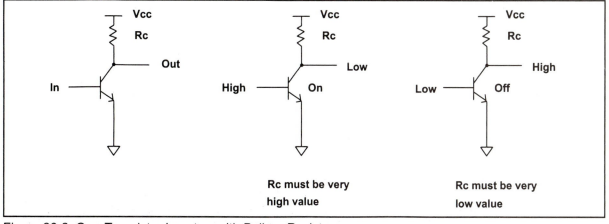

Figure 26-2. One-Transistor Inverter with Pull-up Resistor

The case of a TTL inverter with totem pole output is shown in Figure 26-3. In Figure 26-3, Q3 plays role of a pull-up resistor.

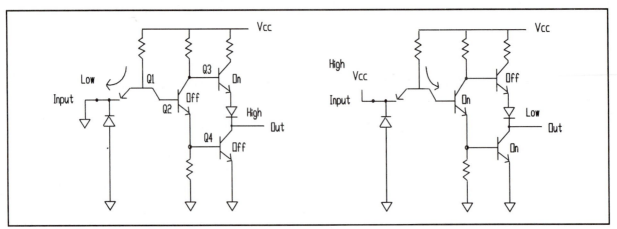

Figure 26-3. TTL Inverter with Totem-Pole Output

CMOS inverter

In the case of CMOS-based logic gates, PMOS and NMOS are used to construct a CMOS (complementary MOS) inverter as shown in Figure 26-4. In CMOS inverters, when the PMOS transistor is off, it provides a very high impedance path, making leakage current almost zero (about 10 nA); when the PMOS is on, it provides a low resistance on the path of V_{DD} to load. Since the speed of the hole is slower than the electron, the PMOS transistor is wider to compensate for this disparity; therefore, PMOS transistors take more space than NMOS.

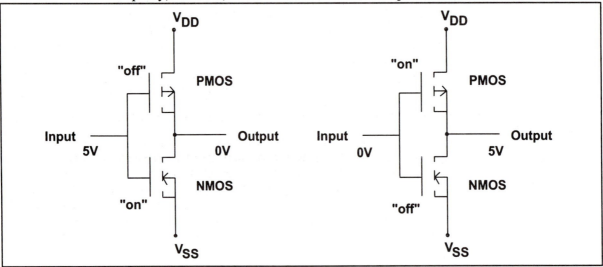

Figure 26-4. CMOS Inverter

Input, output characteristics of some logic families

In 1968 the first logic family made of bipolar transistors was marketed. It was commonly referred to as the standard TTL (transistor-transistor logic) family. The first MOS-based logic family, the CD4000/74C series, was marketed in 1970. The addition of the Schottky diode to the base-collector of bipolar transistors in the early 1970s gave rise to the S family. The Schottky diode shortens the propagation delay of the TTL family by preventing the collector from going into what is called *deep saturation*. Table 26-1 lists major characteristics of some logic families. In Table 26-1, note that as the CMOS circuit's operating frequency rises, the power dissipation also increases. This is not the case for bipolar-based TTL.

Table 26-1: Characteristics of Some Logic Families

Characteristic	Std TTL	LSTTL	ALSTTL	HCMOS
V_{CC}	5 V	5 V	5 V	5 V
V_{IH}	2.0 V	2.0 V	2.0 V	3.15 V
V_{IL}	0.8 V	0.8 V	0.8 V	1.1 V
V_{OH}	2.4 V	2.7 V	2.7 V	3.7 V
V_{OL}	0.4 V	0.5 V	0.4 V	0.4 V
I_{IL}	−1.6 mA	−0.36 mA	−0.2 mA	−1 μA
I_{IH}	40 μA	20 μA	20 μA	1 μA
I_{OL}	16 mA	8 mA	4 mA	4 mA
I_{OH}	−400 μA	−400 μA	−400 μA	4 mA
Propagation delay	10 ns	9.5 ns	4 ns	9 ns
Static power dissipation (F=0)	10 mW	2 mW	1 mW	0.0025 nW
Dynamic power dissipation at F=100 KHz	10 mW	2 mW	1 mW	0.17 mW

History of logic families

Early logic families and microprocessors required both positive and negative power voltages. In the mid-1970s, 5V V_{CC} became standard. For example, Intel's 4004, 8008, and 8080 all used negative and positive voltages for the power supply. In the late 1970s, advances in IC technology allowed combining the speed and drive of the S family with the lower power of LS to form a new logic family called FAST (Fairchild Advanced Schottky TTL). In 1985, AC/ACT (Advanced CMOS Technology), a much higher speed version of HCMOS, was introduced. With the introduction of FCT (Fast CMOS Technology) in 1986, at last the speed gap between CMOS and TTL was closed. Since FCT is the CMOS version of FAST, it has the low power consumption of CMOS but the speed is comparable with TTL. Table 26-2 provides an overview of logic families up to FCT.

Table 26-2: Logic Family Overview

Product	Year Introduced	Speed (ns)	Static Supply Current (mA)	High/Low Family Drive (mA)
Std TTL	1968	40	30	−2/32
CD4K/74C	1970	70	0.3	−0.48/6.4
LS/S	1971	18	54	−15/24
HC/HCT	1977	25	0.08	−6/−6
FAST	1978	6.5	90	−15/64
AS	1980	6.2	90	−15/64
ALS	1980	10	27	−15/64
AC/ACT	1985	10	0.08	−24/24
FCT	1986	6.5	1.5	−15/64

Reprinted by permission of Electronic Design Magazine, c. 1991.

Recent advances in logic families

As the speed of high-performance microprocessors such as the 386 and 486 reached 25 MHz, it shortened the CPU's cycle time, leaving less time for the path delay. Designers normally allocate no more than 25% of a CPU's cycle time budget to path delay. Following this rule means that there must be a corresponding decline in the propagation delay of logic families used in the address and data path as the system frequency is increased. In recent years, many semiconductor manufacturers have responded to this need by providing logic families that have high speed, low noise, and high drive. Table 26-3 provides the characteristics of high-performance logic families introduced in recent years. ACQ/ACTQ are the second-generation advanced CMOS (ACMOS) with much lower noise. While ACQ has the CMOS input level, ACQT is equipped with TTL-level input. The FCTx and FCTx-T are second-generation FCT with much higher speed. The x in the FCTx and FCTx-T refers to various speed grades, such as A, B, and C, where A designation means low speed and C means high speed. For designers who are well versed in using the FAST logic family, the use of FASTr is an ideal choice since it is faster than FAST, has higher driving capability (I_{OL}, I_{OH}), and produces much lower noise than FAST. At the time of this writing, next to ECL and gallium arsenide logic gates, FASTr is the fastest logic family in the market (with the 5V V_{CC}), but the power consumption is high relative to other logic families, as shown in Table 26-3. Recently, a 3.3V V_{CC} with higher speed and lower power consumption is starting to appear. The combining of high-speed bipolar TTL and low power consumption of CMOS has given birth to what is called BICMOS. Although BICMOS seems to be the future trend in IC design, at this time it is expensive due to extra steps required in BICMOS IC fabrication, but in some cases there is no other choice. For example, Intel's Pentium microprocessor, a BICMOS product, had to use high-speed bipolar transistors to speed up some of the internal functions in order to keep up with RISC processor performance. Table 26-3 provides advanced logic characteristics. Table 26-4 shows logic families used in systems with different speeds. The x is for different speed where A, B, and C are used for designation. A is the slower one while C is the fastest one. The above data is for the '244 buffer.

Table 26-3: Advanced Logic General Characteristics

Family	Year	Number Suppliers	Tech Base	I/O Level	Speed	Static Current	I_{OH}/I_{OL}
ACQ	1989	2	CMOS	CMOS/CMOS	6.0 ns	80 μA	−24/24 mA
ACQ	1989	2	CMOS	TTL/CMOS	7.5 ns	80 μA	−24/24 mA
FCTx	1987	3	CMOS	TTL/CMOS	4.1-4.8 ns	1.5 mA	−15/64 mA
FCTxT	1990	2	CMOS	TTL/TTL	4.1-4.8 ns	1.5 mA	−15/64 mA
FASTr	1990	1	Bipolar	TTL/TTL	3.9 ns	50 mA	−15/64 mA
BCT	1987	2	BICMOS	TTL/TTL	5.5 ns	10 mA	−15/64 mA

Reprinted by permission of Electronic Design Magazine, c. 1991.

Table 26-4: Importance of Speed

System Clock Speed (MHz)	Clock Period (ns)	Predominant Logic for Path
2-10	100-500	HC, LS
10-30	33-100	ALS, AS, FAST, FACT
30-66	15-33	FASTr, BCT, FCTA

Reprinted by permission of Electronic Design Magazine, c. 1991.

Evolution of IC technology in Intel's 80x86 microprocessors

Since 1971, when Intel introduced the first microprocessor, the 4004, until the introduction of the Pentium microprocessor, IC technology has gone through some massive changes. The early processors (4004 and 8008) used PMOS. The 8080, 8085, 8088, 8086, and 80286 all used NMOS when first introduced. In recent years, CMOS versions of the 8088, 8086, and 286 have been introduced for power-efficient systems. Currently, CMOS is the universal technology in the design of microprocessors. Only CMOS could allow designers to put over 3 million transistors on a single chip, make it work at 100 MHz, and consume around 10 watts of power. There has been a steady decline in the transistor's dimension throughout the 1970s, 1980s, and 1990s. The design rule, the thickness of the lines inside the IC, has come down from a few microns to a fraction of a micron during this time. See Table 26-5.

The early microprocessors used power supplies with negative (−) and positive (+) voltages. For example, the 4004 used −10 and +5 V. The 8008 used −9 and +5 V, and the 8080 used −5, +5, +12 V. Since the introduction of the 8085, the use of a +5 V power supply has become standard in all microprocessors. To reduce power consumption, 3.3V V_{CC} is being embraced by many designers. The lowering of V_{CC} to 3.3 V has two major advantages: (1) it lowers the power consumption, resulting in prolonging the life of the battery in systems such as a laptop PC or hand-held personal digital assistant, and (2) it allows a further reduction of line size (design rule) to submicron dimensions. This reduction results in putting more transistors in a given die size. The decline in the line size is expected to reach 0.1 µm by the year 2000 and transistor density per chip will reach 100 million transistors.

Table 26-5: Intel Microprocessor Evolution

Microprocessor	Year	IC Tech	Line Thickness (µm)	Power Supply (V)	Number of Transistors
8086	1978	NMOS	3.0	5	29,000
80286	1982	NMOS	1.5	5	130,000
80386	1985	CMOS	1.5	5	275,000
80486	1989	CMOS	1.0	5	1.2 million
Pentium	1992	BICMOS	0.8	5	3.1 million
Pentium II	1993	BICMOS	0.6	3.3	3.1 million

Review Questions

1. State the main advantages of MOS and bipolar transistors.
2. True or false. In logic families, the higher the noise margin, the better.
3. True or false. Generally, high-speed logic consumes more power.
4. Power dissipation increases linearly with the increase in frequency in _____ (CMOS, TTL).
5. In a CMOS inverter, indicate which transistor is on when the input is high.
6. For system frequencies of 10 - 30 MHz, which logic families are used for the address and data path?

SECTION 26.2: IC INTERFACING AND SYSTEM DESIGN CONSIDERATIONS

There are several issues to be considered in designing a microprocessor-based system. They are IC fan-out, capacitance derating, ground bounce, V_{CC} bounce, crosstalk, and transmission lines. This section provides an overview of these design issues in order to provide a sampling of what is involved in high-performance system design.

IC fan-out

In interfacing IC, fan-out/fan-in is a major issue. How many inputs can an output signal drive? This question must be addressed for both logic "0" and logic "1" outputs. Fan-out for low and fan-out for high are as follows:

$$\text{fan-out (of low)} = \frac{I_{OL}}{I_{IL}} \qquad \text{fan-out (of high)} = \frac{I_{OH}}{I_{IH}}$$

Of the above two values the lower number is used to ensure the proper noise margin. Figure 26-5 shows the sinking and sourcing of current when ICs are connected.

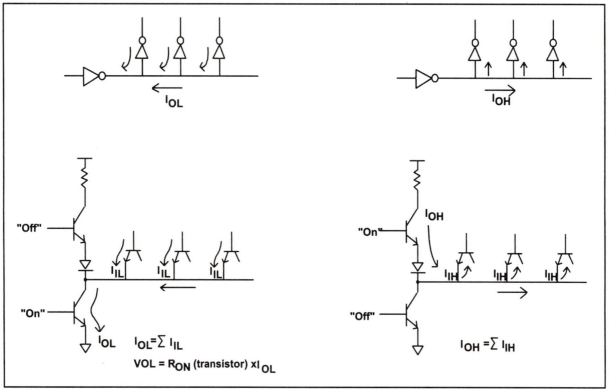

Figure 26-5. Current Sinking and Sourcing in TTL

In Figure 26-5, as the number of inputs connected to the output increases, I_{OL} rises, which causes V_{OL} to rise. If this continues, the rise of V_{OL} makes the noise margin smaller and this results in the occurrence of false logic due to the slightest noise.

In designing the system, very often an output is connected to various kinds of inputs. See Example 26-2.

The total I_{IL} and I_{IH} requirement of all the loads on a given output must be less than the driver's maximum I_{OL} and I_{OH}. This is shown in Example 26-3.

Example 26-1

Find how many unit loads (UL) can be driven by the output of the LS logic family.

Solution:

The unit load is defined as $I_{IL} = 1.6$ mA and $I_{IH} = 40\,\mu A$. Table 26-1 shows $I_{OH} = 400\,\mu A$ and $I_{OL} = 8$ mA for the LS family. Therefore, we have

$$\text{fan-out (low)} = \frac{I_{OL}}{I_{IL}} = \frac{8\ mA}{1.6\ mA} = 5$$

$$\text{fan-out (high)} = \frac{I_{OH}}{I_{IH}} = \frac{400\,\mu A}{40\,\mu A} = 10$$

This means that the fan-out is 5. In other words, the LS output must not be connected to more than 5 inputs with unit load characteristics.

Example 26-2

An address pin needs to drive 5 standard TTL loads in addition to 10 CMOS inputs of DRAM chips. Calculate the minimum current to drive these inputs for both logic "0" and "1".

Solution:

The standard load for TTL is $I_{IH} = 40\,\mu A$ and $I_{IL} = 1.6$ mA and for CMOS $I_{IL} = I_{IH} = 10\,\mu A$. minimum current for "0" = total of all $I_{IL} = 5 \times 1.6$ mA $+ 10 \times 10\,\mu A = 8.1$ mA
minimum current for "1" = total of all $I_{IH} = 5 \times 40\,\mu A + 10 \times 10\,\mu A = 300\,\mu A$

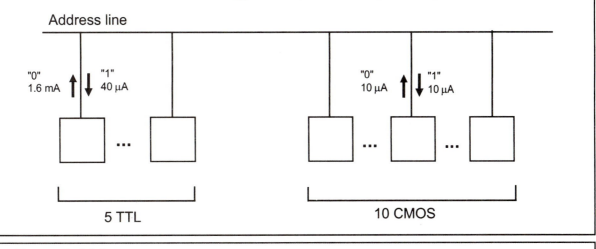

Example 26-3

Assume that the microprocessor address pin in Example 26-2 has specifications $I_{OH} = 400\,\mu A$ and $I_{OL} = 2$ mA. Do the input and output current needs match?

Solution:

For a high output state, there is no problem since $I_{OH} > I_{IH}$. However, the number of inputs exceeds the limit for I_{OL} since an I_{IL} of 8.1 mA is much larger than the maximum I_{OL} allowed by the microprocessor.

In cases such as Example 26-3 where the receiver current requirements exceed the drivers' capability, we must use a buffer (booster), such as the 74xx245 and 74xx244. The 74xx245 is used for bidirectional and the 74xx244 for unidirectional signals. See current 74LS244 and 74LS245 characteristics in Table 26-6.

Table 26-6: Electrical Specifications for Buffers

Buffer	I$_{OH}$ (mA)	I$_{OL}$ (mA)	I$_{IH}$ (μA)	I$_{IL}$ (mA)
74LS244	3	12	20	0.2
74LS245	3	12	20	0.2

Note: V$_{OL}$ = 0.4 V and V$_{OH}$ = 2.4 V are assumed.

Capacitance derating

Next we study what is called *capacitance derating* and its impact in system design. A pin of an IC has an input capacitance of 5 to 7 pF. This means that a single output that drives many inputs sees a large capacitance load since the inputs are in parallel and therefore added together. Look at the following equations.

$$Q = CV \tag{26–1}$$

$$\frac{Q}{T} = \frac{CV}{T} \tag{26–2}$$

$$F = \frac{1}{T} \tag{26–3}$$

$$I = CVF \tag{26–4}$$

In Equation (26-4), I is the driving capability of the output pin, C is C$_{IN}$ as seen by the output, and V is the voltage. The equation indicates that as the number of C$_{IN}$ loads goes up, there must be a corresponding increase in I$_O$, the driving capability of the output. In other words, outputs with high values of I$_{OL}$ and I$_{OH}$ are desirable. Although recently there have been some logic families with I$_{OL}$ = 64 mA and I$_{OH}$ = 15 mA, their power consumption is high. Equation (26-4) indicates that if I = constant, as C goes up, F must come down, resulting in lower speed. The most widely accepted solution is the use of a large number of drivers to reduce the load capacitance seen by a given output. Assume that we have 16 address bus lines A0 - A15 driving 4 banks of 32-bit-wide memory. Each bank has 32 chips of 64Kx1 organization, which results in 128 memory chips. Depending on how many 244s are used to drive the memory addresses, the delay due to the address path varies substantially. To understand this we examine four cases.

Case 1: Two 244 drivers

This option uses two 244 drivers, one for A0 - A7 and one for A8 - A15. An output of the 244 drives 4 banks of memory, each with 32 inputs. Assuming that each memory input has 5 pF capacitance, this results in a total of $128 \times 5 = 640$ pF capacitance load seen by the 244 output. However, the 244 output can handle no more than 50 pF. As a result, the delay due to this extra capacitance must be added to the address path delay. For each 50 to 100 pF capacitor, an extra 3 ns delay is added to the address path delay. In our calculation, we use 3 ns for each 100 pF of capacitance. Figure 26-6 shows driving memory inputs by two 244 chips. See Example 26-4.

Case 2: Doubling the number of 244 buffers

Doubling the number of 244 buffers will reduce the address path delay. A single 244 drives only two banks, or a total of 64 inputs, since there are 32 inputs in each bank. As a result, a 244 output will see a capacitance load of $64 \times 5 = 320$ pF. In this case, we use only four 244 buffer chips, as shown in Figure 26-7 and Example 26-5.

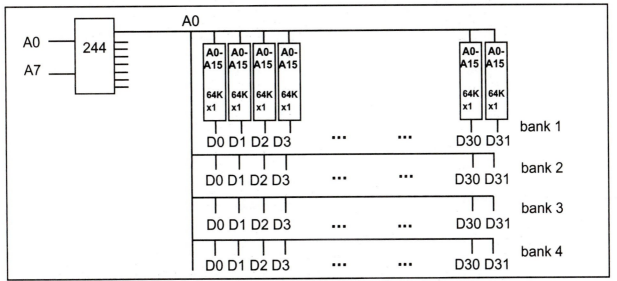

Figure 26-6. Case 1: Two 244 Address Drivers (the second 244 for A8 - A15 is not shown)

Example 26-4

Calculate the following for Figure 26-6, assuming a memory access time of 25 ns and a propagation delay of 10 ns for the 244.
(a) delay due to capacitance derating on the address path
(b) the total address path delay for case 1

Solution:

(a) Of the 640 pF capacitance seen by the 244, only 50 pF is taken care of; the rest, which is 590 (640 - 50 = 590), causes a delay. Since there are 3 ns for each extra 100 pF, we have the following delay due to capacitance derating, $(590/100) \times 3$ ns = 17.7 ns.

(b) Address path delay = 244 buffer propagation delay + capacitance derating delay + memory access time   = 10 ns + 17.7 ns + 25 ns = 52.7 ns.

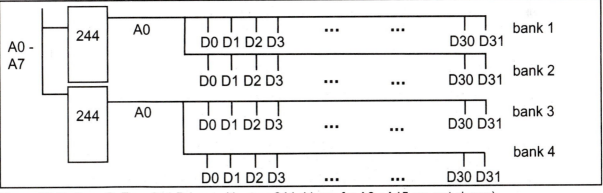

Figure 26-7. Case 2: Four 244 Drivers (the two 244 drivers for A8 - A15 are not shown)

Example 26-5

Calculate (a) delay due to capacitance derating on the address path, and (b) total address path delay for case 2. Assume a memory access time of 25 ns and a propagation delay of 10 ns for the 244.

Solution:

(a) Of the 320pF capacitance seen by the 244, only 50 pF is taken care of; the rest, which is 270 pF, causes a delay. Since there are 3 ns for each extra 100 pF, we have $(270/100) \times 3$ ns = 8.1 ns delay due to capacitance derating.

(b) The address path delay = 244 buffer propagation delay + capacitance derating delay + memory access time   = 10 ns + 8.1 ns +25 ns = 43.1 ns.

SECTION 26.2: IC INTERFACING AND SYSTEM DESIGN CONSIDERATIONS

Case 3: Doubling again

In this case, we double the number of 244 buffers again, so that an output of the 244 drives one bank, each with 32 inputs. This results in a total capacitance load of $32 \times 5 = 160$ pF. Only 50 pF of it is taken care of by the 244, leaving 110 pF, causing a delay. In this case, we use total of eight 244 buffer chips, as shown in Figure 26-8 and Example 26-6.

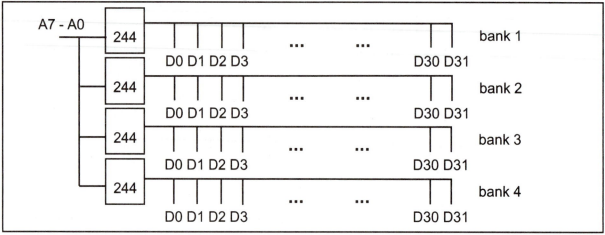

Figure 26-8. Case 3: A Single 244 for Each Bank (A8 - A15 not shown)

Example 26-6

Calculate (a) the delay due to capacitance derating on the address path, and (b) the total address path delay for case 3. Assume memory access time of 25 ns and propagation delay of 10 ns for the 244.

Solution:
(a) Of the 160 pF capacitance seen by the 244, only 50 pF is taken care of; the rest, which is 110 pF, causes a delay. Since there are 3 ns for each extra 100 pF, we have $(110/100) \times 3$ ns = 3.1 ns delay, due to capacitance derating.
(b) The address path delay = 244 buffer propagation delay + capacitance derating delay + memory access time   = 10 ns + 3.1 ns +25 ns = 38.1 ns.

Case 4: Doubling again

Doubling the number of 244 chips again allows each buffer to drive 16 inputs, or one half of the 32-bit bank. This eliminates (almost) the capacitance derating all together. As a result, the address path delay consists of a 244 and memory access time of 35 ns (10 ns + 25 ns = 35 ns) and no capacitance derating. In this case we must use sixteen 244 buffers.

Examining cases 1 through 4 shows that for high-speed system design we must accept a higher cost due to extra parts and higher power consumption.

Power dissipation considerations

Power dissipation of a system is a major concern of system designers, especially for laptop and hand-held systems such as personal digital assistants (PDA). Although power dissipation is a function of the total current consumption of all components of a system, the impact of V_{CC} is much more pronounced, as shown next. Earlier we showed in Equation (26-4) that $I = CFV$. Substituting this in equation $P = VI$ yields the following:

$$P = VI = CFV^2 \qquad\qquad (26\text{--}5)$$

In Equation (26-5), the effects of frequency and V_{CC} voltage should be noted. While the power dissipation goes up linearly with frequency, the impact of the power supply voltage is much more pronounced (squared).

Dynamic and static currents

There are two major types of currents flowing through an IC: dynamic and static. A dynamic current is a function of the frequency under which the component is working, as seen in Equation (26-4). This means that as the frequency goes up, the dynamic current and power dissipation go up. The static current, also called dc, is the current consumption of the component when it is inactive (not selected).

Power-down option and Intel's SL series

The popularity of notebook and laptop PCs led Intel to market a series of 386, 486, and Pentium processors called the SL series. Intel originally designed SL microprocessor technology for mobile computers that needed to conserve battery power. This technology will appear in all enhanced Intel 486 and Pentium processors. These processors have what is called *system management mode* (SMM), which reduces energy consumption by turning off peripherals or the entire system when not in use. According to Intel, SMM can put the entire system, including the monitor, into sleep mode during periods of inactivity, thereby reducing "power from 250 watts to less than 30 watts." The effects on the 3.3 V power supply alone translates into a power savings of up to 56% over systems with a 5 V power supply, as shown in Example 26-7.

Example 26-7

Prove that a system 3.3 V system consumes 56% less power than a system with a 5 V power supply.

Solution:
Since $P = VI$, by substituting $I = V/R$, we have $P = V^2/R$. Assuming that $R = 1$, we have $P (3.3)^2 = 10.89$ W and $P = (5)^2 = 25$ W. This results in using 14.11 W less (25 - 10.89 = 14.11), which means a 56% power saving (14.11 W/25 W $\times$ 100 = 56%).

In addition to V_{CC}, the system design can also have a great impact on power dissipation. As an example of the impact of system design on power consumption, look at Example 26-8, where 64K bytes of memory are designed using two different memory organizations.

Ground bounce

One of the major issues that designers of high-frequency systems must grapple with is *ground bounce*. Before we define ground bounce, we will discuss lead inductance of IC pins. There is a certain amount of capacitance, resistance, and inductance associated with each pin of the IC. The size of these elements varies depending on many factors such as length, area, and so on. Figure 26-9 shows the lead inductance and capacitance of the 24 pins of a DIP IC.

The inductance of the pins is commonly referred to as *self-inductance* since there is also what is called *mutual inductance*, as we will show below. Of the three components of capacitor, resistor, and inductor, self-inductance is the one that causes the most problems in high-frequency system design since it can result in ground bounce. Ground bounce is caused when a massive amount of current flows through the ground pin when a multiple of outputs change from high to low all at the same time. The voltage relation to the inductance of the ground lead follows:

$$V = L \frac{di}{dt} \qquad (26-6)$$

As we increase the system frequency, the rate of dynamic current, *di/dt*, is also increased, resulting in an increase in the inductance voltage *L (di/dt)* of the ground pin. Since the low state (ground) has a small noise margin, any extra voltage due to the inductance voltage can cause a false signal. To reduce the effect of ground bounce, the following steps must be taken where possible.

Example 26-8

Calculate and compare the power consumption of 64K bytes of memory if we use the following organizations. Assume that the active and standby current consumptions are the same for both memory chips: $I_{active} = 180$ mA and $I_{standby} = 30$ mA
(a) 64Kx1   (b) 8Kx8 memory chips

Solution:

(a) Using 64Kx1 memory chips, we have one bank, where it is selected by a decoder output: (A0 - A15 go to all 8 memory chips)

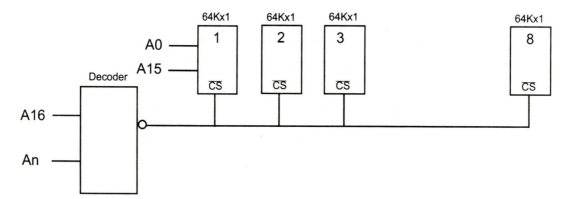

Therefore, the current dissipation is $P = [N \times I_{act} + M \times I_{sb}] V_{CC}$, where N =number of active devices, M the number of standby devices, and I_{sb} and I_{act} are standby and active currents, respectively. $P = [8 \times 180$ mA $+ 0] \times 5$ V $= 7.2$ W.

(b) When using the 8Kx8 devices we have 8 banks each with 8K bytes of memory but only one of them active, with the other 7 in standby mode.

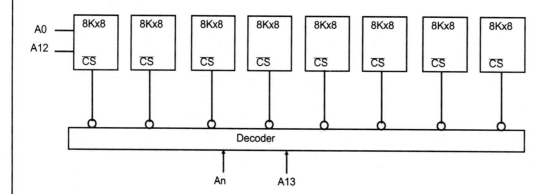

Therefore, $P = [1 \times 180$ mA $+ 7 \times 30$ mA$] \times 5$ V $= 1.95$ W.

Comparing (a) and (b), we conclude that with an extra decoder (e.g.,74xx138), a substantial amount of power is saved. However, using the memory organizations x8, x16, and x32 wide can cause other kinds of problems, such as ground bounce.

1. The V_{CC} and ground pins of the chip must be located in the middle rather than at the opposite ends of the IC chip (the 14-pin TTL logic IC uses pins 14 and 7 for ground and V_{CC}). This is exactly what we see in high-performance logic gates such as Texas Instrument's advanced logic AC11000 and ACT11000 families. For example, the ACT11013 is a 14-pin DIP chip where pin numbers 4 and 11 are used for the ground and V_{CC} instead of 7 and 14 as in the TTL. We can also use the SOIC packages instead of DIP. The self-inductance of the leads is shown in Table 26-7.

Table 26-7: 20-Pin DIP and SOIC Lead Inductance

Pins	DIP (nH)	SOIC (nH)
1,10,11,20	13.7	4.2
2,9,12,19	11.1	3.8
3,8,13,18	8.6	3.3
4,7,14,17	6.0	2.9
5,6,15,16	3.4	2.4

Courtesy of Texas Instruments

2. Use logics with a minimum number of outputs. For example, a 4-output is preferable to an 8-output. This explains why many designers of high-performance systems avoid using memory chips or the drivers and buffers of 16- or 32-bit wide outputs since all the outputs switching at the same time will cause a massive flow of current in the ground pin, and hence cause ground bounce (see Figure 26-10).

3. Use as many pins for the ground and V_{CC} as possible to reduce the lead length, since the self-inductance of a wire with length l and a cross section of $B \times C$ is:

$$L = 0.002 \, l \ln\left(\frac{2l}{B+C} + \frac{l}{2}\right) \qquad (26\text{–}7)$$

As seen in Equation (26-7), the wire length, l, contributes more to self-inductance than does the cross section. This explains why all high-performance microprocessors and logic families use several pins for the V_{CC} and ground. For example, in the case of Intel's Pentium processor there are over 50 pins for the ground and another 50 pins for the V_{CC}.

The discussion of ground bounce is also applicable to V_{CC} when a large number of outputs changes from the low to high state and is referred to as V_{CC} bounce. However, the effect of V_{CC} bounce is not as severe as ground bounce since the high ("1") state has wider noise margin than the low ("0") state.

Pin	Self-inductance	Capacitance
1	15.10 nH	1.86 pF
2	12.20 nH	1.70 pF
3	9.54 nH	1.29 pF
4	7.44 nH	0.95 pF
5	5.31 nH	0.61 pF
6	3.73 nH	0.43 pF
7	3.41 nH	0.43 pF
8	4.66 nH	0.61 pF
9	6.95 nH	0.95 pF
10	8.96 nH	1.29 pF
11	11.70 nH	1.70 pF
12	14.50 nH	1.86 pF
13	14.50 nH	1.86 pF
14	11.70 nH	1.70 pF
15	8.96 nH	1.29 pF
16	6.95 nH	0.95 pF
17	4.66 nH	0.61 pF
18	3.41 nH	0.43 pF
19	3.73 nH	0.43 pF
20	5.31 nH	0.61 pF
21	7.44 nH	0.95 pF
22	9.54 nH	1.29 pF
23	12.20 nH	1.70 pF
24	15.10 nH	1.86 pF

Figure 26-9. Inductance and Capacitance of 24-pin DIP
Reprinted by permission of Electronic Design Magazine, c. 1992.

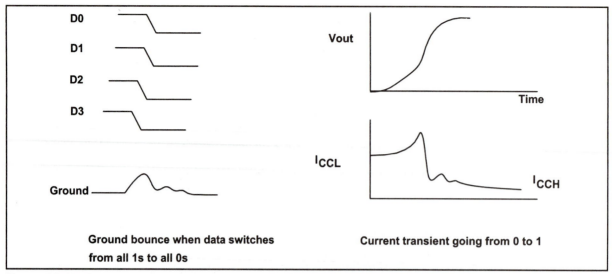

Figure 26-10. Ground Bounce and Transient Current for Output Transitions

Filtering the transient currents using decoupling capacitors

In the TTL family, the change of the output from low to high can cause what is called *transient current*. In totem-pole output when the output is low, Q4 is on and saturated, whereas Q3 is off. By changing the output from the low to high state, Q3 becomes on and Q4 becomes off. This means that there is a time that both transistors are on and drawing currents from the V_{CC}. The amount of current depends on the R_{ON} values of the two transistors and that in turn depends on internal parameters of the transistors. However, the net effect of this is a large amount of current in the form of a spike for the output current, as shown in Figure 26-10. To filter the transient current, a 0.01 µF or 0.1 µF ceramic disk capacitor can be placed between the V_{CC} and ground for each TTL IC. However, the lead for this capacitor should be as small as possible since a long lead results in a large self-inductance and that results in a spike on the V_{CC} line [$V = L\ (di/dt)$]. This is also called V_{CC} bounce. The ceramic capacitor for each IC is referred to as a *decoupling capacitor*. There is also a bulk decoupling capacitor, as described next.

Bulk decoupling capacitor

As many IC chips change state at the same time, the combined currents drawn from the board's V_{CC} power supply can be massive and cause a fluctuation of V_{CC} on the board where all the ICs are mounted. To eliminate this, a relatively large (relative to an IC decoupling capacitor) tantalum capacitor is placed between the V_{CC} and ground lines. The size and location of this tantalum capacitor varies depending on the number of ICs on the board and the amount of current drawn by each IC, but it is common to have a single 22 µF to 47 µF capacitor for each of the 16 devices, placed between the V_{CC} and ground lines.

Crosstalk

Crosstalk is due to mutual inductance. See Figure 26-11. Previously, we discussed self-inductance, which is inherent in a piece of conductor. *Mutual inductance* is caused by two electric lines running parallel to each other. It is calculated as follows:

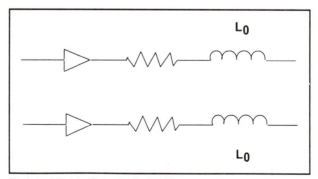

Figure 26-11. Crosstalk (EMI)

CHAPTER 26: IC TECHNOLOGY AND SYSTEM DESIGN

$$M = 0.002 \, l \ln \frac{2l}{d} - \ln (K-1 + \frac{d}{l} - \frac{d}{2l}) \, 2 \qquad\qquad (26\text{–}8)$$

where l is the length of two conductors running in parallel, and d is the distance between them, and the medium material placed in between affects K. Equation (26-8) indicates that the effect of crosstalk can be reduced by increasing the distance between the parallel or adjacent lines (in printed circuit boards, it will be traces). In many cases, such as printer and disk drive cables, there is a dedicated ground for each signal. Placing ground lines (traces) between signal lines reduces the effect of crosstalk. This method is used even in some ACT logic families where there is a V_{CC} and GND pin next to each other. Crosstalk is also called EMI (electromagnetic interference). This is in contrast to ESI (electrostatic interference), which is caused by capacitive coupling between two adjacent conductors.

Transmission line ringing

The square wave used in digital circuits is in reality made of a single fundamental pulse and many harmonics of various amplitudes. When this signal travels on the line, not all the harmonics respond the same way to capacitance, inductance, and resistance of the line. This causes what is called *ringing*, which depends on the thickness and the length of the line driver, among other factors. To reduce the effect of ringing, the line drivers are terminated by putting a resistor at the end of the line. See Figure 26-12. There are three major methods of line driver termination: parallel, serial, and Thevenin. We saw the case of serial termination for the DRAM connection of the IBM PC/XT in Chapter 11, where resistors of 30 - 50 ohms are used to terminate the line. The parallel and Thevenin methods are used in cases where there is a need to match the impedance of the line with the load impedance. This requires a detailed analysis of the signal traces and load impedance, which is beyond the scope of this volume. In high-frequency systems, wire traces on the printed circuit board (PCB) behave like transmission lines, causing ringing. The severity of this ringing depends on the speed and the logic family used. Table 26-8 provides the length of the traces, beyond which the traces must be looked at as transmission lines.

Table 26-8: Line Length Beyond Which Traces Behave Like Transmission Lines

Logic Family	Signal Line Length (in.)
LS	25
S, AS	11
F, ACT	8
AS, ECL	6
FCT, FCTA	5

(Reprinted by permission of Integrated Device Technology, c. IDT 1991)

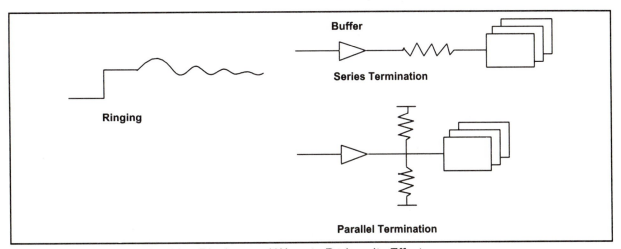

Figure 26-12. Transmission Line Ringing and Ways to Reduce Its Effects

1. What is the fan-out of "0" state?
2. If the fan-out of "low" and "high" are 10 and 15, respectively, what is the fan-out?
3. If I_{OL} = 12 mA, I_{OH} = 3 mA for the driver, and I_{IL} = 1.6 mA, I_{IH}=40 µA for the load, find the fan-out.
4. Why do I_{IL} and I_{OH} have negative signs in many TTL books?
5. What are the 74xx244 and 74xx245 used for?
6. What is capacitive derating?
7. Ground bounce happens when the output makes a transition from _____ to _____.
8. Give one way to reduce ground bounce.
9. Transient current is due to transition of output from _____ to _____.
10. Why do high-speed logic gates using DIP packaging put the V_{CC} and ground pins in the middle instead of the corners?

SECTION 26.3: DATA INTEGRITY AND ERROR DETECTION IN DRAM

As advanced operating systems, such as MS Windows NT, IBM OS/2, Sun Micro Solaris, and Nextstep from Next Corp., become the dominant operating systems in microprocessor-based computers, the need for large amounts of DRAM in such systems is inevitable. This is especially true for multiuser and multitasking systems. However, in designing a system with a large amount of DRAM, we must make sure that data integrity is maintained. As we discussed in Chapter 11, the parity bit method is used for the data integrity of PC DRAM, but the parity method cannot correct the error. It simply indicates that an error has occurred. With system memory reaching 64 megabytes, the idea of incorporating an EDC (error detection and correction) circuitry on the system board is becoming popular among designers of microprocessor-based high-performance computers. To understand the need for EDC, we first discuss the probability of error caused by soft error and hard error.

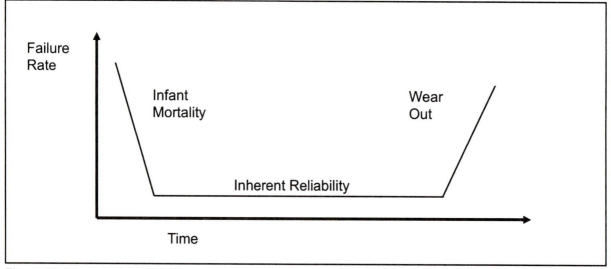

Figure 26-13. Bathtub Failure Rate (Courtesy of Texas Instruments)

Soft error and hard error

In DRAM there are two kinds of errors that can cause a bit to change: soft error and hard error. If the cell bit gets stuck permanently in a "high" or "low" state, this is referred to as a *hard error*. Hard error is due to deterioration of the cell caused by wear-out (see Figure 26-13). There is no remedy for hard error except to replace the defective DRAM chip since the damage is permanent. The other kind of error, a *soft error*, alters the cell bit from 1 to 0 or from 0 to 1, even though the cell is perfectly fine (no hard error). Soft error is caused by radiation of alpha particles and power surges. The sources of the alpha particles are the radiation in the air or the materials in the plastic package enclosing the DRAM die. The occurrence of a soft error as a result of the alpha particle ionizing the charges in a DRAM cell is a greater source of concern since it is 5 times more likely to happen than a hard error. As the density of DRAM chips increases and the size of the DRAM cell goes down, the probability of a soft error for a given cell goes up, but the relation is not linear. DRAM manufacturers provide a parameter called FIT (failure in time) to measure the soft error rate for a single DRAM chip. The FIT of a single DRAM chip is the number of expected errors in 10^9 hours of operation. Few DRAM manufacturers provide the FIT data in their DRAM data books. Even when they do provide FIT data, sometimes it is not very clear if it is for soft error or hard error or both, since the numbers vary wildly. In this section we use a FIT of 252 for Toshiba's 1Mx1 DRAM, used widely in many application notes.

Mean time between failure (MTBF) and FIT for DRAM

The mean time between failure for a single chip is calculated using the failure-in-time (FIT) as follows for a single DRAM chip

$$MTBF = \frac{1,000,000,000 \text{ hours}}{FIT} \qquad (26-9)$$

To get the MTBF rate for the system we must divide the single chip MTBF by the number of DRAM chips in the system.

$$MTBF \text{ of system memory} = \frac{MTBF \text{ of one DRAM chip}}{\text{number of DRAM chips}} \qquad (26-10)$$

The higher FIT number increases the probability of a system memory failure, as shown in Examples 26-9 and 26-10.

Example 26-9

Assuming that the FIT for 1Mx1 DRAM is 252, calculate the MTBF for:
(a) a single DRAM chip
(b) a system with 64 megabytes of DRAM memory

Solution:

(a) The MTBF for a single 1Mx1 is as follows
   MTBF for 1Mx1 = 1,000,000,000 hr / 252 = 3,968,254 hr = 453 years

(b) To implement the 64M of system memory using the 1Mx1 DRAM chips, we need (64M × 8) / 1Mx1 = 512 chips. The MTBF for the system is as follows.
   MTBF for 64M of system memory =453 years / 512 chips = 0.884 year = 323 days

---

Example 26-10

Calculate the system MTBF for the system in Example 26-9 if FIT =745.

Solution:

MTBF for a single DRAM chip = 10^9 / 745 hrs. = 153 years.
For the system it is 153 years / 512 = 109 days.

Error detection and correction

The use of EDC (error detection and correction) in failproof systems with large amounts of DRAM is widespread. EDC not only detects the error but also corrects it, in contrast to parity bit, which only detects the error. EDC stores an error correction code for each word, but the number of bits for the error code varies depending on the word size, as shown in Table 26-9.

Table 26-9: Bits Required for Parity vs. EDC

Data Word Size	Parity Bits	EDC Bits
8	1	5
16	2	6
32	4	7
64	8	8
128	16	9

As seen from Table 26-9, the overhead for EDC is much higher, especially in small words such as 16- and 32-bit data buses. Implementing EDC can reduce the probability of memory error drastically if we can justify the extra cost associated with it. Many semiconductor companies, including Integrated Device Technology (IDT) and Advanced Micro Device (AMD), market EDC chips. Some of the concepts in this section are based on application notes from these two companies. Although EDC detects and corrects a single-bit error occurring in a word, it is unable to do anything about any 2-bit errors occurring in the same word. The probability of a 2-bit error happening in the same word is very low, as shown in Equation (26-11).

$$\text{MTBF with EDC} = \text{MTBF without EDC} \times \sqrt{\frac{\pi \times \text{number of memory words}}{2}} \quad (26\text{--}11)$$

Notice in Equation (26-11) that the MTBF with EDC depends on the word size and the number of words in the system.

Example 26-11

Calculate the MTBF for the Example 26-9 using EDC. Assume that the system board uses 32-bit words (80386, 80486 PC).

Solution:

64M of memory using 32-bit words has 16 banks each with 32 1Mx1 memory chip since 512/32 = 16 banks. This means that we have 16M words of 32-bit size. Since each 32-bit word requires an extra 7 chips (see Table 26-9) for EDC, a total of (64Mx8/32Mx1) × 7 =16 × 7 = 112 DRAM chips are used by EDC. This results in a total chip count of 624 (512 + 112) of 1Mx1 chips. Since MTBF for a single chip is 453, we have:

$$\text{MTBF of 624 DRAM} = \frac{453\,\text{years}}{624} = 0.726 \text{ year} = 265 \text{ days}$$

$$\text{MTBF with EDC} = 265 \text{ days} \times \sqrt{\frac{\pi \times 16M}{2}} = 3726 \text{ years}$$

This means that by detecting and correcting single-bit errors, the probability that a soft error will alter 2 bits in the same word is once in 3726 years.

Comparing Examples 26-9 and 26-11 shows that with 112 extra DRAMs of 1Mx1, or a total of 14M of DRAM (112/8 = 14M) and EDC circuitry, MTBF becomes 3726 years instead of 323 days as we saw in Example 26-9. Example 26-12 shows what happens when the word size is increased in systems such as Pentiums. Example 26-13 shows the impact of DRAM size on the MTBF.

Example 26-12

Calculate the MTBF for 64M of memory if the memory words are 64-bit and 1Mx1 DRAM chips of FIT=252 are used (a) without EDC, and (b) with EDC. Notice that this is the same as Example 26-11 except that the word size is 64-bit (Pentium PC).

Solution:

Since the words size is 64-bit, we have 8 megawords of 64-bit size where 64Mx8/64 × 1Mx1 =8. This means 8 banks, each with 64 1Mx1 DRAM chips. Therefore, the chip count is 8 × 64 = 512, which is the same as Example 26-11. Since the MTBF for a single chip is 453 years,

$$MTBF = \frac{453 \text{ years}}{512 \text{ chips}} = 0.884 \text{ year} \quad \text{(without EDC)}$$

Each 64-bit word requires 8 bits to store EDC error codes according to Table 26-9. Therefore, each bank has an extra 8 1Mx1 DRAM chips for EDC, or a total of 64 DRAMs for EDC. The total chip count is 576 (8 × 64 +8 × 8 = 576), where 8 × 64 is for the memory and 8 × 8 is for EDC. The MTBF for 576 chips, where EDC chips are included, is 453/576 = 0.786 year.

$$MTBF = 0.786 \text{ years} \sqrt{\frac{\pi \times 8,388,608 \text{ words}}{2}} = 2852 \text{ years}$$

We conclude that the wider the word size, the higher the probability of 2-bit errors. Notice in this example that there are 8 banks of 64-bit wide data, or 8,388,608 words.

Example 26-13

Calculate the MTBF for 64M of memory on the Pentium-based system using 4Mx1 DRAM chips where FIT=381 (a) without EDC, and (b) with EDC.

Solution:

(a) The total chip count is 64Mx8/4Mx1 =128 4Mx1 DRAM chips. The MTBF for one DRAM is 1,000,000,000 hours/381=299.6 years and the MTBF for all 128 DRAM chips is

$$MTBF = \frac{299.6}{128} = 2.34 \text{ years} \quad \text{(without EDC)}$$

(b) Since the Pentium word size is 64 bits wide and we are using 4Mx1 chips, the banks are 4Mx64. Each 64-bit words needs 8 extra bits of memory for the EDC, or an extra 8 memory chips for each bank. Organizing 64M of system memory for 64-bit words using 4Mx1 DRAM chips results in a total of 2 banks, each taking 32M. Each bank uses 64 4Mx1 chips. Therefore, the chip count is 144 (2 × 64 + 2 × 8) with 128 (2 × 64) for memory and 16 (2 × 8) for EDC. MTBF for the 144 chips is 299.6 / 144 =2.08 years and since each bank has 8M (8,388,608) memory words (2 banks each with 4M words), the MTBF is as follows:

$$MTBF \text{ (with EDC)} = 2.08 \sqrt{\frac{3.14 \times 8,388,608}{2}} = 7548.4 \text{ years} \quad \text{(with EDC)}$$

Table 26-10, from an AMD application note, provides a comparison of memory size and MTBF. For the same memory systems, if we use FIT = 1000, we have the results shown in Table 26-11 from the AMD data book.

For more discussion of the use of EDC chips, see the "Error detection and correction with IDT 49C466" Application Note 94 by Anupama Hedge, published in IDT high-speed CMOS logic design from Integrated Device Technology Inc.

Table 26-10: MTBF for 32-bit System with 1Mx1 DRAM   (FIT = 252)

Memory Size	Without EDC		With EDC	
	No. of DRAMs	MTBF	No. of DRAMs	MTBF
4M	32	14.1 yr	39	14,907 yr
8M	64	7.1 yr	78	10,541 yr
12M	96	4.7 yr	117	8,607 yr
16M	128	3.5 yr	156	7,454 yr
24M	192	2.4 yr	234	6,086 yr

(Reprinted by permission of Advanced Micro Devices, Copyright AMD Corp. 1990)

Table 26-11: MTBF for 32-bit System with 1Mx1 DRAM   (FIT = 1000)

Memory Size	Without EDC		With EDC	
	No. of DRAMs	MTBF	No. of DRAMs	MTBF
4M	32	3.6 yr	39	3,757 yr
8M	64	1.8 yr	78	2,656 yr
12M	96	1.2 yr	117	2,168 yr
16M	128	326 days	156	1,878 yr
24M	192	217 days	234	1,534 yr

(Reprinted by permission of Advanced Micro Devices, Copyright AMD Corp. 1990)

ECL and gallium arsenide (GaAs) chips

The use of secondary cache (L2: level 2 cache, as many call it) and EDC in systems with speeds of 66 MHz and higher is adding to the data and address path delay. This is forcing designers to resort to using ECL and GaAs chips. Due to the fact that ECL chips have a very high power dissipation, they are not used in PC/Workstation design. However, GaAs chips are showing up in high-speed Pentium and RISC-based computers. This is especially the case for the GaAs EDC and cache controller chips. The mass of electrons in GaAs is lighter than silicon, due to its quantum mechanics structure. As a result, the electrons in GaAs have a much higher speed. This means that GaAs chips can achieve a much higher speed than silicon. The power dissipation of the GaAs transistor is comparable to the silicon-based MOS transistor. Therefore, GaAs technology might appear to provide the ideal chip since it has the speed of ECL (it is even faster than ECL) and the power dissipation of CMOS. However, it has the following disadvantages.

1. Unlike silicon, of which there is a plentiful supply in nature in the form of sand, GaAs is a rare commodity, and therefore more expensive.
2. GaAs is a compound made of two materials, Ga and As, and therefore is unstable at high temperatures.
3. It is very brittle, making it impossible to have large wafers. As a consequence, at this time no more than 100,000 transistors can be placed on a single chip. Contrast this to the millions of transistors for silicon-based chips.
4. The GaAs yields are much lower than silicon, making the cost per chip much more expensive than silicon chips.

These problems make the building of an entire computer based on GaAs a visionary product, if not an impossible one. This is now the case for the CRAY III supercomputer, which is based on GaAs and runs at speeds of 1 GHz but is also several years behind and millions of dollars over budget. We expect to see more GaAs chip usage for time-critical parts of the PC/Workstation such as the cache controller and EDC as microprocessor speed reaches 100 MHz and beyond.

Review Questions

1. True or false. Soft error is permanent.
2. True or false. Hard error is permanent.
3. Alpha particle radiation causes _____ (soft, hard) errors.
4. FIT is in _____ (hours, months, years) of device operation.
5. What is the MTBF for 4 megabytes of memory if DRAM chips used are 1Mx1 with FIT = 252?
6. What is the MTBF for 4 megabytes of memory if DRAM chips used are 1Mx1 with FIT = 1000?
7. Find the MTBF in Question 5 with EDC if the word size is 32-bit.
8. Find the MTBF in Question 6 with EDC if the word size is 32-bit.
9. Compare the number of extra bits required for a 32-bit word using the following.
   (a) parity   (b) EDC
10. EDC detects and corrects the occurrence of _____ (1-bit, 2-bit) errors in the same word.

SUMMARY

The first section of this chapter provided an overview of IC technology. Transistors form the building blocks of modern electronics. The two types of transistors are bipolar and MOS. Bipolar transistors are faster, but MOS transistors are widely used because they have low power dissipation. Logic families are judged according to (1) speed, (2) power dissipation, (3) noise immunity, (4) input/output interface compatibility, and (5) cost. The IC technology used in Intel microprocessors has included PMOS in the 8008, NMOS in the 8086 and 80286, CMOS in the 80386 and 80486, and BICMOS in the Pentium.

The second section of this chapter examined several issues that must be considered in designing a microprocessor-based system: IC fan-out, capacitance derating, ground bounce, V_{CC} bounce, crosstalk, and transmission line ringing. Fan-out is a measure of how many inputs an output signal can drive. Capacitance derating refers to the problem of increased capacitance as more inputs are connected to an output. Buffers (boosters) are implemented in designs to compensate for capacitance derating. Ground bounce is caused by a massive amount of current flowing through the ground pin when many outputs change from high to low at the same time, which can lead to false signals. The effects of ground bounce can be reduced by placing V_{CC} and ground pins in the middle of the chip. Crosstalk is caused by mutual inductance of parallel lines. It can be reduced by placing ground lines between signal lines. Transmission line ringing is caused by the varied responses of harmonics to capacitance, inductance, and resistance of the line. Its effect can be reduced by placing resistors at the ends of lines.

The third section of this chapter discussed data integrity in DRAM. Errors in DRAM can be classified as soft and hard errors. A hard error occurs when a cell (bit) gets permanently stuck to 0 or 1. A soft error is not permanent but is caused by radiation and power surges. FIT (failure in time) is a measure of the occurrence of soft errors in a chip. MTBF (mean time between failures) is a measure of system memory's reliability. EDC (error detection and correction) is a means to detect and correct errors. EDC chips use Hamming code or some variation.

PROBLEMS

SECTION 26.1: OVERVIEW OF IC TECHNOLOGY

1. Why do bipolar transistors dissipate more power?
2. Why is the MOS transistor slower than the bipolar?
3. Why has the use of NMOS replaced PMOS?
4. For a TTL inverter indicate which transistors are "on" and "off" for the following.
   (a) input = high   (b) input = low
5. Repeat Problem 4 for CMOS.
6. Why in CMOS does the current dissipation rise as the frequency goes up?
7. What is the purpose of the Schottky diode in the 74LSxx family?
8. What is the noise margin for "0" and "1" in the LS family?
9. What is the noise margin for "0" and "1" in the HCMOS family?
10. Which one uses more static current, LS or HCMOS?
11. Which one is faster, LS or ALS?
12. Which one is more power efficient, LS or ALS?
13. Which one is faster, AC/ACT or FCT? Which one is more power efficient?
14. What is the FCT logic family?
15. What is the advantage of FASTr over FAST logic?
16. What is the BCT logic family?
17. True or false. The LS family is used for system frequency of less than 10 MHz.
18. True or false. The BCT family is used for system frequency of less than 15 MHz.
19. Pentium uses a line size of _____ micron.
20. Why is CMOS the technology of choice in microprocessor design?

SECTION 26.2: IC INTERFACING AND SYSTEM DESIGN CONSIDERATIONS

21. Calculate the fan-out if LS drives ALS.
22. Calculate the fan-out for LS driving unit loads.
23. Calculate the fan-out for ALS driving unit loads.
24. Calculate the number of LS that the 74LS244 can drive,
25. Find I_{OL} and I_{OH} needed to drive 10 LS and 20 CMOS input loads.
26. True or false. Capacitance derating is a function of frequency.
27. To minimize capacitance derating, use _____ (high, low) drive capability logics.
28. Calculate the path delay if one 244 is driving 2 banks each with 32 inputs and C_{in} for each input is 7 pF. Assume that the 244 delay is 8 ns and memory access time = 15 ns.
29. Repeat Problem 28 where the number of 244s is doubled.
30. Repeat Problem 29 where the number of 244s is doubled again.
31. Which current is a function of frequency, dynamic or static?
32. Give the advantages of lower V_{CC}.
33. If V_{CC} = 3.7, compare the power dissipation in comparison with V_{CC} = 5V.
34. Compare the power dissipation of memory system using the following memory organizations. Assume I_{act} =230 mA and I_{sb} = 50 mA. V_{CC}=5 V.
   (a) 256Kx1   (b) 32Kx8
35. Repeat Problem 34 using V_{CC} = 3.3 V.
36. Contrast the following 14-pin ACT chip pin-out from TI with the pin-out of TTL. Discuss the effect of V_{CC} and GND pin locations.

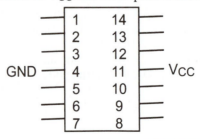

---

37. Discuss the causes and cures for ground bounce.
38. Why is the effect of V_{CC} bounce less severe than ground bounce?
39. Discuss the cause of transient current and ways to reduce its effects.
40. Discuss why many 245 drivers are used for the system data bus as shown below.

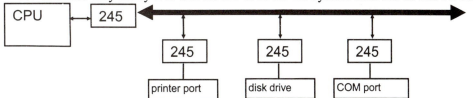

41. Discuss the cause and methods to reduce crosstalk.
42. What is the cause of ringing?
43. PCB traces behave like transmission lines most in the _____ logic family and least in the _____ logic family.
44. What is the purpose of 30 - 50 ohms resistance at the end of lines driving the DRAM arrays?

SECTION 26.3: DATA INTEGRITY AND ERROR DETECTION IN DRAM

45. Discuss the difference between soft error and hard error.
46. Give the main causes of soft error and hard error.
47. Calculate the MTBF for 16M of memory using 1Mx1 and FIT = 252.
48. Calculate the MTBF for Problem 47 with EDC if the word size is 32-bit.
49. Calculate the MTBF for 32M of memory using 1Mx1 and FIT = 680.
50. Calculate the MTBF for Problem 49 with EDC if the word size is 32-bit.
51. Calculate the MTBF for 48M of memory using 1Mx1 and FIT=252 for each of the following.
    (a) without EDC   (b) with EDC
52. In which word size is the overhead for parity and EDC the same?
53. Verify the MTBF for the 24M memory shown in Tables 26-10 and 26-11 for both cases, with and without EDC.
54. Wider data word size _____ (increases, decreases) the MTBF.
55. Higher DRAM density _____ (increases, decreases) the MTBF.

ANSWERS TO REVIEW QUESTIONS

SECTION 26.1: OVERVIEW OF IC TECHNOLOGY
1. MOS is more power efficient, while bipolar is faster.
2. true   3. true
4. CMOS   5. NMOS
6. in the lower end, ALS, and in the higher end, FAST

SECTION 26.2: IC INTERFACING AND SYSTEM DESIGN CONSIDERATIONS
1. It is the number of loads that the driver can support and it is calculated by I_{OL}/I_{IL}.
2. 10
3. I_{OL}/I_{IL}=12 mA/1.6 mA=7 and I_{OH}/I_{IH}=3 mA/40 uA=75. Fan-out is 7, a lower number.
4. The negative sign indicates that these currents are flowing out of the IC (conventional current flow).
5. They are used for the line driver: the 74xx244 for unidirectional and 74xx245 for bidirectional lines.
6. It is signal delay caused by excessive load capacitance.
7. high, low
8. Make the ground pin length as small and short as possible.
9. low, high.
10. to make the self-inductance of pins V_{CC} and GND small in order to reduce the ground and V_{CC} bounce

SECTION 26.3: DATA INTEGRITY AND ERROR DETECTION IN DRAM
1. false   2. true
3. soft   4. hours
5. 453/32=14.1 yr   6. 3.56 years (114.15 years for one DRAM divided by 32 chips)
7. 39 (32+7) chips including EDC. MTBF for 39 chips is 11.6 yr and for the system with EDC is 14,883 yr.
8. 39 (32+7) chips including EDC. MTBF for 39 chips is 2.92 yr and for the system with EDC is 3746 yr.
9. (a) extra 4 bits for parity   (b) extra 7 bits for EDC
10. 1-bit

CHAPTER 27

ISA, PCI , AND USB BUSES

OBJECTIVES

Upon completion of this chapter, you will be able to:

» **Define the meaning of the terms** *master*, *slave*, *bus arbitration*, *bus protocol*, **and** *bus bandwidth* **and describe their importance in PC design**
» **Describe the evolution of bus architecture from ISA to PCI and USB**
» **List the limitations of the ISA bus**
» **List the major characteristics of PCI architecture**
» **List the enhancements of the PCI bus over the ISA bus**
» **Contrast and compare ISA, PCI, and USB buses in terms of bus bandwidth**
» **Define the term** *local bus* **and describe its merits**
» **List the major characteristics of the PCI local bus**
» **List the major characteristics of the USB bus**
» **Contrast and compare the performance of the ISA, PCI, and USB buses**

In a given system board, the microprocessor and peripherals are connected to each other through traces of wire called *buses*. The system board also provides a path to microprocessor and peripheral signals via the expansion slot, making them accessible to add-in cards. The term *bus standard* refers to the layout, availability of system signals and resources, and signal characteristics present in the expansion slot. In this chapter we discuss the bus standards, ISA, its enhanced form EISA, PCI, and the USB bus. In Section 27.1, we look at the evolution of ISA and EISA buses. In Section 27.2, we present the merits of local buses and examine VESA and PCI local buses. Finally, Section 27.3 covers USB buses.

SECTION 27.1: ISA BUSES

Before delving into a discussion of ISA buses, we define some widely used bus terminology.

Master and slave

Many devices connected together communicate with each other through address, data, and control buses. When one device wishes to communicate with another, it sends an address to distinguish it from others since each device is assigned a unique address. It also sends a read or write signal to indicate its intention. The *master* device is the one that initiates and controls the communication while the responding device is called the *slave*. In 80x86-based PCs, the CPU is an example of a master and memory is an example of a slave.

Bus arbitration

There is only one set of global address, data, and control buses available in a given system. This means that requests by more than one master to use the buses must be arbitrated in an orderly fashion, since no bus can serve two masters at the same time. For a master to access the buses, it must ask permission from the central bus arbitrator and wait for a response before it proceeds. Depending on the system design, the central arbitrator can assign access to each master according to a priority scheme or on a first-come-first-served basis.

Bus protocol

To coordinate activity among various parts of the system, buses must follow a strict set of timing and signal specifications. The term *bus protocol* refers to these specifications for a given bus. The two major bus protocols are *synchronous* and *asynchronous*. In synchronous protocol, bus activity is synchronized according to a central frequency, the system frequency. In the IBM PC, the CPU accesses memory using synchronous protocol since memory cannot deviate from the timing specifications of the central clock oscillator. Asynchronous protocol obeys its own timing in that it decides when it is ready and does not operate according to the central clock frequency. Printer interfacing in the IBM PC is an example of asynchronous bus protocol. As discussed in Chapter 18, if the CPU is to send data to the printer, it must continuously monitor the printer's busy signal; only when the printer is not busy (ready) can it issue data to the printer's data bus. The CPU must also signal the availability of the data to the printer by the strobe signal and wait for its acknowledgment. In the asynchronous method of CPU-printer communication, the CPU is the master and the printer is the slave. The slave (printer) obeys its own timing for the acknowledge signal, independent of the system frequency. However, in CPU-memory communication, memory timing specifications are according to the system frequency and the CPU does not poll memory to see if it is ready to accept data. Asynchronous protocol is used when there is a mismatch between the bus timing of the master and slave. Normally, the slave is slower than the master and has self-timing, whereas in synchronous protocol, the timing of the master and slave match. Synchronous protocol generally has a higher rate of data transfer than asynchronous protocol.

Bus bandwidth

The rate at which a bus can transfer data from master to slave is called the *bus bandwidth* or *bus throughput*. It is measured in megabytes per second. The bus bandwidth depends on the bus speed, width, and protocol. Higher clock frequencies allow a higher bus bandwidth. In microprocessors up to the 386, bus throughput was low due to the fact that the microprocessor transferred one word per memory cycle. In 486 and Pentium microprocessors, burst mode data transfers allow much higher bus bandwidth by transferring 4 words of 32-bit data in only 5 clocks instead of the normal 8 clocks (see Chapter 23). Another way to increase bus bandwidth is to increase the width of the data bus. The Intel Pentium has done this by using a 64-bit external data bus. The single data line in serial data communication provides the lowest bus throughput, whereas a 64-bit data bus provides the highest.

ISA buses

The ISA (Industry Standard Architecture) bus is the IBM PC AT bus. Although some refer to it as an AT bus, the makers of 80x86-based IBM PC compatibles call it an ISA bus since AT is a trademark of IBM. The origin (and the limitations) of the ISA bus go back to the 8088-based IBM PC introduced in 1981. As discussed in Chapters 9 and 10, the PC/XT bus has a 62-pin connector, providing add-in boards a path to CPU and peripheral signals. It includes the D0 - D7 data bus, A0 - A19 address bus, IOR, IOW, MEMR, and MEMW control signals, and some control signals belonging to the system board and peripherals such as DMA and the interrupt controller.

With the introduction of the IBM PC AT in 1984, another 36 pins were added to the PC/XT bus to accommodate the 16-bit data bus and 24-bit address bus of the 80286 microprocessor. The extra 36 pins of the PC AT are used for the D8 - D15 data signals, A21 - A23 address signals, new DMA channel control signals, and some other system control signals. The following is an overview of the ISA bus signals. A detailed description of many of these signals was provided in Chapters 9, 10, 14, and 15. See Figure 27-1 for the layout of the ISA bus.

A0 - A19

A0 - A19 (also referred to as SA0 - SA19, where SA is for system address) provides the 20-bit address for accessing memory or I/O. These addresses are provided by the microprocessor to the expansion slot by first going through a latch (such as the 373). The latch is activated by the positive edge of the ALE signal and the address is latched into the 373 on the negative edge of ALE, thereby providing a valid stable address throughout the bus cycle. For I/O, only A0 - A15 are used.

AEN

AEN (address enable) is used to indicate who has control over the buses, the microprocessor or the DMA. When AEN = 1, the DMA has control over address, data, IOR, IOW, MEMR, and MEMW buses. When AEN = 0, the microprocessor controls the buses.

ALE

ALE (address latch enable) is also referred to as BALE (buffered ALE). When ALE goes from low to high, it indicates the presence of a valid address on pins A0 - A19. The falling edge of ALE can be used to latch addresses from the microprocessor.

CLK

Clock is the system frequency with which all memory and I/O read and write operations are synchronized. In the PC/XT, CLK = 4.7 MHz, but in the PC/AT, CLK = 6 MHz. In later PCs and compatibles, this frequency is higher.

D0 - D7

The 8-bit bidirectional data bus is the data path for the microprocessor, memory, and I/O, and is used by 8-bit I/O devices connected to an expansion slot.

DRQ1, DRQ2, DRQ3, DACK1, DACK2, DACK3

DMA request and DMA acknowledge are used to get DMA service. DRQ must be held high until the corresponding DACK goes active. BIOS is programmed for DACK to be active low. In the event that more than one of DRQ1, DRQ2, and DRQ3 is activated, DRQ1 has the highest priority and DRQ3 the lowest.

IOCHCHK

IO channel check is an active-low input signal that indicates to the motherboard the occurrence of an error on an add-in card plugged into an expansion slot (I/O channel). The term *I/O channel* is used by IBM to designate any card connected to the motherboard via the expansion slot. If there is a memory card plugged into the expansion slot, this pin is used to indicate parity error detection. Internally, the IOCHCHK pin is connected to NMI (nonmaskable interrupt) to indicate noncorrectable system errors, as was discussed in Chapter 14.

IOCHRDY

IO channel ready, an active-low input signal, is used by any I/O device or memory plugged into an expansion slot to insert wait states into the memory (or I/O) cycle. When it is pulled low (not ready), it inserts WS to prolong the memory or I/O cycle. How long this signal can be low varies among systems. For example, in the 6-MHz IBM PC/AT, this signal should not be held low more than 2.5 μs. IOCHRDY allows the interfacing of slow devices or memory to a PC.

IOR, IOW

The I/O read and I/O write control signals are both active low. IOW instructs the I/O device to grab the data off the data bus. IOR instructs the I/O device to provide (drive) its data onto the data bus.

IRQ3 - IRQ7 and IRQ9

The interrupt request (IRQ) line is used by an I/O device to signal the CPU that it needs its attention. When the IRQ signal goes from low to high (edge triggered), an interrupt request is generated. If more than one is activated, the priority is given to IRQ9, then IRQ3, IRQ4, and so on. IRQ7 has the lowest priority and IRQ9 the highest (see Chapter 14 for more discussion of these concepts).

OSC

The oscillator is an output signal with a frequency of 14.31818 MHz. It has 50% duty cycle. Some early video boards use this frequency. OSC is not synchronized with the system clock (CLK).

REFRESH

REFRESH is an active-low signal. When it is an output, it indicates that a refresh cycle is in progress. It can be an input signal driven by the device on the expansion slot to indicate a refresh cycle.

RESET DRV

Reset drive is an output signal and is active high. Internally, it is used by the motherboard to reset or initialize peripheral devices at power-up time before they are programmed by BIOS. It can be used for the same purpose by devices mounted on add-in cards plugged into expansion slots. This is the system's main reset signal and is generated by the power supply.

SMEMR, SMEMW

Both are active-low output signals. SMEMR (memory read) instructs memory to provide (drive) its data onto the data bus. SMEMW (memory write) instructs memory to grab the data off the data bus. SMEMR and SMEMW are activated only when the memory address is in the range 00000 to FFFFFH. These two signals are provided by the motherboard to distinguish between memory within 1M and memory spaces beyond 1M. As a consequence, any EEPROM mapped into the upper memory area (UMA) will use SMEMR for the memory read signal. Similarly, video RAM uses SMEMR and SMEMW for memory read and memory write signals.

TC

Terminal count, an active-high output signal, goes high when any DMA channel reaches its terminal count (see Chapter 15 for further discussion).

0WS

Zero wait state is an input signal and is active low. The wait-state generator circuitry on the motherboard

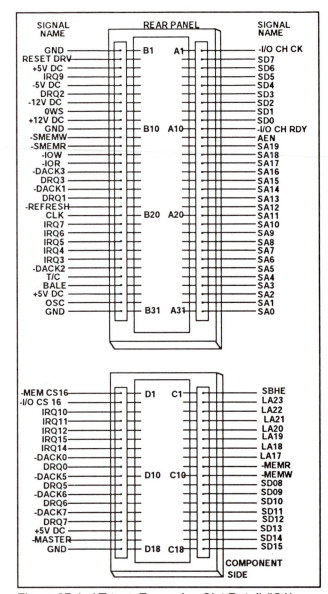

Figure 27-1. AT-type Expansion Slot Detail (ISA)
(Reprinted by permission from "IBM Technical Reference" c. 1985 by International Business Machines Corporation)

automatically inserts 1 wait state to prolong memory and I/O read and write cycles. However, by using this pin, an I/O device or memory on an add-in card can inform the wait-state generator circuitry that it can complete the present read or write cycle without a wait state. According to IBM PC AT *Technical Reference*, the 0WS input "should be driven with an open collector or tri-state driver capable of sinking 20 mA." Notice that this pin does not exist on the PC/XT bus. On ISA computers based on 286 and higher CPUs, the motherboard circuitry inserts 1 WS to make the 16-bit bus cycle a 3-clock affair. This lowers the speed to maintain compatibility with slower boards.

The 0WS pin is used to instruct the motherboard not to insert WS and let the cycle be completed in 2 clocks, the normal cycle time of 286/386/486/Pentium processors. See Figures 27-2 and 27-3.

+5V, - 5V, +12V, - 12V, GND, GND

A total of 6 pins are set aside for the power supply voltages and ground. Notice that there are only 2 ground pins in the entire 62-pin ISA bus. As we will discuss soon, this is a major obstacle in raising the ISA bus speed beyond 8 MHz.

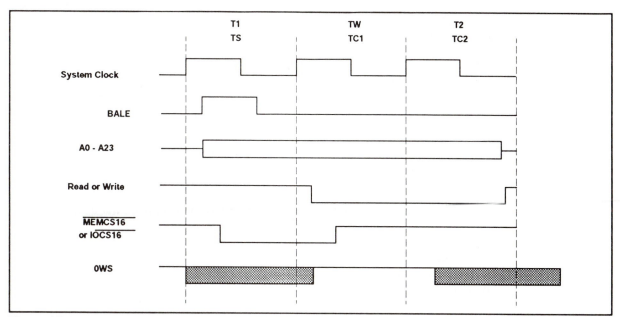

Figure 27-2. Use of 0WS in 286/386/486/Pentium ISA Bus for 3-Clock Bus Cycle
(Reprinted by permission of Intel Corporation, Copyright Intel, 1992)

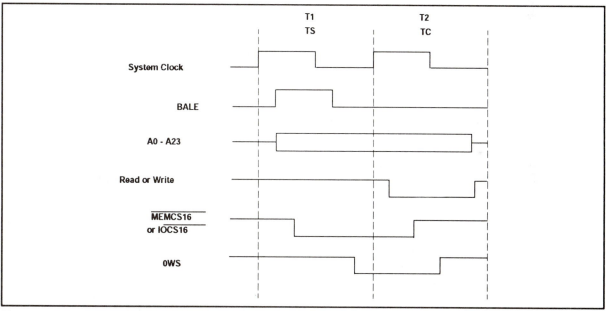

Figure 27-3. Use of 0WS in 286/386/486/Pentium ISA Bus for 2-Clock Bus Cycle
(Reprinted by permission of Intel Corporation, Copyright Intel, 1992)

36-pin part of the ISA bus

The following describes the signals of the 36-pin part of the ISA bus.

A17 - A23 (LA17 - LA23)

A17 - A23 is also called LA17 - LA23 for latchable address. We expect to have A20 - A23, but IBM also provided a duplicate of A17 - A19. Why are they duplicated? As we saw in Chapters 9 and 10, A0 - A19 in the PC/XT/AT are latched. This process of latching adds to the address path delay, leaving less time for the decoding circuitry of add-in cards. This is particularly a problem for the 286/386/486/Pentium processor since in these processors the memory cycle is only 2 clocks, unlike the 8088 memory cycle, which is 4 clocks. For this reason, LA17 - LA23 are not latched and come directly from the microprocessor. These signals are provided by the CPU the moment ALE goes from low to high and can be latched on the negative edge of ALE (going from high to low). Since these signals are not latched, they are not stable during the entire microprocessor read and write cycle. For this reason, they should be latched by the add-in card to provide stable addresses. According to IBM AT *Technical Reference*, signals LA17 - LA23 are used "to generate memory decodes for 16-bit, 1 wait-state, memory cycle." Remember that for the I/O address space, only addresses A0 - A15 are used, since we can have a maximum of 65,536 I/O ports in 80x86 microprocessors.

D8 - D15

The upper byte of the 16-bit data bus extends the 8-bit data bus of the PC/XT, making the ISA a 16-bit bus. The 16-bit data path of D0 - D15 allows the CPU to communicate with memory or I/O devices with a 16-bit data path. I/O devices can be either 8-bit or 16-bit. If the I/O device on the expansion slot is an 8-bit device, the D0 - D7 data bus must be used. D0 - D15 should be used for 16-bit I/O devices. As discussed in Chapter 10, internal circuitry exists on the motherboard called the Hi/Lo byte copier that copies data from the upper byte to D0 - D7 when a 16-bit microprocessor operation transfers data to an 8-bit device. This means that any 16-bit data transfer by the CPU to an 8-bit device is performed in two consecutive cycles where one byte is transferred in each cycle.

DRQ0, DRQ5, DRQ6, DRQ7, DACK0, DACK5, DACK6, DACK7

The 4 additional DMA channels are accessed by these signals. DRQ0 is an 8-bit DMA channel, whereas DRQ5 - DRQ7 are 16-bit channels. In the PC/XT, DRQ0 was used for DRAM refreshing, but the need for more 8-bit channels made IBM release this channel and perform DRAM refreshing by other means.

IOCS16

The I/O 16-bit chip select signal is an active-low input signal. To maintain compatibility with the 8-bit PC/XT system, the PC AT bus assumes that all data transfers are 8-bit unless it is informed otherwise. The IOCS16 input pin is used to tell the motherboard circuitry that the present I/O cycle is a 16-bit data transfer. When the data transfer is a 16-bit transfer to an 8-bit peripheral, if the IOCS16 pin is not pulled low the data transfer is performed in two consecutive I/O cycles, each transferring one byte at a time, which requires more time. Add-in cards with a 16-bit data path use the IOCS16 pin to instruct the motherboard not to convert a word transfer into a byte transfer (see Hi/Lo byte copier discussion in Chapter 10). This pin must be driven with an open collector or tri-state driver capable of sinking 20 mA. Bus data transfers are summarized as shown in Table 27-1.

Table 27-1: ISA Bus Data Transfer Summary

Data Transfer Type (From Source to Destination)	Number of WS	Number of Clocks per Cycle
From 16-bit to 16-bit	1	3
From 8-bit to 8-bit	4	6
From 16-bit to 8-bit	10	12

Reminder: In 286/386/486/Pentium machines, the memory (or I/O) cycle consists of 2 clocks when it is a zero-wait-state cycle.

MEMCS16

The memory 16-bit chip select signal is an active-low input signal. To maintain compatibility with the 8-bit PC/XT system, the PC AT bus assumes that all data transfers are 8-bit unless it is informed otherwise. This is the function of MEMCS16 input pin. It tells the motherboard circuitry that the present memory cycle is a 16-bit data transfer. In the absence of pulling low the MEMCS16 pin, a

16-bit data transfer is performed in two consecutive memory cycles, each transferring one byte at a time. This pin must be driven with an open-collector or tri-state driver capable of sinking 20 mA and must be driven by the LA17 - LA23 decoder. Note that $\overline{MEMCS16}$ must be used along with the LA17 - LA23 signals.

IRQ10, IRQ11, IRQ12, IRQ14, IRQ15

These are additional interrupt requests introduced in the PC AT system. These IRQs have a higher priority than IRQ3 - IRQ7 but lower priority than IRQ9. When the IRQ signal goes from low to high (edge triggered), an interrupt request is generated. For more discussion of IRQ, see Chapter 14.

MASTER

This is an input signal and is active low. It allows an add-in card, plugged into the expansion slot, to become the master and gain control over the system buses. In doing so, the add-in card must release the system buses in order for DRAM to be refreshed. This means that the MASTER input must not be held low for more than 15 μs. If the new master holds this pin low for more than 15 μs, the contents of the system DRAM will be lost due to lack of refreshing and the system will crash. Note that this is a major addition to the PC AT system. In the PC/XT, there was no way for an add-in card to become the bus master. The problem with MASTER in the ISA-type PC is that it must work with the DMA channels. In other words, it cannot be an independent master accessing the system bus. This is especially a problem in multiprocessor systems where many microprocessors (independent of DMA) want to access the system bus. Micro Channel and EISA buses have solved this problem. In the ISA bus, prior to assertion of the MASTER input, DRQn of the DMA channel must be asserted and DACKn must be received.

$\overline{MEMR}$, $\overline{MEMW}$

The memory read and memory write control signals are both active low. $\overline{MEMR}$ instructs the memory to provide (drive) its data onto the data bus. $\overline{MEMW}$ instructs the memory to grab the data off the data bus. $\overline{MEMR}$ and $\overline{MEMW}$ are active on all memory operations regardless of the address locations, in contrast to the $\overline{SMEMR}$ and $\overline{SMEMW}$ signals on the 62-pin section that are activated only when the memory address is in the range 00000 - FFFFFH (1M). Note that these two pins are bidirectional, allowing a master DMA from the expansion slot also to activate it for accessing memory space of the PC.

BHE

BHE, also called SBHE, is the system bus high enable. It is an active-high signal. It indicates the transfer of data on D8 - D15, the upper byte of the 16-bit data bus D0 - D15. When 16-bit devices are used in an add-in card, BHE must also be used in addition to A0 to activate the 245 data transceiver. For odd/even byte selection using A0 and BHE on 16-bit data bus systems, see Chapter 10.

+5V, GND

These two are the only power and ground pins provided in the 36-pin section of the ISA bus.

Limitations of the ISA bus

In 1984 IBM extended the life of the PC/XT bus by adding an extra 36 pins. Although this made the AT bus a 16-bit bus, it did not solve some other problems associated with the AT bus. In 1985 with the introduction of the 386 chip, a microprocessor with a 32-bit data bus, it was obvious that something had to be done about the limitations of the AT bus. The limitation of the ISA (or AT bus) are as follows.

1. The data path is limited to 16 bits; therefore, it is unable to accommodate the 32-bit data bus of the 386/486/Pentium microprocessors.
2. The 24-bit address bus limits the maximum memory accessible through the expansion slot to 16M. Therefore, it is unable to accommodate the 32-bit address bus (4 gigabyte address space) of the 386, 486, and Pentium.
3. In the ISA motherboard, there could be up to 8 ISA expansion slots. The expansion slot is bulky and has a large surface contact, resulting in a massive amount of capacitance and inductance load on each signal. The accumulated capacitance and inductance associated with all the slots, plus the problem of crosstalk, limits the working frequency of the expansion slot of the ISA bus to 8 MHz. That means that the CPU can be 20, or 33, or even 50 MHz, but when it is communicating with the expansion slot it must slow down to 8 MHz. The absence of extra ground pins to reduce the effects of crosstalk and radio-frequency emissions makes the ISA bus irredeemable for good.
4. Since the interrupts (IRQs) are edge triggered, each can be assigned only to a single device and there cannot be any sharing of the interrupt between two or more devices. In high-frequency systems, the edge-triggered interrupt can also result in false activation of the interrupt due to a spike or noise on the IRQ input.
5. The PC/XT had 3 8-bit channels (channels 1 - 3) for DMA as shown in Chapter 15. Channel 0 was used for DRAM refreshing. The IBM PC AT released channel 0 from the task of refreshing the DRAM and added three more DMA channels, all with 16-bit data transfer capability. This made the AT bus capable of handling a total of 7 DMA channels, 4 8-bit channels and 3 16-bit channels. However, the problem is that DOS cannot handle 16-bit DMA channels. This is due to the fact that a 16-bit DMA channel requires that data be aligned on even addresses, but DOS will transfer data from RAM locations with odd or even addresses. The inability of DOS to use the 16-bit DMA channels was a major dilemma for designers of the PC AT hard disk controller. This, plus the fact that the data transfer rate of 8-bit DMA channels is too low for the hard disk, made the designers of the PC AT search for a novel solution. Thanks to string instructions of the 80x86 microprocessors, the data transfer of the hard disk in all PC AT machines is performed by the CPU and not DMA. The absence of enough 8-bit DMA channels and the inability of DOS to use 16-bit DMA channels puts the burden of any mass data transfer on the CPU. This explains why the use of DMA to transfer a mass of data to devices such as a laser printer is rare. Although some devices use the 16-bit DMA, bypassing DOS, they are not industry standards. Another major problem of DMA channels is the 16M address space limitation, due to the availability of the A0 - A23 address bus. This means that 386/486/Pentium machines, with their 4 gigabyte memory space, cannot be used for DMA bus activity to transfer data to memory space located beyond 16M.
6. In the PC/XT motherboard, users had to use DIP switches to inform BIOS about the presence of various disk options, video boards, and the amount of DRAM memory installed. The PC AT motherboard replaced the PC/XT DIP switches with a real-time clock CMOS (RTCMOS) RAM chip, Motorola MC146818, and allowed the user to program options via the keyboard rather than opening the PC case and setting DIP switches. This was a major improvement, but not sufficient. In PC AT and ISA computers, DIP switches and jumpers on the add-in card still must be used to assign a unique address and IRQ to each add-in card. This makes it more complicated for users who want to add a new card to a PC. The wrong selection of DIP switches and jumpers can shut down the system. Setting them properly requires extreme attention, a set of manuals, and sometimes trial and error to get it right.

The combined effects of the above limitations means that the performance of a system with a powerful and fast microprocessor such as the 386/486/Pentium is limited by its expansion slot and system design. This fact led IBM and other PC makers to search for a solution. While IBM decided to design a whole new bus standard, radically different from the AT bus, called IBM Micro Channel, clone makers decided to go for a local bus or extending and improving the ISA bus, which they called the EISA bus. IBM Micro Channel was not made an open architecture by IBM. The industry developed other, more powerful buses; consequently, Micro Channel is no longer used.

EISA bus

The EISA bus is the PC-compatible makers' enhancement to the ISA bus. Although many of the limitations of the ISA bus discussed earlier are removed from the EISA bus, there remains one major one in that the EISA bus is limited to 8 MHz, just like the ISA and PC/XT buses. This is due to the fact that the PC industry wanted to keep the EISA bus AT-bus compatible, down to the smallest detail. The EISA bus is in reality an upgrade of ISA and consequently carries many of its limitations, but it has the A23 - A31 address lines and D16 - D31 data lines to accommodate the 386/486/Pentium. Without making the expansion board longer, the additional signals were placed in between the ISA pins as shown in Figure 27-4. Moving from the ISA to EISA, the following signals were added.

LA2 - LA16

Latchable A2 - A16 provide the address directly from the microprocessor. These are unlatched to reduce the path delay (see the explanation for LA17 - LA23 in the ISA section). Note that in ISA there are A0 - A19, but they are latched. LA2 - LA16 are provided to the expansion slots on the positive edge of ALE and can be latched on the negative edge of ALE. Since LA17 - LA23 are provided by ISA, EISA provides the latchable version of LA2 - LA16 only.

LA24 - LA31

Latchable A24 - A31. These address buses allow EISA to accommodate the 386/486/Pentium microprocessors with 32-bit address pins.

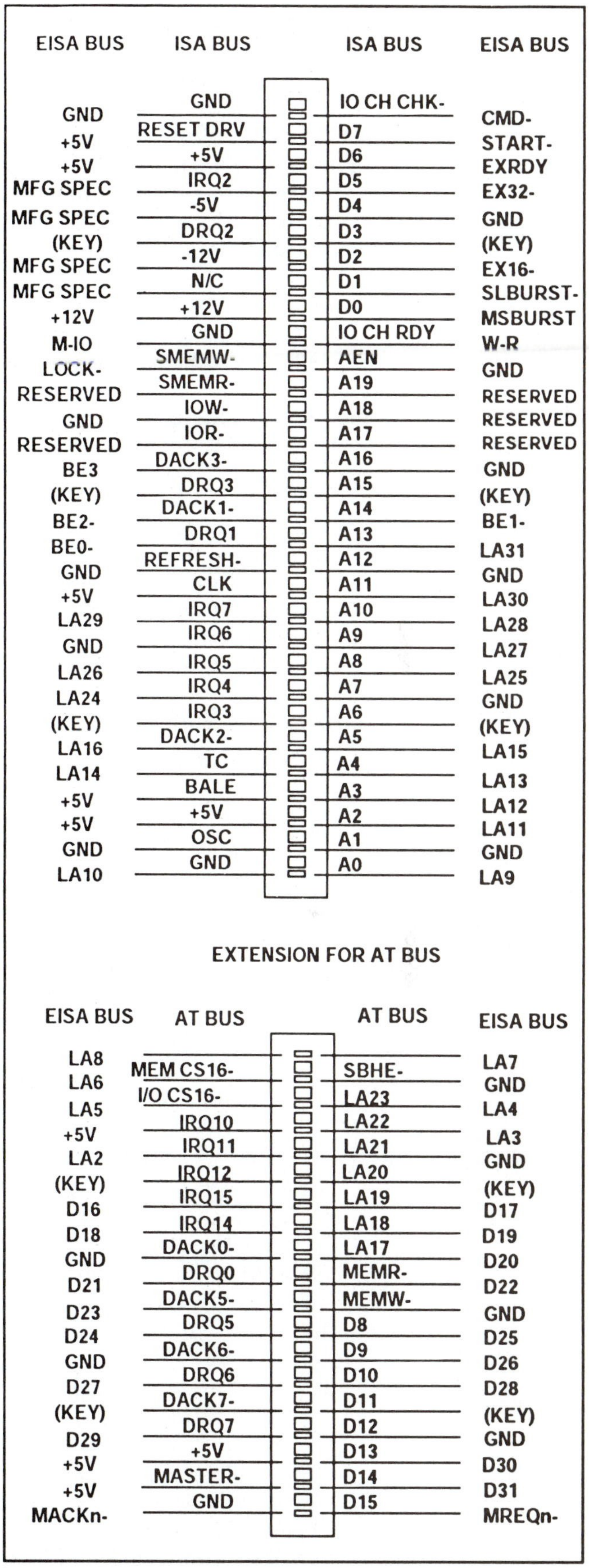

EISA BUS	ISA BUS		ISA BUS	EISA BUS
GND	GND		IO CH CHK-	CMD-
+5V	RESET DRV		D7	START-
+5V	+5V		D6	EXRDY
MFG SPEC	IRQ2		D5	EX32-
MFG SPEC	-5V		D4	GND
(KEY)	DRQ2		D3	(KEY)
MFG SPEC	-12V		D2	EX16-
MFG SPEC	N/C		D1	SLBURST-
+12V	+12V		D0	MSBURST
M-IO	GND		IO CH RDY	W-R
LOCK-	SMEMW-		AEN	GND
RESERVED	SMEMR-		A19	RESERVED
GND	IOW-		A18	RESERVED
RESERVED	IOR-		A17	RESERVED
BE3	DACK3-		A16	GND
(KEY)	DRQ3		A15	(KEY)
BE2-	DACK1-		A14	BE1-
BE0-	DRQ1		A13	LA31
GND	REFRESH-		A12	GND
+5V	CLK		A11	LA30
LA29	IRQ7		A10	LA28
GND	IRQ6		A9	LA27
LA26	IRQ5		A8	LA25
LA24	IRQ4		A7	GND
(KEY)	IRQ3		A6	(KEY)
LA16	DACK2-		A5	LA15
LA14	TC		A4	LA13
+5V	BALE		A3	LA12
+5V	+5V		A2	LA11
GND	OSC		A1	GND
LA10	GND		A0	LA9

EXTENSION FOR AT BUS

EISA BUS	AT BUS		AT BUS	EISA BUS
LA8			SBHE-	LA7
LA6	MEM CS16-		LA23	GND
LA5	I/O CS16-		LA22	LA4
+5V	IRQ10		LA21	LA3
LA2	IRQ11		LA20	GND
(KEY)	IRQ12		LA19	(KEY)
D16	IRQ15		LA18	D17
D18	IRQ14		LA17	D19
GND	DACK0-		MEMR-	D20
D21	DRQ0		MEMW-	D22
D23	DACK5-		D8	GND
D24	DRQ5		D9	D25
GND	DACK6-		D10	D26
D27	DRQ6		D11	D28
(KEY)	DACK7-		D12	(KEY)
D29	DRQ7		D13	GND
+5V	+5V		D14	D30
+5V	MASTER-		D15	D31
MACKn-	GND			MREQn-

Figure 27-4. EISA Bus Layout

D16 - D31

The data bus is extended to 32-bit for 386/486 32-bit microprocessors.

BE0 - BE3

Bus high enable are the pins with the same name and function as the 386/486 microprocessor. It allows selection of any of the four bytes of D0 - D7, D8 - D15, D16 - D23, or D24 - D31. Notice that there is no LA0 and LA1 on the EISA bus since we have BE0 - BE3.

CMD

Command is an active-low signal used by the EISA motherboard to provide the timing reference for valid data read or write.

EXE16

This signal is used to indicate that the current bus cycle is a 16-bit data transfer.

EXE32

This signal is used to indicate that the current bus cycle is a 32-bit data transfer. By default EISA will act like an ISA bus and it assumes that data transfers are 8-bit unless EXE16 or EXE32 is activated to indicate that the current bus cycle can support 16- or 32-bit data transfers.

EXRDY

This is used by add-in cards to request wait-state insertion.

MAKn

Master acknowledge n is an output signal indicating that master number n has been acknowledged. This is in response to MREQn, explained next.

MREQn

Master request is an input to bus arbitration circuitry signaling request for use of the buses by master number n.

MSBURST and SLBURST

These are two signals used by the master or slave to indicate that they support burst mode data transfer. These two signals are especially useful for the 486 and higher microprocessors that support burst mode data transfer.

M-IO

This is used by the master to indicate if the EISA bus cycle is a memory cycle or an I/O cycle. When M-IO = 0 it is an IO cycle, and if M-IO = 1 the cycle is a memory cycle.

START

This is used by the master to indicate the start of the EISA bus cycle.

To prevent mixup between the add-in cards and the system boards there are keys in the expansion slot which effectively block insertion of the wrong card. See Figures 27-5 and 27-6. ISA cards can be plugged into an EISA expansion slot, but an EISA card should not be plugged into an ISA expansion slot.

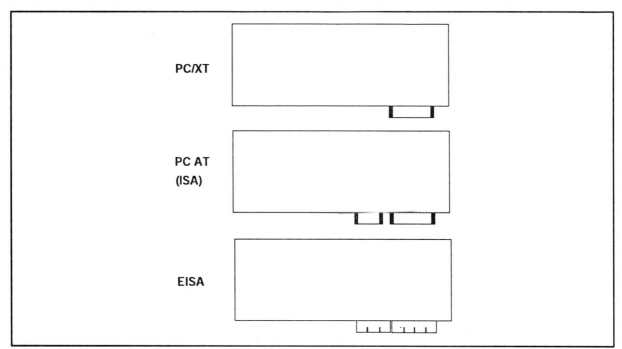

Figure 27-5. PC/XT, PC AT (ISA), and EISA Add-In Cards

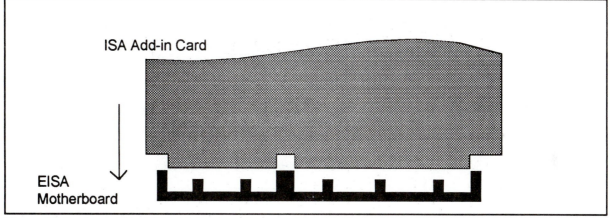

Figure 27-6. Keys of the EISA Motherboard

EISA slot numbering

A major improvement of the EISA bus over the ISA bus is its automatic configuration. For an ISA-based PC to install an add-in card into an expansion slot requires setting switches and jumpers. EISA-based PCs do not have this limitation. An ID number is assigned to each slot and its BIOS is modified to work with the setup software in configuring the system and resolving any conflicts. Every EISA motherboard has NVRAM to store the system configuration setup, where it keeps all the information about the IRQ number, the I/O port address, the memory address, and DMA channels used by each add-in card. This information is provided to BIOS when the system is turned on. When you buy an EISA card to be plugged into an expansion slot of the EISA motherboard, you also get the configuration file for that card. After installing the card, you run the configuration file to inform the NVRAM of the new card. It is the job of the EISA motherboard and EISA card to work together in configuring the system to remove any user involvement in assigning resources or in removing any conflict. This means that in designing any EISA-based PC or an EISA add-in card, you must abide by the rules set by the EISA consortium. One of these rules is the way an ID number is assigned to each slot. The slot numbering of the EISA is described as follows.

1. The system board (motherboard) is assigned slot number 0.
2. The physical slots are numbered sequentially from 1 to a maximum of 15.
3. The embedded devices, which would have otherwise used a slot, are numbered sequentially from one more than the highest physical slot to a maximum of 63.

Example 27-1

In a given EISA motherboard, the video board is embedded into the motherboard and there are 3 physical expansion slots available on the motherboard, which can be used for plug-in cards. What ID number is assigned to the permanently installed video board?

Solution:

Since the motherboard has slot 0, the three expansion slots are numbered 1, 2, and 3 and finally, the embedded video card is assigned ID number 4.

Bus performance comparison

Since the EISA bus is an extension of the ISA bus and since ISA in turn is an extension of the PC/XT bus, EISA has many of the performance limitations of the early PCs. See Example 27-2. A comparison of ISA and EISA bus performance is shown in Table 27-2.

Example 27-2

Calculate and compare the fastest possible bus bandwidth for (a) ISA and (b) EISA bus expansion slots. In EISA, assume that the transfer is nonburst.

Solution:

The fastest bus bandwidth uses 2 clocks per cycle for memory or I/O cycles. Since the frequency is limited to 8 MHz, each clock is 125 ns (1/8 MHz =125 ns).

(a) $2 \times 125 = 250$ ns per cycle and each cycle transfers 2 bytes since the data path is 16-bit. Therefore, ISA bus bandwidth = (1/250 ns) × 2 bytes = 8 megabytes per second
(b) $2 \times 125 = 250$ ns per cycle and each cycle transfers 4 bytes since the data path is 32-bit. Therefore, EISA bus bandwidth = (1/250 ns) × 4 bytes = 16 megabytes per second

---

Table 27-2: ISA and EISA Bus Bandwidth

	ISA	EISA
Maximum data path	16-bit	32-bit
Expansion slot bus speed (MHz)	8	8
Max. bandwidth (megabytes/second)	8	16

In the EISA documentation, often 8.3 MHz is mentioned instead of 8 MHz, where the 33 MHz microprocessor frequency is divided by 4 to get 8.3-MHz EISA bus speed.

It must be noted that EISA has burst mode data transfer, but since EISA specifications were cemented before introduction of the 486 microprocessor, EISA burst mode is not compatible with 486 burst data transfer of 2-1-1-1. See Chapter 23 for a discussion of 2-1-1-1 burst mode in the 486/Pentium. In the next section we discuss how the local bus overcomes the performance limitations of ISA and EISA.

Review Questions

1. The device that initiates the communication is the _____ (master, slave).
2. True or false. In synchronous protocol, the bus activity is according to a central clock frequency.
3. The CPU-printer communication is of _____ (asynchronous, synchronous) protocol.
4. How many ground pins exist in the 62-pin section of ISA?
5. The ISA is a(n) _____ (16-bit, 8-bit) data bus.
6. Can the MASTER pin in the ISA bus be used by a microprocessor without the involvement of the DMA channel?
7. Does EISA support automatic configuration?
8. Does ISA support automatic configuration?
9. Although the ISA and EISA buses have the same frequency, why is the bus bandwidth of EISA twice that of ISA?

SECTION 27.2: PCI LOCAL BUSES

Just as a high-performance car needs high-performance roads (no bumps, no speed limit) to explore its full potential, high-performance CPUs also require high-performance buses. While microprocessor performance is rapidly rising, buses are not keeping up. Many high-performance systems such as CRAY supercomputers and mainframes use their own proprietary buses, but their limited use makes them nonstandard and consequently expensive. When 286 microprocessors of 10 to 16 MHz were used, many manufacturers resorted to proprietary buses to overcome the 8-MHz limitation associated with the ISA bus. This was especially the case where memory was concerned. In 80286/386 systems with 16- or 20-MHz speed, memory boards plugged into expansion slots could be accessed no faster than 8 MHz. This fact led manufacturers such as Compaq to have their own memory expansion modules. In such systems, while ISA expansion slots are used for peripheral boards such as video, hard disk, or network cards, memory expansion was done by a specially designed slot on the motherboard used only for memory modules. These memory modules work at the same speed as the CPU, or close to it. These systems were often advertised as *dual-bus systems*. One bus was for the ISA cards and another one was for the memory modules. In the late 1980s with the widespread adaptation of SIP (single in-line pin) and SIMM (single in-line memory module), this problem was resolved. However, the lack of a bus standard for video and other adapter cards such as disk controllers forced PC board designers to come up with what is called a *local bus*.

Definition and merits of local bus

The idea of a local bus is to access the system buses at the same speed as the microprocessor, or close to it. In a 33-MHz microprocessor system with both ISA and local buses, the speed of the ISA bus signals is limited to 8 MHz, but the local bus signals are accessed at the same speed as the CPU, 33 MHz. In PC/XT systems of 4.7 MHz, the XT buses were accessed at the same speed as the 8088 microprocessor. The gap between CPU speed and expansion slot speed started to develop when the 80286 speed exceeded 8 MHz. In those days, there were not many devices that needed speed beyond 8 MHz. This changed with the introduction of graphical user interface (GUI) software such as Microsoft Windows. In ISA bus systems, even the 16-bit video card plugged into the ISA expansion slot was not fast enough to keep up with the demand of the graphics software. This led some PC manufacturers to embed the video card into the motherboard and bypass the use of an ISA expansion slot for the video board. The problem with this option is that if the video section of the motherboard goes bad, one must either discard the motherboard or connect a video card to the expansion slot, depending on how the system board is designed. To solve the problem of slow video speed in ISA systems, some video board makers used a graphics processor to relieve the main CPU, the 386/486, from the burden of data manipulation of graphic data stored in video RAM. In the absence of a graphic processor on the video board, the main CPU is responsible for graphic data manipulation, which means that it must go through the slow ISA bus to access the data since the 80x86 CPU is connected to the video RAM through the ISA bus. The use of a specially designed processor called a *graphic processor* on the video board with the sole responsibility of taking care of the calculation-intensive work of graphics provided a major improvement in video systems of the PC. However, it has one limitation. If the graphic data needs to be transferred from the disk to video RAM (or vice versa), it must still go through the slow ISA bus. Table 27-3 shows the bus bandwidth requirements for graphics and real-time video.

Table 27-3: Bus Bandwidth Requirements for Graphics and Real-Time Video

Graphics			
Resolution (pixels)	**Colors (bits/pixels)**	**Redraw Rate (update/s)**	**Bandwidth (bytes/s)**
640x480	8	10	2.9M
1024x768	16	10	15M
1280x1024	24	10	37.5M
Real-Time Video			
Frame Size (pixels)	**Colors (bits/s)**	**Frame Rate (frames/s)**	**Bandwidth (bytes/s)**
160x120	8	15	288K
320x240	24	15	3.5M
640x480	24	30	26.3M
1024x768	24	30	67.5M

Example 27-3

Verify the bus bandwidth requirement for each of the following.
(a) 1024x768 resolution, 16 colors, 10 redraw rate
(b) 640x480 resolution, 24 colors, 30 frames per second

Solution:

(a) Bus bandwidth $= 1024 \times 768 \times 16 \times 10 = 125,829,120$ bits/second $=15$ megabytes/second
(b) Bus bandwidth $= 640 \times 480 \times 24 \times 30 = 221,184,000$ bits/second $=26.3$ megabytes/second

---

Table 27-3 and Example 27-3 explain why the urge for a standard local bus was initiated by group called Video Electronics Standards Association (VESA). This widely used local bus standard in today's PC is called *VESA local bus,* or *VL bus* for short. Although the VL bus was created through the efforts of video board makers, most PC makers have adapted it out of necessity and use it even for many other peripherals, such as disk controllers. With the introduction of high-performance microprocessors such as the Pentium, Intel had to do something about bus performance lest their processor be buried under slow buses. For this reason Intel introduced a new local bus standard called PCI (peripheral component interconnect). Due to the superior performance and characteristics of the PCI local bus, we describe briefly some aspects of the VL bus and concentrate on the PCI for the remaining part of this section.

VL bus (VESA local bus) characteristics

The following are the major characteristics of the VL bus.
1. VL bus version 1.0 is a 32-bit bus. However, the 64-bit version is under development, referred to as version 2.0.
2. The VL bus can work up to a maximum of 33 MHz clock frequency with a maximum of 3 expansion slots. It can go to 40 MHz if the number of slots is reduced to 2 and to 50 MHz if there is only one VL bus slot.
3. It does not specify the automatic configuration standard. Unlike EISA, automatic configuration is not an integral part of the VL bus.

PCI local bus

High-performance microprocessors such as the 486 and Pentium require a high bus bandwidth to take advantage of their full potential. Therefore, it is not surprising that Intel became involved in defining a new bus standard. Although Intel came up with the specifications of the PCI local bus, it has become available free of charge to all PC and add-in board manufacturers. PCI was conceived as a specification standard for peripheral connections for Intel's high-performance microprocessors such as the 80486 and Pentium. Later, with encouragement and input from the PC industry, it became a local bus standard with the pin-out for expansion slot connections. It has incorporated the following major characteristics: (a) burst mode data transfer, (b) level-triggered interrupts, (c) bus mastering, (d) automatic configuration, and (e) high bus bandwidth. More important, it has a bridge, which allows any kind of add-in card based on ISA or EISA, to be plugged into the PCI local bus. PCI local bus characteristics are listed next.

PCI local bus characteristics

1. It has a maximum speed of 33 MHz.
2. It has 32- and 64-bit data paths.
3. It supports burst mode data transfer of 2-1-1-1 used by microprocessors such as the 486 and Pentium.
4. It supports bus mastering, allowing the implementation of multiprocessors where any number of microprocessors can become master and take control of the buses.
5. It is compatible with ISA and EISA. With implementation of a bus bridge, it supports the slow ISA and EISA buses as shown in Figure 27-7. Buffers in the bridge allow the microprocessor to write into the buffer and go about its own business, leaving the task of handling the slow ISA/EISA to the bridge.
6. The PCI local bus is processor independent. It can be used with any microprocessor, not just Intel 80x86. For this reason, companies such as Digital Equipment and Apple have also announced support for the PCI to be used with their non-80x86 microprocessors. This feature ensures that future changes in the 80x86 family will not make the PCI an obsolete bus.
7. It supports both 5- and 3.3-V expansion cards, allowing smooth transition from 5- to 3.3-V systems. The placing of small cutouts (keys) prevents users from plugging a card with one voltage into a motherboard with a different voltage.
8. It provides autoconfiguration capability, where a user can install a new add-in card without setting DIP switches, jumpers, and selecting the interrupt. Configuration software automatically selects an unused address and interrupt to resolve conflicts.
9. It has a ground or V_{CC} pin between every two signals to reduce crosstalk and radio-frequency emissions.
10. It implements level-triggered interrupts, which support interrupt sharing.
11. It supports up to 10 peripherals. Some of the peripherals must be embedded into the motherboard.
12. The maximum number of expansion slots working at 33 MHz varies, depending on the 5-V or the 3.3-V implementation. The increase in the number of expansion slots beyond 5 means a speed lower than 33 MHz. The use of a highly refined connector with a small area of contact makes the PCI bus a high-frequency bus.

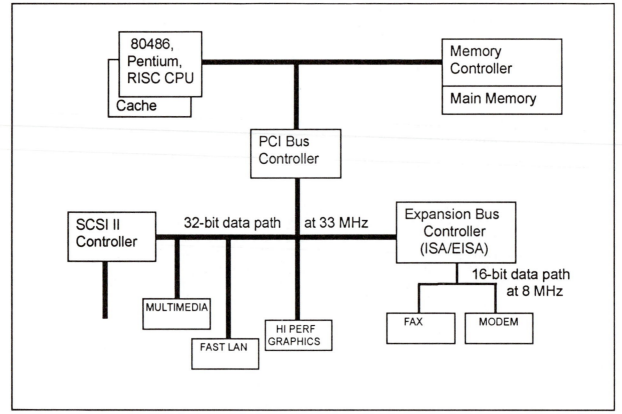

Figure 27-7. PCI Local Bus Architecture
(Reprinted by permission of Intel Corporation, Copyright Intel, 1993)

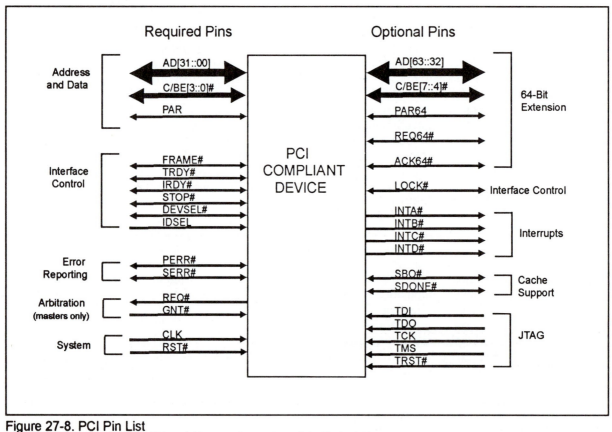

Figure 27-8. PCI Pin List
(Reprinted by permission of PCI Special Interest Group, Copyright 1992, 1993)

Pin	5V Environment Side B	5V Environment Side A	3.3V Environment Side B	3.3V Environment Side A	Pin	5V Environment Side B	5V Environment Side A	3.3V Environment Side B	3.3V Environment Side A
1	-12V	TRST#	-12V	TRST#	50	CONNECTOR KEY		Ground	Ground
2	TCK	+12V	TCK	+12V	51	CONNECTOR KEY		Ground	Ground
3	Ground	TMS	Ground	TMS	52	AD[08]	C/BE[0]#	AD[08]	C/BE[0]#
4	TDO	TDI	TDO	TDI	53	AD[07]	+3.3V	AD[07]	+3.3V
5	+5V	+5V	+5V	+5V	54	+3.3V	AD[06]	+3.3V	AD[06]
6	+5V	INTA#	+5V	INTA#	55	AD[05]	AD[04]	AD[05]	AD[04]
7	INTB#	INTC#	INTB#	INTC#	56	AD[03]	Ground	AD[03]	Ground
8	INTD#	+5V	INTD#	+5V	57	Ground	AD[02]	Ground	AD[02]
9	PRSNT1#	Reserved	PRSNT1#	Reserved	58	AD[01]	AD[00]	AD[01]	AD[00]
10	Reserved	+5V(I/O)	Reserved	+3.3V(I/O)	59	+5V(I/O)	+5V(I/O)	+3.3V(I/O)	+3.3V(I/O)
11	PRSNT2#	Reserved	PRSNT2#	Reserved	60	ACK64#	REQ64#	ACK64#	REQ64#
12	Ground	Ground	CONNECTOR KEY		61	+5V	+5V	+5V	+5V
13	Ground	Ground	CONNECTOR KEY		62	+5V	+5V	+5V	+5V
14	Reserved	Reserved	Reserved	Reserved		CONNECTOR KEY		CONNECTOR KEY	
15	Ground	RST#	Ground	RST#		CONNECTOR KEY		CONNECTOR KEY	
16	CLK	+5V(I/O)	CLK	+3.3V(I/O)	63	Reserved	Ground	Reserved	Ground
17	Ground	GNT#	Ground	GNT#	64	Ground	C/BE[7]#	Ground	C/BE[7]#
18	REQ#	Ground	REQ#	Ground	65	C/BE[6]#	C/BE[5]#	C/BE[6]#	C/BE[5]#
19	+5V(I/O)	Reserved	+3.3V(I/O)	Reserved	66	C/BE[4]#	+5V(I/O)	C/BE[4]#	+3.3V(I/O)
20	AD[31]	AD[30]	AD[31]	AD[30]	67	Ground	PAR64	Ground	PAR64
21	AD[29]	+3.3V	AD[29]	+3.3V	68	AD[63]	AD[62]	AD[63]	AD[62]
22	Ground	AD[28]	Ground	AD[28]	69	AD[61]	Ground	AD[61]	Ground
23	AD[27]	AD[26]	AD[27]	AD[26]	70	+5V(I/O)	AD[60]	+3.3V(I/O)	AD[60]
24	AD[25]	Ground	AD[25]	Ground	71	AD[59]	AD[58]	AD[59]	AD[58]
25	+3.3V	AD[24]	+3.3V	AD[24]	72	AD[57]	Ground	AD[57]	Ground
26	C/BE[3]#	IDSEL	C/BE[3]#	IDSEL	73	Ground	AD[56]	Ground	AD[56]
27	AD[23]	+3.3V	AD[23]	+3.3V	74	AD[55]	AD[54]	AD[55]	AD[54]
28	Ground	AD[22]	Ground	AD[22]	75	AD[53]	+5V(I/O)	AD[53]	+3.3V(I/O)
29	AD[21]	AD[20]	AD[21]	AD[20]	76	Ground	AD[52]	Ground	AD[52]
30	AD[19]	Ground	AD[19]	Ground	77	AD[51]	AD[50]	AD[51]	AD[50]
31	+3.3V	AD[18]	+3.3V	AD[18]	78	AD[49]	Ground	AD[49]	Ground
32	AD[17]	AD[16]	AD[17]	AD[16]	79	+5V(I/O)	AD[48]	+3.3V(I/O)	AD[48]
33	C/BE[2]#	+3.3V	C/BE[2]#	+3.3V	80	AD[47]	AD[46]	AD[47]	AD[46]
34	Ground	FRAME#	Ground	FRAME#	81	AD[45]	Ground	AD[45]	Ground
35	IRDY#	Ground	IRDY#	Ground	82	Ground	AD[44]	Ground	AD[44]
36	+3.3V	TRDY#	+3.3V	TRDY#	83	AD[43]	AD[42]	AD[43]	AD[42]
37	DEVSEL#	Ground	DEVSEL#	Ground	84	AD[41]	+5V(I/O)	AD[41]	+3.3V(I/O)
38	Ground	STOP#	Ground	STOP#	85	Ground	AD[40]	Ground	AD[40]
39	LOCK#	+3.3V	LOCK#	+3.3V	86	AD[39]	AD[38]	AD[39]	AD[38]
40	PERR#	SDONE	PERR#	SDONE	87	AD[37]	Ground	AD[37]	Ground
41	+3.3V	SBO#	+3.3V	SBO#	88	+5V(I/O)	AD[36]	+3.3V(I/O)	AD[36]
42	SERR#	Ground	SERR#	Ground	89	AD[35]	AD[34]	AD[35]	AD[34]
43	+3.3V	PAR	+3.3V	PAR	90	AD[33]	Ground	AD[33]	Ground
44	C/BE[1]#	AD[15]	C/BE[1]#	AD[15]	91	Ground	AD[32]	Ground	AD[32]
45	AD[14]	+3.3V	AD[14]	+3.3V	92	Reserved	Reserved	Reserved	Reserved
46	Ground	AD[13]	Ground	AD[13]	93	Reserved	Ground	Reserved	Ground
47	AD[12]	AD[11]	AD[12]	AD[11]	94	Ground	Reserved	Ground	Reserved
48	AD[10]	Ground	AD[10]	Ground					
49	Ground	AD[09]	Ground	AD[09]					

Figure 27-9. Pinout of the PCI Connector
(Reprinted by permission of PCI Special Interest Group, Copyright 1992, 1993)

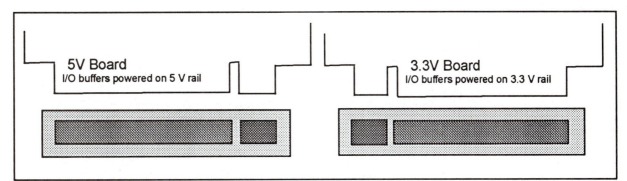

Figure 27-10. PCI Board Connectors
(Reprinted by permission of PCI Special Interest Group, Copyright 1992, 1993)

Plug and play feature

The PCI is equipped with the autoconfiguration feature but at the same time it has a slot for the ISA bus, in which the autoconfiguration is not supported. How can this work? This lack of autoconfiguration is a major headache for computer users and network managers. This led Microsoft and Intel to work together to equip the ISA bus with the autoconfiguration feature. This feature is often referred to as *plug and play*. The PCI autoconfiguration feature can work completely only after the ISA cards and BIOS are equipped with autoconfiguration (plug and play), since EISA is already equipped with this important feature. Plug and play falls into the following three categories.

1. Neither the motherboard BIOS nor the add-in card is equipped with the plug-and-play feature. This is sometimes called "plug and pray." You may get it to work by trial and error.
2. The motherboard BIOS is equipped with plug and play, but the add-in card is not. In this case, setup software will help you to assign the I/O addresses, IRQs, and DMA channels.
3. Both the motherboard BIOS and the add-in card are equipped for plug and play. In this case, autoconfiguration will take care of everything. It will assign I/O addresses, IRQs, and DMA channels without any user involvement.

PCI connector

A few points must be noted about the PCI connector. First, notice in Figure 27-8 that very few PCI signals match the signals of 80x86 microprocessors. The reason is that PCI is a mezzanine bus, meaning that the PCI controller sits between the CPU and the external bus connection. In this way, any CPU can be used with the PCI bus. Standardizing the bus connection frees the CPU buses from any restriction. This is in contrast to the VL bus (version 1.0), in which signals come directly from the 486 pins and have the same name. When new signals were added to the Pentium, the VL bus had to be upgraded (version 2.0). PCI solves this problem by being microprocessor independent.

Another point to be noted is the multiplexing of address and data on the PCI bus, since the same pins are used for address and data. In the first clock, the address is provided and in the second clock, the data is provided. Therefore, the PCI bus has a cycle time of 2 clocks in nonburst mode, just like the 386/486/Pentium. For burst mode, in the first clock the address is provided and in each subsequent clock a word (32-bit) of data is provided.

Another point to be noted is the 64-bit extension for the PCI bus. The PCI bus can be implemented for a 32-bit data bus or a 64-bit data bus. The 32-bit sections end at pin 62. The pinout is shown in Figure 27-9. Pins 63 through 94 are used for 64-bit data/address extension only.

Notice also in Figure 27-9 that every third pin is dedicated to ground or V_{CC}. This eliminates the crosstalk problem and allows the bus to be used for frequencies up to 33 MHz.

Figure 27-10 illustrates the PCI board connectors.

PCI performance

The PCI local bus supports both single memory cycle and burst mode. In the single cycle, it takes 2 clocks to read or write a word of data. In the first clock, the address is provided and in each subsequent clock, the data is accessed. This makes it 2-1-1-1-1-1.... Example 27-4 calculates the bus bandwidth for the PCI.

Table 27-4 provides the performance comparison of all the buses for non-burst mode data transfer.

Example 27-4

Calculate the bus bandwidth of PCI for (a) single and (b) burst transfer, both on a 32-bit data path.

Solution:

PCI can work up to a maximum of 33 MHz. The clock period is 30 ns.

(a) For the single transfer, each transfer takes 2 clocks or a total of 60 ns to transfer 4 bytes (32 bits) of data. Therefore, bus bandwidth = (1/60 ns) × 4 bytes = 66.6 megabytes/second

(b) In burst mode, ignoring the overhead of the first clock for the address, it takes 1 clock or 30 ns to transfer 32-bit data. Therefore, bus bandwidth = (1/30 ns) × 4 bytes = 133 megabytes/second.

Table 27-4: ISA, EISA, VL, and PCI Local Bus Bandwidth Comparison

	ISA	EISA	VL Bus	PCI	PCI
Data path (bits)	16	32	32	32	64
Bus speed (MHz)	8	8.3	33	33	33
Bandwidth (megabytes/s)	8	16	66	66	133

Note: In the bus bandwidth calculation, 2 clocks per memory cycle are assumed.

Review Questions

1. What is the local bus?
2. The memory expansion connections to 16-MHz CPUs are through the _____ (ISA bus, local bus).
3. What is a dual bus system?
4. Which needs a local bus, the modem or the hard disk controller?
5. True or false. PCI is a 32- and a 64-bit bus.
6. How has PCI reduced the effects of crosstalk (EMI) for high-frequency systems?

SECTION 27.3: USB PORT

USB stands for *universal serial bus*. Next to PCI, it is one of the most important additions to the PC system in recent years. In this section we provide an overview of the USB bus. To see the need for the USB bus, we first review the limitations and benefits of ISA and PCI buses, as well as serial and parallel ports.

ISA and PCI buses provide a high rate of data transfer between the CPU and the outside world. This is because they have a wide data path, high frequency bus speed, and communicate directly with the CPU. As mentioned earlier, the ISA bus is a 16-bit bus and has a speed of 8 MHz. For the PCI bus, the data path is 64 bits wide and it has a maximum speed of 64 MHz.

The following are some of the major limitations of the ISA and PCI buses.
1. Both ISA and PCI are inside the PC; therefore, to access them you need to open the PC's case and plug the card into an expansion slot.
2. The PCI and ISA expansion slots take too much physical space on the motherboard. This limits the number of expansion slots that are available on a given motherboard.
3. Both ISA and PCI buses require too much power. For every ISA expansion slot, an extra 25 watts must be incorporated into the PC's power supply; every PCI slot requires an extra 10 watts. As a result, a motherboard with 2 ISA and 3 PCI expansion slots has burdened the PC power supply with an additional 80 watts of power (2 × 25 + 3 × 10 = 80). This can make a significant difference in handheld and laptop systems, and that is the reason that neither of these devices have any expansion slots on their boards.

The most important disadvantage of serial and parallel ports is the limit of 4 of each in a given motherboard. The PC BIOS limits the number of serial (COM) and parallel (LPT) ports to 4. Of course this is a theoretical limitation imposed by the BIOS. Practically speaking, there are a limited number of IRQs to assign to all of these LPT and COM ports. This is the reason that there are no more than 2 COM ports and 1 LPT port on motherboards. There are other limitations associated with the COM and LPT ports that come to light only when compared with the major features of the USB port.

Major features of USB

Here are some of the most important features of the USB. They demonstrate why this is one of the most important additions to PC architecture in recent years.

1. A single USB port on the back of the PC can accommodate up to 127 devices such as a mouse, scanner, printer, modem, etc. The devices are daisy chained together with the help of external hubs. The devices are recognized automatically by the PC. Many devices such as a printer and monitor can be equipped with a hub, thereby saving the additional expense of buying a separate hub. More importantly, daisy chaining the devices via hubs requires no opening of the PC case when connecting additional devices to a PC.
2. The data transfer rate is between 1.5 to 12 megabits per second (Mbs) at this time. Intel is discussing with the industry to raise the maximum limit to 200 Mbs.
3. USB is hot-pluggable. This means that new devices can be connected to the USB port without first turning off the PC. Remember that was not the case with ISA, PCI, LPT, and COM ports. In all these devices, the PC had to be turned off prior to installing the device and configuring the system. The hot-pluggability of USB is one of its most important features.
4. USB does not have to burden the system's power supply. Unlike ISA and PCI, connecting additional USB devices to a PC does not require an exorbitant amount of power from the PC power supply. Each new device requires no more than a maximum of 500 mW from the PC power supply. More importantly, the USB hubs sitting outside the PC can have their own power supply, thereby relieving the motherboard's power supply of the burden of providing power to every device. For example, a printer or monitor that is equipped with a USB hub can provide power to all external USB devices. The USB is also equipped with power managing capability. This allows a device that is not being used for a period of time to be powered down into sleep mode.

Software compatibility

It must be noted that Microsoft's Windows 95 does not come with USB drivers and other support software. However, you can download them from the Microsoft Web site. The same is true for Windows NT 4.00. Starting with Windows 98, USB is part of the operating system; the same is true for Windows NT 5.00.

Bus comparison

Table 27-5 shows the comparison of ISA, PCI, COM, LPT, and USB. Notice in the calculation of data transfer rate (bus bandwidth) for ISA and PCI buses that a 2-clock read and write cycle is assumed. For the LPT port, the 2 microsecond timing for the parallel port is assumed. Also notice that in serial COM and USB ports, there is only a single wire for transfer (RxD) and another single wire for receive (TxD).

Both Microsoft and Intel are working to eliminate ISA bus, LPT, and COM ports from PC motherboards. Early 2000 may bring PCs with PCI and USB ports only.

Table 27-5: Date Transfer Rate for Buses and Ports

Bus/Port	Data Path bit	Maximum Bus Bandwidth
ISA (8 MHz)	16	8 M bytes/second
PCI (33 MHz)	32	66 M bytes/second
PCI (33 MHz)	64	133 M bytes/second
PCI (66 MHz)	64	266 M bytes/second
USB	1	12 M bits/second
LPT	8	500 K bytes/second
COM port	1	56 K bits/second

Note: In the bus bandwidth calculation, 2 clocks per memory cycle are assumed.

Review Questions

1. True or false. Both PCI and USB are hot-pluggable.
2. What advantage do USB devices have over other devices in terms of system power requirements?
3. True or false. Each USB device must have its own USB connection to the system via an expansion slot.

SUMMARY

This chapter began with an overview of bus terminology. A *master* device initiates and controls communication over buses with a *slave* device. *Bus arbitration* is the method of deciding which device can control the buses at a given point in time. *Bus protocol* refers to timing specifications and signal directions. The two protocols are *synchronous*, where bus activity is synchronized according to a central clock, and *asynchronous*, where bus activity proceeds according to the timing of the devices involved. *Bus bandwidth* is a measure of the rate at which a bus can transfer data from master to slave or vice versa. The ISA (Industry Standard Architecture) bus is the IBM PC AT bus. It consists of two physical parts, a 62-pin portion, which is the PC/XT bus, and a 36-pin portion, which was added to accommodate the 16-bit data bus and 24-bit address bus of the 80286, and other advanced features. Manufacturers developed an enhanced ISA bus, called EISA, which featured 32-bit address and data paths, plus other features. ISA cards can be plugged into an EISA expansion slot, but EISA cards cannot be plugged into an ISA slot.

The second section of this chapter described the local bus, which was developed so that the system buses could be accessed at speeds close to the speed of the microprocessor. The VL bus, or VESA bus, is a 32-bit bus that can work up to a maximum frequency of 33 MHz with 3 expansion slots. The PCI local bus has both 32- and 64-bit data paths and a maximum speed of 33 MHz. PCI supports many advanced features critical to 486 and higher CPUs, such as burst mode data transfer and bus mastering, and allows use of ISA and EISA cards.

The third section of this chapter covered a recent addition to PC systems: USB ports. USB stands for *universal serial bus*. USB ports can accommodate numerous USB devices which are daisy-chained together. Devices that contain USB hubs can have other devices connected to them via the hub, and can provide power supply to those devices. USB is hot-pluggable, meaning that new devices can be connected via a hub without shutting down the system and opening the PC case.

PROBLEMS

SECTION 27.1: ISA BUSES

1. Explain the difference between master and slave devices.
2. Why is there a need for bus arbitration?
3. What is the function of central bus arbitration?
4. The CPU/memory relation is an example of _____ (asynchronous, sychnronous) protocol.
5. The CPU/printer relation is an example of _____ (asynchronous, sychnronous) protocol.
6. The CPU/modem relation is an example of _____ (asynchronous, sychnronous) protocol.
7. True or false. The higher the bus frequency, the higher the bus bandwidth.
8. True or false. The wider the data bus, the lower the bus bandwidth.
9. Generally, which bus protocol has the highest bus bandwidth?
10. True or false. The PC AT bus is an extension of the PC/XT bus.
11. Why is the AT bus called ISA?
12. What is the function of the AEN signal?
13. Is AEN an output signal or an input signal?
14. What is the purpose of IOCHRDY? Is it an input or an output signal?
15. What is the signal direction for signals DREQs, IRQs, and DACKs?
16. Explain the direction and function of the RESET DRV signal.
17. Explain the difference between signals SMEMR/SMEMW and MEMR/MEMW.
18. Does the 0WS signal exist on the PC/XT bus?
19. What is the function of the 0WS signal?
20. How many pins on the 62-pin section of the ISA are allocated to GND and V_{CC}?
21. Which address lines are duplicated on the 62- and 36-pin parts of the ISA bus?
22. Give the number of clocks it takes to transfer between the following devices.
   (a) 16-bit to 8-bit device   (b) 16-bit to 16-bit device   (c) 8-bit to 8-bit device
23. If we are connecting an 8255 to an ISA bus PC, which portion of the data bus must be connected to the data bus of the 8255?
24. What is the function of MEMCS16 and IOCS16 pins? Are they input or output signals?
25. Explain the limitation of the MASTER signal in the ISA bus.
26. In the ISA bus, which of the following cases is (are) permitted?
   (a) an external microprocessor from the expansion slot as master
   (b) DMA from the expansion slot as master
27. What is the bus frequency in ISA?
28. What are the limitations of the 16-bit DMA channels in the ISA bus?
29. Explain the difference between edge and level triggering of an interrupt.
30. What is the disadvantage of edge-triggered interrupts?
31. Do we have any access to NMI of the CPU through the expansion slot of ISA?
32. What is the difference between the PC/XT and PC AT as far as system configurations are concerned?
33. EISA is which type of bus?
   (a) 32-bit   (b) 16-bit   (c) both (a) and (b)
34. Why is the EISA bus speed limited to 8 MHz (or some say 8.3 MHz)?
35. Where does the 8.3 MHz figure come from?
36. What is autoconfiguration, and why it is highly desired?
37. True or false. EISA has the autoconfiguration feature.
38. A given EISA with 5 expansion slots has embedded the video card and hard disk controller card into the motherboard. Give the ID number assigned to each of the cards and slots.
39. Give the following for ISA and EISA buses.
   (a) bus size   (b) bus speed   (c) maximum bus bandwidth

40. EISA can transfer data to which type of devices?
(a) 8-bit   (b) 16-bit   (c) 32-bit   (d) all of the above
41. Why in the EISA do we have only LA2 - LA31 buses? What happens to LA0, LA1 for byte selection?

SECTION 27.2: PCI LOCAL BUSES

42. What kind of bus is designated as a local bus?
43. Why is a PC working with 12 MHz frequency and higher sometimes referred to as a dual-bus PC?
44. If the memory of a given 486 ISA PC is designed with 0 WS, give the cycle time for memory and I/O accessed from the ISA bus.
45. In Problem 44, calculate the bus bandwidth for both memory and I/O.
46. Give three major desired characteristics of a bus.
47. The _____ (PCI, VL bus) is processor independent.
48. Calculate the required bus bandwidth for graphics of 1280x1024 resolution, 24 bits for color, and redraw rate of 10.
49. Calculate the required bus bandwidth for real-time video of 640x480 resolution, 24 bits for color, and frame rate of 30.
50. The VL bus version 1.0 is a _____ (32-bit, 64-bit) bus.
51. What is the maximum speed of the VL bus for 3 expansion slots?
52. The PCI has a maximum speed of ____ MHz.
53. True or false. The PCI bus can accommodate 64-bit data buses of the Pentium.
54. True or false. The PCI bus supports autoconfiguration.
55. True or false. Interrupt sharing is not allowed in PCI.
56. Calculate and compare the maximum bus bandwidth for the following. Assume that all are non-burst mode.
(a) 32-bit EISA   (b) 32-bit PCI
57. Calculate the maximum bus bandwidth for the following. Assume burst mode.
(a) 32-bit PCI   (b) 64-bit PCI

SECTION 27.3: PCI LOCAL BUSES

58. What does hot-pluggable mean?
59. True or false. One disadvantage of USB is that USB devices create an exorbitant demand on the system's power supply.
60. List the maximum bus bandwidth for the following.
(a) USB   (b) COM port

ANSWERS TO REVIEW QUESTIONS

SECTION 27.1: ISA BUSES
1. master   2. true   3. asynchronous   4. two
5. 16-bit, but it can also be used for 8-bit peripherals
6. No; the Master pin must be activated after the DRQ has been acknowledged.
7. yes
8. Not at this time. Microsoft and Intel are working together to equip the ISA bus with autoconfiguration. This feature is often referred to as *plug and play*.
9. It is because EISA has a 32-bit data path.

SECTION 27.2: PCI LOCAL BUSES
1. It is the bus that is closely attached to the CPU and works with the same frequency as the CPU (or close to it).
2. local bus
3. in 286/386/486/Pentium systems, where there is an ISA (or EISA) bus for the peripheral connection and there are very refined buses for the SIMM memory connection using the surface mount technology.
4. hard disk controller   5. true
6. by placing a ground or V_{CC} pin between every 2 signal lines

SECTION 27.3: USB PORTS
1. false   2. Devices connected together via a USB hub can share a power supply.
3. false

CHAPTER 28

PROGRAMMING DOS, BIOS, HARDWARE WITH C/C++

OBJECTIVES

Upon completion of this chapter, you will be able to:

» **Use C/C++ language structures to access 80x86 register values**

» **Program DOS INT21H function calls with the intdos function in C/C++**

» **Program BIOS INT10H function calls with the int86 function in C/C++**

» **Program C/C++ functions intdosx and int86x to access segment registers DS and ES**

» **Access memory address space of the 80x86 in C/C++**

» **Access the BIOS data area with C/C++**

» **Access and program CMOS RAM with C/C++**

» **Program 80x86 I/O ports with C/C++**

» **Program the 8253/54 timer with C/C++**

» **Access and program CMOS RAM with Assembly instructions**

Today, C/C++ is the language of choice among developers of application software. Although C is a high-level language, it has strong bit manipulation capability. For this reason, some programmers refer to C as a "high-level assembly" language. Both Microsoft and Borland provide a rich library of functions designed to be used for accessing hardware and software of the PC. In this chapter we discuss programming of DOS, BIOS, and PC hardware with C. In Section 28.1, BIOS interrupt and DOS function call programming with C/C++ is explored. In Section 28.2, the use of C/C++ in accessing PC hardware and I/O ports is discussed.

SECTION 28.1: BIOS AND DOS INTERRUPT PROGRAMMING WITH C

For C/C++ programmers who do not have detailed knowledge of 80x86 Assembly language programming but want to write programs using DOS function calls INT 21H and BIOS interrupts, there is help from compilers in the form of *int86* and *intdos* functions. The int86 function is used for calling any of the PC's interrupts, while the intdos function is used only for the INT 21H DOS function calls. We look first at int86.

Programming BIOS interrupts with C/C++

To use the int86 function, we must set the registers to desired values and then call int86. Upon return from int86, we can access the 80x86 registers. In this regard, int86 is just like the "INT #" instruction in 80x86 Assembly language. To access the 80x86 registers, we use the union of the REGS structure already defined by the C compiler. It has the following format, where regin and regout are variable names:

union REGS regin,regout;

The union of the REGS structure allows 80x86 registers to be accessed in either their 16- or 8-bit format. The 16-bit registers are referred to as x and 8-bit registers are referred to as h (for "halfword"). This is shown in Table 28-1.

Table 28-1: REGS Union Elements and Assembly Equivalent

16-bit		8-bit	
C language	**Assembly Language**	**C Language**	**Assembly Language**
regin.x.ax	AX	regin.h.al	AL
		regin.h.ah	AH
regin.x.bx	BX	regin.h.bl	BL
		regin.h.bh	BH
regin.x.cx	CX	regin.h.cl	CL
		regin.h.ch	CH
regin.x.dx	DX	regin.h.dl	DL
		regin.h.dh	DH
regin.x.si	SI		
regin.x.di	DI		
regin.x.cflag	CY		

The following code compares loading the registers and invoking the interrupt in C and Assembly language.

```
/*     C language                          Assembly language      */
union REGS regin,regout;
regin.h.ah=0x25;                   /* mov ah,25h  ;AH=25H    */
regin.x.dx=0x4567;                 /* mov dx,4567h ;DX=4567H */
regin.x.si=0x1290;                 /* mov si,1290h ;SI=1290H   */
int86(interrupt#,&regin,&regout);  /* int #                  */
```

In the code above, interrupt # is a value from 00 to 255 (or 0x00 to 0xFF in hex, using the C syntax for hexadecimal numbers), and &regin and &regout are the addresses of the REGS variables. Upon returning from the int86 function, we can access the contents of registers just as in 80x86 Assembly language programs. This is shown as follows:

```
mydata=regout.h.ah;      /* mov mydata,ah  ;assign AH to mydata */
myvalu=regout.x.bx;      /* mov myvalu,bx  ;assign BX to myvalu */
```

Example 28-1 demonstrates how int86 is used in C programming. Example 28-2 shows how to access registers upon returning from int86.

Example 28-1

Use the int86 function to clear the screen. Show the equivalent INT 10 instruction.

Solution:

```
/* example 28-1A using 16-bit registers */
#include <dos.h>               /* int86 is part of this library */
main()
{
union REGS regin,regout;
regin.x.ax=0x0600;             /* MOV AX,0600H      */
regin.h.bh=0x07;               /* MOV BH,07H        */
regin.x.cx=0;                  /* MOV CX,0          */
regin.x.dx=0x184F;             /* MOV DX,184FH      */
int86(0x10,&regin,&regout);    /* INT 10H           */
}
```

We can mix 8- and 16-bit registers as shown next:

```
/* example 28-1B using 8-bit registers */
#include <dos.h>               /* int86 is part of this library */
main()
{
union REGS regin,regout;
regin.h.ah=6;                  /* MOV AH,6          */
regin.h.al=0;                  /* MOV AL,0          */
regin.h.bh=07                  /* MOV BH,07         */
regin.x.cx=0;                  /* MOV CX,0          */
regin.h.dl=0x4F;               /* MOV DL=4FH        */
regin.h.dh=0x18;               /* MOV DH=18H        */
int86(0x10,&regin,&regout);    /* INT 10H           */
}
```

Example 28-2

Use the int86 function to perform the following functions.
(a) Save the current cursor position
(b) Set the cursor to row 12, column 8
(c) Display the message "Hello" using the printf function.

Solution:

```
/* example 28-2 */
#include <stdio.h>
#include <dos.h>
main()
{
unsigned char oldrow;
unsigned char oldcol;
union REGS regin,regout;
regin.h.ah=3;                    /* MOV AH,3 ;option 3 INT 10H   */
regin.h.bh=0;                    /* MOV BH,0 ;page 0                 */
int86(0x10,&regin,&regout);      /* INT 10H ;video INT              */
oldrow=regout.h.dh;              /* MOV oldrow,DH ;save row         */
oldcol=regout.h.dl;              /* MOV oldcol,DL ;save col         */
printf("Cursor was at row=%d,column=%d \n",oldrow,oldcol);
regin.h.ah=2;                    /* MOV AH,2  ;option 2 of int 10H  */
regin.h.bh=0;                    /* MOV BH,0  ;Page zero            */
regin.h.dl=8;                    /* MOV DL,8  ;col location         */
regin.h.dh=12;                   /* MOV DH,12 ;rol location         */
int86(0x10,&regin,&regout);      /* INT 10H                         */
printf("Hello\n");
}
```

Finding the conventional memory size with INT 12H

As shown in Chapter 14, BIOS INT 12H provides the size of conventional memory installed in the PC. Example 28-3 reports its size.

Example 28-3

Use function int86 with INT 12H to find the size of conventional memory installed on a given PC.

Solution:

INT 12H provides the size of conventional memory in register AX, as was shown in Chapter 14.

```
/* example 28-3 */
#include <stdio.h>
#include <dos.h>
main()
{
unsigned int convmem;
union REGS regin,regout;
int86(0x12,&regin,&regout);
convmem=regout.x.ax;
printf("This PC has %dKB of Conventional memory\n" ,convmem);
}
```

INT 16H and keyboard access

In Chapter 18 we discussed how to access the keyboard through INT 16H. Example 28-4 shows how to access INT 16H with the int86 function.

Example 28-4

Using function int86 with INT 16H option 0, write a program to indicate the key activated, its scan code, and its ASCII code.

Solution:

As discussed in Chapter 18, option AH=0 of INT 16H waits for a keyboard entry and returns the scan code and ASCII code in registers AH and AL, respectively.

```
#include <stdio.h>
#include <dos.h>
main()
{
unsigned char scancode;
unsigned char ascicode;
union REGS regin,regout;
regin.h.ah=0;
int86(0x16,&regin,&regout);
scancode=regout.h.ah;
ascicode=regout.h.al;
printf("The %c has scan code %X and ASCII code of %X\n" ,ascicode,
    scancode,ascicode);
}
```

Programming INT 21H DOS function calls with C/C++

Although we can use the int86 function for INT 21H DOS function calls, there is a specially designated function, intdos, that can be used for DOS function calls. The format of intdos is as follows. Example 28-5 shows how to use intdos.

```
intdos(&regin,&regout);  /* to be used for INT 21H only */
```

Accessing segment registers

Both int86 and intdos allow access to registers AX, BX, CX, DX, SI, and DI, but not segment registers CS, DS, SS, and ES. In some of the interrupt services, we need access to the segment registers, as well. In such cases we must use int86x instead of int86, and intdosx instead of intdos. In using int86x and intdosx, we must also pass the argument SREGS. Functions int86x and intdosx have the following formats. See the SREGS structure at the end of this section.

```
int86x(interrupt #,&regin,&regout,&regseg);
intdosx(&regin,&regout,&regseg);
struct SREGS regseg;
```

Functions int86x and intdosx provide access only to registers ES and DS and not the segment registers CS and SS. The contents of SS and CS cannot be altered since their alteration will cause the program to crash. Fortunately, BIOS and DOS function calls that use segment registers do not request the alteration of CS and SS. Example 28-6 shows how to get the values of interrupt vector tables.

Example 28-5

Use INT 21H option 2AH to display the date in the form dd-mm-yy on the screen.
(a) Use intdos functions.   (b) Use the int86 function.

Solution:

Upon returning from the INT 21H function 2AH, DL contains the day, DH the month, CX the year.

(a) This program uses intdos.

```
#include <stdio.h>
#include <dos.h>
main()
{
unsigned Int year;
unsigned char month;
unsigned char day;
union REGS regin,regout;
regin.h.ah=0x2A;
intdos(&regin,&regout);
day=regout.h.dl;
month=regout.h.dh;
year=regout.x.cx;
printf("Today's date is %d-%d-%d\n",month,day,year);
}
```

(b) In this program we can replace the intdos statement with

```
int86(0x21,&regin,&regout)
```

Example 28-6

Using INT 21 option 35H, get the CS:IP in the interrupt vector table for INT 10H.

Solution:

From Appendix D, we have INT 21H, AH=35, and AL=interrupt number. Upon return, ES contains the code segment (CS) value and BX register has the instruction pointer (IP) value from the vector table.

```
#include <stdio.h>
#include <dos.h>
main()
{
unsigned int ipvalu;
unsigned int csvalu;
union REGS regin,regout;
struct SREGS regseg;
regin.h.ah=0x35;                /* MOV AH,35H */
regin.h.al=0x10;                /* MOV AL,10H */
int86x(0x21,&regin,&regout,&regseg);
/* or we can use intdosx(&regin,&regout,&regseg) */
ipvalu=regout.x.bx;             /* MOV ipvalu,BX */
csvalu=regseg.es;               /* MOV csvalu,ES */
printf("The CS:IP of INT 10H is %X:%X \n " ,csvalu,ipvalu);
}
```

Accessing the carry flag in int86 and intdos functions

Upon returning from many of the interrupt functions, we need to examine the carry flag. Functions int86, intdos, int86x, and intdosx allow us to examine the carry flag bit only, and no other flag bits are available through these functions. To access the carry flag bit we write

```
if(regout.x.cflag)
```

The structures of word registers, byte registers, and segment registers are shown below.

```
union REGS {
    struct WORDREGS {
        unsigned int ax;
        unsigned int bx;
        unsigned int cx;
        unsigned int dx;
        unsigned int si;
        unsigned int di;
        unsigned int cflag;
        } x;

    struct BYTEREGS {
        unsigned char al,ah;
        unsigned char bl,bh;
        unsigned char cl,ch;
        unsigned char dl,dh;
        } h;
    } *inregs;
union REGS *outregs;

struct SREGS {
    unsigned int es;
    unsigned int cs;
    unsigned int ss;
    unsigned int ds;
    } *seregs;
```

Review Questions

1. True or false. Function int86 can be used for any interrupt number.
2. True or false. Function intdos can be used for any interrupt number.
3. The int86 has_____ arguments, whereas intdos has _____.
4. True or false. Operand regin.h.al accesses the 16-bit register.
5. Is the following code correct?
   ```
   union REGS rin,rout;
   rin.x.ax=0x1250;
   ```
6. To access segment registers we use _____(int86x, int86, intdos).
7. The int86x function has _____ arguments and they are _____.
8. True or false. In the int86x and intdosx functions, only the ES and DS registers are accessible.

SECTION 28.2: PROGRAMMING PC HARDWARE WITH C/C++

In addition to accessing the CPU's registers, we can also access memory and input/output ports with C/C++.

Accessing 80x86 SEGMENT:OFFSET memory addresses in C

The 1M memory space of the 80x86 microprocessor is accessed by addresses in the form of seg:offset. For example, the logical address B000:0000 represents the physical address B0000H. If the address is within the same segment, it is a 16-bit address and uses a near pointer. If the address is outside the segment, it is in the 32-bit form of seg:offset and uses a far pointer. In C compilers for the 80x86 PC, the physical address of B8000H is represented in its seg:offset form of B800:0000H and is declared as 0xB8000000, where "0x" indicates a hex number.

```
unsigned int far *pter;        /*declare a far pointer */
pter = (unsigned int far*) 0xB8000000;
/*assigned the hex B8000000 address to pter pointer */
```

The code "unsigned int far *" is written to typecast address 0xB8000000 since pter=0xB8000000 will cause the compiler to generate a warning message. This typecasting informs the compiler that 0xB0000000 (a long integer) is a far pointer pointing to an unsigned integer. Next we show how the BIOS data area can be accessed to examine the PC hardware configurations and its devices.

Accessing BIOS data area with C

As we have seen throughout many of the chapters in this book, the physical memory locations 00400H to 004FFH, commonly referred to as the BIOS data area, hold some very important information about PC hardware configuration and the status of many of its devices, such as printer and COM ports. The address for the BIOS data area in seg:offset format is 0000:0400H - 0000:04FFH. Example 28-7 uses C to detect the installation of LPT1 and display the port address assigned to it. Example 28-8 shows how the COM2 port is detected and reported.

Example 28-7

Write a C program to detect the installation of LPT1 and report the I/O port address assigned to it.

Solution:

BIOS detects all the LPTs installed on the PC and reports the I/O port addresses to BIOS memory locations 00408H - 40FH, where 408H and 409H hold the I/O port address for LPT1, 40AH and 40BH for LPT2, and so on. If no LPT is installed, zeros are found in these memory locations.

```
/* this program detects the installation of LPT1 and reports the I/O port address
assigned to it. */
#include <stdio.h>
#include <dos.h>
main()
{
unsigned int far *xptr;           /* a far pointer */
xptr=(unsigned int far *) 0x00000408;  /* assign address */
if(*xptr >0)
printf("I/O base address assigned to LPT1 is %X \n",*xptr);
else printf("LPT1 = None found");
}
```

Example 28-8

Write a C program to detect the presence of COM2 and report the I/O port address assigned to it.

Solution:

As was discussed in Chapter 17, BIOS detects all the COM ports installed on the PC and reports the I/O port addresses to BIOS data area memory locations 00400H - 407H, where 400H and 401H hold the I/O port address for COM1, 402H and 403H for COM2, and so on. If no COM is installed, zeros are found in these memory locations. Therefore, we have the following C program.

```
/* this program detects the installation of COM2 and reports the I/O port address
assigned to it. */
        #include <stdio.h>
        #include <dos.h>
        main()
        {
        unsigned int far *xptr;              /* a far pointer */
        xptr=(unsigned int far *) 0x00000402;  /* assign address */
        if(*xptr >0)
        printf("I/O base address assigned to COM2 is %X \n",*xptr);
        else printf("COM2 = None found");
        }
```

Programming input/output ports with C/C++

All major C compilers provide functions to access I/O ports of the 80x86 microprocessor. Table 28-2 shows accessing the 8-bit ports of the 80x86.

Table 28-2: Accessing 8-bit I/O Ports with Assembly, C

80x86 Assembly	Microsoft C	Borland C*	Borland C (native)
OUT port#,AL	_outp(port#,byte)	outp(port#,byte)	outportb(port#,byte)
OUT DX,AL	_outp(port#,byte)	outp(port#,byte)	outportb(port#,byte)
IN AL,port#	_inp(port#)	inp(port#)	inportb(port#)
IN AL,DX	inp(port#)	inp(port#)	inportb(port#)

Note: For both Microsoft and Borland compilers, port# is an unsigned integer data type for port address (0000 to 0xFFFF) and byte is an unsigned char data type (00 - 0xFF) for data to be sent to the port. For inp, the data received is of unsigned char type (00 - 0xFF).

* Indicates Borland C code that is Microsoft-compatible.

Next we use the input/output port in C to play musical notes.

Revisiting playing music

In Chapter 13 we studied the 8253/54 timer of the 80x86 PC and discussed how counter 2 is programmed in Assembly language to play music. We repeat the same concept except that we use C language. Example 28-9 is a C version of Example 13-6.

Accessing parallel printer's (LPT1) data bus with C

As another example of accessing I/O ports with C, in Example 28-10 we program the LPT1 parallel printer port to display binary counts through its D0 - D7 data bus. This illustrates the concept of data acquisition through the PC parallel port used so widely by today's peripheral devices, such as CD-ROM and data acquisition boards.

Example 28-9

Rewrite Example 13-6 in C to play notes D3, A3, A4 for 250, 500, and 500 ms, respectively.

Solution:

```
/* For time delay generation we are using the delay (unsigned milliseconds) func-
tion from Borland C/C++ dos.h library. */
        #include <dos.h>
        main()
        {
        unsigned char orgbits;    /* for orginal status of port B              */
        unsigned char bits;       /* new status of port B             */
        outp(0x43, 0xB6);         /* 8253/54 control byte             */
/* For D3 note 1.1931MHz/147Hz=8116 =1FB4 Hex                */
        outp(0x42,0xB4);          /* send the low byte to port 0x42            */
        outp(0x42,0x1F);          /* send the high byte to port 0x42           */
        orgbits=inp(0x61);        /* get the original status of port B         */
        bits=orgbits|3;           /* make d0=1,d1=1 to turn the speaker on       */
        outp(0x61,bits);          /* speaker is on and note is playing         */
        delay(250);               /* wait for 250 milliseconds        */
        outp(0x61,orgbits);       /* turn the speaker off             */
        delay(100);               /* wait for 100 milliseconds        */
/* Repeat for A3 note where 1.1931MHz/220Hz=5423 =152F Hex */
        outp(0x42,0x2F);          /* low byte                         */
        outp(0x42,0x15);          /* high byte                        */
        orgbits=inp(0x61);
        bits=orgbits|3;
        outp(0x61,bits);
        delay(500);
        outp(0x61,orgbits);
        delay(100);
/* Repeat for A4 note where 1.1931MHz/440Hz=2711=0A97 Hex */
        outp(0x42,0x97);          /* low byte */
        outp(0x42,0x0A);          /* high byte */
        orgbits=inp(0x61);
        bits=orgbits|3;
        outp(0x61,bits);
        delay(500);
        outp(0x61,orgbits);
        }
```

```
LPT1 Connector                          Digital Trainer

        D0  ──────────────►  LED1
        D1  ──────────────►  LED2

        ...      ...    ...
        D6  ──────────────►  LED7
        D7  ──────────────►  LED8

        BUSY ◄─────────────  SW1
        ACK  ◄─────────────  SW2

        GND  ──────────────  GND
```

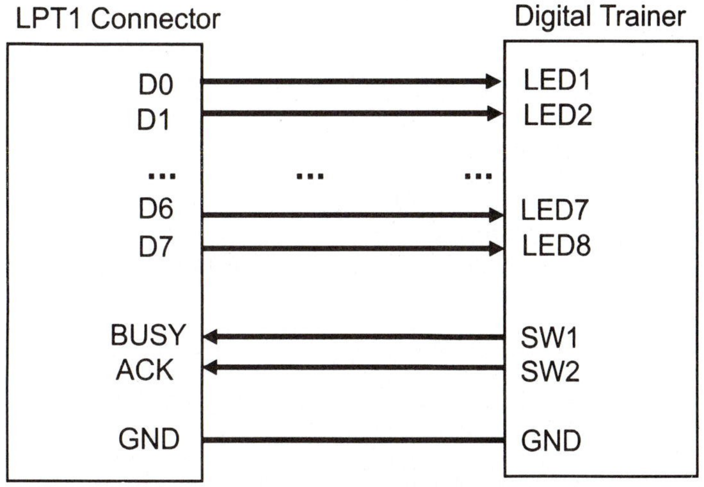

Figure 28-1. Switches and LED Connections to LPT1 for Example 28-10

Example 28-10

Some devices use the D0 - D7 data bus of a parallel port to access the IBM PC system board data bus instead of D0 - D7 of the expansion slot. Connect the LPT1 D0 - D7 data bus to LEDs and write a C program to perform a binary count.

Solution:

After refreshing your memory on the role of the ACK and BUSY printer's status signals, discussed in Chapter 18, follow these steps.

1. Connect the LPT1 data port to LEDs of your digital I/O trainer.
2. Connect SW1 of your digital I/O trainer to the BUSY status signal of printer. BUSY is bit 7 of the status port. See Figure 28-1.
3. Connect the printer's ACK status signal to SW2 of your digital I/O trainer. ACK is bit 6 of the status port.
4. Write a C program to perform the following objectives.
    (a) If SW1=0, it counts.
    (b) If SW1=1, it quits.
    (c) If SW2=0, the LEDs should flash.
    (d) If SW2=1, the LEDs will show the binary count with some time off in between each count. If the time-off is too short, the LEDs will be on all the time instead of showing the count.

```c
/* SW1=BUSY: When SW1=0 run. When SW1=1 quit */
/* SW2=ACK: When SW2=0 flash. When SW2=1 show the count */
        #include <stdio.h>
        #include <dos.h>
        #define datalpt1  0x3BC          /* LPT1's I/O port address for D0-D7*/
        #define statlpt1 0x3BD   /* status port address for LPT1 */

/* Notice: As shown in Chapter 18, in some PCs the port addresses of
378H,379H,and 37AH are assigned to LPT1.  Modify      the port addresses for your
PC before you run it */
        main()
        {
        int i;
        while(inp(statlpt1) & 0x80)         /* keep monitoring bit 7 for BUSY */
           {
           if(inp(statlpt1)&0x40)                   /* if ACK bit (bit 6) is high*/
              for(i=0; i <= 0xFF;i++)   /* then count up*/
                    {
                    outp(datalpt1,i); /* and send it to LEDs*/
                    delay(200);               /*and wait in between */
                    }
           else                   /* otherwise flash since ACK bit 6 is low */
              {
              outp(datalpt1,0xFF);    /* Turn on all LEDs */
              delay(100);             /* Wait */
              outp(datalpt1,0x00);    /* Now turn off all LEDs */
              delay(100);             /* wait. The flash rate is 100 ms */
              }
           }
        }


/* This example is adapted from a fine book, Technical C Programming  by    Vin-
cent Kassab, published by Prentice Hall */
```

Before ending the discussion about input and output ports, it is interesting to note that C compilers also provide a means by which one can access 16-bit devices where the data bus is 16-bit (word size). This is shown in Table 28-3. Since the peripheral chips used on the PC motherboard, such as the 8253/54, 8259, 8237, are all 8-bit, we do not use these functions to access them. However, these functions are useful for data acquisition board interfacing with a 16-bit data path.

Table 28-3: Accessing 16-bit Devices

80x86 Assembly	Microsoft	Borland
OUT DX,AX	outpw(port#,word)	outport(port#,word)
IN AX,DX	inpw(port#)	inport(port#)

Note: Variable port# is an integer, word is an integer. See Chapter 12 for more on this topic.

Finding memory above 1MB: the extended memory size

In the today's 286/386/486/Pentium PC, the DRAM memory installed is often more than 1M. As discussed in Chapter 25, memory above 1M is referred to as extended memory. To find out how much extended memory is available, we can use INT 15H option 88H. After executing the function 88H of BIOS INT 15H, the total available extended memory is reported in the AX register. Notice that it is the total available extended memory and not the total installed extended memory. The difference between them will be discussed soon. Example 28-11 shows how to get the size of available extended memory.

Running Example 28-11 on your PC might give you 0K extended memory even though you are absolutely sure (by using the MEM command) that you have extended memory. This is due to the fact that the DOS memory manager (or any third-party memory manager) will take over all the memory management of the PC, including extended memory, and provides 0 value to INT 15H option 88H. The zero amount of extended memory available does not mean that zero amount is installed. To find the amount of extended memory installed, we must get its size from the CMOS RAM of the 286/386/486/Pentium PC. The CMOS RAM is described next.

Example 28-11

Use the int86 function to find the available memory above 1MB (extended memory) in your PC. Use BIOS INT 15H function 88H.

Solution:

```
/* The BIOS INT 15 option 88H provides the total available extended
memory in KBytes size in AX register.  This program must be used only on 286
and higher PCs */

#include <stdio.h>
#include <dos.h>
main()
{
int extmem;
union REGS regin,regout;
regin.h.ah=0x88;        /* option 88H */
int86(0x15,&regin,&regout); /* of BIOS INT 15H */
extmem=regout.x.ax;
printf("The available extended memory is %d KB \n",extmem);
}
```

Programming the CMOS RAM real-time clock (RTC)

The PC/XT used DIP switches on the motherboard to set the hardware configuration of the PC. Starting with the 80286 PC/AT, all 80x86 PCs use Motorola's real-time clock MC146818 chip (or a compatible one) to store the PC configuration. In these types of PCs, the configuration is performed by a setup program and eliminates any need to open the PC in order to set the DIP switches. The CMOS RAM is powered by a small nickel-cadmium or lithium battery when the PC is off. If the battery runs down, the contents of CMOS RAM are erased and you will not be able to access your hard disk since the CMOS RAM holds the table for hard disk drive type. Without access to the drive type number, BIOS will not be able to recognize the hard drive. For this reason you must keep a copy of the CMOS RAM contents somewhere safe in case you need it. The CMOS RAM has a capacity of 64 bytes and is assigned the port addresses of 70H and 71H in the 286/386/486/Pentium PC. Table 28-4 shows the assignment of the 64 bytes of CMOS RAM.

Much of the information in CMOS RAM is accessible by means of BIOS INT 1AH or INT 21H DOS function calls. We must never write into CMOS RAM unless we know what we are doing, and even in that case we must be extremely careful since CMOS RAM holds some very critical information. The contents of CMOS RAM are accessible only one byte at a time. Next we explain why.

Accessing the CMOS RAM bytes

To access any bytes of the CMOS RAM, the following procedures must be performed.

1. Send the location of the desired byte to port address 70H.
2. Read its contents by way of port address 71H.

This way of accessing CMOS RAM avoids assigning any of the 80x86 memory space to the MC146818 RTC (real-time clock) chip. This method prevents memory fragmentation of RAM space in the PC. This means that although CMOS RAM is RAM memory, it is not taking any of the 80x86 memory space. Instead, it is mapped into the I/O space.

Example 28-12

Show how to read the contents of CMOS RAM locations 30H and 31H in Assembly language.

Solution:

```
            MOV    AL,30H        ;location 30H of CMOS RAM
            OUT    70H,AL        ;to be accessed
            IN     AL,71H        ;read its contents
            MOV    CL,AL         ;save it in CL
            MOV    AL,31H        ;location 31H of CMOS RAM
            OUT    70H,AL        ;to be accessed
            IN     AL,71H        ;read its contents
            MOV    CH,AL         ;save it in CH
;now CX has the size of memory above 1MB (extended memory)
;installed in this PC, the size is in K bytes
```

Of the CMOS RAM contents, we are interested in the size of extended memory. BIOS of the 286/386/486/Pentium PC determines the amount of installed memory above 1M and reports its size (in kilobytes) to CMOS RAM locations 30H and 31H (see Table 28-4). Location 30H holds the low byte, while location 31H holds the high byte. This was shown in Example 28-11 for Assembly language.

Table 28-4: CMOS RAM Information

Address	Description
00	* Seconds
01	* Second alarm
02	* Minutes
03	* Minutes alarm
04	* Hours
05	* Hour alarm
06	* Day of week
07	* Day of month
08	* Month
09	* Year
0A	* Status register A
0B	* Status register B
0C	* Status register C
0D	* Status register D
0E	* Diagnostic status byte
0F	* Shutdown Status byte
10	Diskette drive type byte (drives A and B)
11	Reserved
12	Fixed disk type byte (types 1 - 14)
13	Reserved
14	Equipment byte
15	Low base memory byte (conventional memory size is set during setup)
16	High base memory byte (conventional memory size is set during setup)
17	Low expansion memory byte (extended memory size is set during setup)
18	High expansion memory byte (extended memory size is set during setup)
19	Disk C extended byte
1A	Disk D extended byte
1B - 2D	Reserved
2E - 2F	2-byte CMOS checksum
30	* Low expansion memory byte (set during POST)
31	* High expansion memory byte (set during POST)
32	* Date century byte
33	* Information flags (set during power on)
34 - 3F	Reserved

* These bytes are not included in the checksum calculation and are not part of the configuration record.
(Reprinted by permission from "IBM Technical Reference" c. 1988 by International Business Machines Corporation)

Programming CMOS RAM with C/C++

Locations 30H and 31H hold the size of installed extended memory. CMOS RAM locations can be accessed a byte at a time. In other words, they cannot be accessed as words or doublewords since they are not part of the 80x86 memory space. The values held by locations 30H and 31H are in hexadecimal. Therefore, when we get the second byte it must be shifted left 2 digits in hex to make it the most significant byte of word size data and then added to the first byte to get the size of installed extended memory. Examples 28-13 and 28-14 find the size of installed extended memory in C language, using two different methods.

Example 28-13

Write a C program to display the size of extended memory installed on a 286/386/486/Pentium PC.

Solution:

```
/*This C program displays the size of installed extended memory as
reported by BIOS to CMOS RAM locations 30H and 31H during power-up. */
#include <stdio.h>
#include <dos.h>
main()
{
unsigned char b1,b2;
int extmem;
int w;        /* upper byte of word size extended memory */
outp(0x70,0x30);       /* get the low byte */
b1=inp(0x71);
outp(0x70,0x31);       /* get the high byte */
b2=inp(0x71);
w=0x100*b2;  /* shift left 2 hex digits to make it upper byte */
extmem=w+b1; /* add it to lower byte and display the size */
printf("The extended memory installed is: %d KB \n",extmem);
}
```

Example 28-14

This example shows another version of the program in Example 28-13. Notice how the upper byte is generated.

```
#include <stdio.h>
#include <dos.h>
main()
{
unsigned char b1,b2;
int extmem;
int w;             /* upper byte of word size extended memory */
outp(0x70,0x30);       /* get the low byte */
b1=inp(0x71);
outp(0x70,0x31);       /* get the high byte */
b2 = inp(0x71);
w = b2<<8;  /* shift left 8 digits (in binary) to make it upper byte */
extmem=w+b1; /* add it to lower byte and display the size */
printf("The extended memory installed is: %d KB \n",extmem);
}
```

Review Questions

1. Show how the starting address of video RAM in an MDA card is represented in C/C++.
2. Repeat Question 1 for the graphics mode address of a VGA card.
3. Declare the pointer and assign the address for Question 1.
4. Declare the pointer and assign the address for Question 2.
5. To access memory space within the segment the _____ (near, far) pointer is used.
6. In function "outp (port#,data)", the port address can be _____ or _____ bits.
7. What is the size of data in Question 6?
8. The function "inp(port#)" returns a value of _____ -bit size.
9. Put "INT 3" at the end of Example 28-12 and run it in DEBUG. What is the extended memory in your PC?
10. If CMOS RAM locations 30H and 31H have the values 00 and 04, respectively, calculate the size of installed extended memory. Show your calculation.

SUMMARY

The first section of this chapter shows how to access registers, as well as DOS and BIOS function calls, with C/C++. The C structure REGS can be used to access both 8- and 16-bit 80x86 general-purpose registers, plus SI, DI, and the carry flag. The int86 function can be used to call any of the 80x86 PC's interrupts, including BIOS INT 10H function calls. The intdos function is used for INT 21H DOS function calls. Segment registers CS and SS cannot be accessed through C, but the intdosx and int86x functions can be used to access segment registers ES and DS.

The second section demonstrated PC hardware programming with C/C++. The memory address space of the 80x86 can be accessed through C/C++ by initializing a pointer to a logical address. The BIOS data area of the PC and CMOS RAM can be accessed by this means, as well. C can be used to program I/O ports such as COM and LPT ports.

PROBLEMS

SECTION 28.1: BIOS AND DOS INTERRUPT PROGRAMMING WITH C

1. Show how to use the union REGS to set AX =9878H, BH=90H, and CL=F4H.
2. Write a C function to set the cursor using int86. Then use it to set the cursor to row=10, col=20 and display the message "HELLO".
3. Write a C function for changing the cursor shape. The prompt should ask for the start and end lines of the cursor. See the end of Chapter 16 for a discussion of video cursor shape.
4. Write a C program with the following objectives.
   (a) Clear the screen. Use int86.
   (b) Set the cursor to somewhere around the middle of the screen. Use int86.
   (c) Display the date and time continuously in the following format. Use intdos.
          Time: hr:min:sec
          Date: mon/day/yr
   (d) A prompt should ask for "Q" to quit. Use C functions.
   (e) When the user types in Q, it should quit displaying time and date and go back to DOS. Use C functions.
5. A programmer has declared the REGS union as follows. Would this work?
          union REGS inregs,outregs;
6. In Problem 5, write a program to clear the screen using int86. Use 16-bit registers.

---

7. A memory location in the BIOS data area holds the size of the conventional memory. Write a program in C to get the value and display it. This problem is just like Example 28-3 except that the memory size comes from the BIOS data area.

8. Write a C program to display the I/O port address assigned to all the LPTs. If a given LPT is not installed, it should display none.

9. Repeat Problem 8 for the COM ports.

10. Rewrite Example 28-9 for the first line of the song "Mary Had a Little Lamb." See Chapter 13 for the notes.

11. Write an Assembly program to access memory locations 17H and 18H of the CMOS RAM and put it in CX. Show the run in DEBUG.

12. Write an Assembly program to access memory locations 15H and 16H of the CMOS RAM and put it in CX. Show the run in DEBUG.

13. If, in a given PC, CMOS RAM locations 30H and 31H have the values of 00 and 1D, respectively, calculate the size of the extended memory for this PC. Is this the size of available or installed extended memory?

14. Write and run a C program to display the memory size in bytes for the following cases.
    (a) The size of conventional memory as indicated by locations 15H and 16 of the CMOS RAM.
    (b) The size of extended memory as indicated by locations 17H and 18H of the CMOS RAM.
    (c) The size of extended memory as indicated by locations 30H and 31H of the CMOS RAM.

15. Repeat Problem 14, parts (b) and (c), to display the memory in kilobytes and megabytes. Use the float data type to get the decimal points for megabytes.

ANSWERS TO REVIEW QUESTIONS

SECTION 28.1: BIOS AND DOS INTERRUPT PROGRAMMING WITH C
1. true
2. false; only for the INT 21H
3. 3, 2
4. false
5. Yes; we can use any name. Other commonly used names are inregs,outregs, and r1,r2.
6. int86x
7. four: INT #, &regin, &regout, &regseg
8. true

SECTION 28.2: PROGRAMMING PC HARDWARE WITH C/C++
1. 0xB0000000
2. 0xA0000000
3. unsigned int far *mdaptr;
   mdaptr = (unsigned int far*) 0xB0000000;
4. unsigned int far *vgaptr;
   vgaptr = (unsigned int far*) 0xA0000000;
5. near
6. 8, 16
7. 8 bits
8. 8-bits
9. CX=0C00, size=3072KB
10. The memory size (Kbytes) is 0400 in hex or 4 x 256 = 1024K in decimal.

APPENDIX A: DEBUG PROGRAMMING

DEBUG is a program included in the MS-DOS and PC-DOS operating systems that allows the programmer to monitor a program's execution closely for debugging purposes. Specifically, it can be used to examine and alter the contents of memory, to enter and run programs, and to stop programs at certain points in order to check or even change data. This appendix provides a tutorial introduction to the DEBUG program. You will learn how to enter and exit DEBUG, how to enter, run, and debug programs, how to examine and alter the contents of registers and memory, plus some additional features of DEBUG that prove useful in program development. Numerous examples of Assembly language programming in DEBUG are given throughout and the appendix closes with a quick reference summary of the DEBUG commands.

First, a word should be said about the examples in this appendix. Within examples, what you should type in will be represented in italic caps:

ITALICS CAPS REPRESENT WHAT THE USER TYPES IN

and the response of the DEBUG program will be in bold caps:

BOLD CAPS REPRESENT THE COMPUTER RESPONSE

The examples in this appendix assume that the DEBUG program is in drive A and that your programs are on drive B. If your system is set up differently, you will need to keep this in mind when typing in drive specifications (such as "B:"). It is strongly suggested that you type in the examples in DEBUG and try them for yourself. The best way to learn is by doing!

SECTION A.1: ENTERING AND EXITING DEBUG

To enter the DEBUG program, simply type its name at the DOS level:

A>*DEBUG <return>*
-

"DEBUG" may be typed in either uppercase or lowercase. Again let us note that this example assumes that the DEBUG program is on the diskette in drive A. After "DEBUG" and the carriage return (or enter key) is typed in, the DEBUG prompt "-" will appear on the following line. DEBUG is now waiting for you to type in a command.

Now that you know how to enter DEBUG, you are ready to learn the DEBUG commands. The first command to learn is the quit command, to exit DEBUG.

The quit command, Q, may be typed in either uppercase or lowercase. This is true for all DEBUG commands. After the Q and carriage return have been entered, DEBUG will return you to the DOS level. This is shown in Example A-1, on the following page.

```
┌─────────────────────────────────────────────────────────────────────────┐
│  Example A-1: Entering and Exiting DEBUG                                  │
├─────────────────────────────────────────────────────────────────────────┤
│  A>DEBUG <return>                                                         │
│  -Q  <return>                                                             │
│  A>                                                                       │
│                                                                           │
└─────────────────────────────────────────────────────────────────────────┘
```

SECTION A.2: EXAMINING AND ALTERING THE CONTENTS OF REGISTERS

The register command allows you to examine and/or alter the contents of the internal registers of the CPU. The R command has the following syntax:

R   <register name >

The R command will display all registers unless the optional <register name> field is entered, in which case only the register named will be displayed.

```
┌─────────────────────────────────────────────────────────────────────────┐
│  Example A-2: Using the R Command to Display All Registers                │
├─────────────────────────────────────────────────────────────────────────┤
│  A>DEBUG  <return>                                                        │
│  -R <return>                                                              │
│                                                                           │
│  AX=0000 BX=0000 CX=0000 DX=0000 SP=FFEE BP=0000 SI=0000 DI=0000          │
│  DS=0C44 ES=0C44 SS=0C44 CS=0C44 IP=0100   NV UP DI PL NZ NA PO NC         │
│  0C44:0100 0000    ADD    [BX+SI],AL            DS:0000=CD                 │
│  -                                                                        │
│                                                                           │
└─────────────────────────────────────────────────────────────────────────┘
```

After the R and carriage return are typed in, DEBUG responds with three lines of information. The first line displays the general-purpose, pointer, and index registers' contents. The second line displays the segment registers' contents, the instruction pointer's current value, and the flag register bits. The codes at the end of line two, "NV UP DI ... NC", indicate the status of eight of the bits of the flag register. The flag register and its representation in DEBUG are discussed in Section A.6. The third line shows some information useful when you are programming in DEBUG. It shows the instruction pointed at by CS:IP. The third line on your system will vary from what is shown above. For the purpose at hand, concentrate on the first two lines. The explanation of the third line will be postponed until later in this appendix.

When you enter DEBUG initially, the general-purpose registers are set to zero and the flag bits are all reset. The contents of the segment registers will vary depending on the system you are using, but all segment registers will have the same value, which is decided by the DOS operating system. For instance, notice in Example A-2 above that all segment registers contain 0C44H. It is strongly recommended not to change the contents of the segment registers since these values have been set by the operating system. *Note:* In a later section of this appendix we show how to load an Assembly language program into DEBUG. In that case the segment registers are set according to the program parameters and registers BX and CX will contain the size of the program in bytes.

If the optional register name field is specified in the R command, DEBUG will display the contents of that register and give you an opportunity to change its value. This is seen next in Example A-3.

Example A-3: Using the R Command to Display/Modify Register

(a) Modifying the contents of a register

```
-R CX
CX 0000
:FFFF
-R CX
CX FFFF
:
```

(b) DEBUG pads values on the left with zero

```
-R AX
AX 0000
:1
-R AX
AX 0001
:21
-R AX
AX 0021
:321
-R AX
AX 0321
:4321
-R AX
AX 4321
:54321
     ^ Error
```

(c) Entering data into the upper byte

```
-R DH
BR Error
-R DX
DX 0000
:4C00
-
```

Part (a) of Example A-3 first showed the R command followed by register name CX. DEBUG then displayed the contents of CX, which were 0000, and then displayed a colon ":". At this point a new value was typed in, and DEBUG prompted for another command with the "-" prompt. The next command verified that CX was indeed altered as requested. This time a carriage return was entered at the ":" prompt so that the value of CX was not changed.

Part (b) of Example A-3 showed that if fewer than four digits are typed in, DEBUG will pad on the left with zeros. Part (c) showed that you cannot access the upper and lower bytes separately with the R command. If you type in any digit other than 0 through F (such as in "2F0G"), DEBUG will display an error message and the register value will remain unchanged.

See Section A.6 for a discussion of how to use the R command to change the contents of the flag register.

SECTION A.3: CODING AND RUNNING PROGRAMS IN DEBUG

In the next few topics we explore how to enter simple Assembly language instructions, and assemble and run them. The purpose of this section is to familiarize the reader with using DEBUG, not to explain the Assembly language instructions found in the examples.

A, the assemble command

The assemble command is used to enter Assembly language instructions into memory.

    A   \<starting address>

The starting address may be given as an offset number, in which case it is assumed to be an offset into the code segment, or the segment register can be specified explicitly. In other words, "A 100" and "A CS:100" will achieve the same results. When this command is entered at the command prompt "-", DEBUG will begin prompting you to enter Assembly language instructions. After an instruction is typed in and followed by \<return>, DEBUG will prompt for the next instruction. This process is repeated until you type a \<return> at the address prompt, at which time DEBUG will return you to the command prompt level. This is shown in part (a) of Example A-4.

Before you type in the commands of Example A-4, be aware that one important difference between DEBUG programming and Assembly language programming is that DEBUG assumes that all numbers are in hex, whereas most assemblers assume that numbers are in decimal unless they are followed by "H". Therefore, the Assembly language instruction examples in this section do not have "H" after the numbers as they would if an assembler were to be used. For example, you might enter an instruction such as "MOV AL,3F". In an Assembly language program written for MASM, for example, this would have been typed as "MOV AL,3FH".

Example A-4: Assemble, Unassemble, and Go Commands

(a) Assemble command

```
-A 100
103D:0100 MOV AX,1
103D:0103 MOV BX,2
103D:0106 MOV CX,3
103D:0109 ADD AX,BX
103D:010B ADD AX,CX
103D:010D INT 3
103D:010E
-
```

(b) Unassemble command

```
-U 100 10D
103D:0100 B80100        MOV    AX,0001
103D:0103 BB0200        MOV    BX,0002
103D:0106 B90300        MOV    CX,0003
103D:0109 01D8          ADD    AX,BX
103D:010B 01C8          ADD    AX,CX
103D:010D CC            INT    3
-
```

(c) Go command

```
-R
AX=0000 BX=0000 CX=0000 DX=0000 SP=CFDE BP=0000 SI=0000 DI=0000
DS=103D ES=103D SS=103D CS=103D IP=0100  NV UP DI PL NZ NA PO NC
103D:0100 B80100        MOV    AX,0001
-G
AX=0006 BX=0002 CX=0003 DX=0000 SP=CFDE BP=0000 SI=0000 DI=0000
DS=103D ES=103D SS=103D CS=103D IP=010D  NV UP DI PL NZ NA PE NC
103D:010D CC            INT    3
-
```

As you type the instructions, DEBUG converts them to machine code. If you type an instruction incorrectly such that DEBUG cannot assemble it, DEBUG will give you an error message and prompt you to try again. Again, keep in mind that the value for the code segment may be different on your machine when you run Example A-4. Notice that each time DEBUG prompts for the next instruction, the offset has been updated to the next available location. For example, after you typed the first instruction at offset 0100, DEBUG converted this to machine language, stored it in bytes 0100 to 0102, and prompted you for the next instruction, which will be stored at offset 0103. *Note:* Do not assemble beginning at an offset lower than 100. The first 100H (256) bytes are reserved by DOS and should not be used by your programs. This is the reason that examples in this book use "A 100" to start assembling instructions after the first 100H bytes.

U, the unassemble command: looking at machine code

The unassemble command displays the machine code in memory along with their equivalent Assembly language instructions. The command can be given in either format shown below.

U   \<starting address \> \<ending address\>
U   \<starting address \> \< L number of bytes\>

Whereas the assemble instruction takes Assembly language instructions from the keyboard and converts them to machine code, which it stores in memory, the unassemble instruction does the opposite. Unassemble takes machine code stored in memory and converts it back to Assembly language instructions to be displayed on the monitor. Look at part (b) of Example A-4 on the preceding page. The unassemble command was used to unassemble the code that was entered in part (a) with the assemble command. Notice that both the machine code and Assembly instructions are displayed. The command can be entered either with starting and ending addresses, as was shown in Example A-4: "U 100 10D", or it can be entered with a starting address and a number of bytes in hex. The same command in the second format would be "U 100 LD", which tells DEBUG to start unassembling at CS:100 for D bytes. If the U command is entered with no addresses after it: "U \<return\>", then DEBUG will display 32 bytes beginning at CS:IP. Successively entering "U \<return\>" commands will cause DEBUG to display consecutive bytes of the program, 32 bytes at a time. This is an easy way to look through a large program.

G, the go command

The go command instructs DEBUG to execute the instructions found between the two given addresses. Its format is

G   \< = starting address\> \<stop address(es)\>

If no addresses are given, DEBUG begins executing instructions at CS:IP until a breakpoint is reached. This was done in part (c) of Example A-4 on the preceding page. Before the instructions were executed, the R command was used to check the values of the registers. Since CS:IP pointed to the first instruction, the G command was entered, which caused execution of instructions up until "INT 3", which terminated execution. After a breakpoint is reached, DEBUG displays the register contents and returns you to the command prompt "-". Up to 10 stop addresses can be entered. DEBUG will stop execution at the first of these breakpoints that it reaches. This can be useful for programs that could take several different paths.

The program is first assembled:

```
-A 100
103D:0100 MOV AX,1
103D:0103 MOV BX,2
103D:0106 MOV CX,3
103D:0109 ADD AX,BX
103D:010B ADD AX,CX
103D:010D INT 3
103D:010E
```

(a) Go command in form "G"

```
-G
AX=0006 BX=0002 CX=0003 DX=0000 SP=CFDE BP=0000 SI=0000 DI=0000
DS=103D ES=103D SS=103D CS=103D IP=010D  NV UP DI PL NZ NA PE NC
103D:010D CC       INT   3
-
```

(b) Go command in form "G = start address"

```
-G =100
AX=0006 BX=0002 CX=0003 DX=0000 SP=CFDE BP=0000 SI=0000 DI=0000
DS=103D ES=103D SS=103D CS=103D IP=010D  NV UP DI PL NZ NA PE NC
103D:010D CC       INT   3
-
```

(c) Go command form "G = start address ending address"

```
-G =100 109
AX=0001 BX=0002 CX=0003 DX=0000 SP=CFDE BP=0000 SI=0000 DI=0000
DS=103D ES=103D SS=103D CS=103D IP=0109  NV UP DI PL NZ NA PE NC
103D:0109 01D8      ADD   AX,BX
-
```

(d) Go command format "G   address"

```
-R IP
IP 0109
 :0100
-G 109
AX=0001 BX=0002 CX=0003 DX=0000 SP=CFDE BP=0000 SI=0000 DI=0000
DS=103D ES=103D SS=103D CS=103D IP=0109  NV UP DI PL NZ NA PE NC
103D:0109 01D8      ADD   AX,BX
-
```

At this point the third line of the register dump has become useful. The purpose of the third line is to show the location, machine code, and Assembly code of the next instruction to be executed. In Example A-5, look at the last line in the register dump given after the G command. Notice at the leftmost part of line three, the value CS:IP. The values for CS and IP match those given in lines one and two. After CS:IP is the machine code, and after the machine code is the Assembly language instruction.

Part (a) of Example A-5 is the same as part (c) of Example A-4. The go command started at CS:IP and executed instructions until it reached instruction "INT 3". Part (b) gave a starting address but no ending address; therefore, DEBUG executed instructions from offset 100 until "INT 3" was reached. This could also have been typed in as "G =CS:100". Part (c) gave both starting and ending addresses. We can see from the register results that it did execute from offset 100 to 109. Part (d) gave only the ending address. When the start address is not given explicitly, DEBUG uses the value in register IP. Be sure to check that value with the register command before issuing the go command without a start address.

T, the trace command: a powerful debugging tool

The trace command allows you to trace through the execution of your programs one or more instructions at a time to verify the effect of the programs on registers and/or data.

```
T        <= starting address>  <number of instructions>
```

This tells DEBUG to begin executing instructions at the starting address. DEBUG will execute however many instructions have been requested in the second field. The default value is 1 if no second field is given. The trace command functions similarly to the go command in that if no starting address is specified, it starts at CS:IP. The difference between this command and the go command is that trace will display the register contents after each instruction, whereas the go command does not display them until after termination of the program. Another difference is that the last field of the go command is the stop address, whereas the last field of the trace command is the number of instructions to execute.

Example A-6 shows a trace of the instructions entered in part (a) of Example A-4. Notice the way that register IP is updated after each instruction to point to the next instruction. The third line of the register display shows the instruction pointed at by IP, that is, the next instruction to be executed. Tracing through a program allows you to examine what is happening in each instruction of the program. Notice the value of AX after each instruction in Example A-6.

Example A-6: Trace Command

```
-T=100 5
  AX=0001 BX=0000 CX=0000 DX=0000 SP=CFDE BP=0000 SI=0000 DI=0000
  DS=103D ES=103D SS=103D CS=103D IP=0103  NV UP DI PL NZ NA PO NC
  103D:0103 BB0200      MOV    BX,0002

  AX=0001 BX=0002 CX=0000 DX=0000 SP=CFDE BP=0000 SI=0000 DI=0000
  DS=103D ES=103D SS=103D CS=103D IP=0106  NV UP DI PL NZ NA PO NC
  103D:0106 B90200      MOV    CX,0003

  AX=0001 BX=0002 CX=0003 DX=0000 SP=CFDE BP=0000 SI=0000 DI=0000
  DS=103D ES=103D SS=103D CS=103D IP=0109  NV UP DI PL NZ NA PO NC
  103D:0109 01D8        ADD    AX,BX

  AX=0003 BX=0002 CX=0003 DX=0000 SP=CFDE BP=0000 SI=0000 DI=0000
  DS=103D ES=103D SS=103D CS=103D IP=010B  NV UP DI PL NZ NA PE NC
  103D:010B 01C8        ADD    AX,CX

  AX=0006 BX=0002 CX=0003 DX=0000 SP=CFDE BP=0000 SI=0000 DI=0000
  DS=103D ES=103D SS=103D CS=103D IP=010D  NV UP DI PL NZ NA PE NC
  103D:010D CC          INT    3
  -
```

The same trace as shown in Example A-6 could have been achieved with the command "-T 5", assuming that IP = 0100. Experiment with the various forms of the trace command. "T" with no starting or count fields will execute one instruction starting at CS:IP. If no first field is given, CS:IP is assumed. If no second field is given, 1 is assumed.

If you trace a large number of instructions, they may scroll upward off the screen faster than you can read them. <Ctrl-num lock > can be used to stop the scrolling temporarily. To resume the scrolling, enter any key. This works not only on the trace command, but for any command that displays information to the screen.

Example A-7: Moving Data into 8- and 16-bit Registers

```
A>DEBUG
-R
AX=0000  BX=0000  CX=0000  DX=0000  SP=CFDE  BP=0000  SI=0000  DI=0000
DS=103D  ES=103D  SS=103D  CS=103D  IP=0100   NV UP DI PL NZ NA PO NC
103D:0100 B664        MOV     DH,64
-A 100

103D:0100 MOV AL,3F
103D:0102 MOV BH,04
103D:0104 MOV CX,FFFF
103D:0107 MOV CL,BH
103D:0109 MOV CX,1
103D:010C INT 3
103D:010D
-T =100 5
AX=003F  BX=0000  CX=0000  DX=0000  SP=CFDE  BP=0000  SI=0000  DI=0000
DS=103D  ES=103D  SS=103D  CS=103D  IP=0102   NV UP DI PL NZ NA PO NC
103D:0102 B704        MOV     BH,04

AX=003F  BX=0400  CX=0000  DX=0000  SP=CFDE  BP=0000  SI=0000  DI=0000
DS=103D  ES=103D  SS=103D  CS=103D  IP=0104   NV UP DI PL NZ NA PO NC
103D:0104 B9FFFF      MOV     CX,FFFF

AX=003F  BX=0400  CX=FFFF  DX=0000  SP=CFDE  BP=0000  SI=0000  DI=0000
DS=103D  ES=103D  SS=103D  CS=103D  IP=0107   NV UP DI PL NZ NA PO NC
103D:0107 88F9        MOV     CL,BH

AX=003F  BX=0400  CX=FF04  DX=0000  SP=CFDE  BP=0000  SI=0000  DI=0000
DS=103D  ES=103D  SS=103D  CS=103D  IP=0109   NV UP DI PL NZ NA PO NC
103D:0109 B90100      MOV     CX,0001

AX=003F  BX=0400  CX=0001  DX=0000  SP=CFDE  BP=0000  SI=0000  DI=0000
DS=103D  ES=103D  SS=103D  CS=103D  IP=010C   NV UP DI PL NZ NA PO NC
103D:010C CC          INT     3
-
```

Example A-8 shows some common programming errors in moving data into registers. The DEBUG assemble command catches this type of error when an instruction is entered and it tries to assemble it. In instruction "MOV DS,1200" the error is that immediate data cannot be moved into a segment register. The other errors involve move instructions, where the first and second operands do not match in size.

Example A-8: Common Errors in Register Usage

```
A>DEBUG
-A 100
103D:0100 MOV AL,FF3
              ^ Error
103D:0100 MOV AX,12345
              ^ Error
103D:0100 MOV DS,1200
              ^ Error
103D:0100 MOV SI,DH
              ^ Error
103D:0100 MOV AX,BH
              ^ Error
103D:0100 MOV AL,BX
              ^ Error
103D:0100
-Q
A>
```

```
Example A-9: Assembling and Unassembling a Program

A>DEBUG
-R
AX=0000  BX=0000  CX=0000  DX=0000  SP=CFDE  BP=0000  SI=0000  DI=0000
DS=1132  ES=1132  SS=1132  CS=1132  IP=0100   NV UP DI PL NZ NA PO NC
1132:0100 BED548     MOV    SI,48D5
-A 100
1132:0100 MOV AL,57
1132:0102 MOV DH,86
1132:0104 MOV DL,72
1132:0106 MOV CX,DX
1132:0108 MOV BH,AL
1132:010A MOV BL,9F
1132:010C MOV AH,20
1132:010E ADD AX,DX
1132:0110 ADD CX,BX
1132:0112 ADD AX,1F35
1132:0115
-U 100 112
1132:0100 B057      MOV      AL,57
1132:0102 B686      MOV      DH,86
1132:0104 B272      MOV      DL,72
1132:0106 89D1      MOV      CX,DX
1132:0108 88C7      MOV      BH,AL
1132:010A B39F      MOV      BL,9F
1132:010C B420      MOV      AH,20
1132:010E 01D0      ADD      AX,DX
1132:0110 01D9      ADD      CX,BX
1132:0112 05351F    ADD      AX,1F35
-
```

The program above is stored starting at CS:IP of 1132:0100. This logical address corresponds to physical address 11420 (11320 + 0100).

SECTION A.4: DATA MANIPULATION IN DEBUG

Next are described three DEBUG commands that are used to examine or alter the contents of memory.

F   the fill command fills a block of memory with data
D   the dump command displays contents of memory to the screen
E   the enter command examines/alters the contents of memory

F, the fill command: filling memory with data

The fill command is used to fill an area of memory with a data item. The syntax of the F command is as follows:

F   <starting address > <ending address> <data>
F   <starting address > < L number of bytes > <data>

This command is useful for filling a block of memory with data, for example to initialize an area of memory with zeros. Normally, you will want to use this command to fill areas of the data segment, in which case the starting and ending addresses would be offset addresses into the data segment. To fill another segment, the register should precede the offset. For example, the first command below would fill 16 bytes, from DS:100 to DS:10F with FF. The second command would fill a 256-byte block of the code segment, from CS:100 to CS:1FF with ASCII 20 (space).

F 100 10F FF
F CS:100 1FF 20

SECTION A.4: DATA MANIPULATION IN DEBUG   833

Example A-10 demonstrates the use of the F command. The data can be a series of items, in which case DEBUG will fill the area of memory with that pattern of data, repeating the pattern over and over. For example:

F 100 L20 00 FF

The command above would cause 20 hex bytes (32 decimal) starting at DS:100 to be filled alternately with 00 and FF.

D, the dump command: examining the contents of memory

The dump command is used to examine the contents of memory. The syntax of the D command is as follows:

D   <start address > <end address>
D   <start address > < L number of bytes>

The D command can be entered with a starting and ending address, in which case it will display all the bytes between those locations. It can also be entered with a starting address and a number of bytes (in hex), in which case it will display from the starting address for that number of bytes. If the address is an offset, DS is assumed. The D command can also be entered by itself, in which case DEBUG will display 128 consecutive bytes beginning at DS:100. The next time "D" is entered by itself, DEBUG will display 128 bytes beginning at wherever the last display command left off. In this way, one can easily look through a large area of memory, 128 bytes at a time.

Example A-10: Filling and Dumping a Block of Memory

(a) Fill and dump commands

```
A>DEBUG
-F 100 14F 20
-F 150 19F 00
-D 100 19F
103D:0100  20 20 20 20 20 20 20 20-20 20 20 20 20 20 20 20
103D:0110  20 20 20 20 20 20 20 20-20 20 20 20 20 20 20 20
103D:0120  20 20 20 20 20 20 20 20-20 20 20 20 20 20 20 20
103D:0130  20 20 20 20 20 20 20 20-20 20 20 20 20 20 20 20
103D:0140  20 20 20 20 20 20 20 20-20 20 20 20 20 20 20 20
103D:0150  00 00 00 00 00 00 00 00-00 00 00 00 00 00 00 00   ...............
103D:0160  00 00 00 00 00 00 00 00-00 00 00 00 00 00 00 00   ...............
103D:0170  00 00 00 00 00 00 00 00-00 00 00 00 00 00 00 00   ...............
103D:0180  00 00 00 00 00 00 00 00-00 00 00 00 00 00 00 00   ...............
103D:0190  00 00 00 00 00 00 00 00-00 00 00 00 00 00 00 00   ...............
-
```

(b) Filling and dumping selected memory locations

```
-F 104 10A FF
-D 104 10A
103D:0104  FF FF FF FF-FF FF FF                             .......
-D 100 10F
103D:0100  20 20 20 20 FF FF FF FF-FF FF FF 20 20 20 20 20     .......
-
```

(c) Filling and dumping code segment memory

```
-F CS:100 12F 20
-D CS:100 12F
103D:0100  20 20 20 20 20 20 20 20-20 20 20 20 20 20 20 20
103D:0110  20 20 20 20 20 20 20 20-20 20 20 20 20 20 20 20
103D:0120  20 20 20 20 20 20 20 20-20 20 20 20 20 20 20 20
-
```

                           APPENDIX A: DEBUG PROGRAMMING

Example A-10, on the preceding page, demonstrates use of the fill and dump commands. Part (a) shows two fill commands to fill areas of the data segment, which are then dumped. Part (b) was included to show that small areas of memory can be filled and dumped. Part (c) shows how to fill and dump to memory from other segments. Keep in mind that the values for DS and CS may be different on your machine.

It is important to become thoroughly familiar with the format in which DEBUG dumps memory. Example A-11 provides further practice in dumping areas of memory.

Example A-11: Using the Dump Command to Examine Machine Code

```
-U 100 112
1132:0100 B057    MOV    AL,57
1132:0102 B686    MOV    DH,86
1132:0104 B272    MOV    DL,72
1132:0106 89D1    MOV    BX,DX
1132:0108 88C7    MOV    BH,AL
1132:010A B39F    MOV    BL,9F
1132:010C B420    MOV    AH,20
1132:010E 01D0    ADD    AX,DX
1132:0110 01D9    ADD    CX,BX
1132:0112 05351F         ADD    AX,1F35
-D CS:100 11F
1132:0100 B0 57 B6 86 B2 72 89 D1-88 C7 B3 9F B4 20 01 D0  0W6.2r.Q.G3.4 .P
1132:0110 01 D9 05 35 1F 19 83 3E-E3 45 00 74 12 53 56 BB  .Y.5t..cE.t.SV;
-
```

Example A-11 shows a program being unassembled that had been loaded previously into DEBUG. Below that, the portion of the code segment containing the program is dumped. Notice that the machine codes are stored one after another continuously. It is important to become thoroughly familiar with the way DEBUG dumps memory. The following is one line from Example A-11:

```
1132:0100  B0 57 B6 86 B2 72 89 D1-88 C7 B3 9F B4 20 01 D0  0W6.2r.Q.G3.4 .P
```

The line begins with the address of the first byte displayed on that line, in this case 1132:0100, with 1132 representing the contents of CS and 0100 being the offset into the code segment. After the address, 16 bytes of data are displayed followed by a display of those items. Bytes that contain ASCII characters will display the characters. If the contents of a byte are not an ASCII code, it is not displayable and will be represented by ".". The first byte displayed above is offset 0100, the second offset 0101, the third offset 0102, and so on, until the last byte on that line, which is offset 010F.

```
OFFSET:    100 101 102 103 104 105 106 107-108 109 10A 10B 10C 10D 10E 10F

1132:0100  B0  57  B6  86  B2  72  89  D1- 88  C7  B3  9F  B4  20  01  D0
```

The addresses displayed are logical addresses. The logical address of 1132:0100 above would correspond to physical address 11420 (11320 + 0100 = 11420).

E, the enter command: entering data into memory

The fill command was used to fill a block with the same data item. The enter command can be used to enter a list of data into a certain portion of memory. The syntax of the E command is as follows:

```
E        <address > <data list>
E        <address>
```

SECTION A.4: DATA MANIPULATION IN DEBUG   **835**

Example A-12: Using the E Command to Enter Data into Memory

(a) Entering data with the E command

```
-E 100 'John Snith'
-D 100 10F
103D:0100  4A 6F 68 6E 20 53 6E 69-74 68 20 20 20 20 20 20   John Snith
```

(b) Altering data with the E command

```
-E 106
103D:0106  6E.6D
-D 100 10F
103D:0100  4A 6F 68 6E 20 53 6D 69-74 68 20 20 20 20 20 20   John Smith
```

(c) Another way to alter data with the E command,
     hitting the space bar to go through the data a byte at a time

```
-E 100
103D:0100 4A.  6F.  68.  6E.  20.  53.  6E.6D
-D 100 10F
103D:0100  4A 6F 68 6E 20 53 6D 69-74 68 20 20 20 20 20 20   John Smith
-
```

(d) Another way to alter data with the E command

```
-E 107
103D:0107 69.-
103D:0106 6E.6D
-
```

Part (a) of Example A-12 showed the simplest use of the E command, entering the starting address, followed by the data. That example showed how to enter ASCII data, which can be enclosed in either single or double quotes. The E command has another powerful feature: the ability to examine and alter memory byte by byte. If the E command is entered with a specific address and no data list, DEBUG assumes that you wish to examine that byte of memory and possibly alter it. After that byte is displayed, you have four options:

1. You can enter a new data item for that byte. DEBUG will replace the old contents with the new value you typed in.
2. You can press <return>, which indicates that you do not wish to change the value.
3. You can press the space bar, which will leave the displayed byte unchanged but will display the next byte and give you a chance to change that if you wish.
4. You can enter a minus sign, "-", which will leave the displayed byte unchanged but will display the previous byte and give you a chance to change it.

Look at part (b) in Example A-12. The user wants to change "Snith" to "Smith". After the user typed in "E 106", DEBUG responded with the contents of that byte, 6E, which is ASCII for n, and prompted with a ".". Then the user typed in the ASCII code for "m", 6D, entered a carriage return, and then dumped the data to see if the correction was made. Part (c) of Example A-12 showed another way to make the same correction. The user started at memory offset 100 and pressed the space bar continuously until the desired location was reached. Then he made the correction and pressed carriage return.

Finally, part (d) showed a third way the same correction could have been made. In this example, the user accidentally entered the wrong address. The address was one byte past the one that needed correction. The user entered a minus sign, which caused DEBUG to display the previous byte on the next line. Then the correction was made to that byte. Try these examples yourself since the E command will prove very useful in debugging your future programs.

The E command can be used to enter numerical data as well:

E 100 23 B4 02 4F

Example A-13 gives an example of entering code with the assemble command, entering data with the enter command, and running the program. This use of the E command is common in program testing and debugging. Example A-14 shows the little endian storage convention of 80x86 microprocessors.

Example A-13: Entering Data and Code and Running a Program

```
A>DEBUG
-A 100
103D:0100 MOV AL,00
103D:0102 ADD AL,[0200]
103D:0106 ADD AL,[0201]
103D:010A ADD AL,[0202]
103D:010E ADD AL,[0203]
103D:0112 ADD AL,[0204]
103D:0116 INT 3
103D:0117
-E DS:0200 25 12 15 1F 2B
-D DS:0200 020F
103D:0200  25 12 15 1F 2B 02 00 E8-51 FF C3 E8 1E F6 74 03   %...+..hQ.Ch.vt.
-G =100 116
AX=0096  BX=0000  CX=0000  DX=0000  SP=CFDE  BP=0000  SI=0000  DI=0000
CS=103D  ES=103D  SS=103D  CS=103D  IP=0116   OV UP DI NG NZ AC PE NC
103D:0116 CC          INT     3
-
```

Example A-14: How the 80x86 Stores Words: Little Endian

(a) Moving a word from memory into a register

```
A>DEBUG
-D 6820 LF
103D:6820  26 00 EA 27 CF 5B 48 22-0D 00 B8 15 45 00 EA 20   &.j'O[H"..8.E.j
-A 100
103D:0100 MOV BX,[6826]
103D:0104 INT 3
103D:0105
-T
AX=0000  BX=2248  CX=0000  DX=0000  SP=CFDE  BP=0000  SI=0000  DI=0000
DS=103D  ES=103D  SS=103D  CS=103D  IP=0104   NV UP DI PL NZ NA PO NC
103D:0104 CC          INT     3
-
```

(b) Moving a word from a register into memory

```
-D 200 20F
103D:0200  F2 FF 0E 0B 37 B8 FF FF-50 E8 B3 08 83 C4 02 20   R...78..Ph3..D.
-A 100
103D:0100 MOV BX,1234
103D:0103 MOV [200],BX
103D:0107 INT 3
103D:0108
-G =100
AX=0000  BX=1234  CX=0000  DX=0000  SP=CFDE  BP=0000  SI=0000  DI=0000
DS=103D  ES=103D  SS=103D  CS=103D  IP=0107   NV UP DI PL NZ NA PO NC
103D:0107 CC          INT     3
-D 200 LF
103D:0200  34 12 0E 0B 37 B8 FF FF-50 E8 B3 08 83 C4 02 20   4...78..Ph3..D.
-
```

In Example A-14, part (a), the direct addressing mode was used to move the two bytes beginning at offset 6826 to register BX. Looking at the dump shows that location 6826 contains 48 and the following byte at offset 6827 contains 22. These bytes were moved into register BX in low byte to low byte, high byte to high byte order. The contents of lower memory location 6826, which were 48, were moved to the low byte, BL. The contents of higher memory location 6827, which were 22, were moved to the high byte, BH. In part (b), value 1234H was moved into register BX and then stored at offset 200. Notice that offset address 200 contains the lower byte 34 and the higher offset address 201 contains the upper byte 12.

SECTION A.5: EXAMINING THE STACK IN DEBUG

In this sections we explore the implementation of the stack in 80x86 Assembly language programming and how the stack can be examined through DEBUG.

Pushing onto the stack

Example A-15 demonstrates how the stack is affected by PUSH instructions. First the assemble command is used to enter instructions that load three registers with 16-bit data, initialize the stack pointer to 1236H, and push the three registers onto the stack. Then the instructions are executed with the go command and the contents of the stack examined with the dump command. The following shows the contents of the stack area after each push instruction, assuming that SP = 1236 before the first push. Notice that the stack grows "upward" from higher memory locations toward lower memory locations. After each push, the stack pointer is decremented by 2.

```
SP = 1236
After "PUSH AX"
        103D:1230  00 00 00 00 B6 24 00 00-00 00 00 00 00 00 00 00
SP = 1234
After "PUSH DI"
        103D:1230  00 00 C2 85 B6 24 00 00-00 00 00 00 00 00 00 00
SP = 1232
After "PUSH DX"
        103D:1230  93 5F C2 85 B6 24 00 00-00 00 00 00 00 00 00 00
SP = 1230
```

Example A-15: Pushing Onto the Stack

```
A>DEBUG
-A 100
103D:0100 MOV AX,24B6
103D:0103 MOV DI,85C2
103D:0106 MOV DX,5F93
103D:0109 MOV SP,1236
103D:010C PUSH AX
103D:010D PUSH DI
103D:010E PUSH DX
103D:010F INT 3
103D:0110
-F 1230 123F 00
-D 1230 LF
103D:1230  00 00 00 00 00 00 00 00-00 00 00 00 00 00 00 00  ................
-G =100
AX=24B6 BX=0000 CX=0000 DX=5F93 SP=1230 BP=0000 SI=0000 DI=85C2
DS=103D ES=103D SS=103D CS=103D IP=010F   NV UP DI PL NZ NA PO NC
103D:010F CC        INT   3
-D 1230 123F
103D:1230  93 5F C2 85 B6 24 00 00-00 00 00 00 00 00 00 00  ._B.6$.........
```

Popping the stack

Example A-16 demonstrates the effect of pop instructions on the stack. The trace shows that after each pop is executed, the stack pointer SP is incremented by 2. As the stack is popped, it shrinks "downward" toward the higher memory addresses.

```
Example A-16: Popping the Stack Contents into Registers

-A 100
103D:0100 MOV SP,18FA
103D:0103 POP CX
103D:0104 POP DX
103D:0105 POP BX
103D:0106 INT 3
103D:0107
-E SS:18FA 23 14 6B 2C 91 F6
-D 18FA 18FF
103D:18FA  23 14 6B 2C 91 F6                    #.k,.v
-R IP
IP 010F
:0100
-T
AX=24B6 BX=0000 CX=0000 DX=5F93 SP=18FA BP=0000 SI=0000 DI=85C2
DS=103D ES=103D SS=103D CS=103D IP=0103  NV UP DI PL NZ NA PO NC
103D:0103 59        POP    CX
-T
AX=24B6 BX=0000 CX=1423 DX=5F93 SP=18FC BP=0000 SI=0000 DI=85C2
DS=103D ES=103D SS=103D CS=103D IP=0104  NV UP DI PL NZ NA PO NC
103D:0104 5A        POP    DX
-T
AX=24B6 BX=0000 CX=1423 DX=2C6B SP=18FE BP=0000 SI=0000 DI=85C2
DS=103D ES=103D SS=103D CS=103D IP=0105  NV UP DI PL NZ NA PO NC
103D:0105 5B        POP    BX
-T
AX=24B6 BX=F691 CX=1423 DX=2C6B SP=1900 BP=0000 SI=0000 DI=85C2
DS=103D ES=103D SS=103D CS=103D IP=0106  NV UP DI PL NZ NA PO NC
103D:0106 CC        INT    3
-
```

SECTION A.6: EXAMINING/ALTERING THE FLAG REGISTER IN DEBUG

The discussion of how to use the R command to examine/alter the contents of the flag register was postponed until this section, so that program examples that affect the flag bits could be included. Table A-1, on the following page, gives the codes for 8 bits of the flag register which are displayed whenever a G, T, or R DEBUG command is given.

If all the bits of the flag register were reset to zero, as is the case when DEBUG is first entered, the following would be displayed for the flag register:

NV UP DI PL NZ NA PO NC

Similarly, if all the flag bits were set to 1, the following would be seen:

OV DN EI NG ZR AC PE CY

Example A-17 shows how to use the R command to change the setting of the flag register.

```
Example A-17: Changing the Flag Register Contents

-R F
NV UP DI PL NZ NA PO NC -DN OV NG
-R F
OV DN DI NG NZ NA PO NC -

-
```

Example A-17 on the preceding page showed how the flag register can be examined, or examined and then altered. When the R command is followed by "F", this tells DEBUG to display the contents of the flag register. After DEBUG displays the flag register codes, it prompts with another "-" at the end of the line of register codes. At this point, flag register codes may be typed in to alter the flag register, or a simple carriage return may be typed in if no changes are needed. The register codes may be typed in any order.

Table A-1: Codes for the Flag Register

Flag	Code When Set (=1)	Code When Reset (= 0)
OF overflow flag	OV (overflow)	NV (no overflow)
DF direction flag	DN (down)	UP (up)
IF interrupt flag	EI (enable interrupt)	DI (disable interrupt)
SF sign flag	NG (negative)	PL (plus, or positive)
ZF zero flag	ZR (zero)	NZ (not zero)
AF auxiliary carry flag	AC (auxiliary carry)	NA (no auxiliary carry)
PF parity flag	PE (parity even)	PO (parity odd)
CF carry flag	CY (carry)	NC (no carry)

Impact of instructions on the flag bits

Example A-18, on the following page, shows the effect of ADD instructions on the flag register. The ADD in part (a) involved byte addition. Adding 9C and 64 results in 00 with a carry out. The flag bits indicate that this was the result. Notice the zero flag is now ZR, indicating that the result is zero. In addition, the carry flag was set, indicating the carry out. The ADD in part (b) involves word addition. Notice that the sign flag was set to NG after the ADD instruction was executed. This is because the result, CAE0, in its binary form will have a 1 in bit 15, the sign bit. Since we are dealing with unsigned addition, we interpret this number to be positive CAE0H, not a negative number. This points out the fact that the microprocessor treats all data the same. It is up to the programmer to interpret the meaning of the data. Finally, look at the ADD in part (c). Adding AAAAH and 5556H gives 10000H, which results in BX = 0000 with a carry out. The zero flag indicates the zero result (BX = 0000), while the carry flag indicates that a carry out occurred.

Hexarithmetic command

This command is like an on-line hex calculator that performs hex addition and subtractions. Its format is

H   <number 1>   <number 2>

When this command is entered at the DEBUG "-" prompt, DEBUG will display their sum followed by their difference (number 1 − number 2).

Procedure command

This command has a syntax similar to the trace command:

P   < = start address>   <number of instructions>

It is used to execute a loop, call, interrupt or repeat string operation as if it were a single instruction instead of tracing through every instruction in that procedure.

Example A-18: Observing Changes in the Flag Register

(a)

```
A>DEBUG
-A 100
103D:0100 MOV AL,9C
103D:0102 MOV DH,64
103D:0104 ADD AL,DH
103D:0106 INT 3
103D:0107
-T 3
AX=009C BX=0000 CX=0000 DX=0000 SP=CFDE BP=0000 SI=0000 DI=0000
DS=103D ES=103D SS=103D CS=103D IP=0102   NV UP DI PL NZ NA PO NC
103D:0102 B664      MOV     DH,64

AX=009C BX=0000 CX=0000 DX=6400 SP=CFDE BP=0000 SI=0000 DI=0000
DS=103D ES=103D SS=103D CS=103D IP=0104   NV UP DI PL NZ NA PO NC
103D:0104 00F0      ADD     AL,DH

AX=0000 BX=0000 CX=0000 DX=6400 SP=CFDE BP=0000 SI=0000 DI=0000
DS=103D ES=103D SS=103D CS=103D IP=0106   NV UP DI PL ZR AC PE CY
103D:0106 CC        INT     3
-
```

(b)

```
-A 100
103D:0100 MOV AX,34F5
103D:0103 ADD AX,95EB
103D:0106 INT 3
103D:0107
-T =100 2
AX=34F5 BX=0000 CX=0000 DX=6400 SP=CFDE BP=0000 SI=0000 DI=0000
DS=103D ES=103D SS=103D CS=103D IP=0103   NV UP DI PL NZ NA PO NC
103D:0103 05EB95    ADD     AX,95EB

AX=CAE0 BX=0000 CX=0000 DX=6400 SP=CFDE BP=0000 SI=0000 DI=0000
DS=103D ES=103D SS=103D CS=103D IP=0106   NV UP DI NG NZ AC PO NC
103D:0106 CC        INT     3
-
```

(c)

```
-A 100
103D:0100 MOV BX,AAAA
103D:0103 ADD BX,5556
103D:0107 INT 3
103D:0108
-G =100 107
AX=34F5 BX=0000 CX=0000 DX=6400 SP=CFDE BP=0000 SI=0000 DI=0000
DS=103D ES=103D SS=103D CS=103D IP=0107   NV UP DI PL ZR AC PE CY
103D:0107 CC        INT     3
-
```

Example A-19, on the following page, shows how to code a simple program in DEBUG, set up the desired data, and execute the program. This program includes a conditional jump that will decide whether to jump based on the value of the zero flag. This example also points out some important differences between coding a program in DEBUG and coding a program for an assembler such as MASM. First notice the JNZ instruction. If this were an Assembly language program, the instruction might be "JNZ LOOP_ADD", where the label LOOP_ADD refers to a line of code. In DEBUG we simply JNZ to the address. Another important difference is that an Assembly language program would have separate data and code segments. In Example A-19, the test data was entered at offset 0200, and consequently, BX was set to 0200 since it is being used as a pointer to the data. In an Assembly language program, the data would have been set up in the data segment and the instruction might have been "MOV BX,OFFSET DATA1" where DATA1 is the label associated with the data directive that stored the data.

SECTION A.6: EXAMINING/ALTERING THE FLAG REGISTER IN DEBUG

Example A-19: Tracing through a Program to Add 5 Bytes

```
A>DEBUG
-A 100
103D:0100 MOV CX,05
103D:0103 MOV BX,0200
103D:0106 MOV AL,0
103D:0108 ADD AL,[BX]
103D:010A INC BX
103D:010B DEC CX
103D:010C JNZ 0108
103D:010E MOV [0205],AL
103D:0111 INT 3
103D:0112
-E 0200 25 12 15 1F 2B
-D 0200 020F
103D:0200  25 12 15 1F 2B 9A DE CE-1E F3 20 20 20 20 20 20   %...+.^n.
-G =100 111
AX=0096  BX=0205  CX=0000  DX=0000  SP=CFDE  BP=0000  SI=0000  DI=0000
DS=103D  ES=103D  SS=103D  CS=103D  IP=0111   NV UP DI PL ZR NA PE NC
103D:0111 CC       INT   3

-D 0200 020F
103D:0200  25 12 15 1F 2B 96 DE CE-1E F3 20 20 20 20 20 20   %...+.^n.
```

Example A-20: Data Transfer Program in DEBUG

```
A>DEBUG
-A 100
103D:0100 MOV SI,0210
103D:0103 MOV DI,0228
103D:0106 MOV CX,6
103D:0109 MOV AL,[SI]
103D:010B MOV [DI],AL
103D:010D INC SI
103D:010E INC DI
103D:010F DEC CX
103D:0110 JNZ 0109
103D:0112 INT 3
103D:0113
-E 0210 25 4F 85 1F 2B C4
-D 0210 022F
103D:0210  25 4F 85 1F 2B C4 43 0C-01 01 01 00 02 FF FF FF   %O..+DC........
103D:0220  FF FF FF FF FF FF FF FF-FF FF FF FF 45 0D CA 2A   ............E.J*
-G =100
AX=00C4  BX=0000  CX=0000  DX=0000  SP=CFDE  BP=0000  SI=0216  DI=022E
DS=103D  ES=103D  SS=103D  CS=103D  IP=0112   NV UP DI PL ZR NA PE NC
103D:0112 CC       INT   3
-D 0210 022F
103D:0210  25 4F 85 1F 2B C4 43 0C-01 01 01 00 02 FF FF FF   %O..+DC........
103D:0220  FF FF FF FF FF FF FF FF-25 4F 85 1F 2B C4 CA 2A   ........%O..+DJ*
-
```

SECTION A.7: ADDITIONAL DEBUG DATA MANIPULATION COMMANDS

The following commands are often useful in manipulating the data in your programs.

M, the move command: copying data from one location to another

The move command is used to copy data from one location to another. The original location will remain unchanged. The syntax of this command is

---

M   <starting address> <ending address> <destination address>
M   <starting address> <L number of bytes> <destination address>

In other words, this command will place a copy of the data found from starting address to ending address at the destination address. Part (a) in Example A-21 gives an example of using the move command. This command copied the data found in locations 130 to 13F to location 140. The same result could have been achieved by "M 130 LF 140".

C, the compare command: checking blocks of data for differences

The compare command is used to check two areas of memory and display bytes that contain different data. If the two blocks are identical, DEBUG will simply display the command prompt "-". The syntax of the command is

C   <starting address> <ending address> <compare address>
C   <starting address> < L number of bytes> <compare address>

In other words, this command will compare the data found from the starting address to the ending address with the data found beginning at the compare address and will display any bytes that differ. Part (b) in Example A-21 contains examples of using the compare command. The first command compared from offsets 130 to 134 with memory beginning at location 140. Since no differences were found, DEBUG responded with the command prompt "-". That command could also have been entered as "C 130 L5 140". The next command compared from offsets 130 to 134 with memory beginning at location 150 and printed all five locations since all of them differed.

Example A-21: Move, Search, and Compare Commands

(a) Move command

```
-F 130 13F FF
-D 130 15F
103D:0130  FF FF FF FF FF FF FF FF-FF FF FF FF FF FF FF FF ................
103D:0140  00 00 00 00 00 00 00 00-00 00 00 00 00 00 00 00 ...............
103D:0150  00 00 00 00 00 00 00 00-00 00 00 00 00 00 00 00 ...............
-M 130 13F 140
-D 130 15F
103D:0130  FF FF FF FF FF FF FF FF-FF FF FF FF FF FF FF FF ...............
103D:0140  00 00 00 00 00 00 00 00-00 00 00 00 00 00 00 00 ...............
103D:0150  00 00 00 00 00 00 00 00-00 00 00 00 00 00 00 00 ...............
```

(b) Compare command

```
-C 130 134 140
-C 130 134 150
103D:0130  FF  00  103D:0150
103D:0131  FF  00  103D:0151
103D:0132  FF  00  103D:0152
103D:0133  FF  00  103D:0153
103D:0134  FF  00  103D:0154
```

(c) Search command

```
-S 150 15F FF
-S 130 133 FF
103D:0130
103D:0131
103D:0132
103D:0133
-
```

S, the search command: search a block of memory for a data item

The search command is used to search a block of data for a specific data value. If the item is not found, DEBUG simply displays the command prompt "-". Otherwise, all locations where the data item was found will be displayed. The syntax is:

```
S       <starting address>  <ending address>  <data>
S       <starting address>  <L number of bytes>  <data>
```

DEBUG will search from the starting to the ending address to find data. Look at part (c) of Example A-21. This example searched from locations 150 to 15F for FF and did not find it. The next command searched from 130 to 133 for FF and printed all four addresses since all four contained FF. The following command would have achieved the same result: "S 130 L4 FF". The data may be a list of data items, in which case DEBUG will search for that pattern of data.

SECTION A.8: LOADING AND WRITING PROGRAMS

The write and load commands below are used to save a program onto disk and load a previously saved program from disk into DEBUG. They both require a thorough familiarity with advanced DOS concepts.

W, the write command: saving instructions on disk

The write command is used to save instructions onto a disk. Its format is

```
W       <starting address> <drive number> <starting sector> <sectors>
```

Writing to specific sectors is not recommended since a thorough familiarity with the way information is stored on drives is needed. Writing to the wrong sector could damage the disk's directory, rendering it useless. Example A-23 shows how to use the W command without any parameters to save code on a disk, after the N command had been used to set up a filename.

L, the load command: loading instructions from disk

The load command performs the opposite function of the write command: It loads from disk into memory starting at the specified address. Its syntax is

```
L       <starting address> <drive number> <starting sector> <sectors>
```

After the load, registers BX CX will hold the size of the program in bytes. Using the load command with all its options requires a thorough understanding of disk storage and it is not recommended for beginning students. However, the L command may be used after the name command in a simple format, shown below.

N, the name command: used to load a file from disk

The name command can be used with the load command to load a program into DEBUG.

```
-N <filename>
-L
```

The name command above sets up the filename to be loaded by the load command. An alternative way to load a program into DEBUG is when you initially enter the DEBUG program:

```
-DEBUG <filename>
```

Example A-22: Loading an Assembled Program into DEBUG

(a) Loading with the name and load commands

```
-N B:\PROGRAMS\PROG1.EXE
-L
-
```

(b) Loading on entering DEBUG

```
A>DEBUG B:\PROGRAMS\PROG1.EXE
-
```

Example A-23 first shows how to save code that has been entered in DEBUG. The code entered is the code for Example A-19. After the code has been entered with the A command, registers BX and CX must be set up to contain the number of bytes to be saved. CX is set to 12 to save 12 bytes, BX is the high word and in this case should be zero. The N command sets up the drive and filename to be used, then the W command writes the code to disk. Note that the filename extension must be "com" because "exe" files cannot be saved in this manner. The rest of Example A-23 shows how to load saved code into DEBUG. The N command sets up the file reference and the L command loads the code into DEBUG. Table A-2 provides a summary of DEBUG commands.

Example A-23: Saving and Loading Code

```
C>DEBUG
-A 100
12B0:0100 MOV CX,05
12B0:0103 MOV BX,0200
12B0:0106 MOV AL,0
12B0:0108 ADD AL,[BX]
12B0:010A INC BX
12B0:010B DEC CX
12B0:010C JNZ 0108
12B0:010E MOV [0205],AL
12B0:0111 INT 3
12B0:0112
-R CX
CX 0000
:12
-r
AX=0000  BX=0000  CX=0012  DX=0000  SP=CFDE  BP=0000  SI=0000  DI=0000
DS=12B0  ES=12B0  SS=12B0  CS=12B0  IP=0100    NV UP DI PL NZ NA PO NC
12B0:0100 B90500      MOV    CX,0005
-N B:EX19.COM
-W
Writing 0012 bytes
-Q

C>DEBUG
-N B:EX19.COM
-L
-R
AX=0000  BX=0000  CX=0012  DX=0000  SP=FFFE  BP=0000  SI=0000  DI=0000
DS=12CC  ES=12CC  SS=12CC  CS=12CC  IP=0100   NV UP DI PL NZ NA PO NC
12CC:0100 B90500      MOV    CX,0005
-u cs:100 111
12CC:0100 B90500              MOV       CX,0005
12CC:0103 BB0002              MOV       BX,0200
12CC:0106 B000                MOV       AL,00
12CC:0108 0207                ADD       AL,[BX]
12CC:010A 43                  INC       BX
12CC:010B 49                  DEC       CX
12CC:010C 75FA                JNZ       0108
12CC:010E A20502              MOV       [0205],AL
12CC:0111 CC                  INT       3
-
```

Table A-2: Summary of DEBUG Commands

Function	Command Options
Assemble	A \<starting address\>
Compare	C \<start address\> \<end address\> \<compare address\>
	C \<start address\> \< L number of bytes\> \<compare address\>
Dump	D \<start address\> \<end address\>
	D \<start address\> \< L number of bytes\>
Enter	E \<address\> \<data list\>
	E \<address\>
Fill	F \<start address\> \<end address\> \<data\>
	F \<start address\> \< L number of bytes\> \<data\>
Go	G \< = start address\> \<end address(es)\>
Hexarith	H \<number 1\> \<number 2\>
Load	L \<start address\> \<drive\> \<start sector\> \<sectors\>
Move	M \<start address\> \<end address\> \<destination\>
	M \<start address\> \< L number of bytes\> \<destination\>
Name	N \<filename\>
Procedure	P \< = start address\> \<number of instructions\>
Register	R \<register name\>
Search	S \<start address\> \<end address\> \<data\>
	S \<start address\> \< L number of bytes\> \<data\>
Trace	T \< = start address\> \<number of instructions\>
Unassemble	U \<start address\> \<end address\>
	U \<start address\> \< L number of bytes\>
Write	W \<start address\> \<drive\> \<start sector\> \<sectors\>

Notes:
1. All addresses and numbers are given in hex.
2. Commands may be entered in lowercase or uppercase, or a combination.
3. Ctrl-c will stop any command.
4. Ctrl-numlock will stop scrolling of command output. To resume scrolling, enter any key.

APPENDIX B: 80x86 INSTRUCTIONS AND TIMING

In the first section of this appendix, we list the instructions of the 8086, give their format and expected operands, and describe the function of each instruction. Where pertinent, programming examples have been given. These instructions will operate on any 8086 or higher IBM-compatible computer. There are additional instructions for higher microprocessors (80186 and above); however, these instructions are not given in this list. The second section is a list of clock counts for each instruction across the 80x86 family.

SECTION B.1: THE 8086 INSTRUCTION SET

AAA   ASCII Adjust after Addition

Flags: Affected: AF and CF. Unpredictable: OF, SF, ZF, PF.
Format: AAA
Function: This instruction is used after an ADD instruction has added two digits in ASCII code. This makes it possible to add ASCII numbers without masking off the upper nibble "3". The result will be unpacked BCD in AL with carry flag set if needed. This instruction adjusts only on the AL register. AH is incremented if the carry flag is set.

Example 1:
```
MOV   AL,31H       ;AL=31 THE ASCII CODE FOR 1
ADD   AL,37H       ;ADD 37 (ASCII FOR 7) TO AL; AL=68H
AAA                ;AL=08 AND CF=0
```

In the example above, ASCII 1 (31H) is added to ASCII 7 (37H). After the AAA instruction, AL will contain 8 in BCD and CF = 0. The following example shows another ASCII addition and then the adjustment:

Example 2:
```
MOV   AL,'9'       ;AL=39 ASCII FOR 9
ADD   AL,'5'       ;ADD 35 (ASCII FOR 5) TO AL THEN AL=6EH
AAA                ;NOW AL=04 CF=1
OR    AL,30H       ;converts result to ASCII
```

AAD   ASCII Adjust before Division

Flags: Affected: SF, ZF, PF. Unpredictable: OF, AF, CF.
Format: AAD
Function: Used before the DIV instruction to convert two unpacked BCD digits in AL and AH to binary. A better name for this would be BCD to binary conversion before division. This allows division of ASCII numbers. Before the AAD instruction is executed, the ASCII tag of 3 must be masked from the upper nibble of AH and AL.

Example:

MOV	AX,3435H	;AX=3435 THE ASCII FOR 45
AND	AX,0F0FH	;AX=0405H UNPACKED BCD FOR 45
AAD		;AX=002DH HEX FOR 45
MOV	DL,07	;DL=07
DIV	DL	;2DH DIV BY 07 GIVES AL=06,AH=03
OR	AX,3030H	;AL=36=QUOTIENT AND AH=33=REMAINDER

AAM   ASCII Adjust after Multiplication

Flags: Affected: AF, CF. Unpredictable: OF, SF, ZF, PF.
Format: AAM
Function: Again, a better name would have been BCD adjust after multiplication. It is used after the MUL instruction has multiplied two unpacked BCD numbers. It converts AX from binary to unpacked BCD. AAM adjusts only AL, and any digits greater than 9 are stored in AH.

Example:

MOV	AL,'5'	;AL=35
AND	AL,0FH	;AL=05 UNPACKED BCD FOR 5
MOV	BL,'4'	;BL=34
AND	BL,0FH	;BL=04 UNPACKED BCD FOR 4
MUL	BL	;AX=0014H=20 DECIMAL
AAM		;AX=0200
OR	AX,3030H	;AX=3230 ASCII FOR 20

AAS   ASCII Adjust after Subtraction

Flags: Affected: AF, CF. Unpredictable: OF, SF, ZF, PF.
Format: AAS
Function: After the subtraction of two ASCII digits, this instruction is used to convert the result in AL to packed BCD. Only AL is adjusted; the value in AH will be decremented if the carry flag is set.

Example:

MOV	AL,32H	;AL=32 ASCII FOR 2
MOV	DH,37H	;DH=37 ASCII FOR 7
SUB	AL,DH	;AL-DH=32-37=FBH WHICH IS -5 IN 2'S COMP
		;CF=1 INDICATING A BORROW
AAS		;NOW AL=05 AND CF=1

ADC   Add with Carry

Flags: Affected: OF, SF, ZF, AF, PF, CF.
Format: ADC dest,source ;dest = dest + source + CF
Function: If CF=1 prior to this instruction, then after execution of this instruction, source is added to destination plus 1. If CF = 0, source is added to destination plus 0. Used widely in multibyte and multiword additions.

ADD   Signed or Unsigned ADD

Flags: Affected: OF, SF, ZF, AF, PF, CF.
Format: ADD dest,source ;dest = dest + source
Function: Adds source operand to destination operand and places the result in destination. Both source and destination operands must match (e.g., both byte size or word size) and only one of them can be in memory.

Unsigned addition:

In addition of unsigned numbers, the status of CF, ZF, SF, AF, and PF may change, but only CF, ZF, and AF are of any use to programmers. The most important of these flags is CF. It becomes 1 when there is carry from D7 out in 8-bit (D0 - D7) operations, or a carry from D15 out in 16-bit (D0 - D15) operations.

Example 1:
```
MOV   BH,45H              ;BH=45H
ADD   BH,4FH              ;BH=94H (45H+4FH=94H)
                         ;CF=0,ZF=0,SF=1,AF=1,and PF=0
```

Example 2:
```
MOV   AL,FEH              ;AL=FEH
MOV   DL,75H              ;DL=75H
ADD   AL,DL              ;AL=FE+75=73H
                         ;CF=1,ZF=0,AF=0,SF=0,PF=0
```

Example 3:
```
MOV   DX,126FH           ;DX=126FH
ADD   DX,3465H           ;DX=46D4H (126F=3465=46D4H)
                         ;CF=0,ZF=0,AF=1,SF=0,PF=1

MOV   BX,0FFFFH
ADD   BX,1               ;BX=0000 (FFFFH+1=0000)
                         ;AND CF=1,ZF=1,AF=1,SF=0,PF=1
```

Signed addition:

In addition of signed numbers, the status of OF, ZF, and SF must be noted. Special attention should be given to the overflow flag (OF) since this indicates if there is an error in the result of the addition. There are two rules for setting OF in signed number operation. The overflow flag is set to 1:

1. If there is a carry from D6 to D7 and no carry from D7 out in an 8-bit operation or a carry from D14 to D15 and no carry from D15 out in a 16-bit operation
2. If there is a carry from D7 out and no carry from D6 to D7 in an 8-bit operation or a carry from D15 out but no carry from D14 to D15 in a 16-bit operation

Notice that if there is a carry both from D7 out and from D6 to D7, then OF = 0 in 8-bit operations. In 16-bit operations, OF = 0 if there is both a carry out from D15 and a carry from D14 to D15.

Example 4:
```
MOV   BL,+8     ;BL=0000 1000
MOV   DH,+4     ;DH=0000 0100
ADD   BL,DH     ;BL=0000 1100 SF=0,ZF=0,OF=0,CF=0
```
Notice SF = D7 = 0 since the result is positive and OF = 0 since there is neither a carry from D6 to D7 nor any carry beyond D7. Since OF = 0, the result is correct [(+8) + (+4) = (+12)].

Example 5:
```
MOV   AL,+66    ;AL=0100 0010
MOV   CL,+69    ;CL=0100 0101
ADD   CL,AL     ;CL=1000 0111 = -121 (INCORRECT)
                ;CF=0,SF=1,ZF=0, AND OF=1
```
In Example 5, the correct result is +135 [(+66) + (+69) = (+135)], but the result was −121. The OF = 1 is an indication of this error. Notice that SF = D7 = 1 since the result is negative; OF = 1 since there is a carry from D6 to D7 and CF=0.

Example 6:
```
MOV    AL,-12        ;AL=1111 0100
MOV    BL,+18        ;BL=0001 0010
ADD    BL,AL         ;BL=0000 0110 (WHICH IS +6 )
                     ;SF=0,ZF=0,OF=0, AND CF=1
```

Notice above that OF = 0 since there is a carry from D6 to D7 and a carry from D7 out.

Example 7:
```
MOV    AH,-30        ;AH=1110 0010
MOV    DL,+14        ;DL=0000 1110
ADD    DL,AH         ;DL=1111 0000 (WHICH IS -16 AND CORRECT)
                     ;AND SF=1,ZF=0,OF=0, AND CF=0
```
OF = 0 since there is no carry from D7 out nor any carry from D6 to D7.

Example 8:
```
MOV    AL,-126       ;AL=1000 0010
MOV    BH,-127       ;BH=1000 0001
ADD    AL,BH         ;AL=0000 0011 (WHICH IS +3 AND WRONG)
                     ;AND SF=0,ZF=0 AND OF=1
```
OF = 1 since there is carry from D7 out but no carry from D6 to D7.

AND   Logical AND

Flags: Affected: CF = 0, OF = 0, SF, ZF, PF.
           Unpredictable: AF.
Format: AND dest,source
Function: Performs logical AND on the operands, bit by bit, storing the result in the destination.

X Y	X AND Y
0 0	0
0 1	0
1 0	0
1 1	1

Example:
```
MOV    BL,39H ;BL=39
AND    BL,09H ;BL=09
;39 0011 1001
;09 0000 1001
;-- ---------
;09 0000 1001
```

CALL   Call a Procedure

Flags: Unchanged.
Format: CALL proc ;transfer control to procedure
Function: Transfers control to a procedure. RET is used to return control to the instruction after the call. There are two types of CALLs: NEAR and FAR. If the target address is within the same code segment, it is a NEAR call. If the target address is outside the current code segment, it is a FAR CALL. Each is described below.

NEAR CALL: If calling a near procedure (the procedure is in the same code segment as the CALL instruction) then the content of the IP register (which is the address of the instruction after the CALL) is pushed onto the stack and SP is decremented by 2. Then IP is loaded with the new value, which is the offset of the procedure. At the end of the procedure when the RET is executed, IP is popped off the stack, which returns control to the instruction after the CALL. There are three ways to code the address of the called NEAR procedure:

1. Direct:

```
                    CALL    proc1
                    ...
        proc1       PROC NEAR
                    ...
                    RET
        proc1       ENDP
```

2. Register indirect:

```
                    CALL    [SI]        ;transfer control to address in SI
```

3. Memory indirect:

```
                    CALL    WORD PTR [DI]  ;DI points to the address that
                                           ;contains IP address of proc
```

FAR CALL: When calling a far procedure (the procedure is in a different segment from the CALL instruction), the SP is decremented by 4 after CS:IP of the instruction following the CALL is pushed onto the stack. CS:IP is then loaded with the segment and offset address of the called procedure. In pushing CS:IP onto the stack, CS is pushed first and then IP. When the RETF is executed, CS and IP are restored from the stack and execution continues with the instruction following the CALL. The following addressing modes are supported:

1. Direct (but outside the present segment):

```
                    CALL    proc1
                    ...
        proc1       PROC FAR
                    ...
                    RETF
        proc1       ENDP
```

2. Memory indirect:

```
                    CALL    DWORD PTR [DI]  ;transfer control to CS:IP where
                                            ;DI and DI+1 point to location of CS and
                                            ;DI+2 and DI+4 point to location of IP
```

CBW   Convert Byte to Word

Flags: Unchanged.
Format: CBW
Function: Copies D7 (the sign flag) to all bits of AH. Used widely to convert a signed byte in AL into a signed word to avoid the overflow problem in signed number arithmetic.

Example:
```
MOV    AX,0
MOV    AL,-5       ;AL=(-5)=FB in 2's complement
                   ;AX = 0000 0000 1111 1011
CBW                ;now AX=FFFB
                   ;AX = 1111 1111 1111 1011
```

CLC   Clear Carry Flag

Flags: Affected: CF.
Format: CLC
Function: Resets CF to zero (CF = 0).

CLD   Clear Direction Flag

Flags: Affected: DF.
Format: CLD
Function: Resets DF to zero (DF = 0). In string instructions if DF = 0, the pointers are incremented with each execution of the instruction. If DF = 1, the pointers are decremented. Therefore, CLD is used before string instructions to make the pointers increment.

CLI   Clear Interrupt Flag

Flags: Affected: IF.
Format: CLI
Function: Resets IF to zero, thereby masking external interrupts received on INTR input. Interrupts received on NMI input are not blocked by this instruction.

CMC   Complement Carry Flag

Flags: Affected: CF.
Format: CMC
Function: Changes CF from 0 to 1 or from 1 to 0.

CMP   Compare Operands

Flags: Affected: OF, SF, ZF, AF, PF, CF.
Format: CMP dest,source ;sets flags as if "SUB dest,source"
Function: Compares two operands of the same size. The source and destination operands are not altered. Performs comparison by subtracting the source operand from the destination and sets flags as if SUB were performed. The relevant flags are as follows:

	CF	ZF	SF	OF
dest > source	0	0	0	SF
dest = source	0	1	0	SF
dest < source	1	0	1	inverse of SF

CMPS/CMPSB/CMPSW   Compare Byte or Word String

Flags: Affected: OF, SF, ZF, AF, PF, CF.
Format: CMPSx
Function: Compares strings a byte or word at a time. DS:SI is used to address the first operand; ES:DI is used to address the second. If DF = 0, it increments the pointers SI and DI. If DF = 1, it decrements the pointers. It can be used with prefix REPE or REPNE to compare strings of any length. The comparison is done by subtracting the source operand from the destination and sets flags as if SUB were performed.

CWD   Convert Word to Doubleword

Flags: Unchanged.
Format: CWD
Function: Converts a signed word in AX into a signed doubleword by copying the sign bit of AX into all the bits of DX. Often used to avoid the overflow problem in signed number arithmetic.

Example:
```
MOV   DX,0
MOV   AX,-5          ;AX=(-5)=FFFB in 2's complement
;DX = 0000H
CWD
;DX = FFFFH
```

DAA   Decimal Adjust after Addition

Flags: Affected: SF, ZF, AF, PF, CF, OF.
Format: DAA
Function: This instruction is used after addition of BCD numbers to convert the result back to BCD. It adds 6 to the lower 4 bits of AL if it is greater than 9 or if AF = 1. Then it adds 6 to the upper 4 bits of AL if it is greater than 9 or if CF = 1.

Example 1:
```
MOV   AL,47H        ;AL=0100 0111
ADD   AL,38H        ;AL=47H+38H=7FH.   invalid BCD
DAA                 ;NOW AL=1000 0101 (85H IS VALID BCD)
```

In this example, since the lower nibble was larger than 9, DAA added 6 to AL. If the lower nibble is smaller than 9 but AF = 1, it also adds 6 to the lower nibble.

Example 2:
```
MOV   AL,29H        ;AL=0010 1001
ADD   AL,18H        ;AL=0100 0001 INCORRECT RESULT
DAA                 ;AL=0100 0111 A VALID BCD FOR 47H.
```

The same thing can happen for the upper nibble.

Example 3:
```
MOV   AL,52H        ;AL=0101 0010
ADD   AL,91H        ;AL=1110 0011 AN INVALID BCD
DAA                 ;AL=0100 0011 AND CF=1
```
Again the upper nibble can be smaller than 9 but because CF = 1, it must be corrected.

Example 4:
```
MOV   AL,94H        ;AL=1001 0100
ADD   AL,91H        ;AL=0010 0101 INCORRECT RESULT
DAA                 ;AL=1000 0101 A VALID BCD FOR 85 AND CF=1
```
It is entirely possible that 6 is added to both the high and low nibbles.

Example 5:
```
MOV   AL,54H        ;AL=0101 0100
ADD   AL,87H        ;AL=1101 1011 INVALID BCD
DAA                 ;AL=0100 0001 AND CF=1 (141 IN BCD)
```

DAS   Decimal Adjust after Subtraction

Flags: Affected: SF, ZF, AF, PF, CF. Unpredictable: OF.
Format: DAS
Function: This instruction is used after subtraction of BCD numbers to convert the result to BCD. If the lower 4 bits of AL represent a number greater than 9 or if AF = 1, then 6 is subtracted from the lower nibble. If the upper 4 bits of AL is now greater than 9 or if CF = 1, 6 is subtracted from the upper nibble.

Example:

```
MOV   AL,45H      ;AL=0100 0101 BCD for 45
SUB   AL,17H      ;AL=0010 1110 AN INVALID BCD
DAS               ;AL=0010 1000 BCD FOR 28(45-17=28)
```

For more examples of problems associated with BCD arithmetic, see DAA.

DEC   Decrement

Flags: Affected: OF, SF, ZF, AF, PF. Unchanged: CF.
Format: DEC dest ;dest = dest - 1
Function: Subtracts 1 from the destination operand. Note that CF (carry/borrow) is unchanged even if a value 0000 is decremented and becomes FFFF.

DIV   Unsigned Division

Flags: Unpredictable: OF, SF, ZF, AF, PF, CF.
Format: DIV source ;divide AX or DX:AX by source
Function: Divides either an unsigned word (AX) by a byte or an unsigned doubleword (DX:AX) by a word. If dividing a word by a byte, the quotient will be in AL and the remainder in AH. If dividing a doubleword by a word, the quotient will be in AX and the remainder in DX. Divide by zero causes interrupt type 0.

ESC   Escape

Flags: Unchanged.
Format: ESC
Function: This instruction facilitates the use of math coprocessors (such as the 8087), which share data and address buses with the microprocessor. ESC is used to pass an instruction to a coprocessor and is usually treated as NOP (no operation) by the main processor.

HLT   Halt

Flags: Unchanged.
Format: HLT
Function: Causes the microprocessor to halt execution of instructions. To get out of the halt state, activate an interrupt (NMI or INTR) or RESET.

IDIV   Signed Number Division

Flags: Unpredictable: OF, SF, ZF, AF, PF, CF.
Format: IDIV source ;divide AX or DX:AX by source
Function: This division function divides either a signed word (AX) by a byte or a signed doubleword (DX:AX) by a word. If dividing a word by a byte, the signed quotient will be in AL and the signed remainder in AH. If dividing a doubleword by a word, the signed quotient will be in AX and the signed remainder in DX. Divide by zero causes interrupt type 0.

IMUL   Signed Number Multiplication

Flags: Affected: OF, CF. Unpredictable: SF, ZF, AF, PF.
Format: IMUL source ;AX =source x AL or DX:AX =source x AX
Function: Multiplies a signed byte or word source operand by a signed byte or word in AL or AX with the result placed in AX or DX:AX.

IN   Input Data from Port

Flags: Unchanged.
Format: IN accumulator,port ;input byte or word into AL or AX
Function: Transfers a byte or word to AL or AX from an input port specified by the second operand. The port address can be direct or register indirect:

1. Direct: port address is specified directly and cannot be larger than FFH.

Example 1:
```
IN   AL,99H              ;BRING A BYTE INTO AL FROM PORT 99H
```

Example 2:
```
IN   AX,78H              ;BRING A WORD FROM PORT ADDRESSES 78H
                         ;AND 79H. THE BYTE FROM PORT 78 GOES
                         ;TO AL AND BYTE FROM PORT 79H TO AH.
```

2. Register indirect: the port address is kept by the DX register. Therefore, it can be as high as FFFFH.

Example 3:
```
MOV   DX,481H        ;DX=481H
IN    AL,DX          ;BRING THE BYTE TO AL FROM THE PORT
                     ;WHOSE ADDRESS IS POINTED BY DX
```

Example 4:
```
IN    AX,DX          ;BRING A WORD FROM PORT ADDRESS OF
                     ;POINTED BY DX.  THE BYTE FROM PORT
                     ;DX GOES TO AL AND BYTE FROM PORT
                     ;DX+1 TO AH.
```

INC   Increment

Flags: Affected: OF, SF, ZF, AF, PF. Unchanged: CF.
Format: INC destination ;dest = dest + 1
Function: Adds 1 to the register or memory location specified by the operand. Note that CF is not affected even if a value FFFF is incremented to 0000.

INT   Interrupt

Flags: Affected: IF, TF.
Format: INT type ;transfer control to INT type
Function: Transfers execution to one of the 256 interrupts. The vector address is specified by the type number, which cannot be greater than FFH (0 to FF = 256 interrupts).

The following steps are performed for the interrupt:

1. SP is decremented by 2 and the flags are pushed onto the stack.
2. SP is decremented by 2 and CS is pushed onto the stack.
3. SP is decremented by 2 and the IP of the next instruction after the interrupt is pushed onto the stack.
4. Multiplies the type number by 4 to get the address of the vector table. Starting at this address, the first 2 bytes are the value of IP and the next 2 bytes are the value for CS of the interrupt handler (interrupt handler is also called interrupt service routine).
5. Resets IF and TF.

SECTION B.1: THE 8086 INSTRUCTION SET

Interrupts are used to get the attention of the microprocessor. In the 8086/88 there are a total of 256 interrupts: INT 00, INT 01, INT 02, ... , INT FF. As mentioned above, the address that an interrupt jumps to is always four times the value of the interrupt number. For example, INT 03 will jump to memory address 0000CH ($4 \times 03 = 12 = 0$CH). Table B-1 is a partial list of the interrupts, commonly referred to as the interrupt vector table.

Table B-1: Interrupt Vector Table

INT # (hex)	Physical Address	Logical Address
INT 00	00000	0000:0000
INT 01	00004	0000:0004
INT 02	00008	0000:0008
INT 03	0000C	0000:000C
INT 04	00010	0000:0010
INT 05	00014	0000:0014
...	...	...
INT FF	003FC	0000:03FC

Every interrupt has a program associated with it called the interrupt service routine (ISR). When an interrupt is invoked, the CS:IP address of its ISR is retrieved from the vector table (shown above). The lowest 1024 bytes ($256 \times 4 = 1024$) of RAM are set aside for the interrupt vector table and must not be used for any other function.

Example: Find the physical and logical addresses of the vector table associated with (a) INT 14H and (b) INT 38H.
Solution:
(a) The physical address for INT 14H is 00050H - 00053H
(4×14H = 50H). That gives the logical address of 0000:0050H - 0000:0053H.
(b) The physical address for INT 38H is 000E0H - 000E3H, making the physical address 0000:00E0H - 0000:00E3H.

The difference between INTerrupt and CALL instructions

The following are some of the differences between the INT and CALL FAR instructions:

1. While a CALL can jump to any location within the 1-megabyte address range (00000 - FFFFF) of the 8088/86 CPU, "INT nn" jumps to a fixed location in the vector table as discussed earlier.
2. While the CALL is used by the programmer at a predetermined point in a program, a hardware interrupt can come in at any time.
3. A CALL cannot be masked (disabled), but "INT nn" can be masked.
4. While a "CALL FAR" automatically saves on the stack only the CS:IP of the next instruction, "INT nn" saves the FR (flag register) in addition to the CS:IP.
5. While at the end of the procedure that has been CALLed the RETF (return FAR) is used, for "INT nn" the instruction IRET (interrupt return) is used.

The 256 interrupts can be categorized into two different groups: hardware and software interrupts.

Hardware interrupts

The 8086/88 microprocessors have two pins set aside for inputting hardware interrupts. They are INTR (interrupt request) and NMI (nonmaskable interrupt). Although INTR can be ignored through the use of software masking, NMI cannot be masked using software. These interrupts are activated externally by putting 5 volts on the hardware pins of NMI or INTR. Intel has assigned INT 02 to NMI. When it is activated it will jump to memory location 00008 to get the address (CS:IP) of the interrupt service routine (ISR). Memory locations 00008, 00009, 0000A, and 0000B contain the 4-byte CS:IP. There is no specific location in the vector table assigned to INTR because INTR is used to expand the number of hardware interrupts and should be allowed to use any "INT nn" instruction that has not been assigned previously. In the IBM PC, one Intel 8259 PIC (programmable interrupt controller) chip is connected to INTR to add a total of eight hardware interrupts to the microprocessor. IBM PC AT, PS/2 80286, 80386, and 80486 computers use two 8259 chips to allow up to 15 hardware interrupts.

Table B-2: IBM PC Interrupt System

Interrupt	Logical Address	Physical Address	Purpose
0	00E3:3072	03EA2	Divide error
1	0600:08ED	068ED	Single step (trace command in DEBUG)
2	F000:E2C3	FE2C3	Nonmaskable interrupt
3	0600:08E6	068E6	Breakpoint
4	0700:0147	00847	Signed number arithmetic overflow
5	F000:FF54	FFF54	Print screen (BIOS)
10	F000:F065	FF065	Video I/O (BIOS)
...	...	...	...
21	relocatable	---	DOS function calls
...	...	...	...

Software interrupts

These interrupts are called software interrupts since they are invoked as a result of the execution of an instruction and no external hardware is involved. In other words, these interrupts are invoked by executing an "INT nn" instruction such as the DOS function call "INT 21H" or video interrupt "INT 10H". These interrupts can be invoked by a program at any time, the same as any other instruction. Many of the interrupts in this category are used by the DOS operating system and IBM BIOS to perform the essential tasks that every computer must provide to the system and the user. Also within this group of interrupts are predefined functions associated with some of the interrupts. They are "INT 00" (divide error), "INT 01" (single step), "INT 03" (breakpoint), and "INT 04" (signed number overflow). Each one is described below. These interrupts are shown in Table B-2. Looking at Table B-2, one can say that aside from "INT 00" to "INT 04", which have predefined functions, the rest of the interrupts, from "INT 05" to "INT FF", can be used to implement either software or hardware interrupts.

Functions associated with "INT 00" to "INT 03"

As mentioned earlier, interrupts "INT 00" to "INT 03" have predefined functions and cannot be used in any other way. The function of each is described next.

INT 00 (divide error)

This interrupt, sometimes referred to as a conditional or exception interrupt, is invoked by the microprocessor whenever there is a condition that it cannot take care of, such as an attempt to divide a number by zero. "INT 00" is invoked by the microprocessor whenever there is an attempt to divide a number by zero. In the IBM PC and compatibles, the service subroutine for this interrupt is responsible for displaying the message "DIVIDE ERROR" on the screen if a program such as the following is executed:

```
MOV   AL,25   ; put 25 into AL
MOV   BL,00   ; put 00 into BL
DIV   BL      ; divide 25 by 00
```

This interrupt is also invoked if the quotient is too large to fit into the assigned register when executing a DIV instruction.

INT 01 (single step)

There is often a need to execute a given program one instruction at a time and then inspect the registers (possibly memory as well) to see what is happening inside the CPU. This is commonly referred to as single-stepping. IBM and Microsoft call it TRACE in the DEBUG program. To allow the implementation of single-stepping, Intel has set aside "INT 01" specifically for that purpose. For the Trace command in DEBUG after execution of each instruction, the CPU jumps automatically to physical location 00004 to fetch the 4 bytes for CS:IP of the interrupt service routine. One of the functions of this ISR is to dump the contents of the registers onto the screen.

INT 02 (nonmaskable interrupt)

This interrupt is used in the PC to indicate memory errors, among other problems.

INT 03 (breakpoint)

While in single-step mode, one can inspect the CPU and system memory after execution of each instruction. A breakpoint allows one to do the same thing, after execution of a group of instructions rather than after each instruction. Breakpoints are put in at certain points of a program to monitor the flow of the program and to inspect the results after certain instructions. The CPU executes the program to the breakpoint and stops. One can proceed from breakpoint to breakpoint until the program is complete. With the help of single-step and breakpoints, programs can be debugged and tested more easily. The Intel 8086/88 CPUs have set aside "INT 03" for the sole purpose of implementing breakpoints. When the instruction "INT 03" is placed in a program the CPU will execute the program until it encounters "INT 03", and then it stops. One interesting point about this interrupt is that it is a one-byte instruction, in contrast to all other interrupt instructions, "INT nn", which are two-byte instructions. This allows the user to insert 1 byte of code and remove it to proceed with the execution of the program. The opcode for INT 03 is "CC".

IBM PC and DOS assignment of interrupts

When the IBM PC was being developed, the designers at IBM had to coordinate the assignment of the 256 available interrupts for the 8086/88 with Microsoft, the developer of the DOS operating system, lest a conflict occur between the BIOS and DOS interrupt designations. The result of cooperation in assigning interrupts to IBM BIOS subroutines and DOS function calls is shown in Table B-2. The table gives a partial listing of interrupt numbers from 00 to FF, the logical address of the service subroutine for each interrupt, their physical addresses, and the purpose of each interrupt. It must be mentioned that depending on the computer and the DOS version, some of the logical addresses could be different from Table B-2.

How to get the vector table of any PC

One can get the vector table of any IBM PC/XT, PC AT, PS/2, PS/1, or any 80x86 IBM-compatible computer and inspect the logical address assigned to each interrupt. To do that use DEBUG's DUMP command "-D 0000:0000", as shown next.

```
A>debug
-D 0000:0000
0000:0000  E8 56 2B 02 56 07 70 00-C3 E2 00 F0 56 07 70 00  ..........
0000:0010  56 07 70 00 54 FF 00 F0-47 FF 00 F0 47 FF 00 F0  ..........
```

Note: The contents of the memory locations could be different, depending on the DOS version.

Example: From the dump above, find the CS:IP of the service routine associated with INT 5.
Solution: To get the address of "INT 5", calculate the physical address of 00014H ($5 \times 4 = 00014$H). The contents of these locations are 00014 = 54, 00015 = FF, 00016 = 00, and 00016 = F0. This gives CS = F000 and IP = FF54.

INTO   Interrupt on Overflow

Flags: Affected: IF, TF.
Format: INTO
Function: Transfers execution to an interrupt handler written for overflow if OF (overflow flag) has been set. Intel has set aside INT 4 for this purpose. Therefore, if OF = 1 when INTO is executed, the CPU jumps to memory location 00010H ($4 \times 4 = 16 = 10$H). The contents of memory locations 10H, 11H, 12H, and 13H are used as IP and CS of the interrupt handler procedure. This instruction is widely used to detect overflow in signed number addition. In signed number operations, OF becomes 1 in two cases:

1. Whenever there is a carry from d6 to d7 in 8-bit operations and no carry from D7 out (or in 16-bit operations when there is carry from d14 to d15 and CF = 0)
2. When there is carry from from D7 out and no carry from D6 to D7 (or in the case of 16-bit operation when there is a carry from D15 out and no carry from D14 to D15)

IRET   Interrupt Return

Flags: Affected: OF, DF, IF, TF, SF, ZF, AF, PF, CF.
Format: IRET
Function: Used at the end of an interrupt service routine (interrupt handler), this instruction restores all flags, CS, and IP to the values they had before the interrupt so that execution may continue at the next instruction following the INT instruction. While the RET instruction is used at the end of the subroutine associated

---

with the CALL instruction, IRET must be used for the subroutine associated with the "INT XX" instruction or the hardware interrupt handler.

JUMP Instructions

The following instructions are associated with jumps (both conditional and unconditional). They are categorized according to their usage rather than alphabetically.

J condition

Flags: Unchanged.
Format: Jxx target ;jump to target upon condition
Function: Used to jump to a target address if certain conditions are met. The target address cannot be more than −128 to +127 bytes away. The conditions are indicated by the flag register. The conditions that determine whether the jump takes place can be categorized into three groups,

(1) flag values,
(2) the comparison of unsigned numbers, and
(3) the comparison of signed numbers.
Each is explained next.

1. "J condition" where the condition refers to flag values. The status of each bit of the flag register has been decided by execution of instructions prior to the jump. The following "J condition" instructions check if a certain flag bit is raised or not.

JC	Jump Carry	jump if CF=1
JNC	Jump No Carry	jump if CF=0
JP	Jump Parity	jump if PF=1
JNP	Jump No Parity	jump if PF=0
JZ	Jump Zero	jump if ZF=1
JNZ	Jump No Zero	jump if ZF=0
JS	Jump Sign	jump if SF=1
JNS	Jump No Sign	jump if SF=0
JO	Jump Overflow	jump if OF=1
JNO	Jump No Overflow	jump if OF=0

Notice that there is no "J condition" instruction for AF.

2. "J condition" where the condition refers to the comparison of unsigned numbers. After a compare (CMP dest,source) instruction is executed, CF and ZF indicate the result of the comparison, as follows:

	CF	ZF
destination > source	0	0
destination = source	0	1
destination < source	1	0

Since the operands compared are viewed as unsigned numbers, the following "J condition" instructions are used.

JA	Jump Above	jump if CF=0 and ZF=0
JAE	Jump Above or Equal	jump if CF=0
JB	Jump Below	jump if CF=1
JBE	Jump Below or Equal	jump if CF=1 or ZF=1
JE	Jump Equal	jump if ZF=1
JNE	Jump Not Equal	jump if ZF=0

3. "J condition" where the condition refers to the comparison of signed numbers. In the case of the signed number comparison, although the same instruction, "CMP destination,source", is used, the flags used to check the result are as follows:

destination > source	OF=SF or ZF=0
destination = source	ZF=1
destination < source	OF inverse of SF

Consequently, the "J condition" instructions used are different. They are as follows:

JG	Jump Greater	jump if ZF=0 or OF=SF
JGE	Jump Greater or Equal	jump if OF=SF
JL	Jump Less	jump if OF≠SF
JLE	Jump Less or Equal	jump if ZF=1 or OF≠SF
JE	Jump if Equal	jump if ZF = 1

There is one more "J condition" instruction:
JCXZ   ;Jump if CX is Zero. ZF is ignored.

All "J condition" instructions are short jumps, meaning that the target address cannot be more than −128 bytes backward or +127 bytes forward from the IP of the instruction following the jump. What happens if a programmer needs to use a "J condition" to go to a target address beyond the −128 to +127 range? The solution is to use the "J condition" along with the unconditional JMP instruction, as shown next.

```
          ADD       BX,[SI]
          JNC       NEXT
          JMP       TARGET1
NEXT:     ....
          ...
TARGET1:  ADD       DI,10
          ...
```

JMP   Unconditional Jump

Flags: Unchanged.
Format: JMP [directives] target ;jump to target address
Function: This instruction is used to transfer control unconditionally to a new address. The difference between JMP and CALL is that the CALL instruction will return and continue execution with the instruction following the CALL, whereas JMP will not return. The target address could be within the current code segment, which is called a near jump, or outside the current code segment, which is called a far jump. Within each category there are many ways to code the target address, as shown next.

1. Near jump

(a) direct short jump: In this jump the target address must be within −128 to +127 bytes of the IP of the instruction after the JMP. This is a 2-byte instruction. The first byte is the opcode EBH and the second byte is the signed number displacement, which is added to the IP of the instruction following the JMP to get the target address. The directive SHORT must be coded, as shown next:
```
          JMP  SHORT OVER
          ...
OVER:     ...
```

If the target address is beyond the −128 to +127 byte range and the SHORT directive is coded, the assembler gives an error.

(b) Direct jump: This is a 3-byte instruction. The first byte is the opcode E9H and the next two bytes are the signed number displacement value. The displacement is added to the IP of the instruction following the JMP to get the target address. The displacement can be in the range −32,768 to +32,767. In the absence of the SHORT directive, the assembler in its first pass always uses this kind of JMP, and then in the second pass if the target address is within the −128 and +127 byte range, it uses the NOP opcode 90H for the third byte. This is the reason to code the directive SHORT if it is known that the target address of the JMP is within the short range.

(c) Register indirect jump: In this jump the target address is in a register as shown next:

```
JMP     DI  ;jump to the address found in DI
```

Any nonsegment register can be used for this purpose.

(d) Memory indirect jump: In this jump the target address is in a memory location whose address is pointed at by a register:

```
JMP     WORD PTR [SI] ;jump to the address found at the address in SI
```

The directive WORD PTR must be coded to indicate this is a near jump.

2. Far jump

In a far JMP, the target address is outside the present code segment; therefore, not only the offset value but also the segment value of the target address must be given. A far jump is a 5-byte instruction: the opcode EAH and 4 bytes for the offset and segment of the target address. The following shows the two methods of coding the far jump.

(a) Direct far jump: This requires that both CS and IP be updated. One way to do that is to use the LABEL directive:

```
        JMP         TARGET2
        ...
TARGET2 LABEL       FAR
ENTRY:  ...
```

This is exactly what IBM has done in BIOS of the IBM PC/XT when the computer is booted. When the power to the PC is turned on, the 8088/86 CPU begins to execute at address FFFF:0000H. IBM uses a FAR jump to make it go to location F000:E05BH, as shown next:

```
                    ;CS=FFFF and IP=0000
0000 EA5BE000F0     JMP     RESET
                    ;CS=F000
                    ORG     0E05BH
E05B    RESET       LABEL   FAR
E05B    START:
E05B                CLI
E05C                ...
```

The EXTRN and PUBLIC directives also can be used for the same purpose.

(b) Memory indirect far jump: The target address (both CS:IP) is in a memory location pointed to by the register:

```
JMP    DWORD PTR [BX]
```

The DWORD and PTR directives must be used to indicate that it is a far jump.

LAHF   Load AH from Flags

Flags: Unchanged.
Format: LAHF
Function: Loads the lower 8 bits of the flag register into AH.

LDS   Load Data Segment Register

Flags: Unchanged.
Format: LDS dest,source ;load dest and DS starting at source
Function: Loads into destination (which is a register) the contents of two memory locations indicated by source and loads DS with the contents of the two succeeding memory locations. This is useful for accessing a new data segment and its offset.

Example: Assume the following memory locations with the contents:
```
;DS:1200=(46)
;DS:1201=(10)
;DS:1202=(38)
;DS:1203=(82)
LDS    DI,[1200]        ;now DI=1046 and DS=8238.
```

EA   Load Effective Address

Flags: Unchanged.
Format: LEA dest,source ;dest = OFFSET source
Function: Loads into the destination (a 16-bit register) the effective address of a direct memory operand.

Example 1:
```
       ORG    0100H
DATA   DB     34,56,87,90,76,54,13,29
       ...
       ;to access the sixth element:
       LEA    SI,DATA+5    ;SI=100H+5=105 THE EFFECTIVE ADDRESS
       MOV    AL,[SI]         ;GET THE SIXTH ELEMENT
```

Example 2:
```
       ;if BX=2000H and SI=3500H
       LEA    DX,[BX][SI]+100H
       ;DX=effective address=2000+3500+100=5600H
```

The following two instructions show two different ways to accomplish the same thing:

```
       MOV    SI,OFFSET DATA      ;advantage: executes faster
       LEA    SI,DATA
```

LES   Load Extra Segment Register

Flags: Unchanged.
Format: LES dest,source ;load dest and ES starting at source
Function: Loads into destination (a register) the contents of two memory locations indicated by the source and loads ES with the contents of the two succeeding memory locations. Useful for accessing a new extra segment and its offset. This instruction is similar to LDS except that the ES and its offset are being loaded.

LOCK Lock System Bus Prefix

Flags: Unchanged.
Format: LOCK ;used as a prefix before instructions
Function: Used in microcomputer systems with more than one processor to prevent another processor from gaining control over the system bus during execution of an instruction.

LODS/LODSB/LODSW   Load Byte or Word String

Flags: Unchanged.
Format: LODSx
Function: Loads AL or AX with a byte or word from the memory location pointed to by DS:SI. If DF = 0, SI will be incremented to point to the next location. If DF = 1, SI will be decremented to point to the next location. SI is incremented/decremented by 1 or 2, depending on whether it is a byte or word string.

LOOP Loop until CX=0

Flags: Unchanged.
Format: LOOP target ;DEC CX, then jump to target if CX not 0
Function: Decrements CX by 1, then jumps to the offset indicated by the operand if CX not zero, otherwise continues with the next instruction below the LOOP. This instruction is equivalent to

```
DEC  CX
JNZ  target
```

LOOPE/LOOPZ   LOOP if Equal / Loop if Zero

Flags: Unchanged.
Format: LOOPx target ;DEC CX, jump to target if CX ≠ 0 and ZF=1
Function: Decrements CX by 1, then jumps to location indicated by the operand if CX is not zero and ZF is 1, otherwise continues with the next instruction after the LOOP. In other words, it gets out of the loop only when CX becomes zero or when ZF = 0.

Example:
Assume that 200H memory locations from offset 1680H should contain 55H. LOOPE can be used to see if any of these locations does not contain 55H:

```
            MOV    CX,200        ;SET UP THE COUNTER
            MOV    SI,1680H      ;SET UP THE POINTER
BACK:  CMP    [SI],55H      ;COMPARE THE 55H WITH MEM LOCATION
                                    ;POINTED AT BY SI
            INC     SI              ;INCREMENT THE POINTER
            LOOPE  BACK        ;CONTINUE THE PROCESS UNTIL CX=0 OR
                                    ;ZF=0. IN OTHER WORDS EXIT IF ONE
                                    ;LOCATION DOES NOT HAVE 55H
```

LOOPNE/LOOPNZ   LOOP While CF Not Zero and ZF Equal Zero

Flags: Unchanged.
Format: LOOPxx target ;DEC CX, then jump if CX and ZF not zero
Function: Decrements CX by 1, then jumps to location indicated by the operand if CX and ZF are not zero, otherwise continues with the next instruction below the LOOP. In other words it will exit the loop if CX becomes 0 or ZF = 1.

Example:
Assume that the daily temperatures for the last 30 days have been stored starting at memory location with offset 1200H. LOOPE can be used to find the first day that had a 90-degree temperature.

```
            MOV     CX,30         ;SET UP THE COUNTER
            MOV     DI,1200H      ;SET UP THE POINTER
AGAIN:      CMP     [DI],90
            INC     DI
            LOOPNE  AGAIN
```

MOV   Move

Flags: Unchanged.
Format: MOV dest,source ;copy source to dest
Function: Copies a word or byte from a register, memory location, or immediate number to a register or memory location. Source and destination must be of the same size and cannot both be memory locations.

MOVS/MOVSB/MOVSW   Move Byte or Word String

Flags: Unchanged.
Format: MOVSx
Function: Moves byte or word from memory location pointed to by DS:SI to memory location pointed to by ES:DI. If DF = 0, both pointers are incremented; otherwise, they are decremented. SI and DI are incremented/decremented by 1 or 2 depending on whether it is a byte or word string. When used with the REP prefix, CX is decremented each time until CX is zero.

MUL   Unsigned Multiplication

Flags: Affected: OF, CF. Unpredictable: SF, ZF, AF, PF.
Format: MUL source ;AX = source × AL or DX:AX = source × AX
Function: Multiplies an unsigned byte or word indicated by the operand by a unsigned byte or word in AL or AX with the result placed in AX or DX:AX.

NEG   Negate

Flags: Affected: OF, SF, ZF, AF, PF, CF.
Format: NEG dest ;negates operand
Function: Performs 2's complement of operand. Effectively reverses the sign bit of the operand. This instruction should only be used on signed numbers.

NOP   No Operation

Flags: Unchanged.
Format: NOP
Function: Performs no operation. Sometimes used for timing delays to waste clock cycles. Updates IP to point to next instruction following NOP.

SECTION B.1: THE 8086 INSTRUCTION SET   **865**

NOT   Logical NOT

Flags: Unchanged.
Format: NOT dest ;dest = 1's complement of dest
Function: Replaces the operand with its negation (the 1's complement). Each bit is inverted.

OR   Logical OR

X Y	X OR Y
0 0	0
0 1	1
1 0	1
1 1	1

Flags: Affected: CF=0, OF=0, SF, ZF, PF. Unpredictable: AF.
Format: OR dest,source ;dest= dest OR source
Function: Performs logical OR on the bits of two operands, replacing the destination operand with the result. Often used to turn a bit on.

OUT   Output Byte or Word

Flags: Unchanged.
Format: OUT dest,acc ;transfer acc to port dest
Function: Transfers a byte or word from AL or AX to an output port specified by the first operand.   Port address can be direct or register indirect as shown next:
1. Direct: port address is specified directly and cannot be larger than FFH.

Example 1:
```
OUT    68H,AL          ;SEND OUT A BYTE FROM AL TO PORT 68H
or
OUT    34H,AX ;SEND OUT A WORD FROM AX TO PORT
                       ;ADDRESSES 34H AND 35H.  THE BYTE
                       ;FROM AL GOES TO PORT 34H AND
                       ;THE BYTE FROM AH GOES TO PORT 35H
```

2. Register indirect: port address is kept by the DX register.   Therefore, it can be as high as FFFFH.

Example 2:
```
MOV    DX,64B1H        ;DX=64B1H
OUT    DX,AL           ;SENT OUT THE BYTE IN AL TO THE PORT
                       ;WHOSE ADDRESS IS POINTED TO BY DX
or
OUT    DX,AX           ;SEND OUT A WORD FROM AX TO PORT
                       ;ADDRESS POINTED TO DX.   THE BYTE
                       ;FROM AL GOES TO PORT DX AND AND BYTE
                       ;FROM AH GOES TO PORT DX+1.
```

POP   POP Word

Flags: Unchanged.
Format: POP dest ;dest = word off top of stack
Function: Copies the word pointed to by the stack pointer to the register or memory location indicated by the operand and increments the SP by 2.

POPF   POP Flags off Stack

Flags: OF, DF, IF, TF, SF, ZF, AF, PF, CF.
Format: POPF
Function: Copies bits previously pushed onto the stack with the PUSHF instruction into the flag register. The stack pointer is then incremented by 2.

PUSH   PUSH Word

Flags: Unchanged.
Format: PUSH source ;PUSH source onto stack
Function: Copies the source word to the stack and decrements SP by 2.

PUSHF   PUSH Flags onto stack

Flags: Unchanged.
Format: PUSHF
Function: Decrements SP by 2 and copies the contents of the flag register to the stack.

RCL/RCR Rotate Left through Carry and Rotate Right through Carry

Flags: Affected: OF, CF.
Format: RCx dest,n ;dest = dest rotate right/left n bit positions
Function: Rotates the bits of the operand right or left. The bits rotated out of the operand are rotated into the CF and the CF is rotated into the opposite end of the word or byte. *Note*: "n" must be 1 or CL.

RET   Return from a Procedure

Flags: Unchanged.
Format: RET [n] ;return from procedure
Function: Used to return from a procedure previously entered by a CALL instruction. The IP is restored from the stack and the SP is incremented by 2. If the procedure was FAR, then RETF (return FAR) is used, and in addition to restoring the IP, the CS is restored from the stack and SP is again incremented by 2. The RET instruction may be followed by a number that will be added to the SP after the SP has been incremented. This is done to skip over any parameters being passed back to the calling program segment.

ROL/ROR Rotate Left and Rotate Right

Flags: Affected: OF, CF.
Format: ROx dest,n ;rotate dest right/left n bit positions
Function: Rotates the bits of a word or byte indicated by the second operand right or left. The bits rotated out of the word or byte are rotated back into the word or byte at the opposite end. *Note*: "n" must be 1 or CL.

SAHF   Store AH in Flag Register

Flags: Affected: SF, ZF, AF, PF, CF.
Format: SAHF
Function: Copies AH to the lower 8 bits of the flag register.

SAL/SAR Shift Arithmetic Left/ Shift Arithmetic Right

Flags: Affected: OF, SF, ZF, PF, CF. Unpredictable: AF.
Format: SAx dest,n ;shift signed dest left/right n bit positions
Function: Shifts a word or byte left /right. SAR/ SAL arithmetic shifts are used for signed number shifting. In SAL, as the operand is shifted left bit by bit, the LSB is filled with 0s and the MSB is copied to CF. In SAR, as each bit is shifted right, the LSB is copied to CF and the empty bits filled with the sign bit (the MSB). SAL/SAR essentially multiply/divide destination by a power of 2 for each bit shift. *Note*: "n" must be 1 or CL.

SBB   Subtract with Borrow

Flags: Affected: OF, SF, ZF, AF, PF, CF.
Format: SBB dest,source ;dest = dest - CF - source
Function: Subtracts source operand from destination, replacing destination. If CF =1, it subtracts 1 from the result; otherwise, it executes like SUB.

SCAS/SCASB/SCASW   Scan Byte or Word String

Flags: Affected: OF, SF, ZF, AF, PF, CF.
Format: SCASx
Function: Scans a string of data pointed by ES:DI for a value that is in AL or AX. Often used with the REPE/REPNE prefix. If DF is zero, the address is incremented; otherwise, it is decremented.

SHL/SHR Shift Left/Shift Right

Flags: Affected: OF, SF, ZF, PF, CF. Unpredictable: AF.
Format: SHx dest,n ;shift unsigned dest left/right n bit positions
Function: These are logical shifts used for unsigned numbers, meaning that the sign bit is treated as data. In SHR, as the operand is shifted right bit by bit and copied into CF, the empty bits are filled with 0s instead of the sign bit as is the case for SAR. In the case of SHL, as the bits are shifted left, the MSB is copied to CF and empty bits are filled with 0, which is exactly the same as SAL. In reality, SAL and SHL are two different mnemonics for the same opcode. SHL/SHR essentially multiply/divide the destination by a power of 2 for each bit position shifted. *Note*: "n" must be 1 or CL.

STC   Set Carry Flag

Flags: Affected: CF.
Format: STC
Function: Sets CF to 1.

STD   Set Direction Flag

Flags: Affected: DF.
Format: STD
Function: Sets DF to 1. Used widely with string instructions. As explained in the string instructions, if DF = 1, the pointers are decremented.

STI   Set Interrupt Flag

Flags: Affected: IF.
Format: STI
Function: Sets IF to 1, allowing the hardware interrupt to be recognized through the INTR pin of the CPU.

STOS/STOSB/STOSW   Store Byte or Word String

Flags: Unchanged.
Format: STOSx
Function: Copies a byte or word from AX or AL to a location pointed to by ES:DI and updates DI to point to the next string element. The pointer DI is incremented if DF is zero; otherwise, it is decremented.

SUB   Subtract

Flags: Affected: OF, SF, ZF, AF, PF, CF.
Format: SUB dest,source ;dest = dest - source
Function: Subtracts source from destination and puts the result in the destination. Sets the carry and zero flag according to the following:

	CF	ZF	
dest >source	0	0	the result is positive
dest=source	0	1	the result is 0
dest < source	1	0	the result is negative in 2's comp

The steps for subtraction performed by the internal hardware of the CPU are as follows:

1. Takes the 2's complement of the source
2. Adds this to the destination
3. Inverts the carry and changes the flags accordingly
The source operand remains unchanged by this instruction.

TEST   Test Bits

Flags: Affected: OF, SF, ZF, PF, CF. Unpredictable: AF.
Format: TEST dest,source ;performs dest AND source
Function: Performs a logical AND on two operands, setting flags but leaving the contents of both source and destination unchanged. While the AND instruction changes the contents of the destination and the flag bits, the TEST instruction changes only the flag bits.

Example:
Assume that D0 and D1 of port 27 indicate conditions A and B, respectively, if they are high and only one of them can be high at a given time. The TEST instruction can be used as follows:

```
            IN    AL,PORT_27
            TEST AL,0000 0001B      ;CHECK THE CONDITION A
            JNZ   CASE_A            ;JUMP TO INDICATE CONDITION A
            TEST AL,0000 0010B      ;CHECK FOR CONDITION B
            JNZ   CASE_B            ;JUMP TO INDICATE CONDITION B
            ....                    ;THERE IS AN ERROR SINCE NEITHER
            ....                    ;   A OR B HAS OCCURRED.
CASE_A:  ....
            ....
CASE_B:  ....
```

WAIT   Puts Processor in WAIT State

Flags: Unchanged.
Format: WAIT
Function: Causes the microprocessor to enter an idle state until an external interrupt occurs. This is often done to synchronize it with another processor or with an external device.

XCHG   Exchange

Flags: Unchanged.
Format: XCHG dest,source ;swaps dest and source
Function: Exchanges the contents of two registers or a register and a memory location.

XLAT   Translate

Flags: Unchanged.
Format: XLAT
Function: Replaces contents of AL with the contents of a look-up table whose address is specified by AL. BX must be loaded with the start address of the look-up table and the element to be translated must be in AL prior to the execution of this instruction. AL is used as an offset within the conversion table. Often used to translate data from one format to another, such as ASCII to EBCDIC.

XOR   Exclusive OR

Flags: Affected: CF = 0, OF = 0, SF, ZF, PF.
Unpredictable: AF.
Format: XOR dest,source
Function: Performs a logical exclusive OR on the bits of two operands and puts the result in the destination. "XOR AX,AX" can be used to clear AX.

X Y	X XOR Y
0 0	0
0 1	1
1 0	1
1 1	0

SECTION B.2: INSTRUCTION TIMING

In this section of the appendix we provide clock counts for all the instructions of Intel's 8086, 286, 386, and 486 microprocessors. The clock count is the number of clocks that it takes the instruction to execute. They are extracted from Intel's reference manuals on these microprocessors. The number of clocks for each instruction is given with the assumption that the instruction is already fetched into the CPU. The actual clock count can vary depending on the memory hardware design of the system. Note the following points when calculating the clock counts for a given CPU.

1. In calculating the total clock cycles for the 8086/88, one must add the extra clocks associated with the effective address (EA) provided in Table B-3.

2. In calculating the time required for the 8086, 286, and 386SX microprocessors, its 16-bit external data bus must be taken into consideration. In addition, whether the operand address is odd or even must be considered. To reduce the time required to fetch data from memory, these CPUs require that the data be aligned on even address boundaries. If addresses are not on even boundaries, an extra 4 clock-cycle penalty is added when fetching a 16-bit operand. Look at the following examples, assuming that DS = 2500H and BX = 3000H.

MOV   AX,[BX] ;total clocks = 10 + 5

Since the physical address is 25000H+3000H=28000H, an even address, and the data bus in the 8086 is 16 bits wide, the contents of memory locations 28000H and 28001H will be fetched into the CPU in one memory cycle. In all 80x86 microprocessors, the low byte goes to the low address and the high byte to the high address. In the example above, the contents of memory location 28000H will go to register AL and 28001H to AH. If BX = 3005H, the physical address would be 25000H + 3005H = 28005H, an odd location, and the clocks required would be as follows due to the extra 4 clock penalty for nonaligned data.

MOV   AX,[BX]   ;total clocks=10+4+5

In the instruction above the contents of memory location 28005 are moved to AL and 28006H to AH. In actuality, the way the 8086 accesses memory is that in the first memory cycle, the 16-bit data from 28004H and 28005H is accessed on the D0 - D15 data bus and then the 16-bit data of memory locations 28006H and 28007H is fetched in the second memory cycle using the 16-bit data bus. In other words, although memory locations 28004H, 28005H, 28006H, and 28007H were addressed by the 8086 in two consecutive memory cycles, only the contents of 28005H and 28006H are used; the contents of memory locations 28004H and 28007H are discarded. For this reason the data must be word (16-bit) aligned in the 16-bit data bus microprocessors. What happens if an odd address is accessed in the 8086? It still will take only one memory cycle consisting of 4 clocks. For example, in "MOV AH,[BX]" with BX = 3005H and DS = 25000H, the contents of memory locations 28004H and 28005H both are accessed with one memory cycle, but only the contents of address 28005H are fetched into register AH.

3. In the 8088 microprocessor, the time required to execute an instruction can vary from the 8086 since the data bus is only 8 bits in the 8088. A 16-bit operand would require two memory cycles (each consisting of 4 clock cycles) to move the operand in or out of the microprocessor: for example,

MOV   AL,[BX]   ;total clocks = 10 + 5

MOV   AX,[BX]   ;total clocks = 10 + 4 + 5

4. For conditional jumps and LOOP instructions, the first number is the number of clocks if the jump is successful (jump is taken) and the second number is for when the jump is not taken (noj = no jump). For example, the 8086 column for the JNZ instruction has "16,noj 4" for the clock count. The 16 is the clock cycle for the case when the jump is taken. If there is no jump, the clock is 4.

5. The clock number for the 80386SX is the same as the 80386, except for accessing 32-bit operands, for which an extra 2 clocks should be added since the data bus in the 80386SX is 16-bit and the memory cycle time of the 80386SX is 2 clocks.

6. An extra 2 clocks must be added for the 80286 and 80386SX if a 16-bit word operand is not aligned and also for the 386 if a 32-bit operand is not aligned at the 32-bit boundary.   See the discussion above in point 2.

7. The number of clocks given for the 80486 microprocessor is for situations when the operand is in the cache memory of the 486 chip; otherwise, extra clocks should be added for the cache miss penalty. For the list of the cache miss penalties, refer to Intel's "i486 Microprocessor Programming Reference Manual."

8. PM (privilege mode) instruction timings are for situations when the CPU is switched to protected mode.

SECTION B.2: INSTRUCTION TIMING   871

9. The "m" (often seen in 286 and 386 instructions) represents the number of components associated with the next instruction to be executed. The value of m varies because the size of the instruction located at the target address can vary. Generally, m can be averaged to 2.

10. The "n" represents the number of repetitions of a given instruction.

11. Due to the ever-advancing architectural design of the 286/386/486 microprocessors, the total clock count for a given program cannot be 100% correct. For this reason a 10% margin of error should be taken into consideration when calculating the total clock count of a given program.

12. With every new generation of 80x86, new instructions are added; therefore, there is no clock count for the prior generations. This indicated by "--".

Table B-3: Clock Cycles for Effective Address

Addressing Mode	Operand	CLK
Direct	label	6
Register indirect	[BX]	5
	[SI]	5
	[DI]	5
	[BP]	5
Based relative	[BX]+disp	9
	[BP]+disp	9
Indexed relative	[DI]+disp	9
	[SI]+disp	9
Based indexed	[BX][SI]	7
	[BX][DI]	7
	[BP][SI]	8
	[BP][DI]	8
Based indexed relative	[BX][SI]+disp	11
	[BX][DI]+disp	11
	[BP][SI]+disp	12
	[BP][DI]+disp	12

Note:
These times assume no segment override. If a segment override is used, 2 clock cycles must be added.

(Reprinted by permission of Intel Corporation, Copyright Intel Corp. 1989)

A summary of the clock cycles for various Intel microprocessors, by instruction, is given in Table B-4.

Table B-4: Clock Cycles for Various Intel Microprocessors by Instruction

Code	Description	8086	80286	80386	80486
AAA	ASCII adjust for addition	8	3	4	3
AAD	ASCII adjust for division	60	14	19	14
AAM	ASCII adjust for multiplication	83	16	17	15
AAS	ASCII adjust for subtraction	8	3	4	3
ADC	Add with carry				
	reg to reg	3	2	2	1
	mem to reg	9+EA	7	6	2
	reg to mem	16+EA	7	7	3
	immed to reg	4	3	2	1
	immed to mem	17+EA	7	7	3
	immed to acc	4	3	2	1
ADD	Addition				
	reg to reg	3	2	2	1
	mem to reg	9+EA	7	6	2
	reg to mem	16+EA	7	7	3
	immed to reg	4	3	2	1
	immed to mem	17+EA	7	7	3
	immed to acc	4	3	2	1
AND	Logical AND				
	reg to reg	3	2	2	1
	mem to reg	9+EA	7	6	2
	reg to mem	16+EA	7	7	3
	immed to reg	4	3	2	1
	immed to mem	17+EA	7	7	3
	immed to acc	4	3	2	1
ARPL	Adjust RPL (requested privilege level)				
	reg to reg	--	10	20	9
	reg to mem	--	11	21	9
BOUND	Check array bounds	--	13noj	10noj	7noj
BSF	Bit scan forward				
	reg to reg	--	--	10+3n	6/42
	mem to reg	--	--	10+3n	7/43
BSR	Bit scan reverse				
	reg to reg	--	--	10+3n	6/103
	mem to reg	--	--	10+3n	7/104
BSWAP	Byte swap	--	--	--	1
BT	Bit test				
	reg to reg	--	--	3	3
	reg to mem	--	--	12	8
	immed to reg	--	--	3	3
	immed to mem	--	--	6	3
BTC/	Bit test complement/				
BTR/	Bit test reset/				
BTS	Bit test set				
	reg to reg	--	--	6	6
	reg to mem	--	--	13	13
	immed to reg	--	--	6	6
	immed to mem	--	--	8	8

Table B-4: Clock Cycles for Various Intel Microprocessors by Instruction (continued)

Code	Description	8086	80286	80386	80486
CALL	Call a procedure				
	intrasegment direct	19	7+m	7+m	3
	intrasegment indirect				
	through register	16	7+m	7+m	5
	instrasegment indirect				
	through memory	21+EA	11+m	10+m	5
	intersegment direct	28	13+m	17+m	18
	486: to same level				20
	486: thru Gate to same level				35
	486: to inner level, no parameters				69
	486: to inner level, x parameter (d) words				77+4x
	486: to TSS				37+TS
	486: thru Task Gate				38+TS
	intersegment direct PM	--	26+m	34+m	
	intersegment indirect	37+EA	16+m	22+m	17
	486: to same level				20
	486: thru Gate to same level				35
	486: to inner level, no parameters				69
	486: to innter level, x parameter (d) words				77+4x
	486: to TSS				37+TS
	486: thru Task Gate				38+TS
	intersegment indirect PM	--	29+m	38+m	
CBW	Convert byte to word	2	2	3	3
CDQ	Convert double to quad	--	--	2	
CLC	Clear carry flag	2	2	2	2
CLD	Clear direction flag	2	2	2	2
CLI	Clear interrupt flag	2	3	3	5
CLTS	Clear task switched flag	--	2	5	7
CMC	Complement carry flag	2	2	2	2
CMP	Compare				
	reg to reg	3	2	2	1
	mem to reg	9+EA	6	6	2
	reg to mem	9+EA	7	5	2
	immed to reg	4	3	2	1
	immed to mem	10+EA	6	5	2
	immed to acc	4	3	2	1
CMPS/	Compare string/				
CMPSB/	Compare byte string/				
CMPSW	Compare word string				
	not repeated	22	8	10	8
	REPE/REPNE CMPS/CMPSB/CMPSW	9+22/rep	5+9/rep	5+9/rep	7+7c
CMPXCHG	Compare and exchange				
	reg with reg	--	--	--	6
	reg with mem	--	--	--	7/10
CWD	Convert word to doubleword	5	2	2	3
CWDE	Convert word to extended double	--	--	3	3
DAA	Decimal adjust for addition	4	3	4	2
DAS	Decimal adjust for subtraction	4	3	4	2

Code	Description	8086	80286	80386	80486
DEC	Decrement by 1				
	16-bit reg	3	2	2	1
	8-bit reg	3	2	2	1
	memory	15+EA	7	6	3
DIV	Unsigned division				
	8-bit reg	80-90	14	14	16
	16-bit reg	144-162	22	22	24
	double	--	--	--	40
	8-bit mem	(86-96)+EA	17	17	16
	16-bit mem	(150-168)+EA	25	25	24
	double	--	--	--	40
ENTER	Make stack frame				
	W,0	--	11	10	14
	W,1	--	15	12	17
	dw,db	--	12+4(n-1)	15+4(n-1)	17+3n
ESC	Escape				
	reg	2	9-20	varies	
	mem	8+EA	9-20	varies	
HLT	Halt	2	2	5	4
IDIV	Integer division				
	8-bit reg	101-112	17	19	19
	16-bit reg	165-184	25	27	27
	32-bit reg	--	--	43	43
	8-bit mem	(107-118) +EA	20	22	20
	16-bit mem	(171-190) +EA	28	30	28
	32-bit reg	--	--	46	44
IMUL	Integer multiplication				
	8-bit reg	80-98	13	9-14	13-18
	16-bit reg	128-154	21	9-22	13-26
	32-bit reg	--	--	9-38	13-42
	8-bit mem	(86-104)+EA	16	12-17	13-18
	16-bit mem	(134-160)+EA	24	12-25	13-26
	32-bit reg	--	--	12-41	13-42
	immed to 16-bit reg	--	21	9-34	13-18
	immed to 32-bit reg	--	21	9-38	13-18
	reg to reg (byte)	--	--	9-38	13-18
	reg to reg (word)	--	--	9-38	13-26
	reg to reg (dword)	--	--	9-38	13-42
	mem to reg (byte)	--	--	12-25	13-18
	mem to reg (word)	--	--	12-25	13-26
	mem to reg (dword)	--	--	12-41	13-42
	reg with imm to reg (byte)	--	--	9-14	13-18
	reg with imm to reg (word)	--	--	9-22	13-26
	reg with imm to reg (dword)	--	--	9-38	13-42
	mem with imm to reg (byte)	--	--	12-17	13-18
	mem with imm to reg (word)	--	--	12-25	13-26
	mem with imm to reg (dword)	--	--	12-41	13-42

Code	Description	8086	80286	80386	80486
IN	Input from I/O port				
	fixed port	10	5	12	14
	variable port through DX	8	5	13	14
INC	Increment by 1				
	16-bit reg	3	2	2	1
	8-bit reg	3	2	2	1
	mem	15+EA	7	6	3
INS/	Input from port to string				
INSB/	Input byte				
INSW/	Input word				
INSD	Input double	--	5	15	17
	PM	--	--	9,29	10-32
	REP INS/INSB/INSW	--	5+4/rep	13+6/rep	
	REP INS/INSB/INSW PM	--	--	(7,27)+6/rep	
INT	Interrupt				
	type=3	52	23+m	33	
	type=3 PM	--	(40,78)+m	59,99	
	type3	51	23+m	37	
	type3 PM	--	(40,78)+m	59,99	
INTO	Interrupt if overflow				
	interrupt taken	53	24+m	35	
	interrupt not taken	4	3	3	
	PM	--	(40,78)+m	59,99	
INVD	Invalidate data cache	--	--	--	4
INVLPG	Invalidate TLB entry	--	--	--	12/11
IRET	Return from interrupt	32	17+m	22	15
	PM	--	(31,55)+m	38,82	36
IRETD	Return from interrupt double	--	--	22	20
	PM	--	--	38,82	36
JA/	Jump if above/	16,noj 4	7+m,noj 3	7+m,noj 3	3,noj 1
JNBE	Jump if not below or equal				
JAE/	Jump if above or equal/	16,noj 4	7+m,noj 3	7+m,noj 3	3,noj 1
JNB	Jump if not below/				
JNA	Jump if not above				
JCXZ	Jump if CX is zero	18,noj 6	8+m,noj 4	9+m,noj 5	8,noj 5
JECXZ	Jump if ECX is zero	--	--	--	8,noj 5
JE/	Jump if equal/	16,noj 4	7+m,noj 3	7+m,noj 3	3,noj 1
JZ	Jump if zero				
JG/	Jump if greater	16,noj 4	7+m,noj 3	7+m,noj 3	3,noj 1
JNLE	Jump if not less, or equal				
JGE/	Jump if greater or equal/	16,noj 4	7+m,noj 3	7+m,noj 3	3,noj 1
JNL	Jump if not less				
JL/	Jump if less/	16,noj 4	7+m,noj 3	7+m,noj 3	3,noj 1
JNGE	Jump if not greater, or equal				
JLE/	Jump if less or equal/	16,noj 4	7+m,noj 3	7+m,noj 3	3,noj 1
JNG	Jump if not greater				

Code	Description	8086	80286	80386	80486
JMP	Jump				
	intrasegment direct short	15	7+m	7+m	3
	intrasegment direct	15	7+m	7+m	3
	intersegment direct	15	11+m	12+m	17
	PM	--	23+m	27+m	18
	intrasegment indirect				
	through memory	18+EA	11+m	10+m	5
	intrasegment indirect				
	through register	11	7+m	7+m	5
	intersegment indirect	24+EA	15+m	12+m	8
	PM	--	26+m	27+m	18
	direct intersegment				17
	486: to same level				19
	486: thru call gate to same level				32
	486: thru TSS				42+TS
	486: thr Task Gate				43+TS
	indirect intersegment				13
	486: to same level				18
	486: thru call gate to same level				31
	486: thru TSS				41+TS
	486: thr Task Gate				42+TS
JNE/	Jump if not equal/	16,noj 4	7+m,noj 3	7+m,noj 3	3,noj 1
JNZ	Jump if not zero				
JNO	Jump if not overflow	16,noj 4	7+m,noj 3	7+m,noj 3	3,noj 1
JNP/	Jump if not parity/	16,noj 4	7+m,noj 3	7+m,noj 3	3,noj 1
JPO	Jump if parity odd				
JNS	Jump if not sign	16,noj 4	7+m,noj 3	7+m,noj 3	3,noj 1
JO	Jump if overflow	16,noj 4	7+m,noj 3	7+m,noj 3	3,noj 1
JP/	Jump if parity/	16,noj 4	7+m,noj 3	7+m,noj 3	3,noj 1
JPE	Jump if parity even				
JS	Jump if sign	16,noj 4	7+m,noj 3	7+m,noj 3	3,noj 1
LAHF	Load AH from flags	4	2	2	3
LAR	Load access rights				
	reg to reg	--	14	15	11
	mem to reg	--	16	16	11
LDS/	Load pointer using DS/				
LES	Load pointer using ES	16+EA	7	7	6
	PM	--	21	22	12
LFS/	Load far pointer				
LGS/					
LSS		--	--	7	6/12
	PM	--	--	22-25	
LEA	Load effective address	2+EA	3	2	2,noj 1
LEAVE	High level procedure exit	--	5	4	5
LGDT	Load global descriptor table	--	11	11	11
LIDT	Load interrupt desc. table	--	12	11	11
LLDT	Load local desc. table				
	reg	--	17	20	11
	mem	--	19	24	11
LMSW	Load machine status word				
	reg	--	3	10	13
	mem	--	6	13	13

Code	Description	8086	80286	80386	80486
LOCK	Lock bus	2	0	0	1
LODS/	Load string/				
LODSB/	Load byte string/				
LODSW	Load word string				
	not repeated	12	5	5	5
	repeated	9+13/rep			7+4c
LOOP	Loop	17,noj 5	8+m,noj 4	11+m	7,noj 6
LOOPE/	Loop if equal/				
LOOPZ	Loop if zero	18,noj 6	8+m,noj 4	11+m	9,noj 6
LOOPNE/	Loop if not equal/				
LOOPNZ	Loop if not zero	19,noj 5	8+m,noj 4	11+m	9,noj 6
LSL	Load segment limit				
	reg to reg	--	14	20,25	10
	mem to reg	--	16	21,26	10
LTR	Load task register				
	reg	--	17	23	20
	mem	--	19	27	20
MOV	Move				
	acc to mem	10	3	2	1
	mem to acc	10	5	4	1
	reg to reg	2	2	2	1
	mem to reg	8+EA	5	4	1
	reg to mem	9+EA	3	2	1
	immed to reg	4	2	2	1
	immed to mem	10+EA	3	2	1
	reg to SS/DS/ES	2	2	2	3/9
	reg to SS/DS/ES PM	--	17	18	
	mem to SS/DS/ES	8+EA	5	5	3/9
	mem to SS/DS/ES PM	--	19	19	3
	segment reg to reg	2	2	2	3
	segment reg to mem	9+EA	3	2	3
	control reg to reg	--	--	6	4
	reg to control reg 0	--	--	10	16
	reg to control reg 2	--	--	4	4
	reg to control reg 3	--	--	5	4
	debug reg 0-3 to reg	--	--	22	10
	debug reg 6-7 to reg	--	--	14	10
	reg to debug reg 0-3	--	--	22	11
	reg to debug reg 6-7	--	--	16	11
	test reg to reg	--	--	12	3,4
	reg to test reg	--	--	12	6,4
MOVS/	Move string/				
MOVSB/	Move byte string/				
MOVSW	Move word string				
	not repeated	18	5	7	7
	REP MOVS/MOVSB/MOVSW	9+17/rep	5+4/rep	8+4/rep	12+3/rep
MOVSX	Move with sign-extend				
	reg to reg	--	--	3	3
	mem to reg	--	--	6	3
MOVZX	Move with zero-extend				
	reg to reg	--	--	3	3
	mem to reg	--	--	6	3

Table B-4: Clock Cycles for Various Intel Microprocessors by Instruction (continued)

Code	Description	8086	80286	80386	80486
MUL	Unsigned multiplication				
	8-bit reg	70-77	13	9-14	13/18
	16-bit reg	118-133	21	9-22	13/26
	double	--	--	9-38	13/42
	8-bit mem	(76-83)+EA	16	12-17	13/18
	16-bit mem	(124-139)+EA	24	12-25	13/26
	double	--	--	12-41	13/42
NEG	Negate				
	reg	3	2	2	1
	mem	16+EA	7	6	3
NOP	No operation	3	3	3	3
NOT	Logical NOT				
	reg	3	2	2	1
	mem	16+EA	7	6	3
OR	Logical OR				
	reg to reg	3	2	2	1
	mem to reg	9+EA	7	6	2
	reg to mem	16+EA	7	7	3
	immed to acc	4	3	2	1
	immed to reg	4	3	2	1
	immed to mem	17+EA	7	7	3
OUT	Output to I/O port				
	fixed port	10	3	10	16
	fixed port PM	--	--	4,24	11,31
	variable port	8	3	11	16
	variable port PM	--	--	5,25	10,30
OUTS/	Output string to port/				
OUTSB/	Output byte				
OUTSW/	Output word				
OUTSD	Output double	--	5	14	17
	PM	--	--	8,28	10,32
	REP OUTS/OUTSB/OUTSW	--	5+4/rep	12+5/rep	
	REP OUTS/OUTSB/OUTSW PM	--	--	(6,26)+5/rep	
POP	Pop word off stack				
	reg	8	5	4	4
	segment reg	8	5	7	3/9
	segment reg PM	--	20	21	9
	memory	17+EA	5	5	6
POPA/	Pop all			9	
POPAD	Pop all double	--	19	24	9
POPF	Pop flags off stack	8	5	5	9
POPFD	Pop flags off stack double	--	--	5	
PUSH	Push word onto stack				
	reg	11	3	2	4
	segment reg: ES/SS/CS	10	3	2	3
	segment reg: FS/GS	--	--	2	3
	memory	16+EA	5	5	4
	immed	--	3	2	1
PUSHA	Push All	--	17	18	11
PUSHF/	Push flags onto stack				
PUSHD	Push double flag onto stack	10	3	4	4

Table B-4: Clock Cycles for Various Intel Microprocessors by Instruction (continued)

Code	Description	8086	80286	80386	80486
RCL/	Rotate left through carry/				
RCR	Rotate right through carry/				
	reg with single-shift	2	2	9	3
	reg with variable-shift	8+4/bit	5+n	9	8/30
	mem with single-shift	15+EA	7	10	4
	mem with variable-shift	20+EA+4/bit	8+n	10	9/31
	immed to reg	--	5+n	9	8/30
	immed to mem	--	8+n	10	9/31
RET/	Return from procedure/				
RETF/	Return far/				
RETN	Return near				
	intrasegment	16	11+m	10+m	5
	intrasegment with constant	20	11+m	10+m	5
	intersegment	26	15+m	18+m	18
	intersegment PM	--	25+m,55	32+m,62	13
	intersegment with constant	25	15+m	18+m	33
	intersegment w/constant PM	--	25+m,55	32+m,68	17
	486: imm. to SP	--	--	--	14
	486: to same level	--	--	--	17
	486: to outer level	--	--	--	33
ROL/	Rotate left				
ROR	Rotate right				
	reg with single-shift	2	2	3	3
	reg with variable-shift	8+4/bit	5+n	3	3
	mem with single-shift	15+EA	7	7	4
	mem with variable-shift	20+EA+4/bit	8+n	7	4
	immed to reg	--	5+n	3	2
	immed to mem	--	8+n	7	4
SAHF	Store AH into flags	4	2	3	2
SAL/	Shift arithmetic left/				
SAR/	Shift arighmetic right/				
SHL/	Shift logical left/				
SHR	Shift logical right				
	reg with single-shift	2	2	3	3
	reg with variable-shift	8+4/bit	5+n	3	3
	mem with single-shift	15+EA	7	7	4
	mem with variable-shift	20+EA +4/bit	8+n	7	4
	immed to reg	--	5+n	3	2
	immed to mem	--	8+n	7	4
SBB	Subtract with borrow				
	reg from reg	3	2	2	1
	mem from reg	9+EA	7	7	2
	reg from mem	16+EA	7	6	3
	immed from acc	4	3	2	1
	immed from reg	4	3	2	1
	immed from mem	17+EA	7	7	3
SCAS/	Scan string/				
SCASB/	Scan byte string/				

Table B-4: Clock Cycles for Various Intel Microprocessors by Instruction (continued)

Code	Description	8086	80286	80386	80486
SCASW	Scan word string				
	not repeated	15	7	7	6
	REPE/REPNE SCAS/SCASB/SCASW	9+15/rep	5+8/rep	5+8/rep	7+5/rep
SET	Set conditionally				
	reg	--	--	4	4 or 3
	mem	--	--	5	3 or 4
SGDT	Store global descript. table	--	11	9	10
SIDT	Store interrupt desc. table	--	12	9	10
SLDT	Store local desc. table				
	reg	--	2	2	2
	mem	--	3	2	3
SHLD/	Shift left double precision/				
SHRD	Shift right double				
	reg to reg	--	--	3	2
	mem to mem	--	--	7	3
	reg by CL	--	--	3	3
	mem by CL	--	--	7	4
SMSW	Store machine status word				
	reg	--	2	10	2
	mem	--	3	3	3
	mem PM	--	--	2	
STC	Set carry flag	2	2	2	2
STD	Set direction flag	2	2	2	2
STI	Set interrupt flag	2	2	3	5
STOS/	Store string/				
STOSB/	Store byte string/				
STOSW	Store word string				
	not repeated	11	3	4	5
	REP STOS/STOSB/STOSW	9+10/rep	4+3/rep	5+5/rep	7+4/rep
STR	Store task register				
	reg	--	2	2	2
	mem	--	3	2	3
SUB	Subtraction				
	reg from reg	3	2	2	1
	mem from reg	9+EA	7	7	2
	reg from mem	16+EA	7	6	3
	immed from acc	4	3	2	1
	immed from reg	4	3	2	1
	immed from mem	17+EA	7	7	3
TEST	Test				
	reg with reg	3	2	2	1
	mem with reg	9+EA	6	5	2
	immed with acc	4	3	2	1
	immed with reg	5	3	2	1
	immed with mem	11+EA	6	5	2
VERR	Verify read				
	reg	--	14	10	11
	mem	--	16	11	11
VERW	Verify write				
	reg	--	14	15	11
	mem	--	16	16	11

Table B-4: Clock Cycles for Various Intel Microprocessors by Instruction (continued)

Code	Description	8086	80286	80386	80486
WAIT	Wait while TEST pin				
	not asserted	4	3	6	1-3
WBINVD	Write-back invalid data cache	--	--	--	5
XADD	Exchange and add				
	reg with reg	--	--	--	3
	reg with mem	--	--	--	4
XCHG	Exchange				
	reg with acc	3	3	3	3
	reg wtih mem	17+EA	5	5	5
	reg with reg	4	3	3	3
XLAT/	Translate	11	5	5	4
XLATB					
XOR	Logical exclusive OR				
	reg with reg	3	2	2	1
	mem with reg	9+EA	7	7	2
	reg wtih mem	16+EA	7	6	3
	immed with acc	4	3	2	1
	immed with reg	4	3	2	1
	immed with mem	17+EA	7	7	3

APPENDIX C: ASSEMBLER DIRECTIVES AND NAMING RULES

This appendix consists of two sections. The first section describes some of the most widely used directives in 80x86 Assembly language programming. In the second section Assembly language rules and restrictions for names and labels are discussed and a list of reserved words is provided.

SECTION C.1: 80x86 ASSEMBLER DIRECTIVES

Directives, or as they are sometimes called, pseudo-ops or pseudo-instructions, are used by the assembler to help it translate Assembly language programs into machine language. Unlike the microprocessor's instructions, directives do not generate any opcode; therefore, no memory locations are occupied by directives in the final ready-to-run (exe) version of the assembly program. To summarize, directives give directions to the assembler program to tell it how to generate the machine code; instructions are assembled into machine code to give directions to the CPU at execution time. The following are descriptions of the some of the most widely used directives for the 80x86 assembler. They are given in alphabetical order for ease of reference.

ASSUME

The ASSUME directive is used by the assembler to associate a given segment's name with a segment register. This is needed for instructions that must compute an address by combining an offset with a segment register. One ASSUME directive can be used to associate all the segment registers. For example:

        ASSUME   CS:name1,DS:name2,SS:name3,ES:name4

where name1, name2, and so on, are the names of the segments. The same result can be achieved by having one ASSUME for each register:

        ASSUME   CS:name1
        ASSUME   DS:name2
        ASSUME   SS:name3
        ASSUME   ES:nothing
        ASSUME   nothing

The key word "nothing" can be used to cancel a previous ASSUME directive.

DB   (Define Byte)

The DB directive is used to allocate memory in byte-sized increments. Look at the following examples:

```
DATA1      DB        23
DATA2      DB        45,97H,10000011B
DATA3      DB        'The planet Earth'
```

In DATA1 a single byte is defined with initial value 23. DATA2 consists of several values in decimal (45), hex (97H), and binary (10000011B). Finally, in DATA3, the DB directive is used to define ASCII characters. The DB directive is normally used to define ASCII data. In all the examples above, the address location for each value is assigned by the assembler. We can assigned a specific offset address by the use of the ORG directive.

DD   (Define Doubleword)

To allocate memory in 4-byte (32-bit) increments, the DD directive is used. Since word-sized operands are 16 bits wide (2 bytes) in 80x86 assemblers, a doubleword is 4 bytes.

```
VALUE1     DD        4563F57H
RESULT     DD        ?              ;RESERVE  4-BYTE LOCATION
DAT4       DD        25000000
```

It must be noted that the values defined using the DD directive are placed in memory by the assembler in low byte to low address and high byte to high address order. This convention is referred to as little endian. For example, assuming that offset address 0020 is assigned to VALUE1 in the example above, each byte will reside in memory as follows:

```
DS:20=(57)
DS:21=(3F)
DS:22=(56)
DS:23=(04)
```

DQ   (Define Quadword)

To allocate memory in 8-byte increments, the DQ directive is used. In the 80x86 a word is defined as 2 bytes; therefore, a quadword is 8 bytes.

```
DAT_64B    DQ        5677DD4EE4FF45AH
DAT8       DQ        10000000000000
```

DT   (Define Tenbytes)

To allocate packed BCD data, 10 bytes at a time, the DT directive is used. This is widely used for memory allocation associated with BCD numbers.

```
DATA       DT        399977653419974
```

Notice there is no H for the hexadecimal identifier following the number. This is a characteristic particular to the DT directive. In the case of other directives (DB, DW, DD, DQ), if there is no H at the end of the number, it is assumed to be in decimal and will be converted to hex by the assembler. Remember that the little endian convention is used to place the bytes in memory, with the least significant byte going to the low address and the most significant byte to the high address. DT can also be used to allocated decimal data if "d" is placed after the number:

```
DATA       DT        65535d ;stores hex FFFF in a 10-byte location
```

DUP (Duplicate)

The DUP directive can be used to duplicate a set of data a certain number of times instead of having to write it over and over.

```
DATA1    DB      20 DUP (99)          ;DUPLICATE 99 20 TIMES
DATA2    DW      6 DUP (5555H)        ;DUPLICATE 5555H 6 TIMES
DATA3    DB      10 DUP (?)           ;RESERVE 10 BYTES
DATA4    DB      5 DUP (5 DUP (0))    ;25 BYTES  INITIALIZED TO ZERO
DATA5    DB      10 DUP (00,FFH)      ;20 BYTES ALTERNATE 00, FF
```

DW (Define Word)

To allocate memory in 2-byte (16-bit) increments, the DW directive is used. In the 80x86 family, a word is defined as 16 bits.

```
DATAW_1  DW      5000
DATAW_2  DW      7F6BH
```

Again, in terms of placing the bytes in memory the little endian convention is used with the least significant byte going to the low address and the most significant byte going to the high address.

END

Every program must have an entry point. To identify that entry point the assembler relies on the END directive. The label for the entry and end point must match.

```
HERE:    MOV      AX,DATASEG   ;ENTRY POINT OF THE PROGRAM
         ...
         ...
         END      HERE          ;EXIT POINT OF THE PROGRAM
```

If there are several modules, only one of them can have the entry point, and the name of that entry point must be the same as the name put for the END directive as shown below:

```
;from the main program:
              EXTRN  PROG1:NEAR
              ...
MAIN_PRO:     MOV    AX,DATASG        ;THE ENTRY POINT
              MOV    DS,AX
              ...
              CALL   PROG1
              ...
              END    MAIN_PRO         ;THE EXIT POINT

;from the module PROG1:
              PUBLIC PROG1
PROG1         PROC
              ...
              RET                     ;RETURN TO THE MAIN MODULE
PROG1         ENDP
              END                     ;NO LABEL IS GIVEN
```

Notice the following points about the above code:

1. The entry point must be identified by a name. In the example above the entry point is identified by the name MAIN_PRO.
2. The exit point must be identified by the same name given to the entry point, MAIN_PRO.
3. Since a given program can have only one entry point and exit point, all modules called (either from main or from the submodules) must have directive END with nothing after it.

ENDP (see the PROC directive)

ENDS (see the SEGMENT and STRUCT directives)

EQU (Equate)

To assign a fixed value to a name, one uses the EQU directive. The assembler will replace each occurrence of the name with the value assigned to it.

```
FIX_VALU   EQU        1200
PORT_A     EQU        60H
COUNT      EQU        100
MASK_1     EQU        00001111B
```

Unlike data directives such as DB, DW, and so on, EQU does not assign any memory storage; therefore, it can be defined at any time at any place, and can even be used within the code segment.

EVEN

The EVEN directive forces memory allocation to start at an even address. This is useful due to the fact that in 8086, 286, and 386SX microprocessors, accessing a 2-byte operand located at an odd address takes extra time. The use of the EVEN directive directs the assembler to assign an even address to the variable.

```
           ORG        0020H
DATA_1     DB         34H
           EVEN
DATA_2     DW         7F5BH
```

The following shows the contents of memory locations:

```
DS:0020 = (34)
DS:0021 = (? )
DS:0022 = (5B)
DS:0023 = (7F)
```

Notice that the EVEN directive caused memory location DS:0021 to be bypassed, and the value for DATA_2 is placed in memory starting with an even address.

EXTRN   (External)

The EXTRN directive is used to indicate that certain variables and names used in a module are defined by another module. In the absence of the EXTRN directive, the assembler would search for the definition and give an error when it couldn't find it. The format of this directive is

    EXTRN name1:typea [,name2:typeb]

where type will be NEAR or FAR if name refers to a procedure, or will be BYTE, WORD, DWORD, QWORD, TBYTE if name refers to a data variable.

```
;from the main program:
        EXTRN  PROG1:NEAR
        PUBLIC DATA1
        ...
MAIN_PRO MOV      AX,DATASG              ;THE ENTRY POINT
        MOV       DS,AX
        ...
        CALL      PROG1
        ...
        END       MAIN_PRO               ;THE EXIT POINT

;PROG1 is located in a different file:
        EXTRN DATA1:WORD
        PUBLIC PROG1
PROG1   PROC
        ...
        MOV       BX,DATA1
        ...
        RET                              ;RETURN TO THE MAIN MODULE
PROG1   ENDP
        END
```

Notice that the EXTRN directive is used in the main procedure to identify PROG1 as a NEAR procedure. This is needed because PROG1 is not defined in that module. Correspondingly, PROG1 is defined as PUBLIC in the module where it is defined. EXTRN is used in the PROG1 module to declare that operand DATA1, of size WORD, has been defined in another module. Correspondingly, DATA1 is declared as PUBLIC in the calling module.

GROUP

The GROUP directive causes the named segments to be linked into the same 64K byte segment. All segments listed in the GROUP directive must fit into 64K bytes. This can be used to combine segements of the same type, or different classes of segments. An example follows:

```
SMALL_SYS   GROUP                DTSEG,STSEG,CDSEG
```

The ASSUME directive must be changed to make the segment registers point to the group:

```
ASSUME   CS:SMALL_SYS,DS:SMALL_SYS,SS:SMALL_SYS
```

The group will be listed in the list file, as shown below:

Segments and Groups:

Name	Length	Align	Combine Class
SMALL_SYS	GROUP		
STSEG	0040	PARA	NONE
DTSEG	0024	PARA	NONE
CDSEG	005A	PARA	NONE

INCLUDE

When there is a group of macros written and saved in a separate file, the INCLUDE directive can be used to bring them into another file. In the program listing (.lst file), these macros will be identified by the symbol "C" (or "+" in some versions of MASM) before each instruction to indicate that they are copied to the present file by the INCLUDE directive.

LABEL

The LABEL directive allows a given variable or name to be referred to by multiple names. This is often used for multiple definition of the same variable or name. The format of the LABEL directive is

```
name    LABEL  type
```

where type may be BYTE, WORD, DWORD, QWORD. For example, a variable name DATA1 is defined as a word and also needs to be accessed as 2 bytes, as shown in the following:

```
DATA_B    LABEL    BYTE
DATA1     DW       25F6H

          MOV      AX,DATA1        ;AX=25F6H
          MOV      BL,DATA_B       ;BL=F6H
          MOV      BH,DATA_B +1    ;BH=25H
```

The following shows the LABEL directive being used to allow accessing a 32-bit data item in 16-bit portions.

```
DATA_16   LABEL    WORD
DATDD_4   DD       4387983FH
          ...
          MOV      AX,DATA_16      ;AX=983FH
          MOV      DX,DATA_16 + 2  ;DX=4387H
```

The following shows its use in a JMP instruction to go to a different code segment.

```
          ....
          JMP      PROG_A
          ....
PROG_A    LABEL    FAR
INITI:    MOV      AL,12H
          OUT      PORT,AL
```

In the program above the address assigned to the names "PROG_A" and "INITI" are exactly the same. The same function can be achieved by the following:

```
JMP  FAR PTR INITI
```

LENGTH

The LENGTH operator returns the number of items defined by a DUP operand. See the SIZE directive for an example.

OFFSET

To access the offset address assigned to a variable or a name, one uses the OFFSET directive. For example, the OFFSET directive was used in the following example to get the offset address assigned by the assembler to the variable DATA1:

```
        ORG     5600H
DATA1   DW      2345H
        ...
        MOV     SI,OFFSET DATA1     ;SI=OFFSET OF DATA1 = 5600H
```

Notice that this has the same result as "LEA SI,DATA1".

ORG (Origin)

The ORG directive is used to assign an offset address for a variable or name. For example, to force variable DATA1 to be located starting from offset address 0020, one would write

```
        ORG     0020H
DATA1   DW 41F2H
```

This ensures the offset addresses of 0020 and 0021 with contents 0020H = (F2) and 0021H = (41).

PAGE

The PAGE directive is used to make the ".lst" file print in a specific format. The format of the PAGE directive is

PAGE [lines],[columns]

The default listing (meaning that no PAGE directive is coded) will have 66 lines per page with a maximum of 80 characters per line. This can be changed to 60 and 132 with the directive "PAGE 60,132". The range for number of lines is 10 to 255 and for columns is 60 to 132. A PAGE directive with no numbers will generate a page break.

PROC and ENDP   (Procedure and End Procedure)

Often, a group of Assembly language instructions will be combined into a procedure so that it can be called by another module. The PROC and ENDP directives are used to indicate the beginning and end of the procedure. For a given procedure the name assigned to PROC and ENDP must be exactly the same.

```
name1   PROC    [attribute]
        ...
name1   ENDP
```

There are two choices for the attribute of the PROC: NEAR or FAR. If no attribute is given, the default is NEAR. When a NEAR procedure is called, only IP is saved since CS of the called procedure is the same as the calling program. If a FAR procedure is called, both IP and CS are saved since the code segment of the called procedure is different from the calling program.

PTR (Pointer)

The PTR directive is used to specify the size of the operand. Among the options for size are BYTE, WORD, DWORD, and QWORD. This directive is used in many different ways, the most common of which are explained below.

1. PTR can be used to allow an override of a previously defined data directive.

```
DATA1    DB       23H,7FH,99H,0B2H
DATA2    DW       67F1H
DATA3    DD       22229999H
         ...
         MOV      AX, WORD PTR DATA1          ;AX=7F23
         MOV      BX, WORD PTR DATA1 + 2      ;BX,B299H
```

Although DATA1 was initially defined as DB, it can be accessed using the WORD PTR directive.

```
         MOV      AL, BYTE PTR DATA2   ;AL=F1H
```

In the above code, notice that DATA2 was defined as WORD but it was accessed as BYTE with the help of BYTE PTR. If this had been coded as "MOV AL,DATA2", it would generate an error since the sizes of the operands do not match.

```
         MOV      AX, WORD PTR DATA3          ;AX=9999H
         MOV      DX, WORD PTR DATA3 + 2      ;DX=2222H
```

DATA3 was defined as a 4-byte operand but registers are only 2 bytes wide. The WORD PTR directive solved that problem.

2. The PTR directive can be used to specify the size of a directive in order to help the assembler translate the instruction.

```
         INC      [DI]                    ;will cause an error
```

This instruction was meant to increment the contents of the memory location(s) pointed at by [DI]. How does the assembler know whether it is a byte operand, word operand, or doubleword operand? Since it does not know, it will generate an error. To correct that, use the PTR directive to specify the size of the operand as shown next.

```
         INC      BYTE PTR [SI]        ;increment a byte pointed by SI
or
         INC      WORD PTR [SI]            ;increment a word pointed by SI
or
         INC      DWORD PTR [SI]       ;increment a doubleword pointed by SI
```

3. The PTR directive can be used to specify the distance of a jump. The options for the distance are FAR and NEAR.

```
         JMP      FAR PTR INTI   ;ensures that it will be a 5-byte instruction
         ...
```

INITI: MOV   AX,1200

See the LABEL directive to find out how it can be used to achieve the same result.

PUBLIC

To inform the assembler that a name or symbol will be referenced by other modules, it is marked by the PUBLIC directive. If a module is referencing a variable outside itself, that variable must be declared as EXTRN. Correspondingly, in the module where the variable is defined, that variable must be declared as PUBLIC in order to allow it to be referenced by other modules. See the EXTRN directive for examples of the use of both EXTRN and PUBLIC.

SEG (Segment Address)

The SEG operator is used to access the address of the segment where the name has been defined.

```
DATA1    DW       2341H
         ...
         MOV      AX,SEG DATA1 ;AX=SEGMENT ADDRESS OF DATA1
```

This is in contrast to the OFFSET directive, which accesses the offset address instead of the segment.

SEGMENT and ENDS

In full segment definition these two directives are used to indicate the beginning and the end of the segment. They must have the same name for a given segment definition. See the following example:

```
DATSEG   SEGMENT
DATA1    DB       2FH
DATA2    DW       1200
DATA3    DD       99999999H
DATSEG   ENDS
```

There are several options associated with the SEGMENT directive, as follows:

name1 SEGMENT [align] [combine] [class]

name1 ENDS

ALIGNMENT: When several assembled modules are linked together, this indicates where the segment is to begin. There are many options, including PARA (paragraph = 16 bytes), WORD, and BYTE. If PARA is chosen, the segment starts at a hex address divisible by 10H. PARA is the default alignment. In this alignment, if a segment for a module finished at 00024H, the next segment will start at address 00030H, leaving from 00025 to 0002F unused. If WORD is chosen, the segment is forced to start at a word boundary. In BYTE alignment, the segment starts at the next byte and no memory is wasted. There is also the PAGE option, which aligns segments along the 100H (256) byte boundary. While all these options are supported by many assemblers, such as MASM and TASM, there is another option supported only by assemblers that allow system development. This option is AT. The AT

---

option allows the program to assign a physical address. For example, to burn a program into ROM starting at physical address F0000, code

```
ROM_CODE    SEGMENT    AT F000H
```

Due to the fact that option AT allows the programmer to specify a physical address that conflicts with DOS's memory management responsibility, many assemblers such as MASM will not allow option AT.

COMBINE TYPE: This option is used to merge together all the similar segments to create one large segment. Among the options widely used are PUBLIC and STACK. PUBLIC is widely used in code segment definitions when linking more than one module. This will consolidate all the code segments of the various modules into one large code segment. If there is only one data segment and that belongs to the main module, there is no need to define it as PUBLIC since no other module has any data segment to combine with. However, if other modules have their own data segments, it is recommended that they be made PUBLIC to create a single data segment when they are linked. In the absence of that, the linker would assume that each segment is private and they would not be combined with other similar segments (codes with codes and data with data). Since there is only one stack segment, which belongs to the main module, there is no need to define it as PUBLIC. The STACK option is used only with the stack segment definition and indicates to the linker that it should combine the user's defined stack with the system stack to create a single stack for the entire program. This is the stack that is used at run time (when the CPU is actually executing the program).

CLASS NAME: Indicates to the linker that all segments of the same class should be placed next to each other by the LINKER. Four class names commonly used are 'CODE', 'DATA', 'STACK', and 'EXTRA'. When this attribute is used in the segment definition, it must be enclosed in single apostrophes in order to be recognized by the linker.

SHORT

In a direct jump such as "JMP POINT_A", the assembler has to choose either the 2-byte or 3-byte format. In the 2-byte format, one byte is the opcode and the second byte is the signed number displacement value added to the IP of the instruction immediately following the JMP. This displacement can be anywhere between −128 and +127. A negative number indicates a backward JMP and a positive number a forward JMP. In the 3-byte format the first byte is the opcode and the next two bytes are for the signed number displacement value, which can range from −32,768 to 32,767. When assembling a program, the assembler makes two passes through the program. Certain tasks are done in the first pass and others are left to the second pass to complete. In the first pass the assembler chooses the 3-byte code for the JMP. After the first pass is complete, it will know the target address and fill it in during the second pass. If the target address indicates a short jump (less than 128) bytes away, it fills the last byte with NOP. To inform the assembler that the target address is no more than 128 bytes away, the SHORT directive can be used. Using the SHORT directive makes sure that the JMP is a 2-byte instruction and not 3-byte with 1 byte as NOP code. The 2-byte JMP requires 1 byte less memory and is executed faster.

SIZE

The size operator returns the total number of bytes occupied by a name. The three directives LENGTH, SIZE, and TYPE are somewhat related. Below is a description of each one using the following set of data defined in a data segment:

```
DATA1  DQ    ?
DATA2  DW    ?
DATA3  DB    20 DUP (?)
DATA4  DW    100 DUP (?)
DATA5  DD    10 DUP (?)
```

TYPE allows one to know the storage allocation directive for a given variable by providing the number of bytes according to the following table:

```
bytes
1         DB
2         DW
4         DD
8         DQ
10        DT
```

For example:
```
        MOV      BX,TYPE DATA2     ;BX=2
        MOV      DX,TYPE DATA1     ;DX=8
        MOV      AX,TYPE DATA3     ;AX=1
        MOV      CX,TYPE DATA5     ;CX=4
```

When a DUP is used to define the number of entries for a given variable, the LENGTH directive can be used to get that number.

```
        MOV      CX,LENGTH DATA4   ;CX=64H        (100 DECIMAL)
        MOV      AX,LENGTH DATA3   ;AX=14H        (20 DECIMAL)
        MOV      DX,LENGTH DATA5   ;DX=0A (10 DECIMAL)
```

If the defined variable does not have any DUP in it, the LENGTH is assumed to be 1.

```
        MOV      BX,LENGTH DATA1   ;BX=1
```

SIZE is used to determine the total number of bytes allocated for a variable that has been defined with the DUP directive. In reality the SIZE directive basically provides the product of the TYPE times LENGTH.

```
        MOV      DX, SIZE DATA4    ;DX=C8H=200 (100 x 2=200)
        MOV      CX, SIZE DATA5    ;CX=28H=40 (4 x 10=40)
```

STRUC   (Structure)

The STRUC directive indicates the beginning of a structure definition. It ends with an ENDS directive, whose label matches the STRUC label. Although the same mnemonic ENDS is used for end of segment and end of structure, the assembler knows which is meant by the context. A structure is a collection of data types that can be accessed either collectively by the structure name or individually by the labels of the data types within the structure. A structure type must first be defined and then variables in the data segment may be allocated as that structure type. Looking at the following example, the data directives between STRUC and ENDS declare what structure ASC_AREA looks like. No memory is allocated for such a structure definition. Immediately below the structure definition is the label ASC_INPUT, which is declared to be of type ASC_AREA. Memory is allocated for the variable ASC_INPUT. Notice in the code segment that ASC_INPUT can be accessed either in its entirety or by its component parts. It is accessed as a whole unit in "MOV DX,OFFSET ASC_INPUT". Its component parts are accessed by the variable name followed by a period, then the component's name. For example, "MOV BL,ASC_INPUT.ACT_LEN" accesses the actual length field of ASC_INPUT.

```
;from the data segment:
ASC_AREA      STRUC                    ;defines struc for string input
MAX_LEN       DB    6                  ; maximum length of input string
ACT_LEN       DB    ?                  ; actual length of input string
ASC_NUM       DB    6 DUP (?)          ; input string
ASC_AREA      ENDS                     ;end struc definition
ASC_INPUT     ASC_AREA    <>           ;allocates memory for struc

;from the code segment:
              ...
GET_ASC:      MOV   AH,0AH
              MOV   DX,OFFSET ASC_INPUT
              INT   21H
              ...
              MOV   SI,OFFSET ASC_INPUT.ASC_NUM  ;SI points to ASCII num
              MOV   BL,ASC_INPUT.ACT_LEN         ;BL holds string length
              ...
```

TITLE

The TITLE directive instructs the assembler to print the title of the program on top of each page of the ".lst" file. What comes after the TITLE pseudo-instruction is up to the programmer, but it is common practice to put the name of the program as stored on the disk right after the TITLE pseudo-instruction and then a brief description of the function of the program. Whatever is placed after the TITLE pseudo-instruction cannot be more than 60 ASCII characters (letters, numbers, spaces, punctuation).

TYPE

The TYPE operator returns the number of bytes reserved for the named data object. See the SIZE directive for examples of its use.

SECTION C.2: RULES FOR LABELS AND RESERVED NAMES

Labels in 80x86 Assembly language for MASM 5.1 and higher must follow these rules:

1. Names can be composed of:
   alphabetic characters: A - Z and a - z
   digits: 0 - 9
   special characters: "?" "." "@" "_" "$"

2. Names must begin with an alphabetic or special character. Names cannot begin with a digit.

3. Names can be up to 31 characters long.

4. The special character "." can only be used as the first character.

5. Uppercase and lowercase are treated the same. "NAME1" is treated the same as "Name1" and "name1".

Assembly language programs have five types of labels or names:

1. Code labels, which give symbolic names to instructions so that other instructions (such as jumps) may refer to them

2. Procedure labels, which assign a name to a procedure

3. Segment labels, which assign a name to a segment

4. Data labels, which give names to data items

5. Labels created with the LABEL directive

Code labels

These labels will be followed by a colon and have the type NEAR. This enables other instructions within the code segment to refer to the instruction. The labels can be on the same line as the instruction:

```
          ...
ADD_LP:   ADD  AL,[BX]              ;label is on same line as the instruction
          ...
          ...
          LOOP ADD_LP
```

or on a line by themselves:

```
          ...
ADD_LP:                             ;label is on a line by itself
          ADD  AL,[BX]              ;ADD_LP refers to this instruction
          ...
          ...
          LOOP ADD_LP
```

Procedure labels

These labels assign a symbolic name to a procedure. The label can be NEAR or FAR. When using full segment definition, the default type is NEAR. When using simplified segment definition, the type will be NEAR for compact or small models but will be FAR for medium, large, and huge models. For more information on procedures, see PROC in Section C.1.

Segment labels

These labels give symbolic names to segments. The name must be the same in the SEGMENT and ENDS directives. See SEGMENT in Section C.1 for more information. Example:

```
DAT_SG   SEGMENT
SUM      DW        ?
DAT_SG   ENDS
```

Data labels

These labels give symbolic names to data items. This allows them to be accessed by instructions. Directives DB, DW, DD, DQ, and DT are used to allocate data. Examples:

```
DATA1    DB        43H
DATA2    DB        F2H
SUM      DW        ?
```

Labels defined with the LABEL directive

The LABEL directive can be used to redefine a label. See LABEL in Section C.1 for more information.

Reserved Names

The following is a list of reserved words in 80x86 Assembly language programming. These words cannot be used as user-defined labels or variable names.

Register Names:

AH	AL	AX	BH	BL	BP	BX	CH	CL	CS	CX	DH
DI	DL	DS	DX	ES	SI	SP	SS				

Instructions:

AAA	AAD	AAM	AAS	ADC	ADD
AND	CALL	CBW	CLC	CLD	CLI
CMC	CMP	CMPS	CWD	DAA	DAS
DEC	DIV	ESC	HLT	IDIV	IMUL
IN	INC	INT	INTO	IRET	JA
JAE	JB	JBE	JCXZ	JE	JG
JGE	JL	JLE	JMP	JNA	JNAE
JNB	JNBE	JNE	JNG	JNGE	JNL
JNLE	JNO	JNP	JNS	JNZ	JO
JP	JPE	JPO	JS	JZ	LAHF
LDS	LEA	LES	LOCK	LODS	LOOP
LOOPE	LOOPNE	LOOPNZ	LOOPZ	MOV	MOVS
MUL	NEG	NIL	NOP	NOT	OR
OUT	POP	POPF	PUSH	PUSHF	RCL
RCR	REP	REPE	REPNE	REPNZ	REPZ
RET	ROL	ROR	SAHF	SAL	SAR
SBB	SCAS	SHL	SHR	STC	STD
STI	STOS	SUB	TEST	WAIT	XCHG
XLAT	XOR				

Assembler operators and directives

```
$    *    +    -    .    /    =    ?    [    ]
ALIGN          ASSUME     BYTE       COMM        COMMENT     DB
DD             DF         DOSSEG     DQ          DS          DT
DW             DWORD      DUP        ELSE        END         ENDIF
ENDM           ENDS       EQ         EQU         EVEN        EXITM
EXTRN          FAR        FWORD      GE          GROUP       GT
HIGH           IF         IFB        IFDEF       IFDIF       IFE
IFIDN          IFNB       IFNDEF     IF1         IF2         INCLUDE
INCLUDELIB     IRP        IRPC       LABEL       LE          LENGTH
LINE           LOCAL      LOW        LT          MACRO       MASK
MOD            NAME       NE         NEAR        NOTHING     OFFSET
ORG            PAGE       PROC       PTR         PUBLIC      PURGE
QWORD          RECORD     REPT       REPTRD      SEG         SEGMENT
SHORT          SIZE       STACK      STRUC       SUBTTL      TBYTE
THIS           TITLE      TYPE       WIDTH       WORD
.186           .286       .286P      .287        .386        .386P
.387           .8086      .8087      .ALPHA      .CODE       .CONST
.CREF          .DATA      .DATA?     .ERR        .ERR1       .ERR2
.ERRB          .ERRDEF    .ERRDIF    .ERRE       .ERRIDN     .ERRNB
.ERRNDEF       .ERRNZ     .FARDATA   .FARDATA?   .LALL       .LFCOND
.LIST          .MODEL     %OUT       .RADIX      .SALL       .SEQ
.SFCOND        .STACK     .TFCOND    .TYPE       .XALL       .XCREF
.XLIST
```

APPENDIX D: DOS INTERRUPT 21H AND 33H LISTING

This appendix lists many of the DOS 21H interrupts, which are used primarily for input, output, and file and memory management. In addition, this appendix covers some functions of INT 33H, the mouse handling interrupt. As was mentioned in Chapter 5, this interrupt is not a part of DOS or BIOS, but is part of the mouse driver software.

SECTION D.1: DOS 21H INTERRUPTS

First, before covering the DOS 21H interrupts, a few notes are given about file management under DOS. There are two commonly used ways to access files in DOS. One is through what is called a file handle, the other is through an FCB, or file control block. These terms are defined in detail below. Function calls 0FH through 28H use FCBs to access files. Function calls 39H through 62H use file handles. Handle calls are more powerful and easier to use. However, FCB calls maintain compatibility down to DOS version 1.10. FCB calls have the further limitation that they reference only the files in the current directory, whereas handle calls reference any file in any directory. FCB calls use the file control block to perform any function on a file. Handle calls use an ASCIIZ string (defined below) to open, create, delete, or rename a file and use a file handle for I/O requests. There are some terms used in the interrupt listing that will be unfamiliar to many readers. DOS manuals provide complete coverage of the details of file managment, but a few key terms are defined below.

ASCIIZ string

This is a string composes of any combination of ASCII characters and terminated with one byte of binary zeros (00H). It is frequently used in DOS 21H interrupt calls to specify a filename or path. The following is an example of an ASCIIZ string that was defined in the data segment of a program:

NAME_1   DB 'C:\PROGRAMS\SYSTEM_A\PROGRAM5.ASM',0

Directory

DOS keeps track of where files are located by means of a directory. Each disk can be partitioned into one or more directories. The directory listing lists each file in that directory, the number of bytes in the file, the date and time the file was created, and other information that DOS needs to access that file. The familiar DOS command "DIR" lists the directory of the current drive to the monitor.

DTA Disk transfer area

This is essentially a buffer area that DOS will use to hold data for reads or writes performed with FCB function calls. This area can be set up by your program anywhere in the data segment. Function call 1AH tells DOS the location of the DTA. Only one DTA can be active at a time.

FAT File allocation table

Each disk has a file allocation table that gives information about the clusters on a disk. Each disk is divided into sectors, which are grouped into clusters. The size of sectors and clusters varies among the different disk types. For each cluster in the disk, the FAT has a code indicating whether the cluster is being used by a file, is available, is reserved, or has been marked as a bad cluster. DOS uses this information in storing and retrieving files.

FCB File control block

One FCB is associated with each open file. It is composed of 37 bytes of data that give information about a file, such as drive, filename and extension, size of the file in bytes, and date and time it was created. It also stores the current block and record numbers, which serve as pointers into a file when it is being read or written to. DOS INT 21H function calls 0FH through 28H use FCBs to access files. Function 0FH is used to open a file, 16H to create a new file. Function calls 14H - 28H perform read/write functions on the file, and 16H is used to close the file. Typically, the filename information is set up with function call 29H (Parse Filename), and then the address of the FCB is placed in DS:DX and is used to access the file.

File handle

DOS function calls 3CH through 62H use file handles. When a file or device is created or opened with one of these calls, its file handle is returned. The file handle is used thereafter to refer to that file for input, output, closing the file, and so on. DOS has a few predefined file handles that can be used by any Assembly language program. These do not need to be opened before they are used:

Handle value	Refers to
0000	standard input device (typically, the keyboard)
0001	standard output device (typically, the monitor)
0002	standard error output device (typically, the monitor)
0003	standard auxiliary device (AUX1)
0004	standard printer device (PTR1)

PSP Program segment prefix

The PSP is a 256-byte area of memory reserved by DOS for each program. It provides an area to store shared information between the program and DOS.

<u>AH</u>   <u>Function of INT 21H</u>

00   **Terminate the program**

<u>Additional Call Registers</u>   <u>Result Registers</u>
CS = segment address of   None
PSP (program segment prefix)

Note: Files should be closed previously or data may be lost.

01   **Keyboard input with echo**

<u>Additional Call Registers</u>   <u>Result Registers</u>
None   AL = input character

Note: Checks for ctrl-break.

02   **Output character to monitor**

<u>Additional Call Registers</u>   <u>Result Registers</u>
DL = character to be displayed   None

03   **Asynchronous input from auxiliary device (serial device)**

<u>Additional Call Registers</u>   <u>Result Registers</u>
None   AL = input character

04   **Asynchronous character output**

<u>Additional Call Registers</u>   <u>Result Registers</u>
DL = character to be output   None

05   **Output character to printer**

<u>Additional Call Registers</u>   <u>Result Registers</u>
DL = character to be printed   None

06   **Console I/O**

<u>Additional Call Registers</u>   <u>Result Registers</u>
DL = 0FFH if input   AL = 0H if no character available
or character to be   = character that was input, if
displayed, if output   input successful

Note: If input, ZF is cleared and AL will have the character. ZF is set if input and no character was available.

07   **Keyboard input without echo**

Additional Call Registers	Result Registers
None	AL = input character

Note: Does not check for ctrl-break.

08   **Keyboard input without echo**

Additional Call Registers	Result Registers
None	AL = input character

Note: Checks for ctrl-break.

09   **String output**

Additional Call Registers	Result Registers
DS:DX = string address	None

Note: Displays characters beginning at address until a '$' (ASCII 36) is encountered.

0A   **String input**

Additional Call Registers	Result Registers
DS:DX = address at which to store string	None

Note: Specify the maximum size of the string in byte 1 of the buffer. DOS will place the actual size of the string in byte 2. The string begins in byte 3.

0B   **Get keyboard status**

Additional Call Registers	Result Registers
None	AL = 00 if no character waiting = 0FFH if character waiting

Note: Checks for ctrl-break.

0C   **Reset input buffer and call keyboard input function**

Additional Call Registers	Result Registers
AL = keyboard function number 01H, 06H, 07H, 08H or 0AH	None

Note: This function waits until a character is typed in.

AH   Function of INT 21H

0D   Reset disk

Additional Call Registers
None

Result Registers
None

Note: Flushes DOS file buffers but does not close files.

0E   Set default drive

Additional Call Registers
DL = code for drive
(0=A, 1=B, 2=C, etc.)

Result Registers
AL = number of logical drives
in system

0F   Open file

Additional Call Registers
DS:DX = address of FCB

Result Registers
AL = 00 if successful
   = 0FFH if file not found

Note: Searches current directory for file. If found, FCB is filled.

10   Close file

Additional Call Registers
DS:DX = address of FCB

Result Registers
AL = 00 if successful
   = 0FFH if file not found

Note: Flushes all buffers. Also updates directory if file has been modified.

11   Search for first matching filename

Additional Call Registers
DS:DX = address of FCB

Result Registers
AL = 00 if match is found
   = 0FFH if no match found

Note: Filenames can contain wildcards '?' and '*'.

12   Search for next match

Additional Call Registers
DS:DX = address of FCB

Result Registers
AL = 00 if match found
   = 0FFH if no match found

Note: This call should be used only if previous call to 11H or 12H has been
successful.

13   Delete file(s)

Additional Call Registers	Result Registers
DS:DX = address of FCB	AL = 00 if file(s) deleted = 0FFH if no files deleted

Note: Deletes all files in current directory matching filename, provided that they are not read-only. Files should be closed before deleting.

14   Sequential read

Additional Call Registers	Result Registers

DS:DX = address of opened FCB	AL = 00H if read successful
	    = 01H if end of file and no
	        data is read
	    = 02H if DTA is too small to
	        hold the record
	    = 03H if partial record read and
	        end of file is reached

Note: The file pointer, block pointer, and FCB record pointer are updated automatically by DOS.

15   Sequential write

Additional Call Registers	Result Registers

DS:DX = address of opened FCB	AL = 00H if write successful
	    = 01H if disk is full
	    = 02H if DTA is too small to
	        hold the record

Note: The file pointer, block pointer, and FCB record pointer are updated automatically by DOS. The record may not be written physically until a cluster is full or the file is closed.

16   Create/open a file

Additional Call Registers	Result Registers
DS:DX = addr. of unopened FCB = 0FFH if unsuccessful	AL = 00H if successful

Note: If the file already exists, it will be truncated to length 0.

17   Rename file(s)

Additional Call Registers	Result Registers
DS:DX = address of FCB	AL = 00H if file(s) renamed = 0FFH if file not found or new name already exists

Note: The old name is in the name position of the FCB; the new name is at the size (offset 16H) position.

AH   Function of INT 21H

18   Reserved

19   Get default drive

Additional Call Registers	Result Registers
None	AL = 0H for drive A
	= 1H for drive B
	= 2H for drive C

1A   Specify DTA (disk transfer address)

Additional Call Registers	Result Registers
DS:DX = DTA	None

Note: Only one DTA can be current at a time. This function must be called before FCB reads, writes, and directory searches.

1B   Get FAT (file allocation table) for default drive

Additional Call Registers	Result Registers
None	AL = number of sectors per cluster
	CX = number of bytes per sector
	DX = number of cluster per disk
	DS:BX FAT id

1C   Get FAT (file allocation table) for any drive

Additional Call Registers	Result Registers
DL = drive code	AL = number of sectors per cluster
0 for A	CX = number of bytes per sector
1 for B	DX = number of cluster per disk
2 for C	DS:BX FAT id

1D   Reserved

1E   Reserved

1F   Reserved

20   Reserved

21   Random read

Additional Call Registers	Result Registers
DS:DX = address of opened FCB	AL = 00H if read successful
	= 01H if end of file and no data read
	= 02H if DTA too small for record
	= 03H if end of file and partial read

Note: Reads record pointed at by current block and record fields into DTA.

22   Random write

Additional Call Registers
DS:DX = address of opened FCB
= 01H if disk is full
= 02H if DTA too small for record

Result Registers
AL = 00H if write successful

Note: Writes from DTA to record pointed at by current block and record fields.

23   Get file size

Additional Call Registers
DS:DX = addr. of unopened FCB
  of records is set in FCB random-
  record field (offset 0021H)
= 0FFH if no match found

Result Registers
AL = 00H if file found, number

Note: The FCB should contain the record size before the interrupt.

24   Set random record field

Additional Call Registers
DS:DX = address of opened FCB

Result Registers
None

Note: This sets the random-record field (offset 0021H) in the FCB. It is used prior to switching from sequential to random processing.

25   Set interrupt vector

Additional Call Registers
DS:DX = interrupt handler addr.
AL = machine interrupt number

Result Registers
None

Note: This is used to change the way the system handles interrupts.

26   Create a new PSP (program segment prefix)

Additional Call Registers
DX = segment addr. of new PSP

Result Registers
None

Note: DOS versions 2.0 and higher recommend not using this service, but using service 4B (exec).

27   Random block read

Additional Call Registers
DS:DX = address of opened FCB
CX = number records to be read
= 02H if DTA too small for block
= 03H if EOF and partial block read
CX = number of records actually read

Result Registers
AL = 00H if read successful
= 01H if end of file and no data read

Note: Set the FCB random record and record size fields prior to the interrupt. DOS will update the random record, current block, and current record fields after the read.

28   Random block write

Additional Call Registers
DS:DX = address of opened FCB
CX = number records to write
= 02H if DTA too small for block
CX = number of records actually written

Result Registers
AL = 00H if write successful
= 01H if disk is full

Note: Set the FCB random record and record size fields prior to the interrupt. DOS will update the random record, current block and current record fields after the write. If CX = 0 prior to the interrupt, nothing is written to the file and the file is truncated or extended to the length computed by the random record and record size fields.

29   Parse filename

Additional Call Registers
DS:SI = address of command line
ES:DI = address of FCB
AL = parsing flags in bits 0-3
    Bit 0 = 1 if leading separators are to be ignored; otherwise no scan-off takes place
    Bit 1 = 1 if drive ID in FCB will be changed only if drive was specified in command line
    Bit 2 = 1 if filename will be changed only if filename was specified in command line
    Bit 3 = 1 if extension will be changed only if extension was specified in command line

Result Registers
DS:SI = address of first char after
ES:DI = address of first byte of formatted unopened FCB
AL = 00H if no wildcards were in filename or extension
= 01H if wildcard found
= 0FFH if drive specifier is invalid

Note: The command line is parsed for a filename, then an unopened FCB is created at DS:SI. The command should not be used if path names are specified.

AH   Function of INT 21H

2A   Get system date

Additional Call Registers
None

Result Registers
CX = year (1980-2099)
DH = month (1-12)
DL = day (1-31)
AL = day of week code
             (0 = Sunday, ... , 6 = Saturday)

2B   Set system date

Additional Call Registers
CX = year (1980-2099)
DH = month (1-12)
DL = day (1-31)

Result Registers
AL = 00H if date set
   = 0FFH if date not valid

2C   Get system time

Additional Call Registers
None

Result Registers
CH = hour (0 .. 23)
CL = minute (0 .. 59)
DH = second (0 .. 59)
DL = hundredth of second
        (0 .. 99)

Note: The format returned can be used in calculations but can be converted to a printable format.

2D   Set system time

Additional Call Registers
CH = hour (0 .. 23)
CL = minute
DH = second
DL = hundredth of second

Result Registers
AL = 00H if time set
   = 0FFH if time invalid

2E   Set/reset verify switch

Additional Call Registers
AL = 0 to turn verify off
   = 1 to turn verify on

Result Registers
None

Note: If verify is on, DOS will perform a verify every time data is written to disk. An interrupt call to 54H gets the setting of the verify switch.

2F   Get DTA (disk transfer area)

Additional Call Registers
None

Result Registers
ES:BX = address of DTA

<u>AH</u>   <u>Function of INT 21H</u>

30   Get DOS version number

Additional Call Registers   Result Registers
None   AL = major version number (0,2,3,etc.)
                             AH = minor version number

31   Terminate process and stay resident (KEEP process)

Additional Call Registers   Result Registers
AL = binary return code   None
                             DX = memory size in paragraphs

Note: This interrupt call terminates the current process and attempts to place the memory size in paragraphs in the initial allocation block, but does not release any other allocation blocks. The return code in AL can be retrieved by the parent process using interrupt 21 call 4DH.

32   Reserved

33   Ctrl-break control

Additional Call Registers   Result Registers
AL = 00 to get state of   DL = 00 if ctrl-break check off
   ctrl-break check   = 01 if ctrl-break check on
   = 01 to modify state of
   ctrl-break check
DL = 00 to turn check off
   = 01 to turn check on

Note: When ctrl-break check is set to off, DOS minimizes the times it checks for ctrl-break input. When it is set to on, DOS checks for ctrl-break on most operations.

34   Reserved

35   Get interrupt vector address

Additional Call Registers   Result Registers
AL = interrupt number   ES:BX = address of interrupt handler

36   Get free disk space

Additional Call Registers   Result Registers
DL = drive code   AX = FFFFH if drive code invalid
(0 = default,   = sectors per cluster if valid
   1 = A, 2 = B,etc.)   BX = number of available clusters
                             CX = bytes per sector
                             DX = total clusters per drive

<u>AH</u>   <u>Function of INT 21H</u>

37   **Reserved**

38   **Country dependent information**

Additional Call Registers	Result Registers
DS:DX = address of 32-byte block of memory AL = function code	None

39   **Create subdirectory (MKDIR)**

Additional Call Registers	Result Registers

DS:DX = address of ASCIIZ path
  name of new subdirectory
AX = 3 if path not found
	Carry flag = 0 if successful
	  = 1 if failed
	  = 5 if access denied

3A   **Remove subdirectory (RMDIR)**

Additional Call Registers	Result Registers

DS:DX = address of ASCIIZ path
  name of subdirectory
AX = 3 if path not found
  = 5 if directory not empty
  = 15 if drive invalid
	Carry flag = 0 if successful
	  = 1 if failed

Note: The current directory cannot be removed.

3B   **Change the current subdirectory (CHDIR)**

Additional Call Registers	Result Registers

DS:DX = address of ASCIIZ path
  name of new subdirectory
	Carry flag = 0 if successful
	  = 1 if failed
	AX = 3 if path not found

3C   **Create a file**

Additional Call Registers	Result Registers

DS:DX = address of ASCIIZ path
  and file name
CX = file attribute
	Carry flag = 0 if successful
	  = 1 if failed
	AX = handle if successful
	  = 3 if path not found
	  = 5 if access denied

Note: Creates a new file if filename does not exist, otherwise truncates the file to length zero. Opens the file for reading or writing. A 16-bit handle will be returned in AX if the create was successful.

AH   Function of INT 21H

3D   Open file

Additional Call Registers | Result Registers
DS:DX = addres of ASCIIZ path | Carry flag = 0 if successful
and file name | = 1 if failed
AL = mode flags (see below) | AX = 16-bit file handle if successful
  | = 1 if function number invalid
  | = 2 if file not found
  | = 3 if path not found
  | = 4 if handle not available
  | = 5 if access denied
  | = 0CH if access code invalid

AL mode flag summary:

76543210 (bits)	Result
000	open for read
001	open for write
010	open for read/write
0	reserved
000	give others compatible access
001	read/write access denied to others
010	write access denied to others
011	read access denied to others
100	give full access to others
0	file inherited by child process
1	file private to current process

3E   Close file

Additional Call Registers | Result Registers
BX = file handle | Carry flag = 0 if successful
  | = 1 if failed
  | AX = 6 if invalid handle or file not open

Note: All internal buffers are flushed before the file is closed.

3F   Read from file or device

Additional Call Registers | Result Registers
DS:DX = buffer address | Carry flag = 0 if successful
BX = file handle | = 1 if failed
CX = number of bytes to read | AX = number of bytes actually read,
  | = 5 if access denied
  | = 6 if file not open or invalid handle

Note: When reading from the standard device (keyboard), at most one line of text will be read, regardless of the value of CX.

AH   Function of INT 21H

40   Write to file or device

Additional Call Registers
DS:DX = buffer address
BX = file handle
CX = number of bytes to write

Result Registers
Carry flag = 0 if successful
  = 1 if failed
AX = number of bytes actually
    written if successful
  = 5 if access denied
  = 6 if file not open or invalid handle

Note: If the carry flag is clear and AX is less than CX, a parital record was written or a disk full or other error was encountered.

41   Delete file (UNLINK)

Additional Call Registers
DS:DX = address of ASCIIZ
    file specification

Result Registers
Carry flag = 0 if successful
  = 1 if failed
AX = 2 if file not found
  = 5 if access denied

Note: This function cannot be used to delete a file that is read-only.   First, change the file's attribute to 0 by using interrupt 21 call 43H, then delete the file. No wildcard characters can be used in the filename. This function works by deleting the directory entry for the file.

42   Move file pointer (LSEEK)

Additional Call Registers
BX = file handle
CX:DX = offset
AL = 0 to move pointer offset
bytes from start of file
= 1 to move pointer offset
  bytes from current location
= 2 to move pointer offset
  bytes from end-of-file

Result Registers
Carry flag = 0 if successful
= 1 if fail
AX = 1 if invalid function number
  = 6 if file not open or invalid handle
DX:AX = absolute offset from start of
    file if successful

Note: To determine file size, call with AL = 2 and offset = 0.

43   Get or set file mode (CHMOD)

Additional Call Registers
DS:DX = address of ASCIIZ
    file specifier
AL = 0H to get attribute
= 1H to set attribute
CX = attribute if setting
= attribute codes if
    getting (see below)

Result Registers
Carry flag = 0 if successful
  = 1 if failed
CX = current attribute if set
AX = 1 if invalid function number
  = 2 if file not found
  = 3 if file does not exist or
    path not found
  = 5 if attribute cannot be changed

---

43   Get or set file mode (CHMOD) (continued from previous page)

```
76543210  attribute code bits
0         reserved
 0        reserved
  x       archive
   0      directory  (do not set with 43H; use extended FCB)
    0     volume-label (do not set with 43H; use ext. FCB)
     x    system
      x   hidden
       x  read-only
```

44   I/O device control (IOCTL)

Additional Call Registers	Result Registers
AL = 00H to get device info	AX = number of bytes
= 01H to set device info	transferred if CF=0
= 02H char read device to buffer	otherwise = error code
= 03H char write buffer to device	
= 04H block read device to buffer	
= 05H block write buffer to device	
= 06H check input status	
= 07H check output status	
= 08H test if block device changeable	
= 09H test if drive local or remote	
= 0AH test if handle local or remote	
= 0BH to change sharing retry count	
= 0CH char device I/O control	
= 0DH block device I/O control	
= 0EH get map for logical drive	
= 0FH set map for logical drive	
DS:DX = data buffer	
BX = file handle; CX = number of bytes	

45   Duplicate a file handle (DUP)

Additional Call Registers	Result Registers
BX = opened file handle	Carry flag = 0 if successful
	= 1 if failed
	AX = returned handle if successful
	= 4 if no handle available
	= 6 if handle invalid or not open

Note: The two handles will work in tandem; for example, if the file pointer of one handle is moved, the other will also be moved.

46 **Force a duplicate of a handle (FORCDUP)**

Additional Call Registers	Result Registers
BX = first file handle	Carry flag = 0 if successful
CX = second file handle	= 1 if failed
	AX = 4 if no handles available
	= 6 if handle invalid or not open

Note: If the file referenced by CX is open, it will be closed first. The second file handle will be forced to point identically to the first file handle. The two handles will work in tandem; for example, if the file pointer of one handle is moved, the other will also be moved.

47 **Get current directory**

Additional Call Registers	Result Registers
DL = drive code	Carry flag = 0 if successful
(0 = default,1 = A,...)	= 1 if failed
DS:SI = address of 64-byte buffer	DS:SI = ASCIIZ path specifier
	AX = OFH if drive specifier invalid

Note: The returned pathname does not include drive information or the leading "\".

48 **Allocate memory**

Additional Call Registers	Result Registers
BX = number of paragraphs	Carry flag = 0 if successful
	= 1 if failed
	AX = points to block if successful
	= 7 if memory control blocks destroyed
	= 8 if insufficient memory
	BX = size of largest block available if failed

49 **Free allocated memory**

Additional Call Registers	Result Registers
ES = segment address of block being released	Carry flag = 0 if successful
	= 1 if failed
	AX = 7 if memory control blocks destroyed
	= 9 if invalid memory block addr in ES

Note: Frees memory allocated by 48H.

AH   Function of INT 21H

4A   Modify memory allocation (SETBLOCK)

Additional Call Registers
ES = segment address of block
BX = requested new block size
    in paragraphs

Result Registers
Carry flag = 0 if successful
          = 1 if failed
BX = max available block size
          if failed
AX = 7 if memory control blocks destroyed
   = 8 if insufficient memory
  = 9 if invalid memory block
          address in ES

Note: Dynamically reduces or expands the memory allocated by a previous call to interrupt 21 function 48H.

4B   Load and/or execute program (EXEC)

Additional Call Registers
DS:DX = address of ASCIIZ path
  and filename to load
ES:BX = address of
  parameter block
AL = 0 to load and execute
   = 3 to load, not execute

Result Registers
AX = error code if CF not zero

4C   Terminate a process (EXIT)

Additional Call Registers
AL = binary return code

Result Registers
None

Note: Terminates a process, returning control to parent process or to DOS. A return code can be passed back in AL.

4D   Get return code of a subprocess (WAIT)

Additional Call Registers
None

Result Registers
AL = return code
AH = 00 if normal termination
   = 01 if terminated by ctrl-break
   = 02 if terminated by critical
     device error
   = 03 if terminated by call to
     interrupt 21 function 31H

Note: Returns the code sent via interrupt 21 function 4CH. The code can be returned only once.

<u>AH</u>   Function of INT 21H

4E   Search for first match (FIND FIRST)

Additional Call Registers	Result Registers
DS:DX = address of ASCIIZ file specification	Carry flag = 0 if successful
	= 1 if failed
CX = attribute to use in search	AX = error code

Note: The filename should contain one or more wildcard characters. Before this call, a previous call to interrupt 21 function 1AH must set the address of the DTA. If a matching filename is found, the current DTA will be filled in as follows:

Bytes 0 - 20: reserved by DOS for use on subsequent search calls
    21   · attribute found
    22 - 23: file time
    24 - 25: file date
    26 - 27: file size (least significant word)
    28 - 29: file size (most significant word)
    30 - 42: ASCIIZ file specification

4F   Search for next filename match (FIND NEXT)

Additional Call Registers	Result Registers
None	Carry flag = 0 if successful
	= 1 if failed
	AX = error code

Note: The current DTA must be filled in by a previous interrupt 21 4EH or 4FH call. The DTA will be filled in as outlined on interrupt 21 function 4E.

50   Reserved

51   Reserved

52   Reserved

53   Reserved

54   Get verify state

Additional Call Registers	Result Registers
None	AL = 00 if verify OFF
	= 01 if verify ON

Note: The state of the verify flag is changed via interrupt 21 function 2EH.

55   Reserved

<u>AH</u> Function of INT 21H

56 Rename file

Additional Call Registers	Result Registers
DS:DX = address of old ASCIIZ filename specification	Carry flag = 0 if successful
	= 1 if failed
ES:DI = address of new ASCIIZ	AX = 2 if file not found
filename specification	= 3 if path or file not found
	= 5 if access denied
	= 11H if different device in new name

Note: If a drive specification is used, it must be the same in the old and new filename specifications. However, the directory name may be different, allowing a move and rename in one operation.

57 Get/set file date and time

Additional Call Registers	Result Registers
AL = 00 to get	Carry flag = 0 if successful
= 01 to set	= 1 if failed
BX = file handle	CX = time if getting
CX = time if setting	DX = date if getting
DX = date if setting	AX = 1 if function code invalid
	= 6 if handle invalid

Note: The file must be open before the interrupt. The format of date and time is:

TIME:		DATE:	
Bits	0BH-0FH hours (0-23)	Bits	09H-0FH year (rel.1980)
	05H-0AH minutes (0-59)		05H-08H month (0-12)
	00H-04H number of 2- second increments (0-29)		00H-04H day (0-31)

58 Get/set allocation strategy

Additional Call Registers	Result Registers
AL = 00 to get strategy	Carry flag = 0 if successful
= 01 to set strategy	= 1 if failed
BX = strategy if setting	AX = strategy if getting
00 if first fit	= error code if setting
01 if best fit	
02 if last fit	

59 Get extended error information

Additional Call Registers	Result Registers
BX = 00	AX = extended error code
	(see Table D-1)
	BH = error class
	BL = suggested remedy
	CH = error locus

Warning! This function destroys the contents of registers CL, DX, SI, DI, BP, DS, and ES. Error codes will change with future version of DOS.

<u>AH</u>   Function of INT 21H

5A   **Create temporary file**

Additional Call Registers
DS:DX = address of ASCIIZ path
CX = file attribute
(00 if normal, 01 if read-only,
02 if hidden, 04 if system)

Result Registers
Carry flag = 0 if successful
= 1 if failed
AX = handle if successful
= error code if failed
DS:DX = address of ASCIIZ path
   specification if successful

Note: Files created with this interrupt function are not deleted when the program terminates.

5B   **Create new file**

Additional Call Registers
DS:DX = address of ASCIIZ
file specification
CX = file attribute
00 if normal
01 if read-only
02 if hidden
04 if system

Result Registers
Carry flag = 0 if successful
= 1 if failed
AX = file handle if successful
= error code if failed

Note: This function works similarly to interrupt 21 function 3CH; however, this function fails if the file already exists, whereas function 3CH truncates the file to length zero.

5C   **Control record access**

Additional Call Registers
AL = 00 to lock ,= 01 to unlock
BX = file handle
CX:DX = region offset
SI:DI = region length

Result Registers
Carry flag = 0 if successful
= 1 if failed
AX = error code

Note: Locks or unlocks records in systems that support multitasking or networking.

5D   **Reserved**

AH AL Function of INT 21H

5E 00   Get machine name

Additional Call Registers
DS:DX = address of buffer

Result Registers
Carry flag = 0 if successful
        = 1 if failed
CH = 0 if name undefined
      $\neq$ 0 if name defined
CL = NETBIOS number if successful
DS:DX = address of identifier if successful
AX = error code

Note: Returns a 15-byte ASCIIZ string computer identifier.

5E 02   Set printer setup

Additional Call Registers
BX = redirection list index
CX = setup strength length
DS:SI = address of setup string

Result Registers
Carry flag = 0 if successful
     = 1 if failed
AX = error code

Note: This function specifies a string that will precede all files sent to the network printer from the local node in a LAN. Microsoft Networks must be running in order to use this function.

5E 03   Get printer setup

Additional Call Registers
BX = redirection list index
ES:DI = address of buffer

Result Registers
Carry flag = 0 if successful
     = 1 if failed
AX = error code
CX = length of setup string
ES:DI = setup string if successful

5F 02   Get redirection list

Additional Call Registers
BX = redirection list index
DS:SI = address of 16-byte
  device name buffer
ES:DI = address of 128-byte
  netword name buffer

Result Registers
Carry flag = 0 if successful
= 1 if failed
BH = device status flag
bit 1 = 0 if valid device
      = 1 in invalid device
BL = device type
CX = parameter value
DS:SI = addr. ASCIIZ local device name
ES:DI = addr. ASCIIZ network name
AX = error flag

5F 03   **Redirect device**

Additional Call Registers
BL = device type
  03 printer
  04 drive

Result Registers
Carry flag = 0 if successful
          = 1 if failed
AX = error code
CX = caller value
DS:SI = address of ASCIIZ
  local device name
ES:DI = address of ASCIIZ
  network name

Note: Used when operating under a LAN, this function allows you to add devices to the network redirection list.

5F 04   **Cancel redirection**

Additional Call Registers
DS:SI = address of ASCIIZ
local device name

Result Registers
Carry flag = 0 if successful
          = 1 if fail
AX = error code

Note: Used when operating under a LAN, this function allows you to delete devices from the network redirection list.

60   **Reserved**

61   **Reserved**

62   **Get PSP (program segment prefix) address**

Additional Call Registers
None

Result Registers
BX = address of PSP

A summary of the IBM error codes is given in Table D-1.

TABLE D-1: Extended Error Code Information

Code	Error
1	invalid function number
2	file not found
3	path not found
4	too many open files
5	access denied
6	invalid handle
7	memory control blocks destroyed
8	insufficient memory
9	invalid memory block address
10	invalid environment
11	invalid format
12	invalid access code
13	invalid data
14	unknown unit
15	invalid disk drive
16	attempt to remove current directory
17	not same device
18	no more files
19	attempt to write on write-protected diskette
20	unknown unit
21	drive not ready
22	unknown command
23	data error (CRC)
24	bad request structure length
25	seek error
26	unknown media type
27	sector not found
28	printer out of paper
29	write fault
30	read fault
31	general failure
32	sharing violation
33	lock violation
34	invalid disk change
35	FCB unavailable
36	sharing buffer overflow
37-49	reserved
50	network request not supported
51	remote computer not listening
52	duplicate name on network
53	network name not found
54	network busy
55	network device no longer exists
56	net BIOS command limit exceeded
57	network adapter hardware error
58	incorrect response from network
59	unexpected network error
60	incompatible remote adapter
61	print queue full
62	not enough space for print file
63	print file was deleted
64	network name not found
65	access denied
66	network device type incorrect
67	network name not found
68	network name limit exceeded
69	net BIOS session limit exceeded
70	temporarily paused
71	network request not accepted
72	print or disk redirection is paused
73-79	reserved
80	file exists
81	reserved
82	cannot make directory entry
83	fail on INT 24
84	too many redirections
85	duplicate redirection
86	invalid password
87	invalid parameter
88	network device fault

Reprinted by permission from "IBM Disk Operating System Technical Reference" c. 1987 by International Business Machines Corporation.

SECTION D.2: MOUSE INTERRUPTS 33H

Mouse interrupts are covered in Section 2 of Chapter 5. The following is a partial list of commonly used functions of the mouse handler interrupt, 33H. This interrupt is loaded into the system with the mouse interrupt handler, which is loaded upon reading the "device=" directive in the CONFIG.SYS file.

AX   Function of INT 33H

00   Initialize the mouse

Additional Call Registers   Result Registers
None   AX = 0H if mouse not available
                                      = FFFFH if mouse available
                                   BX = number of mouse buttons

Note: This function is called only once to initialize the mouse. If mouse support is present, AX=FFFFH and the mouse driver is initialized, the mouse pointer is set to the center of the screen and concealed.

01   Display mouse pointer

Additional Call Registers   Result Registers
None   None

Note: This function displays the mouse pointer and cancels any excusion area.

02   Conceal mouse pointer

Additional Call Registers   Result Registers
None   None

Note: This function hides the mouse pointer but the mouse driver monitors its position. Most programs issue this command before they terminate.

03   Get mouse location and button status

Additional Call Registers   Result Registers
None   BX=mouse button status
                                     bit 0 -- left button
                                     bit 1 -- right button
                                     bit 2 -- center button
                                     =0 if up; =1 if down
                                   CX = horizontal position
                                   DX = vertical position

Note: The horizontal and vertical coordinates are returned in pixels.

04   Set mouse pointer location

Additional Call Registers   Result Registers
CX = horizontal position   None
DX = vertical position

Note: The horizontal and vertical coordinates are in pixels. Will display the mouse pointer only within set limits; will not display in exclusion areas.

<u>AX</u>   <u>Function of INT 33H</u>

05   **Get button press information**

<u>Additional Call Registers</u>
BX=button: 0 for left;
    1 for right; 2 for center

<u>Result Registers</u>
AX=button status
  bit 0 -- left button
  bit 1 -- right button
  bit 2 -- center button
  =0 if up; =1 if down
BX = button press count
CX=horizontal position
DX=vertical position

Note: This returns the status of all buttons as well as the number of presses for the button indicated in BX when called. The position of the mouse pointer is given in pixels and represents the position at the last button press.

06   **Get button release information**

<u>Additional Call Registers</u>
BX=button: 0 for left;
    1 for right; 2 for center

<u>Result Registers</u>
AX=button status
  bit 0 -- left button
  bit 1 -- right button
  bit 2 -- center button
  =0 if up; =1 if down
BX = button release count
CX=horizontal position
DX=vertical position

Note: This returns the status of all buttons as well as the number of releases for the button indicated in BX when called. The position of the mouse pointer is given in pixels and represents the position at the last button release.

07   **Set horizontal limits for mouse pointer**

<u>Additional Call Registers</u>
CX=minimum horizontal position
DX=maximum horizontal position

<u>Result Registers</u>
None

Note: This sets the horizontal limits (in pixels) for the mouse pointer. After this call, the mouse will be displayed within these limits.

08   **Set vertical limits for mouse pointer**

<u>Additional Call Registers</u>
CX=minimum vertical position
DX=maximum vertical position

<u>Result Registers</u>
None

Note: This sets the vertical limits (in pixels) for the mouse pointer. After this call, the mouse will be displayed within these limits.

AX   Function of INT 33H

10   **Set mouse pointer exclusion area**

Additional Call Registers | Result Registers
CX=upper left horizontal coordinate | None
DX=upper left vertical coordinate
SI=lower right horizontal coordinate
DI=lower right vertical coordinate

Note: This defines an area in which the mouse pointer will not display. An exclusion area can be cancelled by calling functions 00 or 01.

24   **Get mouse information**

Additional Call Registers | Result Registers
None | BH=major version
 | BL=minor version
 | CH=mouse type
 | CL=IRQ number

Note: This returns the version number (e.g., version 7.5: BH=7,BL=5). Mouse type: 1 for bus; 2 for serial; 3 for InPort; 4 for PS/2; 5 for HP; IRQ=0 for PS/2; otherwise=2, 3, 4, 5 or 7.

APPENDIX E: BIOS INTERRUPTS

This appendix covers the most commonly used BIOS interrupts. INT 10H is used extensively for graphics programming. INT 11H returns the equipment configuration and INT 12H returns the memory size. INT 14H is used for asynchronous communication. Two functions of INT 15H are included: one to initiate a wait and another to return extended memory size. INT 16H is used for the keyboard and INT 17H for the printer. INT 1AH handles the timer and RTC.

SECTION E.1: INT 10H VIDEO FUNCTION CALLS

AH   **Function**

00   **Set video mode**

Additional Call Registers	Result Registers
AL = video mode	None

See Table E-2 for a list of available video modes and their definition.

01   **Set cursor type**

Additional Call Registers	Result Registers
CH = beginning line of cursor (bits 0 - 4)	None
CL = ending line of cursor (bits 0 - 4)	

Note: All other bits should be set to zero. The blinking of the cursor is hardware controlled.

02   **Set cursor position**

Additional Call Registers	Result Registers
BH = page number	None
DH = row	
DL = column	

Note: When using graphics modes, BH must be set to zero. Text coordinates of the upper left-hand corner will be (0,0).

03H   **Read cursor position and size**

Additional Call Registers	Result Registers
BH = page number	CH = beginning line of cursor
	CL = ending line of cursor
	DH = row
	DL = column

Note: When using graphics modes, BH must be set to zero.

AH   Function

04H   Read light pen position

Additional Call Registers
None

Result Registers
AH = 0 if light pen not triggered
 = 1 if light pen triggered
BX = pixel column
CH = pixel row (modes 04H - 06H)
CX = pixel row (modes 0DH - 13H)
DH = character row
DL = character column

05H   Select active display page

Additional Call Registers
AL = page number
(see Table E-1 below)

Result Registers
None

Table E-1: Display Pages for Different Modes and Adapters

Mode	Pages	Adapters			
00H	0 - 7	CGA	EGA	MCGA	VGA
01H	0 - 7	CGA	EGA	MCGA	VGA
02H	0 - 3	CGA			
	0 - 7		EGA	MCGA	VGA
03H	0 - 3	CGA			
	0 - 7		EGA	MCGA	VGA
07H	0 - 7		EGA		VGA
0DH	0 - 7		EGA		VGA
0EH	0 - 3		EGA		VGA
0FH	0 - 1		EGA		VGA
10H	0 - 1		EGA		VGA

All other mode-adapter combinations support only one page.

06   Scroll window up

Additional Call Registers
AL = number of lines to scroll
BH = display attribute
CH = y coordinate of top left
CL = x coordinate of top left
DH = y coordinate of lower right
DL = x coordinate of lower right

Result Registers
None

Note: If AL = 0, the entire window is blank. Otherwise, the screen will be scrolled upward by the number of lines in AL. Lines scrolling off the top of the screen are lost, blank lines are scrolled in at the bottom according to the attribute in BH.

AH	Function

07   Scroll window down

Additional Call Registers	Result Registers
AL = number of lines to scroll	None
BH = display attribute	
CH = y coordinate of top left	
CL = x coordinate of top left	
DH = y coordinate of lower right	
DL = x coordinate of lower right	

Note: If AL = 0, the entire window is blank. Otherwise, the screen will be scrolled down by the number of lines in AL. Lines scrolling off the bottom of the screen are lost, blank lines are scrolled in at the top according to the attribute in BH.

08   Read character and attribute at cursor position

Additional Call Registers	Result Registers
BH = display page	AH = attribute byte
	AL = ASCII character code

09   Write character and attribute at cursor position

Additional Call Registers	Result Registers
AL = ASCII character code	None
BH = display page	
BL = attribute	
CX = number of characters to write	

Note: Does not update cursor position. Use interrupt 10 Function 2 to set cursor position.

0A   Write character at cursor position

Additional Call Registers	Result Registers
AL = ASCII character code	None
BH = display page	
BL = graphic color	
CX = number of characters to write	

Note: Writes character(s) using existing video attribute. Does not update cursor position. Use interrupt 10 Function 2 to set cursor position.

AH   Function

0B   Set color palette

Additional Call Registers | Result Registers
BH = 00H to set border or
background colors
= 01H to set palette
BL = palette/color

None

Note: If BH = 00H and in text mode, this function will set the border color only. If BH = 00H and in graphics mode, this function will set background and border colors. If BH = 01H, this function will select the palette. In 320 x 200 four-color graphics, palettes 0 and 1 are available:

Pixel Colors for Palettes 0 and 1

Pixel	Palette 0	Palette 1
0	background	background
1	green	cyan
2	red	magenta
3	brown/yellow	white

0C   Write pixel

Additional Call Registers | Result Registers
AL = pixel value
CX = pixel column
DX = pixel row
BH = page

None

Note: Coordinates and pixel value depend on the current video mode. Setting bit 7 of AL causes the pixel value in AL to be XORed with the current value of the pixel.

0D   Read pixel

Additional Call Registers | Result Registers
CX = pixel column
DX = pixel row
BH = page

AL = pixel value

0E   TTY character output

Additional Call Registers | Result Registers
AL = character
BH = page
BL = foreground color

None

Note: Writes character to the display and updates cursor position. TTY mode indicates minimal character processing. ASCII codes for bell, backspace, linefeed, and carriage return are translated into the appropriate action.

AH AL Function

0F   **Get video mode**

Additional Call Registers	Result Registers
None	AH = width of screen in characters
	AL = video mode
	BH = active display page

Note: See Table E-2 for a list of possible video modes.

10 00   **SubFunction 00H: set palette register to color correspondence**

Additional Call Registers	Result Registers
AL = 00H	None
BH = color
CL = palette register
(00H to 0FH)

10 01   **SubFunction 01H: set border color**

Additional Call Registers	Result Registers
AL = 01H	None
BH = border color

10 02   **SubFunction 02H: set palette and border**

Additional Call Registers	Result Registers
AL = 02H	None
ES:DX = address of color list

13   **Write String**

Additional Call Registers	Result Registers
AL = write mode	None
=00H, attribute in BL,
cursor not moved
=01H, attribute in BL,
cursor moved
=02H,attributes follow char,
cursor not moved
= 03H, attributes follow char,
cursor moved
ES:BP = address of string
CX = character count
DH = initial row position
DL = intial column position
BH = page

Note: For AL = 00 and 01, the string consists of characters only, which will all be displayed with the attribute in BL. For AL = 02 and 03, the data is stored with the attributes (char, attrib, char, attrib, and so on).

Table E-2: Video Modes and Their Definition

AL	Pixels	Characters			Char box	Text/ graph	Colors		Aadpter	Max pages	Buffer start
00H	320x200	40	x	25	8x8	text	16	*	CGA	8	B8000h
	320x350	40	x	25	8x14	text	16	*	EGA	8	B8000h
	360x400	40	x	25	9x16	text	16	*	VGA	8	B8000h
	320x400	40	x	25	8x16	text	16	*	MCGA	8	B8000h
01H	320x200	40	x	25	8x8	text	16		CGA	8	B8000h
	320x350	40	x	25	8x14	text	16		EGA	8	B8000h
	360x400	40	x	25	9x16	text	16		VGA	8	B8000h
	320x400	40	x	25	8x16	text	16		MCGA	8	B8000h
02H	640x200	80	x	25	8x8	text	16	*	CGA	8	B8000h
	640x350	80	x	25	8x14	text	16	*	EGA	8	B8000h
	720x400	80	x	25	9x16	text	16	*	VGA	8	B8000h
	640x400	80	x	25	8x16	text	16	*	MCGA	8	B8000h
03H	640x200	80	x	25	8x8	text	16		CGA	8	B8000h
	640x350	80	x	25	8x14	text	16		EGA	8	B8000h
	720x400	80	x	25	9x16	text	16		VGA	8	B8000h
	640x400	80	x	25	8x16	text	16		MCGA	8	B8000h
04H	320x200	40	x	25	8x8	graph	4		CGA	1	B8000h
	320x200	40	x	25	8x8	graph	4		EGA	1	B8000h
	320x200	40	x	25	8x8	graph	4		VGA	1	B8000h
	320x200	40	x	25	8x8	graph	4		MCGA	1	B8000h
05H	320x200	40	x	25	8x8	graph	4	*	CGA	1	B8000h
	320x200	40	x	25	8x8	graph	4	*	EGA	1	B8000h
	320x200	40	x	25	8x8	graph	4	*	VGA	1	B8000h
	320x200	40	x	25	8x8	graph	4	*	MCGA	1	B8000h
06H	640x200	80	x	25	8x8	graph	2		CGA	1	B8000h
	640x200	80	x	25	8x8	graph	2		EGA	1	B8000h
	640x200	80	x	25	8x8	graph	2		VGA	1	B8000h
	640x200	80	x	25	8x8	graph	2		MCGA	1	B8000h
07H	720x350	80	x	25	9x14	text	mono		MDA	8	B0000h
	720x350	80	x	25	9x14	text	mono		EGA	4	B0000h
	720x400	80	x	25	9x16	text	mono		VGA	8	B0000h
08H	reserved										
09H	reserved										
0AH	reserved										
0BH	reserved										
0CH	reserved										
0DH	320x200	40	x	25	8x8	graph	16		EGA	2/4	A0000h
	320x200	40	x	25	8x8	graph	16		VGA	8	A0000h
0EH	640x200	80	x	25	8x8	graph	16		EGA	1/2	A0000h
	640x200	80	x	25	8x8	graph	16		VGA	4	A0000h
0FH	640x350	80	x	25	9x14	graph	mono		EGA	1	A0000h
	640x350	80	x	25	8x14	graph	mono		VGA	2	A0000h
10H	640x350	80	x	25	8x14	graph	4		EGA	1/2	A0000h
	640x350	80	x	25	8x14	graph	16		VGA	2	A0000h
11H	640x480	80	x	30	8x16	graph	2		VGA	1	A0000h
	640x480	80	x	30	8x16	graph	2		MCGA	1	A0000h
12H	640x480	80	x	30	8x16	graph	16		VGA	1	A0000h
13H	320x200	40	x	25	8x8	graph	256		VGA	1	A0000h
	320x200	40	x	25	8x8	graph	256		MCGA	1	A0000h

* color burst off

SECTION E.2: INT 11H -- EQUIPMENT DETERMINATION

Get equipment configuration

Call Registers
None

Result Registers
AX = equipment code (see below)

Note: BIOS data area 40:10 is set during POST according to the equipment code word, which shows the optional equipment that is attached to the system.

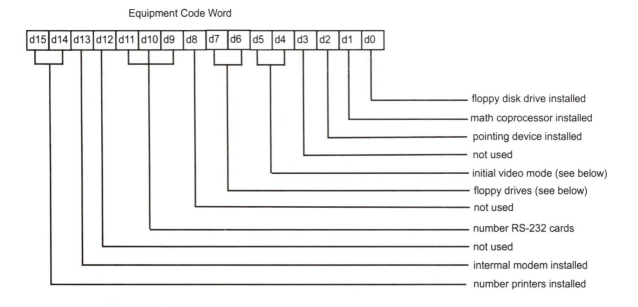

Equipment Code Word

Initial Video Mode
00 = Reserved
01 = 40x25 color
10 = 80x25 color
11 = 80x25 monochrome

Number of Floppy Drives
00 = 1 drive
01 = 2 drives

SECTION E.3: INT 12H -- MEMORY SIZE DETERMINATION

Get Conventional Memory Size

Call Registers
None

Result Registers
AX = memory size (KB)

Note: Returns amount of conventional memory available to DOS and application programs.

AH   **Function**

00   **Initialize COM Port**

Additional Call Registers	Result Registers
AL = parameter (see below)	AH = port status (see below)
DX = port number (0 if COM1, 1 if COM2, etc.)	AL = modem status (see below)

Note 1: The parameter byte in AL is defined as follows

7 6 5 4 3 2 1 0	Indicates
x x x	Baud rate (000=110, 001=150, 010=300, 011=600, 100=1200, 101=2400, 110=4800, 111=9600)
x x	Parity (01=odd, 11=even, x0=none)
x	Stop bits (0 = 1, 1 = 2)
x x	Word length (10=7 bits, 11=8 bits)

Note 2: The port status returned in AH is defined as follows

7 6 5 4 3 2 1 0	Indicates
1	Timed-out
1	Transmit shift register empty
1	Transmit holding register empty
1	Break detected
1	Framing error detected
1	Parity error detected
1	Overrun error detected
1	Received data ready

Note 3: The modem status returned in AL is defined as follows

7 6 5 4 3 2 1 0	Indicates
1	Received line signal detect
1	Ring indicator
1	DSR (data set ready)
1	CTS (clear to send)
1	Change in receive line signal detect
1	Trailing edge ring indicator
1	Change in DSR status
1	Change in CTS status

AH   Function

01   Write character to COM Port

Additional Call Registers | Result Registers
AL = character | AH bit 7 =0 if successful, 1 if not
DX = port number (0 if COM1, | AH bits 0 - 6 = status if successful
    1 if COM2, etc.) | AL = character

Note: The status byte in AH, bits 0 - 6, after the call is as follows

6 5 4 3 2 1 0	Indicates
1	Transmit shift register empty
1	Transmit holding register empty
1	Break detected
1	Framing error detected
1	Parity error detected
1	Overrun error detected
1	Receive data ready

02   Read character from COM Port

Additional Call Registers | Result Registers
DX = port number (0 if COM1, | AH bit 7 =0 if successful, 1 if not
    1 if COM2, etc.) | AH bits 0 - 6 = status if successful
  | AL = character read

Note: The status byte in AH, bits 1 - 4, after the call is as follows

4 3 2 1	Indicates
1	Break detected
1	Framing error detected
1	Parity error detected
1	Overrun error detected

03   Read COM Port Status

Additional Call Registers | Result Registers
DX = port number (0 if COM1, | AH =port status
    1 if COM2, etc.) | AL =modem status

Note: The port status and modem status returned in AH and AL are the same format as INT 14H function 00H, described above.

AH   Function

04   Extended Initialize COM Port

Additional Call Registers

AL = 00H (break), 01H (no break)
DX = port number (0 if COM1,
    1 if COM2, etc.)
BH = parity
    = 00H none
    = 01H odd
    = 02H even
    = 03H stick parity odd
    = 04H stick parity even
BL = stop bits
    = 00H (one stop bit)
    = 01H (1.5 bits for 5-bit word)
    = 01H (2 bits for > 5-bit word)
CH = word length
    = 00H 5-bit
    = 01H 6-bit
    = 02H 7-bit
    = 03H 8-bit
CL = baud rate
    = 00H 110 baud
    = 01H 150 baud
    = 02H 300 baud
    = 03H 600 baud
    = 04H 1200 baud
    = 05H 2400 baud
    = 06H 4800 baud
    = 07H 9600 baud
    = 08H 19200 baud

Result Registers

AH = port status (see function AH=0)
AL = modem status (see function AH=0)

05   Extended COM Port Control

Additional Call Registers

AL = 00H (read control register),
    = 01H (write to control register)
DX = port number (0 if COM1,
    1 if COM2, etc.)
BL = Modem control register
    (see Figure 9-14)
    bits 7 - 5: reserved
    bit 4: loop
    bit 3: out2
    bit 2: out1
    bit 1: RTS
    bit 0: DTR

Result Registers

If read subfunction,
    BL = modem control register
If write subfunction,
    AL = modem status (see Figure 9-16)
    AH = line status (see Figure 9-15)

Note: Subfunction AL = 00H returns the modem control register contents in
BL. Subfunction AL = 01H writes the contents of BL into the modem control
register and returns modem and line status register contents in AL and AH.

SECTION E.5: INT 15H -- SYSTEM SERVICES

AH   Function

86H   **Wait**

Additional Call Registers	Result Registers
CX:DX time to wait in ms	CF = 0 for successful wait
	= 1 if wait not performed

Note: The duration of the wait will always be a multiple of 976 microseconds.

88H   **Extended memory size determination**

Additional Call Registers	Result Registers
None	AX = extended memory (KB)

Note: Returns the amount of installed extended memory in KB, that is, the memory beginning at address 100000H, as determined by the POST. If DOS memory management is in control, 0 will be returned in AX even if you have extended memory.

SECTION E.6: INT 16H -- KEYBOARD

AH   Function

00H   **Keyboard read**

Additional Call Registers	Result Registers
None	AH = key scan code
	AL = ASCII char

Note: Reads one character from the keyboard buffer and updates the head pointer.

01H   **Get keyboard status**

Additional Call Registers	Result Registers
None	If no key waiting,
	ZF = 1.
	If key waiting,
	ZF = 0,
	AH = key scan code,
	AL = ASCII char.

Note: If a key is waiting, the scan code and character are returned in AH and AL, but the head pointer of the keyboard buffer is not updated.

AH   Function

02H   Get shift status

Additional Call Registers	Result Registers
None	AL = status byte

bit 7: Insert pressed
bit 6: Caps Lock pressed
bit 5: Num Lock pressed
bit 4: Scroll Lock pressed
bit 3: Alt pressed
bit 2: Ctrl pressed
bit 1: Left Shift pressed
bit 0: Right Shift pressed

Note: The keyboard status byte returned in AL indicates whether certain keys have been pressed. If the bit = 1, the key has been pressed.

03H   Set typematic rate

Additional Call Registers	Result Registers
AL = 05H	None
BH = repeat delay (see below)
BL = repeat rate (see below)

Note: Sets the rate at which repeated keystrokes are accepted.
The delay value in BH can be 00H (for 250), 01H (for 500), 02H (for 750), or 03H (for 1000). All values are in milliseconds. The repeat rate in BL represents the number of characters per second. Options are:

00H: 30.0	0BH: 10.9	16H: 4.3
01H: 26.7	0CH: 10.0	17H: 4.0
02H: 24.0	0DH: 9.2	18H: 3.7
03H: 21.8	0EH: 8.6	19H: 3.3
04H: 20.0	0FH: 8.0	1AH: 3.0
05H: 18.5	10H: 7.5	1BH: 2.7
06H: 17.1	11H: 6.7	1CH: 2.5
07H: 16.0	12H: 6.0	1DH: 2.3
08H: 15.0	13H: 5.5	1EH: 2.1
09H: 13.3	14H: 5.0	1FH: 2.0
0AH: 12.0	15H: 4.6	20H to FFH - reserved

10H   Extended keyboard read

Additional Call Registers	Result Registers
None	AH = key scan code
	AL = ASCII char

Note: Used in place of INT 16H function 00H to allow program to detect F11, F12, and other keys of the extended keyboard. After the read, the head pointer of the keyboard buffer is updated.

AH   Function

11H   **Extended keyboard status**

Additional Call Registers	Result Registers

None	If no key waiting,
	    ZF = 1.
	If key waiting,
	    ZF = 0,
	    AH = key scan code,
	    AL = ASCII char.

Note: This function is used instead of INT 16H function 01H so that programs can detect keys of the extended keyboard such as F11 and F12. If a key is waiting, the scan code and character are returned in AH and AL, but the head pointer of the keyboard buffer is not updated.

12H   **Extended shift status**

Additional Call Registers	Result Registers

None	AL = shift status
	    bit 7: Insert locked
	    bit 6: Caps Lock locked
	    bit 5: Num Lock locked
	    bit 4: Scroll Lock locked
	    bit 3: Alt pressed
	    bit 2: Ctrl pressed
	    bit 1: Left Shift pressed
	    bit 0: Right Shift pressed
	AH = extended shift status
	    bit 7: SysRq pressed
	    bit 6: Caps Lock pressed
	    bit 5: Num Lock pressed
	    bit 4: Scroll Lock pressed
	    bit 3: Right Alt pressed
	    bit 2: Right Ctrl pressed
	    bit 1: Left Alt pressed
	    bit 0: Left Ctrl pressed

Note: The keyboard status bytes returned in AL and AH indicate whether certain keys have been pressed. If the bit = 1, the key has been pressed.

SECTION E.7: INT 17H -- PRINTER

AH   Function

00H   Print character

Additional Call Registers	Result Registers

AL = character to print
DX = printer (0 for LPT1,
    1 for LPT2, 2 for LPT3)
	AH = status
	  bit 7: printer not busy
	  bit 6: printer acknowledge
	  bit 5: out of paper
	  bit 4: printer selected
	  bit 3: I/O error
	  bits 2-1: reserved
	  bit 0: printer timed-out

01H   Initialize printer port

Additional Call Registers	Result Registers
DX = printer (0 for LPT1, 1 for LPT2, 2 for LPT3)	AH = status (see function 00H)

02H   Read status

Additional Call Registers	Result Registers
DX = printer (0 for LPT1, 1 for LPT2, 2 for LPT3)	AH = status (see function 00H)

SECTION E.8: INT 1AH -- TIMER AND REAL-TIME CLOCK SERVICES

AH   Function

00H   Read system-timer time counter

Additional Call Registers	Result Registers

None	CX = high portion of count
	DX = low portion of count
	AL = 0 if 24 hours has not passed
	            since last read
	    > 0 if 24 has passed since last read

Note: This function returns the number of ticks since midnight. A second is about 18.2 ticks. When the number of ticks indicates that 24 hours has passed, AL is incremented and the tick count is reset to zero. Calling this function resets AL so that whether 24 hours has passed can only be determined once a day.

<u>AH</u>   <u>Function</u>

01H   **Set system-timer time counter**

<u>Additional Call Registers</u>   <u>Result Registers</u>
CX = high portion of tick count   None
DX = low portion of tick count

Note: Calling this function will cause the timer overflow flag to be reset.

02H   **Read real-time clock time**

<u>Additional Call Registers</u>   <u>Result Registers</u>
None   CH = hours
                                         CL = minutes
                                         DH = seconds
                                         DL = 01 for daylight savings option
                                              = 00 for no option
                                         CF = 0 if clock operating, otherwise = 1

Note: Hours, minutes, and seconds are returned in BCD format. This function is used to get the time in the CMOS time/date chip.

03H   **Set real-time clock time**

<u>Additional Call Registers</u>   <u>Result Registers</u>
CH = hours   None
CL = minutes
DH = seconds
DL = 01 for daylight savings option
     = 00 for no option

Note: Hours, minutes, and seconds are in BCD format. This function is used to set the time in the CMOS time/date chip.

04H   **Read real-time clock date**

<u>Additional Call Registers</u>   <u>Result Registers</u>
None   CH = century (19 or 20)
                                         CL = year
                                         DH = month
                                         DL = day
                                         CF = 0 if clock operating, otherwise = 1

Note: Century, year, month, and day are in BCD format. This function is used to get the date in the CMOS time/date chip.

AH   Function

05H   **Set real-time clock date**

Additional Call Registers	Result Registers
CH = century (19 or 20)	None
CL = year	
DH = month	
DL = day	

Note: Century, year, month, and day are in BCD format. This function is used to set the date in the CMOS time/date chip.

APPENDIX F: ASCII CODES

Ctrl	Dec	Hex	Ch	Code	Dec	Hex	Ch	Dec	Hex	Ch	Dec	Hex	Ch
^@	0	00		NUL	32	20		64	40	@	96	60	`
^A	1	01	☺	SOH	33	21	!	65	41	A	97	61	a
^B	2	02	●	STX	34	22	"	66	42	B	98	62	b
^C	3	03	♥	ETX	35	23	#	67	43	C	99	63	c
^D	4	04	♦	EOT	36	24	$	68	44	D	100	64	d
^E	5	05	♣	ENQ	37	25	%	69	45	E	101	65	e
^F	6	06	♠	ACK	38	26	&	70	46	F	102	66	f
^G	7	07	●	BEL	39	27	'	71	47	G	103	67	g
^H	8	08	□	BS	40	28	(	72	48	H	104	68	h
^I	9	09	○	HT	41	29	)	73	49	I	105	69	i
^J	10	0A	◉	LF	42	2A	*	74	4A	J	106	6A	j
^K	11	0B	♂	VT	43	2B	+	75	4B	K	107	6B	k
^L	12	0C	♀	FF	44	2C	,	76	4C	L	108	6C	l
^M	13	0D	♪	CR	45	2D	–	77	4D	M	109	6D	m
^N	14	0E	♫	SO	46	2E	.	78	4E	N	110	6E	n
^O	15	0F	☼	SI	47	2F	/	79	4F	O	111	6F	o
^P	16	10	►	DLE	48	30	0	80	50	P	112	70	p
^Q	17	11	◄	DC1	49	31	1	81	51	Q	113	71	q
^R	18	12	↕	DC2	50	32	2	82	52	R	114	72	r
^S	19	13	‼	DC3	51	33	3	83	53	S	115	73	s
^T	20	14	¶	DC4	52	34	4	84	54	T	116	74	t
^U	21	15	§	NAK	53	35	5	85	55	U	117	75	u
^V	22	16	▬	SYN	54	36	6	86	56	V	118	76	v
^W	23	17	↨	ETB	55	37	7	87	57	W	119	77	w
^X	24	18	↑	CAN	56	38	8	88	58	X	120	78	x
^Y	25	19	↓	EM	57	39	9	89	59	Y	121	79	y
^Z	26	1A	→	SUB	58	3A	:	90	5A	Z	122	7A	z
^[	27	1B	←	ESC	59	3B	;	91	5B	[	123	7B	{
^\	28	1C	∟	FS	60	3C	<	92	5C	\	124	7C		
^]	29	1D	↔	GS	61	3D	=	93	5D	]	125	7D	}
^^	30	1E	▲	RS	62	3E	>	94	5E	^	126	7E	~
^_	31	1F	▼	US	63	3F	?	95	5F	_	127	7F	⌂

APPENDIX G: I/O ADDRESS MAPS

SECTION G.1: IBM PC AT I/O ADDRESS MAP

Hex Range	Device
000 - 01F	DMA controller 1, 8237A-5
020 - 03F	Interrupt controller 1, 8259A, Master
040 - 05F	Timer, 8254-2
060 - 06F	8042 (keyboard)
070 - 07F	Real-time clock, NMI (non-maskable interrupt) mask
080 - 09F	DMA page register, 74LS612
0A0 - 0BF	Interrupt controller 2, 8237A-5
0C0 - 0DF	DMA controller 2, 8237A-5
0F0	Clear math coprocessor busy
0F1	Reset math coprocessor
0F8 - 0FF	Math coprocessor
1F0 - 1F8	Fixed disk
200 - 207	Game I/O
20C - 20D	Reserved
21F	Reserved
278 - 27F	Parallel printer port 2
2B0 - 2DF	Alternate enhanced graphics adapter
2E1	GPIB (adapter 0)
2E2 & 2E3	Data acquisition (adapter 0)
2F8 - 2FF	Serial port 2
300 - 31F	Prototype card
360 - 363	PC network (low address)
364 - 367	Reserved
368 - 36B	PC network (high address)
36C - 36F	Reserved
378 - 37F	Parallel printer port 1
380 - 38F	SDLC, bisynchronous 2
390 - 393	Cluster
3A0 - 3AF	Bisynchronous 1
3B0 - 3BF	Monochrome display and printer adapter
3C0 - 3CF	Enhanced graphics adapter
3D0 - 3DF	Color/graphics monitor adapter
3F0 - 3F7	Diskette controller
3F8 - 3FF	Serial port 1
6E2 & 6E3	Data acquisition (adapter 1)
790 - 793	Cluster (adapter 1)
AE2 & AE3	Data acquisition (adapter 2)
B90 - B93	Cluster (adapter 2)
EE2 & EE3	Data acquisition (adapter 3)
1390 - 1393	Cluster (adapter 3)
22E1	GPIB (adapter 1)
2390 - 2393	Cluster (adapter 4)
42E1	GPIB (adapter 2)
62E1	GPIB (adapter 3)
82E1	GPIB (adapter 4)
A2E1	GPIB (adapter 5)
C2E1	GPIB (adapter 6)
E2E1	GPIB (adapter 7)

Note: I/O address, hex 000 to 0FF, are reserved for the system I/O board. Hex 100 to 3FF are available on the I/O channel.

Reprinted by permission from "IBM Technical Reference Personal Computer AT"

c. 1985 by International Business Machines Corporation.

SECTION G.2: IBM PS/2 I/O ADDRESS MAP

Hex Range	Device
0000 - 001F	DMA controller (0 - 3)
0020, 0021	Interrupt controller (Master)
0040, 0042 - 0044, 0047	System timers
0060	Keyboard, auxiliary device
0061	System control port B
0064	Keyboard, auxiliary device
0070, 0071	RT/CMOS and NMI mask
0081 - 0083, 0087	DMA page registers (0 - 3)
0089 - 008B, 008F	DMA page registers (4 - 7)
0090	Central arbitration control point
0091	Card selected feedback register
0092	System control port A
0094	System board enable/setup register
0096	Adapter enable/setup register
00A0 - 00A1	Interrupt controller (slave)
00C0 - 00DF	DMA controller (4 - 7)
00E0, 00E1	Memory encoding registers
00F0 - 00FF	Math coprocessor
0100 - 0107	Programmable option select
01F0 - 01F8	Fixed disk drive controller
0278 - 027B	Parallel port 3
02F8 - 02FF	Serial port 2 (RS-232C)
0378 - 037B	Parallel port 2
03B4, 03B5, 03BA	Video subsystem
03BC - 03BF	Parallel port 1
03C0 - 03C5	Video sybsystem
03C6 - 03C9	Video DAC
03CA, 03CC, 03CE, 03CF	Video subsystem
03D4, 03D5, 03DA	Video subsystem
03F0 - 03F7	Diskette drive controller
03F8 - 03FF	Serial port 1 (RS-232C)

Reprinted by permission from "IBM Personal System/2 Hardware Interface Technical Reference"

c. 1988 by International Business Machines Corporation.

This section lists the ISA computer I/O address ports as supported by Intel's SL SuperSet chips. The SL SuperSet chips, the Intel 386SL and Intel 486SL, consist of the core processor (386 or 486) and I/O peripheral components. They support ISA standard peripherals.

I/O PORT LIST

The following pages contain descriptions of all the I/O ports supported by the SL SuperSet.

Address	R/W/S	Description
0H	RW	DMA Channel 0 base and current address Loaded with DMA base address and incremented as transfers take place. Low byte accessed after clearing byte pointer, address 000CH. Upper byte loaded or read after low byte access.
1H	RW	DMA Channel 0 base and current word count Low byte of word count. Count = value + 1 transfers. Accessed after byte pointer is cleared (000CH.) Current count is read back. Upper byte loaded or read after low byte access.
2H	RW	DMA Channel 1 base and current address Loaded with DMA base address and incremented as transfers take place. Low byte accessed after clearing byte pointer, address 000CH. Upper byte loaded or read after low byte access.
3H	RW	DMA Channel 1 base and current word count Low byte of word count. Count = value + 1 tranfers. Accessed after byte pointer is cleared (000CH). Current count is read back. Upper byte loaded or read after low byte access.
4H	RW	DMA Channel 2 base and current address Loaded with DMA base address and incremented as tranfers take place. Low byte accessed after clearing byte pointer, address 000CH. Upper byte loaded or read after low byte access.
5H	RW	DMA Channel 2 base and current word count Low byte of word count. Count = value + 1 transfers. Accessed after byte pointer is cleared (000CH). Current count is read back. Upper byte loaded or read after low byte access.
6H	RW	DMA Channel 3 base and current address Loaded with DMA base address and incremented as tranfers take place. Low byte accessed after clearing byte pointer, address 000CH. Upper byte loaded or read after low byte access.

intel.

I/O PORT LIST

Address	R/W/S	Description
0BH	WO	Mode Register, DMA controller 1 Bits [7,6] Mode 0,0 Demand mode 0,1 Signal mode 1,0 Block mode 1,1 Cascade mode Bits [4-5] : Reserved Bits [3,2] Operation 0,0 Verify 0,1 Write 1,0 Read 1,1 Reserved Bits [1,0] channel select 0,0 0 0,1 1 1,0 2 1,1 3
0CH	WO	Clear byte pointer, DMA controller 1
0DH	WO	Master Clear, DMA controller 1
0DH	RO	Temporary Register, DMA controller 1
0EH	WO	Clear Mask Register, DMA controller 1
0FH	WO	Write all Mask Register Bits, DMA controller 1
20H	WO	Initialization Control Word ICW1, Interrupt Controller 1 Bits [5-7] : MCS 80/85 mode only Bit 4 = 1, must be set Bit 3 : Trigger mode, 0-edge, 1-level Bit 2 : Call address interval, 0-8, 1-4 Bit 1 : Mode, 0-cascade, 1-single Bit 0 : 0-no ICW4 needed, 1-ICW4 needed
21H	WO	Initialization Control Word ICW2, Interrupt Controller 1 Bits [3-7] : Interrupt type, MCS 86,88 mode Bits [0-2] : Reserved, MCS 80,85 mode only
21H	WO	Initialization Control Word ICW3 (Master Device) Interrupt Controller 1 Bits [0-7] : 0-IR input has no slave controller 1-IR input has a slave controller

A-3

intel.

I/O PORT LIST

Address	R/W/S	Description
7H	RW	DMA Channel 3 base and current word count Low byte of word count. Count = value + 1 transfers. Accessed after byte pointer is cleared (000CH). Current count is read back. Upper byte loaded or read after low byte access.
8H	WO	Command register, DMA controller 1* Bit 7 : DACK sense, 0-active low, 1-active high Bit 6 : DREQ sense, 0-active high, 1-active low Bit 5 : late write, 1-extended write, x-if B3=1 Bit 4 : 0-fixed priority, 1-rotating priority Bit 3 : 0-normal timing, 1-compressed timing Bit 2 : 0-controller enable, 1-controller disable Bit 1 : Channel 0 address hold, 0-disable, 1-enable Bit 0 : Memory to Memory transfer, 0-disable, 1-enable
8H	RO	Status Register, DMA controller 1 Bit 7 = 1 channel 3 request Bit 6 = 1 channel 2 request Bit 5 = 1 channel 1 request Bit 4 = 1 channel 0 request Bit 3 = 1 channel 3 reached TC Bit 2 = 1 channel 2 reached TC Bit 1 = 1 channel 1 reached TC Bit 0 = 1 channel 0 reached TC
9H	WO	Request Register, DMA controller 1 Bits [3-7] : Reserved Bit 2 : 0-reset request bit, 1-set request bit Bits [1,0] channel select 0,0 0 0,1 1 1,0 2 1,1 3
0AH	WO	Mask Register, DMA controller 1 Bits [3-7] :: Reserved Bit 2 : 0-clear mask bit, 1-set mask bit Bits [1,0] channel select 0,0 0 0,1 1 1,0 2 1,1 3

*Compressed timing and memory to memory transfer functions are not supported.

A-2

APPENDIX G: I/O ADDRESS MAPS

I/O PORT LIST

Address	R/W/S	Description
22H	RW	CPU Power Mode Register Bit 15 : De-Turbo select;IOCFGOPN Bit 14 : 0, Reserved Bits [11-13] : Slow numerics option select Bits [9-10] : Slow CPU clock select Bit 8 : CPUCNFG lock Bit 7 : Flash Disk enable Bit 6 : Halt status Bits [4-5] : Fast CPU clcck select Bits [2-3] : Unit ID Bit 1 : Unit enable Bit 0 : Unlock status
23H	RO	Configuration Space Status Register Bit 7 : 82360SL configuration open bit 0-open, 1-closed Bits [0-6] : Reserved
24H	RW	Configuration Index Register
25H	RW	Configuration Data Register
40H	RW	Timer Counter 1 Channel 0 count. Load with count value and read back current count. High/Low byte selected from Control Register.
41H	RW	Timer counter 1 Channel 1 count. Load with count value and read back current count. High/Low byte selected from Control Register.
42H	RW	Timer Counter 1 Channel 2 count. Load with count value and read back current count. High/Low byte selected from Control Register.

A-5

I/O PORT LIST

Address	R/W/S	Description
21H	WO	Initialization Control Word ICW3 (Slave Device) Interrupt Controller 1 Bits [3-7] : 0 Bits [2,1,0] Slave ID 0,0,0 0 0,0,1 1 0,1,0 2 0,1,1 3 1,0,0 4 1,0,1 5 1,1,0 6 1,1,1 7
21H	WO	Initialization Control Word ICW4, Interrupt Controller 1 Bits [5-7] = 0, Reserved Bit 4 : Fully nested mode 0-not special, 1-special Bits [3,2] Mode 0,0 non-buffered mode 0,1 non-buffered mode 1,0 slave buffered 1,1 master buffered Bit 1 : End of interrupt 0-normal, 1-automatic Bit 0 : 0-MCS 80/85 mode 1-8086/8088 mode
21H	RW	Operation Control Word OCW1, Interrupt Controller 1 Bits [0-7] : Interrupt masks 0-mask reset, 1-mask set
20H	WO	Operation Control Word OCW2, Interrupt Controller 1 Bits [5-7] : EOI command Bits [3-4] : 0 Bits [0-2] : interrupt request to be acted upon
20H	WO	Operation Control Word OCW3, Interrupt Controller 1 Bit 7 = 0, Reserved Bits [5-6] : Special mask Bit 4 - 0, Reserved Bit 3 - 1, Reserved Bit 2 - Command, 0-no poll, 1-poll Bits [1,0] Read register command 1,0 Read interrupt request register 1,1 Read interrupt in-service register

A-4

(Reprinted by permission of Intel Corporation, Copyright Intel Corp. 1992)

SECTION G.3: ISA STANDARD I/O ADDRESS PORTS

Address	R/W/S	Description
43H	WO	Timer Counter 1 Command Register Bits [7,6] Function 0,0 Select counter 0 0,1 Select counter 1 1,0 Select counter 2 1,1 Read-back command Bits [5,4] Function 0,0 Counter latch command 0,1 Read/Write least significant byte only 1,0 Read/Write most significant byte only 1,1 Read/Write least significant byte first, then most significant byte Bits [3,2,1] Function 0,0,0 Select Mode 0 0,0,1 Select Mode 1 X,1,0 Select Mode 2 X,1,1 Select Mode 3 1,0,0 Select Mode 4 1,0,1 Select Mode 5 Bit 0 : 0-Binary counter 16 bits 1-Binary coded decimal counter
48H	RW	Timer Counter 2 Channel 0 count. Load with count value and read back current count. High/Low byte selected from Control Register.
4AH	RW	Timer Counter 2 Channel 2 count. Load with count value and read back current count. High/Low byte selected from Control Register.
4BH	WO	Timer Counter 2 Command Register Bits [7,6] Function 0,0 Select counter 0 0,1 Select counter 1 1,0 Select counter 2 1,1 Read-back command Bits [5,4] Function 0,0 Counter latch command 0,1 Read/Write least significant byte only 1,0 Read/Write most significant byte only 1,1 Read/Write least significant byte first, then most significant byte Bits [3,2,1] Function 0,0,0 Select Mode 0 0,0,1 Select Mode 1 X,1,0 Select Mode 2 X,1,1 Select Mode 3 1,0,0 Select Mode 4 1,0,1 Select Mode 5 Bit 0 - 0-Binary counter 16 bits 1-Binary coded decimal counter

A-6

Address	R/W/S	Description
60H	R	Keyboard Controller data I/O input buffer
60H	W	Keyboard Controller data I/O output buffer
61H	RO	Port 61 Bit 7 : 1-parity error Bit 6 : I/O channel parity error Bit 5 : 82C54 Timer 1 counter 2 output Bit 4 : Refresh detect Bit 3 : I/O parity check enable 0-enable, 1-disable Bit 2 : System parity check enable, 0-enable, 1-disable Bit 1 : Speaker enable, 1-enable Bit 0 : Timer 2 gate to speaker enable 0-disable, 1-enable
61H	WO	Port 61 Bit 7 : Reserved Bit 6 : Reserved Bit 5 : Reserved Bit 4 : Reserved Bit 3 : I/O parity check enable 0-enable, 1-disable Bit 2 : System parity check enable, 0-enable, 1-disable Bit 1 : Speaker enable, 1-enable Bit 0 : Timer 2 gate to speaker enable 0-disable, 1-enable
64H	WO	Keyboard Controller Command
64H	RO	Keyboard Controller Status
70H	WO	CMOS RAM Address port and NMI Mask Bit 7 : NMI mask, 1-disable Bits [0-6] : CMOS RAM Address
71H	RW	RTC CMOS RAM data port
74H	RW	Extended CMOS RAM address port (protected) Bit 7 : Reserved Bits [0-6] : CMOS RAM Address
76H	RW	Extended CMOS RAM data port (protected)
80H	RW	Reserved
81H	RW	DMA Memory Address Mapper Page Register Channel 2
82H	RW	DMA Memory Address Mapper Page Register Channel 3
83H	RW	DMA Memory Address Mapper Page Register Channel 1

A-7

Address	R/W/S	Description
84H	RW	Reserved
85H	RW	Reserved
86H	RW	Reserved
87H	RW	DMA Memory Address Mapper Page Register Channel 0
88H	RW	Reserved
89H	RW	DMA Memory Address Mapper Page Register Channel 6
8AH	RW	DMA Memory Address Mapper Page Register Channel 7
8BH	RW	DMA Memory Address Mapper Page Register Channel 5
8CH	RW	Reserved
8DH	RW	Reserved
8EH	RW	Reserved
8FH	RW	DMA Memory Address Mapper Page Register — Refresh
92H		Port 92
		Bits [4-7] : Reserved
	RO	Bit 3: Password security lock, 1-locked
		Bit 2: Reserved
	RW	Bit 1: A20 gate, 0-disabled, 1-enabled
	RW	Bit 0: Fast CPU Reset, low to high transition
0A0H	RW	Initialization Control Word ICW1, Interrupt Controller 2
		Bits [5-7] : MCS 80/85 mode only
		Bit 4 = 1, must be set
		Bit 3 : Trigger mode, 0-edge, 1-level
		Bit 2 : Call address interval, 0-8, 1-4
		Bit 1 : Mode, 0-cascade, 1-single
		Bit 0 : 0-no ICW4 needed, 1-ICW4 needed
0A1H	RW	Initialization Control Word ICW2, Interrupt Controller 2
		Bits [3-7] : Interrupt type, MCS 86,88 mode
		Bits [0-2] : Reserved, MCS 80,85 mode only
0A1H	RW	Initialization Control Word ICW3 (Master Device) Interrupt Controller 2
		Bits [0-7] : 0-IR input has no slave controller / 1-IR input has a slave controller

A-8

Address	R/W/S	Description
0A1H	RW	Initialization Control Word ICW3 (Slave Device) Interrupt Controller 2
		Bits [3-7] : 0
		Bits [2,1,0]: Slave ID
		0,0,0 0
		0,0,1 1
		0,1,0 2
		0,1,1 3
		1,0,0 4
		1,0,1 5
		1,1,0 6
		1,1,1 7
0A1H	RW	Initialization Control Word ICW4, Interrupt Controller 2
		Bits [5-7. = 0, Reserved
		Bit 4 : Fully nested mode
		0-not special, 1-special
		Bits [3,2. Mode
		0,0 non-buffered mode
		0,1 non-buffered mode
		1,0 slave buffered
		1,1 master buffered
		Bit 1 : End of interrupt
		0-normal, 1-automatic
		Bit 0 : 0-MCS 80/85 mode
		1-8086/8088 mode
0A1H	RW	Operation Control Word OCW1, Interrupt Controller 2
		Bits [0-7] : Interrupt masks
		0-mask reset, 1-mask set
0A0H	RW	Operation Control Word OCW2, Interrupt Controller 2
		Bits [5-7] : EOI command
		Bits [3-4] : 0
		Bits [0-2] : interrupt request to be acted upon
0A0H	RW	Operation Control Word OCW3, Interrupt Controller 2
		Bit 7 = 0, Reserved
		Bits [5-6] : Special mask
		Bit 4 - 0, Reserved
		Bit 3 - 1, Reserved
		Bit 2 - Command, 0-no poll, 1-poll
		Bits [1,0] Read register command
		1,0 Read interrupt request register
		1,1 Read interrupt in-service register

A-9

SECTION G.3: ISA STANDARD I/O ADDRESS PORTS

I/O PORT LIST

Address	R/W/S	Description
0C0H	RW	DMA Channel 4 base and current address Loaded with DMA base address and incremented as transfers take place. Low byte accessed after clearing byte pointer, address 000CH. Upper byte loaded or read after low byte access.
0C2H	RW	DMA Channel 4 base and current word count Low byte of word count. Count = value + 1 transfers. Accessed after byte pointer is cleared (000CH.) Current count is read back. Upper byte loaded or read after low byte access.
0C4H	RW	DMA Channel 5 base and current address Loaded with DMA base address and incremented as transfers take place. Low byte accessed after clearing byte pointer, address 000CH. Upper byte loaded or read after low byte access.
0C6H	RW	DMA Channel 5 base and current word count Low byte of word count. Count = value + 1 tranfers. Accessed after byte pointer is cleared (000CH.) Current count is read back. Upper byte loaded or read after low byte access.
0C8H	RW	DMA Channel 6 base and current address Loaded with DMA base address and incremented as tranfers take place. Low byte accessed after clearing byte pointer, address 000CH. Upper byte loaded or read after low byte access.
0CAH	RW	DMA Channel 6 base and current word count Low byte of word count. Count = value + 1 transfers. Accessed after byte pointer is cleared (000CH). Current count is read back. Upper byte loaded or read after low byte access.
0CCH	RW	DMA Channel 7 base and current address Loaded with DMA base address and incremented as tranfers take place. Low byte accessed after clearing byte pointer, address 000CH. Upper byte loaded or read after low byte access.
0CEH	RW	DMA Channel 7 base and current word count Low byte of word count. Count = value + 1 transfers. Accessed after byte pointer is cleared (000CH). Current count is read back. Upper byte loaded or read after low byte access.

(Reprinted by permission of Intel Corporation, Copyright Intel Corp. 1992)

A-10

I/O PORT LIST

Address	R/W/S	Description
0D0H	WO	Command register, DMA controller 2* Bit 7 : DACK sense, 0-active low, 1-active high Bit 6 : DREQ sense, 0-active high, 1-active low Bit 5 : late write, 1-extended write, x-if B3 = 1 Bit 4 : 0-fixed priority, 1-rotating priority Bit 3 : 0-normal timing, 1-compressed timing Bit 2 : 0-controller enable, 1-controller disable Bit 1 : Channel 0 address hold, 0-disable, 1-enable Bit 0 : Memory to Memory transfer, 0-disable, 1-enable
0D0H	RO	Status Register, DMA controller 2 Bit 7 = 1 channel 7 request Bit 6 = 1 channel 6 request Bit 5 = 1 channel 5 request Bit 4 = 1 channel 4 request Bit 3 = 1 channel 7 reached TC Bit 2 = 1 channel 6 reached TC Bit 1 = 1 channel 5 reached TC Bit 0 = 1 channel 4 reached TC
0D2H	WO	Request Register, DMA controller 2 Bits [3-7] : Reserved Bit 2 : 0-reset request bit, 1-set request bit Bits [1,0] channel select 0,0 4 0,1 5 1,0 6 1,1 7
0D4H	WO	Mask Register, DMA controller 2 Bits [3-7] : Reserved Bit 2 : 0-clear mask bit, 1-set mask bit Bits [1,0] channel select 0,0 4 0,1 5 1,0 6 1,1 7

*Compressed timing and memory to memory transfer functions are not supported.

A-11

Address	R/W/S	Description
0102H	RW	Parallel Port Configuration Bit 7: LPT mode, 0—unidirectional, 1—bi-directional Bits [6,5] LPT Select 0,0 LPT 1 0,1 LPT 2 1,0 LPT 3 1,1 LPT disabled Bits [0-4] : Not used
278H	RW	LPT2 Data Port
279H	RO	LPT2 Status Port Bit 7 : 0-Printer Busy Bit 6 : 0-Acknowledge Bit 5 : 1-Out of paper Bit 4 : 1-Printer is selected Bit 3 : 0-Error Bits[0-2] : Not Used
27AH	RW	LPT2 Control Bits [6-7] : Reserved Bit 5 : Direction, PS/2 mode only Bit 4 : Interrupt enable, 1-enable, 0-disable Bit 3 : Select printer, 1-select Bit 2 : Initialize printer, 0-initialize Bit 1 : Automatic line feed, 1-automatic Bit 0 : Data Strobe
27BH–27FH	RW	Automatic data strobe registers
2F8H	RO	Serial Controller Port B Receiver Buffer Serial Data Receive Buffer when DLAB = 0
2F8H	WO	Serial Controller Port B Transmit holding Buffer Serial Data Transmit holding Buffer when DLAB = 0
2F8H	RW	Serial Controller Port B Divisor Latch Least Significant Byte Clock Divisor for BAUD rate when DLAB = 1
2F9H	RW	Serial Controller Port B Divisor Latch Most Significant Byte Clock Divisor for BAUD rate when DLAB = 1
2F9H	RW	Serial Controller Port B Interrupt Enable Register (DLAB=0) Bits [4-7] = 0 Bit 3 : Modem status interrupt enable Bit 2 : Receiver line status interrupt enable Bit 1 : Transmitter holding register empty interrupt enable Bit 0 : Received data available interrupt enable 1 - enable, 0 - disable

A-3

Address	R/W/S	Description
0D6H	WO	Mode Register, DMA controller 2 Bits [7,6] Mode 0,0 Demand mode 0,1 Signal mode 1,0 Block mode 1,1 Cascade mode Bits [4-5] : Reserved Bits [3,2] Operation 0,0 Verify 0,1 Write 1,0 Read 1,1 Reserved Bits [1,0] channel select 0,0 4 0,1 5 1,0 6 1,1 7
0D8H	WO	Clear byte pointer, DMA controller 2
0DAH	WO	Master Clear, DMA controller 2
0DAH	RO	Temporary Register, DMA controller 2
0DCH	WO	Clear Mask Register, DMA controller 2
0DEH	WO	Write all Mask Register Bits,DMA controller 2
0EEH	RW	Special Feature Set Fast A20 Gate Dummy read — disable Dummy write — enable
0EFH	RO	Special Feature Set Fast CPU Reset A dummy read generates a fast CPU Reset
0F0	WO	MCP Register 0 A dummy write clears the numerics option busy signal.
0F4H	WO	Slow CPU Register A dummy write slows the CPU clock to the rate specified by the De-Turbo bit of the CPUPWRMODE register (22H)
0F5H	WO	Fast CPU Register A dummy write causes the CPU clock to run at the rate specified in the Fast CPU clock field in the CPUPWRMODE register (22H)
0F9H	WO	Special Feature Set Disable A dummy write disables the special feature set
0FBH	WO	Special Feature Set Enable A dummy write enables the special feature set

A-12

SECTION G.3: ISA STANDARD I/O ADDRESS PORTS

I/O PORT LIST

Address	R/W/S	Description
2FAH	RO	Serial Controller Port B Interrupt Id Register Bits [3-7] = 0 Bits [2,1] Interrupt ID 0,0 Modem status 0,1 Transmitter holding register 1,0 Received data available 1,1 Receiver line status Bit 0 : 0 – interrupt pending
2FBH	RW	Serial Controller Port B Line Control Register Bit 7 : 0 - Receiver buffer, transmitter holding, or interrupt enable access 1 - Divisor latch access Bit 6 : Set Break, 1-enable Bit 5 : Stick parity Bit 4 : Even parity select Bit 3 : Parity enable, 1-even, 0-odd Bit 2 : Number of stop bits 0 - 1 stop bit, 1 - if word length is 5 bits then stop bit length is 1½ bit times If word length is 6, 7, or 8, then stop bit length is 2 bit times Bits [1-0] Bits per character 0,0 5 0,1 6 1,0 7 1,1 8
2FCH	RW	Serial Controller Port B MODEM Control Register Bits [5-7] : Reserved Bit 4 : 1-Loopback mode Bit 3 : Out2 interrupt enable, 1-enable Bit 2 : Out1 Active, 1-active Bit 1 : Request to send active, 1-active Bit 0 : Data terminal ready, 1-active
2FDH	RO	Serial Controller Port B Line Status Register* Bit 7 : 0 Bit 6 : Transmitter empty Bit 5 : Transmitter holding register empty Bit 4 : Break interrupt Bit 3 : Framing error Bit 2 : Parity error Bit 1 : Overrun error Bit 0 : Data ready

* Writing to this register is not allowed.

A-14

I/O PORT LIST

Address	R/W/S	Description
2FEH	RO	Serial Controller Port B MODEM Status Register Bit 7 : Data carrier detect Bit 6 : Ring indicator Bit 5 : Data set ready Bit 4 : Clear to send Bit 3 : Delta data carrier detect Bit 2 : Trailing edge ring indicator Bit 1 : Delta data set ready Bit 0 : Delta clear to send
2FFH	RW	Serial Controller Port B Scratch Register Independent Register for General Data
372H	WO	Floppy Disk Controller Digital Output Register Bits [6-7] : 0, Reserved Bit 5 : Motor 1 enable, 1-enable Bit 4 : Motor 0 enable, 1-enable Bit 3 : Floppy disk interrupt and DMA request enable, 0-enable Bit 2 : Floppy disk controller reset, 0-reset Bit 1 : 0, Reserved Bit 0 : Drive select, 0-drive 0, 1-drive 1
374H	WO	Floppy Disk Controller Main Status Register Bit 7 : Data register is ready, 1-ready Bit 6 : Data I/O, 1-transfer is from controller to system Bit 5 : Execution mode, 1-non-DMA mode Bit 4 : FDC busy, 1-busy Bit 3 : Drive 3 busy, 1-busy Bit 2 : Drive 2 busy, 1-busy Bit 1 : Drive 1 busy, 1-busy Bit 0 : Drive 0 busy, 1-busy
375H	RW	Floppy Disk Controller Data Register
376H	WO	Fixed Disk Register Bits [4-7] : Reserved Bit 3 : 0-enables reduced write current Bit 2 : Reset fixed disk, 1-reset Bit 1 : fixed disk interrupts enable, 0-enable Bit 0 : Reserved
377H	RO	Floppy Disk Controller Digital Input Register Bit 7 : Diskette change Bits [1-6] : Reserved bit 0 : High density select

A-15

Address	R/W/S	Description

377H	WO	Floppy Disk Controller Control Register
		
		Bits [2-7] : Reserved
		Bits [0,1] Data Rate
		      0,0 500,000 bps
		      0,1 300,000 bps
		      1,0 250,000 bps
		      1,1 Reserved

378H	RW	LPT1 Data Port
379H	RO	LPT1 Status Bit 7 : 0-Printer Busy Bit 6 : 0-Acknowledge Bit 5 : 1-Out of paper Bit 4 : 1-Printer is selected Bit 3 : 0-Error Bits[0-2] : Not Used
37AH	RW	LPT1 Control Bits [6-7] : Reserved Bit 5 : Direction, PS/2 mode only Bit 4 : Interrupt enable, 1-enable, 0-disable Bit 3 : Select printer, 1-select Bit 2 : Initialize printer, 0-initialize Bit 1 : Automatic line feed, 1-automatic Bit 0 : Data Strobe
37BH–37FH	RW	Automatic data strobe registers
3BCH	RW	LPT3 Data Port
3BDH	RO	LPT3 Status Bit 7 : 0-Printer Busy Bit 6 : 0-Acknowledge Bit 5 : 1-Out of paper Bit 4 : 1-Printer is selected Bit 3 : 0-Error Bits[0-2] : Not Used
3BEH	RW	LPT3 Control Bits [6-7] : Reserved Bit 5 : Direction, PS/2 mode only Bit 4 : Interrupt enable, 1-enable, 0-disable Bit 3 : Select printer, 1-select Bit 2 : Initialize printer, 0-initialize Bit 1 : Automatic line feed, 1-automatic Bit 0 : Data Strobe

A-16

Address	R/W/S	Description
3F0H	RO	Floppy Disk Controller Port Status Register A Bit 7 : Interrupt pending Bit 6 : Second drive, 0-installed Bit 5 : Step Bit 4 : Track 0 Bit 3 : Head 1 select Bit 2 : Index Bit 1 : Write protect Bit 0 : Data received
3F1H	RO	Floppy Disk Controller Port Status Register B Bits [6-7] : Reserved Bit 5 : Select drive Bit 4 : Write data Bit 3 : Read data Bit 2 : Write enable Bit 1 : Drive 1 motor enable, 1-enable Bit 0 : Drive 0 motor enable, 1-enable
3F2H	WO	Floppy Disk Controller Digital Output Register Bits [6-7] : 0, Reserved Bit 5 : Motor 1 enable, 1-enable Bit 4 : Motor 0 enable, 1-enable Bit 3 : Floppy disk interrupt and DMA request enable, 0-enable Bit 2 : Floppy disk controller reset, 0-reset Bit 1 : 0, Reserved Bit 0 : Drive select, 0-drive 0, 1-drive 1
3F4H	WO	Floppy Disk Controller Main Status Register Bit 7 : Data register is ready, 1-ready Bit 6 : Data I/O, 1-transfer is from controller to system Bit 5 : Execution mode, 1-non-DMA mode Bit 4 : FDC busy, 1-busy Bit 3 : Drive 3 busy, 1-busy Bit 2 : Drive 2 busy, 1-busy Bit 1 : Drive 1 busy, 1-busy Bit 0 : Drive 0 busy, 1-busy
3F5H	RW	Floppy Disk Controller Data Register
3F6H	WO	Fixed Disk Register Bits [4-7] : Reserved Bit 3 : 0-enables reduced write current Bit 2 : Reset fixed disk, 1-reset Bit 1 : fixed disk interrupts enable, 0-enable Bit 0 : Reserved
3F7H	RO	Floppy Disk Controller Digital Input Register Bit 7 : Diskette change Bits [1-6] : Reserved bit 0 : High density select

A-17

APPENDIX H: IBM PC/PS BIOS DATA AREA

This appendix lists the BIOS data area contents as provided by IBM PS/2 Technical Reference.

BIOS Data Area

The BIOS Data Area is allocated specifically as a work area for system BIOS and adapter BIOS. The BIOS routines use 256 bytes of memory from absolute address hex 400 to hex 4FF. A description of the BIOS data area follows:

Address	Function	Size
40:00	RS-232-C Communications Line 1 Port Base Address	Word
40:02	RS-232-C Communications Line 2 Port Base Address	Word
40:04	RS-232-C Communications Line 3 Port Base Address	Word
40:06	RS-232-C Communications Line 4 Port Base Address	Word

Note: The RS-232-C communications line port base address fields may be initialized to 0 by the POST if the system configuration contains less than four serial ports. The POST never places 0 in the RS-232-C communications line port base address table between two valid RS-232-C communications line port base addresses.

Figure 3-1. RS-232-C Port Base Address Data Area

Address	Function	Size
40:08	Printer 1 Port Base Address	Word
40:0A	Printer 2 Port Base Address	Word
40:0C	Printer 3 Port Base Address	Word
40:0E	Reserved	Word
Exceptions		
40:0E	Printer 4 Port Base Address (PC, PC XT, AT, and PC Convertible)	Word

Note: The printer port base address fields may be initialized to 0 by the POST if the system configuration contains less than four parallel ports. The POST never places 0 in the printer port base address table between two valid printer port base addresses.

Figure 3-2. Printer Port Base Address Data Area

Reprinted by permission from "IBM Personal System/2 and Personal Computer BIOS Interface Technical Reference" c. 1987 by International Business Machines Corporation.

Figure 3-3. System Equipment Data Area

Address	Function	Size
40:10	Installed Hardware	Word
Bits 15,14	Number of Printer Adapters	
Bit 13	Reserved	
Bit 12	Reserved	
Bits 11,10,9	Number of RS-232-C Adapters	
Bit 8	Reserved	
Bits 7,6	Number of Diskette Drives (0-based)	
Bits 5,4	Video Mode Type (Values are Binary)	
	00 = Reserved	
	01 = 40x25 Color	
	10 = 80x25 Color	
	11 = 80x25 Monochrome	
Bit 3	Reserved	
Bit 2	Pointing Device	
Bit 1	Math Coprocessor	
Bit 0	IPL Diskette	
Exceptions		
Bit 13	Internal Modem (PC Convertible Only)	
Bit 2	Reserved (PC, PC XT, AT, and PC Convertible)	

Note: Refer to INT 11H for equipment return information.

Figure 3-6. Keyboard Data Area 1

Address	Function	Size
40:17	Keyboard Control	Byte
Bit 7	Insert Locked	
Bit 6	Caps Lock Locked	
Bit 5	Num Lock Locked	
Bit 4	Scroll Lock Locked	
Bit 3	Alt Key Pressed	
Bit 2	Ctrl Key Pressed	
Bit 1	Left Shift Key Pressed	
Bit 0	Right Shift Key Pressed	
40:18	Keyboard Control	Byte
Bit 7	Insert Key Pressed	
Bit 6	Caps Lock Key Pressed	
Bit 5	Num Lock Key Pressed	
Bit 4	Scroll Lock Key Pressed	
Bit 3	Pause Locked	
Bit 2	System Request Key Pressed	
Bit 1	Left Alt Key Pressed	
Bit 0	Left Ctrl Key Pressed	
40:19	Alternate Keypad Entry	Byte
40:1A	Keyboard Buffer Head Pointer	Word
40:1C	Keyboard Buffer Tail Pointer	Word
40:1E	Keyboard Buffer	32 Bytes

Figure 3-4. Miscellaneous Data Area 1

Address	Function	Size
40:12	Reserved	Byte
Exceptions		
40:12	Power-On Self-Test Status (PC Convertible Only)	Byte

Figure 3-5. Memory Size Data Area

Address	Function	Size
40:13	Memory Size in Kb (Range 0 to 640)	Word
40:15 to 40:16	Reserved	Byte

Address	Function	Size
40:3E	Recalibrate status	Byte
Bit 7	Interrupt Flag	
Bit 6	Reserved	
Bit 5	Reserved	
Bit 4	Reserved	
Bit 3	Recalibrate Drive 3	
Bit 2	Recalibrate Drive 2	
Bit 1	Recalibrate Drive 1	
Bit 0	Recalibrate Drive 0	
40:3F	Motor Status	Byte
Bit 7	Write/Read Operation	
Bit 6	Reserved	
Bits 5,4	Diskette Drive Select Status (Values in Binary)	
	00 = Diskette Drive 0 Selected	
	01 = Diskette Drive 1 Selected	
	10 = Diskette Drive 2 Selected	
	11 = Diskette Drive 3 Selected	
Bit 3	Diskette Drive 3 Motor On Status	
Bit 2	Diskette Drive 2 Motor On Status	
Bit 1	Diskette Drive 1 Motor On Status	
Bit 0	Diskette Drive 0 Motor On Status	
40:40	Motor off counter	Byte
40:41	Last Diskette Drive Operation Status	Byte
	00H = No Error	
	01H = Invalid Diskette Drive Parameter	
	02H = Address Mark not Found	
	03H = Write-protect Error	
	04H = Requested Sector not Found	
	06H = Diskette Change Line Active	
	08H = DMA Overrun on Operation	
	09H = Attempt to DMA Across a 64Kb Boundary	
	0CH = Media Type not Found	
	10H = CRC Error on Diskette Read	
	20H = General Controller Failure	
	40H = Seek Operation Failed	
	80H = Diskette Drive not Ready	
40:42	Diskette Drive Controller Status Bytes	7 Bytes

Figure 3-7. Diskette Drive Data Area

Address	Function	Size
40:49	Display Mode set	Byte
40:4A	Number of Columns	Word
40:4C	Length of Regen Buffer in Bytes	Word
40:4E	Starting Address in Regen Buffer	Word
40:50	Cursor Position Page 1	Word
40:52	Cursor Position Page 2	Word
40:54	Cursor Position Page 3	Word
40:56	Cursor Position Page 4	Word
40:58	Cursor Position Page 5	Word
40:5A	Cursor Position Page 6	Word
40:5C	Cursor Position Page 7	Word
40:5E	Cursor Position Page 8	Word
40:60	Cursor Type	Word
40:62	Display Page	Byte
40:63	CRT Controller Base Address	Word
40:65	Current Setting of 3x8 Register	Byte
40:66	Current Setting of 3x9 Register	Byte

Figure 3-8. Video Control Data Area 1

Address	Function	Size
40:67	Reserved	DWord
40:6B	Reserved	Byte
Exceptions		
40:67	Pointer to reset code upon system reset with memory preserved (Personal System/2 products except Model 30). Reset Flag at 40:72 = 4321H	DWord

Figure 3-9. System Data Area 1

Address	Function	Size
40:6C	Timer Counter	DWord
40:70	Timer Overflow (If non 0, timer has counted past 24 hours.)	Byte

Figure 3-10. System-Timer Data Area

Address	Function	Size
40:71	Break Key State	Byte
40:72	Reset Flag	Word
	1234H = Bypass Memory Test	
	4321H = Preserve Memory (Personal System/2 products except Model 30)	
	5678H = System Suspended (PC Convertible)	
	9ABCH = Manufacturing Test Mode (PC Convertible)	
	ABCDH = System POST Loop Mode (PC Convertible)	

Figure 3-11. System Data Area 2

Address	Function	Size
40:74	Last Fixed Disk Drive Operation Status	Byte
	00H = No Error	
	01H = Invalid Function Request	
	02H = Address Mark not Found	
	03H = Write Protect Error	
	04H = Sector not Found	
	05H = Reset Failed	
	07H = Drive Parameter Activity Failed	
	08H = DMA Overrun on Operation	
	09H = Data Boundary Error	
	0AH = Bad Sector Flag Detected	
	0BH = Bad Track Detected	
	0DH = Invalid Number of Sectors on Format	
	0EH = Control Data Address Mark Detected	
	0FH = DMA Arbitration Level Out of Range	
	10H = Uncorrectable ECC or CRC Error	
	11H = ECC Corrected Data Error	
	20H = General Controller Failure	
	40H = Seek Operation Failed	
	80H = Time Out	
	AAH = Drive not Ready	
	BBH = Undefined Error Occurred	
	CCH = Write Fault on Selected Drive	
	E0H = Status Error/Error Register is 0	
	FFH = Sense Operation Failed	
40:75	Number of Fixed Disk Drives Attached	Byte
40:76	Reserved	Byte
40:77	Reserved	Byte
Exceptions		
40:74	Reserved (IBM ESDI Fixed Disk Drive Adapter/A)	Byte
40:76	Fixed Disk Drive Control (PC XT)	Byte
40:77	Fixed Disk Drive Controller Port (PC XT)	Byte

Figure 3-12. Fixed Disk Drive Data Area

Address	Function	Size
40:78	Printer 1 Time-out Value	Byte
40:79	Printer 2 Time-out Value	Byte
40:7A	Printer 3 Time-out Value	Byte
40:7B	Reserved	Byte
Exceptions		
40:7B	Printer 4 Time-out Value (PC, PC XT, and AT)	Byte

Figure 3-13. Printer Time-Out Value Data Area

Address	Function	Size
40:7C	RS-232-C Communications Line 1 Time-out Value	Byte
40:7D	RS-232-C Communications Line 2 Time-out Value	Byte
40:7E	RS-232-C Communications Line 3 Time-out Value	Byte
40:7F	RS-232-C Communications Line 4 Time-out Value	Byte

Figure 3-14. RS-232-C Time-Out Value Data Area

Address	Function	Size
40:80	Keyboard Buffer Start Offset Pointer	Word
40:82	Keyboard Buffer End Offset Pointer	Word

Figure 3-15. Keyboard Data Area 2

Address	Function	Size
40:84	Number of Rows on the Screen (Minus 1)	Byte
40:85	Character Height (Bytes/Character)	Word
40:87	Video Control States	Byte
40:88	Video Control States	Byte
40:89	Reserved	Byte
40:8A	Reserved	Byte

Figure 3-16. Video Control Data Area 2

Address	Function	Size
40:96	Keyboard Mode State and Type Flags	Byte
Bit 7	Read ID in Progress	
Bit 6	Last Character was First ID Character	
Bit 5	Force Num Lock if Read ID and KBX	
Bit 4	101/102-Key Keyboard Installed	
Bit 3	Right Alt Key Pressed	
Bit 2	Right Ctrl Key Pressed	
Bit 1	Last Code was E0 Hidden Code	
Bit 0	Last Code was E1 Hidden Code	
40:97	Keyboard LED Flags	Byte
Bit 7	Keyboard Transmit Error Flag	
Bit 6	Mode Indicator Update	
Bit 5	Resend Receive Flag	
Bit 4	Acknowledgment Received	
Bit 3	Reserved (Must be 0)	
Bits 2,1,0	Keyboard LED State Bits	

Figure 3-18. Keyboard Data Area 3

Address	Function	Size
40:98	Offset Address to User Wait Complete Flag	Word
40:9A	Segment Address to User Wait Complete Flag	Word
40:9C	User Wait Count - Low Word (Microseconds)	Word
40:9E	User Wait Count - High Word (Microseconds)	Word
40:A0	Wait Active Flag	Byte
Bit 7	Wait Time Elapsed and Post	
Bits 6 to 1	Reserved	
Bit 0	INT 15H, AH = 86H (Wait) has Occurred	
40:A1 to 40:A7	Reserved	Byte

Figure 3-19. Real-Time Clock Data Area

Address	Function	Size
40:8B	Media Control	Byte
Bits 7,6	Last Diskette Drive Data Rate Selected (Values in Binary)	
	00 = 500Kb Per Second	
	01 = 300Kb Per Second	
	10 = 250Kb Per Second	
	11 = Reserved	
Bits 5,4	Last Diskette Drive Step Rate Selected	
Bit 3	Reserved	
Bit 2	Reserved	
Bit 1	Reserved	
Bit 0	Reserved	
40:8C	Fixed Disk Drive Controller Status	Byte
40:8D	Fixed Disk Drive Controller Error Status	Byte
40:8E	Fixed Disk Drive Interrupt Control	Byte
40:8F	Reserved	Byte
40:90	Drive 0 Media State	Byte
40:91	Drive 1 Media State	Byte
Bits 7,6	Diskette Drive Data Rate (Values in Binary)	
	00 = 500Kb Per Second	
	01 = 300Kb Per Second	
	10 = 250Kb Per Second	
	11 = Reserved	
Bit 5	Double Stepping Required	
Bit 4	Media Established	
Bit 3	Reserved	
Bits 2,1,0	Drive/Media State (Values in Binary)	
	000 = 360Kb Diskette/360Kb Drive not Established	
	001 = 360Kb Diskette/1.2Mb Drive not Established	
	010 = 1.2Mb Diskette/1.2Mb Drive not Established	
	011 = 360Kb Diskette/360Kb Drive Established	
	100 = 360Kb Diskette/1.2Mb Drive Established	
	101 = 1.2Mb Diskette/1.2Mb Drive Established	
	110 = Reserved	
	111 = None of the Above	
40:92	Reserved	Byte
40:93	Reserved	Byte
40:94	Drive 0 Current Cylinder	Byte
40:95	Drive 1 Current Cylinder	Byte
Exceptions		
40:8B to 40:95	Reserved (PC, PCjr, PC XT BIOS Dated 11/8/82, and PC Convertible)	Byte

Figure 3-17. Diskette Drive/Fixed Disk Drive Control Data Area

Reprinted by permission from "IBM Personal System/2 and Personal Computer BIOS Interface Technical Reference"
c. 1987 by International Business Machines Corporation.

For systems with EGA capability and Personal System/2 products, the save pointer table contains pointers that define specific dynamic overrides for the video mode set function, INT 10H, (AH) = 00H.

Address	Function	Size
40:A8	Pointer to Video Parameters and Overrides	DWord
DWord 1	Video Parameter Table Pointer	
	Initialized to the BIOS video parameter table. This value must contain a valid pointer.	
DWord 2	Dynamic Save Area Pointer (except Personal System/2 Model 30)	
	Initialized to 00:00, this value is optional. When non 0, this value points to an area in RAM where certain dynamic values are saved. This area holds the 16 EGA palette register values plus the overscan value in bytes (0-16), respectively. A minimum of 256 bytes must be allocated for this area.	
DWord 3	Alpha Mode Auxiliary Character Generator Pointer	
	Initialized to 00:00, this value is optional. When non 0, this value points to a table that is described as follows:	
	Bytes/Character	Byte
	Block to Load, 0 = Normal Operation	Byte
	Count to Store, 256 = Normal Operation	Word
	Character Offset, 0 = Normal Operation	Word
	Pointer to a Font Table	DWord
	Displayable Rows	Byte
	If 0FFH, the maximum calculated value is used, otherwise this value is used.	
	Consecutive bytes of mode values for this font description. The end of this stream is indicated by a byte code of 0FFH.	Byte
Note:	Use of the DWord 3 pointer may cause unexpected cursor type operation. For an explanation of cursor type, see INT 10H, (AH) = 01H.	
DWord 4	Graphics Mode Auxiliary Character Generator Pointer	
	Initialized to 00:00, this value is optional. When non 0, this value points to a table that is described as follows:	

Figure   3-20 (Part 1 of 2).   Save Pointer Data Area

Address	Function	Size
	Displayable Rows	Byte
	Bytes Per Character	Word
	Pointer to a Font Table	DWord
	Consecutive bytes of mode values for this font description. The end of this stream is indicated by a byte code of 0FFH.	Byte
DWord 5	Secondary Save Pointer (except EGA and Personal System/2 Model 30)	
	Initialized to the BIOS secondary save pointer. This value must contain a valid pointer.	
DWord 6	Reserved and set to 00:00.	
DWord 7	Reserved and set to 00:00.	

Figure   3-20 (Part 2 of 2).   Save Pointer Data Area

Address	Function	Size
Word 1	Table Length	
	Initialized to the BIOS secondary save pointer table length.	
DWord 2	Display Combination Code (DCC) Table Pointer	
	Initialized to ROM DCC table. This value must exist. It points to a table described as follows:	
	Number of Entries in Table	Byte
	DCC Table Version Number	Byte
	Maximum Display Type Code	Byte
	Reserved	Byte
	00,00 Entry 0 No Displays	
	00,01 Entry 1 MDPA	
	00,02 Entry 2 CGA	
	02,01 Entry 3 MDPA + CGA	
	00,04 Entry 4 EGA	
	04,01 Entry 5 EGA + MDPA	
	00,05 Entry 6 MEGA	
	02,05 Entry 7 MEGA + CGA	
	00,06 Entry 8 PGC	
	01,06 Entry 9 PGC + MDPA	
	05,06 Entry 10 PGC + MEGA	
	00,08 Entry 11 CVGA	

Figure   3-21 (Part 1 of 3).   Secondary Save Pointer Data Area

Address	Function	Size
	Internal Palette Index (0-16; 0 = Normal Operation)	Word
	Pointer to Internal Palette	DWord
	External Palette Count (0-256; 256 = Normal Operation)	Word
	External Palette Index (0-255; 0 = Normal Operation)	Word
	Pointer to External Palette	DWord
	Consecutive bytes of mode values for this font description. The end of this stream is indicated by a byte code of 0FFH.	Byte
DWord 5 to DWord 7	Reserved and set to 00:00.	

Figure 3-21 (Part 3 of 3). Secondary Save Pointer Data Area

Address	Function	Size
40:AC to 40:FF	Reserved	Byte
50:00	Print Screen Status Byte (INT 05H Status)	Word

Figure 3-22. Miscellaneous Data Area 2

Extended BIOS Data Area

The Extended BIOS Data Area is supported on Personal System/2 products only. The POST allocates the highest possible (n) Kb of memory below 640Kb to be used as this data area. The word value at 40:13 (memory size), indicating the number of Kb below the 640Kb limit, is decremented by (n). The first byte in the Extended BIOS Data Area is initialized to the length in Kb of the allocated area.

To access the Extended BIOS Data Area segment, issue an INT 15, (AH) = C1H (Return Extended BIOS Data Area Segment Address). To determine if an Extended BIOS Data Area is allocated, use INT 15, (AH) = C0H (Return System Configuration Parameters).

Address	Function	Size
	01,08 Entry 12 CVGA + MDPA 00,07 Entry 13 MVGA 02,07 Entry 14 MVGA + CGA 02,06 Entry 15 MVGA + PGC Abbreviation Meanings: MDPA = Monochrome Display and Printer Adapter CGA = Color/Graphics Monitor Adapter EGA = Enhanced Graphics Adapter MEGA = EGA with monochrome display PGC = Professional Graphics Controller VGA = Video Graphics Array MVGA = VGA based with monochrome display CVGA = VGA based with color display	
DWord 3	Second Alpha Mode Auxiliary Character Generator Pointer Initialized to 00:00, this value is optional. When non 0, this value points to a table that is described as follows:	
	Bytes/Character	Byte
	Block to load, should be non 0 for normal operation.	Byte
	Reserved	Byte
	Pointer to a Font Table	DWord
	Consecutive bytes of mode values for this font description. The end of this stream is indicated by a byte code of 0FFH.	Byte

Note: Attribute bit 3 is used to switch between primary and secondary fonts. It may be desirable to use the user palette profile to define a palette of consistent colors independent of attribute bit 3.

Address	Function	Size
DWord 4	User Palette Profile Table Pointer Initialized to 00:00, this value is optional. When non 0, this value points to a table that is described as follows:	
	Underlining flag (1 = On, 0 = Ignore, -1 = Off; 0 = Normal Operation)	Byte
	Reserved	Byte
	Reserved	Word
	Internal Palette Count (0-17; 17 = Normal Operation)	Word

Figure 3-21 (Part 2 of 3). Secondary Save Pointer Data Area

APPENDIX I: DATA SHEETS

SECTION I.1: NS8250/NS16450 UART CHIP

NS16450/INS8250A/NS16C450/INS82C50A

5.0 Block Diagram

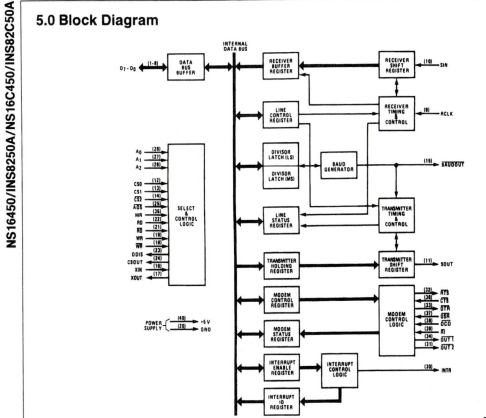

TL/C/8401–10

(Reprinted by permission of National Semiconductor Corporation, Copyright National Semiconductor Corp. 1990)

i486™ MICROPROCESSOR

i486™ Microprocessor Pinout
TOP SIDE VIEW

Figure 1.2

240440-3

i486™ MICROPROCESSOR

Pin Cross Reference by Pin Name

Address		Data		Control		N/C	Vcc	Vss
A2	Q14	D0	P1	A20M#	D15	A3	B7	A7
A3	R15	D1	N2	ADS#	S17	A10	B9	A9
A4	S16	D2	N1	AHOLD	A17	A12	B11	A11
A5	Q12	D3	H2	BE0#	K15	A13	C4	B3
A6	S15	D4	M3	BE1#	J16	A14	C5	B4
A7	Q13	D5	J2	BE2#	J15	B10	E2	B5
A8	R13	D6	L2	BE3#	F17	B12	E16	E1
A9	Q11	D7	L3	BLAST#	R16	B13	G2	E17
A10	S13	D8	F2	BOFF#	D17	B14	G16	G1
A11	R12	D9	D1	BRDY#	H15	B16	H16	G17
A12	S7	D10	E3	BREQ#	Q15	C10	J1	H1
A13	Q10	D11	C1	BS8#	D16	C11	K2	H17
A14	S5	D12	G3	BS16#	C17	C12	K16	K1
A15	R7	D13	D2	CLK	C3	C13	L16	K17
A16	Q9	D14	K3	D/C#	M15	G15	M2	L1
A17	Q3	D15	F3	DP0	N3	R17	M16	L17
A18	R5	D16	J3	DP1	F1	S4	P16	M1
A19	Q4	D17	D3	DP2	H3		R3	M17
A20	Q8	D18	C2	DP3	A5		R6	P17
A21	Q5	D19	B1	EADS#	B17		R8	Q2
A22	Q7	D20	A1	FERR#	C14		R9	R4
A23	S3	D21	B2	FLUSH#	C15		R10	S6
A24	Q6	D22	A2	HLDA	P15		R11	S8
A25	R2	D23	A4	HOLD	E15		R14	S9
A26	S2	D24	A6	IGNNE#	A15			S10
A27	S1	D25	B6	INTR	A16			S11
A28	R1	D26	C7	KEN#	F15			S12
A29	P2	D27	C6	LOCK#	N15			S14
A30	P3	D28	C8	M/IO#	N16			
A31	Q1	D29	A8	NMI	B15			
		D30	C9	PCD	J17			
		D31	B8	PCHK#	Q17			
				PWT	L15			
				PLOCK#	Q16			
				RDY#	F16			
				RESET	C16			
				W/R#	N17			

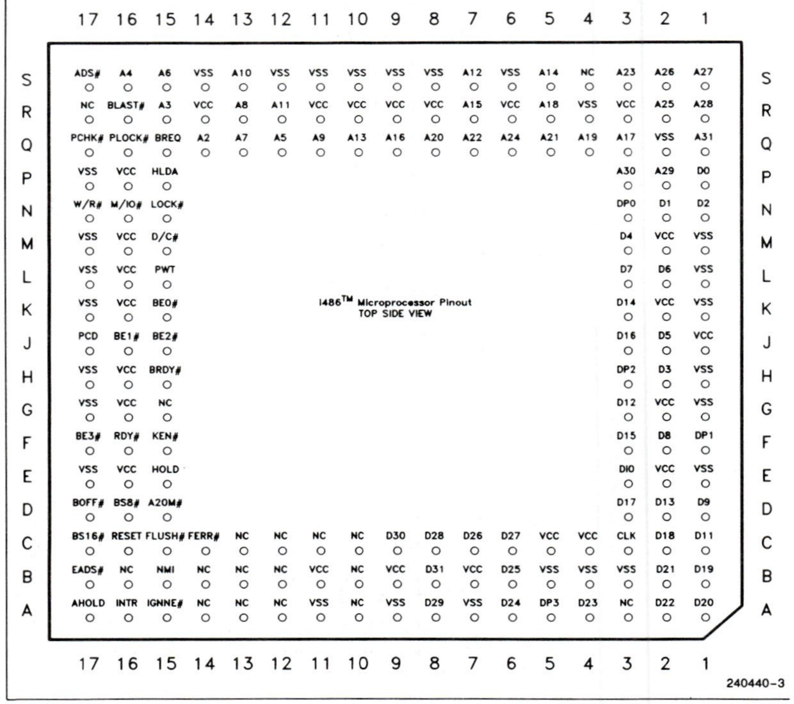

(Reprinted by permission of Intel Corporation, Copyright Intel Corp. 1990)

SECTION I.3: INTEL'S PENTIUM

1.1. PINOUT DIAGRAMS

1.1.1. Pentium™ Processor Pinouts

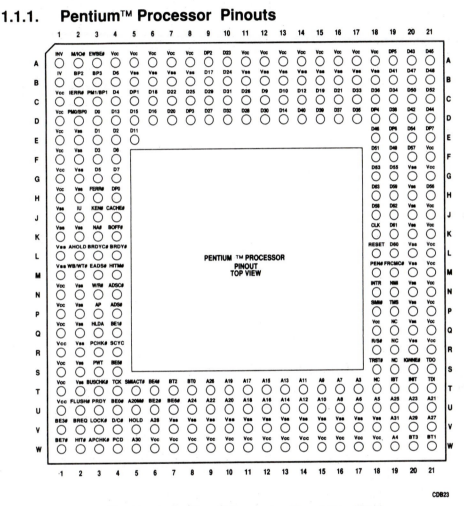

Figure 1-1. Pentium™ Processor Pinout (Top View)

1-1

(Reprinted by permission of Intel Corporation, Copyright Intel Corp. 1993)

1.2. PIN CROSS REFERENCE TABLES

1.2.1. Pentium Processor

Table 1-1. Pentium™ Processor Pin Cross Reference by Pin Name

Address	Pin	Data	Pin	Data	Pin	Control	Pin	Control	Pin
A3	T17	D0	D03	D32	D10	A20M#	U05	FRCMC#	M19
A4	W19	D1	E03	D33	C17	ADS#	P04	HIT#	W02
A5	U18	D2	E04	D34	C19	ADSC#	N04	HITM#	M04
A6	U17	D3	F03	D35	D17	AHOLD	L02	HLDA	Q03
A7	T16	D4	C04	D36	C18	AP	P03	HOLD	V05
A8	U16	D5	G03	D37	D16	APCHK#	W03	IBT	T19
A9	T15	D6	B04	D38	D19	BE0#	U04	IERR#	C02
A10	U15	D7	G04	D39	D15	BE1#	Q04	IGNNE#	S20
A11	T14	D8	F04	D40	D14	BE2#	U06	INIT	T20
A12	U14	D9	C12	D41	B19	BE3#	V01	INTR	N18
A13	T13	D10	C13	D42	D20	BE4#	T06	INV	A01
A14	U13	D11	E05	D43	A20	BE5#	S04	IU	J02
A15	T12	D12	C14	D44	D21	BE6#	U07	IV	B01
A16	U12	D13	D04	D45	A21	BE7#	W01	KEN#	J03
A17	T11	D14	D13	D46	E18	BOFF#	K04	LOCK#	V03
A18	U11	D15	D05	D47	B20	BP2	B02	M/IO#	A02
A19	T10	D16	D06	D48	B21	BP3	B03	NA#	K03
A20	U10	D17	B09	D49	F19	BRDY#	L04	NMI	N19
A21	U21	D18	C06	D50	C20	BRDYC#	L03	PCD	W04
A22	U09	D19	C15	D51	F18	BREQ	V02	PCHK#	R03
A23	U20	D20	D07	D52	C21	BUSCHK#	T03	PEN#	M18
A24	U08	D21	C16	D53	G18	CACHE#	J04	PM0/BP0	D02
A25	U19	D22	C07	D54	E20	CLK	K18	PM1/BP1	C03
A26	T09	D23	A10	D55	G19	D/C#	V04	PRDY	U03

Table 1-1. Pentium™ Processor Pin Cross Reference by Pin Name (Contd.)

Address	Pin	Data	Pin	Data	Pin	Control	Pin	Control	Pin
A27	V21	D24	B10	D56	H21	DP0	H04	PWT	S03
A28	V06	D25	C08	D57	F20	DP1	C05	R/S#	R18
A29	V20	D26	C11	D58	J18	DP2	A09	RESET	L18
A30	W05	D27	D09	D59	H19	DP3	D08	SCYC	R04
A31	V19	D28	D11	D60	L19	DP4	D18	SMI#	P18
		D29	C09	D61	K19	DP5	A19	SMIACT#	T05
BT0	T08	D30	D12	D62	J19	DP6	E19	TCK	T04
BT1	W21	D31	C10	D63	H18	DP7	E21	TDI	T21
BT2	T07					EADS#	M03	TDO	S21
BT3	W20					EWBE#	A03	TMS	P19
						FERR#	H03	TRST#	S18
						FLUSH#	U02	W/R#	N03
								WB/WT#	M02

VCC

A04	C01	N21	B05	W08
A05	D01	P01	B06	W09
A06	E01	P21	B07	W10
A07	F01	Q01	B08	W11
A08	F21	Q18	B11	W12
A11	G01	Q21	B12	W13
A12	G21	R01	B13	W14
A13	H01	R21	B14	W15
A14	J21	S01		W16
A15	K21	T01		W17
A16	L21	U01		W18
A17	M21	W06		
A18	N01	W07		

VSS

B15	H02	L20	Q20	V10
B16	H20	M01	R02	V11
B17	J01	M20	R20	V12
B18	J20	N02	S02	V13
E02	K01	N20	T02	V14
F02	K02	P02		V15
G02	K20	P20		V16
G20	L01	Q02		V17
				V18

NC: Q19, S19, R19, T18

SECTION I.4: INTEL PACKAGING

PACKAGE TYPES

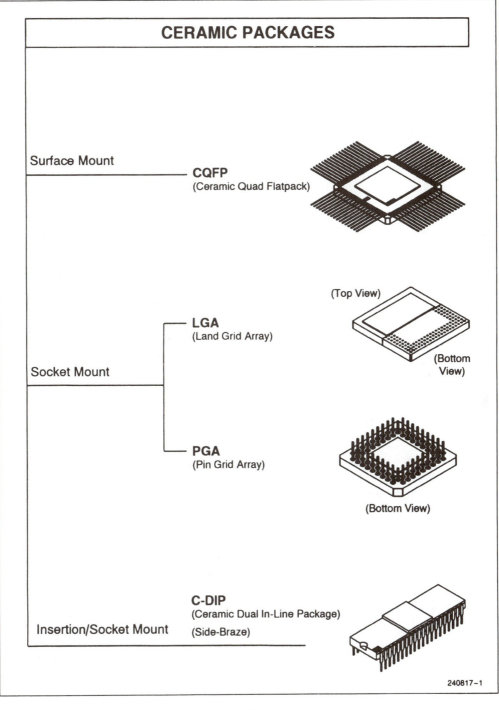

CERAMIC PACKAGES

Surface Mount — **CQFP** (Ceramic Quad Flatpack)

Socket Mount

LGA (Land Grid Array) — (Top View) (Bottom View)

PGA (Pin Grid Array) — (Bottom View)

Insertion/Socket Mount — **C-DIP** (Ceramic Dual In-Line Package) (Side-Braze)

240817-1

1-3

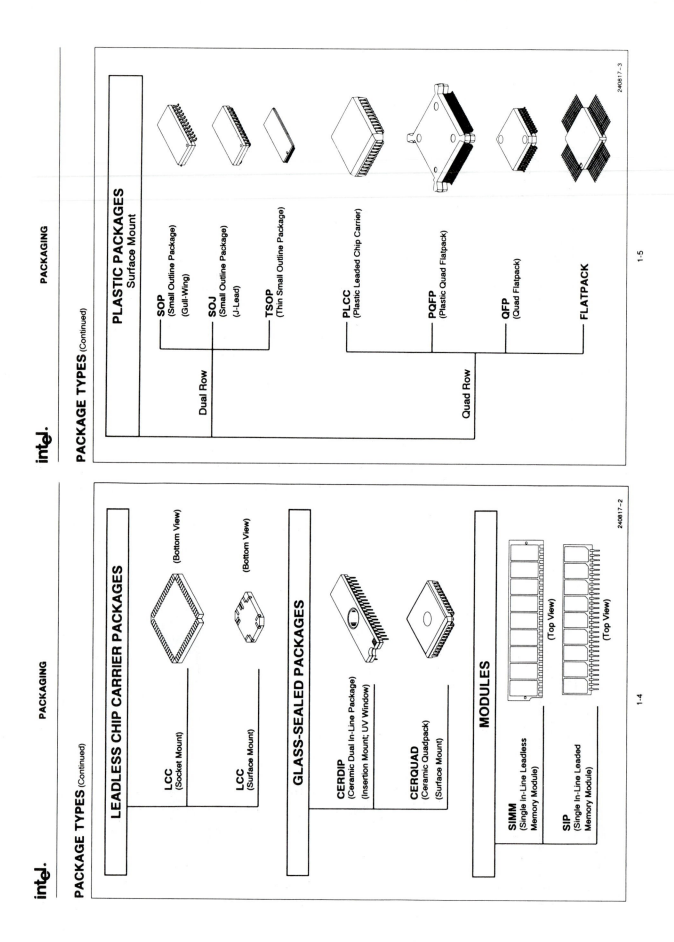

PACKAGE TYPES (Continued)

PLASTIC PACKAGES
Surface Mount

SOP
(Small Outline Package)
(Gull-Wing)

SOJ
(Small Outline Package)
(J-Lead)

Dual Row

TSOP
(Thin Small Outline Package)

PLCC
(Plastic Leaded Chip Carrier)

PQFP
(Plastic Quad Flatpack)

Quad Row

QFP
(Quad Flatpack)

FLATPACK

240817–3

1-5

PACKAGE TYPES (Continued)

LEADLESS CHIP CARRIER PACKAGES

(Bottom View)

LCC
(Socket Mount)

(Bottom View)

LCC
(Surface Mount)

GLASS-SEALED PACKAGES

CERDIP
(Ceramic Dual In-Line Package)
(Insertion Mount; UV Window)

CERQUAD
(Ceramic Quadpack)
(Surface Mount)

MODULES

(Top View)

SIMM
(Single In-Line Leadless
Memory Module)

(Top View)

SIP
(Single In-Line Leaded
Memory Module)

240817–2

1-4

APPENDIX I: DATA SHEETS

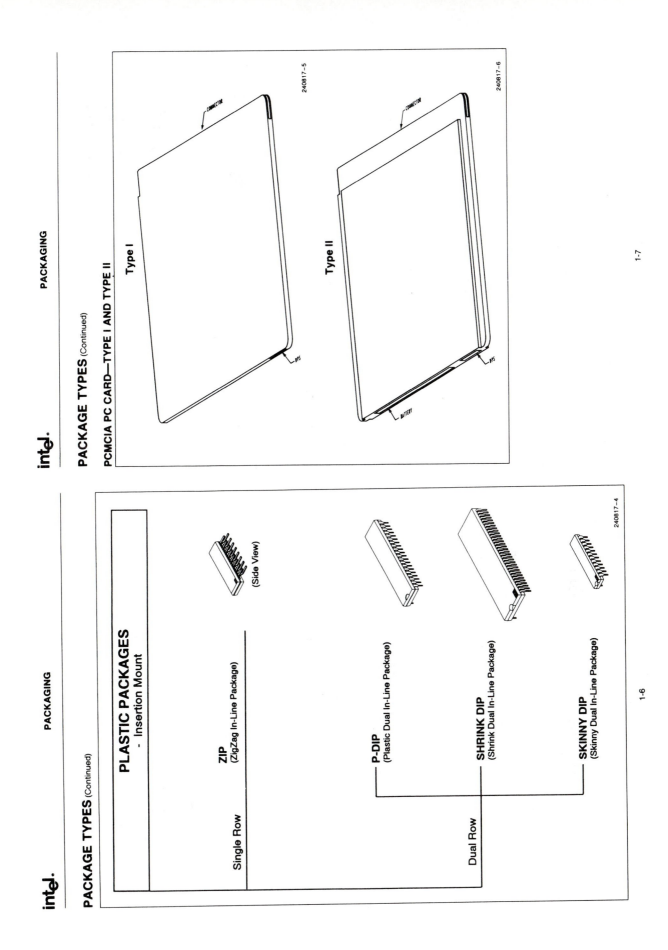

PACKAGE TYPES (Continued)

PCMCIA PC CARD—TYPE I AND TYPE II

Type I

240817-5

Type II

240817-6

PACKAGE TYPES (Continued)

PLASTIC PACKAGES
- Insertion Mount

Single Row

ZIP
(ZigZag In-Line Package)

(Side View)

Dual Row

P-DIP
(Plastic Dual In-Line Package)

SHRINK DIP
(Shrink Dual In-Line Package)

SKINNY DIP
(Skinny Dual In-Line Package)

240817-4

SECTION I.4: INTEL PACKAGING

REFERENCES

Advanced Micro Devices (contact www.amd.com)

Dynamic Memory Design Data Book/Handbook, 1990

Cyrix Corporation (contact www.cyrix.com)

Electronic Design Magazine

Yong-In S. Shin, "Maintain Signal Integrity at High Digital Speeds," May 14, 1992, p. 77.
William Hall, "Avoid Confusion in Choosing Digital Logic," Oct. 24, 1991, pp. 63 - 76.

IBM

IBM Personal System/2 Hardware Interface Technical Reference, 1988
IBM Personal System/2 and Personal Computer BIOS Interface Technical Reference, 1987
IBM Technical Reference Personal Computer AT, 1985
IBM Personal Computer Technical Reference, 1984

IDT

High Performance SRAM Data Book, 1992/1993
IDT Data Book on Logic, 1991, "Characteristics of PCB Traces" by Suren Kodical, AN-49

Intel (contact www.intel.com)

Pentium Processor User's Manual, Volumes 1 - 3, 1993
Pentium Pro Family Developer's Manual, Volumes 1 - 3, 1996
The PCI Local Bus: A Technical Overview, 1993
Packaging Handbook, 1992
Memory Products, 1992
Microprocessors, Volume II, 1992
INTEL386 SL Microprocessor SuperSet Design Guide, 1992
i486 Microprocessor Programmer's Reference Manual, 1990
8086/8088 User's Manual, Programmer's and Hardware Reference, 1989
Memory Components Handbook, 1987
80386 Hardware Reference Manual, 1986
Microprocessor and Peripheral Handbook, 1983

Motorola

Dynamic RAMs and Memory Models, 1993
Microprocessor, Microcontroller and Peripheral Data, Volume II, 1988

National Semiconductor (contact www.national.com)

Data Communications Local Area Networks UARTs Handbook, 1990

PCI Interest Group

PCI Local Bus, Revision 2.0

Seagate

Product Overview, 1990
Product Overview, March 1994

Samsung

MOS Memory Data Book, 1988

Texas Instruments

TTL Logic Data Book, 1988

INDEX

C

N

O

overflow 176 - 179, 181, 416, 639
overscan   479

P

P286 directive 224
P386 directive 224
P486 directive 224
P8086 directive 224
packaging   963 - 965
PAD (end of frame block) character 534
PAGE directive 56, 889
page directory 653
page frames   744, 752, 756
page mode DRAM   665
paging 651 - 653, 655, 745
palettes
        See color palettes
PARA attribute 891
parallel communication 509
parallel execution   702
parity 38, 65, 286 - 288, 692, 700,
                776, 778
        bit generator and checker 286 -
                288
partitioning disks   585
PC Bus Extender   325 - 327
PC DOS operating system   725
PC Interface Trainer   327 - 329
PCI (peripheral component
        interconnect)   798
PCI local bus   799, 802
PCK signal   288
PCLK signal   246
PEACK signal   254
Pentium microprocessor 697 - 704,
                711, 764 - 765, 961 - 962
Pentium Pro microprocessor 710 - 714
PEREQ signal   254
peripheral I/O 314
peripheral interfacing 590
persistence   480
PF (parity flag) 38 - 39, 65
PGA (pin grid array) packaging   634
physical address   647
physical pages 745
pipelining 22, 691, 695 - 698, 701 - 702
pixel   127 - 128, 478
plug and play   802
PMOS logic family   760, 765
pointer registers   23
pointers
        80386 Assembly language
                636, 639
polling devices 337

POP instruction 67, 866
POPA instruction   633
POPF instruction   867
port addressing 316
        COM ports   517, 523
        DMA   450, 452
        LPT ports   557
        of the 8237   460
        of the 8259   430
ports   310
        80386   646
        aliases 316
        interfacing   316 - 319
        of the 8253/54 timer 387 - 388
        of the 8255 320 - 325
        parallel 562 - 564, 566
        See also port addressing
POST (power on self test) 516, 557,
        726 - 727
power supply   765, 771
powers of 2   267
primary memory   266, 647
primary storage 597
        See also RAM
printer
        Centronics specifications 555
        control characters   560
        control signals 556 - 557
        DB-25 connector   557
        interfacing   554, 557
        LPT ports   557
        ports   559, 562 - 564
        programming   559, 561
        status signals 556
        time-out   560
        timing 561
privilege levels 647
PROC directive 52, 889
procedure   52
        See also modules
program counter   11
PROM (programmable ROM)   268
PROMPT command   729
protected mode 20, 251, 260, 301, 440,
        647 - 653, 655, 695, 746
protocol   511, 532, 785
        asynchronous bus   785
        BISYNC   532, 536
        bus   785 - 786
        SDLC   532, 536
        See also serial communication
        synchronous   785
PS/2 port   562 - 563
pseudo-operations
        See directives